Superplasticity in

Advanced Materials

ICSAM-2000

Superplasticity in Advanced Materials ICSAM-2000

Proceedings of the 2000 International Conference on Superplasticity in Advanced Materials (ICSAM-2000) held at Sheraton International Resort, Orlando, USA during August 1-4, 2000.

Editor:

Namas Chandra

TRANS TECH PUBLICATIONS
Switzerland • Germany • UK • USA

Copyright © 2001 Trans Tech Publications Ltd, Switzerland
ISBN 0-87849-874-5

Volumes 357-359 of
Materials Science Forum
ISSN 0255-5476

Distributed in the Americas by

Trans Tech Publications Inc
PO Box 699, May Street
Enfield, New Hampshire 03748
USA

Phone: (603) 632-7377
Fax: (603) 632-5611
e-mail: ttp@ttp.net
Web: http://www.ttp.net

and worldwide by

Trans Tech Publications Ltd
Brandrain 6
CH-8707 Uetikon-Zuerich
Switzerland

Fax: +41 (1) 922 10 33
e-mail: ttp@ttp.net
Web: http://www.ttp.net

Printed in the United Kingdom
by Hobbs the Printers Ltd,
Totton, Hampshire SO40 3WX

INTERNATIONAL CONFERENCE ON SUPERPLASTICITY IN ADVANCED MATERIALS (ICSAM-2000)

Orlando, USA

August 1 to 4, 2000

Conference organized by
Advanced Mechanics and Materials Laboratory
Department of Mechanical Engineering,
FAMU-FSU College of Engineering
Florida State University
Tallahassee, FL 32310

Local Organizing Committee

Namas Chandra (Chairman, Florida State University)
A. Belvin (Florida A&M University)
X. Chen (Florida State University)
B. Cox (Meeting Makers, Inc.)
R. Howell (Meeting Makers, Inc.)
H. Li (Florida State University)
S. Namilae (Florida State University)
C. Shet (Florida State University)
J. Watts (Florida State University)

ICSAM 2000 International Advisory Board

Prof. T. Sakuma (Chairman, Japan)
Prof. T. G. Langdon (University of Southern California, USA)
Prof. N. Furushiro (Osaka University, Japan)
Prof. B. Baudelet (Institut National Polytechnique de Grenoble, France)
Prof. N. Chandra (Florida State University, USA)
Prof. A.H. Chokshi (Indian Institute of Science, India)
Dr. R. Grimes (Buckinghamshire, UK)
Prof Hai Jintao (Beijing Research Institute of Mechanical and Electrical
    Technology,P. R. China)
Prof. K. Higashi (Japan)
Prof. O.A. Kaibyshev (Institute for Metals Superplasticity Problems, Russia)
Prof. M. Kobayashi (Chiba Institute of Technology, Japan)
Prof. A.K. Mukherjee (University of California, Davis, USA)
Dr. N. Ridley (University of Manchester and UMIST, UK)
Prof. T. Sakuma (University of Tokyo, Japan)
Prof. O.M. Smirnov (Moscow Steel and Alloys Institute, Russia)
Dr. J. Wadsworth (Lawrence Livermore National Laboratory, USA)
Dr. F. Wakai (Japan Science and Technology Corporation, Japan)

Preface

The International Conference on Superplasticity in Advanced Materials (ICSAM-2000) represents the seventh in the series of international conferences on this topic held every three years starting with San Diego, USA (1982), then Grenoble, France (1985), Blaine, Washington, USA (1988), Osaka, Japan (1991), Moscow, Former Soviet Union (1994), and Bangalore, India (1997). The interest in superplasticity was confined to the developed nations in the early 70's and mid 80's and has become a subject of intensive study throughout the world from the late 80's onwards. On a similar timeline superplastic forming with and without diffusion bonding has matured from a novel manufacturing process to a well-established advanced production process. Concomitant to that change funding in the area of fundamental aspects of superplasticity has declined significantly in the advanced countries, especially USA. This is very unfortunate since our knowledge even today is not comprehensive enough for us to *design* materials/processes capable of superplasticity in a cost-effective manner.

Once the exclusive domain of space and aeronautical industries, superplastic forming (SPF) has recently made inroads into automotive, rail, architectural, sports, dental and entertainment sectors. However, due to a number of technological issues ranging from die material selection, die design, surface finish (e.g. alpha casing), differential thinning, temperature-pressure-time cycle determination and the overall economy, the application of superplasticity is nowhere near its full potential. Hence in this conference special emphasis was placed on industrial applications. Contrary to the historical precedence, the conference was opened with this topic and ended with the same topic. In all there were six different topical symposiums within the conference.

The meeting returned to United States after 1988 and was well attended by delegation from 22 different countries. The entire meeting was coordinated on the emerging internet with a dedicated website http://icsam.eng.fsu.edu. Extended abstracts were submitted on the web; after a brief review they were compiled and distributed during the conference. The number of countries represents the maximum number of countries participating in the ICSAM series, and perhaps any superplasticity related event in history. This validates the previous observation that superplasticity is alive and thriving in many parts of the world. Scientific research in superplasticity has broadened to include many material systems like metals, intermetallics, ceramics, their composites, nanocrystalline materials and even metallic glasses.

With a total attendance of about 130, many international delegates requested oral presentations, which forced some parallel sessions for the first time. Also for the first time, the entire proceedings were very thoroughly reviewed by experts. Most of the papers were peer-reviewed by at least two referees. Sincere thanks to A.J. Barnes, T. Sakuma, T.G. Nieh, K. Higashi, R. Sadeghi, A.K. Ghosh, H. Iwasaki, T. McNelley, K.A. Padmanabhan, M. Mabuchi, C. Shet, S.C. Rama, J. Humphries, R. Rajagopal, R.S. Misra, O. Ruano, T. Aizawa, A.K. Mukerjee, N. Furushiro, D. Rodriguez, T.G. Langdon, B. Bai, R.Z. Valiev, F. Mohamed, N. Ridley, J. Pimenoff, J.J. Blandin, R. Todd, O.A. Kaibyshev and R. Grimes for their service. The quality of the book is reflected in their timely and careful review; and typically these books (series) remain one of the highly read and referred volumes in superplasticity. There are about 90 papers in this volume, which includes about 15 presented in poster sessions. Only papers presented (oral/poster) at the conference were included in the review process. Most of the authors promptly responded to the corrections and returned the manuscripts in a timely manner. The process is time consuming, but in the interest of quality the whole exercise is worthwhile and should be continued in future meetings.

The meeting was partially funded by FAMU-FSU College of Engineering, Florida State University, MARC Corporation (later MSC-Software), Oakridge National Laboratory and Advanced Mechanics and Materials Laboratory. I would like to acknowledge the participation of the ICSAM Advisory Board, especially Professor Taketo Sakuma. I would like to sincerely thank my students J.D. Watts, A. Belvin, X. Chen, S. Namilae and research associates C. Shet and H. Li. They had dedicated themselves and worked tirelessly towards making the conference a success.

Finally I would like to express my gratitude and appreciation for my wife Usha Chandra and daughters Alli and Kavita Chandra for their patience and encouragement during the long deliberations of the meeting.

Namas Chandra
Chairman, ICSAM-2000
(held in Orlando, Florida, USA)
Tallahassee, Florida, USA

Table of Contents

I. Industrial Applications of Superplastic Forming

Industrial Applications of Superplastic Forming: Trends and Prospects
A.J. Barnes 3

The Current State-of-the-Art and the Future in Airframe Manufacturing Using Superplastic Forming Technologies
D.G. Sanders 17

Generic Ceramic Tooling (GenCerT) for the SPF/DB Process
A. Jocelyn, T. Flower and D. Nash 23

The Design and Manufacture of Superplastic Forming/Diffusion Bonding Presses
R. Whittingham 29

Superplastic Injection Forming of Magnesium Alloys
T. Aizawa and M. Mabuchi 35

Aerospace Part Production Using SP 700
P.N. Comley 41

Dental Implant Superstructures by Superplastic Forming
R.V. Curtis, D. Garriga-Majo, A.S. Juszczyk, S. Soo, D. Pagliaria and J.D. Walter 47

Industrial Applications of Superplastic Forming Technology in China
Z. Li and H. Zhu 53

The Production of Automotive Body Panels in 5083 SPF Aluminium Alloy
B.J. Dunwoody 59

Industrial Applications of the Superplastic Explosive Forming
V.P. Sabelkin and M.A.H. Rojo 65

II. Superplasticity in Metals and Ceramics

Superplasticity in Metals and Ceramics
O.A. Kaibyshev 73

The Role of Impurities in Superplastic Flow and Cavitation
F.A. Mohamed 83

A Microstructural Study of the Effect of Particle Aging on Dynamic Continuous Recrystallization in Al-4Mg-0.3Sc
L.M. Dougherty, I.M. Robertson, J.S. Vetrano and S.M. Bruemmer 93

Investigation of Superplastic Behaviour and Solid State Bonding of Zircaloy-4
P.S. Hill, N. Ridley and R.I. Todd 99

Superplastic Behaviors of a Ti-Alloy with Ultrafine Grain Size
B.Z. Bai, X.J. Sun, J.L. Gu and L.Y. Yang 105

Development of Superplastic AlCuAgMgZr Alloys for Easy Forming
J. Dutkiewicz, J. Kuśnierz and T.G. Nieh 111

Superplastic Deformation Behavior of Undoped and Doped High-Purity 3Y-TZP
T. Satou, F. Hosaka, E. Sato, J. Matsushita, M. Otsuka and K. Kuribayashi 117

Compressive Creep Studies on Alumina Particulate Reinforced Alumino-Silicate Glass
B. Sudhir and A.H. Chokshi 123

Superplastic Characteristics in Germania Based Codoped Y-TZP
K. Sasaki, M. Nakano, J. Mimurada, Y. Ikuhara and T. Sakuma 129

Deformation Characteristics of a 3Y-TZP/20%Al_2O_3 Composite in Tensile Creep
S.S. Sosa and T.G. Langdon 135

Superplastic Flow Stress in Cation-Doped YSZ
M. Nakano, J. Mimurada, K. Sasaki, Yu. Ikuhara and T. Sakuma 141

Effective Diffusivity for Superplastic Flow in Magnesium Alloys
H. Watanabe, H. Tsutsui, T. Mukai, M. Kohzu and K. Higashi 147

Temperature and Strain-Rate Dependence of Elongation in Al - 4,5 Mg Alloy
H. Iwasaki, T. Mori, M. Mabuchi, T. Tagata and K. Higashi 153

Effects of Temperature and Microstructure on the Superplasticity in Microduplex Pb-Sn Alloys
T.K. Ha and Y.W. Chang 159

Low Temperature Superplasticity in a Severely Rolled Sheet of an Al-4.2Mass%Mg Alloy
Y. Takayama, S. Sasaki, H. Kato, H. Watanabe, A. Niikura and Y. Bekki 165

Microstructural Evolution of Quasi-Single Phase Alloy during Superplastic Deformation
W. Bang, T.K. Ha and Y.W. Chang 171

Transformation Superplasticity of Ti-6Al-4V and Ti-6Al-4V/TiC Composites at High Stresses
C. Schuh and D.C. Dunand 177

Superplastic Tensile Behavior of Silicon Nitride
N. Kondo, T. Ohji and Y. Suzuki 183

High Temperature Deformation of a Yttria-Stabilized Tetragonal Zirconia
K. Morita, K. Hiraga and Y. Sakka 187

Cavity Formation and Growth in a Superplastic Alumina Containing Zirconia Particles
K. Hiraga, K. Nakano, T.S. Suzuki and Y. Sakka 193

Superplastic Deformation of a Duplex Stainless Steel
D. Pulino-Sagradi, R.E. Medrano and A.M.M. Nazar 199

Microstructural Evolution in Superplastic Coarse-Grained Fe-27at.%Al: Strain-Rate Effects
J.P. Chu, S.H. Chen, H.Y. Yasuda, Y. Umakoshi and K. Inoue 205

III. Mechanics and Modeling of Superplasticity

Geometric Analysis of Thinning during Superplastic Forming
D. Garriga-Majo and R.V. Curtis 213

Analysis of Superplastic Testing by Using Constant Pressure in Prismatic Die
L. Carrino and G. Giuliano 219

Effect of Hot Deformation under Complex Loading on the Transformation of Lamellar Type Microstructure in Ti-6.5Al-3.5Mo-1.6Zr-0.27Si Alloy
O.A. Kaibyshev, V.K. Berdin, M.V. Karavaeva, R.M. Kashaev and L.A. Syutina 225

High Temperature Load Relaxation Behavior of Superplastic Iron Aluminides
J.H. Song, H.T. Lim, T.K. Ha and Y.W. Chang 231

Glide-Climb Based GBS Model Applied to Submicrocrystalline Superplasticity
A.K. Ghosh 237

Role of Alloying Element in Superplastic Deformation of Aluminum Alloys
N. Furushiro, Y. Umakoshi and K. Warashina 249

Texture, Grain Boundaries and Deformation of Superplastic Aluminum Alloys
M.T. Pérez-Prado, T.R. McNelley, G. González-Doncel and O.A. Ruano 255

A Thermodynamic Framework for the Superplastic Response of Materials
K.R. Rajagopal and N. Chandra 261

An Augmented Rigid Plastic Flow Formulation in Support of Finite Element Modeling of Superplastic Forming
R.S. Sadeghi 273

Effects of the Temperature of Warm Rolling on the Superplastic Behaviour of AA5083 Aluminium Base Alloy
J. Pimenoff, Y. Yagodzinskyy, J. Romu and H. Hänninen 277

Cooperative Processes in Superplastic Forming Under Different Deformation States
M. Zelin, S. Guillard and A.K. Mukherjee 283

Unified Constitutive Equation of CTE-Mismatch Superplasticity Based on Continuum Micromechanics
K. Kitazono, E. Sato and K. Kuribayashi 289

A Coupled Thermo-Viscoplastic Formulation at Finite Strains for the Numerical Simulation of Superplastic Forming
L. Adam and J.P. Ponthot 295

Mechanical Effects of Spatial Distributions of Large Grains in Superplastic Microstructures
E. Bernault, J.J. Blandin and R. Dendievel 301

IV. High Strain Rate Superplasticity

High Strain Rate Superplasticity in Mechanically Alloyed Nickel Aluminides
Y. Doi, K. Matsuki, H. Akimoto and T. Aida 309

Influence of Reversible Hydrogen Alloying on Formation of SMC Structure and Superplasticity of Titanium Alloys
G.A. Salishchev, M.A. Murzinova, S.V. Zherebtsov, D.D. Afonichev and
S.P. Malysheva 315

High Strain Rate Superplasticity in a Zn - 22% Al Alloy after Equal-Channel Angular Pressing
S.M. Lee and T.G. Langdon 321

High Strength and High Strain Rate Superplasticity in Magnesium Alloys
M. Mabuchi, K. Shimojima, Y. Yamada, C.E. Wen, M. Nakamura, T. Asahina,
H. Iwasaki, T. Aizawa and K. Higashi 327

Inhomogeneous Cavity Distribution of Superplastically Deformed AL 7475 Alloy
M.J. Tan and C. Chen 333

The Development of a High Strain Rate Superplastic Al-Mg-Zr Alloy
R.J. Dashwood, R. Grimes, A.W. Harrison and H.M. Flower 339

Recent Advances and Future Directions in Superplasticity
K. Higashi 345

High Strain Rate Superplastic Aluminium Alloys: The Way Forward?
R. Grimes, R.J. Dashwood and H.M. Flower 357

High Strain Rate Superplasticity and Microstructure Study of a Magnesium Alloy
X. Wu, Y. Liu and H. Hao 363

V. Novel Approaches to Superplasticity in Materials

Unified Theory of Deformation for Structural Superplastics, Metallic Glasses and Nanocrystalline Materials
K.A. Padmanabhan and B.S.S. Daniel 371

Characterization of Grain Boundary Properties in Superplastic Al Based Alloys Using EBSD
I.C. Hsiao and J.C. Huang 381

Atomistic Simulation of the Effect of Trace Elements on Grain Boundary of Aluminium
S. Namilae, C. Shet, N. Chandra and T.G. Nieh 387

On the Independent Behavior of Grain Boundary Sliding and Intragranular Slip During Superplasticity
A.D. Sheikh-Ali and H. Garmestani 393

Effect of Chemical Bonding States on the Tensile Ductility in Glass-Doped TZP
A. Kuwabara, S. Yokota, Y. Ikuhara and T. Sakuma 399

Internal Stress Superplasticity in a In-Situ Intermetallic Matrix Composite
R.S. Sundar, K. Kitazono, E. Sato and K. Kuribayashi 405

Analysis of Instability and Strain Concentration during Superplastic Deformation
L.C. Chung and J.H. Cheng 411

Low Temperature and High Strain Rate Superplasticity of Nickel Base Alloys
V.A. Valitov, O.A. Kaibyshev, S.K. Mukhtarov, B.P. Bewlay and M.F.X. Gigliotti 417

VI. Grain Refinement of Materials

Development of Ultrafine Grained Materials Using The MAXStrain® Technology
W.C. Chen, D.E. Ferguson, H.S. Ferguson, R.S. Mishra and Z. Jin 425

Optimization for Superplasticity in Ultrafine-Grained Al-Mg-Sc Alloys Using Equal-Channel Angular Pressing
M. Furukawa, A. Utsunomiya, S. Komura, Z. Horita, M. Nemoto and T.G. Langdon 431

Microstructure and High Temperature Deformation of an ECAE Processed 5083 Al Alloy
L. Dupuy, J.J. Blandin and E.F. Rauch 437

Effect of Temperature on the Microstructure and Superplasticity in a Powder Metallurgy Processed Al-16Si-5Fe-1Cu-0.5Mg-0.9Zr (wt%) Alloy
M.S. Kim, H.S. Cho, J.S. Han, H.G. Jeong and H. Yamagata 443

Grain Refinement and Enhanced Superplasticity in Metallic Materials
R.Z. Valiev, R.K. Islamgaliev and N.F. Yunusova 449

Grain Refinement of a Commercial Magnesium Alloy for Superplastic Forming
T. Mukai, H. Watanabe, K. Moriwaki, K. Ishikawa, Y. Okanda and K. Higashi 459

Improvements in Superplastic Performance of Commercial AA5083 Aluminium Processed by Equal Channel Angular Extrusion
D.R. Herling and M.T. Smith 465

Equal-Channel Angular Pressing as a Production Tool for Superplastic Materials
Z. Horita, S. Lee, S. Ota, K. Neishi and T.G. Langdon 471

Refinement and Stability of Grain Structure
F.J. Humphreys and P.S. Bate 477

Using Severe Plastic Deformation for Grain Refinement and Superplasticity
T.G. Langdon, M. Furukawa, M. Nemoto and Z. Horita 489

Tensile Superplasticity in Nanomaterials - Some Observations and Reflections
S.X. McFadden, A.V. Sergueeva and A.K. Mukherjee 499

Friction Stir Processing: A New Grain Refinement Technique to Achieve High Strain Rate Superplasticity in Commercial Alloys
R.S. Mishra and M.W. Mahoney 507

VII. Poster Presentations

High Strain Rate Superplastic Deformation in 6061 Alloy with 1% SiO_2 Nano-Particles
T.D. Wang and J.C. Huang 515

An Investigation of Cavity Development in a Superplastic Aluminium Alloy Prepared by Equal-Channel Angular Pressing
C. Xu, S. Lee and T.G. Langdon 521

Industrial Superplastic Forming Research and Application for Commercial Aircraft Components at Israel Aircraft Industries
B. Gershon, I. Arbel, S. Hevlin, Y. Milo and D. Saltoun 527

Superplastic Alumina Ceramics Dispersed with Zirconia and Spinel Particles
B.-N. Kim, K. Morita, K. Hiraga and Y. Sakka 533

Grain Refinement and Superplasticity of Reaction Sintered TiC Dispersed Ti Alloy Composites Using Hydrogenation Treatment
N. Machida, K. Funami and M. Kobayashi 539

On the Activation Energies Observed in Al-based Materials Deformed at Ultrahigh Temperatures
B.Y. Lou, T.D. Wang, J.C. Huang and T.G. Langdon 545

Cooperative Grain Boundary Sliding at Room Temperature of a Zn-20.2%Al-1.8%Cu Superplastic Alloy
J.D. Muñoz-Andrade, A. Mendoza-Allende, G. Torres-Villaseñor and
J.A. Montemayor-Aldrete 551

Superplastic Deformation Behavior of the Y-TZP Tested in Torsion
Y. Motohashi, S. Sanada and T. Sakuma 559

Control of Thickness Distribution in a Al7475 Axi-Symmetric Cup
N. Suzuki, M. Kohzu, S. Tanabe and K. Higashi 565

A Molecular Dynamics Study of Large Deformation of Nanocrystalline Materials
H. Ogawa, N. Sawaguchi and F. Wakai 571

Interface Models for GB Sliding and Migration
C. Shet, H. Li and N. Chandra 577

Evaluation of Superplastic Material Characteristics Using Multi-Dome Forming Test
A. El-Morsy, N. Akkus, K. Manabe and H. Nishimura 587

Superplastic Deformation Mechanism of ZrO_2 and Al_2O_3 with Additives Studied by Electron Microscopy
A. Kumao and Y. Okamoto 593

Temperature Effects on the Localization and Mode of Failure of Al 5083
J.D. Watts, X. Chen, A. Belvin, Z. Chen and N. Chandra 599

Superplastic Diffusion Bonding of a Ti-24Al-14Nb-3V-0.5Mo Intermetallic Alloy
H. Zhu, Z. Li and C. Wang 607

High Strain Rate Superplasticity in Al-16Si-5Fe Based Alloys with and without SiC Particulates
J.S. Han, M.S. Kim, H.G. Jeong and H. Yamagata 613

Author Index 619
Keyword Index 623

Industrial Applications
of Superplastic Forming

Industrial Applications of Superplastic Forming: Trends and Prospects

A.J. Barnes

Superform USA, Riverside CA 92517-5375, USA

Keywords: Commercial Alloy, Competitiveness, Markets, Processing Techniques, R&D Initiatives, Techno-Economics

ABSTRACT

What is the process that can take a flat sheet of metal and with the application of heat and just 100 psi or so, shape it into a myriad of complex shapes using simple tools and equipment? The answer is *superplastic forming* (SPF). Since the early 1970's SPF has been fulfilling its initial promise to create complex sheet metal parts for a wide range of end users; from the functional needs of the specialist automobile market to the creative world of architecture and sculpture. With all these beneficial attributes we might be prompted to ask this question, "Why isn't SPF used everywhere and more often if it is that good?"

This paper examines current trends and the reasons for SPF's successes and what limits its wider application. Factors relating to available materials, their cost and performance, the process' techno-economics, R&D initiatives and the prospects of "breakthrough" developments are discussed.

INTRODUCTION

The prospect of industrial application of superplastic forming was first seriously considered in the mid 1960's. It might be surprising to some to know that IBM was then investigating both sheet and bulk forming of the zinc-aluminum eutectoid (Zn22Al) for computer parts manufacture. As early as April 1965 they applied for their first superplastic forming patent [1], by May 1969 they had created a "Superplasticity Design Guide" for their design teams and by 1971, with a number of applications having been produced, an updated report entitled "The Promise of Superplasticity" was issued [2], This report covered parts and tool design; including the use of ceramic dies processing techniques, as well as, mechanical property data for the Zn22Al alloy. Figures 1 & 2 illustrate two of the many parts made by IBM during this period.

Figure 1. Large experimental box forming made from 0.1" Zn22Al sheet

Figure 2. Complex heat exchanger element made from 0.015" Zn22Al sheet.

The large 'box' and the heat exchanger element shown were soon to be overtaken by the advances in the computer industry (semi-conductors/ miniaturization) and although the possible uses of Zn22Al in the automotive industry were being investigated and the first commercial superplastic forming company (ISC Alloys, Avonmouth UK) had been started in 1971, these pioneering efforts had limited success as industrial applications were restricted by the modest mechanical properties and poor creep resistance of the duplex eutectic and eutectoid alloys then available.

The early 1970's saw the culmination of two separate 'break through' research efforts which have most significantly influenced the direction and industrial application of superplastic forming to date; they were: the development in the UK of **dilute superplastic aluminum alloys** known by the trade name SUPRAL [3], leading to the formation of Superform Metals in late 1973 and in the USA, Rockwell International's development of **SPF Titanium technology** including the remarkable process of concurrent superplastic forming and **diffusion bonding** (SPFDB) first demonstrated in 1973 [4],.

The remainder of this paper will focus on current SPF aluminum technology and it's future direction leaving detailed discussion of titanium SPF and SPFDB technology to those more familiar and expert in the minute particulars' which have made it successful. It is sufficient here to state that the technology has become a well established and cost effective solution in the construction of components and assemblies for military aircraft and has more recently transitioned into commercial aircraft, as well as, finding applications in cookware (frying pans) and sporting goods (golf clubs) [5],

COMPETITIVENESS OF SPF TECHNOLOGY

Since the mid 1970's applications for SPF aluminum have continued to grow in many, but not all, market sectors – why? SPF does not gain acceptance in the market place by right or by virtue of its novelty, but must compete with existing and emerging processes and alternative materials to achieve sound engineering, design and economic benefits.

The main processes and material forms that SPF aluminum competes with are listed below:

METAL PROCESSES	**PLASTIC PROCESSES**
CASTING	UNREINFORCED
Die Casting	Injection Molding
Investment Casting	Thermo-forming
Sand Casting	Resin Injection (RIM)
SHEET METAL	REINFORCED
Stamping	SMC
Stretch Forming	RTM
Hydro Forming	HL GRP
Rubber Pressing	Autoclave
Drop Hammer	
Fabrication – folding, bending, rolling	
Weldments	

Each of these processes has economic and technical viability in specific application situations. Many complex and interrelated factors influence selection. These include; quantity, size, weight, mechanical performance, physical properties, functional needs, environmental constraints, as well as, direct economic factors such as material cost and availability, tooling costs and life expectancy, productivity and equipment capital and running costs.

Of the many factors that lead engineers, designers and buyers to consider SPF the limitations of alternative processes and materials are often an important consideration. These include:

- **Size limitations of available process equipment**: Die Casting, Investment casting, Injection molding, Hydro forming.
- **More expensive tooling and insufficient quantity to justify high tool costs**: Die casting, Stamping, Injection molding, SMC, RTM
- **Heavily skill dependent and labor intensive**: Sandcasting, Drop hammer, Fabrication, Weldments, Hand lay-up GRP
- **Not capable of complexity**: Stretch forming, Hydro forming and Rubber pressing
- **Less accuracy and process variability**: Sandcasting, Stretch forming, Drop Hammer, Hand lay-up GRP
- **Environmentally unfriendly:** Sand casting, Drop Hammer, Hand lay-up GRP
- **Does not have the attributes of aluminum needed:** All plastic processes and materials.

Before examining more closely the techno-economic factors that have led to SPFAl's successes, and its limitations, it is appropriate to review the commercial SP alloys and processing techniques currently available.

COMMERCIALLY AVAILABLE SP ALLOYS

Although over the past thirty years many alloy systems have been studied, ranging from the early duplex eutectics and eutectoids (eg: ZnAl, AlCu, AlSi, AlCa & AlPd) to standard or slightly modified "off the shelf" alloys (eg: 7075, 7050, 2219, 6082 & 5083) only a limited number have become commercially available. Most of these are listed in Table 1 along with their typical mechanical properties. All are dilute alloys, which utilize an ultra fine dispersoid of intermetallic particles to pin and stabilize grain size.

Three of the listed alloys represent more than 90% of current usage; they are 2004 [3],, 7475 [6], & 5083 [7],. 2004 and 7475 are used primarily in the Aerospace industry where 2004's excellent SP formability, weldability and medium strength T6 heat treated properties are used in lightly load situations and 7475, after appropriate processing and heat treatment, for more structural applications. 5083 is used exclusively in the as-formed condition for a variety of end uses where it is excellent corrosion resistance and durability are beneficial (building panels, autobody panels and some aerospace applications)

Table 1: Composition and typical properties of superplastic aluminum alloys currently in use.

Alloy Composition	0.2%PS MPa	UTS MPa	Modulus Gpa	Density Mgm-3
2004 Al-6%Cu-0.4%Zr	300	420	74	2.83
2090 Al-2.5%Cu-2.3%Li-0.1Zr	340	450	79	2.57
2095 Al-4.7%Cu-0.37% Mg-1.3%Li0.4%Ag-.14%Zr	585	620	78	2.7
5083 Al-4.5%Mg-.1%Cr	150	300	72	2.67
7475 A1-5.7%Zr-2.3% Mg-1.5%Cu-.2%Cr	500	550	70	2.8
8090 A1-2.4%Li-1.2% Cu-0.6%Mg-0.1Zr	350	450	78	2.55

The limited number of available alloys is indicative of the challenges of producing 'good' superplastic alloys and the willingness (or otherwise) of the aluminum industry to invest their limited resources in developing new alloys. Exotic alloys for limited markets are no longer attractive to the aluminum industry. The next generation of superplastic aluminum alloys may well be made outside of conventional aluminum industries' practices using "breakthrough technology".

SUPERPLASTIC PROCESSING TECHNIQUES

Five basic forming techniques are in regular use today. It is of interest to note that three of them were anticipated in the IBM 1965 patent, they are:

- **The Simple female forming** in which the heated superplastic sheet is clamped around its edge and stretched into the heated cavity tool using gas pressure.
- **Drape forming** where the heated and clamped superplastic sheet is stretched into a cavity containing one or more male form blocks.
- **Male forming** in which gas pressure and tool movement are combined enabling deeper more uniform thickness parts to be made.
- **Backpressure forming** (BPF) utilizes gas pressure on both sides (front and back) of the deforming superplastic sheet which produces a hydrostatic confining pressure capable of suppressing cavitation. Gas control creates a positive pressure differential enabling forming to be achieved. BPF is usually applied to female and drape forming methods when structural applications demand cavitation levels be contained below 0.5% volume fraction.
- **Diaphragm forming** uses one or two 'slave' sheets of superplastic alloy to urge a smaller unclamped blank to be augmentedly drawn and draped into tool contact. The process can shape non-superplastic metals having limited formability at room temperature and in certain cases superplastic alloys are shaped this way when limited thickness variation is demanded. This process has also been successfully applied to shaping advanced thermoplastic graphite composites. [8]

A number of proprietary variations of these basic processes are in current use involving non-planar clamping combined with stretch-draw techniques

The choice of which forming method should be used for a particular component application is a complex one. Laycocks detailed review of forming methods. [9] helps to delineate the problem and a previous paper by the author [10] outlines the selection criteria for the female, drape and male forming methods. The following figures 3-7 illustrate classic examples of each forming technique.

Figure 3. 8 ft wide truck cab roof from a single sheet of SP 5083. Generous radii and the absence of angular features limit the local thinning often associated with female forming.

Figure 4. Twenty different automotive body panel brackets drape formed in a single pressing from SP5083 illustrates the cost effectiveness of multipart drape forming.

Figure 5. A very deep formed missile housing made from SP2004 alloy. Quite uniform thickness distribution is achieved using Superform's unique male forming technique.

Figure 6. Back pressure formed (BPF) structural inner door panel made from SP7475. This aircraft part has strain levels approaching 200%. BPF eliminates the cavitation which otherwise impair it's structural integrity

Figure 7. This PEEK/graphite thermoplastic composite aircraft assembly is 'diaphragm formed' between thin superplastic aluminum sheets.

FORMING EQUIPMENT & TOOLING

Specialized forming equipment (SPF Presses) is needed if superplastic forming is to be successful. The presses used must be capable of uniformly heating tools and the superplastic sheet, accurately controlling gas pressure and in the case of male forming tool velocity and position, to achieve appropriate superplastic strain rates and safety features to prevent press opening under pressure.

The press capacity (tonnage) needed is directly related to component size, the superplastic sheet thickness and the flow stress associated with the superplastic strain rates(s) for the alloy being processed. Commercially available superplastic aluminum alloys (2004, 5083, and 7475) have relatively low flow stresses <15 Mpa and the sheet thickness' used do not usually exceed 4.0 mm. Accordingly gas pressures used do not normally exceed 300 psi. Component geometry, size, thickness, and strain rate dictate the actual gas pressure profile required for 'optimized forming conditions.' Figures 8 and 9 show computer controlled male and female press installations.

Figure 8. One of Superform's unique male forming machine installations in the UK.

Figure 9. Large modern female drape forming press installation producing non-planar clampline aerospace panels from SP 2004 alloy.

Fundamental to the success of most component manufacturing processes is tooling, **tooling** and **TOOLING!** Superplastic forming is no exception, in fact, it is at the heart of the economic viability for commercial superplastic aluminum forming.

Tooling for the lifetime of the product being made, needs to be:
- **Durable;** able to continuously operate at elevated temperature without degrading and safely contain the gas pressures and mechanical forces applied.
- **Accurate;** to consistently produce formings to the required dimensions and surface quality.
- **Productive;** yielding maximum production output commensurate with the superplastic strain rate limits of the alloy being formed and the pressure containment capacity of the forming equipment (forming presses) being used.
- **Cost effective;** manufactured at minimum cost for 'optimized forming conditions'.

With the advent of computer aided design (CAD) almost all SPF tooling is now NC machined from electronic data. However, much proprietary know-how and 'tricks of the trade' are incorporated at all stages of tooling from tool design, material selection, manufacturing methods, through to their mode of operation. Integrating these into a coherent computer based expert system is a key task still needing to be tackled.

Figure 11. This sports car's external body panels are all superplastically formed from 5083, as well as, a number of inner structural panels.

Figure 12. This one piece 1.6m diameter forming is used in cryogenic magnets associated with MRI scanners worldwide. Superplastic forming gave the accuracy and stiffness needed in this exacting application

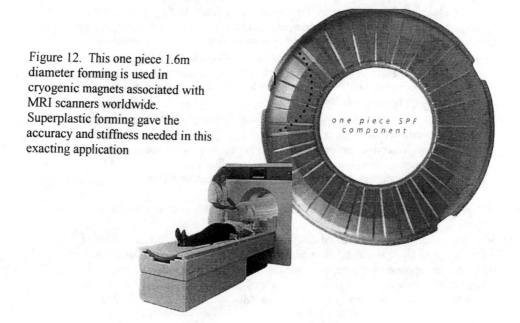

CURRENT MARKETS FOR SPF ALUMINUM

Figure 10. Shows the estimated percentage breakdown, by market segments for superplastic aluminum formed components. Successful markets characteristically have a common profile, which includes:

High added value end products ($K to $M)
- Medium to low volume production (10's to low 1000's)
- Need for the properties and attributes of aluminum alloys
- Innovative design philosophy
  - part designs incorporating thin gages (.020" to .160")
  - complex shape (not simple)
  - medium size parts (1 to 40 sq ft)

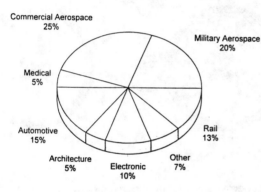

Figure 10. Market segment breakdown for superplastic aluminum formed components.

Of the market segments shown in Figure 10, aerospace is and has been for the last twenty years the most dominant one, steadily growing as designers and engineers have become 'comfortable' with the technology, and procurement and quality personnel have realized the full benefits of part count reduction and the part-to-part consistency characteristic of SPF. The specialty automotive segment also continues to grow as the attributes of SPF aluminum over reinforced plastic body panels (weight-reduction and superior surface quality) are appreciated by both the car builder and the ultimate end user.

Successful innovators of superplastic aluminum forming gain both technical and economic benefits compared to alternative processes and materials through:

- reduced costs associated with tooling and fabrication
- increased design flexibility
- weight savings and reduced part count
- increased structural integrity
- improved consistency/interchangeability
- shorter lead time from design to first off part production

The following examples Figures 11-14 illustrate successful cost saving solutions realized in specific niche markets:

Figure 13. This unconventional Ron Arad designed chair was formed from 5083. This illustrates the unique design freedom offered by SPF.

Figure 14. Probably the world's first superplastic aluminum intensive bicycle! The bonded frame assembly is made from SP5083 and well illustrates the styling possible with SPF.

TECHNO-ECONOMICS OF SPF ALUMINUM

An oversimplified view of the superplastic aluminum parts manufacturing process is:
- it takes a relatively expensive sheet of material
- forms it relatively slowly into a complex shape
- using relatively inexpensive tooling.

These conditions define the viable cost-effective niche shown for SPF aluminum in Figure 15 Material cost and forming time offset the benefit of low cost tooling as annual production quantity increases. To overcome this techno-economic barrier to higher production quantities either material costs and/or forming times must be reduced. This is shown schematically in Figure 16 . The impact of metal cost and forming times on breakeven quantity for autobody panels has been previously reviewed by the author [11] but it is worthwhile to review this again using the graph shown in Figure 17 . Current forming technology and material costs are compared with targets for fast forming (high strain rate) alloys; if realized annual breakeven forming quantities, compared to steel stampings could increase ten fold! Such a quantum change would transform the SPF aluminum business from its present specialized niche position into the main stream of competitive manufacturing. For this transformation to take place it will require radical innovation(s)

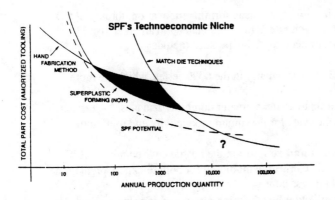

Figure 15. Economics of aluminum sheet forming.

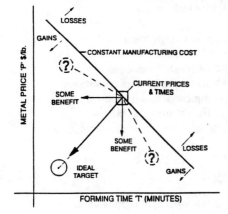

For Constant Manufacturing Cost 'C'

'C' = PW + RT

& P = $\frac{-R}{W}$ T + $\frac{C}{W}$

Where W = Weight of SP Blank (lbs.)
R = Conversion Cost ($/minute)

Figure 16. Metal price: forming time relationship

Figure 17. Steel stamping versus superplastic aluminum forming-influences of material cost and forming time on breakeven economics.

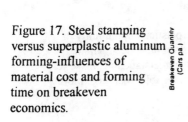
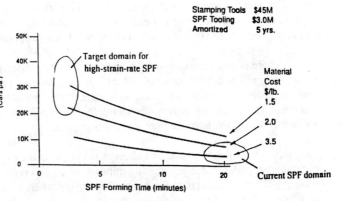

INNOVATION AND BREAKTHROUGH TECHNOLOGY

Innovation draws on a wide spectrum of scientific and technical endeavor, from esoteric 'blue sky's' research of academia through to ready made solutions like *"If it ain't broke don't fix it"* sometimes adopted by industry. The business impact of successful innovation can range from just surviving through to a transforming quantum leap associated with 'break-through' technology.

In the field of superplasticity and superplastic forming innovation, through R&D initiatives, is essential if a healthy growth rate is to be maintained in an otherwise maturing technology. Areas being, or needing to be, pursued include:

- **New SP aluminum alloys**, particularly in the 6000 series, which are more easily heat-treated.
- Better **forming lubricants** to achieve greater durability and class "A" finish.
- **Optimized forming schedules** for maximum productivity and minimum overall cost.
- Greater use of **automation and robotics** in the hostile environment of SPF.
- Better and easier to handle **computer modeling** for both thickness prediction at the design stage and process control.
- The realization of cost effective **fast forming alloys** and forming methods using the emerging high-strain-rate superplasticity technology.

Most of the above initiatives are related to productivity improvements but probably only fast forming via high-strain-rate, if successful, would represent a transforming "break through" taking SPF from its present specialized niche into the mainstream of competitive manufacturing. The technical challenges to be overcome are undoubtedly great but so are the potential rewards. The commercial significance of success is, in the author's opinion, sufficiently great to justify pursuing this goal. Education, training, technology and information transfer are also important elements in the future of SPF.

THE FUTURE PROSPECTS FOR SPF ALUMINUM

The outlook for SPF aluminum has never looked better with new applications coming on stream at an increasing pace. The remarkable formability of these materials combined with the low cost/short lead-time tooling continues to create unique opportunities in markets where the attributes of aluminum are needed and where low to medium quantities of complex shaped components are required.

There is an awareness and anticipation within the superplasticity community in regard to significant developments that may be on-going but as yet have been pursued under a shroud of secrecy. The need for secure development is understandable, however, it is the authors hope that those undertaking this work will soon be able to divulge more and acknowledge the heritage they are building upon.

Although there are many challenges ahead, with the expectation of greater technology transfer, more engineering improvements and process optimization, the prospect of 'smart' computer modeling allowing designers to integrate accurate thickness predictions alongside product-design and finally the possibility of achieving cost effective high volume production via fast forming 'high-strain-rate' materials technology, the future is as exciting as it was 25 years ago.

REFERENCES

[1] US Patent Application N° 445188 2nd April 1965.
[2] Davis S Fields, Jr. *The Promise of Superplasticity* IBM Document EN 200276 July 1972.
[3] R. Grimes, M.J. Stowell & B.M. Watts: Aluminium, 51, 720 (1975)
[4] N.E Paton & C.H. Hamilton ed AIME pp 273-289 (1982)
[5] K. Osada & H. Yoshida Materials Science Forum Vols. 170-172 (1994) pp 715-724
[6] G.J. Mahon, D Warrington, R. Grimes & R.G. Butler. Materials Science ForumVols. 170-172 (1994)
[7] G.A. Mahon: Alcan Tut. Report No GT A04-90 (1994)
[8] A.J. Barnes, J.B. Cattanach: *Advances in Thermoplastic Composite Fabrication Technology*, Proceeding 2nd Conf. in Materials Engineering. IN Eng., London. (1985)
[9] D.B Laycock: *Superplastic Forming of Structural Alloys*, N.E. Paton and C.H. Hamilton ed. AIME pp 257-271. (1982).
[10] A.J. Barnes *Design Optimization of Superplasticity* TMS/AIME Conf. San Diego (1982)
[11] A.J. Barnes, Mater. Sci. Fourm 170-172 pp 701-714. (1994)

The Current State-of-the-Art and the Future in Airframe Manufacturing Using Superplastic Forming Technologies

Daniel G. Sanders

The Boeing Company, PO Box 3707, M/S 5K-63, Seattle WA 98124-2207, USA

ABSTRACT

Superplasticity and the evolution of the Superplastic Forming (SPF) manufacturing process is a classic aerospace development case study, having come from the materials test laboratories to the aircraft production lines in a relatively short period of time. The purpose of this paper is to take a quick look back at the history of the SPF process at Boeing, the current state-of-the-art in SPF design and a glimpse into the future direction that superplasticity is expected to take us in over the next five to ten years.

Early testing of Superplastic materials was performed by the academic engineering scientists, since SPF was generally considered to be a laboratory oddity. Persistence within the commercial aerospace metal forming community, material science universities and the aerospace research & development groups around the world have now taken SPF to the forefront of sheet metal processing. SPF is now considered as a standard process for aerospace design and is emerging, with general uses now being found in rail, architecture, maritime and automotive designs.

SPF parts are now used on virtually all of the commercial and military airframes made by Boeing. Many different superplastic alloys are commercially available, including aluminum, CRES, titanium and aluminum-lithium.

The primary advantages of SPF over conventional design are:
- Freedom of design – the ability to produce complex shapes.
- The ability to build large assemblies with fewer pieces. The accompanying reduction in tool families, assembly jigs, welding, fastening, paperwork and schedule tracking.
- Inventory reduction, through-put improvements and cycle time reduction.
- Performance advantages with streamlined designs and fewer joints.
- Cost and weight savings.

New innovations have been devised to build SPF parts faster-better-cheaper. These advances have been made possible by concurrent advances in many other enabling technologies. As we move further into the twenty-first century, the teaming and partnering of government institutions, industry and the universities is vital. To further the mission of *working together*, Boeing advocates the formation of a "U.S. Committee for the Advancement of Superplasticity".

INTRODUCTION TO SPF AND MONOLITHIC STRUCTURE

Superplasticity is the ability of certain metal alloys and other materials to undergo very large tensile strains with little necking. In addition to being limited to only certain alloys, superplasticity requires elevated temperatures and the application of controlled strain rates. Superplastic Forming (SPF) refers to the manufacturing process that takes advantage of superplasticity to form complex and highly contoured sheet metal parts. With SPF, metals can be stretched to elongations that are more than ten times the limit that was previously possible.

During the 1970's, several companies involved in metals processing, forming and aerospace production invested research and development resources into the development of a manufacturing process to take advantage of the superplasticity phenomenon. These companies took the process from the stage of being a laboratory oddity to the point where high quality airplane parts could be mass produced. Aircraft programs such as the B-1 Bomber drove the development of SPF technologies in the United States.

- **Process is similar to vacuum forming of plastics**

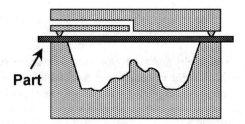

- **Computer controlled gas pressure forms the part into the cavity at a constant strain rate**

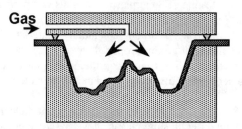

- **Elevated temperature**

- **Fine, equi-axed grain**

An illustration of the basic Superplastic Forming process using a simple tool and gas pressure to form a complex shape of metal part. Many more complex variations of this simple process are now in use.

Simply having the ability to stretch metal alloys with large elongations compared to ordinary materials would have been a valuable advantage for metal forming processes. The process could have evlolved into a method of making existing part shapes with a more economical system. SPF for example, could have simply been a replacement for the antiquated drop hammer forming process that was prevalent in aerospace manufacturing for the previous 50 years.

However, the aerospace industry has now taken advantage of the SPF forming capability to design and build parts that would otherwise be impossible to produce. SPF has been developed as a manufacturing technology and adopted by the Boeing Company to build monolithic structures for commercial and military airframes. The concept of monolithic structure is to eliminate detail parts and assembly by reducing the number of pieces and fasteners that make up a component.

The model 737 Access and Blowout Door was the first part made by Boeing to take advantage of monolithic SPF structure design methodologies.

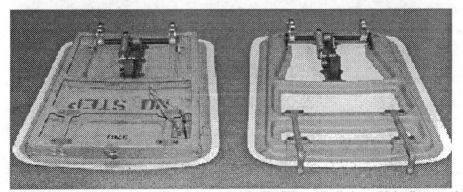

The SPF unit replaced the built up assembly shown at the left. The SPF version eliminates 25 detail parts and 137 rivets. Over 500 of these units are in service. A similar design was implemented on the model Boeing 777.

As the SPF process evolved and the equipment used to make parts became more sophisticated, the aerospace industry capabilities expanded to make larger parts. Since most sheet metal designs used on aircraft are designed for stiffness, the ability to form integrally stiffened panels became a desirable attribute. Aerospace panels could now be built which utilized formed-in stiffening beads and sections. These panels were lighter and less expensive to manufacture than the built-up assemblies which they replaced.

The benefits of part consolidation go way beyond the recurring costs of building and assembling large aerospace structures with multiple pieces. Fewer tools are required to build and assemble the parts. Less tracking paperwork and planning is needed. In summary, there is an accompanying reduction in the initial start-up costs for making SPF parts.

Although the process of forming with SPF is, in itself, an inherently expensive method, the leverage of eliminating detail parts is far greater in magnitude. At Boeing, SPF is now considered as a "baseline design" option for many large assemblies. Many examples of this technology have been proven successful.

The two 6Al-4V titanium formed panels in the front replaced the thirty-two Inconel pieces shown assembled (behind) in the CFM-56 engine Thrust Reverser Assembly. A weight savings of 27 pounds per Boeing 737 aircraft has been realized. The previous used two large Floor Mounted Assembly Jigs (tools) to assemble the pieces together. The new design is made to precision tolerances, which allows it to be assembled without large assembly tooling. Cost was reduced by 40%. Over 2200 of these units are in service. A similar design was implemented on the Boeing 777 and is being considered as a baseline for future designs.

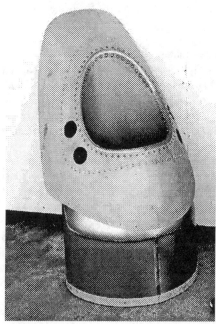
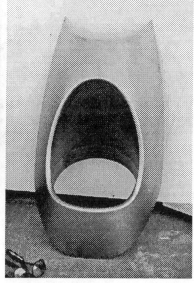

The design shown above on the left is the baseline for the Auxiliary Power Unit Exhaust Fairing that was designed in 1965. The design on the right is the SPF design used for the Boeing Next Generation 737. The old design consisted of 19 detail parts, 126 rivets, 6 weldlines and required a great deal of rework to bring within the required dimensional tolerances. The SPF design consists of 2 parts and one weldline. There are no fasteners for the basic part. The SPF part uses 37 fewer tools to fabricate, 21 hours less labor to install and 38 hours less fabrication time. The SPF part is some 4.7 pounds less in weight.

ACADEMIC AND INDUSTRY TIES

There has been a virtual explosion in academic research applied towards the material science of understanding superplasticity and the advancement of superplastic materials. On the other hand, the comparative use of SPF by industry to produce manufactured goods has been sluggish. The majority of SPF production remain in the aerospace, rail and architectural realms. SPF manufacturing has not yet significantly spilled over into other manufacturing arenas such as household goods. The purpose of this paper is to try and understand why this discrepancy between what seems to be a boom in SPF academic research and a comparatively small growth in SPF industry has occurred.

However, although these economic conditions appear quite commonly in academic literature as being the simple answer to explain the lack of substantial SPF growth in industry, the root cause of the inability to wholesale commercialize SPF really lies in the numerous technical challenges that manufacturers of SPF components have not yet been able to overcome. Simply stated, the economic disadvantage of SPF compared to other metal forming methods lies in the need for state-of-the-art development of the SPF process, not just in the refinement of new or better materials. SPF Material Science, in and of itself, cannot and will not advance SPF into the mainstream of sheet metal fabrication for the world. There is far too much of our worldwide academic resource being used to study the field of SPF materials and the superplastic deformation mechanics. We must now redirect research and begin studying the SPF process as a manufacturing science and make the study of the process issues an equal priority to new SPF materials development. Industry needs some help catching up to the current level of accomplishments made by academia.

SPF metallurgical development activities have focused primarily on exotic new hybrid alloys that are of little value to industry at this time. These alloys show promise for the future (10-20 years) for higher temperature applications or in designs such as engine fan blades where the performance advantage of fiber reinforcement or powder metal alloys can justify their cost. In the interim, the high cost of 6Al-4V titanium sheet metal is a limiting factor for implementing SPF designs in many areas that would otherwise make sense. The basic titanium manufacturing process was devised in the mid 1950's and has not been significantly improved since. The aerospace applications for sheet titanium assemblies could easily be tripled if the cost of raw sheet could be reduced by 33%.

The development of robust manufacturing processes to take advantage of SPF have so far been a low priority for academic research. Very little practical manufacturing development has occurred at the universities. Industry has been left on its own to create production methods. The disconnect between manufacturing and academic experts is unacceptable if we are to take SPF fabrication to its maximum potential use on manufactured goods of all kinds.

The single biggest drawback to SPF, when compared to multiple piece assemblies or alternative forming processes, is the large non-recurring costs, especially tooling. The second largest issue is the prohibitive cost of 6Al-4V titanium sheet. The lead time of producing an initial SPF part is unacceptable. The hot presses and equipment used for SPF are high cost and prone to failure, so few subcontractors have SPF capabilities. If these roadblocks cannot be overcome, there will be little justification to continue with the current level of study of the SPF sciences at the universities.

CONCLUSION

In the SPF Research and Development profession, our horizons should be expanded to include the peripheral process technologies and materials that support SPF manufacturing that have not yet been studied. Better communications, tighter bonds and closer ties between industry and the universities are needed so that we can all work together towards our common goals.

Generic Ceramic Tooling (GenCerT) for the SPF/DB Process

Alan Jocelyn, Terry Flower and Doug Nash

Aerospace Manufacturing Res. Centre, Fac. of Engineering, University of the West of England, DuPont Building, Coldharbour Lane, Bristol BS16 1QY, UK

Keywords: Ceramic, Diffusion Bonding, Low-cost, Novel, Quality, Superplastic, Titanium

Summary
This paper describes the current research being conducted by members of the Aerospace Manufacturing Research Centre (AMRC) at the University of the West of England, Bristol, UK, into improving the Superplastic Forming and Diffusion Bonding (SPF/DB) process. The Centre is primarily interested in novel ways of improving the technology of the process much of which, although now well refined, has changed, conceptually, little in the last twenty-five years. Thus the Centre is conducting research into a new tooling regime the overall objective of which is to reduce the cost of SPF/DB tooling by 50%. This is considered achievable by abandoning the prevailing concept of extending the life of metallic tools, using instead low cost, disposable ceramic liners, containing all component geometry and located within a generic metal support device.

Introduction
Ever since Professor C E Pearson recorded the phenomenon of superplasticity in the 1930s [1], man has sought to harness the discovery within a robust and cost effective technique that industry could use with confidence. In this respect, the process of Superplastic Forming, often combined with diffusion bonding (SPF/DB), has been extensively investigated and the results employed by the automobile, architectural, marine and aerospace industries.

For the aerospace industry, the SPF/DB process has a particular appeal for two reasons. Firstly, the need for highly accurate, complex shapes demanded by aircraft designers can be achieved with relative ease using the process. Secondly, many of the materials used in the industry possess the necessary metallurgical superplastic characteristics, for example 6Al 4V titanium alloy.

Titanium alloys have been used in airframe manufacture for over three decades. They give high strength, lightweight components with exceptional corrosion resistance and enduring fatigue life. Although SPF/DB is now reasonably well established for making complex titanium parts, its use has, nevertheless, been restricted because of difficulties encountered in the manufacturing process. For example, the stainless steel tooling used currently for SPF/DB is expensive, geometrically unstable and a source of component contamination, all of which significantly restrict the widespread use of the process.

Aims and objectives
The prime objective of AMRC's research has been to reduce the high cost of SPF/DB tooling by developing a technique using thin section 'tooling liners' made from suitable ceramic

material. Essentially, expensive bespoke stainless steel tools (Fig 1) would be replaced by generic carriers, or bolsters, into which would fit a purpose-made ceramic liner containing all of the component's geometry (Fig 2).

Characteristics of stainless steel
- High cost.
- Long procurement lead times.
- Cannot be proven as "right first time" until a component is produced.
- Difficult to predict lifecycle.
- Usually difficult, and often impossible, to repair.
- Source of component contamination.
- Progressive, but difficult to predict, deterioration of geometric accuracy from thermal creep and process forming pressures.
- Expensive to maintain.

Fig 1. Conventional SPF/DB tool

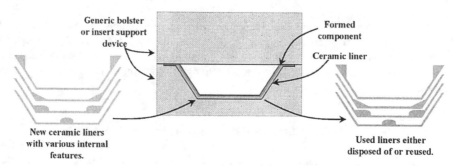

Fig 2. New tooling regimen

The core challenge was to develop a low cost, easily produced ceramic liner, the intrinsic strength of which was sufficient to produce the component whilst enduring the massive loads sustained during the manufacturing process. An additional aim was to find a ceramic that was inert in character, thus eliminating cross-contamination from tool to component and making unnecessary the various coating processes currently employed to prevent contamination. The price of such a liner had to be kept low so that it could be discarded after every production cycle, in marked contrast to prevailing research that aims to extend the life of stainless steel tooling. In order to achieve the 50% cost reduction target set at the beginning of the work, 'success criteria' were established which reflected five major issues considered responsible for the high cost of conventional stainless steel SPF/DB moulds:

1) Reduction in cost of SPF/DB tooling by 30% of current price.
2) Reduction in SPF/DB tooling procurement time by 85% (less than 6 weeks).
3) Increase in SPF/DB press utilisation by 100%.
4) Increased bolster flexibility by 50% improvement.
5) Reduction in tool preparation man-hours by 80%.

Methods
A comprehensive study of ceramics, their uses and methods of casting was undertaken. Properties sought for liners were:

1) Resistance to an internal pressure of at least 450 psi (3.06 MPa) and temperature of 937°C.
2) Readibly castable.
3) Inert in character (no cross-contamination); no stiction, excellent surface finish.
4) Predictable (low) shrinkage during manufacture.
5) Low material cost.
6) Low production cost
7) Re-usable, if possible.

A series of experiments was conducted to select the most appropriate ceramic for SPF/DB application. Test piece tooling was produced and small, 125mm diameter, simple components formed which were tested and evaluated. Subsequently, test components with inward projecting features, based on airframe and gas turbine components, were produced (Fig 3).

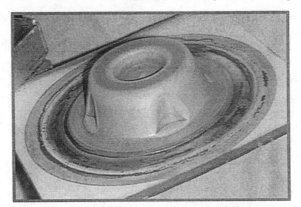

Fig 3. Titanium 6Al4V demonstrator component

Choice of ceramic and production method
Viscous Plastic Process (VPP)
Keele University selected Alumina based feedstock: CT9 + Secar 71(binder) + PVA (Bonding agent) + Glycerol + De-mineralised water which, when processed and fired, would have the potential, it was predicted, to withstand the rigours of the SPF/DB process whilst producing a component of required quality. Furthermore, to produce a liner of maximum strength and integrity, processing of the Alumina 'dough' using the Viscous Plastic Process (VPP) was selected (developed by CERAM Research at Stoke on Trent).

The VPP process eliminates micro-structural defects, the majority of which are caused by agglomeration that occurs spontaneously in fine ceramic powders, by carefully controlling the mixing conditions. In essence, the wetted ceramic powders are mixed into a viscous polymer solution and subjected to very high shear conditions, in this case achieved by feeding the dough through a rolling mill (Fig 4).

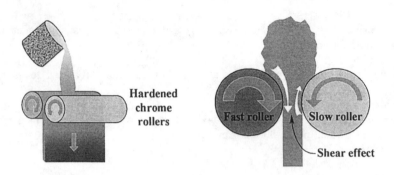

Fig 4. Principle of Viscous Plastic Process for ceramic material

The rolling produced a homogeneous uniform sheet of ceramic dough approximately 1.8 mm thick. Layers of such material were subsequently placed into a mould, cured and then fired.

In addition, the inert character of the Alumina was theoretically analysed recognising, in particular, the highly reactive nature of titanium at high temperatures. The thermodynamic ($\Delta G^\theta - T$) diagram for the formation of oxides was used which indicated that the oxidation of titanium at SPF temperatures (~ 1000° C) would entail the following reactions:

$$\begin{aligned}
Ti + O_2 &= TiO_2 \, ; & \Delta G^\theta &\cong -750 \text{ KJ} \\
{}^2/_3 Al_2 O_3 &= {}^4/_3 Al + O_2 \, ; & \Delta G^\theta &\cong 900 \text{ KJ} \\
\hline
\therefore Ti + {}^2/_3 Al_2 O_3 &= TiO_2 + {}^4/_3 Al \, ; & \Delta G^\theta &\cong +150 \text{ KJ}
\end{aligned}$$

This large positive free energy change implies that spontaneous oxidation (contamination) of Ti would not be expected from the ceramic liner under SPF conditions. However, since the theoretical thermodynamic situation could be altered by the presence of impurities in the ceramic material, it was decided that stiction and contamination tests should be carried out under pressure and temperature conditions similar to those encountered in the SPF process.

Experiments at high temperature and pressures were carried out on small samples of fired ceramic in a controlled inert atmosphere furnace to assess contamination and stiction characteristics. The tests demonstrated that the chosen ceramic neither stuck to, nor contaminated, titanium at high temperature and pressure.

The conclusion, from both theoretical and experimental evaluations of the chemical compatibility of the ceramic material and titanium under SPF conditions, was that the chosen ceramic would satisfy the "no stiction and no contamination" objective of the GenCerT programme. However, subsequent pressure testing showed the strength of liners made using the VPP process to be low and the quality inconsistent, especially at the interface between layers of material.

Freeze casting process (FCP)
In this process various grades of wetted Alumina are injected into a tool cavity and rapidly cooled to -40°C using liquid nitrogen. Both the inner and outer faces of the ceramic freeze first, producing a fine hard shell supported by the mould surfaces (Fig 5). The remaining water within the ceramic mix then freezes, its subsequent volumetric growth being constrained by the frozen shell and mould. As a consequence, substantial internal forces are

generated resulting in high strength atomic bonding between the Alumina particles. Liners made in this way were consequently stronger than those produced by the VPP method.

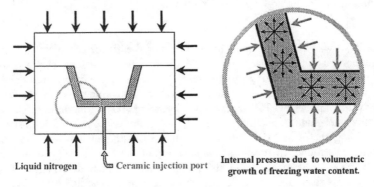

Fig 5. Principles of Freeze Casting Process for ceramic materials

Although the VPP method produced satisfactory properties 3, 4, and 5 (page 3), the liners' low strength and complex manufacturing process and, therefore, high cost, disqualified the process. In marked contrast, the FCP fulfilled all the criteria of properties required of a ceramic liner including satisfactory strength (Fig 6).

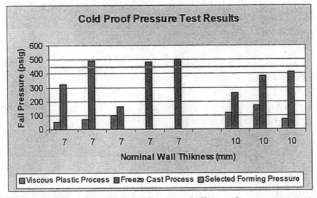

Fig 6. Comparison of VPP and FCP ceramic liners when pressure tested

Test methods
When the various coefficients of expansion of the materials used in the SPF/DB process (titanium/ceramic/steel) were considered, it became clear that the steel bolster would expand away from the ceramic liner leaving it unsupported, whilst pressurised titanium would force the liner into tension. It is well understood that ceramics are weakest when in tension and so some form of external support for the liners was felt to be necessary. Various methods were considered such as gravity fed ceramic powder to fill the gap between the liner and bolster at elevated temperature and externally tapered liners that slid downwards filling the expansion gap. Such liners would have outer fins, the delicate tips of which would crush thus fitting into any geometric inaccuracies in the bolster and preventing the liner sticking when returned to ambient temperature. However, the most realistic solution was felt to be the use of high-pressure argon gas injected progressively into the expansion cavity, thereby offsetting the internal SPF pressure.

However, to establish the level of external support required, it was necessary to know how strong an unsupported liner was. Consequently, a cold proof rig was built in which liners could be pressurised using argon gas to preserve their state of cleanliness.

Results

Unexpectedly, it was found that liners produced using the 'freeze casting' process required no external support so that, in future, the steel bolster may be eliminated. This was surprising as the liners were pressurised during forming to 450 psi causing inevitable high hoop stresses at the top rim, especially as ceramics are at their weakest when under tension. Nevertheless, the liners survived a number of SPF/DB cycles without any apparent harmful effect to either the component or liner. Indeed, one liner survived two cold pressure tests of 500 psi and a further four hot forming cycles of 345, 425, 450 and 463 psi. Although no contamination from the ceramic liner was found, there was some 'sticking' of the test-pieces to liner. Furthermore, geometric accuracy was retained and no thermal creep nor liner deformation was recorded. Thus, components constrained by smaller tolerance envelopes should be possible.

Costs of the proposed GenCerT process have been analysed comparing the liner system with the conventional SPF/DB process of hot platen press production. Capital plant utilisation, effects of tool heating rates, forming cycles, cooling, tool build and breakdown times, cost of tool cleaning and preparation, tool maintenance and component selection and quality have been used in the analysis, the results of which indicate that substantial cost reduction is possible. Furthermore, the ceramic liner production costs used have been from a single source supplier with modest facilities. It is, therefore, quite possible that a cheaper provider could be found which would further reduce ceramic liner prices.

Conclusion

SPF/DB is often restricted to the manufacture of specialist items because of the high cost of the process as currently performed. AMRC is committed to reducing this cost through research, the GenCerT project being an initial, but crucial, indication of that commitment. The improvements in component accuracy and reproducibility, together with the possibility of reducing, or even eliminating surface contamination, will produce components with improved performance at an even lower cost, the precise financial analysis of which was beyond the scope of the project. However, the most significant aspect of the research is the realisation that low cost, disposable ceramic liners may be used without a support bolster. Furthermore, if 'optical windows' are provided within the liner, this could lead to novel ways of heating the unformed blank and thus replace the hot platen press. In conclusion, ceramic liners could be the basis of a new, low cost, high quality, energy efficient and flexible process and a major step to realising the full potential of superplasticity.

Funding

Engineering and Physical Sciences Research Council (EPSRC) UK.
British Aerospace, Airbus, plc.
Rolls-Royce, Civil Engines, plc.

Reference

(1) RIDLEY N, 1995, "C.E. Pearson and his Observations of Superplasticity", Superplasticity: 60 Years after Pearson (1995), The Institute of Materials, ISBN 0 901716 77 4, pp. 1-5.

The Design and Manufacture of Superplastic Forming/Diffusion Bonding Presses

Roy Whittingham

Motherwell Bridge Material Handling Systems Ltd,
Rosemount Works, Huddersfield Road, Elland, West Yorkshire HX5 0EE, UK

Keywords: Aluminium Products, Diffusion Bonding, Press Design for Production, Superplastic Forming, Titanium Products

ABSTRACT

INTRODUCTION

The superplastic forming/diffusion bonding (SPF/DB) process is a method of forming alloys in sheet form which have superplastic properties into a wide variety of shapes and hollow sections.

This process offers huge benefits and brings a totally new range of possibilities into the manufacturing arena particularly in the aerospace industry and more recently the automobile industry where aluminium body panels and fitments are replacing those which have conventionally been made in steel and other materials. To facilitate superplastic behaviour in a material it is necessary to prescribe and control the regimes of strain, strain rate and temperature and for ideal superplastic behaviour these maybe rather narrowly defined. The fact that many engineering components have irregular shapes it is also extremely difficult to maintain uniform strain rate across its section. The process therefore calls for a high level of control accuracy as well as flexibility to prevent the thinning of the material and possible rupture during the forming cycle.

Motherwell Bridge Material Handling Systems Ltd (M.B.M.H.S.) have more than 25 years experience in the manufacture of machines for this purpose for use with both alloys of aluminium and titanium. Over the years a wide variety of machines have been made – each to meet a clients needs in terms of power, platen area, stroke and daylight etc. Nearly all of these machines have been fitted with computer based control systems utilising 'state of the art' equipment and technology to achieve the precise control requirements of the process.

EXPERIMENT

M.B.M.H.S. have recently completed a study of the SPF/DB process with a view to creating a new standard form of machine which will incorporate our existing well proven control systems but adaptable to meet our clients needs in terms of power and dimensional requirements. The advantages and disadvantages of either upstroking and downstroking machines have been considered and we have concluded that a 'Pull Down' type machine, see fig. 1, offers the greatest number of advantages for the following reasons:-

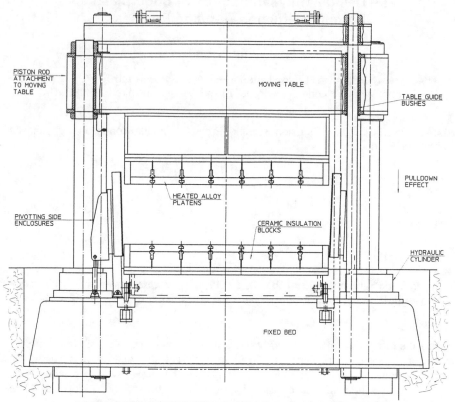

FRONT VIEW OF 'PULL DOWN' TYPE SPF/DB PRESS
FIG. 1

- The tool assembly Upper and Lower sections should be accurately aligned within the machine.

- The tool assembly should be adequately supported with minimum deflection.

- The platen heating system should ensure a uniform tool temperature.

- Tool clamping pressure should vary according to the increase of the gas forming pressure to protect the tool seal and be variable over a wide range.

- The gas forming medium should be finely controlled throughout the forming cycle to avoid the possibility of thinning and component rupture.

- The forming cycle should be fully automatic and without manual inputs. The provision of adequate process data and records is essential.

- The overall height of the machine and floor space occupied should be minimal.

- Fire hazards should be minimised or eliminated.

- The machine incorporates feed/shuttle tables for loading tools etc when required to either the front or rear faces of the machine or both.

RESULTS

MBMHS have now designed and manufactured a machine of this type which has been extremely successful over a period of time. The features we regard as being important for the process have been achieved as follows :-

- The tool assembly is accurately aligned and guided from the external guide system on the downstroking moving slide which locates on columns which link the head & bed sections of the press frame in the cold zone.
- The upper and lower sections of the tools are mounted by various methods (see fig. 3) on the moving slide and press bed respectively, on the platens, insulation blocks and cooling platens. The main load bearing section are the moving slide and bed which are each fabricated from heavy gauge mild steel plates designed to allow only a minimum deflection.

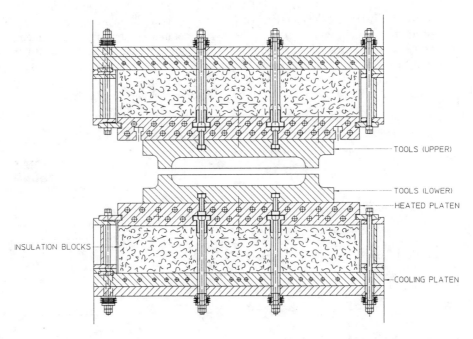

TYPICAL TOOL ASSEMBLY AND PLATEN LAYOUT
FIG. 3

- To ensure that tool assemblies of various shapes and thicknesses are heated to a uniform temperature, the platens are divided into a number of separate temperature control zones. This is achieved by using a multi zone cartridge type heaters in each section of the platen. Typically a 1.5 metre square platen would have 9 control zones, a larger 3 metre by 2 metre platen could have 21 zones. This arrangement enables varying degrees of heat energy to be channelled into each zone according to the thickness and shape of the tools so that tool temperatures can be maintained to within acceptable tolerances, i.e. for titanium 920°C $\pm$ 10°C.

- To ensure that the tool seal does not become damaged it should not be over pressurised. The tool clamping pressure is controlled to a force which is approximately 10% greater than the total tool bursting force being developed within the tool by the gas forming medium. This is achieved by using a motor driven pressure relief valve controlled from a pressure transducer in the gas input box to the tools. This device is also adapted to lock in pressure at the tools in the event of power failure for safety reasons and also to protect the tools and platens against over pressurisation.

- Up to three gas forming lines can be provided to work individually or simultaneously – each line is based on the use of PID Tescom controller with regulator valve sized according to flow and gas pressure requirements. The gas pressure is controlled by preconceived pressure time diagrams using up to 100 contact points during each forming cycle.

- The central computer system is an Allen Bradley type SLC 504 series which is used to manage all the control functions of the press in both automatic production and manual modes, i.e. tool setting. The system embraces the press movements, pressurisation, gas management and platen temperature. Cycle when established are identified by number and stored for future use. A SCADA package provides a display of the press functions typically as shown in fig. 4. Product Data is also provided which may be used as a quality record for each component produced.

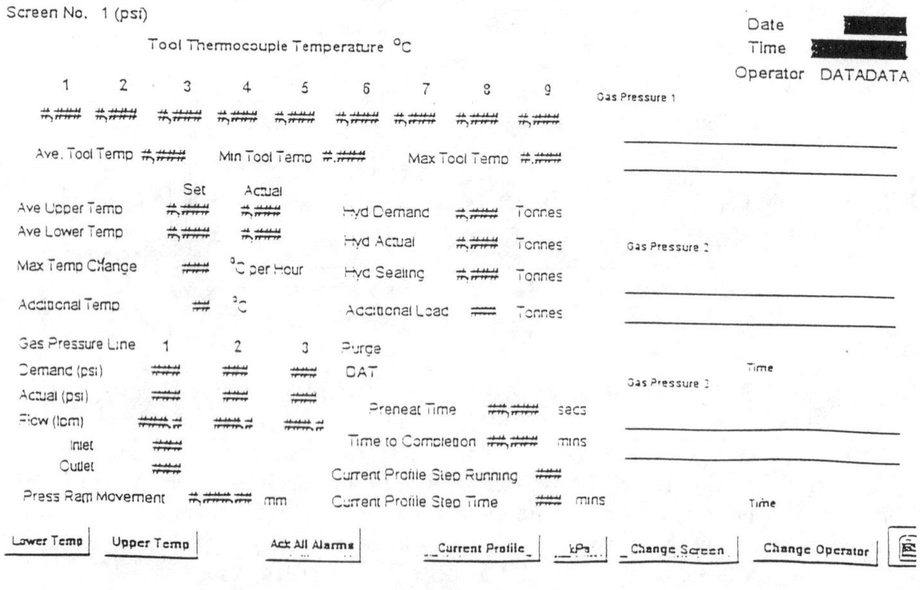

TYPICAL SCREEN FORMAT SHOWING CONTROL SETTINGS
FIG. 4

- The pull down machine is very compact to allow in use the of existing overhead cranes to facilitate tool loading onto the shuttle tables. All ancillary equipment is mounted on a steel gantry positioned on ground level alongside the machine.

- With the 'Pull Down' machine the fire hazard is minimised as all the hydraulic pulldown cylinders are mounted below the level of the hot zone which is enclosed by well insulated guard/enclosure panels.

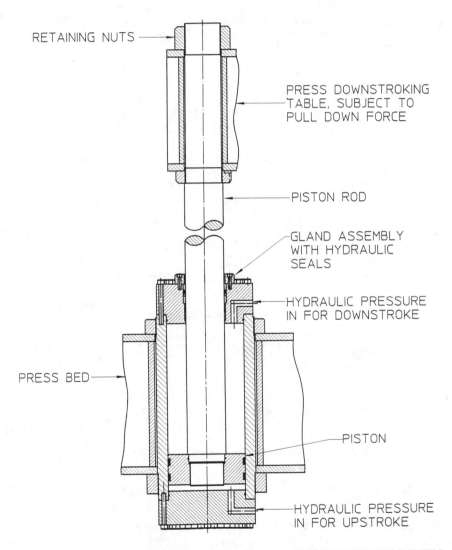

CROSS SECTION THROUGH THE 'PULL DOWN' MAIN CYLINDERS MOUNTED IN THE PRESS BED
FIG. 2

- The machine can be fitted with feed/shuttle tables to either the front or rear of the machine to load/unload the tools or load/unload the component. The shuttle tables are driven through a lead screw mechanism and powered by an hydraulic motor, see fig. 5.

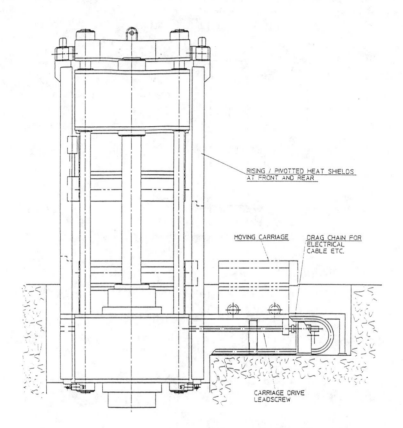

SIDE VIEW OF PRESS SHOWING A SHUTTLE TABLE
FIG. 5

SUMMARY

MBMHS Ltd believe that this style of machine has considerable advantages over others in terms of :
- Access to load/unload machine.
- All hydraulic actuators are below hot zone.
- Ease of assembly.
- Simplified hydraulic system using mineral oil medium.

This design concept will be used on our future machines.

Superplastic Injection Forming of Magnesium Alloys

Tatsuhiko Aizawa[1] and M. Mabuchi[2]

[1] Research Center for Advanced Science and Technology (RCAST), University of Tokyo,
4-6-1 Komaba, Meguro-ward, Tokyo 153-8903, Japan

[2] National Industrial Research Institute of Nagoya,
1-1 Hirate-cho, Kita-ward, Nagoya 462-8510, Japan

Keywords: AZ91, Computer Control, Die-Set Selection, Injection Forming, Magnesium, Near-Net Shaping, Stoke Control, Superplasticity

Abstract: Intrinsic difficulty of metal forming in magnesium alloys must be overcome by new processing. Superplatic injection forming is one of the most promising methodologies. Stroke velocity controlled forming system was developed to evaluate the formability of AZ91 at 573 K both in upsetting and backward-extrusion modes. Uniaxial constitutive equation was used together with consideration of grain growth to predict the stress-strain rate and the stress-stroke relations in upsetting experiments. Double flange, thin-walled cup can be superplastically formed from a cylindrical billet.

Introduction

Magnesium has the highest specific strength and the second highest specific stiffness among various metallic alloys. In addition, large damping capability and electro-magnetic shielding capacity can afford to enhance the practical importance of magnesium alloys [1]. In order to link these intrinsic merits of magnesium alloys with the market demand, high and flexible production technology must be equipped with the IT (Information Technology) - based design tools [2]. Blow forming and mechanical forging are typical superplastic forming methods to realize near-net shaping. Magnesium alloys requires more flexible forming method to fabricate complex-shaped, thin-walled parts even from a bulk preform, or to produce the net-shaped parts from a starting billet through transfer forming. Authors [3-5] have proposed the superplastic injection forming from the starting billet having refined microstructure through hot-extrusion or bulk mechanical alloying. In particular, dynamic recrystallization of magnesium alloys propels refining of microstructure in hot-extrusion. Using the stroke-velocity controlled apparatus, the cylindrical billet can be formed into a double-flange, thin-walled cup by the present method. In the present paper, our superplastic injection forming developed is briefly introduced with some comments on the developing apparatus. In experiments, both upsetting and back-extrusion modes are utilized to evaluate the formability of AZ91 billets. The uniaxial constitutive equation with the grain growth model is used to predict the upsetting behavior. Through discussion over the microstructure control of magnesium via bulk mechanical alloying (BMA), the effect of microstructure refining on the superplastic formability is considered.

Superplastic Injection Forming

In the conventional blow forming, the geometry and dimension of a die is imprinted to the superplastic material plate [6]. Although complex shaping can be done for various panel members, solid parts cannot be formed to shell members. In the superplastic forging, the billet sample is pressed into a closed die-set cavity [7], so that thin-walled parts cannot be forged with sufficient accuracy. Since many targeting parts and members are still difficult to be formed, alternative superplasic forming is necessary to make new ways. The injection molding has succeeded in forming of complex shaped parts for thermoplastic polymers or powder-binder compounds. Due to their sufficient flowability, the closed die-set cavity can be filled and pressurized during the prescribed duration to yield a near-net shaped part. In those cases, the flowability is proved by relatively low shear stress and low viscosity. In the superplasticity, low normal stress is observed all through forming, and sufficient ductility is proved in the relatively wide strain rate range. This intrinsic nature of superplasticity is just adaptive to the injection forming.

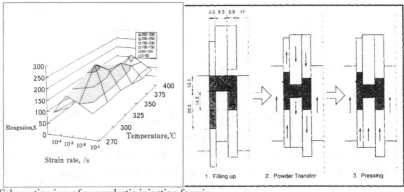

Figure 1: Schematic view of superplastic injection forming.

Figure 1 depicts a concept of superplastic injection forming. The initial preform is subjected to pressing in part and transferred to fill the prescribed die cavity. Different from the conventional forging, the punch and die position is digitally controlled to deform and transfer the materials within the limit of stress, deformation and strain rate for superplasticity. Otherwise, even superplastic materials must suffer form local damage or failure. Key technologies in this superplastic injection forming are consisted in three items. First, the adaptive pass schedule must be designed to fill the final cavity by materials within the duration where no significant grain growth takes place. In second, the digital process control is constructed to feedback the measured strain rate and deformation for process optimization. Finally, preform after preparation must have sufficiently refined

Figure 2: Developing apparatus for superplastic injection forming.

microstructure to reduce the temperature and to widthen the strain rate range. As the first step toward the above ideal processing, stroke-controlled forming apparatus was built to simulate the filling behavior of superplasic preform materials into the designed die cavity. As shown in Figure 2, the whole die set is housed in the inside of furnace with thermal controller. The holding temperature can be controlled within the maximum deviation of 2 K up to 1173 K. The applied load range can be varied in three steps by batch: in the following experiment, maximum pressure is limited by 3 tons. Motion of punches can be controlled by the stroke pass schedule: the stroke velocity is kept constant in the present experiments. Nominal strain rate can be varied from 10^{-6}/s to 10^{1} /s.

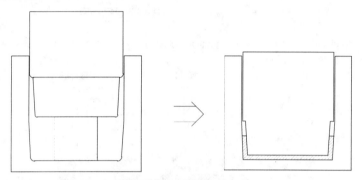

Figure 3: Typical die-set design for superplastic injection forming.

In order to simulate the filling process in the die-punch cavity, two forming modes were selected: upsetting and backward-extrusion modes. As shown in Figure 3, dies and punches were made of autenitic stainless steel of type AIMI 316. In upsetting mode, the stress state becomes near uniaxial compression, so that the measured loading history can be converted to normal stress history. In backward extrusion mode, the compressed materials are further injected through the thin channel between die and punch.

Experimental Results
Two injection modes were selected in this paper to describe the superplastic flow behavior during the injection forming: upsetting mode and back extrusion modes. In the former, the uniaxial constitutive equation was used to predict the flow stress vs. strain rate relation and the flow stress vs. stroke relation with consideration of the grain growth during forming. In the latter, the short-shot samples were compared in series to make evaluation of formability. The hot-extruded AZ91 (Mg-9Al-1Zn-0.2Mn) cylindrical billets were commonly utilized in the following experiments: billet diameter is 18 mm and its height, 20 mm. The holding temperature was varied from 573 K to 727 K. The stroke velocity was employed as a process parameter, and widely varied to control the nominal strain rate. BN powders were only used for lubrication.

<u>Upsetting Mode</u>
The cylindrical billet was uniaxially pressed to investigate the superplastic formability. Table 1 summarizes the experimental conditions and results for upsetting at 573 K. Even when the stroke velocity was so slow as 8.33 x 10^{-4} /s, the grain size was suppressed by 10 μm. This slow grain growth at 573 K is compared to fast grain growth at 673 K. Figure 4 shows the upset specimens at 573 K. Uniform upsetting took place up to the large reduction of 80 to 90 % in thickness. The uniaxial superplastic deformation must be abided by the modified Dorn equation as follows,

$$\varepsilon = [AGb/kT] \, (b/D)^{p} \, (\sigma/G)^{n} \, D_{0} \exp(-Q/RT) . \tag{1}$$

Here, D is the grain size, ε the strain rate, σ the applied stress, T the holding temperature, G the shear modulus, b the burgers vector, Q the activation energy, R the gas constant, and A, D_0 the material constants. P denotes for the exponent for grain growth, n is the stress exponent. Under the constant temperature, this eq. (1) can be simplified into eq. (2):

$$\varepsilon = K D^p \sigma^n \ . \qquad (2)$$

Here K is the materials constant, to which the whole parameters in eq. (1) are reduced. From the uniaxial tensile testing, p = 3 and n = 2 for AZ91. The grain growth is assumed to be in exponential evolution from the initial grain size (D_i) to the final grain size (D_f), where both D_i and D_f were measured by the intersection method from SEM images. D_i was measured to be 3. 62 μm just after raising the temperature to 573 K.

Table 1 Experimental results in the upsetting experiment at 573 K.

Specimen Number	Stroke Velocity (mm/s)	Reduction of Thickness (%)	Average Grain Size D_f (μm)
#1	8. 33 x 10^{-4}	87	10
#2	2. 50 x 10^{-3}	84	8.2
#3	8. 33 x 10^{-3}	80	7.0
#4	2.80 x 10^{-2}	80	6.7

Figure 4: Deformed shapes of AZ91 cylindrical billet specimen by superplastic injection forming in the upsetting mode.

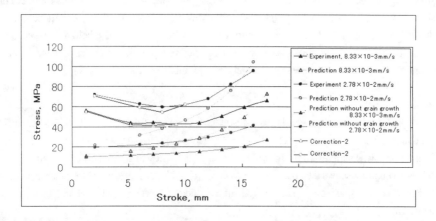

Figure 5: Comparison of theoretically predicted flow-stress vs. stroke relation with the experimental data.

The initial strain rate was defined as the stroke velocity over the preform height. Then, the stoke and stress histories can be calculated by Eqs. (1) and (2). Figure 5 compares the predicted and experimentally measured stress vs. stroke relations. In experiment, interfacial friction between the die/punch surfaces and the sample billet resulted in larger stress value. Hence, the measured stress becomes higher than the predicted value. With higher reduction, both results are in relatively good agreement. This assures that theoretical simulation should be adaptive to superplastic injection forming design.

Backward Extrusion Mode
In the backward extrusion mode, materials were first upset to thin plate with the thickness of 2 mm and driven to fill the cavity of first and second flanges in the successive manner. Figure 6 shows typical short-shot samples to the final product. The side wall was first constructed by back extrusion and two flanges were formed with the imprinted accuracy of die-punch set. From the measured stress vs. stroke history, the final stress state in the formation of two flanges became higher than that for backward extrusion. This might be because of the friction forces working on the die/punch surfaces and the materials in a small die cavity. Lubrication only by using BN at 573 K seemed to be insufficient to reduce friction stresses in practice. Low temperature superplastic flow is favored by injection forming with use of oxide-base lubricants.

Figure 6: Deformed shapes from a cylindrical billet by superplastic injection forming in the backward extrusion mode.

Discussion
The present superplastic injection forming strongly depends on the initial refined microstructure. Remaining the refined microstructure during injection forming, high toughness against plastic localization as well as low stress state can be made full use in the present approach. Reduction of holding temperature and enhancement of allowable strain rate is further necessary to suppress the grain growth and to extend the present research to industrial technology. Figure 7 depicts the refined billet after being subjected to 100 cyclic loading or 800 s duration by the bulk mechanical alloying when starting the magnesium chips. In addition to the fact that this billet has nearly full density, the grain size in the order of 100 nm can be attained only by 800 s processing at room temperature. A had been stressed in Ref. [8], since the magnesium alloy chips or granules with the diameter of 1mm or more can be employed as a starting material, the solid-recycles materials can be upgraded to a preform for superplastic injection forming. This adaptivity to recycling must be indispensable for positive

use of magnesium alloys in the commercial market. With consideration of efficient lubrication for magnesium alloys, the holding temperature might well be lowered to 523 K [9].

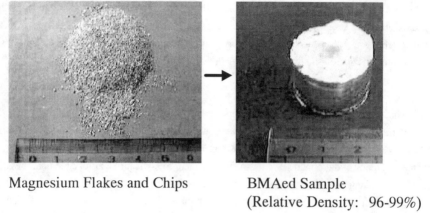

Magnesium Flakes and Chips   BMAed Sample
                                   (Relative Density: 96-99%)

Figure 7: The fabricated magnesium billet by the bulk mechanical alloying starting from the magnesium chips.

Conclusion

The superplastic injection forming was proposed as a new alternative to net-shaping of magnesium alloys from a cylindrical billet or preform. In the upsetting mode experiment, the online-measured stress vs. strain rate and stress vs. stroke relations were theoretically predicted by using the uniaxial constitutive equation with grain growth model. In the backward extrusion mode experiment, double-flange, thin-walled cup was formed in one shot. With aid of further microstructure control by bulk mechanical alloying, the present superplastic injection forming can be operated in lower temperature. The related paper will be reported in near future.

Acknowledgments

Authors would like to express their gratitude to Mr. H. Yamashita for help in experiments. This study is financially supported in part by the Priority group for Innovations in Superplasticity with the contract number of # 0842105.

References

[1] T. Asahina, et al.: Recycling of magnesium alloys and high-strain-rate superplasticity. (2000) NIRIN.
[2] T. Aizawa, et al.: Proc. 50th Japanese Joint Conference for the Technology of Plasticity. (1999, October, Fukuoka) 159-184.
[3] T. Aizawa: Proc. 24th (1999, September, Kyoto) 125-126.
[4] T. Aizawa, et al.: Proc. JSTP (2000, May, Tokyo) 4485-486; 487-488 (in Japanese).
[5] M. Maruyama: J. Nikkei-Mechanical 550 (2000) 68 (in Japanese).
[6] E.N. Paton and C.H. Hamilton: Superplastic Forming of Structural Alloys. Metallurgical Society of AIME (1982).
[7] A.J. Barnes: Materials Science Forum. 306 (1999) 785-796; H. Yamagata: Materials Science Forum. 306 (1999) 785-804.
[8] T. Aizawa, et al.: Proc. COM 2000 (2000, August, Ottawa) (in press).

Aerospace Part Production Using SP700

P.N. Comley

The Boeing Company, PO Box 3707, Seattle WA 98124, USA

Keywords: Alpha Case, Anisotropy, Oxygen Rich Layer, SP 700

Introduction

SP700 is a high strength titanium alloy developed by Nippon Kokon Company (NKK) in Japan, and produced by Reactive Metals Inc. (RMI) in the USA. It has a composition of 4.5%Al, 3%V, 2%Fe and 2%Mo, compared with 6%Al, 4%V for the widely used Ti-6-4 SPF alloy It has been assigned an AMS 4899 spec number. SP700 is promoted as being superplastically formable at 1290°F to 1475°F, with a preferred temperature of 1425°F. The advantages of this alloy include

a) No alpha case formation during SPF, thus chemical milling is not required.
b) As the forming temperature is low, the die surface is less prone to oxidation and corrosion, so they last longer. More importantly, the parts would have a better surface finish and require less handwork on appearance items.
c) Equal or superior properties compared with Ti-6-4 alloy.

The paper examines the SPF properties of SP700 at different temperatures and strain rates, and looks at the comparison between this alloy and Ti-6Al-4V in the production of an aircraft part.

Superplastic properties

NKK[1] had performed some plane strain coupon testing of SP700 at various temperatures. It was found that the stress/strain rate curves are similar to that of Ti-6Al-4V multiplied by a factor that is dependent on temperature (Fig 1).

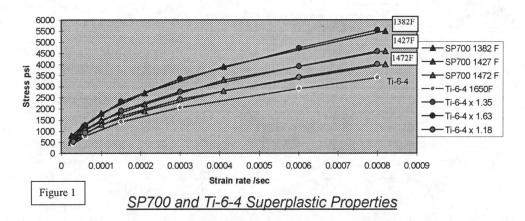

Figure 1

SP700 and Ti-6-4 Superplastic Properties

A multiplication factor of 1.18 applied to the baseline Ti-6Al-4V at 1650F gives a close approximation to the SP700 at 1472F. Similarly multiplication factors of 1.35 and 1.63 give equivalent results to SP700 at 1427F and 1382F respectively. Accordingly, to form a part in SP700 at any particular temperature, the pressure profile of the Ti-6Al-4V part can be used with the appropriate multiplication factor.

Superplastic forming trials

Although coupons had been pulled to obtain the stress/ strain rate curves, little had been done under bi-axial conditions such as is encountered during SPF forming of parts. As a result a series of pans were blown at different temperatures and strain rates to check the published data and to see the effect on the thickness distribution across the pan.

The blank size was 9 x 9", and made a cavity of 2" deep x 6 x 6" with 1" radii on all corners.

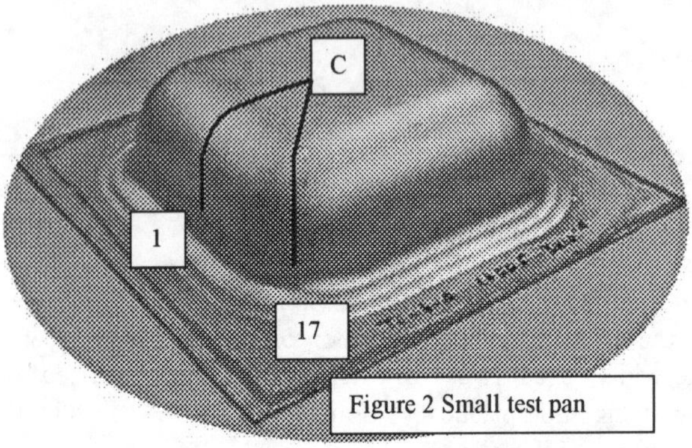

17 measurements were made from the edge to the center and out to the corner as shown (Fig 2). This was repeated for all 4 corners to see if there was any anisotropy between the rolling and transverse directions of the sheet.

Figure 2 Small test pan

The first experiment was to see the effect of forming temperature. 8 pans were formed at a constant strain rate of 3E-4 (derived from a finite element pressure/time cycle), at temperatures of 1380F, 1425F, 1475F and 1525F. At the same time two pans were formed in Ti-6Al-4V at 1650F to see how the thickness compared.

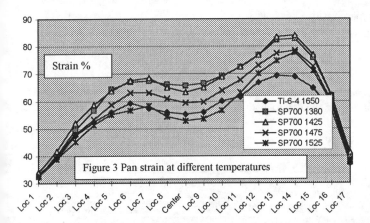

Figure 3 Pan strain at different temperatures

There was no apparent anisotropy evident on any of the pans, and the thickness profiles of pans formed at the same temperature were reasonably consistent.

A material that shows less engineering strain in the graph (Fig 3) would be considered more superplastic for production purposes as it would have less thinnout when formed into a female cavity.

It can be seen that Ti-6Al-4V is better than the SP700 at this strain rate, yielding a lower thickness strain across the pan, especially in the four corners. Of the SP700 data, the best profile is obtained at the highest temperature (1525F) with progressively more strain for lower temperatures. The results are consistent with SPF theory in that a higher temperature gives a better 'm' value and enhanced properties.

From the above it might be concluded that a higher temperature would be better for forming parts. However oxygen contamination, alpha case,

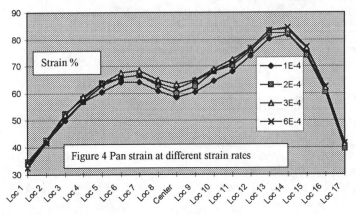

Figure 4 Pan strain at different strain rates

material property degradation and die corrosion all increase at higher temperatures, and it was decided that production parts should be formed at 1425F.

The second experiment was to see the effect of strain rate (Fig 4). Pans were formed at 1425F at strain rates of between 1E-6 to 6E-6 per second. The graph shows that SP700 has little variation in thickness distribution at different strain rates, whereas Ti-6Al-4V has been shown to have significant variation. This property can lead to better thickness across a part due to the fact that during forming, different areas of the sheet form at different rates.

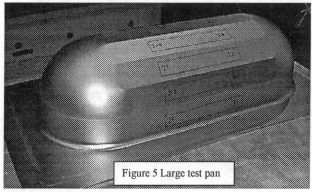

Figure 5 Large test pan

Figure 6 Strain at different strain rates

A larger die was then used to make tensile coupons (Fig 5). Coupons of constant thickness can be extracted from the 9 flat facets of the die. The experiment to examine strain rate dependence on thickness distribution was repeated for this configuration of die shape.

Figure 6 shows the thicknesses obtained from SP700 at 1425F at 3E-4 and 6E-4/sec, and the corresponding values for Ti-6Al-4V at 1650F and the same strain rates. It can be seen that there is only a small difference of strain in the SP700 data, but a larger difference in the Ti-6Al-4V data. This confirms the findings that the SP700 has a more consistent thickness profile over a range of strain rates, which ultimately leads to a more uniform thickness in a production part.

Post SPF properties
Ti-6Al-4V develops alpha case formation on the surface during SPF processing. This is a hard, brittle layer which must be removed by chemical milling otherwise the mechanical properties are severely compromised. Photomicrographs through the etched surface of a formed part (Fig 7) shows that SP700 does not develop alpha case, however there is a oxygen enriched layer that may be detrimental to the properties. Data (Fig 8) shows that the oxygen-enriched layer typically extends 0.003" into the surface and increases the hardness from a baseline of 285 Vickers to 320 near the surface.

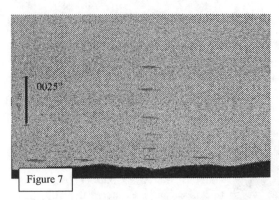
Figure 7

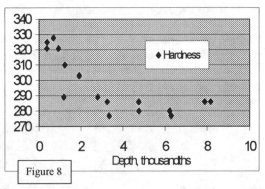
Figure 8

This hardness may have an effect on the tensile properties of SPF'd parts, especially elongation. A design of experiment was devised to test for the effect on properties for various amounts of material removal by pickling. The amount of superplastic elongation and the initial gauge might also influence the tensile properties. Accordingly a test matrix was made for 55 test coupons using five starting gauges (1,2,3,4 and 5mm), three amounts of SPF strain (50, 85 and 115%), and three depths of post SPF pickling (0, 0.5 and 1.0 thousandths per side) to examine these effects. The pickling was performed in a 60% HNO3, 7%HF, 70degF tank, which etches at .003" per hour.

The average yield strength of coupons was 138.8ksi, the ultimate strength 144.3ksi, and the elongation was 10.5%. The yield and ultimate strengths are comparable to Ti-6Al-4V after SPF processing.

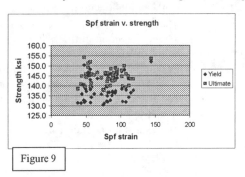
Figure 9

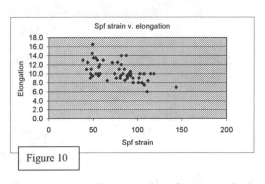
Figure 10

Figures 9 and 10 show the effect of the amount of SPF strain on tensile properties after superplastic deformation. There is little evidence of a trend for the strength, but there is an effect on the elongation. For every 100% of SPF strain there is a reduction of 4.7% in the elongation value. This reduction may be due to an intrinsic property change in the material, or it may be due to the irregularities of surface finish and thickness on the test coupon that are a result of the SPF elongation. These irregularities become worse with increasing SPF elongation.

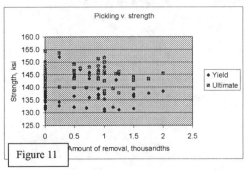
Figure 11

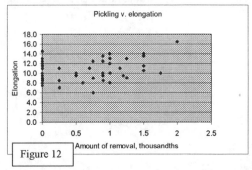
Figure 12

Figures 11 and 12 show the effect of the amount of pickling on tensile properties after superplastic deformation. There is no evidence of a trend for the strength, but there is an effect on the elongation. For every thousandth of inch removed by pickling there is an increase of 0.85% elongation. It is suggested that production parts should have .001" per side pickled after superplastic forming.

Production part

A production part that is currently made in Ti-6Al-4V is a door threshold corner on the Boeing 757 aircraft. Some sample parts were made in SP700 (Fig 13) to compare the thickness profiles of the two materials under production forming conditions. Two gauges of SP700 were tested, .059" and .071", while the initial gauge of the Ti-6Al-4V blank was .071". The thickness was measured at locations shown on the figure.

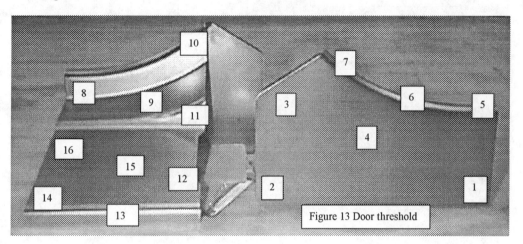

Figure 13 Door threshold

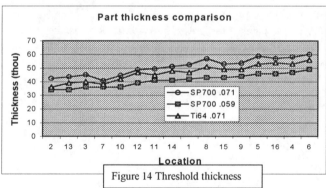

Figure 14 Threshold thickness

Figure 14 shows the variations of thickness of the three parts after all processing, including .003" per side for the Ti-6Al-4V and .001" per side for the SP700. The part drawing calls for certain minimum thickness after processing. The SP700 was thicker than Ti-6Al-4V for the same starting gauge, mainly due to the lesser material removal from chem milling. The .059", although thinner than the other two, would meet drawing requirements. Of interest is the fact that the thinnest areas of the .059" SP700 are only .002" to .004" thinner than the .071" Ti-6Al-4V. It is thought that the thinner starting gauge has less through thickness stress during superplastic forming at critical male bend radii such as is present at the 7, 10 and 13 locations on the part, resulting in less spf deformation at these places.

Following some more pre-production trials a decision was made to implement these thresholds into production. The first parts were made and installed on aircraft in the spring of 2000. The surface finish of the parts was superior to that commonly achieved in Ti-6Al-4V production. The die surface has been examined after several production run cycles, and is has so far been very satisfactory, showing a pale gray oxide with no evidence of pitting or other defects.

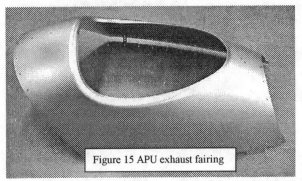

Figure 15 APU exhaust fairing

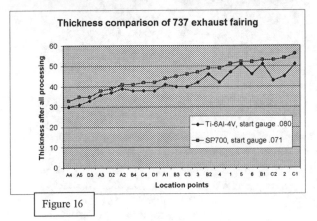

Figure 16

A second part was made to examine the thinnout characteristics of SP700 on a production part. The 737 APU exhaust fairing is currently superplastically formed in Ti-6Al-4V, with a .080" starting gauge. A fairing was made in .071" starting gauge SP700 (Fig 15). A series of points on the fairing were measured and compared after all processing and chem milling operations.

The relative thicknesses are shown on Fig 16. Once again, the final thickness of the SP700 is higher than the Ti-6Al-4V, indicating that a thinner starting gauge could be used to make parts with a similar thickness to Ti-6Al-4V. The part surface finish was comparable or superior.

Conclusion

In conclusion, SP700 can be superplastically formed at 1425F using the same dies and equipment made for Ti-6Al-4V, using the same cycle time and 35% higher pressures. The material does not need to be chemically milled, but it is recommended that .001" per side be pickled to retain the elongation values. The strength of the post formed alloy is about 10% higher than Ti-6Al-4V. Parts were formed into production dies, demonstrating that a thinner starting gauge could be used. The part surface finish quality is generally higher than Ti-6Al-4V. Due to the low temperature it is anticipated that the dies will maintain a better surface for a longer time than the current Ti-6Al-4V production experience.

As a result of the research, one part has gone into production. It is believed this is the first application of SP700 on a commercial aircraft.

---

[1] N Minakawa, NKK Corp

Materials Science Forum Vols. 357-359 (2001) pp. 47-52

Dental Implant Superstructures by Superplastic Forming

R.V. Curtis[1], D. Garriga-Majo[1], A.S. Juszczyk[2], S. Soo[1], D. Pagliaria[1] and J.D. Walter[2]

[1] Dept. of Dental Biomaterials Science, Guy's King's College & St. Thomas's Hospitals Dental Institute, King's College London, London SE1 9RT, UK

[2] Dept. of Prosthetic Dentistry, Guy's King's College & St. Thomas's Hospitals Dental Institute, King's College London, London SE1 9RT, UK

Keywords: Casting Investments, Dental Implant Superstructure, Dental Prostheses, Dies, Induction Heater, Superplasticity, Titanium Alloy

Abstract

A novel application of superplastic forming is described for the production of fixed-bridge dental implant superstructures. Finite element analysis (FEA) has shown that Ti-6Al-4V sheet would be a suitable candidate material for the design of a fixed-bridge dental implant superstucture. Traditionally superstructures are cast in gold alloy onto pre-machined gold alloy cylinders but castings are often quite bulky and 25% of castings do not fit accurately (1) which means that sectioning and soldering is required to obtain a fit that is clinically acceptable and will not prejudice the integrity of the commercially pure cp-titanium implants osseointegrated with the bone. Superplastic forming is shown to be a forming technique that would allow the production of strong, light-weight components of thin section with low residual stress that could be suitable for such applications. Considerable cost savings over traditional dental techniques can be achieved using a low-cost ceramic die material. The properties of these die materials are optimised so that suitable components can be produced. Satisfactory hot strength is demonstrated and thermal properties are matched to those of the titanium alloy for accurate fit of the prosthesis.

Introduction

The positioning of CP-titanium dental implants for a finite element simulation of the stresses in a fixed-bridge dental implant superstructure is based on a study by Hylander (2). Five dental implants are positioned in such a way that a circle of radius 22 mm can be drawn through the centre of the implants. Cantilever's of length 15 mm support posterior teeth. The width of the superstructure should be no less than 6 mm and in cross-section resembles a U-shape (Fig. 1).

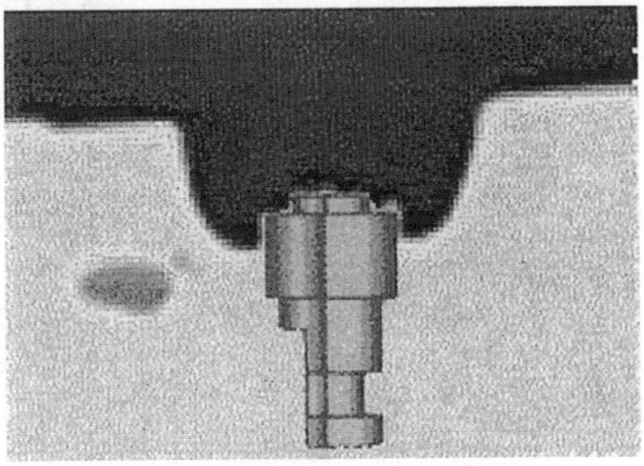

Figure 1. Cross-section of beam through abutments.

Method and Materials

Mean chewing stresses of 50 N and maximum bite forces of 500 N may be expected at the ends of the cantilevers. Stresses in the beam are calculated for increasing wall height for a 100 N load applied at the end of the cantilever.

To ensure reproducibility of shape for quantitative assessment of beam properties and to reduce the time involved in hand-finishing of the superstructures, a simplified beam was designed consisting of 3 implants positioned in a straight line and 8 mm apart. An overall beam width of 10 mm ensures that there are 6 mm of beam in cross-section through an abutment. A master die was produced from which replicas were prepared for superplastic forming. Dies consisted of stainless steel implant abutments mounted in an appropriate casting investment material. The three investment materials used in the study are shown in Table 1. The selection of casting investment is based on strength and control of expansion.

Table 1: Dental casting investments used for superplastic forming dies.

Product Name	Liquid to Powder Ratio	Type of liquid
Croform WB	0.12	Water
Rematitan	0.16	Water to mixing liquid = 0.55
Rema Exakt	0.13	Mixing liquid

Cold strength of the three casting investments using a variety of handling techniques (Table 2) was measured using a four-point bend test. 50 tests for each handling condition and material were performed. Handling techniques that gave highest strengths and lowest strengths were then adopted for the measurement of hot strength. Hot strength was measured using a centrifugal casting machine for which a known weight of alloy was cast against an investment membrane of predetermined thickness. 30 tests were carried out for the two handling conditions for each material. Hot strength testing followed a staircase method whereby a membrane survival resulted in an increased weight of casting alloy for the subsequent test and, likewise, a membrane failure resulted in a decreased weight of casting alloy.

Table 2: Handling techniques for preparing dental casting investments.

Handling technique	Method of mixing	Method of setting
ha	Hand spatulation	In air
hp	Hand spatulation	Under pressure
hva	Hand spatulation in air	In vacuum
ma	Mechanical spatulation in air	In air
mvp	Mechanical spatulation in vacuum	Under pressure
mva	Mechanical spatulation in vacuum	In air

Expansion of the investment was controlled by ratio of manufacturer's special liquid to water ratio. Setting expansion and thermal expansion were measured for the materials shown in Table 1. Thermal expansion of the mill-annealed sheet was also measured to 900°C.

To characterize the flow stress of the titanium alloy (mill-annealed Ti-6Al-4V to BS2TA10 of 2 mm thickness and 140 mm diameter) a 3 cylinder test was developed to allow flow stress to be estimated using one pressing. Flow stress was determined by generating a pressure-time profile based on an initial estimate of the flow stress and on the radius R of a cylinder in the die.

$$P(t) = 4\frac{S_0\sigma}{R} \exp(-1.5\dot{\varepsilon}t)\sqrt{1 - \exp(-\dot{\varepsilon}t)} \quad (1)$$

Where, t = time (s)
S_0 = initial thickness of sheet (mm)
R = radius of cylinder
$\dot{\varepsilon}$ = strain rate (s^{-1})
σ = flow stress (Mpa)

The die was manufactured with 3 cylinders of different radius R_1, R_2 and R_3 (Fig. 2). Three estimates of the flow stress of the material were obtained from one experiment.

Where,

$$\frac{\sigma_1}{R_1} = \frac{\sigma_2}{R_2} = \frac{\sigma_3}{R_3} \qquad (2)$$

$$\sigma_1 = \sigma_2 \times (\frac{R_1}{R_2}) \quad \text{for radius } R_1 \qquad (3)$$
$$\sigma_2 = \sigma_2 \qquad \quad \text{for radius } R_2 \qquad (4)$$
$$\sigma_3 = \sigma_2 (\frac{R_3}{R_2}) \quad \text{for radius } R_3 \qquad (5)$$

For a more reliable estimate of the flow stress of the material for cylinder i, a final depth d_i should be obtained where:

$$d_i^* = \frac{R_i}{\sqrt{3}} \qquad (6)$$

and

$$\frac{d_i}{d_i^*} = 1 \qquad (7)$$

Plotting $\frac{d_i}{d_i^*}$ verses σ_i gives a graph with 3 points that can be interpolated or extrapolated to obtain the better estimate of the flow stress.

An induction heater operating at 415 V using a 3-phase power supply was used to heat a ceramic die contained in a chrome-steel chamber. Heating to 890°C took 90 - 100 minutes. Mill annealed Ti-6Al-4V alloy was inserted into the press at 890°C. To generate a pressure-time profile to form the dental implant superstructure the geometrical model of Garriga-Majo and Curtis was employed (4).

Results and Discussion

A plot of the Von Mises stress verses wall height for a 2 mm thick beam of U-shaped cross-section (Fig. 3) shows that a wall height of 2 mm results in a stress of 178 MPa for a load of 100 N at the end of the cantilever. This equates to a stress of 890 MPa for a 500 N load which is close to the yield stress of Ti-6Al-4V alloy. This suggests that a minimum thickness of 2 mm is required to produce a dental implant superstructures with wall heights of 2 mm. Of course, sheet of thickness greater than this will probably be required given that the cantilever will be subject to fatigue loading. In clinical practice cantilever lengths of 15 mm are unusual and lengths of 10 mm or would be more favourable from the structural point of view.

The expected reduction in thickness during superplastic forming of these superstructures would support mean chewing loads of 50 N but may not support the relatively small number of maximum chewing loads that could reach 500 N.

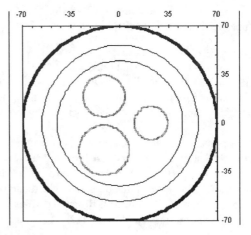

Figure 2. Schematic of the die for the 3 cylinder test for flow stress determination.

Nevertheless, sheet of 2 mm thickness was selected for this study to assess the feasibility of the method to manufacture them. Furthermore, it is assumed that even though sticking friction between die and sheet might reasonably be expected using casting investment materials for dies, complete cessation of flow across the sheet thickness after the point-of-contact is reached is unlikely and, therefore, to test the validity of the model, thinner sheet would probably be more appropriate.

The cumulative distribution function of strengths for Croform WB is shown in Fig. 4. Such plots yield characteristic strengths (i.e. 63.2% of samples failed) that ranged between 4.7 MPa (HA) and 8.1 MPa (MVP) for Croform WB, between 3.7 MPa (HA) and 4.9 MPa (MVP) for Rematitan and between 3.0 MPa (HA) and 4.1 MPa (MVP) for Rema Exakt (Fig. 5). Croform WB was more sensitive to handling technique than Rematitan or Rema Exakt. Strength tests at 900°C for these handling techniques showed that characteristic strengths were 0.37 MPa (HA) and 0.48 MPa (MVP)

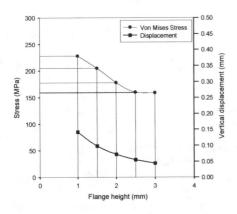

Figure 3. Stress/deflection for a Ti-6Al-4V superstructure by finite element analysis.

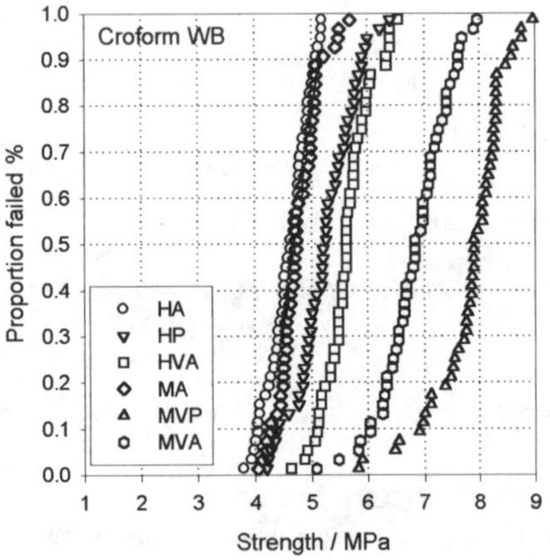

Figure 4. Strengths for Croform investment using Weibull cumulative distribution function.

for Croform WB, 0.39 MPa (HA) and 0.45 MPa (MVP) for Rematitan and 0.77 MPa (HA) and 0.73 MPa (MVP) for Rema Exakt. Differences in hot strengths between handling techniques and between materials (Fig. 6) are less at 900°C than at room temperature, although the strength of Rema Exakt was significantly higher than the other two materials.

Measurements of setting and thermal expansion for the investments and for Ti-6Al-4V sheet (Fig. 7) indicate that Rematitan and Crofom WB are the most likely materials to achieve accurate fit of superstuctures. However, greater control of expansion can be obtained using Rematitan since the manufacturer supplies a mixing liquid. The mixing liquid is diluted with water to achieve the desired setting expansion. Croform WB, on the other hand, is mixed with water so control of setting expansion is not possible. On this basis Rematitan was chosen to manufacture the die for superplastic forming.

The results of the 3 cylinder tests for measurement of flow stress as a function of temperature and strain rate for the mill-annealed Ti-6Al-4V alloy are shown in Fig. 8 from which the strain rate sensitivity index and the constant K in Eq. 8 are obtained:

$$\sigma = K\dot{\varepsilon}^m \tag{8}$$

This linear relationship does not take into account grain growth or hardening/softening during the forming time.

To form the superstructure of cross-section shown in Fig. 1 the geometrical model of Garriga-Majo and Curtis (4) was used and Fig. 9a shows the gas pressure-time profile generated to form the beam

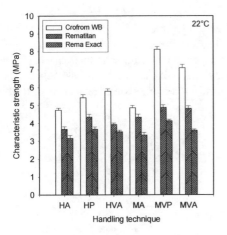

Figure 5. Characteristic strengths at 22°C for Croform WB, Rematitan and Rema Exact.

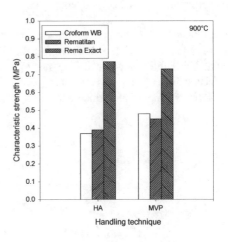

Figure 6. Characteristic strengths for the investments at 900°C for two handling techniques.

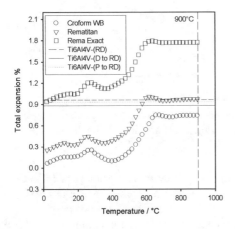

Figure 7. Total expansion for investments and Ti-6Al-4V alloy on heating to 900°C.

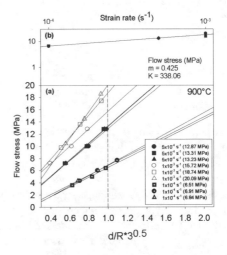

Figure 8. For Ti-6Al-4V sheet (a) flow stresses estimations from 3 cylinder tests (b) flow stress verses strain rate.

at 900°C. This profile assumed a strain rate of 1×10^{-4} s^{-1} and a flow stress of 6.75 MPa. In order to compare the actual degree of adaptation with the geometrical model, the profile was stopped at time = 80 minutes. The formed sheet is shown in Fig. 9b and cross-sections through abutments indicated that thinning has occurred but in a more uniform manner than the geometrical model predicts (Fig. 10). If sticking contact is assumed, and evidence from elsewhere (4) suggests that this is a perfectly valid assumption, then the disparity is probably due to continued flow of material through the thickness of the sheet, normal to the die surface. From the structural and clinical point of view this is more satisfactory since the load applied to the superstructure will be more readily supported by the superplastically formed component.

Three beams were formed in the one pressing representing three different liquid/water ratios. Measurement of beam length showed that a fit with an 11 μm discrepancy occurred at 55% mixing liquid. This compares well with the misfit predicted by measuring the setting and thermal expansions associated with casting investments separately and the shrinkage of the Ti-6Al-4V alloy sheet from the forming temperature which resulted in a predicted misfit of 15 μm. Measurement of fit to this

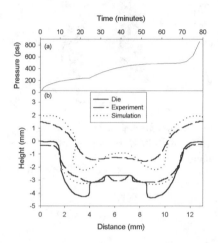

Figure 9. (a) Pressure profile for the 2 mm sheet (b) cross-section of formed component and simulation.

level was achieved using a non-contacting, laser scanning, coordinate measuring device.

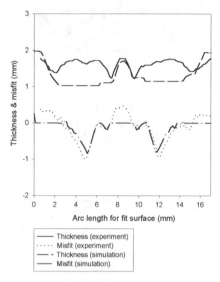

Figure 10. Plot shows variation of sheet thickness across the die for simulation and experiment and misfit with the die surface.

Conclusions
It is feasible for dental implant superstructures to be superplastically formed from Ti-6Al-4V alloy sheet to produce accurately fitting prostheses that are strong in thin section. Material costs are significantly reduced compared to materials used in traditional casting techniques since pre-machined gold alloy cylinders and cast gold alloy used to produce the superstructure are not required.

References
1. G. Goll, Production of accurately fitting full-arch implant frameworks: Part I - clinical procedures. J. Prosthet. Dent. 1991, 66, pp. 377-384.
2. Hylander, W.L., Morphological changes in human teeth and jaws in a high attrition environment. In 'Orofacial growth and Development,' by Albert A Dahlberg. Mouton. 1976. pp. 309-330.
3. RE Goforth, N Chandra, D George. Analysis of the cone test to evaluate superplastic forming characteristics of sheet metal. In "Superplasticity in Aerospace." Eds. HC Heikkenen, TR McNelley. Metallurgical Society of AIME. 1988, pp.149-166.
4. D. Garriga-Majo, R.V. Curtis. Geometric analysis of thinning during superplastic forming. ICSAM 2000, Orlando, Florida, August 1-4, 2000.

Industrial Applications of Superplastic Forming Technology in China

Zhiqiang Li and Hanliang Zhu

Beijing Aeronautical Manufacturing Technology Research Institute,
Dept 106, POB 863, Beijing 100024, China P.R.

Keywords: Industrial Applications, SPF/DB, Superplasticity

Abstract

The superplasticity of materials and related technology have been studied from 1970's in China. More than 50 organizations have been involved in this field and the investigation and the manufacturing center of superplastic forming and diffusion bonding (SPF/DB) technology has been set up. The present paper reviews the industrial application status of SPF/DB technology in China.

More and more materials are developed to exhibit superplasticity in recent years. Some titanium alloys and aluminum alloys have been widely utilized in forming various structural components with complex shapes by SPF/DB technology. More attentions are paid to light-weight and high-temperature materials recently. Single sheet components of Ti_3Al based alloys have been put into use in aerospace and the curve panel with 3-sheets sandwich structure of a Ti_3Al intermetallic has been manufactured successfully by SPF/DB technology. The superplastic characteristics of TiAl, Fe_3Al intermetallics, metal matrix composites and ceramics et al have been studied on a laboratory scale and there is a long way to go to reach the level of industrial application.

With the development of industrial SPF/DB technology, more and more SPF/DB components are manufactured and applied in military and civil industries. The sizes of components produced by the technology become bigger and bigger, and the shapes of components become more complex. At the same time, the components exhibit excellent reliability.

1. Introduction

Superplasticity of materials has been investigated for almost 30 years in China [1]. On the studies of superplasticity, China took her step from the Zn-Al alloys. Today, the studies of superplasticity are concerned with much more materials. The industrial applications increase rapidly in military and civil industry. The investigation and manufacturing center of SPF/DB technology has been set up in China. The combination of fundamental research and technological development as a motive power pushes the application faster.

2. Progress in Industrial Studies and Applications of Superplasticity in China

The studies of superplasticity in China began in early 1970's. During the following ten years, the research increased rapidly and reached a top in the middle of 1980's, which could be reflected from the number of papers published in National Forging Conference and National Superplasticity Symposium (shown in Fig.1). Beijing Aeronautical Manufacturing Technology Research Institute (BAMTRI), Beijing Mechanical and Electrical Research Institute, Harbin Institute of Technology, Beijing Non-Ferrous Metal Research Institute, Institute of Aeronautical Materials, Northeast Polytechnic College, Shanghai Jiaotong University, Changchun Optics and Fine Mechanics Research

Institute and Nanjing University of Aeronautics and Astronautics are some of the earliest groups in the study of superplasticity. On the first National superplasticity symposium held in July 1977, there were only 8 papers given by less than 20 institutes or groups. However, on the second National symposium held in May 1980 in Beijing and the third National symposium held in December 1982 in Shenyang, the numbers of attendees increased 5-6 times as many as the first one and the papers increased as 10 times.

During this period, the research has got great results in the profundity and scope. Most of the work were focused on the fundamental aspects of superplasticity. Superplastic properties of various materials, such as Zn alloys, Al alloys, Ti alloys, Cu alloys, Mg alloys etc, have been studied. The superplasticity mechanisms have also been studied though the observation and analysis of the deformation process and microstructure. The measurement methods of the strain rate sensitivity including section change method, load change method, laxation method, load-specimen elongation method and hardness method have also been investigated. Gas blow forming and superplastic pressing have been developed to manufacture some parts with simple shape, most of which had been put into production. At the same time, superplastic forming and diffusion bonding began to be used to produce two sheet structure.

From the middle of 1980's, the applied research on SPF and SPF/DB technology was widely carried on. 300 ton and 100 ton SPF/DB equipment and gas control systems have been designed and manufactured in BAMTRI. Combined with other research institutes and factories, designing the structures of SPF/DB components, selecting the tooling materials, designing the structure of tooling and manufacturing the tooling, optimizing the SPF/DB parameters by a number of experiments and evaluating the post SPF properties of components, have been carried on.

From the beginning of 1990's, the fundamental research and the technological development were focused on the processing technology of advanced superplastic materials. The development of the SPF/DB technology reached the level of the mass production, which reduce the cost of components. Therefore, more and more SPF/DB components were manufactured and applied in military and civil industry. The structures of parts have been developed from single sheet and two sheets to more complex three sheets and four sheets. The size of components produced by this technology become bigger and bigger, the projectional area of more than $0.5m^2$ can be produced, meanwhile, components exhibit excellent reliability. With development of computer simulation, SPF/DB technology has been found to benefit from modeling of the process, the result of which could not only guide the selection of optimum parameters, predict the thinning characteristics, but could be used to control the blow forming process.

3. Superplastic Materials and Its Industrial Applications

It is well known that a fine grain size is an essential prerequisite for superplasticity. The existing materials must be developed to obtain fine grain and the grain refinement methods have phase separation, phase transformation and mechanical working with recrystallisation. Partial superplastic materials and their superplastic properties are shown in Table 1[2-6].

Zn Alloys
The Zn-Al alloys were firstly developed in China as other countries. The largest elongation of a Zn-Al_5 reach 2400% at temperature of 350°C and strain rate of $8.3 \times 10^{-3} s^{-1}$. Because of its insufficiency in properties and high expense for industrial production, Zn-Al alloys are mainly used to make arts and craft, die for plastic products and electric products.

Al Alloys
Many superplastic aluminum alloys, including Al-Cu, Al-Mg, Al-Li, Al-Ca, Al-Si etc, have been developed to exhibit superplasicity. Rare earth elements have been added to aluminum alloys to enhance their superplastic properties in China. When RE was added to Al-Zn-Mg alloy, a

elongation of 1014% is obtained at 560℃. The elongation is 2 to 5 times as that of alloy without RE. The chemical composition analysis shown that most of rare earth elements concentrated in the second phase. With the addition of RE, the amount and the density of the second phase increase, and the size of the second phase decreases, which result in homogenous distribution of the second phase in the aluminum alloy matrix. At the same time, RE can refine the grain, reduce the rate of grain growth and stabilize the grain structure during superplastic deformation. The above changes of microstructur can reduce the flow stress and enhance the superplastic properties of the alloy[2]. The superplasic aluminum alloys are mainly used in the instrument and electric engineering, aerospace industry. However, because the used properties of superplastic aluminum alloys are still unable to compare with the commercial aluminum alloys. In addition, the material supplies are not interested in small-scale orders, so the price is higher than the conventional aluminum alloys 3-5 times. Therefore, there is a long way to go for the mass commercial production of superplastic aluminum alloys.

Ti Alloys
The superplasticity of Ti alloys has been paid more attention by many researchers from the beginning. Through a large number of studies, Ti alloy products have been widely used in aerospace industry. Ti-6Al-4V has been used to form a mass of components with complex shapes by SPF/DB technology. Ti-15-3 sheet has been rolled and exhibits 300% elongation. However, Ti-15-3 is nearly β Ti alloy, and its superplastic temperature is about 750℃, so SPF/DB can't be finished in one heat cycle. A new process, SPF combined with point welding, has been developed to manufacture structural parts with complex shape in BAMTRI.

Ferrous alloys
The studies of superplasticity of Ferrous alloys in China can be traced back to 1970's. The existing steels were subjected to heat treatment, and obtain fine grain size, which let them to exhibit superplasticity. A large number of experiment datum of several steels, such as structural steels, tool steels, die steels, bearing steels, and so on, have been obtained. The superplastic ferrous alloys are only used in the manufacturing die. It needs more work to improve its usage.

Intermetallics
The superplasticity of Ti_3Al based alloys have been widely studied in recent years in China. A Ti-24Al-14Nb-3V-0.5Mo has been developed to obtain good superplastic properties. The maximum m value of 0.76 is obtained at 960℃, and the strain rate of $1.5 \times 10^{-3} s^{-1}$, while the tensile elongation is 1240%. At optimum strain rate ranging from 3.5×10^{-4} to $4 \times 10^{-3} s^{-1}$, and at temperatures between 940 and 980℃, the m value of the alloy is greater than 0.5, which is the characteristic of superplastic materials. The strain rate at which the m value is greater than 0.5 is higher than previous reported data concerning which is a result of the finer initial grain size for the investigated alloy. Single sheet parts and a curve panel with three sheet sandwich structure have been manufactured successfully by SPF/DB technology. It is expected that the material would be used to fabricate light weight parts, whose serving temperature is 650℃ or so in aerengine.

By hot isostatic pressing and following multi-step thermomechanical processing, a TiAl based alloy obtains about 2μm grain size, and a dramatic change of microstructure. The lamellar structure of ingot has been broken up to form a fine-grained subduplex microstructure with fine equiaxia γ grains, spheroidized α_2 phase and fine lamellar colonies. The microstructure is advantage to the superplastic deformation and diffusion bonding. However, More work is still need to do for fabricating practical parts by SPF/DB technology. The superplastic properties of FeAl and Fe_3Al intermetallics have also been studied on a laboratory scale, there is a long way to reach the level of industrial applications.

Metal Matrix Composites
A foil-fiber-foil technique has been used to fabricate SiC/Ti-15-3 and SiC/Ti$_3$Al composites. The foil-fiber-foil technique used a matrix in the form of rolled foil of Ti-15-3 or Ti3Al. Woven SiC fiber mats are used in alternating layers with the matrix foils. The fibers are held in place in these mats with a stainless steel wire. Consolidation of the composites is achieved by vacuum hot pressing, Which is a SPF/DB process. The interfacial reactions between the fiber and the matrix have been studied. Several methods were used to control the interfacial reactions. Tensile strength of 984MPa is obtained for SiC/Ti-15-3 composites with 15% volume fraction SiC. Based on the studies, a three-sheet sandwich structure has been fabricated. The titanium matrix composites offer great potential for substantial weight savings in advanced engines and airframes.

4. Field of Industrial Applications

The substance of superplasticity is both to select scientifically the deformation conditions and to control seriously the parameters during the SPF process. Under the superplastic conditions, the plasticity of materials is raised up many times, and the deformation resistance is decreased a comparable times, which leads to decrease energy 50%-80% compare with the conventional process, to reduce the consumption of raw materials 20%-50%, and to cut down the expenses of equipment investment 50%-90%. Today, SPF/DB technology has been widely applied in military and civil industry[6-8].

Application in Aerospace Industry
The SPF/DB technology is widely used to manufacture products of military airplane, aeroengine and rocket engine. Fig.2 is a SPF/DB air-conditioner cabin door of titanium alloy. Compared to conventional fabrication methods, the number of parts reduces from 52 to 22, and that of the joint reduces from 840 to 103. The SPF/DB cabin door saving 15% weight and 53% cost. More complex four sheet SPF/DB hollow blades of aeroengine are shown in Fig.3. The blades reduce weight 34.2%. Some single sheet parts of new materials are also formed by gas blow forming (shown in Fig.4 and Fig.5).

Civil Production
The SPF technology has also been applied in manufacturing parts of instrument, electronics engineering, arts and craft, medical equipment, textile machinery, auto, and so on. Fig.6 shows metal drums of various specifications. They can be used for winding cops or cones into cheese or cones of various tapes. Fig.7 is a photo of cover of medical centrifuge. Because of its drum-like shape, a special manipulator is developed to take it off from the tooling after SPF process. Some examples of arts and craft products are shown in Fig.8.

Summary

The SPF technology has been developed for almost 30 years in China. More and more materials are investigated to exhibit superplasticity and the technology has been improved to meet increasingly high requirement for industrial applications in military and civil fields. The industrial applications have been expanded in China as following: from replacing conventional parts to specially designed SPF/DB parts, from manufacturing common parts to load bearing parts. The application area is expending from aircraft to aeroengine, and the material envolved is from titanium alloy, aluminum alloy , Al-Li alloys to metal matrix composites, intermetallics and ceramic composites .

References

[1] J.T.HAI, Z.R.WANG, Y.C.YANG and B.L.ZHANG, Proceeding of 5th National Superplasticity

Symposium, 1992, Beijing, p.1.
[2] H.J.ZHOU, C.Y.ZHAO and A.F.TENG, Proceeding of 5th National Superplasticity Symposium, 1992, Beijing, p.139.
[3] L.Q.WU, S.H.MA and X.L.MA, Proceeding of 4th National Superplasticity Symposium, 1985, Beijing, p.1.
[4] R.Z.LIN, Mechanism and Application of Metal Superplastic Forming, Aviation Industry Publishers, 1989, p.27.
[5] A.Shan, M.Chen, D.Lin and D.Li, Materials Science Forum, Vol.170-172, 1994, p.477.
[6] Development and Application of SPF/DB Technology, China Aviation Industry Corporation, 1995, p.1.
[7] Titanium Science and Engineering, Shanghai Chenguang Publishers, 1993, p.11.
[8] J.T.HAI, Z.R.WANG, CINO-Japan Joint Symposium on Superplasticity, 1985, Beijing, p.1.

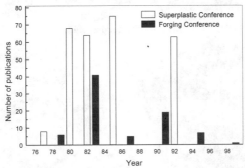

Fig.1 The chronological statistical chart of SP papers on National Forging Conference and National superplasticity symposium in China

Table 1 Partial superplastic materials and their superplastic properties developed in China

	Materials	Grain size μm	Temperature ℃	Strain rate /s^{-1}	m	Elongation ×100
Zn Alloys	Zn-Al5	-	350	8.3×10^{-3}	0.45-0.5	2400
	Zn-Al22	-	200-300	1.6×10^{-3}	0.45-0.6	1000-2500
Al Alloys	8090	20	520	0.01	0.5	820
	2091	-	510-530		0.5	1000
	Al-Zn-Mg-RE	1.2	560	1.48×10^{-3}		1014
	LC4	-	505	6.66×10^{-4}	0.5	400
	Al-Ca-Zn	-	550	(8.33~16.7)×10^{-3}	0.38	900
	Al-Mg	-	560	1×10^{-3}		550-870
	Al-Cu-Mg-Zn	-	430-450	1.67×10^{-3}	0.42-0.47	1290
	LF6	-	435			254
	7475	-	515	0.02	0.6	1200
Ti Alloys	Ti-6Al-4V	10	925	2×10^{-4}	0.6	1590
	Ti-Co-Al	-	700	10^{-2}	0.58	2000
	Ti-15-3	25	750	1×10^{-3}	0.33	300
	Ti55		920		-	510
Steel	3Cr2W8V	-	830	-	0.38	267
	18Cr2Ni4WA	5.92	700	3.33×10^{-4}	-	550
	5Cr-Mn-Mo	-	720	-	0.4	468
Intermetallics	Ti$_3$Al	1	960	1.5 10^{-3}	0.76	1240
	TiAl	1	1075	8 10^{-5}	-	517
	Fe$_3$Al	100	850	1 10^{-3}	0.22-0.42	332.8
	FeAl	300	900-1000	1.39 10^{-2}	-	150-208
other	SiCp/LY12	-	480	1 10^{-3}	0.53	345
	SiCp/2618	5	560	3.3 10^{-3}	-	620
	Y-TZP	-	1450	10^{-4}-10^{-5}	0.42-0.44	240

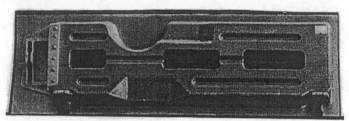

Fig.2 SPF/DB air-conditioner cabin door of Ti-6Al-4V

Fig.3 Four sheet SPF/DB hollow blades of aeroengine

Fig.4 SPF inner stiffening structure of Al-Li alloy   Fig.5 Ti$_3$Al turbine pump shroud of rocket engine

Fig.6 Metal drums of various specifications   Fig.7 Cover of medical centrifuge

Fig.8 Some examples of arts and craft products

The Production of Automotive Body Panels in 5083 SPF Aluminium Alloy

B.J. Dunwoody

AAW Produktions AG, Formtec Dvision,
CH-3965 Chippis, Switzerland

Keywords: Aluminium, Automotive, SPF, Tool Design

ABSTRACT

As aluminium is replacing steel in many automotive applications including body panels, SPF is becoming more and more recognized by car manufacturers as a viable, cost effective alternative to very complex, multi-stage cold pressing tools for prototype parts, pre-series parts and small to medium volume production parts.
This paper describes the design and production of appropriate tooling and the forming by SPF of automotive body panels.

INTRODUCTION

In Europe and the U.S.A., many automobile manufacturers are turning to aluminium body panels to save weight and to improve fuel economy [1,2,3]. For high volume production of panels with less than 35 – 40% equivalent strain, conventional multi-stage cold pressing may be used. However the tooling costs are very high. SPF is being recognized as a viable, cost-effective alternative for prototypes, pre-series parts, small to medium volume production parts (up to 12,000 per year) and for parts with more than 40% equivalent strain. Also SPF tooling costs are often one tenth that of cold press tools. At AAW Formtec we have produced many parts in these categories. This paper describes the design and production of appropriate tooling and SPF forming of two body panels. Tooling and SPF forming were also developed to produce an acceptable surface for body panels with visible areas.

TOOL DESIGN

Two-piece tooling was designed to produce the left and right wheelhousings (shown in Fig. 1) for a utility vehicle.
The requirement was to form left and right wheelhousings together. To form with a conventional SPF tool with a flat blowing plate would have meant a negative tool with a depth of 450 mm. The aspect ratio of the two parts together is 0.46 where:

$$\text{Aspect Ratio (AR)} = \frac{2 \times H}{W+L} \qquad \text{Eq.1}$$

Where H = height; W = width; L = length

For 5083-SPF in a negative tool AR should not exceed 0.3.

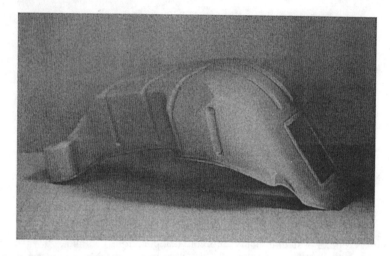

Fig. 1 Superplastically formed utility vehicle wheelhousing in alloy 5083-SPF

One way to achieve this would be to increase W by 1100 mm. But this would give a large area of scrap material and would have been uneconomic.

The solution was to reduce the effective height of the parts (H) by profiling the tool (shown in Fig. 2) and fabricating a shaped blowing plate.

Thus the effective part height (H) is reduced to 250 mm. The length (L) is slightly increased because of the profiling. This gives an aspect ratio of 0.24 with only a narrow strip of scrap material between the parts.

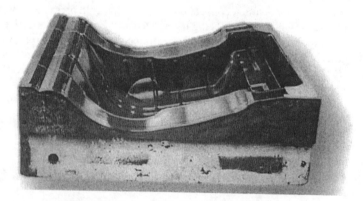

Fig.2 Profiled tool for left and right wheelhousings

The second example is a bulkhead mounting plate for a small hatchback car. This is a structural part, which supports the brake servo unit and other accessories. The finished part is shown in Fig. 3.

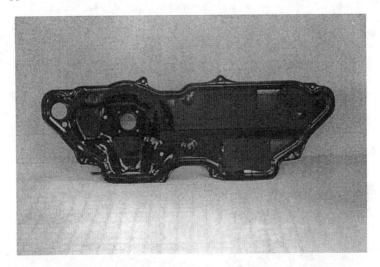

Fig. 3 Superplastically formed bulkhead panel, finished with a cold pressing operation. Alloy 5083-SPF.

The critical areas on this part are a flange with a 3.5 mm radius around the periphery and the deepest area with 70% equivalent strain.
A single impression tool was made to produce prototype parts, complete with the flange.
For series production a 4 position tool was designed without the flange to reduce forming time and to facilitate finishing.

The two internal body panels described above were both formed in negative tools, where the surface finish achieved was acceptable.
For external body panels and internal panels with visible surfaces, the normal practice is to use positive tools where the visible surface is non-tool side.
In order to achieve the required surface finish on automotive panels, tool finish is very critical.
CAD construction of the tool surface is carried out using CATIA data supplied by the customer. This data is then transmitted directly to a CNC machine, which produces a fine milled surface on a tool steel blank. The tool surface is then polished to an acceptable finish, with $R_a < 1\mu m$.

SPF PRODUCTION

The wheelhousing tool has been successfully used to produce pre-series parts in 5083-SPF 1.6 mm sheet supplied by Alusuisse Swiss Aluminium Ltd. On closing the press the sheet is preformed before commencing the forming cycle. The pressure vs. time cycle was calculated, based on an initial strain rate of 3×10^{-3}. Forming temperature was 515°C. The parts were subsequently trimmed by 5-axis laser.

The single impression bulkhead tool was used to produce prototype parts in 5083-SPF as above, which were trimmed and all cut outs made with a 5-axis laser.

Series production was carried out in the 4 position tool. An untrimmed forming from this tool is shown in Fig. 4. The finishing operation was achieved by using a composite 2 part press tool which simultaneously trimmed the parts to size and formed the flange together with the holes.

Fig. 4 Untrimmed forming of 4 bulkhead panels in 5083-SPF

Fig. 5 shows an external hood panel for the utility vehicle mentioned earlier. Figures 6 and 7 show internal door panels, all formed on positive tools.

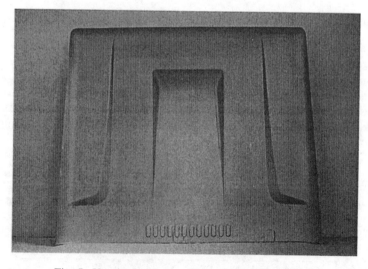

Fig. 5 Hood panel for a utility vehicle in 5083-SPF

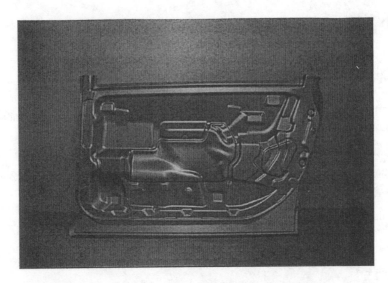

Fig. 6 Door inner panel for a limousine in 5083-SPF

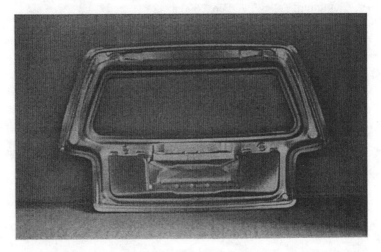

Fig. 7 Superplastically formed rear inner door for a small hatchback car. Alloy 5083-SPF.

The hood panel is highly visible and any surface striations or imperfections are immediately apparent.
Although the internal panels are partially covered by trim, certain areas are painted and are also visible.
It is therefore necessary to monitor surface quality very closely during production runs.

SUMMARY

SPF automotive parts are appropriate for small to medium production runs with currently available SPF Aluminium alloys and where:

1. Deformation is too great for conventional cold pressing.
2. SPF is used for the basic shape followed by a cold pressing/finishing operation.
3. Lower tooling costs are indicated.
4. Positive or negative tooling is chosen to give the required surface finish.

The future may well see more and more SPF aluminium parts in automotive applications in small to medium series using currently available alloys.

If higher strain rate alloys can be developed and produced commercially, SPF may well find its place in higher volume mass produced models.

REFERENCES

[1] The Aluminum Association, Aluminium for Automotive Body Sheet Panels, Publication AT3, December 1998
[2] I. Farmer, Aluminium Today Jan./Feb. 2000 p. 42
[3] W. Kimberley, Automotive Engineer Feb. 2000 p.54

Industrial Applications of the Superplastic Explosive Forming

V.P. Sabelkin[1,2] and Marco A. Hernandez Rojo[2]

[1] Department of Aircraft Engineering, Kharkov State Aerospace University,
17 Chkalov Str., Kharkov 310070, Ukraine

[2] Department of Production Technology, Mexican Petroleum Institute,
152 Eje Central Lazaro Cardenas, Mexico City 07730 DF, México

Keywords: Apparatus for Impulsive Forming, Explosive Forming, Multidimensional Space, Optimal Parameters, Shells, Superplasticity, Titanium Alloys

ABSTRACTS

The impulsive forming is widely used in the modern industry. The superplastic impulsive forming isn't well known, though could have useful applications.

Applied aspects of the superplastic impulsive forming problem, including both control external stress and heat influence parameters are stated in the present research. The problem is reproduced in the vector representation of o timizing functions. In the case of the limited number of measurements the task leads to well known scal r representations with similar physical characteristics.

In the technological processes considered here forming pressures are generated by the impulsive energy sources on the base of high explosives. Preheating of the deformed workpiece is performed from the autonomous heat power source or workpieces are heated during forming process.

THEORETICAL ASPECTS

In the past, explosive forming has been primarily a trial and error process. The application of computer simulation to the process has reduced the trial and error procedures. Further step to refine the technology was made by working out the new method, which permit to determine optimum forming parameters. The method for designing optimum impulsive superplastic forming is developed.

For the first time, the basis of the method for designing optimum technological processes were offered in [1]. The method includes two basic stages: (1) to determine the optimum required characteristics, and (2) to determine the optimum available characteristics. The optimum required characteristics are speeds of workpiece points, final stress-strain state, and part form. The optimum available characteristics are impulsive power fields. Explosive substance charges generate these fields, satisfying these characteristics.

In the present research, additional optimized parameters are temperature field, and boundary conditions. Various heating apparatuses realize the temperature field, and special technological methods realize the boundary conditions.

In this case the optimization of technological process represents vector optimization, leading to the satisfaction to some number of functions. The functions are connected among themselves by means of several beforehand given compromises. Though the quasi-continuous (or continuous) managements, satisfying some initial and boundary conditions, will form set of the allowable management functions.

A workpiece during shaping is described by a surface:

$$Y_3 = S(Y_1, Y_2) \tag{1}$$

Where Y_i (i = 1, 2, 3) = Cartesian coordinates.

A die surface determines the final form of a product:

$$Y_3 = H^*(Y_1, Y_2) \tag{2}$$

And final deformed condition:

$$E_j = E^*_j(Y_1, Y_2). \tag{3}$$

If current deformations are $E_j(Y_\alpha, t)$, and speeds of workpiece points at the moment of the impact between the workpiece and the die are $W(Y_\alpha, t)$, the appropriate relative functions are defined as:

$$I_S = \int_{(F)} (\tilde{S}(Y_\alpha, t_g) - \tilde{H}^*(Y_\alpha))^2 \cdot dF \tag{4}$$

$$I_E = \int_{(F)} (\tilde{E}_j(Y_\alpha, t_g) - \tilde{E}^*(Y_\alpha))^2 \cdot dF \tag{5}$$

$$I_W = \int_{(F)} (\tilde{W}(Y_\alpha, t_g) - \tilde{W}^*(Y_\alpha))^2 \cdot dF \tag{6}$$

Where: I_S = Final detail configuration function
I_E = Deformed state function
I_W = Workpiece-die impact velocity function
α = 1,2.

The required mode of the impulsive superplastic forming should satisfy several optimization criteria equations (4), (5), and (6). The best considered allowable management functions $I_S^1, I_S^2, I_E^1, I_E^2, I_W^1, I_W^2$ are the left-handed functions:

$$I_S^1 < I_S^2, \quad I_E^1 < I_E^2, \quad I_W^1 < I_W^2. \tag{7}$$

The set of management functions, which cannot be improved on three functions simultaneously, makes a subset of allowable functions of workpiece temperature distribution, $T(S, t)$, and impulsive load distribution, $P(S, t)$. The set of optimization criteria will form a vector of criteria, **G**, with components, G_S, G_E, G_W. For searching optimum meaning, G, it is necessary to establish a rule of comparison of two vectors, on which it is possible to make a conclusion about which of the criteria G_S, G_E, G_W to consider the best. In most cases the choice of the relation of criteria, results a vector optimization problem to problem with scalar criterion in result of convolution of the criteria.

In real technological problems there is no sense to achieve an exact optimum of one criterion, because the mathematical model and boundary conditions of a problem aren't always completely adequate to a real existing process. Therefore it is necessary to search for such functions of management, for which the following inequalities are correct:

$$I_S - I_{S\,min} \leq \varepsilon_S, \qquad I_E - I_{E\,min} \leq \varepsilon_E, \qquad I_W - I_{W\,min} \leq \varepsilon_W \qquad (8)$$

The value $\varepsilon = \{\varepsilon_S\ \varepsilon_E\ \varepsilon_W\}$ should be more if initial data fallibility is large. According to the offered method the decision of a vector optimization problem is transformed to the decision of a scalar problem. A goal vector in this case is projected on a chosen direction or given from any practical conditions:

$$G = g_S\, I_S + g_E\, I_E + g_W\, I_W \qquad (9)$$

Where: $0 \leq g_Z$, $\qquad Z = S, E, W$ and

$$g_S + g_E + g_W = 1 \qquad (10)$$

For minimization G the method, developed in [2], is used, where managing parameter was only impulsive force function P (S, t). After determination of functions T (S, t) and P (S, t), minimizing the function G, the parameters of heating and explosive loading are known. Special devices, further described, are used for impulsive superplastic forming of details, satisfying given criteria of the form, thickness, and die lifetime.

Experience has shown that calculated static pressure must be raised at least 5-7 times to arrive at the explosive pressure required to form most materials.

Determination of s ndoff is reached through a compromise between the part's physical shape and the explosive required. For small parts, the explosive charge is placed near the focal point of the part shape or on the part axis; for larger parts, the considerations will be the facility limitations, maximum charge permitted, among others.

INDUSTRIAL APPLICATIONS

Forming with High Explosives

One advantage of superplastic explosive hot forming is the ability to form high strength low ductility materials as aluminum and titanium alloys.

Fig.1 shows the flanged box part stamped from Ti-6Al-2Zr-1Mo-1V flat blank, which has the maximum 120 % deformation in the angular zones. Fig.2 shows the ring frame part, which has the maximum 100% deformation in the flange-wall zone. Fig.3 shows the turbojet engine case explosively shaped from the Ti-5Al-2,5Sn conical preform. Fig.4 shows the ring frame part formed from A2024-T3 high-strength aluminum alloy. Table 1 shows the technological process parameters to shape the specified parts.

Equipment for superplastic impulsive forming of the parts from the flat and shell blanks was developed. Such installations are shown in Fig.5 and Fig.6. The flanged box part (Fig.1) and ring frame (Fig.2) have been formed on the installation shown in Fig.5 and the turbojet engine case (Fig.3) has been formed on the installation shown in Fig.6.

Thick wall tubes with flexible bottom diaphragms are used as explosive chambers to form the flanged box part and ring frame. Explosive chambers with multi-usable flexible thermo-resistant walls are used to form the turbojet engine case. Sometimes single-usable diaphragms have been used to form these parts.

Fig.1. Ti-6Al-2Zr-1Mo-1V flanged box explosively formed from flat square blank

Fig.2. Ti-6Al-2Zr-1Mo-1V explosively formed rocket body ring frame.

Fig.3. Turbojet engine case explosively shaped from a Ti-5Al-2,5Sn conical preform.

Fig.4. 2024-T3 aluminum ring frame explosively formed from conical blank.

Fig.5. Installation for impulsive forming with electric resistant blank heating.

Fig.6. Installation for impulsive forming of shell parts with radiation blank heating.

The ring frame part shown in Fig.4 was stamped without any equipment [3]. The reached maximum deformations are twice more than at static processing methods.

Table 1. Parameters of the technological processes

No	Parameters	Units	Flanged box	Ring frame
1	Dimensions: Width / Length / Height / Diameter	mm	60 / 60 / 8 / -	- / - / 17-26 / 150
2	Material		Ti-6Al-2Zr-1Mo-1V	Ti-6Al-2Zr-1Mo-1V
3	Thickness	mm	1.0/1.2	0.8/1.0
4	Maximum heating temperature	K	1174	1173/1123
5	Charge weight	g	5	6
6	Explosive charge immersing depth	mm	60	70
7	Explosive chamber diameter	mm	120	160
8	Explosive chamber height	mm	220	100
9	Semi-product aperture diameter	mm	33	60/80
10	Forming scheme: Throwing/Direct		No/Yes	Yes/No

These installations consist of horizontal and vertical guide units. The movable plate with mounting means arranged on it is installed on the horizontal guides. The movable plate is displaced from the preparing zone to the technological zone by the hydraulic drive. Forming blank heating is performed in the heat influence zone mainly with electrical contact method in the first installation variant, and with temperature radiation in the second installation variant. Using the method of electrical-contact heating, the installation is provided with compensating readjusting fixing device arranged motionless in respect to the base. A temperature radiation heater is mounted in such a way, that it can move vertically and during explosions, it is protected by shutters.

In the processes considered here high pressures are generated by the impulsive energy sources on the base of the detonating gas mixtures and high explosives. Preheating of the deformed workpiece is performed from the autonomous heat power source or workpieces are heated during forming process.

Part forming is carried out in special explosion chambers with the destructive of singular usage and elastic of multiple usage diaphragms. There are explosion chambers operating in the mode of the transmitting medium throw when the blast pressure, generated by the high explosives, has symmetrical characteristics in the vertical direction. The explosion chambers are installed on the vertically movable plate and the system for vertical displacement is arranged so that the loading of the drive during the explosions is excluded.

The installation (Fig.6) works as follows. In an initial position the pool is below, the die is in a zone of assembly, the heater is in the top position. The workpiece is in the die. A drive moves the die to the working chamber. Simultaneously, the single-useable pool is prepared. After this, a pre-heated heater moves downwards to the die to heat the workpiece. After the heating of the workpiece the heater moves upwards to the initial position. The pool moves to the die cavity. Simultaneously with these shutters the working chamber aperture is closed. The explosive charge is detonated. The die moves to the assembly zone. The stamped detail is taken from the die and the new workpiece and single-useable pool are established. After that the forming cycle is repeated.

Technical data of the installations for superplastic explosion forming from flat and cylindrical blanks are given in Table 2.

Table 2. Technical data of the installations for superplastic explosion forming.

No.	Data name	Unit	Flat Blank	Shell Blank
1	Floor plan dimensions	m	1.5x1.5x2.5	2.4x1.0x1.6
2	Mounting means dimensions	m	0.25x0.25x0.1	0.5x0.5x0.3
3	Maximum charge weight	kg	0.01	0.03
4	Blank heating temperature	K	1273	1440

Detonation Gas Forming

The use of the detonable gaseous mixtures for superplastic forming as the source of high pressures and high temperatures is safeter than the use high explosives. It is possible to mechanize and automates the mixture supply to the process zone and into the blank cavities of the complex configurations, distributed into difficult-to-access places. High duration of the impulse, as compared with high explosives, is favorable for the forming parameters. At present, such installations are used in the industry, and can operate in the open or closed scheme.

The construction of the installation is similar to that in [4] and comprises a supporting base in the form of a frame or bracket, with the movable power supply unit arranged on it. The technological unit ensures the necessary displacement of the blank together with mounting means. Pressurized destructive or reusable elastic diaphragms hold initial excess pressure of the gaseous mixtures. Forming may be performed by direct influence of the gaseous products of the blast or through the combined transmitting medium.

The constructed equipment has the close and open scheme. Table 3 shows the following technical data for the installations.

Table 3. Technical data of the installations for superplastic detonation gas forming.

No.	Data name	Unit	Close Scheme	Open Scheme
1	Floor plan dimensions	m	4.0x3.5x2.5	3.0x4.0x2.5
2	Maximum part dimensions	m	1.2x1.4	3.2x1.4
3	Power supply	kJ	1300	320

CONCLUSIONS

New technological industrial installations for the superplastic impulsive forming have been developed. The technological processes to form some parts from flat and shell blanks are shown. A single operation to shape the required final component is only used.

REFERENCES

[1] V.K. Borisevich, V.P. Sabelkin, and S.N. Solodyankin, About Computer Numerical Simulation of Explosive Forming Process, J. Appl. Mech. and Tech. Phys., No.2, 1979, pp.165-175.
[2] V. K. Borisevich, V. P. Sabelkin, and A. N. Potapenko, Design of Optimal Technological Processes of Hydroexplosive Forming by Computer Simulation, Proc. IX Inf. Conf. on HERF, Novosibirsk, 1986, pp. 209-213.
[3] M. Stuivinga, A. van Doormaal, Vovk V., V. Sabelkin, S. Molodikh, A. Andrienko, H. de Kruijk, and H.D. Groeneveld, Explosive Forming, an Enabling Technology, Proc. VI Int. Conf. on Sheet Metals, Twente, The Netherlands, 1998.
[4] GB Pat. 2 081 630, Int. Cl. B 2ld 26/08. Method, Apparatus and Gas Gun for Forming Articles by Impact Load. V. P. Sabelkin, V. K. Borisevitch, and S. N. Solodyankin, etc., 1984.

Superplasticity in Metals and Ceramics

Superplasticity in Metals and Ceramics

Oscar A. Kaibyshev

Institute for Metals Superplasticity Problems, Russian Academy of Sciences,
Khalturina 39, RU-450001 Ufa, Russia

Keywords: Ceramic, CGBS Bands, Cooperative Grain Boundary Sliding, Liquid Phase, Mechanisms of Deformation, Metals, Superplasticity

Abstract

Generality and distinctions in the mechanism of superplastic (SP) deformation between metals and ceramic materials are analyzed. It is established, that SP deformation depends on properties of grain boundaries and is determined by their structure and long-range area in a polycrystal. Under optimal SP deformation conditions there operates a specific mechanism of deformation, namely, cooperative grain boundary sliding (CGBS). The possible mechanisms of formation of CGBS bands are considered. Difference in SP behavior of metallic and ceramic materials is connected with occurrence of liquid and amorphous phases at grain boundaries during deformation.

Introduction

Superplastic (SP) deformation is observed in metals, intermetallics and ceramics [1,2]. This phenomenon happens to be more general than conventional deformation since under common conditions ceramic materials fail without any visual features of plastic flow and under SP conditions their deformation can be tens and hundreds of percent [3,4,5]. Transformation of materials to a SP state requires refinement of microstructure resulting in an increase of long-range areas of common type grain boundaries. Actually this refinement of microstructure should be $d<10$ μm. Since in this case the volume fraction of grain boundaries increases sharply, their influence on the mechanical properties of a material becomes determinant. Systematic studies conducted in recent years [1,2,6,7,8] have shown that SP deformation depends on properties of grain boundaries and is determined by their structure and long-range area in a polycrystal. This is confirmed by the fact that manifestation of superplasticity does not depend on the crystal lattice type and dislocations present. The influence of temperature-strain rate conditions of deformation on superplasticity is determined by the fact that under optimal SP deformation conditions there operates a specific mechanism of deformation, namely, cooperative grain boundary sliding (CGBS). The operation of this deformation mechanism creates conditions for attaining high strains without failure irrespective of the crystal lattice type of a material.

The aim of the present work is to analyse generalities and distinctions in the mechanism of SP deformation in metallic materials and ceramics.

Material and Experimental Procedure

A number of model metallic and ceramic materials were selected for conducting investigations. For studying the nature of SP deformation of metallic materials the well-known Zn-22%Al alloy, as well as aluminium tricrystals were used. To study the SP behaviour of ceramic materials the single-phase ceramics Bi_2O_3 displaying features of superplasticity at temperatures 600-700°C, ceramics Bi_2O_3 with additions 8.25% B_2O_3 and high temperature superconductive ceramics $YBa_2Cu_3O_{7-x}$ were used.

The main methods of investigations were: TEM, SEM, observation of deformation relief, local analysis – photographs of one and the same area taken before and after deformation.

Results

The consideration of deformation relief in a Zn-22%Al alloy during SP deformation revealed the non-homogeneous character of plastic flow: deformation is localized along surfaces passing through the whole sample cross section and domains are divided by material areas not involved in the deformation. These surfaces were called bands of CGBS (Fig.1, a) [9]. The investigations of mechanisms of SP flow have allowed to establish uniquely that CGBS is the main mechanism of deformation under such conditions. Moreover, formation of CGBS bands is observed not only on the sample surface but also within the material interior (Fig.2, a). To prove this statement the distribution of more than 400 two-side angles at triple junctions arranged along a band (at an

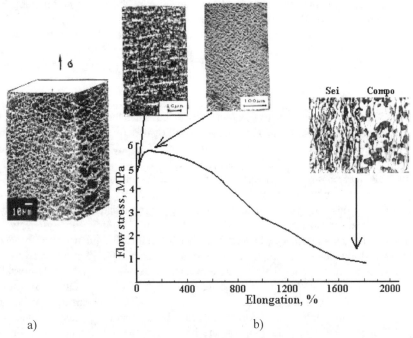

Fig.1. Formation of deformation bands on sample surface (a) and their development during tension (b).

Table 1. Average two-side angles in triple junctions of the Zn-22%Al alloy after tension (T=250°C, $\dot{\varepsilon}=7\times10^{-3}s^{-1}$, e=200%) [10]

Direction (Fig.2, b)	Average two-side angle in a triple junction, deg.
a (along the axis of tension)	122
b (crosswise to the axis of tension)	123
c (at an angle of 45° to the axis of tension)	142

angle of 45° to the axis of tension) and outside of a band (along and crosswise to the axis of tension) (Fig.2, b) [10] were measured in the Zn-22%Al alloy.

As the results of experiments have shown (Table 1), the two-side angles outside the band are close to equilibrium configuration (approximately $2\pi/3$). Triple junctions belonging to the band happen to be "straightened", i.e. their two-side angles display a distinct tendency to open to angles more than $2\pi/3$. So, for realizing cooperative shear along grain boundaries in a material due to "straightening" of a boundary a "smoothed" surface is created for total sliding. "Straightened" boundaries are revealed both on the surface and within the material interior. Thus, it has been shown that bands of CGBS are not the phenomenon connected with a surface but the result of deformation processes occurring within the material interior.

The mechanism of formation of CGBS bands is not yet clear. In this connection, special investigations of aluminium tricrystals were conducted [11]. The samples were produced from aluminium (99.999%) coarse-grained ingots with grain boundaries of general type. Tensile tests were conducted under creep conditions at the temperature 370°C and the stress 2.4 MPa for 7 hours. For local measurements of a horizontal constituent of grain boundary sliding and intragranular strain a thin square grid of marker lines with cell sides of 100 μm was scribed on a sample surface by a diamond point. The application of a system for optical analysis of an image provided an accuracy of measurements ~ 0.3 μm. For determining parameters of migration the angular positions and space paths of grain boundaries were measured before and after deformation after removing a layer, 30 μm thick, by electropolishing. The experiments have shown that there exist at least two possibilities of bands formation: 1) formation of grain boundary dislocation pile-ups and development of further deformation due to movement of intragranular lattice dislocations, fine adjustment of a grain boundary due to migration of a non-

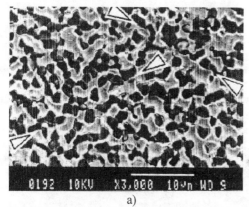
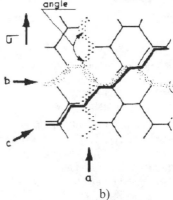

a) b)

Fig. 2. Bands of CGBS within the material interior (Zn-22%Al) (a) and a scheme showing measurements of two-side angles at a triple junction (b). [10].

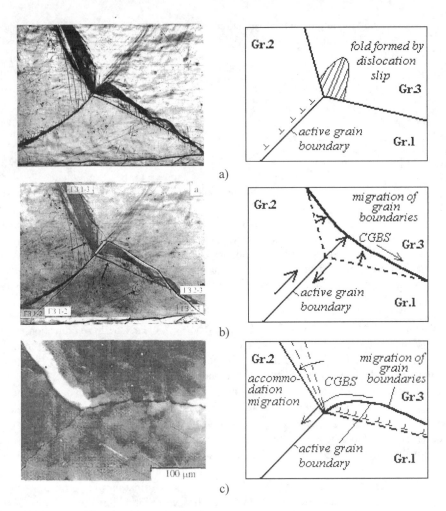

Fig.3. Cooperative grain boundary sliding in Al tricrystals. a) formation of a fold, b) fine adjustment of a grain boundary via migration, c) migration of a grain boundary.

favorably orientated boundary (Fig.3, b), and 2) migration of a triple junction. The investigations conducted on samples of Al tricrystals have shown that it is possible to change the orientation of the common surface of sliding via migration of a triple junction (see Fig.3, c). Due to such migration a pair of grain boundaries is proved to be favorably oriented for sliding.

Thus, generally a CGBS band is a combined physical object incorporating movements of grain boundary and lattice dislocations. The movement of these defects leads to plastic flow and deformation is realized mainly along grain boundaries. So, it is the structure of grain boundary but not the type of crystal lattice that is important for the movement of these defects.

The correlation between a change in mechanical properties and the mechanism of deformation has been established (Fig.1, b). Transition to superplastic deformation is accompanied by occurrence of CGBS bands.

The occurrence of CGBS requires favorable combination of structure conditions (small grain size, common type of grain boundaries) and temperature-strain rate conditions of deformation.

It is clear that such conditions can be realized during deformation of ceramic materials as well. However, the phenomenology of SP flow of ceramic materials can differ to some extent from that of metallic materials that is attributed to peculiarities of grain boundary structure of these materials. The fact is that there can be an amorphous phase or a phase with relatively low melting temperature on grain boundaries of ceramics and at temperatures above which these phases are molten, they can exert a significant influence on the mechanical properties and behaviour of materials.

All ceramic materials can be divided into two groups in respect to their manifestation of SP. The first group has a strain rate sensitivity factor m=0.4-0.5, while the second group has the high strain rate sensitivity m ≈1.0. This division is connected with differences in the structure and state of grain boundaries in ceramics and, consequently, their different deformation behaviour.

The investigations conducted on the model Bi_2O_3 ceramics allowed to reveal the influence of the state of grain boundaries on the SP properties of ceramics [12]. Thus, at temperatures above

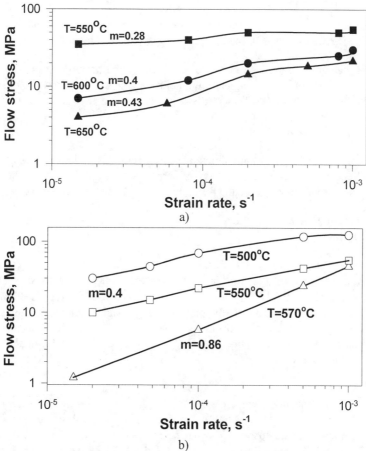

Fig. 4. Strain rate dependences of flow stress for pure Bi_2O_3 (a) and $Bi_2O_3+B_2O_3$-ceramics.

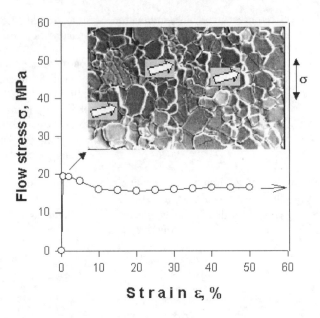

Fig. 5. Flow stress-strain value dependence for pure Bi_2O_3 ceramic and deformation relief after deformation (T=650°C, $\dot{\varepsilon} = 10^{-4} s^{-1}$). Bands of CGBS are indicated by arrows.

650°C pure Bi_2O_3 ceramic becomes ductile and at a temperature of 650°C it transforms to a SP state. With increasing deformation temperature ductility increases abruptly while the flow stress decreases gradually. The strain rate-flow stress dependence (Fig.4, a) has a S-type shape, the maximum value of the strain rate sensitivity coefficient m achieved was 0.43 at $\dot{\varepsilon}=10^{-4}s^{-1}$ and deformation to failure exceeded 70-80%.

Grain deformation and grain growth are almost absent in ceramics under conditions of SP flow. Deformation relief seen during SP deformation of ceramics does not differ from the one typical of metals: after the onset of plastic flow the formation of CGBS bands can be revealed (Fig.5).

Thus, the experimental data testify that for pure Bi_2O_3 ceramic also CGBS is the main mechanism of SP flow.

Addition of 8.25% B_2O_3 oxide to the Bi_2O_3 ceramic abruptly changed the development of deformation. In the temperature interval 500-550°C the mechanical behavior of such ceramics does not differ from that of the Bi_2O_3 ceramic: m ≈ 0.4, ductility achieved 70-80%. However, with increasing temperature up to 570°C the picture changes significantly: the flow stress has a linear strain rate dependence, the value of m approaches 1 (Fig.4, b), ductility decreases sharply, samples fail at small strain values. This is evidently connected with the formation of a liquid phase at grain boundaries (T_{ml} of B_2O_3 =577°C [13]) and the change in the mechanism of deformation.

During deformation of the high temperature superconductive $YBa_2Cu_3O_{7-x}$ ceramic features which are also connected with the structure of grain boundaries are observed [14]. There exist two temperature intervals distinguished by the mechanisms of deformation (Fig.6) and changes in microstructure (Fig.7). In the temperature interval 825-900°C SP deformation is realized via

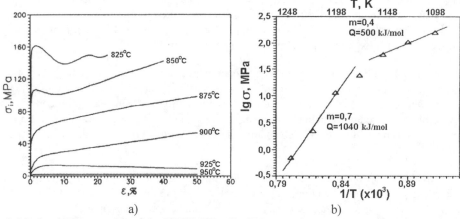

Fig.6. Mechanical properties of the HTSC $YBa_2Cu_3O_{7-x}$ ceramic.

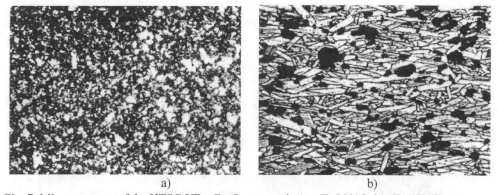

Fig. 7. Microstructure of the HTSC $YBa_2Cu_3O_{7-x}$ ceramic (a – T=850°C, b – T=950°C).

intragranular slip and grain boundary sliding and the superplastic behavior of this ceramics does not differ from that of metallic materials.

HREM studies of grain boundary structure did not reveal the presence of any phases at grain boundaries (Fig.8). The apparent energy of activation was Q=500 kJ/mol and the value of m = 0.4. Above 900°C the value of Q increased up to 1040 kJ/mol, and m reached 0.7. The change in the mechanism of deformation is connected with the formation of a liquid phase at grain boundaries at T > 900°C.

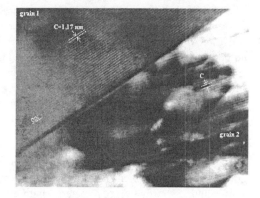

Fig.8. Electron microscopic image of grain boundary in the $YBa_2Cu_3O_{7-x}$ ceramic.

Discussion

For long time it was considered that the presence of the S-type dependence $\sigma=f(\dot{\varepsilon})$ in combination with high values of the coefficient m leads to SP and is its main visible feature. Grain boundary sliding (GBS) between neighbouring grains makes the main contribution to deformation and intragranular slip and diffusion creep provide accommodation of GBS in the vicinity of a triple junction. The notion of "stimulated" GBS seems to have explained SP deformation on a microlevel [15,16].

However, new data have shown that this approach is not sufficient for explaining all features of SP deformation. In particular, high strain rate superplasticity was observed at m < 0.3 [17], and, as shown above, in the Bi_2O_3 ceramic with B_2O_3 oxide additions the increase of m approximately to 1 sharply decreased deformation to failure, i.e. it happens that though low values of m are not impediments to realization of SP deformation, high values of m do not ensure manifestation of SP.

SP is provided by operation of CGBS. On a microscopic scale, it is provided by modification of a sufficient amount of triple junctions being natural retainers to the development of GBS and, on a macroscopic scale, by formation of bands of CGBS within the whole sample cross-section.

Reasoning from the given concept [18,19] it has been shown that SP deformation is a specific phenomenon which cannot be reduced only to favorable combination of operations of different deformation mechanisms. Besides we can explain the $\sigma=f(\varepsilon)$ curve, determine required conditions for average grain size, their dispersion, explain features of SP deformation under conditions of low temperature and high strain rate superplasticity.

The mechanism of CGBS bands formation has its origin in local migration of grain boundaries in the vicinity of a triple junction till such adjustment of its configuration that provides the occurrence of coordinated shear along two adjoining boundaries of the given junction [11]. Migration is possible in case sufficient energy of grain boundary dislocation (GBD) pile-ups is available at the grain boundary under consideration and the appropriate value of its coefficient of surface stress. (Note the analogy between the condition for transfer of deformation of grains in coarse-grained materials as in Hall-Petch stress and the condition for cooperative shear along adjoining boundaries at triple junctions). In this case the grain should not be coarse, otherwise there occurs a transfer to a regime of classic plasticity characterized by a high rate and a low resource of ductility. All these conditions are fulfilled in case of the pure Bi_2O_3 ceramics. However, when B_2O_3 oxide additions are made, a liquid phase is formed at grain boundaries and conditions for local migration disappear, since hardening of such a boundary is impossible (the liquid phase does not support shear stresses). Flow stresses decrease significantly, their strain rate dependence becomes almost linear in compliance with the rheology of the grain boundary phase and since bands of CGBS and not form the strain attained only a value of 20% (the resource connected with the presence of the grain boundary phase was exhausted) and deformation ended with brittle failure typical of ceramics having no bands of cooperative flow: no bands - no SP.

Another consequence of our experiments with ceramics is that we start to understand that SP is a fundamental macrophenomenon distinguished from regimes of classic plasticity of coarse-grained materials which are uniquely determined by mesoscopic scale processes. During conventional deformation of a material with a large average grain size the deformation of grains copies the deformation of the sample as a whole. In other words, the representative volume is given by a size of an average grain and transfer to a macroscopic scale is provided by averaging over a well determined area. In fact, the task of derivation of a physical theory of plasticity for coarse-grained materials is reduced to a correct description of the mesoscopic scale processes by means of Taylor's model.

The study of SP phenomenon involves quite a different situation. In case of a small grain size the densities of lattice dislocations within the grain body do not achieve high values, there is no transfer to a regime of non-linear evolution in the process of deformation and one can consider only interactions of lattice dislocations with grain boundaries. That is, compared to classical plasticity of coarse-grained materials, the mesoscopic scale of SP deformation is characterized by the transfer of the main processes to the grain boundaries. Under these conditions the application of Taylor's model for transfer of regularities of the mesoscopic scale to the microscopic scale becomes impossible: during deformation of a sample by 1000% grains remain almost equiaxed. Under these conditions the volume of the area of averaging for transfer to the macrolevel is determined by a long range area of a main attribute of the deformation mechanism, namely, a band of CGBS.

Thus, we can assume that a SP material is a material with an infinite amount of sliding systems, CGBS bands playing their roles. However, there exists one fundamental and very important distinction between classic sliding systems of a polycrystal and CGBS bands. This distinction forms the basis of an alternative to the Taylor paradigm in SP. The amount and orientation of sliding systems of a monocrystal are uniquely determined by its crystal lattice. Strengthening of the active sliding system in the process of deformation and a low value of the Schmidt factor of other systems can destroy rather rapidly the second term of the Taylor paradigm and thereby the term of compatibility of deformation of a polycrystal as an entity. Failure will take place inevitably. Quite another situation is present in case of CGBS bands. A non-textured material with a small average grain size on the macrolevel is almost isotropic. Due to that the orientation of CGBS bands is determined only by the characteristics of the stress state, namely, CGBS bands are arranged in the plane of action of the maximum shear stresses. Strengthening of bands (for example, due to the action of cavitation effects or rotation during deformation and subsequent elimination of maximum shear stresses from the plane) will block them up. However, changes in the stress state will form the system of CGBS bands with another orientation, since there exists an opportunity to select from a continuum of orientations and not from a finite number of sliding systems. It is more complicated to exhaust the continuum than a fixed number of sliding systems found in conventional plasticity. It is this fact that leads to high strain values during SP deformations. Failure of a SP material can occur only in the following cases: 1) the material has been extremely damaged in the process of deformation (for example cavitation); 2) some portions of the sample have become very thin and effects of fluctuation play a leading role; 3) external conditions have changed and the mechanism of formation of CGBS bands has been destroyed.

Thus, the generality of the mechanism of deformation dictates the generality of SP deformation manifestation in metals and ceramics. Distinctions occur in case of a sharp change in the mechanism of deformation, in particular, due to the formation of a liquid phase at grain boundaries.

Conclusions

1. During SP deformation the CGBS bands are revealed both on the surface and in the entire of the sample bulk in the form of "straightening" of grain boundaries.

2. On experiments with aluminum tricrystals it has been shown that formation of a CGBS band is connected with migration of boundaries near a triple junction.

3. The Bi_2O_3 ceramics at 650°C has all features of SP deformation and during this deformations CGBS occurs.

4. During deformation of the Bi_2O_3 ceramics with B_2O_3 additions at the temperature 570°C an increase in the coefficient m → 1 is observed, the ductility reduces sharply but deformation relief is not observed.

5. Difference in SP behavior of metallic and ceramic materials is connected with occurrence of liquid and amorphous phases at grain boundaries during deformation.

References

1 *Kaibyshev O.A.* Superplasticity of alloys, intermetallides, and ceramics. Berlin; New York: Springer-Verlag,1992. 317 p.
2 *Mucherjee A.K.* Superplasticity in Metals, Ceramic and Intermetallic // in: Plastic Deformation and Fracture of Materials. Ed by Mughrabi H. Materials Science and Technology, v.6, VCR Verlagsgesellschaft mbH, Germany, 1993.
3 *Wakai F., Sakaguchi S. and Murayama M.* Advanced Ceramic Materials, 1986, v.1, pp. 259-263.
4 *Nieh T.G. and Wadsworth J.* Annu. Rev. Mater. Sci., 1990, v.20, pp.117-140.
5 *Maehara Y. and Langdon T.G.* J. Mater. Sci., 1990, v.25, pp. 2275-2286.
6 *Padmanabhan K.A. and Davies J.J.* Superplasticity. Springer-Verlag: Berlin, Germany. 1980, 314 p.
7 *Novikov I.I. and Portnoy V.K.* Superplasticity of alloys with ultrafined grains. Moscow, Metallurgia, 1981, p. 168. (in Russian)
8 *Kaibyshev O.A. and Valiev R.Z.* Grain boundaries and properties of metals. Moscow, Metallurgia, 1987, p. 214. (in Russian).
9 *Astanin V.V., Kaibyshev O.A. and Faizova S.N.* Scripta Met. at Mater., 1991, v.25, No12, pp. 2663-2668.
10 *Astanin V.V., Faizova S.N. and Padmanabhan K.A.* Mater. Sci. Technol., 1996, v.12, June, pp.489-494.
11 *Astanin V.V., Sisanbaev A.V., Pshenichnyuk A.I. and Kaibyshev O.A.* Scripta Met. et Mater., 1997, v.36, No 1, pp.117-122.
12 *Kaibyshev O.A., Zaripov N.G. and Kolnogorov O.M.* Advanced Materials '93, III/B: Composites, Grain Boundaries and Nanophase Materials, Trans. Mat. Res. Soc. Jpn., 1994, 16B, pp. 971-975.
13 *Gogotsi G.A.* J. Less Common Metals, 117 (1986) 225
14 *Imayev M.F., Kaibyshev R.O., Musin F.F., Shagiev M.R.* Materials Forum., 1994, v.170-172, p. 445.
15 *Kaibyshev O.A., Valiev R.Z., Astanin V.V. and Emaletdinov A.K.* Phys. Stat.Sol.(a), 1983, v.78, pp.439-448.
16 *Pshenichnyuk A.I., Astanin V.V. and Kaibyshev O.A.* Phil. Mag. A, 1998, v.77, No 4, pp.1093-1106.
17 *Matsuki K., Iwaki T., Tokizawa M., Murakami Y.* Mater. Sci. and Technology, 1991, v.7, No 6. pp. 513-519.
18 *Kaibyshev O.A., Pshenichnyuk A.I., Astanin V.V.* Acta mater., 1998, v.46, No 14, pp. 4911-4916.
19 *Pshenichnyuk A.I., Kaibyshev O.A. and Astanin V.V.* Phil. Mag. A, 1999, v.79, No 2, pp. 329-338.

The Role of Impurities in Superplastic Flow and Cavitation

Farghalli A. Mohamed

Department of Chemical and Biochemical Engineering and Materials Science,
University of California, Irvine CA 92697-2575, USA

Keywords: Boundary Sliding, Cavitation, Ductility, Former Alpha Boundaries, Impurity Segregation, Micrograin superplasticity

Abstract

Superplasticity refers to the ability of fine-grained materials ($d < 10$ μm, where d is the grain size) to exhibit extensive neck-free elongations during deformation at elevated temperatures ($T > 0.5\ T_m$, where T_m is the melting point). An important characteristic of the deformation behavior of micrograin superplastic alloys is the experimental observation that the relationship between stress and steady-state creep rate is often sigmoidal. Such sigmoidal behavior is characterized by the presence of three regions: region I (the low-stress region), region II (the intermediate-stress region or the superplastic region), and region III (the high-stress region). Recently, the effects of impurity level and type on the sigmoidal relationship reported for superplastic alloys have been studied in detail. Some of the results of these studies are reviewed with particular emphasis on creep behavior, boundary sliding, and cavitation.

1. Introduction

The ability of fine-grained materials ($d < 10$ μm, where d is the grain size) to exhibit extensive plastic deformation at elevated temperatures ($T > 0.5\ T_m$, where T_m is the melting point), often without the formation of a neck prior to fracture, is generally known as micrograin (structural) superplasticity.

It is well-documented that the mechanical behavior of micrograin superplastic alloys at elevated temperatures often exhibits a sigmoidal relationship between stress, σ (tension) or τ (shear), and steady-state creep rate, $\dot{\varepsilon}$ (tension) or $\dot{\gamma}$ (shear). Such a sigmoidal relationship is characterized by the presence of three regions [1-6]: region I (the low-stress region), region II (the intermediate-stress region or the superplastic region), and region III (the high-stress region). In region II, maximum ductility occurs. Because of this characteristic, region II is often referred to as the superplastic region.

Recently, the effects of impurity level and type on the sigmoidal relationship reported for superplastic alloys have been studied in detail [7-17]. In particular, these studies have focused on the origin of the sigmoidal behavior at low stresses (region I), the characteristics of cavitation in superplastic alloys, and the dependence of the contribution of boundary sliding to the total strain on strain rate.

The purpose of this paper is two-fold: (a) to review some of the results obtained on the effects of impurities on superplastic flow, boundary sliding, and cavitation, and (b) to provide evidence in support of the concept of impurity segregation at boundaries during superplastic flow.

2. Discussion
2.1. Deformation behavior

Superplasticity is regarded as a creep phenomenon since it has been observed at temperatures at or above 0.5 of the melting point. Accordingly, in establishing the mechanical behavior of superplastic alloys, investigators extensively studied the following four relationships that define the basic deformation characteristics associated with a creep process: (a) the relationship between stress and strain rate, (b) the relationship between strain rate or stress and temperature, (c) the relationship between strain rate or stress and grain size, and (d) the relationship between strain contributed by boundary sliding and total strain.

As a result of the studies on the aforementioned relationships, three findings are well-documented. First, micrograin superplasticity is a diffusion-controlled process that can be represented by the following dimensionless equation [18]:

$$\frac{\dot{\gamma} k T}{D G b} = A \left(\frac{b}{d}\right)^s \left(\frac{\tau}{G}\right)^n \text{ with } D = D_o \exp\left(-\frac{Q}{RT}\right) \qquad (1)$$

where $\dot{\gamma}$ is the shear creep rate, k is Boltzmann's constant, T is the absolute temperature, D is the diffusion coefficient that characterizes the creep process, G is the shear modulus, b is the Burgers vector, A is a dimensionless constant, d is the grain size, s is the grain size sensitivity, τ is the applied shear stress, n is the stress exponent, Q is the activation energy for the diffusion process that controls the creep behavior, and D_O is the frequency factor for diffusion. Second, the relationship between stress, τ, and strain rate, $\dot{\gamma}$, is often sigmoidal. Under creep testing conditions, this sigmoidal relationship is manifested by the presence of three regions, as illustrated in Figure 1: region I (the low-stress region), region II (the intermediate-stress region or the superplastic region), and region III (the high-stress region). In region III (the high-stress region), the stress exponent, n is higher than 3, the apparent activation energy, Q_a, is higher than that for grain boundary diffusion. Region II (the intermediate-stress region) covers several orders of magnitude of strain rate and is characterized by a stress exponent, n, of 1.5 to 2.5, an apparent activation energy, Q_a, that is close to that for boundary diffusion, and a grain size sensitivity, s of about 2. In this region, maximum ductility occurs. Because of this characteristic, region II is often referred to as the superplastic region. Region I is characterized by a stress exponent of 3 to 5, and an apparent activation energy higher than that for grain boundary diffusion. However, the creep behavior in this region exhibits essentially the same grain size sensitivity noted in region II. Finally, at low elongations (typically of the order of 20-30%), the percentage contribution of boundary sliding to total strain generally ranges from 50-70% in region II but it decreases sharply, to approximately 20-30%, in regions I and III.

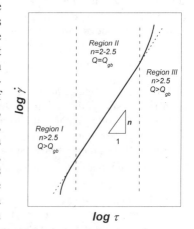

Figure 1: Schematic representation for the three regions associated with superplastic flow.

2.2. Effects of impurities on micrograin superplasticity at low stresses

Recent experimental evidence has revealed that the characteristics of micrograin superplasticity at low stresses are influenced by impurities [7-17]. Some of the details of this evidence are reviewed in the following section.

2.2.1 Creep Characteristics. Data from several creep investigations on micrograin superplasticity have shown that creep characteristics at low stresses are controlled not only by impurity level but also by its type. This finding has been demonstrated by the following experimental observations: (a) Zn-22% Al and Pb-62% Sn do not exhibit region I when the level of impurities in both alloys is reduced to about 6 ppm [7, 10]; (b) for Zn-22% Al, the apparent stress exponent and the apparent activation energy for creep in region I, unlike those in region II, are sensitive to impurity content [7]; and (c) for constant temperature, the transition between region II and region I is transposed to lower strain rates with decreasing the level of Fe [9], and (d) region I is absent in Zn-22% Al when the alloy is doped with Cu [15] whereas this region is observed when the alloy is doped with a comparable atomic level of Fe [12] (Figure 2).

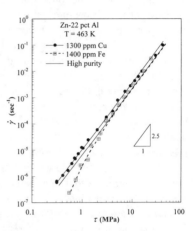

Figure 2: *The stress dependence of creep rate in three grades of Zn-22% Al: High-purity grade, Fe grade, and Cu grade.*

2.2.2. Grain boundary sliding. In a very recent study [19], the effect of Cd on the contribution of boundary sliding to the total strain was examined in the superplastic Pb-62% Sn alloy. The results of that study are briefly presented below.

Figure 3 represents a plot between the contribution of boundary sliding to the total strain, ξ, and initial strain rate, $\dot{\varepsilon}_0$. In this plot, the vertical lines A and B represent the transitions from region I to region II and from region II to region III, respectively, which were determined from earlier data reported for grade 1 (Pb-Sn doped with Cd) [10]. As mentioned earlier, this grade, unlike grade 2 (high purity Pb-62% Sn), exhibits region I that characterizes the sigmoidal relationship between stress and strain rate at low strain rates; for grade 2, region II extends over both intermediate and low strain rates. An examination of such a plot reveals important results that are discussed below in reference to the three deformation regions of the sigmoidal relationship between stress and strain rate reported for superplastic alloys [1-6].

High-strain rate region. For both grade 1 and grade 2, the contribution of boundary sliding to the total strain, ξ, is close to 20% in the high-strain rate region (region III).

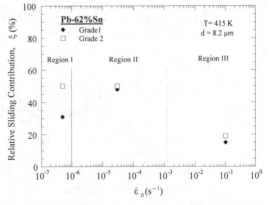

Figure 3: *The contribution of boundary sliding to the total strain, ξ, as a function of strain rate for grade 1 and grade 2 (high purity grade) of Pb-62% Sn.*

This result agrees well with previous measurements documented for a number of superplastic alloys in region III [20]. In addition, the result is consistent with an earlier finding [10] that the advent of region III occurs in both grades of the alloy under about the same stresses (for identical conditions of temperature and grain size).

The intermediate-strain rate region. For grade 1 and grade 2, the contribution of boundary sliding to the total strain, ξ, is essentially the same in region II and is close to a value of 50%. Accordingly, the values of ξ estimated in the present investigation are in good agreement with those reported for superplastic alloys, especially Pb-62% Sn [21]. In addition, it has been reported [16] that three grades of Zn-22% Al containing different levels of various impurities exhibit essentially the same value for ξ in region II (60%). This finding together with the present result signifies that impurities have no noticeable effect on sliding in this region.

Low strain rate region. As plotted in Figure 3, the data on grade 1 and grade 2 reveal the presence of an important difference in the behavior of the two grades with respect to the contribution of boundary sliding to the total strain, ξ. On the one hand, sliding measurements for grade 1 containing 890 ppm of Cd show that the values of ξ in region I (31%) are considerably lower than those in region II (50%). On the other hand, the data for grade 2 indicate that there is no difference between the value of ξ at low strain rates and that at intermediate strain rates (after correction for concurrent grain growth). The behavior of grade 2 (high-purity Pb-62% Sn) resembles that reported for high-purity Zn-22% Al (6 ppm of impurities) [16]. Consideration of the above findings leads to the conclusion that the contribution of boundary sliding to the total strain, ξ, in region I, unlike that in region II, is affected by the presence of impurities.

2.2.3. Cavitation. Experimental results have revealed the following observations: (a) cavities are not observed in high-purity Zn-22% Al [8], (b) the extent of cavitation in Zn-22% Al depends on the impurity content of the alloy [8], and (c) cavitation is nearly absent in Zn-22% Al doped with Cu whereas cavitation is extensive in the alloy when doped with a comparable atomic weight level of Fe [15].

2.2.4. Ductility. The results of a recent investigation have indicated [13] that introducing high Fe levels in Zn-22% Al, while not changing the stress exponent for creep in region II, leads to a loss in the ductility of the alloy in this region. Some of the results of that investigation are shown in Figure 4, where the average elongation to failure estimated from several specimens that were pulled to failure at 1.33×10^{-4} s^{-1} in region II (the superplastic region), where $n = 2.5$, is plotted as a function of Fe content. As shown by the figure, there are two regions of behavior: (a) the average elongation to failure decreases slowly for Fe concentrations less than 125 ppm, and (b) the average elongation to failure decreases rapidly at higher Fe levels.

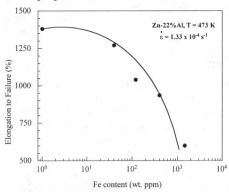

Figure 4: *The average elongation to failure as a function of Fe content.*

2.3. Possible origin of superplastic flow at low stresses

The aforementioned experimental results on superplastic flow in Zn-22% Al and Pb-62% Sn suggest that the origin of the deformation processes controlling region I is related to the presence and activity of impurity atoms. On the basis of this suggestion, it is most likely that region I behavior is related to one of the following processes: (a) viscous glide [7, 22], (b) grain boundary migration controlled by impurities [23], and (c) an impurity-dominated threshold stress [24]. Consideration of the details of processes (a) and (b) has led to the conclusion [7, 9] that they do not provide a satisfactory explanation for all the creep characteristics associated with region I. By contrast, the concept that region I behavior may be a consequence of the operation of a threshold stress process whose origin is related to the segregation of impurity atoms at boundaries and their interaction with boundary dislocations appears to be consistent with present experimental evidence; in this case, the threshold stress, τ_o, is equivalent to the stress that must be exceeded before boundary dislocations can break away from the impurity atmosphere and produce deformation. The consistency between the prediction of the concept of an impurity-dominated threshold stress process and the trend of experimental data is briefly described below.

2.3.1 Threshold stress for superplastic flow. Recent creep data reported for several grades of Zn-22% Al containing different levels of impurities [7, 9], in particular Fe, have revealed the presence of a threshold stress whose characteristics are consistent with various phenomena associated with boundary segregation as summarized below.

i. No threshold stress, τ_o, is observed for superplastic flow in high purity Zn-22% Al [7, 9]. A similar trend for high purity Pb-62% Sn was noted [10]. As reported elsewhere [7, 9, 24], τ_o, is determined from experimental creep data obtained for a superplastic alloy by plotting as $\dot{\gamma}^{1/n}$ ($n = 2.5$) against τ at a single temperature on a double linear scale and extrapolating the resultant line to zero strain rate.

ii. According to the experimental data reported for superplastic alloys [7, 9, 10], the temperature dependence of the threshold stress is described by the following equation, $\tau_o/G = B_o \exp(Q_o/RT)$, where B_o is a constant, Q_o is an activation energy term. The plot of Figure 5, in which experimental data on the threshold stress behavior of several grades of Zn-22% Al doped with Fe are plotted as τ_o/G versus $1/T$ on a logarithmic scale, provides a graphical presentation for the temperature dependence of τ_o. The above equation also resembles in form the following equation [25] that gives, to a first approximation, the concentration of impurity atoms segregated to boundaries, c, as a function of temperature, $c = c_o \exp(W/RT)$, where c_o is the average concentration of impurity, and W is the interaction energy between a boundary and a solute atom.

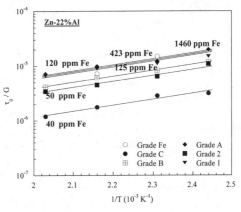

Figure 5: The threshold stress dependence on temperature as a function of Fe content.

iii. The threshold stress appears to approach a limiting value for Fe concentrations above 120 ppm (Figure 5). This Fe level (120 ppm) most likely represents the concentration at which boundary sites available for Fe segregation approach a saturation limit [10].

2.3.2. Segregation of impurities. Observations regarding the effects of impurities (level and type) on cavitation, ductility, and boundary sliding are consistent with the concept of impurity segregation at boundaries in superplastic materials. Given below are four examples for such consistency.

i. The presence of other impurities in addition to Fe in Zn-22% Al results in enhancing cavitation [13]. This observation appears to be consistent with the synergistic effects associated with impurity segregation at boundaries [26].
ii. The observed correlation between the level of impurities and the extent of cavitation in Zn-22% Al is most probably related to effects associated with the presence of excessive impurities at boundaries due to their segregation [8, 10, 27]. As mentioned elsewhere [8, 10, 27], impurity segregation may lead to accelerated cavitation rates through the following processes: (a) reduction of the surface energy, (b) formation of precipitates which serve as cavity nucleation sites, (c) reduction of grain boundary diffusivity, (d) reduction of boundary cohesive strength, and (e) retardation of grain growth which involves boundary migration.
iii. The observation [15] that the presence of Cu in Zn-22% Al does not result in extensive cavitation while the opposite is true in case of the presence of Fe appears to be consistent with the expectation that impurities vary greatly in tendency to segregate at boundaries.
iv. The decrease in the contribution of sliding at low strain rates, where region I is normally observed, is consistent with the expectation that the segregation of impurities at boundaries may influence accommodation processes for boundary sliding, which in general include boundary migration, diffusional flow, dislocation motion, or cavitation. For example, impurities may produce a strong dragging effect on migrating boundaries [28]. As a result, grain boundary migration (GBM), which is a fast process at intermediate strain rates (intermediate stresses, region II) [29], may become too slow to fully accommodate boundary sliding in this range of strain rates, leading to a decrease in the amount of sliding and an increase in the extent of cavitation.

2.3.3. Evidence for impurity segregation. In general, direct examination of boundary segregation requires the application of spectroscopic techniques, such as Auger spectroscopy, in which samples are fractured in vacuum and exposed grain boundaries are examined for segregation. In the case of materials that exhibit a ductile-to-brittle transition, in situ fracture is produced by a combination of sample cooling and impact fracture. However, since superplastic alloys such as Zn-22% Al samples do not exhibit low temperature brittleness, the procedure is not feasible to propagate intergranular failure. Due to the difficulty of obtaining direct information (using conventional methods) regarding boundary segregation in Zn-22% Al, an alternative approach that is based on the presence of former α boundaries ($F\alpha Bs$) has been adopted.

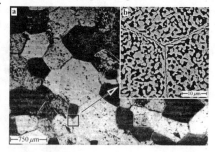

Figure 6: Typical microstructure observed in Zn-22% Al after heat treatment showing $F\alpha Bs$.

$F\alpha Bs$ are residual grain boundaries that develop in Zn-22% Al during a heat treatment which is normally applied to produce the fine structure necessary for micrograin superplasticity. They represent domains consisting of fine elongated α grains, which encompass groups of fine α (Al-rich) and β (Zn-rich) phases (the superplastic microstructure). These characteristics are illustrated in Figure 6. It has recently been demonstrated that the kinetics of $F\alpha B$ domain growth, like that of normal grain growth, is controlled by impurities [15]. This observation which indicates that $F\alpha Bs$, like grain boundaries, serve as favorable sites for impurity segregation, presents an alternative approach to examine the tendency of different impurity atoms to segregate at boundaries as a function of their level and type

Figure 7 show that for five grades of Zn-22% Al, the variation in the *average FαB domain* size, D_α, as a function of annealing time, t_S, exhibits an initial short stage in which the increase is rapid followed by a longer stage in which the increase becomes slow. The annealing curves shown in Figure 7 are similar in trend to those reported for the effect of solute additions on grain growth in zone-refined metals [39]. Consideration of the data of the figure shows the following trends: (a) grade HP (high-purity grade, 6 ppm of impurities) exhibits a much more drastic increase in D_α as compared to grades 1 and 2; (b) D_α in grade 1 (180 ppm of impurities) is always smaller than that in grade 2 (100 ppm of impurities); and (c) the data on grade Cu (1300 ppm), which does not exhibit significant cavitation, fall very close to those on the high-purity grade of the alloy while the data on grade Fe (1400), which exhibits significant cavitation, fall slightly below those on grade 1. These trends imply that the extent of impurity segregation is highest in grade Fe and lowest in grade HP, and that Cu, unlike Fe, has little or no tendency to segregate at boundaries. Such implications regarding the extent of segregation in the five grades are in agreement with the extent of cavitation as shown in Figure 7.

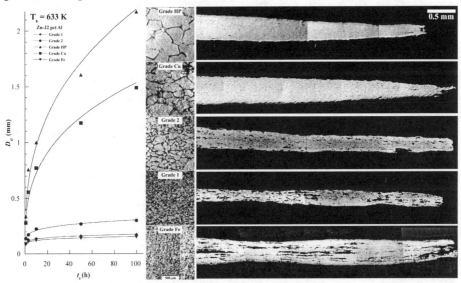

Figure 7: The annealing curves for several grades of Zn-22% Al. The representative micrographs show: (a) the FαB domain size after annealing for 7 h and (b) the extent of cavitation.

In addition to providing indirect evidence for the occurrence of segregation at boundaries in Zn-22% Al, the presence of *FαBs* was utilized to explain the formation of cavity stringers in the alloy. This explanation is based on three primary characteristics of *FαBs*. First, *FαBs*, unlike the fine microduplex structure in Zn-22 pct Al, are elongated rather than equiaxed Accordingly, these boundaries would be resistant to both boundary sliding and rotation. By serving as obstacles to the sliding of either individual grains of α and β or groups of these grains, *FαBs* would play a role similar to that of triple junctions or other discontinuities in the plane of the boundary. In the absence of accommodation by diffusion, boundary migration, and/or grain deformation, the stress concentrations at *FαBs* would be relieved by cavity nucleation. Second, segregation of impurities to *FaBs* is expected to result in nucleating cavities due to effects associated with the presence of excessive impurities at boundaries [28]. Finally, in addition to serving as favorable cavity nucleation sites, recent microstructural observations have shown that, during superplastic deformation, FαBs change their orientation and become aligned with the tensile axis. This finding has not only rationalized the origin of cavity stringers in Zn-22 pct Al but also provided a general explanation for the formation of cavity stringers during superplastic flow [14].

2.4. Correlations between superplastic properties and impurities

In Figure 8, the contribution of boundary sliding together with data on the threshold stress for superplastic flow, the stress exponent for creep at low strain rates, n and the *average FαB domain* size, D_α, are plotted as a function of Fe level in Zn-22% Al. An inspection of this figure reveals the presence of the following three regions of behavior:

i. Region A (Fe content < 15 ppm). In this region, as the Fe content approaches values less than ~15 ppm Fe, the contribution of boundary sliding to the total strain, ξ, approaches the value (60%) measured at intermediate-strain rates (the superplastic region), the threshold stress for superplastic flow, τ_o approaches zero, and the stress exponent for creep, n, at low strain rates approaches 2.5 that characterizes the superplastic behavior at intermediate strain rates (the superplastic region, where maximum ductility occurs), and D_α (which serves as a measure for the extent of segregation) becomes large.

ii. Region B (transition region; 15 ppm < Fe content < 120 ppm). In this region, an increase in the Fe content results in a rapid decrease in both ξ and D_α and parallel increase in both τ_o and n.

iii. Region C (Fe content > 120 ppm). In this region, as the Fe content approaches values greater than 120 ppm Fe, ξ, D_α, τ_o, and n reach a limiting value corresponding to those observed in grade Fe.

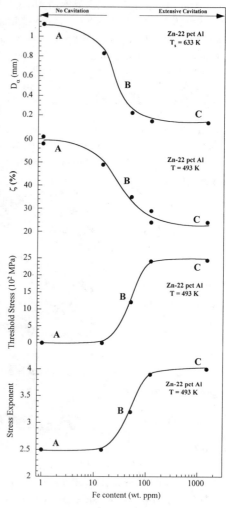

Figure 8: The influence of Fe content in Zn-22%Al on D_α, ξ, τ_o, and n, at low strain rates.

It has been suggested [10] that an Fe level of 120 ppm may represent the concentration at which boundary sites available for Fe segregation approach the saturation limit, or the concentration at which Fe precipitates start to form.

The trends revealed by the aforementioned three regions demonstrate that Fe content influences in *a parallel way* both the *average FαB domain* size, D_α, and the low-stress superplastic characteristics of the alloy (threshold stress for creep, the stress exponent for creep in region I, and the extent of cavitation). This finding provides evidence for the interaction between two roles played by boundaries during superplastic deformation. These roles are: (a) the ability of boundaries to contribute to deformation through the process of boundary sliding [28-33], and (b) the ability of boundaries to serve as favorable sites for the accumulation of impurities [7-17, 24], i.e. boundary segregation.

3. Conclusions

Recent experimental evidence has shown that impurities (level and type) affect the characteristics of micrograin superplasticity at low stresses (region I). These characteristics include the stress exponent, the activation energy for creep, boundary sliding, cavitation, and ductility. The sensitivity of the low-stress superplastic flow and cavitation to impurities have been explained in terms of strong impurity segregation at boundaries. This explanation is supported by: (i) the presence of a threshold stress whose characteristics are consistent with various phenomena associated with boundary segregation, and (ii) the growth kinetics of former alpha boundaries which are formed in Zn-22% Al during heat treatment above the eutectoid temperature.

Acknowledgments

This work was supported by the National Science Foundation under Grant No. DMR-9810422. I am also very thankful to my current and former graduate students for their contributions.

References

1. F. A. Mohamed and T. G. Langdon, *Acta Metall.* **23** (1975) 1443
2. F. A. Mohamed, S.-A. Shei and T. G. Langdon, *Acta Metall.* **23** (1975) 679.
3. F. A. Mohamed and T. G. Langdon, *Phil. Mag.* **32** (1975) 697.
4. S. H. Vale, D. J. Eastgate, and P. M. Hazzledine, *Scipta Metall.* **13** (1979) 1157.
5. D. W. Livesey and N. Ridley: *Scr. Metall.* **16** (1982) 165.
6. A. H. Chokshi and T. G. Langdon, *Metall. Trans. A*, **19A** (1988) 2487.
7. P. K. Chaudhury and F. A. Mohamed, *Acta Metall.* **36** (1988) 1099.
8. K. T. Park and F. A. Mohamed, *Metall. Trans.* **21A,** 2605 (1990).
9. P. K. Chaudhury, K. T. Park and F. A. Mohamed, *Metall. Trans.* **25A** (1994) 2391.
10. S. Yan, J. C. Earthman and F. A. Mohamed, *Phil. Mag. A*, **69** (1994) 1017
11. K. T. Park, S. T. Yang, J. C. Earthman and F. A. Mohamed, *Mat. Sci. Eng.* **A188** (1994) 59
12. S. T. Yang and F. A. Mohamed, *Metall. Mater. Trans. A*, **26A** (1995) 493
13. X. G. Jiang, S. T. Yang, J. C. Earthman and F. A. Mohamed, *Metall. Trans.* **27A** (1996) 863.
14. A.Yousefiani and F. A. Mohamed, *Acta Mater.* **46** (1998) 3557.
15. A.Yousefiani and F. A. Mohamed, *Metall. Mater. Trans. A*, **29A** (1998) 1653.
16. K. Duong and F. A. Mohamed, *Acta Mater.* **46** (1998). 4571
17. A. Yousefiani and F. A. Mohamed, *Metall. Mater. Trans. A*, **31A** (200) 163.
18. J. E. Bird, A. K. Mukherjee and J.E. Dorn, Correlations between high-temperature creep behavior and structure, in *Proceedings of a Symposium on* Quantitative Relation Between Properties and Microstructure, D. G. Brandon and A. Rosen, eds., Israel Univ. Press, Jerusalem, (1969), 22
19. K. Duong and F. A. Mohamed, *Phil. Mag. A* (in press*).*
20. Z.-R. Lin, A. H. Chokshi, and T. G. Langdon, *J. Mater. Sci.* **23** (1988) 2712
21. R. B. Vastava and T. G. Langdon, *Acta Metall.* **27** (1979) 251.
22. T. H.Alden, *J. Aust. Inst. Metals.* 14 (1969) 207.

23. R. C. Gifkins, *in Proceedings of a Symposium on Strength of Metals and Alloys,* Gifkins, ed., Pergamon, Oxford, 1982, p.701.
24. F. A. Mohamed, *J. Mater. Sci.* 18 (1983) 582.
25. H. Gleiter and B. Chalmers, *Prog. Mater. Sci.*, **16** (1973).
26. D. Mclean, *Met. Forum*, 4 (1981) 44.
27. H. Riedel, in *Fracture at High Temperatures*, B. Ilschner and N. Grant, eds., MRE Springer-Verlag, New York, NY, 1986, 116.
28. R. C. Gifkins, in Strength of Metals and Alloys, R. G. Gifkins, ed., Pergamon Press, Oxford, 1982, 1.
29. R. C. Gifkins, *J. Mater. Sci.* **13** (1982) 1926
30. T. G. Langdon, *Mater. Sci. Eng.*, **A174** (1994) 225
31. O.D. Sherby and J. Wadsworth: *Prog. Mater. Sci.*, 33 (1989) 169.
32. A. K. Mukherjee, *Mate. Sci. Eng.* **8** (1971) 83.
33. A. Ball and M. M. Hutchinson, *Metal Sci. J.* **3** (1969) 1.

A Microstructural Study of the Effect of Particle Aging on Dynamic Continuous Recrystallization in Al-4Mg-0.3Sc

L.M. Dougherty[1], I.M. Robertson[1], J.S. Vetrano[2] and S.M. Bruemmer[2]

[1] Department of Materials Science and Engineering, University of Illinois at Urbana-Champaign, Urbana, IL 61801, USA

[2] Pacific Northwest National Laboratory, PO Box 999, Richland, WA 99352, USA

Keywords: Al-Mg-Sc Alloy, Aluminium, Continuous, Dynamic, In-situ TEM, Orientation Imaging Microscopy OIM, Recrystallization, Superplasticity, TEM, Transmission Electron Microscopy TEM

ABSTRACT

Tensile specimens of an Al-4Mg-0.3Sc alloy in the peak-aged (8 hours at 553 K) and the over-aged (96 hours at 623 K) conditions, cold-rolled to a 70% reduction, exhibited dynamic continuous recrystallization during superplastic testing at 733 K and a strain rate of $10^{-3} s^{-1}$. Although the removal of subboundaries and the development of an equiaxed grain structure occurred more rapidly in the over-aged alloy, there was no discernible difference in the microstructures as observed in the transmission electron microscope (TEM). The fundamental mechanisms controlling dynamic continuous recrystallization have been studied by a combination of post-mortem examinations of the developing microstructure frozen at several points during the forming process and dynamic, high-temperature, deformation experiments performed in the TEM. The latter experiments provide direct observation at high spatial resolution of the operating mechanisms and have shown migration, pinning, disintegration and annihilation of subboundaries as well as the incorporation of dislocations into grain boundaries as they occur in real time.

INTRODUCTION

The occurrence of dynamic continuous recrystallization (strain-assisted continuous recrystallization) during superplastic deformation can increase the final strain and also allow higher strain rates to be used during the early stages of the forming process [1-4]. The microstructural changes accompanying dynamic continuous recrystallization produce an equiaxed grain structure that no longer exhibits the rolling texture and contains a relatively low dislocation density. To account for these microstructural changes, a number of mechanisms have been proposed including the migration of subboundaries, which leads to either coalescence or annihilation with other subboundaries [5]; the rotation of subgrains [6]; and the generation of dislocations and their independent migration and absorption into boundaries, which causes incremental increases in misorientation [7]. All of these proposed mechanisms are speculative, however, in that they are based only on post-mortem examinations, which cannot provide information about the route by which the final structure evolved.

In this paper, we describe the microstructures obtained in peak-aged (8 hours at 553 K) and over-aged (96 hours at 623 K) Al-4Mg-0.3Sc material, cold-rolled to a 70% reduction and deformed in tension at 733 K and a strain rate of $10^{-3} s^{-1}$. In addition, we describe the results of in-situ, high-temperature, deformation experiments performed in the TEM, which allow the operative mechanisms to be observed dynamically and at a high spatial resolution. It is important to note that these experiments are performed in thin foils and that the influence of the nearby surfaces on the dislocation dynamics must be considered in the interpretation of all observations. Previous work has, however, shown that similar microstructures are produced in samples deformed both as thin films in the TEM and as bulk specimens in a standard tensile testing apparatus. This provides evidence that similar processes are most likely in operation.

EXPERIMENT

The material used in this study was a cast Al-4Mg-0.3Sc alloy, warm-rolled, solution treated for 4 hours at 848 K, and then aged at temperatures from 523 to 623 K for up to 10^4 seconds. Based on hardness measurements, two aging conditions were selected for further study. These were the peak-aged condition of 8 hours at 553 K and the over-aged condition of 96 hours at 623 K. After aging, the material was cold-rolled to a 70% reduction. Annealing experiments revealed that static recrystallization is completely suppressed in both the peak-aged and the over-aged materials up to at least 733 K. Bulk tensile specimens were fabricated and tensile tested at 733 K and a strain rate of $10^{-3}s^{-1}$. A series of specimens from both the peak-aged and the over-aged conditions were deformed to intermediate true strains of 0.1, 0.2, 0.4 and 0.8. The microstructures at each intermediate strain were preserved by rapidly cooling the samples while under load. The evolution of the grain structure as a function of strain was determined through post-mortem examination of these frozen microstructures using orientation imaging microscopy (OIM). Higher-resolution examination of the microstructures was also performed in the TEM. Tensile TEM samples for in-situ, high-temperature, deformation experiments were cut from the gages of the bulk specimens deformed to 0.2 and 0.4 true strain. These samples had the following dimensions: 11 mm length, 2.5 mm width and 150 µm thickness. The central region of each was thinned to electron transparency using conventional twin jet polishing techniques.

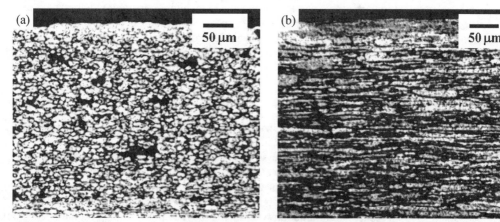

Fig. 1. Recrystallization occurs in the gage of a specimen during superplastic testing but not in the grip. (a) The combination of heat and strain leads to dynamic recrystallization in the gage section. (b) Static recrystallization is suppressed in the grips which are subjected to high-temperature only.

RESULTS AND DISCUSSION

For the peak-aged and the over-aged conditions, there was no difference in the superplastic response of the tensile specimens since both attained an engineering strain at failure of 376%. Comparison of the microstructures in the gage section and grip ends using optical microscopy showed that recrystallization had occurred only in the gage. This is illustrated in Fig. 1 for a specimen aged 64 hours at 593 K, cold-rolled to a 70% reduction and deformed at 733 K and a strain rate of $10^{-3} s^{-1}$. In Fig. 1(a), the gage is mostly recrystallized. In Fig. 1(b), the grip is completely unrecrystallized. Since the gage is the only part of the specimen subjected to both heat and strain, this indicates that dynamic recrystallization occurred in the gage but static recrystallization was suppressed in the grip, which was only subjected to heat. To understand the mechanisms operating during dynamic continuous recrystallization, the microstructural characteristics were determined as a function of strain for the peak-aged and over-aged conditions, and the specimens deformed to intermediate strains were further deformed within the TEM at nominally the superplastic testing temperature. The results of these in-situ TEM experiments were captured on Super VHS video tape.

Examination of the aged microstructures revealed a homogeneous distribution of fine, spherical, coherent, Al_3Sc particles, generally between 12 and 25 nm in diameter. The particles in the over-aged condition were 33% larger than the particles in the peak-aged condition. Superimposed on this distribution of fine particles was a less uniform distribution of large, scandium-rich particles, generally between 125 and 225 nm in diameter. Presumably, these precipitated during casting of the alloy and did not dissolve during the solution treatment at 848 K.

Prior to superplastic testing, the Al-4Mg-0.3Sc specimens were held at the test temperature for 10^3 seconds to allow the system to stabilize. Unstrained specimens, quenched and removed directly after this hold time, were examined using both TEM and OIM to assure that static recrystallization did not occur prior to deformation. The examinations revealed a partially recovered microstructure. In-situ, high-temperature experiments in the TEM were performed on samples prepared from these specimens. The removal of some subboundaries and lattice dislocations was observed, which is consistent with static recovery at this temperature.

Changes in the misorientation distribution of boundaries as a function of strain were studied using OIM. A quicker reduction in the density of low-angle boundaries and a more rapid development of an equiaxed grain structure was observed in the over-aged material as compared to the peak-aged material. However, post-mortem TEM examination at the intermediate strains of 0.1 and 0.4 revealed no discernible differences in subgrain size or lattice dislocation density between the two aging treatments. Considering this observation and the fact that the final strains of the peak-aged and over-aged specimens were identical, it is assumed that the same mechanisms are operating during dynamic continuous recrystallization and that they simply occur more rapidly in the over-aged material. This assumption will be verified in future experiments.

In-situ deformation experiments were performed at a nominal temperature of 733 K on tensile TEM samples fabricated from the peak-aged material that was deformed to 0.2 true strain. Figs. 2-4 illustrate the types of reactions observed and captured on video. Fig. 2 shows two arrays of dislocations that were created from a high-angle grain boundary as the junction between this grain boundary and a subboundary moved due to the migration of the subboundary. The dislocation arrays broke away from the junction once they attained a certain size and migrated as a unit rather than as individual dislocations. The dislocation arrays continued to move as units, which can be seen by comparing the position of array 1 with respect to array 2 in Figs. 2(a) and 2(b). Dislocation array 1 eventually coalesces with array 2 and the new array migrates as a unit as shown in Fig. 2(c). The array eventually intersects large Sc-rich particles, including the one labeled A, and breaks into individual dislocations, which migrate independently. These dislocations are individually incorporated into nearby high-angle grain boundaries.

Two methods by which entire subboundaries can be annihilated are illustrated in Figs. 3 and 4. Fig. 3 illustrates one method in which a migrating subboundary interacts with a large Sc-rich particle (labeled B), eventually breaks free of the particle and immediately disintegrates. The component dislocations from the subboundary migrate independently through the grain and eventually incorporate into a nearby grain boundary. Another method of subboundary annihilation is illustrated in Fig. 4 in which subboundary dislocations are directly incorporated into the grain boundary to which the subboundary is attached. This results in a continual decrease in the length of the subboundary, evidenced by the near-

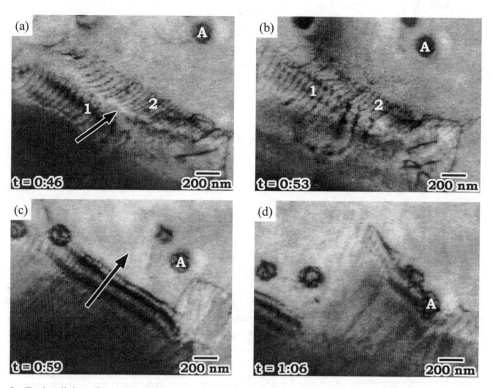

Fig. 2. Entire dislocation arrays migrate as units during a high-temperature deformation experiment in the TEM. (a) Dislocation arrays generated by the breakup of a grain boundary are shown. (b) Dislocation array 1 moves towards and begins to interact with array 2. (c) The two arrays coalesce to form a single array which continues to migrate in the direction indicated. (d) The array interacts with large Sc-rich particles, including particle A, and breaks apart. The units of time, t, in these figures are hours and minutes.

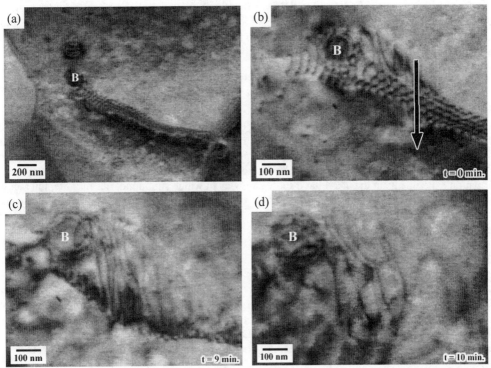

Fig. 3. The interaction between a migrating subboundary and a large Sc-rich particle labeled B leads to subboundary disintegration. (a) The subboundary interacts with particle B. (b) A higher magnification view of this interaction is shown. The arrow indicates the direction of subboundary migration. (c) The subboundary begins to break free of particle B. (d) The subboundary disintegrates and its component dislocations migrate independently and are absorbed in a nearby high-angle grain boundary.

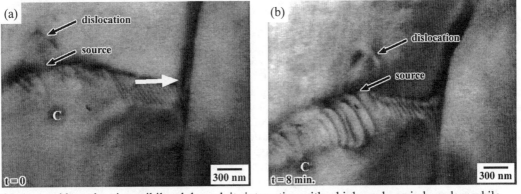

Fig. 4. A subboundary is annihilated through its interaction with a high-angle grain boundary while dislocations are ejected from a source within the subboundary. (a) Subboundary dislocations move in the direction indicated by the white arrow and are incorporated into the attached high-angle grain boundary. Simultaneously, dislocations bow out of the source indicated near a large Sc-rich particle labeled C. (b) The length of the subboundary decreases as dislocations are incorporated into the grain boundary, illustrated by the shift of the dislocation source relative to particle C and the grain boundary.

ing of a dislocation source (indicated in the figure) in the subboundary to the grain boundary. As the subboundary is reduced in length due to the incorporation of subboundary dislocations into the grain boundary, dislocations are also being ejected from the source in the subboundary. These emitted dislocations are shown bowing out from this source in both Figs. 4(a) and 4(b).

CONCLUSIONS

Dynamic continuous recrystallization has been shown to occur in peak-aged and over-aged Al-4Mg-0.3Sc alloy, cold-rolled to a 70% reduction and deformed in tension at 733 K and a strain rate of $10^{-3} s^{-1}$. There are no discernible differences in the microstructures of the peak-aged and over-aged alloys as studied with TEM, although the removal of low-angle boundaries and the development of an equiaxed grain structure appears to occur more rapidly in the over-aged material. In-situ, high-temperature, deformation experiments in the TEM have shown the way in which subboundaries migrate, interact with large Sc-rich particles and annihilate. From these observations, it is clear that the current mechanisms proposed for dynamic continuous recrystallization during superplastic forming are inadequate descriptions of the processes that are occurring.

ACKNOWLEDGEMENTS

The authors gratefully acknowledge sponsorship by the United States Department of Energy and the Office of Basic Energy Sciences through the Pacific Northwest National Laboratory. Additional support is provided by the Center for Excellence in the Synthesis and Processing of Advanced Materials. The use of facilities at PNNL, Argonne National Laboratory and the Center for Microanalysis of Materials in the Frederick Seitz Materials Research Laboratory at the University of Illinois is appreciated.

REFERENCES

[1] K. J. Gardner, P. Griffin, and R. Grimes, "Dynamic recrystallization in an Al-10Zn-0.3Zr superplastic alloy," presented at Recrystallization in the Control of Microstructure, London, UK (1973).
[2] R. Grimes, M. J. Stowell, and B. M. Watts, "Superplastic aluminium-based alloys," Metals Technology 3 (1976), p. 154.
[3] B. M. Watts, M. J. Stowell, B. L. Baikie, and D. G. E. Owen, "Superplasticity in Al-Cu-Zr alloys, Part I: Material preparation and properties," Metal Science 10 (1976), p. 189.
[4] B. M. Watts, M. J. Stowell, B. L. Baikie, and D. G. E. Owen, "Superplasticity in Al-Cu-Zr alloys, Part II: Microstructural study," Metal Science 10 (1976), p. 198.
[5] A. M. Diskin and A. A. Alalykin, "Superplasticity of duralumin and magnalium type alloys with initial non-recrystallized structure," Tsvetnye Metally (English translation) 28 (1987), p. 86.
[6] S. J. Hales and T. R. McNelley, "Microstructural evolution by continuous recrystallization in a superplastic Al-Mg alloy," Acta Metallurgica et Materialia 36 (1988), p. 1229.
[7] X. Zhang and M. J. Tan, "Dislocation model for continuous recrystallisation during initial stage of superplastic deformation," Scripta Metallurgica et Materialia 38 (1998), p. 827.

Investigation of Superplastic Behaviour and Solid State Bonding of Zircaloy-4

P.S. Hill[1], N. Ridley[1] and R.I. Todd[2]

[1] University of Manchester/UMIST, Materials Science Centre,
Grosvenor Street, Manchester M1 7HS, UK

[2] University of Oxford, Department of Materials Science,
Parks Road, Oxford OX1 3PH, UK

Keywords: Diffusion Bonding, Modelling, Solid State Bonding, Superplasticity, Zircaloy 4

Abstract

Superplastic behaviour and isostatic diffusion bonding (DB) have been investigated for as-received annealed Zircaloy-4 sheet of relatively large grain size. The material was superplastic at strain rates up to $5.0 \times 10^{-4} s^{-1}$ at temperatures in the lower part of the $\alpha+\beta$ phase field for β-phase volumes of 10-20%. At higher strain rates and lower temperatures the material deformed by power law creep. The DB studies showed that for intermediate and higher temperatures in the $\alpha+\beta$ phase field, sound bonds could be obtained at relatively low pressures and short times. Bonding in the α-phase field required much higher pressures. Modifications to an earlier model of the DB process involved constraint of collapse by creep of inter-void ligaments at the bond interface. This lead to a dominance of stress directed boundary diffusion and gave improved agreement between prediction and observation for DB of the α- phase.

Introduction

Zirconium alloys are widely used in the nuclear industry but they also find considerable use as structural components in the chemical processing industry because of their high resistance to corrosive attack combined with good mechanical properties at elevated temperatures. Zirconium and titanium are isostructural and appear in the same Group of the Periodic Table. Since the outer electron shells have similar electronic structure, zirconium would be expected to have a similar chemical response and reactivity to titanium. This is significant in the present context since it is well known that titanium alloys may be joined by diffusion bonding (DB), and also that this procedure may be combined with superplastic (SP) forming. The similarities between titanium and zirconium include the tendency of the surface oxides to dissolve at high temperatures, a requirement for the success of DB. Experiment [1] and calculation [2] have shown that the oxide layer on zirconium tends to dissolve rapidly at temperatures of 760°C and above. Sound bonds have been reported for Zircaloy-2 using "pressure bonding" [1] and the feasibility of DB of Zr alloys is reinforced by the success of roll bonding.

Zircaloy-4 has previously been shown to exhibit superplasticity but there has been little work on the diffusion bonding of the material. In the present work both SP behaviour and DB have been examined for temperatures spanning the α and $\alpha+\beta$ phase fields, with particular attention being paid to DB at temperatures in the α-phase field. The DB variables examined also included pressure and time. In parallel with the experimental studies, modelling of the DB of Zircaloy-4 is examined with a model developed previously by the authors being used as a starting point [3].

Material and Superplastic Behaviour

Zircaloy-4 of nominal wt% composition: Zr-1.5Sn-0.22Fe-0.11Cr and 1100ppm O, was received in the form of annealed sheet of 3mm thickness. The material had an equiaxed α-phase microstructure, with a mean linear intercept grain size of ~15μm, and contained a dispersion of intermetallic $Zr(FeCr)_2$ particles. Thermal analysis (DSC) showed that the two-phase α+β structure existed over the temperature range 810°C - 1000°C, which was consistent with the phase diagram. The variation of % β-phase was determined metallographically using small specimens cooled rapidly from temperatures in the 2-phase region.

To characterise flow behaviour, measurements of the stress-strain rate relationships were made for tensile specimens of 20mm gauge length machined parallel to the rolling direction, at temperatures of 770°C - 870°C, using strain rate jump tests. Testing was carried out in an argon atmosphere. At temperatures of 850°C-870°C, when the volume of β-phase varied from 10-20%, the maximum values of strain rate sensitivity of flow stress, m, ranged from 0.3 to >0.5 at strain rates from $5 \times 10^{-5} - 5 \times 10^{-4} s^{-1}$ (Fig.1). This is consistent with SP behaviour. At 870°C, tensile elongations to failure of >350% were recorded during constant strain rate testing, and these could be further enhanced to >450% by constant cross-head velocity straining (Fig.2). Metallographic examination showed that the grains remained equiaxed after straining in the high m region, with the exception of the region near the fracture tip where elongated grains were observed. At higher strain rates, $>10^{-3}s^{-1}$, m was ~0.2.

Insufficient data was available to calculate a reliable activation energy, Q, for the high m region (Region II). At higher strain rates and higher stresses (Region III), values of $Q \sim 310$ kJmol^{-1} were obtained and are consistent with that for power law creep of α-Zr [4]. However, while the α-phase dominates the microstructure, it is difficult to imagine that grain boundary β-phase does not play some role in the accommodation of grain boundary sliding in the SP regime. Reliable diffusion data for β-Zr is not available.

At temperatures below 830°C where the volume fraction of β-phase is very small or zero, values of m over the strain rate range $5 \times 10^{-5} - 5 \times 10^{-1} s^{-1}$ were ~0.2 (Fig.1.) Tensile tests at constant cross-head velocities or constant strain rates in the range $10^{-4} - 10^{-3}s^{-1}$, lead to elongations of 100-120% (Fig.2). Clearly the alloy did not exhibit superplasticity over the range of conditions examined. The activation energy calculated for the temperature range 790-810°C was ~350kJmol^{-1} and is consistent with power law creep in the α-phase as the rate controlling mechanism for plastic flow.

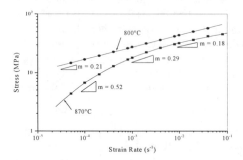

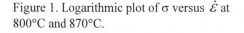

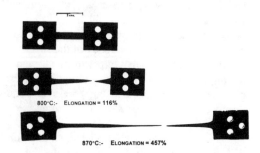

Figure 1. Logarithmic plot of σ versus $\dot{\varepsilon}$ at 800°C and 870°C.

Figure 2. Profile of test pieces strained to failure at $5.0 \times 10^{-4}s^{-1}$.

Isostatic Diffusion Bonding: Experimental

Pairs of rectangular blanks, after preparation to give a P60 grit finish, were electron beam welded around their peripheries to form DB couples. The surface finish was characterised using a Form Talysurf stylus profilometer. Isostatic bonding was carried out in an ABB QIH-9 mini-hipper under predetermined conditions of argon pressure (10 – 100MPa), temperature (750 -1000°C), and time (0 – 6 hours). During bonding the application of pressure brings the two surfaces into intimate contact so creating a planar array of interfacial voids. Time dependent plastic flow and diffusion processes transport atoms to the void surfaces, so progressively reducing interfacial void volume. After the bonding cycle, quantitative optical metallography was used to measure the fractional length of bonded interface, f_b, and also the mean values of spacing, height and width, of the interfacial voids. Compressive lap shear testing was used to determine the strength, τ_b, of the bond in relation to the strength of the parent material, τ_p, subjected to a comparable thermal cycle. For the most part there was a reasonable correlation between metallography and lap shear data. Lap shear fracture surfaces were also examined in a SEM.

Typical results of the many bonding conditions examined are shown in Table 1. At high temperatures e.g. 1000°C (predominantly β-phase) a fully sound bond was obtained. At intermediate temperatures in the α+β region (850°C) an essentially sound bond was formed, while at temperatures in the α-phase field (800°C) high pressures and longer times were required to produce good bonds. Figure 3 shows micrographs of a partly bonded specimen, in which interfacial voids are clearly visible, and a fully bonded specimen, and the corresponding fracture surfaces.

Temp (°C)	Pressure (MPa)	Duration (mins)	Surface Finish	Shear Stress τ_b (MPa)	Parent τ_p Strength (MPa)	τ_b/τ_p	f_b
750	20	60	P60	247.9	424.2	0.58	0.35
850	20	60	P60	325.5	416.3	0.79	0.96
1000	20	60	P60	375.5	403.8	0.93	1.00
800	20	0	P60	180.2	405.2	0.44	0.29
800	20	60	P60	255.3	449.3	0.57	0.53
800	20	120	P60	275.4	449.8	0.61	0.64
800	20	240	P60	293.4	416.1	0.70	0.72
800	20	360	P60	303.7	412.4	0.74	0.82
800	10	120	P60	206.4	420.4	0.49	0.44
800	30	120	P60	308.3	425.8	0.72	0.76
800	50	120	P60	367.2	411.8	0.89	0.93
800	100	120	P60	376.4	424.8	0.89	0.95

Table 1. Results for selected bonding conditions

Examples of the variation of the mean values of void spacing, void width and void height, as a function of fraction bonded, f_b, are seen in Fig. 4. An interesting and important feature of Fig.4 is that both the mean height and mean width of the interfacial voids decrease with fraction bonded while the void spacing (Fig.4c) passes through a minimum before increasing progressively as the bond fraction approaches 1. The shaded area on the LHS of Fig.4(c) represents an estimate of the possible variation of void spacing on initial interface contact based on surface roughness measurements. The initial decrease in the void spacing is attributed to the early formation of void necks as the larger voids collapse plastically leading to a number of smaller voids. As bonding continues, the smaller voids are progressively removed, and this results in a continuing increase in mean intervoid separation.

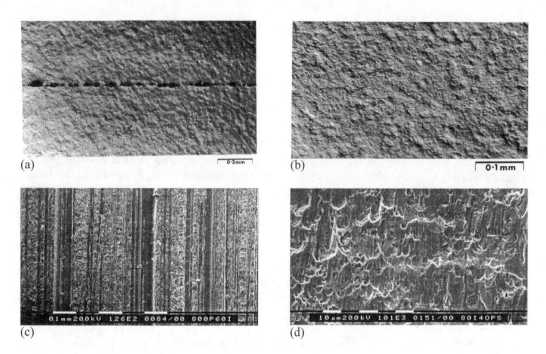

Figure 3. Micrographs of (a) partially bonded (b) fully bonded DB specimens; (c) and (d) corresponding fracture surfaces.

Isostatic Diffusion Bonding: Modelling

The initial stages of modelling applied to bonding at 20 MPa and 800°C with a P60 grit surface finish are reported here. The starting model has been used previously by the authors [3] and considers bonding to take place by both the collapse by creep of the ligaments between the interfacial voids, and by stress directed diffusion of matter along the bond line. Initial results showed that the contribution of ligament collapse led to predicted bonding rates substantially in excess of those observed experimentally.

The most likely explanation for this is the fact that like most other models, it is assumed that the inter-void ligaments can deform freely in plane strain compression. In reality, the expansion of these ligaments into the voids will be constrained by the surrounding bulk of the material, thus reducing the deformation rate. In accounting for this, the original geometry [3], consisting of an array of identical rectangular voids has been retained, but during bonding the width of each ligament has been constrained to remain constant at the point of contact with the bulk of the specimen. This is satisfied by the deformation geometry shown in Figure 5, which can be solved by viewing the deformation as a combination of a plane strain 'forging' of the ligament under the action of the applied gas pressure, with a shear deformation arising from the constraint on the ligament. The displacements due to the 'forging' and shear components are required to be equal and opposite along the line where the ligament meets the bulk of the material, thus satisfying the condition that the net displacement is zero along this line.

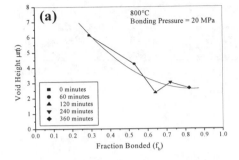
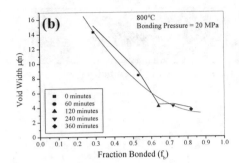
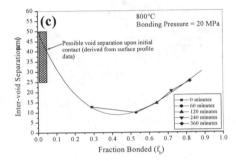

Figure 4. (a) mean void height (b) mean void width, and (c) mean void spacing, of specimens bonded at 800°C.

The width and height of the inter-void ligaments have been included in the model using microstructural measurements such as those used to construct Fig.4. The predicted contributions of ligament creep and diffusion of the modified model are compared in Figure 6. The ligament creep contribution dominates until the fraction bonded, f, exceeds about 0.5, at which point diffusion becomes dominant. This shows that the inclusion of plastic constraint has slowed down the contribution of ligament collapse significantly.

The predicted and measured bonding rates are compared in Figure 7. The two theoretical predictions have been obtained using the lowest and highest reasonable estimates of the grain boundary diffusion coefficient of zirconium available in the literature [5]. These upper and lower bound solutions bracket the experimental results for $f > \approx 0.6$. Agreement is less good for the specimen with $f = 0.53$, with the predicted bonding rate exceeding that measured experimentally by an order of magnitude. Creep collapse of the inter-void ligaments in dominant in this case, and the discrepancy is likely to be a consequence of the following considerations: (i) the degree of constraint depends critically on the dimensions of the features of the interfacial microstructure, and these cannot be measured accurately, and (ii) the effect of (i) and of the approximations made in the model are exaggerated by the high value of the creep exponent, n (≈ 5).

Although model development is at an early stage, some useful conclusions can be drawn about the conditions required to obtain a sound bond. First, the results of Fig. 6 show that under the conditions of moderate temperature (800°C) and pressure (20 MPa) under consideration, the dominant bonding mechanism will be diffusion. The bonding rate varies inversely as the third power of the scale of the interfacial microstructure under diffusional control, so there is considerable scope for reducing the bonding time by improving the surface finish. The dominance of diffusion (bonding rate $\propto$ stress), and the rapid reduction in the contribution of creep collapse as bonding proceeds explain the experimental observation that large increases in stress are required to achieve significant reductions in the bonding time despite the high stress sensitivity of creep in the material.

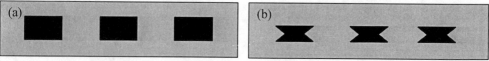

Figure 5. Assumed DB geometry (a) before and (b) during bonding, incorporating plastic constraint.

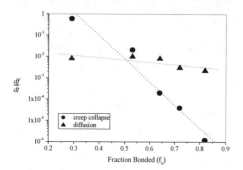

Figure 6. Comparison of predicted contribution to bonding rate of diffusion and creep collapse of inter-void ligaments as a function of fraction bonded at 800°C and 20 MPa pressure.

Figure 7. Predicted and measured bonding rates as a function of fraction bonded at 800°C and 20MPa pressure.

Conclusions

1. As received Zircaloy-4 annealed sheet material of relatively large grain size (~ 15 μm) was shown to exhibit superplasticity at strain rates up to 5.0 x 10^{-4} s^{-1} at temperatures of 850-870°C. The volume % of β-phase at these temperatures was in the range 10-20%.
2. At higher strain rates and/or lower temperatures the material deformed by power law creep consistent with a measured activation energy of ~ 310 – 350 kJmol^{-1}.
3. DB studies showed that at temperatures where β-phase was present >10% by volume, sound bonds could be achieved at relatively low pressures (20 MPa) and short times (60 minutes).
4. At temperatures in the α-phase field e.g. 800°C, the interface void spacing passes through a minimum during bonding, while the height and width of voids progressively decreases.
5. Modifications to an earlier model of DB involved constraint of the plastic collapse of inter-void ligaments at the bond interface. This led to a dominance of stress directed boundary diffusion and gave improved agreement between prediction and observation for DB of the α-phase.

Acknowledgements

The work was funded by the EPSRC under grant no.GR/M43296.

References

(1) J.Weber and W.L.Frankhouser, National Bureau of Standards UC-25, Metals, Ceramics and Materials, Special Distribution Report WAPD-305 (1966).
(2) Z.A.Munir, Welding Journal 62 (1983), p.333.
(3) R.I.Todd, C.S.Hodges, Y.C.Wong, Z.C.Wang and N.Ridley, Mater. Sci. Forum 243-245 (1997), p.675.
(4) P.M.Sargent and M.F.Ashby, Scripta Metall. 16(1982), p.1415.
(5) I.M.Berstein,Trans. TMS-AIME 239 (1967), p.1518.

Superplastic Behaviors of a Ti-Alloy with Ultrafine Grain Size

B.Z. Bai[1], X.J. Sun[1], J.L. Gu[1] and L.Y. Yang[2]

[1] Department of Materials Science and Engineering, Tsinghua University,
Beijing 100084, China P.R.

[2] Beijing Research Institute of Mechanical & Electrical Technology,
No. 18 Xueping Road, Beijing 100083, China P.R.

Keywords: Deformation Mechanisms, Superplasticity, Ti Alloy, Ultrafine Grains

ABSTRACT

The superplastic deformation behavior of a Ti alloy with submicron grain size was investigated. The ultrafine grained (UFG) Ti alloy exhibits good superplasticity at lower temperature (800 °C) and at higher strain rates (not lower than $1.8 \times 10^{-2} s^{-1}$). Severe deformation inhomogeneity(DI) could occur at certain conditions. To lower deformation temperature or to raise strain rate could ease or eliminate inhomogeneity. The influence of deformation conditions and grain size on DI was analyzed. Based on examining on the m value and deformation activation energy of the UFG Ti alloy, SPD mechanisms were proposed to be grain boundary sliding accommodated by dislocation movement at lower temperature and by diffusion at higher temperature.

INTRODUCTION

Maximum ductility of SPD is generally obtained at relatively low strain rates and relatively high temperature [1]. This induced some problems that could make resistance to SPF, such as longer forming time, consumption of energy and oxidization of die.

It has been suggested that superplasticity might be achieved at higher strain rates or lower temperature by grain refinement[2]. As documented in detail elsewhere, most reports on high strain rate superplasticity were confined to a limited range, e.g. metal matrix composites, mechanically-alloyed materials and some metallic alloys through powder metallurgy techniques [3,4]. However, these materials are too expensive for fabrication to be widely used. Recent reports have established that it is possible to attain ultrafine grain sizes in bulk metals(with grain sizes generally in the nanometer and submicrometer ranges) by intensive plastic straining (IPS), e.g. equal-channel angular (ECA) pressing, torsion or accumulative rolling-bonding (ARB) [5-8]. The superlasticity of ultrafine grained Al alloys produced by IPS have been widely explored. It is shown that IPS is capable of producing materials that exhibit superplasticity both at lower temperature and higher strain rates [4]. It is well known that Ti alloy is widely used in aviation or aerospace industries. However, few efforts have been made on the fabrication of ultrafine grained Ti alloy. In the present work, ultrafine grain structure was realized in a Ti alloy and its superplastic behavior was investigated as well.

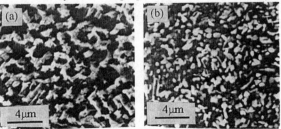

Fig. 1 The microstructures of Ti-6.5Al-3.3Mo-1.8Zr subjected to radial forging at 900 °C (a) and 850 °C (b)

EXPERIMENTAL PROCEDURES

The starting material was water quenched Ti-6.5Al-3.3Mo-1.8Zr alloy with lamellar martensite microstructure. Round rods with about 20mm diameter were muti-stepped radial forged to about 8.5mm diameter at 900 and 850℃, corresponding to the true strain of 1.7. The microstructures after radial forging are shown in Fig. 1. The intercept line grain sizes were measured to be about 0.88μm at 900℃ and about 0.74μm at 850℃ (referred to as A and B hereafter, respectively).

Tensile specimens were machined from the forged samples with a gauge length of 17mm and diameter of 3mm. These specimens were extended to failure in air using Shimadzu testing machine operating at constant strain rates from $5.4 \times 10^{-4} s^{-1}$ to $3.6 \times 10^{-2} s^{-1}$ and at temperatures of 900 and 800 ℃, respectively. Stress-strain rate relationship and m value were examined on Gleeble1500 machine by means of stepped-strain rate test. The stepped-temperature tests were conducted on Gleeble 1500 to determine the deformation activation energy. Microstructures were observed on SEM and OPM.

RESULTS AND DISCUSSION
A. Superplastic tensile properties of the UFG Ti alloy

The σ-$\dot{\varepsilon}$ relationship and m-$\dot{\varepsilon}$ relationship of alloy A and B at 800℃ are shown in Fig. 2. As comparison, Fig. 2 also gives the results of this alloy with grain size of 5μm (referred to as C hereafter). In general, the flow stress decreases and m value increases with refining the grain size. The maximum m values of both alloy A and B are not lower than 0.5, and the maximum m value of alloy C is only about 0.35 at 800℃. Fig.3 shows the stress-strain rate relationship and m-value of alloy A at 900℃. The maximum m value is about 0.85, and it is worthy to be mentioned that the m value is still as high as 0.59 even at the strain rate of $1.6 \times 10^{-2} s^{-1}$.

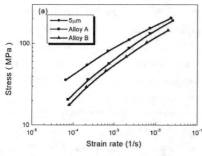 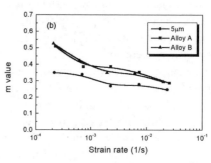

Fig. 2 The σ-$\dot{\varepsilon}$ relationship (a) and m-$\dot{\varepsilon}$ relationship (b) of alloy A and B at relatively low temperature of 800℃

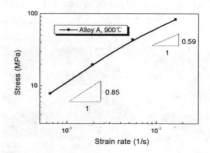

Fig. 3 The stress-strain rate relationship of alloy A at relatively high temperature of 900℃

Table 1 The elongations to failure of the UFG Ti alloys after superplastic tension

Alloy	Deformation conditions	δ(%)
A	800°C, $\dot{\varepsilon}=5.4\times10^{-4}s^{-1}$	488
	800°C, $\dot{\varepsilon}=1.6\times10^{-2}s^{-1}$	263
	900°C, $\dot{\varepsilon}=8.8\times10^{-4}s^{-1}$	412
	900°C, $\dot{\varepsilon}=1.8\times10^{-2}s^{-1}$	471
B	800°C, $\dot{\varepsilon}=5.4\times10^{-4}s^{-1}$	457
	800°C, $\dot{\varepsilon}=1.6\times10^{-3}s^{-1}$	402
	800°C, $\dot{\varepsilon}=1.6\times10^{-2}s^{-1}$	313
	900°C, $\dot{\varepsilon}=8.8\times10^{-4}s^{-1}$	515
	900°C, $\dot{\varepsilon}=1.8\times10^{-2}s^{-1}$	753
	900°C, $\dot{\varepsilon}=3.6\times10^{-2}s^{-1}$	290

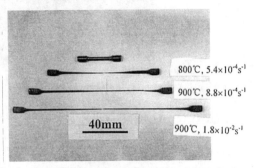

Fig. 4 The examples of tensile specimens

Elongation to failure of alloy A and B at different conditions are listed in Table 1. The examples of tensile specimen B are shown in Fig. 4. It has been reported that the optimum superplasticity for conventional Ti-6.5Al-3.3Mo-1.8Zr alloy (with 5-10μm grain size) are attained at 900-940°C and $10^{-4}s^{-1}$-$10^{-3}s^{-1}$ [9]. However, the UFG Ti alloys can attain good superlasticity at lower temperature 800°C and higher strain rate ($1.8\times10^{-2}s^{-1}$-$3.6\times10^{-2}s^{-1}$).

B. Deformation inhomogeneity of UFG Ti alloy

Inhomogeneity of SPD is frequently found in the materials with non-ideal SP microstructure, i.e. non-equiaxed or non-uniform distributed microstructures [9,10]. However, severe DI can still be observed in the present UFG Ti alloy at certain conditions though the microstructures are equiaxed and uniform. Fig. 5 shows the surface appearances of alloy A at different conditions. The specimens' surface is smooth at higher strain rate or lower temperature. It becomes rough and uneven at lower strain rate and higher temperature. Fig. 6 shows the specimen's shape near the fracture position tensioned at 900°C and $5.4\times10^{-4}s^{-1}$. There are two obvious neckings in the vicinity of fracture position. Fig.5(a) and Fig.6 represent two types of DI.. One is called surface irregularity with size much less than that of specimen(Fig.5a), and the other is called multi-necking with size comparable to that of specimen(Fig.6).

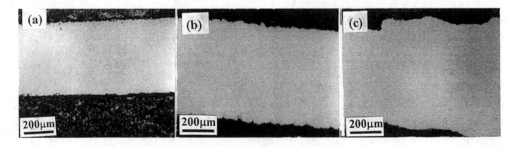

Fig. 5 The surface appearances of alloy A at various deformation conditions
(a) 900°C, $1.8\times10^{-2}s^{-1}$ (b) 800°C, $5.4\times10^{-4}s^{-1}$ (c) 900°C, $8.8\times10^{-4}s^{-1}$

Fig. 6 The specimen's shape near the fracture position tensioned at 900℃ and $5.4\times10^{-4}\text{s}^{-1}$

1 Surface irregularity

It has been established that the formation of surface irregularity is resulted from the non-uniform distribution of microstructure near specimen's surface [9]. Fig. 7(a) shows the near surface appearance of specimen A, where a protuberance with about 100μm height can be clearly observed. Fig. 7(b) shows the detail arrangement of microstructure within the square region marked in Fig. 7 (a). As seen in this figure, stringers of α-α and β-β, oriented at ~45° to tensile axis, were formed near the specimen's surface. These stringers should be formed during SPD since the initial structure is fine, equiaxed and uniform. Similar results were observed with mixture grained microstructure of the same alloy[10]. The phase stringers could be considered as grain groups. They would slide as an entity and gradually float out of the surface of specimen to form the surface irregularity.

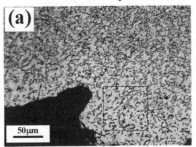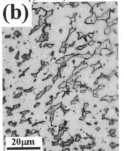

Fig. 7 Near surface appearance of specimen A (a) and detail morphology of the square region marked in it (b).

2 Multi-necking

In general sense, the finer the gains are, the more uniform the deformation is. However, so severe strain localization was found in present UFG material. It brings us to have a brief analysis on the effect of grain size on strain distribution.

Superplastic constitutive relationship can be described by the following general equation [11].

$$\dot{\varepsilon} = K_2 \cdot \frac{\sigma^{n_2}}{d^a} + K_3 \cdot \sigma^{n_3} \tag{1}$$

The letters in it have their usual meanings. First term of the right side represents the contribution of grain boundary sliding, which strongly depends on the grain size. Second term represents the contribution of power law creep, which is independent on the grain size.

Assuming there exists a microstructural non-uniform region with grain size of d' in the material with grain size of d(d>d'). The microstructural non-uniform degree is represented by $f = \frac{d-d'}{d}$. The strain rate difference between the two regions could be obtained according to equation (1):

$$\Delta \dot{\varepsilon} = \dot{\varepsilon} - \dot{\varepsilon}' = K_2 \cdot \sigma^{n_2} (\frac{1}{d'^a} - \frac{1}{d^a}) \qquad (2)$$

Thus, the strain difference at certain strain can be obtained:

$$\Delta \varepsilon \big|_\varepsilon = \Delta \dot{\varepsilon} \frac{\varepsilon}{\dot{\varepsilon}} = \frac{K_2[(\frac{d}{d'})^a - 1]}{K_2 + K_3 \cdot \sigma^{n_3 - n_2} d^a} \cdot \varepsilon = \frac{K_2[(\frac{1}{1-f})^a - 1]}{K_2 + K_3 \cdot \sigma^{n_3 - n_2} d^a} \cdot \varepsilon \qquad (3)$$

The relationship of DI, $\Delta \varepsilon / \varepsilon$, with the grain size and strain rate at certain microstructural inhomogeneity, f=0.1 could be analyzed through (1)~(3). The parameter values used for the calculation were taken from reference [11] (Table 2). As pointed out in this reference, these parameters could be representative of general superplastic alloy.

Table 2 The parameter values taken from [12]

K_2 $((\mu m)^p (MPa)^{-n_1} s^{-1})$	K_3 $((MPa)^{-n} s^{-1})$	a	n_2	n_3
2.26×10^{-4}	4.54×10^{-6}	2	1	5

The calculation results are shown in Fig. 8. The results demonstrate three important aspects: (a) The superplastic DI could be decreased with increasing the strain rate. (b) The DI is more severe with finer grain size at certain strain rate. (c) The DI is very sensitive to grain size at relatively low strain rate, but this is not the case at relatively high strain rate.

Thus, the reasons that severe muti-necking occurs during superplastic deformation for UFG Ti alloy could be qualitatively explained according to Fig. 8.

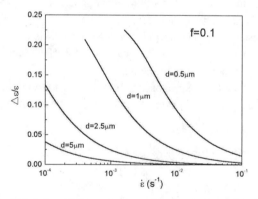

Fig. 8 The relationship of DI, $\Delta \varepsilon / \varepsilon$, with the grain size and strain rate at certain microstructural inhomogeneity, f=0.1

C. Superplastic deformation mechanism

The results of the stepped-temperature tests are given in Fig. 9. Fig. 10 shows the $1/T$ vs. $\ln \sigma$ relationship. Activation energy could be calculated from it. It is evident that Q-value is different in lower temperature region (760-830℃) and in higher region (830-900℃). The former is 85.4KJ/mol, which is close to the activation energy value for grain boundary diffusion of Ti (101KJ/mol for α-Ti and 77KJ/mol for β-Ti). The latter is 183.7KJ/mol, which is similar to the activation energy value for volume diffusion of Ti (169KJ/mol for α-Ti and 131KJ/mol for β-Ti) [13]. From this point of view, it could be primarily supposed that the superplastic deformation mechanisms for the UFG Ti alloy are grain boundary sliding accommodated by slip in relatively low temperature region and grain boundary sliding accommodated by diffusional flow in relatively high temperature region, respectively.

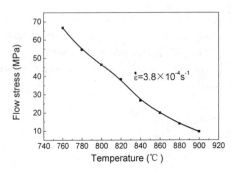

Fig. 9 The relationship between flow stress and deformationtemperatures at the strain rate of $3.1\times10^{-4}s^{-1}$

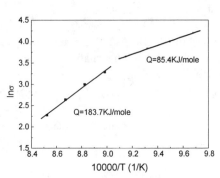

Fig. 10 The relationship between $1/T$ and $\ln\sigma$

CONCLUSIONS

1 The UFG Ti-6.5Al-3.3Mo-1.8Zr alloy can exhibit good superplasticity at relatively low temperature (800°C) and at relatively high strain rates (not lower than $1.8\times10^{-2}s^{-1}$). Compared with the optimum SPD conditions for the conventional Ti-6.5Al-3.3Mo-1.8Zr alloy, the deformation temperature is decreased by about 100°C and the strain rate is increased by one order of magnitude.
2 Severe DI occurs during SPD at relatively high temperature and relatively low strain rate for the UFG Ti alloy, which is manifested as surface irregularity and muti-necking. The surface irregularity is attributed to the grain group movement near the specimen's surface and the muti-necking is resulted from the effect of microstructural ulta-refinement.
3 The SPD mechanisms of UFG Ti alloy are proposed to be grain boundary sliding accommodated by slip at relatively low temperature and by diffusion at relatively high temperature, respectively.

REFERENCES

[1] T.G. Langdon, Metall. Trans. 13A (1982), p. 689.
[2] Y. Ma, M. Furukawa, Z. Horita, et al, Mater. Trans. JIM. 37 (1996), p. 336
[3] M. Mabuchi and K. Higashi. JOM, 6, 34(1999)
[4] T.G Nieh, C.H. Henshall, and J. Wadsworth. Scr. Metall. Mater. 18(1984), p. 1405.
[5] Y. Iwahashi, Z. Horita, M. Nemoto, et al., Acta mater. 46(1998), p. 3317.
[6] Y. Iwahashi, Z. Horita, M. Nemoto, et al., Metall. Mater. Trans. A. 29(1998), p. 2503.
[7] M. Furukawa, P. B. Berbon, Z. Hemoto, et al., Mater. Sci. Forum 233-234(1997), p. 177.
[8] Y. Saito, N. Tsuji, H. Utsunomiya, et al., Scr. Mater. 39(1998), p. 1221.
[9] B.Z. Bai, On the relationship between deformation inhomogeneity and microstructural non-uniformity of titanium and aluminium alloys, Dissertation, Tsinghua University, 1991
[10] B. Z. Bai and J. Fu., Sci. Metal. Mater. 26(1992), p. 743.
[11] C.H. Hamilton, Metall. Trans. A 20A(1989), p. 2783.
[12] J. Warren, L.M. Hsiung and H.N.G. Wadley, Acta Metall. Mater. 43(1995), p. 2773.

Development of Superplastic AlCuAgMgZr Alloys for Easy Forming

J. Dutkiewicz[1], J. Kuśnierz[1] and T.G. Nieh[2]

[1] Institute of Metallurgy and Materials Science, Polish Academy of Sciences,
ul. Reymonta 25, PL-30-059 Krakow, Poland

[2] Lawrence Livermore Natioanl Lab., L-350,
PO Box 808, Livermore CA 94551-9900, USA

Keywords: AlCuAgMgZr Alloys, Grain Refinement, Superplastic Deformation

Abstract: Two aluminium alloys containing up to 6% Cu, 0.8% Ag, 0.7 % Mg and 0.4 % Zr (all in wt %) were continuously cast and hot rolled at 673 K. Thermomechanical treatment involving particle stimulated nucleation (PSN) process was used in order to obtain fine grain size. The microstructure studies after this treatment revealed the presence of subgrains of 0.5-2 µm in size with a misorientation of 3 - 15°. In spite of the presence of zirconium rich and copper rich particles the grain growth up to 10-20 µm was observed after annealing at temperatures above 673K. This was also observed in samples deformed at the temperature of 673K. It affected the superplastic deformation properties of the investigated alloys at higher temperatures in a negative way. The superplastic behaviour was characterised by the **m** value, which was determined by step strain - rate change tests. The tests were conducted at a slow initial strain rate, which increased stepwise to high constant strain rates. The highest **m** value obtained approached 0.7 for the lowest temperature (623K) and the strain rates ($10^{-3} - 10^{-4}$ s^{-1}). A value close to 0.5 was observed for temperatures 623-673K and the same strain rates. The lowest **m** values were found for temperatures T> 623K and higher strain rates. Structure studies of the superplastically deformed samples revealed grain growth and presence of cavities after elongation of a few hundreds percent. Presence of dislocations within grains indicates the participation of the crystallographic slip in the superplastic deformation at 673K.

1. Introduction

Among the aluminium based alloys for superplastic deformation the commercially available SUPRAL type alloys containing up to 6 % Cu and up to 0.5% Zr (all in wt. %) belong to the most important ones [1,2]. Since the necessary condition for superplastic deformation is the small grain size, there are several treatments reported for the grain refinement [1-5]. Two basic modes of recrystallization can be used for grain refinement [4,5], both relying on the presence of particles. In the case of discontinuous recrystallization a high density of nucleation sites can be obtained by a high degree of deformation and presence of large particles [6]. The alternative way is the continuous recrystallization, when subgrain coarsening proceeds until the high angle grain boundaries are formed [2]. The latter method was used in the case of AlCuZr alloys [7].
Depending on the type of SUPRAL alloy, its tensile strength may vary between 420 and 510 MPa [1]. The addition of silver and magnesium to Al-Cu alloys increases its mechanical properties after ageing due to a change in the type of the hardening particles from Θ' to Ω and the habit plane from {100} to {111} [8]. The Ω phase appears to be highly stable at elevated temperatures [9]. These

alloys attain the tensile strength of 600 MPa, higher than that of AlCuZr alloys [2] and possess improved creep resistance at temperatures in the range 150-220°C which is important in aviation applications. Since no investigation was found on superplastic deformation of AlCuAgMg alloys, they has been chosen as the subject of this work.

2. Experimental procedure

Two aluminium alloys containing: 5% Cu, 0.2% Mn, 0.75% Mg, 0.65% Ag, 0.4% Zr (AG1) and 5.6% Cu, 0.16%Mn, 0.44%Mg, 0.43% Ag, 0.4% Zr (AG2), were continuously cast and hot rolled down to 30 mm. In order to obtain a fine grain size, particle stimulated nucleation (PSN) process [3] was used. The alloys were hot rolled down to 15 mm, solutionized at 530°C, water quenched and aged at 523K for 3 hours, then rolled at 573K to the final thickness of 6 mm. The microstructure of the alloys was studied using optical, scanning microscope Philips XL30 and transmission electron microscope (TEM) Philips CM20 operating at 200 kV. Thin foils were obtained by jet electropolishing at subzero temperatures in electrolyte consisting of methanol – 6 % perchloric acid.

3. Results and discussion

Fig.1a shows transmission electron micrograph taken of the alloy AG1 subjected to ageing and next to rolling at 573 K treatment. It shows the presence of subgrains of 0.5-2 µm in size, with a misorientation of 3 - 15° as a result of tilt/contrast change experiments.

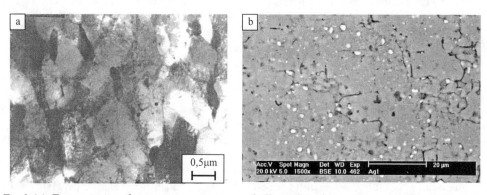

Fig.1 (a) Transmission electron microstructure of Alloy AG1 after rolling at 573K . (b) Scanning electron microstructure of alloy AG1 after tensile deformation of 300% at 400°C.

These grains grow rapidly up to 10-20 µm after annealing at temperatures higher than 673K. This can be also seen in samples deformed by tension at the temperature of 673K (Fig.1b). It affects the superplastic properties at higher temperatures in a negative way. The tensile stress σ is related to the strain rate $\dot{\varepsilon}$ by the equation:

$$\sigma = k\, \dot{\varepsilon}^{\,m} \qquad (1)$$

where **m** = d(ln σ)/d(ln $\dot{\varepsilon}$) is the strain rate sensitivity and **k** is a material constant. The superplastic behaviour is characterised by the **m** value, which was determined by step strain - rate change tests. The tests were conducted at a slow initial strain rate, which increased stepwise to a high constant strain rates. The test results are presented in Fig.2 where it can be seen that the **m** value varies with the strain rate and temperature. The highest **m** value obtained approaches 0.7 for the lowest temperature (623K) and the strain rates ($\dot{\varepsilon} = 10^{-3} - 10^{-4}\ s^{-1}$). A value close to 0.5 was observed for temperatures 623 –673K and the strain rates $10^{-3} - 10^{-4}\ s^{-1}$. The lowest **m** values occur at T> 623K

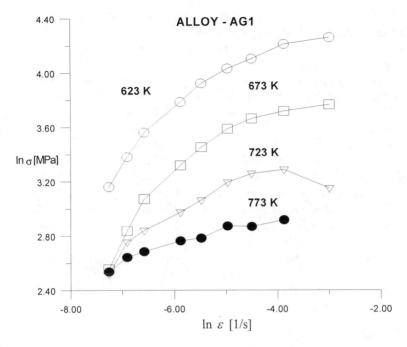

Fig.2 Relationship of tensile stress versus strain rate $\dot{\varepsilon}$ for alloy AG1 at various temperatures.

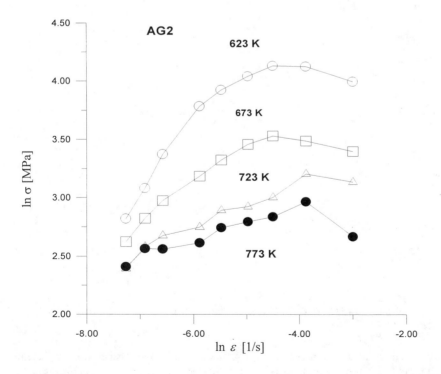

Fig.3 Relationship of tensile stress versus strain rate $\dot{\varepsilon}$ for alloy AG2 at various temperatures.

and the strain rates $> 10^{-3}$. It is to be noted that the flow stress of the present alloys is much more sensitive to the applied strain rate, as compared to other superplastic aluminium alloys. One possible explanation is that zirconium rich precipitates have the form of large particles (1 – 5) μm and are not effective inhibitors for the grain growth. The existing precipitate phases (Al-Cu based) do not prevent the grain boundaries migration. Then, the grain growth occurring during deformation at temperatures above 623K decreases the **m** value. At the interface of large zirconium and copper rich particles the cavities nucleate (as can be seen in Fig.1b) and cause deterioration of superplastic properties.

Fig.3 shows the relationship of tensile stress versus strain rate $\dot{\varepsilon}$ at temperatures 623K, 673K, 723K and 773 K obtained for the alloy AG2. One can see that similarly to the alloy AG1 the curves are more steep for the lower temperatures indicating higher **m** values for these temperatures at low deformation rates. As it results from Fig.3 the **m** value approaches 0.7 at the temperature 623 K and tensile rate $\dot{\varepsilon} = 3 \times 10^{-3} - 7 \times 10^{-4}$ s^{-1}. At 673 K the m value drops to 0.4, independently of $\dot{\varepsilon}$, and at higher temperatures it attains even lower values, close to 0.2 for all rates investigated. This indicates that similarly to the alloy AG2, the grain growth occurs during deformation at higher temperatures influencing the mechanism of deformation and deteriorating the superplastic properties. Apparently, the Ω phase formed during high temperature ageing reported to be stable at higher temperatures [9] does not prevent the grain growth, similarly as large copper rich and zirconium rich particles.

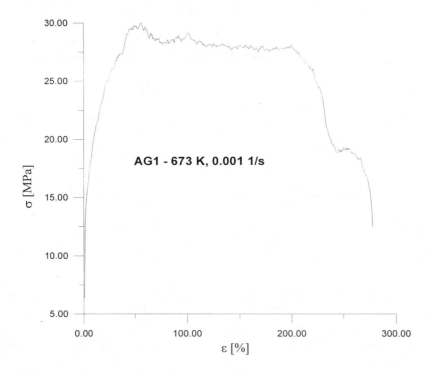

Fig.4 Tensile stress versus strain curve obtained for the alloy AG1 at a constant strain rate $\varepsilon = 10^{-3}$ s^{-1} and temperature of 673 K.

The tensile curve obtained at 673 K presented in Fig. 4 shows that the tensile stress σ increases during the initial period of deformation up to about 28 MPa (the stress values were compensated for a decrease in the cross section of the sample) and then stays constant or even slightly decreases to about 200 % of elongation. This phenomenon may indicate either that there occurs the grain boundaries sliding and consequently no work hardening takes place, or the dislocations contributing to the crystallographic slip are partially annihilated due to a high temperature effect.

In order to clear this problem the structure of the deformed samples was investigated using transmission electron microscopy. Fig.5 shows the transmission electron microstructures of alloy AG1 subjected to elongation of 120% at 673 K. Fig.5a shows a low magnification image where within large grains of several μm in size, much smaller grains in dark contrast can be seen. Additionally visible much finer dark particles are copper rich and zirconium rich phases existing already after casting and partially grown during ageing at 250°C i.e. during PSN treatment. They are often located at grain boundaries, which indicates the grain boundaries pinning mechanism. Apparently, their size is too large and their density too low to prevent the subgrain coarsening as suggested in [2] for continuous recrystallization mechanism.

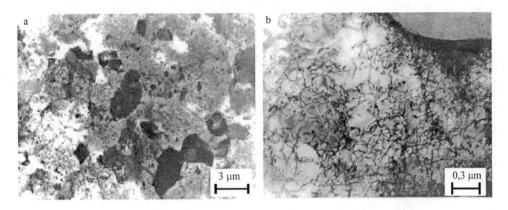

Fig.5 TEM micrographs taken of the sample AG1 deformed isothermally 120% at 673 K

The microstructure given in Fig.5b taken at higher magnification shows two types of dislocations: (i) the grain boundary dislocations causing a jerky shape of the grain boundary and (ii) short dislocation lines within the grain. This microstructure is typical for this stage of deformation and since it shows rather low density of dislocations, one can conclude that dislocation plays a minor role in the deformation process. The presence of grain boundary dislocations and of fine grains after a high degree of deformation suggests that grain boundary sliding is the dominating deformation mechanism.

4. Conclusions

1. The AlCuAgMg alloys continuously cast, aged and deformed at 573 K show good superplastic properties at temperatures below 673 K. The **m** value calculated from step strain tensile tests approaches 0.7 for the lowest temperature (623K) and the strain rates $10^{-3} - 10^{-4}$ s^{-1}. A value close to 0.5 was observed for temperatures 623 - 673K at the strain rates 10^{-3}-10^{-4} s^{-1}.
2. The degradation of superplastic deformation properties at higher temperatures is due to a drastic grain growth occurring at higher temperatures. Its cause is the concentration of Zr rich in large particles and low grain boundary pinning properties of the Ω and Θ phases,
3. Low density of dislocations within grains, presence of grain boundary dislocations and of small grains within larger ones indicate the grain boundary sliding as the dominating deformation mechanism at temperatures below 673K.

5. References

[1] T.G. Nieh, J. Wadsworth and O.D. Sherby, Superplasticity in metals and ceramics, Cambridge University Press, Cambridge 1997, p.64
[2] J.A. Wert, in "Superplastic Forming of Structural Alloys", ed. A.A. Paton and C.H. Hamilton, Metallurgical Society of AIME, Warrendale, Pa, USA, 1982, p.69
[3] H. Ahlborn, E. Hornbogen, and V. Koster, J. Mat. Sci., 4 (1969) 944
[4] H.J. Mc. Queen, Thermomechanical Processing of aluminium Alloys, ed. J.G. Morris, The Metallurgical Society of AIME, Warrendale, PA, 1979, p. 1-24
[6] J.A. Wert, N.E. Paton, C.H. Hamilton, M.W. Mahoney, Metall. Trans., 12A (1981) 1267
[7] B.M. Watts, M.J. Stowell, B.L. Baikie, D.G.E. Owen, Metal Sci., 10, (1976) 189
[8] B.C. Muddle and I.J. Polmear, Acta metall. 37 (1989) 777
[9] I.J. Polmear and M.J. Couper, Met. Trans A., 19A, (1988) 1027

6. Acknowledgements

This work was supported by the Maria-Curie Skłodowska Fund (Contract PAN/NSF -98-331) and the Project supported by the State Committee for Scientific Research No PBZ-12-15.

Superplastic Deformation Behavior of Undoped and Doped High-Purity 3Y-TZP

T. Satou[1], F. Hosaka[2], E. Sato[3], J. Matsushita[1], M. Otsuka[2] and K. Kuribayashi[3]

[1] Department of Applied Chemistry, Tokai University,
1117 Kitakaname, Hiratuka, Kanagawa 259-1292, Japan

[2] Department of Materials Science and Engineering, Shibaura Institute of Technology,
3-9-14, Shibaura, Minatoku, Tokyo 108-8548, Japan

[3] The Institute of Space and Astronautical Science,
3-1-1 Yoshinodai, Sagamihara, Kanagawa 229-8510, Japan

Keywords: Deformation Mechanism Map, Diffusional Creep, Superplasticity, Zirconia

Abstract

Superplastic deformation behavior was examined in undoped and doped high purity 3 mol% yttria-stabilized tetragonal zirconia polycrystals (3Y-TZP) with fine and coarse grains. In undoped 3Y-TZP, fine grain materials below 0.3 μm showed two deformation regions with a stress exponent of two, a grain size exponent of one and an apparent activation energy of 370 kJ/mol at high stress region and three, one and 630 kJ/mol at low stress region, while coarse grain materials above 1 μm showed two deformation regions with a stress exponent of one, a grain size exponent of three and an apparent activation energy of 450 kJ/mol at high stress region and three, one and 550 kJ/mol at low stress region. On the other hand in alumina doped high purity 3Y-TZP, both the fine and coarse grain materials showed two deformation regions with a stress exponent of one, a grain size exponent and an apparent activation energy of 520 kJ/mol at high stress region and two, one and 590 kJ/mol at low stress region. The region of diffusional creep region with a stress exponent of one was first observed in 3Y-TZP. Deformation mechanism maps have been drawn for undoped and doped TZP, and the obtained deformation mechanism maps can predict these deformation regions.

Introduction

Following Wakai's initial report of an elongation of more than 100 % in a 3Y-TZP [1], superplasticity in TZP has been widely studied and the complicated deformation behavior was become clear. Because the deformation behavior is influenced by stress σ, grain sizes d and temperatures T, the constitutive equation is written by

$$\dot{\varepsilon} = \left(\frac{A}{kT}\right)\sigma^n \left(\frac{b}{d}\right)^p D_0 \exp\left(\frac{-Q}{RT}\right) \qquad (1)$$

where $\dot{\varepsilon}$ is the strain rate, A is a dimensionless constant, k is Boltzman's constant, n is the stress exponent, b is the magnitude of Burger's vector, p is the inverse grain size exponent, D_0 is the frequency factor, Q is the apparent activation energy and R is gas constant. Furthermore, the deformation behavior changes largely by trace impurities above 0.1 wt% such as silica and alumina.

We recently showed that high-purity materials possess two deformation regions of stress exponent three at low stress and two at high stress, while low purity materials possess only one deformation region of stress exponent two [2]. The change in the stress exponent in the high-purity materials has been explained in several ways: first, the region with stress exponent two and three are explained to corresponding to independent deformation mechanisms [3]. Second, the region with stress exponent three is explained to be generated by a threshold stress, and a new region below the threshold stress is also claimed [4]. Third, diffusional creep is assumed to be at higher stress region and region with stress exponent two is just a transition from the at low stress region to diffusional creep region [5]. However, diffusional creep has not been observed in 3Y-TZP, and it is observed only in 6Y-PSZ with very coarse grains [6]. In the present study, superplastic deformation behavior is explained using a deformation mechanism map, through examining deformation behavior in a wide range of stress, grain size and temperature for alumina doped and undoped 3Y-TZP. Especially, the present study focused on diffusional creep region with coarses grain at high temperature and high stress region, and low strain rate region with fine grain at high temperature and low stress region, compared to the conventional superplastic region.

Experiment

0.25 wt% alumina doped and undoped high-purity 3Y-TZP powders (Alumina undoped 3Y-TZP : Y_2O_3 5.19, Al_2O_3 0.012, SiO_2 0.007, Fe_2O_3 0.005, Na_2O 0.019, Alumina doped 3Y-TZP : Y_2O_3 5.16, Al_2O_3 0.248, SiO_2 0.006, Fe_2O_3 0.002, Na_2O 0.022) were prepared as raw materials. Dense cylindrical samples of 6 mm diameter and 8 mm length were prepared from these powders thought cold isostatic pressing at 300 MPa for 300 s, air sintering at several temperatures with heating and cooling rates of 10 K/min and mechanical finishing.

Depending on the sintering temperature, we obtained materials with grain sizes of 0.32 (1673 K, 2 h), 0.72 (1873 K, 8 h) and 1.1 µm (1873 K, 24 h) in undoped materials and 0.39 (1673 K, 2 h), 0.60 (1773 K, 6 h) and 2.2 µm (1873 K, 12 h) in alumina doped materials were obtained. By high temperature and long time sintering, we obtained more than 1 µm grain sizes.

High temperature compression tests were performed constant speed test (1573, 1673, 1773 and 1873 K, 10^{-1} to $10^{-6} s^{-1}$) and constant load test with a strain gage (1773 and 1873 K, 10^{-6} to $10^{-7} s^{-1}$) using an Instron type machine equipped with a SiC jig in air.

Results

Figure 1 shows the stress and strain rate relation in undoped TZP at 1873 K with several grain sizes. In fine-grain materials, the stress exponent is three at low stress region and two at high stress region similar to the previous report [2, 3], while in coarse-grain materials, the stress exponent is three at low stress region and one at high stress region. It seems that the region of stress exponent two locates between the region of stress exponent three and one. In fine grain materials, the grain size exponent is one with stress exponent two and three, while in coarse grain materials, grain size exponent is one in the region of stress exponent three and three in the region of one. The apparent activation energy equals 450 kJ/mol in high stress region with stress exponent one observed in coarse grain materials, 370kJ/mol in the intermediate stress region with two observed in fine grain materials, and 550 to 630 kJ/mol in low stress region with three.

Figure 2 shows the stress and strain rate relation in doped TZP at 1773K with several grain sizes. The region of stress exponent two is observed in fine grain materials at low stress region similarly to the previous studies [2], though, at high stress region especially coarse grain materials the region of one is observed. The grain size exponent is one in the region of stress exponent three and three in the region of one. The apparent activation energy is 520kJ/mol in the high stress region with stress exponent one and 590kJ/mol in the low stress region with two. According to the authors's knowledge, this is the first report of diffusional creep in undoped and doped TZP.

Discussion

The parameter n, p and Q obtained in the present study on undoped and doped TZP are compared with those of the recently reported papers [2, 3, 4, 7, 8]. The extra low region proposed by Bravo-Leon et al [4] has not been conformed in preset study. The other regions observed in the previous papers are conformed completely in the present study, i.e., three regions with stress exponent three at low stress region, two at intermediate stress region and one at high stress region for undoped TZP, and two regions with two at low stress region and one at high stress region for doped TZP. The grain size exponent is one for the regions of stress exponent three and two and four for the region of one for undoped TZP, and one for the regions of two and three for the region of one for doped TZP, respectively. These stress exponent and grain size exponent are almost consistent with the present study and the previous studies. The apparent activation energy differs among the studies, but roughly speaking low stress region has high apparent activation energy around 600 kJ/mol, while the high stress region has rather low around 500 kJ/mol. From the above discussion, the present study decide to use values in Table 1 for drawing deformation mechanism maps in undoped and doped TZP.

Figures 3 and 4 are deformation mechanism maps in undoped and doped TZP at 1773 K with line for 0.3, 0.6, 1.2 and 2.4 μm grain sizes. In these maps, stress exponent two and three regions are superplastic regions and one is diffusional creep region. Under the normal superplastic condition, i.e., in fine grains at intermediate stress region and temperatures, only the regions with stress exponent of two and three appear.

In figs. 5 and 6, the previous data are plotted on the deformation mechanism map. Figure 5 is the data by Owen and Chokshi [3] at 1723 K with 0.41, 0.66 and 1.2 μm. The behavior of fine grain materials agrees with our deformation mechanism map. Figure 6 is the data by Bravo-Leon et al. [4] at 1623 K with 0.3 and 0.8 μm. The behavior of fine grain materials is almost consistent with our deformation mechanism map. The extra low stress region claimed by them can be interpreted to be diffusional creep region similar to high stress region. We must notice that the magnitude of the grain sizes slightly differs among the papers according to the measurement methods.

Figure 7 schematically shows the deformation mechanisms of superplasticity in 3Y-TZP. The main deformation process is grain boundary sliding with grain boundary switching, grain boundary sliding requires accommodation process by diffusion, and diffusion is controlled by diffusion itself or interface reaction, which is vacancy formation and absorption at grain boundaries. If deformation is diffusion controlled, the stress exponent becomes one and grain size exponent becomes three. If deformation is interface reaction controlled, assuming that the reaction rate is proportional to the stress and the reaction site density is proportional to stress and stress squared, then the stress exponent two or three appears respectively. This case grain size exponent is one. The obtained deformation mechanism maps can be almost explained by this model.

Conclusion

1) Diffusional creep has been shown with coarse grain materials at high temperatures.
2) Deformation mechanism maps have been drawn for undoped and doped TZP.
3) The obtained deformation mechanism maps are consistent with data in the previous paper of TZP.

References

1. F. Wakai, S. Sakaguchi and Y. Matsumoto, Adv. Ceram. Mater. 1, 1986, 259-263
2. E. Sato, H. Morioka, K. Kuribayashi and D. Sundararaman, J. Mat. Sci. 34, 1999, 4511-4518
3. D. M. Owen and A. H. Chokshi, Acta. Mater. 46, 1998, pp. 667-679
4. A. Bravo-Leon, M. Jimenez-Melendo and A. Dominguez-Rodriguez, Scr. Mater. 34, 1996,1155-1160
5. M. Z. Berbon and T. G. Rangdon, Acta, Mater. 47, 1999, 2485-2495
6. P. E. Evans, J. Am. Ceram. Soc. 53, 1970, 365-369
7. A. Lakki, R. Schaller, M. Nauer and C. Carry, Acta. Metall Mater. 41, 1993, 2845-2853

8. F. Wakai, S. Sakaguchi, N. Murayama, Y. Kodama and N. Kondo, Trans. Mat. Res. Soc. Jpn. 16B, 1994, 947-952

Corresponding author: E.Sato (sato@materials.isas.ac.jp)

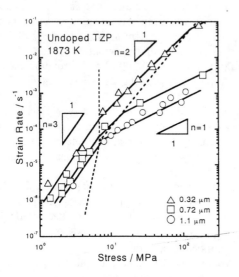

Figure 1. Stress and strain rate relation in undoped TZP at 1873 K at several grain sizes.

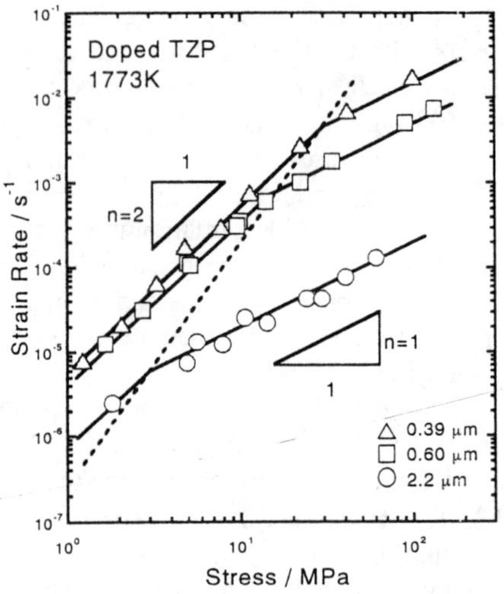

Figure 2. Stress and strain rate relation in doped TZP at 1773 K at several grain sizes.

Table 1. The values of parameters for deformation mechanism maps

Material	Undoped TZP			Doped TZP	
Stress Level	Low	Intermediate	High	Low	High
n	3	2	1	2	1
p	1	1	4	1	3
Q (kJ/mol)	600	500	430	540	510

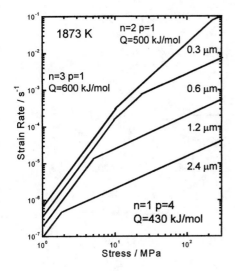

Figure 3. Deformation mechanism map in undoped TZP at 1773 K in several grain sizes

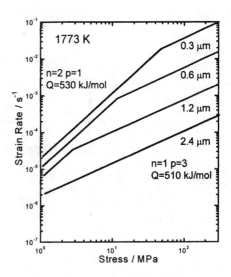

Figure 4. Deformation mechanism map in doped TZP at 1773 K in several grain sizes

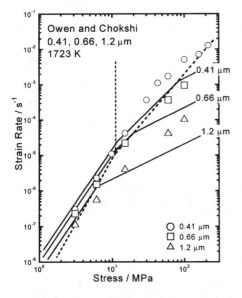

Figure 5. Owen and Chokshi's data plotted on our deformation mechanism map in undoped TZP at 1723K several grain sizes.

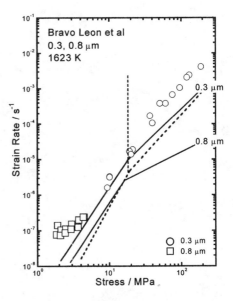

Figure 6. Bravo-Leon et al's data plotted on our deformation mechanism map in doped TZP at 1623K several grain sizes.

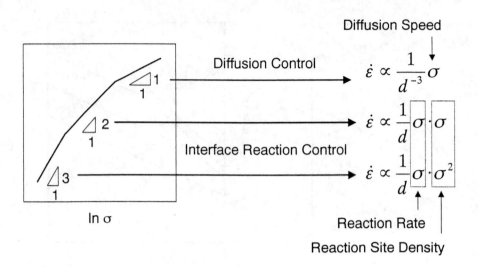

Figure 7. Deformation mechanisms of superplasticity in TZP

Compressive Creep Studies on Alumina Particulate Reinforced Alumino-Silicate Glass

B. Sudhir and Atul H. Chokshi

Department of Metallurgy, Indian Institute of Science, Bangalore 560 012, India

Keywords: Alumina Reinforcement, Aluminosilicate Glass, Compression, Percolation

Abstract The effect of alumina particulate reinforcement on the high temperature compressive creep behaviour of alumino-silicate glass was characterized in the stress and temperature range of 3 to 50 MPa and 1023 to 1123 K, respectively. The resulting strain rates varied from 10^{-6} to 10^{-2} s^{-1}, and the composites exhibited Newtonian viscous flow. The creep activation energies ~ 720 ± 50 kJ mol^{-1}. XRD and density measurements on the sintered and tested samples indicated that there was no crystallization or cavitation in the glass or the composites. It was found that addition of alumina particles moderately increased the creep resistance of the glass and that these materials have the potential for superplasticity.

1. Introduction Glass ceramics are used in a variety of applications such as architectural materials, glass-ceramic to metal seals, fiber optic communications and thermomechanical applications in turbine heat exchangers. An attractive feature of these materials is the ability to form them to the final shape in the glassy form at lower temperatures and subsequently crystallize them for use at higher temperatures. An even more attractive feature of these materials is the ability to vary the properties, such as coefficient of thermal expansion, dielectric constant, toughness and elastic modulus, by addition of a reinforcing phase. Another advantage of studying glass ceramics is that it can help in understanding the behaviour of other glass containing polycrystalline ceramics. For example, the presence of a glassy phase can enhance the ductility of ceramics [1,2]. In fact the largest reported elongation of 1038% in ceramics is in 3mol% yttria stabilized tetragonal zirconia containing 5 wt% SiO_2 [2].

In reinforced glasses a sharp variation in properties with addition of second phase occurs because of percolation threshold, which refers to the specific volume fraction at which the inclusion phase forms a contiguous network in the matrix [3]. The percolation threshold has been shown to be dependent on the size and aspect ratio of the second phase [3,4,5,6]. As the particle size decreases, the volume fraction required for the percolation threshold also decreases. It is logical to expect that use of fine particle sizes would decrease the percolation threshold to still lower volume fractions, and thereby, the mechanical properties are expected to substantially increase even at lower volume fractions. In the present study, the effect of alumina inclusions, of very fine particle size (0.2 µm), on the creep of alumino-silicate glass was characterized at volume fractions of 0, 5 and 10%.

2. Experimental The alumino-silicate glass powder (SiO_2:56.7, Al_2O_3:16.3, BaO: 8.2, CaO:8.1, MgO:5.8 wt%) was procured from Corning, USA (Glass code 1724) and the high purity (99.99%) alumina powder from Taimei, Japan. The mean particle sizes of the glass and alumina powders were 10 µm and 0.2 µm, respectively. The reported densities of the glass and alumina were 2.62 g cc^{-1} and 3.96 g cc^{-1}, respectively, and these were confirmed by pycnometric studies on the powders. The composites were prepared by wet ball milling weighed amounts of glass and alumina powders for 24 hrs using high purity alumina balls in iso-propyl alcohol medium. The powder was then dried at about 353 K for 6 hours and the resulting soft cake crushed and sieved through a 30 µm mesh to ensure uniform granule size.

The powders were cold compacted in a steel die at a stress of about 75 MPa and then free sintered at 1173 K for 1 hour. In the case of 10 vol% composite, free sintering at 1173 K gave compacts with only 95 % density, which did not change even at longer sintering times. Consequently the 10 vol% composites were sinter-forged to full density at the same temperature using a stress of ~ 0.6 MPa. Parallelepiped samples of nominal dimensions 4×4×7 mm were cut from the sintered discs for compression creep studies and the creep behaviour was characterized in the temperature and stress ranges of 1023 to 1123 K and 3 to 50 MPa, respectively. It may be noted that the test temperatures are below the glass transition temperature, which has been reported by Corning as 1199 K. The creep data was analyzed using the equation:

$$\dot{\varepsilon} = A\sigma^n \exp(-Q/RT) \quad (1)$$

where A is a constant, σ is the applied stress, n is the stress exponent, Q is the creep activation energy, R is the gas constant and T is the absolute temperature. XRD and density measurements were done on selected samples to detect any crystallization or cavitation. Also, hardness of the three compositions was determined by Vickers microhardness tester using a load of 1000 g. applied for 10 s.

3. Results and Discussion All samples tested in this study had densities of > 98 %. The hardness, in VHN, of pure glass, 5 vol% and 10 vol% composites are 635 ± 20, 610 ± 15 and 625 ± 15, indicating that there is no significant effect on hardness by the addition of upto 10 vol% alumina. XRD studies on the sintered and tested samples indicated that there was no observable crystallization in the tested glasses. Figure 1 shows a plot of strain rate vs strain for the three materials at a temperature of 1073 K and a stress of 10 MPa, where it can be seen that the strain rate is essentially constant upto strains of 10 %. This indicates that there were no time dependent microstructural changes in the materials. In order to confirm the absence of any microstructural changes in the composites, stress jump tests were performed on them. One such study carried out on 10 vol% composite at a temperature of 1023 K is shown in Fig. 2 as a plot of strain rate vs strain. It is observed that the strain rate returned to the original value when the stress was brought back to the initial stress, confirming that there was no strengthening or degradation such as increased particle-particle contacts [3] or cavitation.

The stress exponent of the composites was determined from the plot of strain rate vs stress and as seen in Fig. 3 the three materials exhibit a stress exponent of unity, indicating Newtonian viscous flow. Further, it can be seen that there is no significant difference between the strain rates of the three materials. Similar, plots at 1073 and 1123 K also gave the stress exponent as unity. Stress exponents close to unity, indicating Newtonian viscous flow, have been reported in other glass ceramics such as β-spodumene [1,7,8], borosilicate glass [4], β-quartz [7] and anorthite [9]. However, in some studies [3,4,5,8], it was observed that at higher reinforcement volume fractions the creep rate was became strain dependent and the flow became non-Newtonian, a feature was not observed in the present study.

The activation energy of the composites was determined from a Arrhenius type plot of strain rate vs inverse temperature (Fig. 4) and it was found that the activation energy for creep is comparable in the three materials and is about 720 ± 50 kJ mol^{-1}. Activation energies around 750 ± 50 kJ mol^{-1} have been reported in other silicate based glasses such as β-spodumene and β-quartz glasses [1,7,8], which are also alumino-silicate glasses with lithium oxide as the major modifier, oxynitride glasses [5] and calcium-alumino-silicate glass [9]. However, it is significantly higher that the value of 540 kJ mol^{-1} reported in borosilicate glass [4] and the activation energy of 530 kJ mol^{-1} in pure fused silica [10]. Raj and Morgan [11] explain such high values of activation energy as a sum of the activation energies associated with viscosity and heat of solution of the crystals in the liquid phase. However, in the present case and in the case of oxynitride glass [5], as there is no evidence for crystallization, this explanation may not be applicable. Moreover, as noted by Wang and Raj [1], the activation energy seems to be similar for a wide variety of compositions and it is unlikely that the heat of dissolution will be similar in different compositions.

The activation energy for creep in glasses, being a measure of the dependence of viscosity on temperature, is expected to depend primarily on the number of non-bridging oxygen ions per glass-forming cation, which is a function of composition [10]. It is thus interesting to note that even though the compositions of the various glasses considered above are different, they have comparable activation energies around 700 kJ mol^{-1}, whereas one would expect glass compositions with higher silica content to exhibit higher activation energies [10]. However, it should be noted that viscosity of most glasses is not linear when plotted over a wider temperature range [12]. In fact, at lower temperatures the apparent activation energy in oxide glasses can be 2 to 3 times the value at higher temperatures [12].

At a volume fraction of 10%, alumina inclusions increased the creep strength of glass by a factor of ~ 2. In comparison, for the same volume fraction, nickel inclusions of 8 μm particle size increased the resistance by a factor of ~ 5 [3], whereas SiC inclusions with particle size of 6 μm did not affect the creep resistance [5]. The difference in the magnitude of the effect between Nickel and SiC particles of comparable size can be explained on the basis of differences in the strength of the contacts between the particles. An increase in viscosity of the composite arises from the ability of the network formed by the inclusion phase to bear load [3]. The strength of the network depends on the strength of the bonds between the contacting particles. We expect the nickel particles to form a stronger bond than the SiC particles, particularly at temperatures below 1273 K, which is the range of temperatures used in the above studies. However, it cannot explain the reason for lack of significant effect on creep by addition of 10 vol% of 0.2 μm alumina particles. Based on the data reported in the previous studies [3,5], the percolation threshold for 0.2 µm sized inclusion phase is expected to be much lower than 10 vol%. In contrast, the present data indicates that percolation threshold was not reached even at 10 vol% of the inclusion phase. Further experiments and analysis are needed to resolve this issue.

3.1. Potential for Superplasticity A stress exponent of unity implies a perfect resistance to strain localization. However, in the case of ceramic materials, the ductility is limited by cavitation [13]. For example, in the case of anorthite glass, even though the stress exponent was 1, there was significant cavitation even under compressive loading [9]. In the present study we did not observe any cavitation even when the samples were tested to strains > 10 %. In particular, a glass sample was tested to 34 % strain at a temperature of 1123 K and a stress of 8 MPa and there was no observable change in its density. However, Wang and Raj [1] found that in the case of β-spodumene glass, even though it showed a potential for infinite ductility under compression, the strains were limited to 135 % under tension because of intergranular fracture. In particle reinforced glasses, failure under tension can be expected at the glass-particulate interfaces, particularly if the glass does not wet the particle or if the glass crystallizes during testing. In the present system, these are not expected to occur because alumino-silicate glass is expected to wet the alumina surface, and also there was no observable crystallization in the glass as determined by XRD studies. Hence it is concluded that the material has potential to exhibit superplasticity.

4. Conclusion The compressive creep behaviour of alumino-silicate glass reinforced with 0.2 µm sized alumina particles was studied in the stress and temperature ranges of 3 to 50 MPa and 1023 to 1123 K. Addition of upto 10 vol% alumina particles was found to moderately improve the creep resistance but had no significant effect on the hardness. Based on data reported in the literature, the percolation threshold for 0.2 µm particles was anticipated to occur much below 10 vol%. However, our data indicates that percolation threshold was not reached even at 10 vol% of the inclusion phase.

In the glass and the reinforced composites the flow behaviour was Newtonian viscous and the activation energy was about 720 ± 50 kJ mol^{-1}. These values are comparable to those reported in other glass ceramics. Our analysis using the XRD, density and stress change experimental data indicates that there was no crystallization or cavitation during creep testing. The Newtonian viscous flow exhibited by these materials coupled with lack of cavitation suggests that they have potential for superplastic flow in tension.

Acknowledgement This work is supported by the Department of Science and Technology under a swarnajayanti presidential young investigator award to Dr. Atul H. Chokshi.

References
[1]   J.G. Wang and R. Raj, J. Am. Ceram. Soc. 67 (1984), p. 399.
[2]   K. Kajihara, Y. Yoshizawa, and T. Sakuma, Scr. Metall. Mater. 28 (1993), p. 559.
[3]   R.E. Dutton and M.N. Rahaman, J. Mater. Sci. Lett. 12 (1993), p.1453.
[4]   V.S.R. Murthy, Santanu Das, G. Banu Prakash and G.S. Murty, Mater. Sci. Forum 243-245 (1997), p. 375.

[5] T. Rouxel, B. Baron, P. Verdier and T. Sakuma, Acta Mater. 17 (1998), p. 6115.
[6] W. Xia and M.F. Thorpe, Physical Rev. A 38 (1988), p. 2650.
[7] S.E. Bold and G.W. Groves, J. Mater. Sci. 13 (1978), p. 611.
[8] J.P. Northover and G.W. Groves, J. Mater. Sci. 16 (1981), p. 1874.
[9] R.F. Mercer and A.H. Chokshi, Scr. Metall. Mater. 28 (1993), p. 1177.
[10] G. Urbain, F. Cambier, M. Deletter, and M.R. Anseau, Trans. J. Br. Ceram. Soc. 80 (1981), p. 139.
[11] R. Raj and P.E.D. Morgan, J. Am. Ceram. Soc. 64 (1981), p. C-143.
[12] G.S. Meiling and D.R. Uhlmann, Phys. Chem. Glasses 8 (1967), p. 62.
[13] A.H. Chokshi, Mater. Sci. Engg. A166 (1993), p. 119.

Corresponding Author
B. Sudhir: sudhir@metalrg.iisc.ernet.in

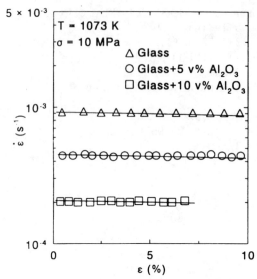

Fig. 1: Plot of strain rate vs strain for the composites at 1073 K and 10 MPa

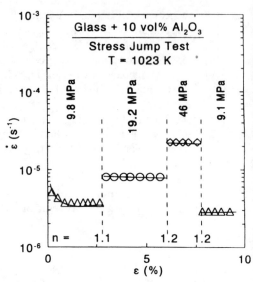

Fig. 2: Stress jump test conducted on 10 vol% composite at 1023 K

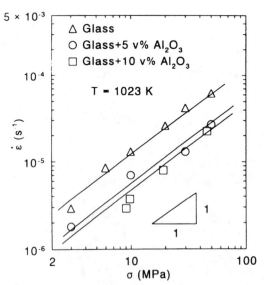

Fig. 3: Strain rate vs stress plot at 1023 K showing n = 1 in the three materials

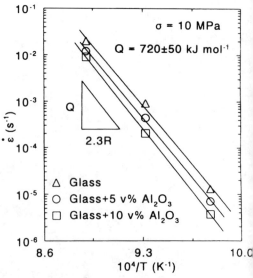

Fig. 4: Arrhenius type plot of strain rate vs inverse temperature at 10 MPa

Superplastic Characteristics in Germania Based Codoped Y-TZP

K. Sasaki, M. Nakano, J. Mimurada, Y. Ikuhara and T. Sakuma

Department of Materials Science, Fauclty of Engineering, University of Tokyo,
7-3-1 Hongo, Bunkyo-ku, Tokyo 113-8656, Japan

Keywords: Co-Doping, Dual Phase, Superplasticity, Zirconia

Abstract

The superplastic characteristics in 3 mol% yttria doped tetragonal zirconia polycrystals (TZP) were drastically improved by the combined addition of germanium oxide and other dopants (titanium oxide, magnesium oxide, and calcium oxide) giving rise to 1000% elongation under optimum conditions. Extensive work hardening of the co-doped compositions for temperatures above 1400°C was observed. These were unique features which could not be explained from the work hardening due to dynamic grain growth. Rather it is proposed that the metastable nature of these co-doped materials, which changes during deformation is the cause of this exaggerated work hardening.

Introduction

ZrO_2 ceramics show excellent mechanical properties by the tailoring of a very refined microstructure consisting of a metastable tetragonal phase retained at room temperature, known as tetragonal zirconia polycrystals (TZP)[1]. TZP ceramics were later found to have the additional benefit of superplastic deformation[2]. Superplasticity occurs under conditions of submicron grain sizes, at temperatures of about 1300°C and higher. Since then a lot of work has been underway for superplasticity in yttria doped tetragonal zirconia polycrystals (Y-TZP), which have shown elongations of up to 800% at 1550°C in 3 mol % Y-TZP[3].

The effect of oxides dopants on the superplastcicity of Y-TZP has been studied extensively in terms of the ionic radius and valence of the dopants in a recent paper[4]. It was found that the change in flow stress due to the addition of metal oxides was strongly dependent on the ionic radius of the cations. While the valence of the dopant cation had a tendency to influence the elongation. Doping of two cations to Y-TZP (co-doping) has shown that greater tensile elongations than single doping in Y-TZP, could be obtained. Titania-calcia doping to Y-TZP gave elongations of about 500%[5] while germanium oxide-neodinium oxide was found to fail at 610%[6] under optimum conditions. Work in co-doped TZP ceramics is however still not clear, therefore the aim of this paper is to clarify some of the important mechanisms that occur in this class of superplastic materials. Compositions with germanium oxide and two divalent oxides were chosen, since their solitary effect on Y-TZP has been examined extensively in the past giving rise to elongations of 1000% under optimum compositions[7].

Experimental Procedure

Commercially available 3 mol% yttria stabilized zirconia high purity powder (Tosoh Co. Ltd.) was used as the base material. The dopants used were germanium oxide (Rare Metallics, 6N), titanium oxide (Sumitomo Osaka Cement Co. Ltd, 3N), calcium oxide (Soekawa Chemical, 3N) and magnesium oxide (Ube Industries Ltd., 3N). The Y-TZP powders were mixed with the metal oxides with the following compositions: 2 mol% Ge-1 mol% Ca (2Ge-1Ca), 2 mol% Ge-1 mol% Mg (2Ge-1Ca) and 2 mol% Ge-2 mol% Ti (2Ge-2Ti) to 3 mol% Y-TZP. The powders were ball milled in ethanol for 24 h. The final slurries were dried in argon and subsequently granulated with a 100 μm sieve. Uniaxial pressing was carried out at 33 MPa, and cold isostatically pressed under a pressure of 100 MPa in a rubber tube. All 4 compositions were sintered at 1400°C in air for 2 h.

Only samples with at least 95% theoretical density were used. High temperature tensile test specimens with an approximate cross section of 2×2 mm^2 and a gage length of 13.4 mm were cut out from the sintered bodies. The tensile tests (Shimazu AG-5000C) were carried out for a temperature range of 1250-1550°C with an initial strain rate of $1.3 \times 10^{-4} s^{-1}$. SEM analysis (JEOL JSM-5200) and optical microscopy (Nikon Optical Microscopes) was carried out for the as sintered deformed and annealed specimens to find the mean grain size using the linear intercept method. XRD analysis (Rigaku RINT 2500V X-Ray Diffractometer), was carried out for the as sintered and elongated samples in order to have qualitative data of the phases present.

Results and Discussion

High elongations were obtained for all compositions as can be clearly seen in Fig. 1, where stress-strain curves are given at a testing temperature of 1400°C. The specimens deformed uniformly with no necking at the fracture tip and in some cases severe deformation was observed on the sample head. Greater elongations are obtained with respect to Y-TZP, when germania is co-doped with either divalent calcium or magnesium, or tetravalent titanium 5 to 9 times. The largest elongation was obtained for the titania-germania combination which shows an elongation of nearly 1000%. Fig. 1 shows for these codoped materials a region of apparent steady state up to 400 % elongation, which is then followed by an increase in flow stress. This phenomenon will be discussed later.

The microstructures of the as sintered specimens were determined using secondary electron imaging by SEM. In all cases, a homogeneous and equiaxed microstructure is observed. No amorphous phase is found by means of conventional electron microscopy. For the compositions investigated the average grain size ranges from about 0.4 to 0.5 μm (3Y-TZP: 0.38μm, 2Ge-1Ca: 0.55μm, 2Ge-1Mg: 0.44μm, 2Ge-2Ti: 0.51μm).

In Fig. 2 the XRD spectra obtained for the 4 compositions in the as sintered state and annealed states are shown. The phases present were examined between 72 and 76 degrees, for the {400} reflections, which are often used to identify tetragonal zirconia and cubic zirconia. In the as sintered state only the undoped Y-TZP and the germania-titania composition do not show any significant amount of cubic phase. Germania-calcia and magnesia show the presence of the cubic {400} reflections, implying that these compositions consist of two phases.

Unlike the as sintered state significant increase in the amount of cubic phase was found for the germania co-doped compositions as shown in Fig. 3. Dynamic grain growth was also observed in all cases (Fig. 3), extensive abnormal grains were observed at large strains for the germania-calcia composition. The abnormal grains were about 20 times larger than the surrounding matrix, when deformed at 1400°C to a strain of 100%. For the calcia-magnesia compositions the grain size of the matrix was employed in the subsequent analysis.

From the current work one striking feature has been found. Extensive work hardening in the stress-strain characteristics (Fig. 1) for the co-doped materials has not been observed in zirconia ceramics to this extent. In Fig. 1 an apparent region of steady state stress is observed followed by a region of constant flow, similar to work hardening. Normally in ceramics such regions of constant stress are a result of dynamic grain growth during deformation[3, 12]. Grain growth causes an increase in stress which is cancelled out by the decrease in strain rate with respect to strain

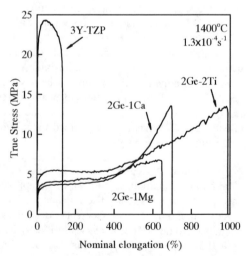

Fig. 1: Stress-strain characteristics of 3Y-TZP and the three co-doped materials deformed at 1400°C.

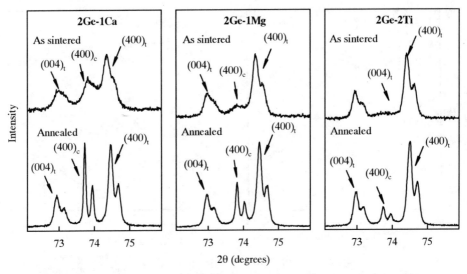

Fig. 2: XRD spectra for the {400} tetragonal and cubic peaks for 3 Y-TZP and the co-doped compositions in the as sintered state (top) and annealed for 15 h at 1400oC (bottom). Note that Cu-Kβ peaks are also shown.

due to constant cross head speed conditions. From such consideration it is necessary to analyse the stress strain curves in terms of the constant strain rate and grain size conditions. The following expression is commonly used to describe the elevated temperature mechanical characteristics of superplastic materials:

$$\dot{\varepsilon} = A \frac{DGb}{kT} \left(\frac{\sigma}{G}\right)^n \left(\frac{b}{d}\right)^p \qquad \text{Eq. 1}$$

Where $\dot{\varepsilon}$ is the steady state creep rate, A is a dimensionless constant, D is the diffusion coefficient, G is the shear modulus, b is the magnitude of the Burgers vector, k is the Boltzmann's constant, T is the absolute temperature, σ is the imposed stress, n is termed the stress exponent ($n=1/m$, m is the strain rate sensitivity), d is the grain size and p is the inverse grain size exponent.

The strain rate and stress dependence for all samples was estimated using a strain rate change method, where the codoped compositions gave values of m ranging from 0.45-0.55, while 3Y-TZP had a value of 0.4. It is clear that the addition of dopants increases the strain rate coefficient, which is likewise observed in the case of impure superplastic zirconia ceramics discussed in many works so far[8-11].

To clarify the effect of grain growth on stress-strain curves, the grain size and flow tress characteristics were examined. According to equation 1, the grain size and flow stress should give a

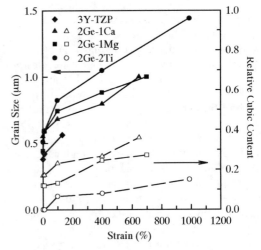

Fig. 3: Grain growth (closed symbols) and cubic content (open symbols) characteristics during deformation for the co-doped samples when deformed at 1400°C at an initial strain rate of $1.3 \times 10^{-4} s^{-1}$.

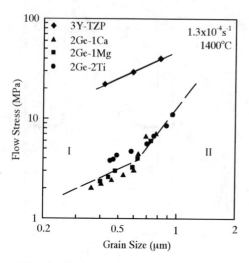

Fig. 4: Grain size dependence of flow stress (10% strain) for the various compositions.

linear expression on a logarithmic plot. For valid flow stress comparisons, specimens with different initial grain sizes were deformed up to 10%, whereby the grain size and flow stress were obtained and arranged in Fig. 4. For the co-doped materials two linear regions are clearly observed, and are depicted as I and II for small grain sizes and larger grains sizes respectively. In region I the dependence of flow stress of the co-doped materials is very similar to that of 3Y-TZP (the slope of the curves mp having values of about 1). On the other hand in region II, the slope of the co-doped samples changes quite drastically showing a greater increase in flow stress with respect to grain size ($mp>2$). For a similar grain size of about 0.5μm, co-doping of any combination has successfully lowered the flow stress. Such decrease in flow stress points directly to an improvement in the accommodation process for grain boundary sliding which in this case is believed to be by diffusional flow. The addition of germania is probably very effective in improving such diffusional accommodation processes together with other dopants.

From the above high temperature characteristics, normalisation of the stress-strain curves was carried out under conditions of constant strain rate and constant grain size. The changes in grain size were taken into account by measuring the

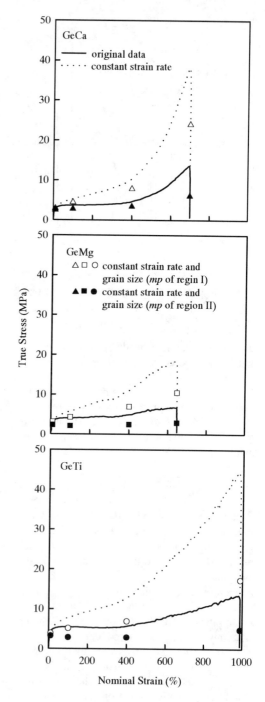

Fig. 5: Modified stress-strain curves under conditions of initial constant strain rate ($1.3\times10^{-4}s^{-1}$) and with an initial constant grain size.

grain sizes at 10%, 100%, 400% and at failure using both *mp* values obtained in Fig. 4. With these grain sizes and equation 1, the assumed flow stress for conditions of constant grain size were obtained and are shown in Fig. 5 as open (*mp* of region I) and closed symbols (*mp* of region II). Fig. 5 shows that even under conditions of constant strain rate and grain size strain hardening at the final stages of deformation are still observed for the co-doped materials. Thus conditions of constant cross head speed and dynamic grain growth are insufficient to explain the work hardening effect observed in the final stages of deformation for the codoped materials.

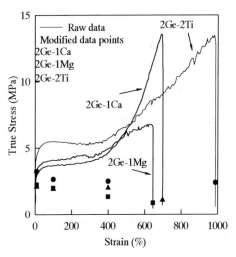

Fig. 6: Modified stress-strain curves for the co-doped compositions. Solid lines and filled symbols indicate the original data and those normalised for constant grain size, strain rate and initial microstructure respectively.

The stress-strain curves were once again calculated, assuming that the deformation of the codoped materials was assumed to obey the grain size dependence in region I through out the tensile test of Fig. 1. This is thought to be reasonable since the slope in region I (Fig. 4), of the co-doped compositions in Fig. 4 was similar to the undoped 3Y-TZP. The "extra" strain hardening term was obtained by extrapolating the difference in stress ($\Delta\sigma$) between region I and II between the two gradients for the specific grain size. The reanalysed data of Fig. 5 are presented in Fig. 6, where it is clear that the exaggerated strain hardening effects are not present.

The abrupt change in flow stress with respect to grain size observed in Fig. 5 may provide some information on the cause of this strain hardening effect. Since the samples were annealed for various times, the results in Fig. 4 imply that changes in microstructure other than grain size must have occurred and somehow influenced the flow stress characteristics of codoped TZP ceramics. A possible explanation for this kind of behaviour may be the presence of the cubic phase, which is present after annealing in the 3 co-doped materials. TZP ceramics lie in the cubic-tetragonal dual phase region. In the as sintered state only a non-equilibrium tetragonal phase is present. This tetragonal phase is known to partition from the metastable tetragonal phase to the equilibrium cubic and tetragonal phases when annealed at elevated temperatures[13]. Part of the original tetragonal phase decomposes into the cubic phase by a diffusion controlled reaction during annealing. Sasaki et. al.[14] reported that this decomposition mechanism occurred under superplastic conditions for temperatures above 1400°C for 4 mol% YSZ.

In metals similar behaviours in the final stages of the stress-strain curves have been reported. Belzunce and Suery[15] reported for α/β brasses having a duplex structure a decrease in stress at ~170%, followed by an increase in stress and eventual failure at ~260%, for a brass containing 63% α phase. The cavity growth behaviour (hence the tensile properties) of this class of superplastic brasses have been known to be very sensitive to the plastic deformation of the most ductile β-phase, and a function of the α-phase volume fraction. The authors argue that it is this difference in deformability between the phases that causes the eventual increase in stress at the final stages of deformation.

As discussed by Wilkinson and Caceres[16] materials exhibiting strain hardening can withstand large amounts of cavitation without fracture, as was observed for the codoped materials in this study. The additional strain hardening mechanism is the likely cause of such large elongations. The exception is obviously the germania-calcia composition, from which one would expect the best strain to failure due to the very large strain hardening. The results obtained may be explained by the presence of abnormal grains, which are ideal sites for cavity nucleation. The relation between cubic content and strain hardening during deformation will be further investigated together with the excellent elongation characteristics shown by 2Ge-2Ti.

Conclusion
1. Superplasticity in Y-TZP can be improved extensively by carefully choosing the type of dopant combination of tetravalent and divalent cations such that elongations of up to 1000% can be obtained.
2. Extensive strain hardening is obtained above 1400°C for all 3 codoped compositions.
3. Strain hardening due to a continuously changing duplex structure is a possible reason for the increase in stress.

Acknowledgement
The authors would like to acknowledge the Ministry of Education, Science and Culture, Japan under Scientific Research B2-10450254 for the support of this research.

Literature
[1] R. C. Garvie, R. H. Hannink and R. T. Pascoe, Nature (London), **258**, 703 (1975).
[2] F. Wakai, S. Sakaguchi and Y. Matsuno, Adv. Ceram. Mater., **1**, 259, (1986).
[3] T. G. Nieh and J. Wadsworth, Acta Metall. Mater., **38**, 1121, (1990).
[4] J. Mimurada, M. Nakano, K. Shimura, K. Sasaki, Y. Ikuhara and T. Sakuma, submitted to J. Am. Cer. Soc.
[5] M. Oka, N. Tabuchi and T. Takashi, Mater. Sci. Forum, 304-306, 451, (1999).
[6] M. Nakano, K. Shimura, J. Mimurada, K. Sasaki, Y. Ikuhara and T. Sakuma, Key Eng. Mater., 171-174, 343, (2000).
[7] M. Nakano, K. Shimura, J. Mimurada, K. Sasaki, Y. Ikuhara and T. Sakuma, to be submitted
[8] M. Jiménez-Melendo, A. Domínguez-Rodríguez and A. Bravo-Leon, J. Am. Ceram. Soc., **81**, 2761, (1998).
[9] D. M. Owen and A. H. Chokshi, Acta Mater., **46**, 667, (1998).
[10] T. G. Langdon, "The Characteristics of Superplastic-like Flow in Ceramics", pp. 251-268, Plastic Deformation of Ceramics. Edited by R. C. Bradt, C. A. Brookes and J. L. Routhort, Plenum Press, New York, 1995.
[11] F. Wakai and T. Nagano, J. Mater. Sci., **26**, 241, (1991).
[12] Y. Yoshizawa and T. Sakuma, Eng. Frac. Mech., **40**, 847, (1991).
[13] Y. Yoshizawa and T. Sakuma, ISIJ International, **29**, 746, (1989).
[14] K. Sasaki, T. Kondo, Y. Ikuhara, and T. Sakuma, 1819, PRICM 3, edited by M. A. Imam, R. DeNale, S. Hanada, Z. Zhong and D. N. Lee, The Minerals, Metals & Materials Society, 1998
[15] J. Belzunce and M. Suery, Acta. Metall. **31**, 1497, (1983).
[16] D. S. Wilkinson and C. H. Caceres, Mater. Sci. and Tech., **2**, 1086, 1996

Deformation Characteristics of a 3Y-TZP/20%Al$_2$O$_3$ Composite in Tensile Creep

Siari S. Sosa and Terence G. Langdon

Departments of Materials Science and Mechanical Eng., University of Southern California, Los Angeles CA 90089-1453, USA

Keywords: Ceramics, Composites, Creep, Diffusion, Yttria-Stabilized Zirconia

Abstract

Creep tests were performed over a range of stresses and temperatures on a 3Y-TZP/20% Al$_2$O$_3$ composite termed 3Y20A. After creep, measurements were undertaken to determine the average shapes and sizes of the zirconia and alumina grains. The experimental data are compared with creep models including interface-controlled diffusion creep. The results suggest that interface-controlled diffusion creep is important at the lower stress levels but grain boundary sliding may be accommodated by intragranular slip within the alumina grains at the higher stresses.

Introduction

Zirconia ceramics are recognized to have superior mechanical properties including a high temperature resistance, excellent wear resistance, high strength and good fracture toughness. As a consequence of these properties, zirconia ceramics are often a good choice for numerous industrial applications. The fracture strength of zirconia ceramics is among the highest of all of the commercial ceramics currently available. In practice, this strength is remarkably high for zirconia-alumina composites, especially the 3 mol % yttria-stabilized tetragonal zirconia with 20% alumina (termed 3Y20A) [1]. Earlier experiments have shown this material is capable of exhibiting superplastic properties at high temperatures [2] and this suggests the potential for developing innovative forming methods to obtain pieces of complex shape, to attain finished surfaces of exceptionally high quality and to achieve improved control over the final dimensions by comparison with traditional procedures [3].

This paper describes experiments designed to investigate the creep properties of a 3Y20A composite. The results include observations of the grain morphology in the samples before and after testing. As will be demonstrated, the experimental evidence supports the advent of an interface-controlled diffusion creep model at low stresses but with the possibility of accommodation by intragranular slip in the alumina grains at higher stresses.

Experimental Materials and Procedures

The material used in this study was a two-phase ceramic composite consisting of a matrix of 3 mol % yttria-stabilized zirconia (termed 3Y-TZP) containing 20 wt % alumina (Al$_2$O$_3$). This material is termed 3Y20A. Fully dense blocks were fabricated from high purity commercial powders of 3Y20A available from the Tosoh Corporation (Nanyo Plant, Shin-nanyo, Yamaguchi-ken, Japan). The chemical composition of the powders (in wt %) was reported as 3.8% Y$_2$O$_3$, 19.79% Al$_2$O$_3$, <0.002% SiO$_2$,

0.002% Fe_2O_3 and 0.02% Na_2O. After fabrication, the alumina grains were uniformly distributed in the 3Y-TZP matrix and both phases had an essentially uniform distribution of grain shapes and sizes.

Tensile samples were machined from the as-sintered blocks and creep tests were performed over a range of temperatures and stresses. Some cyclic temperature creep tests were also performed to determine the apparent activation energy, Q_{app}. A software package named Image Pro Plus™ was used to measure the grain size and the grain aspect ratio (GAR) of the grains in images obtained in a scanning electron microscope (SEM). The GAR was measured as the ratio of the horizontal and vertical dimensions of each grain where the horizontal axis was set parallel to the tensile axis. The mean linear intercept grain size ($\bar{L}$) was determined from the expression $(L_1 L_2^2)^{1/3}$, where L_1 and L_2 were grain size measurements taken in directions parallel and perpendicular to the tensile axis, respectively. The initial mean linear intercept grain size ($\bar{L}_o$) was measured as ~0.4 μm in the as-sintered samples prior to testing. Full details of these experimental procedures were given previously [4].

Results and Discussion

Figure 1 shows the variation of the strain rate with the inverse temperature for 3Y20A samples tested at different temperatures using a stress of σ = 53 MPa: from the slope of the line, the apparent activation energy is estimated as $Q_{app} \approx 575$ kJ/mol. Also included in Fig. 1 are data for 3Y20A tested in tension and compression with a larger grain size of ~0.7 μm and at a slightly higher stress level of 60 MPa [5,6]. These latter results give similar values for Q_{app} but the measured creep rates are lower because of the larger grain size. The results obtained from the cyclic temperature tests were consistent with these data with an average apparent activation energy of ~560 kJ/mol. For low purity 3Y20A samples, Okada et al. [7] reported an apparent activation energy of ~620 kJ/mol and Wakai and Kato [8] obtained an apparent activation energy of ~600 kJ/mol.

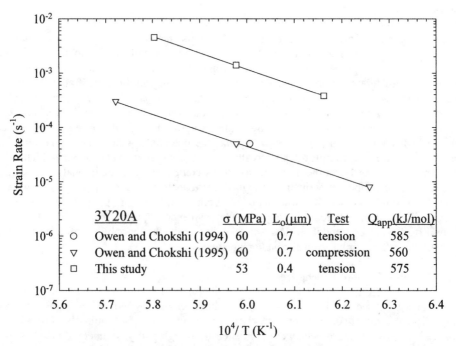

Fig. 1 Variation of strain rate with reciprocal temperature for tests conducted with a stress of 53 MPa: also included are experimental datum points for a stress of 60 MPa [5,6].

Table 1. Grain size and grain shape measurements for 3Y20A samples tested at 1723 K

Stress (MPa)	$\overline{L}$ (Al_2O_3) (μm)	$\overline{L}$ (Y-TZP) (μm)	GAR (Al_2O_3)	GAR (Y-TZP)	Final elongation (%)	$\dfrac{L_{1f}-L_{1o}}{L_{1o}}$ (%) (Al_2O_3)	$\dfrac{L_{1f}-L_{1o}}{L_{1o}}$ (%) (Y-TZP)
0	0.41	0.45	1.12	1.12	0	0	0
12	0.75	1.12	1.12	1.19	10	-	-
33	0.44	0.59	1.60	1.45	125	14	43
53	0.44	0.42	1.51	1.28	130	9	11
57	0.42	0.54	1.32	1.31	55	31	36

As reported earlier [4], the plots of strain rate versus stress showed a decrease in the creep rate at higher strains. This decrease may arise if the material deforms by diffusion creep because the individual grains become elongated at the higher strains [9]. An earlier analysis of the shapes and sizes of the alumina and zirconia grains after creep deformation suggested some evidence for diffusion creep and intragranular slip in the alumina grains [4]. To investigate this trend in more detail, Table 1 summarizes the results obtained for measurements of the mean linear intercept grain sizes, the grain aspect ratios, and the average distortions of the grains along the tensile axis for the 3Y-TZP and Al_2O_3 grains as given by the ratio ($L_{1f} - L_{1o}$)/L_{1o}, where the subscripts o and f denote the initial and final grain size measurements, respectively.

The processes of grain growth, diffusion creep and intragranular slip may affect these results simultaneously. From the results shown in Table 1 for the mean linear intercept grain sizes before and after testing at a stress of 12 MPa, it is evident that very significant grain growth has occurred in this sample and it is therefore not possible to relate these results to any evaluation of the deformation process. For this reason, these results were not used to estimate the average grain distortions as indicated in the last two columns of Table 1. For the samples tested at stresses of 33 and 53 MPa, grain growth in the alumina grains is negligible and the grain elongations may be associated with diffusion creep. The contribution of diffusion creep and intragranular slip is high at a stress of 57 MPa and it represents more than half of the total elongation. This may be a consequence of increasing stress concentrations that activate intragranular slip in the alumina grains at higher stress levels. If this is the case, the estimated contribution to deformation at the lower stresses may be associated exclusively with diffusion creep and under these conditions the stress concentrations are probably insufficient to activate the slip systems in alumina. For the zirconia grains, grain growth tends to obscure all of the estimates of grain strain except at a stress of 53 MPa and for this sample the estimate is almost identical to the result for the alumina grains. If there is a major contribution to deformation from grain boundary sliding, this estimate would be consistent with measurements using atomic force microscopy (AFM) where, for a similar composite with 50% of each phase, the contribution of grain boundary sliding was estimated as ~80% of the total strain [10].

In a previous study, it was suggested that the experimental stresses in the 3Y20A composite tested are comparable to those needed to activate the basal slip system in alumina [4]. Experiments show the stresses needed to activate slip in the prismatic and pyramidal systems in alumina are larger than those needed for basal slip [11,12] and therefore it is necessary only to compare the experimental data with the stresses for the basal system. In practice, the stress concentrations arising at triple junctions from the pile up of grain boundary dislocations may be large enough to activate slip in these other systems.

The solid line in Fig. 2 shows the theoretical stress produced by the pile-up of a group of grain boundary dislocations at a triple point, where this stress is estimated from the relationship ($2L_p\sigma^2/Gb_{gb}$) [13] where L_p is the pile-up length given by the ratio of the grain perimeter divided by the average number of facets in the grain, σ is the applied tensile stress, G is the shear modulus and b_{gb} is the Burgers vector for dislocations in the grain boundary. The shear modulus was calculated as the volume average from the two components in the composite [14-17] and the value of the grain boundary Burgers vector was taken as 2.4 Å, which is two-thirds of the lattice Burgers vector in zirconia (3.6 Å) and one-half of the lattice Burgers vector for basal slip in alumina (4.76 Å). This latter value for b_{gb} is consistent also with observations by Wakai et al. [18] where transmission electron microscopy (TEM) was used to examine the grain boundaries in zirconium-doped alumina at selected points near the triple junctions.

Superimposed on Fig. 2 are the experimental points from the present investigation at a testing temperature of 1723 K together with experimental data from investigations conducted on alumina single crystals [19,20]. It is evident that the stress concentrations at the triple junctions are sufficient to activate the basal slip system in alumina, especially at the faster strain rates when larger tensile stresses are applied. The conclusion from Fig. 2 is that, at a temperature of 1723 K, applied stresses at and above ~30 MPa will be able to activate the basal slip system. It should be noted this is consistent with the earlier observations of the presence of some slip activity within the alumina grains in 3Y20A samples tested at the highest stress levels [4]. There is also a report of dislocation activity occurring in the alumina grains of an Al_2O_3-24 vol % ZrO_2 composite in an area in the immediate vicinity of an indentation introduced at a temperature of 1473 K [21]. A comparative analysis was presented by Wakai et al. [18] documenting the results reported from deformation studies at a temperature of 1523 K in polycrystalline alumina and zirconium-doped alumina. This analysis, together with the results from TEM observations, led to the conclusion that doping decreases the creep rate and supports the possibility of an increase in the pile up stress at triple junctions due to segregation of yttrium at the grain boundaries.

For the case of polycrystalline alumina doped with yttria and magnesia, non-basal slip has been observed after tensile tests at 1523 K when using an initial strain rate of 6×10^{-6} s^{-1} [22]. The TEM observations performed by Lartigue Korinek and Dupau [22] showed evidence for pyramidal slip, an elongation of the grains, the segregation of yttrium at the grain boundaries and the emission of dislocations from triple junctions.

There is a possibility that two deformation mechanisms may act in this composite at low and high stresses, as discussed earlier [4]. Specifically, interface reaction-controlled diffusion creep [23] may occur with accommodation by slip in the alumina grains at the larger stresses. The process of interface reaction-controlled diffusion creep considers grain boundary dislocations as discrete sources and sinks of vacancies, and this process has been suggested as an adequate model for the creep behavior of Y-TZP ceramics [24]. Figure 3 shows the results for the calculations of the interface reaction-controlled diffusion creep model under the experimental conditions used in the present investigation for a testing temperature of 1723 K: the experimental datum points are added for comparison. These calculations were performed following the model developed by Arzt et al. [23] and by estimating the diffusion coefficient for the slowest species along the fastest path, which was taken as Zr^{4+} ions along the grain boundaries. The diffusion data was derived from the experimental report of Sakka et al. [25] for cation grain boundary and lattice diffusion in polycrystalline tetragonal CeO_2-ZrO_2-HfO_2 solid solutions. Following the procedure developed earlier [24], the number of dislocations in a single grain boundary wall was estimated from the expression ($\sigma d/2Gb_{gb}$), where d is the grain size.

It is evident that at low stresses the interface reaction-controlled diffusion creep model fits the experimental data extremely well but at higher stresses, above ~30 MPa, another mechanism may operate. This level of the applied stress is approximately equal to the stress at which the pile ups of grain boundary dislocations may be capable of activating intragranular slip at the triple junctions, as demonstrated in Fig. 2. Thus, the results are mutually consistent and it is apparent that slip deformation in the alumina grains may contribute to flow behavior at the higher stresses. This suggests that the

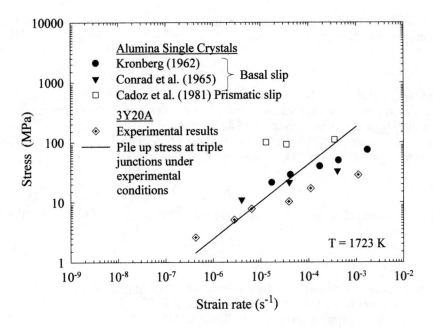

Fig 2 Critical resolved shear stresses for basal and prismatic slip in alumina, theoretical pile up stress at triple junctions and experimental results in a 3Y20A composite.

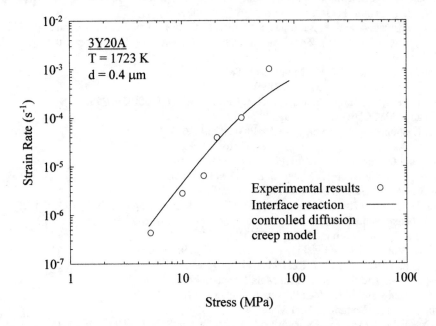

Fig 3 Calculations of the interface reaction-controlled diffusion creep model under experimental conditions compared with the experimental results.

differences in the amounts of deformation in the alumina and zirconia grains may be attributed to the contribution of intragranular slip in the alumina grains.

Conclusions

The creep deformation of the 3Y20A composite under superplastic conditions occurs by grain boundary sliding with the stress concentrations relieved by diffusion creep in both constituent phases and with intragranular slip in the alumina grains at the higher stresses.

There is very good agreement between the interface reaction-controlled diffusion creep model and the experimental data at the low stresses.

Acknowledgements

This work was supported by the United States Department of Energy under Grant No. DE-FG03-92ER45472

References

[1] K. Tsukuma, K. Ueda and M. Shimada, J. Amer. Ceram. Soc. 68 (1985), p. C4.
[2] T.G. Nieh and J. Wadsworth, Acta Metall. Mater. 39 (1991), p. 3037.
[3] I-W. Chen and L.A. Xue, J. Amer. Ceram. Soc. 73 (1990), p. 2585.
[4] S. S. Sosa and T. G. Langdon, Mater. Res. Soc. Symp. Proc. 602 (2000), p. 111.
[5] D. M. Owen and A. H. Chokshi, J. Mater. Sci. 29 (1994), p. 5467.
[6] D.M. Owen and A.H. Chokshi, in "Plastic Deformation of Ceramics" (R.C. Bradt, C.A. Brookes and J.L. Routbort, eds.), Plenum Press, New York (1995), p. 507.
[7] K. Okada, Y. Yoshizawa and T. Sakuma. in "Superplasticity in Advanced Materials" (S. Hori, M. Tokizane and N. Furushiro, eds.), The Japan Society for Research on Superplasticity, Osaka, Japan (1991), p. 227.
[8] F. Wakai and H. Kato, Adv. Ceram. Mater. 3 (1988), p. 71.
[9] H.W. Green, J. Appl. Phys. 41 (1970), p. 3899.
[10] L. Clarisse, F. Petit, J. Crampon and R. Duclos, Ceram. Intl. 26 (2000), p. 295.
[11] D.J. Gooch and G.W. Groves, Phil. Mag. 28 (1973), p. 623.
[12] J. Cadoz, J. Castaing and J. Philibert, Revue Phys. Appl. 16 (1981), p. 135.
[13] J. Friedel, in "Dislocations", Pergamon Press, Oxford, U.K. (1964), p 260.
[14] D.H. Chung and G. Simmons, J. Appl. Phys. 39 (1968), p. 5316.
[15] M. Fukuhara and I. Yamauchi, J. Mater. Sci. 28 (1993), p. 4681.
[16] R.G. Munro, J. Amer. Ceram. Soc. 80 (1997), p. 1919.
[17] M. Hayakawa, H. Miyauchi, A. Ikegami and M. Nishida, Mater. Trans. JIM 39 (1998), p. 268.
[18] F. Wakai, T. Nagano and T. Iga, J. Amer. Ceram. Soc. 80 (1997), p. 2361.
[19] M.L. Kronberg, J. Amer. Ceram. Soc. 45 (1962), p. 274.
[20] H. Conrad, G. Stone and J. Janowski, Trans. AIME 233 (1965), p. 889.
[21] B.-T. Lee and K. Hiraga, J. Mater. Res. 9 (1994), p. 1199.
[22] S. Lartigue Korinek and F. Dupau, Acta Metall. Mater. 42 (1994), p. 293.
[23] E. Arzt, M.F. Ashby and R.A. Verrall, Acta Metall. 31 (1981), p. 1977.
[24] M.Z. Berbon and T.G. Langdon, Acta Mater. 47 (1999), p. 2485.
[25] Y. Sakka, Y. Oishi, K. Ando and S. Morita, J. Amer. Ceram. Soc. 74 (1991), p. 1610.
Corresponding author: Siari S. Sosa (ssosa@usc.edu)

Superplastic Flow Stress in Cation-Doped YSZ

Manabu Nakano, Junpei Mimurada, Kazutaka Sasaki,
Yuichi Ikuhara and Taketo Sakuma

Department of Materials Science, School of Engineering, University of Tokyo,
7-3-1 Hongo, Bunkyo-ku, Tokyo 113-8656, Japan

Keywords: Dopant, Ductility, Superplasticity, YSZ

Abstract
Superplastic behaviors in various cation doped yttria-stabilized tetragonal zirconia polycrystal (Y-TZP) were examined. For 1 mol% cation-doping the flow stress of Y-TZP is very dependent on the ionic radii of the doped cations, for instance, smaller cation radii give rise to lower flow stress when comparing the various compositions for the same grain size, strain rate and testing temperature. The altered flow stress level must be due to the change in diffusivity of the accommodation process for grain boundary sliding caused by the addition of cations in ZrO_2. Meanwhile, the strain to failure of the doped zirconia are affected by both ionic radius and valancy of the dopant cations.

I. Introduction
Yttria-stabilized tetragonal zirconia polycrystal (Y-TZP) consists of a fine-grained microstructure. This material is known to show superplastic flow at elevated temperatures.[1-5] Cation oxides doped into Y-TZP can affect the deformation characteristics of Y-TZP. For example, 10 mol% (5 wt%) SiO_2-doped Y-TZP exhibits an elongation of 1038% at an initial strain rate of $1.3 \times 10^{-4} s^{-1}$ at 1673K.[6,7] TiO_2(5 wt%)-doped Y-TZP,[8] shows larger elongation to failure than Y-TZP below 1773K at an initial strain rate of $1.3 \times 10^{-4} s^{-1}$. In recent study, The addition of 1mol% GeO_2 doped Y-TZP is also known to improve superplastic characteristics throughout whole temperature. However, methodical investigations for the effect of cation-doping on the superplastic flow in Y-TZP have not been made yet. In addition, there are no data available on the effect of dopant cations on the superplastic flow in Y-TZP. Hence, in this paper, we will examine the superplastic behavior in various cation doped Y-TZPs and investigate the relationships between the cation properties and the superplastic characteristiocs.

II. Experimental Procedure
The chemical compositions of the materials prepared in this study were ZrO_2-3 mol%Y_2O_3 (3Y), and 3Y doped with 1 mol% Ba^{2+} (1Ba), Mg^{2+} (1Mg), Nd^{3+} (1Nd), Gd^{3+} (1Gd), Sc^{3+} (1Sc), Ge^{4+} (1Ge), Ti^{4+} (1Ti) and Ce^{4+} (1Ce). Each dopant cation was mixed with 3Y as cation oxides. For 1 mol% these cations are reproted to form a solid solution with ZrO_2[11-17]. The solubility of 1Ba has not been reproted since no information is yet available.
ZrO_2 powders (Tosoh Co. Ltd., TZ-3Y) were mixed with 1 mol% cation oxide powders by a ball-milling method with ZrO_2 balls and ethanol for 24h and dried in Ar atmosphere. Subsequently, the powders were uniaxially pressed under a pressure of 33 MPa and then cold isostatically pressed (CIP) at 100 MPa. The green bodies of these compositions were sintered for 2h in air at the appropriate temperatures to obtain full density. Some of the sintered bodies were further annealed in a temperature range of 1723-1773K for a period of 2 to 4h to control the grain size. Tensile test specimens with a gauge length of 13 mm and a cross sectional area of 2x2 mm were cut and ground from the sintered or annealed bodies. High temperature tensile tests were conducted for these compositions in air at temperatures in the range of 1573−1773K at an initial strain rate of $1.3 \times 10^{-4} s^{-1}$ with a tensile machine (SHIMAZU AG-5000C). Microstructure analysis was carried out with a scanning electron microscope

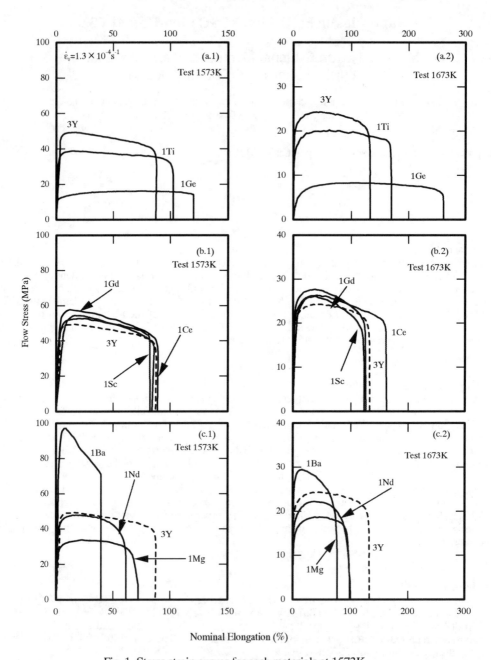

Fig. 1. Stress strain curves for each materials at 1573K

(SEM, JEOL JSM-5200) for the as sintered and annealed samples that were polished and thermally etched. The average grain size was measured directly from SEM micrographs by the linear intercept method. In addition, 1Ge was examined by high resolution electron microscopy (HERM) with a Topcon 002BF (200kV) field emission type high resolution microscope which has a point to point resolution of about 0.17 nm.

III. Results and discussion

(1) High temperature tensile test
High temperature tensile tests were conducted to obtain stress-strain relationships for each specimen at an initial strain rate of $1.3 \times 10^{-4} s^{-1}$ in a temperature range of 1573-1773K. All specimens deformed plastically and there was no necking observed at the time of fracture. The results of the tensile tests are summarized in Fig. 1. The compositions are classified into three groups in terms of their ductilities and compared with 3Y. The groups are namely enhanced ductility, shown in Fig. 1 (a.1) and Fig. 1 (a.2) at 1573K and 1673K respectively unaltered ductility for Fig. 1 (b.1), (b.2) and deteriorated ductility for Fig. 1 (c.1), (c.2). Fig. 1 shows that the ductility behavior of each group is consistent at both 1573K and 1673K.

Fig. 1 (a.1) and Fig. 1 (a.2) are the stress-strain curves of the enhanced group that contains 1Ti and 1Ge. It is considered that the stress reductions in both materials result in enhanced elongations. The unchanged group consists of 1Sc, 1Gd and 1Ce as shown in Fig. 1 (b.1) and Fig. 1 (b.2). These compositions exhibit similar stressstrain curves to that of 3Y in 1573-1673K, which means that the dopings of these cations does not alter effectively the characteristics of the superplastic flow in 3Y. The deteriorated consists of 1Nd, 1Mg and 1Ba in Fig. 1 (c.1) and Fig. 1 (c.2). and 1673K The effects of these cations on the flow stress of 3Y are different. 1Ba exhibits higher stress than that of 3Y while the stress of 1Nd is almost the same as 3Y. Furthermore, 1Mg shows a lower floe stress than 3Y. In case of 1Mg, the ductility of 3Y is lowered by the addition of Mg in spite of the stress decrease. The results in these materials suggest that the decrease in elongation to failure cannot be explained only by the magnitude of flow stress.

(2) Microstructures
XRD analyses were conducted to determine the phase of each composition. Fig. 2 gives the XRD profiles obtained in a diffraction range of 20-80 degree. Judging from these profiles each composition consists of a tetragonal ZrO_2 phase. The spectra are also consistent with the reports on the solid solubility limit of these cations in ZrO_2 (except for 1Ba mentioned above).[9-15] It is difficult to assess whether Ba is in solution in the tetragonal ZrO_2 phase, although the 1Ba spectra indicate a typical zirconia tetragonal phase.

(3) Cation-doping effect on the flow stress
Superplastic flow stress in ceramics is highly dependent on the grain size, for example, a material with larger grain size has a higher flow stress when compared to a smaller grain size at a given strain rate and temperature. High temperature deformations of ceramic materials are described by the following equation

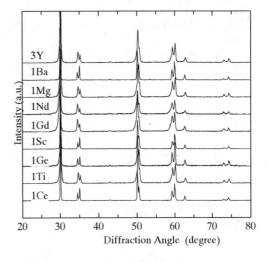

Fig. 2. XRD Profiles

$$\dot{\varepsilon} = A \frac{\sigma^n}{d^p} \exp(-\frac{Q}{RT}) \quad (1)$$

where $\dot{\varepsilon}$ is the strain rate, A is a constant depending on each material, n is the stress exponent, p is the grain size exponent, Q is the activation energy for deformation, R and T have the usual meaning. This equation is considered to hold for the superplastic deformation in Y-TZP.

In this study, to elucidate the effect of cation-doping on the flow stress of 3Y, we carried out the tensile tests for specimens with various average grain sizes for each composition. Fig. 3 gives the log-log relationship between the flow stress and average grain size at 10% deformation at 1673K in the strain rate of $1.1 \times 10^{-4} s^{-1}$. The grain sizes were measured for the specimens deformed to 10% using SEM photographs. In Fig. 3, it is assumed that the plot of each composition is expressed by a straight line with its increment value of 1. This figure enables us to compare the flow stress of these compositions with different average grain sizes. The flow stress of 1Ge is lower than that of 3Y independent of grain size. This indicates the fact that the reduction in flow stress of 3Y with Ge doping cannnot be explained by grain size refinement. Fig. 3 also shows that 1Ti and 1Mg decrease the flow stress when compared at a similar grain size. These cations contribute to the decrease in flow stress of 3Y in the following order Ge>Ti>Mg. This also implies that the stresses of 1Mg in Fig. 1 are lower than that of Ti because of grain size refinement of 1Mg. On the other hand, the stresses of 1Ba, 1Nd and 1Gd are higher than that of 3Y and the effect of Ba is much larger than Nd or Gd. Sc and Ce, have no notable effect on the flow stress of 3Y. As mentioned above, it is concluded that some cations can affect the flow stress of Y-TZP, however, the effect is different for each cation.

Table I gives the flow stresses of these compositions for an average grain size of 0.5μm, which were estimated from Fig. 3. This table also shows the ionic radii of the cations for 8 fold coordination.[17] The change in flow stress due to the addition of these cations is given as:

$$\Delta \sigma = \sigma_{doped} - \sigma_{undoped} \quad (2)$$

where σ_{doped} is the flow stress of the doped Y-TZP and $\sigma_{undoped}$ is for 3Y. A positive value of $\Delta\sigma$ means an increase in flow stress and a negative value means a decrease in flow stress.

Fig. 4 shows the effect of cations on the change in flow stress, $\Delta\sigma$, on the vertical axis and the ionic radii of cations for 8-fold coordination on the horizontal axis. This figure points out specifically the fact that the change in the flow stress is in reports good agreement with the ionic radii of dopant cation. In other words, a smaller cation contributes to a larger stress reduction in Y-TZP and larger cations raise the flow stress. These cations are considered to be in solution with tetragonal zirconia, therefore, the origin of the change in flow stress summarized in Table I is rationalized in terms of the altered diffusivity for the accommodation process required for superplastic deformation caused by the presence of the cations.

The relationship between dopant cation and the diffusivity in Y-TZP is also supported by several reports. Chen et al. studied the effect of cation-doping on the grain growth behavior in Y-TZP and 12mol% CeO$_2$-stabilized TZP.[18] In their study, the grain growth of TZP is slower when doped with larger cations compared to cations of the same valency. On the other hand, Jimenez-Melendo et al.[19] have suggested that the superplastic flow in Y-TZP is dominated by Zr lattice diffusion judging from the activation energy of the superplastic flow in Y-TZP Taking these into consideration, the cation-doping effect in this study can be explained by lattice diffusion.

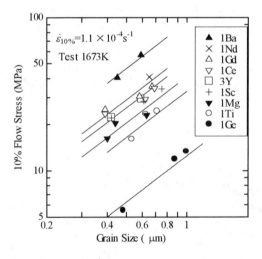

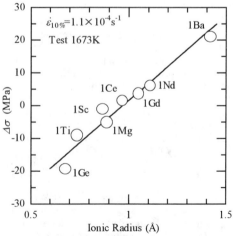

Flow stress at the grain size of 0.5 μm.

Fig. 3 A log-log plot of flow stress against grain size

Fig. 4 A plot of Stress changing

Table I. A list of ionic radii of doped cations, flow stresses and $\Delta\sigma$ for the compositions in this study

Material	Ionic Radius[†] (8-fold, Shannon)[19]	Flow Stress at 0.5μm (MPa)	$\Delta\sigma$ [‡] (MPa)
3Y	-	25.6	-
1Ba	1.42Å (Ba^{2+})	46.4	20.8
1Nd	1.11Å (Nd^{3+})	31.5	5.9
1Gd	1.05Å (Gd^{3+})	29.0	3.4
1Ce	0.97Å (Ce^{4+})	27.0	1.4
1Sc	0.87Å (Sc^{3+})	24.3	-1.3
1Mg	0.89Å (Mg^{2+})	20.3	-5.3
1Ti	0.74Å (Ti^{4+})	16.4	-9.2
1Ge	0.68Å (Ge^{4+})	6.2	-19.4

[†]Reference 17. All radii for 8-fold coordination. [‡] $\Delta\sigma = \sigma_{doped} - \sigma_{undoped}$.

(4) The effect of cation doping on ductility

Fig. 5 is a plot of 10% flow stress of the materials in this study against the elongation to failure in the temperature range 1573-1773K. All the plots show a linear dependence, implying that lower ductilities give rise to higher elongations. The tendency indicates that the dominant factor for enhanced ductility is the decrease in flow stress.

However, 1Mg has deteriorated ductilities in spite of the lower flow stresses when compared to the other compositions as shown in Fig. 1 (c.1), (c.2) and Fig. 4. It implies that another factor independent of flow stress affects ductility. There have been several studies for the dopant effect on the ductility of Y-TZP. One is the study in our group about 2 mol% Nb^{5+}-doped 3Y-TZP that has enhanced ductility.[19] On the other hand, Oka et al. reported deteriorated ductility in 2 mol% Ca^{2+}-doped 3Y-TZP.[4] The effect of these dopants are summarized in Fig. 8 with the results of this study. In this figure, the dopant cations are sorted by the valency and the ionic radius. These cations belong to one of three groups, enhanced, unaltered and deteriorated with respect to 3Y. This figure clearly shows the fact that all the cations of enhanced ductility group are in a region, where the ionic radius is smaller than Zr^{4+} ion and/or has a higher valency than +4. On the other hand, the area between Zr^{4+} and Y^{3+} provides the region for unchanged ductility group. Finally, the region of the deteriorated group corresponds to cations with +2 valency or larger ionic sizes than Y^{3+}. In this way, the effect of cation-doping on the ductility of 3Y is described well by valencies and radii of doped cations. This figure also indicates that the cations with lower valency give rise to lower ductilities. It has been reported that the ductility in Y-TZP-based materials cannot be always described in terms of the flow stress level.[9,10] It is expected

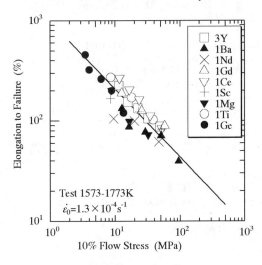

Fig.5 Elongation against flow stress

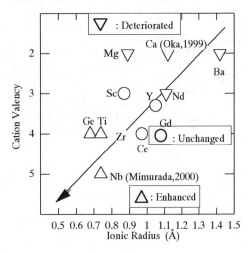

Fig. 6 The effect of cation on the ductility

that the intergranular fracture strength is altered by cation segregation to grain boundaries. The cation dopants in this study are in solid solution, therefore, the flow stress itself is the dominant factor determining the ductility of Y-TZP rather than the strength of intergranular fracture.

IV. Conclusions
   To investigate the relationship between cation properties and the superplastic behavior in 3Y-TZP, high temperature tensile tests and microstructure observation were conducted for various cation doped 3Y-TZP. The conclusions obtained are as follows.
   (1) The superplastic flow of 3Y-TZP can be affected by the small amount of 1 mol% cation addition. The result indicates that the cation-doping method is effective to control the superplastic flow of Y-TZP.
   (2) The effect of cation-doping on the flow stress of 3Y-TZP depends on the ionic radius of each dopant cation. Cations with smaller ionic size decrease the flow stress when compared at the same grain size, strain rate and testing temperature, while larger cations increase flow stress. It is considered that these changes in flow stress must result from the altered diffusivity of the accommodation process for deformation caused by the addition of the dopants in ZrO_2.
   (3) The valency of the dopant cation in 3Y-TZP influences the ductility. Therefore, it is necessary to consider both ionic radius and the valency of the dopant cation. To obtain enhanced ductility, cations with higher valency and smaller ionic size are desirable.

Acknowledgements
Supported by the Ministry of Education, Science and Culture, Japan under Scientific Reserch B2-10450254.

References
[1]   F. Wakai, S. Sakaguchi, and Y. Matsuno, *Adv. Ceram. Mater.*, 1, 259-63 (1986).
[2]   A. H. Chokshi, A. K. Mukherjee, and T. G. Langdon, *Mater. Sci. Eng. R*, 10 [6] 237-74 (1993).
[3]   T. Sakuma, *Mater. Sci. Forum*, 304-306, 3-10 (1999).
[4]   M. Oka, N. Tabuchi, and T. Takashi, *Mater. Sci. Forum*, 304-306, 451-58 (1999).
[5]   J. Mimurada, K. Sasaki, Y. Ikuhara, and T. Sakuma, *Mater. Trans.* JIM, 40 [8] 836-41 (1999).
[6]   K. Kajihara, Y. Yoshizawa and, T. Sakuma, *Acta Metall. Mater.*, 43 [3] 1235-42 (1995).
[7]   P. Thavorniti, Y. Ikuhara, and T. Sakuma, *Acta Metall. Mater.*, 81 [11] 2927-32 (1998).
[8]   K. Tsurui and T. Sakuma, *Scr Maert.*, 34 [3] 443-447(1996)
[9]   C. F. Grain, *J. Am. Ceram. Soc.*, 50 [6] 288-96 (1967).
[10] J. Katamura, T. Seki, and T. Sakuma, *J. Phase Equilibria*, 16 [4] 315-19 (1995).
[11] M. J. Bannister, and J. M. Barnes, *J. Am. Ceram. Soc.*, 69 [11] C269-C271 (1986).
[12] T. S. Sheu, J. Xu, and T. Y. Tien, *J. Am. Ceram. Soc.*, 76 [8] 2027-32 (1993).
[13] D. J. Kim, J. W. Jang, H. J. Jung, J. W. Huh, and I. S. Yang, *J. Mater. Sci. Letters*, 14 [14] 1007-09 (1995).
[14] E. Zshech, P. N. Kountouros, G. Petzow, P. Behrens, A. Lessmann, and R. Frahm, *J. Am. Ceram. Soc.*, 76 [1] 197-201 (1983).
[15] E. Tani, M. Yoshimura, and S. Somiya, *J. Am. Ceram. Soc.*, 66 [7] 506-10 (1983).
[16] A. H. Chokshi, Chokshi, Y. Ikuhara, J. A. Hines, and T. Sakuma, *Acta Mater.*, 46, 5557-67 (1998).
[17] R. D. Shannon, *Acta Crystallogr.*, A32, 751-67 (1976).
[18] S. L. Hwang, and I. W. Chen, *J. Am. Ceram. Soc.*, 73 [11] 3269-77 (1990).
[19] M Jimenez-Melendo, A Dominguez-Rodriguez and A Bravo-Leon, *J. Am. Ceram. Soc.*, 81 [11] 2761-76 (1998).
[20] J. Mimurada; M. Eng. Thesis, The University of Tokyo, Tokyo, Japan, 2000.

Effective Diffusivity for Superplastic Flow in Magnesium Alloys

H. Watanabe[1], H. Tsutsui[2], T. Mukai[1], M. Kohzu[3] and K. Higashi[3]

[1] Osaka Municipal Technical Research Institute,
1-6-50 Morinomiya, Joto-ku, Osaka 536-8553, Japan

[2] Graduate Student, Osaka Prefecture University, 1-1 Gakuen-cho, Sakai 599-8531, Japan

[3] Department of Metallurgy and Materials Science, Osaka Prefecture University,
1-1, Gakuen-cho, Sakai, Osaka 599-8531, Japan

Keywords: Activation Energy, Effective Diffusivity, Grain Size Exponent, Magnesium Alloy, Stress Exponent

Abstract

Effective diffusivity for superplastic flow was investigated using a relatively coarse-grained (17 and 30 μm) Mg–Al–Zn alloy. Tensile tests revealed that the strain rate was inversely proportional to the square of the grain size and to the second power of stress. The activation energy was close to that for grain boundary diffusion at low temperatures, and was close to that for lattice diffusion at high temperatures. From the analysis of stress exponent, grain size exponent and activation energy, it was suggested that the dominant diffusion process was influenced by temperature and grain size. It was demonstrated that the notion of effective diffusivity explained the dominant diffusion process.

Introduction

It has been shown that superplastic behavior of fine-grained metals is well described by a deformation model that grain boundary sliding (GBS) is accommodated by slip [1,2]. This slip accommodation process involves the sequential steps of glide and climb, and the climb is assumed to be the rate-controlling process when the stress exponent is 2 [2]. Accordingly, the rate-controlling process is related to the dominant diffusion process for the climb. Sherby and Wadsworth [2,3] suggested that superplastic flow in fine-grained metals is controlled by grain boundary diffusion or lattice diffusion, and is considered to be controlled by the dominant diffusion process. The dominant diffusion process for superplastic flow may be evaluated through the effective diffusivity [3], although this notion has not been experimentally confirmed. The effective diffusion coefficient, involving the lattice diffusion coefficient, D_L, and grain boundary diffusion coefficient, D_{gb}, is given by [3,4]

$$D_{eff} = D_L + x \frac{\pi \delta}{d} D_{gb} \tag{1}$$

where x is an unknown constant, δ the grain boundary width and d the grain size. The term x is usually considered to be 1×10^{-2} for superplastic flow [3,4]. The critical grain size, d_c, above which a dominant diffusion process is lattice diffusion and below which it is grain boundary diffusion can be analyzed using Eq. 1. The variation in critical grain size as a function of temperature is plotted for magnesium and aluminum in Fig. 1. The critical grain size for magnesium is larger than that for aluminum. This is because the grain boundary diffusion in magnesium is quite rapid. The critical grain size is estimated to be about 10 and 0.1 µm around the superplastic temperature in magnesium and aluminum, respectively. From Fig. 1, magnesium is suggested to be a suitable material to analyze the effective diffusivity, because a grain size of 10 µm can be obtained easily in magnesium alloys.

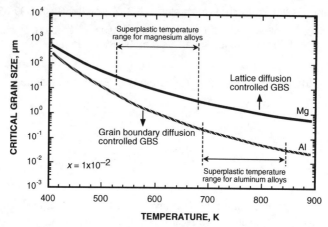

Fig. 1 The variation in critical grain size above which a dominant diffusion process is lattice diffusion and below which it is grain boundary diffusion as a function of temperature in magnesium and aluminum for $x = 1 \times 10^{-2}$.

Experimental

The materials used in the present study were two commercial Mg–Al–Zn alloy (AZ61) sheets. The initial grain size of each alloy was 17 and 30 µm, respectively. The grains were almost equiaxed for both materials. To investigate the mechanical properties, constant strain rate tensile tests were carried out in air.

Results

The variation in (a) flow stress and (b) elongation-to-failure as a function of strain rate is plotted in Fig. 2 for AZ61 alloy with an initial grain size of 30 µm. The strain rate sensitivity exponent, m, exhibited a maximum value of ~ 0.5 in the low strain rate range. This high m-value of 0.5 suggests that GBS could be a dominant deformation process. The strain rate regions exhibiting high m-value were roughly in agreement with the regions where large elongations were attained. However, the elongation value depended on the strain rate, in spite of the range having the same m-value of 0.5. Large elongations were obtained in the low strain rate range of ~ 10^{-5} s^{-1}. This may be associated with cavitation behavior, since the cavity nucleation rate increases rapidly with increasing stress [6]. The maximum elongation of 401% was obtained at 723 K and 1×10^{-5} s^{-1}. Similar results were observed for AZ61 alloy with an initial grain size of 17 µm [5]. The maximum elongation of 461% was obtained at 648 K and 3×10^{-5} s^{-1}, where $m = 0.5$. It was found that the present materials behaved in a superplastic manner even in the relatively coarse-grained materials.

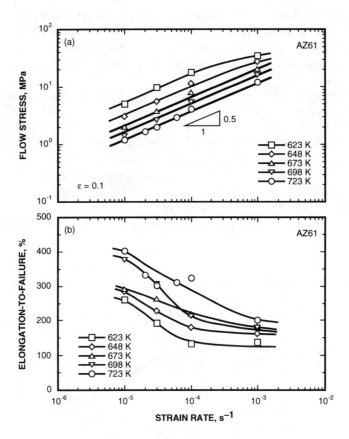

Fig. 2 The variation in (a) flow stress and (b) elongation-to-failure as a function of strain rate in AZ61 with an initial grain size of 30 μm.

The constitutive equation to describe superplasticity is generally expressed as [7]

$$\dot{\varepsilon} = A \frac{D_0 G b}{kT} \left(\frac{b}{d}\right)^p \left(\frac{\sigma - \sigma_0}{G}\right)^n \exp\left(-\frac{Q}{RT}\right) \qquad (2)$$

where $\dot{\varepsilon}$ is the strain rate, A a constant, D_0 the pre-exponential factor for diffusion, σ the stress, σ_0 the threshold stress, G the shear modulus, k the Boltzmann constant, n the stress exponent (=1/m), d the grain size, b the Burgers vector, p the grain size exponent, R the gas constant, T the absolute temperature and Q the activation energy which is dependent on the rate controlling process.

To determine the threshold stress, a plot of $\dot{\varepsilon}^{1/2}$ against σ on a double linear scale was adopted [8]. This is because the use of $n = 2$ gave the best linear fit among the assumed stress exponents ($n = 1, 2, 3, 5$) at all investigated temperatures. The threshold stresses were estimated to be 3.4, 1.0, 0.3, 0, 0 and 0 MPa at 523, 548, 573, 598, 623, 648 and 673 K, respectively, for the material with an initial grain size of 17 μm. The threshold stresses were estimated to be 0 at all temperatures investigated for the material with an initial grain size of 30 μm.

The grain size dependence of the superplastic flow was also characterized. The difference in grain size affected the mechanical properties; the strain rate increased with a decrease in grain size. The p-value was found to be 2 at 648 K in the present materials with relatively coarse grain size [9], contrary to fine-grained magnesium alloys with grain sizes of < 10 μm, whose p-value was estimated to be 3 [10].

The relationship between the normalized strain rate, $\dot{\varepsilon}(T/G)(d/b)^2$, and reciprocal temperature is illustrated in Fig. 3, where $\dot{\varepsilon}(T/G)(d/b)^2$ was determined at a fixed normalized stress of $(\sigma-\sigma_0)/G = 5\times10^{-4}$ for the typical superplastic region. Inspection of the data in Fig. 3 reveals that the behaviors are divided into two regions. The activation energy value at 523 – 573 K is close to that for grain boundary diffusion of magnesium (92 kJ/mol) [11]. However, the activation energy at 598 – 673 K is close to that for lattice diffusion of magnesium (135 kJ/mol) [11]. The transition temperature estimated from the intersection of each line is 585 K. It is suggested that the dominant deformation mechanism is GBS accommodated by slip, which is controlled by grain boundary diffusion at 523 – 573 K, and is controlled by lattice diffusion at 598 – 673 K, respectively.

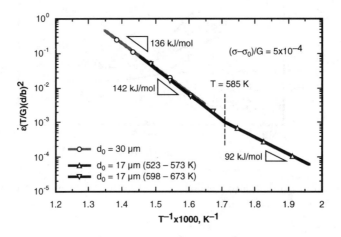

Fig. 3 The variation in $\dot{\varepsilon}(T/G)(d/b)^2$ as a function of $1/T$ in relatively coarse-grained magnesium alloy AZ61.

Discussion

It was demonstrated from the experimental results that the dominant diffusion process varies with temperature. The critical grain size above which a dominant diffusion process is lattice diffusion and below which it is grain boundary diffusion can be analyzed using Eq. 1. In Fig. 4, grain sizes are plotted as a function of temperature for pseudo single phase magnesium alloys, paying attention to the dominant diffusion process. The value of x can be deduced from the Fig. 4 so as to be divided into two regions. The term x is estimated to be 1.7×10^{-2} for the superplasticity in magnesium alloys, which is in agreement with the phenomenological relations in superplastic metals ($x = 1\times10^{-2}$).

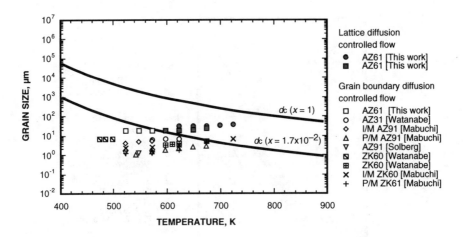

Fig. 4 The relationship between grain size and temperature in AZ61 alloy. Data for grain boundary diffusion controlled GBS in pseudo single phase magnesium alloys are also included. The critical grain size, d_c, above which a dominant diffusion process is lattice diffusion and below which it is grain boundary diffusion is indicated in case for $x = 1$ and 1.7×10^{-2}.

The relationship between $(\dot{\varepsilon}/D_{eff})(kT/Gb)(d/b)^2$ and $(\sigma-\sigma_0)/G$ for AZ61 is shown in Fig. 5, where D_{eff} is taken to be $[D_L+(1.7\times 10^{-2})(\pi\delta/d)D_{gb}]$. The figure also includes the data for superplastic behavior in magnesium alloys, whose deformation mechanisms were considered to be grain boundary diffusion controlled GBS. It is obviously noted that the materials studied behaved identically to other documented magnesium alloys. It was found that the superplastic behavior in magnesium alloys is represented by a single straight line with a slope of 2 in the normalized plot compensated by effective diffusion coefficient. The notion of effective diffusivity satisfactorily explained the dominant diffusion process during superplastic flow.

Summary

Effective diffusivity for superplastic flow was investigated using relatively coarse-grained magnesium alloy AZ61. Effect of temperature and grain size on superplastic flow was characterized for the inclusive understanding of dominant diffusion process. The strain rate was inversely proportional to the square of the grain size and to the second power of stress. The activation energy was close to that for grain boundary diffusion at 523 – 573 K, and was close to that for lattice diffusion at 598 – 673 K. It was suggested that the deformation mechanism was grain boundary sliding accommodated by slip controlled by grain boundary diffusion at low temperatures, and controlled by lattice diffusion at high temperatures. It was demonstrated that the dominant diffusion process was influenced by temperature and grain size according to the effective diffusivity. Finally, superplastic behavior in magnesium alloys is represented by a single straight line with a slope of 2 in the normalized plot compensated by effective diffusion coefficient, involving the lattice diffusion coefficient, D_L, and grain boundary diffusion coefficient, D_{gb}.

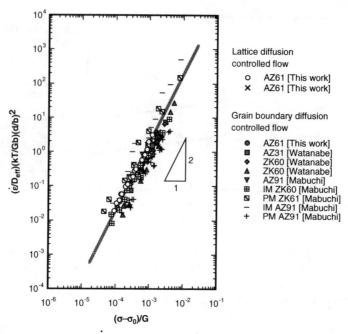

Fig. 5 The relationship between $(\dot{\varepsilon}/D_{\text{eff}})(kT/Gb)(d/b)^2$ and $(\sigma-\sigma_0)/G$ for AZ61, where D_{eff} is taken to be $[D_L+(1.7\times10^{-2})(\pi\delta/d)D_{\text{gb}}]$. Data for grain boundary diffusion controlled GBS in pseudo single phase magnesium alloys are also included.

References

[1] O.A. Ruano and O.D. Sherby, *Mater. Sci. Eng.* 56 (1982), p. 167.
[2] O.D. Sherby, J. Wadsworth, *Prog. Mater. Sci.* 33 (1989), p. 169.
[3] O.D. Sherby and J. Wadsworth, Deformation processing and structure, ASM, Metals Park, Ohio (1982), p. 355.
[4] P. Metenier, G. González-Doncel, O.A. Ruano, J. Wolfenstine and O.D. Sherby, *Mater. Sci. Eng.* A125 (1990), p. 195.
[5] H. Tsutsui, H. Watanabe, T. Mukai, M. Kohzu, S. Tanabe and K. Higashi, Mater. Trans., JIM 40 (1999), p. 931.
[6] R. Raj and M.F. Ashby, Acta Metall. 23 (1975), p. 653.
[7] R.S. Mishra, T.R. Bieler, A.K. Mukherjee, Acta Metall. Mater. 43 (1995), p. 877.
[8] F.A. Mohamed, *J. Mater. Sci.* 18 (1983), p. 582.
[9] H. Watanabe, T. Mukai, M. Kohzu, S. Tanabe and K. Higashi, Acta mater. 47 (1999), p. 3753.
[10] H. Watanabe, H. Hosokawa, T. Mukai and T. Aizawa, Materia Japan 39 (2000), p. 347.
[11] H.J. Frost, M.F. Ashby, *Deformation Mechanism Maps*, Pergamon Press, Oxford (1982), p. 44.

Corresponding author: Hiroyuki Watanabe
E-mail : hwata@omtri.city.osaka.jp

Temperature and Strain-Rate Dependence of Elongation in Al – 4.5 Mg Alloy

H. Iwasaki[1], T. Mori[1], M. Mabuchi[2], T. Tagata[3] and K. Higashi[4]

[1] College of Engineering, Dept. of Materials Science & Engineering, Himeji Institute of Technology, 2167 Shosha, Himeji, Hyogo 671-2201, Japan

[2] National Industrial Res. Institute of Nagoya, 1-1 Hirate-cho, Kita-ku, Nagoya 462-8510, Japan

[3] Sky Aluminium Co., Ltd., 1351, Uwanodai, Fukaya, Saitama, 366-0801, Japan

[4] College of Engineering, Dept. of Metallurgy and Materials Science, Osaka Prefecture University, 1-1, Gakuen-cho, Sakai, Osaka 599-8531, Japan

Keywords: Al-Mg Alloy, Cavitation, Elongation, Grain Boundary Sliding, Strain Rate Sensitivity

ABSTRACT

The effects of temperature and strain rate on elongation of Al - 4.5 Mg alloy were investigated at $10^{-4} \sim 10^{-2}$ s^{-1} and at 573 ~ 773 K. The elongation depended on the temperature and strain rate even in the solute-drag creep region; as a result, a large elongation of 352 % was attained at 653 K and at 10^{-4} s^{-1}. Necking was developed due to a decrease in plastic stability arising from the low strain rate sensitivity. Also, significant cavitation was caused due to grain boundary sliding. The development of necking and cavitation is responsible for the temperature and strain rate dependence of elongation.

INTRODUCTION

Large elongations above 300 % have been obtained at elevated temperatures in coarse-grained Al - high Mg alloys [1~4]. The Al - high Mg alloys exhibit a low value of the stress exponent of 3 or a large value of the strain rate sensitivity of 0.33 for dislocation creep because of viscous dislocation glide [1~11]. Taleff et al. [11] noted that the large elongation for the Al - high Mg alloys is attributed to the high plastic stability arising from the large strain rate sensitivity of about 0.3. In this case, large elongation can be attained under wide conditions of temperature and strain rate, compared to the conditions where a large elongation is attained by superplastic flow. Furthermore, a fine-grained microstructure is not required to attain large elongation for solute-drag creep. However, even in the solute-drag creep region, elongation depends on the temperature and strain rate and large elongation is not attained under all temperature and strain rate conditions [3,11]. This trend is the same as that in superplastic flow where the strain rate sensitivity is greater than 0.3 [12]. These facts suggest that elongation is not determined only by the value of the strain rate sensitivity.

In general, commercial Al - Mg alloys contain impurities of Fe, Si and so on. Recently, Hosokawa et al. [4] investigated the effects of Si content on mechanical properties in the solute-drag creep region for Al - 4.5 Mg alloy, and they revealed that cavities were formed at the impurity particles. Hence, it is suggested that elongation in the solute-drag creep region is affected by cavitation. In the present paper, the elongation of Al - 4.5 Mg alloy is investigated at 573 ~ 773 K and at $10^{-4} \sim 10^{-2}$ s^{-1}. Furthermore, the temperature and strain rate dependence of elongation is investigated from the viewpoint of cavitation and plastic stability due to a strain rate sensitivity.

EXPERIMENTAL PROCEDURE

An Al-4.5Mg alloy with chemical compositions of (wt-%) Al-4.49Mg-0.05Fe-0.01Ti was used. The alloy was homogenized at 803 K for 10 hours, and then it was reheated to 753 K, hot-rolled to a thickness of 6 mm, and cold-rolled to 1 mm. The rolled sheet was annealed at 723 K for 25 s by an immersion molten salt-bath.

Tensile specimens with a gage length of 10 mm and a gage width of 5 mm were prepared with the tensile axes parallel to the rolling direction. Elongation-to-failure tests were conducted at 573 ~ 773 K and at 10^{-4} ~ 10^{-2} s^{-4}. In addition, strain-rate-change tests were carried out at 573 ~ 773 K to investigate the stain rate sensitivity. The specimens were pulled at a series of discrete constant true strain rates ranging from 7×10^{-6} to 7×10^{-2} s^{-1}. The strain rate was changed from a lower to a higher value when a steady flow stress was judged to have been attained. For both types of tests, the specimens were held in the furnace for 1.8×10^3 s before pulling. The temperature variation during the tensile tests was not more than 1 K.

To estimate necking, the local strain was investigated by measurement of the cross-sectional area of the deformed specimen. The local strain is given by

$$\varepsilon_{local} = ln(A_O/A) \quad (1)$$

where ε_{local} is the local strain, A_O is the cross-sectional area of the specimen prior to the tensile test and A is the area of the deformed specimen.

The volume fraction of cavities in the gage length of the deformed specimens was investigated by hydrostatic weighing in water with a corresponding gage head being used as a density standard. Also, cavities were observed by an optical microscope. The grain boundary sliding contribution was investigated by measurement of the transverse offset. The strain due to grain boundary sliding is given by [13]

$$\varepsilon_{GBS} = \varphi w/L \quad (2)$$

where ε_{GBS} is the strain due to grain boundary sliding, w is the average offset, L is the average linear intercept grain size and φ is a geometric constant which is equal to 1.5 for the transverse offset [13] The grain boundary sliding contribution to total strain is given by

$$\xi = \varepsilon_{GBS}/\varepsilon_{tot} \quad (3)$$

where ε_{tot} is the total strain. The grain size was measured by optical microscope. The grain size was given by multiplication of the average linear intercept grain size by a constant (= 1.74). The specimens were polished and then electrolytically etched at 20 V for 90 s.

RESULTS AND DISCUSSION

The variation in elongation to failure as a function of strain rate at 573, 653 and 773 K is shown in Fig. 1. The peak of elongation was attained at 10^{-3} s^{-1} under a constant temperature of 653 K. Also, the peak of elongation was attained at 653 K under a constant strain rate of 10^{-3} s^{-1}. As a result, a maximum elongation of 352 % was attained at 10^{-3} s^{-1} and at 653 K in the strain rate and temperature range investigated. It should be noted that the elongation depended on the strain rate and temperature.

The variation in stress as a function of strain rate was investigated by strain-rate-change tests. The results are shown in Fig. 2. From Fig. 2, the strain rate sensitivity at 653 and 773 K was about 0.3 in the strain rate range of 7×10^{-6} to 7×10^{-2} s^{-1}. On the other hand, the strain rate sensitivity at 573 K was about 0.3 in the strain rate range $< 10^{-4}$ s^{-1}, while it was less than 0.3 in the strain rate range $\geq 10^{-4}$ s^{-1}. The strain rate range $\geq 10^{-4}$ s^{-1} at 573 K is probably in the power-law breakdown region.

The variation in local strain as a function of distance from the center at 10^{-4}, 10^{-3} and 10^{-2} s^{-1} under a constant temperature of 653 K is shown in Fig. 3, where the specimen is deformed to 260 % at 10^{-4} and 10^{-3} s^{-1} and it is deformed to 230 % at 10^{-2} s^{-1}. Inspection of Fig. 3 revealed that necking was significantly developed at 10^{-2} s^{-1}, compared to that at 10^{-4} and 10^{-3} s^{-1}. The variation in local strain at 573, 653 and 773 K under a constant strain rate of 10^{-3} s^{-1} is also shown in Fig. 4, where the specimen is deformed to 200 %. Necking was significantly developed at 573 K, compared to that at 653 and 773 K. From the inspections, it is likely that the decrease in elongation at 10^{-2} s^{-1} at 653 K and at 10^{-3} s^{-1} at 573 K, as shown in Fig. 1, was attributed to significant development of necking. The significant development of necking at at 10^{-2} s^{-1} at 653 K

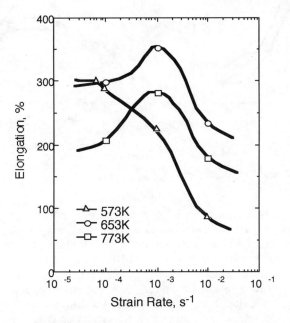

Fig. 1 The variation in elongation to failure as a function of strain rate at 573, 653 and 773 K.

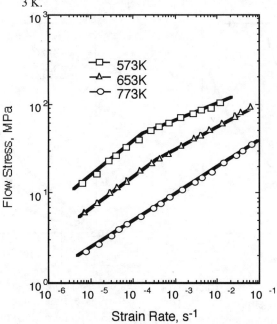

Fig. 2 The variation in stress as a function of strain rate at 573, 653 and 773 K.

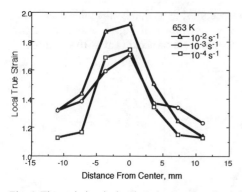
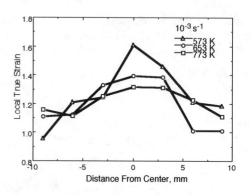

Fig. 3 The variation in local strain as a function of distance from the center at 10^{-4}, 10^{-3} and 10^{-2} s^{-1} under a constant temperature of 653 K, where the specimen is deformed to 260 % at 10^{-4} and 10^{-3} s^{-1} and it is deformed to 230 % at 10^{-2} s^{-1}.

Fig. 4 The variation in local strain as a function of distance from the center at 573, 653 and 773 K under a constant strain rate of 10^{-3} s^{-1}, where the specimen is deformed to 200 %.

and at 10^{-3} s^{-1} at 573 K is likely because of the lower strain rate sensitivity as shown in Fig. 2.

The variation in volume fraction of cavities as a function of strain at 10^{-4}, 10^{-3} and 10^{-2} s^{-1} under a constant temperature of 653 K is shown in Fig. 5. It should be noted that the cavity volume fraction at 10^{-4} s^{-1} was much larger than those at 10^{-3} and 10^{-2} s^{-1}. The variation in volume fraction of cavities as a function of strain at 573, 653 and 773 K under a constant strain rate of 10^{-3} s^{-1} is shown in Fig. 6. The cavity volume fraction at 773 K was much larger than those at 573 and 653 K. As shown in Fig. 2, the elongation at 10^{-4} s^{-1} was lower than that at 10^{-3} s^{-1} under a constant temperature of 653 K. Also, the elongation at 773 K was lower than that at 653 K under a constant strain rate of 10^{-3} s^{-1}. These results are likely to be related to significant development of cavitation formed at grain boundaries.

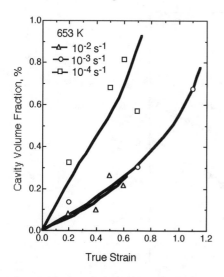
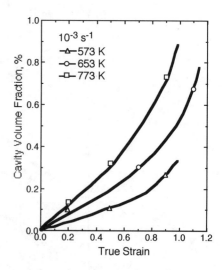

Fig. 5 The variation in volume fraction of cavities as a function of strain at 10^{-4}, 10^{-3} and 10^{-2} s^{-1} under a constant temperature of 653 K.

Fig. 6 The variation in volume fraction of cavities as a function of strain at 573, 653 and 773 K under a constant strain rate of 10^{-3} s^{-1}.

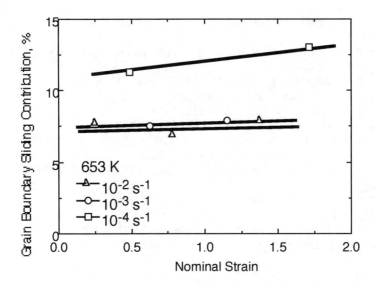

Fig. 7 The variation in grain boundary sliding contribution as a function of strain at 10^{-4}, 10^{-3} and 10^{-2} s^{-1} under a constant temperature of 653 K.

One of the possibilities is due to grain boundary sliding. Grain boundary sliding often occurs even in the dislocation creep region [14]. Ayensu and Langdon [15] showed that the overall level of cavitation increases with strain due to grain boundary sliding during creep. Watanabe and Davies [16] noted that ledges produced by movement of lattice dislocation along the grain boundary could act as nucleation sites of grain boundary cavities. Fig. 7 shows the variation in grain boundary sliding contribution as a function of strain at 10^{-4}, 10^{-3} and 10^{-2} s^{-1} under a constant temperature of 653. It can be seen that grain boundary sliding occurred significantly at 10^{-4} s^{-1}, compared to that at 10^{-3} and 10^{-2} s^{-1}. Therefore, it is suggested that significant development of cavitation at 10^{-4} s^{-1} arises from active grain boundary sliding.

CONCLUSIONS

The effects of temperature and strain rate on elongation of Al - 4.5 Mg alloy were investigated at $10^{-4} \sim 10^{-2}$ s^{-1} and at $573 \sim 773$ K. A maximum elongation of 352 % was attained at 10^{-3} s^{-1} and at 653 K. The temperature and strain-rate dependence of elongation of the alloy was concluded as follows:
1. The lower elongation at 10^{-2} s^{-1} under a constant temperature of 653 K and the lower elongation at 573 K under a constant strain rate of 10^{-3} s^{-1} was attributed to development of necking. The development of necking arose from a decrease in plastic stability due to the lower strain rate sensitivity.
2. The lower elongation at 10^{-4} s^{-1} under a constant temperature of 653 K and the lower elongation at 773 K under a constant strain rate of 10^{-3} s^{-1} was due to significant cavitation. The significant cavitation is likely to be related to grain boundary sliding.

REFERENCES
[1] S.S.Woo, Y.R.Kim, D.H.Shin and W.J.Kim, *Scripta Mater.*, **37** (1997) 1351.
[2] M.Otsuka, S.Shibasaki and M.Kikuchi, *Mater. Sci. Forum*, **233-234** (1997) 193.

[3] E.M.Taleff, G.A.Henshall, T.G.Nieh, D.R.Lesuer and J.Wadswarth, *Metall. Mater. Trans. A*, **29A** (1998) 1081.
[4] H.Hosokawa, H.Iwasaki, T.Mori, M.Mabuchi, T.Tagata and K.Higashi, *Acta Mater.*, **47** (1999) 1859.
[5] K.L.Murty, F.A.Mohamed and J.E.Dorn, *Acta Metall.*, **20** (1972) 1009.
[6] M.S.Mostafa and F.A.Mohamed, *Metall. Trans. A*, **17A** (1986) 365.
[7] W.C.Oliver and W.D.Nix, *Acta Metall.*, **30** (1982) 1335.
[8] P.Yavari, F.A.Mohamed and T.G.Langdon, *Acta Metall.*, **29** (1981) 1495.
[9] O.D.Sherby and P.M.Burke, *Prog. Mater. Sci.*, **13** (1969) 325.
[10] P.Yavari and T.G.Langdon, *Acta Metall.*, **30** (1982) 2181.
[11] E.M.Taleff, D.R.Lesuer and J.Wadsworth, *Metall. Mater. Trans. A*, **27A** (1996) 343.
[12] T.G.Langdon, *Scripta Metall.*, **11** (1977) 997.
[13] K.-T.Park, S.Yan and F.A.Mohamed, *Philos. Mag. A*, **72** (1995) 891.
[14] R.C.Pond, D.A.Smith and P.W.Southerden, *Philos. Mag. A*, **37** (1978) 27.
[15] A.Ayensu and T.G.Langdon, *Metall. Mater. Trans. A*, **27A** (1996) 901.
[16] T.Watanabe and P.W.Davies, *Philos. Mag. A*, **37** (1978) 649.

Effects of Temperature and Microstructure on the Superplasticity in Microduplex Pb-Sn Alloys

Tae Kwon Ha and Young Won Chang

Center for Advanced Aerospace Materials (CAAM), Pohang University of Science and Technology, Pohang, Kyungbuk 790-784, Korea

Keywords: Internal Variable Theory, Microduplex Pb-Sn Alloys, Phase Boundary Sliding, Structural Superplasticity

Abstract

Superplastic deformation behavior of microduplex Pb-Sn alloys has been investigated in this study. The effects of test temperature and microstructure were examined within the framework of the internal variable theory of structural superplasticity. A series of load relaxation and tensile tests were conducted at room temperature for the alloys with various microstructures such as as-cast eutectic, equiaxed eutectic, hypoeutectic and hypereutectic alloys. For eutectic alloys, load relaxation tests were carried out from room temperature to 140 °C. The flow curves obtained from load relaxation tests on the superplastic Pb-Sn alloys were shown to consist of the contributions from interface sliding (IS) and the accommodating plastic deformation. The IS behavior could be described as a viscous flow process characterized by the power index value of $M_g = 0.5$, suggesting the onset of intense phase boundary sliding (PBS). Superplasticity of hypoeutectic alloy was comparable to that of superplastic eutectic alloys and superior to that of hypereutectic alloy. As the test temperature increased, contribution from IS appeared to be exhibited in lower stress and faster strain rate region.

Introduction

It is now well known that structural superplasticity (SSP) is a deformation process that can produce an extraordinary elongation in a fine-grained crystalline material deformed under the optimum conditions of strain rate and temperature [1-3]. Grain boundary sliding (GBS) or phase boundary sliding (PBS) has been widely accepted as the major mechanism of SSP. Recently, PBS has been reported to be different from GBS in nature and the internal variable theory of SSP, which can separately describe GBS or PBS and the accommodation mechanism, has been proposed by the authors [4]. It is attempted, in the present study, to elucidate the effects of test temperature and microstructure on the superplastic deformation behavior of microduplex Pb-Sn alloys within the framework of the internal variable theory of SSP.

Internal Variable Theory of SSP

Figure 1 shows the physical model for SSP employed in this study. In this model, IS is mainly accommodated by a dislocation process, giving rise to an internal strain ($\underline{a}$) and plastic strain ($\underline{\alpha}$). The stress variables σ^I and σ^F represent the internal and the friction [4] stresses, respectively. In the temperature range where SSP can be expected, σ^F is, in general, very small compared to σ^I and the time rate of the internal strain can be neglected if the relaxation test is performed uni-axially at a steady state. So the total inelastic strain rate ($\dot{\varepsilon}$) can be treated to be the sum of non-recoverable plastic strain rate ($\dot{\alpha}$) and the strain rate due to interface sliding ($\dot{g}$);

$$\dot{\varepsilon} = \dot{\alpha} + \dot{g}. \tag{1}$$

We can therefore describe the superplastic deformation with constitutive relations for $\dot{\alpha}$ and $\dot{g}$

elements. The constitutive relations are expressed as follows;

$$(\sigma^*/\sigma^I) = \exp(\dot{\alpha}^*/\dot{\alpha})^p \quad (2)$$

$$(\dot{g}/\dot{g}_o) = (\sigma/\Sigma_g - 1)^{1/M_g} \quad (3)$$

where p and M_g are material constants, σ^* and $\dot{\alpha}^*$ are the internal strength variable and its conjugate reference strain rate as first used by Hart [5], while Σ_g and $\dot{g}_o$ are the static friction stress for IS and its reference rate, respectively. Interface sliding is considered here as a viscous drag process similar to the frictional glide process of dislocations [4].

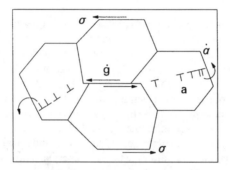

Figure 1. The physical model representing the interface sliding (IS) accommodated by grain matrix deformation (GMD) [4].

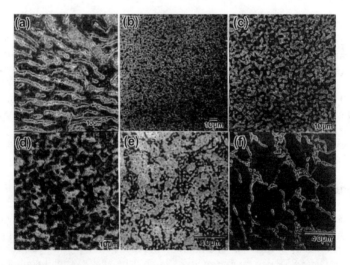

Figure 2. SEM micrographs showing typical microstructures of materials used in this study. (a) AC, (b) E1, (c) E2, (d) E3, (e) AR and (d) BR alloys, respectively.

Experimental procedure

The microduplex Pb-Sn binary alloys with the compositions of 40 wt.%, 62 wt.%, and 80 wt.% Sn were cast by using a 99.99% purity lead and a 99.95% purity tin. The ingots were then extruded into the rods of 12 mm in diameter at room temperature. The specimens with eutectic composition were annealed at 170°C for 2 hrs, 24 hrs, and 168 hrs to obtain the average planar phase size of 2.5 µm, 3.3 µm, and 7.0 µm, respectively. The specimens with hypoeutectic (40 wt.% Sn) and hypereutectic (80 wt.% Sn) compositions were also annealed at 170°C for 2 hrs. For convenience, specimens of as-cast eutectic, fine-grained eutectic, hypoeutectic, and hypereutectic alloys are denoted hereafter as AC, E1(2.5 µm), E2(3.3 µm), E3(7.0 µm), AR, and BR, respectively. The resulting microstructures are summarized in Fig. 2. The Pb-rich α phases are relatively bright in the figure. The ratios between the volume fractions of α and Sn-rich β phases are 3:1, 1:1, and 2:5 in AR, E1, and BR alloys, respectively.

A series of load relaxation tests has been conducted on these materials at various temperatures from 20 to 140 °C in air. Following the usual procedure proposed by Lee and Hart [6], the inelastic strain rate vs. flow stress curves were then deduced. Tensile tests have also been performed under the various strain rates.

Results and discussion

The flow curves obtained from load relaxation tests at room temperature for various microstructures are given in Fig. 3. The flow curves shift into lower stress region as the grain size decreased. The flow curves of specimens of E1, E2, and E3 alloys show the typical sigmoidal shape, a clear manifestation of superplastic characteristics caused by IS process. As noted in the figure, while the flow curve of AR appears to have a high value of the strain rate sensitivity (defined as the slope of flow curves) comparable to that of E1 alloy, those of AC and BR alloys are seemingly less sensitive to strain rate. The flow curves of eutectic alloys obtained at various temperatures are summarized in Fig. 4. As the testing temperature increased, flow curves shift into faster strain rate and lower stress region in AC, E1 and E3 alloys. No change in shape of flow curve of AC specimens is observed even at high temperatures.

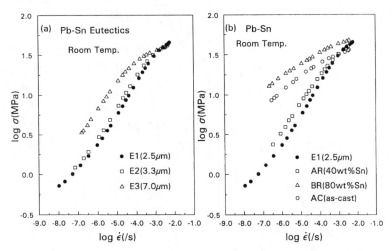

Figure 3. Flow curves obtained at room temperature for various microstructures.

Applying the internal variable theory of SSP, the contributions from IS and accommodating plastic deformation by dislocation process could be separated from the flow curves as given in Fig. 5. The thin solid lines in the figure represent the contributions from the plasticity element described by Eq. 2 and the dotted lines from the interface sliding (IS) described by Eq. 3. The bold lines are the composite curves predicted from Eq. 1 and show good agreements with the experimental data. Only the contribution from plasticity element could be seen in the AC alloy and the flow curve of E1 alloy showed the most predominant effect ($\dot{g}$) of interface sliding process. It is interesting to note that the contribution from PBS process appeared to shift into the faster strain rate and lower stress region as the test temperature increased in E1 alloy. The effect of IS was also exhibited in AR and BR alloys but the contributions were less than those from plasticity element in both cases. The PBS contribution in the BR specimen was exhibited in the slower strain rate range compared to that in the AR specimen and the tensile test results given in Fig. 6 confirm this analysis although indirectly.

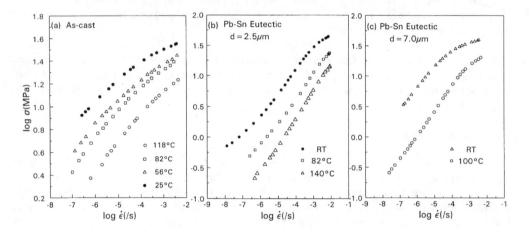

Fig. 4. Flow curves obtained at various temperatures for eutectic alloys.

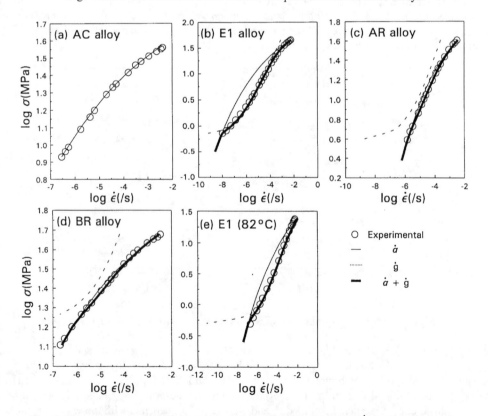

Fig. 5. Flow curves can be divided into the contributions from IS ($\dot{g}$) and plasticity due to dislocation process ($\dot{\alpha}$) based on the internal variable theory of SSP.

The total elongation increased as the initial strain rate decreased in all cases. The specimen E1 and AR, which are expected to show the $\dot{g}$ contribution under faster strain rate in tensile test from Fig. 5, were shown to exhibit much higher elongations than BR specimen.

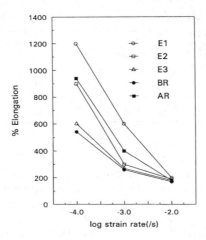

Figure 6. Tensile tests results obtained at room temperature.

The constitutive parameters determined numerically by separating the individual flow curves as explained above are listed in Table 1 and 2. The parameter M_g was determined as 0.5, which is lower than the reported value of $M_g = 1.0$ for various single-phase Al alloys [7-9], suggesting that the IS behavior of microduplex Pb-Sn alloys be different from grain boundary sliding (GBS) of quasi-single phase superplastic materials. In an earlier report [10], the authors examined the surface of a superplastically deformed Pb-Sn eutectic alloy and found the phase boundary sliding (PBS) predominant. The static friction resistance to PBS (Σ_g) can be observed from the tables to decrease with the increase in test temperature and decrease in grain size.

Another important parameter p, denoting the dislocation permeability of barrier such as grain boundaries, was found to be 0.1, the same as that obtained for pure tin [12] with body-centered tetragonal (BCT) structure, except for the AR alloy. It is well known that this parameter depends on the crystal structure [12] and it should be noted that the AR alloy mainly consists of Pb-rich phase with FCC structure. The value of p in FCC metals and alloys has been reported to be 0.15 [4, 8, 9, 11, 13], the same as that for AR alloy in this study.

Table 1. The constitutive parameters determined from the load relaxation tests of Pb-Sn eutectic alloy at room temperature.

Alloy	GMD			IS		
	$\log \sigma^*$	$\log \dot{\alpha}^*$	p	$\log \Sigma_g$	$\log \dot{g}_o$	M_g
AC	1.971	-2.740	0.1	-	-	-
AR	2.069	-2.281	0.15	0.587	-5.550	0.5
E1	2.305	-0.516	0.1	-0.162	-6.701	0.5
E2	2.069	-2.281	0.1	-0.162	-6.905	0.5
E3	2.069	-2.281	0.1	0.206	-6.932	0.5
BR	2.038	-3.339	0.1	1.226	-4.721	0.5

Table 2. The constitutive parameters determined from the load relaxation tests of Pb-Sn eutectic alloy at various temperatures.

Alloy/ Temp. (°C)	GMD			IS		
	$\log \sigma^*$	$\log \dot{\alpha}^*$	p	$\log \Sigma_g$	$\log \dot{g}_o$	M_g
AC/56 °C	1.895	-2.158	0.1	-	-	-
AC/82 °C	1.913	-1.652	0.1	-	-	-
AC/118 °C	1.863	-0.732	0.1	-	-	-
E1/82 °C	2.213	0.675	0.1	-0.302	-5.870	0.5
E1/140 °C	2.243	1.859	0.1	-0.596	-5.819	0.5
E3/100 °C	2.092	0.254	0.1	-0.502	-6.876	0.5

Conclusions

The flow curves of microduplex Pb-Sn alloys could effectively be interpreted as consisting of contributions from PBS and the accommodating plastic deformation due to dislocation process, regardless of microstructures and temperatures. The contribution from PBS was observed in all compositions used in this study except in the AC alloy and was most predominant in the fine-grained eutectic alloy. As the grain size decreased and the test temperature increased, the static friction resistance to PBS decreased and the flow curves shifted into a faster strain rate and lower stress region.

Acknowledgements

The financial support by the Korea Science and Engineering Foundation (KOSEF) is gratefully acknowledged.

References

[1] A. H. Chokshi, A. K. Mukherjee, and T. G. Langdon, Mater. Sci. Eng. R10 (1993), p. 237.
[2] O. D. Sherby, and J. Wadsworth, Prog. Mater. Sci., 33 (1989), p. 169.
[3] K. A. Padmanabhan, and G. J. Davis, "*Superplasticity*", Springer-Verlag, New York, NY (1980) p. 11.
[4] T. K. Ha, and Y. W. Chang, Acta Mater. 46 (1998), p. 2741.
[5] E. W. Hart, "*Stress Relaxation Testing*", A. Fox ed., ASTM, Baltimore, Md. (1979), p. 5.
[6] D. Lee, and E. W. Hart, Metall. Trans., 2A (1971), p. 1245.
[7] T. K. Ha, and Y. W. Chang, Scripta Metall. Mater., 32 (1995), p. 809.
[8] Y.-N. Kwon, and Y. W. Chang, J. Korean Inst. Met. Mater. 32 (1994), p. 878.
[9] W. Bang, T. K. Ha, and Y. W. Chang, Met. & Mater. 6 (2000), p. 203.
[10] T. K. Ha, and Y. W. Chang, Scripta Mater. 37 (1997), p. 1415.
[11] T. K. Ha, H. J. Sung, K. S. Kim, and Y. W. Chang, Mater. Sci. Eng. A271 (1999), p. 160.
[12] T. K. Ha, C. S. Lee, and Y. W. Chang, Scripta Mater. 35 (1996), p. 635.
[13] E. W. Hart, J. Eng. Mater. Tech. 106 (1984), p. 322.

Low Temperature Superplasticity in a Severely Rolled Sheet of an Al-4.2mass%Mg Alloy

Y. Takayama[1], S. Sasaki[1], H. Kato[1], H. Watanabe[1], A. Niikura[2] and Y. Bekki[2]

[1] Department of Mechanical Systems Engineering, Utsunomiya University, Utsunomiya, Tochigi 321-8585, Japan

[2] Furukawa Electric Co., Ltd., Mikuni, Sakai, Fukui 913-8588, Japan

Keywords: Al-Mg Alloy, Fracture Surface, Sever Straining, Superplasticity, Texture

ABSTRACT

Superplasticity at relatively low temperatures around 573K has been investigated a severely rolled sheet of an Al-4.2mass%Mg alloy. It was already found that fine recrystallized grains of about 6μm was gotten by annealing for 120s at 573K in a salt bath for the severely rolled sheet. The maximum elongation to failure of 208% is obtained at 563K and $1.4 \times 10^{-3} s^{-1}$ at a higher heating rate of $5Ks^{-1}$, though the sheet has a small thickness of only 0.1mm. It is suggested from analyses of stress and fracture surface that the elongation for $0.57Ks^{-1}$ and $5Ks^{-1}$ is attained by different mechanisms.

INTRODUCTION

Grain refinements using severe straining techniques known as Equal-channel angular pressing (ECAP) [1-5] and Accumulative roll-bonding (ARB) [6] have been carried out recently to obtain excellent superplasticity, which refers to high strain rate or low temperature superplasticity. A sheet used in the present study has been prepared by severe rolling which has a high possibility of industrial application. It has been already found that fine recrystallized grains of about 6μm is gotten by annealing for 120s at 573K in a salt bath for the severely rolled sheet, which was made from high purity ingots to reduce amount of inclusion and precipitate [7]. It is, therefore, expected that fine-grained superplasticity takes place around the annealing temperature of 573K, which is low temperature superplasticity. In the present study, to investigate possibility of low temperature superplasticity, the severely rolled sheet of an Al-4.2mass%Mg is examined in various conditions of tensile tests.

EXPERIMENTAL

The sample was the thin sheet cold-rolled severely up to reduction of 99.8%. Thickness t of the sheet was 0.1mm. The chemical composition is listed in Table 1. Tensile specimens were machined from the sheet with a 6 mm length and a 4 mm width of the reduced section, and a 6 mm radius of fillets. The elongation to failure was measured from the increase in the gauge length of 5 mm. Gauge marks were put on the reduced section by a Vickers microhardness tester.

Tables 1 Chemical composition of used alloy (mass%)

Si	Fe	Cu	Mn	Mg	Cr	Zn	Zr	Ti	Al
0.02	0.00	0.00	0.00	4.24	0.00	0.01	0.00	0.00	Bal.

Tensile tests were carried out in air in the range of 523-593K using an Instron type testing machine at an initial strain rate of $2.8\times10^{-4}s^{-1} - 1.4\times10^{-2}s^{-1}$. The specimens were tested after heating with an infrared furnace at a rate of $0.57Ks^{-1}$ or $5Ks^{-1}$. Holding time before tensile testing was 0s so as to start the test immediately after fine recrystallized grains were formed. The fracture surfaces of the broken specimens were observed by scanning electron microscopy. Grain size after fracture was evaluated on optical micrographs of some samples. Further, two samples were chosen to analyze grain orientation distribution after fracture using TSL/HITACHI S-3500H Orientation Image Microscopy (OIM) system based on the electron backscatter diffraction pattern (EBSP) technique [8].

RESULTS AND DISCUSSION

Nominal stress - strain curves for the Al-4.2mass%Mg specimens heated at rates of $0.57Ks^{-1}$ and $5Ks^{-1}$ are shown in Figs. 1 and 2, respectively. The curves of $0.57Ks^{-1}$ show a peak of stress at an early stage and gradual decrease with a strain for each strain rate. Maximum stresses for the curves exhibit common dependence of strain rate for superplastic alloys. On the other hand, maximum stress for the curves of $5Ks^{-1}$ is not strongly dependent on strain rate. Curves at lower strain rates show characteristic reduction in the stress after the peak. These facts suggest that deformation at the higher heating rate is accompanied by microstructural change.

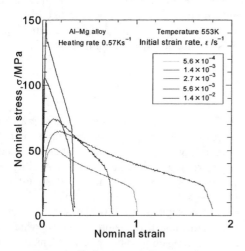 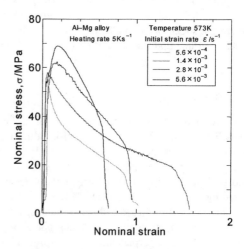

Fig. 1 Nominal stress - strain curves at 553K and a heating rate of $0.57Ks^{-1}$.

Fig. 2 Nominal stress - strain curves at 573K and a heating rate of $5Ks^{-1}$.

Figure 3 shows elongation to failure as a function of strain rate at a heating rate of $0.57Ks^{-1}$. The elongation has a peak value at $1.4\times10^{-3}s^{-1}$ at every temperature. Change in the strain rate for the maximum elongation with increasing temperature is not found in the range tested. The fact means that effect of grain size and temperature on the optimum strain rate counteract each other. The peak elongation at 548K is a fairly large elongation of 189%. Large scatter of the elongation is regarded as effect of thickness of the specimen. Figure 4 shows results at a heating rate of $5Ks^{-1}$ in the same way. The maximum elongation of 208% is obtained at 563K and $1.4\times10^{-3}s^{-1}$. The elongation is markedly large in consideration of specimen's thickness of 0.1mm. Comparing two conditions of heating rates, the scatter of elongation at $5Ks^{-1}$ tends to be smaller than that at $0.57Ks^{-1}$. This is due to the difference in microstructural evolution during heating. The microstructural evolution is more complicated than that at a constant temperature. It is, therefore, understood that precise control of the initial microstructure enables this alloy to exhibit a better superplasticity.

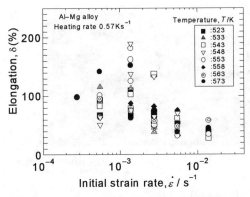
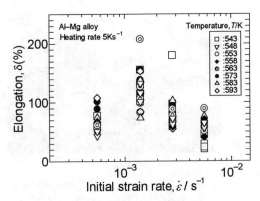

Fig. 3 Relation between elongation to failure and initial strain rate at a heating rate of 0.57Ks^{-1}.

Fig. 4 Relation between elongation to failure and initial strain rate at a heating rate of 5Ks^{-1}.

Relations between maximum nominal stress and initial strain rate at heating rates of 0.57Ks^{-1} and 5Ks^{-1} are plotted in Figs. 5 and 6, respectively. For 0.57Ks^{-1}, strain rate sensitivity coefficient, m is about 1/3 at most of all temperatures, which implies a solute-drag controlled creep mechanism [9,10]. Maximum stress at 573K exhibits a smaller m value of 0.16 at a low strain rate range. This is attributed to grain growth during heating or deformation. On the other hand, m values are fairly small and less than 1/4 for samples of 5Ks^{-1}. These results should be attained by microstructural change during deformation because confused order of the stress on temperature is observed, especially at low strain rates. Though large elongation of about 200% was obtained at both heating rates, strain rate dependence of stress can lead to contribution of different deformation mechanisms to remarkable ductility.

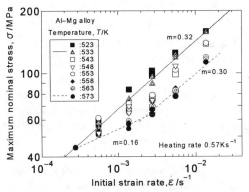
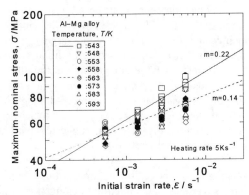

Fig. 5 Relation between maximum nominal stress and initial strain rate at a heating rate of 0.57Ks^{-1}.

Fig. 6 Relation between maximum nominal stress and initial strain rate at a heating rate of 5Ks^{-1}.

SEM micrographs of fracture surfaces are shown for the specimens of 0.57Ks^{-1} with the elongation of 116% and 189% in Fig. 7. While the fracture surface with the smaller elongation exhibits lots of coarsely granular features, the finer features with a size less than 5μm are observed on the fracture surface with a large elongation of 189%. This may suggest that the large elongation is related to fine recrystallized grains.

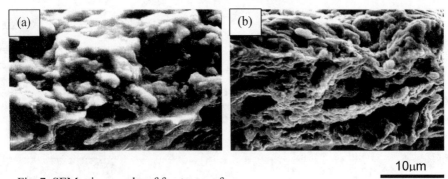

Fig. 7 SEM micrographs of fracture surfaces:
(a) deformed at 533K and $5.6 \times 10^{-4} s^{-1}$ and (b) deformed at 548K and $1.4 \times 10^{-3} s^{-1}$.

In order to discuss grain size effect, mean grain size was evaluated by linear intercept method for some samples after fracture. The mean grain size is plotted in Fig. 8 against testing or holding time at the elevated temperature in consideration of microstructural change during the test. Larger open circles are also attached for large elongation more than 150%. As shown in the figure, grain sizes are about 5µm for the samples with the testing time less than 1500s. Grain size rises up to three times with increasing the testing time. However, the fine grain does not always lead to the large elongation. The larger circles indicating the large elongation concentrate in the testing time range of 1000 to 2000s. Moreover, the circles include a sample of the second largest grain size. This can be understood to be importance of strain rate for large elongation. On the other hand, most of all samples having 5µm grain size did not gain large elongation. Although it should be noted that the grain size is not during deformation but *after fracture*, rapid dynamic growth can not be always supposed immediately before fracture. It is, therefore, suggested that fine grain make less contribution for superplasticity at relatively low temperature.

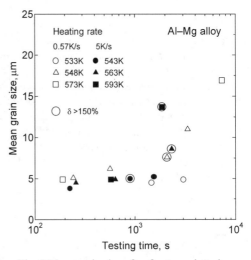

Fig. 8 Mean grain size after fracture plotted against testing time.

As mentioned above, the maximum stress for the curves at $5Ks^{-1}$ is not strongly dependent on strain rate. Nevertheless, the maximum elongation of 208% is obtained at 563K and $1.4 \times 10^{-3} s^{-1}$. Then, two tested samples showing elongation of 62% and 156% was chosen to analyze grain orientation distribution after the fracture using the EBSP technique. Grain boundary sliding reduced the overall level of texture after deformation [11].

Figure 9 shows inverse pole figure maps for the normal direction ND for the two samples. Mean grain sizes are calculated to be 9.5µm and 8.7µm by the planimetric method (equivalent area diameter). Difference in the grain size is fairly small compared with that in elongation. Fractions of low-angle (represented as 'L' in the figure caption), Σ3-Σ29 coincidence site lattice (CSL, 'Σ') and random ('R') boundaries are also calculated. The two samples are not distinguished by grain

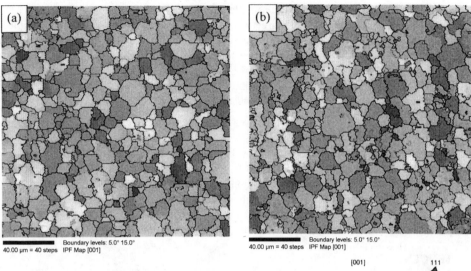

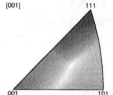

Fig. 9 Inverse pole figure maps in normal direction after fracture:
(a) deformed at 563K and $5.6 \times 10^{-4} s^{-1}$
   δ=62%, GS: 9.5μm, GB/ L: 15% / Σ: 12% / R: 73%
(b) deformed at 573K and $1.4 \times 10^{-3} s^{-1}$
   δ=156%, GS: 8.7μm, GB/ L: 24% / Σ: 11% / R: 65%

boundary character, except for fraction of low-angle boundary. The fraction of low-angle boundary, which is not available for grain boundary sliding, is rather large for the sample showing elongation of 156%.

These EBSP data reveal difference in texture in tensile direction. Figure 10 displays comparison of inverse pole figures in tensile direction between both samples. Because the tensile direction is consistent with rolling one, the observed textures should be originated from severe rolling.

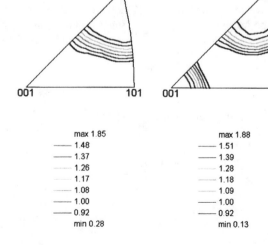

Fig. 10
Inverse pole figures in tensile or rolling direction for samples tested:
(a) at 563K and $5.6 \times 10^{-4} s^{-1}$ (δ=62%) and
(b) at 573K and $1.4 \times 10^{-3} s^{-1}$ (δ=156%).

Certainly, the texture of the sample with the smaller elongation is related to rolling texture, which is corresponding to concentration of [111] direction in Fig. 10(a). In contrast to this, the sample with the larger elongation exhibits concentration to [001] as well as [111]. This is a typical tensile deformation texture. Therefore, multiple slips should take place during deformation to perform concentration of [001] direction. Consequently, contribution of crystallographic slip to gain elongation is confirmed from textural data. Though contribution of grain boundary sliding is not denied, it is not regarded as predominant mechanism for a heating rate of $5Ks^{-1}$.

CONCLUSIONS

Superplasticity at relatively low temperatures has been investigated for a severely rolled sheet of an Al-4.2mass%Mg alloy. The peak elongation of 189% at 548K and $1.4\times10^{-3}s^{-1}$ is found for the sample heated at a rate of $0.57Ks^{-1}$. The maximum elongation of 208% is obtained at 563K and $1.4\times10^{-3}s^{-1}$ for $5Ks^{-1}$, though the tensile specimen is only 0.1mm thick. It is suggested that the elongations for $0.57Ks^{-1}$ and $5Ks^{-1}$ are attained by different mechanisms. Further, contribution of crystallographic slip to gain elongation is confirmed for the sample heated at $5Ks^{-1}$.

REFERENCES

[1] T.G. Langdon, Mater. Sci. Forum 304-306 (1999), p.13.
[2] R.Z. Valiev and R.K. Islamgaliev, Mater. Sci. Forum 304-306 (1999), p.39.
[3] M. Nemoto, Z. Horita, M. Furukawa and T. G. Langdon, Mater. Sci. Forum 304-306 (1999), p.59.
[4] M. Mabuchi, N. Nakamura, K. Ameyama, H. Iwasaki and K. Higashi, Mater. Sci. Forum 304-306 (1999), p.67.
[5] Z. Horita, S. Komura, P.B. Berbon, A. Utsunomiya, M. Furukawa M. Nemoto and T. G. Langdon, Mater. Sci. Forum 304-306 (1999), p.91.
[6] N. Tsuji, K. Shiotsuki, H. Utsunomiya and Y.Saito, Mater. Sci. Forum 304-306 (1999), p.73.
[7] A. Niikura and Y. Bekki, Mater. Sci. Forum 331-337(2000), p.871.
[8] B.L. Adams, S.I. Wright and K. Kunze, Metall. Trans. A, **24A**, (1993), p.819.
[9] E.M.Taleff, D.R.Lesuer and J.Wadsworth: Metall. Mater. Trans., 27A(1996),p.343.
[10] M.Otsuka, S.Shibasaki and M.Kikuchi: Materials Science Forum, 233-234(1997),193.
[11] M.N. Melton, J.W. Edington, J.S. Kallend and C.P. Cutler, Acta Metall., 22(1974),p.165.

ACNOWLEDEMENTS

We wish to thank Mr. H. Ochiai, Mr. H. Takahashi and Mr. Y. Koike for their experimental collaborations. The sample was provided by the Japan national project, "Super Metal Technology ", of the Japan Research and Development Center for Metals. This work was supported by the Grant-in-Aid for Scientific Research, Japan.

Corresponding Author: Y. Takayama
E-mail: takayama@mech.utsunomiya-u.ac.jp
Fax: +81-28-689-6078.

Microstructural Evolution of Quasi-Single Phase Alloy during Superplastic Deformation

W. Bang, T.K. Ha and Y.W. Chang

Center for Advanced Aerospace Materials (CAAM), Pohang University of Science and Technology, San 31, Hyojadong, Nam-ku, Pohang Kyungbook 790-784, Korea

Keywords: Internal Variable Theory, Load Relaxation, Precipitate Free Zone, Superplasticity

Abstract

Flow shifts and changes in microstructure were investigated for superplastic deformation of 7475 Al alloy. The recently proposed internal variable theory of structural superplasticity has been applied to the results of mechanical test. Deformed microstructure was observed using a TEM to check the validity of the constitutive relations. Accommodation mechanism for grain boundary sliding was also examined through the microchemical analysis of deformed microstructures.

INTRODUCTION

A comprehensive understanding for microstructural mechanisms during superplastic deformation has become increasingly important for the optimization of superplastic forming (SPF) process. However, as to the micromechanical origin of structural superplasticity (SSP) in fine-grained materials, no hypothesis has been found capable of accurately describing both the mechanical and microstructural features of deformation. In this study, flow behavior and microstructural changes of a fine-grained 7475 Al alloy during superplastic deformation are investigated. Grain boundary sliding and its accommodation process were figured out schematically by the investigation of precipitate free zone (PFZ), characteristic microstructure of many superplastically deformed aluminum alloys.

EXPERIMENT

Specimens were prepared from thermomechanically treated 7475 Al sheets[1]. Load relaxation tests were carried out at various temperatures ranging from 430℃ to 516℃. In the test, a specimen was first loaded in tension to a certain predetermined extension followed by stopping crosshead motion, and then the load decrement was recorded to a PC for further processing. This load-time (P-t) curve was converted into flow stress-inelastic strain rate (σ-dε/dt) curve applying the method prescribed by Lee and Hart[2]. Constitutive parameters have been determined from load relaxation test by applying the internal variable theory of structural superplasticity (SSP) proposed recently[3]. For microstructural observations, plate type specimens were deformed in tension up to a specific amount of elongation. Deformed specimens were then sliced using an Electric Discharge Machine (EDM) not to cause additional stress during the cutting. Thin foils were prepared by a twin-jet electropolishing after mechanical polishing, and then examined using a transmission electron microscope. TEM-EDS was applied to perform a microchemical analysis of precipitate free zone (PFZ) and grain matrix.

RESULTS

A. Mechanical test results

The relaxation test results are shown in **Fig. 1**. It clearly shows that flow behavior shifts toward the lower stress and faster strain rate region with increasing temperature. Each set of flow data was resolved into what corresponds to grain boundary sliding(GBS) and grain matrix deformation(GMD) as shown in **Fig. 2**. The state variables and material constants defined in constitutive equations (**Eqns. 1 and 2**) were determined by a nonlinear regression and the results are listed in **Table 1**. Detailed description of the internal variable theory and analysis procedure can be found in a referred article[3].

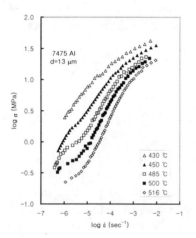
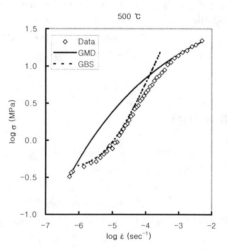

Fig.1. Relaxation Test Results   Fig.2. Resolution and analysis of GMD and GBS

$$(\sigma^*/\sigma') = \exp(\dot{\alpha}^*/\dot{\alpha})^p \qquad \text{Eqn. 1.}$$

$$(\dot{g}/\dot{g}_0) = (\sigma/\Sigma_g - 1)^{1/M_g} \qquad \text{Eqn. 2.}$$

	GMD			
Temp (°C)	log σ^*	log $\dot{\alpha}^*$	p	
430	8.08	−2.09	0.1526	
450	8.03	−1.73	0.1548	
485	7.97	−1.43	0.1470	= 0.15
500	7.94	−1.30	0.1500	
516	7.92	−1.09	0.1501	
	GBS			
Temp (°C)	log Σ_0	log $\dot{g}_0$	M_0	
430	N/A	N/A	N/A	
450	6.18	−5.32	1.0595	
485	5.85	−5.21	0.9982	= 1.0
500	5.60	−5.14	1.0000	
516	5.34	−5.12	1.0000	

Table 1. Constitutive parameters of SSP at various temperatures

B. Observation of deformed microstructure

PFZ formation The GBS of superplastic deformation can generally be confirmed by finding i) the shift of artificial marking along grain boundary(direct method) and ii) the microstructural evolution during accommodation process(indirect method). Since the later method is much more applicable for TEM micrographs, the corresponding deformation mode under a certain condition was determined by its characteristic microstructure. It has been well known that precipitate free zone (PFZ) is created during the superplastic deformation of quasi-single phase Al alloys[4], and the observation was focused on the formation of PFZ. **Fig. 3** shows the grip region of a specimen representing undeformed microstructures. Since no PFZ is found at any condition, the PFZ formation can be thought as a solely deformation induced process in this case. Typical GMD only (non-superplastic) and GBS dominant deformation microstructures are shown in the following **Fig. 4**. Considering that PFZ is the product of GBS accommodation process, PFZ-forming condition shows a good agreement with analytically induced GBS dominant region from the relaxation test results (**Table 2**). These microstructural evidence implies a validity of the proposed constitutive relation based on inelastic deformation theory.

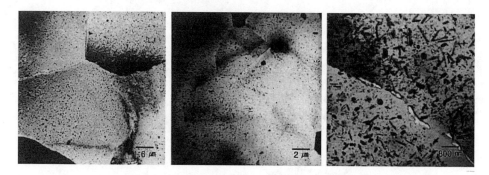

Fig. 3. Grip region after deformation of 50% in tension (450, 485, 516 ℃)

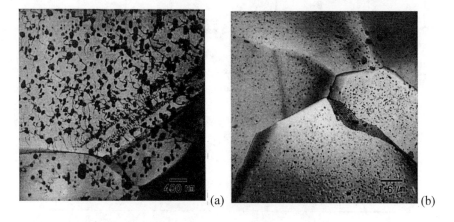

Fig. 4. Typical GMD and GBS dominant microstructures:
(a) 450 ℃, 1×10^{-2}/sec, (b) 516 ℃, 2×10^{-4}/sec

Strain Rate	1×10^{-2}	5×10^{-3}	1×10^{-3}	2×10^{-4}
450°C	No	No	No	No
485°C	No	No	PFZ	PFZ
516°C	No	PFZ	PFZ	PFZ

Table 2. PFZ formation under various deformation conditions

Microchemical analysis Compositional variation through PFZ and grain matrix was analyzed by energy disperse spectroscopy (EDS) attached to a TEM. In **Fig. 5**, EDS spectrum indicates that the fine precipitates in grain interior are Cr-rich particles, and the coarse particles at grain boundary are Fe or Cu rich ones. As to the matrix composition, it shows that the matrix of grain interior is Cr-depleted and lower in solute contents, and PFZ has relatively higher solute contents except Fe. It's quite notable that Cr is found in PFZ, yet only in a very small fraction of the region. Important compositional variation is given in **Table 3**.

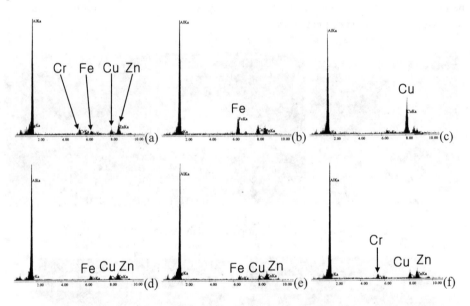

Fig. 5. EDS spectrum of deformed microstructures: (a) fine precipitates at the grain interior, (b)-(c) coarse precipitates at the grain boundary, (d) matrix of the grain interior, (e)-(f) PFZ

weight %	Zn	Mg	Cu	Fe	Cr	Al
PPT	10.7	8.3	4.2	2.3	3.3	Bal.
Matrix	7.6	7.4	4.0	3.7	-	Bal.
PFZ	8.1	9.9	3.4	3.1	-	Bal.

Table 3. Compositional variation of deformed microstructure

DISCUSSIONS

The mechanism of PFZ formation during deformation is a quite disputable subject up to now and is directly related to the accommodation mechanism[5-7]. Accommodation by dislocation slip is applicable to a wide range of strain rates and matches well with the fact that grain shape is remained equiaxed after deformation. GBS localizes plastic strain in the vicinity of grain boundary and this localized deformation develops high energy area along the grain boundaries. If stored strain energy contributes to the resolution of precipitates, The width of PFZ will grow with the amount of inelastic deformation. Since the spherical and rod type fine precipitate are known as $Al_{12}Mg_2Cr$ and η phase $(MgZn_2)$[8], results of the microchemical analysis, showing that PFZ has relatively higher Mg and Zn concentration than in the interior matrix, supports this explanation. Lower contents of Fe and Cu might be due to the presence of coarse Fe or Cu rich precipitates at the grain boundaries.

As for a diffusion accommodation, applied stress acts as a driving force for the diffusion, which delivers atoms from the interface parallel to the tensile direction to the perpendicular interface. But no evidence for the preferential direction of PFZ was found in this study, and diffusional flow is usually too slow to accommodate grain boundary sliding rate of $\sim 10^{-3}$ s^{-1} without loosing compatibility at the microstructural interfaces.

Another explanation based on grain boundary migration [9] is that PFZ is formed behind a moving grain boundary. This movement is to compensate the energy difference between the opposite sides of a grain boundary. In case of a two-phase material under a cyclic loading, stored deformation energy of a phase is usually different from that of adjacent phase, therefore grain boundary migration might be possible to occur continuously. But for a quasi-single phase alloy like 7475 Al, only the boundary curvature contributes to the energy difference. The energy difference is eliminated when boundary is flattened and therefore grain boundary migration can not be continuous up to a large amount of deformation like in SSP. In addition, the PFZ formed by boundary migration should not be observed on both side of a boundary, contrary to the present experimental results.

However, this interpretation should be verified by other methods because EDS spectrum cannot provide a composition spectrum with a sufficient accuracy. An investigation of PFZ width change during superplastic deformation is also necessary for a more comprehensive understanding of PFZ formation and is currently under investigation.

SUMMARY AND CONCLUSIONS

From the mechanical test and the microstructural observation of a 7475 Al alloy, the following important results are obtained: 1) Superplastic flow curves can be resolved quantitatively into GBS and GMD curves using inelastic deformation theory, 2) The proposed constitutive law shows a good agreement with deformed microstructures observed in this study, 3) The evidence from the microstructural investigation and microchemical analysis suggests that PFZ is not likely to form by a grain boundary migration or diffusion in case of structural superplasticity.

ACKNOWLEDGMENT

The financial support by the Korea Science and Engineering Foundation (KOSEF) is gratefully acknowledged. The authors wish to thank Mr. Won Beom Lee at POSTECH for his help with TEM examinations.

REFERENCES

1. J. A. Wert, N. E. Paton, C. H. Hamilton, and M. W. Mahoney, Metall. Trans. 12A, 1267 (1981)
2. D. Lee and E. W. Hart, Metall. Trans. 2A, 1245 (1971)
3. T. K. Ha and Y. W. Chang, Acta Mater., 46, 2741 (1998)
4. U. Koch, W. Bunk and P-J. Winkler: in "Superplasticity in Advanced Materials", (eds. S. Hori et al.), p. 57, Osaka, Japan (1991)
5. J. E. Harris and R. B. Jones, J. Nucl. Mater. 10, 360 (1963)
6. A. Karim, D. L. Holt, and W. A. Backofen, Trans. AIME, 245, 2421(1969)
7. B. Burton and G. L. Reynolds, Mater. Sci. Eng. 191A, 135 (1995)
8. D. H. Shin, K. S. Kim, D. W. Kum, and S. W. Nam, Metall. Trans. 21A, 1267 (1990)
9. R. Maldonado and E. Nembach, Acta Metall. 45, 213 (1997)

CONTACT INFO
Corresponding Author: W. Bang
E-mail: tensile@mail.com
URL: http://tensile.homepage.com/ or http://www.postech.edu/mse/supertop/

Transformation Superplasticity of Ti-6Al-4V and Ti-6Al-4V/TiC Composites at High Stresses

C. Schuh and D.C. Dunand

Department Materials Science and Engineering, Northwestern University,
2225 N. Campus Drive, Evanston IL 60208-3108, USA

Keywords: Phase Transformation, Superplasticity, Titanium Alloy, Titanium Composites

Abstract

Isothermal and thermal cycling tensile creep experiments are reported for Ti-6Al-4V and a Ti-6Al-4V/10 vol% TiC composite, in which the thermal cycles span a broad portion of the α/β phase transformation of Ti-6Al-4V. At low applied stresses, cyclic transformations give rise to *transformation superplasticity*, a deformation mechanism with higher rates of deformation and increased flow stability as compared to isothermal creep. At high stresses this flow stability is lost as a power-law regime is encountered. The experimental results are discussed in light of an analytical model, which considers creep of the weak β-phase under both the internal stresses caused by the transformation and the external stress produced by the applied load.

Introduction

Transformation superplasticity has recently been successful in deforming coarse-grained titanium alloys and titanium matrix composites to very large tensile strains, commonly in excess of 100% [1, 2]. This deformation mechanism relies on thermal cycling, which repeatedly induces the allotropic α/β transformation of titanium, producing internal mismatch strains. A superimposed external stress biases these internal strains, resulting in macroscopic strain increments after each transformation, which can be accumulated to superplastic strains independently of grain size or shape. Recent research has demonstrated that alloys and composites deformed by this method exhibit reasonable room-temperature tensile properties, and that cavitation does not limit the superplastic ductility of Ti-6Al-4V [2].

For implementation of transformation superplasticity as a forming method, it is desirable to maximize the strain rate without inducing cavitation or plastic instabilities which are associated with large applied stresses. Specifically, the linear, Newtonian, flow law commonly observed at low stresses during transformation superplasticity diverges to a stronger stress dependence as the external biasing stress is increased [3, 4]. This non-linear deformation behavior compromises the stability of deformation and thus the achievable forming ductility. The purpose of this work is to present experimental data for Ti-6Al-4V and a Ti-6Al-4V/TiC composite at large applied stresses, where the Newtonian flow model of transformation superplasticity breaks down. A model is presented which captures the deviations to non-linear behavior observed at high stresses.

Experiment

Uniaxial tensile specimens of Ti-6Al-4V and Ti-6Al-4V/10 vol% TiC were machined from billets supplied by Dynamet Technology (Burlington, MA). The materials were prepared by a blended-elemental powder route (the CHIP process [5]) involving cold pressing, vacuum sintering, and containerless hot-isostatic pressing. The microstructures of these materials are typical of powder

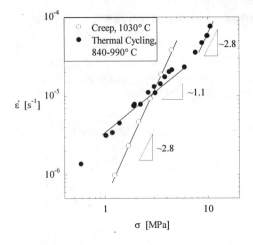 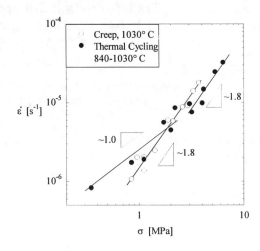

Figure 1: Tensile strain rate of Ti-6Al-4V during isothermal creep and thermal cycling, as a function of applied stress σ.

Figure 2: Stress-dependence of strain rate for Ti-6Al-4V/10 vol% TiC during isothermal creep or thermal cycling.

metallurgy Ti-6Al-4V, with a colony-type microstructure. The TiC reinforcement particles were equiaxed, about 50-100 μm in diameter; typical micrographs of both materials are presented in Ref. [2]. Uniaxial tensile tests were performed in a custom creep frame described in Ref. [6], which allowed for rapid thermal cycling under an applied tensile stress and in an atmosphere of high-purity argon. Isothermal creep data was collected for both materials at 1030° C (in the β-phase field of Ti-6Al-4V) over a narrow range of low stresses. Both materials were thermally cycled with a lower temperature of 840° C, and with upper temperatures of 990° (Ti-6Al-4V) or 1030° C (Ti-6Al-4V/10 vol% TiC), both using triangular 8 minute cycles. These cycles encompass a broad section of the two-phase α+β field of Ti-6Al-4V (β-transus near 1000° C), and involve transformation of about 75 vol% of Ti-6Al-4V [7]. The strain increment after each cycle was measured with a linear voltage-displacement transducer at the cold end of the load train. Superplastic response was investigated over a broad range of experimental stresses (σ = 0.35 – 10.7 MPa).

Results

The isothermal creep rate, $\dot{\varepsilon}$, of Ti-6Al-4V at 1030° C is shown in Fig. 1 as a function of applied stress, σ. The creep rates of the composite at the same temperature (in the β-field of Ti-6Al-4V) are shown in Fig. 2. Over the limited range of stresses investigated here, both materials exhibit power-law type relationships with stress exponents near three and two, for Ti-6Al-4V and the composite, respectively.

After each complete thermal cycle, a strain increment, Δε, is measured; the thermal cycling strain rate is found by dividing Δε by the cycle time. In Figs. 1 and 2, the thermal cycling strain rate is shown as a function of applied tensile stress for Ti-6Al-4V and the Ti-6Al-4V/10 vol% TiC composite, respectively, and compared with the isothermal creep data. During cycling, each of these materials exhibits a linear, Newtonian flow law at low stresses, with stress exponents of 1.1 and 1.0 for Ti-6Al-4V and the composite, respectively. This behavior is characteristic of transformation superplasticity, and is responsible for the enhanced flow stability which allows for accumulation of large tensile strains. Additionally, at low stresses, thermal cycling produces higher rates of deformation as compared to isothermal creep. This is true for both Ti-6Al-4V and Ti-6Al-4V/10 vol% TiC in Figs. 1 and 2, even though the isothermal creep data were acquired at higher temperatures than the average (or effective) temperature during cycling.

At low stresses, provided there is not a significant amount of creep outside of the phase transformation range, the so-called superplastic slope ($\partial\Delta\varepsilon/\partial\sigma$) is a material-specific constant which depends on the internal stress generated during the transformation and the multiaxial creep law which governs the relaxation of that stress. The superplastic slopes found at low stresses in Figs. 1 and 2 ($\partial\Delta\varepsilon/\partial\sigma = 0.21$ GPa^{-1} and 0.11 GPa^{-1} for Ti-6Al-4V and Ti-6Al-4V/10 vol% TiC, respectively) are in agreement with our previous work on similar materials [2, 8].

As shown in Figs. 1 and 2 for Ti-6Al-4V and Ti-6Al-4V/10 vol% TiC, the linear flow law observed at low stresses diverges to a power-law at larger stresses (about 7 and 3 MPa, respectively). For both materials, the stress exponent found at high stresses during thermal cycling is similar to that measured during isothermal creep.

Discussion

Figs. 1 and 2 demonstrate two features of transformation superplasticity which are of interest for shape-forming of advanced structural materials. Provided that the applied external stress is small, thermal cycling results in (i) a stress exponent of unity, leading to exceptional flow stability and (ii) higher deformation rates compared to isothermal creep at similar temperatures. However, both of these advantages are restricted to low stresses; at higher stresses deformation diverges to non-linear behavior which offers no benefit over isothermal deformation. Thus, understanding the transition from Newtonian flow at low stresses to power-law deformation at higher stresses is of practical importance for selecting shape-forming process parameters. Maximizing the applied stress (and deformation rate) without compromising flow stability requires understanding of the micromechanics of deformation.

In this section, we present an analytical model for transformation superplasticity, which is applicable to Ti-6Al-4V and other materials which transform over a broad temperature range, and which is extended to the high stress regime. The model presented here is based on the classic work of Greenwood and Johnson [9], who formulated a successful model of transformation superplasticity at low applied stresses. They consider a material undergoing an isothermal phase transformation, in which there is a volume mismatch $|\Delta V/V|$ between the coexisting phases, which creates internal strains. They assume that one of the phases is substantially weaker than the other, and creeps according to a power-law relation:

$$\dot{\varepsilon}_{ij} = \tfrac{3}{2} \cdot A \cdot \sigma_{eq}^{n-1} \cdot \sigma_{ij} \tag{1}$$

where A is a temperature-dependent constant, σ_{eq} is the equivalent stress, n is the power-law stress exponent, and the subscripted variables are tensors. Considering the combined stress and strain state due to the transformation volume mismatch and an external stress σ applied in the z-direction, Greenwood and Johnson [9] derived the following constitutive law for deformation of the weaker phase in the z-direction:

$$\sigma_{zz} = \frac{\left(\Delta\varepsilon - \left(\tfrac{\Delta V}{V}\right)_{zz}\right)}{\left|\tfrac{\Delta V}{V}\right|} \cdot \left[\frac{3}{2} \cdot \frac{A \cdot \Delta t}{\left|\tfrac{\Delta V}{V}\right|}\right]^{-\tfrac{1}{n}} \cdot \left[1 - \frac{9}{2} \cdot \frac{\Delta\varepsilon \cdot \left(\tfrac{\Delta V}{V}\right)_{zz}}{\left|\tfrac{\Delta V}{V}\right|^2} + \frac{9}{4} \cdot \frac{\Delta\varepsilon^2}{\left|\tfrac{\Delta V}{V}\right|^2}\right]^{\tfrac{1-n}{2n}} \tag{2}$$

in which $\Delta\varepsilon$ is the strain increment in the z-direction after a complete transformation of duration Δt. The term $(\Delta V/V)_{zz}$ is the zz-component of the volume-mismatch tensor, and is a function of the spatial orientation of the transformation front. Greenwood and Johnson [9] assume that all phase orientations are equally likely; a macroscopic σ-$\Delta\varepsilon$ relationship can then be found by averaging Eq. (2) over the surface of a sphere. The exponent $(1-n)/2n$ on the last term of Eq. (2) makes this averaging procedure analytically intractable. Instead, Greenwood and Johnson linearize Eq. (2) by using only the first-order $\Delta\varepsilon$ terms of a Taylor's series expansion. They then average the linearized form of Eq. (2) over the surface of a sphere to obtain a simple constitutive law for transformation superplasticity:

$$\Delta\varepsilon = \frac{2}{3} \cdot \frac{5 \cdot n}{4 \cdot n + 1} \cdot \frac{\Delta V}{V} \cdot \left[\frac{2}{3} \cdot \frac{\Delta V}{V} \cdot \frac{1}{A \cdot \Delta t} \right]^{\frac{1}{n}} \cdot \sigma \qquad (3)$$

Eq. (3) has been successful in predicting the linear stress-dependence of $\Delta\varepsilon$, as well as the numerical value of the superplastic slope ($\partial\Delta\varepsilon/\partial\sigma$), for a number of allotropic metals and polymorphic alloys (see, for example, Refs. [1, 4, 6, 9]). However, Eq. (3) is derived with two assumptions which limit its applicability. First, Greenwood and Johnson [9] explicitly assume that the transformation occurs isothermally, as for allotropic metals. Second, the linearization of Eq. (2) limits the applicable range of Eq. (3) to small strain increments $\Delta\varepsilon$, or small stresses σ. In what follows, we discuss a generalization of the Greenwood and Johnson model which can be applied to alloys which transform over a range of temperatures (as for Ti-6Al-4V), and which is extended to include the non-linear, high-stress deformation regime.

At the outset of their derivation, Greenwood and Johnson [9] assume that the phase transformation occurs isothermally, and use a time-integrated form of Eq. (1) which gives rise to the $A \cdot \Delta t$ terms in Eqs. (2,3). However, their derivation can be followed without this time-integration, and using a time-differential of the volume mismatch tensor, which requires no stipulations regarding the time- or temperature-dependence of the creep parameters A, n, and the external stress σ. This derivation is the subject of a different, forthcoming publication [10], and leads to the following alternate form of Eq. (2):

$$\sigma_{zz} = \frac{(\dot{\varepsilon}_i - \dot{\varepsilon}_{zz}^M)}{\dot{\varepsilon}^M} \cdot \left[\frac{3}{2} \cdot \frac{A}{\dot{\varepsilon}^M} \right]^{-\frac{1}{n}} \cdot \left[1 - \frac{9}{2} \cdot \frac{\dot{\varepsilon}_i \cdot \dot{\varepsilon}_{zz}^M}{(\dot{\varepsilon}^M)^2} + \frac{9}{4} \cdot \frac{\dot{\varepsilon}_i^2}{(\dot{\varepsilon}^M)^2} \right]^{\frac{1-n}{2 \cdot n}} \qquad (4)$$

in which the strain increment $\Delta\varepsilon$ is replaced by an instantaneous strain rate $\dot{\varepsilon}_i$, and the volume mismatch $\Delta V/V$ and its tensor components are replaced by the rate at which the volume-mismatch is developing, $\dot{\varepsilon}^M$. The volume mismatch is assumed to develop in proportion to the fraction of transformed phase, f:

$$\dot{\varepsilon}^M = \dot{f} \cdot \left| \frac{\Delta V}{V} \right| \qquad (5)$$

To determine a macroscopic relationship between $\dot{\varepsilon}_i$ and σ, both sides of Eq. (4) need to be averaged over the surface of a sphere. The zz-component of the mismatch strain-rate tensor has the following orientation dependence [9]:

$$\dot{\varepsilon}_{zz}^M = \tfrac{1}{3} \cdot \dot{\varepsilon}^M \cdot \cos^2(\phi) \cdot \sin^2(\theta) + \tfrac{1}{3} \cdot \dot{\varepsilon}^M \cdot \sin^2(\phi) \cdot \sin^2(\theta) - \tfrac{2}{3} \cdot \dot{\varepsilon}^M \cdot \cos^2(\theta) \qquad (6)$$

Upon introducing Eq. (6), Eq. (4) is averaged over the surface of a sphere using:

$$\overline{X} = \frac{1}{\pi^2 \cdot r^2} \cdot \int_0^{\pi/2} \int_0^{\pi/2} X \cdot 2 \cdot \pi \cdot r^2 \cdot \sin\theta \cdot d\theta \cdot d\varphi \qquad (7)$$

where the quantity to be averaged, X, is the left- or right-hand side of Eq. (4). This procedure yields, after some manipulation:

$$\frac{2}{3} \cdot \sigma = \left[\frac{3}{2} \cdot \frac{A}{\dot{\varepsilon}^M} \right]^{-\frac{1}{n}} \cdot g\left(\frac{\dot{\varepsilon}_i}{\dot{\varepsilon}^M} \right) \qquad (8)$$

in which

$$g\left(\frac{\dot{\varepsilon}_i}{\dot{\varepsilon}^M}\right) = \int_0^{\pi/2} \left(\frac{\dot{\varepsilon}_i}{\dot{\varepsilon}^M} - \frac{1}{3} + \cos^2\theta\right) \cdot \left[1 - \frac{9}{2}\cdot\frac{\dot{\varepsilon}_i}{\dot{\varepsilon}^M}\cdot\left(\frac{1}{3} - \cos^2\theta\right) + \frac{9}{4}\cdot\frac{\dot{\varepsilon}_i^2}{\left(\dot{\varepsilon}^M\right)^2}\right]^{\frac{1-n}{2n}} \cdot \sin\theta \cdot d\theta \qquad (9)$$

By assuming, in analogy with Greenwood and Johnson's [9] original assumption of small strain increments, that the quantity $\dot{\varepsilon}_i/\dot{\varepsilon}^M$ is much less than unity, a Taylor's series expansion can be applied to the terms in the integrand of Eq. (9), and truncated to first-order in $\dot{\varepsilon}_i$. This makes the integral in Eq. (9) soluble in closed form, giving $g(\dot{\varepsilon}_i/\dot{\varepsilon}^M) = (\dot{\varepsilon}_i/\dot{\varepsilon}^M)\cdot(4n+1)/5n$, and yielding a linear constitutive relation between $\dot{\varepsilon}_i$ and σ for the limit of small applied stresses. When averaged over the duration of a cycle, this linear relationship ($\dot{\varepsilon} \propto \sigma$) is indeed observed for Ti-6Al-4V (Fig. 1) and for the Ti-6Al-4V/10 vol% TiC composite (Fig. 2) at low stresses.

Conversely, if we assume that $\dot{\varepsilon}_i/\dot{\varepsilon}^M$ is much larger than unity, Eq. (9) can again be solved in closed form, giving $g(\dot{\varepsilon}_i/\dot{\varepsilon}^M) = ((3\dot{\varepsilon}_i)/(2\dot{\varepsilon}^M))^{1/n}$. Introducing this result into Eq. (8), we recover a simple power-law expression, $\dot{\varepsilon}_i = A\cdot\sigma^n$, corresponding to Eq. (1) in uniaxial form. Thus, when the applied external stress is large and the resulting strain rates are rapid, the material obeys a power-law creep equation, with a power-law exponent identical to that for isothermal creep. This divergence from a stress-exponent of unity to n is indeed observed experimentally for both Ti-6Al-4V and the composite (Figs. 1 and 2).

The full, nonlinear constitutive relationship between $\dot{\varepsilon}_i$ and σ, valid at any applied stress (i.e., any strain rate), is given by Eqs. (8,9), which has the limits described above at small and large stresses. The function $g(\dot{\varepsilon}_i/\dot{\varepsilon}^M)$ can be found by numerically evaluating Eq. (9) for selected values of $\dot{\varepsilon}_i/\dot{\varepsilon}^M$ and a given numerical value of the stress exponent n, as shown in Fig. 3. For $n > 1$, $g(\dot{\varepsilon}_i/\dot{\varepsilon}^M)$ exhibits a clear transition between a linear- and a power-law in $\dot{\varepsilon}_i/\dot{\varepsilon}^M$, as described earlier.

With numerical data such as that in Fig. 3, Eq. (8) can be used to determine the instantaneous strain rate during the phase transformation, for any applied stress. At each moment during thermal cycling the stress σ is known, the mismatch strain-rate $\dot{\varepsilon}^M$ is calculated from Eq. (5) using physical data on the volume mismatch $\Delta V/V$ and the evolution of the phase fraction, f, and the temperature-dependent constant A is given by the isothermal creep law of the weak phase (in this case of Ti-6Al-4V, the high-temperature β phase). Using these parameters in Eq. (8) gives a numerical value for $g(\dot{\varepsilon}_i/\dot{\varepsilon}^M)$, which

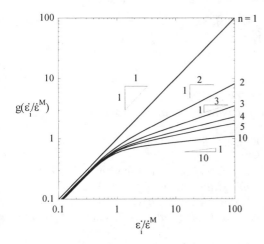

Figure 3: Numerical solutions of Eq. (9) for selected values of the creep stress exponent n.

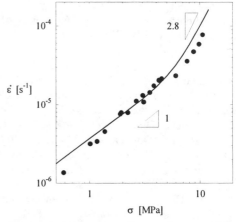

Figure 4: Thermal cycling strain rate of Ti-6Al-4V vs. tensile stress; experimental data points are compared with the model (Eqs. 8-10)

can then be compared with the numerical solutions of Eq. (9) (i.e., Fig. 3) to determine the instantaneous strain rate, $\dot{\varepsilon}_i$. The experimental strain rates (Figs. 1 and 2) represent the average strain rate during thermal cycling; comparing the model to the data thus requires that the instantaneous strain rate be averaged over the cycle duration:

$$\dot{\varepsilon} = \frac{1}{\Delta t} \cdot \int_0^{\Delta t} \dot{\varepsilon}_i \cdot dt \qquad (10)$$

In order to compare with the experimental thermal cycling data for Ti-6Al-4V (Fig. 1), the above model, which is valid at all strain rates and applied stresses, requires several input parameters. The creep law of the weak (β) phase is given by the data in Fig. 1, with stress exponent n = 2.8. The parameter A = K·exp(-Q/RT), where the temperature-independent constant K = 0.72 MPa$^{-2.8}$ is taken from the data in Fig 1, assuming that the creep activation energy is the same as for pure β-Ti, Q = 153 kJ/mol [1]. The parameters R and T are the gas constant and absolute temperature, respectively. The volume fraction of transforming phase, f, and the α/β phase volume mismatch ΔV/V are provided by Ref. [7] as a function of temperature, determined experimentally by high-temperature x-ray diffraction. Finally, the experimental thermal cycles were triangular between 840° and 980° C, providing a relationship between temperature and time.

With the input data described in the preceding paragraph, Eqs. (8,9,10) have been evaluated numerically for Ti-6Al-4V; the results are shown in Fig. 4, compared with the experimental data points. The model successfully predicts the stress dependence of the thermal-cycling strain rate, both in the linear, low-stress regime, and the power-law, high-stress regime. Furthermore, without the use of any adjustable parameters, the model is quite accurate in predicting the absolute value of the strain rate over the full range of stresses, within a factor of ~1.5.

Conclusions

We present experimental data on transformation superplasticity of Ti-6Al-4V and a Ti-6Al-4V/10 vol% TiC composite, acquired during thermal cycling through the α/β phase transformation range of Ti-6Al-4V. The relationship between stress and cycling strain rate is investigated at low stresses, where flow is Newtonian, as well as higher stresses, where non-linear, power-law flow behavior is observed. The power-law observed at high stresses has the same stress exponent as isothermal creep of the β phase of Ti-6Al-4V. We also present an analytical model, based on the work of Greenwood and Johnson [9], which is shown to capture the transition from linear to power-law flow behavior as stress is increased.

<u>Acknowledgements</u>- This study was supported by NSF SBIR #9901850, through a subcontract from Dynamet Technology, and the U.S. Department of Defense, through a graduate fellowship for C.S. Helpful discussions with W. Zimmer of Dynamet Technology are gratefully recognized.

References

1. D.C. Dunand and C.M. Bedell, *Acta Materialia*, 44 (1996), p. 1063.
2. C. Schuh, W. Zimmer, and D.C. Dunand. in *Creep Behavior of Advanced Materials for the 21st Century*: TMS, Warrendale PA. (1999), p. 61.
3. P. Zwigl and D.C. Dunand, *Acta Materialia*, 45 (1997), p. 5285.
4. C. Schuh and D.C. Dunand, *Acta Materialia*, 46 (1998), p. 5663.
5. S. Abkowitz, P.F. Weihrauch, and S.M. Abkowitz, *Industrial Heating*, 12 (September, 1993), p. 32.
6. P. Zwigl and D.C. Dunand, *Metallurgical and Materials Transactions*, 29A (1998), p. 2571.
7. W. Szkliniarz and G. Smolka, *Journal of Materials Processing Technology*, 53 (1995), p. 413.
8. R. Kot, G. Krause, and V. Weiss. in *The Science, Technology and Applications of Titanium*: Pergamon, Oxford. (1970), p. 597.
9. G.W. Greenwood and R.H. Johnson, *Proceedings of the Royal Society of London*, 283A (1965), p. 403.
10. C. Schuh and D.C. Dunand, (unpublished research, 2000).

Superplastic Tensile Behavior of Silicon Nitride

Naoki Kondo, Tatsuki Ohji and Yoshikazu Suzuki

National Industrial Research Institute of Nagoya,
Shimo-Shidami 2268-1, Moriyama-ku, Nagoya-shi 463-8687, Japan

Keywords: Activation Energy, Microstructure, Silicon Nitride, Stress Sensitivity, Superplastic Deformation

Abstract

Microstructural development, stress sensitivity and activation energy for superplastic tensile deformation of silicon nitride consisting of rod-shaped grains were investigated. Rod-shaped grains, which oriented randomly before deformation, aligned along tensile direction. Both the stress sensitivity and the activation energy increased during the deformation.

The principle deformation mechanism is considered to be the Non-Newtonian viscous flow of grain boundary glassy phase at the former and middle stages. At the last stage (more than 200 % elongation), however, superplastic behavior is considered to be mainly controlled by the solution - precipitation with 2-D nucleation.

Introduction

Superplasticity refers to an ability of polycrystalline solids to exhibit extremely large elongation in tension tests at elevated temperatures. Superplastic behaviors of silicon nitride-based ceramics have been reported [1]. At the early period of the research, the microstructure with very fine equiaxed grains was believed to be indispensable for achievement of superplasticity in silicon nitride, because equiaxed grains are preferable for grain boundary sliding which is major mechanism of superplasticity [2]. However, it was demonstrated that silicon nitrides consisting of rod-shaped β-silicon nitride grains can also exhibit a superplastic elongation [3].

In this paper, some additional features for superplasticity of silicon nitrides, i.e., changes in microstructure, stress sensitivity and activation energy against strain, were investigated. Based on the results, deformation mechanisms are discussed

Experimental Procedure

Silicon nitride samples sintered with 5 wt.% Y_2O_3 and 3wt. % Al_2O_3 were used in this study. Dog-bone shaped tensile specimens with a gage of 10 mm length and 3 mm diameter were prepared from the sintered body. The specimens were tensile deformed in 0.1MPa nitrogen atmosphere. Change in stress sensitivity and activation energy against strain was measured by step test. The stress sensitivity, n, was calculated by the following equation,

$$n = \frac{\ln(Cs_1/Cs_2)}{\ln(P_1/P_2)} \quad (1)$$

where P is the nominal stress and Cs is the cross head speed. For the measurement, cross-head speed was alternately changed between 0.013 and 0.018 mm/min at 1630 °C. The activation energy, Q, was calculated by the following equation,

$$Q = nR \frac{\ln P_2 - \ln P_1}{1/T_2 - 1/T_1} \quad (2)$$

where R is the gas constant and T is temperature. Temperature was changed from 1600 to 1630 °C at several strains, and cross-head speed was constant (0.015 mm/min). Microstructural developments was observed by SEM. SEM observation for the tensile deformed specimen was conducted on the cross section parallel to the tensile direction.

Results

Fig. 1 shows the true stress-true strain curve of the deformation. The deformation stress increment during deformation was observed. The specimen deformed under a stress of less than 10 MPa initially, but the stress increased slowly and then sharply, and finally reached ~ 37 MPa just before the fracture.

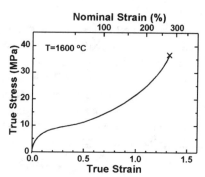

Fig. 1 True stress - strain curve obtained in the tensile deformation. Initial strain rate was 2×10^{-5}/s.

Fig. 2 shows the microstructural development during deformation. The specimen before deformation consisted of randomly oriented rod-shaped β-silicon nitride grains. During deformation, the grains aligned preferentially along the tensile direction according to the strain. Since the grains were elongated and aligned, it is quite difficult to know "real" microstructure by simple observation of the sectioning plane. Therefore, the stereological analysis, which was recently developed by Sato et al. [4], was applied to estimate "real" microstructure. The estimated average radius, aspect ratio and orientation angle are shown in Fig. 3. The average radius increased slightly. The average aspect ratio was almost constant during the former stage (0~140 %) of the

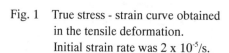

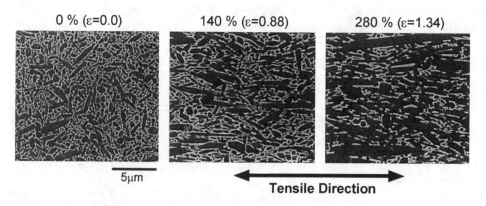

Fig. 2 Microstructural development during tensile deformation.

deformation, however, it increased largely during the latter stage (140~280 %). Decrement of average orientation angle means anisotropic microstructure formation.

Measured stress sensitivity and activation energy are shown in Fig. 4. The stress sensitivity, which was initially about 1.5, first slightly and then largely increased as strain increased, and reached over 2.2 just before the fracture. The activation energy, which was about 630 kJ/mol initially, was almost constant up to 180 % elongation, and then largely increased up to 1000 kJ/mol.

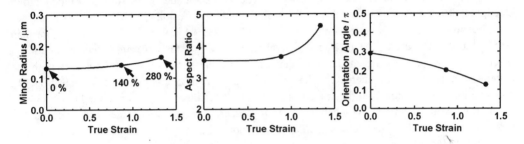

Fig. 3 Change in average radius, aspect ratio and orientation angle against strain.

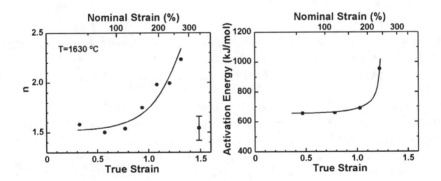

Fig. 4 Measured stress sensitivity (n) and activation energy against the strain.

Discussion

At the former ~ middle stages of the deformation (roughly less than 200 %), stress sensitivity was < 2. Non-Newtonian viscous flow is considered to be consistent with the result. The measured activation energy, 630 kJ·mol^{-1}, is slightly higher than that for the viscous flow (= grain boundary diffusion), 545 kJ·mol^{-1} [5, 6]. This is because additional activation energy for accommodation may be required for the deformation [7]. In addition, as known from Fig. 3, grain growth was not so activated because grain radius and aspect ratio were almost constant during the former stage of the deformation. This fact also supports non-Newtonian viscous flow, since the deformation did not require grain shape change. Therefore, superplastic behavior up to 200 % elongation is considered to be a non-Newtonian viscous flow of grain boundary glassy phase.

As stated previously, silicon nitride grains tend to align along the tensile direction during the deformation and the deformation stress becomes high. The non-Newtonian viscous flow must be suppressed, and instead the solution - precipitation becomes predominant [8].

At the last stage of the deformation (roughly more than 200 %), stress sensitivity exhibits ≥ 2. Wakai proposed a solution - precipitation creep model, where different rate-controlling processes yield different stress sensitivities [9]. According to his model, an interface - reaction controlled creep with two-dimensional (2-D) nucleation results in n ≥ 2. Activation energy also increased more than 1000 kJ·mol^{-1} at this stage. The activation energy for solution - precipitation creep, i.e., the sum of activation energies for grain boundary diffusion (545 kJ·mol^{-1} [5, 6]) and for grain boundary reaction (solution - precipitation) (686 ~ 772 kJ·mol^{-1}) [5 - 7], exhibits more than 1000 kJ·mol^{-1}. Therefore, the measured activation energy is consistent with that of solution - precipitation creep. Wakai's model also mentioned that the stress sensitivity increases with the increment of an activation energy for the solution - precipitation creep with 2-D nucleation [9]. The simultaneous increment of the stress sensitivity and activation energy correspond to his model. In addition, grain growth and elongation noticeably occurred at the latter stage of the deformation as known from Fig. 3. Therefore, superplastic behavior at the last stage is considered to be mainly controlled by the solution - precipitation with 2-D nucleation.

References

[1] F. Wakai, N. Kondo and Y. Shinoda, Current Opinion in Solid State and Materials Science, **4**, 461–465, (1999)
[2] T. G. Nieh, J. Wadsworth and O. D. Sherby, *Superplasticity in Metals and Ceramics*, Cambridge University Press, Cambridge, U.K., (1997)
[3] N. Kondo, F. Wakai, M. Yamagiwa, T. Nishioka and A. Yamakawa, Mat. Sci. Eng., **A206**, 45-48, (1996)
[4] E. Sato, N. Kondo and F. Wakai, Phil. Mag. A, **74**, [1], 215-28, (1996)
[5] M. Kitayama, K. Hirao, M. Toriyama and S. Kanzaki, Acta. Mater., **46**, 6541-50, (1998)
[6] M. Kitayama, K. Hirao, M. Toriyama and S. Kanzaki, Acta. Mater., **46**, 6551-57, (1998)
[7] R. Raj and P. E. D. Morgan, J. Am. Ceram. Soc., **64**, C143-145, (1981)
[8] S. Y. Yoon, T. Akatsu, E. Yasuda, J. Mater. Sci., **32**, 3813-19, (1997)
[9] F. Wakai, Acta Metall. Mater., **42**, 1163-72, (1994)

High Temperature Deformation of a Yttria-Stabilized Tetragonal Zirconia

K. Morita, K. Hiraga and Y. Sakka

National Research Institute for Metals, Sengen 1-2-1, Tsukuba, Ibaraki 305-0047, Japan

Keywords: Creep, Dislocation, Stress Exponent, TEM Observation, Zirconia

Abstract

The creep behavior of a high-purity tetragonal zirconia polycrystal containing 3 mol% yttria (3Y-TZP) is characterized with a stress exponent of $n \sim 2.7$ and a grain size exponent of $p \sim 2.0\text{-}3.0$ in a high stress region, and with $n \sim 1.3$ and $p \sim 2.0$ in a low stress region. These regions are connected with a transition region with $n \geq 5$, and accordingly the creep stress-strain rate relationship as a whole exhibits a sigmoidal feature. The apparent activation energy in the high and the low stress regions is the same, 580 kJ/mol, which corresponds to a value reported for lattice diffusion of cations. In the high stress region, evidence of intragranular dislocations revealed by TEM suggests that the motion of intragranular dislocations would contribute to the accommodation process of grain boundary sliding. In the low stress region where such dislocations were not observed, the combination of $n \sim 1.3$ and $p \sim 2.0$ suggests an intervention of lattice diffusion creep.

INTRODUCTION

High temperature deformation of high-purity yttria-stabilized tetragonal zirconia (Y-TZP) has been characterized by

$$\dot{\varepsilon} = A\,(\sigma^n/d^p)\exp(-Q/RT), \qquad (1)$$

where $\dot{\varepsilon}$ is the steady state creep rate, σ is the applied stress, d is the grain size, n is the stress exponent, p is the grain size exponent, Q is the apparent activation energy, R is the gas constant, T is the absolute temperature and A is a material constant. For fine-grained Y-TZP, creep behavior has been characterized by $n \sim 2.0$ at higher stresses and $n \sim 3.0$ at lower stresses [1-5].

Creep parameters should be determined from the instantaneous relationship among steady state creep rate, stress and grain size in Eq. (1). We have recently noted that both the deformation around the grip portion of tensile specimen and the occurrence of grain growth strongly affect the observed creep behavior, and tends to result in spurious creep parameters [6, 7]. In earlier studies [1-5], however, the creep parameters have been determined without correcting the effects of those factors.

In the present study, therefore, we performed creep test at wide range of experimental condition and determined the creep parameters of n, p and Q for the instantaneous relationship among stress, strain rate and grain size defined in Eq. (1). On the basis of the determined creep parameters and microstructural features observed TEM, we discussed possible deformation mechanisms of high-purity, fine-grained 3Y-TZP.

EXPERIMENTS
Material and Creep Tests

The material used was prepared from a high-purity Y-TZP powder containing 3 mol% yttria (Tosoh Co., Japan: <50 Al_2O_3, 50 SiO_2, <50 Fe_2O_3, 220 Na_2O in wt ppm). The powder was cold-isostatically pressed at 100 MPa followed by sintering at 1673 K in air. The sintering time was adjusted to obtain the grain sizes of 0.35 and 0.65 μm, where the size was determined as 1.56 times the average linear intercept length of grains in SEM micrographs. From the material, dog-bone-shaped flat specimens were machined with a gauge portion of $^t3\text{-}^w3\text{-}^l20$ mm. Tensile creep tests at constant load were performed at 1573-1673 K in an initial stress range of 3-80 MPa. True tensile displacement was measured with an electro-optical extensometer, which can directly monitor the length between targets made at both ends of the gauge portion [6].

To examine microstructure by TEM, conventional tensile creep tests at constant stress were conducted by using an Instron-type machine at 1673 K in a vacuum condition of 2×10^{-3} Pa and dog-bone-shaped flat specimens with a gauge portion of $^t3$-$^w3.5$-$^l10$ mm. In the experimental conditions examined, tensile elongation to failure was nearly the same value as that of earlier study [8]. For TEM observation, tensile specimens were deformed up to desirable strain and rapidly cooled to preserve substructure, which may be developed during deformation. On the Instron-type machine, it makes possible rapid cooling at an initial cooling rate of ~5.0 Ks^{-1} under loading condition. TEM specimens were cut from the deformed gauge portion, mechanically polished less than 100 μm in thickness and further thinned with an Ar ion-milling machine.

Creep Data Analysis

The monitored creep rate, $\dot{\varepsilon}_m$, was corrected for both instantaneous stress and grain size as follows [6, 7]. Assuming uniform deformation and constant volume in the gauge portion, the monitored creep rate $\dot{\varepsilon}_m$ was corrected for instantaneous stress as

$$\dot{\varepsilon}_\sigma = \dot{\varepsilon}_m \exp(-n\varepsilon_a), \qquad (2)$$

where $\dot{\varepsilon}_\sigma$ is the creep rate corrected only for instantaneous stress and the ε_a is the arbitrary strain. If grain growth can not negligible, Eq. (2) can be corrected for instantaneous grain size d_a at ε_a as

$$\dot{\varepsilon}_{\sigma d} = \dot{\varepsilon}_m \exp(-n\varepsilon_a)(d_0/d_a)^p$$
$$= \dot{\varepsilon}_m \exp(-n\varepsilon_a)\{[(d_0^r + kt)^{1/r}\exp(\alpha\varepsilon_a)]/d_0\}^p, \qquad (3)$$

where $\dot{\varepsilon}_{\sigma d}$ is creep rate corrected for both instantaneous stress and grain size, d_0 is the initial grain size, r is the grain growth exponent, k and α are the rate constants for static and strain-induced grain growth [9], respectively.

EXPERIMENTAL RESULTS
Creep Behavior

Figure 1 shows the variation of the corrected creep rate $\dot{\varepsilon}_{\sigma d}$ with true stress σ. The open circles, triangles and squares represent the data obtained at absolute temperatures T of 1673, 1623 and 1573 K, respectively. Although a decrease in the temperature sifts the $\dot{\varepsilon}_{\sigma d}$-$\sigma$ relationship toward lower creep rate and higher stress regions, it does not change the general feature of the $\dot{\varepsilon}_{\sigma d}$-$\sigma$ curves. Inspection of Fig. 1 shows that the creep behavior divides into three distinct regions with n~2.7 in the higher stress region, n≥5.0 in the intermediate stress region and n~1.3 in the lower stress region. Thus, the $\dot{\varepsilon}_{\sigma d}$-$\sigma$ relationship shows a sigmoidal feature similar to superplastic metals.

The value of grain size exponent was determined from the slope of $\ln(\dot{\varepsilon}_\sigma)$-$\ln(d_a)$ plots at T =1673 K. In the lower stress region with n~1.3, the grain size exponent takes a value of p~2.0 at σ= 3-5 MPa, while in the higher stress region with n~2.7, the grain size exponent takes values of p~2.0-3.0 at σ = 15-50 MPa.

Activation energy was determined at the stresses of 3, 5, 30 and 50 MPa. At each stress, the stress exponent takes the same value of ~1.3 for σ=3-5 MPa and ~2.7 for σ=30-50 MPa at whole temperature examined. As shown in Fig. 1, the activation energy determined from the $\ln(\dot{\varepsilon}_{\sigma d})$-$(1/T)$ plot takes a value of ~590 kJ/mol at σ=3-5 MPa and ~570 kJ/mol at σ=30-50 MPa. Considering the scatter of experimental data, it is reasonable to think that the activation energy takes the same value of ~580 kJ/mol in both the higher and the lower stress regions.

The Q-value of ~580 kJ/mol for creep deformation agrees well with the values of static grain growth for 3Y-TZP [10, 11] and of the high temperature deformation in cubic zirconia single crystal containing of 9.4 mol% Yttria [12]. In the earlier studies [11, 12], the Q-values have been ascribed to the lattice diffusion of cations.

As-sintered and Deformed Microstructures

Figure 2 shows the microstructures of (a) the as-sintered and (b)-(d) the deformed materials. Based on the $\dot{\varepsilon}_{\sigma d}$-$\sigma$ relationship shown in Fig. 1, creep test was carried out at 1673 K in the lower stress region with n~1.3 ((b)) and in the higher stress region with n~2.7 ((c) and (d)). Each test was terminated at a true strain of ~0.2, where quasi-steady-state region appears in the creep-rate curves.

Figure 2 clearly shows a difference in the microstructure. For the as-sintered material, dislocation was scarcely observed within most grains. By careful observation under two-beam condition, a very limited number of dislocations were observed in some grains as indicated by arrows. For the material deformed at 5 MPa, dislocation was not found within the grains as shown in Fig. 2(b). In contrast to the materials, a high density of dislocations was observed within the grains of the materials deformed at 30 and 50 MPa.

The intragranular dislocations were frequently observed around the multiple grain junctions. Densely aligned dislocations were observed within the grains labelled as P in Figs. 2(c) and (d). The dislocations are likely to pile-up against the obstacles such as a boundary or a sub-boundary. It should be noted that sub-boundaries were formed in the grain interior labelled as S in Fig. 2 (d).

DISCUSSION
Comparison with Previous studies

The present analysis of creep behavior revealed a sigmoidal feature in the $\dot{\varepsilon}_m$-σ relationship as shown in Fig. 1. The $\dot{\varepsilon}_{od}$-σ relationship is different from earlier one [1-5].

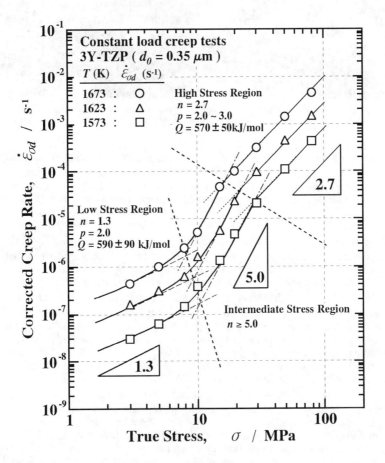

Fig. 1 Steady state creep rate, $\dot{\varepsilon}_{od}$, plotted against true stress for 3Y-TZP tested at 1573-1673 K.

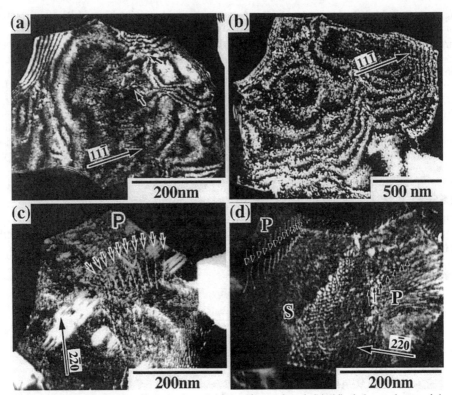

Fig. 2 Dark-field TEM micrographs of (a) as-sintered and (b)-(d) deformed materials. The materials were deformed at 1673 K, at (b) 5 MPa, (c) 30 MPa and (d) 50 MPa.

First, the transition of stress exponent from ~2.0 to ~3.0 reported in earlier studies did not appear at stresses up to 80 MPa. As discussed elsewhere [6, 7], the absence of the transition from n~2.0 to n~3.0 cannot be interpreted in terms of grain growth, because the effect of grain growth on creep behavior decreases with increasing stress. Since the transition stress increases with decreasing grain size in earlier studies [1-5], a n-value of ~2.0 may appear at stresses higher than 80 MPa. To confirm the existence of n~2.0 in 3Y-TZP, additional experiments at σ >80 MPa are necessary.

Second, a large stress exponent of ≥5, which did not appear in earlier studies [1-5], appears in the intermediate stress region. The reason why this region has not appeared in earlier studies is probable that a stress exponent has been determined by ignoring the effects of the deformation around the grip portion of tensile specimen and the occurrence of grain growth.

Third, in earlier studies [1-5], the stress exponent takes a value of ~3.0 at lower stresses, whereas in present study, the stress exponent tends to approach unity with a decrease in stress. As noted in our previous study [6], concurrent grain growth decreases creep rate and this leads to a spuriously larger stress exponent in particular at lower stresses. In the earlier studies [1-5], the stress exponent has been determined without correcting the effect of grain growth. It can be concluded that a part of the stress exponent of n~3.0 in the earlier studies is spurious value and a trend of decreasing n-value toward unity is essential at lower stresses in 3Y-TZP.

With regard to the values of grain size exponent and activation energy, the determination should also be required the instantaneous relationship among steady state creep rate, stress and grain size in Eq. (1). In earlier studies [1-5], however, p- and Q-values have been determined without correcting the effect of grain growth. This may lead to a spurious determination of p- and Q-values as well as n-value. It can be concluded, therefore, that the creep parameters of n, p and Q determined in the instantaneous relationship in Eq. (1) are essential in fine-grained 3Y-TZP: n~2.7, p~2.0-3.0 and Q~580 kJ/mol in the higher stress region, and n~1.3, p~2.0 and Q~580 kJ/mol in the lower stress region.

Deformed Microstructure

Although both the effect of the deformation around the grip portion and accuracy at lower stresses are different between the conventional constant stress creep test and the constant load creep test, a large stress exponent similar to the present study also appeared in the conventional constant stress creep test [13]. Thus, it can be thought that the observed microstructure (Figs. 2(b)-(d)) shows essential deformation substructure.

Densely aligned dislocations and sub-boundaries were observed only in the materials deformed in the higher stress region with $n\sim2.7$. In the as-sintered material, intragranular dislocation was rarely observed, and hence such dislocation substructures as seen in Figs.2 (c) and (d) did not preexist. If the dislocation substructures were due to some artifacts during specimen preparation, such dislocation substructures should also be found in both the as-sintered material and the material deformed at 5 MPa. This is not consistent with the present result of TEM observation (Figs.2 (a) and (b)). It can be concluded that the dislocation substructures observed in the higher stress region were developed during deformation.

For fine-grained Y-TZP, although only a limited number of intragranular dislocations were observed in deformed material [14], there were no experimental data providing significant intragranular dislocations as in the present study. In earlier study, TEM observation has been conducted on the specimens cooled without creep loading [14]. Since dislocation substructure is highly sensitive to unloading and cooling condition after deformation [15], the dislocation substructures developed during deformation may be annealed out during the cooling process in early study. Grain sizes finer than ~0.5 μm may also make it difficult to observe intragranular dislocations under a two-beam condition.

The lower level of flow stresses in fine-grained Y-TZP has also ruled out the possibility of intragranular dislocation activity. The present study, however, revealed the evidence of intragranular dislocation developed during deformation. Although detailed examinations on stress concentration arising from grain boundary sliding (GBS) are necessary, the present result means that stresses much higher than the creep flow stress should be generated in particular at multiple grain junctions during deformation.

Possible Deformation Mechanism

Careful analysis of creep data shows distinct three regions in the creep rate-stress relationship: $n\sim2.7$ in the high stress region, $n\geq5$ in the intermediate stress region and $n\sim1.3$ in the low stress region. To discuss the deformation mechanism of the intermediate stress region with $n\geq5$, additional experiments are necessary. In this study, therefore, we will deal only with the high and the low stress regions.

In the higher stress region with $n\sim2.7$, the present TEM observation revealed the evidence of intragranular dislocations. For fine-grained Y-TZP, only diffusion-related processes have been proposed as deformation mechanisms: e.g., an interface-reaction-controlled process [1, 3-5], a diffusional process incorporated with a threshold stress [2] and a interface-controlled Coble diffusion creep [16]. The observed creep parameters and the intragranular dislocations (Figs.2(c) and (d)), however, do not seem to be compatible with these diffusion-related processes. This suggests that another explanation seems to be necessary to provide for the higher stress region.

The higher n-value of ~3.0 accompanied with intragranular dislocation motion are usually associated with dislocation creep process for metals and some yttria-stabilized cubic zirconia single crystals [17]. Conventional dislocation creep, however, can be excluded as the deformation mechanism, because the creep rates strongly depend on the grain size, i.e., p-value takes a value of 2.0 or 3.0. As mentioned above, the intragranular dislocations were frequently observed around multiple grain junctions as typically shown in Figs. 2(c) and (d). This result suggests that GBS is predominant process of deformation and that the intragranular dislocation motion is likely to contribute to the relaxation of the stress concentration arising from the GBS.

The models involving intragranular dislocation motion were proposed by Ball and Hutchison [18] and Mukherjee [19] for superplastic metals. In these models, a stress concentration was created at multiple grain junctions or at boundary ledges during GBS, and this generates dislocations which pile-up against opposite grain boundaries. The rate-controlling process is the rate of removal of dislocations from the head of the pile-up by climb into and along grain boundary. If this is also the case for fine-grained 3Y-TZP, further GBS cannot occur without a concomitant process of the continuous relaxation of the stress concentration by dislocations and the continuous removal of pile-up dislocations. Unlike the models of Ball-Hutchison [18] and Mukherjee [19], the activation energy for creep deformation corresponds to the value for lattice diffusion of cations in the higher stress region. Present results suggest that the GBS process would be rate-controlled by the intragranular dislocation motion consisting of glide and climb process for fine-grained 3Y-TZP.

In the lower stress region, the stress exponent tends to approach unity. In contrast to the higher stress region, intragranular dislocation was not found in deformed material. The lower n-value of unity is usually associated with diffusion creep process. At very lower stresses, there seems to be no information available for discussing rate-controlling mechanism for fine-grained Y-TZP, but such data for coarse-grained cubic zirconia containing of 25 mol% yttria are available. The experimental result of the cubic zirconia exhibits n~1.2, p~2.2 and Q~550 kJ/mol for grain sizes of 2.6-14.4 μm [20]. Such creep behavior was interpreted in terms of Nabarro-Herring diffusion creep. The creep parameters of n~1.3, p=2.0 and Q~580 kJ/mol is quite similar to those of the coarse-grained cubic zirconia. The combination of the creep parameters suggests that lattice diffusional creep intervenes in the lower stress region in fine-grained 3Y-TZP.

CONCLUSIONS

For fine-grained, high purity 3Y-TZP, the present analysis of creep data led to the creep parameters of n=1.3, p=2 and Q~580 kJ/mol in the lower stress region, and n=2.7, p=2~3 and Q~580 kJ/mol in the higher stress region. These regions were connected with a transition region with $n \geq 5$. A Q-value of ~580 kJ/mol for creep deformation agrees with that of lattice diffusion of cations. In the higher stress region, creep deformation occurs through grain boundary sliding accommodated by intragranular dislocation process. The intragranular dislocation motion consisting of glide and climb would be the rate-controlling mechanism. In the lower stress region, the combination of the creep parameters suggests an intervention of lattice diffusional creep.

REFERENCES

1. A. H. Chokshi, Mat. Sci. Eng., **A166**(1993), pp. 119.
2. M. Jimenez-Melendo, A. Dominguez-Rodoriguez and A. Bravo-Leon, J. Am. Ceram. Soc., **81**(1998), pp. 2761.
3. D. M. Owen and A. H. Chokshi, Acta Metall. Mater., **46**(1998), pp. 667.
4. J. A. Hines, Y. Ikuhara, A. H. Chokshi and T. Sakuma, Acta Metall. Mater., **46**(1998), pp. 5557.
5. E. Sato, H. Morikawa, K. Kuribayashi and D. Sundararaman, J. Mat. Sci., **34**(1999), pp. 4511.
6. K. Morita and K. Hiraga, Scripta Metall., **42**(2000), pp. 183.
7. K. Morita, K. Hiraga and Y. Sakka, "The 8th International Conference on Creep and Fracture of Engineering Materials and Structures", ed T. Sakuma and K. Yagi. (2000), pp. 847.
8. K. Sasaki, K. Shimura, J. Mimurada, Y. Ikuhara and T. Sakuma, *ibid.*, pp. 377.
9. E. Sato and K. Kuribayashi, ISIJ Int., **33**(1993), pp. 825.
10. T. G. Nieh and J. Wadsworth, J. Am. Ceram. Soc., **72**(1989), pp. 1469.
11. J. Zhao, Y. Ikuhara and T. Sakuma, *ibid.*, **81**(1998), pp. 2087.
12. J. Martinez-Fernandez, M. Jimenez-Melendo and A. Dominguez-Rodoriguez, *ibid.*, **73**(1990), pp. 2452.
13. K. Hiraga and K. Nakano, Mater. Sci. Forum, **243-245**(1997), pp. 387.
14. S. Primdahl, A. Tholen and T. G. Langdon, Acta Metall. Mater., **47**(1999), pp. 2485.
15. U. Messerschmidt, D. Baither, B. Baufeld and M. Bartsch, Mat. Sci. Eng., **A233**(1997), pp. 61.
16. M. Z. Berbon and T. G. Langdon, Acta Metall. Mater., **47**(1999), pp. 2485.
17. D. Gomez-Gracia, J. Martinez-Fernandez, A. Dominguez-Rodoriguez, P. Eveno and J. Castaing, *ibid.*, **44**(1996), pp. 991.
18. A. Ball and M. H. Hutchison, J. Met. Sci., **3**(1969), pp. 1.
19. A. K. Mukherjee, Mat. Sci. Eng., **8**(1971), pp. 83.
20. D. Dimos and D. L. Kohlstedt, J. Am. Ceram. Soc., **70**(1987), pp. 531.

Corresponding Author: Koji Morita, E-mail: morita@momokusa.nrim.go.jp

Cavity Formation and Growth in a Superplastic Alumina Containing Zirconia Particles

K. Hiraga, K. Nakano, T.S Suzuki and Y. Sakka

National Research Institute for Metals, Sengen 1-2-1, Tsukuba-shi, Ibaraki 305-0047, Japan

Keywords: Alumina, Cavity Formation, Cavity Growth, Cavity Size Distribution

Abstract The present study on a tensile-deformed superplastic alumina reveals that the growth of cavities finer than the current grain size (fine cavities) into larger ones (large cavities) is constrained from the surrounding matrix. As a result of the constraint, the density of the large cavities becomes noticeably lower than that of the finer ones and shows non-linear dependence on strain, although the total cavity density, i.e., the sum of the fine and large cavity densities depends linearly on strain. Analysis of cavity size distributions also reveals the formation and growth laws of the large cavities.

1. Introduction

Concurrent cavitation is of practical importance in superplastic ceramics, because it degrades post-deformation strength [1] and relates closely to the ability of tensile deformation [2~7]. Earlier studies have reported that tensile failure in fine-grained zirconia and alumina is caused from microcracking occurring through extensive interlinkage among cavities grown to sizes larger than the matrix grains [2-4]. Recent studies [5~7] on alumina-base materials also show that such extensive cavity interlinkage takes place when the cavity separation distance, which is a function of cavity size and density, is decreased to a certain critical value. This means that the formation and growth rates of the cavities control the onset of microcracking and hence the ability of deformation. Irrespective of such importance, however, information is limited on the nature of cavitation in superplastic ceramics. The present study was therefore undertaken to obtain quantitative information on this issue. For this purpose, we examined cavity size distributions in a superplastic zirconia-dispersed alumina [8] as a function of strain.

2. Experimental Procedure

A 10-vol%-zirconia-dispersed alumina was fabricated from high purity α-Al_2O_3 (>99.99%) and 3-mol%-yttria-stabilized tetragonal zirconia (>99.98%) powders. The material had a relative density of 99.7% and equiaxed alumina grains with an average size of 0.45 μm, where the size was defined as 1.56 times the average intercept length of grains corrected for second phase dispersion. It was confirmed that amorphous phases did not coexist along grain boundaries nor at multiple grain junctions. Flat tensile specimens with gauge dimensions of $^l10$-$^t3$-$^w3.5$ mm were loaded to prescribed strains at temperatures (T) between 1400 and 1500 °C, at an initial strain rate ($\dot{\varepsilon}_0$) of 2×10^{-4}~2×10^{-3} s^{-2}. Tensile strain, ε, was monitored between the specimen shoulders. From the deformed specimens, a surface layer with a thickness of ~400 μm was removed for polishing to a 0.25 μm diamond finish. Cavity size distributions, where the size was defined as the area-equivalent circle diameter of each cavity, were measured on the center of the gauge portions with an optical microscope (OM) and a scanning electron microscope (SEM) equipped with an image analyzer. The measurement was performed on the as-polished surfaces for OM and on lightly heat-etched (1250 °C for 1 h) surfaces for SEM with resolutions of 0.8 and 0.02 μm, respectively. The total cavity numbers counted were 2×10^3 before deformation and $(1~10) \times 10^4$ after deformation. The measured cavity densities were corrected against areal expansion due to cavitation damage as $N=n/(1-A_c)$, where n is the cavity density on the examined surface and A_c is the cavitated area fraction of the

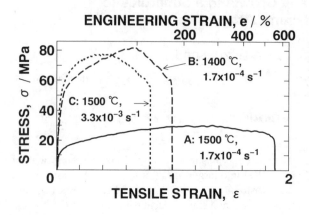

Fig. 1 Tensile flow behavior in the present material. The final grain sizes were 2.1, 0.78 and 0.80 μm under conditions A, B and C, respectively.

same surface. The grain size at each prescribed tensile strain was determined from experiments and interpolation based on a theoretical model for dynamic grain growth [9].

3. Results and Discussion

3.1 Tensile flow behavior

On the basis of tensile data obtained, we chose the conditions presented in Fig. 1. We regarded the deformation at $T=1500$ °C with $\dot{\varepsilon}_0=1.7\times10^{-4}$ s^{-1} (labeled as condition A) as the primary one, since this condition led to a maximum tensile elongation of 550%. To compare the effects of temperature and strain rate on cavitation, examination in the same level of flow stress is desirable. Such a requirement was found to be fulfilled between $T=1400$ °C combined with $\dot{\varepsilon}_0=1.7\times10^{-4}$ s^{-1} (condition B) and $T=1500$ °C combined with $\dot{\varepsilon}_0=3.3\times10^{-3}$ s^{-1} (condition C). Under these conditions, the material showed similar flow behavior. The average flow stress evaluated as $[\int\sigma(\varepsilon)d\varepsilon]/(\varepsilon_f-\varepsilon_y)$, was almost the same, 70±1 MPa, where $\sigma(\varepsilon)$ is the flow stress, ε_f is the failure strain and ε_y is the 0.2% yield strain. We should note, however, that failure strain is larger under condition B than under condition C, irrespective of the similar flow behavior and almost the same final grain size.

3.2 Deformed microstructure and cavity size distribution

As shown in Fig. 2(a), cavities were observed to nucleate mainly at multiple grain junctions and grow toward the matrix grain size. These cavities were also found to grow exceeding the current grain size as seen in Fig. 2(b). Such a sequence of cavity formation and growth is consistent with earlier studies on alumina- and zirconia-base materials [2~7, 10, 11]. At strains up to ~0.85ε_f, more than 95% of the cavities larger than the matrix grains had aspect ratios (=major axis length/minor axis length) smaller than 2.0 and there was a little indication of coalescence (Fig.2 (b)). The onset of extensive cavity interlinkage was found at ~0.9ε_f under each condition. From these observation, we confined the examination at strains <~0.9ε_f. Figure 2(c) shows a typical cavity size distribution after deformation. The numerical density of cavities finer than the current grain size (termed *fine cavities*) is noticeably higher than that of cavities larger than the grain size (termed *large cavities*), and accordingly the size distribution as a whole exhibits a bimodal feature. The feature indicates that the fine cavities do not grow readily to the large ones. This means equivalently that cavity growth exceeding the current grain size is constrained by the surrounding matrix.

3.3 Cavity formation

In the following examination, we used the local true strain, $\varepsilon_l=\ln(S_0/S)$, as the extent of deformation to avoid interference from deformation at the specimen shoulders and from broad necking at large strains. Here, S_0 is the cross-sectional area of the gauge portion before deformation, and S is the cross-sectional area of the gauge portion at which the measurement by both SEM and

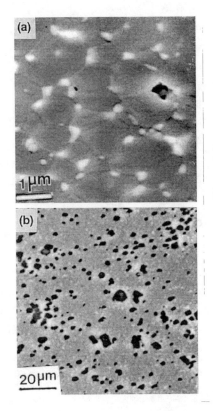
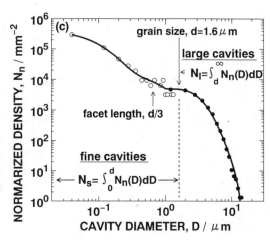

Fig. 2 Microstructure and cavity size distribution in a specimen deformed under condition A to a tensile strain of 1.34. (a) and (b) are obtained by SEM and OM, respectively. The stress axis is horizontal. In (c), D and N_n are the mean diameter and the numerical density of cavities in the size range of $D\pm\Delta/2$, respectively, where Δ is the size interval. The density is normalized against the size interval as $N_n=N(0.8/\Delta)$, where N is the density measured with $\Delta=0.8$ μm for $D\geq0.8$ μm and $\Delta=0.08$ μm for $D<0.8$ μm.

OM was performed. It was confirmed that, although ε_l and ε showed a linear relationship, scattering in the data presented below was minimized by using the former instead of the latter.

Figure 3 represents the total cavity density, N_t, defined by Eq. 1 as a function of strain.

$$N_t=N_s+N_l, \qquad (1)$$

where N_s and N_l are the numerical densities of the fine and the large cavities, respectively. The total density increases linearly with increasing strain under the respective tensile conditions. The rate of cavity nucleation can thus be represented as

$$dN_t/d\varepsilon_l=\alpha, \qquad (2)$$

where α is the rate constant. The data show that the nucleation rate is increased by a decrease in temperature or an increase in strain rate. Almost the same nucleation rate under conditions B and C is consistent with the flow behavior seen in Fig. 1, since existent models [12] predict that the rate of cavity nucleation depends strongly on stress.

In contrast to the total density, it was found that the density of the large cavities is not a linear function of strain (Fig. 4) nor of the density of the fine cavities. This is an additional support to the constraint for the cavity growth exceeding the current grain size. Another point of Fig. 4 is the difference between conditions B and C. The data indicate that the large cavities are formed more slowly under condition B than under condition C, irrespective of nearly the same flow stress level (Fig. 1) and nucleation rate (Fig. 3) under these conditions. Thus, the effects on the formation of the large cavities are different between temperature and strain rate.

To elucidate the origin of such formation behavior, we examined the formation rates of the large cavities. For this purpose, we used an analytical function, $y=a+bx+c\exp(-dx)$, because the function

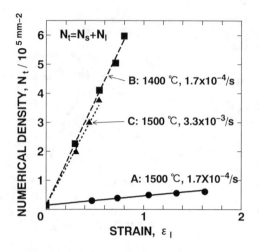
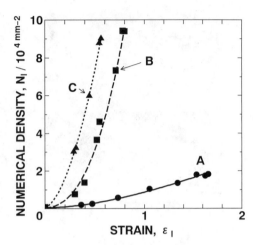

Fig. 3 Total cavity density, N_t, as a function of strain.

Fig. 4 The density of large cavities, N_l, as a function of strain.

was found to give the best fit with r^2-coefficients larger than 0.99 simultaneously to the formation data under conditions A trough C (Fig. 4). A theoretical support to this empirical fitting procedure will later be found in Eq. 4.

In Fig. 5, we find that the formation rate, $dN_l/d\varepsilon_l$, is proportional to the density of the fine cavities, and that the extrapolation of the linear relationship to the N_s-axis yields a positive intercept. This leads to the following formation law for the large cavities:

$$dN_l/d\varepsilon_l = \beta(N_s - \delta), \qquad (3)$$

where β and δ are constants for a given combination of temperature and strain rate. This equation means that the formation of the large cavities from the fine ones has a stochastic nature, where β represents the probability that the fine cavities leave the constraints from the surrounding matrix. The appearance of a positive δ-value indicates that a certain amount of deformation is necessary for cavity nuclei to grow into sizes close to the current grain size. Thus, δ represents the incubation for the appearance of cavities being ready for the growth exceeding the current grain size, and it can be assumed to depend on the growth rates of the fine cavities. Combining Eqs. 1 through 3 and $N_{t0} = N_{s0} + N_{l0}$ at $\varepsilon_l = 0$, we obtain

$$N_l = (N_{t0} - \delta - \alpha/\beta) + \alpha\varepsilon_l + [N_{l0} - (N_{t0} - \delta - \alpha/\beta)]\exp(-\beta\varepsilon_l), \qquad (4)$$

for the formation of the large cavities. Table 1 lists the α-, β- and δ-values evaluated from Figs.3 and 5. From the data, it is found that $(N_{t0} - \delta - \alpha/\beta)$ in Eq. 4 takes a negative value and the other terms depending on strain take positive values. Equations 3 and 4 indicates that the density of the large cavities increases as α or β increases, or as δ decreases.

Equations 2 through 4 and the data listed in Table 1 give the following explanation for the temperature- and strain-rate-dependence of the cavity formation. When a decrease in temperature and an increase in strain rate yield the same increment of flow stress level (Fig. 1), cavity nucleation is accelerated in the same way (Fig. 3), resulting in a general trend of enhanced formation in the large cavities (Fig. 4). The increase in flow stress is also accompanied by an increase in β, i.e., the probability of the formation of the large cavities. This gives an additional contribution to the enhanced cavity formation. However, there is a difference in the β-value between conditions B and C, suggesting that the probability does not depend only on flow stress but also on temperature and

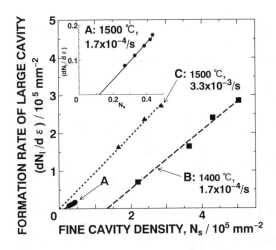
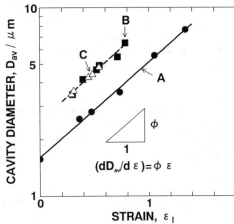

Fig. 5 Formation rate of the large cavities, $dN_l/d\varepsilon$, as a function of the density of the fine cavities, N_s.

Fig. 6 Average cavity diameter, D_{av}, as a function of strain.

/or strain rate itself. For the given increment of flow stress level (Fig. 1), the effects of temperature and strain rate on δ, i.e., the incubation for the formation of the large cavities are opposite: the δ-value increases by a factor of 10 under condition B, whereas it decreases by a factor of 10^3 under condition C. This implies that the growth of cavity nuclei to the current grain size may be controlled by a thermally activated process, i.e., diffusion. As a result of the smaller β- and the larger δ-values under condition B, the large cavities are formed more gradually than under condition C.

3.4 Cavity growth in sizes larger than the current grain size

To get information on the growth of the large cavities without interference from continuous cavity formation, we evaluated the average diameter, D_{lav}, of the largest 500 cavities per square millimeter, which is the density of the large cavities before deformation, using a method described elsewhere [6, 7]. As plotted in Fig. 6, a linear relationship was found between ln (D_{av}) and ε_l under the respective loading conditions. This leads to the following growth law:

$$D_{av}(\varepsilon) = D_{av0} \exp(\phi\varepsilon), \qquad (5)$$

where D_{av0} is the diameter from which cavity growth obeys this equation and ϕ is the growth rate constant. This growth law has also been found in a fine-grained magnesia-doped alumina [6, 7] and a superplastic tetragonal zirconia [13], and is compatible with that for spherical holes in a plastic continuum. This is not contradictory to the fact that the grains of ceramics are rigid, because the aggregate of such rigid grains surrounding the large cavities behaves as a plastic body in superplastic ceramics through enhanced grain boundary sliding accommodated by diffusion. As seen in Fig. 6, D_{av0} takes a larger value under conditions B and C than that under condition A, suggesting that the preexistent large cavities may grow rapidly during initial loading toward the heightened level of flow stress (Fig. 1). As listed in Table 1, however, ϕ takes the same value, 1.2, under conditions A through C. Thus, the growth rate of the large cavities is independent of the deformation conditions.

3.5 Tensile ductility

The close relationship between the cavity separation distance and the onset of extensive cavity interlinkage was also confirmed in the present material: the interlinkage was found to occur when

Table 1 Evaluation of parameters relating to cavity formation and growth.

condition	$T/°C$	$\dot{\varepsilon}_0/s^{-1}$	ε_f	$\alpha/10^4$ mm^{-2}	β	$\delta/10^4$ mm^{-2}	ϕ
A	1500	1.7×10^{-4}	1.88	3.2	0.55	1.3	1.2
B	1400	1.7×10^{-4}	1.00	72	0.77	13	1.2
C	1500	3.3×10^{-3}	0.81	64	0.98	0.001	1.2

the separation distance, of which initial value was 63 μm, decreased to 3~5 μm. This means that the rates of formation and growth of the large cavities strongly control the tensile ductility under the present loading conditions, since the separation distance is a function of cavity size and density, and since the onset strain of extensive cavity interlinkage corresponds to ~$0.9\varepsilon_f$ as described before. From this basis, the decreased ductility under conditions B and C can be attributed mainly to the enhanced formation of the large cavities and subsidiarily to an increase in the D_{av0}-value. The former arises from the increase in the cavity nucleation rate (α) and the probability that the fine cavities leave the constraints from the surrounding matrix (β). The difference between conditions B and C is due to the difference in the probability and in the incubation for cavity nuclei to grow into sizes close to the current grain size (δ). For the same increment of flow stress, the lowered temperature tends to suppress the increase in the β-value and noticeably increases the δ-value. For this reason, the formation rate of the large cavities is lower under condition B than under condition C, and thereby tensile ductility is larger under the former than under the latter.

4. SUMMARY

During tensile deformation in a superplastic alumina, the growth of cavities finer than the current grain size (*fine cavities*) into larger ones (*large cavities*) is constrained from the surrounding matrix. The formation rate of the large cavities is characterized with the rate constant of cavity nucleation, α, the incubation for cavity nuclei to grow into sizes close to the current grain size, δ, and the probability of such cavities leaving the constraint from the matrix, β. The present data show that α is almost determined solely on the level of flow stress, whereas β and δ depend also on temperature, in particular the latter does. The data also show that, since the growth rate is insensitive to loading conditions, the temperature- and the strain-rate-dependence of α, δ and β strongly affects the tensile ductility of the present material.

REFERENCES

[1] F. Wakai and H. Kato and T. Nagano, Adv. Ceram. Mater., 3(1988), p. 71.
[2] D. J. Schissler, A. H. Chokshi, T. G. Nieh and J. Wadsworth, Acta metall., 39(1991), p. 3227.
[3] A. H. Chokshi, T. G. Nieh and J. Wadsworth, J. Am. Ceram. Soc. 74(1991), p. 869.
[4] Y. Yoshizawa and T. Sakuma, Acta metall. mater. 40, (1992), p. 2943.
[5] K. Hiraga, K. Nakano, T.S.Suzuki and Y.Sakka, Scr. Mater, 39(1998), p. 1273.
[6] K. Hiraga, K. Nakano, Y. S. Suzuki and Y. Sakka, Mater. Sci. Forum, 304-306, 431 (1999).
[7] K. Hiraga, Y. Sakka, T. S. Suzuki and K. Nakano, Key Eng. Mater., 171-174, 763 (2000).
[8] K. Nakano, T. S. Suzuki, K. Hiraga and Y. Sakka, Scripta mater., 38(1998), p. 33.
[9] B-N. Kim, K. Hiraga, Y. Sakka and B-W. Ahn, Acta mater., 47(1999), p. 3433.
[10] S. Primdahl, A. Tholen and T. G. Langdon, Acta metall. mater., 43(1995), p.1211.
[11] D. M. Owen, A. H.Chokshi and S. R. Nutt., J. Am. Ceram.Soc., 80(1997), p. 2433.
[12] H. Riedel, Fracture at High Temperatures, Springer-Verlag, Verline, 1988, p. 67.
[13] K.Hiraga and K. Nakano, Mater. Sci. Forum 243-245(1997) p. 387.
e-mail address: hiraga@nrim.go.jp

Superplastic Deformation of a Duplex Stainless Steel

D. Pulino-Sagradi, R.E. Medrano and A.M.M. Nazar

Faculdade de Engenharia Mecanica, Universidade Estadual de Campinas,
Caixa Postal 6122, 13083-970 Campinas SP, Brazil

Keywords: Mechanical Properties, Plastic Deformation, Stainless Steel, Superplasticity

ABSTRACT

Superplastic behavior of a duplex stainless steel has been studied at 1223 K (950°C) and 1253 K (980°C) for initial strain-rates ranging from 2×10^{-4} s^{-1} to 8×10^{-3} s^{-1}. In the given range of temperature, it was observed that the elongation increased as the strain rate decreased. The best superplastic conditions were obtained at 1253 K where the samples presented homogeneous deformation along the gage length. Under this temperature the maximum elongation (900%) was obtained for an initial strain-rate of 4×10^{-4} s^{-1}. Those results were also confirmed by determination of the strain-rate sensitivity coefficient, m, which presented the highest value at 1253 K and clearly increased with decreasing strain-rate. Microstructure studies and previous information about the amount of cavities suggest that the deformation process occurs by a superplastic grain boundary sliding mechanism.

INTRODUCTION

The superplastic behavior in duplex stainless steel has been frequently studied not only because of its useful properties like high strength and superior corrosion resistance, but also by its unusual high ductility at high temperatures, which facilitates deformation processing [1]. The superplastic properties of this material have been extensively reported [2-7] and most of the experiments have been concentrated in the measurement of elongation after previous thermomechanical treatments.

The purpose of this paper is to report a study of the mechanical properties of a duplex stainless steel in order to analyze its superplastic behavior. Efforts were concentrated in the direction of obtaining measurements of elongation, uniformity of deformation throughout the sample, strain-rate sensitivity coefficient, and also in the study of the microstructure.

EXPERIMENTAL

Experiments were performed on samples of a microduplex stainless steel with the following composition: 0.018%C, 0.61%Si, 1.36%Mn, 0.04%P, 0.007%S, 6.42%Ni, 22.48%Cr, 3.05%Mo, 0.16%N. The material for preparation of the samples consisted of cold-rolled sheets of 3.0 mm in thickness. No additional treatment was done before testing. The alloy is characterized by a duplex ferrite (α) - austenite (γ) structure with a phase ratio of approximately unity and with an average phase dimension of 6x20x40 μm.

The samples were machined from the sheets parallel to the rolling direction with a measured gage length, gage width and fillet radius of 5.2, 7.0, and 2.4 mm, respectively. By measured gage length we means the initial length of the smallest cross-section area region of the sample. Although this sample geometry has the inconvenience of producing some deformation in the rest of the sample (grips) [8], it is the most widely used for superplasticity studies (small length-to-width ratio). The reasons for using this geometry are: (i) practical limits on cross-head travel that restrict the specimen length to be pulled to failure, and (ii) combination of flow stress and practical limits on load cell accuracy that can also limit the use of smaller widths.

Tensile tests were conducted to determine elongation to failure and to study deformation homogeneity. Strain-rate change and stress-relaxation tests were also carried out to determine the strain-rate sensitivity coefficient, m. Both tests were performed at 1223 K and 1253 K in a MTS servohydraulic machine, model 810A, using a stroke control mode. Time for temperature stabilization was ten minutes.

In the tensile tests, the samples were pulled to failure with a constant ram speed ranging from 1.04×10^{-3} to 4.16×10^{-2} mm/s. Deformation homogeneity was evaluated from the variation in cross-section area along the gage-length of the fractured samples. After tests, the longitudinal surface of the gage-length of each specimen was polished and etched for metallographic analysis [9].

In the strain-rate change tests and stress relaxation tests, the samples were initially deformed to an elongation of 50% at the lowest strain-rate to obtain a stabilized superplastic microstructure. The m-values from strain-rate change tests were determined resorting to Backofen's method [10]. The stress relaxation tests allow obtaining m-values from stress versus stress-rate curves [9].

In an analysis of superplasticity by using the Finite Element Method [8], it has been shown that plastic deformation occurs also in the gripping region. Therefore, the real gage-length is larger than the measured gage-length as has been defined before. In that paper, it has also been shown that differences between real and measured gage length are important only in the early stages of deformation, a question not studied in the present work. Due to the fact that the real gage-length is undetermined, in the present experiments elongation is used to represent the variable associated with strain. Elongation is defined as the ratio between the ram displacement and measured gage-length, which is well determined.

RESULTS

Tensile tests

Load versus displacement curves depend on deformation conditions such as temperature and strain-rate. Curves of the "apparent" true stress versus elongation are represented in Fig. 1. The apparent true stress is calculated from the measured gage-length. It is not possible to represent the true stress because at early stages of deformation the real gage-length is unknown [8] and at late stages the cross-section of the samples is undetermined due to cavitations [11]. Nevertheless, at intermediate stages, this apparent true stress is close to the real true stress (flow stress).

Data is generally represented in the literature using the so-called strain-rate as a parameter. This strain-rate is the ratio between the ram speed and the measured gage-length. It can be seen in Fig. 1

that most of the stress versus elongation curves shows that the flow stress and its variation decrease when elongation increases. However, in the test with the highest elongation, there is a small and steady increase of the flow stress as deformation proceeds.

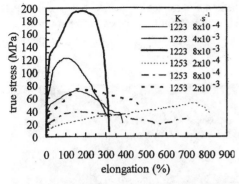 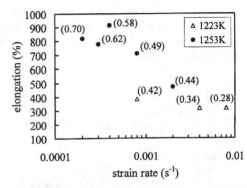

Fig. 1. True stress versus elongation curves for 1223 K and 1253 K at several initial strain-rates.

Fig. 2. Elongation to failure as a function of initial strain-rate for 1223 K and 1253K. The numbers close to the points indicate m-values.

From tensile tests, results of elongation to failure obtained from selected curves are represented in Fig. 2. It can be noticed the effect of strain-rate consisting in that the decrease of strain-rate causes an increase of elongation. At 1253 K the highest value around 900% was attained at 4×10^{-4} s^{-1}. It can be noticed that elongation tends to increase as test temperature increases: this effect is clearly seen comparing the results obtained at the strain-rate of 8×10^{-4}s^{-1} at both temperatures.

Deformation homogeneity

Fig. 3 presents the reduction in area at fracture, along the sample gage-length at different strain-rates and temperatures. Reduction in area is defined as the difference between initial and final areas divided by final area. This definition is not the most common in the literature, but the advantage of using this definition stands in the fact that it permits a direct comparison with elongation (when the volume is constant, both variables are the engineering strain). Generally, the tests performed at 1223 K exhibited a maximum deformation at the fracture surface, indicating necking. On the other hand, necking was not observed at 1253 K, showing that deformation was homogeneous.

Strain-rate sensitivity coefficient, m

The m-values determined by strain-rate change tests increased with the decrease of strain-rate. For strain-rates decreasing from 8×10^{-3}s^{-1} to 2×10^{-4} s^{-1} the m-values vary respectively from 0.25 to 0.70 for 1253 K and from 0.28 to 0.49 for 1223 K. It can be seen in Fig. 2 that the m-values at 1253 K are higher than those obtained at 1223 K. On the other hand, comparing results obtained at both temperatures we find that there are two points with a quite similar elongation (~400%), and m-values (0.42-0.44).

Stress-relaxation tests provided mean m-values of 0.56 for 1253 K and of 0.41 for 1223 K, both for strain-rates lower than 10^{-3}s^{-1}. These values are independent of elongation and of the strain-rate acting during the loading stage.

Microstructure

Metallographic studies of the as-received material showed a microstructure composed by elongated austenite and ferrite phases (Fig. 4). At their boundaries the presence of the sigma phase is observed. These elongated grains remain when the material is heated to test temperatures. As expected [9], there is a decrease in the amount of sigma phase, which is present in boundaries between ferrite and austenite phases, triple points and also inside the austenite phase (Fig. 5). The sigma phase is practically nonexistent at 1253 K.

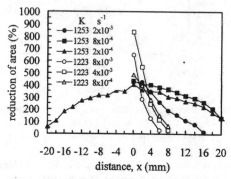

Fig. 3. Deformation along gage-length of tested samples. Position x=0 refers to fracture surface.

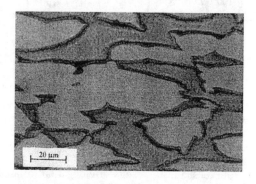

Fig. 4. Microstructure of the as-received material (longitudinal section).

After deformation the microstructure changed to an equiaxial and more refined structure as showed in Fig. 6 for a sample pulled to fracture at 1253 K (strain rate $2 \times 10^{-4} \, s^{-1}$). The microstructure of deformed samples shows cavitations at boundaries between similar and different phases (Fig. 6).

Previous studies [11] indicate that at 1253 K cavitations occurred along the entire gage-length, whereas at 1223 K cavitations spread near the surface of fracture. The amount of cavitations at 1253 K is greater than at 1223 K as shown in Fig. 7 of this reference.

DISCUSSION

The main purpose of this work has been the study of the best condition for superplasticity of the as-received material. Early experiments were performed at 1173 K and 1223 K with an extended range of strain-rates. The choice of these temperatures was based on Maehara´s testing temperatures [2]. Due to the fact that maximum elongation was lower than 400%, the temperature was increased to 1253 K to increase superplastic deformation. In this range of temperature and for the used strain-rate, it was observed a large variation in the degree of superplastic behavior.

Our measurements indicate that there is a maximum in elongation at 1253 K for an initial strain-rate close to $4 \times 10^{-4} \, s^{-1}$. A quadratic fitting of all points of the curve signals a maximum at $3.2 \times 10^{-4} \, s^{-1}$. This maximum is not well defined probably due to deviation of elongation in different samples. It is interesting to compare these elongation results with m values. Strain-rate sensitivity m continuously increases as strain-rate decreases. It seems that there is not a maximum in the m value in this range of strain-rate. Elongation follows this trend for the most part, except in the region close to the apparent maximum at 1253 K. We notice that measurements of m are done after 50% elongation when the structure of the material is not yet totally recrystallized, while in elongation measurements the test goes to fracture when the material is fully recrystallized. This occurrence could explain the slightly decrease of elongation with the increase of *m*-value at strain rates under 4×10^{-2} s^{-1}.

Comparison of measurements of elongation and reduction in area are of a particular interest. The two variables should be equal if there is not deformation of the material in the grip region and the volume remains constant. Since cavitations are observed, the last condition is not verified and the area measurement after deformation is larger than the real one. Thus reduction in area gives an undervalued estimation of the real engineering strains. However, reduction in area is a valuable parameter for characterizing the local deformation on the specimen. Moreover, ram displacement is much larger than change in length in the gage region due to deformation in the grip zone. This fact yields to an overvalued elongation estimative of the real engineering strain, but at large deformations (superplasticity) both variables are practically equal (strain at grips is negligible in comparison with total elongation).

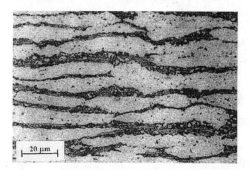

Fig. 5. Microstructure of the material (longitudinal section) before test initiation at 1223 K.

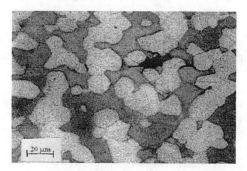

Fig. 6. Microstructure of the material (longitudinal section) pulled to failure with an initial strain-rate 2×10^{-4} s^{-1} at 1253 K.

Measurements of reduction in area along the gage length show that deformation is fairly uniform at 1253 K, but it is not uniform at 1223 K. This result points to the presence of a neck at the lower temperature. This neck becomes less developed as elongation (or parameter *m*) increases for both temperatures, and when *m* is 0.50 or larger the neck practically disappears.

Measurements of elongation, parameter *m*, and reduction in area are good variables to measure the degree of superplasticity. Another variables are necessary to assess deformation mechanisms.

In the recrystallization model [3] the sigma phase favors superplasticity. However, the extension of this phase is larger when the degree of superplasticity is smaller (at 1223 K), contradicting this model. Furthermore, experiments in another duplex stainless steel [12] indicate that the presence of sigma phase reduces the elongation to fracture. The reduction of the degree of superplasticity (elongation) at a lower temperature (1223 K) could be attributed to the presence of this phase at this temperature.

In the case of the best superplastic conditions, the slow increase of the true stress as deformation increases indicates the presence of a mechanism of grain growth instead of recrystallization.

In a similar material Tsuzaki et. al. [7] found that the dislocation density, which is necessary to promote recrystallization, is low during superplastic deformation. Recrystallization occurs in early stages and deformation proceeds by grain boundary sliding.

Cavitations are associated to grain boundary sliding, this being the mechanism producing cavitations at grain boundaries. As reported in reference [11], the presence of this mechanism is supported by the micrograph of cavities shown in its Fig. 7 and also by the observed increase of the degree of superplasticity when the volume of cavities increases.

Finally, the fact that the stress relaxation experiments give a unique curve in stress-rate versus stress diagram indicates that the "hardness" state, as defined by Hart [13] is constant. Following Hart, it means no hardening is produced during deformation in agreement with Tsuzaki et. al.

CONCLUSIONS

The best superplastic behavior was obtained at 1253 K with a strain-rate of 4×10^{-4} s^{-1}, when the material presents a homogeneous deformation throughout the gage length with a maximum elongation of 900%.

The strain-rate sensitivity coefficient constitutes an appropriate parameter to describe the superplastic behavior indicating sensible variations according to the initial strain-rate as well as temperature.

Microstructure studies together with previous reports on the volume-fraction of cavities support the considerations that the mechanism of grain boundary sliding is the main responsible for the superplasticity of duplex stainless steel.

ACKNOWLEDGMENTS

We thank Dr. Oscar Ruano for providing us the material used in our experiments and for helpful discussions, and Dr. Alain J. Isoré of Electrometal SA for providing us the chemical analysis of the material. Financial support from the São Paulo State Foundation (FAPESP) is gratefully acknowledged.

REFERENCES

1. J.O. Nilsson, Mater. Sci. Tech. **8,** (1992) 685-700.
2. Y. Maehara, Trans. ISIJ **25**, (1985) 69-76.
3. Y. Maehara and Y.Ohmori, Metall. Trans.**18A**, (1987) 663-672.
4. Y. Maehara. Trans. ISIJ **27**, (1987) 705-712.
5. K. Osada, S. Ueko and K. Ebato, Trans. ISIJ **28**, (1987) 713-718.
6. K. Osada, S.Ueko, T. Tohge, M. Noda and K. Ebato, Trans. ISIJ **28**, (1988) 16-22.
7. K. Tsuzaki, H. Matsuyama, M. Nagao and T. Maki, Mater. Trans. JIM **31**, (1990) 983-994.
8. M.A. Khaleel, K.I. Johnson, C.A. Lavender, M.T. Smith and C.H. Hamilton, Scripta Materialia **34**, (1996) 1417-1423.
9. D. Pulino-Sagradi, "Superplasticity of a duplex stainless steel"(in portuguese), Ph.D. Thesis, UNICAMP, Campinas, Brazil (1996).
10. W.A. Backofen, I.R.Turner and D.H. Avery, Trans. ASM **57**, (1964) 981-990.
11. D. Pulino-Sagradi, A.M.M. Nazar, J.-J. Amann and R.E. Medrano, Acta Materialia **45**, (1997) 4663-4666.
12. M. Sagradi, D. Pulino-Sagradi and R.E. Medrano, Acta Materialia **46**, (1998) 3857-3862.
13. E.W.Hart, Acta Metallurgica **18** (1970) 599-610.

Microstructural Evolution in Superplastic Coarse-Grained Fe-27at.%Al: Strain-Rate Effects

J.P. Chu[1], S.H. Chen[1], H.Y. Yasuda[2], Y. Umakoshi[2] and K. Inoue[3]

[1] Institute of Materials Engineering, National Taiwan Ocean University, Keelung 202, Taiwan ROC

[2] Department of Materials Science and Engineering, Osaka University, 2-1, Yamada-Oka, Suita, Osaka 565-0871, Japan

[3] Department of Materials Science and Engineering, University of Washington, Seattle WA 98195-2120, USA

Keywords: Dynamic Recovery, Dynamic Recrystallization, Iron Aluminide, Strain Rate, Superplasticity

Abstract

Strain-rate effects on superplastic properties of coarse-grained Fe-27at.%Al based alloys have been examined. The alloy studied has an initial grain size of ~1.6 mm. Tensile tests have been performed in air under initial strain rates of 1×10^{-4}, 1×10^{-2}, and 1×10^{-1} sec^{-1} at 800 and 900° C. Elongation-to-failure obtained are at least 183%, with the maximum of 420% at 800° C, 1×10^{-4} sec^{-1}, confirming the occurrence of superplasticity. At 900° C, the elongation shows a positive strain-rate dependence while a negative strain-rate dependence of elongation is observed at 800° C. At 800° C, the high strain-rate deformation results in a grain-migration structure and poor elongation is obtained. Yet, at 900° C, the grain growth is likely to occur at the low strain rate. A better elongation is thus obtained when the high strain rate is applied to reduce the time for the grain growth. Based on the microstructure observed, it could be stated that the effect of a strain rate increased by two orders of magnitude is approximately equivalent to that of a decrease of 100° C in the testing temperature. To yield a greater elongation and achieve refined grain structures after deformation, the superplastic deformation need to perform at 800° C with a strain rate of 1×10^{-4} sec^{-1} or at 900° C with 1×10^{-2} sec^{-1}.

1. Introduction

One of important characteristics for metals and ceramics exhibiting superplasticity is the fine-grained structure (typically <10 μm) [1]. Yet, Lin et al. and we have reported the superplasticity in Fe-Al-based alloys with much coarser grain structures, >500 μm [2-6]. Fe-Al based alloys are reported to exhibit all the deformation characteristics that conventional fine-grained superplastic materials hold. Using an electron backscattered diffraction technique, our previous works investigated crystallographic features of a superplastic coarse-grained Fe-27at.%Al alloy [4-6]. [Compositions in atomic percent are used throughout this article.] Results showed that the grain structure underwent four major transitions: subgrain-boundary formation, grain-boundary migration, formation and growth of recrystallized grains. Subgrains formed during an initial stage of high-temperature deformation when deformation was conducted at 600° C. Upon further deformation at 700° C, grain boundaries migrated and new grains started to form. When deformation was made at 800° C, dynamic recovery and recrystallization occurred, resulting in grain refinement and hence superplasticity. In addition, refined grains thus formed to maintain crystallographic relationships with parent grains [5]. To establish better understanding of superplastic properties, this study is directed toward investigation of microstructural evolution in superplastic Fe-27Al alloys under various strain rates.

2. Experimental Procedure

Commercial pure iron and aluminum are vacuum induction melted under argon atmosphere. To ensure the homogeneity and cleanliness, alloy are then refined in a consumable vacuum arc remelting system. As determined by an electron probe microanalyzer, the nominal composition of alloy is Fe-26.5Al. After homogenizing for 24 hrs at 1100° C in air, the alloy is hot rolled into ~3 mm sheets at 900-800° C. Tensile test specimens are cut from the sheet by the wire cutting. The tensile test specimen has the gauge section of 6.0x4.4x2.6 mm. All specimens are heat treated at 850° C for 1 hr in air, followed by furnace cooling. The initial grain size is ~1.6 mm. Tensile tests are conducted at 800° C and 900° C in air at initial strain rates of 1×10^{-1}, 1×10^{-2}, 1×10^{-4} sec^{-1}. Immediately after the tensile test, the heating furnace is removed for the specimen to cool. For microstructural examinations, longitudinal sections of specimen are cut and polished to 0.3 μm Al_2O_3 powder without etching. At least three specimens are examined for each testing condition.

Table 1. A summary of tensile properties obtained at 800 and 900° C with various strain rates.

Temperature	Strain Rate (sec^{-1})	Elongation-to-Failure (%)	Ultimate Tensile Strength (MPa)
800	10^{-4}	420	13.1
	10^{-2}	185	42.8
	10^{-1}	229	62.8
900	10^{-4}	183	4.60
	10^{-2}	203	32.1
	10^{-1}	230	31.6

3. Results and Discussion
3.1 Tensile Properties

Table 1 is a summary of tensile properties obtained at 800 and 900° C with various strain rates. Ultimate tensile strengths in this table are of engineering values. It is seen in the tables that at both temperatures with increasing strain rate the strength increases. Elongation-to-failure obtained in the alloys studied are at least 183%, with the maximum of 420% obtained at 800° C at a strain rate of 1×10^{-4} sec^{-1}. Evidently, the large elongation achieved in this study suggests the possible occurrence of superplasticity. Further, distinct strain-rate effects on the elongation are noted for both temperatures. That is in general the elongation shows a positive strain-rate dependence at 900° C whereas a negative strain-rate dependence of elongation is seen at 800° C. Yet, a detailed microstructural examination is needed to understand these distinct strain-rate effects.

3.2 Optical Microstructural Analysis

The microstructural evolution under various strain-rate conditions is studied using longitudinal optical micrographs taken from fractured specimens after deformation. Fig. 1 shows longitudinal optical micrographs from a sample deformed to an elongation of 203% at 900° C, 1×10^{-2} sec^{-1}. It is seen in Fig. 1(a) that a refined-grain structure is obtained in the fracture region after deformation. The refined grains are ranged from 100 to several hundred μm in size, which are in fact smaller than the original grain structure. Away from the fracture area (~5mm), the grain structure appears to be elongated, with some jagged grain boundaries present (Fig. 1(b)). In contrast, when deformed at 1×10^{-4} sec^{-1}, 900° C, the grain structure is completely different. Micrographs shown in Fig. 2 reveal smooth grain boundaries and large grains, comparable to the original grain structure, in areas near and away from fracture region. These characteristics suggest the grains are fully-grown during the deformation. However, this large-grained specimen yields an elongation of 183% at 1×10^{-4} sec^{-1}, which is comparative lower than 203% of the refined specimen at 1×10^{-2} sec^{-1}.

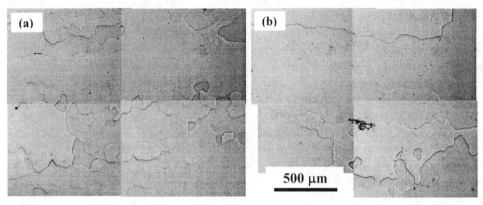

Fig. 1 Longitudinal optical micrographs obtained from a sample deformed to an elongation of 203% at 900° C, 1×10^{-2} sec^{-1}. (a) fracture area and (b) away from fracture area.

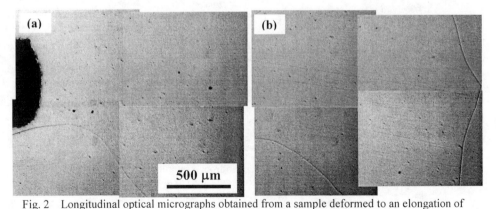

Fig. 2 Longitudinal optical micrographs obtained from a sample deformed to an elongation of 183% at 900° C, 1×10^{-4} sec^{-1}. (a) fracture area and (b) away from fracture area.

During the process of dynamic recovery and recrystallization, subgrain boundaries form initially and then grain boundaries start to migrate, followed by the formation and growth of recrystallized grains. When deformed at high temperatures (such as 900° C) and the low strain rate (1×10^{-4} sec^{-1}), the course of deformation is extended and thus thermally-activated grain growth inevitably takes place thoroughly, resulting in a fully-grown coarse grain structure. Yet, while the course of deformation is also prolonged for the low strain-rate deformation at 800° C, the grains are not thermally activated to grow due to the low testing temperature, yielding a recrystallized, fine structure, as will be shown next in Fig. 3.

At 800° C, the grain structures are also found to be dependent upon the strain rate applied. Nevertheless, at this temperature the low strain-rate deformation gives rise to a better superplastic elongation than the high strain-rate deformation. Fig. 3 shows micrographs obtained from samples deformed to an elongation of 420% at 800° C, 1×10^{-4} sec^{-1}. In the fracture area and the region away from fracture, the dynamic recrystallization results in a refined grain structure. The grains, ~100 to few hundred μm in size, are much smaller than that of the undeformed structure. At the high strain-rate deformation of 1×10^{-2} sec^{-1}, on the other hand, the refined grain structure is not obviously revealed. Instead, the grain-boundary migration is predominately observed, as shown in Fig. 4. In the fracture region (Fig. 4(a)), the elongated grains are found, due to the heavy deformation in the region, and plentiful jagged boundaries resulting from the extensive migration are clearly seen. While the jagged boundaries are present, the grains remain relatively coarse in size even after deformation. As a result, the elongation obtained at this high strain rate of 1×10^{-2} sec^{-1} is rather low, 185%. At the high strain rates (such as 1×10^{-2} sec^{-1}), the time needed for the thermal-activated recrystallization process also plays important roles. At 800° C, the grain-boundary migration is found, attributable to the insufficient time for recrystallization at high strain rate of 1×10^{-2} sec^{-1}. At 900° C and 1×10^{-2} sec^{-1}, on the other hand, the grains are completely recrystallized, but not fully grown because the course of deformation is relatively short (Fig. 2).

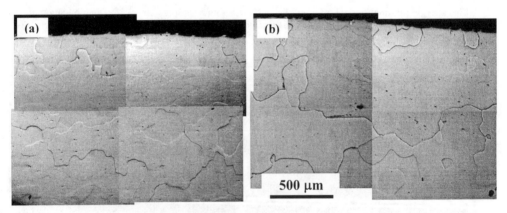

Fig. 3   Longitudinal optical micrographs obtained from a sample deformed to an elongation of 420% at 800° C, 1×10^{-4} sec^{-1}. (a) fracture area and (b) away from fracture area.

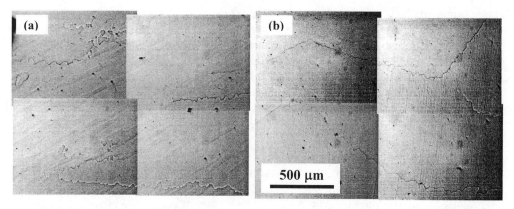

Fig. 4 Longitudinal optical micrographs obtained from a sample deformed to an elongation of 185% at 800° C, 1×10^{-2} sec^{-1}. (a) fracture area and (b) away from fracture area.

Our previous work reported that under a strain rate of 1×10^{-4} sec^{-1} the grain migration occurred at 700° C while the recrystallization took place at 800° C [4,5]. In the present study, the increase in strain rate from 1×10^{-4} to 1×10^{-2} sec^{-1} leads to a grain-migration structure in the samples when deformed at 800° C. Qualitatively, it thus could be stated that the effect of a strain rate increased by two orders of magnitude is approximately equivalent to that of a decrease of 100° C in the test temperature. This is further evidenced by the fact that when the deformation is performed at 900° C. At this temperature, as the strain rate decreases by two orders of magnitude from 1×10^{-2} to 1×10^{-4} sec^{-1}, the grain structure changes considerably from the recrystallization to the grain growth. Such a microstructural change is comparable to that observed in the samples when the testing temperature changes from 800 to 900° C at a strain rate of 1×10^{-4} sec^{-1}. Similar strain-rate dependence results have also been reported in other Fe-Al based alloys such as Fe-28.4Al-2.0Cr and Fe-24.5Al-1.4Ti alloys [7].

5. Summary

We have examined strain-rate effects on superplastic properties of coarse-grained Fe-27Al alloy deformed at 800 and 900° C. Results are summarized as follows:
1. Elongation obtained in the alloys are at least 183%, with the maximum of 420% obtained at 800° C, 1×10^{-4} sec^{-1}, confirming our coarse-grained Fe-Al alloy exhibiting the superplasticity.
2. At 900° C, the elongation shows a positive strain-rate dependence. At this temperature, a better elongation is obtained when the strain rate is high in order to reduce the time avoiding the grain growth. At 800° C, a negative strain-rate dependence of elongation is observed. The high strain-rate deformation results in a grain-migration structure and poor elongation is obtained. To yield a better elongation and achieve refined grain structure, the superplastic deformation needs to be performed at 800° C with a strain rate of 1×10^{-4} sec^{-1}.
3. Based on the microstructure observed, the effect of a strain rate increased by two orders of magnitude is approximately equivalent to that of a decrease of 100° C in the test temperature.

Acknowledgments

Dr. C. Y. Ma and Mr. C. S. Chen of Materials Research and Development Center at Chun Shan Institute of Science and Technology are acknowledged for their preparation of alloy samples. The National Science Council of the Republic of China supports this study under Contract No. NSC 89-2216-E-019-002.

References
1. T. G. Nieh, J. Wadsworth, "Fine Structure Superplastic Intermetallics," International Materials Reviews, 44 (1999), No. 2, p. 59.
2. D. Lin, T.L. Lin, A. Shan and M. Chen, Intermetallics, 4 (1996), p. 489.
3. T.L. Lin, D. Lin, A. Shan and Y. Liu, THERMEC '97, International Conference on Thermomechanical Processing of Steels and Other Materials, Ed. by T. Chandra and T. Sakai, TMS, Warrendal, PA, 1997, p. 19152.
4. J.P. Chu, I.M. Liu, J.H. Wu, W. Kai, J.Y. Wang and K. Inoue, Mat. Sci. Eng. A, A258 (1998), p. 146.
5. J.P. Chu, J. H. Wu, H. Y. Yasuda, Y. Umakoshi and K. Inoue, Intermetallics, 8 (2000), p. 39.
6. J.P. Chu, H.Y. Yasuda, Y. Umakoshi and K. Inoue, Intermetallics, 2000, in press.
7. J.P. Chu, S.H. Chen, W. Kai, H.Y. Yasuda, Y. Umakoshi and K. Inoue, Mat. Sci. Eng. A,, 2000, submitted for publication.

Corresponding Author: Professor Jinn P. Chu
E-mail address: jpchu@mail.ntou.edu.tw

Mechanics and Modeling of Superplasticity

Geometric Analysis of Thinning during Superplastic Forming

D. Garriga-Majo and R.V. Curtis

Dental Biomaterials Science, Guy's King's & St. Thomas Dental Institute,
King's College London, Floor 17, Guys Hospital, London SE1 9RT, UK

Keywords: Complex Analysis, Geometric Simulation, Pressure-Time Profile, Prismatic Die Shape, Sticking Contact, Thickness Profile

Abstract

Two original geometric models applicable to the superplastic forming of prismatic die shapes are presented in this study: the uniform thickness model which determines the average of the final thickness, and the variable thickness model, which gives a first approximation of the thickness distribution by assuming sticking contact with the die. The variable thickness approach demonstrates the important contribution of the die geometry to the thinning process by monitoring the sequence of contact events throughout the process. These two models relate to the limits of the friction regime between the superplastic material and the die, i.e. perfect sliding for the uniform thickness model (lubrication and/or low pressures/low strain rates) and sticking contact for the variable thickness model (no lubricant and/or high pressures/high strain rates). Predictions of the pressure-time profiles required to form the component are also derived from the second model and were successfully applied to the manufacture of dental prostheses in titanium alloy (Ti-6Al-4V). The development and the implementation of this geometric study have been greatly and elegantly simplified by the introduction of complex numbers to represent the different geometric parameters.

Introduction

During superplastic forming, the thinning of the alloy sheet depends on several parameters: the shape of the die, the material's properties, the loading conditions and temperature. The aim of this study is to evaluate separately the contribution of the geometry of the die to the thinning process. Knowing the initial dimensions of the forming sheet and the geometry of the die, good approximations of the final thickness distribution would be desirable. This is particularly important in medical applications where components are usually complex prototypes with strict adaptability requirements and the geometry of the die is derived either from an impression or a scan of the patient's body. Previous geometric approaches developed for industrial applications [1,2], although including frictional effects, are only suitable for specific shapes derived from rectangular box sections with a limited number of geometric parameters. In the current approach, no restriction is made regarding the shape of the die in a specific cross-section but the curvature in the perpendicular direction to the section needs to be small. This can be regarded as a plane strain case in stress analysis or as a two-dimensional case from a geometric point of view.

In addition, complex numbers are an extremely powerful tool with a wide range of practical applications and still remain an active area of mathematical and physical research [3]. Applied to the analysis of superplastic forming they offer an efficient apparatus to visualize and model the thinning process in a more intuitive way. In particular, the representation of geometric quantities by complex numbers combines the advantages of the Cartesian, polar and stereographic coordinates and so reduces dramatically the number of parameters during the analysis.

Computer implementation and experimental procedure

Any language or software supporting complex numbers and conventional functional programming is suitable to implement the model developed in this study. The possibility of defining recursion and nested arrays also simplifies the architecture of the contact algorithm. The dental prostheses have been superplastically formed on a 20-ton press using titanium alloy sheets (Ti-6Al-4V) and ceramic die material (phosphate bonded investment) [4]. The contact curve and the experimental thickness profiles were derived from images scanned using a commercial flatbed scanner with a resolution of 0.085 and 0.064 mm respectively (300 and 400 dpi).

Geometric analysis of SPF

Geometric definitions

The contour of the die, i.e. the contact curve (C) is defined by a list of point coordinates (x,y). The complex analysis is introduced by associating to each point of (C) a complex number such that:

$$z(k) = x_k + i \cdot y_k \qquad k \in [1, K] \qquad \text{Eq. 1}$$

Where, K is the total number of points in (C)

Using that simple parameterization and basic results from elementary differential geometry, the intrinsic features of the contact curve can be evaluated [5]. In particular, the arc-length $l_{a \to b}$ between two points a and b and the total arc-length L of the contact curve are:

$$l_{a \to b} = \int_a^b |z'(u)| \, du \qquad L = l_{z(1) \to z(K)} \qquad \text{Eq. 2, Eq. 3}$$

(the notation $z' = x' + iy'$ represents the first derivative of z)

Regarding the initial geometry of the alloy sheet [M], the initial thickness S is the only input parameter required. In particular, this model does not necessitate any discretization of [M] prior to the simulation. The initial width W of [M] between the two end points of the contact curve (C) is:

$$W = |z(1) - z(K)| \qquad \text{Eq. 4}$$

Uniform thickness model

Most plastic large deformations consider constancy of volume. This assumption supposes also that no voids are created in the material during forming. Hence, at each step of the analysis, the total area of [M] is simply related to its initial dimensions and equal to the product W.S. In particular, when contact with the die is complete, one can evaluate the final average thickness s_{unif} of [M] by the following relation:

$$s_{\text{unif}} = \frac{W \cdot S}{L} \qquad \text{Eq. 5}$$

This value constitutes the first approximation of the final thickness of the sheet after forming.

Variable thickness model

In this geometric approach sticking contact with the die is assumed. The basic elementary shape of the forming sheet between two contact points is defined as an expanding cylinder sector of uniform thickness with two disk sectors at the contact points to accommodate the deformation (Fig. 1.a). With such a shape, thinning is allowed to continue locally even after contact with the die is made. In fact, it carries on at a contact point until both adjacent points also make contact.

During forming the expansion of the cylinder continues until it makes a new contact with the die. The new contact corresponds to the shortest path of the cylinder centre, that is the smallest distance $|\omega_b - \omega_c|$ (Fig. 1.a and b), where ω_c is the centre of the circumcircle of the three points (a,b,c) obtained by:

$$\omega_c = \frac{a\xi - c\overline{\xi}}{\xi - \overline{\xi}} \quad \text{where} \quad \xi = \overline{(a-b)}(c-b) \qquad \text{Eq. 6}$$

(the notation $\overline{z} = x - iy$ represents the complex conjugate of z)

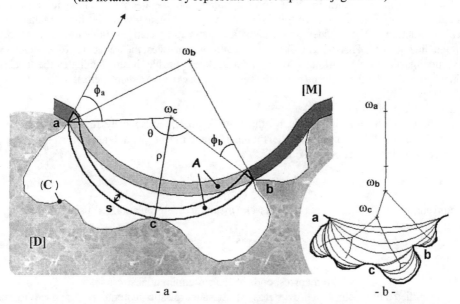

Fig. 1 – Variable thickness model. a - The elementary shape of the geometric model and its evolution between two steps of the analysis. b - The generation of the new elementary shapes during the analysis and the branching structure of the cylinder centre paths.

In complex analysis, the advance of the fit surface can be interpreted as particular mappings of the plane to itself called elliptic Möbius transformations with fixed points a and b [3]. The geometric parameters of the elementary shape (the cylinder radius ρ, the cylinder angle θ, and the two angles ϕ_a and ϕ_b as illustrated in Fig. 1.a) are updated for the new position determined by the contact point c and the cylinder centre ω_c:

$$\rho = |\omega_c - c| \qquad \text{Eq. 7}$$

$$\theta = \text{angle}(a, \omega_c, b) \qquad \phi_a = \text{angle}(\omega_c, a, \omega_a) \qquad \phi_b = \text{angle}(\omega_b, b, \omega_c) \qquad \text{Eq. 8}$$

where the angle function is: $\text{angle}(u, v, w) = \arg(w - v) - \arg(u - v)$

The new thickness s is calculated by assuming that the area A of the elementary shape does not change between the two steps, that is:

$$\frac{1}{2}\phi_a s^2 + \theta\left(\rho - \frac{s}{2}\right)s + \frac{1}{2}\phi_b s^2 = A \qquad \text{Eq. 9}$$

The new thickness s corresponds to the lowest root of this equation of the second degree:

$$s = \frac{2\tau}{1 + \sqrt{1 - 2\delta\frac{\tau}{\rho}}} \quad \text{where} \quad \tau = \frac{A}{\rho\theta} \quad \text{and} \quad \delta = 1 - \frac{\phi_a + \phi_b}{\theta} \qquad \text{Eq. 10}$$

Once the new contact is made, the elementary shape will split into two similar elementary shapes having new areas A_{ac} and A_{cb} defined by:

$$A_{ac} = \frac{1}{2}\phi_a s^2 + \theta_{ac}\left(\rho - \frac{s}{2}\right)s \quad \text{and} \quad A_{cb} = \theta_{cb}\left(\rho - \frac{s}{2}\right)s + \frac{1}{2}\phi_b s^2 \qquad \text{Eq. 11}$$

where $\theta_{ac} = \text{angle}(a,\omega_c,c)$ and $\theta_{cb} = \text{angle}(c,\omega_c,b)$ Eq. 12

At each step in the analysis an algorithm using a recursive function and the set of equations above (Eq. 6 to Eq. 12) establishes the new contact points and the geometric characteristics of each elementary shapes. This process continues until complete contact with the die is achieved. The branching structure of the cylinder centre paths and the step-by-step positions of the fit surface are illustrated in Fig. 1.b. This typical diagram visually summarizes the sequence of contact events during superplastic forming. Along with the final thickness profile, it also facilitates the localisation of critical points on the die surface (minimum thickness, thickness gradients, etc.).

Pressure profile generation
In order to establish a pressure-time profile suitable to superplastically form a component, the process is monitored by controlling the thinning strain rate $\dot{\epsilon}$ in a branch corresponding to a selected point in the contact curve, so that:

$$s(t) = S\,e^{-\dot{\epsilon}\,t} \quad \text{or} \quad t = -\frac{1}{\dot{\epsilon}}\ln\frac{s}{S}$$ Eq. 13

The conventional cylindrical pressure vessel (membrane) formulation is used to estimate the forming pressure:

$$P = \frac{s}{\rho}\sigma$$ Eq. 14

where σ corresponds to the circumferential stress

In this simplified approach, the flow stress of the superplastic material is supposed to remain constant throughout forming at constant strain rate and constant temperature, but any refinement regarding the constitutive equation of the material, as grain growth for instance, can be introduced at this stage. As for the cylinder centre paths, a branching structure of the pressure-time profiles, related to the contact events, is obtained. A suitable pressure profile is then selected at a specific point according to the maximum thinning criteria and/or the maximum pressure required to form the component.

Results and Discussion

Application to basic geometric shapes
The validity of the uniform thickness model has been evaluated using die contact curves of known perimeter such as semi-circles and rectangular boxes with an increasing number of points ($2^3 < K < 2^{12}$). These simulations revealed that a minimum of about 60 points is sufficient to obtain an error of the arc length and the final uniform thickness lower than 1% in both cases.

Regarding the variable thickness model, the important influence of the die geometry on the thinning process is illustrated in Fig. 2 where three rectangular, circular and triangular shapes with the same opening width have been assembled together on the perimeter of a semi-circle. The simulation shows that any deviation from a concave circular shape produces important final thickness variations (Fig. 2.c). In comparison, the relative influence of the initial thickness is much smaller: different values of S produce similar profiles of final normalised thickness. This simulation also illustrates the strong influence of the sequence of contact events and the resulting thickness variations on the pressure-time profiles (Fig. 2.d). In the particular case of rectangular box sections, the model agrees accurately with a previous geometric model developed with the sticking contact assumption [1].

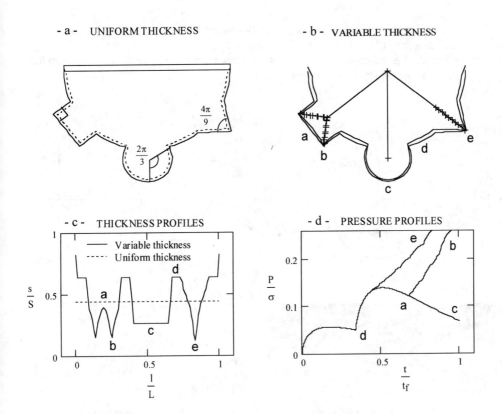

Fig. 2 – Application of the uniform (a) and the variable (b) thickness models to basic shapes showing the influence of the die geometry on the final thickness profile (c) and the pressure profiles generated for each point along the contact curve (d).

Application to the manufacture of dental prostheses
The geometric model has been successfully applied to the manufacture of different kinds of dental prostheses such as conventional denture bases (see Fig. 3) and dental implant superstructures [4]. For these die shapes the curvature of the die in a direction perpendicular to the cross-section is relatively small (< 0.05-0.1 mm^{-1}), so the die surface is well approximated by a prismatic shape of same cross-section. The comparison of the simulated and experimental thickness profiles suggests that the variable thickness model is more appropriate and that the sticking contact assumption for this combination of materials (Ti-6Al-4V/ceramic phosphate bonded investment) is realistic. In particular, a good agreement is obtained in the region of large thickness variations (see sections a-b and d-e in Fig. 3.c). However the contact is not totally complete at the point f where the pressure-time profile was generated. This suggests that the pressure required to form the component was underestimated and that a more refined constitutive equation for the superplastic material (including grain growth kinetics for instance) would be a benefit for the model.

Conclusion

The geometric analysis presented in this study provides good estimations of the uniform (sliding) and variable thinning (sticking) of the alloy sheet during superplastic forming based on the important contribution of die geometry. The contact formulation adopted in the variable thickness model offers the possibility to generate estimations of the pressure-time profiles suitable to form the component. This modelling based on geometrical complex analysis constitutes a fast and efficient tool for the design and the superplastic forming of complex prismatic shapes and was successfully applied to the manufacture of dental prostheses.

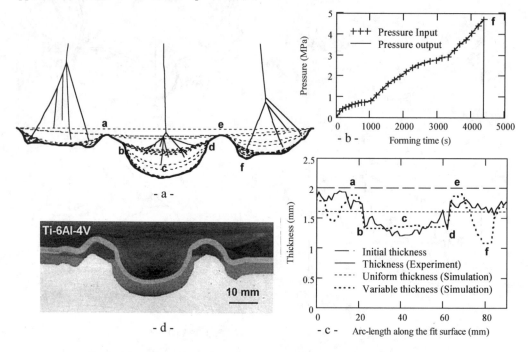

Fig. 3 – Application of the geometric model to the manufacture of a dental prosthesis in Ti-6Al-4V. a - Sequence of contact events and the branching structure of the cylinder centre paths. b - Pressure profile generated at the point f that was used to form the prosthesis ($\dot{\varepsilon} = 1.5\,10^{-4}\ \mathrm{s}^{-1}, \sigma = 8.0\ \mathrm{MPa}, T = 1173\ \mathrm{K}$). c – Experimental and simulated final thickness profiles. d - Scan of the cross-section of the prosthesis after forming.

References

[1] A. Ghosh, C. Hamilton, Proc. of ASM on Process Modeling Fundamentals and Applications to Metals (1978) p. 303-331.

[2] N. Chandra, K. Chandy. J. Mater. Shaping Technol. (1991), vol. 9, p. 27-37.

[3] T. Needham, Visual Complex Analysis, Oxford University Press (1997), chap. 1-3, p. 1-188.

[4] R. Curtis et al., ICSAM2000, August 1-4 2000, Orlando, FA (2000).

[5] B. O'Neill, Elementary Differential Geometry, Academic Press (1997), chap. 2, p. 50-78.

Aknowledgements

The financial support of the Engineering and Physical Science Research Council, UK is gratefully acknowledged.

Corresponding Author Dr. D. Garriga-Majo. E-mail address: denis.garriga-majo@kcl.ac.uk

Analysis of Superplastic Testing by Using Constant Pressure in Prismatic Die

L. Carrino and G. Giuliano

Department of Industrial Engineering, University of Cassino,
Via di Biasio 43, IT-03043 Cassino (FR), Italy

Keywords: Analytical Model, Finite Element Method, Superplastic Forming

Abstract

In the paste, different authors [1,2,3,4,5,6,7,8] have developed analytical process models of superplastic sheets bulged into dies of simple shapes such as cylindrical cavities, parallel walls, conical and prismatic dies. An important data for any type of process model is the constitutive equation of superplastic material. In general, the most common method of characterising a material consists in submitting a specimen, with standardised geometry, to a uniaxial tensile test. For superplastic materials the tests were carried out at different strain-rate in order to determine the flow stress as a function of strain-rate. In alternative, to determine the constitutive equation of superplastic materials, biaxial tests was derived from analytical process models [1,4,5,8].
In this study, the method to determine the strain-rate sensitivity coefficient of superplastic materials, based on the analysis of forming process in a prismatic die, was experimentally validated in the laboratory on specimen of material formed at room temperature. The authors have confirmed, by finite element method, the validity of the material characterisation carried out using a prismatic die.

Analysis of the forming process

The method to determine the strain-rate sensitivity coefficient considers forming as occurring in an indefinite prismatic die. To keep the flow stress dependent only on pressure during the forming process, it is necessary for the die to have an angle of 46.4° at its vertex.
The model assumes a plane strain condition, a cylindrical bulge profile and a uniform thinning in the deforming sheet [7]. It is also necessary to take the following hypotheses into account:
- only the sheet that is not in contact with the die can be deformed (sticking friction);
- all the sheet can be deformed (sliding friction).

The prismatic die forming layout is showed in figure 1.
In the plane strain condition, the value of the flow stress $\bar{\sigma}$ can be expressed as:

$$\bar{\sigma} = \frac{\sqrt{3}}{2} p \frac{r}{s} \qquad \text{Eq. 1}$$

From geometrical conditions is possible to calculate the radius of curvature as:

$$r = r_1 - x \tan\left(\frac{\alpha}{2}\right) \qquad \text{Eq. 2}$$

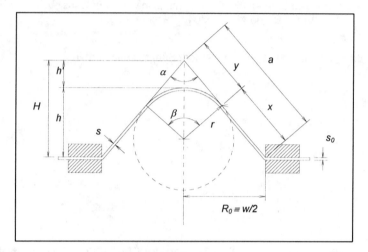

Fig. 1 - Prismatic die forming layout.

Therefore, the distance along the die wall from the top theoretical sharp corner is:

$$x = \frac{s_1 - s}{s} \frac{r_1 \beta_1}{2 - \beta_1 \tan\left(\frac{\alpha}{2}\right)} \qquad \text{(sliding friction)} \qquad \text{Eq. 3}$$

$$x = \frac{r_1}{\tan\left(\frac{\alpha}{2}\right)} \left[1 - \left(\frac{s}{s_1}\right)^{\frac{\beta_1 \tan\left(\frac{\alpha}{2}\right)}{2 - \beta_1 \tan\left(\frac{\alpha}{2}\right)}} \right] \qquad \text{(sticking friction)} \qquad \text{Eq. 4}$$

The index at the down right indicates that the quantities are calculated at the first contact point between sheet and die. Figure 2a shows that the flow stress remains constant, during the forming process, in sticking friction condition while it changes in the sliding friction condition. From figure 2b, it possible to see that, in the first contact point between sheet and die, the strain-rate is independent by friction conditions. In this way, by prismatic die tests, the valuation of the strain-rate in the first contact point between sheet and die permits to calculate the strain-rate sensitivity coefficient.

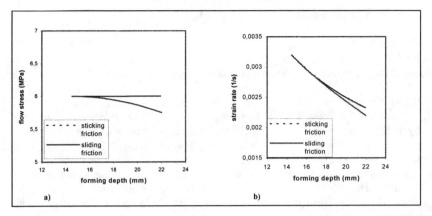

Fig. 2 – a) Flow stress versus forming depth, b) and strain rate versus forming depth, obtained in forming process in prismatic die by using constant pressure. The used constitutive law assumes that the flow stress is dependent on both strain rate and strain.

Forming tests using constant pressure

The superplastic alloy, used for the experimentation in this study, comprised 60% lead and 40% tin. To prepare the sheets in order to reach a grain dimension in the order of microns, a series of lamination cycles [9] were carried out. For the forming a two part lopped-off pyramid-shaped die with rectangular base was used. The lower one is a plate 240mm x 100mm x 5mm in size that has a blank holder, four holes for fixing and a central hole for the removal of air. The upper part is a plate containing the lopped-off pyramid (Fig. 3). The shortest side of the rectangular base is 44mm, and the longest one is circa 200mm. The angle of opening α is 46.4°, so as to assure direct proportion between the value of the flow stress $\bar{\sigma}$ and the pressure applied during forming.

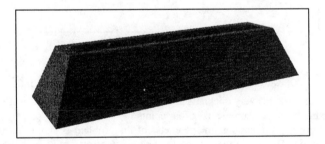

Fig. 3 - Pyramidal die.

The experimental tests have been performed in a random sequence. During each forming test the values of the sheet height has been measured as a function of the processing time by means of the centesimal comparator. A video camera has allows storing the comparator lectures on an hardware support.

It can be affirmed that the trend of $\ln \bar{\sigma}$ in relation to $\ln \dot{\bar{\varepsilon}}$ is expressed by :

$$\ln \bar{\sigma} = 4.46 + 0.42 \ln \dot{\bar{\varepsilon}}$$
Eq. 5

then the parameters of the constitutive equation $\overline{\sigma} = K \cdot \dot{\overline{\varepsilon}}^m$:

$m = 0.42$ e $K = 86.78$. Eq. 6

Validation of the material characterization model

To test the suitability of the characterization method using a prismatic die, the values of K and m, obtained for the PbSn60 alloy, were used to simulate a constant pressure forming process by means of the FEM. Parallel experimental activity was conducted in the laboratory, adopting a forming pressure of 0.15 MPa and using some sheets with initial thickness of 0.320 mm.

The die used comprised an upper part in the shape of a axial symmetric cap shown in cross-section in Fig. 4. The latter is built in Plexiglas and has two air outlet holes drilled in the upper surface. The upper part is anchored to the rest of the die by means of a metallic plate with four holes.

The lower part comprises a metallic base with an extruding ring that acts as blank holder, and a hole that allows compressed air to be withdrawn.

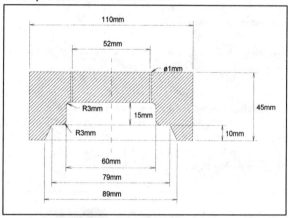

Fig. 4 - Plexiglas die.

Since the thickness of the sheet, at the end of forming, varies along a curvilinear abscissa, it was decided to measure this very thickness at a number of points. In particular it was decided to carry out the thickness measurements in four points.

For closed die forming simulations, the geometry of the problem is simplified by the axial symmetric condition. The nodes that are on the axis of symmetry must be bound in such a way that they cannot move in an orthogonal direction to the axis (Fig. 5). To simulate the action of the blank holder, the outermost node in contact with the die was hinged and the underlying one in contact with the blank holder, was permitted to move only in a direction parallel to the axis of symmetry.

A quadrilateral element specifically dedicated to axial symmetric applications was used. The sheet was discretised with 123 rectangular elements 0.320 mm thick and a total length of 39.5 mm.

On the lower surface of the sheet a uniformly distributed load equal to a pressure of 0.15 *MPa* was applied.
To carry out the simulation the constitutive equation of the material was used:

$$\overline{\sigma} = 86.78 \cdot \dot{\overline{\varepsilon}}^{0.42}.$$ Eq. 7

The output variable of interest chosen was the final thickness of the sheet.

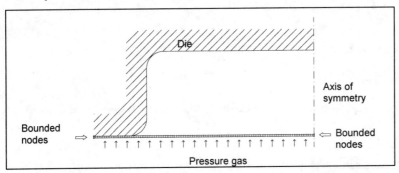

Fig. 5 - Finite element geometry discretisation.

Comparison between the experimentation and FEM simulation

Once the data were gathered, the numerical simulation results were compared to those obtained experimentally.
Table 1 gives the values obtained via the numerical simulation in correspondence to the curvilinear abscissa desired. These values were compared with those obtained by means of the experimentation in every position of the curvilinear abscissa considered. It is possible to notice that the maximum error in predicting thickness is less than 1%.

curvilinear abscissa	Average experimental thickness value	F.E.M. thickness value
x (mm)	s (mm)	s (mm)
0	0.208	0.209
9	0.190	0.191
18	0.166	0.166
36.21	0.212	0.212

Table 1 - Comparison between experimentation and F.E.M. simulation.

Fig. 6 compares numerical activity and experimental data, showing for both cases the trend of the thickness *s* in relation to the curvilinear abscissa *x*.

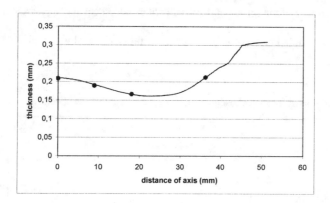

Fig. 6 – Qualitative comparison between simulation results (continuous line) and experimental data (points).

Conclusions

A method to characterise superplastic materials based on an approximated analysis has been presented. It consists in forming the sheet in a quadrangular lopped-off pyramid-shaped die with a characteristic 46.4° angle of opening. The method was validated by comparing the data from both the closed die forming F.E.M. simulation and the experimental activity carried out on the superplastic *PbSn60* alloy. The simulation results corresponded well, in terms of end-of-forming thickness measurement, to the experimental data.

Acknowledgements

The presents work is part of the research project CO.TE.STA., sponsored by Italian Ministry of University and Technological Research (MURST) which is gratefully acknowledged.

References

1. F. Jovane, International Journal of Mechanical Science, **10**, (1968) 403-427.
2. D.L. Holt, International Journal of Mechanical Science, **12**, (1970) 491-497.
3. G.C. Cornfield and R.H. Johnson, International Journal of Mechanical Science, **12**, (1970) 479-490.
4. A.K. Ghosh and C.H. Hamilton, Metallurgical Transactions, **11 A**, (1980) 1915-1918.
5. R.J. Lederich, S.M.L. Sastry, M. Hayaseand and T.L. Mackay, Journal of Metals, (1982) 16-20.
6. A.R. Ragab, Metals Technology, **10**, (1983) 340-348.
7. J.M. Story, *Superplasticity and Superplastic Forming*, edited by C.H. Hamilton and N.C. Paton (The Minerals, Metals & Materials Society 1988) 297-302.
8. R.E. Goforth, N.A. Chandra and D. George, in *Superplasticity in Aerospace*, edited by H.C. Heikkenen and T.R. McNelley (The Metallurgical Society, Inc. 1988) 149-166.
9. L. Carrino and G. Giuliano, Advanced Performance Materials, **6**, (1999) 149-159.

Corresponding Author: Gillo Giuliano, e-mail: giuliano@unicas.it

Effect of Hot Deformation under Complex Loading on the Transformation of Lamellar Type Microstructure in Ti-6.5Al-3.5Mo-1.6Zr-0.27Si Alloy

O.A. Kaibyshev, V.K. Berdin, M.V. Karavaeva,
R.M. Kashaev and L.A. Syutina

Institute for Metals Superplasticity Problems, Russian Academy of Science,
Khalturina 39, RU-450001 Ufa, Russia

Keywords: Complex Loading Conditions, Finite Element Modelling, Grain Refinement, Superplasticity, Titanium Alloy

Abstract

The transformation of the lamellar type microstructure to an equiaxed one in two-phase titanium alloys is determined not only by temperature-strain rate regimes but also by the modes of loading. The experimental results of investigations relating to the influence of one- and two-component monotonous loading on microstructure and mechanical behavior of the two-phase Ti-6.5Al-3.5Mo-1.6Zr-0.27Si titanium alloy are presented in this work. It is shown that under conditions of proportional torsion + tension loading the transformation of the lamellar microstructure occurs more uniformly in the transverse section of a sample and over its length than under conditions of simple torsion or uniaxial tension. The transformation is accompanied by formation of a typical texture based on the strain state. The following uniaxial tension in optimal temperature-strain rate regimes of superplasticity is characterized by a lower value of flow stress, uniform distribution of deformation and further microstructure transformation within the whole sample volume.

Introduction.

Transformation of coarse lamellar type microstructure of two-phase titanium alloys to the equiaxed fine grain one requires intensive plastic deformation in $\alpha+\beta$ region [1,2]. Multiple step forging is a typical method of grain refinement of coarse microstructure [2]. However, multiple step forging is difficult to control, time consumption and power-intensive processing method.

It has been shown [3] that in the two-phase titanium alloy the transformation of the lamellar microstructure to the equiaxed one occurs more efficiently under two-component (tension + torsion) mode of loading. However, the uniformity of deformation in the whole sample volume was ignored and therefore they observed microstructure transformations only within the small sample's region.

It is known [4] that strain is uniform along radius in the gauge portion of a cylindrical sample during uniaxial tension. However, when samples with lamellar type microstructure are deformed, the deformation is localized even at e=0.2-0.4 due to formation of a neck. During torsion tests [6] the deformation is uniformly distributed along sample's length, but not uniform along its radius. So we can expect that during intensive plastic deformation the combination of torsion and tension (in reasonable proportions) should result in more uniform strain distribution and produce uniform and intensive transformation of the lamellar microstructure both over the length and along the radius of the sample.

Therefore, in the present work the influence of one-component either torsion or tension, as well as two-component e.g. torsion + tension proportional loading on microstructure changes in samples with initial lamellar microstructure is investigated. The mechanical behavior of the preliminary deformed samples during additional deformation by uniaxial tension in the optimal temperature-strain rate regime of superplasticity as well as microstructure and texture evolutions is investigated, too.

Material and Experimental Procedure

The two-phase Ti-6.5Al-3.5Mo-1.6Zr-0.27Si(VT9) titanium alloy was used as an initial material (Fig.1). The hot rolled billets of 20.0 mm in diameter were taken as an initial material. The coarse grain microstructure with lamellar morphology of phases was obtained by annealing in β-region (1050 °C - 1 hour, cooling with a furnace). Mechanical tests were conducted under isothermal conditions at temperature T=950 °C on a multi-axial loading machine [6]. Cylindrical samples of 10.0 mm in diameter and 40.0 mm in gauge length were used for testing.

Fig.1. Initial microstructure.

The procedure of mechanical tests included two stages of loading. The first stage is hot plastic deformation by:
- uniaxial tension;
- torsion;
- two - component torsion + tension with torsion to axial strain rate ratio of 10:1.

Tension tests were performed at constant strain rate $1.0 \cdot 10^{-3}$ s^{-1} up to e=0.17 and to failure e=0.65. Torsion tests were conducted at two angular velocities of ω_1=0.1 rev/min and ω_2=1.0 rev/min. The strain rate values at the periphery of samples were ξ_1=1.3 10^{-3} s^{-1} and ξ_2=1.3 10^{-3} s^{-1}, respectively. The angle of twisting in both cases was φ=1080 degrees, and it corresponded to the true strain value e=1.1. Parameters of two-component loading were the same as for one-component one. The total value of strain and strain rate calculated according to the technique described in [7] was equal to e=1.11 and ξ=1.3 10^{-2} s^{-1}. The ratio of torque to axial component 10:1 was selected in order to prevent formation of neck in the gauge portion of samples.

During the second stage of the mechanical tests the samples were subjected to secondary deformation by uniaxial tension in the optimal temperature-strain rate regime of superplasticity in the alloy under study [1,2]. The loading diagrams for uniaxial tension, torsion and torsion+tension were calculated according to the technique [7].

The textures of the material were characterized by means of inverse pole figures using the Kα Cu-radiation for the direction of the specimen axis. X-ray diffractometer DRON-4 was used for these purposes.

The microstructure evolution was studied in the cross section of samples in the periphery and in the center (maximum and minimum torsion strain) by scanning electron microscope JSM 840. The following parameters were estimated: the mean value and distribution of the α-particle shape coefficient Kα= l / t, where l is the length, t is the thickness of α-particle, as well as volume fraction of α-phase Vα:Vβ. The microscope Axiatech 100 and PC program KS300 were used for quantitative analysis of microstructure.

Results of Experiments.
Initial material.

The size of β-transformed grains at the beginning was 350.0 μm. Mean plate thickness was 10.0 μm, the volume fraction of α-phase was Vα:Vβ = 90:10 (Fig.1). Distribution of the α-phase particle shape coefficient is characterized by blur of maximum values at 6-9 that testifies that in the initial state the microstructure is non-uniform. During heating, prior to deformation in the two-phase region, the volume fraction of the α-phase decreases to 50%. It is connected with the fact that the initial microstructure is in a non-equilibrium state at T=950°C of deformation. During heating the excessive α phase is dissolved that leads to a significant reduction of the α-plate thickness from 10.0 μm to 3.0 μm. Kα has finally the value 9.5.

Uniaxial tension.

The process of deformation of the studied alloy is characterized by a dropping diagram attaining the maximum flow stress at low true strain values of 0.05 (Fig.2). After that the flow stress decreases sharply with increasing e. After the strain value of e=0.4, a bend on the diagram is observed. This points to the beginning of strain localization which then leads to failure at e=0.7. The shape of the sample also demonstrates the non-uniformity of deformation along the length - formation of a neck.

The microstructure in the gauge zone of the samples after tension was not distributed uniformly too. In the areas far from the neck α-phase particles retain their lamellar type. The Kα coefficient decreases slightly from 9.5 to 6.7. In the necking zone a significant change in the microstructure was observed. Kα became equal to 2.0.

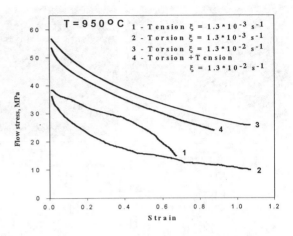

Fig.2. Stress versus strain for different modes of deformation.

Torsion.

Similar to tension, significant softening also occurs during torsion: with increasing the strain value up to e=1.0 the flow stress decreases by almost a factor of 3.0. However, localization of deformation in the gauge zone of the sample was not observed. The value of flow stress significantly depends on a strain rate (curves 2 and 3 Fig.2).

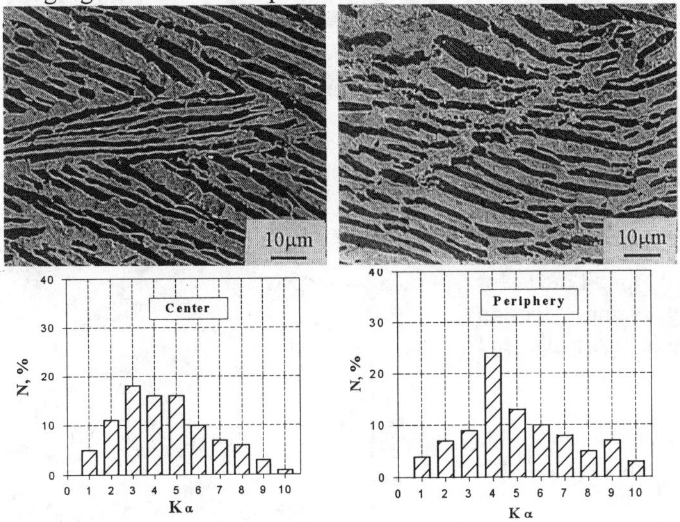

After torsion the microstructure transforms uniformly over the length of the sample and non-uniformly over its diameter. After torsion the α-phase plates were bent, and were divided into fragments. Some portions of fragments were shifted relative to each other. In the β-phase the areas with different orientation from that of the initial structure became visible. It testifies that the process of recrystallization occurs there. In the periphery zone the value of strain was the highest one, a mean value of Kα became equal to 5.0. The thickness of plates was t=3.0 μm there. In the central zone Kα was 6.5 and t=2.5μm, respectively.

Fig.3. Evolution of microstructure after torsion test.

Microstructures of the samples after torsion and distribution of Kα are shown in Fig.3.

Two – component (Torsion + Tension) loading.

Similar to torsion, deformation under two-component (torsion + tension) loading occurs without localization. Significant softening in this case was observed during simple torsion too. With increasing the true strain value up to e=1.0 the flow stress decreased by a factor of three.

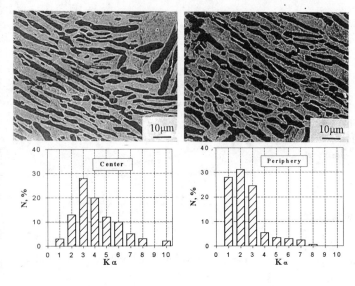

Fig.4. Evolution of microstructure after torsion + tension test.

The character of microstructure evolution during two-component loading was the same as during simple torsion (Fig.4). However, degree and uniformity of microstructure transformation in the whole sample volume had the highest values. In the periphery zone the microstructure was almost equiaxial and was characterized by the value of $K\alpha=2.1$. In the central zone $K\alpha$ was 4.0. It is significantly lower than after torsion when $K\alpha$ was 5.5. As compared to torsion, the uniformity of the parameter $K\alpha$ in the cross section is also higher (histograms at Fig.4.). Moreover, the residual matrix type structure observed after torsion in the central zone of the samples was absent. The thickness of α-particles in the central zone of samples was smaller than in the periphery: 2.8 μm and 3.4 μm, respectively.

Evolution of texture.

The α-textures in the initial state and after deformation by tension, torsion and torsion + tension are shown in Figure 5 a, b, c and d, respectively, by means of inverse pole figures. Before deformation the texture was characterized by low (0001) and $(10\bar{1}0)$ intensity. Intensity of $(11\bar{2}0)$ was the highest one. After deformation intensities of (0001) and $(10\bar{1}0)$ were increased. But the intensity of $(11\bar{2}0)$ for all modes of deformation was decreased. During torsion + tension the lowest intensity values for basic, prismatic and pyramidal plates were registered.

Secondary deformation by tension

The second stage of mechanical tests was uniaxial tension in superplastic conditions for the studied material. In the sample, preliminary subjected to tension up to e=0.17, softening is observed already in the initial stage of tension as compared to the sample not subjected to preliminary treatment (curves 1 and 5 in Fig.7). However, as compared to annealed samples a decrease in the flow stress is not so intensive. No neck was formed in this case.

In spite of the non-uniform distribution of microstructure parameters in the cross section of the sample after preliminary one-component torsion and two-component torsion + tension the mechanical behavior was similar to the one during subsequent

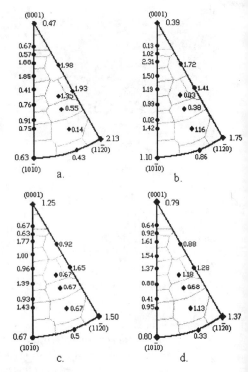

Fig.6. Inverse pole figures.
a - original state.
b - after tension, e = 0.7.
c - after torsion, e = 1.1.
d - after torsion + tension, e = 1.11

tension at temperature and strain rate of testing.

The main distinctions from the non-treated samples: low flow stress σ = 18.0-20.0 MPa and ability to be deformed up to strain values more than 100% without formation of neck. This can be seen both from the tensile diagrams (curves 2,3 and 4, Fig.7) and the sample shapes after deformation.

In the samples, deformed by torsion and torsion + tension and subsequent uniaxial tension in the temperature-strain rate regime of superplasticity the differences in the microstructure parameters between central and periphery zones of the sample disappeared. This effect is most prominent in the case of preliminary deformation according to the mode of proportional two component

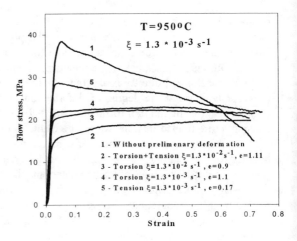

Fig.7. Stress versus strain for different modes of deformation.

(torsion + tension) loading. In particular, the value of Kα in the central zone of the sample was 3.3, in the periphery zone - 2.1. So, the deformation was not concentrated in the periphery layers of the sample with the preliminary processed microstructure. Due to additional superplastic deformation further transformation of the microstructure at the central zone occurred and uniformity of structure within the whole sample volume was increased. The largest volume fraction of the material (about 85%) had a globular type microstructure after two step treatment (Fig.8).

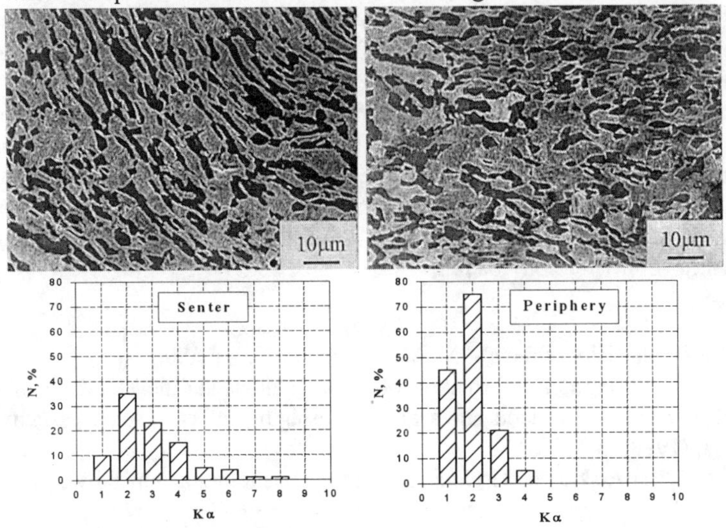

Discussion.

The mechanism of transformation of the lamellar microstructure in two-phase titanium alloys has been studied very well [1,2 etc]. It is known that both the dynamic recrystallization and the phase transformations play an important role in grain refinement processes. The occurrence of dynamic recrystallization in the deformed material requires some definite density of dislocations which are stable enough during high temperature deformation. As noted above, the mode of uniaxial tension is most favorable in terms of uniformity of the stress - strain state. However, during tension significant microstructure transformation occurs in a local part of the sample. As a result, deformation is concentrated in a non-significant volume of the sample where complete transformation of the microstructure takes place.

Fig.8. Evolution of microstructure after torsion + tension.

The experimental data show that in case of the titanium alloy under study with coarse grain lamellar type microstructure the uniform tensile deformation does not exceed 20% and no noticeable microstructure transformations are observed after that. Torsion deformation provides large strain before sample's failure. Therefore, an opportunity to accumulate large densities of crystal lattice defects in the material occurs there. The latter increases the probability of the onset of recrystallization processes. However, the strain value distribution along the radius of the sample is not uniform in case of torsion (Fig.3). During proportional two component (torsion + tension) loading the microstructure of central layers is transformed due to axial strain. At the same time the torque component makes it possible to increase significantly the degree of deformation uniformity in the periphery, and consequently, strain value to failure. In addition to torsion, the deformation by tension increases the rate and uniformity of microstructure transformations within the whole sample volume that is probably due to the formation of stable to dynamic recovery dislocation structure formed during operation of different slip systems typical both for torsion and uniaxial tension. Indirectly it could be confirmed by the results of texture evolution (Fig7). The different modes of loading determine the different type of crystallographic texture.

The tensile diagrams after preliminary treatments show that the alloy can display superplastic flow, though metallographic investigations testify that the microstructure has Non-equilibrium State and its matrix type structure is still preserved. During preliminary treatment a sub-microstructure is evidently formed. After formation of low angle grain boundaries the grain boundary sliding takes place. The latter contributes to increasing the rate and uniformity of further grain refinement processes.

Conclusions
1. During deformation under monotonic loading a significant softening occurs. However, in the case of tension the localization of deformation takes place rather rapidly (e=0.4) which ends in failure at e=0.7. In the case of torsion and proportional (torsion + tension) loading under analogous temperature-strain rate conditions the strain is uniform and localization of deformation is absent up to e>1.1.
2. Structural refinements occurs non-uniformly within the sample volume: in tension the most significant structure transformations take place in the zone of neck formation; in the case of modes containing torsion - in the peripheral zones of samples. α - plates subdivisions both along radius and along sample's length are most uniform under conditions of proportional (torsion + tension) loading.
3. In case of secondary loading by uniaxial tension under superplastic conditions a significant change in mechanical behavior of all preliminary deformed samples is observed. The flow stress is decreased by more than twice and localization of deformation is absent even for preliminary tension up to 0.17.
4. Microstructure refinements after secondary loading takes place more efficiently and uniformly in case of preliminary proportional (torsion + tension) loading.

References.
1. O.A. Kaibyshev, Plasticity and superplasticity of metals. M.:Metallurgy, (1975), p.280.
2. G.A. Salishchev, Structural changes during hot deformation of titanium alloys and development of a method for processing high qualitative semi-products and articles by using the effect of superplasticity. Thesis of doctor of science, Ufa, (1990), p.475.
3. A.A.Korshunov, F.U.Enikeev, M.I. Mazurski et al. Grain-structure refinement in titanium alloy under different loading schedules. Journal of Materials Science 31 (1996) 4635-4639.
4. V.L.Kolmogorov, Mechanics of plastic metal working, M.: Metallurgy, 1986, 688 p.
5. Y.Combres, C.Levaillant, F.Montheillet, Superplastic behaviour of Ti-6Al-4V alloy investigated by Torsion Testing Titanium Prod. and Appl. Proc. Techn. Programm International Conference, Dayton (1986), V2, p.1163-1174.
6. O.A.Kaibyshev, R.A.Vasin, V.K.Berdin, R.M. Kashaev, Device for study of large plastic deformations of materials under conditions of complex loading. KSHP, №4, (1999), p.8-11.
7. Masataka Tokuda, Yoshiya Inagaki, Hiroaki Yoshida. Semimicroscopic study of non-elastic behaviour of polycristalline metal under varying temperature conditions// ISME International Journal. Series A. (1994), V.3. № 2. P.117-123.

High Temperature Load Relaxation Behavior of Superplastic Iron Aluminides

Jin Hwa Song[1], Hyun Tae Lim[2], Tae Kwon Ha[2] and Young Won Chang[2]

[1] Research Institute of Industrial Science and Engineering, Pohang, Kyungbuk 790-600, Korea

[2] Center for Advanced Aerospace Materials (CAAM), Pohang University of Science and Technology, Pohang, Kyungbuk 790-784, Korea

Keywords: Coarse-Grained Iron Aluminides, Dynamic Recrystallization, Load Relaxation Test, Structural Superplasticity

Abstract

Superplastic deformation and high temperature load relaxation behavior of coarse-grained iron aluminides have been investigated in this study. Iron aluminides with compositions of Fe-28 at.% Al and Fe-28 at.% Al-5 at.% Cr were prepared and thermomechanically treated to obtain an average grain size of about 500 μm. A series of load relaxation tests was conducted at temperatures ranging from 600 to 850°C. The flow curves obtained from load relaxation tests were found to have a sigmoidal shape and to exhibit stress vs. strain rate data in a very wide strain rate range from 10^{-7}/s to 10^{-2}/s. Tensile tests have been conducted at 850°C and at various initial strain rates ranging from 3×10^{-5}/s to 1×10^{-2}/s to investigate the superplastic deformation behavior. Maximum elongation of ~500 % was obtained at the initial strain rate of 3×10^{-5}/s and the maximum strain rate sensitivity was found to be 0.68. Microstructural observation through the optical microscopy has been carried out on the deformed specimens and it has revealed the evidences for grain boundary migration and grain refinement to occur during deformation, suggesting the dynamic recrystallization mechanism as reported earlier. The activation energy was evaluated as 387 kJ/mol in the strain rate range from 10^{-4}/s to 10^{-2}/s, which is very close to that for creep deformation in Fe_3Al alloys.

Introduction

Structural superplasticity (SSP) has been usually observed in fine-grained materials (~ 10 μm) under proper conditions of strain rate and test temperature [1-3]. Some fine-grained intermetallic compounds, such as Ni_3Al, Ni_3Si, Ti_3Al, and TiAl, have been reported to show excellent structural superplasticity [4]. Interestingly, the coarse-grained superplasticity has also been reported in Fe-Al based alloys with a grain size of 100-350 μm [5, 6] or even 700-800 μm [7, 8]. All the deformation characteristics such as a large value of strain rate sensitivity, a low flow stresses independent of strain, and high ductility have been exhibited in large-grained Fe-Al based alloys. The mechanism of large-grained superplasticity has, however, been reported and thought to be different from that of the conventional fine-grained superplasticity.

To understand the exact mechanism of high temperature deformation behavior such as creep and superplasticity, wide range of flow data is necessitated and in this regard, load relaxation test has been strongly recommended. It is well known that load relaxation test can provide a much wider range of strain rates, applying only a little amount of plastic strain to the specimen without an appreciable change in microstructures [9]. It has been attempted, in the present study, to establish a better understanding of the mechanism for the large-grained superplasticity in Fe-Al based alloys by employing the high temperature load relaxation test and by applying the recently proposed internal variable theory of structural superplasticity [10].

Experimental procedure

Iron aluminides with compositions of Fe-28 at.% Al and Fe-28 at.% Al-5 at.% Cr were prepared by vacuum induction melting using the 99.99% purity electrolytic Iron and 99.99% purity Aluminum. The ingots were homogenized at 1000°C for 5 hrs and then rolled from 30 mm to 9 mm in thickness starting at 1000°C and finishing at 800°C. Rod type specimens with the gauge dimensions of 4 mm in diameter and 27 mm in length for load relaxation tests and plate type specimens with 5 mm in length and 3 mm in thickness for tensile tests, respectively, were machined from the hot rolled plates. These specimens were then recrystallized at 857°C for 1 hr followed by oil quenching.

Load relaxation tests were carried out at the temperatures from 600 to 850°C by using a computer controlled electro-mechanical testing machine (Instron 1361 model) attached with a furnace capable of maintaining the temperature fluctuation within ± 1°C. It is well known that load relaxation test can provide a much wider range of strain rates, applying a little amount of plastic strain, 1.5 % in this study, to the specimen without an appreciable change in microstructure [9]. Loading strain rate was 5×10^{-2}/s in all cases. In the load relaxation tests employed in this study, the variation of load with time during the tests was monitored through a DVM and stored in a personal computer for the subsequent analysis. The flow stresses as a function of the inelastic strain rate were determined by following the usual procedure described in the literature [11]. A series of tensile tests were also carried out under the various strain rates ranging from 3×10^{-5}/s to 10^{-2}/s to examine superplastic deformation behavior. Deformed microstructure was observed using the conventional optical microscopy.

Results and discussion

Figure 1 shows the typical microstructure of Fe-28 at.% Al alloy prepared and used in this study. The grain size of the thermomechanically-processed materials used in this study was large around 500 μm and the appearance of each grain was an equiaxed shape. The flow curves obtained from load relaxation tests conducted on Fe-28 at.% Al alloy at temperatures ranging from 600 to 850°C are summarized in Fig. 2. It is noted that flow data evaluated are ranging from near 10^{-7}/s to 10^{-2}/s in strain rate. The flow curves are of the sigmoidal shape with maximum strain rate sensitivity, m, in the intermediate strain rate range, which appeared to shift into faster region with increase in testing temperature. Depending on test temperatures, m values varied from 0.28 to 0.68 in lower strain rate range ($\dot{\varepsilon} \leq 10^{-4}$/s) and from 0.18 to 0.22 in higher strain rate range ($\dot{\varepsilon} \geq 10^{-4}$/s). The m value in higher strain rate range obtained in this study is somewhat lower than that (m ≈ 0.3) obtained by Chu et al. in Fe-27 at.% Al at 800°C [7]. The difference in m value is thought to be attributed to the difference in grain size and test method. Maximum value of m has been obtained as 0.68 in Fe-28 at.% Al alloy, which is the largest ever reported for binary Fe-Al alloys. Close examination of the flow curves reveals that there exist rate-insensitive portions in flow curves except at 850°C, at very low strain rate region around 10^{-7}/s. From the viewpoint of the internal variable theory of inelastic deformation [10], frictional resistance to dislocation motion due to lattices is considered to be predominant mechanism in this strain rate region.

Tensile test results of Fe-28 at.% Al alloy conducted at 850°C under the initial strain rates from 3×10^{-5}/s to 10^{-2}/s are summarized in Fig. 3(a) as engineering stress vs. engineering strain curves. With a decrease in the initial strain rate, elongation to failure increased up to ~500 %, which is the largest elongation ever reported for binary Fe-Al alloys. As noted in Fig. 3(b), the whole gauge section of specimens appeared to be stretched and necking was developed severely.

In order to examine the microstructure evolution, longitudinal optical micrographs were taken from fractured specimens along the gauge section and they are shown in Fig. 4. Some grain boundary cavities were found in the specimen tested at the initial strain rate of 10^{-2}/s. Despite large elongations to failure, elongated grain structure is not prominent but irregularly curved grain boundaries, evidence for possible grain boundary migration, were observed at the strain rates of 10^{-2}/s and 10^{-3}/s. Partially and fully recrystallized structures can be noted in the specimens deformed at 10^{-3}/s and 10^{-4}/s, respectively. Grain refinement can be observed in all cases at the tip region of fractured specimens. As the strain rate decreased, grain refinement appeared to occur over the whole gauge region. The aspects of grain boundary migration and dynamic recrystallization during superplastic deformation in iron aluminides are consistent with other researchers' observations [5-8]. Recently, for example, Chu et al. [8] have summarized the microstructure evolution during superplastic deformation of coarse-grained iron aluminides as the sequence of three stages, i.e. subgrain-boundary formation, grain-boundary migration, and formation of recrystallized grains. In the strain rate range of high strain rate sensitivity, however, neither the irregularly curved grain boundaries nor recrystallized grains could be observed as given in Fig. 5, suggesting that a new deformation mechanism could be operated. Further research is necessitated in this regard. The micrographs were taken from the specimen deformed by 100 % strain at the strain rate of 3×10^{-5}/s at 850 °C. For the comparison's purpose, micrographs taken from the specimen deformed at the strain rate of 10^{-3}/s were shown together.

From the flow data given in Fig. 2, activation energy Q can be calculated by plotting the ln σ against 1/T at a given strain rate [6]. The strain rate of 10^{-4}/s was chosen and the Q value was obtained as 387 kJ/mol in this study as illustrated in Fig. 6. It is interesting to note that the activation energy obtained in the present study is very similar to that for creep deformation of Fe-27 at.% Al alloy [12], suggesting that the mechanism operating during the superplastic deformation of Fe-28 at.% Al alloy in the strain rate range higher than 10^{-4}/s is likely controlled by the lattice diffusion.

The deformation behavior of Fe-28 at.% Al-5 at.% Cr alloy was very similar to that of Fe-28 at.% Al alloy. The flow curves obtained at the temperature range from 700 to 850 °C are shown in Fig. 7 and the maximum value of strain rate sensitivity was evaluated as 0.48, which is somewhat lower than binary alloy. Elongation to failure was also found to be comparable to binary alloy. More detailed description of deformation behavior of Fe-28 at.% Al-5 at.% Cr alloy will be published elsewhere.

Summary

The flow curves of Fe-Al based alloys were sigmoidal shape and the strain rate range of maximum m value, appeared to shift into faster region with increase in testing temperature. A large elongation of about 500 % was obtained in Fe-28 at.% Al alloy with the maximum m value of about 0.68, which is the largest ever reported for binary Fe-Al alloys. Microstructure observation revealed the evidences for the strain-induced grain boundary migration and dynamic recrystallization during the deformation of Fe-28 at.% Al alloy at 850°C in the strain rate range higher than 10^{-4}/s, while a new mechanism is expected to occur in the lower strain rate range. Activation energy Q of 387 kJ/mol suggested that the mechanism operating during the superplastic deformation of Fe-28 at.% Al alloy in the strain rate range higher than 10^{-4}/s is likely controlled by the lattice diffusion.

References

[1] A. H. Chokshi, A. K. Mukherjee, and T. G. Langdon, Mater. Sci. Eng. R10 (1993), p. 237.
[2] O. D. Sherby, and J. Wadsworth, Prog. Mater. Sci., 33 (1989), p. 169.
[3] K. A. Padmanabhan, and G. J. Davis, "*Superplasticity*", Springer-Verlag, New York, NY (1980) p. 11.
[4] T. G. Nieh, J. Wadsworth, and O. D. Sherby, "*Superplasticity in Metals and Ceramics*", Cambridge Univ. Press, Cambridge, (1997), p. 125.
[5] D. Lin, T. L. Lin, A. Shan and M. Chen, Intermetallics 4 (1996) p. 489.
[6] D. Lin, D. Li, and Y. Liu, Intermetallics 6 (1998) , p. 243.
[7] J. P. Chu, I. M. Liu, J. H. Wu, W. Kai, J. Y. Wang, and K. Inoue, Mater. Sci. Eng. A258 (1998) p. 236.
[8] J. P. Chu, H. Y. Yasda, Y. Umakoshi, and K. Inoue, Intermetallics 8 (2000) p. 39.
[9] E. W. Hart, "*Stress Relaxation Testing*", A. Fox ed., ASTM, Baltimore, Md. (1979), p. 5.
[10] T. K. Ha, and Y. W. Chang, Acta Mater. 46 (1998), p. 2741.

[11] D. Lee, and E. W. Hart, Metall. Trans., 2A (1971), p. 1245.
[12] C. G. McKamey, P. J. Masiasz, J. W. Jones, J. Mater. Res. 7 (1992), p. 2089.

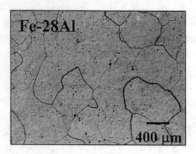

Fig. 1. Typical microstructure of a Fe-28 at.% Al alloy used in this study.

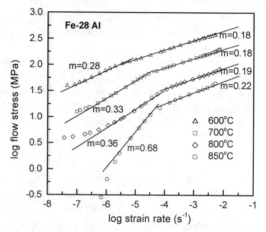

Fig. 2. Flow curves obtained from load relaxation tests on Fe-28 at.% Al alloy conducted at various temperatures.

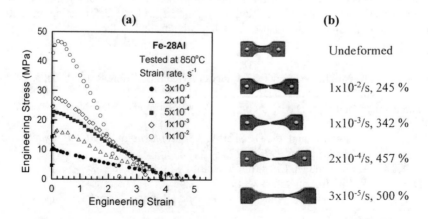

Fig. 3. Tensile test results of Fe-28 at.% Al alloy obtained at 850°C. (a) Engineering stress vs. engineering strain curves. (b) Appearances of some selected tensile specimens.

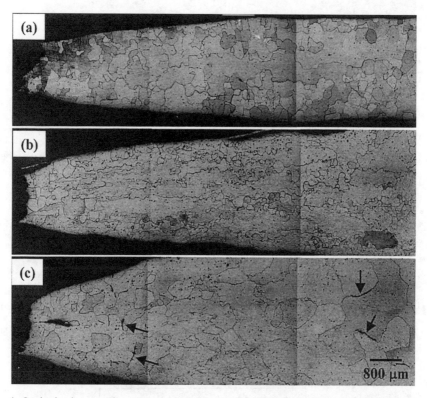

Fig. 4. Optical micrographs taken from surfaces of specimens fractured at 850°C under the strain rate of (a) 2×10^{-4}/s, (b) 1×10^{-3}/s, and (c) 1×10^{-2}/s, respectively.

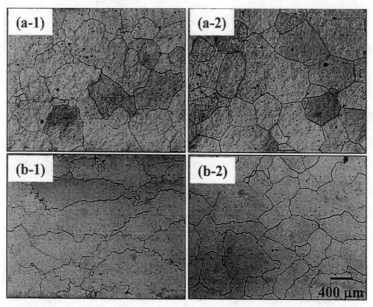

Fig. 5. Optical micrographs taken from the longitudinal cross-sections of specimens deformed at 850°C by 100% strain. (a-1) $\dot{\varepsilon} = 3\times10^{-5}$/s; gauge part, (a-2) $\dot{\varepsilon} = 3\times10^{-5}$/s; grip part, (b-1) $\dot{\varepsilon} = 1\times10^{-3}$/s; gauge part, and (b-2) $\dot{\varepsilon} = 1\times10^{-3}$/s; grip part.

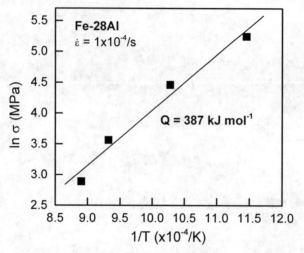

Fig. 6. The plot of ln σ vs. 1/T at the initial strain rate of $\dot{\varepsilon} = 1\times10^{-4}$/s, providing a means to determine the activation energy.

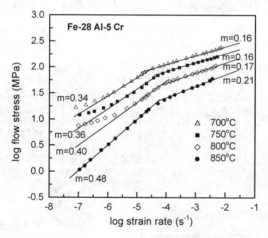

Fig. 7. Flow curves obtained from load relaxation tests on Fe-28 at.% Al-5 at.% Cr alloy conducted at various temperatures.

Glide-Climb Based GBS Model Applied to Submicrocrystalline Superplasticity

A.K. Ghosh

Department of Materials Science and Engineering, University of Michigan,
Ann Arbor MI 48109-2136, USA

Abstract

Research on superplasticity in metals have clearly established the importance of microstructural evolution combined with several phenomena: (i) dislocation emission and absorption at and near grain boundaries (ii) concurrent grain growth and (iii) grain boundary sliding (GBS) during deformation[1]. The combined evolutionary aspects of these processes is essential to superplasticity mechanism. With current interest in submicrocrystalline metallic alloys produced by severe deformation processes, and their reported large degree of strain hardening under the condition of low temperature superplasticity, it is necessary to critically examine the understanding and modeling of their response in these materials containing subgrains and ultrafine grains. After examining several steady state and evolutionary hardening-recovery models, it has become clear that an existing superplasticity model involving a mix of two mechanisms: glide-climb based deformation in the grain mantle and general dislocation creep within the grain core (proposed in 1994) appear to capture all essential features of the experimental behavior. Data for both microcrystalline and submicrocrystalline materials have been reviewed relative to our model and presented here.

Introduction

Ultrafine grain alloys show bimodal distribution of low and high angle grain boundaries and often exhibit a high degree of superplasticity at temperatures below that for conventional micrograin superplasticity. Severely deformed (SD) nanocrystalline alloys show extremely high strength at room temperature and enhancement of strain hardening and strain rate sensitivity at moderate temperatures. It is also reported that their behavior is marked by strong strain hardening response[2] during superplastic flow at low test temperatures, which appears to be poorly understood. Strain hardening during microcrystalline superplasticity is related to grain growth occurring during deformation. Grain growth leads to linear strain hardening at constant strain rate, which also seems to be the case for nanocrystalline alloys[2]. Fig. 1 shows the nature of grain boundaries in the SD nanocrystalline metals from the work of Valiev[3], exhibiting wavy and corrugated grain boundaries even in well annealed condition. It is anticipated from this micrograph that sliding at grain boundaries would be resisted at the grain boundary corrugations and stress concentration effects would arise. In addition, most superplastic alloys contain grain boundary particles which are non-deformable. Thus shear stresses developed at grain boundaries are never uniform, since stress concentration arises at grain corners, particles and other corrugations. Several approaches have been examined to understand the source of strain hardening, by (i) focusing on a combined strain hardening and recovery response and their variations with temperature and strain rate [4], (ii) examining the grain growth induced hardening effect and models which directly incorporate such effect and (iii) analyzing experimental observations relative to the theoretical predictions.

Recovery-Hardening Model

A generalized deformation model[4] has been considered which includes dislocation storage, substructural recovery kinetics, and the variations in the interfacial energies of subgrains and grains during their coarsening. Friedel's network coarsening model[5] is utilized for estimating the recovery element. The nature of the structure of severely deformed nanocrystalline materials may be described by one containing a very high dislocation density, organized either in cells or subgrains and also as additional intragranular dislocations. The high level of stored energy in such structures can be expressed in terms of a rather large prestrain term, ε_0, in the familiar strain hardening equation:

$$\sigma = K (\varepsilon_0 + \varepsilon)^n \qquad (1)$$

During high temperature deformation, as strain (ε) accumulates in the material, the stored energy in the cell walls is minimized by subgrain coarsening thereby altering ε_0. To describe the incremental changes, the differential form of Eq. (1) can then be used:

$$\dot{\varepsilon} + \dot{\varepsilon}_0 = [\sigma^{(1/n-1)} / (n K^{1/n})] \dot{\sigma} \qquad (2)$$

Because subgrain coarsening continues by minimizing subgrain surface area per unit volume of the solid, the resulting removal of stored energy can be described by

$$\dot{\varepsilon}_0 = - (\gamma_s / d_s \sigma) \, d \, (\mathit{ln} \, d_s) / dt \qquad (3)$$

where γ_s = subgrain surface energy and d_s = instantaneous subgrain size. This softening contribution is combined with strain hardening for the intrasubgranular deformation to obtain the overall deformation behavior. Key observations successfully simulated by this combined model include strain-induced hardening over a wide range of strain rate and temperature, specifically for non steady state process such as primary creep in metals. It predicts a hardening curve with diminishing work hardening rate with strain, a saturation stress at large strain, and a strain hardening exponent which increases with increasing deformation rate. These characteristics are observed during warm forming of metals in the dynamic recovery range[6], but do not match the observations cited in Ref[2] for low temperature superplasticity of nanocrystalline metals. After determining that the characteristics of low temperature superplasticity is essentially free from primary creep effects, reexamination of grain growth related hardening and other mechanical characteristics of superplasticity are considered for possible understanding and correlation.

Mantle Creep Based Model involving Glide and Climb

The model presented here was discussed at ICSAM conference in Moscow in 1994 [7], and it is explained here for further clarity. At elevated temperature, the grain boundary bonds are weakened and yet shear and normal stresses must be transferred across them. In these regions, stress transfer gradient and complex loading is set up, and displacements become greater (due to local weakness) than those within the grain interior. This view is essentially the same as the grain mantle vs. core deformation model proposed by Gifkins [8]. While diffusion of atoms is critical to the deformation process, we do not believe that large scale atom transport [9,10] occurs to any appreciable extent in metallic alloys. The details of atom transport, however, involves both glide and climb, as explained below. Figure 2 shows how grain boundary sliding may be resisted at grain boundary steps, ledges, particles (not shown). Discontinuities such as grain corners, ledges, non-deformable particles, and dispersoids on the grain boundaries, capable of creating stress concentrations act as dislocation sources.

Glide is thermally activated stepwise restarting the jumping process of dislocation segments to their next equilibrium position. At low stresses, the glide velocities of the dislocations are low and therefore, under shear + compressive stress field, piled-up dislocations climb out of their planes toward grain boundaries (having lower chemical potential) before gliding through the entire grain core. This local glide-climb process is akin to an accommodating eddy of atoms in the neighborhood of the grain boundary, initiated by a glide threshold, rather than by diffusion. The rows of atoms arriving at the grain boundary, can plate on the boundary as well as travel along it away from the ledge which produced them. The rise of the subsurface atoms to grain boundary surface leads to boundary migration and enhanced concurrent grain growth. Grain boundary sliding and grain growth are thus tied to the same mantle deformation process. The sliding strain is directly proportional to the extension of the grains. When applied stresses are high, some dislocations can still climb to the nearest grain boundaries but many travel across the grain to cause grain core deformation via conventional climb-glide creep. Kinetics of these processes are given below.

Low Stress Behavior
Stress concentration at grain boundary discontinuities is a function of their size and spacing. The largest discontinuities having the largest incompatibility initiate slip first at the lowest stresses. A popular view of source activation is by the statistics of a successful jump of bowed dislocation segment from the grain boundary or a kink segment, onto the adjoining favored slip plane under local state of stress. This glide threshold is a kinetic threshold, τ_0, to produce yield strain at a small quasistatic rate and given by,

$$\tau_0 = v^{*-1} \{ Q - kT \ln [(\dot{\gamma}_s / \dot{\gamma}_0) \cdot (d/d_0)^{-p1}] \} \qquad (4)$$

where v^* is the average activation volume for the successful jump, Q = activation energy, $\dot{\gamma}_s / \dot{\gamma}_0$ is the ratio of quasistatic to reference shear strain rates, d/d_0 is the normalized grain size. The grain size enters into this expression via the probability of finding an activation site on the grain boundary, and the value of exponent p1 can vary between 1 and 3 depending on whether the number of initiation sites is proportional to the available grain boundary surface, or to the number of grain corners per unit volume. If there is rapid grain growth during heating or initial straining of the sample to cause yield, it is clear that threshold stress would rise. This is a major problem when microstructure is not completely stable, as in submicrocrystalline alloys.

The number of activated sources, N, per grain boundary as a function of $(\tau - \tau_0)$ will have a decaying slope toward eventual saturation, approximated by a parabolic function

$$N \propto (\tau - \tau_0)^q \qquad (5)$$

where q = source activation exponent (0.1-1 depending on the alloy, and grain boundary character). The shear displacement rate, $\dot{x}_m$ in the mantle may be expressed as

$$\dot{\gamma}_m \propto b.N.v \qquad (6)$$

where b = Burgers vector and v = glide velocity. As commonly assumed, v is proportional to climb velocity which in turn is related to the volume transport rate of atoms per unit area of the grain boundary driven by the effective stress. It can be shown then that

$$\dot{\gamma}_m \propto \dot{x}_m / d \propto (b\, D_m / d^2)(w/\delta)\,[(\tau - \tau_0)\,\Omega / kT]\,(\tau - \tau_0)^q \qquad (7)$$

where D_m = effective diffusivity in the mantle region, which we believe has the same activation energy as dislocation climb creep, but a much larger preexponential term due to the large vacancy concentration in the mantle, d = grain size, δ = mantle width, w = thickness of climb pipe (dimension ~ 5b), Ω = atomic volume, k = Boltzmann's constant, T = absolute temperature and $(\tau - \tau_0)\,\Omega$ represents chemical potential driving diffusional climb. Since this strain rate directly leads to tensile strain rate, the mantle's contribution to the tensile strain rate can be written as

$$\dot{\varepsilon}_m = (A/d^2)\,(\sigma - \sigma_0)^{1+q} \qquad (8)$$

where $A \propto (b\,\Omega\,D_m)(w/\delta)/kT$ is constant for a fixed temperature, σ is tensile stress and σ_0 is the tensile equivalent of τ_0. The value of q may vary between 0.1 - 0.4 for a well recrystallized grain structure containing a small volume fraction of dispersoids, but for a recovered subgrain containing material with a high volume of fine dispersoids, the stress dependence of source activation can be greater (0.3 < q < 1).

Concurrent Grain Growth
Dynamic grain growth in this model is directly related to the transport occurring from strain contribution due to the mantle region. This could occur in addition to the surface energy driven static grain growth. Thus, instantaneous grain size, d, may be given by

$$d = do + a\,t^p + b\,\varepsilon_m \qquad (9)$$

where do = initial grain size, t = time, a = constant for static grain growth, p = static grain growth exponent (typically 0.5), ε_m = strain contributed from mantle deformation and b = proportionality constant dependent on temperature, alloy chemistry and dispersoid volume fraction.

High Stress Behavior
At high stresses, climb transport is too slow to fully relieve stress concentration at grain boundaries. Dislocations, injected through the grain core, participate in the normal glide-climb creep process with a creep law similar to

$$\dot{\varepsilon}_c = K_1\,\sigma^n\,e^{-Q/kT} \qquad (10)$$

where $\dot{\varepsilon}_c$ = core creep rate, K_1 = preexponential term, Q = activation energy for dislocation creep. Stress concentration at grain corners cannot, however, be relieved by the general dislocation creep alone, and additional local deformation must occur at all grain corners where major slip incompatibility develops. An estimate of this effect has been made in a previous paper [11] which suggests

$$\dot{\varepsilon}_c = (K + A_1/d^3)\sigma^n \qquad (11)$$

The overall superplastic creep rate is then obtained by assuming that both mantle and core deformation contribute to the overall deformation; that is

$$\dot{\varepsilon} = \dot{\varepsilon}_m + \dot{\varepsilon}_c \qquad (12)$$

This two-mechanism summation model has been sucessfully applied to many micrograin alloys, precisely describing the sigmoidal stress-strain rate curves as well as grain growth hardening behavior. The materials include various Al alloys[7], MA21, Ti-6Al-4V, TiAl, Ti_3Al and Ni_3Si[12]. To this author's knowledge no existing model simulates as precisely the observed features of superplastic flow in metal alloys, a basis for serious current application by practitioners. The universality of this model is appreciated by examining the predicted schematic log-log plot in Fig. 3 explaining how changes in the values of q, σ_0, and the relative position of the dislocation creep curve produce values of peak strain rate sensitivity (m) over a wide strain rate range. These plots are shown only as illustration (actual data appears in Ref. 7). It is found that for a low value of σ_0, changing q from 0.1 to 0.9 can change the m value from 0.8 to 0.55, as encountered in micrograin (m~0.8) and submicrograin (~m~0.5) alloys respectively. The range of strain rates over which a high m occurs is also varied by varying the relative position of the dislocation creep curve. No single mechanism-based model or two-mechanism based model involving diffusional creep has such versatility. In the current model, the primary difference among the various materials is controlled by the following physical characteristics: (i) the nature of grain boundary structure, composition, particles and grain size which affect q and σ_0, and (ii) the relative strength difference for core creep compared to σ_0 which is affected by grain size, alloy chemistry and diffusivity. For dynamic grain growth, its kinetics is related to grain boundary character, diffusivity and grain size. The relationship between flow stress and instantaneous grain size is related by a power of magnitude between 2 and 3, divided by (1+q), since these are the exponents for mantle and core creep respectively. Its magnitude can lie between 1 and 3, however, it is not simple to verify its value without first separating threshold stress from Eq. (8) and (12). With this background on the universal constitutive model, attempt is now made to apply it to the behavior of submicrocrytalline alloys.

Review of Related Experimental Data

The following mechanical data relative to both micrograin and submicrograin superplasticity should be considered for a perspective on submicrograin superplasticity. Valiev[3] has demonstrated that room temperature yield strength of 1-1.2 GPa is attainable in Al 1420 (Al-5.5%Mg-2.2%Li-.12%Zr) at 70 nm grain size. However, the microstructure in the SD state is unstable. For superplastic deformation at 300-350C, the relevant grain size is around 200nm - 1 μm [2,3], and tensile elongations are around 850 - 1100% at strain rates in the range of 0.01 -0.1 s^{-1}. Ghosh and Huang[13] have shown that the threshold stress for Al-4.6%Mg-0.7%Mn-0.2%Cr alloy (5083 Al) vary dramatically over a wide range of temperature for different grain sizes down to submicrocrystalline sizes, but in a fairly systematic manner (see Fig. 4). Near room temperature, threshold stress is essentially the same as the yield strength measured in a quasistatic test. On this plot, results of fine grain PM base 7000 series Al alloys (1 and 2), processed conventionally by cold rolling and recrystallization, is also shown[13,14]. At approximately 1μm grain size, these P/M alloys exhibit superplastic elongation values around 1500-1700% at low temperature (460C) and high strain rate (>0.01s^{-1}), far in excess of its ingot metallurgy counterpart (7475 Al at 8-10 μm shows 1000% elongation at 516C and 0.0004s^{-1}). These PM alloys have a room temperature yield strength of about 600 MPa which is similar to or greater than that of the 1420 alloy in the SD state. Fig. 5 (a) shows images of tested specimens of the PM1 alloy, while Fig. 5(b) compares the elongation of this alloy with other alloys currently in use. While 300C data is not available for these alloys, the above results suggest that elongations similar to the 1420 alloy might be possible at 300C. Thus as long as alloy composition and processing method are properly controlled, superplasticity at low temperature or high strain rate may be possible for grain sizes ~ 1μm without using complex SD approaches such as ECAE or torsion under high pressure which are difficult to scale up.

Table I. Chemical composition (wt%), Processing and Grain Size of Al Alloys

Alloy #	Mg	Zn	Cu	Mn	Cr	Zr	Sc	Al	Process	Grain Size(μm)
1	4.6	-	-	0.7	0.2	-	-	bal.	IM / MCF	0.4 - 1.5
2	3.5	4.2	1.6	0.8	0.4	0.3	0.4	bal.	IM-RST / MCF	0.2 - 1.2
3-1	2.5	7.2	2.0	-	0.3	0.3	-	bal.	PM/3axisFORGE	0.6 - 2.5
3-2	2.5	7.2	2.0	-	0.3	0.3	-	bal.	PM / TMT	0.7 - 3.0
4	2.6	7.3	2.1	0.9	-	0.3	-	bal.	PM/3axisFORGE	0.6 - 2.0

IM = Ingot metallurgy source, PM = atomized powder source, TMT = Thermomechanical processing involving overaging, rolling reduction > 90% and recrystallization steps [14], MCF = Multistep Coining-Forging process[13]

Now returning to Fig. 4, examination of the limited data in Refs. 2 and 3 permits addition of a few approximate strength values for Al 1420. Considering the order of grain fineness, from 5083, to dispersoid containing 5083, to PM 7000 alloy and Al 1420, it is clear that the shape of the threshold stress vs. 1000/T curve changes systematically by raising its low temperature strength with concomitant drop in the high temperature glide threshold via decreasing grain size. This temperature dependence of glide resistance, more clearly depicted in the schematic illustration (Fig.6) is expected naturally from an activated flow mechanism. At lower test temperature, yield stress is inversely proportional the barrier spacing, and thus it bears an inverse relationship with grain or subgrain size. The temperature dependence of yield stress in this region is weak but its slope increases with decreasing grain size. Based on the data shown in Fig.4, and Eq. (4), an approximate value of activation energy is found to be around 150 KJ/ mole, close to the value reported for dislocation creep in aluminum alloys. The determination of activation energy is not straightforward because of the associated grain size and strain rate terms involved in Eq. (4).

Fig. 6 shows how the steeper portion of these curves move toward lower temperatures for decreasing grain size, and climb near grain mantle begins at locations with decreasing potency. It is this grain boundary climb creep which is believed to lead to increased strain rate sensitivity and large superplastic elongation [7]. Because the temperature for the onset of mantle climb creep decreases with finer grain size, superplasticity in nanograin materials start at much lower test temperature. High test temperature cause rapid grain growth in such material, thereby rendering the material less superplastic at higher temperature or lower strain rate. Thus a peak in superplastic response in submicrocrystalline materials would occur at temperatues well below that for micrograin alloys.

Next, existing data on grain growth and flow stress for both micrograin and submicrograin alloys are examined to develop an understanding of the grain growth hardening behavior. Table 2 lists data from the literature for a variety of materials with different grain sizes.

Table 2. Grain Size Dependence of Flow Stress during dynamic grain growth

Material (g.s. μm)	Temp °C	Strain Rate, s^{-1}	Ref.	(d_f / d_0)	(σ_f / σ_0)	p'
7475 Al (10)	482	0.001	15	1.6	3.3	2.5
	516	0.0001		1.6	3.5	2.6
PM 7000 -1 Al (1)	460	0.01	14	2.0	2.6	1.4
		0.001	+ this work	2.44	2.5	1.03
Ti-6Al-4V (6.4)	927	0.001	16	1.32	3.0	3.9
		0.0002		1.8	4.0	2.3
Ni$_3$Al (0.08->0.5)	650	0.001	2	2.2	3.8	1.7
Al 1420 (0.1-> 0.8)	300	0.5	2	2.7	4.2	1.4

The approximate value of p' defined as $\log(\sigma_f / \sigma_0) / \log((d_f / d_0))$ ranges from 1.03 to 3.9, with lower values typically seen for finer grain materials and lower test temperatures, but no unusual hardening response of the submicron grain materials is observed. Based on the mantle glide-climb creep model, the value of p' cannot be easily checked, but only through a numerical solution of data utilizing the correct grain growth kinetics. This has been performed for many materials and grain growth hardening has been predicted correctly[7,12] based on the model discussed. An additional factor must be remembered. Due to their ultrafine grain size, SD materials possess very high strength at low temperature. The high threshold stress affects the low temperature flow stress in a significant way which also shows large values; however on a normalized basis the hadening rate of submicron grain materials do not differ from their microcrystalline counterpart. The effect of this on the stress-strain rate behavior is schematically shown in Fig. 7. Compared to the microcrystalline material at high temperature, going to submicron material at low temperature causes a rise in σ_0 as well as a rise in core creep stress, which could shift the transition stress to higher strain rate, i.e. low temperature superplasticity would be obtained at a high strain rate, with a very high stress level. The upward vertical arrows indicate possible strain hardening during grain growth. For the low temperature case, equal size increment on a log stress scale might mean a significantly higher absolute amount of strain hardening, although the behavior pattern fundamentally remains the same. For the high temperature case, submicron grain material shows lower flow stress, but the transition stress shifts to a lower strain rate, indicating the inavailability of high strain rate superplasticity under this condition.

Summary
Experimental results on nanocrystalline materials were considered from the work of Misra et al [2], Valiev[3], Ghosh and Huang[13] for Ni$_3$Al, Alloy 1420, 5083 Al and PM7000 series Al. In all

cases, materials which are essentially brittle at room temperature were found to exhibit extended ductility, or a reasonable degree of superplasticity at mildly elevated temperature, with increasing degree of strain hardening as the applied strain rate increased. This low temperature superplasticity at high strain rate was found to be not related to conventional strain hardening and stress-assisted recovery process responsible for primary creep. It is directly related to conventional superplasticity effects previously explained on the basis of a mantle-core model of superplasticity in which dislocation sources are activated at the grain boundaries, and glide and climb in the grain mantle region at low stresses control both grain boundary sliding and concurrent grain growth[7]. The model accurately predicts stress-strain rate characteristics of microcrystalline and submicrocrystalline metals with fully and partially recrystallized microstructures and their strain hardening characteristics. While the submicron grain materials are very strong at low temperatures, climb creep can also begin at a moderately elevated temperature, leading to a high enough strain rate sensitivity and superplasticity at high strain rate and low temperature. In spite of the high strength and high strain hardening rate, on a normalized basis submicron superplasticity is fundamentally the same as micrograin superplasticity, except for lower strain rate sensitivity values and reduced cavitation.

References

[1] D.H.Bae and A.K.Ghosh, *Acta Mater.*, v. 48, April 2000, p. 1207-1224.
[2] R.S.Misra, S.X. McFadden and A.K.Mukherjee, *Investigations and Applications of Severe Plastic Deformation*, Lowe and Valiev eds., Klewer Academic Publishers, 2000, p.231-240.
[3] R.Z.Valiev, *Investigations and Applications of Severe Plastic Deformation*, Lowe and Valiev eds., Klewer Academic Publishers, 2000, p.221-230.
[4] A.K.Ghosh, *Acta Met.*, v. 28, 1980, p.1443-1465
[5] J. Friedel, *Dislocations*, p.278, Pergamon Press, Oxford (1956)
[6] A. Riposan and A.K.Ghosh, unpublished research, University of Michigan, 2000.
[7] A.K.Ghosh, *Mat. Sci. Forum*, v. 170-172 , 1994, p. 39-46
[8] R.C. Gifkins, Met. Trans. A, 7A, 1225 (1976), also R.C. Gifkins, Met. Trans. A, 8A, 1507 (1977).
[9] R.L. Coble, J. Appl. Phys., 34, 1679 (1963).
[10] M.F. Ashby and R.A. Verrall, Acta Met., 21, 149 (1973).
[11] A.K. Ghosh and A. Basu in Critical Issues in the Development of High Temperature Structural Materials, Edited by Stoloff, Duquette, and Giamei (TMS Publication, 1993) p. 291.
[12] F. Enikeev, to be published in this volume, 2000
[13] A.K.Ghosh and W. Huang, *Ultrafine Grain Materials*, R.S.Misra, S.L.Semiatin...eds, TMS, 2000, p.173-184.
[14] A.K. Ghosh and C. Gandhi, U.S. Patent 4,770,848, September 1988.
[15] A.K. Ghosh in Deformation of Polycrystals: Mechanics and Microstructures, (Proc. 2nd Riso Intl. Symp., Roskilde, Denmark, 1981) p. 277. also A.K. Ghosh and C.H. Hamilton, Met. Trans. A., 13A, 1955 (1982).
[16] A.K. Ghosh and C.H. Hamilton, Met. Trans. A., 10A, 699 (1979).

Acknowledgements

This work was performed under support from US Dept of Energy under grant FG02-96ER45608-A000, and Century Aluminum Co.

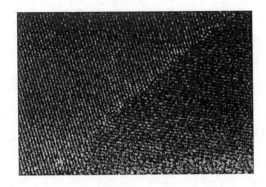

Fig. 1. High Resolution TEM image of grain boundary in nanograin Al 1420 (from Valiev, Ref. 3), showing distorted region along grain boundary

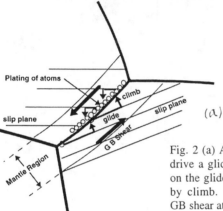

Fig. 2 (a) An overall schematic showing how grain boundary shear drive a glide step into the grain. At low stresses the packed atoms on the glide plane waiting for the next glide jump are carried away by climb. (b) details showing stress concentration arising during GB shear at the ledges.

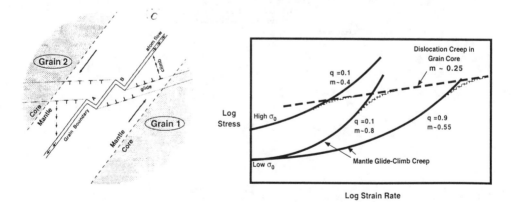

Fig.3 A representative diagram to illustrate the effects of prediction based on key parameters in the glide-climb mantle + core creep model over a wide range of strain rates.

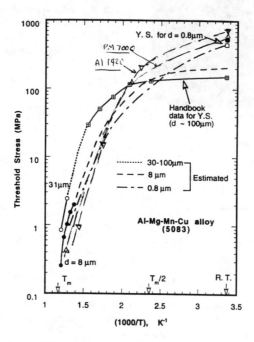

Fig. 4. Threshold stress and / or yield strength determined from several high temperature and room temperature tests 5083 Al, possessing a variety of grain sizes. Intermediate temperature data are from Aluminum Handbook, and the lines labeled "estimated" mean expected interpolation between measured data points.

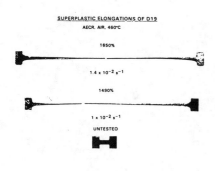
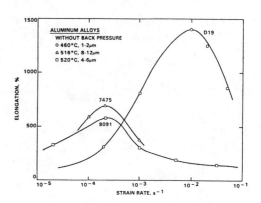

Fig. 5 (a) UFG PM 7000 alloy processed by conventional TMT shows low temperature and high strain rate superplasticity. (b) This alloy exhibits superior tensile elongation at higher strain rate in comparison to micrograin alloys.

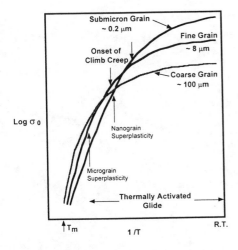
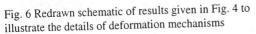

Fig. 6 Redrawn schematic of results given in Fig. 4 to illustrate the details of deformation mechanisms

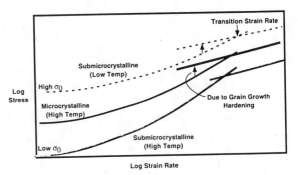

Fig. 7 Schematic illustration of possible predictions of Mantle glide-climb GBS model, for microcrystalline and submicrocrystalline alloys

Role of Alloying Element in Superplastic Deformation of Aluminum Alloys

N. Furushiro[1], Y. Umakoshi[2] and K. Warashina[3]

[1] International Student Center and Department of Materials Science and Engineering, Graduate School of Eng., Osaka University, 2-7, Yamadaoka, Suita, Osaka 565-0871, Japan

[2] Department of Materials Science and Engineering, Graduate School of Eng., Osaka University, 2-1, Yamadaoka, Suita, Osaka 565-0871, Japan

[2] Graduate Student of Osaka Univ., Presently: Tomioka Works, Tanaka Kikinzoku Kogyo Co. Ltd., Ichinomiya, Tomioka Gunma 370-2300, Japan

Keywords: Al-Cu-Zr Alloy, Grain Boundary Sliding, Misorientation of Grain Boundaries, Role of Solute Atoms

Abstract

It is well known that metallic materials which exhibit typical superplasticity are generally not pure metals but alloys. The alloying elements exist in different phases or as solute atoms. Role of the solute atoms during the deformation has been a great interest not only in making clear the deformation mechanism but also in designing new materials. The objective of the present study is to examine the role of copper addition in Al-Cu-Zr alloys. Alloys of Al-Cu-0.3%Zr were prepared for the specimens, copper contents of which were 0, 0.5, 0.9, 1.6, 3.1, 4.9 and 6.4%. Tensile specimens were cut from the sheets warm-rolled at 623K. Tensile tests were carried out at 623-773K and at 1.7×10^{-4}-1.7×10^{-2} s^{-1}. Microstructures were observed on the specimens of deformed or non-deformed condition by TEM. In all specimens, Zr atoms are found to be precipitated as fine particles of metastable Al_3Zr. The large elongations were obtained in specimens including particles of the θ (Al_2Cu) phase. Even in alloys of single phase (solid solution), the elongation increased with an increases of copper content. That implies that solute copper also has a significant effect on superplasticity. Deformation characteristics of the Al-x%Cu-0.3%Zr alloys considerably depend on the copper content. That means the existence of the optimum copper content for superplasticity of the alloys. The role of solute copper is to increase flow stress of the matrix, while the fine particles of the θ phase are useful to both form and hold the fine grain structure.
Using the present results, the minimum misorientation at grain boundary sliding in superplastic aluminum alloys will be also discussed.

Introduction

It is well known that metallic materials which exhibit typical superplasticity are generally not pure metals but alloys. The alloying elements exist in different phases or as solute atoms. Role of the solute atoms during the deformation has been a great interest not only in making clear the deformation mechanism but also in designing new superplastic materials. However, there has been few investigations on the role of the alloying elements in the view of both second phases and solutes. The objective of the present study, therefore, is to examine the role of solute copper and the θ phase in superplastic deformation of Al-Cu-Zr alloys known as one of typical superplastic aluminum alloys [1,2].

Experimental

Alloys of Al-Cu-0.3%Zr were prepared for specimens in the present study, copper contents of which were 0, 0.5, 0.9, 1.6, 3.1, 4.9 and 6.4mass% as shown Table 1. The specimens are called simply by each copper content such as 0Cu, 0.5Cu and so on. These alloys were warm-rolled at 623K with a reduction of 80%. Tensile specimens were cut from the sheets, tensile direction of which are parallel to the rolling direction. Tensile tests were carried out at 623-773K and at $1.7 \times 10^{-4} - 1.7 \times 10^{-2}$ s^{-1}. To evaluate of grain size of the specimens, microstructures were observed on the specimens of deformed or non-deformed condition by TEM. The Kikuchi pattern analysis was carried out to examine misorientation of grain boundaries in these specimens.

Table 1 Chemical compositions of the Al-Cu-Zr alloys used (mass %)

Specimens	0Cu	0.5Cu	0.9Cu	1.6Cu	3.1Cu	4.9Cu	6.4Cu
Cu	<0.01	0.51	0.94	1.59	3.10	4.85	6.41
Zr	0.25	0.32	0.32	0.28	0.32	0.29	0.27

Results and Discussion

Microstructure of all specimens here revealed equiaxed fine grain strucuture. Since Zr atoms are considered to be precipitated as fine particles of metastable Al$_3$Zr, the effect of Zr addition on their deformation behavior is able to be considered constant in all specimens here. In specimens including Cu of more than 1.6%, the second phase of θ (Al$_2$Cu) is observed. This is consistent to the expected result from the equilibrium phase diagram of the Al-Cu system. The θ phase exists in specimens of 3.1%Cu or more.

Grain size distribution of all specimens just before starting tests at 673K is shown in Fig. 1. Very fine

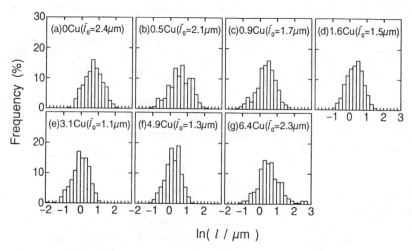

Fig. 1 Grain size distribution of Al-x%Cu-0.3%Zr alloys begore loading at 673K.

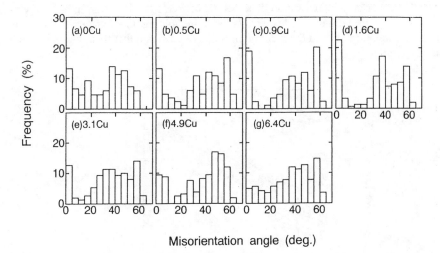

Fig. 2 Distribution of misorientation angle at grain boundaries in Al-x%Cu-0.3%Zr alloys begore loading at 673K.

grain structures of 1.1 and 1.3 μm were obtained in specimens of 3.1Cu and 4.9Cu, respectively, both of which include the θ phase. As shown in Fig. 2, misorientation distribution of grain boundaries in all specimens is also estimated. There are more higher angle grain boundaries in specimens of smaller content of Cu. It has been still unknown of the minimum grain boundary misorientaion angle, at which grain boundaru sliding will be take place in superplastic deformation [3].

Elongations of specimens are found to be much influenced by Cu content and the strain rate, as shown in Fig. 3. The large elongations were obtained in specimens including particles of the θ (Al$_2$Cu) phase. However, it should be emphasized that, even in the solid solution alloys, the elongation increased with an increases of copper content, although the amount of elongations is not so large. This implies that solute copper also has a significant effect on superplasticity.

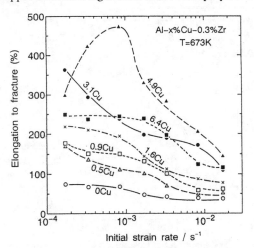

Fig. 3 Effect of the initial strain rate on elongation To fracuture of Al-x%Cu-0.3%Zr alloys deformed at 673K.

The role of solute copper is to increase flow stress of the matrix. In addition, grain boundary migration will be suppressed by the solutes. This results both in holding the fine grain structure and in accelerating the deformation at or in the vicinity of grain boundaries rather than within the grains, such as grain boundary sliding [4].

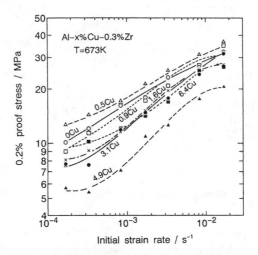

Fig. 4 Effect of the initial strain rate on the 0.2% proof stress of Al-x%Cu-0.3%Zr alloys deformed at 673K.

In Fig. 4, strain rate dependence of 0.2% flow stress is shown for various specimens with different contents of copper. For specimens containing copper, the flow stress is the minimum for 4.9Cu. decreases with an increase of copper content. This may be opposite tendency to that in the copper content dependence of elongation. Strain rate sensitivities of flow stresses [5], m, changed with the

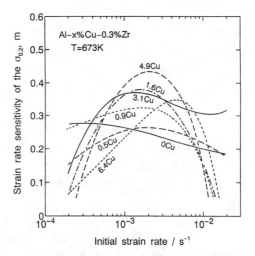

Fig. 5 Effect of the initial strain rate on the strain rate sennsitivity of flow stress, m, of Al-x%Cu-0.3%Zr alloys deformaed at 673K.

strain rate, as shown in Fig. 5. The maximum sensitivity for each specimen was 0.28, 0.33, 0.37. 0.44 and 0.35 for 0.5Cu, 1.6Cu, 3.1Cu, 4.9Cu and 6.4Cu, respectively.
The fine particles of the θ phase are useful to both form and hold the fine grain structure. During the deformation, dynamic recrystallization will take place more easily around the particles. This also means these particles will act an important role in superplasticity. It should be also pointed out that the volume fraction and size dstribution will also have a siginificant effct on superplastic behavior. That means the existence of the optimum copper content for superplasticity of the alloys.
Systematic researchs on the optimum composition of the superplastic Al-Cu-Zr alloy have been carried out by manay wokers [6,7]. The objective of this paper, however, is not to find a optimum composition of the superplasric Al-Cu-Zr alloy, but to discuss the role of the alloying elements.

Conclusions

Deformation characteristics of the Al-x%Cu-0.3%Zr alloys considerably depend on the copper content. That means the existence of the optimum copper content for superplasticity of the alloys.
The role of solute copper is to increase flow stress of the matrix. In addition, grain boundary migration will be suppressed by the solutes. This results both in holding the fine grain structure and in accelerating the deformation at or in the vicinity of grain boundaries, such as grain boundary sliding.
The fine particles of the θ phase are useful to both form and hold the fine grain structure. During the deformation, dynamic recrystallization will take place more easily around the particles. This also means these particles will act an important role in superplasticity.

Acknowledgements
The authors wish to thank Mr. W. Fujitani for his experimental collaboration.

References
[1] B.M. Watts, M.J. Stowell. B.L. Baikie and D.G. Owen, Metal Sci. 10 (1976), p. 189.
[2] B.M. Watts, M.J. Stowell. B.L. Baikie and D.G. Owen, Metal Sci. 10 (1976), p. 198.
[3] F. Weiberg, Trans. AIME 212 (1958) p. 808.
[4] N. Furushiro, M. Toyoda abd S. Hori, Acta Metall. 36 (1988) p. 523.
[5] W.A. Backofen, I.R. Turner and D.H. Avery, Trans. ASM 57 (1964) p. 980.
[6] R. Grimes, M.J.Stowell and Watts, "*Superpalsticity in Aerospace*", p.97-113, ed. C.Heikkenen and T.R.McNelly, TMS-AIME, Warrendale, PA, 1988.
[7] A.J.Barnes, *ibid*, p. 301-313.

Texture, Grain Boundaries and Deformation of Superplastic Aluminum Alloys

M.T. Pérez-Prado[1,3], T.R. McNelley[2], G. González-Doncel[1] and O.A. Ruano[1]

[1] Department of Physical Metallurgy, Centro Nacional de Investigaciones Metalúrgicas, CSIC, Avda. de Gregorio del Amo, 8, ES-28040 Madrid, Spain

[2] Department of Mechanical Engineering, Naval Postgradute School, 700 Dyer Road, Monterey, CA 93943-5146, USA

[3] Present address: Department of Mechanical and Aerospace Engineering, University of California, 9500 Gilman Dr., La Jolla, CA 92093-0411, USA

Keywords: Aluminum Alloys, Deformation Banding, Grain Boundaries, Texture

Abstract

Two distinct microstructural transformation processes have been observed to enable superplastic response in aluminum alloys, depending of composition and thermomechanical processing (TMP). Texture observations, including microtexture data acquired by newly developed electron backcsatter diffraction (EBSD) analysis methods, and mechanical property results are summarized in this work. These data reveal a clear dependence of superplastic response on the mechanism of microstructural transformation that leads to a superplastically enabled condition. Transformation by continuous recrystallization involves recovery-dominated changes that occur gradually and homogeneously throughout the deformation microstructure in the absence of high angle boundary formation and migration. Important features of microstructural evolution include deformation banding leading to grain boundaries that are interfaces between symmetric variants of the main texture component, and the presence of a fine, cellular structure within the bands. Both grain boundary sliding and dislocation creep operate in the superplastic regime and alloys such as Al-5%Ca-5%Zn and Supral 2004 respond in this manner. Alternatively, transformation via primary recrystallization processes involves the heterogeneous formation and growth of grains by the migration of high-angle grain boundaries from the deformation zones surrounding coarse precipitate particles. This leads to essentially random grain orientations and a predominance of high-angle boundaries in the microstructure. The mechanical behavior of these materials is governed by grain boundary sliding in the superplastic range and alloys such as 5083 and 7475 behave in this way. The implications of these results for aluminum alloy development and the design of deformation processing procedures for grain refinement and superplasticity will be discussed. Finally, we present a table describing the superplastic behavior of a number of aluminum alloys according to the two patterns mentioned above.

Introduction

Two different microstructural transformation processes are currently utilized to enable superplasticity in aluminum alloys, depending on alloy composition and processing route. Discontinuous, primary recrystallization results in equiaxed, randomly oriented grains. The mechanical behavior of these alloys can be described by current, well-accepted models of superplastic deformation. The second microstructural transformation, continuous recrystallization, is a recovery-dominated process that takes place homogeneously throughout the microstructure [1,2]. Primary, or discontinuous recrystallization is suppressed by the addition of fine, second-phase particles or by alloying additions. During thermomechanical processing (TMP) of alloys that transform via continuous recrystallization, deformation banding takes place. The as-received material is formed by grains elongated along the rolling direction (RD) and oriented as alternating variants of the main texture component. The deformation behavior of these alloys can not be accurately described by current models of superplastic deformation. These alloys seem promising since they are capable of achieving superplastic deformation at higher strain rates.

In this work, a model for the deformation behavior of alloys that transform by continuous recrystallization will be presented. It will be seen that both GBS and dislocation creep contribute to deformation of these alloys over a wide range of deformation conditions. The implications of this model for the thermomechanical processing (TMP) of aluminum alloys will be discussed. A review of available data regarding texture analysis and superplastic behavior of aluminum alloys is presented. These alloys include Al-5%Ca-5%Zn, Al-6%Cu-0.4%Zr (Supral 2004), Al-Li (8090 and 2090), Al 5083, Al 7475 and Al 2519 that have been evaluated by us and also Al-33Cu and other Al-Cu-Zr alloys for which results are available in the literature.

Experimental procedure

The model presented in this work has been developed by analyzing mechanical test data as well as texture measurements reported in previous papers corresponding to several superplastic aluminum alloys that transform via continuous recrystallization [3-9]. Experimental results will be presented occasionally to illustrate the different aspects of the model. In particular, data from the alloy Supral 2004 [4,5] will be utilized.

Results and discussion

A deformation banding model for alloys that transform via continuous recrystallization

In previous studies [4, 5] it has been shown that the as-received microstructure of Supral 2004 is formed by grains elongated along RD (bands) and oriented as alternating variants of the main texture component. Figure 1 is a schematic of one such microstructure. For Supral 2004, the main texture component was B, {011}<211> (the corresponding symmetric variants are B_1 (110)[1-12], and B_2 (011)[2-11]). In general, these alloys exhibit typical rolling textures and the main texture components belong to the β-fiber in Euler space [7,8]. An additional characteristic feature of this type of alloy is a bimodal distribution of grain-to-grain disorientation angles [6]. This is consistent with the development, during processing, of a cellular substructure within the variant grains. The high-angle boundary peak corresponds to boundaries between symmetric variants, and the low angle boundary peak, of misorientation ranging from 5° to 15°, corresponds to boundaries between cells. Figure 2 shows a disorientation distribution histogram corresponding to the alloy Supral 2004 in the as-processed condition. Annealing of this microstructure at high temperature results in the retention of the initial texture. Simultaneously, gradual sharpening of the cell walls and grain growth take place.

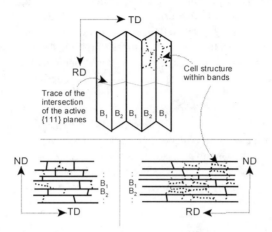

Figure 1. Schematic showing the as-received microstructure of alloys that transform via continuous recrystallization. View along the three orthogonal planes RD-TD, TD-ND, and RD-ND.

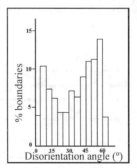

Figure 2. Bimodal disorientation distribution corresponding to Supral 2004 in the as-received condition.

The texture evolution of these alloys during deformation within and without the superplastic regime has been recently reported [6-9]. After deformation under non-superplastic conditions either the initial texture is retained and sharpened, or grain rotation takes place until other β-fiber orientations are stabilized [7-9]. During deformation under highly superplastic conditions texture weakening takes place, but still either the as-received texture or, correspondingly, other β-fiber orientations are retained after elongations as high as 600%. The disorientation distribution resembles that of a random distribution of cubes [10]. It was concluded [9] that both grain boundary sliding and crystallographic slip operate within such a structure during deformation in the superplastic regime.

The model proposed in this paper to describe the microstructural evolution during superplastic deformation of alloys that transform via continuous recrystallization is depicted in Figure 3. As described before, at the onset of deformation (Fig. 3a), cell walls disoriented mainly between 5° and 15° exist within the bands. These boundaries are not suitable for sliding, and therefore the material starts deforming predominantly by crystallographic slip, which leads either to the retention of the initial texture or to the stabilization of other orientations of the β-fiber in Euler space. During the first stages of deformation (Fig. 3b), crystallographic slip coupled with temperature lead to increasing sharpening of the cell walls by climb and cross slip of dislocations. It is not clear how crystallographic slip contributes to accelerate cell wall sharpening. After a certain *"critical deformation (ε_c)"* (Fig. 3c) a large number of cell walls are well defined and, thus, suitable for sliding. At this point grain boundary sliding takes over and predominates during the following stages of deformation (Fig. 3d). For the strain interval during which the cell boundaries are evolving both crystallographic slip and GBS contribute simultaneously - but the relative contributions are changing. The critical deformation is alloy-dependent. Although this would be the subject of further work, preliminary evidence is showing that, for Al-5Ca-5Zn, ε_c >200%; for Al-Cu-Zr, ε_c =70-130%; for 8090, ε_c =50-100%; for Al-33Cu, ε_c ~100%. Alloys that transform via primary recrystallization would be equivalent to alloys that transform via continuous recrystallization having a critical deformation of 0%.

The microstructural evolution depicted in Figure 3 has been observed by the present authors in Al-5%Ca-5%Zn [9], Supral 2004, 2090 and 8090 [4,7-8]. However, this behavior seems to be more general since there are several other alloys described in the literature that appear to behave in a similar way. Table 1 summarizes available data regarding texture analysis and deformation behavior of several aluminum alloys.

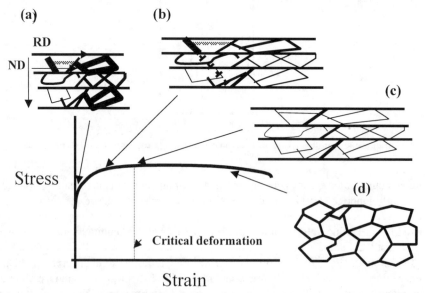

Figure 3. Schematic showing the microstructural evolution during superplastic deformation of alloys that transform via continuous recrystallization. (a) Onset of crystallographic slip; (b) cell wall sharpening; (c) critical deformation-onset of grain boundary sliding; (d) predominance of GBS.

Implications of this model for the design of thermomechanical processing routes

The model presented here for the microstructural evolution during superplastic deformation of alloys that transform via continuous recrystallization has important implications for the design of thermomechanical processing routes. This model suggests that improved superplastic response would be achieved if a large number of small, highly disoriented cells, were present in the microstructure after the TMP. The cell walls would rapidly evolve into well defined boundaries, capable of sliding, during the first stages of deformation. "In-situ" grain refinement would take place, thus, during the course of superplastic deformation.

Conclusions

1. A model is proposed for the microstructural evolution during superplasticity of alloys that transform via continuous recrystallization. At the onset of deformation, the microstructure of these alloys comprises bands elongated along the rolling direction and oriented as alternating symmetric variants of the main texture components. Within the bands, a cellular substructure develops. During the first stages of deformation crystallographic slip predominates and cell wall sharpening takes place. After a "critical deformation", when a sufficient number of cell walls are well defined and thus capable of sliding, grain boundary sliding takes over and predominates until the failure of the material. While cells are evolving, both crystallographic slip and grain boundary sliding operate simultaneously. This pattern of behavior seems to be quite general. Several examples of alloys whose behavior conforms to this pattern are presented.

2. Important implications for the design of thermomechanical processing routes can be inferred from this model. In particular, improved superplastic response in alloys that transform via continuous recrystallization would be achieved if smaller, highly disoriented cells were present in the microstructure as a result of the thermomechanical processing.

References

[1] F.J. Humphreys, Acta Metall., 25 (1977), p. 1323.
[2] R.D. Doherty, D.A. Hughes, F.J. Humphreys, J.J. Jonas, D. Juul-Jensen, M.E. Kassner, W.E. King, T.R. McNelley, H.J. McQueen and A.D. Rollet, Mater. Sci. Eng. A, A238 (1997), p. 219.
[3] M.T. Pérez-Prado, M.C. Cristina, O.A. Ruano and G. González-Doncel, J. Mater. Sci. 32 (1997) p.1313.
[4] T.R. McNelley, M.E. McMahon and M. T. Pérez-Prado, Phil. Trans. R. Soc. Lond. A 357 (1999) p.1683.
[5] T.R. McNelley and M.E. McMahon, Metall. Mater. Trans. A 28 (1997) p.1879.
[6] M.T. Pérez-Prado, T.R. McNelley, O.A. Ruano and G. González-Doncel, Metall. Mater. Trans. A 29 (1998) p.485.
[7] M.T. Pérez-Prado, M.C. Cristina, O.A. Ruano and G. González-Doncel, Textures and Microstructures, in press.
[8] M.T. Pérez-Prado, M.C. Cristina, O.A. Ruano and G. González-Doncel, Mater. Trans. JIM, in press.
[9] M.T. Pérez-Prado, M.C. Cristina, O.A. Ruano and G. González-Doncel, Mater. Sci. Eng. 244 (1998) p.216.
[10] J.K. MacKenzie, Acta Metall. 12 (1964), p.223.
[11] M.T. Pérez-Prado, Ph.D. thesis, Universidad Complutense de Madrid, Madrid, Spain, 1998.
[12] M.T. Pérez-Prado and G. González-Doncel, Textures and Microstructures 34 (2000), p.34.
[13] C.P. Cutler, J.W. Edington, J.S. Kallend and K.N. Melton, Acta Metall. 22 (1974), p.665.
[14] M.E. McMahon, Ph.D. thesis, Naval Postgraduate School, Monterey, CA, 1996.
[15] P. Partridge, A.W. Bowen, C.D. Ingelbrecht and D.S. MacDarmaid, Proc. Int. Conf. Superplasticity, B.Baudelet and M. Suery (eds.), Editions du CNRS, Paris (1985), p.10.1.
[16] K. Matsuki, Y. Uetani, M. Yamada and Y. Murakami, Met. Sci. 10 (1976), p.235.
[17] P.L. Blackwell and P.S. Bate, Metall. Trans. 24A (1983), p.1085.

Alloy	Initial texture (after TMP)	Initial disorientation (after TMP)	Annealing texture	Annealing disorientation distribution	Deformation texture	Disorientation distribution after deformation	Reference
Al-5Ca-5Zn	Near C	Random	Sharp C	Bimodal	SP-L: Weak Cu SP-T: Weak B SLIP-L: Sharp C SLIP-T: Sharp B	SP: Random SLIP: Bimodal	3,6,9
Al-6Cu-0.4Zr Supral 2004	α fiber, close to B.	Bimodal	Sharper initial texture	Bimodal	SP: weaker ε<1: GBS+SLIP ε>1: only GBS SLIP: Sharp B	SP: ε<1: Bimodal ε>1: Random SLIP: Bimodal	4,5
Al-2.4Li-1.2Cu Al 8090 (mid-layer)	β-fiber, close to {156}<877>				SP-L: near C, weaker SP-T: B, weaker SLIP-L: C SLIP-T: B		8,11
Al-2.3Li-2.7Cu Al 2090 (mid-layer)	B				SP-L: near C, weaker SP-T: B, weaker SLIP-L: C SLIP-T: B		7,11
Al-5.7Zn-2.1Mg Al 7475	Along α and β fibers-Weak				SLIP: Retention and strengthening of α and β fiber orientations		12
Al-33Cu (Al-phase)			{103}<301> Skeleton tube <0,-1,0>		SP: ε=50%: Weaker skel. tube, sharper <0,-1,0> ε=200%: all weaker SLIP: <111>fiber		13
Al 5083	Weak Goss	Random	Weaker Goss and {100}<0vw	Random	SP: Random SLIP: <111> fiber	SP:Random SLIP: 0-10°+ random	11
Al-10Mg-0.1Zr			B, S	Bimodal (TMP6) Random(TMP2)	SP: Weaker	Random	5
Al 7475 Al5.5Zn2.3Mg	Cube	Near random	Weaker	Bimodal	SP: Weaker		14
Al 7475 Al5.5Zn2.3Mg	{112}<111>		Weaker		SP: Weaker but rotation: <111>//TA, (111)<110>slip		15
Al-6Mg-0.4Zr			(013)[-100] (115)[-150]		SP: Weaker		16
Al 8090 Al-2.4Li-1.2Cu			B		SP: Weaker for ε>50%		17
Al-8090	{110}<112>		Sharper initial texture		SP: Weaker but rotation about ND <110>		15

Table 1. Review of available data regarding texture analysis and superplastic behavior of aluminum alloys.
Key: S= {123}<634>, Cube = {100}<001>, Cu= {112}<111>, B= {011}<211>, Goss = {011}<100>;TA= Tensile axis, SP= superplastic deformation region, SLIP= slip creep region, L=longitudinal test; T=transverse test.

A Thermodynamic Framework for the Superplastic Response of Materials

K.R. Rajagopal[1] and N. Chandra[2]

[1] Texas A & M University, 701 H R Bright Building, College Station TX 77843-3141, USA

[2] Department of Mechanical Engineering, Florida State University, Tallahassee FL 32310, USA

Keywords: Constitutive Equation, Natural Configurations, Superplastic Response, Thermodynamics

ABSTRACT

Over the years, a number of constitutive equations have been proposed to model the thermomechanical behaviour of superplastic materials. Most of those relations were motivated by phenemenological observations and restricted to uniaxial forms at (high homologous) superplastic temperatures. Not only do they lack the rigor of mechanics (e.g. finite deformation, frame invariance, three dimensional form) but also do not potray the post-formed room temperature response. The latter is an important criterion in determining if a superplastically formed component can be used in a critical industrial application. In this work, we attempt to address these concerns by developing superplastic constitutive relationships within the frame work of nonlinear mechanics and thermodynamics by invoking the concept of natural configurations. This is an imposing but important task. This work (the first in a series of papers) outlines the basic principles of such a formulation and enumerates the restrictions arising from the structural and mechanical observations of superplastic materials in different temperature regimes.

INTRODUCTION

The ability of certain class of solid materials to exhibit very large tensile elongations under some specific conditions of microstructure, temperature and rates of deformation has been the focus of intense research over many years.The mechanical response characteristic of solid materials depends on a variety of factors: the strain rate, temperature, the microstructures and possibly other fields such as elctrical and magnetic fields. In the absence of electrical and magnetic fields, and at a fixed strain rate, certain classes of materials can undergo very large tensile elongations, without any structural damage. Such a response is referred to as "superplastic response". Below this range they behave essentially as strain rate independent or strain rate dependent viscoplastic material. The temperature at which such superplastic response starts manifesting itself is approximately 70% of the homologous melting temperature of the material. Above the temperature interval wherein we observe superplastic response, the materials behaves like a viscous fluid.

Superplastic response is a consequence of the increased mobility of the grain boundaries due to the increase in temperrature. For the superplastic response to be structurally damage free, it is essential that certain accommodation mechanism such as dislocations, diffusion of vacancies, motion of atoms perpendicular to boundary surface or even partial melting come into play. Such accommodating mechanisms, while present at all temperatures, acquire different levels of significance as the temperaute changes. At the appropriate temperature, which could vary from one class of materials to another, the different accommodation mechanisms conjoin in an appropriate manner to engender superplastic response. Materials that are capable of such superplastic response are referred to as superplastic materials in the technical literature. Superplastic bodies can be unloaded to a stress-free state and we shall see later that these stress free states denote different natural configurations, and play an important role in determining the response of such bodies. Thus superplastic materials, at sufficiently low temperature are solid-like while at higher temperatures they behave in a fluid-like manner.In general, most bodies are capable of existing stress free in more than one configuration, and their response characteristics depend on the natural configuration to which the body would return on the removal of the loads. Traditional metal plasticity, twinning and solid to solid phase transitions are examples of the above notion. Moreover, the material can possess different material symmetry with regard to these configurations.

The notion of a configuration is a local notion (see Truesdell and Noll (1992), Rajagopal (2000)). However, within the context of homogeneous bodies we can talk about the configuration as a global quantity. When a body is subject to an inhomogeneous deformation that leads to inelastic response, on being unloaded to a traction free state it cannot also be stress free and geometrically compatible in an euclidean space. However, it is possible to construct a non-euclidean geometry in which it will be stress-free (see Eckart (1948), Rajagopal & Srinivasa (1998)). The various natural configurations that a body can exist in are a part of the constitutive specification of the body, and we need to know them *apriori* to complete the description of its thermomechanical behavior. In a process that a body is subject to, it will access certain natural configurations. For example, in a restorable process (by this we mean a non-dissipative process), a metallic body behaves as an elastic body, and has one natural configuration (modulo rigid motion) associated with it. On the removal of the load it returns to this natural configuration. A metal that is *plastically deformed* in the traditional sense, has an infinity of natural configurations associated with the process, while in a process such as twinning, there are as many natural configurations as the parent and the variants of the twin. In general, the natural configurations that are accessed depend on the process. In different processes the natural configurations evolve differently and their evolution is determined by some thermodynamic criterion. For many processes this could be the maximization of the rate of dissipation, which in turn could depend on the temperature, strain, the symmetric part of the velocity gradient, and also some measures of the microstructure of the body in question. In order to complete the constitutive specification of the body, we have to provide constitutive relationships for the internal energy, entropy, rate of dissipation, heat flux vector, etc, in terms of the kinematical, thermal and microstructural fields. Different constitutive specifications lead to different types of responses. In the choice of such constitutive specifications, one is guided by experimental evidence and experience.

Many types, but not all, of responses can be modeled as essentially elastic response, but an evolving set of natural configurations. Using such an approach a large class of material responses have been modeled recently: multi-network polymers (Rajagopal & Wineman (1990), Wineman & Rajagopal (1992)), traditional plasticity (Rajagopal & Srinivasa (1998), (1998a), (1998b)), twinning (Rajagopal & Srinivasa (1995), (1997), Srinivasa, Rajagopal & Armstrong (1998)), solid to solid phase transitions (Rajagopal & Srinivasa (1999), viscoelasticity (Rajagopal & Srinivasa (1999)), crystallization of polymers (Rao & Rajagopal (2000a), (2000b)), deformation of ligaments and tendons (Johnson, Livesay, Rajagopal & Woo (1996), Muralikrishnan & Rajagopal (2000)), and growth of biological materials (Rajagopal (2000c)). A key aspect of the theory is the determination of the evolution of the natural configurations and the response functions associated with natural configuration as a function of loading and temperature history.

MECHANICAL AND STRUCTURAL CHARACTERISICS OF SUPERPLASTIC MATERIALS

Superplasticity is an intriguing inelastic process in solid materials with structural damage-free deformation up to several hundred percent. Forming sheet and bulk materials using superplastic forming has become an established manufacturing method in aerospace and recently in other industries. It is estimated that there are many thousands of materials that exhibit superplasticity ranging from metals, ceramics, metallic/intermetallic/ceramic matrix composites, and nanocrystalline materials. Thermomechanical constitutive equations describing the relationship between flow stress, strain, strain-rate, temperature and other microstructural quantities have been developed based on heuristics or phemenological observations.

As theoretical and computational mechanicians, we all seek the Holy Grail of an ideal constitutive equation that can describe the material behavior at all ranges of thermomechanical loading conditions, knowing fully well that the microstructure strongly influences the inelastic behavior and that the microstructure is a product of the initial chemistry and the history of the processing conditions beginning with the melt. In the case of superplastic material we seek such a relations involving stress, strain, rate of deformation and structure, taking into account the effects of

- Temperature

- Strain hardening/softening

- Grain growth (static and dynamic)

- Cavitation (initiation, growth and coalescence)

- Deterioration in post-deformed thermomechanical properties

The model should be flexible enough to allow minor changes in the chemical composition of the ingot, as well as small changes in the thermo-mechanical primary and secondary processing of the melt into the product form, and allow for variations in the deformation history (temporal and spatial variations of process parameters) during superplastic forming. Additionally the constitutive equation should to facilitate the experimental evaluation of material constants and model parameters with the least number of tests and high degree of accuracy and reliability. Finally the constitutive equation should be numerically simple to be implemented in a computational model.

The ability of the constitutive equations to accurately model the deformation conditions is critical in superplasticity. For example, the temperature has a profound effect on the behavior, e.g. transition from plastic to superplastic behavior, as shown in Fig. 1 (Chandra (2000f)). Since we are interested in modeling the behavior not only at the superplastic temperature but the behavior at lower temperatures prior and subsequent to superplastic deformation, it is imperative to appreciate the difference between classical plasticity and superplasticity. Table 1 below outlines some of the basic differences.

Superplasticity	Plasticity
Superplasticity represents an inelastic behavior with high strain rate sensitivity.	Plasticity represents inelastic behavior with no rate dependence.
The effect of strain hardening is secondary.	Flow stress primarily increased due to strain hardening.
Grain switching, grain boundary sliding (GBS) are the primary mechanisms.	Grain neighbours remain as such at all times.
Texture decreases (grain orientation becomes random) increasing with strain	Texture decreases (preferred orientation in the principle plastic strain direction).
Deformation primarily due to GBS with diffusion and dislocation as the accommodating mechanisms.	Deformation primarily due to dislocational activities.
Deformation reduces the initial anisotropy.	Deformation induces a strong anisotropy.
Failure due to cavity initiation, growth and finally by geometric instability.	Failure due to material and geometric instability.

Table 1: Plasticity vs. Superplasticity

Inelastic deformations as described by the theory of plasticity are generally confined to the lattice planes that have the highest density of atoms per unit area (low index planes). Dislocations that are always present (their number increasing with absolute temperature), glide along the slip planes, climb over obstacles and end up on the grain boundary or the surface producing the inelastic strain concomitant with the applied stress. Grain boundaries hinder the motion of the dislocations, causing them to pile up leading to the macroscopically observed back stress and the so called Baushinger effect. Thus grain boundaries harden the material, as described by the Hall-Petch relations. Thus fine grains not only increase the yield stress but also the post-yield hardening behavior.

Within the theory of superplasticity, inelastic strain is produced by the motion (sliding and migration) of grain boundaries themselves. Dislocations act as accommodation mechanisms to maintain the continuity of materials during deformations. As noted earlier there are other mechanisms (e.g. diffusion) which play a major role in this accommodation process. Thus fine (equiaxed) grain enhances grain boundary sliding and hence superplasticity. Hardening in superplasticity is minimal and occurs mostly due to grain growth (consequently lower grain boundary area).

A NATURAL CONFIGURATION APPROACH TO SUPERPLASTIC MATERIALS

In this work, we propose to discuss the superplastic response of materials within the multi-configurational approach developed for a range of materials cited above. In order to do so, we need to incorporate the following characteristics in the model. At normal temperatures, the body behaves very much like a stain-rate independent inelastic solid, and thus our choice for the internal energy, rate of dissipation, etc., should describe the behavior of strain-rate independent inelastic solids. The transition from the rate independent to a rate dependent behavior occurs in a range of high homologous temperature. The rate of dissipation at these higher temperatures is primarily due to grain boundary deformation (sliding, rotation and migration, and thus the rate of dissipation should reflect these mechanisms). The global rate of dissipation associated with the grain boundary deformation is usually assumed to be similar to that in power-law viscous fluids (Chandra (2000e)). While the dissipation in the grains themselves can be neglected, the grain size can affect the rate of dissipation and thus has to be included while describing the rate of dissipation. Unlike a granular materials that is deforming in which voids act as self accommodating mechanisms with the grains themselves suffering little or no deformation, in superplastic response we have to avoid the formation of cavities though the grains are once again not significantly deformed. Also, the dissipation in granular materials is due to solid friction of the Coloumb type while in superplastic response the dissipation is of a viscous nature. In both cases, the overall deformation is due to the relative motion between grains. The evolution of the natural configuration is determined by the of the rate of dissipation.

For our purpose here, it is sufficient to think of the reference configuration $\kappa_R(\mathcal{B})$ of the body $\mathcal{B}$ as a three dimensional differentiable manifold. Let $\kappa_t(\mathcal{B})$ denote the configuration of the body at time t. Then, by a motion of the body, we mean a mapping χ_{κ_R} which assigns to each point $\mathbf{X} \subset \kappa_R(\mathcal{B})$, a point $\mathbf{x} \subset \kappa_t(\mathcal{B})$ at each t, i.e.,

$$\mathbf{x} = \chi_{\kappa_R}(\mathbf{X}, t) \qquad (1)$$

The deformation gradient $\mathbf{F}_{\kappa_R}$ is defined through

$$\mathbf{F}_{\kappa_R} := \frac{\partial \mathbf{x}_{\kappa_R}}{\partial \mathbf{X}} \qquad (2)$$

For any motion, $\Phi = \hat{\Phi}(\mathbf{X}, t) = \tilde{\Phi}(\mathbf{x}, t)$, we use the standard notation

$$\frac{d\Phi}{dt} := \frac{\partial \hat{\Phi}(\mathbf{X}, t)}{\partial t}, \qquad \frac{\partial \Phi}{\partial t} := \frac{\partial \tilde{\Phi}}{\partial t}, \qquad \text{Grad}\, \Phi := \frac{\partial \hat{\Phi}}{\partial \mathbf{X}}, \qquad \text{grad}\, \Phi := \frac{\partial \tilde{\Phi}}{\partial \mathbf{x}}$$

Then, the velocity $\mathbf{v}$ is given by

$$\mathbf{v} := \frac{\partial \chi_{\kappa_R}}{\partial t} \qquad (3)$$

and the velocity gradient

$$\mathbf{L} := \text{grad}\, \mathbf{v} \qquad (4)$$

The Cauchy-Green stretch tensors $\mathbf{C}_{\kappa_R}$ and $\mathbf{B}_{\kappa_R}$ are defined through

$$\mathbf{C}_{\kappa_R} := \mathbf{F}_{\kappa_R}^T \mathbf{F}_{\kappa_R} \qquad (5)$$

and

$$\mathbf{B}_{\kappa_R} := \mathbf{F}_{\kappa_R} \mathbf{F}_{\kappa_R}^T \qquad (6)$$

As the body is deformed, its natural configuration can change (see Rajagopal (1995), Rajagopal and Srinivasa (1997)). Suppose $\kappa_{p(t)}$ is the natural configuration corresponding to the deformed body at time t (see Fig. 2), then we shall define the deformation gradient from the current natural configuration as $\mathbf{F}_{\kappa_{p(t)}}$. In general $\mathbf{F}_{\kappa_{p(t)}}$ is not the gradient of a motion. However, if the deformation is homogeneous, it is. We define the mapping $\mathbf{G}$ through

$$\mathbf{G} = \mathbf{F}_{\kappa_R}(\mathbf{F}_{\kappa_{p(t)}})^{-1} \qquad (7)$$

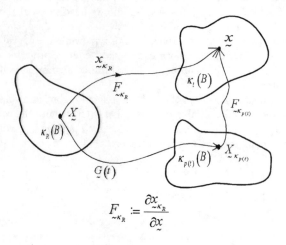

Figure 1: Evolution of natural configurations during a deformation process.

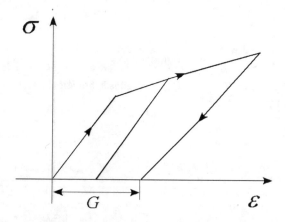

Figure 2: Uniaxial response functions and the evolution of **G**.

Again, the tensor **G** is in general not the gradient of any mapping, but in the case of a homogeneous deformation it can be thought of as the gradient of a mapping. **G** is usually referred to as the plastic deformation gradient (see Figure 3).

Next, we define the Green- St. Venant strain tensor $\mathbf{E}_{\kappa_R}$ through

$$\mathbf{E}_{\kappa_R} = \frac{1}{2}(\mathbf{C}_{\kappa_R} - \mathbf{1}) \tag{8}$$

We define the *"plastic velocity gradient"* through

$$\mathbf{L}_p := \dot{\mathbf{G}}\mathbf{G}^{-1} \tag{9}$$

in an analogous manner with the usual velocity gradient

$$\mathbf{L} := \dot{\mathbf{F}}_{\kappa_R}\mathbf{F}_{\kappa_R}^{-1} \tag{10}$$

The symmetric part of $\mathbf{L}_P$ and $\mathbf{L}$ are denoted respectively through

$$\mathbf{D}_P := \frac{1}{2}\left[\mathbf{L}_P + \mathbf{L}_P^T\right] \tag{11}$$

$$\mathbf{D} := \frac{1}{2}\left[\mathbf{L} + \mathbf{L}^T\right] \tag{12}$$

Next, we introduce the *upper-convected* time derivative (Oldroyd (1950)) through

$$\overset{\triangledown}{\mathbf{A}} = \frac{d\mathbf{A}}{dt} - \mathbf{A}^T\mathbf{L} - \mathbf{L}\mathbf{A} \tag{13}$$

where **A** is any linear transformation. Thus we note that

$$\overset{\triangledown}{\mathbf{1}} = -2\mathbf{D} \tag{14}$$

The principal invariants of a tensor **A** are defined through

$$I_{\mathbf{A}} = tr\,\mathbf{A}, \quad II_{\mathbf{A}} = \frac{1}{2}\left[(tr\,\mathbf{A})^2 - tr\,\mathbf{A}^2\right], \quad III_{\mathbf{A}} = \det\mathbf{A} \tag{15}$$

These minimal kinematical definitions suffice for our illustration. We now introduce the thermodynamic variables. In marked contrast to usual studies in continuum thermodynamic that use some form of the second law, say the Clausius- Duhem inequality to obtain restrictions on constitutive relations, we make a constitutive choice for the rate of dissipation that automatically guarantees that the second law is met (see Green and Naghdi (1977), Rajagopal and Srinivasa (2000a)).

Let ϵ denote the specific internal energy and η the specific entropy of the materials. Then the Helmholtz potential Ψ can be introduced through

$$\Psi := \epsilon - \theta\eta \tag{16}$$

Below the temperature at which superplastic response is apparent, we shall model the response as that of elastic-plastic response which is temperature dependent. Let us consider the body in a state below the temparaure at which superplasic response commences. Then it would be reasonable to assume that the Helmholtz potential ψ is given by

$$\psi = \psi_1(\mathbf{E}_{\kappa_R}, \mathbf{G}, \theta) \qquad \forall \theta < \theta_{cr} \tag{17}$$

and thus we notice that when the natural configuration remains fixed, i.e., $\mathbf{G}$ does not change, the Helmholtz potential reduces to the form for classical elasticity. When the natural configurations changes, then the above form plays the role as is usual in traditional plasticity, provided we are below the critical transition temperature θ_{cr}. Next we provide the appropriate from for rate of dissipation below θ_{cr}. First we note that

$$\xi = \xi_1 \left(\mathbf{B}_{\kappa_{p(t)}}, \mathbf{D}_{\kappa_{p(t)}}, \theta \right) \tag{18}$$

Thus, once again the dissipation is that in a traditional elastic-plastic material, which is temperature dependent. Let us now increase the temperature. We can then provide the forms for Ψ and ξ above θ_{cr} :

$$\Psi = \Psi_2 \left(\mathbf{E}_{\kappa_R}, \theta \right), \ \theta > \theta_{cr} \tag{19}$$
$$\xi = \xi_2 \left(\mathbf{D}_{\kappa_{p(t)}}, \theta \right), \ \theta > \theta_{cr}$$

This choice implies that the behavior of the material is akin to viscoelastic response in that the material has a Helmholtz potential which determines its elastic response and the rate of dissipation determines its viscous response. It is usual to assume the viscous dissipation to have a power-law structure (Chandra (2000e)). Thus

$$\Psi = \begin{Bmatrix} \Psi_1 \left(\mathbf{E}_{\kappa_R}, \mathbf{G}, \theta \right), \ \theta < \theta_{cr} \\ \Psi_2 \left(\mathbf{E}_{\kappa_R}, \theta \right), \ \theta \geqslant \theta_{cr} \end{Bmatrix} \tag{20}$$

$$\xi = \begin{Bmatrix} \xi_1 \left(\mathbf{B}_{\kappa_{p(t)}}, \mathbf{D}_{\kappa_{p(t)}}, \theta \right), \ \theta < \theta_{cr} \\ \xi_2 \left(\mathbf{D}_{\kappa_{p(t)}}, \theta \right), \ \theta \geqslant \theta_{cr} \end{Bmatrix} \tag{21}$$

We shall assume that during superplastic response, i.e. $\theta \geqslant \theta_{cr}$, the deformation is isochoric, thus

$$\det \mathbf{F}_{\kappa_R} = 1 \tag{22}$$

or

$$\operatorname{div} \mathbf{V} = 0 \tag{23}$$

It is possible that the Helmholtz potential and the rate of dissipation can depend on the grain size and this can be introduced into the theory by incorporating a variable to reflect this aspect, similar to the case of the phenomenon of twinning which is also influenced by the grain size. However, just introducing such a variable into the theory is insufficient unless we also provide an equation that governs it.

It is imperative to introduce a method to describe the possibility of cavitation above θ_{cr} as the material is deformed. One way to do this is to assess the rate of dissipation associated with the growth of the cavity as energy is required to produce the increased surface associated with the cavity. The formation of the cavity could be due to one of the following mechanisms: (1) The condensation of vacancies due to the thermal field (2) The enlargement of the cavity due to the stress field and (3) accommodation during grain boundary sliding. Also, during the deformation cavities may come together and coalesce. We need to incorporate the rate of dissipation associated with all the above possiblilities. Thus, we can as a first attempt express the rate of dissipation above θ_{cr} through

$$\xi = \xi_2 \left(\mathbf{D}_{\kappa_{p(t)}}, \theta\right) + \xi_3 \left(d, m\right) \qquad (24)$$

where d denotes the grain size and m is a measure of the grain boundary morphology (energy associated with the boundary, orientation mismatch, and the symmetry structure of the grain boundary). We have tacitly assumed that no energy is required for nucleating the cavities and that they preexist.

Let us consider decreasing the temperature below θ_{cr} and unloading the body subsequent to superplastic response of the material. In this state the body will once again behave like an elasti-plastic thermally dependant material. However, its response will be different from that before its superplastic response. Hence, we need to model this aspect of the response through response functions of the form

$$\Psi = \Psi_4 \left(\mathbf{E}_{\kappa_R}, \mathbf{G}, \theta, d, m\right) \qquad (25)$$

$$\xi = \xi_4 \left(\mathbf{B}_{\kappa_{p(t)}}, \mathbf{D}_{\kappa_{p(t)}}, \theta, d, m\right) \qquad (26)$$

In fact, the above formalism allow us to model the symmetry changes that are engendered by the superplastic response, i.e., a material that initially had a specific anisotropy or texture could after superplastic response become isotropic. To describe the elastic-plastic response post superplasticity below θ_{cr}, it is necessary to allow the response characteristics to change when a certain threshold is reached. Thus, we need a *"yield"* condition which leads to traditional plastic response. The yield surface is represented by means of the equation

$$g(\mathbf{E}_{\kappa_R}, \mathbf{G}, \theta) = 0 \qquad (27)$$

and the elastic domain is characterized by $g(\mathbf{E}_{\kappa_R}, \mathbf{G}, \theta) \leq 0$. The loading criteion etc follow from traditional plasticity.

The rate of dissipation ξ is defined through

$$\xi := -\rho r + div\mathbf{q} + \rho\theta = -\rho\left(\dot{\Psi} + \dot{\theta}\eta\right) - \mathbf{T} \cdot \mathbf{D} \qquad (28)$$

In traditional plasticity ξ depends on $\mathbf{G}$ and $\mathbf{L}_p$ and represents the rate of dissipation in the crystal. Here, in addition to these variables, the rate of dissipation depends on the temperature, as well as on the frictional forces that manifest themselves due to grain boundary sliding, rotation and migration. It might also be necessary to incorporate the effect of the grain size on the rate of dissipation. Thus, we need to assume at the very least that $\xi = \xi(\mathbf{G}, \mathbf{L_p}, \theta)$. Frame indifference implies that.

We introduce the constraint that the bodymapping between the natural configurations is aisochoric and thus the mapping $\mathbf{G}$ satisfies the constraint.

$$\det \mathbf{G} = 1 \qquad (29)$$

We shall assume that amongst the many process that are available to the body, it chooses that which maximizes the rate of dissipation. Thus, we have a constrained maximization problem, one in which we maximize ξ subject to the constraints (28) and (29). We note that the equation (28) is treated as a constraint as the dissipation that is picked has to always satisfy that equation. Thus, we maximize

$$\Phi := \xi + \lambda_1 \left[\xi - \rho\left(\dot{\Psi} + \dot{\theta}\eta\right) + \mathbf{T} \cdot \mathbf{D}\right] + \lambda_2 \left[\det \mathbf{F}_{\kappa_R}\right], \theta \geqslant \theta_{cr} \qquad (30)$$

where λ_1 and λ_2 are Lagrange multiplier, and $\xi = \xi_2 + \xi_3, \Psi = \Psi_2$. Later, to describe the elastic-plastic response below $\theta = \theta_{cr}$, we have to maximize

$$\sum := \xi_4 + \beta_1 \left[\xi_4 - \rho\left(\dot{\Psi}_4 + \dot{\theta}\eta_4\right) + \mathbf{T} \cdot \mathbf{D}\right] + \beta_2 \det \mathbf{G} \qquad (31)$$

We expect, based on the past success of the methodology with respect to material behavior, the procedure will lead to the constitutive theory that is appropriate for superplasticity. The exact structure of the representation for the stress and its evolution, and as well as the evolution of the natural configuration will depend on the specific forms chosen for ψ and ξ. The exact forms of the various functions presented here will be developed in forthcoming papers on this topic.

SUMMARY

Superplasticity in certain class of materials is exhibited in a narrow range of high homologous temperatures and rates of deformation. Contrary to the numerous forms of constitutive relations proposed over the years, in this work we propose to describe superplastic behavior of materials in terms of their natural configurations. Such relationships will not only comply with the rigors of nonlinear mechanics but also describe their characteristics at superplastic temperatures and subsequent room temperature behavior. This wide temperature range description is critical for designing components formed at superplastic temperatures but used at relatively lower temperatures.

Acknowledgement: Rajagopal thanks the National Science Foundation and Chandra, the U.S. Army Research Office for their support.

References

[1] (1949) C. Eckart, Thermodynamics of irreversible processes-Part IV-Theory of elastisicity and anelasticiity, Physical Review, 73, 373-38

[2] (1950) J. G. Oldroyd, On the formulation of a rheological state, Proceedings of Royal Society of London, A200, 523-591.

[3] (1990) A.S.Wineman and K.R.Rajagopal, On a constitutive theory for materials undergoing microstructural changes, Archives of Mechanics, 42, 53-75.

[4] (1992) K.R.Rajagopal and A.S.Wineman, A constitutive equation for nonlinear solids which undergo deformation induced micro-structural changes, International Journal of Plasticity, 8, 385-395.

[5] (1992) Truesdell and Noll, Non-linear Field Theories of Mechanics, Second Edition, Springer Verlag, Berlin-Hydelberg.

[6] (1995) K.R.Rajagopal and A.R.Srinivasa, On the inelastic behavior of solids -Part I:Twinning, International Journal of Plasticity, 6, 653-678.

[7] (1995) K.R.Rajagopal, Multiple configurations in continuum mechanics, Institute for Computational and Applied Mechanics, University of Pittsburgh, Report 6, Pittsburgh, Pa.

[8] (1997) K.R.Rajagopal and A.R.Srinivasa, Inelastic behavior of materials -Part I: Energetics associated with discontinuous deformation twinning, International Journal of Plasticity, 13, 1-35.

[9] (1998)A.R.Srinivasa, K.R.Rajagopal and R.W.Armstrong, A phenomenological model of twinning based on dual reference structures, Acta Materialia, 46, 1235-1248.

[10] (1999)K.R.Rajagopal and A.R.Srinivasa, On the thermodynamics of shape memory wires, ZAMP, 50, 459-496.

[11] (2000a) K.R.Rajagopal and A.R.Srinivasa, A thermodynamic framework for rate type fluid models, Journal of Non-Newtonian Fluid Mechanics, 88, 207-227.

[12] (2000b) I.J.Rao and K.R.Rajagopal, Phenomenological modeling of polymer crystallization using the notion of multiple natural configurations, Interfaces and free boundaries, 2, 73-94.

[13] (2000c) I.J.Rao and K.R.Rajagopal, A study of strain-induced crystallization of polymers, International Journal of Solids and Structures, In Press.

[14] (2000d) K.R.Rajagopal, On growth of biomaterials, To appear.

[15] (2000e) N.Chandra, Constitutive behavior of superplastic materials, To appear.

[16] (2000f) N. Chandra, The micromechanics of inelastic processes in superplastic materials, Recent Advances in the Mechanics of Structed Media, ASME AMD-Vol 244, 5-11.

[17] (2000g) Murlikrishnan and Rajagopal, A Thermodynamic Framework for the constitutive modeling of asphalt concrete: Part I-Theory, To appear.

An Augmented Rigid Plastic Flow Formulation in Support of Finite Element Modeling of Superplastic Forming

R.S. Sadeghi

MSC Software Corporation,
4330 La Jolla Village Dr. Suite 320, San Diego CA 92122, USA

Keywords: Finite Element Modelling, Rigid-Plastic, Simulation, Volume Control

Abstract

The rigid-plastic flow analysis is a solution approach to large deformation analysis, which is well suited for simulation of high temperature metal forming. Two formulations are available: an Eulerian (steady state) and Lagrangian (transient) approach. The effects of elasticity are not included. If these effects are important, this approach should not be employed. However simulation of metal forming applications such as superplastic forming does not require inclusion of elasticity. In the steady state approach, the velocity field (and stress field) is obtained as the solution of a steady-state flow analysis. The time period is considered as 1.0 and, hence, the velocity is equal to the deformation. In the transient formulation, the incremental displacement is calculated. This procedure needs to enforce the incompressibility condition, which is inherent to the strictly plastic type of material response being considered.

Introduction

The Rigid-plastic flow analysis requires an accurate imposition of the incompressibility condition. The following approaches were developed and added to the general purpose MSC.Marc solver in support of simulation of superplastic forming:

1. by means of Lagrange multipliers. Such procedure requires Herrmann elements, which have a pressure variable as the Lagrange multiplier.
2. by means of penalty functions. This procedure uses regular solid elements, and adds penalty terms to any volumetric strain rate that develops. It is observed that the constant dilatation formulation will improve the results.
3. in plane stress analysis (shell and membrane elements), the incompressibility constraint is satisfied exactly by updating the thickness.

This analysis method is very cost effective for SPF simulation. For the case of deformation controlled convergence checking, it is shown that stress recovery is only necessary once an increment has converged. In such cases, considerable savings in execution time are achieved.

Approach
The steady state R-P flow formulation is based on an Eulerian reference system. For problems in which a steady-state solution is not appropriate, an alternative method is available to update the coordinates. The time step should be selected such that the strain increment is never more than one percent. In the transient procedure, there is an automatic updating of the mesh at the end of each increment. During the analysis, the updated mesh can exhibit severe distortion. To overcome the mesh difficulties, mesh rezoning can be employed.

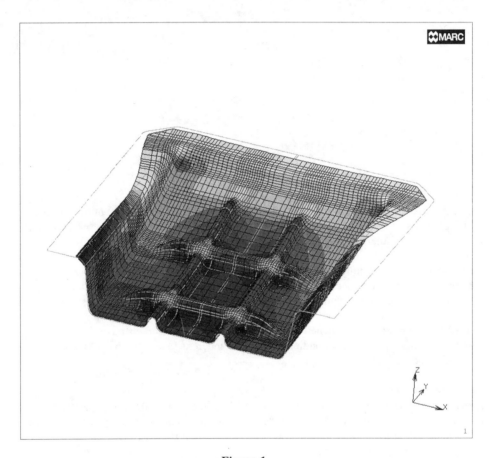

Figure 1.

Results
A new capability based on automatic adjustment of pressure has been introduced to be able to model forming of superplastic materials. These materials are highly rate dependent and show a significant elongation without breaking when subjected to loading. The functionality allows for adaptive meshing, multistage forming and strain-rate control as well as prediction of grain and void growth. The simulation provides post forming

material properties such as final grain size distribution as well as presence and size of voids. Pressure time schedules can be optimized to control grain and void growth.

Vacuum furnace based SPF is modeled using the above methodology. This process has shown great success in SPF of Titanium and Inconel sheet metal components. It offers a more precisely controlled environment. This process allows for material to draw in and hence minimize the overall thinning of the formed part.

Target component shown in Figure 1. was designed to deliver a "Highly Integrated Stiffening". Simulation results and test results showed a 99% agreement in all areas of the formed part. The full-scale component used .080 inch Titanium 6-4 stock. Final component met the .06 inch thickness requirement after SPF operations with close to zero Alpha case from forming.

References
[1] Zienkiewicz, O. C. and R. L. Taylor. *The Finite Element Method (4th ed.) Vol. 1. Basic Formulation and Linear Problems* (1989),*) Vol. 2. Solid and Fluid Mechanics, Dynamics, and Nonlinearity* (1991) McGraw-Hill Book Co., London, U. K.
[2] Bathe, K. J. *Finite Element Procedure*s, Prentice-Hall, Englewood Cliffs, NJ, 1995.
[3] Hughes, T. J. R. *The Finite Element Method–Linear Static and Dynamic Finite Element Analysi*s, Prentice-Hall, Englewood Cliffs, NJ. 1987.
[4] Cook, R. D., D. S. Malkus, and M. E. Plesha, *Concepts and Applications of Finite Element Analysis* (3rd ed.), John Wiley & Sons, New York, NY, 1989.

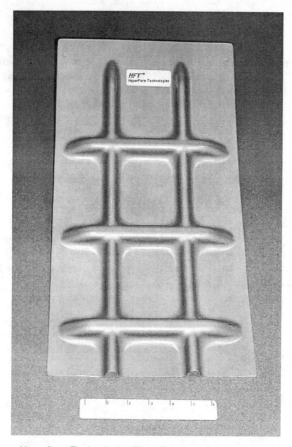

Hyperform Technologies Test Component,
Thickness and Forming Results Verified MARC Modeling

HyperForm Technologies, Inc.

Figure 2.

Effects of the Temperature of Warm Rolling on the Superplastic Behaviour of AA5083 Aluminium Base Alloy

J. Pimenoff[1], Y. Yagodzinskyy[3,4], J. Romu[1] and H. Hänninen[1,2]

[1] Laboratory of Engineering Materials, Helsinki University of Technology,
PO Box 4200, FI-02015 Hut, Finland

[2] Currently on leave at the Institute for Advanced Materials, Joint Research Centre,
Petten Site, P.O.Box 2, NL-1755 ZG Petten, The Netherlands

[3] Institute for Metal Physics, National Academy of Science of Ukraine,
36, Vernadsky Str., Kiev 03142, Ukraine

[4] At Present: Laboratory of Engineering Materials, Helsinki University of Technology,
PO Box 4200, FI-02015 Hut, Finland

Keywords: Aluminium Alloy 5083, Intermetallic Particles, Superplastic Forming, Warm Rolling

ABSTRACT

The effects of the temperature of preliminary thermomechanical treatment (TMT) on the superplastic forming parameters of commercial grade aluminium base alloy AA5083 have been studied. The TMT performed included multiple-pass warm rolling followed by cold rolling with a reduction ratio per pass approximately equal to two. The temperature of the warm rolling, T_{WR}, was varied between 250 and 500 °C. The dependence of superplastic elongation on the temperature of warm rolling during TMT was investigated by tensile testing dog-bone samples at 350 and 500 °C. The research shows that material warm rolled at temperatures close to 350 °C yields maximum values of elongation to fracture. Refinement and change of distribution of intermetallic particles occur in the process of warm rolling AA5083. Pycnometric measurements show that the porosity of the material after a full cycle of TMT, encompassing warm and cold rolling, increases monotonically with the rolling temperature. SEM/EDS observations additionally show that warm rolling at lower temperatures is accompanied by the formation of Si- and Mg-rich segregations at the interfaces between the intermetallic particles and matrix. Both these phenomena lead to a significant decrease in the SPF properties of AA5083.

INTRODUCTION

The need for inexpensive, yet light-weight, structures in the transport industry has led to an increased interest in the use of commercial grade aluminium alloys in superplastic forming (SPF). Recent extensive studies have demonstrated the potential of SPF of AA5083-type aluminium alloys. Alloys of this type are promising in large-scale production, in particular for manufacturing car frame parts and light panel structures for ship-building, due to their fortunate combination of cost, mechanical properties, weldability and corrosion resistance [1].

Large-scale SPF applications set additional requirements on lowering the specific cost of both the mechanical and thermal treatment cycles necessary to prepare the material for superplastic deformation, as also in making the SPF process cheaper. The latter may be achieved by lowering the SPF temperature and increasing the forming strain rate. Formerly, it was considered that the optimal temperature for AA5083-type aluminium alloy lies within the range of 500-550 °C. Nonetheless, elongations up to 400 % have been obtained at 250 °C and a strain rate of 10^{-3} s^{-1}, mainly due to appropriate thermal and mechanical treatments (including warm rolling) [2]. Thus, a warm rolling (WR) stage, may and should be included into the TMT scheme in the preparation of AA5083 for SPF.

The WR stage affects significantly the microstructural state of AA5083. Coarse intermetallic particles, which are present in the initial microstructure of the alloy, are refined during the WR process, hence contributing to developing a beneficial microstructure for SPF. On the other hand, plastic deformation of a heterophase alloy may be accompanied by the nucleation and development of different flaws (*e.g.*, voids and discontinuities) at the particle/matrix interfaces that lead unavoidably to premature cavitation and lowering of the SPF properties of the material.

The present study is an attempt to reveal the effect of the temperature of warm rolling (T_{WR}) on the superplastic behaviour of AA5083 alloy. Main attention was paid to the structural state of the alloy and its porosity after warm rolling at different temperatures.

EXPERIMENTAL

Material and thermal treatment. The studied alloy was AA5083 aluminium alloy from Alcoa Inc.. The chemical composition of the alloy is given in Table 1.

Table 1. Chemical composition of the studied AA5083 wrought aluminium alloy (wt-%).

Alloy	Si	Fe	Cu	Mn	Mg	Zn	Cr	Ti	Al
AA5083	0.08	0.13	0.06	0.58	4.4	0.04	0.07	0.01	bal.

The material was supplied as a 42 mm thick billet in an annealed (O) state. Annealed state AA5083 possesses a coarse grain structure with an average grain size of 23 × 23 × 50 µm and contains equi-axed inclusions of intermetallic particles (Fig. 2A) with an average size of 10 µm. The billet was cut into 8 mm thick plates. The plates were first subjected to warm multiple-pass rolling (WMR) and then to cold multiple-pass rolling (CMR). The WMR and CMR reduction ratios were held close to 2, giving a final plate thickness of 2 mm. The WMR temperatures, which were 250, 350 and 500 °C, were kept constant between passes.

Mechanical testing. Dog-bone type samples with a 10 mm gauge length for tensile testing were water-jet cut from the rolled plates. Hot tensile testing was performed on a MTS-810 tensile testing machine equipped with a split tube furnace. Superplastic uniaxial tensile testing was carried out at 350 and 500 °C, at an initial strain rate of 10^{-3} s^{-1}. Prior to testing, the sample installed into the testing machine was heated to the testing temperature at a heating rate of 25 °C/min, without any preliminary recrystallisation annealing [2].

SEM and EDS observations. Microstructural observations of the material in the as-supplied state and after hot rolling at different temperatures were made on a Zeiss DSM 962 scanning electron microscope. EDS element maps were provided by a LINK Oxford EDS analyser.

Pycnometric measurements. Measuring the density of the material in the as-supplied state and after hot rolling at different temperatures was carried out on an AccuPyc 1330 pycnometer. Porosity, being determined as a change of alloy density after certain mechanical and thermal treatment relative to the as-supplied state, was chosen as a measure of discontinuity.

RESULTS AND DISCUSSION

Figure 1 shows how the tensile elongation of AA5083, tested at 350 and 500 °C, depends on the warm rolling temperature of TMT. It can be seen, that the dependence exhibits a non-monotonic behaviour for both testing temperatures with a maximum elongation in the vicinity of a warm rolling temperature of 350 °C. The maximum elongation at testing temperature 500 °C was 276 %. A similar result was achieved by Vetrano et al. [3] for cold rolled AA5083 tested at 510 °C. The material in Vetrano's work was rolled to a reduction ratio 5:1 and given a recrystallisation anneal at 410 °C for 10 min prior to testing.

Data on porosity development in AA5083 after different temperatures of preliminary warm rolling are also shown in Fig. 1 for comparison with the tensile test results. According to the results, porosity increases monotonically with the rolling temperature. Density of the alloy after rolling at 250 °C is somewhat less than that in the as-supplied state. This indicates both the presence of porosity in the initial annealed state of the alloy, and the absence of its increase during two-stage deformation, i.e., warm rolling at 250 °C and cold rolling. The latter, however, is not true for warm rolling temperatures 350 and 500 °C. Instead, it may be supposed that the significant decrease of elongation during superplastic deformation of AA5083 after warm rolling at 500 °C is due to early cavitation. Initial voids, existing in the alloy after preliminary mechanical and thermal treatment, act as sites for void nucleation and further cavitation [4].

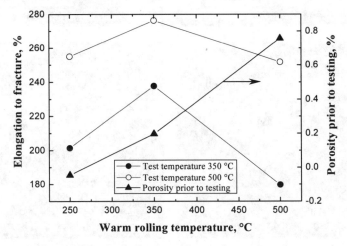

Figure 1. Dependence of superplastic elongation and porosity prior to testing on warm rolling temperature of AA5083 (hot tensile tested at 350 and 500 °C).

The increase of porosity with warm rolling temperature is, apparently, related to the influence that warm rolling has on the distribution of intermetallic phases in the alloy. Figure 2 shows backscattered electron images of microstructures of AA5083 in the as-supplied state and after different TMT. The morphology of intermetallic particles (light contrast) changes with the warm rolling temperature from uniform distribution of fine particles (T_{WR} = 350°C) to localised clusters (T_{WR} = 500 °C). In the former case the particles are, apparently, situated where coarse intermetallic particles were located in the initial annealed material state. In other words, coarse initial particles are fragmented and distributed uniformly in the matrix. At higher temperatures, fragmentation of the initial particles does take place, but their uniform distribution in the matrix is prevented by the relatively low flow stress of the alloy. During the second stage of TMT, i.e., cold rolling, again in the former case a dispersed and more homogeneous distribution of intermetallic particles is developed. This structure is also more stable against porosity formation.

The effective stress for fracture is larger for fine particles embedded into a plastically deformed matrix than for localised clusters. The last statement is a supposition and requires a separate, more detailed study. However, it should be noted that a local cluster of particles may be considered as a large particle with certain values of elastic constants, and the fracture stress of a particle decreases with an increase in its size [5].

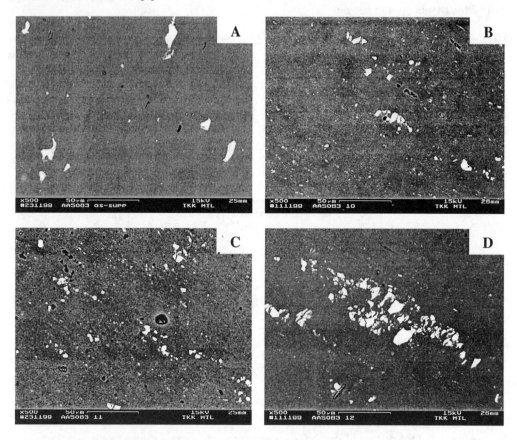

Figure 2. SEM backscattered electron images of AA5083: A) as-supplied state, and after TMT including hot rolling at B) 250 °C, C) 350 °C and D) 500 °C.

The decrease in elongation of AA5083 with decreasing warm rolling temperature (see Fig. 1) is not, seemingly, related to initial porosity. Such behaviour is more likely due to a redistribution of alloying elements, silicon in particular, and the formation of segregations or precipitations of brittle phases at the particle/matrix interfaces. The data on distribution of elements in Figs 3 and 4 obtained by EDS analysis strongly substantiate this statement. Figure 3 shows six images of AA5083 alloy after warm rolling at 250 °C. The backscattered electron image shows an intermetallic particle (light contrast) spalled in the process of preparation of the metallographic sample, revealing the particle/matrix interface. The cavity (dark contrast) left by the split part of the particle was confirmed by secondary electron observation. The Mg and Si element maps (third and fourth images) unequivocally display that the remaining section of the intermetallic particle is free of Si and that the particle/matrix interface is covered with Mg or Si, or a compound containing them.

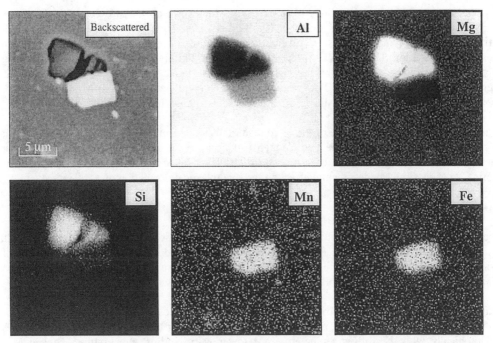

Figure 3. EDS element maps of AA5083 after TMT including warm rolling at 250 °C.

Meanwhile, the analogous images for AA5083 after warm rolling at 500 °C (Fig. 4) show that Si is distributed uniformly in intermetallic particles of a complex composition (Al-Mn-Fe-Si), and that Mg is distributed uniformly in solid solution in the alloy matrix.

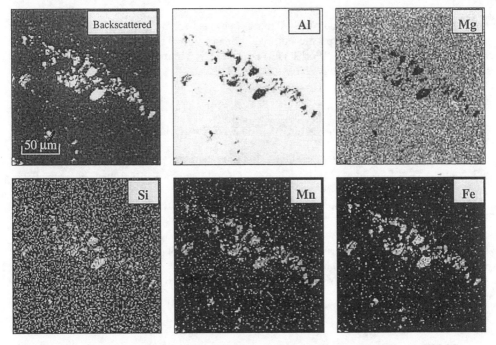

Figure 4. EDS element maps of AA5083 after TMT including warm rolling at 500 °C.

It would appear that the segregations of Si and Mg, which form at the particle/matrix interfaces during warm rolling of AA5083, are of an equilibrium nature and are driven by the chemical affinity of these alloying elements. Warm plastic deformation promotes the diffusive process, and facilitates segregation. The principal difference between the equilibrium segregations of Si and Mg of this study and the non-equilibrium segregations of Si found in earlier research [6], needs therefore to be acknowledged. The non-equilibrium Si-segregations in Al-Mg-Si alloys are promoted specifically by superplastic deformation and form along the grain boundaries.

It has been shown that an increase of the Si-content in AA5083 promotes cavity formation, and that microcavities are nucleated at the particle/matrix interfaces [7]. It seems, based on the present study, that Si- and Mg-segregations, which form efficiently at lower warm rolling temperatures (250 °C), have a similarly detrimental effect on the superplastic deformation of AA5083.

CONCLUSIONS

The stage of warm rolling in the process of TMT prior to SPF significantly influences the parameters of alloy forming, particularly the elongation in uniaxial tensile deformation. The dependence of elongation to fracture on warm rolling temperature of TMT is non-monotonic. Maximum elongation to fracture is achieved for warm rolling temperatures in the vicinity of 350 °C.

Fragmentation and change of distribution of intermetallic particles occur during warm rolling of AA5083. The porosity of AA5083 after completed TMT, including the final cold rolling, increases monotonically with the rolling temperature. At lower warm rolling temperatures, Si and Mg-segregations form at the particle/matrix interfaces. Increased porosity and Si- and Mg-segregations, both being material specific mechanisms, impair the material superplastic properties and hence, lead to a significant decrease in the SPF properties of AA5083.

ACKNOWLEDGEMENTS

The present study was carried out within the framework of the project "Manufacturing of Lightweight Structures with Superplastic Forming", financed by Tekes, the National Technology Agency (Finland), Valmet Turku Works Oy (Finland), Outokumpu Polarit Oy (Finland) and Kvaerner Masa-Yards Oy Technology (Finland).

REFERENCES

[1] P. Impiö, J. Pimenoff, H. Hänninen and M. Heinäkari, Proc. Materials Research Society, Vol. 601 (2000), p. 223.
[2] I.C. Hsiao and J.C. Huang, Scr. Mater., Vol. 40 (1999), p. 697.
[3] J.S. Vetrano, C.A. Lavender, C.H. Hamilton, M.T. Smith and S.M. Bruemmer, Scr. Metall. et Mater., Vol. 30 (1994), p. 565.
[4] H. Iwasaki, T. Mori, T. Tagata, M. Matsuo and K. Higashi, Mater. Sci. Forum 233-234 (1997), p. 81.
[5] A. Needleman and J.R. Rice, Acta Metall., Vol. 28 (1980), p. 1315.
[6] J.S. Vetrano, C.A. Lavender and S.M. Bruemmer, Proc. Superplasticity and Superplastic Forming, Minerals, Metals and Materials Society, (1995), p. 49.
[7] H. Hosokawa, H. Iwasaki, T. Mori, M. Mabuchi, T. Tagata and K. Higashi, Acta Mater., Vol. 47 (1999), p. 1859.

Cooperative Processes in Superplastic Forming under Different Deformation States

Michael Zelin[1], Stephane Guillard[1] and Amiya Mukherjee[2]

[1] Concurrent Technologies Corporation, 100 CTC Drive, Johnstown PA 15904, USA

[2] University of California Davis, Department of Chemical Engineering & Materials Science, Davis CA 95616, USA

Keywords: Cooperative Processes, Deformation States, Superplasticity

Abstract

Evidence on cooperative phenomena including sliding and rotation of grain groups and correlated grain boundary migration during superplastic deformation under different strain states is reported. The pattern of macroscopic surfaces at which grain blocks slide as an entity is consistent with that of slip lines predicted by the slip band field theory. Sliding of grains at these shear surfaces occurs in a sequential manner. Increasing strain and applied stress (strain rate) result in decrease in spacing of the shear surfaces. Cooperative grain rotation results in a change in the orientation of the macroscopic shear surfaces and provides accommodation sliding of grain group. Correlated grain boundary migration is due to the coupling of grain boundary sliding and migration caused by characteristics of dislocation movement at grain boundaries.

1. Introduction

Cooperative grain boundary processes (CGBP) during superplastic (SP) deformation have recently been reported in a number of superplastic materials [1-3]. These processes include cooperative grain boundary sliding (CGBS) and rotation (CGBR), i.e. sliding and rotation of grain groups as an entity, and cooperative grain boundary migration (CGBM), a correlated migration of grain boundaries. These processes play an important role in superplastic flow and microstructural evolution, particularly, grain growth, cavity formation, and phase morphology changes [1, 2]. Studies of cooperative processes during SP deformation were performed primarily under uni-axial tension. In this work, cooperative phenomena during SP deformation were studied under uni-axial tension and compression, shear, and bi-axial gas forming and bulging. Following the approach developed in previous publications [1, 2], cooperative grain boundary phenomena were studied on different microstructural scales.

2. Microstructural scales of CGBP under different strain-stress states

2.1. Macro-scopic scale. It has been demonstrated [2] that the pattern of CGBS surfaces is consistent with that of slip lines predicted by the slip band field theory (SBFT) [4]. This is related to the fact that orientation of both maximum shear stress and zero non-shear strain components coincide with that of slip bands [5]. As an example, Figs. 1a-c illustrate slip band field patterns and CGBS patterns observed at the surface of shapes produced by sheet stretching with punches of a cylindrical, hemispherical,

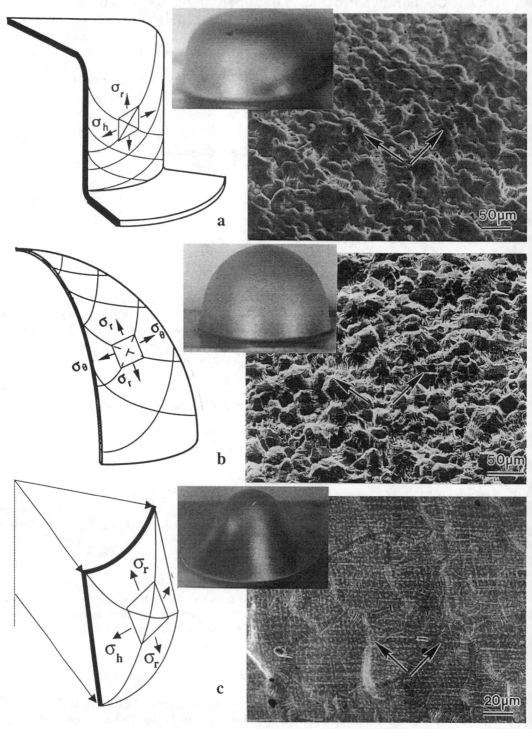

Fig. 1 A pattern of macro-scopic shear surfaces predicted by the slip band field theory and CGBS surfaces (arrowed) observed in superplastically formed AA7475 aluminum alloy cups of (a) cylindrical, (b) hemi-spherical, and (c) conical shape.

cylindrical shape (shown in inserts). Offset of marker lines observed at high magnifications [6] clearly indicates that the shear surfaces originated from grain boundary sliding. The same directionality in offsets of marker lines spaced apart by more than ten grains indicates cooperative manner of grain boundary sliding. The fact that marker line segments are undisturbed between the sliding grain boundaries also suggests grain group sliding.

Theoretical predictions of slip bands pattern are also in accord with experimentally observed CGBS surfaces for the case of a plane strain state. In this case, surfaces of maximum shear stress are inclined under 45 degrees with respect to the sheet mid-plane (Fig. 2). Such shear surfaces were observed in the cross section of a wall of a long rectangular pan gas formed from an AA7475 aluminum alloy sheet [7]. Cavities formed at the intersections of the CGBS surfaces, dispersion free zones, deeper etching, and alignment of sliding grain boundaries along the direction of maximum shear delineates these surfaces (Fig. 2). Thus, the pattern of CGBS surfaces is consistent with that of slip lines that determine a specific arrangement of CGBS surfaces for each stress-strain state.

2.2. Meso-scopic scale. Observations performed at the scale of grain groups provided information on the following important aspects of CGBS:

a. <u>Spacing of CGBS surfaces</u>. Spacing of CGBS surfaces decreases with increase in strain. The number of grains between sliding grain boundaries decreases from 4...8 grain sizes at initial strains to 1...3 grains at high strains [1-3, 6]. Spacing of CGBS surfaces also decreases with increasing strain rate or applied stress [3, 6]. Figure 3 demonstrates results of measurement of the length of marker line segments for the case of conical cups formed under different strain rates and illustrates marker line

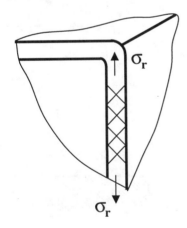
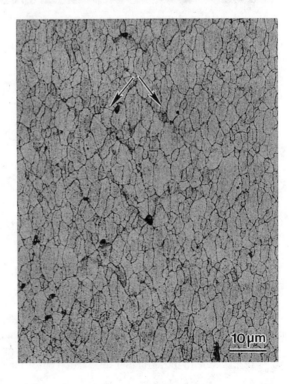

Fig. 2 Pattern of macroscopic shear surfaces and grain structure observed in the cross section of a wall of a superplastically formed AA7475 pan. Arrows indicate shear surfaces revealed by a deeper etching and grain boundary alignment.

distortion due to CGBS. An average size of sliding grain groups in a cup formed with a slow strain rate is approximately six to eight grain diameters. Sliding grain groups of this size are clearly seen in Fig. 3a: they are outlined by the micro-fibers formed at sliding grain boundaries [8]. An average size of sliding grain groups is approximately one and a half of grain diameter in the case of a cup formed under high strain rate. Shear surfaces in this case, propagate not only through grain boundaries, but also through the grain interior (arrowed in Fig. 3b). Directionality of dislocation slip lines inside grains coincides with that of grain boundary displacements caused by CGBS.

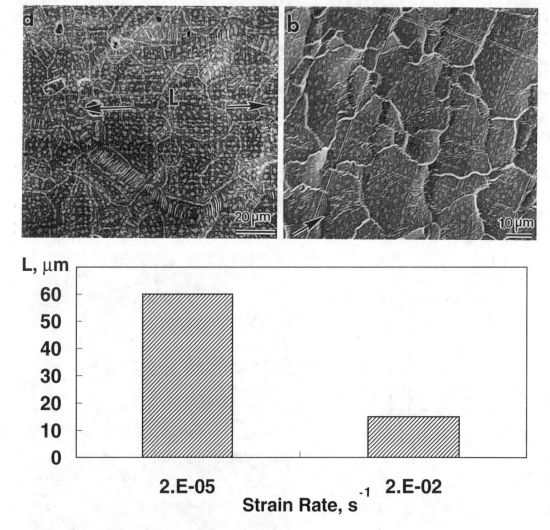

Fig. 3 (a), (b) – Distortion of marker lines due to cooperative grain boundary sliding observed in AA7475 aluminum alloy cups of conical shape deformed at low strain rate and high strain rate, respectively. c- Length of marker line segments as a function of strain rate and distortion of marker lines. Letter L shows a size of sliding grain groups; an arrow in (b) indicates slip lines formed due to intragranular dislocation deformation.

b. <u>Sequential manner of CGBS</u>. In-situ SEM observations of CGBS performed on Pb-62%Sn alloy specimens deformed in shear and in tension [1, 2] showed that grains slide at CGBS surfaces one after another. Such a sequential grain shear at CGBS surfaces is analogous to the dislocation movement at a slip plane, and it can be modeled in terms of cellular dislocations (CDs). Morral and Ashby [8] and later Sherwood and Hamilton [9] employed concept of cellular dislocation for analysis of sliding of grain blocks. Using the typical for dislocation analysis approach, strain rate due to movement of cellular dislocations can be presented as: $\dot{\varepsilon}$=MBv, where M is density of CD sources, B is Burgers vector defined by the grain size, and v is velocity of a CD. This approach allows one to utilize a well developed theory of regular dislocations for analysis of processes occurring on the meso-scopic level.

c. Grain group rotation. Rotation of grain groups can be caused by the change in the orientation of CGBS surfaces, as well as accommodation deformation providing compatibility of strain of sliding grain groups. In the former case, grain groups monotonically rotate in one direction defined by the change in the pattern of slip lines. In the last case, direction of grain group rotation alternates as dictated by the local character of CGBS, for instance, intersection of CGBS surfaces [1, 2].

2.3. Micro-scopic scale. Coupling between CGBS and cooperated grain boundary migration can be explained based on a dislocation model of grain boundary sliding [2]. Since core of a grain boundary dislocation is associated with a grain boundary ledge, their movement results in a grain sliding accompanied by a correlated grain boundary migration. As an example, Fig. 4a demonstrates simultaneous sliding and grain boundary migration for the case of a bi-axially stretched AA5182/AA6090-25%SiC$_p$ laminated composite sheet blank into a conical cup. Light etching revealed a higher dislocation density at, or near, grain boundaries (Fig. 4b).

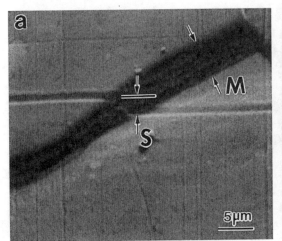
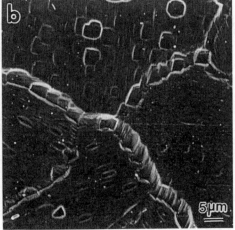

Fig. 4 (a) Coupling of grain boundary migration and grain boundary sliding and (b) etch pits in a superplastically formed AA5182/AA6090-25%SiC$_p$ laminated composite conical cup. Letters S and M show grain boundary sliding and migration, respectively.

3. Summary

Cooperative manner of GBS during SP deformation has been observed for different strain states. Pattern of macroscopic surfaces of CGBS is consistent with that of slip lines predicted by the slip band field theory. Grains slide at CGBS surfaces in a progressive manner, i.e. one after the other. Spacing of CGBS surfaces decreases with increase in accumulated strain and applied stress. Cooperated grain boundary sliding is accompanied by the rotation of grain groups and correlated grain boundary migration.

References

1. M.G. Zelin and A.K. Mukherjee, Acta Metallurgica et Materialia, v. 43, (1995), p. 2359-2372.
2. M. G. Zelin, and A.K. Mukherjee, Mater. Sci. & Eng., v. A208, (1996), p. 210-225.
3. V.V. Astanin, O.A. Kaibyshev, and S.N. Faizova, Acta Metals Mater., v. 42, (1994), p. 2617-2622.
4. O. Hoffman and G. Sachs, Introduction to the Theory of Plasticity for Engineers, McGraw-Hill, New York, (1953), p. 276.
5. A. Nadai, Theory of Flow and Fracture of Solids, 2nd ed., McGraw-Hill, New York, (1950).
6. M. G. Zelin, Materials Characterization, v. 37, (1996), p. 311-330.
7. M. G. Zelin and S. Guillard, Mater. Sci. & Techn., v. 15, (1999), p. 309-315.
8. J.E. Morral and M.F. Ashby, Acta Metall., v. 22, (1974), p. 567.
9. D.J. Sherwood and C.H. Hamilton, Phil. Mag. A, v. 70, (1994), p.109-143.

Corresponding Author: Michael Zelin, email address: adan@sssnet.com

Unified Constitutive Equation of CTE-Mismatch Superplasticity Based on Continuum Micromechanics

K. Kitazono, E. Sato and K. Kuribayashi

The Institute of Space and Astronautical Science,
3-1-1 Yoshinodai, Sagamihara, Kanagawa 229-8510, Japan

Keywords: Internal Stress, Metal Matrix Composite, Monolithic Polycrystal, Thermal Cycle

Abstract

Two types of micromechanics models for CTE-mismatch superplasticity have been proposed independently. They are developed about metal matrix composites and monolithic polycrystals. The present study compares and unifies these two models. Using a newly proposed geometrical factor, similar constitutive equations at low and high applied stresses are derived quantitatively. The unified model can be applicable to explain the CTE-mismatch superplastic behavior in any metal matrix composites and monolithic polycrystals.

Introduction

Internal stress superplasticity appears in some kinds of materials subjected to a cyclic temperature change and a small external applied stress at elevated temperature. The plastic strain per cycle is proportional to the applied stress, i.e., the material shows Newtonian viscous flow, and thus it shows a large ductility without failure through accumulating the uniform plastic strain. This phenomenon is not due to the grain boundary sliding but the generated high internal stress.

Internal stress is induced by transformation mismatch and CTE (Coefficient of Thermal Expansion) mismatch during the temperature change. The former is observed in the materials obeying an allotropic phase transformation such as iron [1]. The CTE-mismatch occurs in composite materials and anisotropic polycrystalline materials. In the case of metal matrix composites such as Al-SiC [2], the CTE-mismatch strain is induced by the difference in CTEs between the matrix and the inclusion. In the case of monolithic polycrystals such as zinc [3], the CTE-mismatch strain is induced by the anisotropic CTEs between the neighboring grains.

For transformation superplasticity, Greenwood and Johnson [4] developed the internal stress model and revealed that the transformation mismatch strain caused linear creep at low stresses. Their model was recently extended to high stress region by Dunand et al. [5] and was applied to several materials [6,7]. However, it is not rigorous to apply Greenwood and Johnson's model to CTE-mismatch superplasticity because the material geometry is quite different from each other. On the other hand, a simple uniaxial model for CTE-mismatch superplasticity was proposed by Sherby and Wadsworth [8]. However, as discussed in the following section, their model could not estimate the quantitative thermal cycling creep rate.

Recently, the authors [9,10] proposed two theoretical models for composite and anisotropic CTE-mismatch superplasticity based on continuum micromechanics independently. Each model enables to estimate the thermal cycling creep rates in metal matrix composites and monolithic polycrystals quantitatively. In the present study, the unified model is proposed to compile the mechanism of CTE-mismatch superplasticity theoretically and experimentally.

Theoretical Models

2.1. One-dimensional qualitative model

The theoretical analysis of CTE-mismatch superplasticity by Sherby and Wadsworth [8] was based on simple uniaxial consideration without material geometries. Under the isothermal condition, they assumed the material obeying steady-state power-law creep expressed as

$$\dot{\varepsilon}^{ISO} = B\sigma^n, \tag{1}$$

where σ is the flow stress, n is the stress exponent and B is the constant. When the cyclic temperature change is applied to the material, they assumed that half of the dislocations experience a stress which is increased by the internal stress $|\sigma^0+\sigma_i|$, and the other half experience a stress which is reduced by the internal stress $|\sigma^0-\sigma_i|$. The thermal cycling creep rate is described as

$$\dot{\varepsilon}^C = \frac{1}{2}B\left\{\left|\sigma^0+\sigma_i\right|^n + \frac{\left|\sigma^0-\sigma_i\right|}{(\sigma^0-\sigma_i)}\left|\sigma^0-\sigma_i\right|^n\right\}. \tag{2}$$

Comparing the magnitude of the generated internal stress and the applied stress, the thermal cycling creep rate becomes at low applied stresses is described as

$$\dot{\varepsilon}^C = nB(\sigma_i)^{n-1}\sigma^0. \qquad (\sigma^0 \ll \sigma_i) \tag{3}$$

On the other hand, the effect of internal stress is negligible at high applied stresses. Then, the thermal cycling creep rate becomes

$$\dot{\varepsilon}^C = B(\sigma^0)^n. \qquad (\sigma_i \ll \sigma^0) \tag{4}$$

Their model simply explained that the generated internal stress caused linear creep at low applied stresses. However, the model could not explain the effect of the material geometry and the accommodation process on the generated internal stress. Therefore, the quantitative estimation of the thermal cycling creep rate cannot be carried out using the above model.

2.2 Micromechanics model
2.2.1 Composite CTE-mismatch superplasticity

The micromechanics model for composite CTE-mismatch superplasticity was first proposed by the present authors [9]. The model assumes a continuum metal matrix containing spherical inclusions. When the uniaxial external stress $\sigma_{ij}^0 = \sigma^0\delta_{i3}\delta_{j3}$ is applied to the composite under the isothermal condition, the mismatch of the plastic strain is generated at the interface between the matrix and the inclusions. Assuming the complete relaxation by the interface diffusion between the matrix and the inclusions, the isothermal creep rate is described as

$$\dot{\varepsilon}_{ij}^{ISO} = \frac{3}{2}K\exp(-\frac{Q_C}{RT})\left(\frac{\sigma_e^0}{1-f}\right)^{n-1}\left(\frac{\sigma^0}{1-f}\right)(\delta_{i3}\delta_{j3}-\frac{1}{3}\delta_{ij}), \tag{5}$$

where K is the constant, Q_C is the activation energy of isothermal creep, σ_e^0 is the equivalent stress of σ_{ij}^0 and f is the volume fraction of the inclusions. The summation convention for the repeated indices is employed throughout the present study.

Under the small temperature change of $\dot{T}\delta t$, the thermal strain ε_{ij}^T is induced in the inclusion which is expressed as

$$\varepsilon_{ij}^T = (\alpha^I - \alpha^M)\delta_{ij}\dot{T}\Delta t = \Delta\alpha^C\delta_{ij}\dot{T}\delta t, \tag{6}$$

where α^I and α^M are CTEs of the inclusion and the matrix, respectively. In order to achieve the steady state deformation, the thermal strain must be accommodated by the plastic deformation of the matrix immediately. Then, the internal stress induced in the material is expressed as

$$\sigma_e^I = K^{-1/n}\exp(\frac{Q_C}{nRT})\left[\left|\Delta\alpha^C\dot{T}\right|\frac{2r^3}{x^3}\right]^{1/n}, \tag{7}$$

where r is the radius of the inclusion and x is the distance from the center of the inclusion.

Under the thermal cycling creep condition, the external and internal stresses occur in the material simultaneously. Then, the local strain rate around a inclusion is described as

$$\dot{\varepsilon}_{ij}^{Local} = \frac{3}{2} K \exp(-\frac{Q_C}{RT})[(\sigma_e^0 + \sigma_e^1)]^{n-1}(\sigma_{ij}^{'0} + \sigma_{ij}^{'1}), \qquad (8)$$

where $\sigma_{ij}^{'0}$ and $\sigma_{ij}^{'1}$ are the deviatoric stresses of external and internal stresses, respectively. According to the two calculations, that is, volume averaging in the matrix and time averaging per cycle, the final formula of the constitutive equation at low applied stress becomes

$$\bar{\dot{\varepsilon}}_{33}^D = K^{1/n} \exp(-\frac{Q_C}{nRT_{eq}^L})(\frac{\sigma^0}{1-f})\left|\Delta\alpha^C \bar{\dot{T}}\right|^{1-1/n} F^C, \qquad (9)$$

where T_{eq}^L is the equivalent temperature at low stresses and $\bar{\dot{T}}$ is the average heating or cooling rate. The definitions of T_{eq}^L and $\bar{\dot{T}}$ is described in the section 2.4. Equation 9 describes that the average strain rate is proportional to the applied stress.

The term F^C in Eq. 9 is a geometrical factor, depending only on the stress exponent and the volume fraction of the inclusions, which is expressed as

$$F^C = \frac{(1-f^{1/n})f}{(1-f)f^{1/n}} \frac{2n(n+4)}{2^{1/n}5}. \qquad (10)$$

According to Eqs. 9 and 10, we can quantitatively estimate the average strain rate at low stresses in metal matrix composites.

2.2.2 Anisotropic CTE-mismatch superplasticity

The model of anisotropic CTE-mismatch superplasticity was first proposed by the present authors [10]. We consider a monolithic polycrystal D consisting of a number of crystal grains, which are equiaxial in shape. Each grain is assumed to be sufficiently large to ignore the effects of grain boundary sliding and diffusion. The material D has an anisotropic crystal structure such as hexagonal and tetragonal structures and the CTEs in the direction of the crystal axes a, b and c result in $\alpha_a = \alpha_b \neq \alpha_c$. When a uniaxial external stress $\sigma_{ij}^0 = \sigma^0 \delta_{i3}\delta_{j3}$ is applied parallel to the x_3-axis, the material D deforms in a multiaxial power-law equation, expressed as

$$\dot{\varepsilon}_{ij}^{ISO} = \frac{3}{2} K \exp(-\frac{Q_C}{RT})(\sigma_e^0)^{n-1}\sigma^0(\delta_{i3}\delta_{j3} - \frac{1}{3}\delta_{ij}). \qquad (11)$$

Above equation is identical to Eq. 5 when the volume fraction f becomes zero in metal matrix composites.

Under the small temperature change of $\dot{T}\delta t$, the thermal strain is induced in a crystal expressed as

$$\varepsilon_{ij}^T = (\alpha_{ij}^\Omega - \alpha_{ij}^{D-\Omega})\dot{T}\delta t. \qquad (12)$$

Here, we consider that the monolithic polycrystal consists of a continuum matrix $D-\Omega$ containing a spherical grain Ω. Using the polar density $P(\phi,\psi)$ of the basal plane, the CTEs are expressed as

$$\alpha_{ij}^\Omega = \begin{pmatrix} \alpha_a + \Delta\alpha^P \cos^2\psi \sin^2\phi & -\Delta\alpha^P \sin\psi\cos\psi\sin^2\phi & \Delta\alpha^P \cos\psi\sin\phi\cos\phi \\ & \alpha_a + \Delta\alpha^P \sin^2\psi\sin^2\phi & -\Delta\alpha^P \sin\psi\sin\phi\cos\phi \\ \text{Sym.} & & \alpha_a + \Delta\alpha^P \cos^2\phi \end{pmatrix}, \text{(in } \Omega\text{)} \qquad (13)$$

$$\alpha_{ij}^{D-\Omega} = \alpha_{ij}^D = \frac{1}{2\pi}\int_0^{\pi/2} d\phi \int_0^{2\pi} \alpha_{ij}^{\Omega_m} P(\phi,\psi)\sin\phi \, d\psi \qquad \text{(in } D-\Omega\text{)} \quad (14)$$

where $\Delta\alpha^P$ is the anisotropic CTE-mismatch expressed as $\Delta\alpha^P = \alpha_c - \alpha_a$.

In order to achieve the steady state deformation, the thermal strain must be accommodated by the plastic deformation in D. Then, the internal stress in Ω is calculated as

$$\sigma_e^1 = K^{-1/n} \exp(\frac{Q_C}{nRT})\left[|\Delta\alpha^P \dot{T}|\beta(\phi,\psi)\right]^{1/n}, \qquad (15)$$

where $\beta(\phi,\psi)$ is a function depending on the texture.

Under the thermal cycling creep condition, the external and internal stresses occur in the material at the same time. Then, the local strain rate around an inclusion is described as

$$\dot{\varepsilon}_{ij}^{Local} = \frac{3}{2} K \exp(-\frac{Q_C}{RT})[(\sigma_e^0 + \sigma_e^1)]^{n-1}(\sigma_{ij}'^0 + \sigma_{ij}'^1). \qquad (16)$$

According to the two calculations, that is, volume averaging in D and time averaging per cycle, the final formula of the constitutive equation at low applied stress becomes

$$\bar{\dot{\varepsilon}}_{33}^D = K^{1/n} \exp(-\frac{Q_C}{nRT_{eq}^L})\sigma^0 \left| \Delta\alpha^P \bar{\bar{T}} \right|^{1-1/n} F^P, \qquad (17)$$

where T_{eq}^L and $\bar{\bar{T}}$ are the same parameters shown in Eq. 9. Equation 17 also describes that the average strain rate is proportional to the applied stress.

The term F^P in Eq. 17 is the geometrical factor depending only on the stress exponent and polycrystalline texture, which is expressed as

$$F^P = \frac{1}{2\pi} \int_0^{\pi/2} d\phi \int_0^{2\pi} \left[\{\beta(\phi,\psi)\}^{1-1/n} + \frac{9(n-1)}{25} \{\beta(\phi,\psi)\}^{-1-1/n} (\cos^2\phi - \xi_3)^2 \right] P(\phi,\psi) \sin\phi \, d\psi. \qquad (18)$$

According to Eqs. 17 and 18, we can quantitatively estimate the average strain rate at low stresses in monolithic polycrystals.

2.3 Geometrical factors

To examine the contribution of the geometrical factor to the thermal cycling creep rate, the relation between F and n is plotted in Fig. 1 for the respective models. The factor in metal matrix composite increased with increasing the volume fraction. The factor in monolithic polycrystals decreased with aligning the crystallographic orientation. The plots of NT, WF and SF are for no-texture, weak fiber texture and strong fiber texture, respectively. From Fig. 1, the two geometrical factors are nearly the same.

2.4 Equivalent temperatures

The equivalent temperatures shown in Eqs. 9 and 17 are the important factors to discuss CTE-mismatch superplasticity. Though Sherby et al. [8] used the similar parameter in their previous report [3], their definition was incorrect. According to the present analysis, two equivalent

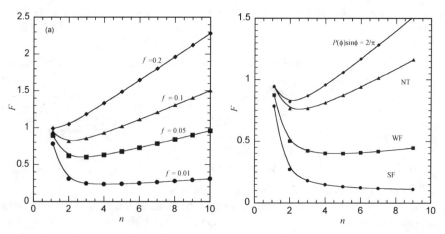

Fig. 1 Geometrical factor is plotted against the stress exponent of isothermal power-law creep. The factors depend on the volume fraction of the inclusions (a) and the crystallographic texture (b).

temperatures are defined at high and low stresses. Under the thermal cycling condition at high stresses, the equivalent temperature T_{eq}^{H} is defined as

$$\exp(-\frac{Q_C}{RT_{eq}^H}) = \frac{1}{\Delta t}\int_0^{\Delta t} \exp(-\frac{Q_C}{RT(t)})dt, \quad (19)$$

where Δt is the cycle period and $T(t)$ is the temperature profile during the thermal cycle. The thermal cycling creep equation at high applied stresses is identical to the isothermal power-law creep equation at T_{eq}^H.

Under the thermal cycling condition at low applied stresses, the equivalent temperature T_{eq}^L is defined as

$$\exp(-\frac{Q_C}{nRT_{eq}^L}) = \frac{1}{|\bar{T}|^{1-1/n} \Delta t}\int_0^{\Delta t} |\dot{T}(t)|^{1-1/n} \exp(-\frac{Q_C}{nRT(t)})dt, \quad (20)$$

where $\bar{\dot{T}}$ is the average heating or cooling rate. The parameter $\bar{\dot{T}}$ which is shown in Eqs. 9, 17 and 20 is defined as

$$|\bar{\dot{T}}| = \frac{2(T_{max} - T_{min})}{\Delta t}, \quad (21)$$

where T_{max} and T_{min} are the maximum and minimum temperatures through the thermal cycle, respectively. The equivalent temperature depends on the activation energy of isothermal creep and the temperature profile of the thermal cycle.

Discussion

The obtained constitutive equations are listed in Table 1. It is found that the constitutive equations in each material are similar to each other except for the geometrical factor and the CTE-mismatch. These equations do not contain any unknown parameters. Therefore, we can estimate CTE-mismatch superplastic behavior in all materials using Table 1.

The obtained micromechanics models were verified using typical materials. As a metal matrix composite, Al-Be eutectic alloy was selected [11]. As an anisotropic monolithic polycrystal, pure zinc was selected [10]. The results of isothermal and thermal cycling creep tests were plotted in Fig. 2. All creep tests were performed under compressive load with fixed heating and cooling rates. Both thermal cycling creep curves showed linear creep at low stresses and power-law creep at high stresses. The linear creep rates were much higher than the isothermal creep rates.

The solid lines show theoretically calculated average strain rates using the present micromechanics models. The geometrical factors become $F^C = 0.22$ for Al-Be alloy and $F^P = 0.56$ for pure zinc. It was found that the experimental values agreed well with the theoretical ones.

Table 1 Constitutive equations in CTE-mismatch superplasticity.

	Low applied stress	High applied stress
Metal matrix composite	$\bar{\dot{\varepsilon}} = K^{1/n} \exp(-\frac{Q_C}{nRT_{eq}^L})(\frac{\sigma^0}{1-f})|\Delta\alpha^C \bar{\dot{T}}|^{1-1/n} F^C$	$\bar{\dot{\varepsilon}} = K \exp(-\frac{Q_C}{RT_{eq}^H})(\frac{\sigma^0}{1-f})^n$
Monolithic polycrystal	$\bar{\dot{\varepsilon}} = K^{1/n} \exp(-\frac{Q_C}{nRT_{eq}^L}) \sigma^0 |\Delta\alpha^P \bar{\dot{T}}|^{1-1/n} F^P$	$\bar{\dot{\varepsilon}} = K \exp(-\frac{Q_C}{RT_{eq}^H})(\sigma^0)^n$
Equivalent temperature	$\exp(-\frac{Q_C}{nRT_{eq}^L}) = \frac{1}{|\bar{\dot{T}}|^{1-1/n} \Delta t}\int_0^{\Delta t} |\dot{T}(t)|^{1-1/n} \exp(-\frac{Q_C}{nRT(t)})dt$	$\exp(-\frac{Q_C}{RT_{eq}^H}) = \frac{1}{\Delta t}\int_0^{\Delta t} \exp(-\frac{Q_C}{RT(t)})dt$

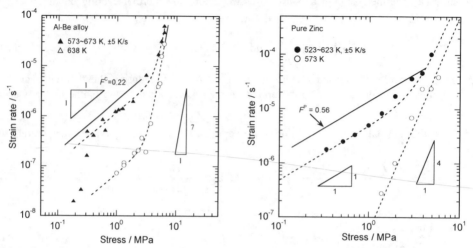

Fig. 2 Thermal cycling creep rates in Al-Be eutectic alloy (a) and pure zinc (b) are plotted against the uniaxial applied stress. The solid lines show theoretically calculated thermal cycling creep rates.

Summary

Constitutive equations of thermal cycling creep in metal matrix composite and monolithic polycrystal were deduced using continuum micromechanics. From the isothermal power-law creep equations, linear creep at low stresses and power-law creep at high stresses under the condition of the thermal cycle were quantitatively obtained for two materials. The linear creep equations correspond to CTE-mismatch superplasticity. Using the newly proposed geometrical factor, two models are unified to a frame to discuss the CTE-mismatch superplastic behavior for any metal matrix composites and monolithic polycrystals.

Acknowledgements

This work was financially supported by Research Fellowships of the Japan Society for the Promotion of Science and Amada Foundation for Metal Work Technology.

References

[1] M. de Jong and G. W. Rathenau, Acta metall. 9 (1961), p. 714.
[2] M. Y. Wu and O. D. Sherby, Scripta metall. 18 (1984), p. 773.
[3] M. Y. Wu, J. Wadsworth and O. D. Sherby, Metall. Trans. 18A (1987), p. 451.
[4] G. W. Greenwood and R. H. Johnson, Proc. Roy. Soc. London 283A (1965), p. 403.
[5] P. Zwigl and D. C. Dunand, Acta mater. 45 (1997), p. 5285.
[6] P. Zwigl and D. C. Dunand, Metall. Trans. 29A (1998), p. 2571.
[7] C. Schuh and D. C. Dunand, Acta mater. 46 (1998), p. 5663.
[8] O. D. Sherby and J. Wadsworth, Mater. Sci. Tech. 1 (1985), p. 925.
[9] E. Sato and K. Kuribayashi, Acta metall. mater. 41 (1993), p. 1759.
[10] K. Kitazono, R. Hirasaka, E. Sato, K. Kuribatashi and T. Motegi, Mat. Res. Soc. Symp. Proc. 601 (2000), p. 199.
[11] K. Kitazono and E. Sato, Acta mater. 46 (1998), p. 207.

All correspondence should be addressed to K. Kitazono (kitazono@materials.isas.ac.jp).

A Coupled Thermo-Viscoplastic Formulation at Finite Strains for the Numerical Simulation of Superplastic Forming

Laurent Adam and Jean-Philippe Ponthot

LTAS - Milieux Continus et Thermomécanique, Université de Liège,
1, Chemin des Chevreuils, BE-4000 Liège 1, Belgium

Keywords: Contact, Finite Element, Numerical Simulation, Thermomechanics, Viscoplasticity

Abstract: This paper is concerned with the numerical simulation of hot metal forming, especially superplastic forming. A complete thermo-viscoplastic formulation at finite strains is derived and a unified stress update algorithms for thermo-elastoplastic and thermo-elasto-viscoplastic constitutive equations is obtained. The resulting unified implicit algorithm is both efficient and very inexpensive. Finally, numerical simulations of superplastic forming are exposed.

1 Introduction

Our goal in this paper is to present a complete thermo-viscoplastic formulation at finite strains and its implementation in the finite element code METAFOR [5]. The formulation is derived from the classical J2-plasticity and extended to take into account thermal and rate dependent effects. A stagerred scheme will be used for the resolution of the thermomechanical problem as well as an extension of the radial return algorithm for the integration of the constitutive law and a penalty approach to manage contact interactions. Finally, some results of superplastic forming will be presented and discussed.

2 Non-isothermal large strains formulation

2.1 Constitutive equation

In the present model we use an additive decomposition of the rate of deformation [9, 10] and we assume an hypoelastic stress-strain relation of the form :

$$\overset{\triangledown}{\sigma}_{ij} = H(T)_{ijkl}(D_{kl} - D^{vp}_{kl} - D^{th}_{kl}) \qquad (1)$$

where :
- $H(T)$ is the temperature dependent Hooke stress-strain tensor,
- T is the temperature,
- $\overset{\triangledown}{\sigma}$ is an objective rate of Cauchy stress tensor,
- D is the rate of deformation,
- D^{vp} is the viscoplastic part of D,
- D^{th} is the thermal part of D.

In the case of viscoplastic deformation we use a classical flow rule for associative viscoplasticity given by

$$D^{vp} = \Lambda N \quad \text{where} \quad N_{ij} = \frac{s_{ij}}{\sqrt{s_{kl}s_{kl}}} \qquad (2)$$

is the unit outward normal ($N : N = 1$) to the yield surface and Λ is a positive parameter called the *consistency parameter*. In a viscoplastic formulation, Λ cannot be determined, as in J2-plasticity, by expressing the so-called consistency condition. So one more equation is needed to be able to express that consistency parameter, and thus the viscoplastic part of the rate of deformation. Various models were proposed to express Λ in terms of the current stress level. For example, classical viscoplastic models of the Perzyna type [3, 4] use a consistency parameter of the form :

$$\Lambda = \sqrt{\frac{3}{2}} \left\langle \frac{\bar{\sigma} - \sigma^v(T)}{\eta(T)(\bar{\epsilon}^{\text{vp}})^{1/n(T)}} \right\rangle^{m(T)} \tag{3}$$

where :
- $\bar{\sigma}$ is the effective stress, i.e. $\bar{\sigma} = \sqrt{\frac{3}{2} s : s}$,
- s is the deviator of the stress tensor,
- $\sigma^v(T)$ is the current yield stress,
- $n(T)$ is a hardening exponent,
- $m(T)$ is a rate sensitivity parameter,
- $\eta(T)$ is a viscosity parameter.

The resulting yield function is the viscoplastic extension of the classical von Mises criterion [7] and is given by

$$\begin{aligned} \bar{f} &= \bar{\sigma} - \sigma^v(T) - \eta(\bar{\epsilon}^{\text{vp}})^{1/n}(\dot{\bar{\epsilon}}^{\text{vp}})^{1/m} \\ &= 0 \end{aligned} \tag{4}$$

The equations governing the evolution of the thermal part of the tensor of the rate of deformation is a generalization of the equation used in infinitesimal strain theory, which is given by :

$$D_{ij}^{th} = \beta \, \dot{T} \, \delta_{ij} \tag{5}$$

where β is the linear thermal expansion coefficient.

2.2 Integration procedure

Let us consider $[t_n, t_{n+1}]$ a sub-interval of the global time interval of interest, and ϕ the vector of primary variables defined as :

$$\phi = (x, v, T) \tag{6}$$

where :
- x is the current nodal coordinates vector,
- v is the current nodal velocities vector,
- T is the current nodal temperature.

We want to determine ϕ at time t_{n+1} so we need to integrate the constitutive law and to update the vector of internal parameters.

The governing equations can be formulated as a first order problem of evolution. In our model we have decided to use the backward-Euler formula for the approximation of the time derivatives, and the finite element technique for the spatial discretization. For the time integration we use a staggered scheme, based on a split of the first order operator, which, in comparison with a monolithic scheme, considerably reduces the computational cost of such resolution [1].

As far as the integration of the constitutive law is concerned we have to find the new values of the stress tensor, viscoplastic strains and other internal parameters given the incremental strain history. To compute this part of the formulation we rely on the general methodology of elastic-predictor/plastic-corrector (return mapping algorithm), as synthesized in Simo & Hughes [8] but here, we use an extension of this methodology to the time-dependent case, see [7, 6] for details.

3 Numerical illustrations
3.1 Superplastic forming of an hemisphere

This case was proposed by Bellet & Chenot [2] and consists in the forming of a Ti-6Al-4V sheet into an hemispherical rigid die. The dimensions of the sheet and its die are given in Figure 1. The periphery of the sheet has fixed radial displacements. The material rheology is given in table 1 and the optimum strain rate to induce superplastic deformations is $3.10^{-4} s^{-1}$. Bellet & Chenot use pressure boundary elements to compute the sheet forming. In our case we will combine pressure and punch forming. So, the pressure is automatically updated to keep a maximum value for the equivalent viscoplastic strain rate close to the optimum strain rate. The pressure adaptation algorithm is an explicit one, based both on membrane theory and empirical considerations. For more complex shapes (wich need more important pressure variations within a time step) we can use an implicit algorithm which ensure an unconditionnal stability and an error limit on the maximum equivalent viscoplastic strain rate.

In this computation we have used two different meshes. The coarser one is made of 30x1 quadrilateral axisymmetric elements (with constant pressure) and the finer one is composed of 100x4 elements. Most of the numerical results are issued from the finer mesh. Figure 2 shows the final thicknesses of the sheet using the two different meshes in the case of pressure forming and are compared to the results of Bellet & Chenot [2]. The coarser mesh was choosen to have rather the same mesh density as those of Bellet & Chenot.

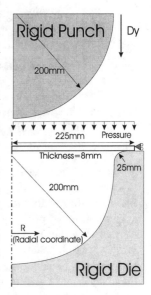

Figure 1: Superplastic forming of an hemisphere

We observe a good agreement between the results. The finer mesh improves the results in the outer part of the sheet contrary to the coarser one which does not contain enough elements in this region to correctly simulate the forming of the transition radius. We also plot on this figure the results obtained if we use a perfect slip condition for the contact interactions between the sheet and the die. Computational features of this numerical simulation are given in table 1.

Young Modulus	$E = 15000 N/mm^2$
Poisson ratio	$\nu = 0.33$
Viscoplastic behaviour	$\bar{\sigma} = 460(\dot{\bar{\epsilon}}^{vp})^{0.5} N/mm^2$
Friction coefficient (μ)	0.2
Forming Time	4650 sec
CPU (finer mesh)	1 min 30 sec
Number of time steps	251
Number of iterations	571

Table 1: Material properties for Ti-6Al-4V and computational features of the hemisphere forming simulation

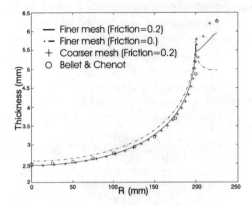
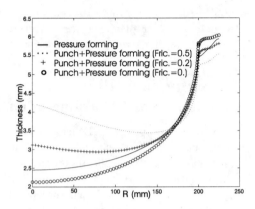

Figure 2: Thickness distributions versus radial coordinates

Figure 3: Thickness distributions in the case of punch+pressure forming

The final thickness of the sheet can be modulated using a rigid punch to partially form the sheet instead of applying a pressure during the whole process. The process setup is exposed in Figure 1. In the following results the vertical displacement of the punch, Dy, is 170mm and next a controlled pressure is applied to complete the forming (note that initially the punch is in contact with the sheet). Since the forming pressure is controlled, the displacement of the punch also needs to be controlled. Figure 3 shows the final thickness for different friction coefficients of the punch (keeping the friction coefficient between the die and the sheet at 0.2). We observe that the lowest value of the thickness can be improved using this type of process. It is also evident that Dy and the friction coefficients can be optimized to produce the ideal thickness distribution.

3.2 Forming of a quarter cylinder.

This numerical test aims to show especially the importance of the thermal spring-back, and so of the unloading phase, at different forming temperatures. The material parameters are typical of an aluminium alloy which has a superplastic behaviour at high temperature.

The test conditions are represented in Figure 4. The right end of the sheet is clamped and the rigid die has a fixed position. The pressure and the temperature are, in a first phase, increased from their room values to their forming values, then they are maintained during the stress relaxation phase, and finaly lowered to the room values. In the results exposed below, the thickness to die radius ratio is $\frac{1}{50\pi}$. The forming time is 3100s (300s for the loading, 2700s for the stress relaxation and 100s for the unloading). The mesh is composed of 50x4 quadrilateral finite elements with constant pressure.

Figure 5 shows final configurations for the different forming temperatures. We observe that the cold forming of the sheet induces few irreversible deformations and so the spring-back is almost total. On the other hand, for the forming at high temperature, there is almost no spring back. The final stress level is a decreasing function of the forming temperature. For example the final maximum of the $J2$ stress is $26.5 MPa$ at $493°K$, $4.5 MPa$ at $793°K$ and $1.6 MPa$ at $993°K$. This numerical simulation needs 250 time steps and 657 iterations, leading to 1min 47s of CPU time.

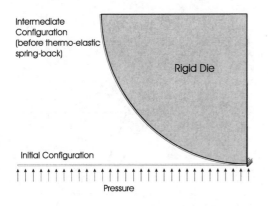
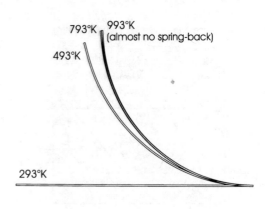

Figure 4: Description of the process setup

Figure 5: Final configurations for different forming temperatures

3.3 SPF/DB forming of multi-cell sandwich sheet

This numerical example is also inspired by one proposed in Bellet & Chenot [2]. This example simulates a superplastic forming-diffusion bonding (SPF-DB) process which is advantageously used to produce light sandwich panels. The process consists of 3 superposed sheets, each 1mm thick, of titanium alloy bonded together over a part of the total area by diffusion bonding and formed by injection of gas within the cells. The material behaviour is the same as in the hemisphere forming simulation (see Table 1). Figure 6 shows the process setup. The upper side of the panel has fixed vertical displacements, and the left end of the panel is clamped. In this case the diffusion bonding is simulated by a perfect sticking contact condition. Different computed configurations are represented in Figure 7. Computational features of this numerical simulation are given in table 2.

Forming Time	1500 sec
CPU	4 min 03 sec
Number of time steps	299
Number of iterations	752

Table 2: Computational features of the SPF-DB process

4 Conclusions

A complete thermo-elasto-viscoplastic formulation was derived and implemented in a finite element code. This formulation is able to simulate non-isothermal metal forming involving viscous behaviour. Further developments will have the aim of simulating more accurately the microscopic evolution of the material through a non-isothermal process and its influence over macroscopic constitutive coefficients. Heat transfer computation during contact interactions is also in progress. These two advances will enable us to simulate more precisely and completly a superplastic forming-diffusion bonding process.

References

[1] F. Armero and J.C. Simo. A new unconditionnaly stable fractionnal step method for non-linear coupled thermomechanical problem. *International Journal of Numerical Methods in Engineering*, 35:737–766, 1992.

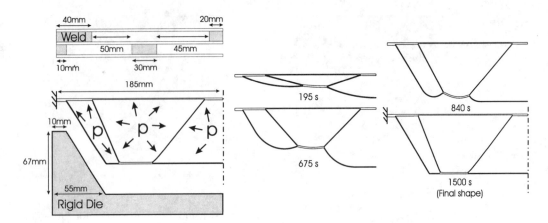

Figure 6: Process set up of a SPF-DB of a sandwich panel

Figure 7: Different computed configurations of a SPF-DB of a sandwich panel

[2] M. Bellet and J.L. Chenot. Numerical modelling of thin sheet superplastic forming. In *NUMIFORM*, pages 401–406, 1989.

[3] P. Perzyna. Fundamental problems in visco-plasticity. In G. Kuerti, editor, *Advances in Applied Mechanics*, volume 9, pages 243–377. Academic Press, 1966.

[4] P. Perzyna. Thermodynamic theory of plasticity. In Chia-Shun Yih, editor, *Advances in Applied Mechanics*, volume 11, pages 313–355. Academic Press, 1971.

[5] J.P. Ponthot. *Traitement unifié de la Mécanique des Milieux Continus solides en grandes transformations par la méthode des éléments finis*. PhD thesis, University of Liège, Liège, Belgium, 1995.

[6] J.P. Ponthot. An extension of the radial return algorithm to account for rate-dependent effects in frictional contact and visco-plasticity. *Journal of Materials Processing Technology*, 80-81:628–634, 1998.

[7] J.P. Ponthot. Unified stress update algorithms for the numerical simulation of large deformation elasto-plastic and elasto-viscoplastic processes. *International Journal of Plasticity*, Accepted for publication.

[8] J.C. Simo and T.J.R Hughes. General return mapping algorithms for rate-independent plasticity. In *Constitutive Laws for Engineering materials : Theory and Applications*. Elsevier Science Publishing Co, 1987.

[9] T.B. Whertheimer. Thermal mechanically analysis in metal forming processes. In Pittman Wood Alexander Zienkiewicz, editor, *Numerical Methods in Industrial Forming Processes*, pages 425–434, 1982.

[10] P. Wriggers, C. Miehe, M. Kleiber, and J.C. Simo. On the coupled thermo-mechanical treatment of necking problems via FEM. In D.R.J. Owen, E. Hinton, and E. Oñate, editors, *International Conference on Computational Plasticity (COMPLAS2)*, pages 527–542, Barcelona, Spain, 1989. Pineridge Press.

Mechanical Effects of Spatial Distributions of Large Grains in Superplastic Microstructures

E. Bernault, J.J. Blandin and R. Dendievel

Institut National Polytechnique de Grenoble (INPG), Génie Physique et Mécanique des Matériaux (GPM2), ENSPG-UJF, BP 46, FR-38402 Saint-Martin d'Hères Cedex, France

Keywords: Cooperative GBS, Creep, Modeling

Abstract

The effect of some microstructural heterogeneities on the superplastic behaviour is studied. Grain size distributions are taken into account with a particular attention given to the spatial distribution of the grains through the microstructure. The predictions concern both the macroscopic behaviour and the extent of strain localisation. It is shown that when a superplastic microstructure contains large grains, the parameters which can influence the mechanical behaviour are not only the proportion of large grains or their size but also their degree of agglomeration or even the spatial distribution of these agglomerates.

Introduction

To obtain superplastic properties, significant efforts are frequently made to generate fine microstructures not only with small mean grain sizes but also with narrow distributions around this mean value. However, the effect on superplastic properties of microstructural heterogeneities remains poorly documented. In particular, very few models deal with grain size distribution or spatial distribution of large grains through a fine grained microstructure and therefore can be used as predictive tools for superplastic alloy design.

Investigating the effect of such microstructural heterogeneities requires a description of the microstructure at a mesoscopic level. A large population of grains with different sizes and a specific spatial distribution must be taken into account. The aim of this paper is to present some predictions concerning the superplastic responses in relation to such a mesoscopic approach.

Model basis

The foundations of the model have been detailed elsewhere [1]. All the studied microstructures are derived from numerically generated Voronoi cells. For each generated microstructure subjected to a macroscopic stress, the macroscopic deformation is supposed to result from shear along percolating paths through the sample, which are preferentially oriented with respect to the stress axis. For a given applied shear stress $\bar{\tau}$, the shear rates are calculated along all these paths and the macroscopic behaviour $\bar{\tau} = f(\bar{\dot{\gamma}})$ is derived. From a local point of view, information can also be obtained from the possible difference between the shear rates associated with each path. In superplasticity, grain boundaries are the regions where strain preferentially occurs. In an homogeneous microstructure, the mean distance between the paths is then expected to be roughly equal to the grain size. Moreover, in this case, no significant differences will be detected between all the shear rates : the deformation will be considered homogeneous. Conversely, if along some paths, the associated shear rates are particularly large, it suggests that strain is preferentially concentrated in these regions and deformation is then considered heterogeneous. Such a framework can be related to the occurrence of cooperative grain boundary sliding (CGBS), which was recently developed in superplasticity [2,3].

It is worth noticing that the concept of percolating paths developed in this approach does not imply a systematic localisation of deformation through the sample, since the strain heterogeneity will arise only when significant differences between the behaviours of the paths appear.

For each grain (i) along a path $P^{(j)}$, grain boundary sliding (GBS) and dislocation creep (DC) can take place simultaneously, resulting in a displacement rate $\dot{U}_i^{(j)}$ given by :

$$\dot{U}_i^{(j)} = \dot{U}_{i,GBS}^{(j)} + \dot{U}_{i,DC}^{(j)} \qquad (1)$$

where $\dot{U}_{i,GBS}^{(j)}$ is the sliding rate resulting from accommodated GBS and $\dot{U}_{i,DC}^{(j)}$ is the displacement rate resulting from DC. Along a path $P^{(j)}$, the displacement rates $\dot{U}_i^{(j)}$ are supposed constant, noted in the following $\dot{U}^{(j)}$. As already mentionned, the predictions concern firstly the macroscopic behaviour of the microstructure (typically the curve $(\bar{\tau}, \bar{\dot{\gamma}})$ but also the contribution of GBS to total strain and the value of the strain rate sensitivity parameter). Secondly, it deals with the possible localisation of strain along preferential paths.

The relation between the local stress $\tau_i^{(j)}$ and the sliding rate $\dot{U}_{i,GBS}^{(j)}$ is derived from the sliding of a sinusoidal grain boundary with an amplitude h [4] :

$$\dot{U}_{i,GBS}^{(j)} = \frac{8\Omega\delta D_{GB}}{kT} \cdot \frac{\tau_i^{(j)} - \tau^*}{\left(h_i^{(j)}\right)^2} \qquad (2)$$

with D_{GB} the diffusion coefficient along the boundary, δ the thickness of the boundary, Ω the atomic volume and τ^* a threshold stress. Along a path $P^{(j)}$, each grain is associated with a specific value of $h = h_i^{(j)}$. In the case of dislocation creep, the dependence of the displacement rate $\dot{U}_{i,DC}^{(j)}$ on the local stress $\tau_i^{(j)}$ is given by :

$$\dot{U}_{i,DC}^{(j)} = \frac{\Delta^{(j)}}{K_{DC}} \left(\frac{\tau_i^{(j)}}{G}\right)^{n_{DC}} \qquad (3)$$

where $\Delta^{(j)}$ is the half distance between the two paths $P^{(j-1)}$ and $P^{(j+1)}$. By combining these equations, one obtains an implicit relation between $\bar{\tau}$ and the displacement rate $\dot{U}_i^{(j)}$. For a given applied $\bar{\tau}$, it is then possible, thanks to a numerical inversion of this implicit relation, to determine the shear rate $\dot{\gamma}^{(j)} = \dot{U}^{(j)} / \Delta^{(j)}$ on each path. The macroscopic shear $\bar{\dot{\gamma}}$ rate is defined as the average of these local shear rates $\dot{\gamma}^{(j)}$. The numerical values used in this work are also given in [1].

A particular procedure has been developed to control the generation of a population of large grains more or less agglomerated in a fine grained microstructure [1].

Predictions

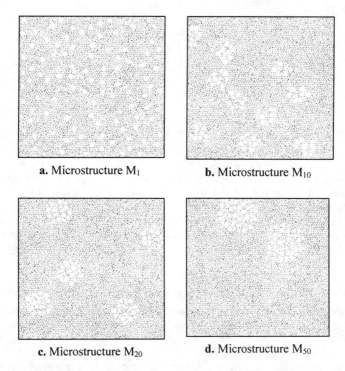

a. Microstructure M_1 **b.** Microstructure M_{10}

c. Microstructure M_{20} **d.** Microstructure M_{50}

Fig. 1 : Examples of microstructures with different $N_{G/A}$

Figure 1a-d displays generated microstructures of about 4000 grains with the same mean grain size (≈ 10 μm), the same size ratio between the large and small grains (= 3) and the same concentration in large grains $C_{LG} \approx 0.18$. The main differences between these microstructures deal with the degree of agglomeration of the large grains. A way to quantify this degree of agglomeration is to define the number of grains per agglomerate $N_{G/A}$. It is assumed that $N_{G/A}$ is constant for a given microstructure. In microstructure M_1, the large grains are dispersed through the microstructure ($N_{G/A} \approx 1$, no agglomerate) whereas in microstructure M_{50}, all the large grains are concentrated in only two agglomerates ($N_{G/A} \approx 50$). Microstructures M_{10} ($N_{G/A} \approx 10$) and M_{20} ($N_{G/A} \approx 20$) display intermediate configurations for which several agglomerates are present through the microstructure.

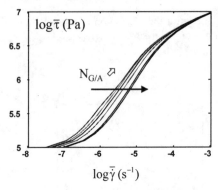 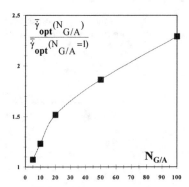

Fig. 2: Macroscopic behaviours for the investigated microstructures

Fig. 3: Effect of $N_{G/A}$ on the macroscopic shear rates

Figure 2 shows the macroscopic behaviour obtained for different microstructures with $N_{G/A} = 1, 5, 10, 20, 50$ and 100. It can be seen that the slopes of the curves are quite similar which indicates that the effect of $N_{G/A}$ on the value of m remains limited. Nevertheless, these curves suggest that the degree of agglomeration of large grains can affect the strain rate for a given macroscopic stress in the superplastic domain : when $N_{G/A}$ is increased, the strain rate is also increased. In other words, from a rheological point of view, it is better to deal with a microstructure with a large agglomerate rather than with dispersed large grains (of course, for the same concentration of large grains).

Figure 3 shows the effect of $N_{G/A}$ on the value of the optimal macroscopic shear rate $\bar{\dot{\gamma}}_{opt}(N_{G/A})$ for a given $N_{G/A}$ normalised by the corresponding one for the microstructure with homogeneously dispersed large grains ($N_{G/A} = 1$). It is shown that the optimum strain rate is increased when the large grains are concentrated in a limited number of agglomerates, despite the fact that these agglomerates contain thus a quite important number of large grains. Indeed, when only few agglomerates are present through the microstructure, the agglomerate free domain, for which GBS is promoted, remains quite large.

Figures 4 and 5 show the local behaviour (i.e. the variations with the applied stress $\bar{\tau}$ of the shear rates $\dot{\gamma}^{(j)}$ for all the paths) for two degrees of agglomeration ($N_{G/A} = 1$ and 50). When the large grains are homogeneously dispersed (Figure 4), the difference between the maximum and the minimum shear rates is very limited, suggesting a homogeneous strain. Conversely, when the large grains are agglomerated (Figure 5), this difference is significantly increased, indicating that some paths are promoted and consequently suggesting heterogeneous deformation. This is illustrated by Figure 6, which shows the paths obtained for the microstructure M_{20} ($N_{G/A} = 20$). Arrows indicate the paths for which the shear rates are larger than the macroscopic shear rate. These paths do not intercept any large grains. More generally, for a given concentration of large grains, the probability of intercepting large grains decreases when $N_{G/A}$ is increased. In consequence, the deformation will be more heterogeneous.

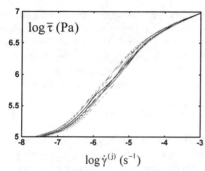
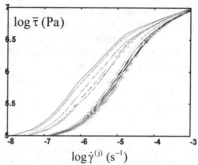

Fig. 4 : Local behaviour for a homogeneous dispersion of large grains ($N_{G/A} = 1$) leading to homogeneous strain through the sample

Fig. 5 : Local behaviour for agglomerations of large grains ($N_{G/A} = 50$) leading to strain localisation through the sample

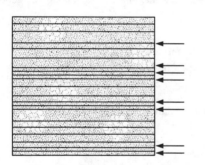
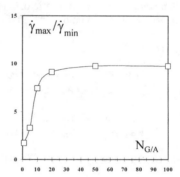

Fig. 6 : Paths operating for a microstructure with $N_{G/A} \approx 20$

Fig. 7 : Effect of $N_{G/A}$ on strain heterogeneity

Figure 7 displays the effect of $N_{G/A}$ on the ratio between the maximum and minimum shear rates obtained along the paths through the sample. The extent of strain heterogeneity is increased with $N_{G/A}$. This increase concerns essentially $N_{G/A}$ values lower than 20.

These results indicate that the heterogeneity of deformation is strongly related to the fraction of paths intersecting agglomerates. If all paths intercept agglomerates, strain localisation will be limited whereas if only some paths intercept agglomerates, strain localisation will be expected. However, for a given degree of agglomeration, the ability to obtain a path avoiding the large grains, will depend on the concentration C_{LG} of large grains. Moreover, for a given population of agglomerates, their spatial distribution may also influence the superplastic properties.

This study supports the idea that the characterisation of the grain size distribution is unable to give appropriate predictions of the superplastic response of a microstructure (in terms of macroscopic behaviour as well as strain homogeneity). The spatial distribution of the grains must be taken into account since it was demonstrated that for the same mean grain size and the same grain size distribution, the degree of agglomeration of large grains in a fine grained microstructure can affect significantly superplastic properties.

Conclusions

A 2D model describes superplastic deformation at a mesoscopic scale in order to predict the effect on the mechanical behaviour of microstructural heterogeneities, like grain size distribution or spatial distribution of grains. Predictions deal with the macroscopic rheology and the extent of localisation of deformation through the sample. Macroscopic deformation is assumed to result from preferential shear along paths through the sample. Microstructures with various microstructural heterogeneities were studied and it is shown that the degree of agglomeration of the large grains and the spatial distribution of these agglomerates through the microstructure are key parameters to obtain a good homogeneity of deformation through the microstructure.

References

[1] J.J. Blandin, R. Dendievel, Acta Mater., 48 (2000), p. 1541.
[2] V.V. Astanin, O.A. Kaybishev, S.N. Faizova, Scripta Metall. Mater, 25 (1191) p.2663.
[3] M.G. Zelin, T.R. Bieler, A.K. Mukherjee, Metall. Trans., 24A (1993) p.1208.
[4] R. Raj, M.F. Ashby, Metall. Trans., 2A (1971) p. 1113.

Corresponding author : Jean-Jacques Blandin, e-mail : Jean-Jacques.Blandin@inpg.fr

High Strain Rate Superplasticity

High Strain Rate Superplasticity in Mechanically Alloyed Nickel Aluminides

Y. Doi[1], K. Matsuki[2], H. Akimoto[3] and T. Aida[2]

[1] Toyama Industrial Technology Center, Futagami 150, Takaoka, Toyama 933-0981, Japan

[2] Department of Mechanical & Intelligent Systems Engineering, Toyama University, Faculty of Engineering, Gofuku 3190, Toyama 930-8555, Japan

[3] Graduate School, Toyama University, 3190 Gofuku, Toyama, 930-8555, Japan

Keywords: Grain Refining, High Strain Rate Superplasticity, Nickel Aluminides, Wet Mechanical Alloying

Abstract

Grain refining for three nickel aluminides such as γ'-Ni$_3$Al (25Al) and β-NiAl (49Al) single-phase intermetallics, and (γ'+β) two-phase intermetallic (34Al) has been attained by using a wet mechanical alloying and vacuum hot pressing process. The values of average grain size of the three compacted intermetallics were very close and as small as about 1.0 μm. The superplastic flow behaviors of the fine grain intermetallics were examined mainly by hot compression tests, together with some tensile tests. The compression tests were performed at the temperatures of 1073K-1273K and at an initial strain rate range from $1.4 \times 10^{-4} s^{-1}$ to $5.6 \times 10^{-2} s^{-1}$. The results obtained from the compression and tensile tests revealed that γ' single-phase intermetallic (25Al) and (γ'+β) two-phase intermetallic (34Al) can be deformed superplastically at 1273K and at a high strain rate range of about $4 \times 10^{-3} s^{-1} \sim 5.6 \times 10^{-2} s^{-1}$. However, β (49Al) single-phase intermetallics showed poor ductility in tensile tests, probably due to a cavity formation during the deformation. Grain boundary sliding accommodated by slip was considered to contribute much to the high strain rate superplasticity in the intermetallics including γ' phase.

Introduction

Recently, several intermetallics such as titanium aluminides, and nickel aluminides and silicides have received considerable interest as important high-temperature structural materials. In the nickel aluminides, L1$_2$ structural Ni$_3$Al (γ') is known to show anomalously positive temperature dependence of the yield stress, and B2 structural NiAl (β) has excellent oxidation resistance, low density and high melting points.

Furthermore, in the recent years, some superplastic studies of γ' single phase [1]-[4] and (γ'+β) two phase [5] nickel aluminides have been reported. However, the optimum superplastic strain rates for these nickel aluminides intermetallics were relatively low and ranging of $10^{-5} \sim 10^{-4} s^{-1}$ [1]-[5].

High strain rate superplasticity is considered interesting and potentially useful in the manufacturing of complex parts from the hard-to fabricate materials, such as the intermetallics. Grain refining is a principal method to shift the optimum superplastic strain rates to higher levels.

In this study, the possibility of grain refining by a wet mechanical alloying process [6] and high strain rate superplasticity has been investigated for nickel aluminides such as γ' and β single phase intermetallics, and (γ'+β) two-phase intermetallic.

Experimental

Nickel aluminide intermetallics, of which nominal compositions are Ni-24.9Al-0.1at%B (γ' phase: 25Al), Ni-33.9Al-0.1at%B ((γ'+β) phase: 34Al), and Ni-48.9Al-0.1at%B (β phase: 49Al), were fabricated from Ni, NiAl and B powders by a wet mechanical alloying process [6]. In this process, a large quantity of hexane enough to separate the powders was used as the organic agent. The chemical compositions of these intermetallics are shown in Table 1. These MA powders were consolidated by a vacuum hot pressing at 1423K.

The true stress, σ, - true strain, ε, and $\log\sigma$ -$\log\dot{\varepsilon}$ curves for the three intermetallics after compaction were obtained mainly by compression tests using an instron type testing machine. The compression tests were performed at the temperatures of 1073K-1273K and at an initial strain rate range from $1.4\times 10^{-4} s^{-1}$ to $5.6\times 10^{-2} s^{-1}$. The specimen size for the test was 3 mm$\times$3 mm$\times$6 mm.

The microstructure change with the compression deformation was examined by SEM and TEM. Specimens for TEM observation were thinned by an ion milling machine.

Some tensile tests at the selected test conditions were carried out in order to evaluate the ductility for the compacted intermetallics. The specimen dimension for the tensile test is 8 mm in gage length and 3 mm in gage width.

Table 1 The chemical composition of three intermetallics. (at%)

	Ni	Al	Fe	Cr	Ti	W	O	C	N	B
25Al (γ')	73.94	23.06	0.22	0.05	0.08	0.04	0.94	1.61	0.01	<0.04
34Al ($\gamma'+\beta$)	65.01	32.17	0.17	0.03	0.07	-	1.09	1.40	0.02	<0.04
49Al (β)	52.02	45.11	0.15	0.03	0.10	-	1.66	0.87	0.02	<0.04

Results and Discussion

Microstructure before test

SEM and TEM microstructures for 25Al, 34Al and 49Al intermetallic specimens before compression tests are shown in Fig.1 (a)-(f), respectively. In Fig.1 (b), dark contrast phases are β phase. By the wet mechanical alloying followed by a vacuum hot pressing, equiaxial and fine grain structures were formed in the three intermetallics. These intermetallics after consolidation were almost free from pores, as in Fig.1 (a)-(c).

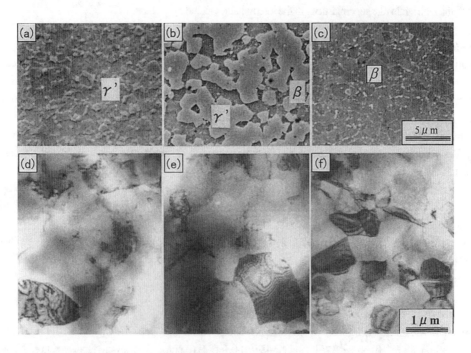

Fig.1 SEM and TEM micrographs for 25Al, 34Al and 49Al before compression test.

Table 2 shows the average grain sizes of these intermetallics evaluated from TEM microstructures. Very small average grain size of about 1.0 μm was obtained in each as-compacted intermetallic. The grains in these intermetallics are seen to be almost free from dislocations.

Table 2 Average grain sizes of three intermetallics before test.

	Average grain size (μm)		
	25Al (γ)	34Al (γ'+β)	49Al (β)
As compacted	1.0	1.2	1.0

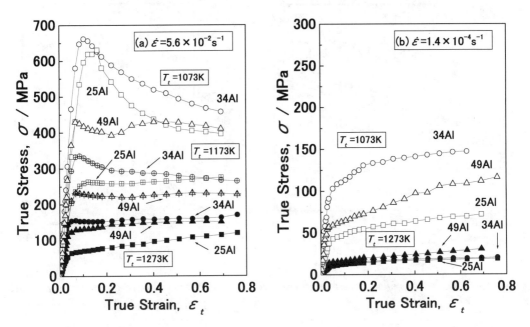

Fig.2 Typical true stress-true strain curves at the initial strain rates of $5.6 \times 10^{-2} s^{-1}$ (a) and $1.4 \times 10^{-4} s^{-1}$ (b).

Comparison of flow behavior

The high temperature compression tests for the fine-grained intermetallics were carried out in the wide deformation condition described above. Typical true stress–true strain curves at the initial strain rates of $5.6 \times 10^{-2} s^{-1}$ and $1.4 \times 10^{-4} s^{-1}$ are shown in Fig.2 (a) and (b), respectively, for the three intermetallics. At a lower strain rate range less than about $10^{-3} s^{-1}$, relatively steady state flow behavior or gradual strain hardening behavior were observed in these alloys irrespective of temperatures. However, at a higher strain rate range larger than about $10^{-2} s^{-1}$, a strain softening behavior, as in Fig.2 (a), were observed at a 1073K in 25Al and at temperatures lower than 1173K in 34Al. However, 49Al did not show this behavior at the all temperatures. Thus, this strain softening behavior implies the occurrence of a dynamic recrystallization in γ' phase which ordered structure is $L1_2$ [4].

Fig.3 shows the comparison of log σ- log $\dot{\varepsilon}$ curves of the three intermetallic specimens tested at 1073K and 1273K. At a strain rate range higher than about $2 \times 10^{-3} s^{-1}$, the strain rate sensitivity, m, is higher than 0.35 in 25Al at 1073K and in the three intermetallics at 1273K. This result suggests the possibility of high strain rate superplasticity in these intermetallics. However, low strain rate region showing lower m value than about 0.15 are observed in 25Al and 49Al intermetallics, suggesting the existence of a threshold stress. It is also recognized in Fig.3 that the flow stress for 25Al is remarkably lower than those for 34Al and 49Al across a wide strain rate range.

Fig.4 shows a comparison of total elongation at 1273K for the three intermetallics. In 25Al and 34Al intermetallics, maximum elongations of as high as about 250% were obtained at an initial strain rate of 4.2×10^{-3} s^{-1} and higher than about 180% elongations were obtained even at a high strain rate of 4.2×10^{-2} s^{-1}. However, the total elongation was relatively low for 49Al intermetallic composed of β single phase. Thus, 25Al and 34Al intermetallics, which are including γ' phase, can be deformed superplastically at the high strain rate range. Further experiments on cavitation formation are necessary to explain the fracture mechanisms of them.

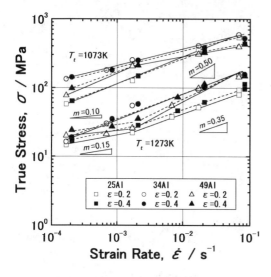

Fig.3 log σ - log $\dot{\varepsilon}$ relationships of the three intermetallic.

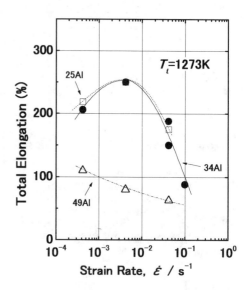

Fig.4 The comparison of total elongation at 1273K for the three intermetallics.

Microstructure after deformation

Fig.5 shows a TEM microstructure of 25Al after the compression to about 30% strain at 1073K and at a high strain rate of 5.6×10^{-2} s^{-1}. At this deformation condition, rapid strain softening was observed in 25Al, as shown in Fig.2(a). Compared with Fig.1 (d), the grain structure is seen to be refined by the deformation. Many grains contain a considerable amount of dislocations. These dislocations are relatively homogeneously distributed within the grains without forming cell boundaries. However, some small grains (arrowed) containing few dislocations are also observed in Fig.5. This was proved by careful tilting experiments in TEM. Thus, these small grains regarded as new grains produced by a dynamic recrystallization.

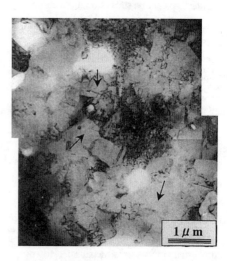

Fig.5 TEM micrograph of 25Al after the compression to about 30% strain. $\dot{\varepsilon} = 5.6 \times 10^{-2}$ s^{-1}, $T_t = 1073$K

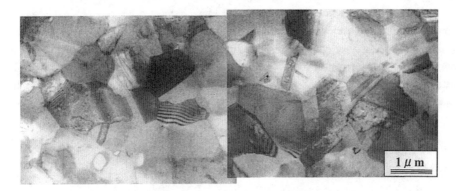

Fig.6 TEM micrograph of 25Al after the compression to about 50% strain. $\dot{\varepsilon}=5.6\times10^{-2}\text{s}^{-1}$, $T_t=1273\text{K}$

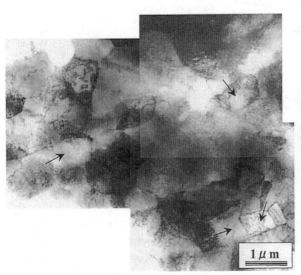

Fig.7 TEM micrograph of 34Al after the compression to about 50% strain. $\dot{\varepsilon}=5.6\times10^{-2}\text{s}^{-1}$, $T_t=1273\text{K}$

Fig.8 TEM micrograph of 49Al after the compression to about 50% strain. $\dot{\varepsilon}=5.6\times10^{-2}\text{s}^{-1}$, $T_t=1273\text{K}$

Fig.6-Fig.8 show TEM microstructures of 25Al, 34Al and 49Al, respectively, after the compression to about 50% strain at 1273K and at a high strain rate of $5.6\times10^{-2}\text{s}^{-1}$. Equiaxial and fine grain structures were observed in three intermetallics after the compression. This observation implies the occurrence of grain boundary sliding during the deformation at the high strain rate. In Fig.9, the change of average grain size with the compression deformation was shown for the three intermetallics. In the both intermetallics of 25Al and 34Al containing γ′ phase, the grain structures are very stable during the deformation even at 1273K. The dynamic recrystallization and the grain growth might be occurred concurrently during the deformation in both intermetallics. However, the different dislocation structures are clearly observed between 25Al(γ′) and 49Al(β). In 25Al(γ′), many grains are almost dislocation free, while some grains contain a relatively parallel arrayed dislocation structure. These dislocations seemed to accommodate the grain boundary sliding and to be easily absorbed into boundaries. In contrast, most of grains in 49Al(β) contain a considerable amount of dislocation distributed homogeneously. No sub-boundary formation has been observed in the deformation microstructure. The dislocation glide and climb process seemed to be more difficult in β phase in comparison with γ' phase. Dislocation structure of 34Al (γ'+β) is rather a similar case with 49Al(β) intermetallics, except the formation of small grains including a few dislocation (arrowed). Thus, the superplastic deformation mechanism at the high strain rate for the intermetallics is considered to be grain boundary sliding accommodated by slip.

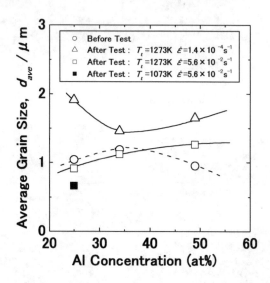

Fig.9 Change of average grain size with the compression deformation.

Conclusions

1) Grain refining for three nickel aluminides such as γ' (25Al) and β (49Al) single phase intermetallics, and (γ'+β) two-phase intermetallic (34Al) has been attained by using a wet mechanical alloying and vacuum hot pressing process. The values of average grain size of the three compacted intermetallics were very close and as small as about 1.0 μm.

2) The intermetallics composed of γ' single-phase (25Al) and (γ'+β) two-phases (34Al) can be deformed superplastically at 1273K and at a high strain rate range of about $4 \times 10^{-3} s^{-1} \sim 4 \times 10^{-2} s^{-1}$.

3) Grain boundary sliding accommodated by slip was considered to contribute much to the high strain rate superplasticity.

References

[1] A.K.Mukherjee and R.S.Mishra, Mater. Sci.Forum243-245(1997),p.609.
[2] M.S.Kim, S.Hanada, S.Watanabe and O.Izumi, Mater. Trans. *JIM*, 30(1989),p.77.
[3] J.Mukhopadhyay, G.Kascner and A.K.Mukherjee, Scripta Met. Mater.,24(1990),p.857.
[4] M.Oikawa, S.Hanada, T.Sakai and S.Watanabe, Mater. Trans. *JIM*, 36(1995),p.1140.
[5] S.Ochiai, Y.Doi, I.Yamada and Y.Kojima, J.Japan Inst. of Metals,57(1993),p.214.
[6] Y.Doi, H.Akimoto, K.Matsuki, T.Aida and S.Ochiai, J.Japan Inst. of Metals,64(2000),p.518.

Influence of Reversible Hydrogen Alloying on Formation of SMC Structure and Superplasticity of Titanium Alloys

G.A. Salishchev, M.A. Murzinova, S.V. Zherebtsov,
D.D. Afonichev and S.P. Malysheva

Institute for Metals Superplasticity Problems, Russian Academy of Sciences,
Khalturina 39, RU-450001 Ufa, Russia

Keywords: Dynamic Recrystallization, Hydrogen Alloying, Submicrocrystalline Structure, Superplasticity, Titanium Alloy

Abstract
A method of production of a submicrocrystalline (SMC) structure (d<1µm) in large workpieces of the commercial pure titanium and the two-phase titanium alloys based on initiation of dynamic recrystallization (DRX) during hot working and reversible hydrogen alloying has been developed. The method involves continuous grain refinement due to DRX at decreasing temperature. Hydrogen alloying decreases deformation stresses and increases ductility at lower temperatures than in hydrogen-free alloy and promotes the formation of the finest grain structure. The grain sizes of 0.1-0.2 µm and 0.04 µm have been produced in the hydrogenated commercial pure Ti and Ti-11.4Al-1.7Mo-0.88Zr (at.%) based alloys, respectively. An oxidized layer retards the hydrogen removal in these alloys at temperatures lower than 500°C. Therefore the SMC structure during degassing can be retained only in complex alloyed two-phase titanium alloys. After degassing the grain size as low as 0.04 µm has been obtained. The hydrogen alloying of commercial pure Ti prevents observing of low temperature superplasticity (SP), but sharply increases superplastic properties at higher temperatures. The low temperature SP at 550°C occurs in the two-phase titanium alloy with the grain size 0.04 µm that has been obtained by using reversible hydrogen alloying.

1. Introduction
Materials with nano- and submicrocrystalline structure exhibit unique mechanical properties [1]. Strength of such materials is sharply increased and temperature of superplastic deformation significantly decreases [2]. That is why the development of an efficient method for obtaining SMC materials and the study of their properties is a very urgent problem. One of them is severe plastic deformation providing the intensive occurrence of DRX [2,3]. It is based on the dependence of the size (d) of recrystallized grains on the steady flow stress

$$\sigma_s = kd^{-n} \quad (Eq.\ 1),$$

where k is an empirical constant and the grain size exponent n≅0.5 [4]. Therefore, the size of recrystallized grains can be decreased via decreasing the temperature or increasing the strain rate since the change of these parameters leads to an increase in the flow stress. Grains produced during DR are almost the same in size as subgrains, while the subgrain size can be as less as several ten nanometers. Hence, a SMC structure is expected to be produced in materials by DRX. The SMC structure with d as low as 0.06 µm has been formed in Ti-10.30Al-1.55Mo-0.33Fe (at.%) (α+β)-titanium alloy at 550°C [5]. It should be noted that a critical strain for DRX initiation increases considerably when the deformation temperature decreases and the high strains are required to produce fully recrystallized SMC structure. Besides, high stresses and low ductility at lower temperatures lead to the crack formation in the billet [5]. However, the lower is the deformation temperature, the finer is the size of recrystallized grains.
Application of reversible hydrogen alloying for formation of SMC structure in titanium alloys seems

to be very interesting. First, hydrogen reduces the temperature of $(\alpha+\beta)\Leftrightarrow\beta$ transformation and improves the workability of titanium alloys [6]. Second, hydrogen facilitates DRX of α-phase [7] and the formation of a finer structure in two-phase alloys during heat treatment. This leads to the increase in the fraction of recrystallized grains and the decrease in their size [6,8]. Third, one can expect for an additional effect of $\beta\rightarrow\alpha$ transformation stresses on grain refinement during vacuum annealing for hydrogen removal [6]. On the other hand, decreases in the temperature of $(\alpha+\beta)\Leftrightarrow\beta$ transformation during hydrogen alloying and in the grain size assume a significant decrease in the temperature of SP of titanium alloys. Moreover, hydrogen alloying in the α titanium alloys suppresses dynamic-strain aging (DSA) [6], which occurs in the temperature interval close to the temperature interval of superplastic deformation [9]. Consequently, one can expect higher SP properties in hydrogen alloyed α-titanium alloys. And finally, hydrogen alloying provides transformation of single phase alloys to two-phase ones that also may improve their SP characteristics. The present work was undertaken to answer these questions.

2. Experimental

The commercial pure titanium Ti with the chemical composition Ti-0.56Al-0.08Fe-0.12Si-0.20C-0.1H-0.3O and the two-phase titanium alloy marked as ATi with the chemical composition Ti-11.4Al-1.7Mo-0.88Zr-0.09Fe-0.45Si (both at.%) were used. The forged and then rolled materials produced by VSMPO were machined into 20 mm long by 20 mm diameter rods. Hydrogenation for these specimens was carried out in hydrogen atmosphere at 650-700°C. The commercial pure titanium was alloyed by hydrogen up to 5 and 16 at.%, the ATi alloy up to 14 at.%. These alloys are marked as Ti-5H, Ti-16H and ATi-14H, respectively. The ATi-14H alloy after hydrogen removal is marked as ATiR. The Ti-5H alloy is an α- alloy, the Ti-16H and ATi-14H alloys are $(\alpha+\beta)$-alloys with a close volume fraction of phases at 550-650°C. The samples of the commercial pure titanium were preliminary annealed at the temperature of a single phase β-region, thus, after saturation by hydrogen they had a similar grain size of 270±30 μm. The ATi and ATi-14H alloy samples were quenched from the temperature of β-region, after that their grain sizes were 180±20 and 30±3 μm, respectively. Compression and tension mechanical tests of the samples were conducted at 20-800°C and the strain rates 5×10^{-5}-$10^{-2}s^{-1}$. The structure of the samples was studied by optical and electron microscopy. The SMC structure was obtained in billets out of Ti and titanium alloy ATi, 20 mm in diameter and 40 mm in length, by 2 stage a-b-c-forging at 380-420 and 530-570°C, respectively.

3. Results

3.1. Formation of SMC structure

In the initial condition the alloys had microstructure of two types: coarse grained single phase - in Ti and its alloy Ti-5H and coarse lamellar two-phase- in alloys Ti-16H, ATi and ATi-14H. During compression tests at 400-700°C and $\dot{\varepsilon}=5\times10^{-4}$ s^{-1} the occurrence of DRX and the refinement of the initial microstructure were observed in all alloys. A decrease in the deformation temperature leads to decrease in the size of recrystallized grains (Fig.1). The mean grain size in Ti, Ti-5H and Ti-16H alloys at final testing temperatures was 0.30±0.05, 0.20±0.03 and 0.10±0.02, respectively and in the ATi and ATi-14H alloys - 0.1±0.02 and 0.04 μm. Unlike the dependence corresponding to the two-phase alloys, in the dependence for the single phase Ti and Ti-5H alloys there are two parts with different slopes with a point of a bend at 650°C and 550°C, respectively.

The dependence of recrystallized grain size on flow stress at $\varepsilon=50\%$ is plotted according to the (Eq. 1) and is shown in Fig.2. The dependence for the two-phase condition of the alloys (Ti-16H, ATi, ATi-14H) is approximated by straight lines with exponents varied for different compositions from $\mathbf{n}=0.7$ to 1.6. The curves corresponding to the single phase Ti and Ti-5H alloys were found two stages. Similar to the temperature dependence the slope of curves corresponds to the same temperature points. The values of n for Ti and Ti-5H alloys are equal to 0.3 and 0.2 in the first part and 1.1 and 0.9 in the second part. The presence of bends on the curve probably testifies to a change in the recrystallization mechanism. A decrease in the temperature of deformation leads to a significant growth of flow stress. Hydrogen alloying reduces a value of flow stresses required for

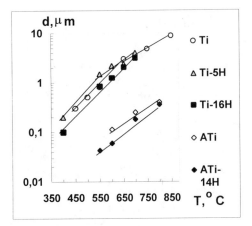

Fig.1. Dependence of deformation temperature on size recrystallized grains (d) in titanium alloys.

attaining the similar grain size.

A decrease in the testing temperature reduces ductility. For example, the ATi alloy samples failed already after compression by 20% at 580°C and $\dot{\varepsilon}=5\times10^{-4}s^{-1}$. At lower deformation temperatures the ductility can increase due to a decrease in the initial grain size. Hydrogen alloying of alloys provides it retarding the grain growth [7]. The size of β-grains in the ATi-14H alloy after β-region quenching is three times less as compared to the basic ATi alloy. Hydrogen alloying provides the increase in dispersity of metastable phase decomposition (for instance, martensite) that contributes to formation of the finest grain size during DRX. In particular, after holding of β-quenched samples at 600^0C for 20 min a mean thickness of precipitates in the ATi-14H alloy was 0.05 μm and 0.15 μm in the ATi alloy. The occurrence of phase transformation along with DRX should contribute to formation of a SMC structure [2]. In this condition the samples were deformed by 80% at 525°C without crack formation. As a result, after deformation of the ATI-14H alloy at 530°C a microstructure with a grain size of 0.04 μm was formed.

The billets with dimensions of 30x20x10 mm with SMC structure were produced by using multiple forging. Depending on alloy composition, the grain sizes corresponding to the smallest ones in Fig.1 were formed in the billets. The microstructure in the SMC condition is typical for alloys produced by severe plastic deformation: the presence of internal elastic stresses caused by high density of dislocations at grain boundaries was observed (Fig.3). In the SMC titanium alloys Ti-5H and Ti-16H the features of microstructure caused by formation of hydrides have been revealed. As known [6], in hydrogenated titanium alloys hydride plates like of large size up to several micrometers precipitate during cooling from the temperature higher than eutectoid (328°C). These hydrides make the alloys brittle. Fine hydrides of a globular shape mainly are formed in the SMC alloy that probably assumes the influence of internal elastic stresses on their formation.

One of the important aspect of applying hydrogen alloying for refinement of microstructure is retaining of the SMC condition after hydrogen removal. The presence of an oxidized layer in the titanium does not allow to remove hydrogen at temperatures up to 500°C at which retaining of SMC structure is possible. At the same time the structure of ATi alloy is more stable. A vacuum annealing at 550°C of ATi-14H alloy specimens decreases hydrogen concentration until safety level like in basic alloy and retains SMC structure.

3.2. Mechanical properties at room and elevated temperatures
3.2.1. Ti, Ti-5H, Ti-16H

The tensile properties were determined at 20-600°C (Fig.4 a). At room temperature the SMC titanium both in basic and hydrogen alloyed conditions displays high values of yield strength. The larger is the hydrogen content, the higher is these values. The later is connected with strengthening of titanium by fine hydrides, whose volume fraction increases with increasing hydrogen concentration. It seems to be interesting that a significant elongation (17-20%) was obtained in hydrogen alloyed alloys. In the coarse grained condition the hydrogen alloyed titanium is brittle. An increase in temperature up to 400°C leads to a sharp drop of a yield strength while a relative elongation changes slightly. However, the ductility of hydrogen alloyed samples at 400°C is higher than in Ti. An increase in temperature up to 450°C leads to further decrease in yield strength and the strain rate sensitivity coefficient, m and δ are 0.3 and 117%, respectively. A value of m for the SMC titanium with hydrogen at 450°C does not exceed 0.25 though elongation is higher than 140%.

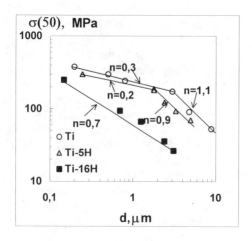
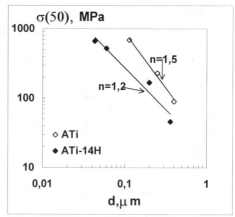

Fig.2. Size of recrystallized grains (d) versus steady state stress (σ_{50}) in titanium alloys

Fig.3. Microstrucrure of titanium alloys obtained by a-b-c-forging; (a)- Ti-16H, (b)- ATiR after hydrogen removal.

Further increase in temperature leads to slight growth of ductility in basic titanium and hydrogen alloyed Ti-5H, though in the later at 600°C and $\dot{\epsilon}=4.2\times10^{-4}$ s^{-1} m=0.48 and relative elongation is 220% at flow stress 45 MPa. At the same time, in the Ti-16H alloy deformed in ($\alpha+\beta$) region at 600°C and $\dot{\epsilon}=4.2\times10^{-4}$ s^{-1}, a considerable increase in elongation up to 640% takes place and m=0.48 at flow stress 20 MPa.

The estimation of grain growth has shown that the grain size becomes coarser at T>400°C and the strain rate 4.2×10^{-4} s^{-1} in all conditions of titanium but hydrogen alloying retards grain growth. So, the grain size in the Ti-5H alloy (deformation in α-field) exceeds 1 μm after testing at 500°C, in the alloy Ti-16H (($\alpha+\beta$) deformation) - at 600°C, whereas in Ti the grain growth up to 1 μm is observed already at 450°C.

The temperature interval of elevated ductility of titanium coincides with the range of DSA. The estimation of a strain-hardening rate showed that DSA was observed in the basic SMC titanium and was suppressed in hydrogen alloyed alloys (Fig.5) and, consequently, can effect on its ductility.

3.2.2. ATiR, ATi-14H

The influence of hydrogen on deformation behavior of the complex alloyed SMC alloy is ambiguous too. The temperature dependence of strength and ductility is shown in Fig.4 b. In the temperature interval 20-350°C both alloys are characterized by low ductility and high strength. In the temperature interval 500-550°C the ductility of SMC alloys is increased and their strength is decreased. The presence of hydrogen suppresses the low temperature SP: m=0.24 and maximum

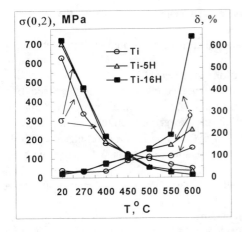 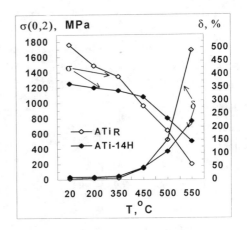

Fig.4. Dependence of yield stress (σ_{02}) and elongation (δ) on deformation temperature at d $\dot\varepsilon = 5\times10^{-4}$ s^{-1} in SMC titanium alloys: a - for Ti, Ti-5H, Ti-16H; b - for ATiR and ATi-14-H

elongation is 240% (T=550°C, d $\dot\varepsilon = 4.2\times10^{-4}$ c^{-1}). At the same time dehydrogenated ATiR alloy is superplastic at 550°C and at d $\dot\varepsilon = 2\times10^{-4}$ s^{-1}: m= 0.52, relative elongation achieves 550%.

4. Discussion

The results of the present study testify the favourable influence of hydrogen alloying on formation of SMC structure in titanium and two-phase titanium alloy and its ambiguous influence on SP. Actually hydrogen alloying decreases flow stress during processing for an ultrafine grain size and increases maximum acceptable strain values. A number of features contribute to formation of microstructure with fine grain size. First, in the hydrogen containing alloys the rate of grain growth is decreased. In single phase condition it may be connected with increasing concentration of impurities (O,C,N,Fe) at grain boundaries, which are replaced by hydrogen from intergranular defects [6]. In case of two-phase structure formation the grain growth is retarded by second phase particles. Another feature that can be applied for refining microstructure is the formation of finer precipitates of phases in the hydrogen saturated alloy as compared to the basic alloy. That occurs due to quenching and subsequent decomposition of metastable phase. However, the potentialities of hydrogen application along with the retention of the SMC structure during vacuum annealing are determined by the final temperature of treatment. The necessity of hydrogen removal at the temperature higher than dissolution of an oxidized layer (500°C) does not allow to retain it in titanium, but it becomes possible in complex alloyed titanium alloys with a more thermal stable structure. No grain growth was observed after vacuum annealing of the ATi alloy with initial hydrogen content of 14 at.% that is possibly due to transformation stresses and recrystallization in hydrogen free condition. At elevated temperatures (450°C) the SMC titanium displays poor SP properties. This is probably connected with occurrence of DSA (Fig.5) in the temperature interval of 350-450°C and with intensive grain growth at higher temperatures. Hydrogen suppresses DSA and retards grain growth. Due to that the ductility of Ti-5H and Ti-16H alloys is increased already at 400°C. But SP is observed only at T=600°C. SP properties are the most prominent in the two-phase condition when α and β phase proportion is approximately equal. It can be explained by increase of the quantity of impurities at grain boundaries that results in retarding grain boundary sliding. It is confirmed indirectly by the fact that in the hydrogen alloyed titanium the growth of ductility coincides with the increase in the fraction of β-phase and, consequently, the decrease of the hydrogen concentration in it. Due to such softening effect the negative influence of impurities on ductility can be reduced. The sharp growth of superplastic properties in the two-phase titanium alloy

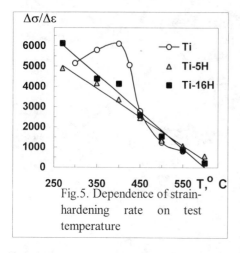

Fig.5. Dependence of strain-hardening rate on test temperature

after its degassing can be explained by the same way. However, in this case the fraction of β-phase decreases which is accompanied with the decrease in strengthening with hydrogen and the intensification of diffusion processes [6].

The results obtained by testing the degassed two-phase titanium alloy with the grain size 0.04 μm show that the lower temperature limit of SP is observed at 550°C, and it almost coincides with the temperature of alloy aging. In VT8 alloy with the grain size 0,06 μm that possess lower high temperature strength SP was observed at the same testing temperature [5].

5. Conclusions

A new method of obtaining a submicrocrystalline structure in workpieces made of titanium and two-phase titanium alloy by employing dynamic recrystallization and reversible hydrogen alloying was developed. At the same deformation temperature and strain rate hydrogen alloying decreases flow stress during processing for an ultrafine grain size and increases maximum acceptable strains. Grain sizes of 0,1-0.2 μm and 0.04 μm were produced in hydrogenated Ti and two-phase Ti-11.4Al-1.7Mo-0.88Zr (at.%) based alloys, respectively. Retention of SMC structure during hydrogen removal can be employed only in two-phase alloy because of an oxidized layer, which does not allow to remove hydrogen at temperatures lower than 500°C. Hygroden suppresses the low temperature superplasticity in SMC alloys, but allows to observe prominent superplastic properties at higher temperatures in hydrogenated Ti with equal proportion α- and β-phases. Degassing of hydrogenated two-phase alloy does not change the grain size and the low temperature superplasticity takes place at 550°C and $\dot{\varepsilon}=2\times10$ s^{-1} at that m=0.52, relative elongation achieves 550%.

6. References

1. H. Gleiter, Nanostructured Materials. (1992), Vol.1, p.1.
2. O.A. Kaibyshev, Superplasticity of alloys, intermetallides and ceramics. - Berlin, Springer-Verlag. (1992), pp.317.
3. N.A.Smirnova, V.I. Levit, V.I. Pilyugin. and et., Fis. Met. Metalloved. (1986), Vol. 61, p. 127.
4. G. Glovers and C.M. Sellars, Met. Trans. (1973), A4, p.765.
5. G.A. Salishchev, O.R. Valiakhmetov, R.M. Galeev, J. of Materials Science (1993), 28, p.2898.
6. O.N. Senkov, F.H. Froes. Int.J. of Hydrogen Energy (1999), Vol. 24, p.565.
7. M.A. Murzinova, G.A. Salishchev, D.D. Afonichev. Metalls (2000), to be published(in Russian).
8. M.A. Murzinova, M.I. Mazursky, G.A. Salishchev, D.D. Afonichev. Intern. J. Hydrogen Energy (1997), Vol. 22, No. 2/3. p. 201.
9. H. Conrad, Progress in Materials Science, (1981), Vol. 26, p.123.

High Strain Rate Superplasticity in a Zn - 22% Al Alloy after Equal-Channel Angular Pressing

Sang-Mok Lee and Terence G. Langdon

Departments of Materials Science and Mechanical Eng., University of Southern California,
Los Angeles CA 90089-1453, USA

Keywords: Equal-Channel Angular Pressing, High Strain Rate Superplasticity, Ultrafine Grains, Zn-Al Alloy

Abstract

It is well established that superplasticity requires a small grain size and therefore there is an interest in developing processing methods having the capability of reducing the grain size below that generally produced in conventional thermo-mechanical processing. Experiments were conducted on a superplastic Zn-22%Al eutectoid alloy where the material was subjected to equal-channel angular pressing (ECAP) through 4, 8 or 12 passes at a temperature of 373 K. It is shown that this processing procedure leads to additional grain refinement such that the grain size was reduced from an initial value of ~1.8 μm to values as low as ~0.6 μm. Tensile testing after ECA pressing revealed a significant enhancement of the superplastic properties including both increases in the total elongations to failure and the occurrence of these high elongations at very rapid strain rates. Elongations were achieved of up to >2380 % at a strain rate of 1 s^{-1}. The results suggest that ECAP may be a very useful procedure for attaining superplasticity at high strain rates in commercial alloys.

Introduction

A very small grain size, typically in the range ~1 - 10 μm, is an important requirement for attaining high superplastic elongations [1]. Generally, these small grain sizes are achieved either by using two-phase eutectic or eutectoid alloys where grain growth is limited by the presence of two separate phases or by incorporating a very fine dispersion of a second phase into an alloy to act as a grain refiner. The Zn-22%Al alloy is a classic example of a two-phase superplastic material that is capable of exhibiting exceptionally high tensile ductilities (up to >1000%) over a limited range of strain rates and temperatures [2,3]. Typically, high elongations are attained in this alloy at strain rates in the vicinity of ~10^{-3} - 10^{-2} s^{-1}. This alloy also has significant commercial applications because it has been used in the superplastic forming industry for the fabrication of parts ranging from office equipment to instrument covers and car body panels [4-6].

Very recently, interest has developed in a processing method known as Equal-Channel Angular Pressing (ECAP) in which, by using a process of repeated simple shear, a material is subjected to a very large plastic strain without any concomitant change in the cross-sectional dimensions of the sample [7,8]. This simple processing procedure is an example of processing through Severe Plastic Deformation (SPD) [9] and ECAP is especially attractive by comparison with only SPD procedures, such as High-Pressure Torsion (HPT), because it can be used to produce large bulk samples that may have commercial applications.

In ECAP, a sample is pressed through a die containing two channels, equal in cross-section, which form an L-shaped configuration. Two angles characterize the die and determine the strain imposed on the sample during a single pass: there is an angle φ between the two channels within the die and an additional angle Ψ denoting the outer arc of curvature at the point where the two channels intersect. Experiments show that the processing route is an important variable during ECAP, where these routes denote the potential for rotating the sample between consecutive passes through the die [10-13]. It has been shown for pure aluminum that, when φ = 90°, route B_C is the most effective pressing route for rapidly achieving a homogeneous microstructure of equiaxed grains separated by boundaries having high angles of misorientation [12], where route B_C denotes the procedure of rotating the billet in the same sense by 90° between each separate pass. Since high angle boundaries are a prerequisite for the occurrence of grain boundary sliding in superplasticity, it is reasonable to anticipate that subjecting a sample to ECAP through route B_C may give a microstructure suitable for attaining high superplastic ductilities.

The present investigation was motivated by the prossibility that additional grain refinement of the commercial Zn-22% Al alloy through ECAP may lead to enhanced superplastic properties. It should be noted that some earlier experiments revealed only limited superplastic elongations in this alloy after ECAP due, it was suggested, to the agglomeration of grains of similar phases after pressing [14]. The present investigation was designed to evaluate this problem in more detail. As will be demonstrated, these experiments show that ECAP is capable of producing remarkably high ductilities in this alloy under optimum conditions and, furthermore, these high ductilities occur at very rapid strain rates. Thus the results demonstrate that ECAP may be useful procedure for achieving high strain rate superplasticity (HSR SP) in the commercial Zn-22% Al alloy.

Experimental Material and Procedures

The material used in this investigation was a commercial Zn-22%Al eutectoid alloy obtained in a superplastic condition. Measurement showed that the initial grain size before ECA pressing was ~1.8 μm. Cylindrically shaped billets with a diameter of 9.3 mm and total lengths of ~63 mm were machined for ECAP.

The principle of ECAP is that a sample is pressed through the die with a plunger and it experiences simple shear at the intersection of the two channels. The present experiments were conducted using a solid die with an internal angle of φ = 90° and an arc of curvature at the point of intersection given by ψ ≈ 45°. It can be shown that these values of φ and ψ lead to an imposed strain of ~1 on each pass through the die [15]. In the present experiments, samples were pressed for totals of 4, 8 and 12 passes corresponding to total imposed strains of ~4, ~8 and ~12, respectively. All samples were pressed at a temperature of 373 K with the temperature held constant to within ±5°. To allow the samples to reach thermal equilibrium prior to testing, each sample was held in the die for ~2 minutes before pressing. Unlike the earlier experiments [14], all samples were rotated in the same sense by 90° between each pressing in the procedure designated as route B_C.

After completion of ECAP, the pressed billets were machined into tensile specimens with gauge lengths of 4 mm and cross-sections of 2×3 mm^2. All samples were oriented with the tensile axes lying parallel to the longitudinal axes of the ECAP billets. The samples were pulled in tension in air using an Instron testing machine operating at a constant rate of cross-head displacement. Testing temperatures were in the range from 423 to 533 K. Most samples were pulled to failure in order to determine the maximum elongations as a function of strain rate.

Experimental Results and Discussion

Measurements showed the average grain sizes after ECA pressing were ~1.0, ~0.8 and ~0.6 µm for totals of 4, 8 and 12 passes through the ECAP die, respectively. Tensile testing of the as-pressed samples revealed very high elongations to failure, with elongations up to and exceeding 2000%. The optimum superplastic conditions are summarized in Fig.1 where (a) shows the maximum elongation to failure recorded at each testing temperature from 423 to 533 K for samples pressed through 4, 8 and 12 passes and (b) shows the optimum testing strain rate associated with each temperature and pressing condition.

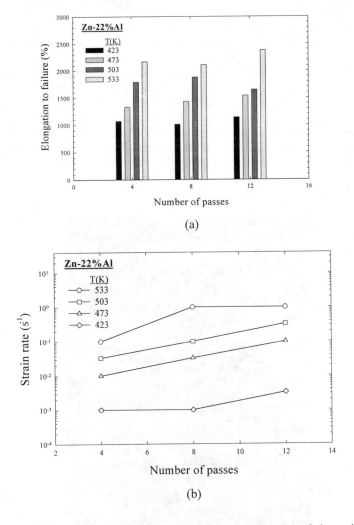

Fig. 1 The optimum superplastic conditions after ECAP: (a) variation of elongation to failure with number of passes through the die for four different testing temperatures, (b) strain rate associated with maximum elongations at each temperature as a function of the number of passes.

Inspection of Fig. 1 reveals several important trends. First, it is apparent from Fig. 1(a) that the elongations to failure increase as the testing temperature is increased, with the highest elongations recorded at the highest testing temperature of 533 K. The appearance of the specimens after pressing through 12 passes and testing at 533 K is shown in Fig. 2 together with an untested sample. Second, it is also apparent from Fig. 1(a) that higher elongations tend to occur after larger numbers of passes. This latter result is consistent with earlier observations on a commercial Al-Mg-Li-Zr alloy [16,17]. Third, inspection of Fig. 1(b) shows the optimum superplastic conditions occur at higher strain rates as the number of passes is increased. Again, this result is consistent with earlier reports [16,17] and all of these trends confirm the increase in microstructural evolution which occurs when pressing is continued to larger numbers of passes. In Fig. 2, the appearance of the samples after testing reveals uniform deformation within the gauge lengths with no evidence for any necking and with the samples pulling down to fine points at failure: all of these characteristics are consistent with conventional superplasticity [18]. It is important to note also the exceptionally high elongations achieved in this alloy at 533 K. At a strain rate of $1\ s^{-1}$, the test was terminated at an elongation of 2380% without failure of the sample. Furthermore, an elongation of 1630% was achieved at a strain rate of $2\ s^{-1}$ and the elongation was 1910% at $10^{-1} s^{-1}$. All of these results confirm the potential reported earlier for achieving HSR SP after processing by ECAP [19,20].

Further information is given in Fig. 3 where the elongations to failure are plotted against the strain rate for samples pressed through 12 passes. This plot clearly reveals the tendency for the optimum strain rate to increase with increasing temperature. Figure 4 shows the corresponding plot of the measured flow stress versus the strain rate and these results suggest a division into two regions with a strain rate sensitivity close to ~0.4 in the region of maximum superplasticity.

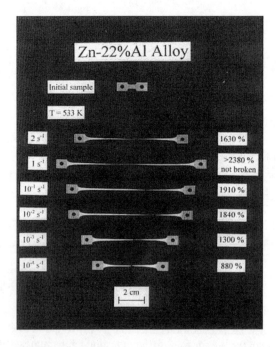

Fig. 2 Appearance of the specimens pressed through 12 passes and tested at 533 K.

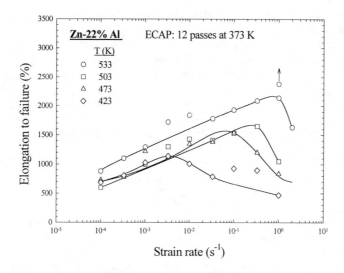

Fig. 3 Elongation to failure versus strain rate for samples pressed through 12 passes.

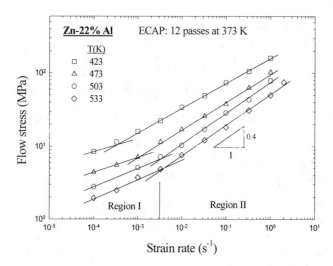

Fig. 4 Flow stress versus strain rate for samples pressed through 12 passes.

Summary and Conclusions

1. The grain size of a Zn-22% Al eutectoid alloy was reduced from ~1.8 μm to a value as low as ~0.6 μm through ECAP.
2. Tensile testing after ECAP revealed very high elongations to failure, with the elongations increasing with (a) the number of passes through the die and (b) the testing temperature. Elongations were recorded of up to >2300% at a temperature of 533 K with a strain rate of 1 s^{-1}.

Acknowledgements

This work was supported in part by the U.S. Army Research Office under Grant No. DAAD19-00-1-0488.

References

1. T.G. Langdon, Metall. Trans. 13A (1982) p. 689.
2. H. Ishikawa, F.A. Mohamed and T.G. Langdon, Phil. Mag. 32 (1975) p.1269.
3. F. A. Mohamed, M. M. I. Ahmed, and T. G. Langdon, Met. Trans. 8A (1977) p. 933
4. D.S. Fields, IBM J. 9 (1965) 134.
5. L.T. Feng and D.T. Camp, SAE Paper #700133, Society of Automotive Engineers, Warrendale, PA (1970).
6. T.G. Langdon, in *Recent Advances in Science, Technology and Applications of Zn-Al Alloys* (edited by G. Torres-Villaseñor, Y. Zhu and C. Piña-Barba), Instituto de Investigaciones en Materiales, Universidad Nacional Autónoma de México, Mexico City (1994), p.177.
7. V. M. Segal, Mat. Sci. Eng. A197 (1995) p.157.
8. T.C. Lowe and R.Z. Valiev, JOM 50 (4) (2000) p.27.
9. T.C. Lowe and R.Z. Valiev (eds.), *Investigations and Applications of Severe Plastic Deformation*, Kluwer, Dordrecht, The Netherlands (2000).
10. Y. Iwahashi, Z. Horita, M. Nemoto and T. G. Langdon, Acta Mater. 45 (1997) p.733.
11. Y. Iwahashi, Z. Horita, M. Nemoto and T. G. Langdon, Acta Mater. 46 (1998) p.3317.
12. M. Furukawa, Y. Iwahashi, Z. Horita, M. Nemoto and T.G. Langdon, Mater. Sci. Eng. A257 (1998) p.328.
13. K. Oh-ishi, Z. Horita, M. Furukawa, M. Nemoto and T. G. Langdon, Metall. Mater. Trans. 29A (1998) p.2011.
14. M. Furukawa, Y. Ma, Z. Horita, M. Nemoto, R. Z. Valiev and T. G. Langdon, Mater. Sci. Eng. A241 (1998) p.122.
15. Y. Iwahashi, J. Wang, Z. Horita. M. Nemoto and T. G. Langdon, Scripta Mater. 35 (1996) p. 143.
16. P.B. Berbon, N.K. Tsenev, R.Z. Valiev, M. Furukawa, Z. Horita, M. Nemoto and T.G. Langdon, Metall. Mater. Trans. 29A (1998) p.2237.
17. S. Lee, P. B. Berbon, M. Furukawa, Z. Horita, M. Nemoto, N. K. Tsenev, R. Z. Valiev and T. G. Langdon, Mater. Sci. Eng., A272 (1999) p.63.
18. T.G. Langdon, Metal Sci. 16 (1982) p.175.
19. R.Z. Valiev, D.A. Salimonenko, N. K. Tsenev, P. B. Berbon and T. G. Langdon, Scripta Mater. 37 (1997) p.1945.
20. P. B. Berbon, M. Furukawa, Z. Horita, M. Nemoto, N. K. Tsenev, R. Z. Valiev and T. G. Langdon, Phil. Mag. Lett. 78 (1998) p.313.

Corresponding author: Sang-Mok Lee, e-mail: sangmokl@usc.edu

High Strength and High Strain Rate Superplasticity in Magnesium Alloys

M. Mabuchi[1], K. Shimojima[1], Y. Yamada[1], C.E. Wen[1], M. Nakamura[1], T. Asahina[1], H. Iwasaki[2], T. Aizawa[3] and K. Higashi[4]

[1] Materials Processing Department, National Industrial Research Institute of Nagoya, 1-1 Hirate-cho, Kita-ku, Nagoya 462-8510, Japan

[2] College of Engineering, Dept. of Materials Science & Engineering, Himeji Institute of Technology, Faculty of Engineering, 2167 Shosha, Himeji, Hyogo 671-2201, Japan

[3] Research Center for Advanced Science and Technology (RCAST), University of Tokyo, 4-6-1 Komaba, Meguro-ku, Tokyo 153-8904, Japan

[4] College of Engineering, Dept. of Metallurgy and Materials Science, Osaka Prefecture University, 1-1, Gakuen-cho, Sakai, Osaka 599-8531, Japan

Keywords: High Strain Rate Superplasticity, Magnesium Alloy, Powder Metallurgy

Abstract

Mechanical properties of fine-grained Mg alloys processed by powder metallurgy (P/M) and ingot metallurgy (I/M) routes have been investigated by tensile tests at room temperature and 573 K. The superplastic strain rate range for the P/M Mg alloys was higher than that for the I/M Mg alloys. Furthermore, the P/M Mg alloys exhibited higher room temperature strength than the I/M Mg alloys. These excellent mechanical properties of the P/M Mg alloys are attributed to a very small grain size of 0.5 ~ 1 µm.

Introduction

Magnesium alloys have high potential for use in the structural components in aerospace and outerspace applications due to their low density. In general, however, magnesium alloys have poor workability because of the h.c.p. structure. Therefore, it is desirable to improve the poor workability and superplastic forming is expected to be applied in the processing in practical applications.

It has been reported [1,2] that aluminum alloys exhibit superplastic behavior at high strain rates above 10^{-2} s^{-1}. High strain rate superplasticity is very attractive for commercial applications because one of the current drawbacks in superplastic forming technology is a slow forming rate, typically ~ 10^{-4} s^{-1}. Because the superplastic strain rate range increases with a decrease in grain size, a very small grain size is required to attain high strain rate superplasticity.

Recently, it was shown that strength of Mg alloys strongly depends on the grain size, indicating that high strength can be attained in fine-grained Mg alloys. Hence, grain refinement leads to not only high strain rate superplasticity but also high strength.

A powder metallurgy (P/M) process can give rise to a smaller grain size, compared to an ingot metallurgy (I/M) process. Therefore P/M alloys may show high strain rate superplasticity and high strength. This paper describes superplastic behavior at 573 K and tensile properties at room temperature of Mg-Al-Zn (AZ91) and Mg-Zn-Zr (ZK60 and ZK61) alloys processed by P/M and I/M routes.

Experimental Procedure

P/M AZ91 (Mg-9wt.%Al-1wt.%Zn-0.2wt.%Mn) and P/M ZK61 (Mg-6wt.%Zn-0.8wt.%Zr) were processed with rapidly solidified powders. The powders were sintered at 523 K at a pressure of 500 MPa in vacuum, respectively. The sintered billets were extruded at 553 K with a reduction ratio of 100 : 1. The grains of the P/M Mg alloys were equiaxed and recrystallization was almost complete. The grain size of the as-extruded bar was 0.5 µm for the P/M AZ91 and 0.6 µm for the P/M ZK61, respectively. A very small grain size of 0.5 ~ 0.6 µm was attained by the P/M method. Fine-grained I/M AZ91 was processed from machined chips. The details of the processing from machined chips are shown in the previous paper [3]. Briefly, machined chips were produced by machining an as-cast ingot. The machined chips were sintered at 573 K at a pressure of 100 MPa in air. The sintered billet was extruded at 553 K with a reduction ratio of 100 : 1. In addition, fine-grained I/M ZK60 (Mg-6wt.%Zn-0.5wt.%Zr) was processed directly from an as-cast ingot. The ingot was extruded at 553 K with a reduction ratio of 100 : 1. The grains were equiaxed and recrystallization was complete for the I/M Mg alloys as well as for the P/M Mg alloys. The grain size of the as-extruded bar was 1.4 µm for the I/M AZ91 and 1.2 µm for the I/M ZK60, respectively.

Tensile specimens with 5 mm gauge length and 2.5 mm gauge diameter were machined from the as-extruded bars. Tensile tests were carried out at 573 K to investigate superplastic properties. The flow stress was determined at a small strain of 0.1. The samples required about 1.8 ks to equilibrate at the given testing temperatures prior to initiation of straining. In addition, tensile tests were conducted at room temperature. The tensile axis was selected to be parallel to the extrusion direction for all the tests.

Results and Discussion

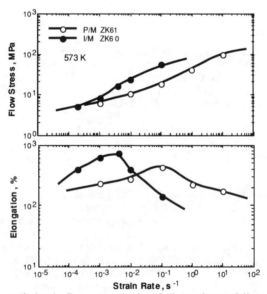

Fig. 1 The variation in flow stress (top) and elongation to failure (bottom) as a function of strain rate for P/M ZK61 and I/M ZK60.

The variation in flow stress (top) and elongation to failure (bottom) at 573 K as a function of strain rate is shown in Fig. 1 for P/M ZK61 and I/M ZK60 and in Fig. 2 for P/M AZ91 and I/M AZ91, respectively. In general, the logarithmic stress - logarithmic strain rate relation is sigmoidal for superplastic metals. The strain rate sensitivity of stress is high (> 0.3) and large elongations are attained in an intermediate strain rate region, which is a superplastic region. However, in both low and high strain rate regions, the strain rate sensitivity is low and large elongations are not attained. For the Mg alloys, the logarithmic stress - logarithmic strain rate curves were sigmoidal, as has been observed for superplastic metals. The high strain rate sensitivity of about 0.5 was attained in an intermediate strain rate region. The strain rate regions exhibiting the high strain rate sensitivity were roughly in agreement with the regions where large elongations were attained.

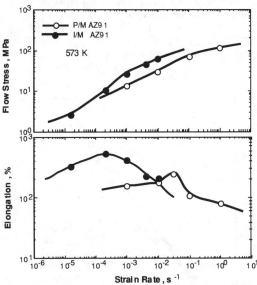

Fig. 2 The variation in flow stress (top) and elongation to failure (bottom) as a function of strain rate for P/M AZ91 and I/M AZ91.

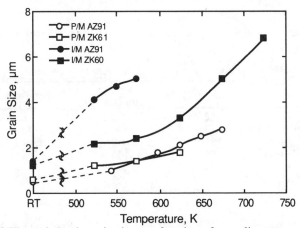

Fig. 3 The variation in grain size as a function of annealing temperature.

A maximum elongation of 432 % was obtained at a strain rate of 10^{-1} s^{-1} for the P/M ZK61, on the other hand, a maximum elongation of 730 % was obtained at a strain rate of 4×10^{-3} s^{-1} for the I/M ZK60. Thus, the P/M ZK61 exhibited superplastic behavior at a higher strain rate than the I/M ZK60. The similar result was obtained for the P/M and I/M AZ91. It should be noted that the P/M Mg alloys showed superplasticity at a high strain rate range of $10^{-2} \sim 10^{-1}$ s^{-1}.

The grain size was measured from the samples annealed for 1.8 ks at the given temperature. The variation in grain size as a function of annealing temperature is shown in Fig. 3. Grain growth was not large and the grain size at 573 K was 1.4 µm for both P/M Mg alloys. On the other hand, grain growth was relatively large for the I/M Mg alloys and the grain size at 573 K was 2.4 µm for the I/M ZK60 and 5.0 µm for the I/M AZ91, respectively.

As shown in Fig. 1 and Fig. 2, the P/M Mg alloys showed superplastic behavior at higher strain rates than the I/M Mg alloys. In particular, the P/M ZK61 alloy exhibited superplasticity at a high strain rate of 10^{-1} s^{-1}. However, superplastic behavior is attained in a low strain rate range ~ 10^{-4} s^{-1} for 7475 Al [4] and 5083 Al [5], which are typical of superplastic aluminum alloys. The P/M Mg alloys showed superplastic behavior at much higher strain rates, compared to the typical superplastic aluminum alloys. The variation in superplastic strain rate as a function of the inverse of the grain size is shown in Fig. 4. For comparison, the data of 7475 Al [4] and 5083 Al [5] are plotted in the figure. Clearly, high strain rate superplasticity for the P/M Mg alloys results from the very small grain sizes.

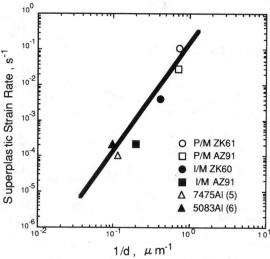

Fig. 4 The variation in superplastic strain rate as a function of the inverse of the grain size.

Table 1 The results of tensile tests at room temperature.

	UTS (MPa)	0.2 % Proof Stress (MPa)	Elongation (%)
P/M AZ91	**432**	**376**	**6.0**
I/M AZ91(extruded)	341	244	12.9
I/M AZ91(as-ingot)	131	72	1-3
I/M AZ91 ingot (T6)	235	108	3
P/M ZK61	**400**	**383**	**7.3**
I/M ZK60(extruded)	371	288	18.3

The results of tensile tests at room temperature are shown in Table 1. The P/M AZ91 and P/M ZK61 showed high strength of 432 and 400 MPa, respectively. The strengths of the P/M Mg alloys are higher than those of the I/M Mg alloys. In general, the yield strength as a function of grain size can be represented as Hall-Petch equation

$$\sigma = \sigma_o + Kd^{-1/2} \tag{1}$$

where σ is the yield stress, σ_o is the yield stress of a single crystal, K is a constant and d is the grain size. A value of K increases with increasing the Taylor factor [6]. The Taylor factor generally depends on the number of the slip systems. Because the slip systems are limited and the Taylor factor is larger for $h.c.p.$ metals than for $f.c.c.$ and $b.c.c.$ metals, $h.c.p.$ metals exhibit the strong influence of grain size on strength. Therefore, it is suggested that high strength can be attained in fine-grained Mg alloys.

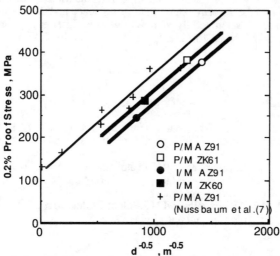

Fig. 5 The relationship between the yield stress and the grain size for Mg alloys.

The relationship between the yield stress and the grain size is shown in Fig. 5 for Mg alloys. It can be seen from Fig. 5 that the high strength of the P/M Mg alloys is attributed to the small grain size.

The experimental results in the present investigation showed that the P/M Mg alloys exhibited high strain rate superplasticity in a strain rate range of $10^{-2} \sim 10^{-1}$ s^{-1} and high strength of more than 400 MPa. These excellent properties of the P/M Mg alloys are attributed to the small grain size.

The tensile properties of the P/M Mg alloys are compared with those of the other alloys in Table 2. It can be seen that the specific strengths of the P/M Mg alloys are higher than those of the A7075 Al alloy.

Table 2 Comparison of tensile properties of the P/M Mg alloys with A7475 and Ti-6Al-4V.

Materials	Ultimate Tensile Strength, MPa	0.2% Proof Stress, MPa	Elongation to Failure, %	Specific Strength, MPa	Specific Proof Stress, MPa
P/M AZ91	432	376	6	240	209
P/M ZK61	400	383	7	222	213
A7075(T6)	573	505	11	205	180
Ti-6Al-4V(T6)	1166	1030	7	261	233

Summary

Mechanical properties of fine-grained Mg alloys processed by powder metallurgy (P/M) and ingot metallurgy (I/M) routes were investigated by tensile tests at room temperature and 573 K. The results are summarized as follows;
(1) The superplastic strain rate range for the P/M Mg alloys was higher than that for the I/M Mg alloys.

(2) Furthermore, the P/M Mg alloys exhibited higher room temperature strength than the I/M Mg alloys. These excellent mechanical properties of the P/M Mg alloys are attributed to a very small grain size of 0.5 ~ 1 µm.

Acknowledgment

The authors gratefully acknowledge the financial support of the Ministry of Education, Science, Culture and Sports, Japan as the Priority Area "Platform Science and Technology for Advanced Magnesium Alloys". Also, M.M. gratefully acknowledges the financial support from the project "Barrier-Free Processing of Materials for Life-Cycle Design for Environment" by Science and Technology Agency.

References

[1] O.D.Sherby and J.Wadsworth, Prog. Mater. Sci., 33 (1989), p. 169.
[2] K.Higashi, Sci. Forum, 170-172 (1994), p. 131.
[3] M.Mabuchi, K.Kubota and K.Higashi, Mater. Trans. JIM, 36 (1995), p. 1249.
[4] H.Iwasaki, S.Hayami, K.Higashi, T.Ito and S.Tanimura, J. Japan Soc. Tech. Plasticity, 32 (1991), p. 359.
[5] H.Iwasaki, K.Higashi, S.Tanimura, T.Komatubara and S.Hayami, in Proc. Conf. Superplasticity in Advanced Materials, edited by S.Hori, M.Tokizane and N.Furushiro, pp. 447-452, 1991, Jpn. Soc. Res. Superplasticity.
[6] R.Armstrong, I.Codd, R.M.Douthwaite and N.J.Petch, Philos. Mag., 7 (1962), p. 45.
[7] G.Nussbaum, P.Sainfort, G.Regazzoni and H.Gjestland, Scripta Metall., 23 (1989), p. 1079.

Inhomogeneous Cavity Distribution of Superplastically Deformed AL 7475 Alloy

Ming Jen Tan and Chilong Chen

School of Mechanical & Production Engineering, Nanyang Technological University, 50 Nanyang Avenue, Nanyang 639798, Singapore

Keywords: Al Alloy 7475, Grain Boundary Character Distribution (GBCD), Inhomogeneous Cavitation

ABSTRACT

In this work, uniaxial superplastical deformation of Al7475 alloys was performed along the rolling and transverse directions. Samples in rolling direction that were deformed showed inhomogeneous cavity distribution along the thickness direction. Samples deformed along the transverse direction showed inhomogeneous cavity distribution not only just along thickness direction but also along stress direction, which resulted in the relatively low elongation. Inhomogeneous cavity distribution along sheet thickness is related to particle distribution variation and microtexture gradient along the thickness. The particle distribution and grain boundary character distribution (GBCD) of the as-received materials played an important role on the inhomogeneous cavitation along stress direction. The inhomogeneous cavitation distribution is closely related to the cavity stringer formation. The cavity distribution can be improved by a homogenization treatment.

INTRODUCTION

The cavitation in a wide range of the superplastic materials after superplastic forming (SPF) causes the degradation of overall post-SPF properties of the materials [1,2]. Large volume fraction cavity inside materials, especially when cavities are inhomogeneously distributed, will cause serious reduction of mechanical properties and even premature failure during deformation [3,4,5]. Therefore, investigations on the formation mechanism and control method of inhomogeneous cavity distribution of structural materials have great engineering significance [4,5,6,7].

The formation of inhomogeneous cavitation is closely related to the characteristerics of superplastic material microstructure. For example, precipitated particles were used for refining grain during thermomechanical treatment (TMT) and pinning grain boundaries during SPF in quasi-single-phase (QSP) superplastic materials. For Al7475, a typical QSP superplastic material, small chemical composition differences or subtle differences in solidification and TMT processing condition between batches can result in large difference in cavitation behaviours [8]. This suggested that, besides the modification of TMT processing employed to achieve fine grain structure [8,9,10], homogenization treatment could be a potential method to alleviate inhomogeneous cavitation. There are numerous investigations that have been carried out on the cavitation nucleation, growth, coalescence and interlinkage mechanisms, and on the cavitation controlling or alleviation methods [11,12,13,14]. Little information on the formation mechanism of inhomogeneous cavitation is available. The current study, therefore, aimed to investigate the inhomogeneous cavitation formation and the effect of homogenization treatment.

MATERIALS AND EXPERIMENTAl PROCEDURE

Commercial Al 7475 sheet of 2.0mm thickness was used, with chemical composition of (wt%) 5.64 Zn, 2.34 Mg, 1.58 Cu, 0.19 Cr, 0.08 Fe, 0.05 Si, 0.02 Ti, 0.01 Mn and Al balance. The as-received sheet was in T4 condition. The tensile test specimens were of 4mm width and 15mm length in gauge parts, with cutting direction parallel (R-sample) or perpendicular (T-sample) to the rolling direction of the sheet. The surfaces of the tensile sample gauge part were polished up to 1μm diamond paste.

Uniaxial superplastic tensile test was performed at 480°C to 516°C at constant initial strain rate of $3.3 \times 10^{-4} s^{-1}$ to $10^{-3} s^{-1}$. The test parameters used were close to the optimum superplastic deformation condition of Al 7475 [9,12,15]. Cavitation ratio measurements were performed on the stress direction-short transverse (ST) section for both R- and T-samples. To characterise the distribution of cavities along thickness, the whole section was arbitrary defined as three regions, as shown in Fig. 1. According to its distance from the surface, regions were defined as region I (0~1/6t), II (1/6~2/6t) and III (2/6~3/6t). These measurements were carried out with the aid of an image analyser. Cavitation ratios along the stress direction were measured on the same section, with the each datum covering an area of the whole thickness × 0.9mm (Fig. 1). SEM and TEM were used for sample surface morphology and microstructure observations.

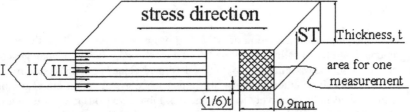

Fig. 1 Illustration of the definition of three regions along thickness and the area for one measurement of cavitation ratio along stress direction. Not drawn in to scale.

RESULTS AND DISCUSSIONS

Cavities of R-samples were basically uniformly distributed along gauge length direction while they were inhomogeneously distributed along thickness (Fig. 2). It is interesting to note that the region II of the sheet has higher resistance to cavitation at all test conditions (Fig. 3). The cavitation ratio and distribution inhomogeneity decreased as the test temperature increased. In all of the test conditions used in this work, R-samples were deformed uniformly. Cavities of T-samples were distributed inhomogeneously along length (Fig. 4). The profile of the sample was not uniform after deformation; there were many small neckings in the gauge part after straining to fracture. Though inhomogeneous cavity distribution along the thickness existed in T-samples, its deleterious effect might not be as important compared with the inhomogeneous cavitation along stress direction. With the increase of strain, cavities in some region grow more quickly than others and thus result in premature failure.

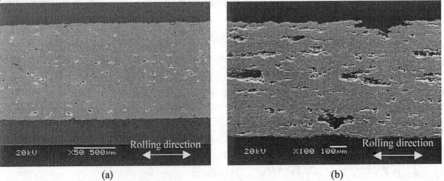

Fig. 2 Inhomogeneous distribution of cavities along thickness direction of R-samples; at 516°C at initial strain rate $10^{-3} s^{-1}$, (a) 300%, (b) fracture; R-ST section, stressed horizontally.

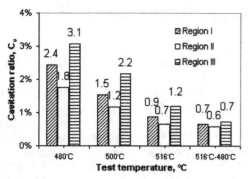
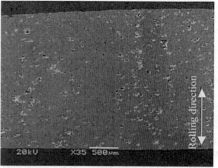

Fig. 3 Inhomogeneous distribution of cavities of R-samples at three regions, at 200%, at strain rate $10^{-3} s^{-1}$.

Fig. 4 Inhomogeneous distribution of cavities along length direction, T-sample, at 516°C, 300%, $10^{-3} s^{-1}$, R-T section, stressed horizontally.

1. Inhomogeneous cavity distribution along thickness

The inhomogeneous cavitation is closely related to the microstructure of the materials. To achieve fine grain structures for superplasticity of quasi-single phase (QSP) superplastic materials like Al7475, thermomechanical treatments (TMT) processing is usually used, which mainly consists of solution heat treatment, precipitation (by overaging), warm or cold rolling and recrystallization [9,16]. Large particles precipitated during TMT processing, which are needed for grain refinement [16], are possible cavity nucleation sites during superplastic deformation. The distribution of large particles may then influence the density of pre-existing cavities and cavity nucleation sites. In the present work, most serious cavitation occurred at the center of the sheet. This phenomenon is related to the rolling process of the materials, since the material at the center part experienced less shear stress and some large particles might still exist after rolling, and then these particles could serve as the nucleation sites for cavitation. This is evidenced by the observation of large particles (>1μm in diameter) in the region III (center part) of the materials (Fig. 5), while large particles are rarely found in the region I and II.

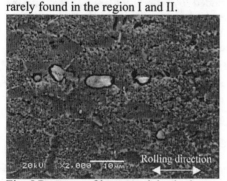
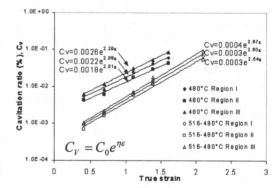

Fig. 5 Presence of large particles in the materials center after strained to fracture, at initial strain rate $10^{-3} s^{-1}$, at 480°C.

Fig. 6 Cavitation growth of different regions, stressed in rolling direction.

For plasticity-controlled cavity growth, Hancock's model [17] predicts a linear relationship between logarithmic cavitation ratio, C_v and true strain, ε. The results of three regions fit this relationship well for both as-received and homogenized samples (Fig. 6). Detailed cavitation growth results were presented elsewhere [18]. It is noted that the three regions have the similar cavity growth parameters, η, while the initial cavitation ratios, C_0, have a larger difference. As can be seen from

Fig. 6, region III has the highest C_0 and region II has the lowest one. This shows that the cavity growth mechanism of the three regions is the same and the inhomogeneous cavity distribution is closely related to the initial deformation state, i.e., the initial cavity ratio and / or sites for cavity nucleation. In addition, the special condition of the surface layer may contribute to the relatively high cavitation ratio in region I compared to region II, since surface open cavities cannot be filled with surround materials.

The inhomogeneous cavitation along the thickness also relate to the microtexture of the materials. The local texture of the as-received sheet was found to vary along the thickness. The texture gradient along thickness resulted from the material processing. It is noted that the misorientation trace [21] measured along longitudinal or transverse direction was anisotropic and varied along thickness. There are more low angle grain boundaries in the center part of the material. The distribution of low angle grain boundaries will influence the grain boundary sliding (GBS) [19]. The higher the ratio of low angle grain boundaries, the higher stress needed to generate GBS, and further, the higher ratio of cavitation nucleation and growth.

2. Inhomogeneous cavitation distribution along stress direction

At the same deformation condition, the amount of cavitation of T-samples is much higher than that of R-samples, as shown in Fig. 7. As can be seen from Fig. 7, serious inhomogeneous cavitation of the T-sample along stress direction is evidenced by the large scattered data of cavitation ratios. In this work, the rate of the maximum cavitation ratio to the minimum one among the ten measurements on different areas, β, was used to describe the inhomogeneous degree of the samples, see Table 1. For T-samples, β is around 3.5. Cavitation of the R-sample along stress direction is relatively uniform; β is about 2.2. It needs to be pointed out that this rate is not an absolute evaluation of the inhomogeneous degree and this rate will change depending on the size of the area of each measurement.

The difference between the cavity growth rate of T- and R- samples is not very large, while C_0 of T-sample is about 4 times large than that of the R-sample (Fig. 7). This indicated that, besides for the possible different cavity growth mechanisms, the T-sample has more pre-existed cavities or nucleation sites than R-sample. Table 1 also presents the cavitation ratio at fracture, C_f, which was measured from gauge with the final fracture region excluded. C_f reflects the cavity level that the material can withstand, beyond which unstable traverse cavity interlink may take place and result in the final fracture. It is noted that the C_f is not obviously influenced by the sample cutting directions. C_f seems more closely relate to the deformation temperature. This suggested that the ability of materials to tolerate cavitation at superplastic condition is an important parameter which control the elongation of the materials [20].

Table 1. List of cavitation ratio C_f (%)fracture and inhomogeneous degree β along stress direction at 200%, at an initial strain rate of $10^{-3}s^{-1}$.

Sample	As-received				516°C, 15mins, homogenized			
	R-sample		T-sample		R-sample		T-sample	
Test temperature	480°C	516°C	480°C	516°C	480°C	516°C	480°C	516°C
C_f (%)	6.0	20.4	6.6	22.9	6.8	19.4	6.4	23.8
$\beta = \dfrac{C_{max}}{C_{min}}$ (at 200%)	2.4	2.1	3.8	3.3	2.1	2.0	3.2	2.9

The higher pre-existing cavities or nucleation sites of T-samples may be related to the anisotropic grain boundary character distribution (GBCD) of the material. For the material used in this work, there are more low-angle grain boundaries presented parallel to the transverse direction than to the rolling direction. Low angle grain boundaries are not as active to GBS as high angle boundaries. Large particles were present more frequently on the grain boundaries parallel to the rolling direction. This may result in the different GBS and cavitation nucleation patterns when a sample strained along T or L directions. In the early stage of superplastic flow, these anisotropic characteristics may dominate the pattern of the cavity nucleation and further influence the growth, coalescence and interlinkage of cavities. For the QSP materials, cavities nucleate frequently at particles on grain boundaries, triple junctions of grain boundaries, and grow, coalesce and interlink intergranularily [13], though some small cavities can nucleate within grains [21]. This suggests that GBCD of the material has influence on the cavity distribution. In this work, cavity stringer formation along rolling direction was observed, regardless of the stress direction, as shown in Fig. 8. The formation of cavity stringers along stress direction is not as bad as the formation of stringers perpendicular to stress direction. The formation of cavity stringers perpendicular to stress direction, as frequently observed in the case of T-sample, causes cavitation concentrated in certain area, and thus results in the serious inhomogeneous cavitation along stress direction and leads to premature failure. Details of the effect of GBCD on cavity stringer formation were discussed elsewhere [21].

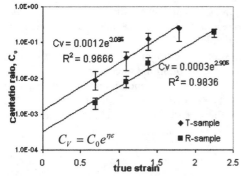

Fig. 7 Variation of cavitation ratio with strain of samples deformed along longitudinal and transverse directions, at 516°C, at $10^{-3}s^{-1}$.

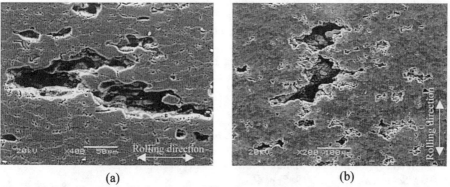

Fig. 8 Cavity stringers in (a) longitudinal direction of R-sample, (b) transverse direction of T-sample, after superplastic deformation, 516°C, $10^{-3}s^{-1}$, at fracture, stressed horizontally.

3. Homogenization treatment

To investigate the effect of homogenization treatment, 15mins high temperature exposure at 516°C was used before later deformed at 480°C or 516°C. Results can be found in Fig. 3, Fig. 6 (labelled as 516-480), and Table 1. R-samples exposed at 516°C for 15mins before straining at 480°C has relatively low C_v and quite uniform cavity distribution along thickness compared with samples deformed 480°C or 516°C, see Fig. 3. The decrease of C_v and inhomogeneous cavitation may be related to the closing-up of the small pre-existing cavities and the dissolution or partial dissolution of some particles during the 516°C exposure. Homogenization treatment also improved the elongation to failure [18]. However, after a 15mins 516°C exposure, η is nearly 30% higher. The increase of η is related to the variation of the liquid phase along grain boundaries [18].

Homogenization treatment also ameliorates the inhomogeneous cavitation degree along the stress direction (Table 1). Inhomogeneous cavitation of T- sample is still much higher than that of R- samples. The decrease of β is only about 10% for both T- and R- samples. Since the inhomogeneous degree β depends on the area size used for each measurement, β of value 2 seems represent a relative uniform cavity distribution. Further decrease of β seems limited. However, longer homogenization time may be beneficial for further alleviation of the inhomogeneous cavitation.

CONCLUSIONS

1. Inhomogeneous cavity distributions along sample thickness and stress direction were observed. The serious inhomogeneous cavitation along stress direction of samples deformed along transverse direction leads to the relatively low elongation.
2. Cavity distribution along sample thickness was quite inhomogeneous, and it is related to precipitate distribution variations and microtexture gradient along the thickness. Inhomogeneous cavitation along stress direction was closely related to the grain boundary character distribution of the as-received materials and the formation of cavity stringers along rolling direction.
3. Homogenization treatment before deformation can alleviate the inhomogeneous cavitation both along thickness and stress direction, although the effect on the latter seems limited.

REFERENCES

1. Schelleng, R.D. and Reynolds, G. H., *Metall. Trans.*, **4** (1973), p.2199.
2. Ridley, N., Livesey, D.W. and Mukherjee, A.K., *Metall. Trans.* **A, 15A** (1984), p.1443.
3. Ph. Bompard, J. Y Lacroix and A. Varloteaux, *Aluminium* **64** (1988), p.162.
4. K. Kannan and C.H. Hamilton, *Scripta Mater.* **37** (1997), p. 455.
5. K. Kannan and C.H. Hamilton, *Scripta Mater.* **38** (1998), p.299.
6. M.J. Stowell, *Metal Sci.* **17** (1983), p.92.
7. K. Kannan and H.Hamilton, *Acta mater.* **46** (1998), p.5533.
8. C.C. Bamption and J.W. Edington, *Metall. Trans. A*, **13A** (1982), p. 1721.
9. J. A. Wert, JOM, Sept. (1982), p.35.
10. G. J. Mahon, D. Warrington, R. G. Butler and R. Grimes, *Mat. Sci. Forum*, **170-172** (1994), p. 187.
11. E. Tanaka, S. Murakami and H. Ishikawa, *Mat. Sci. Forum*, **233-234** (1997) p.21.
12. Dong Hyuk Shin, Chong Soo Lee, Woo-Jin Kim, *Acta mater.* **45**, n12 (1997), p. 5195
13. C.C. Bamption and J.W. Edington, *Metall. Trans. A*, **13A** (1982), p. 1721.
14. J.Pilling, *Mat. Sci. Tech.*, **1**, June, (1985) p. 461.
15. R.K. Mahidhara, *J. of Mat. Eng. & Perf.*, **4** (6), Dec., (1995), p.674
16. H. Yoshida, *Materials Science Forum*, **204-206** (1996), p.657.
17. J.W. Hancock, *Metal Sci.* **10** (1976), p.319.
18. C.L. Chen and M.J. Tan, *Mat. Sci. Engg.* **A**, to be published.
19. T. Watanabe, *Mat. Sci. Forum*, **233-234** (1997), p.375.
20. R.K. Mahidhara, *J. of Mat. Engg. & Perf.*, **4** (6) Dec., (1995), p.674.
21. C.L. Chen and M.J. Tan, to be submitted.

The Development of a High Strain Rate Superplastic Al-Mg-Zr Alloy

R.J. Dashwood, R. Grimes, A.W. Harrison and H.M. Flower

Department of Materials, Imperial College of Sicence Technology and Medicine,
London SW7 2BP, UK

Keywords: Al-Mg-Zr, High Strain Rate Superplasticity, Rapid Solidification

ABSTRACT

In order for superplastic forming of aluminium to break out of the niche market low cost alloys are required that exhibit higher strain rate capability that are capable of volume production. This paper describes an investigation into the feasibility of producing such an alloy. A series of Al-4Mg alloys with 0, 0.25, 0.5, 0.75 & 1 % Zr additions was prepared using a cheap particulate casting route, in an attempt to achieve higher levels of Zr supersaturation than are possible with conventional casting. The particulate was processed into a sheet product via hot extrusion followed by cold rolling and the effect of a number of process variables on the SPF performance of the sheet was investigated. It was found that increasing the Zr content, and manipulation of the thermomechanical processing conditions improved the SPF performance. Ductilities in excess of 600% have been achieved at a strain rate of 0.01 s^{-1}, together with flow stresses less than 15MPa.

INTRODUCTION

There are now increasing pressures to make motor vehicles less environmentally damaging and, in consequence, most car builders are, at least, experimenting with the construction of body structure from aluminium alloy sheet. It is generally agreed that aluminium vehicle body sheet has a somewhat lower forming capability than deep drawing quality steel strip and, in consequence, sheet aluminium can have difficulties in forming the more complex inner panels. With the advent of aluminium alloys capable of being superplastically formed at relatively high strain rates[1-3] the possibility arises for superplastic forming of aluminium to break out of the niche market that it currently occupies. To achieve success in the volume car market simply producing alloys with a higher strain rate capability is insufficient. Appropriate in-service properties, significantly lower price than existing superplastic aluminium alloys and capability of volume production are also essential. This paper will describe an exploratory programme that attempts to address these fundamental requirements.

EXPERIMENTAL

As the development was concerned with automotive structural applications an Al-4% Mg matrix was chosen in order to confer the necessary strength level in combination with good corrosion resistance, subsequent cold forming capability and weldability. Zirconium was selected as the grain control addition and prior work had suggested that a relatively high level of zirconium would be required. In order to minimise the separation of primary Al$_3$Zr particles and maximise the retention of zirconium in super saturated solid solution a casting technique with a higher solidification rate than conventional, semi-continuous, direct chill casting was selected. A proprietary particulate casting process with a solidification rate of ~10^2 °C.s^{-1}, developed by ALPOCO (UK), was judged capable of achieving the required structure, while having the potential for low cost volume production. A series of alloys with zirconium contents of 0, 0.25, 0.5, 0.75, and 1.0 wt.%. was prepared using this technique. Cold compacted billets were heated to an initial temperature ranging from 400-550°C, extruded with a ram speed of 3mm.s^{-1} and an extrusion ratio of 18:1 to yield 8mm thick bar, prior to cold rolling to 1.5 mm thick sheet. Various heat treatments were applied to the material before and after the extrusion process. Isothermal, constant strain rate hot tensile tests were performed on a servo hydraulic testing frame employing a 12.5 mm gauge length sample. Tests were performed over a strain rate range from 10^{-3}-10^0 s^{-1}. Microstructural analysis (SEM, FEGSEM, TEM) was performed on the materials at various stages in the production route. In order to assess the hot forming behaviour of the 0.5, and 1.0 % Zr variants a series of 450 x 170 x 1.5 mm sheets was prepared for cone test experiments by cross rolling 40 x 18mm extruded bar. The bar was extruded from 100mm diameter billets at 500°C, with a ram speed of 3mm.s^{-1} and an extrusion ratio of 11:1. Biaxial testing was performed at Superform Aluminium (UK); complementary uniaxial SPF tests were performed under the same conditions as previously stated.

RESULTS AND DISCUSSION

Firstly the effectiveness of the particulate casting process in achieving a significant level of zirconium supersaturation had to be assessed. Figure 1 shows a backscattered electron FEGSEM image of the Al-4Mg-1Zr alloy in the as cast condition, together with the results of X-EDS analysis. A number of sub 5µm primary Al$_3$Zr particles were clearly observed, however the size and quantity of such particles are indicative of a high degree of supersaturation being achieved. Moreover the X-EDS results clearly show a significant zirconium content within the cells.

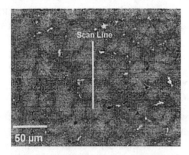 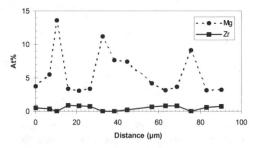

Figure 1: FEGSEM image of Al-4Mg-1Zr in the as cast condition together with the corresponding X-EDS analysis

All the alloys discussed in this paper were processed over a range of extrusion temperatures, subjected to a variety of heat treatments, and uniaxially tested between 400 and 550°C. For simplification purposes only the best observed ductilities, irrespective of the test temperature and condition, are reported in this paper. It is assumed that this is the optimum temperature and condition for SPF and will hence forth be referred to as the optimum SPF ductility. In order to assess the level of zirconium required to achieve a superplastic material a series of alloys was prepared with increasing levels of zirconium (0-1wt.% Zr). The results of this investigation are summarised in figure 2 where the optimum SPF ductility (achieved at a strain rate of 0.01s^{-1}) is plotted as a function of zirconium content. The 0 and 0.25 wt.% alloys did not exhibit SPF characteristics, whilst alloys with 0.5 wt% Zr or more, were superplastic. For this reason all subsequent work was concentrated on alloys with 0.5 to 1 wt.% Zr.

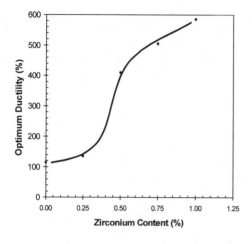

Figure 2: Effect of zirconium content on SPF performance (strain rate of 0.01 s^{-1})

Figure 3: Effect of extrusion temperature and heat treatment on SPF performance (strain rate of 0.01 s^{-1})

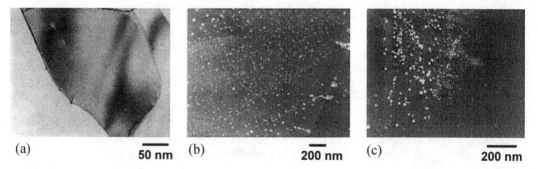

Figure 4: TEM showing Al$_3$Zr distribution in Al-4Mg-0.5Zr (a) as extruded at 400°C (b) extruded at 400°C and heat treated at 360°C for 100 hours (c) as extruded at 550°C

As previously stated the alloys were subjected to hot deformation, in the form of extrusion, over a range of extrusion temperatures, varying from 400 to 550°C and were also subjected to various pre and post extrusion heat treatments. Having achieved a supersaturation of zirconium during solidification the desire is to achieve a controlled decomposition of the solid solution to yield a fine dispersion of grain growth restricting metastable cubic Al$_3$Zr particles. Potentially this can be achieved at various stages during the thermomechanical processing cycle. Experience with AA2004 showed that heat treatment prior to hot working was very effective in achieving this objective [4]. Concerns over particle oxidation made this an undesirable route, leaving extrusion and/or post extrusion heat treatment as the possible methods for manipulating the Al$_3$Zr distribution. Prior work [5] involving extrusion consolidation of a rapidly solidified Al-7Mg-1Zr alloy showed that minimal decomposition of the zirconium solid solution takes place during extrusion consolidation at temperatures less than 400C, permitting post extrusion heat treatment. Figure 3 summarises the results of trials conducted to study the effects of these different heat treatments. Extrusion at 400°C yields a material, that exhibits poor SPF ductility in the untreated condition, but that can be significantly improved by the application of a post extrusion treatment (e.g. 100 hours at 360°C). TEM examination (figure 4) of material extruded at 400°C before and after heat treatment shows that extrusion alone is insufficient to cause significant decomposition whereas after 100 hours at 360°C a fine dispersion of particles of ~10nm particles was obtained. However extrusion alone at higher temperatures produced similar dispersions, and near identical SPF response. Even after extrusion at as high a temperature as 550°C improvement in SPF performance by further heat treatment was observed in the 0.5 % Zr alloy. However, in the 1.0 % Zr variant optimum SPF performance was achieved without recourse to any heat treatments.

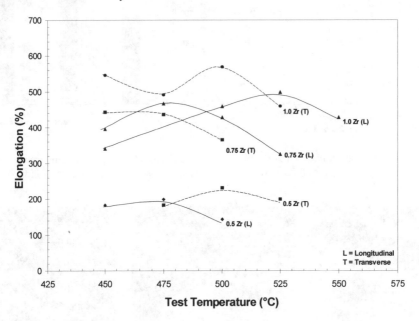

Figure 5: Effect of zirconium addition and test temperature on SPF ductility at for the scaled up material (extruded at 500°C, no heat treatment, strain rate 0.01 s^{-1})

The above results were all achieved with rolled product too narrow to allow biaxial testing. Scaling up of the production route demonstrated comparable (if not improved) SPF performance, and reasonable freedom from anisotropy (figure 5). The results of the biaxial forming trials are given in table 1 and figure 6. The 0.75% and 1% Zr alloys produced the best results achieving greater deformation, at much lower pressures and forming times, than SPF5083. This is further emphasised from the uniaxial flow stress plots in figure 7.

Table 1:Result of biaxial testing trials

Alloy	Pressure (psi)	Time to Failure (s)	Cone Height (mm)
Al-4Mg -0.5Zr	30	243	52
Al-4Mg -0.75Zr	30	194	59
Al-4Mg -1Zr	30	190	60
SPF5083	70	340	50

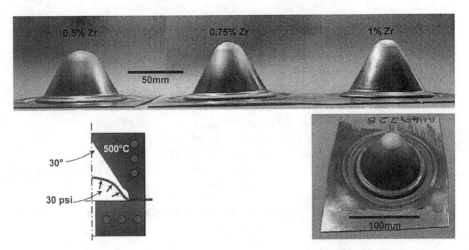

Figure 6: Effect of zirconium content on cone height during after biaxial testing

The biaxial testing procedure results in extremely rapid heating rates (>500 C.min^{-1}) being experienced by the sheet prior to testing. As this was significantly greater than the standard heating rate employed during uniaxial testing, a biaxial test was performed in which the heating rate was crudely reduced, resulting in modest improvement in the cone test. In consequence the effect of heating rate on SPF ductility was investigated by uniaxial testing the results of which are given in figure 8. It is apparent that the as rolled sheet performs better with a relatively slow heating rate. Since commercial forming would normally employ heating rates comparable with those in the cone test this implies that improved performance should be achieved by a controlled final anneal before superplastic forming. While significant strain-induced recrystallisation is occurring during SPF (figure 9), it is clear that some static recrystallisation is also involved. The fact that a lower heating rate results in improved SPF behaviour suggests that the balance between strain induced and static recrystallisation is being tipped in favour of the strain induced mechanism. More metallographic evidence is required to clarify this and other associated issues.

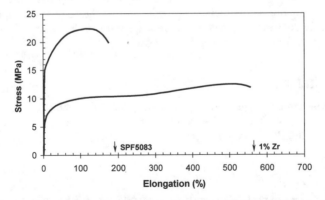

Figure7: Comparison of SPF flow stresses of commercial SPF 5083 and Al-4Mg-1Zr
(tested at 500°C, and 0.01s^{-1})

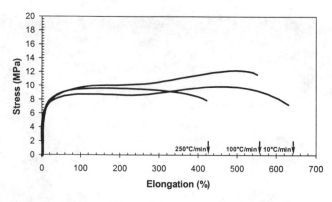

Figure8: Effect of heating rate on SPF ductility of Al-4Mg-1Zr (tested at 500°C, and 0.01s^{-1})

Figure 9: FEGSEM showing microstructure of Al-4Mg-1Zr deformed to 600% (tested at 525°C, 0.01s^{-1})

CONCLUSIONS

The feasibility of producing a simple, but novel, alloy capable of superplastic performance appreciably exceeding that of available commercial competitors for the envisaged applications has been demonstrated. Not only does the new material possess the ability to be deformed to greater strains, but it also exhibits significantly lower concomitant flow stresses.

ACKNOWLEDGEMENTS

Funding for the programme is gratefully acknowledged under the EPSRC/MOD/DERA scheme (GR/L43848). Support and valuable discussions have taken place with DERA, LSM, ALPOCO, Superform Aluminium and British Aluminium Sheet.

REFERENCES

[1] T.G. Nieh, C.A. Henshall, J. Wadsworth, *Scr. Metall.*, **18**, 1405 (1984)
[2] T.R. Bieler, T.G. Nieh; J. Wadsworth, A.K. Mukherjee, *Scr. Metall.*, **22**, 81 (1988)
[3] K. Higashi, M. Mabuchi, *Materials Science Forum*, **304-306**, 209 (1999)
[4] R. Grimes, *AGARD Lecture Series*, **154**, 8/1 (1987)
[5] R.J. Dashwood, T. Sheppard, *Materials Science and Technology*, **9**, 483 (1993)

Recent Advances and Future Directions in Superplasticity

Kenji Higashi

Department of Metallurgy and Materials Science, College of Eng., Osaka Prefecture University,
1-1, Gakuen-cho, Sakai, Osaka 599-8531, Japan

Keywords: Accommodation, Cavitation, Fine-Grained Structure, Grain Boundary Plasticity, Grain Boundary Sliding, High Strain Rate Superplasticity, Low-Temperature Superplasticity

Abstract

Superplasticity is generally associated with fine grains, grain boundary sliding, accommodation, high tensile ductility and high strain-rate-sensitivity at elevated temperatures. This paper reviews some of the recent important findings in very fine-grained superplastic materials, including the areas of high-strain-rate superplasticity in aluminum based materials and of low-temperature superplasticity in magnesium based materials. Deformation mechanism maps are shown to be powerful methods for predicting the conditions where high-strain-rate superplasticity and/or low-temperature superplasticity can be expected. Ultrafine grained materials, processed economically, remain an important future objective in achieving high-strain-rate superplasticity and/or low-temperature superplasticity, which have considerable promise for net-shape isothermal forming of sheet and bulk components in industry scale.

1. Introduction

It is noted that the research on superplasticity has been advanced in recent ten years by the discovery of superplasticity in new materials such as high-performance ceramics or intermetallics, and also by the challenge to timely topics such as high-strain-rate superplasticity [1,2]. Grain size refinement is the most important factor to obtain superplasticity. In respect to the mechanical properties of the very fine grained materials one should hope to obtain extremely high strength states that can be expected in the case of strong grain refinement according to the Hall-Petch relationship. Another very important aspect is superplasticity in fine-grained materials. It can be expected from the existing models that superplasticity could be obtained at high strain rates and/or relatively low temperatures by reducing grain size of the materials to less than one micrometer. There are obvious technological advantages of high-strain-rate superplasticity and low-temperature superplasticity. High-strain-rate superplasticity could make superplastic forming very attractive for bulk component production, and low-temperature superplasticity could allow the use of the conventional tooling materials and energy savings. The discovery of new phenomena in superplasticity has been particularly important to keep the activity as high as possible.

A pair of the key subjects is to develop bulk nanocrystalline materials and to examine their flow behavior. Another principal subject is to analyze the elementary process of grain boundary sliding and its accommodation. Grain boundary structure analysis has highly been developed by the advancement of modern research facilities such as high-resolution electron microscopy and related analytical techniques. However, the grain boundary sliding has seldom been discussed from the knowledge of structure and atomic bonding state in grain boundaries. The analysis from atomistic or electronic level has given us the physical image of grain boundary sliding, and been effective to break through the limitation of phenomenological analysis. It is so successful approach that the plastic flow caused by grain boundary sliding has been termed "grain boundary plasticity". The superplastic deformation can be studied from different viewpoints, which may be divided into two general categories emerging from the vast content of superplasticity. One is the phenomenological approach with particular emphasis on the mathematical formulation and practical applications. The other is rather physically motivated as to deal with the physical background of materials subject to superplastic deformation, especially devoted to the actual deformation mechanisms and their interrelationship with the evolution of microstructure during superplastic flow.

Now the essential knowledge of micro-superplasticity, including an atomistic mechanism of grain boundary sliding, is in great demand not only from scientists but also from engineers. As a central link between micro-and macro-scales, however, meso-superplasticity aims at introducing the essential micro-superplasticity concepts to various intermediate scales where the quantitative theory of continuum mechanics is still applicable in describing the evolution of material microstructures during the superplastic flow. Meso-superplasticity represents an important connection between the continuum-based macro-superplasticity and the atomistic physical theory of micro-superplasticity. The connection here should be in both length scales and the method of investigations. Author believes that a new area created by meso-superplasticity stimulates a close workmanship among metallurgists, mechanists and engineers, and promotes a combined approach of solid mechanics and material science. In this paper, therefore, the previously reported results will be summarized from mesoscopic viewpoint, especially with focusing to high-strain-rate superplasticity and low-temperature superplasticity.

2. Processing of Grain Size Refinement for Superplastic Structure

Grain size refinement is related to high-strain-rate superplasticity and low-temperature superplasticity in materials because an optimum state of strain rate and temperature for superplasticity is a function of grain size. The deformation behavior of a material at elevated temperatures can be represented by a constitutive equation, which incorporates an activation energy term as given by [3,4]

$$\dot{\varepsilon} = \frac{AGb}{kT}\left(\frac{b}{d}\right)^p \frac{\sigma - \sigma_0}{G}^n D_0 exp\left(\frac{-Q}{RT}\right) \quad (1)$$

where $\dot{\varepsilon}$ is the strain rate, b is the Burgers vector, σ is the flow stress, σ_0 is the threshold stress, G is the shear modulus. D_0 is the pre-exponential factor for diffusion, Q is the activation energy for superplastic flow (presumably $Q_{superplastic} = Q_{diffusion}$), R is the gas constant, T is the temperature in degrees Kelvin, n is the stress exponent at typically 2 for superplastic flow, p is the grain size exponent at 2~3 depending on the dominant diffusion mechanism, k is Boltzmann's constant, and A is a constant which is principally a function of the deformation mechanism. From the grain size dependence of the superplastic strain rate and the superplastic temperature from eq. (1), it is clearly concluded that high-strain-rate superplasticity or low-temperature superplasticity requires a very small grain size. The possibility of high-strain-rate superplasticity was first recognized over 10 years ago in a series of experiments conducted on a metal matrix composite where high elongations were achieved at strain rates above ~ 10^{-2} s^{-1} [5]. Subsequently, high-strain-rate superplasticity was defined as superplasticity occurring at strain rates at or above 10^{-2} s^{-1} and there have been numerous reports of the occurrence of high-strain-rate superplasticity in a limited range of metal matrix composites, mechanically alloyed materials and alloys fabricated using powder metallurgy (PM) procedures [6].

The microstructural features in high-strain-rate superplastic materials are as follows [6,7]; a typical microstructure of the mechanically alloyed materials consisted of very fine grains with a 350 ~ 500 nm mean size. Carbon and oxygen were present as fine (<30 nm) carbide (Al_4C_3) and oxide (Al_2O_3, MgO, or LiO_2) particles with a volume fraction of about 5 vol%. The structures of the mechanically alloyed materials were very stable at high temperatures for long times. The bulk materials consolidated only by extrusion from amorphous or nanocrystalline powders consisted of nano or near-nano scale structures from 50 to 100 nm in grain size. It was noted that grain growth of these materials depends on both the annealing temperature and the holding time, and also heating rate up to the annealing temperature. For example, a typical microstructure of the Al-Ni-Mm-Zr alloy, heated up to an annealing temperature of 873 K with a very rapid rate and then annealed for a holding time of 30 s, consisted of grains in a large size of about 1 μm and particulates 0.5 μm in size and 30~40 % in volume fraction, but nano or near-nano scale grained structures and particulates less than 200 nm in size remained even after annealing for 30 s at 773 K with a rapid heating rate. A typical microstructure of superplastic vapor quenched (VQ Al-7wt%Cr-1wt%Fe) alloy consisted of a very fine grained structure Al-

Cr matrix about 500 nm in size and a uniform dispersion of iron-rich precipitates, $(Cr,Fe)Al_7$, 500 nm in diameter (30~40 % volume fraction). It was found that all the grains of the high-strain-rate superplastic composites were equiaxed and the sizes were about 1 to 3 µm [6,8]. A group of the PM processed aluminum alloys with a large amount of Cr, Mn or Zr was made to a fine-grained superplastic structure of less than 5 µm by optimum powder metallurgical processings. It is noted that the size of the dispersed particles of Al_3Zr is one order of magnitude finer than that of precipitates with Cr or Mn. These reports on high-strain-rate superplasticity have prompted much interest because of the potential for using the high-strain-rate superplasticity phenomenon to achieve a substantial reduction in the processing time associated with commercial superplastic forming operations.

It was noted that ultrafine grain sizes and high-strain-rate superplasticity might be achieved in materials produced by a combination of powder metallurgy and/or advanced processing methods. In the recent works, a variety of processes such as new or improved thermomechanical processing [9,10], accumulative roll-bonding [11], torsion straining, equal-channel-angular extrusion (ECAE) etc. have been examined. For example an Al-Mg alloy with a grain size of about 90 nm is produced by torsion straining [12]. Especially it is noteworthy that very recent investigations reveal the possibility of achieving high-strain-rate superplasticity in ingot metallurgically processed (IM) alloys at relatively low temperatures (below the melting point) [13]. An occurrence of high-strain-rate superplasticity was found in a commercial cast light-weight alloy, exhibiting exceptional elongations including >1180 % without failure at 623 K with a strain rate of 10^{-2} s^{-1} and 340 % at the same temperature with a strain rate of 1 s^{-1} [14,15]. The metallic alloys were subjected to intense plastic straining through a processing procedure known as ECAE in order to attain a grain size of about 1 µm. The result suggests no clear demarcation between the IM and PM processing routes but rather it is apparent that ECAE is capable of achieving both the ultrafine grain sizes and the superplastic ductilities at high strain rates that are a characteristic feature of some PM alloys, as shown in **Fig.1**.

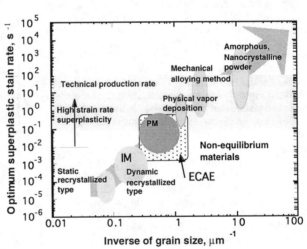

Fig.1; The change of the optimum superplastic strain rate for several aluminum alloys with various grain sizes produced by different processing routes, including ECAE.

3. High-Strain-Rate Superplasticity

In conventional superplastic materials, grain boundary sliding is a dominant deformation process, and the grain compatibility during grain boundary sliding is maintained by concurrent accommodation processes which may involve grain boundary migration, grain rotation, diffusion or dislocation motion. If the sliding displacements are too large to be accommodated elastically, the sliding must be accommodated by diffusional or plastic flow. Therefore, the shorter distance with the refinement of grain size can enhance the accommodation by diffusional or plastic flow. The origin of high-strain-rate superplasticity is associated with ultra-fine grain sizes of the materials. In order to obtain large superplastic elongations to failure for ultra-fine grained superplastic materials with many pinning particles, including the metal matrix composites, special mechanisms are required for the accommodation process to relax stress concentrations near interfaces as well as grain boundary triple junctions. Especially in metal matrix composites having a high volume fraction of hard reinforcements, high-strain-rate superplasticity is not necessarily attained only by ultra-fine grain sizes because excessive cavitation is considered to be caused due to the high stress concentrations around the reinforcements intersected by grain boundaries. In particular, it appears to be difficult to accommodate grain boundary sliding at high strain rates by diffusion processes, also including diffusion-controlled dislocation movement, because the times are too short [7]. High-strain-rate superplasticity seems to be critically controlled by the accommodation process to relax the stress concentration resulting from sliding at grain boundaries and/or interfaces, involving an accommodation helper such as a liquid phase [8,16].

It was reported very recently [8] that if the above view is correct, the accommodation helper such as a liquid phase is required as an additional accommodation process for superplastic flow when the stress concentrations are not sufficiently relaxed by a conventional accommodation process. They proposed the concept of critical strain rate, $\dot{\varepsilon}_c$, i.e., the accommodation mechanism is diffusional flow and/or diffusion-controlled dislocation movement in a strain rate range below the critical strain rate, on the other hand, a special accommodation process by an accommodation helper such as a liquid phase is required in a strain rate range above the critical strain rate because stress concentrations are caused around reinforcements [17]. The variation in $\dot{\varepsilon}_c$ as a function of the value of T/T_m in the aluminum alloy matrix composites having a mean grain size of 1 μm and a volume fraction of the reinforcements of 20 % is shown in **Fig. 2** for reinforcement particle sizes of 0.2 and 1 μm in the case of lattice diffusion-controlled and grain boundary diffusion-controlled superplastic flow respectively. The T_m value is taken to be the absolute melting point of pure aluminum. The critical strain rate strongly depends on the particle size of the reinforcements, and increases with increasing temperature and with decreasing reinforcement particle size, d_p. Furthermore, the critical strain rate is dependent on the dominant diffusion process, and especially in the higher temperature range the critical strain rate in the grain boundary diffusion-controlled superplastic flow is higher than in the lattice diffusion-controlled superplastic flow. It is therefore suggested that it is important to use reinforcements having small size in order to limit cavity formation caused by the stress concentrations around reinforcements for the given constant matrix grain size and constant volume fraction of the reinforcements. It is noted that the predicted critical strain rate obtained in the analysis consisted with the experimental results for the high-strain-rate superplastic aluminum matrix composites.

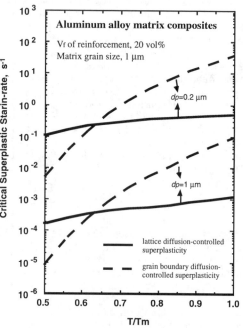

Fig. 2; The variation in $\dot{\varepsilon}_c$ as a function of the value of T/T_m in aluminum alloy matrix composites having a mean grain size of 1 μm and a volume fraction of the reinforcements of 20 % for the reinforcement particle sizes of 0.2 and 1 μm in the case of lattice diffusion-controlled and grain boundary diffusion-controlled superplastic flow, respectively.

4. Local Melting in High-Strain-Rate Superplastic Composites

The detailed mechanistic origins of high-strain-rate superplasticity are not yet fully understood. However it has been suggested for the composites exhibiting high-strain-rate superplasticity that the presence of a small amount of liquid phase at grain boundaries and/or interfaces plays an essential role in the deformation mechanisms during superplastic flow, and not only enhances the strain rate but also assists strain accommodation and thus delays the fracture process [6]. Clearly, one of the striking facts is that the presence of partial melting along the interfaces at superplastic temperatures in many aluminum alloy matrix composites and its significance to the high-strain-rate superplasticity were experimentally confirmed by in-situ TEM and HREM [18]. Investigation with an analytical TEM also revealed that solute segregation took place along the matrix-reinforcement interfaces during processing by hot extrusion and was attributed to partial melting at the enriched regions with solute elements of Mg, Si and Cu along the interfaces because an addition of some solute elements decreases the solidus temperature of the aluminum alloys. So Mg-free in chemical composition of the matrix alloys is expected to decrease the efficiency in segregating along the interfaces and then in reducing the partial melting temperature. Therefore, two composites with Si_3N_4/Al-Cu-Mg and Mg-free Si_3N_4/Al-Cu were prepared by a complete same thermomechanical processing route to understand the mechanical properties in superplasticity, then the second phase formation in these composites was studied to clarify in what form solute atoms exist and to examine the validity of segregation as an origin for partial melting. Comparison of typical superplastic properties between Mg-free Si_3N_4/Al-Cu and Si_3N_4/Al-Cu-Mg composites are summarized in **Table 1**, where T_{op} represents the optimum superplastic temperature, T_i the incipient melting point, Ts the solidus temperature and d the grain size. The results indicate that the Mg-free Si_3N_4/Al-Cu composite can not exhibit a large superplastic elongation in a similar deformation condition to that for the Si_3N_4/Al-Cu-Mg composite although the grain sizes in both composites are almost same to be very small at 2 μm. Also it is noted that the incipient melting point could not observe in the Mg-free Si_3N_4/Al-Cu composite by the DSC investigation.

Table 1 Comparison of the experimental data between Mg-free Si_3N_4/Al-Cu and Si_3N_4/Al-Cu-Mg composites.

Material	T_{op} (K)	T_i (K)	T_s (K)	Strain rate (s^{-1})	Stress (MPa)	m value	Elong. (%)	d (μm)
Mg-free Si_3N_4/Al-Cu	838	-	838	0.1	2	0.4	93	2.0
Si_3N_4/Al-Cu-Mg	773	784	853	0.1	5	0.3	640	2.0

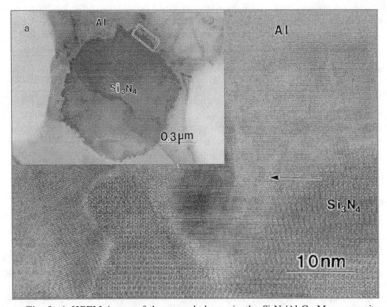

Fig. 3; A HREM image of the second phases in the Si_3N_4/Al-Cu-Mg composite exhibiting high-strain-rate superplasticity.

The HREM investigation reveals the absence of the new second phases as an evidence of partial melting along the interfaces and grain boundaries in the as-processed Mg-free Si_3N_4/Al-Cu composite hardly exhibiting high-strain-rate superplasticity. Whereas a distribution of the second phases, as shown in **Fig. 3**, is clearly found in the Si_3N_4/Al-Cu-Mg composite exhibiting high-strain-rate superplasticity. The EDS analysis shows no segregation of special elements at the interfaces in the Al-Cu alloy composite, whereas magnesium and silicon are enriched in the reaction phases around the interfaces in the Al-Cu-Mg alloy composites. It is suggested that magnesium in the matrix and precipitates moves to the interfaces during the hot extrusion and superplastic deformation and is magnesium enriched at the interfaces. The enriched magnesium serves the reaction to the Si_3N_4 crystals. The reaction phases in the Al-Cu-Mg alloy composites are an Al-based solid solution supersaturated by magnesium and silicon, and consequently the melting point of the reaction phase is lower than that of the matrix. Therefore, the reaction phase causes partial melting at the interfaces at the tensile testing temperature. This conclusion can be consistent with the argument that partial melting is caused by solute segregation and that the raised stress concentration around hard phases by grain boundary sliding can be relaxed by accommodation helper such a liquid on a localized scale, causing partial melting. Also the partial melting at the interfaces can not only relax the stress concentration but also suppress the extensive development of microcracks and cavities at the interfaces during superplastic flow. It is also important to note that the tendency of melting was found to depend on the nature of the grain boundaries, probably because of the observed dependence of segregation on the grain boundary structure, and the results obtained were explained by a decrease of the solidus temperature due to segregation whose extent depends on the type of the grain boundary structure. The tendency of melting appears to be dependent on the type of grain boundary structure characterized by the misorientation between neighboring grains [19].

5. Cavitation Fracture in High-Strain-Rate Superplasticity

It was reported recently that grain boundary microstructures of Al-Li based alloy were analyzed quantitatively to obtain the fundamental knowledge for optimizing superplastic microstructure [20]. The results are summarized in **Table 2**, showing the relationship between grain boundary connectivity at triple junctions and the frequency of cavitation. It is evident from Table 2 that cavitation preferably occurred at grain boundary triple junctions where two or more random grain boundaries were connected. Therefore, it is concluded that the connectivity of random grain boundaries is of great importance as well as introduction of large amount of random grain boundaries for development of superplasticity.

Table 2; Relationship between grain boundary connectivity at triple junction and cavity nucleation.

GB Connectivity at Triple Junction	Total Number of Triple Junction	The Number of Cavity Nucleation
L-L-L	8	0
L-L-Σ	0	0
L-L-R	7	1
L-Σ-Σ	0	0
L-Σ-R	6	1
L-R-R	35	12
Σ-Σ-Σ	0	0
Σ-Σ-R	8	1
Σ-R-R	8	4
R-R-R	28	16
Total	100	35

L; low-angle grain boundaries, Σ; CSL grain boundaries, R; random grain boundaries.

On the cavitation behavior for a high-strain-rate superplastic composite, it was reported [21] that the rate of increase in the cavity volume at a temperature slightly above the onset temperature for partial melting was significantly lower than at a temperature below the onset temperature for partial melting. The variation in cavity growth rate as a function of cavity radius for the above cavity growth mechanisms in the cases of consideration for stress concentrations, that is, $\sigma = 108$ MPa and of no consideration for stress concentrations, that is, $\sigma = 8$ MPa wad reported [22]. Here the local tensile stress is calculated to be 108 MPa at the testing conditions for the high-strain-rate superplastic Si_3N_{4p}/Al-Mg-Si composite. The diffusion growth rate is expected to be low because of short times for high-strain-rate deformation. However, when the stress concentrations are cause, the diffusion growth rate is very fast in a small cavity radius range < about 0.2 µm because mass transfer due to diffusion is accelerated by high stress. On the other hand, when the stress concentrations are relaxed by the presence of liquid, the diffusion growth is significantly slow and consequently the diffusion growth is negligible even in a small cavity size range. These results revealed that plastic growth was also significantly limited in a cavity radius range ≤ about 0.5 µm by relaxation of stress concentrations. The experimental growth rates were in good agreement with the rate predicted by plasticity-controlled growth mechanism in the case of consideration for relaxation of the stress concentrations. The fact that the testing temperature is slightly above the partial melting temperature indicates that a liquid phase due to partial melting plays an important role in relaxing the stress concentrations, that is, as accommodation helper [8,16], and consequently cavitation is limited.

The optimum amount of liquid phase may depend upon the precise material composition and the precise nature of a grain boundary or interface, such as local chemistry which determines the chemical interactions between atoms in the liquid phase and atoms in the neighboring grains, and also the magnitude and the type of the misorientation. Only a small amount of liquid phase may be present at temperatures close to the incipient melting point, and it would be expected to segregate to grain boundaries and particularly at grain triple junctions. However, larger volumes of a liquid, or a continuous liquid layer, can not support normal tractions, and therefore can not contribute to large elongations. Thus, intergranular decohesion at a liquid grain boundary leads to intergranular fracture and very limited elongation. This increase in liquid explains the drop in elongation observed at the highest temperatures above the melting point in many high-strain-rate superplastic materials [6,7,8]. There apparently exists a critical amount of liquid phase for the optimization of grain/interface boundary sliding during superplastic deformation. The HREM investigations revealed that the thickness of the liquid phases at the optimum superplastic temperature is about less than 30 nm, and the distribution is discontinuous. The estimated volume fraction of a liquid phase at the optimum superplastic condition is probably about 1 % or less. Thus, the volume fraction of a liquid phase is required to be very small in order to attain superplasticity.

The optimum superplastic strain rates strongly depend on the refinement of grain structure. Also the high-strain-rate superplasticity is critically controlled by the accommodation helper to relax the stress concentration resulting from grain boundary sliding. Optimum superplasticity is obtained at a temperature close to the partial melting point or solidus. Control of the distribution of a liquid phase is the most important to limit decohesion at liquid phase in tensile stress field.

6. Low-Temperature Superplasticity

High-strain-rate superplasticity is of great interest because it is expected to result in economically viable, near-net-shape forming techniques. In general, High-strain-rate superplasticity is observed at relatively high temperatures of ~ 0.8 T_m, where T_m is the melting temperature of the materials. Therefore, one of the subject is to lower the superplastic temperature. Grain size refinement is related to low-temperature superplasticity in materials because an optimum state of temperature for superplasticity is a function of grain size. Documented low-temperature superplastic behavior in aluminum-based and magnesium-based materials are listed in **Table 3**. It should be noted that the processed materials exhibit superplastic behavior at low temperatures of about $0.5T_m$.

The low temperature superplasticity is attributed to a very small grain size in materials. Although the superplastic temperature is very low, large elongations are obtained only in the low strain rate range from 10^{-5} to 10^{-3} s^{-1}. One of the recent notable topics in superplasticity research is on the combination of high-strain-rate superplasticity and low-temperature superplasticity [23]. This is significant, and may have technological implications for superplastic forming. A lower forming temperature enables reduction in energy cost, reducing surface oxidation and preventing selective depletion of the alloy addition [24]. Magnesium alloys have a high potential for superplasticity at lower temperature compared with aluminum alloys, which have similar melting point with magnesium [25]. This is because the pre-exponential factor for grain boundary diffusion, δD_{gb}, for magnesium is two orders of magnitude larger than that for aluminum, though the activation energy is close to each other

[26]. Especially, low-temperature superplasticity would be beneficial for magnesium alloys. Since many magnesium alloys have low formability near the room temperature because of the h.c.p. structure, and magnesium is susceptible to surface oxidation at high temperatures.

Table 3. Low-temperature superplastic behavior ($T \leq \sim 0.5\, T_m$) in aluminum and magnesium-based materials.

Material	d, μm	T, K	T/T_m [†]	$\dot{\varepsilon}$, s^{-1}	σ, MPa	δ, %	m-value
Aluminum based materials							
Al-3Mg	0.2	403	0.43	1.3×10^{-5}	105	180	–
Al-5.5Mg-2.2Li-0.12Zr (1420)	1.2	523	0.56	1.0×10^{-3}	44	620	0.5
Al-5.5-Mg-2.2Li-0.1Zr (1420)	5	593	0.64	7.0×10^{-4}	15	1000	0.5
Al-5.5Mg-2.2Li-0.1Zr (1420)	50	593	0.64	7.0×10^{-4}	52	400	0.3
Al-10.2Mg-0.52Mn	0.2–0.5	573	0.61	1.4×10^{-3}	32	400	0.45
Al-10Mg-0.1Zr	2	573	0.61	6.7×10^{-4}	29	580	0.5
Al-4.8Mg-0.07Mn-0.06Cr (5056)	0.3	548	0.59	1.0×10^{-5}	30	200	0.3
Al-4.7Mg-0.7Mn (5083)	0.5	523	0.56	1.0×10^{-3}	95	400	0.55
Al-4.5Mg-0.6Mn-0.06Cr (5083)	0.28	473	0.51	1.7×10^{-3}	190	220	0.37
Al-4Mg-0.5Sc	1–2	589	0.63	2.0×10^{-3}	–	511	–
Al-3Mg-0.2Sc	0.2	573	0.61	3.3×10^{-4}	–	720	–
Al-2.4Li-1.3Cu-0.63Mg-0.11Zr (8090)	1	623	0.67	8.0×10^{-4}	14	700	0.3
Al-4Cu-1.5Mg-1.1C-0.8O (IN9021)	0.5	573	0.61	1.0×10^{-2}	80	250	0.15
Al-4Cu-0.5Zr	0.3	493	0.53	3.0×10^{-4}	23	> 250	0.48
Al-6Al-0.4Zr (Supral 100)	0.5	573	0.61	1.0×10^{-2}	–	970	–
Magnesium based materials							
Mg-1.5Mn-0.3Ce (MA8)	0.3	453	0.49	5.0×10^{-4}	33	> 150	0.38
Mg-5Al-5Zn-5Nd-0.1Mn (EA55RS)	–	473	0.51	1.0×10^{-4}	–	270	–
Mg-9Al-1Zn-0.2Mn (AZ91)	0.5	473	0.51	6.2×10^{-5}	25	661	0.5
Mg-9Al-1Zn-0.2Mn (AZ91)	1	543	0.59	1.0×10^{-3}	16.9	190	0.5
Mg-9Al-0.7Zn-0.15Mn (AZ91)	1.2	523	0.57	3.3×10^{-3}	–	> 500	0.52
Mg-9Al-1Zn-0.2Mn (AZ91)	4.1	523	0.57	3.2×10^{-3}	11.5	425	0.5
Mg-6Zn-0.5Zr (ZK60)	2.2	523	0.57	1.0×10^{-3}	31.4	450	0.5
Mg-6Zn-0.5Zr (ZK60)	3.4	423	0.46	1.0×10^{-5}	66	340	0.3
Mg-0.58Zn-0.65Zr (ZK60)	3.7	523	0.57	1.1×10^{-4}	8.4	680	0.55
Mg-6Zn-0.5Zr (ZK60)	6.5	498	0.54	1.0×10^{-5}	15	449	0.5
Mg-6Zn-0.8Zr (ZK61)	0.65	473	0.51	1.0×10^{-3}	21.7	659	0.5
Mg-6Zn-0.8Zr (ZK61)	1.2	523	0.57	1.0×10^{-2}	24.7	350	0.5
ZK60/SiC/17p	1.7	463	0.50	1.0×10^{-4}	29.1	337	0.38

[†] T_m: Melting point of the material (T_m = 933K for aluminum, 924 K for magnesium)

The constitutive equation to describe the superplastic flow of the magnesium alloys is generally can be unified by introducing a notion of effective diffusion coefficient, D_{eff}, and given by [27]

$$\dot{\varepsilon}_{sp} = 1.8 \times 10^6 \left(\frac{Gb}{kT}\right)\left(\frac{\sigma - \sigma_0}{G}\right)^2 \left(\frac{b}{d}\right)^2 D_{eff} \tag{2}$$

The effective diffusion coefficient is described by a combination of grain boundary diffusion, D_{gb}, coefficient and lattice diffusion coefficient, D_L, [28]:

$$D_{eff} = D_L + x\left(\frac{\pi}{d}\right)\delta D_{gb} \tag{3}$$

where $\dot{\varepsilon}_{sp}$ is the strain rate, D_{gb} is the grain boundary diffusion coefficient, D_L the lattice diffusion coefficient, δ the grain boundary width, x is an unknown constant and estimated to be 1.7×10^{-2} for superplastic flow in magnesium alloys [27]. The effects of decrease in grain size expected from the constitutive equation for superplastic flow are (i) to increase the strain rate and/or (ii) to decrease the temperature for optimum superplastic flow. In **Fig. 4**, low-temperature superplastic behavior in magnesium alloys and aluminum alloys is correlated in the relationship between effective-diffusion-compensated strain rate and the reciprocal grain size. The figure also includes the data for conventional and high-strain-rate superplasticity in magnesium alloys, and the relationship derived from equation (2) at a normalized stress of $(\sigma - \sigma_0)/G = 5 \times 10^{-4}$. From this figure, it is suggested that the observation of low-temperature superplasticity is directly attributed to the fine grain size. This figure also indicates the possibility that the combination of high-strain-rate superplasticity and low-temperature superplasticity can be attained by extreme grain refinement.

Therefore, the desired grain size in pseudo-single phase magnesium alloys is estimated to be less than ~ 0.4 μm for obtaining high-strain-rate superplasticity ($\dot{\varepsilon} \geq 1 \times 10^{-2} \, s^{-1}$) and low-temperature superplasticity ($T =$ ~ 473 K (~ 0.5 T_m)). A typical microstructure of the as-extruded PM ZK61 alloy is shown in **Fig. 5**. The grain size is estimated to be 0.65 μm, which is near the target value in grain size. The grains were almost equiaxed. Typical microstructural feature of alloy produced by PM route, that is distinctly different from the IM processed alloy, is that the PM alloy contains a significant volume fraction of small second phase particles. In addition to the precipitates, oxide particles, which was probably introduced during atomization process, are invariably present. The dispersions were observed to reside in both the grain boundaries and the grain interiors. The particles were almost spherical with an average size of ~ 25 nm. On the other hand, it is evident that only the extremely fine precipitates of less than 10 nm are present in IM alloy. These precipitates are suggested to be fully taken into solution by the annealing treatment at above 573 K.

It has been suggested that the grain size is much finer and stable in PM alloys as compared with IM alloys, since the particles can pin the grain boundary. By the investigation of grain growth between IM and PM magnesium alloys at elevated temperatures, it is obvious that not only the grain size of PM alloys are finer than that of IM alloys, but also the PM alloys show higher grain size stability than IM alloys at elevated temperatures. The grain sizes of IM alloys slightly coarsened below ~ 600 K, but increased rapidly at higher temperatures. On the other hand, fine grain sizes are retained in PM alloys at higher temperatures.

The variation in flow stress as a function of strain rate is plotted in **Fig. 6** (a) for PM ZK61. The flow stress for each strain rate was determined at a small strain of 0.1. The *m*-value

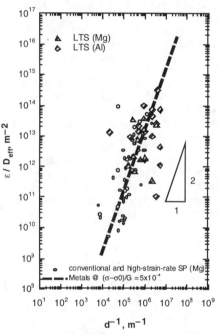

Fig. 4; The variation in effective-diffusion-compensated strain rate as a function of the reciprocal grain size in low-temperature superplastic magnesium alloys and aluminum alloys. The data for the conventional and high-strain-rate superplastic magnesium alloys are also included in the figure. A dotted line indicates the phenomenological relation derived from equation (2) at a normalized stress of $(\sigma - \sigma_0)/G = 5 \times 10^{-4}$.

Fig. 5; TEM microstructure of the as-extruded PM ZK61 alloy.

exhibited a maximum value of about 0.5 at all investigated temperatures. This high m-value of 0.5 suggests that grain boundary sliding could be a dominant deformation process. This was confirmed from a scanning electron microscopic (SEM) observation of surface appearance, which showed direct evidence for grain boundary sliding. The effective grain sizes (grain group sizes) for active sliding process inspected by SEM observations agree well with those by TEM observations. The m-value decreased to a small value in the low strain rate range at all temperatures. Low m-value in the low strain rate range is often observed in PM alloys; such values have been associated with the existence of a threshold stress.

The variation in elongation-to-failure as a function of strain rate is plotted in Fig. 6 (b) for PM ZK61. The strain rate regions exhibiting high m-value were roughly in agreement with the regions where large elongations were attained. The maximum elongation of 659% was obtained at 473 K at 10^{-3} s^{-1}. However, it should be noted that a large elongation of 283% was attained even at a high strain rate of 10^{-2} s^{-1}. It was demonstrated that the high-strain-rate superplasticity was attained even at low temperatures of ~ 0.5 T_m.

7. Commercial Superplastic Forming

Now superplastic forming has become a well established technology for specific niche market applications that benefit from the attributes of current superplastic materials and processes [29]; specifically; (1) greater design freedom afforded by high superplastic ductility allowing the manufacture of complex single piece components in one operation which might otherwise require expensive fabrication and/or the assembly of separately made parts, and (2) lower overall manufacturing cost for medium/small production quantities; basically as a result of lower tooling costs and, in some instances, less expensive capital equipment.

As shown in **Fig. 7**, high-strain-rate superplastic forging technology has been developed to mass-produce high-performance-engine pistons with near-net-shape of the Al-Si based PM alloys [30,31]. Also the new wider applications in Japan using commercially produced superplastic 5083 alloy have been found in some diverse markets, *i.e.*, the output port of a food ticket vending machine, the gate, the window panel of a car of the train, the top tablet frame of an automatic drug packager for medical purposes,

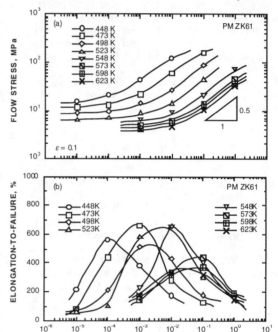

Fig. 6; The variation in (a) flow stress and (b) elongation-to-failure as a function of strain rate in PM ZK61 with an initial grain size of 0.65 μm.

Fig. 7; The high-performance-engine pistons with near-net-shape of the PM Al-Si based alloy.

the tire housing for a recreational vehicle, the fuel tank for a motorcycle, the architectural panel and the door sill of the Chrysler Dodge Viper [32]. The positive attributes of aluminum over plastics (*e.g.* fire resistance, specific strength and easy recycling) will give superplastic forming aluminum advantages in transportation industries. Very recently, HONDA has produced the roof top component of Sport 2000 car using IM processed 5083 alloy sheet exhibiting high-strain-rate superplasticity, as shown in **Fig. 8**. This material shows large elongations at very high strain rates around 10^{-1} s^{-1} [33]. Further refinements to currently available materials, tooling methods and equipment will allow for better optimization of forming conditions and facilitate expansion of the techno-economic niche. It will require break-through technology capable of producing superplastic aluminum sheet having ultra fine grain size at low cost to transform superplastic forming in aluminum alloys from its present specialized niche into the mainstream of competitive manufacturing. Achieving this break-through will require much effort and requires the combined commitment and resources of material producers, fabricators, user industries and government.

Fig.8; An application for the automobile by HONDA. The used material processed by IM, is the high-strain-rate superplastic 5083 alloy sheet, exhibits about 500 % elongations at very high strain rates around 10^{-1} s^{-1}.

8. Concluding Remarks

Possibility of combination of high-strain-rate superplasticity and low-temperature superplasticity was experimentally confirmed using extremely fine-grained magnesium alloy. It was indicated that the decrease in grain size increases the strain rate and/or decreases the temperature for the optimum superplastic flow. Low-temperature superplastic behavior in magnesium and aluminum alloys was well described by the relationship between effective-diffusion-compensated-strain rate and the reciprocal grain size. Based on the phenomenological relation for superplastic flow, target grain size for the present purpose was estimated to be $\leq \sim 0.4$ μm. It is concluded that the extremely fine grain sizes of < 1 μm is required in order to attain the combination of high-strain-rate superplasticity and low-temperature superplasticity.

It is clear that the study of sciences in superplasticity has been narrowly limited to the phenomenon of superplasticity and new developing materials, and not to the manufacturing tools, presses and processes that take advantage of it. In the research and development profession for superplasticity, the horizons should be expanded to include the peripheral process technologies and materials that support the manufacturing of superplasticity that have not yet been studied. Better communication and a tighter bond between industry and the universities is needed. It is desired that fruitful collaboration between scientists and industrialists will highly be anticipated.

ACKNOWLEDGMENT
The authors express their thanks to the financial support of the Priority Area "Innovation in Superplasticity" from the Ministry of Education, Science, Culture and Sports, Japan.

REFERENCES
(1) Towards innovation in superplasticity I, Ed. by T. Sakuma, T. Aizawa, and K. Higashi : Materials Science Forum, Vols. 233-234, Trans Tech Publications, Hampshire, (1997).
(2) Towards innovation in superplasticity II, Ed. by T. Sakuma, T. Aizawa, and K. Higashi, Materials Science Forum, Vols. 304-306, Trans Tech Publications, Hampshire, (1999).
(3) A.K. Mukherjee, J.E. Bird and J.E. Dorn: Trans. ASM, **62** (1969), 155-179.
(4) J.E. Bird, A.K. Mukherjee and J.F. Dorn: Quantitative Relation Between Properties and Microstructure, Ed. by D.G.Brandon and A.Rosen, Israel Universities Press, Jerusalem, (1969), 255-342.
(5) T.G. Nieh, C.A. Henshall and J. Wadsworth: Scripta Metall., **18** (1984), 1405-1408.
(6) K. Higashi, M. Mabuchi and T. G. Langdon: ISIJ Int., **36** (1996), 1423-1438.
(7) K. Higashi, T.G. Nieh and J. Wadsworth: Acta Metall. Mater., **43** (1995), 3275-3282.
(8) K. Higashi: Mater. Sci. Forum, **170-172** (1994), 131-140.
(9) E. Sato, S. Furimoto, T. Furuhara, K. Tsuzaki and T. Maki: Mater. Sci. Forum, **304-306** (1999), 133-138.
(10) N. Tsuji, K. Shiotsuki, H. Utsunomiya and Y. Saito: Mater. Sci. Forum, **304-306** (1999), 73-78.
(11) Y. Saito, N. Tsuji, H. Utsunomiya, T. Sakai and R.G. Hong,: Scripta Mater. **39** (1998), 1221-1227.
(12) M. Furukawa, Z. Horita, M. Nemoto, R.Z. Valiev and T.G. Langdon, Acta Mater., **44** (1996), 4619-4629.
(13) T.G. Langdon, M. Furukawa, Z. Horita and M. Nemoto: JOM, **50** (1998), 41-45.
(14) P.B. Berbon, M. Furukawa, Z. Horita, M. Nemoto, N.K. Tsenev, R.Z. Valiev and T.G. Langdon: Phil. Mag. Lett., **78** (1998), 313-318.
(15) T.G. Langdon: Mater. Sci. Forum, **304-306** (1999), 13-20.
(16) M. Mabuchi and K. Higashi: Acta.Mater., **47** (1999), 1915-1922.
(17) M. Mabuchi and K. Higashi: Mater. Tans. JIM., **40** (1999), 787-793.
(18) H-G.Jeong, K. Hiraga, M. Mabuchi and K. Higashi: Acta Mater., 46 (1998), 6009-6020.
(19) J. Koike, M. Mabuchi and K. Higashi: J. Mater. Res., **10** (1995), 133-138.
(20) S. Kobayashi, T. Yoshimura, S. Tsurekawa and T. Watanabe: Mater. Sci. Forum, **304-306** (1999), 591-596.
(21) M. Mabuchi, H. Iwasaki, K. Higashi and T.G. Langdon: Mater. Sci. Tech., **11** (1995), 1295-1299.
(22) H. Iwasaki, M. Mabuchi and K. Higashi: Acta Mater., **45** (1997), 2759-2766.
(23) S.X. McFadden, R.S. Mishra, R.Z. Valiev, A.P. Zhilyaev and A.K. Mukherjee: Nature, **398** (1999), 684-686.
(24) H.P. Pu, F.C. Liu and J.C. Huang: Metall. Mater. Trans. A, **26A** (1995), 1153-1160.
(25) H. Watanabe, T. Mukai and K. Higashi: Scripta Mater., **40** (1999), 477-483.
(26) H.J. Frost and M.F. Ashby, Deformation-mechanism maps, Pergamon Press, Oxford, 1982, p. 44.
(27) H. Watanabe, T. Mukai, M. Kohzu, S. Tanabe and K. Higashi, Acta mater., **47** (1999), 3753–3758.
(28) O.D. Sherby and J. Wadsworth, Development and Characterization of Fine-Grain Superplastic Materials, ASM, Metals Park, OH, 1982, p. 355–389.
(29) A.J. Barnes: Mater. Sci. Forum, **304-306** (1999), 785-796.
(30) H. Yamagata: Mater. Sci. Forum, **304-306** (1999), 797-804.
(31) N. Kuroishi, S. Tsuboi, S. Fujino, Y. Shiomi and K. Matsuki: Mater. Sci. Forum, **304-306** (1999), 837-842.
(32) Y. Onishi: Mater. Sci. Forum, **304-306** (1999), 819-824.
(33) H. Uchida, M.Asano and H.Yoshida: Mater. Sci. Forum, **304-306** (1999), 309-314.

High Strain Rate Superplastic Aluminium Alloys: The Way Forward?

R. Grimes, R.J. Dashwood and H.M. Flower

Department of Materials, Imperial College of Sicence Technology and Medicine,
London SW7 2BP, UK

Keywords: Alloying Additions, Alloying Costs, High Strain Rate Superplasticity, Production Costs, Scandium, Zirconium

ABSTRACT

The technical and commercial barriers to the development and successful exploitation of a high strain rate superplastically deformable aluminium alloy for use in the automotive industry are considered in this paper. Batch processing routes, such as mechanical alloying or equal channel angular extrusion, employed to deliver appropriate chemistry and structure, are inherently costly and unlikely to deliver either the quantity or the size of strip required commercially. There is evidence that there is still scope for development of conventional casting and rolling routes, but a particulate casting route combined with roll consolidation offers the prospect of a commercially viable Al-Mg-Zr product. The use of alloying additions, including zirconium, is also discussed and comparative costs are presented: on this basis the use of scandium appears economically prohibitive.

INTRODUCTION

Superplastic aluminium alloys have been commercially exploited for rather more than twenty years. From a very early stage some of the applications were in motor cars. However, the combination of inherently slow forming with a significantly higher raw material price has resulted in the automotive applications being confined to the exotic or specialist markets.

In the meantime, many millions of dollars have been spent by the world's major aluminium companies to persuade the world's major car builders to switch from drawing quality steel sheet to aluminium. In turn, the car builders have been under growing pressure (e.g. in the USA, the CAFE regulations and the PNGV) to make serious efforts to reduce the environmental damage for which the motor car is responsible. Thus, it can be argued that the climate has never been more favourable for sheet aluminium to make the breakthrough as a significant material in body structure. The technical feasibility of producing vehicles with all the major structure in aluminium and with no sacrifice in safety or comfort has been demonstrated beyond doubt in a number of concept vehicles.

Since the demonstration, in metal matrix composites, that superplastic behaviour could be obtained at much higher strain rates considerable effort has been devoted to developing the phenomenon in aluminium. It has been suggested that this would then make possible the mass production of automotive components. Unfortunately most of the alloys that have, so far, demonstrated high strain rate capability have either been made by commercially impractical routes, and/or they contain exotic and prohibitively expensive additions.

It should be borne in mind that there are two compelling reasons for employing superplastic forming and they both, more than anything else, come down to economics:

1. relatively cheap, single sided tooling is used and
2. part consolidation can, frequently, be achieved

Against these advantages, the forming rates are slow – even a simple, "rapidly formed" SPF component takes of the order of five minutes while complex components can take over 60 minutes and available commercial SPF aluminium sheet is expensive (up to 5x conventional automotive aluminium sheet).

The purpose of this paper is to provide some indications of the way that the price of aluminium semi fabricated products builds up, in the hope that it will encourage researchers to pursue alloys and production routes that stand some chance of achieving commercial success. It will also endeavour to criticise constructively the routes and alloys that appear to be under current development.

DISCUSSION

Today, the dominant material in the construction of motor car body/chassis structure (the body-in-white and its associated closure panels) remains steel. In modern plants steel is produced in lot sizes of the order of 300 tonnes (produced by continuous casting), using high speed rolling mills and taking full advantage of the economies of scale. Thus, not only is the raw material, relatively, cheap but also processing costs are modest and a high quality, consistent product with extremely good cold formability results with a price not greatly more than $600 per tonne.

Conventional Aluminium Strip Production: In a modern aluminium rolling mill the ingot size can be up to 25 tonnes (produced by vertical, semi-continuous, direct chill casting) and this is then hot rolled on a reversing mill before passing through a multi stand tandem hot mill from which it emerges at about 275 metres/minute to be coiled for cold rolling. Where there are no over-riding metallurgical considerations, the gauge at the end of hot rolling will be chosen to minimise the amount of the more expensive cold rolling, but, typically, would probably be between 6.5mm and 3.5mm. A variety of single stand or multi stand cold mills may be employed, but in the most modern the strip may emerge at up to 1200 metres per minute. The capital involved in such a rolling system is obviously very large, but when operated at the intended rate, the entire rolling operation would add about $550 per tonne of product. With basic aluminium costing about $1500 per tonne, this leads to an aluminium alloy strip price of around $2500 - $3000 per tonne depending on alloy, quantity, condition and gauge. While there are still many factories operating with relatively small ingots and slower, less modern mills than those outlined, there is no doubt that if aluminium makes the breakthrough into major structural application in volume motor cars, then it will have to be produced in the largest, most efficient mill systems. Thus, it is these general levels of price with which those who hope to see superplastic forming of aluminium expand into the general automotive market will have to compete.

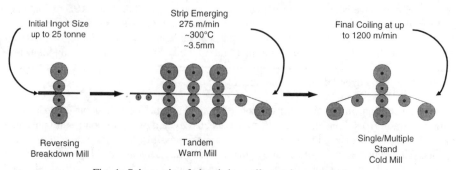

Fig. 1: Schematic of aluminium alloy strip manufacture

Current commercial superplastic alloys are priced at levels up to an order of magnitude higher than those for conventional strip, but neither the original superplastic aluminium alloy, 2004, nor the SPF version of 7475 that succeeded it, has service properties appropriate for the majority of structural automotive applications. The SPF variants of 5083 have suitable service properties but relatively modest SPF performance. Nevertheless, they carry a price premium of 2-3x that of conventional aluminium alloy strip.

The target, therefore, should be for an alloy with service properties similar to 5083 but exhibiting good superplastic behaviour at strain rates up to 10^{-1} s^{-1}. It is doubtful whether practical use could be made of strain rates higher than this for sheet forming because the times involved in pressurising and de-pressurising the forming machine then dominate the cycle and further reduction in the actual time of forming becomes insignificant. The following paragraphs consider some of the methods employed to achieve high strain rate superplasticity and their prospects for eventual commercialisation.

Routes Investigated to Develop High Strain Rate Superplasticity:

Metal Matrix Composites: High strain rate superplasticity was first observed in metal matrix composites but, for automotive applications, it seems very doubtful whether a potential volume production route exists that could possibly be economically competitive and, in most cases, the service properties of the resultant strip would not be suitable.

Mechanical Alloying: Excellent high superplastic strain rate behaviour has been observed in material made via mechanical alloying. However, as a slow, inherently expensive, batch process, with very few opportunities to achieve economies of scale there seem even fewer possibilities for an economic volume process.

Equal Channel Angular Extrusion (ECAE): This seems to have become the favourite tool for producing the very fine grain structures necessary for achieving high strain rate superplasticity. While the ability to tailor microstructures using this process is extremely impressive and very valuable in leading to further understanding of superplasticity and other microstructural phenomena, there seems very little prospect of developing the process into a route for commercial sheet production.

While it is true that, in the past, some aluminium strip – and particularly that intended for very high quality trim applications – was made by initially extruding a flat "plank" as a pre-cursor for subsequent rolling, developments in rolling mills have rendered this route hopelessly uneconomic and, in the UK, it was discontinued some thirty years ago. It is the present authors belief that any route for aluminium strip production that incorporates any extrusion step would be destined to fail economically, but there are, perhaps, two possibilities that could be considered. While the normal extrusion press has a cylindrical container, some have been built with rectangular containers in order to allow the extrusion of wide, thin sections, particularly for applications such as rail vehicles where significant numbers of long, wide, integrally stiffened panels are required. With such a press it would be possible to extrude a flat blank 800 mm wide and 8 mm thick. Starting from a 1000 kg billet the strip, weighing about 700 kg could, in theory, be coiled (Figure 2) and subsequently cold rolled. Unless cross rolling of the blank were to be employed, the extruded blank width would be too narrow for many SPF applications. In any case, there are few such presses in the world and the cost would be appreciably greater than that involved in rolled strip. Using ECAE, in which several passes through the press would be required, where a far smaller billet would have to be employed at the outset and where, because the extruded product would not be wide and thin, far more complex cold rolling procedures would be required, the route would be even more un-economic.

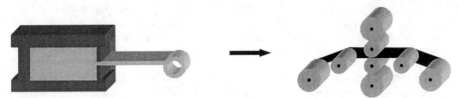

Fig. 2: Schematic of strip production from a press with a rectangular container

The second possibility would involve extrusion of a tube and the use of the "expansion technology" described by Michisaka, Matsumura and Hattori [1] to produce relatively wide and thin flat blank suitable for further working (Figure 3). They claim that using a 440mm diameter billet it is possible, after splitting, to produce a 1200 mm wide blank up to 15 metres long. The advantages of this process would include the fact that there are numerous extrusion presses with containers of about 400 mm diameter. Nevertheless, the piece weight would still be very small compared with conventional strip rolling.

Fig. 3: Use of "expansion technology" to produce flat blank for cold rolling

Conventional Casting and Rolling: Although conventional semi-continuous, direct chill casting imposes limitations including a restricted ability to take alloying elements, such as zirconium, into super saturated solid solution, work in a number of organisations suggests that there is still scope for developing improved superplastic performance in conventionally processed material. Uchida, Asano and Yoshida [2] have developed an Al-Mg alloy that exhibits reasonable superplastic behaviour at strain rates of 10^{-2} to 10^{-1} s^{-1}. Work at the Pacific North West Laboratories/University of Washington has demonstrated that the superplastic performance of conventionally manufactured Al-Mg alloys can be improved by optimising manganese content together with additions of zirconium [3] and they have also investigated the addition of scandium to Al-Mg alloys [4]. This, inevitably, raises the issue of the cost/availability of alloying additions and this is, briefly, addressed below. Perhaps most interestingly, work at the Manchester Materials Science Centre [5] has demonstrated the possibility of achieving geometric dynamic recrystallisation during hot rolling thus, potentially, avoiding the needs for the exceptionally large strains involved in ECAE.

Unconventional Casting and Rolling: The present authors have conducted an exploratory programme employing a particulate casting route to enable larger than normal additions of zirconium to an Al-Mg alloy. The experimental processing has employed extrusion as a convenient method of consolidation. However, considerably earlier work [6] has demonstrated that large volume sheet production is feasible by hot rolling particles directly to strip (Figure 4) for subsequent cold rolling and, should the development proceed further, some such route is envisaged. Good superplastic behaviour has been achieved at strain rates of 10^{-2} to 10^{-1} s^{-1}; details of which are reported elsewhere in this Conference [7]. A quite different approach has been employed by Tsuji, Shiotsuki, Utsunomiya and Saito [8] who overcome the limitations on strain imposed by the conventional rolling route by means of the "accumulative roll-bonding" sketched in Figure 5. While technically ingenious, from a cost point of view the amount of handling involved appears to be a major disadvantage. Finally, while considering the possibilities of unconventional casting routes for the production of superplastic sheet, it should be remembered that there have been many advances in strip casting systems over the last two decades. This raises the question of whether the previously perceived problems of centre line segregation and difficulties with wide freezing range alloys could now be overcome.

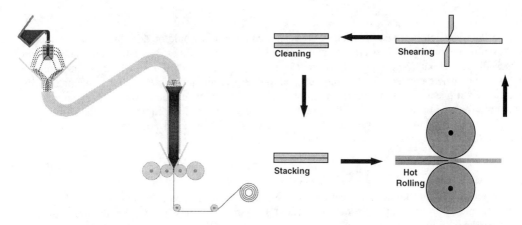

Fig. 4: Schematic of direct particle rolling [6]   Fig. 5: Schematic "accumulative roll bonding" [8]

Alloying Additions: Most of the alloying additions commonly made to aluminium only contribute modestly to the basic cost of the raw material. Table 1 indicates the cost of one tonne of a variety of notional alloys, together with the cost per kilogram recovered of each alloying addition. With the exceptions of lithium and scandium, the range from cheapest to most expensive is quite small. However, while it is extremely unlikely that lithium would be included in any automotive alloy, the possibility of scandium is frequently mentioned. The addition works, in many respects, in a manner similar to zirconium and the Al_3Sc phase has a lattice parameter (a=0.410 nm), like that of Al_3Zr (a=0.4009 nm), very close to that of aluminium (a=0.40494 nm). The further benefit is that the aluminium-scandium phase diagram is a eutectic system and the solubility of scandium in aluminium at high temperatures is appreciably greater than that of zirconium so that it is considerably easier to make reasonably large additions to conventionally cast ingots. Against this, any coherent precipitated particles of Al_3Sc coarsen rapidly when exposed to thermal treatments, such as ingot homogenisation, so that additions of (Sc+Zr) seem to have become the favoured route to obtain the best technical results [9].

Table 1: Indicative Costs of a Variety of Aluminium Alloys and Alloying Additions

Alloy	Cost per tonne of alloy ($)	$ per kg of addition recovered
Al-4%Mg	1530	3
Al-6%Cu	1608	4
Al-6%Zn	1491	2
Al-0.75%Mn	1506	3
Al-1%Zr	1613	17
Al-2.5%Li	3525	110
Al-0.25%Sc	23063	8626*

(* Price taken from Stanford Materials Web Site www.stanfordmaterials.com/sc.html on 26 July 00)

The problem with scandium, as far as potential use in any automotive application is concerned lies in its cost and/or availability. In geological circles there is general agreement that scandium comprises about

0.003% of the earths crust [10,11]. However, despite this, relative, abundance the element is highly dispersive, forming solid solutions in a large number of rocks and minerals, but, in general, only being present in very small concentrations [12]. Most existing scandium has been a by-product of uranium enrichment and although the possibility exists for its extraction as a by-product of other extractive operations (e.g. the red mud residues from Jamaican bauxite) a continuing (very) high price seems inevitable.

The further question is raised of whether the scandium resources would be adequate for a commercially successful, automotive, high strain rate alloy. The US Bureau of Mines, has estimated global resources of scandium at about 770 tonnes. If one envisaged an eventual market for a high strain rate aluminium alloy as somewhere between 50 000 and 100 000 tonnes per annum and if that alloy contained 0.25%Sc and further, if one assumes that major automobile manufacturers would need to be convinced of security of supply for, say, twenty five years, the exploitable reserves would have to be no less than 5 000 tonnes of scandium. Even if it is assumed that the US Bureau of Mines did not adequately survey eastern sources, this seems an insurmountable barrier.

CONCLUSIONS

The opportunities for aluminium sheet to achieve a major role in motor vehicle body construction are probably greater now than ever before. Hand in hand with this should go the opportunity for superplastically formed aluminium parts to achieve a significant role in mass produced vehicles. However, this will only be achieved if alloys are developed that not only possess the required superplastic performance but also can be made in volume at prices not substantially greater than that of conventional aluminium alloy sheet. Too much effort has been, and is being, devoted to techniques that have little, or no, prospect of utilisation for the production of commercial material, while it appears that scope still exists for improving the performance of sheet of relatively normal composition and manufacturing route.

ACKNOWLEDGEMENTS

Funding for the associated research under the joint EPSRC/MOD/DERA scheme (GR/L43848) is gratefully acknowledged. Support and valuable discussions have taken place with DERA, LSM, ALPOCO, Superform Aluminium and British Aluminium Sheet.

REFERENCES

[1] H. Michisaka, M. Matsumara, and M. Hattori, Procedings Seventh International Aluminium Extrusion Technology Seminar, Vol II, (2000), p489
[2] H. Uchida, M. Asano and H. Yoshida, Mater. Sci. Forum, 304-306, (1999), p309
[3] C. A. Lavender, J. S. Vetrano, M. T. Smith, S. J. Bruemmer and C. H. Hamilton, ibid, 170-172, (1994), p279
[4] J. S. Vetrano, C. H. Henager, S. M. Bruemmer, Y. Ge and C. H. Hamilton, Superplasticity and Superplastic Forming, The Minerals, Metals & Materials Society, (1998), p89
[5] A. Gholinia, P. B. Prangnell and F. J. Humphreys, Materials Congress 2000, to be published.
[6] T. Stevens Daugherty, Powder Metallurgy,11,(1968),p342
[7] R. J. Dashwood, R. Grimes, A. W. Harrison and H. M. Flower, this volume.
[8] N. Tsuji, K. Shiotsuki, H. Utsunomiya and Y. Saito, Mater. Sci. Forum,304-306, (1999), p73
[9] Y. Filatov, V. Yelagin and V. Zakharov, Materials Science and Engineering, A280, (2000), p97
[10] V. C. Fryklund and M. Fleischer, Geochim.Cosmochim. Acta, 27, (1963), p643
[11] S. R. Taylor, ibid, 28, (1964), p1273
[12] H. H. Schock, "Scandium; its occurrence, chemistry, physics, metallurgy, biology and technology", London Academic Press, (1975), p50

High Strain Rate Superplasticity and Microstructure Study of a Magnesium Alloy

Xin Wu*, Yi Liu and Hongqi Hao

Department of Mechanical Engineering, Wayne State University, Detroit, MI 48202, USA

Keywords: Dynamic Recrystallization, Mg Alloy, Microstructure, Superplasticity

Abstract

Superplastic-like behavior was found in a commercial large-grained Mg-3Al-1Zn alloys (AZ31), at high temperature and high strain rate regime. The creep behavior of this material was studied from RT to 550°C and the strain rates of 0.001 s^{-1} ~1 s^{-1}, and the maximum elongation to failure of 170% was obtained at 0.01 s^{-1} and at 500°C (or 0.84 T_m), which is much higher than the equivalent temperature of most known superplastic materials. Optical microscopy, transmission electron microscopy (TEM) and scanning electron microscopy with electron backscattering diffraction technique (SEM-EBSD) were used to investigate the microstructure and deformation mechanisms involved. It is evident that high temperature and high rate dislocation creep played a critical role at the early stage deformation in breaking down the initial large grains without fracture, and combined GBS and dislocation creep define the steady-state grain size and control the quasi-superplastic behavior.

1. Introduction

Magnesium is the lightest of all structural metals. As such, it forms the basis for commercial alloys that have found successful use in a wide variety of applications. However, due to very poor ductility at room temperature defined by its (hcp) crystal structure, the net shape forming of Mg alloys through deformation process are very limited, mainly in primary hot working (e.g. rolling, forging and extrusion), and under compressive stress states. As of today the principal fabrication technique of Mg alloys is die casting, which is relatively slow and costly for many part geometry if compared to plastic forming. Now it is necessary to develop a low-cost thermal forming technique as a new manufacturing route, so the relatively high material cost (over steel and aluminum alloys) can be compensated by the reduced manufacturing cost and improved functionality.

Superplastic forming is of special importance for Mg manufacturing, because low temperature deformation process is not feasible for Mg. However, conventional superplastic forming relies on a fine and stable microstructure, which is usually obtained through a thermal-mechanical cycling and recrystallization process. This technique is not applicable for Mg alloys in general, due to the difficulty in cold-working Mg to >50%. Dynamic recrystallization process with mechanical alloys and metal matrix composites (MMC) had great success in obtaining superplasticity, mostly in Al-based MMCs, in particular in the directions of high strain rate or low temperature superplasticity, although fabricating MMC involves additional investment. One work [1] proposed to use hot forging for grain refinement by dynamic recrystallization. One report [2] showed that through extrusion and thermal treatment ZC63-T6 exhibited superplasticity at a very low strain rate of 10^{-4} to 10^{-5} s^{-1} at 400°C, with elongation to failure of 100%-350%. Obviously, if feasible it is highly desired to obtain superplasticity from commercial Mg alloys that may have large grains. Very recently one of such large grained monolithic Mg alloys, AZ31, has been reported to have quasi-superplastic behavior at high strain rate and high temperature by the

---

* Contact: xwu@eng.wayne.edu, (313) 577-3882

current authors [3]. The current paper is to further explore the detailed deformation process and microstructure evolution involved.

2. Experimental Procedures

Material used in this study is a Commercial AZ31 magnesium alloy (containing 3 wt.% aluminum and 1 wt.% zinc), in the form of hot-rolled 5mm thick plate. Tensile specimens were machined from the plate with the rolling direction parallel to the tensile axis, and the gage dimension of $10 \times 5 \times 0.8$mm (length × width × thickness).

Uniaxial tensile test was conducted on Instron machine (model 8801) equipped with a digital controller and a resistance furnace. Load and displacement signals were obtained from a load cell of 10 kN capacity and LVDT sensor, and recorded with a computer data acquisition system. The tensile tests were performed under constant crosshead speeds from 0.01mm/sec to 100mm/sec, corresponding to initial strain rates from 1×10^{-3} s^{-1} to 1×10^{1} s^{-1}. Mounted specimens were heated up to certain desired testing temperatures at 20°C/min heating rate, and held for 20 min for stabilizing, followed by uniaxial tension to failure. Immediately after the tensile test was completed (detected from load drop to zero), the specimen was quickly removed from the furnace and cooled down to room temperature for further metallographic examination.

Microstructure examination was performed after tensile tests, with the help of optical microscopy, scanning electron microscopy with electron backscattering diffraction technique (SEM-EBSD), and with transmission electron microscopy (TEM). SEM-EBSD can distinguish grains based on their crystal orientations, even with a very small misorientation angle (sub-grains). The operating voltage is 20kV. TEM samples were sliced along the tensile direction on the specimen flat surfaces, and then mechanically thinned to below 100μm. TEM foils were prepared by conventional twin jet polishing technique using an electrolyte of 10% perchloric acid in methanol at -40°C. The foils were examined in a JEM-2000FX transmission electron microscope that operates at 200kV.

3. Results and Analyses

3.1 Effect of temperature and strain rate on elongation: At the initial strain rates of 1×10^{-2} s^{-1}, 5×10^{-3} s^{-1} and 1×10^{-3} s^{-1}, the elongation to failure as a function of temperature is shown in Fig 1(a). Despite the scattering of the elongation data, the general trend is that, for these strain rates applied and the temperature above 150°C, the elongation increases gradually with temperature up to 500°C, and then decreases, with the peak value occurred at 500°C. Note that at room temperature only 2-6% of elongation were obtained. This further confirms that without thermal activation Mg alloys are not suitable for plastic forming as a manufacturing route.

The effect of strain rate on the total elongation at 500°C is shown in Fig 1(b). At the strain rate of 10^{-3} s^{-1} to 10 s^{-1} elongation increases with strain rate, and then it decreases with strain rate, with the peak elongation of 189% obtained at 1×10^{-2} s^{-1}. At the high strain rate of 10 s^{-1} the alloy lost its ductility and behaviors similar to that at room temperature. Note that at the high strain rates of 10^{-1} to 1 s^{-1} this material can still get about 80% elongation, which is already much superior to the formability of today's most engineering metals at room temperature. On the other hand, if compared to conventional superplastic forming, the best elongation obtained is still relatively low. Nevertheless, if consider that this alloy is from commercial off-shelf without complex grain refinement processing, and consider the high strain rate applied, the results obtained here are very attractive for industry application.

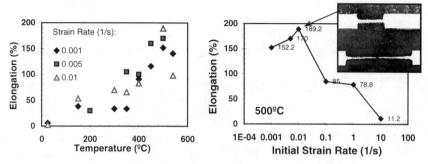

Figure 1. Effect of temperature and strain rate on elongation

3.2 Stress-strain curves: The tensile engineering stress-strain curves at different temperatures are shown in Fig 2. With the increase of testing temperature the flow stress decreases. At room temperature, the material yields at about 200 MPa, and it fractured at about 250 MPa with limited plastic strain. At the temperature from 150-300°C the material exhibits strain-hardening behavior, with an elongation to failure of 40%, a very high necking strain, and a high necking stress to yield stress ratio. Such a yielding and necking behavior of this (hcp) alloy at the high strain rate and elevated temperatures (0.45-0.6 T_m) are much alike that in common (bcc) or (fcc). metals at room temperature (e.g. mild steel and aluminum alloys). Further increasing the temperature (above 350°C) a steady state flow behavior was reached, suggesting a new deformation mechanism took place.

Figure 2(b) shows the engineering stress-strain curves at different initial strain rates and at 500°C (0.84 T_m). At high strain rate (1 s^{-1}) the material exhibits strain-hardening at beginning, and then strain-softening and localized necking, similar to that at lower temperature and lower strain rate (i.e. 150-300°C at 0.01 s^{-1}, see Fig 4a). With the decrease of strain rate at $10^{-2} \sim 10^{-3}$ s^{-1}, after initial yielding at a peak stress the material shows strain softening, with diffused necking and extended elongation, similar to conventional superplastic behavior. The decrease of flow stress is not only associated with the effect of decreased true strain rate during the constant speed test, but due to microstructure evolution during deformation, to be further discussed in the next section.

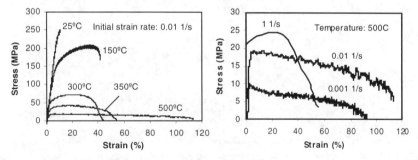

Figure 2. Tensile stress-strain curves at different temperatures

3.3 Microstructure evolution during deformation: Static grain growth occurred at elevated temperature, and it became more severe at above 500°C. Figure 3 shows the initial grain size and static growth at 500°C after 20 min. and after strain to 1.2 (local) at strain rate of 1 s^{-1} (in 10 second). At 500°C there was significant static grain growth, including abnormal grain growth. However, with deformation the grain size decreased significantly, indicating that both dynamic grain refinement and static grain growth took place. The grain shapes were almost equiaxed, suggesting the deformation be dominated by grain boundary sliding and diffusion (because otherwise grains tend to form elongated shape in power law creep or by volume diffusion).

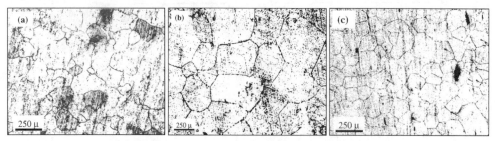

Figure 3. Microstructures of (a) as-received material, and (b) a specimen isothermal heating at 500° for 20 min (shoulder area) and (c) then strained at strain rate of 1 s^{-1} with local strain of 1.2 (gage area).

Grain refinement during deformation at different temperatures was studied with SEM-EBSD, and the results are shown in Figure 4. The initial grain size is further confirmed to be about 250μm, as shown in Figure 4(a). The grain size was reduced to about 30, 50 and 100 μm after deformation at 350, 400, and 500°C, respectively. See Figure 4(b), (c) and (d). This results indicates that dynamic recrystallization is temperature dependent. No obvious microstructure texture was found.

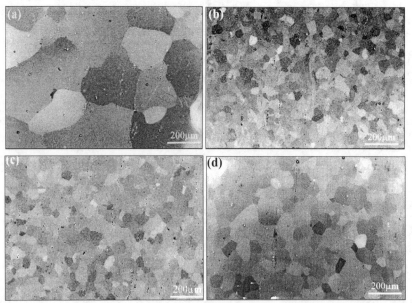

Figure 4. SEM-EBSD observations on the microstructures of Mg alloy (AZ31) as received and after deformation at various temperatures. At strain rate = 0.005 s^{-1} (a) As-received microstructure, (b) Deformed at 300 C, δ =105%, (c) Deformed at 400°C, δ =160%, (d) Deformed at 500°C, δ =170%. (Where δ is elongation)

3.4. TEM study on deformation mechanisms. TEM analysis on the sample after deformation indicates that the density of dislocation is very low. From diffraction patterns, it is known that most grain boundaries are large angle ones, although small angle grain boundaries are occasionally observed. Figure 5(a) shows the grain boundaries in the sample after deformation at 400°C. Boundaries 2 and 3 are large angle GB, while boundary 1 is small angle GB. The boundary marked "A" is sub-grain boundary that is composed of dislocations. On the boundaries 2 and 3, dislocations can be seen (marked with arrows), indicating that grain boundary sliding is an important mechanism. Figure 5(b) shows a dislocation array piled up at the boundary. This indicates that slip is also an important mechanism.

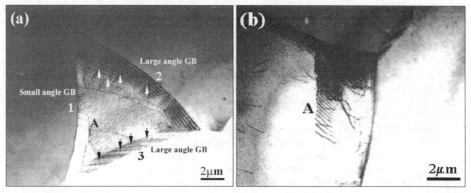

Figure 5. TEM observations of grain boundaries in Mg alloy (ZA31) after deformed at 400°C.

Figure 6 shows the interaction between grain boundaries and dislocations, where the dislocation motion direction can be determine from the bowling direction of dislocation, based on force equilibrium. In Fig 6(a) dislocations were moving towards the grain boundary, and in Fig 6(b), dislocations were moving away from the boundary, indicating that dislocation absorption and emission occurred at grain boundaries.

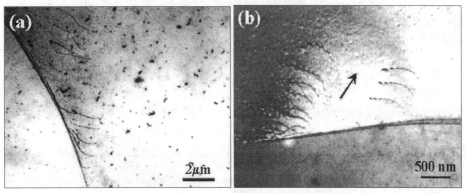

Figure 6. Interaction between grain boundaries and dislocations. (By TEM).
(a) Dislocation absorption, (b) dislocation emission.

4. Discussion

4.1. Dynamic recovery and recrystallization: During deformation at elevated temperatures a high stress concentration will develop at grain boundaries and triple-junction points, and it becomes more severe for large grained material. Continuation of deformation requires stress relaxation, which can be achieved through either recovery or recrystallization. It is generally agreed that dynamic recrystallization occurs most readily in those materials where dynamic recovery is slow, e. g. in materials with medium or low stacking fault energy such as Copper, Nickel and austenitic steels [4,5,6]. But for Mg, the reported values of stacking fault energy are high, similar to that of aluminum [7]. Thus, it is expected that Mg be softened by dynamic recovery, rather than recrystallization. An early review by Sellars [8] indicates that there is little or no evidence for dynamic recrystallization in aluminum, zinc, magnesium, tin, or ferritic steels. In controversy, later work [9, 10], as well as the present work, shows that dynamic recrystallization indeed exists and plays important role during the high temperature high-speed deformation. This is probably related to the constraints imposed by the lack of easily activated slip systems of this (hcp) material rather than to the effect of stacking fault energy. Factors such as relatively high flow stresses observed at very high temperature (see Fig 2), very large grain size, and high grain boundary strength before fracture may also provide chance for recrystallization.

4.2. *Microstructure evolution process:* Based on this study, a possible microstructure evolution process can be described by 3 steps:
(1) Break-down of initial large grains: involving GBS at high angle GB and stress concentration build-up, dislocation emission and absorption at BG and triple junctions;
(2) Subgrain formation, involving the formation of dislocation networks and substructures, and formation of low-angle subgrains;
(3) Formation of small grains, involving subgrain rotation to form large angle GB, superplastic flow by GBS, and dynamic (as well as static) grain growth to large grains. At a critical grain size the process returns to (1) and (2), and a steady-state microstructure and superplastic flow will be established after certain amount of strain, which is temperature and strain rate dependent.

Obviously, this process is very different from conventional recrystallization process involving nucleation and growth of new grains.

5. Conclusions

(1) Commercial large-grained Mg alloy AZ31 was found to have quasi-plastic behavior at high strain rates and high temperature, and 189% elongation was achieved at $1 \times 10^{-2} s^{-1}$ and 500°C.

(2) Optical and SEM-EBSD study reveals that dynamic grain refinement and growth are competing processes determining steady-state grain size and it is train-rate and temperature dependent;

(3) SEM and TEM study suggests that dislocation creep plays a critical role to break down the initial large grains, and the combination of dislocation creep and grain boundary sliding are responsible for large elongation at high temperature and strain rate;

(4) The results show attractive potential for net shape high-rate superplastic forming of Mg alloys.

Acknowledgment

Funding of this study is provided by US National Science Foundation under grant #9970053.

References

1. H. Takuda, S. Kikuchi, N. Hatta, "Possibility of grain refinement for superplasticity of a Mg-Al-Zn alloy by pre-deformation", J. Mat. Sci, 27, 937-940 (1992).
2. K. U. Kainer, "Superplasticity in extruded magnesium alloys," Proceedings of the 3rd International Magnesium Conference, ed. G. W. Lorimer, 10-12 April 1996, Manchester, UK (published by the Institute of Materials), p533-543.
3. X. Wu, Y. Liu and H. Hao, Submitted to Scripta Mat. July 2000.
4. T.Sakai and J. J. Jonas, Acta Metall., 32 (1984) 189-209.
5. R. D. Doherty, D. A. Hughes, F. J. Humphreys, J. J. Jonas, D. Juul Jensen, M. E. Kassner, W. E. King, T. R. McNelley, H. J. McQueen and A. D. Rollett, Mater. Sci. Eng., A238 (1997) 219.
6. F. Gao, Y. Xu, B. Song and K. Xia, Metall. Trans. A, 31A (2000) 21-27.
7. B. Sestak, Proc. 5th Int. Conf. On Strength of Metals and Alloys, Acchen(edited by P. Haasen, V. Gerold and G. Kostorz), 3 (1979) 1461.
8. M. Sellars, Phil. Trans. R. Soc. A288 (1978) 147
9. S. E. Ion, F. J. Humphreys and S. H. White, Acta Metall., 30 (1982) 1909-1919.
10. Mwembela, E. B. Konopleva and H. J. McQueen, Scripta Metall. Mater., 37 (1997) 1789-1795.

Novel Approaches
to Superplasticity in Materials

Unified Theory of Deformation for Structural Superplastics, Metallic Glasses and Nanocrystalline Materials

K.A. Padmanabhan[1] and B.S.S. Daniel[2]

[1] Dept of Materials & Metallurgical, Indian Institute of Technology, Kanpur 208 016, India

[2] Present address: IFW Dresden, Postfach 270016, DE-01171 Dresden, Germany

Keywords: Atomistic Mechanisms, Cooperative Grain Boundary Sliding, Grain Boundary Sliding, Interphase Boundary Sliding, Mesoscopic Grain Boundary Sliding, Metallic Glass, Nanostructured Material, Rate Controlling Process, Structural Superplastics

Abstract

A rate equation for grain/interphase boundary sliding is developed which is able to accurately account for the deformation of structural superplastics, metallic glasses and nanostructured materials on a common physical basis. In some structural superplastics, however, at the highest strain rates dislocation climb controlled creep becomes important. In its present state of development, the model for the optimal range is able to predict all the three constants of the rate equation *ab initio* if the grain size is uniform and constant, the grain shape is simple, e.g., rhombic dodecahedron and the number of grain boundaries that participate in a mesoscopic boundary sliding event is known from experiments. When a grain size distribution is present and the grain shape is not regular, the grain size exponent in the rate equation will have to be obtained empirically (in addition to the number of boundaries involved in a mesoscopic sliding event). Understanding of behaviour in the region where grain deformation co-exits with grain boundary flow is phenomenological at present.

1. Introduction

It has been accepted for a long time that grain / interphase boundary sliding is dominant during structurally superplastic flow. But the belief that this process could control the rate of deformation was confined to a minority, see, for example, [1-3]. Only in the 1990s it was conceded that nearly 100% of the observed strain during optimal superplastic flow arose from grain/interphase boundary sliding [4]. In our view, structural superplasticity at least up to the point of inflection in the log tensile stress (σ) – log tensile strain rate ($\dot{\varepsilon}$) curve (the optimal range) is rate controlled by grain/interphase boundary sliding, the dominant process for whose presence there is very clear experimental evidence. Local boundary migration and intragranular dislocation activity for which also there is experimental evidence are regarded as faster accommodation processes that do not enter the rate equation. In a series of recent publications this idea has been shown to be very viable not only for understanding optimal structural superplasticity but also superplastic flow in metallic glasses and deformation in nanostructured materials [5-15].

2. Crucial Experiment

Optical or electron microscopy can not help decide if boundary migration or intragranular dislocation motion is rate controlling or not. Quantitative assessment of texture changes allows such a deduction as these changes represent the entire thermomechanical history of a material [16, 17]. And so, in a crucial experiment three dimensional orientation distribution functions (ODFs), corrected for the so-called "ghost errors", were determined for Al-6Cu-0.4Zr (SUPRAL) alloy sheets of 1.6mm thickness in the starting, room temperature deformed and superplastically elongated conditions [18]. The orientations g were characterised by the Euler angles ϕ_1, ϕ and ϕ_2 and the ODFs were represented by plotting iso-intensity lines in sections of ϕ_2 = constant ($\Delta\phi_2$ = 5°) through the Euler angle space (Fig.1 a-d). (This rigorous, three dimensional analysis applicable to face centered cubic materials uses the series expansion method in which generalised polynomial functions involving up to 22 coefficients are employed.) Time at test temperature had a negligible effect on texture (Fig.1a). During *room temperature deformation* the initial texture peak of the orientation {001} <110> at the Euler angles $(\phi_1,\phi,\phi_2) = (45°, 0°, 0°)$ readily changed towards the orientation {011}<111> at ~ (55°,45°,0°) (Fig.1b). This orientation change of the tensile direction from <110> towards a direction close to <111> suggests deformation by dislocation slip on crystallographic planes in accordance with the predictions of Taylor poly-slip models for uniaxial stressing [18].

During *superplastic flow* at 730 K, in contrast, the position of the texture peaks practically did not change, but the texture intensity had decreased substantially (Fig. 1c – 175% elongation, true strain $\varepsilon \sim 1$; Fig.1d-660% elongation, $\varepsilon \sim 2$). The observed texture changes clearly indicated that texture annihilation during superplastic flow is mainly the result of grain rotations accompanying *grain boundary sliding* as a *primary deformation mode*. To the best of our understanding, significant grain rotations of the kind observed in these experiments can result *only when* boundary sliding is the *rate controlling process* and not when it is a rapid accommodation step as suggested in some diffusion models including that of Ashby and Verrall [19]. Intragranular dislocation slip control, on the other hand, will predict totally different texture changes. In fact the observed texture annihilation has been successfully modeled at the level of phenomenology by treating it as a *random walk problem* [18]. The texture results also revealed that in some superplastic alloys grain boundary sliding control may persist even beyond the point of inflection in the sigmoidal log σ – log $\dot{\varepsilon}$ curve.

3. An Atomistic Model for Grain/Interphase Boundary Sliding Controlled Flow

In this model a network of deforming grain and interphase boundaries surround grains that do not deform except for what is required to ensure strain and geometric compatibility [5, 6]. Boundaries of high viscosity can be bypassed by grain rotations [18]. A boundary is divided into a number of atomic scale ensembles which contain free volume [20]. Each of these ensembles constitutes a basic unit of sliding. Shear in a basic sliding unit is independent of that present in its neighbours. For calculations, the basic unit located on either side of a boundary is assumed to be an oblate spheroid of ground area πW^2 in the boundary plane and radius on either side of the boundary of (W/2), where W is the grain boundary width. To achieve unit sliding displacement, a short range stress, τ_i, has to be overcome before a neighboring topologically equivalent configuration can be reached by thermally assisted shear. As the basic unit is embedded in a solid matrix, concomitant with the internal shear a transient volume expansion will also occur which, however, will vanish immediately on reaching the new equilibrium configuration. By contrast, the release of the induced shear component will depend on the progress of shear in the vicinity of the basic unit under consideration. It is only after the surroundings have also been sheared by the same amount that the induced shear will be released completely. The free energy of activation for this process will be the sum of the elastic energies of distortion due to shear in the basic unit and the concomitant transient volume expansion. Steric hindrance, e.g., due to a triple junction, will make this kind of sliding at a boundary rather ineffective. This is why during high temperature creep grain boundary sliding leads to stress concentration at triple junctions, cavitation and premature failure. It is suggested that superplastic flow results from *boundary/interphase sliding developing to a mesoscopic scale* (defined to be of the order of a grain diameter or more). Then, two

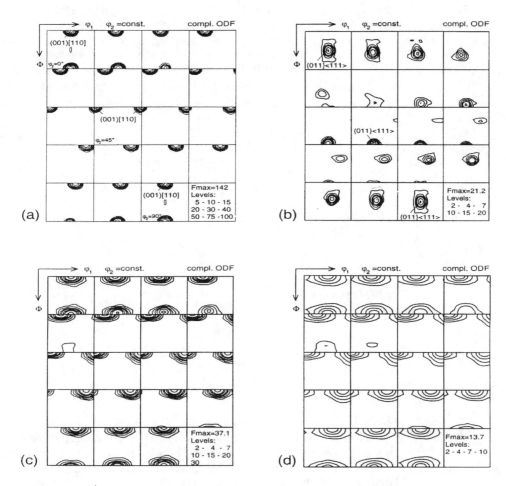

Fig.1. Example of the texture at different states of deformation of an Al-6Cu-0.4 Zr alloy. (a) undeformed state (grip section, soaked at 730K), (b) deformed by 9.2% at room temperature, (c) deformed by 175% at 730K ($\varepsilon \sim 1$), (d) deformed by 660% at 730K ($\varepsilon \sim 2$).

or more boundaries will cooperate to form a plane interface by boundary migration. (By interconnection with other plane interfaces, such a process will lead to long range sliding and superplasticity.) This requires energy expenditure and will give rise to a threshold stress.

3.1. Threshold Stress

To analyse mesoscopic boundary sliding, the polycrystalline structure has been described as a cubic dense packing of equisized spheres [5, 6]. The walls of the Wigner-Seitz cell containing one sphere represent the grain boundaries. This corresponds to a FCC structure on a mesoscopic scale. Since roughness is the least on the {111} planes, they are assumed to be the glide planes for mesoscopic sliding. The grain shape has been assumed to be rhombic dodecahedron. Then, the surfaces of a mesoscopic {111} plane will consist of a regular arrangement of triangular pyramids and pits of height $h = (6^{0.5}/12) L$ and ground area $A = (3^{0.5}/4) L^2$, where L is the grain size. Levelling off the pyramids and pits by grain boundary migration results in an enlargement, ΔA, perpendicular to the boundary, of the grain boundary area per grain. A corresponding (larger) decrease in the grain boundary area will be present along the plane interfaces and this is the driving force for this mesoscopic boundary sliding process [21]. If N is the number of grain boundaries participating in a mesoscopic sliding event, γ_B the specific grain boundary energy, G the shear modulus and α_f is a form factor of the order of unity, τ_o the long range threshold shear stress can be obtained as [6, 9]

$$\tau_o = \left[\frac{2G\gamma_B \overline{\Delta A / A}}{\alpha_f (NA)^{0.5}}\right]^{0.5} \quad (1)$$

with

$$\overline{\Delta A} \text{ (for a grain shape of rhombic dodecahedron)} = \frac{2^{0.5}}{8} L^2 \left(1 - \frac{2 \times 6^{0.5} W}{L}\right) \quad (2)$$

Thus, in a single phase material $\overline{\Delta A}$ (and the resistance to form a plane interface) will vanish at $L = L_o = 2 \times 6^{0.5} W$. Introducing the abbreviation

$$L_1 = 2^{1.5} \times 3^{-0.75} N^{-0.5} \frac{\gamma_B}{G\alpha_f} \quad (3)$$

one obtains $\tau_o = G\left[\frac{L_1}{L}\left(1 - \frac{L_o}{L}\right)\right]^{0.5} \quad ; \quad L \geq L_o \quad (4)$

$$\tau_o = 0; \quad L < L_o \quad (5)$$

$$\tau_o = G\left(\frac{L_1}{L}\right)^{0.5} \quad ; L >> L_o \quad (6)$$

Eq. (4) (or its approximate form Eq. (6)) is the mathematical form of the *inverse Hall – Petch effect,* often discussed in the literature on *nanostructured materials.* In a two phase material at interphase boundaries, when a large difference in chemical composition is present between adjacent grains, dislocation emission from boundary obstacles only can lead to boundary migration. Then, the exponent 0.5 in Eqs. (4) and (6) should be replaced by a constant 'a'; 0 < a < 0.5. The actual value of 'a' will depend on the ratio between the number of intercrystalline and interphase boundaries and their nature. N and 'a' are determined experimentally. While the operation of such a dislocation process at interphase interfaces is possible in a superplastic alloy, in the low nanometer grain size range it will not be possible to nucleate a lattice dislocation from an interphase boundary corner or obstacle. Such boundaries will have to be bypassed by grain rotation [18]. This would imply that when the grain size is of the order of only a few nanometers, boundary sliding will be present at less number of boundaries in a nanocrystalline composite compared with a single phase nanostructured material. This will lead to a decrease in the strain rate sensitivity index, m. Experimental evidence for this prediction exists [22-24].

3.2. Rate Equation

The threshold stress in tension for the occurrence of mesoscopic sliding is $\sigma_o = \sqrt{3}\tau_o$ (von Mises yield behaviour is assumed). Then, an effective stress $\sigma'(=\sigma - \sigma_o)$, where σ is the externally applied tensile stress, would be available for the atomic scale sliding process at a boundary described at the beginning of this section. σ_o for a pseudo-single phase superplastic alloy will be obtained from Eq. (6) and that for a nanocrystalline material from Eq. (4). (When grain size and shape distributions are present the exponent 0.5 in these equations will be replaced by 'a' –see previous section.) The isothermal, isostructural tensile strain rate of flow, $\dot{\varepsilon}$, for both classes of materials will be given by [5,6,10-12]

$$\dot{\varepsilon} = \begin{cases} 0 & ; \sigma \leq \sigma_o \\ A \sinh[B(\sigma - \sigma_o)]; & \sigma > \sigma_o \end{cases} \quad (7)$$

where $A = \left(\dfrac{10 b \varepsilon_o v}{3^{0.5} L}\right) \exp\left(\dfrac{-\Delta F_o}{kT}\right)$ and $B = \dfrac{125 \pi b^3 \varepsilon_o}{12 \times 3^{0.5} \, kT}$.

The other symbols stand for the following: b is the nearest neighbour interatomic spacing in the grain boundary region, v the thermal vibration frequency, ΔF_o the free energy of activation for the rate controlling process, k the Boltzmann constant, T the absolute temperature of deformation and ε_o is the mean tensile strain associated with unit sliding displacement at a boundary. The free energy of activation, in turn, can be expressed as

$$\Delta F_o = \dfrac{1}{2}\left(\beta_1 \gamma^2_o + \beta_2 \varepsilon_o^2\right) G V_o \quad (8)$$

where $\beta_1 = 0.470(1.590 - p)/(1 - p)$, $\beta_2 = 4(1+p)/[9(1-p)]$, with p the Poisson ratio and V_o the volume of the basic unit equal to $2/3\pi W^3$. In the calculations, based on field ion microscopic results W is taken as equal to 2.5 b. From bubble raft experiments, $\varepsilon_o = (1/2)\gamma_o = 0.05$. [6,9]. When the applied stresses are small (sinh x~x), Eq. (7) reduces to

$$\dot{\varepsilon} = \begin{cases} 0 & ; \sigma \leq \sigma_o \\ A'(\sigma - \sigma_o) ; & \sigma > \sigma_o \end{cases} \quad (9)$$

where $A' = A.B$. From elementary thermodynamics $\Delta F_o = Q - T\Delta S$, with Q the activation energy for the atomic scale unit sliding process and ΔS the entropy of the basic sliding transformation.

3.3. Grain Size Dependence of Superplastic Flow

From Eq.(6), for a material in which grain size and shape distributions are present, $\sigma_o = \dfrac{K_1}{L^a}$, where K_1 is a constant equal to $\left[6^{0.5}\left(\dfrac{2}{3}\right)^{0.5a}(G\gamma_B)^{0.5}\left(\dfrac{1}{\alpha_f N^{0.5} 3^{0.25}}\right)\right]^a$ [6,9,15]. Substituting this grain size dependence in Eq. (9), taking logarithms on both sides and truncating the expansion of the resultant logarithmic expression on the right hand side at the second degree, one obtains

$$\ln \dot{\varepsilon}_\sigma \sim A_1 - \ln L - \dfrac{A_2}{L^a} + \dfrac{A_3}{L^{2a}} \quad ; \sigma > \sigma_o \quad (10)$$

Where A_1, A_2, A_3 are constants that have absorbed all other constants. The suffix (σ) indicates that this variable is kept constant. Eq. (10) should account for the grain size dependence of strain rate over a fairly wide range of strain rate and grain size (in the region of optimal superplastic flow).

3.4. Grain Rotation

It can be shown at the level of phenomenology [18] that

$$\phi' = (4S\varepsilon)^{0.5} \tag{11}$$

where ϕ' is the orientation change of a grain, S is the 'orientation diffusion coefficient' and ε is the external tensile superplastic strain. Experimental support for this prediction exists in the case of superplastic flow [18].

3.5. Deformation of Superplastics at High Strain Rates

It will be shown below that for some superplastic alloys Eq. (7) is sufficient to describe the entire sigmoidal $\log \sigma - \log \dot{\varepsilon}$ curve. In contrast, for some systems, where testing has been over a wider range, at the highest strain rates a rate equation describing multiple dislocation slip, e.g., Weertman creep, will take over from Eq.(7) as the rate controlling process. The strain rate due to this process is given by [25]

$$\dot{\varepsilon}_{dis} = \frac{\alpha_1 D \Omega}{G^{3.5} kT} \sigma^{4.5} \quad \text{(applicable only at large stresses)} \tag{12}$$

where $\dot{\varepsilon}_{dis}$ is the tensile creep rate due to climb controlled multiple dislocation slip, α_1 is a constant of value $5 \times 10^{20} cm^{-2}$ when the stress exponent, n(=1/m), is 4.5, D is the bulk diffusion coefficient and Ω is the atomic volume.

4. Superplastic Deformation of Metallic Glasses

Superplastic flow has also been reported in metallic glasses [26, 27]. Here too the steady state log stress- log strain rate curve is sigmoidal. It is presently suggested that the entire $\log \sigma - \log \dot{\varepsilon}$ plot in this case can be accounted for using Eq. (7). The regions of relatively lower density where deformation is concentrated in metallic glasses are analogous to the grain / interphase boundaries. The islands that are surrounded by these regions of intense deformation and within which there is negligible deformation are comparable to the grains of a superplastic alloy. The nature of development of the shear process described in this model, which leads to Eq. (7), only involves shear stress aided redistribution of free volume and such a process can apply equally to micro / nanocrystalline materials as well as to metallic glasses. In truly nanocrystalline materials too if the $\log \sigma - \log \dot{\varepsilon}$ curve extends beyond the point of inflection, the entire sigmoidal curve will have to be accounted for using Eq. (7).

5. Validation of Ideas

Computer simulation based on molecular dynamics has revealed that during the deformation of a nanocrystalline material both grain boundary sliding and grain rotation are present. The deformation features have been considered to be similar to what is described in the present grain boundary sliding controlled flow model [28, 29].

5.1 Experimental Validation
When the method of least squares is used in the form

$$\sum_{i=1}^{N_1} (Y_{iobs} - Y_{ipred})^2 = \min. \tag{13}$$

where N_1 is the number of observations, an implicit assumption is that the variance is constant for all the observations. If such a scheme were to be used over a wide range, a greater weight gets attached to observations of larger magnitudes compared with those of lower values. Then, even when Eq. (13) is satisfied, very poor fit will be obtained in the lower ranges of the domain of interest. This problem is overcome by dispensing with the assumption of constant variance and minimising the function [30]

$$\sum_{i=1}^{N_1}\left(\frac{Yiobs - Yipred}{Yiobs}\right)^2 = min. \qquad (14)$$

Experimental data available in the literature were analysed using Eq. (7) and the minimization scheme described by Eq. (14). Eq. (7) described accurately the entire sigmoidal curve (all three regions of superplastic flow) in case of some alloy systems. Fig. 2 is an example. It can be seen that the fit in case of this Al – 33Cu – 0.4Zr alloy [31] is excellent : standard error 0.004; coefficient of correlation 99.68%. However, in some other systems, e.g., Al – 12Si eutectic alloy [32], Eq. (7) accounted for data well beyond the point of inflection, but not the entire curve. The applicability of Eq. (7) beyond the point of inflection as seen presently is consistent with the texture annihilation results (see Section 2) [18] which suggest that in this strain rate range too grain / interphase boundary sliding is rate controlling. However, at still higher stresses / strain rates multiple dislocation slip emerges as a dominant process. The above points are illustrated in Fig. 3 where at the highest strain rates the grain boundary sliding based model predicts strain rates faster than those given by the dislocation climb controlled model (Eq. 12) with a stress exponent of 4.5. Evidently, in this highest strain rate range, Eq.(12) is rate controlling. It is seen that the fit is excellent in this case also. As mentioned in Section 4, Eq. (7) can also be used to predict superplastic flow in metallic glasses. Fig. 4 pertaining to metallic glass Zr65 Al10 Ni10 Cu15 [29] is an example. Once again the fit is very good. In a series of publications [11-14] it has been demonstrated already that grain / interphase boundary sliding controlled flow (Eq. (7)) is very useful in explaining the deformation of nanostructured materials. The grain size dependence of superplastic flow, as predicted by Eq. (10), is verified in Fig. 5 right into the sub-micrometer range. This result reveals that the expression for the grain size dependence of superplastic flow rate is rather complicated.

Finally, it is essential to show that the best fit values of A, B, σ_o and ΔF_o in Eqs. (4), (6), (7) and (8) obtained by fitting the experimental data in terms of these equations are similar to those calculated *ab initio* using the values from literature for the micro-level constants that are included in A, B, σ_o and ΔF_o. This has been done successfully [9, 10, 11, 12, 14, 15].

6. Concluding Remarks

In region III of superplastic flow (e.g., the highest strain rates in Fig.3 at which Eq.(12) has been found to apply), depending on alloy composition and test conditions dislocation arrangement could be in the form of single pile-ups, tangles, networks or cells [3]. As pointed out by Weertman [25], his analysis applies to sub-grains as well as to ordinary grains, provided that everywhere along their lengths the dislocations are good vacancy sources or sinks. But when grains contain sub-grains, the size of the latter should be used in the analysis for deriving strain rates instead of the former. This analysis is partly phenomenological and with suitable assumptions, a stress exponent (n) anywhere in the range of 3 to 6 can be obtained. The case n = 4.5, $\alpha_1 = 5\times10^{20}$ cm^{-2} was obtained empirically by examining the experimental data pertaining to a number of metals. So, the fit obtained at the highest stresses in Fig.3 is empirical and the conclusion is that the superplastic response here is similar to what was seen earlier by Weertman in case of some metals.

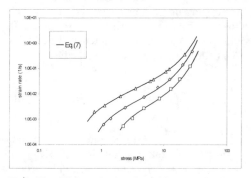
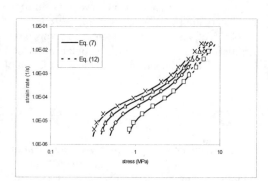

Fig.2. Sigmoidal steady state log strain rate – log stress relationship in an Al–33Cu – 0.4Zr alloy. Experimental points are after Matsuki et al. [31]. The theoretical prediction (full curves) is based on grain /interphase boundary sliding process control Eq. (7).

Fig.3. Sigmoidal steady state log strain rate- log stress relationship in an Al – 12 Si alloy. Experimental points are after Chung and Cahoon [32]. The theoretical prediction (full curves) is based on Eq. (7) as the rate controlling process at all strain rates except the highest ones where the flow rate is governed by Eq. (12).

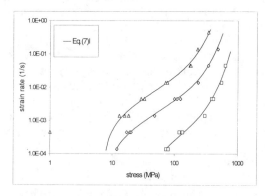
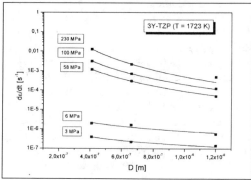

Fig.4. Sigmoidal steady state log strain rate – log stress relationship in a metallic glass Zr65Al10Ni10Cu15. Experimental points are after Kawamura et al. [27]. The theoretical prediction is based on Eq. (7).

Fig.5. Logarithmic strain rate versus grain size data from [33] for 3Y – TZP at 1723 K and different stress levels. The data were fitted using Eq. (10) (solid lines).

Strictly speaking, grain deformation and grain boundary sliding are processes that are independent of each other and so both will contribute to the external strain rate, once the stress level beyond which dislocation sources within the grains start to operate is crossed. However, to ensure strain compatibility, the strain rates arising out of both grain and grain boundary deformation should be matched. (This may require a modification of both Eqs.(7) and (12).) This is yet to be done. As grain boundary sliding predicts a grain size dependence, such an approach will also be able to account for the grain size dependent mechanical response seen in region III.

Materials scientists/physicists capture the correct physical details in a model at least qualitatively and when some of the predictions agree with experimental observations at the macrolevel, the model is considered to be accurate. In the present case, the derivations concern a uniaxial stress system. Generalising this to cover the multi-axial case is non-trivial. Often, the scalar properties of a constitutive equation for the 3D case are obtained through some of the tensor invariants. A common assumption is to regard the superplastic alloy to be a von Mises solid. The vector properties are obtained through St. Venant type of equations. It is not clear how relevant the proportionality between a stress and a strain rate vector component assumed in such equations is for superplastic flow. Also, even hydrostatic stresses of the order of 0.5 times the (rather low) flow stress affects superplastic flow while the deformation of conventional materials is not affected by hydrostatic stresses of large magnitude. Then, the assumption of a von Mises solid, whose scalar properties depend only upon the second tensor invariant, becomes questionable. An added complication is that most constitutive equations are written for infinitesimally small strains, but superplastic flow is associated with finite strain behaviour. In the latter case these are no unique measures for the stress and the strain rate tensors. All these aspects are discussed in a forthcoming book [34].

An examination of Eqs.(7) and (9) reveals that they describe the flow behaviour of *a solid* and not a non-newtonian viscous liquid, as is erroneously suggested by the equation $\sigma = K\dot{\varepsilon}^m$.

The hyperbolic sine function for the strain rate tensor obtained in Eq.(7) may be regarded by some as 'artificial'. However, it is emphasised that *all* theories of high temperature deformation based on the kinetic theory of solids/transition state theory(which has a firm basis in statistical mechanics) lead to such a form for the rate equation. At low stresses, it reduces to the linear Eq.(9). A quadratic approximation, which has been demonstrated to be superior, has also been obtained [6,10]. How these latter two cases can be generalised for the 3D case and the transient regions of flow are discussed elsewhere [34]. But here the hyperbolic sine function has been found to describe the entire sigmoidal curve in many cases and hence this form is very useful. How to generalise this type of equation for the 3D case will form a part of a future investigation.

For generalising the rate equation for the 3D case, the strain rate is identified with its deviatoric part. (Volume constancy is assumed.) How such equations are to be handled for both infinitesimally small strains and finite strain behaviour is discussed in the book mentioned above.[34].

Finally, experimental determination of texture corresponding to different deformed states is based on the series expansion method [18] which takes into account the evolution of the symmetry group. But the model for texture annihilation based on an analogy with the random walk problem is phenomenological[18] and the details of the physical picture are yet to be presented.

Acknowledgements

This work was supported by the Indo-US project No. N00014-97-1-0993 sponsored by the Department of Science and Technology, Government of India and the Office of Naval Research in the Naval Research Laboratory, Washington, D. C., USA.

References

[1]. K. A. Padmanabhan, Ph.D. thesis, University of Cambridge, UK (1971).
[2]. K. A. Padmanabhan, Mater. Sci. Eng. 29 (1977) p.1.
[3]. K. A. Padmanabhan and G. J. Davies, Superplasticity, Springer Verlag, Heidelberg-Berlin- New York (1980).
[4]. R. Z. Valiev and T. G. Langdon, Acta Metall. Mater. 41 (1993) p. 949.
[5]. K. A. Padmanabhan and J. Schlipf. Proc. First Int. Conf. 'Transport Phenomena in Processing', Technomic Publishing Co., Lancaster, USA (1993) p. 491.
[6]. K. A. Padmanabhan and J. Schlipf, Mater. Sci. Technol. 12 (1996) p. 391.
[7]. V. V. Astanin, S. N. Faizova and K. A. Padmanabhan, Mater. Sci. Technol. 12 (1996) p. 489.
[8]. V. V. Astanin, K. A. Padmanabhan and S. S. Bhattacharya, Mater. Sci. Technol. 12 (1996) p.545.

[9]. T. A. Verkatesh, S. S. Bhattacharya, K. A. Padmanabhan and J. Schlipf, Mater. Sci. Technol. 12 (1996) p. 635.
[10]. F. U. Enikeev, K. A. Padmanabhan and S. S. Bhattacharya, Mater. Sci. Technol. 15 (1999) p. 673.
[11]. K. A. Padmanabhan, R. Nitsche and H. Hahn, Fourth Euro. Conf. Advanced Materials and Processes (Euromat 95), Symp. G: Special and Functional Materials, Associazone Italiana Di Metallurgia, Milano (1995) p. 289.
[12]. H. Hahn and K. A. Padmanabhan, Philos. Mag. 76B (1997) p.559.
[13]. H. Hahn, P. Mondal and K. A. Padmanabhan, Nanostructured Materials 9(1997) p. 603.
[14]. K. A. Padmanabhan, J. Metastable and Nanocrystalline Materials 8 (2000) p.753.
[15]. U. Betz, H. Hahn and K. A. Padmanabhan, unpublished work.
[16]. J. W. Edington, K. N. Melton and C. P. Cutler, Prog. Mater. Sci. 21(1976) p.61.
[17]. K. A. Padmanabhan and K. Luecke, Z. Metallk. 77 (1986) p. 765.
[18]. O. Engler, K. A. Padmanabhan and K. Luecke, Modelling and Simulation in Mater. Sci. Eng. 8 (2000) 477.
[19]. M. F. Ashby and R. A. Verrall, Acta Metall. 21 (1973) p. 149.
[20]. D. Wolf, Acta Metall. 38 (1990) pp 781; 791.
[21]. K. A. Padmanabhan, Int. Conf. Superplasticity in Advanced Materials (ICSAM-97) Trans. Tech. Publ., Aedermannsdorf, Switzerland (1997) p.1.
[22]. H. Hahn and K. A. Padmanabhan, Nanostructured Materials 6 (1995) p. 191.
[23]. H. Hahn and K. A. Padmanabhan, 'Advanced Materials Processing', PRICM-2, The Korean Institute of Metals and Materials, Kyangju, Korea, 3(1995) p. 2119.
[24]. K. A. Padmanabhan and H. Hahn, Proc. Symp. 'Synthesis and Processing of Nanocrystalline Powder', The Minerals, Metals and Materials Society, Pittsburgh, USA (1996) p.21.
[25]. J. Weertman, Trans. ASM 61 (1968) 681.
[26]. Y. Kawamura, T. Shibata, A. Inoue and T. Masumoto, Appl. Phys. Lett. 69(1996) p.1208.
[27]. Y. Kawamura, T. Shibata, A. Inoue and T. Masumoto, Scripta Mater. 37(1997) p.431.
[28]. H. Van Swygenhoven and A. Caro, Nanostructured Mater. 9 (1997) p. 669.
[29]. H. Van Swygenhoven and A. Caro, ' Nanophase and Nanocomposite Materials II', Mater. Res. Soc., Pittsburgh, USA (1997) p. 457.
[30] N. R. Draper and H. Smith, Applied Regression Analysis, Second Edition, John Wiley, New York (1981) pp. 108 – 116.
[31]. K. Matsuki, K. Minami, M. Tokizawa and Y. Murakami, Met. Sci. 13 (1979) p. 619.
[32]. D. W. Chung and J. R. Cahoon, Met. Sci. 13 (1979) p. 635.
[33]. D. M. Owen and A. H. Chokshi, Acta Mater. 46 (1998) p.667.
[34]. K. A. Padmanabhan, R. A. Vasin and F. U. Enikeev, Superplastic Flow: Phenomenology and Mechanics, Springer Verlag, Heidelberg-Berlin- NewYork (in press).

Corresponding Author: K. A. Padmanabhan, Director@iitk.ac.in, kap@iitk.ac.in

Characterization of Grain Boundary Properties in Superplastic Al Based Alloys Using EBSD

I.C. Hsiao and J.C. Huang

Institute of Materials Science and Engineering, National Sun Yat-Sen University,
Kaohsiung 804, Taiwan ROC

Keywords: Al Alloy, CSL Boundaries, EBSD, Grain Boundary, Low Temperature Superplasticity, Misorientation Angle

Abstract

The grain boundary characteristics in the 1050, 5052, 5083, 6061 and 1420 LTSP Al alloys were examined using EBSD as a function of TMT and LTSP deformation true strain from 1.5-12.0. The grain boundary character was observed to evolve continuously during TMT and LTSP deformation. With increasing accumulated deformation strain, the bimodal grain boundary misorientation distribution gradually transformed into semi-random behavior. For different alloys under cold or warm working, the minimum working strain to achieve a semi-random grain boundary misorientation distribution favorable for GBS and LTSP varied accordingly. Only minor variation was observed for the CSL boundary fraction during TMT and LTSP deformation. Still over 80% of the MAB+HAB belonged to random higher angle boundaries that could proceed GBS. No apparent relationship was seen for the CSL population and LTSP performance.

Introduction

The evolution of micro-texture and grain boundary misorientation in superplastic or non-superplastic aluminum or copper alloys have been examined by X-ray, transmission electron microscopy or electron backscattered diffraction (EBSD) associated with scanning electron microscopy [e.g. 1-3]. Numerous interesting observations on the grain orientation evolution as a function of deformation strain level at room or elevated temperatures have provided important evidence for acting deformation mechanisms during thermomechanical treatment (TMT) and tensile loading.

Recently, there have been a number of aluminum alloys that were processed to exhibit low temperature and/or high strain rate superplasticity (LTSP and/or HRSP). For example, the 1420 Al-Mg-Li alloy was processed via warm equal channel angular pressing (ECAP), resulting in LTSP and HRSP of ~910% at 350 °C and 1×10^{-1} s^{-1} [4]. Also, the 5083 alloy has been processed via warm rolling-type TMT and exhibited LTSP of 511% at 230 °C and 2×10^{-3} s^{-1} and 300% at 230 °C and 1×10^{-2} s^{-1} [5,6]. Such LTSP Al materials can be formed through deep drawing or hydroforming using hot silicon oil at 200-300 °C, rather than using the more expensive Ar-gas blow forming for high temperature superplastic (HTSP) Al sheets at 450-550 °C. While the texture and grain boundary evolution during HTSP has been studied by a number of investigators, the evolution during LTSP receives much less attention. This paper compares the variation of grain boundary characteristics in such LTSP aluminum alloys, and the results can be compared with those obtained from the HTSP ones.

Experimental

The grain boundary characteristics in various commercial Al alloys subjected to severe TMT or ECAP to a true strain level of 1.5-12.0 have been examined using EBSD. The alloys included the industry-pure Al 1050, Al-2.5Mg 5052 and Al-4.5Mg-0.7Mn 5083, Al-1Mg-0.6Si 6061 and Al-5.2Mg-2.0Li 1420 alloys. The details of the TMT or ECAP processing have been presented elsewhere [5,7]. Published results on the 1420 alloy [8] are included in this paper for overall comparison. The grain structure and grain-orientation distribution were examined using a JEOL 6400 SEM, equipped with an Oxford Link Opal EBSD system. The spatial resolution limit of the EBSD was around 0.3 μm. The data were gathered from at least three separate scanned areas measuring ~40x40 μm, each including 500-1000 grain boundaries. The reported results were the average values. The boundaries with angles <10° are termed as low angle boundaries, LAB, unlikely to motivate grain boundary sliding (GBS). Those >30° are classified as high angle boundaries, HAB, most favorable for GBS. The rest within 10-30° are called as medium angle boundaries, MAB, which can be viewed as an index for the severity of bimodal behavior.

Results

(A) Grain boundary misorientation distribution

During the initial stage of TMT or ECAP processing, the Al alloys, independent of cold or warm working, all exhibited apparent bimodal grain boundary misorientation distribution (Fig. 1a). With increasing deformation strain, the bimodal behavior gradually diminished and the distribution was replaced by a semi-random appearance (Fig. 1b), with LAB decreasing and approaching ~2%, HAB increasing and approaching ~79% and MAB approaching a stable level of ~19%, characteristic of complete random boundary misorientation distribution. Cold deformation usually induced a severer bimodal distribution than the warm counterpart to the same strain. An example is presented in Table 1 for the 5083 alloy to a true working strain of 2.7. Generally, cold working would produce a higher fraction of LAB and a lower fraction of HAB, as well as a lower MAB fraction reflecting the severer bimodal behavior.

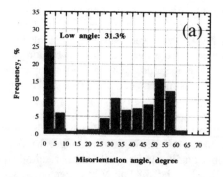 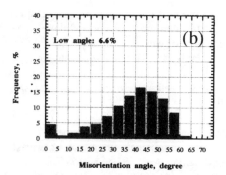

Fig. 1 Representative grain boundary misorientation distributions observed in Al alloys subjected to (a) TMT true strain ~2.5, and (b) TMT strain ~5.6.

Based on the experimental results obtained from the cold or warm worked aluminum alloys, it was found that the grain boundary evolution continuously proceeded during TMT processing and subsequent LTSP tensile loading. Figure 2 shows the variation of LAB, MAB and HAB fractions in the 5083 alloy as a function of warm-TMT (Fig. 2a) and LTSP (Fig. 2b) true strain, both at similar deformation temperatures ~200-250 °C. If the accumulated true strains received during TMT and LTSP are combined, it can be seen from Fig. 3 that the evolution follows a continuous trend. This behavior was observed in a variety of aluminum alloys during TMT, ECAP and LTSP at 150-300 °C without rapid recrystallization, as depicted in Fig. 4.

Table 1 Comparison between the grain boundary misorientation distribution in the 5083 alloys after cold or warm TMT strain of 2.7.

TMT type	0°-15°	15°-65°	LAB	MAB	HAB
Cold rolling	37.5%	62.5%	36.2%	6.4%	57.4%
Warm rolling	30.1%	69.9%	28.0%	9.5%	62.5%

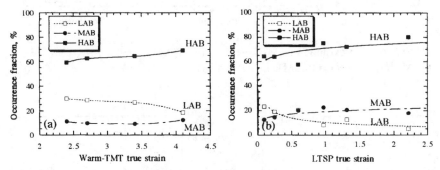

Fig. 2 Variation of LAB, MAB and HAB occurrence fractions as a function of (a) warm-TMT and (b) LTSP true strain observed in the 5083 alloy.

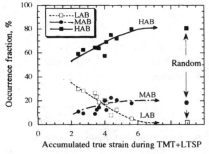

Fig. 3 Illustration of the continuous evolution trend of LAB, MAB and HAB when the accumulated true strains from TMT and LTSP are added.

On the base of massive experimental data, the fitted lines for the overall evolution trends can be drawn, as the examples shown in Fig. 5 for LAB and HAB. From experience, it was known that once the materials possessed a semi-random grain boundary misorientation distribution, GBS could proceed more effectively and the materials could exhibit reasonable LTSP behavior. Thus we can set a level for semi-random distribution, as the dashed horizontal lines in Fig. 5a for LAB and Fig. 5b for HAB. From these dashed lines, the minimum working true strain needed for that specific Al alloy could be roughly determined. For cold deformation at room temperature, the 1050 Al needed a true strain level ε~7.5 to reach semi-random distribution, while the 5052, 5083 and 6061 Al required ε~9.0, 9.5 and 12.5, respectively, as summarized in Table 2. The purer and simpler aluminum alloys would evolve more rapidly. And warm deformation at 200-400 °C would proceed even faster. In Table 2, it can be seen that, under the optimum condition, most alloys required a warm working true strain ε~5.0± 0.5 to reach semi-random grain boundary characteristics so as to result in satisfactory LTSP elongations. The EBSD results might be used as one of the indexes for LTSP performance.

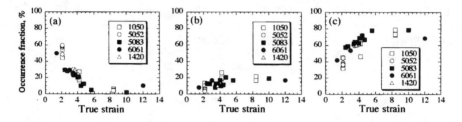

Fig. 4 Summary of evolution trends for (a) LAB, (b) MAB, and (c) HAB observed in various Al alloys subjected to TMT, ECAP and LTSP.

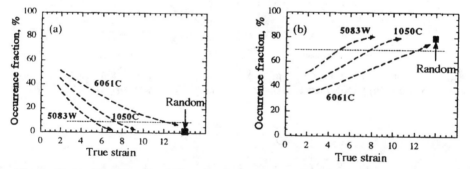

Fig. 5 Examples of the fitted lines for the evolution of (a) LAB and (b) HAB in cold deformed 1050 and 6061 alloys and warm deformed 5083 alloy. The dashed horizontal lines represent the semi-random situation.

Table 2 Minimum deformation true strain levels required for different Al alloys to reach the semi-random grain boundary misorientation distribution

Alloy	Cold deformation	Warm deformation
1050	~7.5	--
5052	~9.0	~5.5 (at 200 °C)
5083	~9.5	~5.0 (at 250 °C)
6061	~12.5	~4.5 (at 300 °C)
1420	--	~5.7 (at 400 °C)

The gradual evolution of the grain micro-texture and grain boundary misorientation distribution during LTSP of Al alloys at 200-300 °C was similar to that occurred during HTSP at 450-550 °C [1-3]. This suggests that, even though dislocation activity should have been involved during LTSP, the dominant deformation mechanism which contributes most of the LTSP strain is still GBS. The combined effects from GBS, grain rotation and grain boundary migration lead to the disappearance of strong β-fiber texture and bimodal grain boundary misorientation distribution in the as-TMT specimens into much more random orientation distribution at large strains.

(B) Special coincidence site lattice boundary

Among the MAB and HAB, some of the boundaries belong to special coincidence site lattice (CSL) boundary, or called the Σ boundaries. These special boundaries possess special matching relationship between neighboring grains, and are not considered to be the boundaries favorable for GBS even though the misorientation angles are greater than 10°. Thus the characterization and quantification of the CSL boundaries becomes necessary.

There were only minor overall variations of the total Σ boundary population as a function of alloy chemical composition or deformation condition, with an average fraction of 11.5%±2.5% of the total grain boundary population. Some of the representative data are listed in Table 3. Among all of the CSL boundaries, $\Sigma=3$ was always the most prevalent one, the next could be $\Sigma=9$, followed by $\Sigma=11$, 7, or 17b. Around one-half of the CSL boundaries belonged to the $\Sigma=3^n$ twin-related categories, accounting for about 5.8%±1.5%.

The Σ/(MAB+HAB) ratios remained relatively fixed irrespective of alloy type or deformation strain level, scattering around 16.6%±3.1%. In other words, still over 80% of the MAB+HAB belonged to random high angle boundaries along which GBS could proceed. The overall average ratio of $\Sigma=3^n$/(MAB+HAB) was 7.3%±2.8%, and this average ratio seemed to increase slightly from 6.1 to 7.9 with decreasing stacking fault energy in various Al alloys. For example, the overall average ratio for the 1050 alloy was 6.1, and the average values for the 5052, 5083 and 6061 alloys were 7.2, 7.6 and 7.9, respectively. Finally, the minor variations in total Σ and $\Sigma=3^n$ boundaries in these alloys did not seem to correspond to any direct relationship with the low temperature superplastic elongation performance.

Summary

The grain boundary characteristics in various Al alloys were seen to evolve continuously during TMT and LTSP deformation. With increasing accumulated deformation strain, the bimodal grin boundary misorientation distribution gradually transformed into semi-random behavior, with LAB, MAB and HAB fractions approaching ~2%, 19% and 79%, respectively. For different alloys under cold or warm working, the minimum working strain to achieve a semi-random grain boundary misorientation distribution favorable for GBS and LTSP varied accordingly. The pure and simpler Al alloys would evolve more rapidly. Only minor variation was observed for the CSL boundary fraction during TMT and LTSP deformation. Still over 80% of the MAB+HAB belonged to random higher angle boundaries that could proceed GBS. No apparent relationship was seen for the CSL population and LTSP performance.

References

[1] M. T. Perez-Prado, M. C. Cristina, M. Torralba, O. A. Ruano, and G. Gonzalez-Doncel, Scripta Mater., 35 (1996), p. 1455.
[2] J. Liu and D. J. Chakrabarti, Acta Mater., 44 (1996), p. 4647.
[3] T. R. McNelley and M. E. McMahon, Metall. Mater. Trans., 27A (1996), p. 2252.
[4] R. Z. Valiev, D. A. Salimonenko, N. K. Tsenev, P. B. Berbon, and T. G. Langdon, Scripta Mater., 37 (1997), p. 1945.
[5] I. C. Hsiao and J. C. Huang, Scripta Mater., 40 (1999), p. 697.
[6] I. C. Hsiao and J. C. Huang, on-going unpublished research, National Sun Yat-Sen University, Taiwan, (2000).
[7] P. L. Sun, P. W. Kao, and C. P. Chang, Proc. 1999 Annual Conf. Chinese Society for Mater. Sci., Tsinchu, Taiwan, (1999), p. B-05-1.
[8] M. Furakawa, Y. Iwahashi, Z. Horita, M. Nemoto, N. K. Tsenev, P. Z. Valiev, and T. G. Langdon, Acta Mater., 45 (1997), p. 4751.

Table 3 Some representative measurements of the CSL boundaries for various Al alloys. All values given are expressed in %.

Alloy	1050*	1050*	5052	5083	5083	5083	6061	6061
d (μm)	~0.7	~0.4	~0.6	~0.5	~0.5	~0.5	~1.0	~0.4
Accumulated strain	ε=4.2	ε=8.4	ε=3.1	ε=3.4	ε=4.7	ε=5.6	ε=1.4	ε=12.0
Σ=3	4.6	4.8	1.9	4.2	4.0	4.3	2.2	5.1
Σ=5	0.8	0.6	0.1	0.7	1.5	0.6	0.5	0.8
Σ=7	0.4	1.1	0.2	0.6	1.0	1.0	0.3	0.9
Σ=9	1.2	0.6	1.7	2.3	1.3	0.6	2.2	1.1
Σ=11	0.2	0.4	1.6	1.9	--	1.3	1.3	0.8
Σ=13a	0.4	0.3	0.1	0.1	0.5	0.1	--	--
Σ=13b	0.4	0.3	0.7	0.1	0.3	0.1	0.5	0.7
Σ=15	0.2	0.4	0.1	0.3	0.5	0.6	--	0.1
Σ=17a	--	0.1	0.1	0.1	--	0.1	--	--
Σ=17b	0.3	0.1	0.7	0.8	0.8	0.3	0.8	0.9
All Σ	10.4	11.6	8.9	12.5	12.4	10.7	8.6	13.5
Σ=3^n	6.0	5.6	3.6	6.7	5.3	4.9	4.4	6.3
(Σ=3^n)/(All Σ)	57.7	48.3	41.0	53.6	42.7	45.8	51.2	46.7
MAB+HAB (10°-65°)	72.9	95.0	48.1	60.0	87.5	95.0	53.0	85.7
(All Σ)/(MAB+HAB)	14.3	12.2	18.5	20.8	14.2	17.8	16.2	15.8
(Σ=3^n)/(MAB+HAB)	8.2	5.9	7.5	11.2	6.1	5.2	8.3	7.4

* Data from Ref. 7.

Atomistic Simulation of the Effect of Trace Elements on Grain Boundary of Aluminum

S. Namilae[1], C. Shet[1], N. Chandra[1] and T.G. Nieh[2]

[1] Department of Mechanical Engineering, FAMU-FSU College of Eng., Florida State University, Tallahassee FL 32310, USA

[2] Lawrence Livermore Natioanl Laboratory, L-350, PO Box 808, Livermore CA 94551-9900, USA

Keywords: Grain Boundary Energy, Hydrostatic Stress, Segregation

Abstract
Molecular statics simulations are used in conjunction with the Embedded Atom Method to study the segregation tendency of magnesium and the effect of doping on aluminum symmetric tilt grain boundaries (STGB). It is observed that a state of hydrostatic stress exists predominantly near the grain boundary and influences the segregation energy of magnesium. The origin of hydrostatic stress is explained in terms of local geometrical arrangement of atoms at the grain boundary. The energy of a given grain boundary is found to increase with addition of Mg atoms and the increase depends on the normal distance of the atom from the grain boundary.

Introduction
Impurities in minor quantities are known to affect the deformation behavior in superplastic materials [1]. It has also been observed through experiments and simulations that grain boundary sliding is affected by the geometry and orientation of grain boundaries [2,3]. It is well known that the certain trace elements have a propensity to segregate to the grain boundaries. This segregation might affect the deformation behavior of grain boundaries, however the effect of structure and energy of grain boundary on the segregation tendencies is not well understood. The purpose of the paper is to examine the segregation behavior and its effects by performing atomic simulations.

Atomic simulations used along with experimental techniques such as HRTEM and STEM can provide a better insight into grain boundary processes like segregation and sliding. In this paper, molecular statics using the Embedded Atom Method (EAM) potentials [4] is employed to study the effect of magnesium on various symmetric tilt grain boundaries of aluminum.

Simulation procedure
Embedded atom method (EAM) [4] potentials implicitly include many body interactions and have been proven more reliable than conventional pair potentials in representing the atomic interactions in metals. The EAM potential for single atom is given by

$$E_i = F(\rho_i) + \tfrac{1}{2}\sum_{j \neq i} \phi(r_{ij}) \tag{1}$$

The first term in Eq.1 represents the energy required to embed an atom in a cloud of electron density, and the second term is pair interactions. In molecular statics, the total energy is minimized to give the equilibrium position of all the atoms in the given crystal. In the present work, bicrystals with about 5000 atoms were modeled with one Mg atom placed at various positions near the grain boundary. For each of the bicrystal studied, a grain is first generated based on its orientation. The grain boundary in the bicrystals was generated by the reflection operation in half of the simulation block It is necessary to remove or add atoms at the grain boundary because, multiple energy minima may exist with a similar structure [5]. The resulting configuration is equilibrated by molecular statics. The crystals are periodic in X and Z directions, free in Y direction. To avoid interference of free surface effects, crystals are designed to have an adequate number of atomic layers in Y direction. EAM 0functions developed by Oh and Johnson [6] are used in all the simulations. A molecular statics-dynamics code DYNAMO [7] developed by Sandia National Laboratory, Livermore is used. Four frequently studied low Σ grain boundaries [110]Σ3 (1,1,1), [001]Σ5(2,1,0), [110]Σ9(2,21) and [110]Σ11(1,1,3) are considered in this work. Fig 1 illustrates schematic of a typical grain boundary. Fig 2 shows the equilibrium structures of the grain boundaries studied.

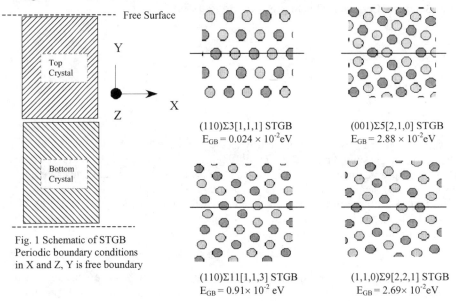

Fig. 1 Schematic of STGB Periodic boundary conditions in X and Z, Y is free boundary

Fig. 2 Equilibrium structures of grain boundaries studied

Mg atoms are substituted in the place of Al atoms at various locations near the grain boundary. Segregation energy (E_S) for Mg at these positions was determined as:

$$E_S = E_B - E_0 \qquad (2)$$

where E_0 is energy of atom in bulk and E_B is energy of atom at grain boundary. Hydrostatic stress state of atoms at the grain boundary was calculated from the local stress tensor.

Grain boundary energy (E_{GB}) is calculated as the difference between energy of all the atoms in the grain boundary region and a perfect crystal containing same number of atoms, divided by the area of the grain boundary. Grain boundary energies of pure Al grain boundaries (See Fig.2) calculated in this work compare well with earlier simulations and experiments [3].

Hydrostatic Stress and Segregation Energy

The actual positions of Mg atoms influence the magnitude of the segregation energy irrespective of the type of STGB. This motivated us to look into other factors affecting it like stress fields, especially the hydrostatic stress field. Variation of hydrostatic stress with distance from grain boundary is shown in Fig.3. It can be observed that in general there is alternating tension and compression along the boundary. Origin of hydrostatic stress can be explained by considering the geometrical distribution of atoms around the stressed atoms. Fig. 4 shows radial distribution of various atoms at the grain boundaries and in bulk. It is known that an atom in FCC crystal structure is surrounded by twelve nearest neighbors and six next nearest neighbors. Atom 1a (shown in Fig.5) in Σ5 grain boundary is in a state of high compressive stress (-0.93 eV/Å^3) surrounded by 14 atoms instead of 12. Similarly, atom 2a in Σ9 boundary is also surrounded by 14 nearest neighbors and is in a state of compression. Atoms 2b and 2a, which are in tensile state are surrounded by 10 nearest neighbors.

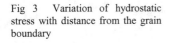

Fig 3 Variation of hydrostatic stress with distance from the grain boundary

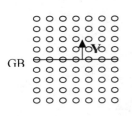

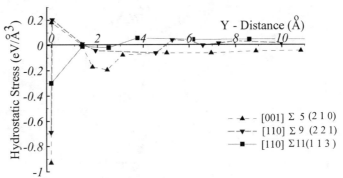

In the case of Σ11 boundary, atom 1c is in a state of compression and atoms in the adjacent layers are in tensile state. In general, grain boundary atom experiences compression if there is higher density of neighbors surrounding it. In addition, there is a tensile state when it has lesser number of atoms surrounding it. Hydrostatic stress fluctuates and tends to zero as we move across the boundary. Similar oscillatory behavior of interplanar spacing at free surface and grain boundary has been reported by Chen and coworkers [8].

Substitution of Mg atom in Al position away from the grain boundary results in hydrostatic tension at the atom position because Mg atom is about 15% larger then Al atom. It has been observed that Mg atom tends to segregate to positions where there is in hydrostatic compression. In Σ9 grain boundary the most preferred site is 1b, which is under highest hydrostatic compression, the segregation energy here is –0.18 eV. However site 2b experiences hydrostatic tension and Mg atom is strongly repelled from this site. Similar behavior is observed for Σ5 and Σ11 boundaries. In Σ5 Mg atom prefers to sit at 1a but is repelled from position 2a and 3a In Σ11 the most preferred site is 1c, the symmetric center of the grain boundary. These results match with earlier work on Mg segregation in Al [9]. Similar studies on Cu segregation in Al show that Cu atoms behave in an exactly opposite manner to that of Mg [10,11]. This can be explained from the fact that Cu is smaller than Al where as Mg is larger than Al.

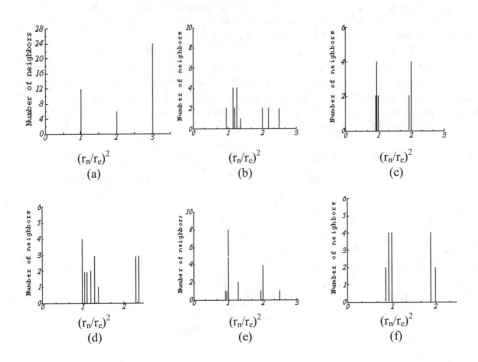

Fig 4 Distribution of atoms around the stressed atom (a) Ideal FCC, (b) 1a, (c) 2a, (d) 1b (e) 3a (f) 2b

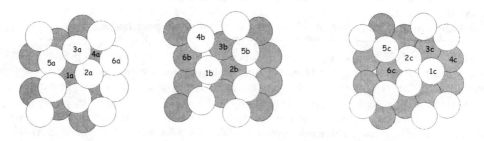

Fig 5 Schematic of atomic sites at [001] Σ 5, Σ 9 and Σ 11 grain boundaries respectively

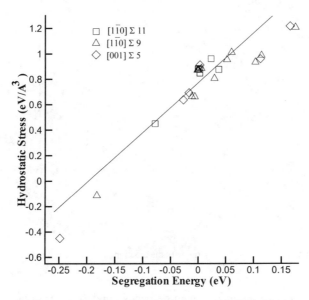

Fig. 6 Variation of segregation energy with hydrostatic stress

Fig 6 shows variation of segregation energy of Mg with hydrostatic stress. There seems to be nearly one to one correspondence between hydrostatic stress and segregation energy. This indicates that hydrostatic stress is one of the important factors, which influences segregation to particular positions at grain boundary.

A simple explanation of the above interesting behavior can be made by relating the segregation energy E_S (eV) to pressure ΔP (ev/A^3) and volume V^* through $E_S = \Delta P \cdot V^*$. Here V^* is difference between volumes of Al and Mg, but the calculated value of $V^* = 39.9$ A^3 does not correlate with the value predicted in the graph. This opens up the possibility that the effect observed in Fig.6 may be due to factors other than $\Delta P \cdot V^*$ effect.

Grain boundary Energy

The effect of Mg doping on grain boundary energy was studied. Doped bicrystal have a higher grain boundary energy then pure bicrystal. Presence of dopant atom in the bulk also raises the grain boundary energy, however this increase in grain boundary depends on the position of Mg atom and the structure of the boundary. In case of Σ5 grain boundary there is a comparative decrease in grain boundary energy as the dopant atom moves closer to the Σ5 grain boundary (Fig7b). For Σ3 boundary there is no variation in grain boundary energy irrespective of the position of dopant atom as shown in Fig 7a.

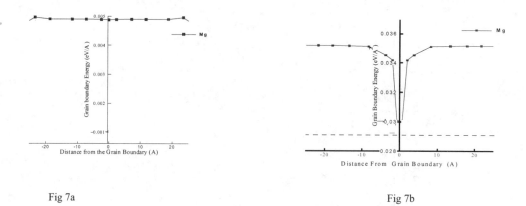

Fig 7a and b. Variation of grain boundary energy with the position of dopant atom in [001] Σ5 and [110] Σ3

This can be explained by the fact that there is no difference in the energy or stress of the atom across Σ3 boundary.

Summary

Mg segregation in Al bicrystals was studied using molecular statics. It is found that impurity segregation strongly depends on hydrostatic stress field at the grain boundary. Mg atoms tend to segregate to the positions of high compressive hydrostatic stress. Origin of hydrostatic stress can be explained in terms of density of atoms around the stressed atom. Grain boundary energy was found to increase with addition of Mg atoms, however the extent of this increase in grain boundary energy depends on the position of Mg atom and the structure of grain boundary.

References
[1] A.H.Chokshi, Mater. Sci. and eng. A166, 119-133, (1993)
[2] A.D. Sheikh Ali, F.F.Lavrentyev and YU.G.Kazarov Acta Metall. 45 (1997) 4505
[3] N. Chandra and P. Dang, J.of Mat. Sci. 34 (1999), 655-666.
[4] M.S. Daw and I. Baskes, Phys. Rev. Letters, 50 (1983), 1285.
[5] G.J. Wang, A.P.Sutton and V.Vitek, Acta metal. 32 (1994) 1093
[6] D .J.Oh and R.A.Johnson ,"Atomistic simulation of Material: Beyond Pair Potentials" edited by V.Vitek and D.J.Srolovitz (Plenum Press, 1989,p.233)
[7] S. F. Foiles , Private communication 1999
[8] S.P.Chen, D.J.Srolovitz and A.F.Voter J.Mater. Res. 4,(1989) 62
[9] X-Y Liu and J.B.Adams , Acta metal. 46 (1998) 3467
[10] X.-Y.Liu, Wei Xu, S.M.Foiles and J.B Adams App. Phy. Letters, 72 (1998) 1578
[11] Hanchen Huang, T.Diaz de la Rubia and M.J.Fluss Mat. Res. Soc.Proc.Vol 428 (1996) 177

Corresponding Author : N. Chandra

On the Independent Behavior of Grain Boundary Sliding and Intragranular Slip During Superplasticity

A.D. Sheikh-Ali and H. Garmestani

Laboratory for Micromechanics of Materials, National High Magnetic Field Laboratory,
1800 E. Paul Dirac Drive, Tallahassee FL 32310, USA

Keywords: Grain Boundary Sliding, Intragranular Slip, Lattice Dislocations, Superplasticity

ABSTRACT

Operation of grain boundary sliding is examined for conditions of plastic strain incompatibility that is the most frequent case for deformation of polycrystals. Two coexisting components of grain boundary sliding: dependent and independent on intragranular slip are distinguished. Theoretical estimate of a ratio between slip induced sliding and intragranular slip is obtained. It is concluded that at the beginning of deformation intragranular slip and grain boundary sliding behave independently. However, at high strains they can be viewed to some extent as interdependent processes.

Introduction

Usual high-temperature deformation is associated with accumulation of lattice dislocations in grains that often induces processes of dynamic recovery and recrystallization. Structural superplasticity is a specific type of high-temperature deformation accompanied by insignificant accumulation of dislocations thus produces minor changes in microstructure. Such behavior is primarily attributed to the process of grain boundary sliding (GBS) that operates with intragranular (crystallographic) slip and diffusional creep in a favorable combination. Sliding and slip make their maximum contributions to total strain at the optimum and high strain rates respectively. There is a link between these processes. Under deformation with a constant strain rate the accumulation of lattice dislocations in grains is decreased by operation of GBS. Lattice dislocations entered grain boundaries are dissociated into extrinsic grain boundary dislocations (GBDs) that participating in the process of cooperative GBS [1, 2] can be carried out to the specimen surface. This makes possible further development of slip. At the same time giving a significant contribution to overall strain, GBS decreases the contribution of slip. Such complex interconnection between sliding and slip results in appearance of two different concepts of structural superplasticity. According to the first one [3] slip and sliding are interconnected in the optimum superplastic region: facilitation of intragranular slip increases the rate grain boundary sliding. Following the other concept [4] sliding and slip are independent and competing processes. In this paper, results that can support these different concepts are analyzed and relationship between slip and sliding is determined. To elucidate the problem experimental results obtained by different researchers on bicrystals and polycrystalline materials are considered.

Concept of Sliding Operating under Conditions of Incompatibility

Let us consider a hypothetical two-dimensional bicrystal containing a finite boundary with the ends designated as O and C (Fig. 1 (a)). If the applied shear stress is sufficient to initiate GBS but

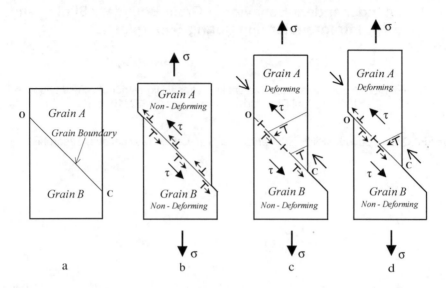

Fig. 1. (a) Original hypothetical bicrystal; (b) pure GBS provided by GBDs generated from boundary sources; (c) slip induced sliding provided entirely by GBDs generated as a result of interaction of lattice dislocations with boundary; (d) combination of pure sliding and slip induced sliding.

not enough to deform grains this is the case of pure sliding (Fig. 1(b)). It is worth noting that Horton et al. [5] used the term of pure sliding when it was accommodated by a small slip. At microscopic scale pure sliding can be considered as the motion of glissile GBDs of opposite signs generated in the boundary plane by numerous grain boundary sources distributed along the whole boundary including points O and C (Fig. 1(b)). When initiation of grain boundary shear or generation of GBDs by grain boundary sources is inhibited, sliding can be produced purely by the difference of deformation between neighboring grains [6]. In the case of incompatible deformation shown in Fig. 1(c) the upper grain A deforms and the lower grain B does not. Only GBDs of one sign are produced as a result of interaction of lattice dislocations with the boundary. If motion of GBDs is governed mainly by applied stresses the amount of sliding increases from zero at point O to some value at point C. Often, sliding is a combination of pure sliding and sliding induced by intragranular deformation (Fig. 1(d)). Assuming that these processes are independent the total amount of sliding is a sum of two kinds of sliding:

$$S = S_{pure} + S_{SIS}, \qquad (1)$$

where S is total amount of sliding, S_{pure} is a value of pure sliding and S_{SIS} is a value of slip induced sliding. Due to non-uniform distribution of sliding induced by slip the overall sliding is also non-uniform. However in this case, some amount of sliding is reached at point O and this is a pure sliding whereas sliding at point C is a sum of the amounts of pure sliding and sliding induced by slip. At dislocation level the independence of different types of sliding means the non-interacting and independent operation of two different kinds of GBD sources: lattice dislocations impinged into the boundary and grain boundary sources of GBDs. The interaction between GBDs of different origin results in remaining of glissile GBDs of one sign (Fig. 1(d)).

Interaction Between Sliding and Slip in Incompatible Bicrystals

In the case of incompatible bicrystals the activation of intragranular slip results in gradual changing of the amount of GBS from one end of the boundary to the other [7]. It is shown that at respectively small intragranular strain (less than 1%), intragranular slip increases the magnitude of sliding more than twice in comparison with the case when intragranular slip is negligible. At the same time, the amount of sliding at point O remains the same [8]. Therefore, according to the concept described in previous section lattice dislocations and/or GBDs appeared as a result of LD-GB interaction obstruct neither the generation of glissile GBDs from boundary sources nor their motion along the boundary. Thus, a combination of two types of sliding obeys an additive (linear) law at respectively small strains. It is important noting that incompatible deformation is the most frequent case in polycrystalline materials.

Relation Between Slip and Sliding under Superplastic Conditions

Experimental Results

The study of anisotropic behavior of Zn-0.4%Al sheet alloy during superplastic flow have shown that straining in the direction favorable for basal slip reduces the strain-rate sensitivity exponent m [9] (Fig. 2). The analogous results have been obtained for fine-grained cadmium and magnesium alloy [10, 11]. The value of m correlates with contribution of sliding to overall strain [12, 13]. Therefore, these results may indicate on concurrent and independent character of sliding and slip during superplastic deformation: the greater contribution of GBS to total strain, the lower contribution of slip and vice versa.

The notion of the two independent processes contradicts to the results obtained by Matsuki et al. [14, 15] on Al-10.8%Zn-0.85%Mg-0.29%Zr and Zn-0.92%Cu-0.60%Mn alloys. For these alloys the stress exponents for GBS and slip were measured separately. Fig. 3 demonstrates that the stress exponents for sliding and slip are similar in the optimum region of superplasticity (n_{GBS}=2.0 and n_{IS}=2.2 for both alloys) and dissimilar in the region of the higher strain rates (n_{GBS}=2.0 and n_{IS}=4.9 for aluminum alloy, n_{GBS}=2.0 and n_{IS}=4.2 for zinc alloy). These results show that in the optimum region of superplasticity GBS and slip are interdependent processes whereas in the region of high strain rates they can be considered as independent.

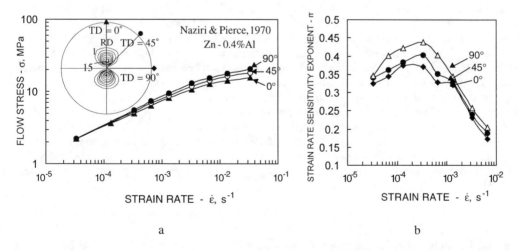

Fig. 2. Anisotropic superplasticity in Zn-0.4%Al [9]. (a) Flow stress-strain rate relations for specimens cut at 0, 45 and 90° to the rolling direction and (b) variation of m with strain-rate and with direction of straining.

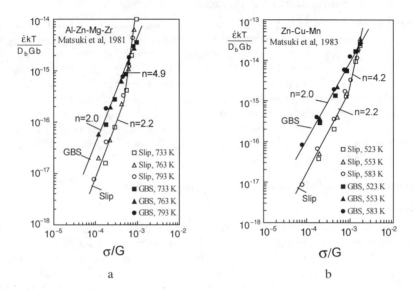

Fig. 3. Relation between individual strain rates of grain boundary sliding and intragranular slip and flow stress in dimensionless plots: $D_b = D_o \exp(-Q/RT)$ for Al (a) and Zn (b) alloys [14, 15].

Theoretical Consideration

Consideration of sliding in a hypothetical bicrystal shows that there is a component of sliding dependent on intragranular slip called slip induced sliding. Contribution of slip induced sliding to total strain without operation of pure GBS can be assessed assuming that accommodation processes are rapid in triple boundary junctions and glissile GBDs disappear easily. Fig. 4 (a) illustrates two hexagonal grains having a common boundary. The upper grain is subjected to uniform elongation, the other is non-deforming (Fig. 4 (b, c)). The upper grain is allowed to slide when it deforms. In general, there are two different cases of sliding induced by slip. In the first case illustrated in Fig. 4 (b), slip induced sliding accommodates grain deformation without contribution to total strain and all strain is provided by intragranular deformation. It happens when there is no applied shear stress along boundary or its value is small. Under internal stresses grain boundary dislocations move in the opposite directions and appear in both triple boundary junctions creating facets. In the second case, slip induced sliding makes a direct contribution to strain and simultaneously accommodates intragranular deformation (Fig. 4 (c)). In this case, applied shear stress is sufficient to effect on slip induced sliding pushing GBDs in one direction and creating facet in the triple junction. Apparently, this type of sliding occurs in the optimal superplastic region being a part of cooperative GBS. The difference between strains in two different cases gives the strain produced purely by slip induced sliding:

$$\varepsilon_{SIS} = \varepsilon_{TOT} - \varepsilon_{IS} = \frac{l_2 - l_1}{l_0}, \qquad (2)$$

Here l_0 - length between marker points in grains before deformation, l_1, l_2 - distances between the same points in two different cases of deformation.

Dividing equation (2) on ε_{TOT} we obtain the contribution of slip induced sliding to overall strain:

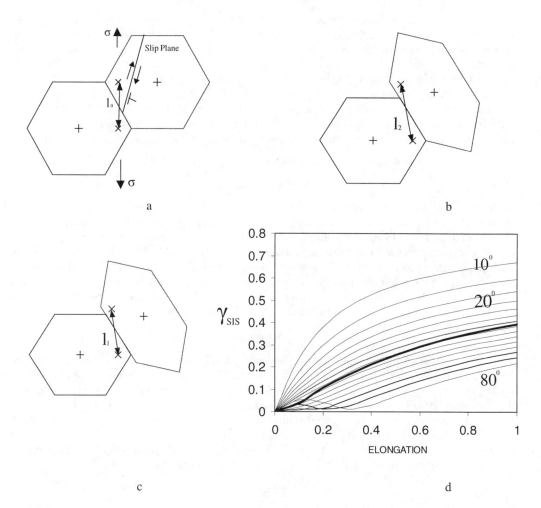

Fig. 4. Contribution of slip induced sliding to total elongation. Initial hexagonal grains (a). Slip induced sliding (b) with and (c) without contribution to total elongation at two-dimensional deformation of the upper grain by shear along slip planes. Contribution of slip induced sliding to total elongation as a function of strain for different inclination of slip planes to tensile axis (d). Thick curve designate average contribution of slip induced sliding.

$$\gamma_{SIS} = 1 - \gamma_{IS} = 1 - \frac{l_1}{l_2}, \qquad (3)$$

where γ_{SIS} and γ_{IS} is the contributions of slip induced sliding and intragranular slip to total strain, respectively.

Fig. 4 (d) represents the dependencies of γ_{SIS} on total strain at different angles between direction of tension and slip planes. With increasing strain it increases from zero up to significant values. Depending on inclination of slip planes to deformation axis γ_{SIS} can vary in a wide range.

From comparison of theory and experimental results obtained on bicrystals, it follows that pure and slip induced sliding obey an additive law. In other words, these two types of sliding can operate as independent and concurrent processes at the same boundary. In polycrystalline materials pure sliding is a rare event. Nevertheless, we use the term "pure" to designate the component of sliding that is produced from grain boundary dislocation sources. This component is independent on slip. Slip induced sliding is directly connected with intragranular deformation. The latter makes a contribution to overall strain if motion of GBDs is governed by applied stresses. Hence, GBS along the same boundary can be separated into two components: dependent and independent on slip, each of them can make a contribution to total strain. We denote processes which are interdependent by including the appropriate strains in square brackets and processes which are independent, including the resultants of sequential pairs, just by addition signs. Thus, the total strain ε_{TOT} can be expressed in terms of strains due to pure GBS ε_{GBS}, intragranular slip ε_{IS} and slip induced sliding ε_{SIS}:

$$\varepsilon_{TOT} = \varepsilon_{GBS} + [\varepsilon_{IS} + \varepsilon_{SIS}] \tag{4}$$

With changing strain and strain rate, the contributions of pure GBS, intragranular slip and slip induced sliding are varied. The increase of intragranular strain always results in decrease of total sliding composed of pure and slip induced sliding and vice versa. It means that competing character of slip and sliding does not mean that these processes are entirely independent. Fig. 4 (d) shows that with straining the contribution of slip induced sliding can exceed the contribution of intragranular slip. However, these results obtained from simplistic model. All experimental results on contribution of intragranular slip to total strain have been obtained at respectively small strains. Therefore, the exact relationship between slip and sliding at high strains is still unclear.

SUMMARY

Grain boundary sliding along the same boundary can be separated into two components that are dependent and independent on intragranular slip. Each of them can make own contribution to total strain during superplastic deformation in the optimum superplastic region. Theoretical assessment of ratio between intragranular slip and slip induced sliding shows that straining can increase the contribution of slip induced sliding to total strain from zero to more than half of that. Therefore, at small strains intragranular slip and grain boundary sliding can be considered as independent processes. At high strains the relationship between these processes is complex due to coexistence of slip dependent and slip independent components of sliding.

REFERENCES

1. V.V. Astanin, O.A. Kaibyshev and S.N. Faizova, *Acta Metal. Mater.*, **42** (1994), p. 2663.
2. M.G. Zelin and A.K. Mukherjee, *Acta Metal. Mater.*, **43**, (1994), p. 2359.
3. O.A. Kaibyshev, Superplasticity of Alloys, *Superplasticity Intermetallides, and Ceramics*, Springer-Verlag, Berlin, 1992, 317 p.
4. T.G. Nieh, J. Wadsworth and O.D. Sherby, *Superplasticity in Metals and Ceramics*, Cambridge University Press, 1996, 273 p.
5. A.P. Horton, N.B.W. Thompson and C.J. Beevers, *Metal Sci. J.*, **2** (1968), p. 19.
6. A.D. Sheikh-Ali, F.F. Lavrentyev and Yu.G. Kazarov, *Acta Mater.*, **45** (1997), p. 4505.
7. A.D. Sheikh-Ali and R.Z. Valiev, *Scripta Metall. Mater.*, **31** (1994), p. 1705.
8. A.D. Sheikh-Ali, J. Szpunar and H. Garmestani, *Mater. Sci. Eng.*, (submitted).
9. H. Naziri and R. Pierce, *J. Inst. Metals*, **98** (1970), p. 71.
10. Shu-En Hsu, G.R. Edwards and O.D. Sherby, *Acta Metall.*, **31** (1983), p. 763.
11. O.A. Kaibyshev, I.V. Kazachkov and N.G. Zaripov, *J. Mater. Sci.*, **23** (1988), p. 4369.
12. D. Lee, *Acta Metall.*, **17** (1969), p. 1057.
13. R.B. Vastava and T.G. Langdon, *Acta Metall.*, **27** (1979), p. 251.
14. K. Matsuki, N. Hariyama and M. Tokizawa, *J. Japan Inst. Metals*, **45** (1981), p. 935.
15. K. Matsuki, N. Hariyama, M. Tokizawa and Y. Murakami, *Metal Sci.*, **17** (1983), p. 503.

Effect of Chemical Bonding States on the Tensile Ductility in Glass-doped TZP

A. Kuwabara[1], S. Yokota[1], Y. Ikuhara[1] and T. Sakuma[2]

[1] Department of Materials Science, Faculty of Engineering, University of Tokyo,
7-3-1 Hongo, Bunkyo-ku, Tokyo 113-8656, Japan

[2] Department of Advanced Materials Science, Graduate School of Frontier Sciences,
University of Tokyo, 7-3-1 Hongo, Bunkyo-ku, Tokyo 113-8656, Japan

Keywords: Chemical Bonding State, Grain Boundary, Molecular Orbital Calculations, Superplasticity, TZP

ABSTRACT

The chemical bonding states of silica-doped YTZP are computed by the DV-Xα molecular orbital calculation. It is clarified that silicon ions substituted into the lattice of TZP have extremely high covalent bonding with oxygen ions, while other cations, magnesium and titanium ions reduce the covalency. The increment of covalency of bondings around doped cations seems to have a critical role to exhibit thee improved tensile ductility in silica-doped YTZP.

INTRODUCTION

Tetragonal zirconia polycrystal (TZP) with fine grain size shows superplasticity at high temperatures. In particular, SiO_2-doped YTZP (Si-YTZP) exhibits large tensile elongation in excess of 1000%[1]. However, the extensive ductility in Si-YTZP is suppressed by an addition of a small amount of metal oxide such as MgO, TiO_2, and so on, as shown in Fig. 1[2-4]. According to our experimental data, the ductility in glass-doped YTZP is caused by cations segregation in grain boundaries[2-4]. The present paper aims to report that the chemical bonding states in or near grain boundaries are likely to change with dopant segregation, and such a change affects very seriously the tensile ductility in glass-doped TZP.

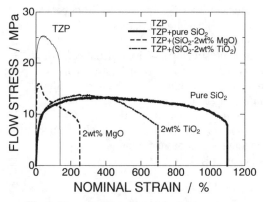

Fig. 1. Stress-Strain curve obtained for pure SiO_2-doped TZP, (SiO_2-2wt%MgO)-doped TZP, and (SiO_2-2wt%TiO_2)-doped TZP at a strain rate of 1.3×10^{-4} s^{-1} at 1400°C in air.

CALCULATION

The DV-Xα molecular orbital calculation[5] is made to estimate the chemical bonding state. The $(Y_4Zr_{18}O_{86})^{88-}$ cluster used for the calculation of 3Y-TZP is shown in Fig. 2. This cluster is composed of 108 ions and includes up to the twelfth nearest neighbor from the Zr ions at the sites of C(1) or C(2) indicated by arrows. Two Y^{3+} ions are substituted with Zr^{4+} ions at the third nearest neighbor from $Zr_{C(1)}$. An oxygen vacancy for maintaining electric neutrality is introduced at the mutual second nearest neighbor site from the two Y^{3+} ions[6]. Similarly, two Y^{3+} ions are located at the third nearest neighbor sites from $Zr_{C(2)}$ and an oxygen vacancy is produced at the second nearest neighbor from these two Y^{3+} ions. $(SiY_4Zr_{17}O_{86})^{88-}$ and $(Si_2Y_4Zr_{16}O_{86})^{88-}$ clusters are constructed for the calculation of Si-YTZP, and $(SiMgY_4Zr_{16}O_{85})^{88-}$ and $(SiTiY_4Zr_{16}O_{85})^{88-}$ clusters are built for the calculation of Si-TZP with MgO and with TiO_2, respectively. The structure of each cluster is demonstrated in Table 1. All of these clusters are put into a field of Madelung potential composed of about 6000 point charges. From the calculation, we can obtain the information of bonding states such as bond overlap population (BOP) and net charge (NC). The BOP indicates the degree of covalency of a bonding, while the NC corresponds to the effective ionicity of an ion in a crystal. The product of NCs is expected to describe the strength of an ionic bonding. In DV-Xα molecular orbital calculation, the electronic structures of ions at the edge of a cluster are very much affected by dangling bondings. To exclude the influence, it is necessary to examine the electronic structure near the center of a cluster. In this paper, we analyze the bonding states around ions at the site of C(1) or C(2)

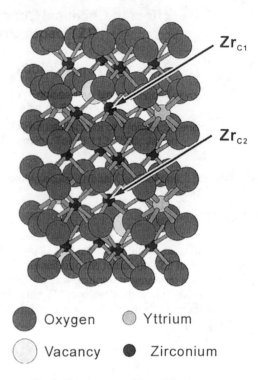

Fig. 2. The cluster model used for the calculation of 3Y-TZP.

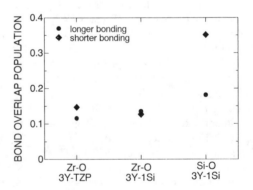

Fig. 3. The BOP of Zr-O in 3Y-TZp and 3Y-1Si. The BOP of Si-O in 3Y-1Si are indicated together.

RESULTS

In t-ZrO_2, Zr-O bondings can be classified into two types in terms of the

Table 1. The structure of each cluster model.

	$(Y_4Zr_{18}O_{86})^{88-}$	$(SiY_4Zr_{17}O_{86})^{88-}$	$(Si_2Y_4Zr_{16}O_{86})^{88-}$	$(SiMgY_4Zr_{16}O_{85})^{88-}$	$(SiTiY_4Zr_{16}O_{85})^{88-}$
C1	Zr	Si	Si	Mg	Ti
C2	Zr	Zr	Si	Si	Si
Name	3Y-TZP	3Y-1Si	3Y-2Si	3Y-MgSi	3Y-TiSi

length of the bondings. The longer Zr-O bonding and the shorter one are named as Zr-O(l) and Zr-O(s), respectively. Other bondings between cation and oxygen ion are expressed in a similar way, for example Si-O(s) and Si-O(l).

Fig. 3 shows the BOPs of Si-O and Zr-O bondings in 3Y-1Si. For comparison, the BOPs of Zr-O bondings in 3Y-TZP are also shown. The BOPs of Zr-O bondings in 3Y-TZP and 3Y-1Si are not so different but the BOP of Si-O(l) are slightly larger, and of Si-O(s) is about three times as large as those of Zr-O(s) or Zr-O(l). This result means that Si ions in Si-YTZP have a very high covalency with O ions in comparison with Zr ones.

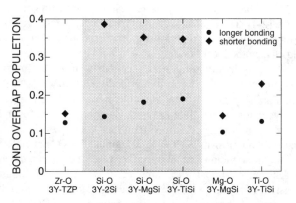

Fig. 4. The BOP of bondings around doped cation in 3Y-2Si, 3Y-Mg-Si, and 3Y-TiSi. The BOP of Zr-O in 3Y-TZp is indicated together.

Fig. 4 indicates the BOPs of Mg-O and Si-O in 3Y-MgSi and those of Ti-O and Si-O in 3Y-TiSi. The BOPs of Si-O in 3Y-2Si and those of Zr-O in 3Y-TZP are also demonstrated in the same figure. The BOPs of Si-O don't change remarkably among 3Y-MgSi, 3Y-TiSi, and 3Y-2Si. This result would suggest that an additional dopant scarcely influence on the bonding state of Si-O in Si-YTZP. The BOPs of Mg-O bondings have almost same value as those of Zr-O in 3Y-TZP. On the other hands, the BOPs of Ti-O(s) is larger than those of Zr-O in 3Y-TZP and is less than those of Si-O(s).

Discussion

The change of ductility of Si-YTZP cannot be explained only from the viewpoint of the flow stress[7]. Our previous reports reveal that glassy phase exists only at the multiple junctions of grain boundaries and neighboring grains contact directly in atomic level without amorphous film[2-4]. Any glassy film or second phase is not observed at grain boundaries. EDS analyses of

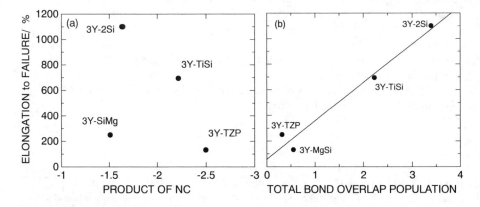

Fig. 5. The dependence of the elongation to failure in Si-YTZP on ionicity(a) and covalency of bondings.

Si-YTZP nearby grain boundaries show the segregation of doped cations in two-grain junctions[2-4]. The elongation to failure is greatly enhanced by the segregation of Si^{4+} ion, but is deteriorated by the segregation of other codoped cations such as Mg^{2+}. The co-segregation would change the local bonding state around grain boundaries and influence on the ductility.

SiO_2 is hardly soluble in the matrix of ZrO_2. In contrast, our previous experimental results prove that Si^{4+} ion is likely to dissolve in ZrO_2 near grain boundaries[2-4]. From the calculation using DV-Xα method, silicon ions in YTZP have extremely high covalent bonding with oxygen ions rather than Zr ones. The grain boundaries in Si-YTZP must be strengthened by the increase of covalency with silicon ions which will lead to the resistance against the fracture at grain boundaries. In Si-YTZP containing MgO, not only silicon ions but magnesium ones segregate in the vicinity of grain boundaries. The present calculation shows that Mg-O bondings in YTZP have almost the same BOP as Zr-O bondings. Because of co-segregation of magnesium ions, the covalency of bondings around the grain boundaries in Si-YTZP containing MgO does not increase so much as in pure Si-YTZP. Doping of MgO to Si-YTZP would suppress the increment of resistibility against the fracture at grain boundaries and reduce its ductility. The introduction of vacancies by MgO doping, i.e. the increase of dangling bonding must be an origin of the reduction in tensile ductility with the co-doping of MgO

Fig. 5 shows the relationship between the tensile ductility and the chemical bonding sates in Si-YTZP. The elongation to failure cannot be described by the ionicity of bondings, while, the ductility has a good relationship with the covalency of bondings. If the bondings between dopant and oxygen have higher covalency, Si-YTZP exhibits larger tensile elongation. From these results it is concluded that the covalency of bondings is one of the major origins which decide the tensile ductility in Si-YTZP.

CONCLUSION

We examined the chemical bonding states of pure Si-YTZP, Si-YTZP with MgO, or with TiO_2 using DV-Xα method. Silicon ions dissolved in YTZP induce high covalent bonding with oxygen ones. Ti-O bondings are also high covalent bonding but the covalency of Mg-O bondings is much weaker than Si-O bondings and is nearly the same with that of Zr-O bondings. These dopant cations affect the electronic structures of grain boundaries in YTZP and hence contribute toelongation. It is concluded that the elongation to failure is enhanced as the covalency of bondings around dopant cations increases. The covalency of bondings is one of the most important factors for the exhibition of remarkable ductility in Si-YTZP.

Acknowledgment

The authors wish to express their gratitude to the Ministry of Education, Science and Culture, Japan, for the financial support by a Grant-in-Aid for Developmental Scientific Research --1135 5028 for Fundamental Scientific Research. We also wish to express our thanks to Dr P. Thavorniti for her experimental assistance.

Reference

1. K. Kajihara, Y. Yoshizawa, and T. Sakuma, Acta. Metal Mater., 1995, **43**[3], pp. 1235-1242.
2. Y. Ikuhara, K. Sasaki, P. Thavorniti, and T. Sakuma, The Third Pacific Rim International Conference on Advanced Materials and Proceeding (PRICM3) Edited by M. A. Imam, R. denale, S. Hanada, Z. Zhong, and D. N. Lee, 1998, pp.1747-1754.
3. P. Thavorniti, Y. Ikuhara, and T. Sakuma, Acta. Mater., **45**[12], 1997, pp. 5275-5284.
4. Y. Ikuhara, P. Thavorniti, and T. Sakuma, Mater. Sci. Forum, **243-245**, 1997, pp. 345-350.

5. H. Adachi, M. Tsukada, and C. Satoko, J. Phys. Soc. Jpn., **45** [3] pp. 875-883 (1978).
6. P. Li, I-W. Chen, and J.E.P. Hahn, J. Am. Ceram. Soc., **77** [1] pp. 118-128 (1994).
7. T. Sakuma and Y. Yoshizawa, Mater. Sci. Forum, **170-172**, 1994, pp. 369-378

Internal Stress Superplasticity in an In-Situ Intermetallic Matrix Composite

R.S. Sundar, K. Kitazono, E. Sato and K. Kuribayashi

The Institute of Space and Astronautical Science,
3-1-1 Yoshinodai, Sagamihara, Kanagawa 229-8510, Japan

Keywords: Creep, In-situ Composite, Intermetallics, Internal Stress Superplasticity, NiAl

Abstract

Processing of NiAl-Cr based eutectic in-situ composite via internal stress superplasticity has been explored. The alloy is heat treated to produce spherodized Cr particles in NiAl matrix. Results of thermal cycling creep experiments indicate that internal stress of considerable magnitude are generated between the NiAl matrix and Cr particles, which aids the external applied stress in deforming the alloy in viscous manner. The origin of internal stress is attributed to the difference in thermal expansion behavior of the constituent phases. At low stresses, the thermal cycling creep rates are higher than the isothermal creep rates. However, the observed thermal cycling creep rates are an order of magnitude lower than values predicted by an internal stress superplasticity model based on micromechanics. The possible reasons for this discrepancy are discussed.

Introduction

Intermetallic compounds are gaining importance as high temperature structural materials. However, successful application of these materials depends critically on identification of processing routes, which will produce a defect free material in the required shape at a competitive cost. Forming of these materials by internal stress superplasticity (ISS) is promising, as these materials are difficult to form through conventional processing routes. Further, processing through ISS is attractive, as it does not require fine grain size as in conventional superplastic forming. Recently, internal stress superplasticity has been reported in super α_2 alloy [1] and NiAl-ZrO$_2$ composite [2]. Apart from the above two studies, there are no other studies on internal stress superplasticity of intermetallic compounds available in the literature.

The aim of the present investigation is to study the ISS in NiAl-Cr based eutectic alloy. This two-phase in-situ composite has shown high thermal stability, improved room temperature toughness and attractive tensile strength and stress-rupture properties [3–6]. The thermal expansion coefficient (α) of NiAl is $15.1 \times 10^{-6} K^{-1}$ and that of Cr is 9.4×10^{-6} K^{-1}. The difference in thermal expansion behavior of the constituent phases can be exploited in processing the alloys via ISS, by thermally cycling them under a small external applied stress. In this report, some of the preliminary results on thermal cycling creep behavior of NiAl-Cr alloy are presented.

Experimental

(1) Materials:

Near eutectic NiAl-Cr (NiAl-33.7at%Cr) alloy was prepared by induction melting and casting under vacuum. The as-cast microstructure of the alloy exhibited fine fibrous eutectic structure. It has been shown earlier [7] that diffusional accommodation around precipitates during thermal cycling is more difficult for elongated particle than in globular one. Hence, an additional heat treatment was given at 1673 K for 15h to produce globular eutectic structure. As cast and heat-treated samples were observed in SEM, after etching with a solution containing 10vol.% HNO_3 + 10vol.% HF + 80vol.% H_2O.

(2) Creep tests:

Hollow compression samples of 10 mm outer diameter and 8 mm inner diameter and 15 mm length were machined from the heat treated blocks by electro discharge machining. Hollow compression samples were favored for thermal cycling creep experiments to minimize the difference in temperature and difference in cooling rate between surface and inside of the samples. Compression creep tests under thermal cycling and isothermal conditions were performed using a servo hydraulic testing machine equipped with an induction heating facility. The specimens were heated by a high frequency induction coil and cooled by atmospheric air, with a temperature control through a R-type thermocouple welded on them. The maximum controlling error, which occurred at points where heating was reversed to cooling, was less than ±3 K. An example of the temperature profile and the associated specimen length change are shown in Fig. 1. Broken line in the figure indicates an average strain rate of thermal cycling creep.

Initial thermal cycling creep tests were done at 1223-1423 K temperature range. Additional tests were done at 1248-1448 K and 1273-1473 K (at low stress levels) to study the temperature dependence of ISS. In all the above experiments, the thermal cycling was triangular wave, whose temperature amplitude and heating/cooling rates of thermal cycling were fixed at 200 K and 10 Ks^{-1}, respectively. To compare the thermal cycling creep with isothermal creep behavior, the equivalent temperature T_{eq} [8] is defined as :

$$\exp(-\frac{Q}{nRT_{eq}}) = \frac{1}{T_{max}-T_{min}} \int_{T_{min}}^{T_{max}} \exp(-\frac{Q}{nRT}) dT \qquad (Eq.\ 1)$$

where T_{max} and T_{min} are maximum and minimum temperature during thermal cycling and Q is the apparent activation energy for creep of NiAl-Cr alloy under isothermal condition (300 $kJmol^{-1}$). The T_{eq} for the 1223-1443 K thermal cycling temperature range is calculated as 1326 K. Isothermal creep experiments were done at 1323 K, which is close to the equivalent temperature (T_{eq}).

Results

The alloy is hypo eutectic and contains fine fibrous eutectic mixture and primary NiAl phase (Fig. 2a). The amount of primary NiAl phase is about 5vol.%. The average grain size of the matrix is about 200 μm, which is too large for conventional fine grain superplasticity. The microstructure of the heat-treated alloy is shown in Fig 2b. Annealing at 1673 K for 15 hrs spheroidized most of the Cr fibers in the eutectic mixture. The size and volume fraction of the Cr particle is about $a = 1.5$ μm and $f = 0.32$ respectively. The Cr particles were seen both at the grain boundaries as well as inside the grains. The ratio of volume fraction of Cr within the grain and at the grain boundary is 15:1. The internal stress generated by particles at the grain boundaries is easily accommodated by fast grain boundary diffusion [8], hence, volume fraction of Cr particles within the grain (0.3) is utilized for theoretical calculation.

The thermal cycling and isothermal creep results are presented in Fig. 3. Under isothermal condition, the stress exponent for steady state creep rate is about 6.1, which indicates the operation of a dislocation creep mechanism. At low stresses, thermal cycling creep rates are more than several orders of magnitude higher than that of isothermal creep rates. More over the value of stress exponent at low stress condition is about 1. This region is considered as internal stress superplastic region. With increasing stress, the stress exponent for thermal cycling creep rate tends to depart from one and the thermal cycling creep rates tend to approach the isothermal creep rates. The above trends are agreeing well with the other studies [1,7,8] of ISS.

Temperature dependence of ISS is studied by carrying out thermal cycling creep rate at 3 different temperature range at constant heating and cooling rate of 10 Ks^{-1}. The results are presented in Fig. 4. From the above plot, activation energy (Q_{ISS}) for ISS is calculated by Arrhenius plot (Fig. 5) at constant stress levels. From the slope of the lines at different stress levels, the value of Q_{ISS} is deduced and is about 245 kJmol^{-1}. The Q_{ISS} is less than that of isothermal creep deformation (300 kJmol^{-1}).

Discussion

The thermal cycling strain rates of NiAl-Cr alloy at low stresses are higher (more than several orders of magnitude) than those of isothermal creep rates. However at high stresses, thermal cycling creep rates are similar to isothermal creep rates. These results imply that enhancement of creep rate by thermal cycling takes place at low stresses, where the internal stress caused by the difference in thermal expansion of the constituent phases are relatively larger than the applied stresses.

Based on micromechanics, Sato et al. [9] have proposed a model for internal stress superplasticity in a creeping matrix with rigid inclusion. According to the model, the average creep rate during thermal cycling is given by:

$$\bar{\varepsilon} = |\Delta \alpha \, \dot{T}|^{1-1/n} B^{1/n} \left(\frac{\sigma_A}{1-f} \right) \frac{2n(n+4)}{2^{1/n}5} \frac{(1-f^{1/n})f}{(1-f)f^{1/n}} \quad \text{(low } \sigma_A\text{)} \quad \text{(Eq. 2)}$$

and
$$\bar{\varepsilon} = B \left(\frac{\sigma_A}{1-f} \right)^n \quad \text{(high } \sigma_A\text{)} \quad \text{(Eq. 3)}$$

where $\Delta \alpha$ is difference in thermal expansion co-efficient between two phases, $\dot{T}$ is rate of temperature change, B is constant of proportionality in the matrix power law creep equation and σ_A is applied stress. The model predicts that under an applied stress much higher than the internal stress, the thermal cycling creep rate coincides with the isothermal creep rate, and that at low stresses, it is proportional to the applied stress. The theoretical prediction by equation (2) for the present alloy is shown in Fig 3 as broken line. Following values are utilized for the calculation: $f = 0.3$, $n = 6.1$, $B = 3.6 \times 10^{-18}$ MPa^{-1}s^{-1}, $\Delta \alpha = 5.7 \times 10^{-6}$K^{-1}. An order of magnitude difference in creep rate is observed between the experimental creep rates and model prediction. In addition, according to a recent analysis [10], Q_{ISS} should be equal to Q/n. However, the experimental activation energy for thermal cycling creep is about 4 to 5 times higher than the expected value ($Q_{ISS} = 240$ kJmol^{-1} vs. $Q_{Theoritical} = 50$ kJmol^{-1}). The possible reasons for the deviations are discussed below:

The reason for the discrepancy between the theoretical and experimental rate may be due to dominance of transient processes associated with temperature change. The model treats only the quasi-steady state stress distribution during heating and cooling, but in actual experiment, a stationary internal stress distribution is achieved after a certain period during each temperature reversal [8]. It was shown in Al-Be eutectic [8] that, the difference between the theoretical creep rate and experimental rate was reduced, when thermal cycling creep experiments were carried out at large temperature amplitude. Under this condition, there is sufficient time to achieve stationary state of internal stress distribution. Hence the effect of transient process on over all deformation is reduced. We are planning to carry out thermal cycling creep experiments at different temperature amplitude at constant T_{eq}, constant stress and constant heating rate, to bring out the effect of transient process on the deformation.

The difference between the experimental and theoretical creep rates may also arise due to plastic deformation of Cr particles during thermal cycling. In the model [9], the inclusions are treated as rigid. Plastic deformation of Cr particles may reduce some portion of the internal stress, and the alloy is expected to deform at lower strain rate. However, the above possibility can be ruled out based on the following discussion. During thermal cycling, the spherical inclusions are subjected to pure hydrostatic state of stress (σ^T). However, when the particles are slightly distorted from spherical shape, shear component of stress is created in them. The magnitude of shear stress will be proportional to the σ^T, with the proportionality constant less than one. The following equation [9] is utilized to calculate the magnitude of σ^T in Cr particles during thermal cycling creep condition:

$$\sigma^T = \frac{2}{3} \frac{\Delta\alpha \dot{T}}{|\Delta\alpha \dot{T}|} \left(\frac{2\Delta\alpha \dot{T}}{(1-f)^n B_o} \right)^{1/n} \quad \text{(Eq. 4)}$$

The estimated magnitude of σ^T in Cr particle is equal to 120 MPa, and the magnitude of shear component in non-spherical Cr particle is less than 120 MPa. At these stress levels, dislocations may not be activated in Cr particles [11]. Hence, the plastic deformation of Cr particles during thermal cycling condition can be ruled out.

Summary

Internal stress superplasticity regime, induced due to difference in thermal expansion behavior of constituent phases in NiAl-Cr in-situ composites is identified. The ISS region is characterized by higher creep rates when compared to isothermal creep rates and n value close to 1. Deforming the alloy under this viscous flow region is expected to yield large tensile elongation. Further studies are needed to clarify the differences between the prediction of the theoretical model and experimental results.

Acknowledgement – The authors thank Dr. S. Kuramoto and Prof. M. Kanno of The University of Tokyo for their help in alloy preparation.

References

1. C. Schuh and D.C. Dunand, Acta. Mater. 46 (1998), p. 5663.
2. P. Zwigl and D.C. Dunand, in "Superplasticity and Superplastic Forming 1998", Supplemental volume, ed. A.K.Ghosh and T.R.Bieler, TMS, 1998, p. 40.
3. J.L. Walter and H.E. Cline, Met. Trans. 1 (1970), p. 1221.
4. J.L.Walter and H.E. Cline, Met Trans. 4(1973), p. 33.
5. D.R. Johnson, X.F. Chen, B.F. Oliver, R.D. Noebe and J.D. Whittenberger, Intermetallics 3 (1995), p.99.
6. F.E.Heredia, M.Y.He, G.E.Lucas, A.G.Evans, H.E.Deve, D.Konitzer, Acta Metall. 41(1993), p.

505.
7. K.Kitazono and E.Sato, Acta Mater. 47 (1999), p. 135
8. K.Kitazono and E.Sato, Acta Mater. 46 (1998), p. 207.
9. E.Sato and K.Kuribayashi, Acta Metall. Mater. 41 (1993), p. 1759.
10. K.Kitazono, E.Sato and K.Kuribayashi, Acta Mater.47 (1999), p. 1653.
11. V.K.Dakshinamurthy, PhD. Thesis, Carnegie Mellon University, (1997).

Corresponding Author : E.Sato(sato@materials.isas.ac.jp)

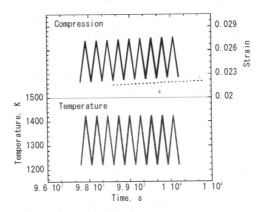

Fig. 1 Typical result of thermal cycling compression creep test.

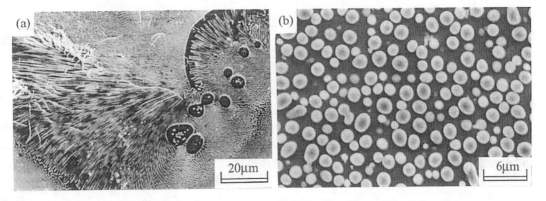

Fig.2 SEM photographs of (a) as-cast and (b) heat-treated NiAl-Cr alloy.

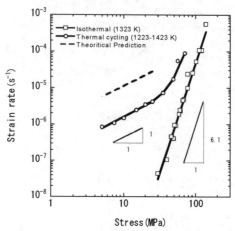

Fig. 3 Thermal cycling and isothermal creep rates as a function of applied stress in NiAl-Cr alloy.

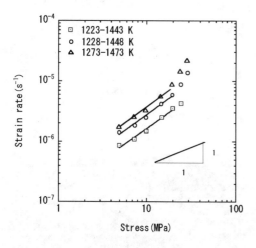

Fig. 4 Thermal cycling creep rate vs stress at different temperature range.

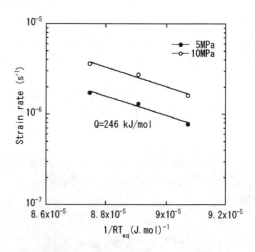

Fig. 5 Equivalent temperature of dependence thermal cycling creep rate.

Analysis of Instability and Strain Concentration during Superplastic Deformation

Li Chuan Chung and Jung-Ho Cheng

Department of Mechanical Engineering, National Taiwan University,
Taipei 106, Taiwan ROC

Keywords: Finite Element Model, Flow Localization Factor, Instability, Stability Criterion, Strain Concentration, Superplasticity, Unstable Plastic Flow

Abstract

The instability and the localization process of unstable plastic flow during constant pressure bulging of superplastic Ti-6Al-4V sheets are analyzed. A generalized stability criterion is developed by generalizing Hart's stability criterion[1] in terms of strain and strain rate. According to the stability criterion, a new concept of the "flow localization factor (FLF)," which enables quantitative description of the localization process of unstable plastic flow, is presented. A finite element model that simulates experiments on constant pressure bulging is applied to compute the FLF at the fracture point. It is found that the localization process during constant pressure bulging can be divided into three stages: (1) the developing period of initial localized flow, (2) the steady stage of strain concentration, and (3) the accelerating stage of strain concentration, which results in the final fracture.

1 Introduction

Interest in superplastic forming (SPF) has been considerable due to the large amounts of deformation under low levels of stress. To take advantage of SPF, improve productivity, and at the same time avoid fracturing during the forming process, further understanding of the fracture mechanisms of superplastic deformation is required. Among them, instability is one of the most significant factors [2-4].

The failure of superplastic sheet metals is a result of the combination or interaction of two processes: unstable plastic flow and internal cavity evolution [5]. Zhou et al. [6] has shown that for materials that are less sensitive to cavity growth, the fracture mode is dominated by unstable plastic flow, and the ductility is limited by the flow localization process beyond instability.

The instability of superplastic deformation has been the subject of several studies [1],[7-10]. These existing approaches are only able to predict the instability of materials or simply to assume that forming limits are reached as long as instability occurs. However, for superplastic materials, there is still large amounts of post-uniform deformation beyond instability, and it is too conservative to take instability as the forming limit. The purpose of the present work is to propose a parameter which can be applied more generally to predict instability and describe strain concentration beyond instability.

2 Stability Analysis

2.1 Hart's Stability Criterion[1]

Hart assumed $\sigma = \sigma(\varepsilon, \dot{\varepsilon})$, which combined with plastic incompressibility leads to

$$\delta\sigma = \frac{\partial\sigma}{\partial\varepsilon}\delta\varepsilon + \frac{\partial\sigma}{\partial\dot{\varepsilon}}\delta\dot{\varepsilon}, \tag{1}$$

$$\delta\varepsilon = -\frac{\delta A}{A},\qquad(2)$$

$$\delta\dot{\varepsilon} = -\frac{\delta\dot{A}}{A} + \frac{\dot{A}}{A}\left(\frac{\delta A}{A}\right).\qquad(3)$$

Here A is the cross-sectional area, and δ signifies a variation between the nominal and a local nonuniformity. Defining the strain-hardening coefficient (γ) and the strain-rate sensitivity index (m) to be

$$\gamma = \frac{1}{\sigma}\left(\frac{\partial\sigma}{\partial\varepsilon}\right),\text{ and } m = \frac{\dot{\varepsilon}}{\sigma}\left(\frac{\partial\sigma}{\partial\dot{\varepsilon}}\right),\qquad(4)$$

it is obtained that

$$\frac{\delta\ln\dot{A}}{\delta\ln A} = -\frac{1-\gamma-m}{m}.\qquad(5)$$

Hart's criterion states that if the magnitude of the cross-section difference does not increase, then the deformation is stable. Thus, the stability criterion can be written as follows. *The deformation is stable if*

$$\frac{\delta\dot{A}}{\delta A} \leq 0.\qquad(6)$$

Since $\dot{A}/A$ is negative in tension, this condition can be restated as *the deformation is stable in tension if*

$$\gamma + m \geq 1.\qquad(7)$$

2.2 Re-expressing Hart's Stability Criterion

Eq.3 can be expressed as

$$A\frac{\delta\dot{\varepsilon}}{\delta A} = -\frac{\delta\dot{A}}{\delta A} + \frac{\dot{A}}{A},\qquad(8)$$

which combined with Hart's stability criterion, Eq.6, deduces that

$$\frac{\delta\dot{\varepsilon}}{\delta A} \geq \frac{1}{A}\frac{\dot{A}}{A}.\qquad(9)$$

Since $\dot{\varepsilon} = -\frac{\dot{A}}{A}$, and $\delta\varepsilon = -\frac{\delta A}{A}$, the above criterion can be expressed as

$$\frac{\delta\dot{\varepsilon}}{\delta\varepsilon} \leq \dot{\varepsilon}.\qquad(10)$$

Thus, we obtain our stability criterion in terms of strain and strain rate.

2.3 Flow Localization Factor (FLF)

The discussion in this section is to derive a parameter that characterizes the degree of flow localization, which we name the "flow localization factor (FLF)."

2.3.1 Uniaxial Tension

For uniaxial tension, the FLF ξ_I can be defined according to Eq.10 as

$$\xi_I = \begin{cases} \dfrac{\delta\dot{\varepsilon}/\delta\varepsilon - \dot{\varepsilon}}{\dot{\varepsilon}} = \dfrac{\delta\dot{\varepsilon} - \dot{\varepsilon}\delta\varepsilon}{\dot{\varepsilon}\delta\varepsilon} & \text{if } \dfrac{\delta\dot{\varepsilon} - \dot{\varepsilon}\delta\varepsilon}{\dot{\varepsilon}\delta\varepsilon} > 0 \\ 0 & \text{if } \dfrac{\delta\dot{\varepsilon} - \dot{\varepsilon}\delta\varepsilon}{\dot{\varepsilon}\delta\varepsilon} \leq 0 \end{cases},\qquad(11)$$

From Eq.1,
$$\delta\dot{\varepsilon} = (\delta\sigma - \frac{\partial\sigma}{\partial\varepsilon}\delta\varepsilon)\Big/\frac{\partial\sigma}{\partial\dot{\varepsilon}}, \qquad (12)$$

which divided by $\dot{\varepsilon}\delta\varepsilon$ gives
$$\frac{\delta\dot{\varepsilon}}{\dot{\varepsilon}\delta\varepsilon} = \left(\frac{\delta\sigma}{\delta\varepsilon} - \frac{\partial\sigma}{\partial\varepsilon}\right)\Big/\dot{\varepsilon}\frac{\partial\sigma}{\partial\dot{\varepsilon}}. \qquad (13)$$

Since $\delta\sigma = -\dfrac{\sigma\delta A}{A} = \sigma\delta\varepsilon$, $\delta\sigma/\delta\varepsilon = \sigma$, and Eq.13 gives
$$\frac{\delta\dot{\varepsilon}}{\dot{\varepsilon}\delta\varepsilon} = \left(1 - \frac{1}{\sigma}\frac{\partial\sigma}{\partial\varepsilon}\right)\Big/\frac{\dot{\varepsilon}}{\sigma}\frac{\partial\sigma}{\partial\dot{\varepsilon}} = \frac{1-\gamma}{m}. \qquad (14)$$

Therefore, the FLF is expressed as
$$\xi_I = \begin{cases} \dfrac{1-\gamma}{m} - 1 & \text{if } 1-\gamma-m > 0 \\ 0 & \text{if } 1-\gamma-m \leq 0 \end{cases}, \qquad (15)$$

in which the stability condition, $\xi_I = 0$, is equivalent to Hart's stability criterion, $\gamma + m \geq 1$ (Eq.7). The definition of the FLF for uniaxial tension is the same as that of the "instability parameter" defined by Cáceres [11] except that the former is defined to be zero when the deformation is stable. The instability parameter defined by Cáceres does not have a clear physical meaning, as the FLF does, and it is only valid for uniaxial tension. Nevertheless, the FLF can be generalized to biaxial stretching conditions, which will be done in the next section.

2.3.2 Biaxial Stretching

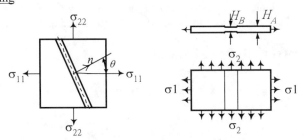

Fig. 1. Biaxial stretching of a thin sheet [8]

The localization factor for biaxial stretching is derived via the stability analysis for thin sheets under biaxial stretching studied in [8]. It is assumed that when the neck develops, it is normal to the maximum principal stress (σ_1), and $\Delta\dot{\varepsilon}_2 = 0$ (see Fig.1 [8]). Combining the assumptions with incompressibility and equilibrium equation produces the following equation:
$$\sigma_1 = \frac{\delta\sigma_1}{\delta\varepsilon_1}. \qquad (16)$$

The equivalent strain, equivalent strain rate, and equivalent stress can be expressed as
$$\bar{\varepsilon} = a\varepsilon_1, \quad \bar{\dot{\varepsilon}} = a\dot{\varepsilon}_1, \text{ and } \bar{\sigma} = b\sigma_1, \qquad (17)$$

where

$$a = \sqrt{\frac{4}{3}(1+\rho+\rho^2)}, \text{ and } b = \frac{1}{2+\rho}\sqrt{3(1+\rho+\rho^2)}, \text{ with } \rho = \frac{\varepsilon_2}{\varepsilon_1}. \tag{18}$$

The formulas of thin sheets under biaxial stretching presented so far were developed by Ding[8], and now our analysis is developed based on them.

Since the neck is normal to direction 1, and $\Delta\dot{\varepsilon}_2 = 0$, the FLF under these conditions should be

$$\xi_{II} = \begin{cases} \dfrac{\delta\dot{\varepsilon}_1/\delta\varepsilon_1 - \dot{\varepsilon}_1}{\dot{\varepsilon}_1} = \dfrac{\delta\dot{\varepsilon}_1 - \dot{\varepsilon}_1\delta\varepsilon_1}{\dot{\varepsilon}_1\delta\varepsilon_1} & \text{if } \dfrac{\delta\dot{\varepsilon}_1 - \dot{\varepsilon}_1\delta\varepsilon_1}{\dot{\varepsilon}_1\delta\varepsilon_1} > 0 \\ 0 & \text{if } \dfrac{\delta\dot{\varepsilon}_1 - \dot{\varepsilon}_1\delta\varepsilon_1}{\dot{\varepsilon}_1\delta\varepsilon_1} \leq 0 \end{cases} \tag{19}$$

Under biaxial stretching, Eq.13 should be modified to

$$\frac{\delta\bar{\varepsilon}}{\dot{\bar{\varepsilon}}\delta\bar{\varepsilon}} = \left(\frac{\delta\bar{\sigma}}{\delta\bar{\varepsilon}} - \frac{\partial\bar{\sigma}}{\partial\bar{\varepsilon}}\right) \bigg/ \left(\dot{\bar{\varepsilon}}\frac{\partial\bar{\sigma}}{\partial\dot{\bar{\varepsilon}}}\right). \tag{20}$$

From Eq.16, 17, 20, and defining

$$\gamma = \frac{1}{\bar{\sigma}}\left(\frac{\partial\bar{\sigma}}{\partial\bar{\varepsilon}}\right), \text{ and } m = \frac{\dot{\bar{\varepsilon}}}{\bar{\sigma}}\left(\frac{\partial\bar{\sigma}}{\partial\dot{\bar{\varepsilon}}}\right), \tag{21}$$

we obtain the FLF under biaxial stretching as

$$\xi_{II} = \begin{cases} \dfrac{1-a\gamma}{m} - 1 & \text{if } 1-a\gamma-m > 0, \\ 0 & \text{if } 1-a\gamma-m \leq 0. \end{cases} \tag{22}$$

Under stable conditions, Eq.22 is reduced to the stability criterion of Ding[8]. Under uniaxial tension, $\rho = -0.5$, and Eq.22 is reduced to Eq.15. If $\rho = 1$ (balanced biaxial stretching), then $a = 2$.

3 Superplastic Fracture Bulging Experiments

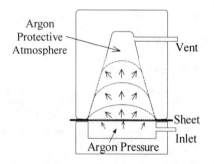

Fig. 2. Schematic diagram for superplastic bulging   Fig. 3. Fracture bulged part of Ti-6Al-4V sheet

The relation between the "FLF" and the forming limits requires further investigation through experimentation. Thus, we designed a series of experiments, a design accomplished by placing a 2.0mm-thick Ti-6Al-4V sheet over a cone-shaped die and blowing high purity argon gas over the sheet at 900°C with constant pressures until fracturing occurs. Fig.2 shows the schematic diagram for the superplastic bulging of a cone.

The picture of a typical formed part is shown in Fig.3. The fracture time (the time required to fracture) and fracture height for various forming pressures are shown in Table 1. The experimental results show that lower forming pressures lead to longer fracture times and greater forming heights.

Table 1. Experimental and simulation results of Ti-6Al-4V sheets bulged at 900°C

Forming pressure [MPa]	2.94	2.548	2.254	1.96
Fracture height [mm]	45.4	48.1	49.0	51.1
Fracture time [s]	591	748	883	1133
Simulation time to reach the fracture height [s]	607	775	921	1175
Error (%)	2.7	3.6	4.3	3.7

4 Finite Element Analysis

4.1 Basic Assumptions and Mesh

A commercial finite-element package ABAQUS is used to perform the modeling and analysis for the constant pressure bulging experiments. Since the Ti-6Al-4V alloy is less sensitive to void growth, the effects of void growth can be ignored.

Two-node linear axisymmetric membrane elements are used for the sheet. The sheet and die model for the superplastic bulging experiments is shown in Fig.4.

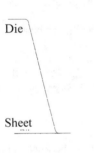

Fig. 4. Sheet and die model

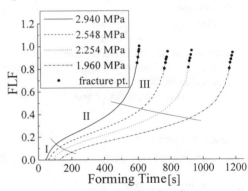

Fig. 5. FLF vs. forming time

4.2 Material Properties

In order to simplify the complexity of analysis, the behavior of superplastic materials is often characterized by the relation $\sigma = K\dot{\varepsilon}^m$; i.e. strain rate hardening is considered, but strain hardening is not. However, the Ti-6Al-4V alloy possesses a stronger strain hardening effect then other superplastic materials [12]; consequently, the constitutive equation $\sigma = K\dot{\varepsilon}^m \varepsilon^n$, which is more representative of the Ti-6Al-4V alloy, is used in this study, and the coefficients m and n are taken as functions of strain rate. The material coefficients at 900°C are obtained and converted from [7] and [12].

4.3 Simulation Results and Discussions

The simulation results for the experiments described in Sec.3 are also summarized in Table 1, where the errors of the forming time are less than 5 %. The critical point of the observation is the fracture point, i.e. the top of the blow-formed cone, which is in the balanced biaxial condition. As derived in Sec. 2.3.2, the FLF in this condition is $\xi_{II} = [(1-2\gamma)/m] - 1$. Fig.5 shows the curves of FLF at the fracture point versus forming time. According to the curves, the localization process can be divided into three stages:

I. The developing period of initial localized flow: The FLF somehow grows at a higher rate, with a noticeable increase in the growth rate as the forming pressure increases. This, however, is only a short duration in the entire forming process, and the second stage appears soon after.

II. The steady stage of strain concentration: The FLF grows slowly and steadily, and thus the slopes of the curves are nearly constant. This stage is the major part of the forming process.

III. The accelerating stage of strain concentration: The formed part approaches its forming limit in this stage, where the FLF again grows fast, and strain concentration is apparent throughout the stage until the end of the forming process, when fracturing occurs.

According to the above observations, an effective superplastic forming process can be designed by controlling the bulging pressure to keep it within the second stage in order to avoid the third stage.

5 Conclusions

The relation between superplastic failure behavior and strain concentration is studied. Since SPF is a slow process, there is a strong incentive to increase the efficiency of production while maintaining the required quality; techniques to achieve this goal can be developed by exploring new control schemes [7]. The quantitative analysis of the localization process of unstable plastic flow proposed in this paper can be applied as an aid in this work, to which we will devote ourselves in the future. From the analysis and experimental results, we conclude the following.

1. Hart's stability criterion is re-expressed in terms of strain and strain rate, and the FLF is defined according to the expression.
2. The proposed FLF stands for the degree of flow localization, and it can be applied for both uniaxial tension and biaxial stretching conditions.
3. The flow localization process during constant pressure bulging can be divided into three stages. The second stage takes the most of the forming time, and the fracturing occurs in the third stage.
4. The presented quantitative analysis of the flow localization process can be applied to develop SPF processes that achieve both high productivity and high quality.

Acknowledgement

This research was sponsored by the National Science Council of the Republic of China.

References

[1] E. W. Hart, Acta Metall., Vol.15 (1967), p.351.
[2] A. R. Tayupov, Scr. Metall. Mater., Vol.30, No.11 (1994), p.1387.
[3] N. Q. Chinh, G. Kapovics, P. Szommer, and I. Kovacs, Mater. Sci. Forum 217-222 (1996), p.1455.
[4] E. Sato and K. Kuribayashi, ISIJ International, Vol.33, No.8 (1993), p.825.
[5] J. Pilling and N. Ridley, Superplasticity in Crystalline Solids, Institute of Metals (1989), p.102-158.
[6] D.-J. Zhou, J. Lian and M. Suery, Mater. Sci. Technol., Vol.4 (1988), p.348.
[7] X. D. Ding, H. M. Zbib, C. H. Hamilton and A. E. Bayoumi, J. Mater. Eng. Perform., Vol.4, No.4 (1995), p.474.
[8] X. D. Ding, H. M. Zbib, C. H. Hamilton and A. E. Bayoumi, J. Eng. Mater. Technol., ASME, Vol.119 (1997), p.26.
[9] Du Zhixiao, Li Miaoquan, Liu Mabao and Wu Shichun, Appl. Math. Mech., V.17, N.2 (1996), p.133.
[10] W. A. Spizig, R. E. Smelser, and O. Richmond, Acta Metall., Vol.36, No.5 (1988), p.1201.
[11] C. H. Cáceres and D. S. Wilkinson, Acta Metall., Vol.32, No.3 (1984), p.415.
[12] C. J. Lin, Principles and Applications of Superplastic Metal Forming (in Chinese), 1st Ed., Aero-industry Publisher, Beijing (1990), p.51-52.

Low Temperature and High Strain Rate Superplasticity of Nickel Base Alloys

V.A. Valitov[1], O.A. Kaibyshev[1], Sh.Kh. Mukhtarov[1],
B.P. Bewlay[2] and M.F.X. Gigliotti[2]

[1] Institute for Metals Superplasticity Problems, Russian Academy of Sciences,
Khalturina 39, RU-450001 Ufa, Russia

[2] General Electric Corporate Research and Development,
PO Box 8, Schenectady NY 12301, USA

Keywords: Nanocrystalline Structure, Nickel Based Alloy, Submicrocrystalline, Superplasticity

ABSTRACT

This paper will describe features of micro-, submicro-, and nano-crystalline structure formation under severe plastic deformation and the influence of structure on superplastic (SP) behavior of high alloyed nickel base alloys with a range of hardened precipitates (γ'; $\gamma''+\delta$, $\gamma'+Y_2O_3$). It has been shown that severe plastic deformation over a wide range of homologous temperatures (0.9-0.2T_m) can refine the microstructure to a size of several tens of nanometers. In comparison with the microcrystalline (MC) state, the submicrocrystalline (SMC) structure in dispersion hardened alloys reduces the optimal temperature of SP deformation from 0.9-0.8T_m to 0.7-0.6T_m. The PDS alloy combines precipitation and dispersion hardening ($\gamma'+Y_2O_3$), and, in the SMC state, it can display both low temperature and high strain rate superplasticity. Features of microstructure transformations and failure of samples during SP deformation are considered.

1. INTRODUCTION

Broad application of SP forming of nickel base alloys with MC structures is limited by the fact that superplasticity only occurs at high homologous temperatures (0.8-0.85T_m) and low strain rates (10^{-3}-$10^{-4}s^{-1}$) [1-3]. SMC and nanocrystalline (NC) materials [4, 5] display SP behavior at higher strain rates and lower working temperatures, and they have the potential to expand the application of SP forming. There is presently substantial interest in structure-property relationships in materials with SMC and NC structures because these materials display unique physical and mechanical properties [6, 7]. However, there have been few previous studies of the formation of SMC and NC structures in complex nickel base alloys and the effect of these structures on properties. The goal of the present paper is to describe processing regimes for generating MC and SMC states in nickel base alloys with different chemical and phase compositions, and to evaluate the effect of these microstructures on SP properties.

2. EXPERIMENTAL PROCEDURE

The chemical compositions of the nickel base alloys that were studied are shown in Table 1.

Table 1. Compositions of the alloys that were studied

Alloy	Concentrations (wt. %)										
	Cr	Co	W	Mo	Fe	Al	Ti	Nb	C	Ni	Other
Inconel 718	19	-	-	3.1	18	0.5	1.0	5.1	0.05	base	
EP962	13.1	10.7	2.8	4.6	0.6	3.2	2.6	3.4	0.1	base	0.01 B
PDS	10.1	-	7.5	-	-	6.8	-	-	0.1	base	1.2 Y_2O_3

The materials selected for this study were considered to be typical examples of precipitation hardened nickel base alloys with a range of strengthening phases: Inconel 718 -$\gamma''+\delta$, and EP962 - γ'; PDS alloy -$\gamma'+Y_2O_3$. Samples of Inconel 718 and EP962 alloys were machined from 200 mm diameter billets. The powder PDS alloy was produced by mechanical alloying and was in the form of a 22 mm diameter rod. Compression and tension tests were conducted over a wide range of temperatures (500-1270°C) and strain rates ($1.33 \cdot 10^{-4}$-$1.7 \cdot 10^{-1} s^{-1}$). Cylindrical samples, 10 mm in diameter and 15 mm in height, were compressed in a SCHENK-PSA testing machine. Larger samples (40-80 mm in diameter and 10-12 mm tall), with SMC structures, were generated on 100 and 630 ton presses using processing conditions derived from the results of the smaller samples. SP characteristics were measured using tension tests on flat samples, 10x5x2mm in gauge dimension, and cylindrical samples, 5 mm in diameter and 25 mm in gauge length. Transmission electron microscope (TEM) investigations were performed using a JEM-2000EX.

3. RESULTS

Initial state

The microstructure of hot-deformed precipitation-hardened alloys Inconel 718 and EP962 in the initial state is completely recrystallized coarse-grained with a mean grain size of 40-80 μm [4,9]. The TEM studies have shown that within the Inconel 718 grains there are coarse δ-phase plates between which there is a uniform dispersion of coherent disk-type γ''- precipitates. In the EP962, partially coherent precipitates of γ'-phase, 0.5 μm in size, are observed within the grains.

The powder dispersion-strengthened nickel base alloy, PDS, was produced by mechanical alloying. This generated an SMC structure with an initial grain size of 0.8 μm. The SMC microstructure is characterized by high density of dislocations both within the grains and along grain boundaries [4].

Formation of MC and SMC structure of Inconel 718 and EP962 alloys

In Inconel 718 and EP962, the microduplex type MC structures were formed using high strain thermo-mechanical processing. During processing intense dynamic recrystallization was accompanied by significant changes in morphology of second phase precipitates and their distribution in the matrix. Analysis of the evolution of the microstructure during hot deformation allowed establishment of the bases of transformation of the coarse-grained structure to the MC structure [2, 4, 8, 9].

High strain thermo-mechanical processing was used for formation of SMC and NC structures in the alloys under study. This process includes isothermal deformation with intermediate annealing and a gradual step by step reduction in the temperature of processing. Although there are significant differences in the chemical compositions and constituent phases of the alloys studied, the formation of the SMC and NC structures possess certain similarities.

High-strain thermo-mechanical processing of Inconel 718 and EP962 in the temperature regimes 0.6-0.78 T_m generated refinement of the MC structure to a grain size of 0.5-0.3 μm. In both Inconel 718 and EP962, some coarse second phase particles (up to 1 μm) were still present. This suggests localization of the deformation in the softer matrix, which was probably due to insufficient accumulated strain. The noteworthy feature of the SMC structure of the alloys under study is the non-equilibrium nature of the structure as described by the high dislocation density both within the grains and along the grain boundaries. Grain boundary contrast typical of high-angle grain boundaries was not observed (Fig. 1a, b). Post-deformation annealing contributes to formation of the SMC structure with equilibrium grain boundaries; however, elastic stresses have not been completely removed.

In order to achieve additional grain refinement, thermo-mechanical processing of the PDS alloy was performed at temperatures of 950-900°C. This processing refined the initial γ and γ'-phase grains to a size of ~0.5 μm (Fig. 1c).

Additional deformation at much lower temperatures of 0.18-0.64 T_m generated NC in all alloys with grain sizes of 0.02-0.05 μm. TEM analysis of images of alloys in the NC state revealed that the microstructures are similar to those observed in other NC materials.

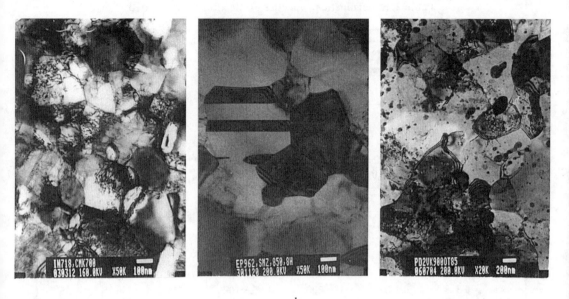

Fig.1. Microstructures after thermomechanical processing at temperatures in the range 0.6-0.78 T_m
a- Inconel 718; b - EP962; c - PDS

Microstructure and mechanical properties of alloys during SP deformation

Though structural states with approximately similar grain size can be formed in the alloys under study the differences in chemical composition and constituent phases should evidently exert a significant influence on their SP behavior. In this connection, the comparative analysis of SP behavior of nickel base alloys with SMC and MC structures has been conducted in the given work.

Comparison of the results of mechanical test data for MC and SMC samples (see Table 2) indicates a strong sensitivity of the mechanical behavior of the Inconel 718 to grain size. Table 2 indicates that a decrease in the grain size from 6 to 1-2 µm reduces the lower temperature limit of the effect of superplasticity. The mechanical test data show that additional refinement of the microstructure to SMC sizes further reduces the low temperature limit for superplasticity. The Inconel 718 with the SMC structure displays SP behavior even at 650°C. In addition, in Inconel 718 the lowest level of flow stress is observed for the SMC structure.

Calculations of the activation energy for SP deformation in Inconel 718 with MC and SMC structures (Table 3) have been performed using the procedure described previously [12]. These values correspond to the activation energy of grain boundary diffusion in Ni and Ni base alloys. The activation energy for SP deformation of Inconel 718 is in the range of 255-180 kJ/mol. The activation energies for grain boundary diffusion in Ni and Ni base alloys. The activation energy for deformation of the SMC structure is significantly lower than that for the MC structure. Heat treated samples that were deformed at 925°C revealed deformation bands which indicate cooperative grain boundary sliding during SP deformation [16].

Investigation of the microstructure of the deformed SMC Inconel 718 samples revealed that, after SP deformation, the grains remain equiaxed. The SP deformation also provided an equilibrium structure. The density of dislocations decreased, some grain growth was observed and non-equilibrium grain boundaries were converted to equilibrium boundaries, as indicated by the band contrast. At temperatures above 900°C the grain size exceeded 1 µm.

Table 2. Superplastic properties of the alloys investigated.

Alloy	Grain size (particles), μm		Temperature °C	Strain rate, s^{-1}	σ_{40}, MPa	δ, %	m
	γ- phase	δ- phase					
Inconel 718	6-12	-	930-980				0.7-0.9 [10]
	6	-	980	5·10^{-4}	70	514	0.5 [11]
	1-2	0.15-0.6	950	5.5·10^{-4}	63	660	0.6
			900	5.5·10^{-3}	226	215	0.31
			900	5.5·10^{-4}	96	480	0.4
			800	5.5·10^{-3}	205	140	0.2
			800	1.5·10^{-4}	134	390	0.33
			700	3·10^{-4}	563	170	0.2
			650	5.5·10^{-4}	742	85	0.1
	0.3	0.1-0.6	900	5.5·10^{-4}	65	790	0.6
			800	5.5·10^{-3}	292	270	0.3
			800	5.5·10^{-4}	123	430	0.5
			800	1.5·10^{-4}	87	1095	0.54
			700	5.5·10^{-3}	540	195	0.29
			700	5.5·10^{-4}	306	440	0.35
			700	3·10^{-4}	224	700	0.41
			650	5.5·10^{-3}	833	130	0.2
			650	5.5·10^{-4}	514	370	0.3
			600	5.5·10^{-4}	838	110	0.15
			600	1.5·10^{-4}	662	150	0.2
	0.08 -0.1		600	3·10^{-4}	573	154	0.31
			600	1.5·10^{-4}	414	350	0.37
EP962	γ-phase	γ'-phase					
	5.5	2.5	1075	1.33·10^{-3}	50	>550	0.6
			950	1.33·10^{-3}	120	>500	0.3
			950	1.33·10^{-2}	551	76	0.16
	2.5	1.3	875	1.33·10^{-3}	349	116	0.25
	0.25		950	1.33·10^{-2}	144	252	0.43
			875	1.33·10^{-3}	187	337	0.39
			800	1.33·10^{-3}	435	98	0.21
			800	3.3·10^{-4}	350	>550	0.3
			800	6.6·10^{-5}	170	>450	0.35
PDS	0.5		800	3.3·10^{-4}	71	240	0.35
			1100	8.3·10^{-2}	52	125	0.37
			1200	8.3·10^{-2}	33	190	0.48
			1250	8.3·10^{-2}	19	200	0.5

Analogous results were obtained in the EP962 (Table 2). Formation of the SMC structure reduces the lower temperature limit for superplasticity. In addition, for the SMC structure the strain rate interval for SP behavior is expanded, as shown for the data at 950°C.

TEM analysis of the SMC EP962 deformed at t=800°C and strain rate $3.3·10^{-4}s^{-1}$ by 550% showed no significant grain growth during deformation for grain sizes up to 0.4 μm. As in the case of Inconel 718, in the EP962 pore formation occurred during SP deformation, and it increased towards the failure zone. Away from the failure zone, individual pores less than 1 μm in size were observed. Near the failure zone, pores were aligned with the direction of deformation, and pore coalescence was observed. During SP deformation of alloys both in the SMC state and in the microduplex condition, pore formation developed mainly in the vicinity of carbide particles. However, in the SMC state, the mean pore sizes were 2-3 times smaller.

Table 3. Activation energies (kJ·mol^{-1}) for SP deformation of Ni and Ni-based alloys.

Structure	CoarseC	MC	SMC
Activation energies for diffusion of Inconel 718 : at temperature 675-725°C at temperature 875-900°C.	480$^+$-30	489$^+$-12 235$^+$-35	255$^+$-20 180$^+$-30
Activation energies for volume diffusion of Ni and Nickel-based alloys of IN-100	278-293 [13]	307-483 [14]	
Activation energies for grain boundary diffusion of Ni and Nickel-based alloys of IN-100	115 [13]	153-241 [14]	
Activation energies of Inconel 718	400 [15]		
Activation energy for dislocation pipes diffusion of Ni and Nickel-based alloys	170 [13]		

Analysis of the mechanical test data of the PDS alloy with SMC structure showed that it can display both low-temperature and high-strain-rate superplasticity (HSRSP) in a wider temperature-strain rate interval (Table 2) than for Inconel 718 and EP962. In the SMC PDS, HSRSP was observed at strain rates of $8.3 \cdot 10^{-2} s^{-1}$ - $1.7 \cdot 10^{-1} s^{-1}$. These are an order of magnitude higher than the strain rates observed during usual SP deformation of nickel base alloys with the MC structure.

Investigations of microstructure of tensile samples at both low (800°C) and high (1250°C) temperatures showed equiaxed grains of γ and γ'-phases (less than 0.9 μm in size), and the yttria particles were decorated by dislocations. During SP deformation pore formation also occurred. However, unlike the dispersion-hardened alloys, failure of the PDS samples during SP deformation occurred due to coalescence of micropores at yttria particles on transverse boundaries.

4. DISCUSSION

These results show that in nickel base alloys with different chemical compositions and constituent phases the microstructure can be refined to a scale of several ten nanometers by high strain thermo-mechanical processing. The analysis of SMC and NC states formed in nickel base alloys suggests that they are analogous to other materials with equivalent structures [4-5]. These states are characterized, especially for the NC state, by a high level of residual stress both within the grains and at the grain boundaries, as indicated by the high dislocation density (non-equilibrium state) in these regions.

Formation of deformation bands observed during deformation of Inconel 718 is consistent with the mechanism of cooperative grain boundary sliding during SP deformation. The following deformation mechanisms operate in the deformation band: GBS, dislocation slip, and diffusion [16]. Analysis of the structural transformations observed during SP deformation of Inconel 718 suggest that common deformation mechanisms control SP behavior of these different alloys.

The analysis of the results of activation energy calculations indicates that SP deformation in SMC Inconel 718 is controlled by grain boundary diffusion. However, values of activation energy of SP deformation of Inconel 718 are close to the values of activation energy of diffusion along the dislocation pipes in nickel base alloys [13]. Thus, it is possible that grain boundary sliding in the SMC material with non-equilibrium boundaries may also be controlled by diffusion along the dislocation pipes.

Comparison of the behavior of nickel base alloys with MC and SMC structures suggests that analogous microstructural transformations occur during SP deformation; these include retention of equiaxed grain morphologies, grain growth, and pore formation. At the same time, the increase in the total area of intergranular and interphase boundaries, and the increase in the dislocation density at the grain boundaries, lead to both low temperature SP behavior (in all the alloys that were studied) and high-strain-rate superplasticity at high homologous temperatures in the PDS alloy. Finally, SP deformation generated a structural condition closer to equilibrium as a result of the transformation of "non-equilibrium grain boundaries" to lower energy boundaries similar to those usually observed after annealing of SMC and NC materials [6].

5. CONCLUSIONS

1. High strain thermo-mechanical processing with a gradual reduction in the deformation temperature generates MC, SMC and NC structures in nickel bases alloys with different chemical composition and constituent phases.
2. During SP deformation of dispersion hardened alloys both with MC and SMC structures, similar microstructural transformations are observed. This suggests the operation of one and the same mechanism of SP deformation. The formation of deformation bands suggests that the operative deformation mechanism is cooperative grain boundary sliding.
3. The activation energy for SP in the alloy Inconel 718 was estimated as 255-180 kJ/mol, suggesting that grain boundary diffusion was the rate controlling mechanism for deformation.
4. Low-temperature superplasticity was observed in SMC dispersion-hardened alloys with different chemical compositions and constituent phases. SP deformation allows formation of SMC structures with equilibrium grain boundaries.
5. High thermal stability of the SMC structure in the PDS alloy is provided by the Y_2O_3 particles. This structure displays SP behavior at low strain rates and low temperatures ($<0.65T_m$), and high strain rates at high homologous temperatures ($0.92\ T_m$).

REFERENCES

[1] O.A. Kaibyshev, F.Z. Utyashev, V.A. Valitov, Metal. Term. Obr. Met., N7 (1989), pp. 40-44.
[2] O.A. Kaibyshev, Superplasticity of alloys, Intermetallides and Ceramics, Springer Verlag, Berlin (1992), 316.
[3] V.A. Valitov, F.Z. Utyashev, Sh. Kh. Mukhtarov, Mat. Sci. For., 304-306 (1999), pp. 79-84.
[4] V.A Valitov, G.A. Salishchev, Sh.Kh. Mukhtarov, Mat. Sci. For, ICSAM97, 243-245, 557(1997).
[5] V.M. Imaev, et al., Scripta Mater., Vol. 40, No 2 (1999), pp.183-190.
[6] R.Z. Valiev, A.V. Korznikov, R.R. Mulyukov, Mater. Sci. Eng. A168, 141 (1992).
[7] G.A. Salishchev, O.R. Valiakhmetov, V.A. Valitov, Sh. Kh. Mukhtarov, Mat. Sci. For. 170-172, 121 (1994).
[8] O.A. Kaibyshev, V.A. Valitov, G.A. Salishchev, Fiz. Met. Metal., Vol.75,4 (1993), pp.110-117.
[9] O.A. Kaibyshev, V.A. Valitov, Sh. Kh. Mukhtarov, B.P. Bewlay and M.F.X. Gigliotti, MRS Fall Proceedings, Vol. 601, pp. 43-48.
[10] Superplastic Forming of Structural Alloys, Edited by N.T. Paton and C.H. Hamilton, The Metallurgical Society of AIME (1982).
[11] M.W. Mahoney, The Minerals, Metals & Materials Society (1989), pp. 391-405.
[12] P. Poirier. Plasticitte a haute temperature des solides cristallins (1976) Paris.
[13] H.F. Frost and M.F. Ashby, Pergamon Press, 1982.
[14] R.G. Menzies, J.W. Edington and G.J. Davies, Metal Science Vol.15 (1981), pp. 210-216.
[15] S.C. Medieros, et al., Scripta Mater., 42 (2000), pp. 17-23.
[16] O.A.Kaibyshev, A.I. Pshenichniuk, V.V. Astanin. Acta Mater. Vol. 46, No. 14 (1998) pp. 4911-4916.

Grain Refinement of Materials

Development of Ultrafine Grained Materials Using the MAXStrain® Technology

W.C. Chen[1], D.E. Ferguson[1], H.S. Ferguson[1], R.S. Mishra[2] and Z. Jin[3]

[1] Dynamic Systems Inc., Poestenkill NY 12140, USA

[2] University of Missouri, Department of Mechanical Engineering, Rolla MO 65409-0340, USA

[3] Reynolds Metals Company, Corporate Research & Development,
13203 N. Enon Church Road, Chester VA 23836-3122, USA

Keywords: MAXStrain Technology, Severe Plastic Deformation, Ultra-Fine Grain

ABSTRACT

Ultrafine-grained structures, having grain size of 1μm or less in diameter have shown high rate superplasticity at lower temperatures compared to the conventional superplastic forming conditions. Many techniques have been developed to produce the ultrafine-grained structures primarily by severe plastic deformation, which include equal channel angular pressing (ECAP), 3D forging, high pressure torsion (HPT), and accumulative roll bonding (ARB), etc. A multi-axis-restraint deformation technique (MAXStrain® Technology) was recently developed to obtain ultrafine-grained structures by performing extremely large strains with accurate control of strain, strain rate and temperature. The technology has demonstrated a potential to produce industrial-size ultrafine-grained bulk materials. In this paper, an application of the MAXStrain Technology for developing ultrafine grain structures of a commercial aluminum alloy is presented. The results show that the technology is a promising technique to produce ultrafine-grained structures of aluminum alloys for the high rate superplastic forming.

INTRODUCTION

High rate superpasticity has become more and more attractive to both academia and industries because it promises that complex engineering products can be formed more rapidly than the conventional superplastic forming. High rate superplasticity has been observed in many alloy systems [1-4] that have very fine grain structures. It has been also demonstrated that the superplasticity temperatures can be reduced by reducing the grain size to sub-micronmeter in diameter (or ultrafine grained structures) [1-4]. Several techniques have been developed to produce ultrafine grain structures primarily using severe-plastic-deformation methods, such as, equal channel angular pressing

---

® MAXStrain is a registered trade mark of Dynamic Systems Inc,, NY, USA.

(ECAP), 3D forging, high pressure torsion (HPT), and accumulative roll bonding (ARB). A multi-axis restraint deformation technique (Maxstrain® Technology) was developed to achieve extremely large strains such that it allows production of ultrafine grained structures with controllable conditions of strain, strain rate, and temperature. The sample size produced by this technique to date is large enough for subsequent tensile mechanical testing. This technique has a great potential to be scaled-up for the production of large bulk superplastic materials for real engineering components.

Superplasticity in some ultrafine grained materials has been shown to occur at temperatures from 200°C to 250°C below those in ultrafine grain materials [3]. For AA5083, superplasticity has been observed at 480°C to 550°C at an elongation of 670% [4]. In this paper, a commercial 5083 aluminum alloy was deformed at room temperature using the Maxstrain Technology. The tensile mechanical properties were measured at room temperature for as-deformed and annealed samples to examine the effect of grain refinement and annealing on the mechanical properties of the processed commercial 5083 aluminum alloy. A preliminary study on possible superplastic conditions of the processed samples was also pursued.

THE MAXSTRAIN TECHNOLOGY

Severe-plastic deformation may apply to a strain of at least 4 or more. Often used deformation strains, such as in rolling, are the order of 0.3 to 0.4 per roll stand. 10 or more such deformations are required to achieve a strain of 4. A strain of 4 may be the lower limit required to achieve ultrafine grains in many materials [5, 6]. Since during deformation material flows in all unrestricted directions, most deformation processes excepting the above mentioned methods have very little material left in the work zone after a strain of 4 or more has been introduced. If the deformation is plane strain type, less than 2% of the thickness remains after a strain of 4 is introduced. The MAXStrain Technology applies strains only in two axes with little strain in the third axis. With this procedure strains of 10 or more are possible before material failure. For a strain of 10, more than 80% of the original section remains and can be used for subsequent test work.

The system used to obtain very large strains is comprised of a rotation assembly as shown in Fig. 1 mounted in a thermomechanical system. The rotation assembly is comprised of a very rigid frame in which the material is mounted for rotation around and perpendicular to the axis of the thermomechanical single axis machine. The grips at each end of the specimen are electrically insulated from one another for high electric current passing through the specimen. The current through the specimen is controlled in a thermal servo system that permits rapid and accurate temperature control. The thermomechanical system provides accurate servo hydraulic control of two independent rams, one on each side of the specimen as shown by the arrows depicting the deformation direction (Fig.1). Deformation in the middle of the specimen cannot cause elongation of the specimen because of the restraint at the ends of the specimen. Fig. 2 shows two specimens, where one has been deformed without restraint and the other with full restraint in the longitudinal direction. Without restraint the specimen elongates when deformed at the midspan (Fig. 2A). With full restraint the specimen increases in width with no increase in length (Fig. 2B). During deformation the middle section of the specimen flows sideways such that the middle portion of the specimen becomes wider and thinner. The specimen is deformed repeatedly with 90-degree rotation along specimen longitudinal axis between deformations such that the deformation flow becomes perpendicular between two sequential hits. The repeatedly deformation with 90-degree rotation will lose little volume of material in the deformed zone.

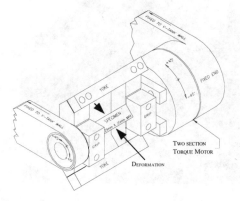

Figure 1 Schematic drawing of the rotation assembly of the MAXStrain System

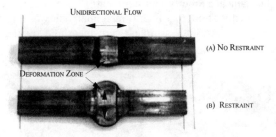

Figure 2 Unidirectional metal flow with no restraint (A) and with restraint (B)

The two-axis Maxstrain system has a maximum specimen size of 25 mm square and 195 mm in length. Consecutive deformation cycles may be as short as 0.5 seconds. After inducing severe strain in the midspan of the specimen, mechanical test specimens are machined for subsequent mechanical tests. The deformation can be done at any programmed temperature and strain rate. Controlled heating and cooling rates are part of the program. The time between deformations is also programmed along with the strain and strain rate of each deformation. Some aluminum materials are deformed at room temperature, but the majority of the high strain work is done at elevated temperatures.

EXPERIMENTAL

A commercial aluminum alloy AA5083 was deformed at different strains up to 9 using the Maxstrain system. The alloy was homogenized before deformation. Specimens were 15 mm by 15 mm in cross section and 195 mm in length. They were deformed at a strain rate of 2/s at room temperature with a different number of hits (strains). Each hit was programmed with 30% reduction. A specimen after 21 hits was shown in Fig. 3 on the Maxstain system.

Figure 3 An aluminum specimen deformed after 21 hits

Round tensile test samples (Fig.4) were machined from deformed specimens for room temperature mechanical property measurement. A small diameter of the reduced section was used since the strength of deformed material was higher. Mechanical properties of the deformed AA5083 samples were measured on the Gleeble thermal mechanical simulator.

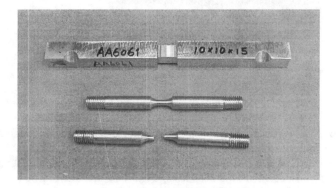

Figure 4 Tensile testing samples machined out of MAXStrain-deformed specimens

The effect of annealing temperature was also studied with a cold deformation strain of 5. The deformed specimens were annealed at temperatures of 100 °C, 150 °C, 200 °C and 250 °C for 1 hour using an electric box furnace. The tensile samples were machined after annealing and tested at room temperature.

RESULTS

The room temperature mechanical properties of the aluminum alloy AA5083 deformed using the MAXStrain system are shown in Fig. 5 as a function of cold deformation strain. The tensile strength increases from 290MPa for non pre-deformed to 560MPa for the deformed to strain of 5. However, the elongation decreases from 22% to 6.7% at the same strain.

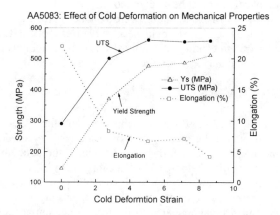

Figure 5 Effect of cold deformation strain on mechanical properties of a commercial AA5083 at room temperature

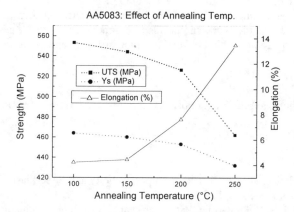

Figure 6 Effect of annealing temperature on room temperature mechanical properties of a commercial aluminum alloy AA5083 deformed to strain of 5

The effect of annealing temperature on mechanical properties is shown in Fig. 6 for the specimen with a prior strain of 5. Both the yield strength and the tensile strength decrease as the annealing temperature increases. The decrease is more dramatic when the temperature is above 200 °C. The elongation increases to 13.4% at 250°C. This implies that the temperature at 200°C may be close to the temperature where the ultrafine grains start to grow significantly.

A preliminary superplasticity test at 10^{-3}/s was also conducted at 500°C. The sample was annealed at 200°C for 1 hour after a strain of 5. An elongation of 151% with the final reduction of area of 70.3% was obtained. The lower elongation of superplasticity with a relatively low reduction of area could be related to the minor defects or micro-cracks that were observed under SEM in the deformation zone at high strains. Also, a percussion-welded thermocouple junction on the 3mm diameter reduced section gauge may also have contributed to the lower elongation because it had ruined the microstructure at the junction causing early fracture at the thermocouple junction. The grain stability during superplastic forming needs to be investigated and the effect of grain orientations on superplasticity must be studied too.

Both optical and SEM examination were conducted for the deformed samples. No clear grain/sub-grain boundaries could be observed except for tangled flow lines. Further work is under way to examine the grain size and orientation using both TEM and OIM for both deformed and annealed samples.

CONCLUDING REMARKS

A high strain of 10 or more can be achieved using the Maxstrain system. The bulk deformed specimens can be machined into tensile testing samples for subsequent property measurement. The cold deformation can double the strength of the AA5083, although it drops the elongation below 10%. Annealing below a certain temperature may recover the elongation significantly without a big loss of the strength.

Fracture and cracks start to occur when the strain gets higher at room temperature. This may have contributed in some extent to the low elongation at high strains. Future work should focus on deformation of the alloy at warm temperature such as $100^{\circ}C$ to $200^{\circ}C$ to improve the workability. Processing at warm temperatures promotes conversion of low angle boundaries to high angle boundaries which are beneficial for superplasticity. Hot furnace testing system should be used to conduct superplastic tensile tests to avoid possible microstructure damage due to percussion thermocouple welding. More detailed microscopic work should also be conducted on the Maxstrain deformed alloys to explore the relationship between the microstructure and the properties.

ACKNOWLEDGEMENT

The authors appreciate supply of the test materials from Reynolds Metals Company.

REFERENCES

1. Amit K. Ghosh and Thomas R. Bieler, Superplasticity and Superplastic Forming, TMS, 1998, p. 43, 89, 155, 227, 297, 305, 321.
2. Rajiv S. Mishra, S.L. Semiatin, C. Suryanarayana, Naresh N. Thadhani, and Terry C. Lowe, Ultrafine Grained Materials, TMS, 2000, p.381, 421.
3. R.Z. Valiev, A. V. Korznikov, and R.R. Mulyukov, Materials Science and Engineering, A168 (1993) 141-148.
4. P G Partridge; D S McDarmaid; I Bottomley; D Common, NATO, Agard Lecture Series, p 6.1 - 6.23
5. W.Y. Choo and S. W. Lee, "High Performance Steels for Structural Application", Cleveland, OH, ASM Annual Meeting, (1995), 117.
6. H.R. Hou, Q.Y. Liu, Q.A. Chen, and H.Dong, "Grain Refinement of Microalloyed Steel Through Heavy Hot Deformation and Controlled Cooling", ACTA Metallurgica SINICA, Vol. 13, No. 2, pp.508-513, April 2000.

Optimization for Superplasticity in Ultrafine-Grained Al-Mg-Sc Alloys Using Equal-Channel Angular Pressing

M. Furukawa[1], A. Utsunomiya[2], S. Komura[2], Z. Horita[2], M. Nemoto[3] and T.G. Langdon[4]

[1] Department of Technology, Fukuoka University of Education, Munakata, Fukuoka 811-4192, Japan

[2] Department of Materials Science and Engineering, Kyushu University, Faculty of Engineering, 6-10-1 Hakozaki, Higashiku, Fukuoka 812-8581, Japan

[3] Sasebo National College of Technology, 1-1 Okishin-cho, Sasebo 857-1193, Japan

[4] Departments of Materials Science and Mechanical Eng., University of Southern California, Los Angeles CA 90089-1453, USA

Keywords: Al-Mg-Sc Alloy, Equal-Channel Angular Pressing, High Strain Rate Superplasticity, Severe Plastic Deformation, Ultrafine Grain Size

Abstract
Ultrafine grain sizes may be achieved in materials subjected to severe plastic deformation through processing procedures such as equal-channel angular pressing (ECAP). Experiments were conducted to evaluate the factors influencing the superplastic characteristics of Al-Mg-Sc alloys. Specifically, detailed testing was undertaken to determine the influence of (i) the processing route, (ii) the strain introduced by ECAP and (iii) the Mg content.

1. Introduction
There is considerable current interest in using severe plastic deformation (SPD) in order to reduce the grain size of materials to the submicrometer or nanometer level [1-3]. Several SPD techniques are available but the most promising procedure appears to be equal-channel angular pressing (ECAP) where a sample is pressed repetitively through a die containing a channel which is bent through an angle close to 90° [4-6]. The process of ECAP has the advantage that it may be used for relatively large samples and there is a potential for scaling-up the procedure by developing multi-pass facilities in which high total strains are achieved on a single passage through the die [7].

Earlier experiments showed it was possible to use ECAP to achieve an ultrafine submicrometer grain size in Al-Mg-Sc alloys [8] and subsequent tensile testing after ECAP yielded elongations to failure of up to >1500% under optimum testing conditions [9]. The present experiments were conducted to evaluate three factors which influence the measured superplastic ductilities. Since ECAP is generally conducted by pressing a sample repetitively through a die, there is an opportunity to rotate the sample between consecutive passes. Four distinct processing routes may be identified and these are termed route A where there is no rotation of the sample, route B_A where the sample is rotated by $\pm 90°$ in alternate directions between each pass, route B_C where the sample is rotated by $+90°$ in the same direction between each pass and route C where the sample is rotated by $+180°$ between each pass [10]. In the present work, the effects of processing through routes A, B_C and C were investigated since it was shown earlier that route B_C is preferable to route B_A in producing an array of grains separated by high-angle boundaries [11]. In addition, experiments were conducted to determine the influence of the total strain imposed by ECAP and the effect of using different concentrations of Mg.

2. Experimental materials and procedures

The experiments to investigate the processing route and the total imposed strain were conducted using an Al-3% Mg-0.2% Sc alloy. This material was produced by casting, homogenizing at 753 K for 24 hours, swaging into rods with a diameter of 10 mm and then cutting to lengths of ~60 mm. Prior to ECAP, all samples were solution treated for 1 hour at 883 K to give an initial grain size of ~200 μm. The ECAP was conducted using a solid die with the internal channel bent into an L-shaped configuration with an angle, Φ, of 90° between the two segments of the channel and an angle, ψ, of 45° representing the outer arc of curvature where the two channels intersect. These values of Φ and ψ lead to an imposed strain of ~1 on each passage of the sample through the die. All pressings were conducted at room temperature using a facility in which the pressing speed was ~19 mm s^{-1}.

To investigate the effect of the processing route, samples were pressed using routes A (no rotation), B$_C$ (rotations of +90°) and C (rotations of +180°) to a total of 8 passes and an equivalent strain of ~8. Tensile specimens were then machined parallel to the pressing direction with gauge lengths of 5 mm and cross-sections of 2 × 3 mm^2 and these samples were pulled to failure at a temperature of 673 K using a machine operating at a constant rate of cross-head displacement with an initial strain rate of 3.3 × 10^{-2} s^{-1}.

To investigate the effect of the imposed strain, all samples were pressed using route B$_C$ but with imposed strains in the range from ~1 to ~12. These samples were also pulled to failure at 673 K using the same testing conditions as described above.

To investigate the influence of the Mg content, additional materials were prepared to give alloys of Al-0.2% Sc, Al-0.5% Mg-0.2% Sc and Al-1% Mg-0.2% Sc. Thus, it was possible to evaluate the effect of the Mg content in the range from 0% to 3%. These other alloys were subjected to the same homogenizing and solution treatments as for the alloy containing 3% Mg. All of these samples were pressed for 8 passes using route B$_C$, machined into tensile samples parallel to the longitudinal axes of the cylinders and then pulled to failure at 673 K using initial strain rates in the range from 1.0 × 10^{-4} to 3.3 s^{-1}.

Samples were examined by transmission electron microscopy (TEM) after ECAP with the foils prepared using the procedure described earlier [12]. All TEM observations were made on the Y planes of the samples, defined as the plane parallel to the side face of the sample at the point of exit from the die [10]. Selected area electron diffraction (SAED) patterns were recorded for each experimental condition using an aperture size of 12.3 μm.

3. Experimental results

Figure 1 shows examples of the microstructures of the Al-3% Mg-0.2% Sc alloy observed by TEM after pressing through 8 passes using (a) route A, (b) route B$_C$ and (c) route C, together with the appropriate SAED patterns. Inspection shows the grains tend to be elongated after processing through routes A and C and the SAED patterns reveal [110] zone-axis nets. By contrast, there is a reasonably homogeneous microstructure of predominantly equiaxed grains after processing through route B$_C$ and the diffracted beams in the SAED pattern then form rings indicative of grain boundaries having high angles of misorientation. The measured grain size for each of these conditions was ~0.2 μm.

Each sample was pulled to failure at 673 K using the same initial strain rate of 3.3 × 10^{-2} s^{-1}. The results are shown in Fig. 2 where it is apparent that each sample exhibits exceptional superplastic properties. However, the sample pressed using processing route B$_C$ shows the highest elongation to failure of 2280%. This result provides a very clear demonstration that the occurrence of high tensile ductilities requires an array of ultrafine grains separated by high angle boundaries. This is consistent with the proposal that superplasticity occurs almost exclusively by grain boundary sliding [13] since sliding is difficult or impossible in boundaries having low misorientation angles.

Figure 3 illustrates the effect of the pressing strain for the same alloy subjected to ECAP to (a) 1 pass, (b) 4 passes and (c) 8 passes using the optimum route B$_C$. There is a well-defined banded structure visible after 1 pass, with these bands oriented at approximately 45° to the upper and lower surfaces of the sample. The width of these bands was measured as ~0.2 μm which is equal to the grain size recorded in Fig. 1. It is apparent that the SAED pattern reveals a [110] zone-axis net after 1 pass so that the bands consist of subgrains separated by low angle boundaries. There is also some evidence for the banded structure after 4 passes but the microstructure is more homogeneous and the SAED pattern reveals a [110] zone-axis net but the diffracted beams are now diffuse. These observations show there is an increase in the boundary misorientations although the grain size remains at ~0.2 μm. These results may be compared with Fig. 3(c) which shows a reasonably homogeneous microstructure after 8 passes with a grain size of ~0.2 μm and an SAED pattern which confirms the presence of high angle boundaries. Careful measurements showed the area fractions of subgrains were ~20-30% after 6 passes but <10% after 8, 10 and 12 passes.

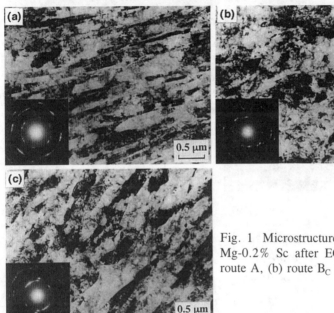

Fig. 1 Microstructures in the Y plane for Al-3% Mg-0.2% Sc after ECAP for 8 passes using (a) route A, (b) route B_C and (c) route C.

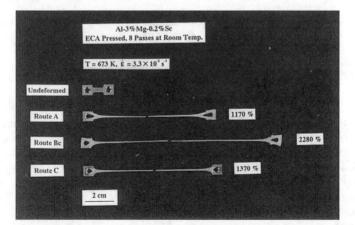

Fig. 2 Appearance of Al-3% Mg-0.2% Sc samples after pulling to failure at 673 K with an initial strain rate of 3.3×10^{-2} s^{-1}: samples were pressed through 8 passes using three different processing routes.

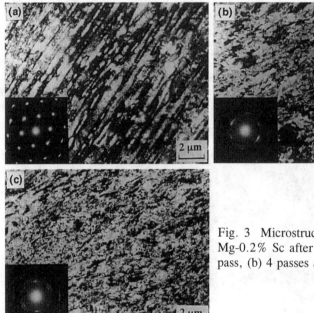

Fig. 3 Microstructures in the Y plane for Al-3% Mg-0.2% Sc after ECAP using route B_C for (a) 1 pass, (b) 4 passes and (c) 8 passes.

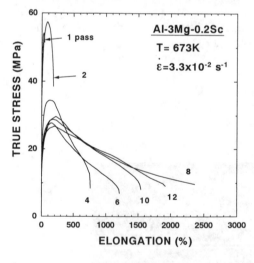

Fig. 4 Stress versus elongation for Al-3% Mg-0.2% Sc after ECAP using route B_C for 1-12 passes and then testing at 673 K with an initial strain rate of 3.3×10^{-2} s^{-1}.

Each of these samples was pulled in tension at 673 K using an initial strain rate of 3.3×10^{-2} s^{-1}. The resultant plots of true stress versus elongation are shown in Fig. 4 where it is apparent that the flow stress increases rapidly after 1 or 2 passes and with elongations as low as ~20% after a single pass. There are extended periods of strain softening after pressing through 4 to 12 passes with a maximum elongation of 2280% recorded after 8 passes. Thus, it is concluded that the strain imposed in ECAP must be sufficient for the microstructure to attain a reasonably homogeneous array of grains separated by high angle boundaries.

The effect of changing the Mg content is illustrated in Fig. 5 for samples pressed through 8 passes using route B_C and with 0%, 0.5%, 1% and 3% Mg, respectively. Inspection shows the microstructure depends markedly upon the amount of Mg contained within the matrix. In the absence of Mg in the Al-0.2% Sc alloy, there is a reasonably homogeneous microstructure of equiaxed grains having high angle boundaries but the grain size is ~0.70 μm. The addition of 0.5% Mg reduces the grain size to ~0.48 μm and this average size is further reduced to ~0.36 μm and ~0.20 μm with the additions of 1% Mg and 3% Mg, respectively. However, whereas the microstructure consisted almost exclusively of equiaxed grains in the Al-0.2% Sc alloy, close inspection revealed area fractions of ~50% of elongated grains in the three alloys containing Mg. In all four materials, these ultrafine grain sizes were retained up to temperatures of ~750 K.

Figure 6 shows the results obtained by pulling samples of these four materials to failure at 673

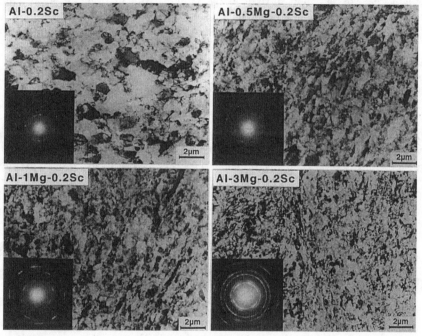

Fig. 5 Microstructures in the Y plane for Al-X% Mg-0.2% Sc alloys after ECAP for 8 passes using route B_C and with Mg contents corresponding to values of X of 0%, 0.5%, 1% and 3%, respectively.

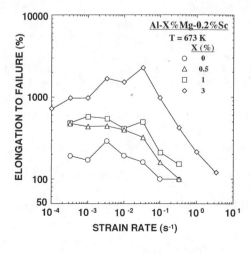

Fig. 6 Elongation to failure versus strain rate for samples tested at 673 K with different amounts of Mg after ECAP using route B_C.

K over a range of strain rates from 1.0×10^{-4} to 3.3 s^{-1}. Thus, the measured elongations to failure in Al-Sc alloys increase with increasing additions of Mg. In this investigation, maximum elongations are achieved with the addition of 3% Mg and the minimum elongations occur in the absence of Mg in the Al-0.2% Sc alloy. This trend is consistent with the grain sizes recorded after ECAP where, as shown in Fig. 5, the average grain size decreases with increasing additions of Mg. It is reasonable to suggest that even higher superplastic elongations may be attained with the addition of 5% Mg. However, it was found in practice that the introduction of 5% Mg led to a melting of the alloy during the solution treatment at 883 K. Conversely, if the solution treatment temperature is reduced to prevent incipient melting, it is no longer possible to dissolve the 0.2% Sc in the Al matrix and this gives a reduction in the formation of Al$_3$Sc particles and a consequent decrease in the stability of these ultrafine grains at high temperatures. It is concluded therefore that optimum superplasticity is attained with the addition of 3% Mg.

4. Summary and conclusions

1. Very high superplastic ductilities may be attained in an Al-3% Mg-0.2% Sc alloy after processing by equal-channel angular pressing at room temperature. In the present experiments, elongations to failure were achieved of up to 2280% when testing at 673 K under an initial strain rate of 3.3×10^{-2} s^{-1}.

2. When ECAP is conducted to a total of 8 passes, maximum elongation occur when using processing route B$_C$ in which the sample is rotated by +90° between each pass through the die. At least 8 passes are needed in this alloy to attain high ductilities when testing in tension.

3. The grain size after pressing of Al-0.2% Sc alloys decreases with the addition of Mg up to a maximum of ~3% Mg and there is an increase in the elongation to failure with increasing Mg.

Acknowledgements

This work was supported in part by the Light Metals Educational Foundation of Japan, in part by a Grant-in-Aid for Scientific Research from the Ministry of Education, Science, Sports and Culture of Japan, in part by the Japan Society for the Promotion of Science and in part by the National Science Foundation of the United States under Grants No. DMR-9625969 and INT-9602919.

References

1. T.C. Lowe and R.Z. Valiev, JOM 52 (4) (2000) p.27.
2. R.Z. Valiev, R.K. Islamgaliev and I.V. Alexandrov, Prog. Mater. Sci. 45 (2000) p.103.
3. T.C. Lowe and R.Z. Valiev (eds.), *Investigations and Applications of Severe Plastic Deformation*, Kluwer, Dordrecht, The Netherlands (2000).
4. V.M. Segal, Mater. Sci. Eng. A197 (1995) p.157.
5. Y. Iwahashi, Z. Horita, M. Nemoto and T.G. Langdon, Acta Mater. 46 (1998) p.3317.
6. V.M. Segal, Mater. Sci. Eng. A271 (1999) p.322.
7. K. Nakashima, Z. Horita, M. Nemoto and T.G. Langdon, Mater. Sci. Eng. A281 (2000) p.82.
8. S. Komura, P.B. Berbon, M. Furukawa, Z. Horita, M. Nemoto and T.G. Langdon, Scripta Mater. 38 (1998) p.1851.
9. P.B. Berbon, S. Komura, A. Utsunomiya, Z. Horita, M. Furukawa, M. Nemoto and T.G. Langdon, Mater. Trans. JIM 40 (1999) p.772.
10. M. Furukawa, Y. Iwahashi, Z. Horita, M. Nemoto and T.G. Langdon, Mater. Sci. Eng. A257 (1998) p.328.
11. K. Oh-ishi, Z. Horita, M. Furukawa, M. Nemoto and T.G. Langdon, Metall. Mater. Trans. 29A (1998) 2011.
12. J. Wang, Y. Iwahashi, Z. Horita, M. Furukawa, M. Nemoto, R.Z. Valiev and T.G. Langdon, Acta Mater. 44 (1996) p.2973.
13. T.G. Langdon, Mater. Sci. Eng. A174 (1994) p.225.

Corresponding author: M. Furukawa, e-mail: furukawm@fukuoka-edu.ac.jp

Microstructure and High Temperature Deformation of an ECAE Processed 5083 Al Alloy

L. Dupuy, J.J. Blandin and E.F. Rauch

Institut National Polytechnique de Grenoble (INPG), Génie Physique et Mécanique des Matériaux (GPM2), ENSPG-UJF-UMR CNRS 5010, BP 46, FR-38402 Saint-Martin d'Hères Cedex, France

Keywords: Aluminium Alloys, Damage, ECAE, Severe Plastic Deformation

Abstract

An industrial Al-Mg alloy was processed by Equal Channel Angular Extrusion. The resulting microstructures were studied and the mean orientation between cells were measured by transmission electron microscopy. The as-processed materials were deformed at high temperature. The effects of the number of extrusions on the mean grain size, the rheology at high temperature and the associated damage behaviour of the as-processed microstructures were investigated. It is shown that some damage occurs nearby second phase particles and the effect of ECAE processing on the size and the spatial distribution of these particles is discussed.

Introduction

Ultra fine grained microstructures (mean grain size less than 1 µm) in aluminium alloys can be obtained by severe plastic deformation [1-3], allowing to obtain superplastic properties at lower temperatures or higher strain rates than those conventionally used in superplastic forming. Among the available procedures, Equal Channel Angular Extrusion (ECAE) is the most popular one. The key parameters in this process are mainly the geometry of the device (in particular the channel angle ϕ), the temperature, the number of extrusions and the type of route (related to the possible rotations of the sample between each pass). This process has been extensively applied in the case of pure Al or binary Al-Mg alloys. The produced microstructures are generally characterised by TEM observations associated with selected area diffractions [2] and more recently by electron backscattered diffraction techniques carried out with field emission gun microprobe [3].

The aim of this paper is to investigate the effect of the number of ECAE extrusions on both the resulting microstructures and the deformability at high temperature of an industrial Al-Mg alloy.

Experimental procedure

The studied alloy is a 5083 Al-Mg alloy, of which the composition is given in table I, with an initial grain size close to 40 µm. It was provided as 10 mm thick hot-rolled plate.

Mg	Mn	Fe	Cr	Al
4.7	0.8	0.2	0.1	bal.

Table I: Composition of the investigated alloy (wt %)

The ECAE device used in this study, admits a channel angle equal to 90° (resulting in an equivalent von Mises strain of 1.15 after each pass) and molybdenum disulphide (MoS_2) was used as lubricant. The results presented in this work were obtained for extrusions performed at 150°C, according to a B_C route : between each extrusion, the sample is rotated from 90°. The number of extrusions N_E varied from 1 to 8. The resulting microstructures were characterised by TEM observations, performed on thin foil discs which plane contains the channel axes. The orientations of the cells were individually determined in indexing spots diffraction patterns since the degree of such misorientations between adjacent cells remains poorly characterised by selected area diffractions. To measure the misorientations between the cells, a procedure was developed [4] : diffraction patterns recorded with a small electron beam ($\approx$ 5 nm) for individual subgrains aligned on a straight lines over a distance of about 20 µm were directly analysed with a dedicated software which leads to orientation measurements with an accuracy of 1°. The threshold angle between low (LAB) and high angle boundaries (HAB) was arbitrarily fixed to 15°

High temperature mechanical properties were studied by uniaxial tensile tests, performed on specimens cut parallel to the extrusion direction and with a gauge region of 8x2x1.5 mm^3. To preserve the fine grain size when deformation starts, tests were carried out in a lamp furnace, which allows an heating rate of about 1°C/sec. The strain-induced damage behaviour was mainly studied by SEM observations on polished sections of the deformed samples.

Extruded microstructures

After processing, average structure sizes in the range 0.25-0.50 µm were measured, as illustrated by figure 1, which shows a TEM micrograph of the as-processed microstructures after 8 extrusions. TEM observations carried out after 1,2, and 4 extrusions, suggested that this average structural size was reached after 2 passes and then remained roughly constant in the further course of processing [4].

A typical line measurement of misorientation between cells is shown in figure 2. Each dot corresponds to a local misorientation between adjacent cells. Even if it was demonstrated that the proportion of HAB continuously increases with N_E [4], figure 2 shows that after 8 extrusions (which corresponds to a total strain of about 9), a significant proportion of LAB are still present. A mean grain size, defined in this case as the mean distance between two adjacent HAB, of about 1 µm is measured whereas after one extrusion, it was close to 3 µm. This difference between the structural size and the effective grain size confirms the difficulty to deduce directly a mean grain size from TEM observations of ECAE processed microstructures. Data shown in figure 2, indicate also that the microstructure is quite heterogeneous since grains of about 3 µm size can be detected.

The effect of N_E on the mechanical properties at room temperature of the studied alloy has been reported elsewhere [5] : the maximum stress increases and the ductility decreases with increasing N_E.

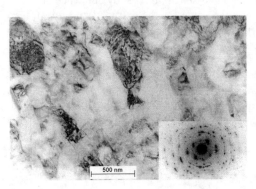

Fig. 1 : Typical TEM micrograph of the alloy microstructure after 8 extrusions

Fig. 2 : Line measurement of misorientation between cells after 8 extrusions.

Mechanical properties at high temperature

The stability of the processed microstructures was studied in static conditions. Thermal treatments were carried out between 150°C and 350°C and both hardness measurements and TEM observations were performed. Table II summaries the variations of the hardness after a treatment of 1 hour at different temperatures in the case of an alloy extruded 4 times, for which the hardness was 130 HV after processing.

T (°C)	160	200	240	260	280	300	350
HV	130	131	122	110	83	85	84

Table II : Variation of hardness with the temperature of a maintain of 1 hour (4 extrusions)

Important recovery takes place when temperature exceeds 280°C. At 260°C, recovery was more sluggish since after one hour, the hardness decreased only from 130 HV to 110 HV. In consequence, mechanical properties were preferentially investigated at 260°C. Strain-rate jump tests were carried out for strain rates between 10^{-4} s^{-1} and 10^{-2} s^{-1}. Figure 3 displays the resulting variation with strain rate of the m values at 260°C as a function of N_E. For a given strain rate, m increases with N_E and for a given N_E, m continuously decreases when strain rate is increased.

Figure 4 shows the effect of N_E on the curves (σ,ε) when the microstructures are deformed at 10^{-3} s^{-1}. In the case of the as-received alloy, high stresses (> 140 MPa) and limited elongations to fracture (< 100 %) are obtained. As long as N_E increases, the flow stress is reduced and the elongation to fracture is increased. A maximum elongation of about 300 % ($\varepsilon \approx 1.4$), associated with a flow stress of around 70 MPa are obtained after 8 extrusions. These results are correlated to the data shown in figure 3, where m increases from a value close to 0.2 after one extrusion to 0.35 after 8 extrusions.

The increase of both deformability and m values with N_E can be attributed to the concomitant transformation of LAB to HAB, which is expected to promote grain boundary sliding (GBS) during deformation. The contribution of such a grain size dependent mechanism of deformation is confirmed by the important sensitivity of the flow stress to N_E. These results differ from those obtained in the case of an ECAE processed 5056 Al alloy deformed at 473 K and 10^{-3} s^{-1} [6]. In this case, the flow stress was the same for the extruded and unextruded alloy despite different grain sizes which suggested that the contribution of GBS to macroscopic strain was negligible.

It can be noted that larger values of m are obtained for strain rates lower than 10^{-3} s^{-1}. For instance, at 2×10^{-4} s^{-1}, values of m $\approx$ 0.45 are measured after 8 extrusions. However, tests performed in such conditions led to a reduction of the fracture strain of the alloy, which can be attributed to the longer times required to reach large strains whereas the stability of the microstructure remains limited at 260°C.

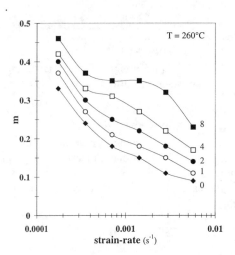
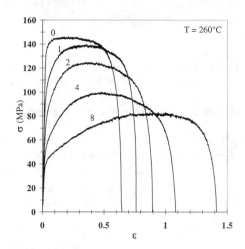

Fig. 3 : Effect of N_E on the variation of m with strain rate

Fig. 4 : Effect of N_E on the curves (σ,ε) when the alloy is deformed at 10^{-3} s^{-1}.

Damage behaviour

In the case of conventional superplastic Al alloys, it has been demonstrated that the fracture was not controlled only by the rheology (m value) of the alloy but also by the population of second phase particles, since these particles can act as preferential nucleation sites for strain-induced damage.

The effect of the population of second phase particles in the superplastic properties of a conventional 5083 alloy was studied by Chanda *et al.* [7]. These authors produced, thanks to various thermomechanical treatments, microstructures with similar mean grain sizes of about 6 µm and they demonstrated that the superplastic properties, and particularly the cavitation behaviour, were affected by the characteristics of the population of intermetallic particles. The role of the second phase particles is particularly important in the case of 5083 alloy, since this material can contain a large variety of particles, like $(Cr,Fe)Al_7$, $(Mn,Fe)Al_6$ or $Mg_3Mn_2Al_{12}$. Figure 5 shows a SEM micrograph of the investigated alloy in the as-received conditions.

Large inclusions can be detected with size larger than 10 µm and the particles are preferentially distributed along the hot rolling direction. In the case of conventional superplastic alloys, it is known that these particles can crack during further rolling and that these cracks may be preexisting nuclei for subsequent cavities.

Despite the expected role of the second phase particles on the damage behaviour of the extruded microstructures, the effect of ECAE processing on the size and the spatial distribution of these particles remains poorly documented.

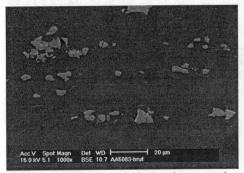

Fig. 5 : SEM micrograph showing second phase particles in the as-received alloy (the hot rolling direction is horizontal)

Fig. 6 : SEM micrograph of broken particles after 2 extrusions

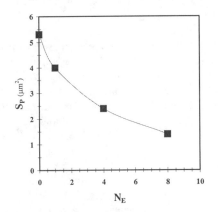

Fig. 7 : Effect of N_E on the mean surface of second phase particles

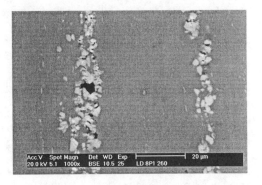

Fig. 8 : SEM micrograph showing strain-induced cavities developed near particles ($N_E = 8$ and the stress axis is vertical).

Figure 6 displays a SEM micrograph of severely broken particles detected in the alloy after only 2 extrusions. Due to the intense shear during each extrusion, an increasing fraction of particles are broken when N_E is increased. Figure 7 shows the effect of N_E on the mean surface of the particles : a significant reduction of this surface is obtained when N_E is increased. Such a reduction is expected to decrease cavity nucleation at the matrix/particle interface, due to the increase of the probability to get particles with size smaller than the critical diameter needed for nucleation. Moreover, it must be kept in mind that when N_E is increased, the flow stress is also reduced.

Nevertheless, as shown in figure 6, cracks are also induced by the extrusions and these cracks can accelerate the damage behaviour during the further deformation at high temperature.

The impact of the modification during ECAE of the population of second phases on strain-induced damage is not straightforward since the spatial distribution of the particles is also affected by the extrusions. As long as N_E increases, particles concentrate in stringers which maximum length is also increased. During deformation, these particle stringers are aligned with the stress axis. In consequence, the local strain rate in the matrix between the cavities is likely to be significantly higher than the macroscopic one, resulting in an increase of the cavity growth rate. Moreover, due to the proximity of the growing cavities, coalescence process may also be promoted.

From these comments, it appears that the effect of the ECAE process on the population of second phase particles affects the ability to reach large elongations to fracture with the extruded microstructures. Work is now in progress to get a more detailed description (for instance, by the use of high resolution micro-tomography techniques) of the effect of the successive extrusions on the spatial distribution of the second phase particles and its further connection with the cavities induced by high temperature deformation.

Conclusions

An industrial 5083 Al alloy has been processed by ECAE at 150°C. An average structure size of about 0.25-0.50 µm is rapidly developed but from the measurements of local misorientations between cells, it was shown that even after 8 extrusions, an important fraction of the boundaries were LAB, the mean grain size remaining close to 1 µm.

At 260°C, the strain rate sensitivity parameter and the deformability are increased with the number of extrusions but the elongation to fracture is also connected to the population of second phase particles through the microstructures. The ECAE process modifies the size as well as the spatial distribution of the particles. These changes can affect the nucleation, the growth as well as the coalescence of the cavities and consequently, they must be taken into account in the study of the superplastic properties of ECAE processed industrial alloys.

References

[1] S. Ferrasse, V.M. Segal, K.T. Hartwig, R.E. Goforth, Metall. Mater. Trans., 28A (1997), p. 1047.
[2] Y. Iwahashi, Z. Hojita, M. Nemoto, T.G. Langdon, Acta Mater., 46 (1998), p. 3317.
[3] A. Gholinia, P.B. Prangnell, M.V. Markushev, Acta Mater., 48 (2000), p. 1115.
[4] L. Dupuy, E.F. Rauch, J.J. Blandin, NATO Advanced Research Workshop "Investigations and Applications of Severe Plastic Deformation", (Moscow, Russia), T.C. Lowe et R.Z. Valiev Eds., (1999), p. 189.
[5] L. Dupuy, J.J. Blandin, E.F. Rauch, Materials Congress 2000 (Cirencester, GB), to be published in Mater. Sc. Tech.
[6] M. Kawazoe, T. Shibata, T. Mukai, K. Higashi, Scripta Mater., 36 (1997), p. 699.
[7] T. Chanda, A.K Ghosh, C. Lavender, Superplasticity and Superplastic Forming, Ed. A.K. Ghosh and T.R. Bieler, TMS, (1995), p. 41.

Effect of Temperature on the Microstructure and Superplasticity in a Powder Metallurgy Processed Al-16Si-5Fe-1Cu-0.5Mg-0.9Zr (wt%) Alloy

M.S. Kim[1], H.S. Cho[1], J.S. Han[1], H.G. Jeong[2] and H. Yamagata[3]

[1] School of Materials Science and Engineering, Inha University,
253 Yonghyun-dong, Nam-gu, Inchon 402-751, Korea

[2] Institute of Materials Research, Tohoku University,
Katahira 2-1-1, Aoba-ku, Sendai 980-8577, Japan

[3] R&D Division, Yamaha Motor Co. Ltd,
2500 Shingai, Iwata, Shizuoka 438-8501, Japan

Keywords: Al-Si Alloy, Grain Boundary Sliding, Grain Growth, High Strain Rate, Liquid Phase, Microstructure, Powder Metallurgy, Superplasticity, Temperature

Abstract The effect of temperature on microstructure and tensile deformation behavior of an Al-16Si-5Fe-1Cu-0.5Mg-0.9Zr (wt%) alloy fabricated by a powder metallurgy processing was investigated in temperature ranges from 673 to 813K under a high strain rate of $1.4 \times 10^{-1} s^{-1}$. The elongation increased with increasing temperature up to 793K, and then decreased with further increase in temperature. A superplastic elongation over 300% was achieved in the optimum temperature range of 783-803K owing to the occurrence of grain boundary sliding, a limited grain growth and the formation of an optimum amount of a liquid phase in the vicinity of grain boundary. On the other hand, above the optimum temperature, a rapid grain growth and an excessive amount of a liquid phase were suggested to cause the rapid drop of elongation.

1. Introduction

A hyper-eutectic Al-Si alloy has a low density, low thermal expansion coefficient and good wear resistance, which are suitable for light-weight engine part application. Recently developed powder metallurgy (PM) processed hyper-eutectic Al-Si alloy containing Fe, Cu, Mg and Zr had submicron sized grain structure and enhanced tensile strength [1]. In addition, the PM Al-Si based alloy exhibited superplastic deformation with a large elongation of about 400% at 793K and at a high strain rate of $1.4 \times 10^{-1} s^{-1}$ [2]. In the present paper, temperature dependence of the microstructure and tensile deformation behavior including superplasticity is demonstrated for the PM Al-Si based alloy over a wide range of temperature from 673 to 813K at the strain rate of $1.4 \times 10^{-1} s^{-1}$.

2. Experimentals

The rapidly solidified Al-16Si-5Fe-1Cu-0.5Mg-0.9Zr (wt%) alloy powders were produced by an air atomization method. The powders were then degassed, canned and extruded with extrusion ratio of 18 to a rod shape of 60mm in diameter. Tensile test specimens with a gage diameter of 3mm and a gage length of 12mm were prepared from the as-extruded bar with the tensile axis parallel to the extrusion direction. Tensile tests were conducted in air at temperatures from 673 to 813K under the strain rate of

Fig. 1 TEM micrograph of as-extruded alloy

$1.4 \times 10^{-1} s^{-1}$. Prior to testing, each specimen was heated up to each test temperature with the heating rate of 10K/min, and held at each temperature for about 20minutes. The specimens were cooled by air soon after failure. Microstruture was observed by scanning electron microscope (SEM) and transmission electron microscope (TEM). Microchemical analysis was carried out using energy dispersive spectrometer (EDS). Also, differential scanning calorimetry (DSC) experimentation was done at temperatures up to 873K using the same heating rate of 10K/min as the case of tensile tests. A typical microstructure of as-extruded specimen is shown in Fig.1. The grains are almost equiaxed, and the mean grain size is about 800nm. In addition, very fine particles (100-400 nm) are exhibited on grain boundaries and in the matrix. In previous work [1], the fine particles were identified as an intermetallic Al_5FeSi.

3. Results and Discussion

3.1 Tensile properties

Figure 2 shows the true stress-true strain curves of the specimens tensile-deformed at several testing temperatures with the high strain rate of $1.4 \times 10^{-1} s^{-1}$. It can be shown that the flow stress tends to decrease continuously with increasing strain when specimen is deformed at and below 723K. In contrast, when specimen is deformed at and above 773K, a steady-state flow or a slight strain hardening during deformation occurs up to a

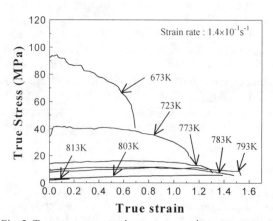

Fig. 2. True stress-true strain curves at various temperatures

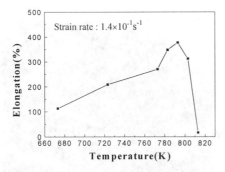

Fig. 3. Temperature dependence of elongation.

Fig. 4. Surface morphology of the specimen deformed superplastically at 793K.

high strain or up to the final stage of deformation. At the given strain of 0.1, the flow stress decreases rapidly with increasing temperature in the temperature range of 673-773K. A variation of elongation with test temperature is plotted in Fig. 3. As shown in this figure, elongation increases with increasing temperature up to 793K, and then decreases with further increase in temperature. A superplastic elongation over 300% is achieved at temperatures between 783 and 803K. The elongation of the specimen deformed at 813K, however, rapidly drops to the order of 25%.

3.2 Microstructures
It has been demonstrated that superplasticity at high strain rate of more than $10^{-2}s^{-1}$ can be obtained in various aluminum alloys by successful refinement in microstructure through PM process [3, 4]. It has also been demonstrated that grain boundary sliding is the dominant flow mechanism in the high strain rate superplasticity for fine-grained aluminum alloys [5, 6]. In the present investigation, the inspection of specimen surfaces revealed the occurrence of grain boundary sliding in the specimens deformed superplastically. One example is presented in Fig. 4 for the specimen deformed at 793K, showing a striation-like pattern formed in parallel to the tensile direction. Such a pattern is considered to be the trace of plastic flow at the newly developed surface of the grain or grain cluster caused by grain boundary sliding [6].

It is recognized that grain size is one of controlling factors to maintain superplasticity, and a stable microcrystalline structure during deformation is an essential condition for superplasticity [7]. In order to examine microstructural changes during deformation, TEM observations were performed in both specimens deformed at 793K and at 813K, and the results are shown in Fig. 5. It can be shown from Fig. 5(a) that the grain size is still small (about 1.2 μm) in the specimen deformed superplastically at the temperature of 793K. On the other hand, Fig. 5(b) of the specimen deformed at 813K with a limited elongation shows a considerable increase in grain size (3-4μm) compared to the case of the as-extruded specimen (Fig. 1). Therefore, it is likely that the high strain rate superplasticity occurs at the optimum temperature range where a refined grain structure is maintained, and that the superplastic property reduces rapidly above the optimum temperature where the rapid grain growth occurs.

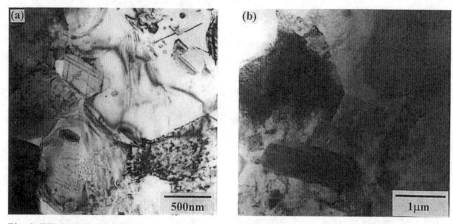

Fig. 5. TEM micrograph of the specimen deformed superplastically at (a) 793K and (b) 803K.

SEM fracture surface observations reveal that a number of cavities (Fig. 6(a)) are visible on the fracture surface in the specimens deformed at and below 773K. Interestingly, in case of the specimens deformed superplastically at temperatures from 783 to 803K, a filament-like phase is formed in the vicinity of grain boundary, as illustrated in Fig. 6(b). Similar formations of the filament-like phase have been observed in superplastically deformed metal matrix composites and considered to a result of the occurrence of a partial melting [8-10]. Using the as-extruded specimen, DSC experimentation was performed with the same heating rate of 10K/min as the case of the tensile tests. The result in Fig. 7 clearly indicates the onset of incipient melting at 781K, which is slightly lower temperature at which the filament-like phase appears. EDS analysis data in Fig. 8 reveals that the filament-like phase contains Cu and Mg in addition to Al, Si and Fe, whereas the matrix contains mainly Al, Si and Fe. According to Al-Cu-Mg ternary phase diagram [11], the eutectic reaction occurs at 781K as follows:

$$\text{Liquid} \rightarrow \alpha\text{-Al} + CuAl_2 + CuMgAl_2$$

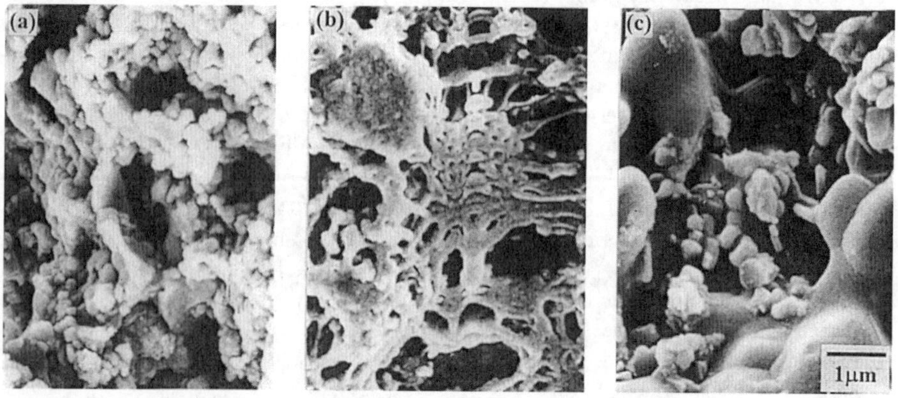

Fig. 6. Fracture surface of the specimens deformed at (a) 723K, (b) 793K and (c) 813K.

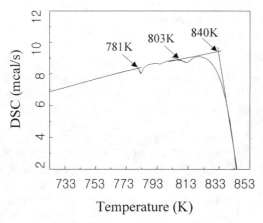

Fig. 7. DSC data of as-extruded alloy.

Therefore, it is thinkable that the filament-like phase exhibited in the vicinity of grain boundary in the specimens deformed superplastically at temperatures from 783 to 803K results from a liquid phase formed by the partial melting attributed to the above-mentioned eutectic reaction in connection with the segregation of Cu and Mg at grain boundary. When deformation temperature increases up to 813K, the fracture surface contains a lot of a smooth grain boundary facet as shown in Fig. 6(c), which may indicate an increased amount of a liquid phase.

Higashi et al. [12] considered that an optimum amount of a liquid phase formed in the vicinity of grain boundary contributes to an accommodation helper, whereas an excessive amount of a liquid phase does not make stress to transfer effectively to adjacent grains owing to a thick intergranular layer of a liquid phase. If such is the case, it can be concluded that the superplastically large elongation (>300%) developed at the optimum temperature range of 783-803K is related to an easy relaxation of the stress concentration arising from grain boundary sliding owing to the formation of an optimum amount of a liquid phase in the vicinity of grain boundary, whereas the rapid

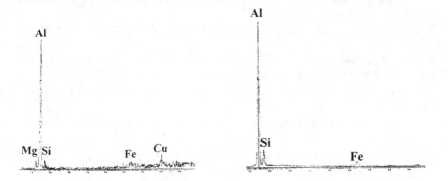

Fig. 8. EDS analysis data from (a) filament-like phase and (b) matrix of the fracture surface in Fig. 6(b).

drop of elongation above the optimum temperature is related to a thick intergranular layer (an excessive amount) of a liquid phase.

4. Conclusions

Temperature dependence of microstructure and tensile deformation behavior of the PM processed Al-16Si-5Fe-1Cu-0.5Mg-0.9Zr (wt%) alloy was investigated in a range of temperature from 673 to 813K under a high strain rate of $1.4\times10^{-1}s^{-1}$. The following conclusions were obtained.

(1) Elongation increased with increasing temperature up to 793K, and then decreased with further increase in temperature. A superplastic elongation over 300% was achieved at temperatures between 783 and 803K. On the other hand, the elongation rapidly reduced to 25% when temperature reaches at 813K.

(2) In the specimens deformed superplastically at the optimum temperature range of 783-803K, occurrence of grain boundary sliding, a retarded grain growth and the formation of a filament-like phase in the vicinity of grain boundary were observed. Above the optimum temperature, a rapid grain growth and an excessive amount of a liquid phase were considered to cause the rapid reduction of elongation.

References

[1] H.S. Cho and M.S. Kim, J. Kor. Inst. Met. & Mater. 37 (1999), p. 1191.
[2] H. S. Cho, H.G. Jeong, M. S. Kim and H. Yamagata, Scripta Mater. 42 (2000), p. 221.
[3] T.R. Bieler and A.K. Mukherjee, Mater. Sci. Eng. A128 (1990), p. 171.
[4] K. Higashi, T. Okada, T. Mukai and S. Tanimura, Scripta Metall. 25 (1991), p. 2053.
[5] T. G. Nieh, J. Wadsworth and T. Imai, Scripta Metall. 26 (1992), p. 703.
[6] M. Mabuchi, K. Higashi and T.G. Langdon, Acta Mater. 42 (1994), p.1739.
[7] O.A. Kaibyshev, "Superplasticity of Alloys, Intermetallics and Ceramics", Springer-Verlag, (1992), p.8.
[8] T. Imai, G.L. Esperance and B.D. Hong, Scripta Metall. 31 (1994), p. 321.
[9] J. Koike, M. Mabuchi and K. Higashi, J. Mater. Res. 10 (1995), p.133.
[10] H.G. Jeong, K. Hiraga, M. Mabuchi and K. Higashi, Acta Mater. 46 (1998), p.6009.
[11] "Metals Handbook", ASM, 8 (1973).
[12] K. Higashi, T.G. Nieh and J. Wadsworth, , Acta Mater. 43 (1995), p.3275.

Corresponding Author

M.S.Kim : mskim@inha.ac.kr

Grain Refinement and Enhanced Superplasticity in Metallic Materials

R.Z. Valiev, R.K. Islamgaliev and N.F. Yunusova

Institute of Physics of Advanced Materials, Ufa State Aviaton Technical University,
RU-450000 Ufa, Russia

Keywords: Deformation Mechanisms, High Strain Rate Superplasticity, Low Temperature Superplasticity, Severe Plastic Deformation, Ultrafine Grained Materials

Abstract

Recent works on fabrication of bulk ultrafine-grained (UFG) metals and alloys with a grain size less than 1 μm have provided new opportunities to considerably enhance superplastic properties through attaining superplasticity at low temperatures and high strain rates. Among different techniques for processing UFG materials severe plastic deformation (SPD) has a special interest because it enables the refinement of microstructure till submicron and nanometer range in different metallic materials, including commercial cast alloys and intermetallics.

The present paper overviews recent achievements in this field and focuses on requirements to microstructure in order to obtain enhanced superplasticity in UFG materials.

Introduction

One of the main requirements of superplastic metallic materials is the existence of fine grains ranging in microns which are traditionally processed by various methods of thermal and mechanical treatments [1,2]. During recent years the active development of new techniques allowing to attain ultrafine-grained metals and alloys with grain size less than 1 μm, i.e. within submicron (100 nm-1000 nm) and nanometer (less than 100 nm) range have been ensured [3-5]. Among these new techniques methods of advanced powder metallurgy (gas condensation, ball milling) and severe plastic deformation processing (high pressure torsion, equal-channel angular pressing and so on) draw special attention.

Based on the constitutive law of superplastic flow one can expect that a decrease in grain size should provide a sharp increase in superplasticity at relatively low temperatures and/or high strain rates. Therefore, expectations to attain enhanced superplasticity defined a great interest in development of techniques to fabricate UFG materials.

The present paper considers recent achievements in developing grain refinement techniques with an accent on severe plastic deformation (SPD) methods which allow to fabricate UFG structure in various metallic materials, including commercial cast alloys and intermetallics. Special attention is paid to recent studies in this field and requirements to microstructure (an optimal grain size, phase composition, types of grain boundaries) in order to obtain enhanced superplasticity in UFG materials.

Methods of Grain Refinement

Methods of advanced powder metallurgy (gas condensation, ball milling/mechanical alloying) and SPD processing (high pressure torsion, equal channel angular pressing and others) are alternative techniques for grain refinement. In the first case one produces nanostructured powders with further consolidation. In the second case, UFG structures are fabricated from microstructural refinement in bulk samples.

For example, in IN90211 and IN905XL Al-based alloys UFG structure with a mean grain size of about 0.5 μm was fabricated by mechanical alloying with further consolidation through extrusion. This method enabled to attain high strain rate superplasticity which was observed at temperatures close to melting [2,6]. At the same time, the factors causing enhanced superplasticity in these materials: a fine grain size or the appearance of liquid phases on grain boundaries, remain still unclear. Moreover, there still exist problems associated with these techniques including the retention of some residual porosity and the difficulty of fabricating large samples with UFG structure for subsequent tensile testing. As a result there have been no experimental works up till now on superplasticity in metallic materials produced by these techniques and having a grain size of about 100 nm or less.

In this connection, a high interest for investigations on superplasticity has arisen by our proposed and developed procedure for fabrication of UFG structures by severe plastic deformation, i.e intense plastic straining under high imposed pressure [5,7]. Two SPD techniques: high pressure torsion (HPT) (Fig. 1a) and equal-channel angular pressing (Fig.1b) became very well known [8]. For consideration of superplasticity it is highly important that a change in a phase composition during SPD-processing can occur along with grain refinement, and formation of high angle grain boundaries occurs only when a certain strain is attained [5]. Therefore differences of microstructures in fabricating samples are associated with deformation regimes and routes of SPD technique employed [9].

As a rule, HPT technique enables essential microstructure refinement till a nanometer grain size (less than 100 nm) and ECA pressing provides formation of submicrometer grains with a size range 0.3 – 1 μm. Therefore, below we consider superplastic behavior of nano- and submicrocrystalline alloys produced by both the above described techniques.

Nanocrystalline alloys

Superplastic behavior was observed in several nanocrystalline alloys and intermetallics, processed by HPT. Nanostructures are characterized usually by low stability at elevated temperatures and, in fact, nanocrystalline pure metals are unstable often even at room temperature. However, nanostructures in two-phase alloys and intermetallics are more stable. Such structures were fabricated in the boron doped intermetallic Ni_3Al (Ni-3.5%Al-7.8%Cr-0.6%Zr-0.02%B) [10,11], the Russian light weight aluminum alloy 1420 (Al-5.5%Mg-2.2%Li-0.12%Zr) [12] and the Zn-22%Al eutectoid alloy by means of HPT at room temperature. This technique (Fig. 1a) has the advantage of producing small discs (12 mm in diameter and 0.5 mm in thickness) having uniform nanoscale microstructures. Tensile specimens of 1 mm gauge length were electro discharge machined from torsion straining processed discs. Tensile tests were performed on a custom-built computer-controlled constant strain rate testing machine [10,12].

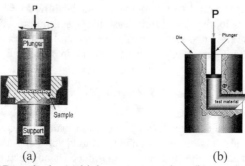

Fig. 1. Principles of SPD methods: (a) high pressure torsion; (b) equal channel angular pressing.

Ni₃Al intermetallic. After HPT at room temperature a strong refinement of structure with a mean grain size of about 50 nm was found in a Ni₃Al intermetallic material. This value remained almost unchanged after annealing at T=650°C, and even after heating to 750°C the grain size did not exceed 100 nm [10,13]. Transmission electron microscopy and high resolution electron microscopy (TEM/HREM) observations indicated that the grains after such heating are almost free of lattice dislocations (Fig. 2a), but the grain boundaries still retained their non-equilibrium character because they are not atomically flat and usually wavy and contained many defects, such as facets, steps and grain boundary dislocations (Fig. 2b). These are similar to the HREM observations for other metals after HPT [14].

The nanostructured Ni₃Al revealed superplastic behavior at temperatures as low as 650°C. It is apparent from Fig. 3 that the specimens exhibit very high tensile ductility (several hundred percent elongation to failure) without visible macroscopic necking. However, such low temperature superplastic behavior has several unusual features, associated with an extensive region of strain hardening and low strain rate sensitivity less than 0.3 [10]. Special TEM/HREM investigations were performed on the thin foil prepared from a gauge section of Ni₃Al sample strained about 300% (see Fig. 3). Although there is some grain growth during superplastic deformation, the grain size remained less than 100 nm.

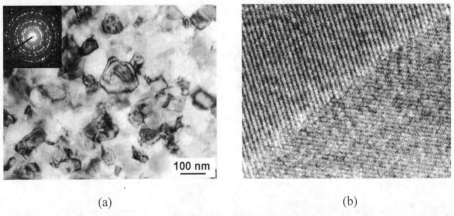

Fig. 2. Microstructure of Ni₃Al intermetallic after high pressure torsion and annealing at 650 °C: (a) TEM micrograph; (b) HREM micrograph of a typical grain boundary.

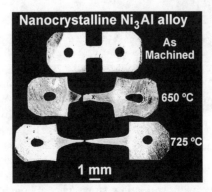

Fig. 3. Appearance of nanostructured Ni₃Al samples prior and after tension at temperature 650°C, strain rate 1×10^{-3}, elongation 390% (a cross indicates the place where the foil was cut out for HREM/TEM) and at temperature 725°C and strain rate 1×10^{-3}, elongation 560%.

Fig. 4. HREM micrograph of a typical grain boundary in Ni_3Al from a gage section of the specimen superplastically elongated at 650°C.

Grains were not elongated and we could not find any evidence of large dislocation activity inside the grains. From the HREM observations it appears that grain boundaries still retain their narrow width and they are still atomically not flat (Fig. 4). There was also some evidence for grain boundary dislocations.

<u>Al 1420 alloy</u>. After quenching and HPT (5 turns, P = 6 GPa) at room temperature the typical microstructure in Al 1420 alloy had the average grain size measured from the dark field image less than 70 nm [12]. After heating up to 300°C there was some grain growth, but the mean grain size was still less than 300 nm (Fig. 5).

The specimens after testing at various temperatures in the interval of 250-350°C show neck free elongation that is characteristic of superplastic flow. The alloy demonstrates high strain rate superplasticity (HSRS) at a rather low temperature of 300°C (Fig. 6). The maximum elongation of 900% was observed at a strain rate of 1×10^{-2} s^{-1} with a flow stress of 40 MPa.

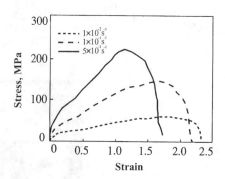

Fig. 5. Microstructure of the Al 1420 alloy after HPT and annealing at 300°C for 5 min.

Fig. 6. The variation of true stress with strain as a function of strain rate for the HPT Al 1420 alloy.

Elongation decreases slightly with increasing strain rate although it remains quite high and significant strain hardening takes place. Increasing in testing temperature up to 350°C at a strain rate of 1×10^{-1} s^{-1} leads to a decrease in the flow stress to 16 MPa and the sample demonstrates high elongation of 890% (Fig. 7).

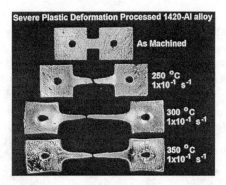

Fig. 7. View of the Al 1420 alloy samples after tensile tests.

<u>Zn-22%Al alloy.</u> Another investigated material is the Zn-22%Al eutectoid alloy. It is a classic two-phase superplastic alloy capable of exhibiting tensile elongation exceeding 2000% under optimal conditions: temperature of 250°C and strain rate of $10^{-3} - 10^{-2}$ s^{-1} [1]. Superplasticity here was achieved in specimens with a grain size in the range 2-6 µm. In the present work, samples from Zn-22%Al were quenched and subjected to HPT. This processing led to the formation of a two-phase nanoduplex structure with a mean grain size of ~80 nm (Fig. 8) [15].

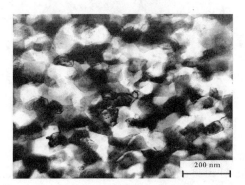

Fig. 8. TEM micrograph of the Zn-22%Al alloy, processed by HPT.

Superplasticity in this Zn-Al alloy subjected to HPT was observed at 120°C and a strain rate of 1×10^{-4} s^{-1}, thus demonstrating superplasticity at a relatively low temperature. However the elongation to failure was equal to 280% only. It is significantly less than for the ECAP samples with a mean grain size of 0.4 µm where elongation to failure was about of 600% at 120°C and a strain rate of 1×10^{-4} s^{-1} [9].

Submicrocrystalline alloys

Recent investigations have shown that the application of another SPD method, i.e. ECA pressing not only decreases significantly the temperature of superplastic flow but also produces high strain rate superplasticity in UFG alloys having a grain size of 0.5-1.0 µm [16,17]. It was shown [12] that the presence of mainly high angle grain boundaries is one of the most important requirements to obtain high strain rate superplasticity in aluminum alloys.

After ECA pressing at temperatures 370°C the microstructure of the Al 1420 alloy was characterized by a mean grain size of about 1 μm (Fig. 9). Inspection of the microstructures reveals also an essential change in a phase composition, because the specimens contain particles of the T-Al_2LiMg phase, with a volume fraction of 10 - 20% and 0.3-0.4 μm in size, though there was no evidence of the presence of this phase in the initial specimens. It should be noted that these particles were also observed in the HPT samples (Fig. 5) which demonstrated high strain rate superplasticity (HSRS).

The samples after ECA pressing were pulled to failure at different testing temperatures (Fig. 10). It is apparent that the samples exhibit the fundamental characteristics of superplastic flow: uniform deformation within the gauge length and an absence of visible localized necking. The optimum level of superplasticity at 400°C was found at a high strain rate of 1.2×10^{-2} s^{-1}, which is two orders of magnitude higher that in a specimen with a grain size of about 5 μm [1]. Moreover an elongation to failure of 1500% recorded in this material (Fig. 10) is the highest elongation ever reported for this commercial Al 1420 alloy.

Under these conditions the flow stress is not high and it reaches 20 MPa. While increasing the strain rate up to $1,2\times10^{-1}$ s^{-1} elongation to failure also exceeded a value of 1000%, being 1200%, and even at the highest strain rate 1.2 s^{-1} it amounted to 340%. The flow stress in this case increased noticeably; however, even at the highest strain rate of 1.2 s^{-1} it did not exceed 90 MPa (Fig. 11).

The strain rate sensitivity of the flow stress, which was measured by a strain rate jump, is quite interesting (Fig. 12). It is seen that the strain rate sensitivity at 400°C is not high and at the optimal strain rate reaches 0.40 - 0.45 depending on the strain in the sample.

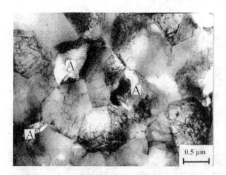

Fig. 9. Microstructure of the Al 1420 alloy after ECA pressing. The particles (grains) of T-Al_2LiMg phase are marked by symbol A.

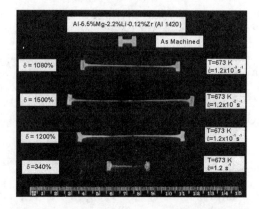

Fig. 10. Appearance of specimens processed by ECAP and pulled to failure.

At higher strain rates the value of the strain rate sensitivity is less and does not exceed 0.25 whereas elongation to failure is high. The obtained values of the strain rate sensitivity in the UFG 1420 alloy produced by ECAP during high strain rate superplasticity are considerably lower compared with conventional superplasticity in the given alloy with a grain size of 5 µm, where the strain rate sensitivity was equal to 0.5 for superplastic flow at 450°C and strain rate 4×10^{-4} s^{-1} [1].

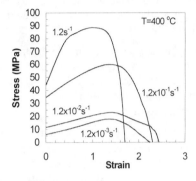

Fig. 11. The variation in true stress with strain as a function of strain rate for the ECAP Al 1420 alloy.

Fig. 12. The dependence of the strain rate sensitivity of flow stress in the ECAP Al 1420 alloy on strain.

The calculations of apparent activation energy revealed that in the steady-state of plastic flow it is equal to 148 kJ/mol (at 170%) which corresponds to the activation energy of bulk self-diffusion in pure aluminum.

TEM replica studies of deformation relief of the samples surface after tensile tests of the ECA-processed Al 1420 alloy allowed to reveal the formation of deformation zones which are caused by grain boundary sliding [1,18]. For example, Fig 13 shows a TEM picture prepared by the replica method from sample of Al 1420 alloy strained by 60% at 350°C using a strain rate 10^{-2} s^{-1}. Here one can observe the formation of typical grain boundary sliding deformation zones and the absence of visible slip lines in grains. It should be noted that the zones are formed mostly perpendicular to the tensile axis and banded. At the same time, the spacing between the zones was 3 - 4 µm, i.e. essentially larger than the grain size suggesting a cooperative character of grain boundary sliding during HSRS in Al 1420 alloy.

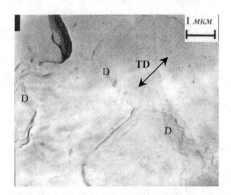

Fig. 13. TEM photograph from the surface of the UFG Al 1420 alloy (pulled at 60%) obtained by replica method. **D** – deformation zones; TD - tensile direction.

Discussion

It is known that all the three investigated alloys demonstrate superplastic behavior when they have a grain size of a few microns [1,2,19]. Further decrease in grain size to the submicron or nanocrystalline range in the alloys by SPD enhances their superplastic properties resulting in a decrease in temperature and/or increase in strain rate of superplastic deformation. A first manifestation of enhanced superplastic properties in SPD materials was demonstrated in the Al-based alloy from the Al-Cu-Zr-system [20]. Up to present time the effect has been revealed in a number of alloys and intermetallics [5,11,16,21-25 and other].

The results of this work exhibit that enhancement of superplasticity and attaining high strain rate superplasticity in the Al 1420 alloy can be provided both by ECA pressing and HPT at room temperature. HPT leads to formation of a very fine grain size of about 70 nm. However, due to grain growth during heating the samples produced by both techniques have similar microstructures and, therefore, it is clear why their properties at 350°C were very similar.

It is very important that the observed effect of high strain rate superplasticity in this case is conditioned by both the fine grain size (about 1 μm) and the presence of particles of the T-phase. As it is shown in [9], the dissolution of these particles during ECA pressing at an elevated temperature, 420°C, results in a sharp decrease in superplastic properties. The positive influence of a two-phase micro-duplex structure on superplasticity is well known in literature [1,2,26]. The data in the present work demonstrate that the submicron duplex structure contributes considerably to attaining high strain rate superplasticity in the Al1420 alloy.

Recent investigations have showed that a very important requirement for producing high strain rate superplasticity is not only the strong refinement of microstructure by SPD techniques, but the formation of ultrafine-grained structure with high angle grain boundaries [17]. The latter can be achieved by an increase in the number of passes during ECA pressing. According to the data obtained it may be assumed that another requirement for enhancement of superplasticity is the optimization of the alloy phase composition, and this can be attained by optimal temperature and strain rate conditions in SPD processing.

Discussing the requirements of microstructure, let us consider in more detail the role of grain refinement. The results of the present paper as well as previous works [16,17,23] indicate that grain refinement till a submicron range definitely contributes to the enhancement of superplastic properties and attaining low temperature and/or high strain rate superplasticity. However, a further decrease in grain size to the nanometer range does not necessarily lead to an additional enhancement of superplasticity. Moreover, as it is seen from the results of the Zn-22%Al alloy, the elongation to failure, in the case of a grain size of 80 nm, is even less than in an alloy with a grain size of 0.5 μm. Obviously, there exists a critical grain size within the range of 100-300 nm, and below this range any further grain refinement does not lead to additional enhancement of superplastic properties. The reasons for the possible difficulty in obtaining superplasticity in nanocrystalline materials was recently discussed in [27], where it was supposed that this behavior is related to the difficulty in generation and motion of dislocations inside the very fine grains. The data on the microstructure of superplastically deformed Ni_3Al presented above also confirm this assumption. The decrease in the strain rate sensitivity in UFG alloys to values of 0.3 - 0.4, and even less, as compared to values of 0.5 - 0.6, which are typical for conventional superplasticity, indicate a possible change in the deformation processes in UFG materials. In addition, it is important to note that in SPD alloys one can observe visible strain hardening. Taking into account the mechanical approach of Hart [26] it may be assumed that enhanced ductility and good stability of superplastic flow under conditions of low temperature and/or high strain rate superplasticity in UFG alloys result from both the strain rate sensitivity of the flow stresses and strain hardening. This points to the importance of further investigations of the phenomenology and nature of unusual superplasticity in ultrafine-grained materials.

Summary

Grain refinement, in particular when using innovative SPD techniques, enables to enhance essentially superplastic properties in metallic materials, i.e., through a manifestation of low temperature and high strain rate superplasticity. At the same time, there is the relationship between SPD processing, UFG structure formation, and enhanced superplasticity, as it is shown in this paper for the Al 1420 and Zn-22%Al alloys as well as a Ni_3Al intermetallic. It was found that enhanced superplasticity in the Al 1420 alloy has important microstructural requirements, including both an ultrafine grain size and the formation of an optimal phase composition. Most significant superplastic properties can be observed in Al 1420 UFG duplex-type structures. There is also probably a "critical" grain size of 100-300 nm, below which the enhancement of superplastic properties becomes less because of a possible change of the deformation mechanisms.

Acknowledgment.

The authors acknowledge the cooperation with Profs. T.G. Langdon and A.K. Mukherjee in the field of superplasticity of ultrafine-grained materials and thank the support of the U.S. NSF-International Program grant # NSF-DMR-960881, the Russian Foundation for Basic Research (grant # 00-02-16583) and the INTAS grant #97-1243.

References

[1] O.A. Kaibyshev, Superplasticity in Metals, Intermetallics and Ceramics, Springer, Frankfurt (1992), p.280.
[2] T.G. Nieh, J. Wadsworth and O.D. Sherby. Superplasticity in Metals and Ceramics, Cambridge Univer. Press. (1997), p. 290.
[3] H.Gleiter, Prog. Mater. Sci. 33 (1989), p. 223.
[4] D.G. Morris, Trans Tech. Publ. (1998), p. 85.
[5] R.Z. Valiev, R.K. Islamgaliev, I.V. Alexandrov, Prog. Mat. Sci. 45 (2000), p. 103.
[6] A.K. Mukherjee in H. Mughrabi (ed), Mater. Sci. And Tech. 6 (1993), p. 339.
[7] R. Z. Valiev, A. V. Korznikov and R. R. Mulyukov, Mater. Sci. Eng. 168 (1993), p. 141.
[8] Proceedings of the NATO ARW on Investigations and Applications of Severe Plastic Deformation (Moscow, Russia), NATO Sci. Series, eds. Lowe, T.C. and Valiev, R.Z. Kluwer Publ., 80, (2000).
[9] R.Z. Valiev, R.K. Islamgaliev, Mat. Res. Soc. 601 (2000), p.335.
[10] R.S. Mishra, R.Z. Valiev, S.X. McFadden and A.K. Mukherjee, Mater. Sci. Eng. A252 (1998), p.174.
[11] S.X. McFadden, R.S. Mishra, R.Z. Valiev, A.P. Zhilyaev and A.K. Mukherjee, Neture 396 (1999), p.684.
[12] R. S. Mishra, R. Z. Valiev, S. X. McFadden, R. K. Islamgaliev and A. K. Mukherjee Phil.Mag., to be published.
[13] R.Z. Valiev, T. Son, S. X. McFadden, R. S. Mishra, A. K. Mukherjee, Phil. Mag. (in press)
[14] Z. Horita, D.J. Smith, M. Nemoto, R.Z. Valiev and T.G. Langdon, J.Mater. Res. 13 (1998), p. 446.
[15] R.Z. Valiev, R.K. Islamgaliev, V.V. Stolyarov, R.S. Mishra and A.K. Mukherjee, Mater. Sci. Forum 269-272 (1998), p.969.
[16] R. Z. Valiev, A. D. Salimonenko, N. K. Tsenev, P. B. Berbon and T. G. Langdon, Scripta Mater. 37 (1997), p. 724.
[17] P.B. Berbon, M. Furukawa, Z. Horita, M. Nemoto, N. K. Tsenev, R. Z. Valiev and T. G. Langdon, Phil. Mag. Lett. 4, V. 78 (1998), p. 313.

[18] I.I. Novikov, V.K. Portnoy, Superplasticity of ultrafine grained alloys. M.: Metallurgy (1981), p. 168.
[19] R.Z. Valiev, R.M. Gayanov, H.S. Yang and A.K. Mukherjee, Scripta Met. Mater. 25 (1991), p. 1945.
[20] R.Z. Valiev, O.A. Kaibyshev, R.I. Kuznetsov, R.Sh. Musalimov, N.K. Tsenev, DAN SSSR 301, 4 (1988), p. 864.
[21] M. Mabuchi, H. Iwasaki, K. Yanase and K. Higashi, Scripta Mater. 36, 6 (1997), p. 681.
[22] R.Z. Valiev, Mater. Sci. Forum 243-245 (1997), p. 207.
[23] R.Z. Valiev, R.K. Islamgaliev, Superplasticity and Superplastic Forming, edited by A.K. Ghosh and T.R. Bieler (The Minerals, Metals & Materials Society, 1998), p.117.
[24] G.A.Salishchev, R.M.Galeev, S.P.Malysheva and O.R.Valiakhmetov, Mater. Sci. Forum 243-245 (1997), p. 585.
[25] S. Lee, P.B. Berbon, M. Furukawa, Z. Horita, M. Nemoto, N.K. Tsenev, R.Z. Valiev, T.G. Langdon, Mater. Sci. Eng. 272, 1 (1999), p. 63.
[26] J.Pilling and N.Ridley, Superplasticity in Crystalline Solids (Manchester, Inst. of Metals, 1992), p.214.
[27] R. S. Mishra, A. K. Mukherjee, Superplasticity and Superplastic Forming, edited by A.K. Ghosh and T.R. Bieler (The Minerals, Metals & Materials Society, 1998), p.109.

Grain Refinement of a Commercial Magnesium Alloy for Superplastic Forming

T. Mukai[1], H. Watanabe[1], K. Moriwaki[2], K. Ishikawa[1], Y. Okanda[1] and K. Higashi[3]

[1] Osaka Municipal Technical Research Institute,
1-6-50 Morinomiya Joto-ku, Osaka 536-8553, Japan

[2] Graduate Student, Osaka Prefecture University

[3] Department of Metallurgy and Materials Science, Osaka Prefecture University,
1-1, Gakuen-cho, Sakai, Osaka 599-8531, Japan

Keywords: Grain Size, High Strain Rate Superplasticity, Low Temperature Superplasticity, Magnesium Alloy, Superplastic Forging

Abstract

Superplastic forming was performed at a high strain rate of 10^{-2} s^{-1} for a commercial ZK60 magnesium. The relatively high forming temperature of 673 K was decided from the superplastic characteristics owing to its grain size ~ 5μm. In order to prevent the grain growth during the high temperature superplasticity, the possibility of low temperature superplasticity(LTSP) in magnesium was derived. The required grain size for high strain rate superplasticity(HSRS) at the low homologous temperature was estimated, and the fabrication techniques to develop the fine-grained structure were discussed.

Introduction

Recent activities in the research of magnesium become higher in order to reduce the weight of components such as motor vehicles from the economical and ecological point of view. There are many potential opportunities for the use of magnesium alloys in motor vehicle components [1]. This is not only a result of magnesium's relatively low density, which can directly and substantially reduce vehicle weight, but is also a result of its good damping characteristics, dimensional stability, machinability, and low casting cost. Despite these advantages, magnesium alloys normally exhibit limited ductility (~ 7%). Therefore, recent fabrications of magnesium-based material were mainly performed by die casting. However, the microstructure of the fabricated products contains cast structures [2], thus the strength and ductility are poor. In order to use for structural components, which require high toughness, their microstructures must be modified. Wrought magnesium is noted to have a higher strength and ductility than cast magnesium. This primarily result from the fact that the grain size of wrought magnesium is normally finer than that of the cast material. Therefore, there is a strong need to develop fine-grained magnesium for structural applications. In order to exploit the benefits of fine-grained magnesium, it is important to develop secondary processing which can effectively produce complex engineering components directly from wrought products. Superplastic forming is a viable technique that can be used to fabricate magnesium into complex shapes. Superplasticity usually occurs when grain size is small, typically less than ~10 μm. At this small grain size, grain boundary sliding is readily taking place at elevated temperatures. The deformation mechanisms at elevated temperatures in magnesium alloys with various grain sizes are discussed. From the discussion, the required grain size for high strain rate superplasticity($\sim 10^{-2}$ s^{-1}) and low temperature superplasticity(~ 0.5 T_m, T_m is the melting temperature of magnesium) is estimated.

Superplastic Forming of a Commercially Extruded ZK60

The material used for the experimental study on superplastic forming is a commercially extruded ZK60 alloy. The chemical composition is Mg-6Zn-0.5Zr (by wt. %). The microscopic observation revealed that the microstructure consisted of fine-grains with the sizes of 2 ~ 5 μm. The alloy is expected to be deformed superplastically at a reasonable high strain rate of 10^{-2} s^{-1} owing to its fine-grained structure.

The mechanical properties at elevated temperatures were characterized before the superplastic forming in order to decide the fabrication condition [3]. The variations of flow stress and elongation-to-failure against the temperature are shown in Fig. 1 at a target strain rate of 10^{-2} s^{-1}. The elongation-to-failure increases with temperature and reaches almost a constant value of ~ 300 % above the temperature of 673 K, while the flow stress decreases with increasing temperature monotonically. From the results of tensile test, the fabrication temperature was selected to 673 K.

The cylindrical specimen of ZK60 was superplastically forged to a square box [3]. The schematic configuration of the die is illustrated in Fig. 2. The forging was conducted at a fixed speed of 0.5 mm/s corresponding to a reasonable high-strain rate of ~ 10^{-2} s^{-1}. The experimental forming was carried out at a temperature of 673 K, a better condition for superplastic manner as shown in Fig. 1. The appearance of the superplastically forged specimen is shown in Fig. 3. There can be hardly seen macroscopic cracks and/or defects on the surface. Inspection of the microstructure of the post deformed specimen revealed that there could be hardly seen any cracks or cavities and its microstructure was consisted of equi-axed and fine grains(~ 6 μm). The structure grew a little comparing with the initial grain structure for the elevated temperature. Since the grain growth occurs, the strength of the post deformed product is expected to be relatively lower than that of the as-received(as-extruded) alloy. In order to prevent the grain growth, superplastic forming in the magnesium alloy should be conducted at relatively low homologous temperatures.

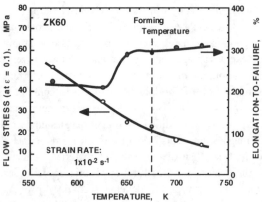 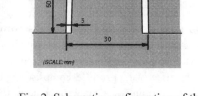

Fig. 1 The variations of flow stress and elongation-to-failure against the temperature.

Fig. 2 Schematic configuration of the die.

Fig. 3 Appearance of the superplastically forged specimen.

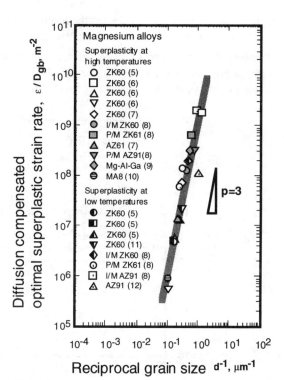

Fig.4 Optimal superplastic strain rate compensated by grain boundary diffusivity against the inverse of grain size for fine-grained magnesium alloys [5-12].

Required Grain Size for Low Temperature Superplastic Forming

The deformation characteristics in HSRS and LTSP in several magnesium alloys are compared to investigate the dominant deformation mechanism as follows. In general, the constitutive equation to describe the superplasticity of magnesium alloys can be expressed by [4]

$$\dot{\varepsilon} = A \frac{D_0 G b}{kT} \left(\frac{b}{d}\right)^p \left(\frac{\sigma - \sigma_0}{G}\right)^n \exp\left(-\frac{Q}{RT}\right) \quad (1)$$

where $\dot{\varepsilon}$ is the strain rate, A a constant, D_0 the pre-exponential factor for diffusion, σ the stress, σ_0 the threshold stress, G the shear modulus, k the Boltzmann's constant, n the stress exponent ($=1/m$), d the grain size, b the Burgers vector, p the grain size exponent, R the gas constant, T the absolute temperature and Q the activation energy which is dependent on the rate controlling process. The three variables of n, p and Q in equation (1) are often used to identify the deformation mechanism. The deformation characteristics of magnesium alloys with various grain sizes are summarized as follows; (1) grain size refinement is essentially to effectively promote grain boundary sliding in magnesium, and (2) superplastic behavior exhibits a strong dependence on grain size in fine-grained magnesium alloys [3]. Figure 4 shows the optimal superplastic strain rate compensated by grain boundary diffusivity against the inverse of grain size for fine-grained magnesium alloys [5-12]. The relations can be well described by a single straight line. The slope of the line represents the grain size exponent, p which is about 3. The relationship in Fig. 4 can be represented by the relations between the reciprocal grain size and optimum temperature as shown in Fig. 5. As can be seen in this figure, the grain size decreases with decreasing temperature and increasing strain rate. For a target strain rate of 10^{-2} s^{-1}, the grain size must be developed about 0.5 µm at a low homologous temperature of 473 K, as indicated in Fig. 5. Thus, microstructure of magnesium alloys should be refined to be ~ 0.5 µm.

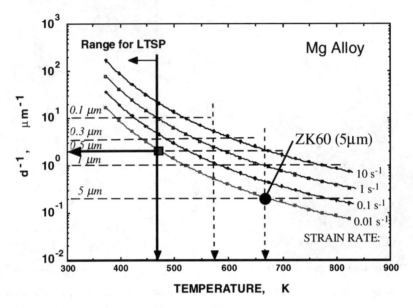

Fig. 5 Relations between the reciprocal grain size and optimum temperature.

The values of p=3 and Q=Q$_{gb}$ in LTSP is the same as those in high temperature superplasticity (HTSP) [5]. The variation of normalized strain rate, $\dot{\varepsilon}(kT/D_{gb}Gb)(d/b)3$, as a function of normalized stress, $(\sigma-\sigma_o)/G$, is plotted in Fig. 6 [5,6,8,10,13]. Since the normalized relations in LTSP is essentially similar to that in HSRS, the deformation mechanism in LTSP is the same as that in HTSP.

Grain Refinement of Magnesium Alloys

Kaibyshev et al. have demonstrated that grain refinement could be achieved for Mg-5.8 Zn-0.65 Zr alloy with increasing Zener-Hollomon parameter, i.e. $\varepsilon exp(Q/RT)$ [14]. Mabuchi et al. have also reported that grain size in AZ91 machined-chip increased with increasing extrusion temperature, and grain refinement could be achieved with an increase in Zener-Hollomon parameter [15]. The experimental evidence provides useful guides for the grain-refinement in magnesium: (1) increase the strain rate (or stress) in the process, or (2) decrease the temperature near the recrystallization temperature. Following these guides, three thermomechanical processes were subsequently selected for refining the grain structures; (a) extruding cast alloys with a high ratio, (b) severe plastic deformation, e.g., Equal-Channel-Angular-Extrusion (ECAE), (c) powder-metallurgy (P/M) processing of rapidly solidified powders. Experiments of refining grain size were conducted for commercial magnesium alloys of AZ31, ZK60 and ZK61 [3]. The developed grain sizes for the above three processes were summarized as follows; (a) 2 ~ 5 µm, (b) 0.5 ~ 1 µm, (c) 0.5 ~ 1 µm.

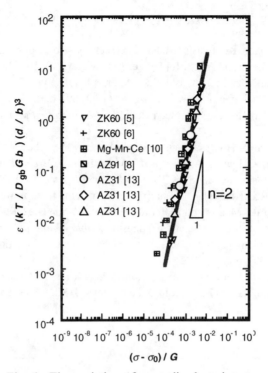

Fig. 6 The variation of normalized strain rate, $\dot{\varepsilon}(kT/D_{gb}Gb)(d/b)3$, as a function of normalized stress, $(\sigma-\sigma_o)/G$ [5,6,8,10,13]

Therefore the target grained-structures ($d \sim 0.5$ μm) can be obtained by (b)severe plastic deformation and (c) P/M of rapidly solidified powders. Routes (b) and (c), however, have drawbacks of quantitative and economical problem, respectively. Further grain refinement should be performed by the thermomechanical processing with conventional extrusion for the commercial application of magnesium alloys.

Summary

Superplastic forming was performed at a high strain rate of 10^{-2} s^{-1} for a commercial ZK60 magnesium at a high temperature of 673 K. In order to prevent the grain growth during the high temperature superplasticity, the possibility of low temperature superplasticity(LTSP) in magnesium was estimated. The required grain size for high strain rate superplasticity($\sim 10^{-2}$ s^{-1}) at the low homologous temperature(~ 0.5 T_m) was found to be ~ 0.5 μm. The fine-grained structures could be produced via severe plastic deformation of ECAE or P/M route. From the economical point of view, further grain refinement should be performed by the thermomechanical processing with conventional extrusion for the commercial application.

References

[1] J. Davis: Technical Paper No. 910551, (SAE International, Warrendale, PA, 1985).

[2] ASM Speciality Handbook, 'Magnesium and Mangesium Alloys', ASM International, Materials park, Ohio, 1999.
[3] T. Mukai, H. Watanabe, K. Higashi: Mater. Sci. Technol., -to be published.
[4] R.S. Mishra, T.R. Bieler and A.K. Mukherjee: Acta Mater., 43, 1995, p. 877.
[5] H. Watanabe, T. Mukai and K. Higashi, Scr. Mater. 40, (1999) p.477.
[6] A.Karim and W.A.Backofen, *Mater. Sci. Eng.* 3, (1968-69), p.306.
[7] M.M.Tilman, R.L.Crosby and L.A.Neumeier, *RI8662, Bureau of Mines, U.S. Department of the interior* (1979).
[8] M.Mabuchi, T.Asahina, H.Iwasaki and K.Higashi, *Mater. Sci. Tech.* 13, (1997) p.825.
[9] A.Uoya, T.Shibata, K.Higashi, A.Inoue and T.Masumoto, *J. Mater. Res.* 11, (1996) p.2731.
[10] R.Z.Valiev, N.A.Krasilnikov and N.K.Tsenev, *Mater. Sci. Eng. A* A137, (1991) p.35.
[11] A.M.Galiyev and R.O.Kaibyshev, A supplemental volume to superplasticity and superplastic forming, The Minerals, Metals and Materials Society, Warrendale (1998), p. 20.
[12] M.Mabuchi, H.Iwasaki, K.Yanase and K.Higashi, *Scripta Metall. Mater.* 36, (1997) p.699.
[13] H. Watanabe, T. Mukai, K. Ishikawa, Y. Okanda, M. Kohzu and k. Higashi, J. Japan Inst. Light Metals, 49 (8), (1999), p.401.
[14] R.O. Kaibyshev, A.M. Galiev and B.K. Sokolov: Phys. Met. Metallogr., 78 (1994), p.209.
[15] M. Mabuchi, K. Kubota and K. Higashi: Mater. Trans. JIM., 36, (1995), p.1249.

Corresponding author: Toshiji Mukai
    E-mail : toshiji@pp.iij4u.or.jp

Improvements in Superplastic Performance of Commercial AA5083 Aluminium Processed by Equal Channel Angular Extrusion

Darrell R. Herling and Mark T. Smith

Materials Processing, Pacific Northwest National Laboratory,
902 Battelle Blvd, MSIN: P8-35, Richland WA 99352, USA

Keywords: Aluminium Alloy 5083, ECAE, Equal-Channel-Angular-Extrusion, SPF, Superplasticity

ABSTRACT

An improvement in superplastic performance, as measured by elongation to failure, was achieved in an AA5083 aluminum alloy. Equal channel angular extrusion (ECAE) processing was performed on commercially available hot rolled AA5083 material, in order to refine the microstructure to achieve superplastic characteristics. The ECAE process offered several potential advantages in the processing of an SPF-grade aluminum alloy. The first was the ability to introduce a high level of energy into the structure through localized shearing, which developed a well-defined sub-grain structure and ultimately a refined microstructure upon recrystallization. Secondly, with ECAE there is the unique ability to achieve the desirable microstructure in bulk form without reducing the dimensions of the starting material, as is the case in conventional roll processing of SPF sheet materials.

Total elongation of >350% was measured at a temperature of 550°C and strain rate of 5×10^{-4}. This was an increase in elongation of over 150% compared to the same material processed by convention thermal mechanical processing (TMP) techniques, designed to produce a fine grain microstructure in sheet material. Past work on the same commercial AA5083 material using an appropriate TMP schedule, revealed <200% elongation was achievable at the same temperature and strain rate. Through conventional thermal mechanical processing (a series of heat treatments and hot and cold rolling steps), the minimum grain size achieved was typically limited to 10 µm, but with ECAE an average grain size <2 µm was produced, after the recovery heat treatment step.

INTRODUCTION

There has been a lot of interest in aluminum 5083 alloy for superplastic forming, due to its good strength, weldability, and corrosion resistance. Aluminum alloys that exhibit superplastic properties require appropriate processing schedules that are successful in producing an equiaxed fine grain microstructure. Typical processing techniques to achieve the desired microstructure are based on conventional hot and cold rolling processes. However, this processing method limits the final material to thin gauge sheet metal. It may be desirable in some cases to be superplastically form materials other than in sheet form. Such application would include the production of complex components, such as powertrain gear sets or other bulk piece that require a large degree of deformation, that may not be possible with conventional metal forging or closed die forming techniques.

Equal channel angular extrusion (ECAE) has the potential to produce a superplastic grade of aluminum alloy in bulk form. The objective of this research work was to produce a fine grain microstructure in a commercially available 5083 aluminum alloy, that is in bulk form and has improved superplastic properties over conventionally processed sheet materials.

BACKGROUND

Previous research work by Smith and colleges [1] investigated the possibility of producing a microstructure in an off-the-shelf 5083 aluminum alloy that is conducive to superplastic forming. The purpose of the work was to apply an appropriate thermal mechanical processing (TMP) schedule to the preparation 5083 sheet, from a commercially and readily available 5083 plate. The proper TMP applied needed to produce an equiaxed fine-gained microstructure that is typically associated with good SPF materials. One of the implications of success to such an approach was that a more conventional production-oriented route for sheet production could be taken for the production of cheaper SPF sheet material.

The starting material used in the work by Smith et al, was selected for its content of Mg and Mn levels and was supplied as a commercial H321 temper plate. To provide a baseline for comparison of a superplastic microstructure and properties, a sample of the 5083 plate was processed to sheet form using hot and cold rolling steps. The material was initially given a standard annealing treatment to obtain an initial O temper, then hot rolled at 510°C to a thickness of 12.7 mm. The following thermomechanical processing utilized a warm rolling step to 7.6 mm followed by a cold rolling step to 2 mm (cold reduction ratio of 4:1), and finally a flash anneal at 510°C for 10 minutes. Metallographic studies of the microstructure revealed a fairly equiaxed microstructure, with an average grain size of 10 μm.

Tensile tests were conducted for the samples, at various temperatures and strain rates typical to the superplastic deformation regime. Tensile test specimens utilized for the constant strain rate tests had a gauge length of 25.4 mm, and a gauge length to width ratio of 4:1. Strain rates selected for the testing included 5×10^{-4} s^{-1}. More details of the specimen design, testing procedures and results are described in previously published research [1].

EXPERIMENT SETUP AND MATERIAL PROCESSING

The purpose of deforming these materials through the ECAE process was to create an equiaxed grain size on the order of 1 μm. With ECAE, materials can be highly strained by multiple passes through the extrusion die. Each consecutive pass through a 90-degree die arrangement increases the equivalent strain intensity by 1.15. It has been shown in the literature, that a minimum of four passes is required to create a homogeneous microstructure [2,3].

There are many combinations of billet orientation schedules associated with equal channel angular extrusion that can be executed in order to develop different microstructures. However, the most typical ECAE billet orientation schedules are refereed to as Route A, Route B, and Route C. A detailed description of other routes and the associated microstructures that can be developed can be found in the literature [4]. With Route A, the billet orientation is constant for each pass, or no rotation of the ECAE billet. Each successive pass increases the amount of deformation in the same direction. This develops a strong texture, with a laminar grain structure along the extrusion axis. For Route B, the billet orientation is rotated back-and-forth 90 degrees about the extrusion axis on each successive pass. As a result, the material is deformed on two orthogonal directions. This creates a long fibrous structure in the material in a similar direction as Route A. Route C consists of rotating the ECAE billet 180 degrees about the extrusion axis after each even numbered pass. On odd numbered passes, the material is deformed as in

Route A, and restored to a similar original form at each even numbered pass. The result is a uniform, heavily deformed, equiaxed microstructure, with many high angled grain boundaries.

The material used in this study was the same as that used previously, which was an AA5083 aluminum alloy in an H321 temper and purchased as an off-the-self, 25.4 mm thick aluminum plate, from a metal supplier. Appropriately sized ECAE billets were machined from the plate and subsequently annealed at 415°C to give an O temper. The ECAE processing was performed under two conditions as outlined in Table 1.

Specimen #	Alloy ID	ECAE Route	Total # Passes	Axial Rotation Angle Between Passes	Process Temp	Process Speed
A500-4B	ALC-5083	B	4	90°	200°C	10 mm/min
A501-4C	ALC-5083	C	4	180°	200°C	10 mm/min

Table 1: ECAE processing conditions for AA5083 material

Based on work by Segal and others [4,5], ECAE Route C was selected as the primary mode for processing for the 5083 material. This was thought to be the optimum process route to directly obtain a fine-grain equiaxed microstructure, required for SPF materials. With the ECAE Route C, every odd numbered pass produces an elongated structure similar to that of route A. When a subsequent pass is conducted and the work piece is rotated 180° about it's extrusion axis, such as on even numbered passes, the elongated structure is reversed and a more equiaxed type structure is produced. Route B was also selected as an alternative ECAE processing route that potentially could also produce the desired microstructure. This was done in an attempt to produce a more homogeneous microstructure that had undergone deformation on multiple planes, versus just one, as with Route C. A total of four extrusion passes were used with each route.

Initially, the 5083 billets were to be extruded at room temperature, in order to achieve maximum cold work with the minimum extrusion passes. However, due to material strength, limited ductility at room temperature, and extrusion load limits, it was determined that this extrusion condition was not possible with the available setup. Therefore, a series of elevated temperature ECAE runs were attempted to determine the lowest possible extrusion temperature, without adverse effects such as billet cracking and die failure. Multiple passes were successful at an extrusion temperature and rate of 200°C and 10 mm/min, respectively. All materials were processed through an ECAE die that was designed and fabricated at the Idaho National Environmental and Engineering Laboratory (INEEL), in Idaho Falls, Idaho. The die set was designed to minimize the interaction of the work piece with the die channels. This insured that friction due to the two surfaces sliding past one another was as low as possible, so there would not be any adverse surface effects. The billet dimensions for this die were 22.3 mm square and 100 mm long. Low viscosity oil was used during the ECAE process to further reduce friction and required extrusion loads. The ECAE die included insertion heaters in order to extrude materials at elevated temperatures. Billets to be processed were preheated in-situ in the die entry channel and then held at a predetermined temperature for ECAE processing.

After ECAE processing, the billets received a flash anneal heat treatment at 510°C for 10 minutes, as a recovery/recrystallization step. Flat dog-bone style tensile specimens were sliced from each billet for elevated temperature uniaxial tensile testing. The specific specimen geometry was the same as that used in previous superplastic tensile testing on 5083 sheet material [1]. Testing temperatures ranged from 450°C to 550°C and constant strain rates from 1×10^{-3} s^{-1} to 5×10^{-5} s^{-1}.

RESULTS

Maximum total elongation to failure was used as the metric of superplastic performance and comparison to previous results. Elongations of >350% were measured at a temperature of 550°C and strain rate of 5×10^{-4} s^{-1}, with the AA5083 material processed via both ECAE routes 4B and 4C. However, route 4B, which produced a finer grain structure, yielded a slight increase in total elongation at 385%, with an m-value of 0.65. Likewise, the material processed by route 4C failed at an elongation of 351%, and had an m-value of 0.58 associated with it. These results accounted for an increase in elongations of over 150% compared to the same material processed by convention thermal mechanical processing techniques, designed to produce a fine grain microstructure in sheet material.

Past work on the same commercial AA5083 material using an appropriate TMP schedule, revealed that <200% elongations were achievable, under the best test conditions of 500°C temperature and 5×10^{-4} s^{-1} strain rate [1]. At the equivalent temperature and strain rate where maximum elongation occurred with the ECAE processed material, the elongation to failure was <100%. The TMP schedule use prior, consisted of a series of heat treatments and hot and cold rolling steps, which was thought to produce the necessary deformation, strain induced energy and thermal recovery to produce a refined microstructure. However, due to physical limitations, such as edge cracking of the sheet metal during cold rolling and limited maximum final gauge thickness (limited cold work), rolling by itself was not able to produce as fine of a grain structure that would otherwise have been desired. The minimum average grain size achieved with rolled from a 1-inch plate to sheet material was limited to 10 μm.

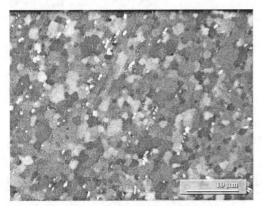 

Figure 1a: Microstructure of 5083 processed by ECAE Route B, plus flash anneal at 510°C for 10 minutes. Average grain size is 1 μm.

Figure 1b: Microstructure of 5083 processed by ECAE Route C, plus flash anneal at 510°C for 10 minutes. Average grain size is 2 μm.

Since a billet cross section dimensional change does not occur with ECAE, it was possible to extrude the material multiple times through the same die. With each pass through an ECAE die, with a 90° included channel angle, a strain of 1.15 is effectively produced. Therefore, the four passes used during processing produced a total equivalent strain on the order of 4.6, or about a 100:1 reduction ratio. The tri-axial stress that occurs in the billet while extruding helped resist the tendency for cracking at high effective strain levels, therefore allowing for more cold work. With the ECAE process an average grain size <2 μm was produced, when coupled with a recovery heat treatment step. Optical and transmission electron microscopy (TEM) images are shown in Figures 1 and 2, for 5083 aluminum processed by ECAE. The micrographs show that the recovered microstructure is indeed equiaxed for both cases processed by Routes B and C, and the respective average grain sizes are 1 μm and 2 μm. Based on the common deformation modes of superplastic materials (grain boundary sliding), a smaller grain structure is more desirable. This was shown in the comparison of results with those from the same material, but processed by conventional rolling practices, which had a larger average grain size.

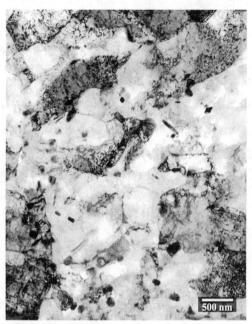

Figure 2a: TEM micrograph of 5083 aluminum processed by ECAE Route B, total of 4 passes at 200C.

Figure 2b: TEM micrograph of 5083 aluminum processed by ECAE Route C and a total of 4 passes at 200C.

CONCLUSIONS

An improvement in superplastic performance, as measured by elongation to failure, was achieved in an AA5083 aluminum alloy. Equal channel angular extrusion processing was performed on commercially available hot rolled AA5083 material, in order to refine the microstructure to achieve superplastic characteristics. The ECAE process offered several potential advantages in the processing of a SPF grade aluminum alloy. The first was the ability to introduce a high level of energy into the structure through localized shearing, which developed a well-defined sub-grain structure and ultimately a refined microstructure upon recovery. Secondly, with ECAE there is the unique ability to achieve the desirable microstructure in bulk form without reducing the dimensions of the starting material, as is the case in conventional roll processing of superplastic aluminum sheet materials.

ACKNOWLEDGMENTS

This research work was funded by the DOE Office of Transportation Technology – Office of Heavy Vehicle Technology. The authors would like to thank Dr. Gary Korth of the Idaho Environmental and Engineering Laboratory and Dr. Jenya Macheret of the DOE Idaho Operations Office in Idaho Falls, Idaho for their support in ECAE processing. The authors would also like to thank Karl Mattlin, Val Guertsman and Konrad Lasota of the Pacific Northwest National Laboratory for their assistance in materials processing and metallographic preparation.

This work was conducted at Pacific Northwest National Laboratory, which is operated by Battelle for the U.S. Department of Energy under contract DE-AC06-76RLO 1830.

REFERENCES

[1] M.T. Smith, J.S. Vetrano, E.A. Nyberg, D.R. Herling, "Effects of Mg and Mn Content on the Superplastic Deformation of 5000-Series Alloys," TMS, Superplasticity and Superplastic Forming, (1998), 99-108.

[2] M. Kawazoe, T. Shibata, T. Mukai and K. Higashi, "Elevated Temperature Mechanical Properties of A 5056 Al-Mg Alloy Processed by Equal Channel Angular Extrusion," Scripta Materialia, vol. 36, no. 6 (1997), 699-705.

[3] M. Mabuchi, H. Iwasaki and K. Higashi, "Microstructure and Mechanical Properties of 5056 Al Alloy Processed by Equal Channel Angular Extrusion," Nanostructured Materials, vol. 8, no. 8 (1997), 1105-1111.

[4] M. Furukawa, Y. Iwahashi, Z. Horita, M. Nemoto and T.G. Langdon, "The Shearing Characteristics Associated with Equal Channel Angular Pressing," Materials Science and Engineering, A257 (1998), 328-332.

[5] V.M. Segal, "Materials Processing by Simple Shear," Materials Science and Engineering, A197 (1995), 157-164.

Equal-Channel Angular Pressing as a Production Tool for Superplastic Materials

Z. Horita[1], S. Lee[2], S. Ota[1], K. Neishi[1] and T.G. Langdon[2]

[1] Department of Materials Science and Engineering, Kyushu University, Faculty of Engineering, 6-10-1 Hakozaki, Higashiku, Fukuoka 812-8581, Japan

[2] Departments of Materials Science and Mechanical Eng., University of Southern California, Los Angeles CA 90089-1453, USA

Keywords: Aluminium Alloys, Equal-Channel Angular Pressing, Severe Plastic Deformation, Ultrafine Grain Size

Abstract

Equal-channel angular pressing (ECAP) was applied for grain refinement of Al-3%Mg-0.2%Fe and Al-3%Mg-0.2%Ti alloys and also for a commercial Al 2024 alloy. The grain sizes of the alloys were reduced to ~0.3 µm. The stability of the fine-grained structures were examined and it was found that the small grains remained stable up to the temperatures of ~250°C for the Al-3%Mg-0.2%Fe and Al-3%Mg-0.2%Ti alloys and ~400°C for the Al 2024 alloy. Tensile tests revealed maximum elongations of ~370% and ~180% in the Al-3%Mg-0.2%Fe and Al-3%Mg-0.2%Ti alloys, respectively, at a temperature of 250°C with an initial strain rate of 3.3×10^{-4} s^{-1}. There is some evidence for low temperature superplasticity in the Al-3%Mg-0.2%Fe alloy. A maximum elongation of ~460% was attained in the Al 2024 alloy at 400°C with an initial strain rate of 10^{-3} s^{-1}. It is demonstrated that the ECAP can be effective in producing superplastic materials.

1. Introduction

Equal-channel angular pressing (ECAP) is a processing procedure capable of producing fine grains in metallic materials [1]. The advantages of the ECAP technique are that it is possible to produce a large bulk sample without introducing residual porosity and the fine grains can be produced in a wide range of materials without relying on complex thermomechanical heat treatments. Because the grain sizes are reduced to ~1 µm or less, it is possible to anticipate superplasticity at higher strain rates and/or lower temperatures [2]. However, when considering the attainment of superplasticity in the ECAP materials, it is important that the fine grains should be stable at testing temperatures where diffusion is rapid. Such a condition may be achieved by dispersing fine particles in a sample so that they inhibit significant grain growth.

It was shown that an Al-3%Mg alloy containing 0.2% Sc exhibits superplasticity with an elongation of greater than 2000% at an initial strain rate of 3.3×10^{-2} s^{-1} and at a testing temperature of 673 K [3]. For this material, Sc is effective in retaining the small grain size as it forms a fine dispersion of Al_3Sc particles [4]. However, Sc is an expensive element and it is desirable to replace it with other common elements or alternatively to achieve superplasticity in commercially available Al alloys. In this study, 0.2%Fe or 0.2%Ti were added to Al-3%Mg prior to processing by ECAP and the alloys were then examined for the occurrence of superplasticity at high strain rates or low temperatures. In addition, the grain refinement of a commercial Al alloy, Al 2024, was examined using ECAP and the potential for achieving superplasticity in this alloy was evaluated.

2. Experimental

Al-3%Mg alloys containing 0.2%Ti or 0.2%Fe were melted and cast into a steel molt. Ingots with dimensions of 17x55x120 mm^3 were homogenized at 480°C for 24 hours and then swaged to a diameter of 10 mm. Rods having lengths of 60 mm were annealed at 600°C for 1 hour prior to ECAP. Rods of the same size were also made from a billet of a fully annealed Al 2024 alloy. ECAP was conducted using a solid die having a channel angle of 90° and the samples were subjected to 8 pressings at room temperature with a rotation of 90° in the same sense between each pressing (ie., route B_C [5]). Tensile specimens were machined with the tensile axes parallel to the longitudinal directions of the ECAP rods. Tensile tests were conducted at temperatures in the range of 250 to 450°C with initial strain rates in the range of 10^{-4} to 1 s^{-1}. Microstructures were observed using transmission electron microscopy for samples after ECAP, after static annealing and following tensile testing. For comparison purposes, tensile tests were also conducted on Al-3%Mg and Al-3%Mg-0.2%Sc alloys prepared in earlier studies [3,6].

3. Results and Discussion

Figure 1 shows the microstructure of the as-pressed Al-3%Mg-0.2%Fe alloy observed parallel to the longitudinal axis (ie., the y plane defined earlier [5]) and the microstructures after annealing at 200, 250 and 300°C for 1 hour. Selected area electron diffraction (SAED) patterns recorded from an area of 12.3 μm in diameter are included in the first three micrographs. The grain size of the as-pressed sample is ~0.3 μm and subsequent annealing leads to gradual grain growth to a grain size of ~0.7 μm at 250°C while the grain boundaries become progressively better defined. The SAED patterns exhibit many diffracted beams forming rings, demonstrating that the grains are separated by high angle grain boundaries. Significant grain growth occurs at the annealing temperature of 300°C. A similar grain growth behavior was observed also in the Al-3%Mg-0.2%Ti alloy.

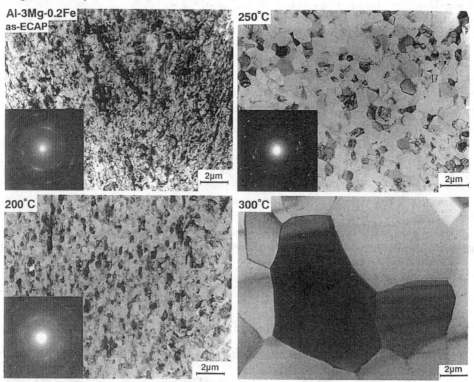

Fig.1 Microstructures of as-pressed Al-3%Mg-0.2%Fe alloy and after annealing at 200, 250 and 300°C for 1 hour.

Figure 2 shows the microstructure of the Al 2024 alloy after ECAP including microstructures after annealing at 200, 400 and 450°C for 1 hour following ECAP. A grain size of ~1 μm is retained even after annealing at 400°C but annealing at 450°C leads to significant grain growth. The diffracted beams in the SAED patterns are scattered around rings, indicating that the structure again consists of grains with high angle grain boundaries. There are also many particles visible within the grains and on the grain boundaries. These particles are considered to inhibit grain growth and the thermal stability of the fine-grain structure is due to the presence of such particles.

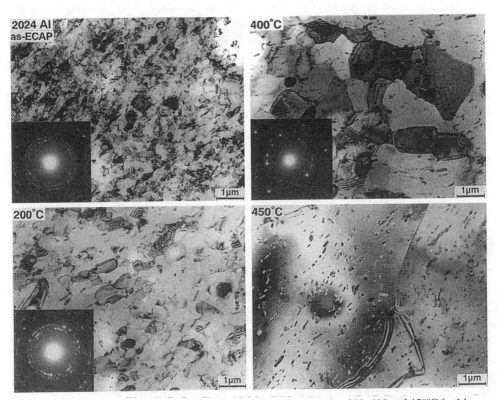

Fig.2   Microstructures of as-pressed 2024 alloy and after annealing at 200, 400 and 450°C for 1 hour.

The variation of the grain size with the annealing temperature is shown in Fig.3 for the three alloys plus Al-3%Mg and Al-3%Mg-0.2%Sc alloys examined in an earlier study [4]. For the Al-3%Mg-0.2%Fe and Al-3%Mg-0.2%Ti alloys, it is confirmed that fine grains having a grain size less than ~0.7 μm are retained up to an annealing temperature of 250°C. When compared with the results for the Al-3%Mg alloy, the stability of the fine grains is improved by 50°. This improvement is due to the presence of the particles formed by the addition of 0.2%Fe or 0.2%Ti. The stability is further improved for the Al 2024 alloy and, as confirmed in Fig.2, the fine-grained structure is stable up to 400°C.

Figure 4 plots the elongation to failure against the initial strain rate at testing temperatures of (a) 250°C and (b) 400°C for the Al-3%Mg-0.2%Fe and Al-3%Mg-0.2%Ti alloys. The results of the Al-3%Mg and Al-3%Mg-0.2%Sc alloys obtained in this study are also included for comparison. At a testing temperature of 250°C, corresponding to the temperature where significant grain growth begins to occur in the Al-3%Mg-0.2%Fe and Al-3%Mg-0.2%Ti alloys, the elongation increases as the initial strain rate decreases. The Al-3%Mg-0.2%Fe alloy exhibits the elongation of ~370% at the strain rate of 3.3×10^{-4} s^{-1}. This elongation is smaller than the elongation of ~640% attained in the Al-3%Mg-0.2%Sc alloy but it is large by comparison with the elongation of ~200% in the Al-3%Mg alloy without Fe. It should be noted that the testing temperature of 250°C is equivalent to ~0.6Tm where Tm is the absolute melting temperature of Al-3%Mg, and therefore the elongation of ~370% attained in the Al-3%Mg-0.2%Fe alloy is close to the range of low temperature superplasticity.

The elongation to failure obtained in the Al-3%Mg-0.2%Ti alloy is ~180% which is almost the same as in the Al-3%Mg alloy. It is considered that deformation occurred in the Al-3%Mg-0.2%Ti alloy through dislocation glide as in the Al-3%Mg alloy whereas in the Al-3%Mg-0.2%Fe alloy there is some contribution to the deformation from grain boundary sliding. It is probable that the difference is due to a more homogeneous distribution of fine particles in Al-3%Mg-0.2%Fe than in Al-3%Mg-0.2%Ti.

For a testing temperature of 400°C, both the Al-3%Mg-0.2%Fe and Al-3%Mg-0.2%Ti alloys give elongations similar to the Al-3%Mg alloy or even less. Since the grain sizes are no longer small in both alloys at 400°C and yet particles are present, it is concluded that deformation occurs through dislocation glide with some resistance due to the particles.

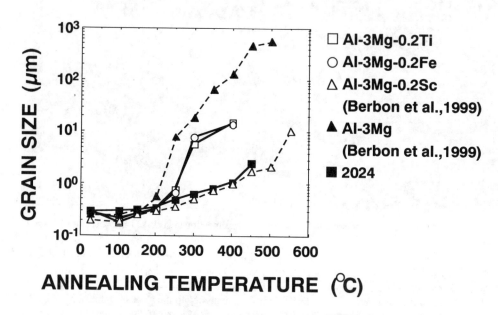

Fig.3   Grain size versus annealing temperature for the different alloys.

Figure 5 plots the elongations to failure against the initial strain rate at the testing temperature of 400°C in the Al 2024 alloy. The results for specimens without ECAP are also included for comparison. There is a significant improvement of the elongation when the alloy is processed by ECAP. A maximum elongation of ~460% was attained at a strain rate of 10^{-3} s^{-1}. The elongations to failure are further plotted in Fig.6 against the testing temperature for a strain rate of 10^{-3} s^{-1}. It is evident that there is a clear difference between the elongations above or below ~400°C: the elongation is large provided the small grains are retained but, when the grain size is large due to significant grain growth, the elongation becomes similar in magnitude to that obtained on the sample without ECAP. It is confirmed that a large elongation is attained when the small grains produced by ECAP are stable at higher temperatures.

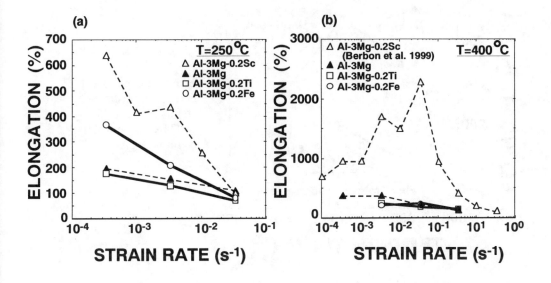

Fig.4 Elongations to failure versus initial strain rate at (a) 250°C and (b) 400°C.

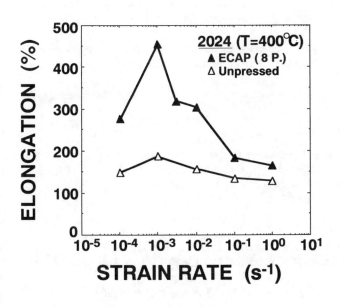

Fig.5 Elongations to failure versus initial strain rate for the Al 2024 alloy with and without ECAP.

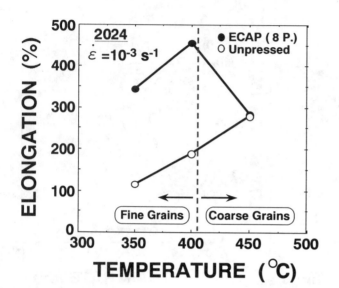

Fig.6 Elongations to failure versus testing temperature for the Al 2024 alloy with and without ECAP.

4. Summary and Conclusions

1. Fine grains with a size of ~0.3 μm ware produced using the technique of equal-channel angular pressing (ECAP) in Al-3%Mg alloys containing 0.2%Fe or 0.2%Ti and in an Al 2024 alloy.
2. The Al-3%Mg-0.2%Fe alloy after processing by ECAP exhibited a maximum elongation of ~370% at 250°C with an initial strain rate of 3.3×10^{-4} s^{-1}. This elongation is significantly larger than in the coarse-grained Al-3%Mg alloy. There is evidence for possible low temperature superplasticity in the Al-3%Mg-0.2%Fe alloy.
3. An elongation of ~460% was obtained in a fully annealed Al 2024 alloy after processing by ECAP. It is demonstrated that superplasticity is attained in a commercially available alloy when ECAP was applied for grain refinement.
4. The ECAP technique is capable of producing superplasticic materials provided it is possible to stabilize the fine-grained structure at high temperatures.

Acknowledgements

This work was supported in part by the Light Metals Educational Foundation of Japan, in part by a Grant-in-Aid for Scientific Research from the Ministry of Education, Science, Sports and Culture of Japan, in part by the Japan Society for the Promotion of Science and in part by the National Science Foundation of the United States under Grants No. DMR-925969 and INT-9602919.

References
1. R. Z. Valiev, N. A. Krasilnikov and N. K. Tsenev, Mater. Sci. Eng. A137 (1991) p.35.
2. Y. Ma, M. Furukawa, Z. Horita, M. Nemoto, R.Z. Valiev and T.G. Langdon, Mater. Trans. JIM 37 (1996) p.336.
3. Z. Horita, M. Furukawa, M. Nemoto, A.J. Barnes and T.G. Langdon, Acta Mater. (in press).
4. P.B. Berbon, S. Komura, A. Utsunomiya, Z. Horita, M. Furukawa, M. Nemoto and T.G. Langdon, Mater. Trans. JIM 40 (1999) P.772.
5. M. Furukawa, Y. Iwahashi, Z. Horita, M. Nemoto and T.G. Langdon, Mater. Sci. Eng. A257 (1998) p.328.
6. Y. Iwahashi, Z.Horita, M.Nemoto and T.G.Langdon, Metall. Mater. Trans. 29A (1998) p.2503.

Corresponding author: Zenji Horita, horita@zaiko.kyushu-u.ac.jp

Refinement and Stability of Grain Structure

F.John Humphreys and Peter S. Bate

Manchester Materials Science Centre
Grosvenor Street, Manchester M1 7HS, UK

Keywords: Grain Growth, Grain Refinement, Grain Size, Large Strain, Recrystallization, Stability

Abstract
The various methods of producing fine-grained alloys are discussed and it is concluded that thermomechanical processing routes are most suitable for the economic production of the large quantities of material required for structural applications. The limits of grain refinement by conventional discontinuous recrystallization are considered, and the production of micron-grained alloys by continuous recrystallization processes during or after large strain deformation is discussed. The stability of highly deformed microstructures against recrystallization is analysed, and the effect of second-phase particles on grain growth is discussed. It is shown that perturbations of the Zener drag during high temperature deformation may lead to dynamic grain growth in two-phase alloys.

1. INTRODUCTION
Fine-grained alloys are of industrial significance not only for their superplastic properties, which are of particular interest to this conference, but also because of the effect of the grain size on the mechanical properties, e.g. improvements in strength and fracture properties. In this paper we will examine the various methods available for refining the grain size, and the various factors which control the stability of a fine-grained material at high temperatures. There is often a conflict between the grain size which can be produced in a material, the amount or the size of material which can be so processed, and the cost of processing. This paper is focussed on fine-grained materials for structural applications and particular attention will be given to materials processed by routes which can in principle produce reasonably large quantities of material economically.

2. OVERVIEW OF METHODS OF GRAIN REFINEMENT
Conventional processing of alloys generally results in grain sizes of the order of 50-250μm. Smaller grain sizes may be produced from the melt, vapour, powders or thermomechanical processing.

Solidification. The cast grain size decreases as the solidification rate increases and very small grains or even amorphous metals may be produced under extreme conditions. However, because the size of the product is small – powder, wire or ribbon, further consolidation and processing is required to produce a bulk material.

Vapour deposition. There is a growing interest in the field of nanophase materials and the production of grains which are only a few atoms in diameter [1], generally by vapour deposition. Such materials often have impressive physical and mechanical properties, but it is not usually possible to produce material in the size and quantity required for structural applications.

Powder Processing. Powder metallurgy is a well established technology. In its simplest form, metal powders are mixed with ceramic or intermetallic powders and consolidated by hot pressing to produce dispersion strengthened products. The ceramic or intermetallic may provide both strengthening and also resistance to grain growth. High energy ball milling of the powder mixtures provides further microstructural refinement and reactive processing in which the metal powder reacts with an additive to produce a second-phase has been widely investigated. Such routes are currently the most suitable for production of nanostructured materials on a reasonable scale, but in addition to cost implications there are potential problems of contamination, gas entrapment and densification with such routes.

Thermomechanical processing. The deformation and annealing of bulk alloys is generally the optimum method in terms of cost, quantity and size of material, and the thermomechanical routes to producing fine grained alloys will be discussed in detail in the following sections.

3. THE LIMITS OF CONVENTIONAL THERMOMECHANICAL PROCESSING ROUTES

Conventional thermomechanical processing of alloys generally results in grain sizes of the order of 30-250µm [2], and the production of finer grain structures requires careful control of the material and the processing parameters [2,3,4] as is discussed in this paper.

3.1 Discontinuous recrystallization

Fine-grain microstructures can be produced by the discontinuous recrystallization of a cold worked metal. A small grain size is promoted by a large stored energy resulting from deformation and a large density of sites for nucleating recrystallization [2]. The smallest grain sizes in aluminium are typically achieved by deformation to large strains of alloys containing second-phase particles larger than ~1µm, and subsequent annealing to stimulate recrystallization (PSN) [2,5]. The commercial Rockwell process which produces ~10µm grains in AA7xxx alloys for superplastic applications is based on this principle [6], and similar grain sizes may be achieved by large strain deformation and annealing of commercial Al-Mg-Fe (AA5xxx) alloys [e.g. 2]. In order to achieve even smaller grain sizes, the number of large second-phase particles needs to be increased, for example by the use of particulate metal-matrix composites [2,7], in which grain sizes of less than 5µm can be produced. Ferrite has a high stacking fault energy, so that recovery processes precede and accompany recrystallization of cold rolled ferrite and consequently it is difficult to obtain grain sizes less than about 10µm in ferritic steels by discontinuous recrystallization. However, it has recently been shown [8] that cold rolling and annealing of steel with a pre-refined grain size of ~3µm resulted in a recrystallized grain size of less than 1µm. This is thought to be attributable to the large number of recrystallization nucleation sites available at pre-existing grain boundaries.

3.2 Continuous recrystallization

If a metal is given a large strain at intermediate or high temperatures, a microstructure containing predominantly high angle grain boundaries may evolve with little or no further annealing. Such a microstructure is virtually indistinguishable from one which has been conventionally recrystallized, but because it has evolved gradually and uniformly throughout the material, this phenomenon is generally known as **continuous recrystallization**. An example of a continuous recrystallization process in a compressed metal is shown schematically in Fig.1. During hot deformation of a polycrystalline metal, subgrains develop within the grains and the high angle boundaries become corrugated (Fig 1a). At larger strains (Fig 1b) the high angle boundaries are pushed together and, because the subgrain size remains approximately constant during deformation, they are separated by fewer subgrains. Eventually, the separation of the high angle boundaries is equal to the subgrain size, the high angle boundaries impinge and a microstructure of small and almost equiaxed grains is formed (Fig 1c). This process, known as **geometric dynamic recrystallization** [9] has been extensively studied in aluminium alloys [2].

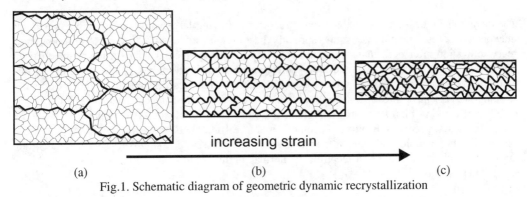

Fig.1. Schematic diagram of geometric dynamic recrystallization

The conditions under which geometric dynamic recrystallization can occur are a function of the initial grain size and the deformation strain, temperature and strain rate and are shown in Fig.2 in which the deformation temperature (T) and strain rate ($\dot{\varepsilon}$) have been combined into a single term, the Zener-Hollomon parameter (Z)

$$Z = \dot{\varepsilon}.\exp\left(\frac{Q}{RT}\right) \tag{1}$$

The subgrain size is inversely related to Z, being small at low deformation temperatures and high strain rates (large Z) and vice versa. It is seen from Fig.2 that if an alloy with a large initial grain size (D_1) is deformed under conditions of low Z (subgrain size d_1), then the grain boundary spacing decreases with strain until impingement occurs at A. The size of the grains so formed (D*) is approximately equal to the subgrain size and the strain required is ε_A. If the initial grain size were smaller – D_2, then geometric recrystallization would occur at a lower strain (ε_B) but would result in the same grain size. A smaller final grain size (D**) can be achieved by deformation at lower temperatures and strain rates (larger Z), but this will require a larger strain (ε_C). Therefore in order to achieve a small grain size at reasonable strains, a smaller initial grain size (D_2) is required, together with a larger Z so that the grain boundary spacing and subgrain size intersect at ε_D in Fig.2.

The conditions for high angle boundary impingement may be estimated on simple geometric criteria. The critical strain in plane strain compression (ε_c) is given [2] by

$$\varepsilon_c = \ln(Z^{1/m}D_0) + k_1 \tag{2}$$

where D_0 is the initial grain size and m and k_1 are constants, and the "recrystallized" grain size (D) [10, 11] is

$$D = k_2(1 - k_3 \ln Z) \tag{3}$$

where k_2 and k_3 are constants

It should be noted that the value of ε_c obtained from equation 2 is for plane strain compression, and different relationships will hold for the more complicated deformation paths discussed in section 4. Also, the above analysis assumes that no "new" high angle grain boundaries are created during deformation. The formation of new high angle boundaries by grain subdivision, which is well documented [e.g. 3,12] and particularly important for large initial grain sizes, will reduce the value of εc below that of equation 2.

The development of relatively stable grain structures following large strain deformation, is perhaps unexpected because a microstructure normally becomes unstable with respect to discontinuous recrystallization after deformation. It is not thought that any special microstructural mechanism operates during geometric dynamic recrystallization, but that the microstructural stability is due solely to the increased ratio of high angle to low angle boundaries illustrated in Fig.1. Computer simulations [13] have shown that such structures are stable, and a recent analysis [10,11] of the stability of cellular microstructures (subgrains and grains) has confirmed that as the HAGB/LAGB ratio increases, recrystallization becomes more difficult to nucleate and predicts that recrystallization should not occur if the fraction of HAGB is larger than ~0.7 [3]. Such fine grain structures will of course be susceptible to normal grain growth, and the prevention of this is discussed in section 6.

It has long being known that the development of fine-grain superplastic microstructures in Al-Cu-Zr (Supral) alloys occurred by continuous recrystallization [14]. Earlier work suggested that this process involved progressive rotation of subgrains during deformation [see 2], but recent research [15,16] in which the grain boundaries have been characterised by electron backscatter diffraction (EBSD), has shown that the mechanism is essentially one of high angle boundary accumulation of the type shown in figure 1.

Very large strain deformation of Al-Fe-Si and similar alloys at room temperature has been shown to result in sub-micron grain structures [3, 17]. On low temperature annealing, a sub-micron grain structure evolves by a continuous process and the structures are resistant to discontinuous recrystallization. It is thought that their formation is very similar to the process of geometric dynamic recrystallization discussed above.

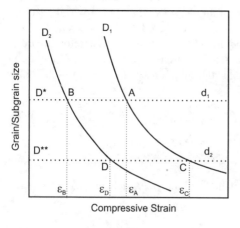
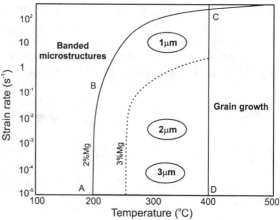

Fig.2. Processing conditions for the formation of a stable grain structure during hot deformation.

Fig.3. The process window for producing an ultra-fine grain structure in Al-2%Mg by plane strain compression to a strain of 3. (after ref 18)

An important difference between deformation at low and elevated temperatures is that at lower temperatures, grains, particularly large ones, will subdivide or fragment by inhomogeneous deformation [2,3,12], thus producing a larger amount of high angle grain boundary than the simple geometric increase discussed above. Additionally, during low temperature deformation, the large second-phase particles present in most commercial aluminium alloys will lead to inhomogeneous local deformation [3] which will assist the break-up of the grains structure.

The use of continuous recrystallization to produce fine grain structures is of considerable interest because it appears to overcome the grain size limits imposed by discontinuous recrystallization, which were discussed in section 3.1. In addition, it uses conventional rolling technology and in principle is capable of economically producing large quantities of sheet. However, the conditions under which very fine-grained material can be produced by such a process may be restricted, and this has been demonstrated in a recent experimental investigation [18] of Al-Mg-Cr alloys deformed by plane strain deformation.

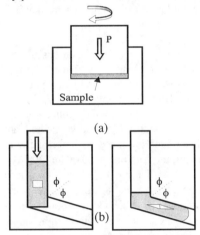
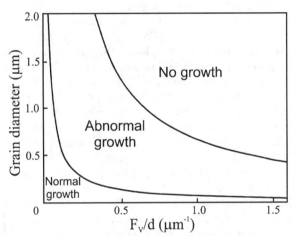

Fig.4. Redundant strain processing
a) Torsion under hydrostatic pressure.
b) Equal Channel Angular Extrusion.

Fig.5. The predicted regimes of no growth, normal and abnormal growth for fine-grained alloys with a volume fraction F_V of particles of diameter d.

The critical strain required for continuous recrystallization is given approximately by equation 2, and depends on the initial grain size and processing conditions (Z). In this investigation an initial grain size of ~12μm was used, and in order to produce micron sized grains in 1mm sheet, the maximum strain was ~3. The process window for the Al-Mg-Cr alloys deformed in plane strain compression to a von Mises strain of 3 is shown in Fig 3. For the Al-2%Mg alloys, the smallest grains will be formed at low temperatures and large strain rates, i.e. large Z (equation 3). However, there is a limit (A-B-C) imposed by the condition for geometric dynamic recrystallization as given by equation 2, and deformation at larger Z results in banded microstructures rather than new grains. For the Al-3%Mg alloy, restricted boundary mobility at low temperatures moves the limit to A^I, B^I, C^I. As discussed in section 6, a dispersion of small second phase particles is required to prevent grain growth of the small grains formed by continuous recrystallization. However, the presence of too many small particles will impede the boundary migration required for continuous recrystallization [19] and in the present case the moderate dispersion of Al-Cr particles was insufficient to prevent grain growth above processing temperatures of 400°C. The processing window is therefore defined between the limits ABC for continuous recrystallization and for grain growth CD. The grain size is determined by Z (equation 3) and it may be seen that in this investigation, grain sizes of 1-3µm could be produced. However, the small process window means that it would be difficult to process such a material to a 1µm grain size in a commercial rolling mill.

4. SEVERE DEFORMATION PROCESSING
The production of micron-grained alloys by conventional thermomechanical processing technology is seen from the above considerations to be limited by the achievable strains, which are typically ~3-4 if the resultant material is to have a minimum thickness of ~1mm. Several processing routes which allow larger strains have been developed and are discussed below. The mechanisms by which the original grain structure is broken up are thought to be similar to those discussed in the previous section although the more complicated deformation geometries have a significant effect.

A technique which increases the available strain is **accumulative roll bonding (ARB)** in which the material is rolled, stacked and re-rolled, thus maintaining the sample thickness, and this technique has been applied to aluminium alloys [20] and steels [21] with total strains of up to 8, resulting in sub-micron grain sizes. Critical factors in successful accumulated roll bonding are surface preparation and cleaning, the deformation temperature, and the amount of strain.

There has been a substantial recent research effort in developing laboratory scale methods which, by imposing complex and redundant strains, can achieve very large total strains whilst retaining a specimen of reasonable dimensions. One method which has been used with a wide range of materials is that of **torsion under hydrostatic pressure**, adapted from the Bridgeman anvil [22,23]. In this technique (Fig.4a), a thin disc is deformed in torsion using the friction provided by the application of a large hydrostatic pressure. The equivalent strains that have been induced with this method are typically of the order of 7, and grain sizes as fine as 0.2µm have been produced by deformation at room temperature. An alternative method is that of **Equal Channel Angular Extrusion (ECAE)** developed in the former Soviet Union by Segal and colleagues [24]. During ECAE the sample is extruded in a closed die which has two intersecting channels of equal size (Fig.4b) offset at an angle 2φ. Assuming ideal conditions and a sharp die corner, the sample will be subjected to a homogenous shear of $2\cot\varphi$, and the process may be repeated, thus giving a large cumulative strain to the sample. There are a number of factors including friction and die shape which make ECAE deformation more complicated than indicated above, and the deformation behaviour in the die has been numerically modelled [25], and the effects of friction and back pressure on the homogeneity of shear have been investigated [26]. Total strains of ~10 are readily achievable and because the sample is constrained, the process can even be used for less ductile materials.

The die angle is an important factor in ECAE processing as not only does it determine the strain per pass, but it also affects the geometry of deformation [27]. In ECAE it is not necessary to keep the work piece orientation the same for each repeated pass, and several permutations are possible, the most common being either no sample rotation or a rotation of 90° between passes.

An understanding of the formation of sub-micron grain structures during processing requires detailed quantitative microstructural characterisation. In many investigations [e.g. 27] the arcing of diffraction patterns in the TEM has been used to obtain a qualitative assessment of the grain structure. However, more detailed characterisation such as the determination of grain boundary character and the percentage of high angle boundaries etc, is only obtainable with high resolution Electron Backscatter Diffraction (EBSD) using a Field Emission SEM [e.g. 2-4].

5. GRAIN REFINEMENT UTILISING PHASE TRANSFORMATIONS
High temperature phase transformations such as those occurring in iron and titanium alloys may be used in conjunction with deformation and recrystallization to refine the grain structure, and this has

been extensively exploited for steels. Deformation during the α–γ transformation will affect the phase transformation and may also result in dynamic or static recrystallization of the austenite and ferrite. A summary of recent research which has concentrated on controlling the phase transformation and recrystallization processes so as to produce ultra-fine grained ferrite is given by Priestner and Ibraheem [8]. Hodgson and colleagues [28, 29] have produced 1μm ferrite in the surface layers of steel sheet by utilising the large surface strains resulting from roll friction, together with roll chilling, to intergranularly nucleate a high density of ferrite grains during hot rolling of austenite. Priestner and Ibraheem [8] have shown that a refined starting austenite grain size, combined with accelerated cooling, can also produce a 300μm layer of grains less than 1μm in a Nb-microalloyed steel.

6. STABILITY OF FINE GRAIN STRUCTURES DURING STATIC ANNEALING

The potential applications of very fine-grained alloys are dependent on the stability of the microstructures, particularly at high temperatures. Fine-grain microstructures have a large area of grain boundary and therefore a large stored energy, and are intrinsically unstable with respect to grain growth during high temperature annealing or even during the high temperature deformation operation during which they are formed. A dispersion of second-phase particles is generally required to prevent grain growth, although abnormal (discontinuous) grain growth may still be possible. The theory of the stability of cellular microstructures [17] has been extended to particle-containing alloys [30] and the predictions of the effect of boundary pinning pressure (F_V/d) on the stability of a fine grain structure are shown in Fig.5. It is seen that although only moderate pinning pressure is required to prevent grain growth, abnormal grain growth will readily occur. Thus to completely stabilise a 1μm grain size at elevated temperature, it is predicted that F_V/d should be larger than ~0.7μm^{-1} [3].

7. STABILITY OF FINE GRAIN STRUCTURES DURING DYNAMIC ANNEALING

Deformation at high temperatures further increases the instability of fine-grained structures, and this is of particular interest during superplastic deformation. Dynamic grain growth is a ubiquitous feature in superplasticity, and even the rate of growth with straining is similar in diverse materials [31].

Dynamic grain growth in superplastic deformation- at least in simple tension- is, to a good approximation, a simple function of strain. In rate form, this is:

$$\frac{\dot{D}}{D} = \lambda \dot{\varepsilon} \tag{4}$$

-where D is the current mean grain size, ε the tensile strain and the superimposed dot indicates time derivative. Because time derivatives occur on both sides of the equation, the natural time effectively cancels and growth depends only on strain. The constant, λ, usually has a value between 0.1 and 1. There has been some confusion about this value caused by the use of initial or final grain size rather than the current value. Further confusion is caused by the fact that static growth could take place at the temperatures involved, and this is simply subtracted from the dynamic growth in some reports. Both of those factors contribute to the range in reported values of λ.

Several theories for dynamic grain growth in superplasticity have been proposed [32-35]. These tend to be based on the usual assumptions that some combination of boundary sliding and diffusional creep is the principal deformation mechanism. The fact that deformation will produce a geometric change in the grain structure, which will affect the total boundary energy and the balance of surface tensions and so grain growth, has been noted by Rabinovich and Trivonof [36]. This type of geometric effect is likely to be very significant in Zener pinned grain structures, and will lead to dynamic grain growth as shown in a recent analysis of this effect [37], which is discussed below.

Simple models of the limiting grain size in Zener pinned systems assume a proportionality between the grain size and the boundary curvature which drives migration. If a Zener pinned system is deformed in a homogeneous manner, initially 'spherical' grains will distort to ellipsoids. The effective curvature will then change. If the assumed proportionality between grain size and boundary curvature is extended to a proportionality between the deformed ellipsoid and boundary curvatures, then in the absence of boundary migration the maximum curvature will increase with tensile strain. The curvature pressure will then exceed that in balance with the Zener pressure, and growth would be expected. A simple analysis shows that if boundary migration perpendicular to the tensile axis occurs to maintain the maximum curvature at the original level, then $\lambda = 0.75$ is predicted.

Further investigation using computer simulation of dynamic grain growth shows that simple result to be rather high. Combining a two-dimensional network model of grain growth with homogeneous deformation in the presence of a spatially uniform Zener drag leads to a prediction of $\lambda \approx 0.4$. This increases somewhat, to $\lambda \approx 0.5$, when inhomogeneous deformation, of the type which would result from different grain orientations having different plastic responses, is included in the model.

One factor which that modelling revealed was the sensitivity of the Zener pinned state to geometric perturbation. Any deviation from spatial uniformity of the Zener pressure would increase that sensitivity, and give more rapid dynamic growth. Such heterogeneity will exist in real alloys, and an extreme example has been investigated.

The eutectic alloy Al-6%Ni can be processed to give a distribution of nearly spherical Al_3Ni particles in an aluminium matrix, with a volume fraction of about 0.1. Extended annealing followed by rolling and recrystallization leads to a situation where there is approximately one pinning particle per grain [38]. Static annealing at 500 °C for 24hrs gives a mean particle section diameter of 2 μm and the grain section diameter is stable at 5.7 μm.

Tensile testing of this material was carried out at 10^{-3} s^{-1} with temperatures of 500 °C and 550 °C. The ductility was limited, as will be discussed below. Dynamic grain growth was observed, and a section of specimen gauge strained to 0.3 at 500 °C gave a mean grain section diameter of 7.3 μm. This corresponds to $\lambda \approx 0.85$, which though high is not exceptional. There were, however, indications that the growth was becoming discontinuous i.e. the grain size distribution did not remain self-similar. The standard deviation of the linear intercept distribution divided by the mean intercept after a strain of 0.3 was 0.15, compared to 0.11 with no deformation. Normal, continuous, growth would give similar values. Higher strains were only accessible in necked regions of the specimen, but the grain structure in those regions reinforced the view that some form of discontinuous dynamic growth was taking place, and typical grain structures are shown in Fig. 6.

Although the dynamic grain growth model mentioned above [37] does not deal with discrete particles, it is possible to introduce spatial heterogeneity in the potential for Zener drag. To do this, a set of Cauchy- type distributions is distributed in the model domain, and the drag on a boundary is given by the integral of the resulting potential along its length. The actual drag, as well as boundary energy and mobility, also depends on misorientation. Two cases have been investigated, in which the centres of the distributions were arranged with either moderate or weak uniformity. The diagrams shown in fig. 7 demonstrate the potential drag fields with associated, stagnant, static grain structures superimposed.

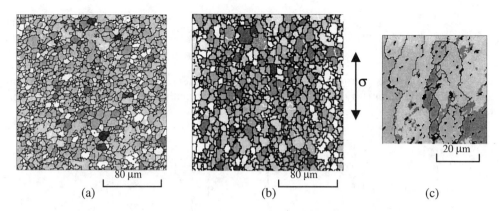

Fig. 6. Orientation image maps (EBSD) showing the grain structures in an Al-6Ni tensile specimen; in (a) the grip region (unstrained), in (b) a section of gauge strained to 0.3 and in (c) the local neck. Boundaries with misorientations ω >5° are shown. Note the change of scale in (c).

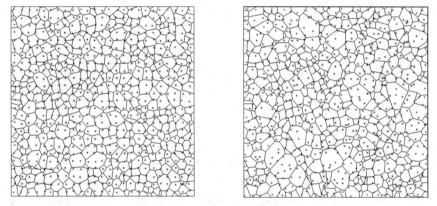

Fig. 7. Maps showing inhomogeneous drag potential fields, with intensity indicated by grey scale, for cases with moderate (left) and weak (right) regularity. Grain structures, which are at their Zener limit in the fields, are superimposed.

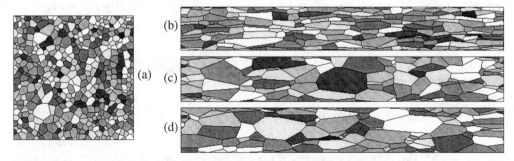

Fig. 8. Grain structures predicted by the dynamic grain growth model with homogeneous deformation. A typical, statically stagnant, structure prior to deformation is shown in (a), and structures are shown following a strain of 1 for (b) a spatially uniform drag potential, for (c) a non-uniform potential with moderate regularity and (d) for a non-uniform potential with weak regularity.

Deforming those domains results in much faster dynamic growth than was predicted with spatially uniform drag: in both cases, $\lambda \approx 0.7$. Examples of grain structures predicted at a strain of unity are shown in Fig. 8

As well as this increase in growth rate, there was a marked tendency for the grain size distribution to deviate from self- similarity when the drag potential was highly inhomogeneous. Predicted grain size distributions are shown in Fig. 9.

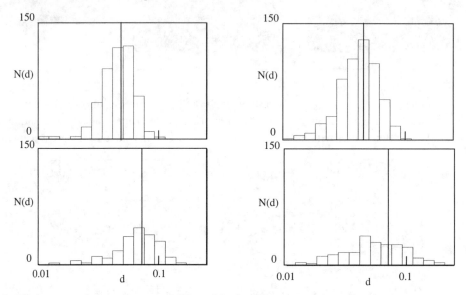

Fig. 9. Histograms of grain diameters, d (logarithmic diameter scale) predicted using homogeneous drag (left side) and inhomogeneous drag (right side). The upper figures are the initial, static, structures with about 500 total grains. The lower figures are when dynamic growth has reduced that number to about 230: that corresponds to a strain of 1 for the homogeneous case, and 0.5 for the inhomogeneous one.

Even at half the strain, the logarithmic standard deviation with inhomogeneous drag has clearly increased compared with the homogeneous drag case, which has remained about the same as the initial structure. There is, however, no indication of strongly discontinuous growth in the model predictions.

Both the growth rate and increase in spread of the size distribution are in general agreement with experimental observations. It is clear that homogeneously dispersed pinning particles which are significantly smaller than the grain size are more likely to give self-similar, continuous, grain growth than inhomogeneously dispersed ones or, equivalently, large individual particles.

As noted above, the Al-6Ni alloy was not superplastic- with total elongations of 50% at 500 °C and 65% at 550 °C. This is despite a fine initial grain size and low flow stresses: peak values were 13 MPa at 500 °C and 8 MPa at 550 °C. Work some years ago on this alloy [39] concluded that dynamic grain growth was too rapid. It could be that discontinuous dynamic growth is occurring, and that this causes the lack of ductility. Discontinuous growth has been observed previously during isothermal forging of $\alpha+\beta$ titanium alloys when the α fraction was relatively small [41], a phenomenon known as "β breakout". There may be other reasons for the lack of superplasticity in

Al-6%Ni compared with, say Al-5%Ca-5%Zn [40], but investigating materials which are **not** superplastic under conditions where similar alloys are is potentially useful, and may help to clarify the role of dynamic grain growth in superplastic deformation.

8. CONCLUSIONS

1. Grain refinement by discontinuous static recrystallization can be achieved by utilising particle stimulated nucleation, and the minimum grain size achievable in conventional aluminium alloys is ~10μm. However, in a Nb-steel which a 3μm grain size was produced by controlled rolling, further cold rolling followed by static recrystallization resulted in a grain size of <1μm.

2. Micron sized grains can be produced in aluminium alloys by continuous recrystallization during or following large strain rolling deformation. Although this can readily be achieved in low solute alloys of moderate initial grain size by rolling at room temperature, in Al-Mg alloys, the process window for continuous recrystallization is very restricted.

3. Continuous recrystallization to sub-micron grain sizes may readily be achieved using very large shear strains in techniques such ECAE. However, such techniques are generally limited to the processing of small amounts of material.

4. The conditions under which normal and abnormal grain growth may be prevented by dispersions of second-phase particles may be predicted using a theory of the stability of cellular microstructures.

5. A recent model of dynamic grain growth based on the effects of the changes in grain geometry during deformation on Zener pinning, is shown to give realistic predictions of the rate of dynamic grain growth.

6. Modelling of dynamic grain growth in microstructures with large particle volume fractions, in which discrete pinning events need to be considered, shows, in agreement with experiment, that inhomogeneous particle distribution may lead to abnormal grain growth during high temperature deformation.

ACKNOWLEDGEMENTS

The authors gratefully acknowledge the support of Alcan International, Banbury Laboratories, and the EPSRC, for this research.

REFERENCES

[1] Gleiter, H. Acta Mater. 2000. **48,** p1.
[2] Humphreys, F.J. and Hatherly, M. **Recrystallization and related annealing phenomena**. 1995. Pergamon Press, Oxford.
[3] Humphreys, F.J., Prangnell, P.B., Bowen, J.R., Gholinia, A. and Harris, C. Phil. Trans. Royal Society. 1999. **A357**, p1663.
[4] Humphreys, F.J., Prangnell, P.B. and Priestner, R. Proc. Rex99, Tsukuba. Eds. T.Sakai and H.Suzuki. 1999. p69.
[5] Humphreys, F.J. Acta Met. 1977. **25,** p1323.

[6] Wert, J.A., Paton, N.E., Hamilton, C.H. and Mahoney, M.W. Metall. Trans. 1981. **12A**, p1267.
[7] Humphreys, F.J., Miller, W.S. and Djazeb, M.R. Mats. Sci. & Tech. **6**, 1990. p1157.
[8] Priestner, R. and Ibraheem, A.K. Mats. Sci. and Tech. 2000. (in press).
[9] McQueen, H.J., Knustad, O., Ryum, N. and Solberg, J.K. Scr. Met. 1985. **19**, p73.
[10] Humphreys, F.J. Acta Mater. 1997. **45**, p4235.
[11] Humphreys, F.J. Mats Sci and Tech. 1999. **15**, p37.
[12] Hughes, D.A. and Hansen, N. Acta Mater. 1997. **45**, p3871.
[13] Oscarsson, A., Hutchinson, W.B., Nicol, B., Bate, P.S. and Ekstrom, H-E. Mats. Sci. Forum. 1994. **157-162**, p1271.
[14] Watts, B.M., Stowell, M.J., Baikie, B.L. and Owen, D.G.E. Met. Sci. J. 1976. **10**, p189.
[15] McNelley, T.R., NcMahon, M,E. and Perez-Prado, M.T. Phil. Trans. Royal. Soc. Lond. 1998. **A357**, p1683.
[16] Ridley, N., Cullen, E.M. and Humphreys, F.J. Mats. Sci & Tech. 2000. **16**, p117.
[17] Oscarsson, A., Ekstrom, H.E. and Hutchinson, W.B. Materials Science Forum. 1992. **113-115**. p177.
[18] Gholinia, A., Humphreys, F.J. and Prangnell, P.B. Mats. Sci. & Tech 2000, (in press).
[19] Gholinia, A., Sarkar, J., Withers, P.J. and Prangnell, P.B. Mats. Sci. and Tech. 1999. **15**, p605.
[20] Saito, Y., Utsunomiya, H., Tsuji, N. and Sakai, T. Acta Mater. 1999. **47**, p579.
[21] Tsuji, N., Saito, Y., Utsunomiya, H. and Tanigawa, S. Scripta Mater. 1999. **40**, p795.
[22] Valiev, R.Z., Korznikov,A.V. and Mulyukov, R.R. The Physics of Metals and Metallography. 1992. **4**, 70.
[23] Horita, Z., Smith, D.J., Furukawa, M., Nemoto, M., Valiev, R.Z. and Langdon,T.G. Proc. Thermec'97, ed. T. Chandra and T. Sakai. TMS. 1997. p1937.
[24] Segal, V.M. Mat. Sci. Eng. 1995. **A 197**, p157.
[25] Prangnell, P.B., Harris, C. and Roberts, S.M. Scripta Materialia, 1997. **37**, p983.
[26] Bowen, J.R., Gholinia, A., Roberts, S.M. and Prangnell, P.B. Mats. Sci. & Eng. 2000, (In press).
[27] Nakashima, K., Horita, Z., Nemoto, M and Langdon, T.G. Acta Mater. 1998. **46**, p1589.
[28] Hodgson, P.D., Hickson, M.R. and Gibbs, R.K. Scripta Mater. 1999. **40**, p1179.
[29] Hurley, P.J., Kelly, G.L. and Hodgson, P.D. Materials Sci. and Tech. 2000. (in press).
[30] Humphreys, F.J. Acta Mater. 1997. **45**, p5031.
[31] Seidensticker, J. R. and Mayo, M. J. Scripta mater., 1998, **38**, p1091.
[32] Clark, M. A. and Alden, T. H. Acta Metall., 1973, **21**, p1195.
[33] Wilkinson, D. S. and Cáceres, C. H. Acta Metall., 1984, **32**, p1335.
[34] Kim, B. N. , Hiraga, K., Sakka, Y. and Ahn, B. W. Acta Mater., 1999, **47**, p3433.
[35] Seidensticker, J. R. and Mayo, M. J. Acta Mater., 1998, **46**, p4883.
[36] Rabinovich, M. Kh. and Trifonov. V. G. Acta Mater. , 1996, **44**, p2073.
[37] Bate, P. (to be published)
[38] Humphreys, F. J. and Chan, H. M. Mats. Sci. and Tech., 1996, **12**, p143.
[39] Lloyd, D. J. , Private communication
[40] Lloyd, D. J. and Moore, D. M. in" Superplastic Forming of Structural Alloys", eds. Paton, N.E. and Hamilton, C. H., The Metallurgical Society of AIME, Warrendale, 1982, p147.
[41] Bate, P. S., Blackwell, P. L. and Brooks, J. W. 6th world conference on titanium, 1988, p287

Using Severe Plastic Deformation for Grain Refinement and Superplasticity

Terence G. Langdon[1], Minoru Furukawa[2], Minoru Nemoto[3] and Zenji Horita[4]

[1] Departments of Materials Science and Mechanical Engineering, University of Southern California, Los Angeles CA 90089-1453, USA

[2] Department of Technology, Fukuoka University of Education Munakata, Fukuoka 811-4192, Japan

[3] Sasebo National College of Technology, 1-1 Okishin-cho, Sasebo 857-1193, Japan

[4] Department of Materials Science and Engineering, Kyushu University, Faculty of Engineering, 6-10-1 Hakozaki, Higashiku, Fukuoka 812-8581, Japan

Keywords: Equal-Channel Angular Pressing, High Strain Rate Superplasticity, Severe Plastic Deformation, Superplastic Forming, Ultrafine Grain Size

Abstract

The grain size of many metals may be reduced to the submicrometer or nanometer level using various procedures incorporating the application of severe plastic deformation. This paper examines the procedure and potential of equal-channel angular pressing (ECAP) in which a material is pressed through a die and a high strain is imposed without any change in the cross-sectional dimensions of the workpiece. Various factors influence the ultrafine grains attained using ECAP and examples are presented showing the application of ECAP in fabricating materials exhibiting high strain rate superplasticity and rapid forming capabilities.

1. Introduction

Superplasticity occurs over a limited range of strain rates, typically in the vicinity of $\sim 10^{-3}$ s^{-1}, in materials having grain sizes below ~ 10 μm [1]. However, there is experimental evidence showing the superplastic regime is displaced to faster strain rates when the grain size is decreased [2] and this has led to considerable interest in developing procedures for refining the grain sizes of polycrystalline metals to within the submicrometer or nanometer range [3].

Thermomechanical processing is generally limited to producing materials with grain sizes no smaller than ~ 1 μm but techniques have been developed in which grain sizes may be reduced to values which are often significantly smaller than 1 μm using procedures involving the introduction of high dislocation densities through severe plastic deformation (SPD) [4-6]. Two examples of these procedures are (i) Equal-Channel Angular Pressing (ECAP) in which a bulk sample, typically in rod form, is pressed through a die to introduce a high strain and (ii) high-pressure torsion (HPT) in which a sample, in the form of a disc, is subjected to torsion straining under a high pressure. Experiments suggest that HPT may be more effective than ECAP in refining the grain size of a material to a size below 100 nm but, nevertheless, ECAP has the advantage that it may be used to produce relatively large bulk samples and there is a potential for scaling-up the process by using multi-pass facilities in which the die is designed so that it introduces a very large strain during a single pressing [7]. The objectives of this paper are to present a brief overview of the factors affecting the refinement of grains using ECAP and to describe results illustrating the subsequent superplastic properties after ECAP.

2. The principle of ECAP

The principle of ECAP is illustrated schematically in Fig. 1. A die is fabricated containing two channels of equal cross-section intersecting near the center of the die at an angle of Φ which is generally close to, or equal to, 90°. The test sample is machined so that it fits within the channel and it is then pressed through the die using a plunger. Thus, the sample is subjected to straining by shear as it passes from one channel to the other and it emerges from the die without experiencing any change in the cross-sectional dimensions. This processing method is therefore distinct from the conventional metal working processes of extrusion and rolling where the cross-sectional dimensions of the work-piece are reduced in each pass. In practice, the strain imposed on the sample during a single passage through the die is dependent both upon the angle Φ between the two channels and upon an additional angle Ψ defining the outer arc of curvature where the two channels intersect. It can be shown that the total strain, ε_N, accumulated by repeated passages through the die is given by [8]

$$\varepsilon_N = \frac{N}{\sqrt{3}} \left[2 \cot\left(\frac{\Phi}{2} + \frac{\Psi}{2}\right) + \Psi \csc\left(\frac{\Phi}{2} + \frac{\Psi}{2}\right) \right] \quad (1)$$

where N is the number of passes through the die. The validity of equation (1) has been confirmed in model experiments [9].

Equation (1) shows that the total strain is directly proportional to the number of passes through the die, N. Since the cross-sectional dimensions of the sample remain unchanged in ECAP, it is possible to use repetitive pressings to impose very high strains and this provides an opportunity to introduce different shearing systems by rotating the sample between each separate pass [10]. In practice, it is possible to define four separate and distinct processing routes.

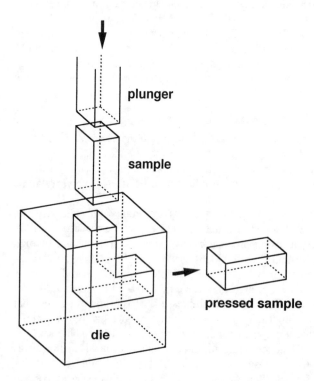

Fig. 1 Schematic illustration showing the principle of ECAP for imposing a high total strain without any change in the cross-sectional dimensions of the sample.

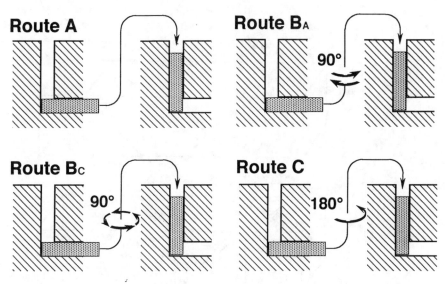

Fig. 2 The four different processing routes for ECAP.

The different processing routes are illustrated schematically in Fig. 2. In route A, the sample is pressed repetitively without any rotation between each pass. In route B, the sample is rotated by 90° between each pass, with this procedure subdivided into route B_A where the sample is rotated by ±90° so that the rotations are in alternate directions between each pass and route B_C where the sample is rotated by +90° in the same direction between each pass. In route C, the sample is rotated by 180° between each pass.

In order to understand the shearing systems introduced by each of these processing routes, it is instructive to consider the internal shearing associated with the use of these four routes when the die has an internal angle of $\Phi = 90°$. These shearing patterns are illustrated in Fig. 3 for three orthogonal planes labelled X, Y and Z, where these planes are defined as the plane perpendicular to the pressing direction, the plane parallel to the side face at the point of exit from the die and the plane parallel to the top face at the point of exit from the die, respectively.

Inspection shows that processing by route A leads to shearing on two separate planes and in practice this leads to increasing deformation on the X and Y planes but with no deformation on the Z plane [11]. Route C is very simple because it leads to a shearing on a single plane with the direction of shear reversed in each consecutive cycle. This means there is a redundant strain in route C and a restoration of the original structure every 2n passes, where n is an integer. The introduction of a redundant strain suggests that route C has similarities to procedures such as reciprocating extrusion where large and unlimited strains may be introduced by repetitively extruding a sample between two closed dies [12]. Route B_C also involves redundant strains because, as is evident in Fig. 3, the third pass reverses the shearing on the first pass and the fourth pass reverses the shearing on the second pass so that the original element is restored after 4n passes. However, it is found in practice that route B_C is not similar to reciprocating extrusion because the redundant strains do not occur in consecutive passes and, as will be demonstrated in the following section, this has important implications in the development of a fine and homogeneous microstructure. Finally, route B_A also involves no redundant strains but, as with route A, there is a progressive deformation with increasing numbers of passes through the die although for this route the deformation occurs on all three orthogonal planes. An important requirement in ECAP is to determine the optimum processing route.

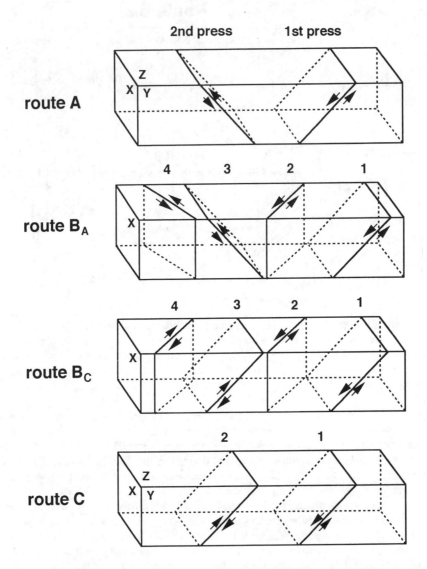

Fig. 3 The shearing patterns associated with the four processing routes in ECAP.

3. Optimizing the pressing conditions for achieving ultrafine grain sizes

There have been several detailed investigations in which samples were subjected to ECAP using different processing routes [13-19]. Figure 4 shows representative photomicrographs taken on the X plane in samples of pure Al after ECAP for a total of 4 passes using a die with an internal angle of $\Phi = 90°$ after processing using routes A, B_A, B_C and C, respectively, together with the selected area electron diffraction (SAED) patterns taken using a diameter of 12.3 μm. The material used for these experiments had an initial grain size of ~1.0 mm prior to ECAP and it is therefore apparent from inspection of Fig. 4 that ECAP to a strain of ~4 leads to remarkable grain refinement with an average grain size after pressing of ~1.3 μm.

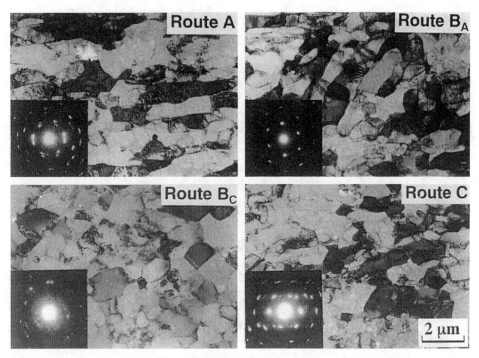

Fig. 4 Microstructures and SAED patterns on the X plane for pure Al after ECAP to a total of 4 passes using different processing routes at room temperature.

Although the grain size in each photomicrograph in Fig. 4 is very small and close to ~1 μm, nevertheless it is apparent that the grains tend to be elongated after processing through routes A, B_A and C whereas there is an essentially equiaxed grain configuration after processing through route B_C. It is also apparent from the SAED patterns that the diffracted beams form net patterns for routes A, B_A and C indicating the presence of many subgrain boundaries with low misorientation angles whereas the SAED pattern for route B_C shows diffracted beams lying randomly around rings indicating the presence of boundaries with high angles of misorientation.

Observations of this type have led to the conclusion that processing route B_C is probably the optimum procedure for producing a homogeneous microstructure of equiaxed grains separated by high angle grain boundaries [15]. However, it should be noted that an interpretation of the boundary misorientations from SAED patterns taken in a transmission electron microscope provides only a qualitative indication of the nature of the microstructure. Recent quantitative measurements employing electron back-scatter diffraction (EBSD) methods have suggested instead that, with Al-Mg and Al-Mn alloys and a die having an internal angle of $\Phi = 120°$, a higher fraction of high angle boundaries are generated using processing route A [18]. The reason for this apparent dichotomy may lie in the use of an ECAP die having a different internal angle Φ but, nevertheless, these experiments found also that the grains were more elongated when using route A by comparison with route B_C. Since an equiaxed grain structure with high angle boundaries is an important prerequisite for the occurrence of grain boundary sliding in superplasticity, it is reasonable to conclude that route B_C is the optimum procedure for fabricating a microstructure suitable for exhibiting superplastic elongations.

It is often difficult to perform ECAP successfully at room temperature because of limitations on sample ductility and/or a requirement for very high pressures. It is therefore instructive to consider the effect of conducting the ECAP at different temperatures.

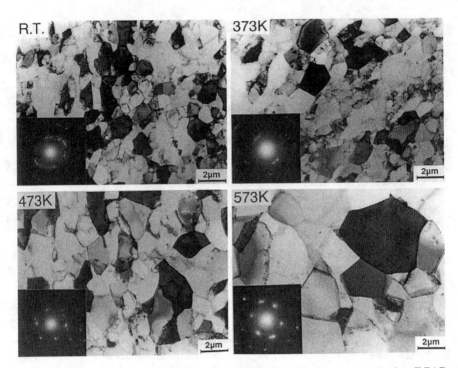

Fig. 5 Microstructures and SAED patterns on the X plane for pure Al after ECAP to a total of 6 passes using processing route B_C at different temperatures.

Figure 5 shows photomicrographs and SAED patterns recorded on the X plane in samples of pure Al after ECAP through 6 passes using route B_C with a die having an angle of $\Phi = 90°$ and with pressing temperatures in the range from room temperature to 573 K [20]: as previously, the SAED patterns were recorded using a diameter of 12.3 μm. It is apparent from Fig. 5 that an array of equiaxed grains is achieved at each of the pressing temperatures but the measured average grain size increases from ~1.3 μm at room temperature to ~1.5 μm at 373 K, ~2.0 μm at 473 K and ~4.2 μm at 573 K. Furthermore, there is a transition in the SAED patterns at pressing temperatures at and above 473 K to well-defined nets indicative of the presence of many low angle boundaries. These results show that an increase in the ECAP temperature leads to a larger average grain size and a higher fraction of boundaries with low angles of misorientation. It is reasonable to conclude that ECAP should be conducted at the lowest possible temperature which is consistent with the production of unfractured billets.

Several other factors may influence the nature of the microstructure generated in ECAP. For example, it has been shown that the magnitude of the internal angle Φ is important such that optimum microstructures are produced when the sample is subjected to an intense plastic strain using a value of Φ close to 90° [21]. Conversely, the pressing speed has only a minor influence although fewer extrinsic dislocations are present in the microstructures when using slower speeds [22].

4. Application of ECAP to superplasticity

The preceding section demonstrates the potential for using ECAP to fabricate materials with ultrafine grain sizes and with microstructures which are reasonably equiaxed and where the grain boundaries have high angles of misorientation. It is reasonable to anticipate these materials may exhibit exceptional superplastic properties. This possibility is now examined.

Figure 6 provides a demonstration of the exceptionally high superplastic ductilities which may be realized in conventional commercial alloys through the use of ECAP [23]. A light-weight high-strength alloy, known commercially as 1420 and with a composition of Al-5.5% Mg-2.2% Li-0.12% Zr, was subjected to ECAP using route B_C through a die with an internal angle of $\Phi = 90°$. Each sample was pressed for 8 passes at 673 K and a further 4 passes at 473 K to a total strain of ~12. The plot shows the variation of the elongation to failure with the imposed initial strain rate for tensile samples tested at temperatures within the range from 573 to 723 K. Also included in Fig. 6 are three experimental points for the same alloy in an unpressed condition and pulled to failure at 603 K. Inspection shows that ECAP leads to exceptionally high superplastic elongations up to and exceeding 1000%. By contrast, the unpressed alloy exhibits elongations of ~200% and lower. This difference is attributed to the grain refinement due to ECAP since the initial grain size in the unpressed non-superplastic condition was ~400 μm and the grain size after ECAP was ~0.8 μm. Furthermore, the high elongations after ECAP occur at very rapid strain rates typically in the vicinity of ~10^{-1} s^{-1} at the two highest testing temperatures and with an elongation of ~950% at a strain rate of 1 s^{-1} at 673 K. These results provide unambiguous confirmation that ECAP is capable of introducing a microstructure leading to high strain rate superplasticity in commercial alloys where the tensile elongations are generally fairly limited.

A further example of superplastic elongations after ECAP is shown in Fig. 7 where samples of an Al-3% Mg-0.2% Sc alloy were pulled to failure at 673 K [24]. This alloy was prepared by arc melting and casting with a solution treatment of 1 hour at 883 K to maximize the scandium in solid solution and giving an initial grain size of ~200 μm in the unpressed condition. The ECAP was conducted at room temperature with a die having an internal angle of $\Phi = 90°$ and with the samples pressed using route B_C for 8 passes to a strain of ~8. In Fig. 7, the upper sample is untested and the other samples were pulled to failure at decreasing strain rates from 3.3 s^{-1} where the elongation was 120% to 1.0×10^{-4} s^{-1} giving an elongation of 710%. A maximum elongation of 2280% was recorded under these testing conditions at a strain rate of 3.3×10^{-2} s^{-1}.

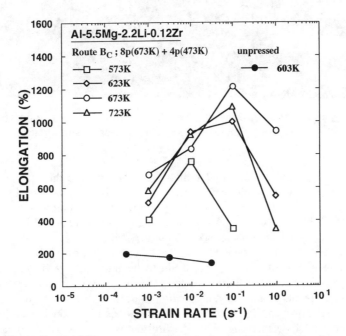

Fig. 6 Elongation to failure versus strain rate for an Al-Mg-Li-Zr alloy after ECAP and testing from 573 to 723 K: also shown are results for the unpressed alloy.

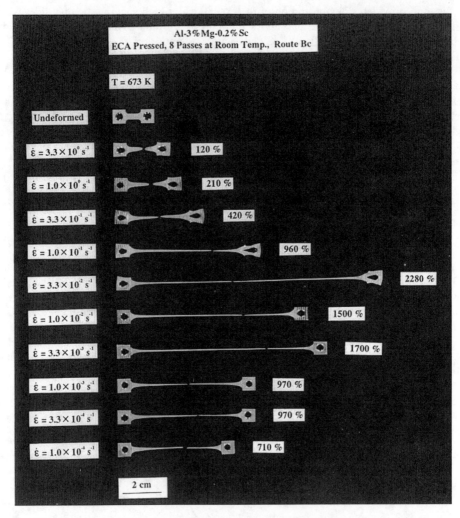

Fig. 7 Samples of an Al-Mg-Sc alloy subjected to ECAP to a strain of ~8 and then pulled to failure at 673 K using different strain rates.

The high elongations recorded in Fig. 7 at high strain rates suggest this material can be used for rapid superplastic forming. An example of this effect is shown in Fig. 8 where three discs were cut from a rod after ECAP to a strain of ~8 at room temperature: the disc on the left is untested and the other discs were inserted into a biaxial gas-pressure forming facility, held at 673 K, and then subjected to a constant gas pressure of 10 atmospheres, equivalent to 1 MPa, for periods of 30 s for the disc in the center and 60 s for the disc on the right. It is evident the discs have formed smooth domes despite the short time and the relatively low pressure. To evaluate the uniformity of thinning, the domes in the center and on the right were carefully cut into two parts and measurements were taken to determine the local thickness at angular points around the domes with respect to the centers of the undeformed discs. These measurements are shown in Fig. 9 where 90° corresponds to the thickness of the undeformed sheet. The measurements show there has been uniform deformation about the pole at 0° and this uniformity in thinning is consistent with a material having a high strain rate sensitivity. Thus, ECAP can produce materials suitable for use in rapid superplastic forming operations.

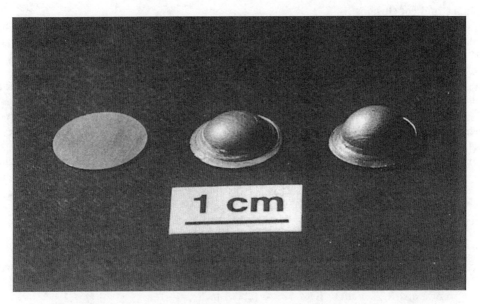

Fig. 8 Three discs of an Al-Mg-Sc alloy after ECAP to a strain of ~8: disc is untested on left and other discs subjected to a gas pressure of 10 atmospheres at 673 K for 30 s for disc in the center and 60 s for disc on the right.

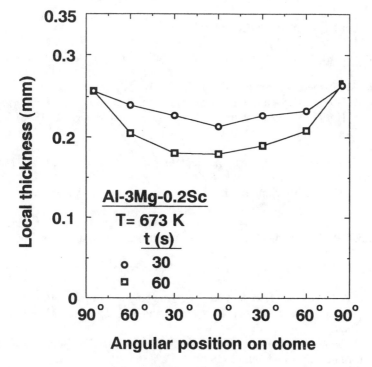

Fig. 9 Variation of local thickness with angular position around dome for domes shown in Fig. 8 after forming for 30 and 60 s.

5. Summary and conclusions

1. Equal-channel angular pressing (ECAP) is a processing method involving the introduction of severe plastic deformation and the consequent refinement of the grain size of metals to values not generally attained using conventional thermomechanical treatments.

2. Materials subjected to ECAP are capable of exhibiting high superplastic elongations at very rapid strain rates in the vicinity of $\sim 10^{-1}$ s^{-1}. In addition, there is a potential for using these materials for rapid superplastic forming operations.

Acknowledgements

This work was supported in part by the Light Metals Educational Foundation of Japan, in part by a Grant-in-Aid for Scientific Research from the Ministry of Education, Science, Sports and Culture of Japan, in part by the Japan Society for the Promotion of Science and in part by the National Science Foundation of the United States under Grants No. DMR-9625969 and INT-9602919.

References

1. T.G. Langdon, Metall. Trans. 13A (1982) p.689.
2. F.A. Mohamed, M.M.I. Ahmed and T.G. Langdon, Metall. Trans. 8A (1977) p.933.
3. Y. Ma, M. Furukawa, Z. Horita, M. Nemoto, R.Z. Valiev and T.G. Langdon, Mater. Trans. JIM 37 (1996) p.336.
4. T.C. Lowe and R.Z. Valiev, JOM 52 (4) (2000) p.27.
5. R.Z. Valiev, R.K. Islamgaliev and I.V. Alexandrov, Prog. Mater. Sci. 45 (2000) p.103.
6. T.C. Lowe and R.Z. Valiev (eds.), *Investigations and Applications of Severe Plastic Deformation*, Kluwer, Dordrecht, The Netherlands (2000).
7. K. Nakashima, Z. Horita, M. Nemoto and T.G. Langdon, Mater. Sci. Eng. A281 (2000) p.82.
8. Y. Iwahashi, J. Wang, Z. Horita, M. Nemoto and T.G. Langdon, Scripta Mater. 35 (1996) p.143.
9. Y. Wu and I. Baker, Scripta Mater. 37 (1997) p.437.
10. V.M. Segal, Mater. Sci. Eng. A197 (1995) p.157.
11. M. Furukawa, Y. Iwahashi, Z. Horita, M. Nemoto and T.G. Langdon, Mater. Sci. Eng. A257 (1998) p.328.
12. J. Richert and M. Richert, Aluminium 62 (1986) p.604.
13. Y. Iwahashi, Z. Horita, M. Nemoto and T.G. Langdon, Acta Mater. 45 (1997) p.4733.
14. Y. Iwahashi, Z. Horita, M. Nemoto and T.G. Langdon, Acta Mater. 46 (1998) p.3317.
15. K. Oh-ishi, Z. Horita, M. Furukawa, M. Nemoto and T.G. Langdon, Metall. Mater. Trans. 29A (1998) p.2011.
16. Y. Iwahashi, M. Furukawa, Z. Horita, M. Nemoto and T.G. Langdon, Metall. Mater. Trans. 29A (1998) p.2245.
17. Y. Iwahashi, Z. Horita, M. Nemoto and T.G. Langdon, Metall. Mater. Trans. 29A (1998) p.2503.
18. A. Gholinia, P.B. Prangnell and M.V. Markushev, Acta Mater. 48 (2000) p.1115.
19. P.B. Prangnell, J.R. Bowen, A. Gholinia and M.V. Markushev, Mater. Res. Soc. Symp. Proc. 601 (2000) p.323.
20. A. Yamashita, D. Yamaguchi, Z. Horita and T.G. Langdon, Mater. Sci. Eng. A287 (2000) p.100.
21. K. Nakashima, Z. Horita, M. Nemoto and T.G. Langdon, Acta Mater. 46 (1998) p.1589.
22. P.B. Berbon, M. Furukawa, Z. Horita, M. Nemoto and T.G. Langdon, Metall. Mater. Trans. 30A (1999) p.1989.
23. S. Lee, P.B. Berbon, M. Furukawa, Z. Horita, M. Nemoto, N.K. Tsenev, R.Z. Valiev and T.G. Langdon, Mater. Sci. Eng. A272 (1999) p.63.
24. Z. Horita, M. Furukawa, M. Nemoto, A.J. Barnes and T.G. Langdon, Acta Mater. (in press).

Corresponding author: Terence G. Langdon, e-mail: langdon@usc.edu

Tensile Superplasticity in Nanomaterials - Some Observations and Reflections

S.X. McFadden, A.V. Sergueeva and A.K. Mukherjee

Department of Chemical Engineering & Materials Science, University of California,
One Shields Avenue, Davis CA 95616, USA

Keywords: Aluminium, Nanocrystalline, Nanomaterials, Nickel, Severe Plastic Deformation, Superplasticity, Titanium

Abstract

The synthesis of nanocrystalline materials has provided new opportunities to explore grain size dependent phenomenon to a much finer scale. Superplasticity is a well-established grain size dependent phenomenon. In this paper, we analyze some of the tensile superplasticity data obtained in the last few years on Ti-6Al-4V, Ni_3Al, and 1420-Al alloy processed by severe plastic deformation (SePD). The experimental results show higher flow stresses for superplasticity in nanocrystalline materials than in microcrystalline materials. It is suggested that the conventional slip accommodated grain boundary sliding is likely to be difficult in nanomaterials

Introduction

The synthesis of nanocrystalline materials has provided new opportunities to explore grain size dependent phenomenon to a much finer scale. Superplasticity is a well-established grain size dependent phenomenon. The generalized constitutive equation for superplasticity is given by the Mukherjee-Bird-Dorn equation [1]. In generalized form, this equation can be written

$$\dot{\varepsilon} = A \frac{DGb}{kT} \left(\frac{b}{d}\right)^p \left(\frac{\sigma}{G}\right)^n \qquad (1)$$

where $\dot{\varepsilon}$ is the strain rate, D is the appropriate diffusivity (lattice or grain boundary), G is the shear modulus, b is the Burgers vector, k is the Boltzmann constant, T is the test temperature, d is the grain size, p is the grain size exponent, σ is the applied stress and n is the strain rate sensitivity of the flow stress. A large body of data for microcrystalline superplasticity in metals, intermetallics, and ceramics, has shown the grain size exponent p to be 2 in the case of lattice diffusion control, or 3 in the case of grain boundary diffusion control [2]. Consequently, a reduction in grain size can lead to a reduction in the superplastic temperature at constant strain rate, or an increase in the superplastic strain rate at constant temperature. This grain size dependence has led to the interest in superplasticity of nanocrystalline materials. A basic question is, "does the superplastic mechanism scale with grain size to the nanocrystalline range or is there a transition in superplastic behavior at such fine grain size?"

There are two major processing routes used in the synthesis of nanocrystalline materials: (a) consolidation of powders, and (b) severe plastic deformation (SePD) of bulk materials. Although a grain size of <100 nm can be obtained by both these methods in a variety of materials, the microstructural details vary. Some early investigations of nanocrystalline materials produced by consolidation of powders provided misleading results in terms of values for diffusivity and mechanical behavior. These results were influenced primarily by the presence of porosity in the consolidated samples. Grain size refinement of bulk materials by SePD eliminates problems associated with consolidation of powders as well as providing a means to study commercially important alloys that would be difficult to produce from powders. In this paper, we analyze some of the tensile superplasticity data obtained in the last few years on materials processed by SePD.

Experimental Procedure

Samples of Ti-6Al-4V, Ni_3Al, and 1420-Al alloy were processed by SePD to produce disks having nominal dimensions of 12mm diameter and 0.3mm thickness. A discussion of SePD processing can be found elsewhere [3]. The Ni_3Al was obtained from a bar of extruded alloy having the composition Ni-18 at.% Al-8 at. % Cr-at.% Zr-0.15 at. % B. The 1420-Al samples were processed from the Russian commercial alloy having the composition Al-5 wt% Mg-2 wt% Li-0.1 wt% Zr. Constant strain rate tensile testing was performed on specimens cut from the processed disks using electro-discharge machining. The tensile specimens had a gage length and width of 1mm, and were ground and polished to a thickness of around 0.2mm prior to testing. The tensile machine had a displacement resolution of 5μm and a load resolution of 0.1N. Testing was done with the specimens enclosed in a clam-shell furnace, in air.

Transmission electron microscopy (TEM) was used to observe the as-processed microstructures and to estimate grain size. Grain growth experiments were conducted by TEM *in situ* heating, and by annealing bulk specimens prior to preparing thin foils. Thin foils were prepared by electrochemical jet polishing. Differential scanning calorimetry (DSC) was used to study microstructural changes with heating. The DSC results were correlated with TEM and tensile testing data in an effort to understand the effect of heating on microstructural evolution and the effect of microstructure on superplastic properties.

Results

The experimental observations of tensile superplasticity in some nanocrystalline materials are summarized in Table 1. Reduction in the superplastic temperature was observed for all of the nanocrystalline alloys, but the temperatures for Ni_3Al and Ti-6Al-4V were particularly important in terms of practical superplastic forming operations. For example, in previous work [4], 550% elongation of 6μm grain size Ni_3Al was obtained at 1100°C. The temperature limit for conventional superplastic tooling is about 900°C.

Table 1. A Summary of Experimental Data for Tensile Superplasticity in Nanocrystalline Alloys

Material	Grain size (nm)	Strain Rate (s^{-1})	Temperature (°C)	Ductility (%)	Reference
Ti-6Al-4V	100	10^{-3}	725	740	Sergueeva et al. [5]
Ni_3Al	50	10^{-3}	650	560	Mishra et al. [6]
1420-Al alloy	100	10^{-1}	300	750	McFadden et al. [7]

Deformed tensile specimens of Ti-6Al-4V, Ni_3Al, and 1420-Al alloy are shown in Fig. 1. The as-processed microstructures of these materials are shown in Fig. 2. The structures were typical of materials processed by

SePD. The indistinct grain boundaries and non-uniform contrast in the bright-field images indicated the presence of a large defect density and associated internal strains. These features are less apparent in the 1420-Al alloy, presumably due to recovery at room temperature.

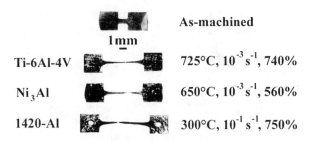

Fig. 1. Tensile specimens of Ti-6Al-4V, Ni$_3$Al, and 1420-Al.

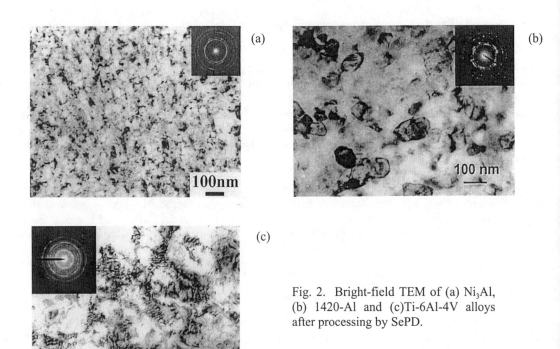

Fig. 2. Bright-field TEM of (a) Ni$_3$Al, (b) 1420-Al and (c) Ti-6Al-4V alloys after processing by SePD.

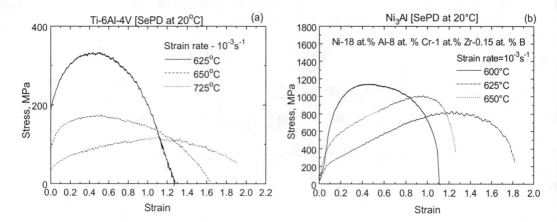

Fig. 3. Superplastic stress–strain curves for (a)Ti-6Al-4V and (b)Ni$_3$Al processed by SePD. Note the extensive strain hardening and high flow stresses.

Typical superplastic stress-strain curves for materials processed by SePD are shown in Fig. 3. The tensile data showed large flow stresses and significant strain hardening. Fig. 4 shows superplastic flow stress as a function of strain rate normalized by grain size and diffusivity for these materials. Even on a normalized plot, the nanocrystalline samples had much higher flow stresses than microcrystalline samples. This is opposite to some of the earlier expectations of enhanced superplasticity in nanocrystalline materials. Interestingly, Fig. 4 shows that at 573°C grain growth in 1420-Al resulted in the data merging with that for microcrystalline materials.

Grain growth investigations using TEM *in situ* heating and annealing of bulk specimens has shown that the onset of superplasticity coincides with the onset of grain growth. For Ni$_3$Al, grain growth was shown to be sluggish enough to maintain a grain size of 100nm or less throughout the duration of a tensile test at 650°C. Fig. 5(a) shows a TEM image taken from the deformed gage section of a Ni$_3$Al specimen deformed at 650°C. For 1420-Al alloy, TEM *in situ* heating showed that the microstructure was less resistant to grain growth at superplastic temperatures, and that a sub-microcrystalline grain size evolved at the lowest superplastic temperature of 250°C. Fig. 5(b) shows a TEM image taken from the gage of a 1420-Al specimen after superplastic deformation at 250°C.

The correlation between the onset of superplasticity and grain growth for Ni$_3$Al is shown in Fig. 6. Specimens of Ni$_3$Al processed by SePD were heated in the DSC to temperatures of 300°C, 375°C, 500°C, and 725°C. Fig. 6(a) the DSC results are overlaid with elongation as a function of temperature. Thin foils were prepared from the DSC specimens and examined by TEM. The corresponding bright-field TEM images are shown in Fig. 6(b). The TEM images show that recovery and some grain growth were associated with the large exothermic peak in the DSC curve. Also apparent from Fig. 6 is that the onset of superplastic deformation coincided with the onset of grain growth. A similar correlation between DSC results and superplasticity is shown in Fig. 7 for Ti-6Al-4V and 1420-Al alloys.

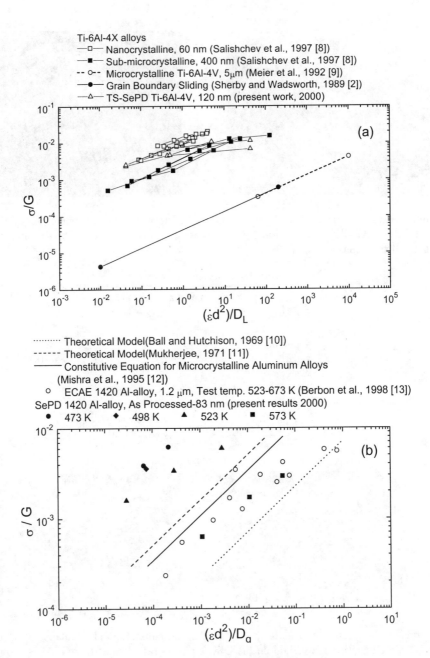

Fig. 4. Superplastic flow stress as a function of strain rate normalized by grain size and diffusivity for (a) Ti-6Al-4V alloys and (b) 1420-Al alloy. The nanocrystalline material processed by SePD showed much higher flow stresses than microcrystalline materials, and higher flow stresses than predicted by accepted models for superplasticity.

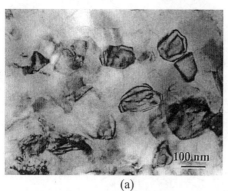 

Fig. 5. Microstructure in the deformed gage section taken from tensile specimens of (a) Ni$_3$Al deformed at 650°C, and (b) 1420-Al alloy deformed at 250°C. Image (a) is bright-field TEM and image (b) is dark-field TEM. Note the dislocations visible in the bright grain of 1420-Al (b).

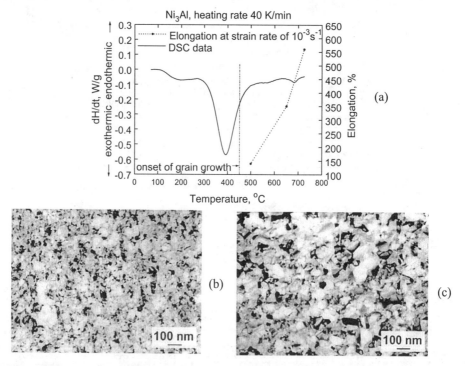

Fig. 6. Correlation between DSC results, superplastic deformation, and TEM images of specimens heated in the DSC. Tensile elongation is overlaid on the DSC curve in (a). Bright-field TEM is shown for specimens heated to (b) 300°C and (c) 500°C in the DSC. From the TEM images, recovery and the onset of grain growth were associated with the large exothermic peak in the DSC curve. Superplastic deformation began around 600°C.

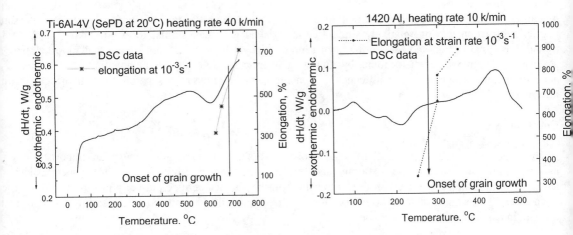

Fig. 7. Correlation between DSC data and superplastic elongation for Ti-6Al-4V and 1420-Al alloys. The onset of grain growth was confirmed by TEM *in situ* heating.

Discussion

The driving force for grain growth is significantly higher in nanocrystalline materials than in microcrystalline materials. Consequently, significant grain growth can occur even at the lower superplastic temperatures found for nanocrystalline materials. In addition to limiting the amount of superplastic deformation that may actually occur while the grain size is in the nanocrystalline range, this introduces additional complications in the analysis of data and interpretation of superplastic mechanisms. One of the assumptions of Eq. 1 is that the deformation is steady-state. Steady-state superplasticity is commonly identified by constant flow stress, which gives rise to classic flat-top flow curves in constant strain rate tensile tests. The flow curves for nanocrystalline materials do not exhibit regions of constant flow stress. However, the critical criterion for steady-state in terms of Eq. 1 is that the relevant mechanistic details of deformation are in a steady-state. Consequently, the shape of the flow curve alone is not a sufficient indicator of steady-state. At this point it is not clear whether or not the relevant mechanistic details of deformation are constant during superplastic flow in this system. That does not preclude the assumption of steady-state and an attempt to evaluate the parameters of Eq. 1 to further understand the observed behavior.

One way to explain the origin of higher flow stresses for superplasticity in nanocrystalline materials is to consider the possible influence of grain size on slip accommodation during grain boundary sliding. We suggest that dislocation nucleation is difficult in nanocrystalline materials. Although the details are not clear at this stage, based on the limited data and present analysis, we propose that the kinetics of superplastic flow would undergo a transition at a critical grain size. Another issue we would like to discuss is the issue of microstructural instability. As pointed out earlier, some grain growth does occur at temperatures where superplasticity has been reported. The conventional thinking based on constitutive equation is that the grain growth would lead to increase in flow stress. However, we have presented arguments that slip accommodation is difficult in nanocrystalline materials. From this it is imperative that grain growth in nanocrystalline materials should make the slip accommodation step easier. Does this imply that grain growth would enhance superplasticity when the starting grain size is in the nanocrystalline range? We believe the answer is "Yes". The reason might lie in two basic considerations. First, as discussed above, the stress required for s accommodation will come down. Second, grain growth during testing would provide a metastab microstructure, which might be conducive for fast recovery processes. A simple reasoning would be th migrating boundaries during concurrent grain growth would sweep the material which has undergone son plastic deformation leaving relatively strain free material in its wake, similar to events taking place durir dynamic recrystallization and transformational superplasticity where the microstructure undergoes continuo cleansing (i.e. removal of dislocations by migrating interfaces). We are conducting further experiments develop an understanding of the influence of concurrent grain growth on superplasticity in nanocrystalli materials.

Conclusions

Analysis of the tensile superplasticity data shows that slip accommodation plays an important role nanocrystalline material. The experimental results show higher flow stresses for superplasticity nanocrystalline materials. Based on this, a transition in the kinetics of superplasticity is expected at a critic grain size. It is suggested that the conventional slip accommodated grain boundary sliding is likely to k difficult in nanomaterials.

Acknowledgements

This work was supported by a grant from the US National Science Foundation: NSF-DMR-9903321. Tl authors are grateful to Professor Ruslan Valiev at the Institute of Physics and Advanced Materials, Ufa Sta Aviation Technical University, Russia, for processing the materials by SePD. We would also like to thank tl National Center for Electron Microscopy, Lawrence Berkeley National Laboratory, USA, for use of the TEM *situ* heating facility.

References

[1] A. K. Mukherjee, J. E. Bird and J. E. Dorn, Trans. ASM., 62 (1969), p.155.
[2] O. D. Sherby, J. Wadsworth, Progress in Materials Science, 33 (1989), p. 169.
[3] R. Z. Valiev, R. K. Islamgaliev, I. V. Alexandrov, Prog. Mater.Sci. 45 (2000), p. 103.
[4] J. Mukhopadhyay, G. Kaschner, A. K. Mukherjee, Scripta Materialia, 24 (1989), p. 857.
[5] A. V. Sergueeva, V. V. Stolyarov, R. Z. Valiev, and A. K. Mukherjee, "Enhanced Superplasticity in Ti-6A 4V Alloy Processed by Severe Plastic Deformation" submitted to Scripta Materialia.
[6] R. S. Mishra, R. Z. Valiev, S. X. McFadden and A. K. Mukherjee, Mat. Sci. Eng., A252 (1998), p. 174.
[7] S. X. McFadden, R. S. Mishra, R. Z. Valiev, A. P. Zhilyaev and A. K. Mukherjee, Nature 398 (1999), 684.
[8] G. A., Salishchev, R. M. Galeyev, S. P. Malisheva, O.R. Valiakhmetov, Mater. Sci. Forum, 243-24 (1997), p. 585.
[9] M. L. Meier, D. R. Lesuer, A. K. Mukherjee, Mater. Sci. Eng., A154 (1992), p.165.
[10] A. Ball, M. M. Hutchison, Metal. Sci. J., 3 (1969), p. 1.
[11] A. K. Mukherjee, Mater. Sci. Eng. 8 (1971), p. 83.
[12] R. S. Mishra, T. R. Bieler, A. K. Mukherjee, Acta. Metal. Mater., 43 (1995), p. 877.
[13] P. B Berbon, N. K. Tsenev, R. Z. Valiev, M. Furukawa, Z. Horita, M. Nemoto, T. G. Langdon, Metal Mater. Trans., 29A (1998), p. 2237.

Friction Stir Processing: A New Grain Refinement Technique to Achieve High Strain Rate Superplasticity in Commercial Alloys

R.S. Mishra[1] and M.W. Mahoney[2]

[1] Department of Metallurgical Engineering, University of Missouri,
1870 Miner Circle, Rolla MO 65409-0340, USA

[2] Rockwell Science Center, Thousand Oaks CA 91360, USA

Keywords: Aluminium Alloys, Friction Stir Processing, Grain Misorientation, Grain Refinement, High Strain Rate Superplasticity

Abstract

Friction stir processing is a new thermo-mechanical processing technique that leads to a microstructure amenable for high strain rate superplasticity in commercial aluminum alloys. Friction stirring produces a combination of very fine grain size and high grain boundary misorientation angles. Preliminary results on a 7075 Al demonstrate high strain rate superplasticity in the temperature range of 430-510 °C. For example, an elongation of >1000 % was observed at 490 °C and 1×10^{-2} s^{-1}. This demonstrates a new possibility to economically obtain a superplastic microstructure in commercial aluminum alloys. Based on these results, a three-step manufacturing process to fabricate complex shaped components can be envisaged: cast sheet or hot-pressed powder metallurgy sheet + friction stir processing + superplastic forging or forming.

Introduction

In spite of several advantages, the widespread use of superplastic forming has been hampered by slow strain rate for optimum superplasticity. This is a critical issue for the high volume manufacturing sector. The conventional superplastic forming rate of 10^{-4}-10^{-3} s^{-1} is too slow. High strain rate superplasticity, optimum strain rate $>10^{-2}$ s^{-1}, has been demonstrated in many materials, but requires either modification of the alloy chemistry or extensive thermo-mechanical processing [1,2]. Extensive thermo-mechanical processing or use of other processing routes for microstructural modification, such as powder processing, result in higher material cost. There is a clear need to find economical processing techniques that can produce microstructures amenable for high strain rate superplasticity in commercial alloys. Only then can superplastic forming be economically feasible for a high volume manufacturing sector such as automotive.

A new processing technique, Friction Stir Processing (FSP), has been developed by adapting the concepts of friction stir welding [3]. In the last ten years, Friction Stir Welding (FSW) has emerged as an exciting solid state joining technique for aluminum alloys. This technique, developed by The Welding Institute (TWI), involves traversing a rotating tool that produces intense plastic deformation through a 'stirring' action [3]. The localized heating is produced by friction between the rotating tool shoulder and the sheet top surface, as well as plastic deformation of the material in contact with the tool. This results in a stirred zone with a very fine grain size in a single pass [4]. Fine grain size is one of the requirements for superplasticity. Very recently, Mishra et al. [5] have reported high strain rate superplasticity in FSP 7075 Al, with optimum superplastic strain rate of 10^{-2} s^{-1} at 490 °C. These results suggest that a simple FSP technique can be used to obtain high strain rate superplasticity in commercial aluminum alloys.

Experimental

Commercial aluminum alloys, 7075 Al and 5083 Al, were selected for FSP and superplastic evaluation. A single pass friction stir processed zone of 0.3 m length was made using conventional friction stir welding practices, the difference being that the starting material was single piece (i.e. no joint). Tensile specimens with 1mm gage length were machined in the transverse direction, as shown in Fig. 1(a), where the gage length was centered in the fine grain FSP nugget. All the tensile tests were carried out using a custom-built, computer-controlled bench-top tensile testing machine at constant strain rate. The microstructure of friction stir processed specimens was examined by optical and transmission electron microscopy. In addition, orientation imaging microscopy was used to evaluate the nature of grain boundaries after the FSP process.

Results and Discussion

Fig. 1(a) presents a montage of optical micrographs showing a transverse cross section of the as processed friction stir region. The flow lines within the deformed region (nugget) and the elliptical shape are typical of the friction stir process for some aluminum alloys. The friction stir nugget experiences extensive plastic deformation, producing a very fine grain size. Fig. 1(b) shows a bright field transmission electron micrograph of the nugget region. The average grain size of 3.3±0.4 μm is considerably smaller than what is achieved by conventional thermo-mechanical processing (TMP). Fig. 1(c) shows the grain boundary misorientation in the center of the friction stir nugget. The distribution of grain boundaries from 20-60° misorientation suggests the FSP leads to high angle grain boundaries. The combination of very fine grain size and high angle grain boundaries is ideal for superplastic behavior.

The effectiveness of friction stir processing for grain refinement is very good. Table 1 summarizes some examples of the range of grain sizes reported in aluminum alloys. It should be noted that most of these results were obtained during metal joining and all data are first attempts, i.e. no process optimization was performed to study the minimum grain size possible by this technique. The grain size range of 0.8-12.1 μm is comparable to the best rolling based thermo-mechanical process schedule currently used to manufacture commercial superplastic aluminum alloy sheets. A particularly noteworthy aspect of friction stir processing is that it produces the refined microstructure in a single step. We note that the most recent patent of Brown [10] describes a rolling-based method comprises of the following steps, (a) homogenizing the alloy, (b) hot rolling, (c) over ageing, holding at a temperature and time period sufficient to create precipitates of diameters ranging from about 0.5 to 10 microns, (d) hot rolling to an exit temperature ranging from 70 °F to 650 °F, (e) controlled cold rolling, and (f) recrystallization anneal. These steps are typical in the rolling-based TMP, often referred to as the Rockwell process. Although this six-step procedure leads to a superplastic microstructure, it has two important implications:
  i.   the starting material is costly, and
  ii. more importantly, true strain exceeding 2.3 is required and from a practical standpoint, this results in sheets having a thickness less than 3 mm [11].

In addition, rolling develops different textures and anisotropic properties. By eliminating most of these steps, friction stir processing has the potential to economically produce superplastic microstructures in commercial aluminum alloys.

Another important aspect of friction stir processing is the grain boundary misorientation achieved after single pass. As illustrated in 1(c), FSP of 7075 Al produces a large majority of high angle grain boundaries. Currently, a number of severe plastic deformation processing routes are being developed to enhance superplasticity [12]. In terms of manufacturing pieces of significant size, equal channel angle extrusion (ECAE) offers maximum potential for scaling. ECAE can reduce the grain size to 0.5-2 μm in aluminum alloys, but requires a true strain of >4.0 [12,13]. It generally takes >8 ECAE passes to achieve very fine grain size with high grain boundary misorientation [14].

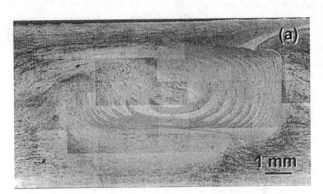

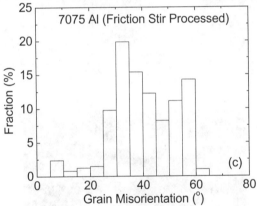

Fig. 1. (a) A montage of optical micrographs showing the 'nugget' region of very fine grain size produced by friction stir processing with superimposed miniature tensile specimen used in this study. (b) A bright field transmission electron micrograph from the center of the 'nugget', representing the microstructure in the gage region of the tensile specimen. (c) The grain boundary misorientation distribution in the friction stir zone indicates formation of high angle grain boundaries in this region.

Table 1. Some examples of the grain refinement observed in friction stir welded/processed aluminum alloys.

Alloy	Grain Size, μm	Reference	Remark
7075 Al	3-4	Rhodes et al., 1997, [4]	Joining
1100 Al	4.0	Murr et al., 1997, [6]	Joining
6061 Al	10.0	Liu et al., 1997, [7]	Joining
2024 Al	0.8	Benavides et al., 1999, [8]	Joining
Al-Li-Cu	9.0	Jata and Semiatin, 2000, [9]	Joining
7075 Al	3.3±0.4	*Present study*	Processing
5083 Al	12.1±3.6	*Present study*	Processing

The FSP 7075 Al exhibits superplasticity in a wide range of strain rates and temperatures. The variation of ductility with strain rate is illustrated in Fig. 2(a). The ductility at 1×10^{-2} s^{-1} and 490 °C was >1000%. The change in ductility with temperature at 1×10^{-2} s^{-1} is shown in Fig. 2(b). It can be noted that the specimens show high strain rate superplasticity at all test temperatures (430° to 510°C). The deformed specimens are shown in Fig. 2(c). The uniform elongation in the gage region is typical of superplastic flow. To compare the effectiveness of FSP with a conventional TMP route, the data of Xinggang et al. [15] are included in Fig. 2(a). The data of Xinggang et al. [15] represent the previous best TMP effort on 7XXX aluminum alloy. The increase in optimum strain rate due to FSP by more than an order of magnitude is noteworthy. The ductility at 1×10^{-1} s^{-1} and 490 °C was 318 %. For most forming operations, required ductility is less than 200%. Thus, the high strain rate ductility demonstrated herein is highly desirable because it leads to rapid forming. This work shows the effectiveness of FSP in evolving a microstructure in one step that is amenable to high strain rate superplasticity.

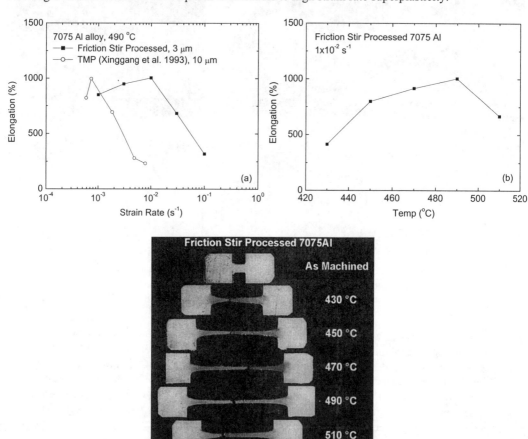

Fig. 2. Variation of elongation with (a) strain rate at 490 °C, and (b) temperature at 1×10^{-2} s^{-1}. (c) The appearance of specimens before and after superplastic deformation at various temperatures and at a strain rate of 1×10^{-2} s^{-1}. Note an order of magnitude improvement in optimum strain rate for superplasticity over the best results reported in literature [15].

In this initial evaluation of superplasticity in FSP 7075-Al, we took the mini-tensile specimens from the center of the nugget. This region displays microstructural stability in the temperature range of 430-490 °C. However, following heat treatment of the entire friction stir region a thin surface layer (0.13mm) exhibits secondary recrystallization and formation of large grains (Fig. 3). It is critical for superplastic forming to establish the microstructural stability at elevated temperatures.

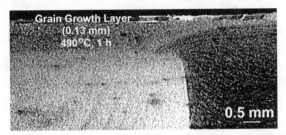

Fig. 3. Microstructural change after 1h heat treatment at 490 °C. Note that grain growth is limited to a thin surface layer.

For commercial viability, entire sheets or specific locations within a sheet will need to be rastered with the friction stir process. An example of overlapping passes is shown in Fig. 4. Preliminary results show grain refinement during both passes is similar and the overlap region does not have any adverse microstructural change. Some of the critical issues being pursued are; (a) microstructural stability at elevated temperature, (b) influence of microstructural gradients on superplasticity, and (c) optimization of friction stir processing parameters. The overall success of this process will depend on the outcome of these studies. From the initial studies, friction stir processing is likely to provide several technological breakthroughs in the area of superplastic forming.

Fig. 4. An example of overlapping nuggets using two passes. A continuous raster pattern to cover the desired area for superplastic forming would lead to overlapping regions.

Concluding Remarks

The overall implication of the present results is significant. Early studies show the possibility of using a simple FSP to produce a microstructure conducive to high strain rate superplasticity in commercial aluminum alloys. Further, superplasticity may now be achievable in thick section aluminum sheet via FSP. In conventional thermo-mechanical processing involving sheet rolling, the sheet thickness is reduced with each pass. To give sufficient total strain for grain refinement, a number of passes are applied resulting in sheet generally <2.5mm thick. On the other hand, in FSP the thickness of the sheet does not change. The entire sheet can be FSP in an overlapping sequence to obtain the desired microstructure. It should also be possible to use this technique to achieve a very fine grain size directly from a cast structure in commercial alloys. Based on these results, a three-step manufacturing of components can be envisaged: cast sheet or hot-pressed powder metallurgy sheet + friction stir process + superplastic forging or forming.

Acknowledgements

One of the authors (RSM) gratefully acknowledges the support of the University of Missouri Research Board and help of Siddharth Sharma with experiments at UMR.

References

1. A. K. Mukherjee, R. S. Mishra and T. R. Bieler, Materials Science Forum, 233-234 (1997) 217.
2. K. Higashi and M. Mabuchi, Materials Science Forum, 243-245 (1997) 267.
3. W. M. Thomas, E. D. Nicholas, J. C. Needham, M. G. Murch, P. Templesmith and C. J. Dawes, "Friction Stir Butt Welding", G.B. Patent Application No. 9125978.8, Dec. 1991; U.S. Patent No. 5460317, Oct. 1995.
4. C. G. Rhodes, M. W. Mahoney, W. H. Bingel, R. A. Spurling and C. C. Bampton, Scripta Mater., 36 (1997) 69.
5. R. S. Mishra, M. W. Mahoney, S. X. McFadden, N. A. Mara, and A. K. Mukherjee, Scripta Materialia, 42 (2000) 163.
6. L. E. Murr, G. Liu and J. C. McClure, J. Mat. Sci. Lett., 16 (1997) 1801.
7. G. Liu, L. E. Murr, C-S. Niou, J. C. McClure and F. R. Vega, Scripta Mater., 37 (1997) 355.
8. S. Benavides, Y Li, L. E. Murr, D. Brown and J. C. McClure, Scripta Mater., 41 (1999) 809.
9. K. V. Jata and S. L. Semiatin, Scripta Mater., (2000) in press.
10. K. R. Brown, Method of producing aluminum alloys having superplastic properties, U.S. Patent, 5,772,804, June 30, 1998.
11. R. Grimes and R. G. Butler, in Superplasticity in Advanced Materials, Edited by S Hori, M Tokizane and N Furushiro, Japan Society for Research on Superplasticity 1991, p771.
12. R. Z. Valiev, R. K. Islamgaliev and I. V. Alexandrov, Progress in Materials Science, 45 (2000) 103.
13. M. Furukawa, Z. Horita, M. Nemoto and T. G. Langdon, in Ultrafine Grained Materials, Edited by R. S. Mishra, S. L. Semiatin, C. Suryanarayana, N. N. Thadhani and T. C. Lowe, The Minerals, Metals and Materials Society, 2000, p125.
14. S. D. Terhune, Z. Horita, M. Nemoto, Y. Li, T. G. Langdon and T. R. McNelley, in Proceedings of the Fourth International Conference on Recrystallization and Related Phenomena, Edited by T. Sakai and H. G. Suzuki, Japan Inst. of Metals, Sendai, Japan, 1999, p. 515.
15. J. Xinggang, C. Jianzhong and M. Longxiang, Acta Metall. Mater., 41 (1993) 2721.

Poster Presentations

High Strain Rate Superplastic Deformation in 6061 Alloy with 1% SiO₂ Nano-Particles

T.D. Wang and J.C. Huang

Institute of Materials Science and Engineering, National Sun Yat-Sen University,
Kaohsiung 804, Taiwan ROC

Keywords: 6061 Alloy, HSRS, Nano-particle, Powder Metallurgy, Thermomechanical Treatment

Abstract

Four 6061 aluminum systems were prepared, including the cast 6061 alloy processed by TMT or ECAP, the PM 6061 alloy, and the modified PM 6061 alloy added with 1 vol% nano-SiO₂. The modified 6061/1%nano-SiO₂ alloy possessed a grain size ~0.5 μm and maintained fine grain size upon loading at high temperatures, resulting in HSRS over 300% at 550-590 °C and 1-5×10^{-1} s^{-1}. The grain boundary misorientation distribution in the modified 6061/1%nano-SiO₂ alloy was also most random compared with other three unmodified 6061 alloys. The current results suggest that the addition of a small amount (1 vol%) of cheaper SiO₂ or Al₂O₃ nano-particles into commercial Al alloys can effectively suppress grain growth at high temperatures and enhance HSRS, similar to the effects by adding 15-25 vol% micro-sized SiC or Si₃N₄ reinforcing particulates or whiskers.

Introduction

Commercial 6061 Al alloys after severe thermomechanical treatments (TMT) or equal channel angular pressing (ECAP) usually do not exhibit high strain rate superplasticity (HSRS) at temperatures greater than 500 °C due to extensive grain growth [1]. Previous efforts were made by adding 15-25 vol% Si₃N₄ or SiC reinforcing particles or whiskers measuring around 0.2-1 μm [2-4]. The inter-particle spacing L_s can be roughly estimated from $L_s = <r>(2\pi/3V_f)^{1/2}$ [5], where $<r>$ is the average particle radius and V_f is the particle volume fraction. With reinforcing particles of V_f=20% and $<r>$=0.3 μm in typical HSRS aluminum base composites, L_s will be 1.0 μm. The resulting grain size in HSRS composites would be limited by L_s, and was usually around this range (0.5-2 μm).

It is well known that the HSRS aluminum composites are highly expensive and usually exhibit unsatisfactory tensile ductility less than 3% at room temperature, both limiting their industrial applications. Meanwhile, recent studies on a commercial HSRS 6061/SiC$_w$ composites, a product of Advance Materials Corp of USA, also showed that the HSRS composite was difficult to fusion welded [6]. Therefore, it was inspired to study the feasibility in producing HSRS 6061 alloy which is reinforced by 1 vol% SiO₂, Al₂O₃ or SiC nano-particles measuring around 30-50 nm. The material is called as the modified 6061 alloy in comparison with the unmodified 6061 cast alloy. According to the above equation, L_s in the modified alloy with V_f=1% and $<r>$=20 nm will be 0.3 μm. This suggests that the addition of 1% nano-powders could stabilize the grain size below 1 μm and enhance HSRS in the modified alloys. The modified alloy may be fabricated via powder metallurgy (PM) or ingot-casting route, and the product might be cheaper and weldable.

Experimental

Four Al-1%Mg-0.6%Si 6061 aluminum systems were prepared. The first two were commercial cast 6061 alloys, either processed by TMT or room temperature ECAP to a true strain level of 2.0-12.0. The other two were the PM 6061 alloys, including the unmodified PM 6061, and the modified 6061 alloys added with 1 vol% nano-SiO_2. The average particle sizes of the 6061 and SiO_2 powders, purchased from Valimet of USA and Plasmachen GmbH of Germany, were 20 μm and 30 nm, respectively. The mixed powders were first vacuum hot pressed at 600 °C under a pressure around 120 MPa, followed by one-step hot extrusion at 500 °C or two-step TMT comprising hot extrusion plus cold or warm rolling to a total TMT true strain of 3.0-5.0.

The tensile loading axis was consistently parallel to the extrusion or rolling axes. Instron 5582 universal testing machine, equipped with a three-zone furnace allowing a temperature control within ±2°, was used. Microstructural characterization was done using JEOL 200CX TEM or JEOL 6400 SEM. The grain-orientation distribution were examined using an Oxford Link Opal EBSD system. The spatial resolution limit of the EBSD was around 0.3 μm. Around 500-1000 grain boundaries were examined for each distribution histogram. The boundaries with angles <10° are termed as low angle boundaries, LAB, unlikely to motivate grain boundary sliding (GBS). Those >30° are classified as high angle boundaries, HAB, most favorable for GBS. The rest within 10-30° are called as medium angle boundaries, MAB.

Results

Figure 1a shows the mixture of the 6061 powders surrounded by the nano-SiO_2 particles under the as-hot-pressed condition. Upon further hot extrusion and rolling, the 1 vol% nano-SiO_2 particles finally dispersed semi-uniformly in the 6061 matrix, as shown in Fig. 1b. The resulting grain sizes under the as-TMT or as-ECAP condition for all of the four 6061 systems were ~0.5±0.2 μm. The ECAP processed 6061 alloy possessed a slightly smaller grain size measuring ~0.3 μm. Table 1 lists the measured grain sizes for the four materials under various conditions. Upon static annealing at 300 and 500 °C for 30 min, the grain size increased to ~1.5 μm and 2.3 μm respectively (Fig. 2) in the modified 6061/1%nano-SiO_2 materials, compared with ~2.5 μm and over 10 μm for the unmodified TMT or ECAP cast 6061 alloy exposed at the same temperatures.

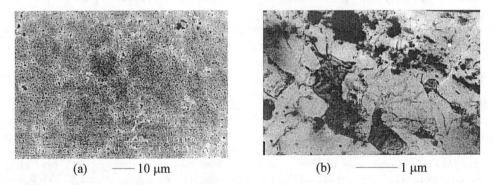

Fig. 1 (a) SEM micrograph showing the hot-pressed 6061 powders surrounded by the nano-SiO_2 particles, and (b) TEM micrograph showing the microstructure after further hot extrusion and rolling.

Table 1 Average grain size in the four materials measured from optical or TEM micrographs

Condition	TMT-6061 d, μm	ECAP-6061 d, μm	PM-6061 d, μm	6061/1%nano-SiO$_2$ d, μm
As-TMT	0.7	0.3	0.7	0.5
Annealed at 300 °C for 30 min	2.5	1.5	1.5	1.5
Annealed at 500 °C for 30 min	20	7	10	2.3
Annealed at 550 °C for 30 min	40	12	20	2.5
Annealed at 590 °C for 30 min	50	15	20	3.0

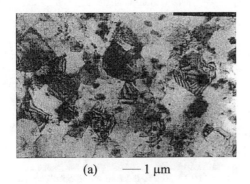
(a)   — 1 μm

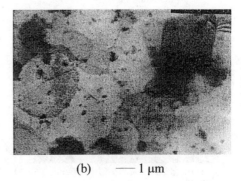
(b)   — 1 μm

Fig. 2   TEM micrographs showing the microstructures in the rolled sheets after further annealing at (a) 300 °C and (b) 500 °C for 1 h.

The grain orientation and grain boundary mutual misorientation angle distribution has been characterized using EBSD. Under the as-TMT or as-ECAP condition for all of the four materials, a bimodal distribution was seen, with more LAB or HAB and less MAB fractions, as shown in Fig. 3. The occurrence fractions for LAB, MAB and HAB for four systems are compared in Table 2. The PM 6061/1%nano-SiO$_2$ possessed the highest HAB after processing, implying a better superplastic performance during tensile loading. After HSRS at 570±20 °C for around 100%, these materials all possessed a higher fraction of HAB and a much lower population of LAB, as listed in the bottom row in Table 2. The bimodal behavior disappeared and a semi-random distribution took over.

The high strain rate tensile elongations at 1×10^{-1} s^{-1} for the unmodified 6061 alloy after TMT or ECAP processing were ~35%, 160% and 180% at 300, 550 and 590 °C, respectively, with a maximum flow stress of ~100, 18 and 14 MPa, as listed in Table 3. In comparison, the modified 6061 with 1 vol% nano-SiO$_2$ exhibited ~100%, 270% and 310% at the same temperatures, with a lower maximum flow stress of ~80, 10 and 4 MPa. The higher tensile elongation and the lower flow stress at high temperatures and high strain rates suggested the operation of GBS in the modified 6061 alloy under such conditions. It was noted that the maximum stress of 4 MPa for the 6061 with 1% nano-particles at ~590 °C and 1×10^{-1} s^{-1} was similar to the observed stresses for most HSRS Al composites with 15-25% micro-sized particles.

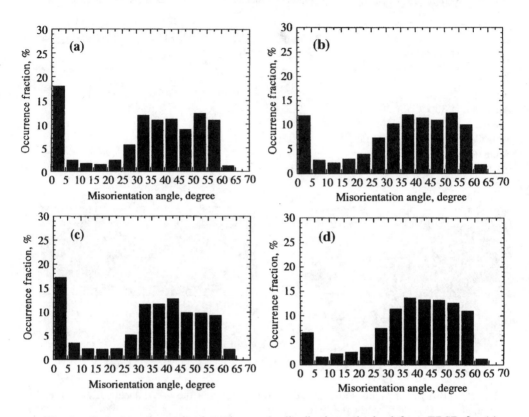

Fig. 3 Grain boundary misorientation angle distributions obtained from EBSD for (a) as-TMT cast 6061 alloy, (b) as-ECAP 6061 cast alloy, (c) as-TMT PM 6061 alloy, and (d) as-TMT PM 6061/1%nano-SiO$_2$.

Table 2 Comparison of the relative populations of LAB, MAB and HAB in the four 6061 base materials under different conditions

Condition	LAB, %	MAB, %	HAB, %
Cast 6061 alloy, as-TMT	20.3	11.3	68.4
Cast 6061 alloy, as-ECAP	14.3	17.1	68.6
PM 6061 alloy, as-TMT	20.7	12.0	67.3
PM 6061/1%nano-SiO$_2$, as-TMT	8.2	15.9	75.9
PM 6061/1%nano-SiO$_2$, HSRS to ~100%	4.3	15.3	80.4

At the optimum HSRS temperatures around 570-590 °C, partial melting at grain or interface boundaries is expected since the incipient partial melting temperature determined from DSC curves for the 6061 alloys was determined to be ~580 °C. No liquid filament was observed on the fracture surface of the failed tensile specimens loaded at 550 °C and 1×10^{-1} s^{-1}, as shown in Fig. 4a. In contrast, fine filaments were frequently seen in samples loaded at 590 °C and 1×10^{-1} s^{-1}, as shown in Fig. 4b. The basic deformation mechanisms in the modified 6061/1%nano-SiO$_2$ alloy is considered to be similar to that occurred in HSRS aluminum composites.

The apparent m-value increased from 0.15-0.20 over 570-590 °C and 1×10^{-2}-5×10^{-1} s^{-1} for the unmodified alloy to 0.3-0.35 for the 1 vol% modified one. Preliminary calculations resulted in an activation energy Q of ~190-300 kJ/mol for the unmodified alloy over 500-590 °C and 10^{-2}-10^{-1} s^{-1}, compared with ~160-500 kJ/mol for the modified counterpart. After the threshold stresses were considered, the true m was around 0.5. The details of controlling deformation mechanisms in the unmodified and modified 6061 alloys will be analyzed and presented elsewhere.

Table 3   Representative tensile loading data on the unmodified cast 6061 alloy and modified PM 6061/1vol% nano-SiO_2

Material	T, °C	$\dot{\varepsilon}$, s^{-1}	σ, MPa	e, %	m	Q_a, kJ/mol
Cast 6061	25	1×10^{-3}	300	12	--	--
	300	1×10^{-1}	100	35	--	--
	500	1×10^{-1}	26	150	~0.2	--
	550	1×10^{-1}	18	160	~0.2	~190
	570	1×10^{-1}	15	175	~0.2	--
	590	1×10^{-1}	14	180	~0.2	~300
	590	5×10^{-1}	18	180	~0.2	--
6061/1%SiO_2	25	1×10^{-3}	343	25	--	--
	300	1×10^{-1}	80	100	--	--
	500	1×10^{-1}	21	180	~0.3	~160
	550	1×10^{-1}	10	270	~0.3	--
	570	1×10^{-1}	8	240	~0.3	~210
	590	1×10^{-1}	4	310	~0.3	~500
	590	5×10^{-1}	10	297	~0.3	--

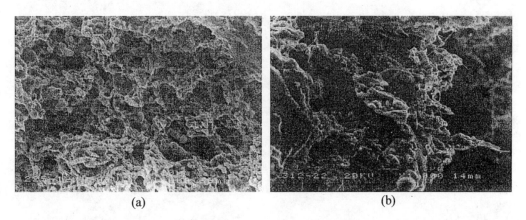

(a)   (b)

Fig. 4   SEM fractographs of the 6061/1%nano-SiO_2 alloy loaded at 1×10^{-1} s^{-1} and a temperature of (a) 550 °C and (b) 590 °C. Fine liquid filaments were only observed in (b).

The addition of Al_2O_3, SiO_2 or SiC nano-particles to commercial Al alloys can be done not only via the PM method, but also via the cast route by adding into molten Al during casting. It appears that a small amount of nano-particles of ~1 vol% can effectively suppress grain growth at high temperatures and enhance HSRS, similar to the effects by adding 15-25 vol% micro-sized SiC or Si_3N_4 reinforcing particulates or whiskers. The modified aluminum alloys with 1 vol% 30 nm Al_2O_3 or SiO_2 (which are lower-priced than SiC) nano-particles might have potential in industrial applications.

Summary

(1) All four 6061 aluminum systems, after proper TMT or ECAP processing, possessed fine grain sizes around 0.5±0.2 μm. But only the modified PM 6061 alloy added with 1 vol% nano-SiO_2 could maintain fine grain size ~1-3 μm upon loading at high temperatures, resulting in HSRS over 300% at 550-590 °C and $1-5 \times 10^{-1}$ s^{-1}.
(2) The grain boundary misorientation distribution in the modified 6061/1%nano-SiO_2 alloy after TMT was also most random compared with other three unmodified 6061 alloys.
(3) The low flow stress level of only 4 MPa for the 6061 with 1% nano-particles at ~590 °C and 1×10^{-1} s^{-1}, similar to the observed stresses for most HSRS Al composites with 15-25% micro-sized reinforcements, suggested the smooth operation of GBS in this modified alloy at high temperatures.
(4) The current results demonstrate that the addition of a small amount (1 vol%) of cheaper SiO_2 or Al_2O_3 nano-particles into commercial Al alloys can effectively suppress grain growth at high temperatures and enhance HSRS, similar to the effects by adding 15-25 vol% micro-sized SiC or Si_3N_4 particulates or whiskers. The modified aluminum alloys with 1 vol% 30 nm SiO_2 or Al_2O_3 nano-particles might have potential in industrial applications.

References

[1] T. G. Nieh, R. Kaibyshev, F. Musin and D. R. Lesuer, Superplasticity and Superplastic Forming, ed. A. K. Ghosh and T. R. Bieler, TMS, Warrendale, PA, 1998, p. 137.
[2] M. Mabuchi and K. Higashi, Mater. Trans. JIM, 35 (1994), p. 399.
[3] T. Imai, S. Kojima, G. L'Esperance, B. Hong and D. Jiang, Scripta Mater., 35 (1996), p. 1199.
[4] B. Q. Ham and K. C. Chan, Scripta Mater., 36 (1997), p. 593.
[5] A. J. Ardell, Metall. Trans., 16A (1985), p. 2131.
[6] R. Y. Huang, S. C. Chen, and J. C. Huang, unpublished research, National Sun Yat-Sen University, Taiwan, (2000).

An Investigation of Cavity Development in a Superplastic Aluminium Alloy Prepared by Equal-Channel Angular Pressing

Cheng Xu, Sungwon Lee and Terence G. Langdon

Departments of Materials Science and Mechanical Engineering, University of Southern California, Los Angeles CA 90089-1453, USA

Keywords: Aluminium Alloys, Cavitation, Equal-Channel Angular Pressing, High Strain Rate Superplasticity

Abstract

Experiments were conducted on a commercial aluminum Al-2004 alloy known as Supral 100 and containing 6% Cu and 0.4% Zr. This material was subjected to Equal-Channel Angular Pressing (ECAP) over a range of experimental conditions and the mechanical properties of the as-pressed samples were determined using standard tensile testing. Samples were pulled to failure and the level of internal cavitation was then recorded quantitatively using metallographic techniques. The results show this alloy exhibits good superplastic properties after ECAP with total tensile elongations up to >1000%. As in conventional superplasticity, there is the development of extensive internal cavitation in this alloy in the as-pressed condition. Measurements are reported for the average sizes and shapes of these cavities.

1. Introduction

There is considerable interest in investigating the occurrence of superplasticity at very high strain rates and at low temperatures in commercial aluminum-based alloys. This interest arises because superplastic forming is currently conducted at relatively slow forming rates and high forming temperatures and this tends to limit the potential for fabricating a large number of identical components. It is well established that superplasticity depends upon the grain size of the material and there is evidence that a reduction in the grain size has the potential of both decreasing the temperature and increasing the strain rate associated with optimum superplastic flow [1,2]. The Equal-Channel Angular Pressing (ECAP) process makes use of an essentially pure shear to subject a material to intense plastic straining and it has been shown that this procedure is effective in achieving an ultrafine-grained structure [3-6]. For example, the grain size of a material is often reduced to the submicrometer level when using ECAP.

The development of internal cavitation is generally important in conventional superplasticity and there have been numerous reports of the cavitation characteristics associated with these materials [7]. Although it is well known that ECAP is capable of producing microstructures that exhibit very extensive superplasticity in tensile testing at high temperatures [8-10], there is no information currently available concerning the role of cavity development in these materials. Accordingly, the present investigation was initiated to evaluate, through quantitative measurements, the significance of internal cavitation in a commercial aluminum-based alloy after ECAP.

A commercial alloy, Supral 100, was selected for this investigation. This alloy was chosen for two reasons. First, it is a conventional superplastic alloy [11,12] with applications in the superplastic forming industry [13]. Second, there are earlier quantitative measurements of cavity development in this alloy when testing under superplastic conditions [14-16]. The objectives of the research were therefore two-fold: (i) to investigate the mechanical properties of the Supral 100 alloy in tensile testing after processing by ECAP and (ii) to examine the extent of any cavity development in this alloy.

2. Experimental material and procedures

The experiments were conducted using an aluminum Al-2004 alloy having a chemical composition, in weight %, of Al-6% Cu-0.4% Zr and with the commercial designation of Supral 100. The alloy was prepared by casting, homogenizing at 648 K for 5 hours and then hot rolling. The initial grain size of the alloy was ~100 μm. All of the ECAP tests were performed using cylindrical samples with diameters of ~10 mm and lengths of ~70 mm. Samples were subjected to ECAP using processing route B_C where the sample is rotated by 90° in the same direction between each pass [17]. The pressing was conducted under two different conditions: (i) for a total of 12 passes at 573 K and (ii) for a total of 6 passes at 673 K. The ECAP die had an internal channel angle of 90° so that each separate pass corresponded to a strain of ~1.

After ECAP, tensile samples were machined parallel to the pressing direction with gauge lengths of 4 mm and cross-sectional areas of 3×2 mm^2. All samples were tested to failure at temperatures from 623 to 723 K over a range of initial strain rates from 1.0×10^{-3} to 1.0×10^{-1} s^{-1}. Following tensile testing, the samples were sectioned and polished for detailed metallographic examination. Quantitative measurements were taken of the cavity morphology using an optical microscope and a video camera connected through a computer with appropriate software. Measurements were taken to determine the sizes and shapes of all cavities having areas of at least 1 μm^2 within a window having an area of 0.018 mm^2. A total of 40 windows was measured on each sample, with the measurements taken to within a distance of ~5 μm of the fracture tip.

3. Experimental results and discussion

Inspection after ECAP revealed average grain sizes of ~0.9 and ~2 μm after pressing through 12 passes at 573 K and 6 passes at 673 K, respectively. These measurements confirm, therefore, that processing by ECAP leads to very significant grain refinement.

Specimens were tested in tension after ECAP and the results of these tests are summarized in Table 1, including the ultimate tensile stress (UTS) and the measured elongation to failure. By comparison, the unpressed material gave a maximum elongation of ~450% at a temperature of 773 K when using a strain rate of 3.3×10^{-4} s^{-1} and the elongation was reduced to ~260% at the same temperature when using a strain rate of 1.0×10^{-2} s^{-1} [18]. As a consequence of processing by ECAP, the elongations to failure are therefore very significantly increased despite conducting the tensile tests at a lower temperature. The occurrence of elongations up to >1000% at an imposed strain rate of 1.0×10^{-2} s^{-1} demonstrates the occurrence of high strain rate superplasticity (HSR SP) in this alloy [19].

Figure 1 shows typical curves of stress, σ, versus strain, ε, obtained on samples of the alloy at a temperature of 723 K after ECAP at 673 K for a total of 6 passes. Inspection shows these curves exhibit a short period of strain hardening and then extensive strain softening up to the point of failure. There appears to be a reasonably well-defined steady-state region in the sample tested at 1.0×10^{-2} s^{-1}.

Table 1. Strength and elongation after ECAP.

Processing condition	Test temp. (K)	Strain rate (s^{-1})	UTS (MPa)	Elongation (%)
12 passes at 573 K	623	1.0×10^{-2}	48	420
	673	1.0×10^{-1}	63	330
	673	3.3×10^{-2}	42	530
	673	1.0×10^{-2}	25	1070
	673	1.0×10^{-3}	13	730
6 passes at 673 K	623	1.0×10^{-2}	54	250
	673	1.0×10^{-2}	31	580
	698	1.0×10^{-2}	27	650
	723	1.0×10^{-1}	47	380
	723	3.3×10^{-2}	31	550
	723	1.0×10^{-2}	22	1100
	723	1.0×10^{-3}	9	520

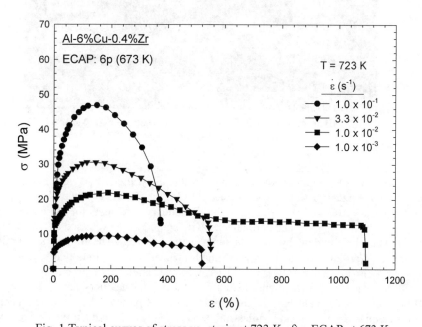

Fig. 1 Typical curves of stress vs. strain at 723 K after ECAP at 673 K for a total of 6 passes.

The appearance of these four specimens after failure is shown in Fig. 2 where it is apparent the deformation is reasonably uniform within the gauge lengths. There is no evidence for the occurrence of any visible necking in these samples but failure occurred ultimately without pulling to a fine point. This type of fracture is consistent with materials that fail because of the nucleation, growth and interlinkage of internal cavities [7].

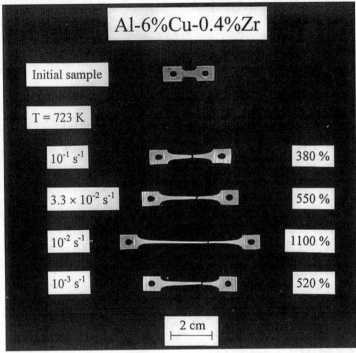

Fig. 2 Examples of tensile ductility in samples after ECAP at 673 K and tensile testing at 723 K.

The occurrence of cavitation was examined in this alloy after ECAP and Fig. 3 shows the distribution of cavity sizes for samples subjected to ECAP at 673 K and then tested at 723 K at strain rates from 1.0×10^{-3} to 1.0×10^{-1} s^{-1}. The roundness coefficients for the cavities in these three samples are given in Fig. 4. The normalized cavity numbers are expressed as a percentage and they were estimated by dividing by the total number of cavities. The roundness coefficient is defined as $4\pi \times$Area/(perimeter)^2 giving a maximum value of 1 for a circle and values of <1 for deviations from a circle. The results show that cavity nucleation is important in this alloy after ECAP and subsequent tensile testing and there are many cavities having sizes in the range of 1-5 µm^2 which are comparable with the specimen grain size. The plots of the roundness coefficients shows that the cavities have elliptical profiles and in practice it was found that the cavities tended to be elongated preferentially in the direction of the tensile axis.

4. Summary and conclusions

1. The Supral 100 alloy (Al-6% Cu-0.4% Zr) was subjected to ECAP and then tested in tension at elevated temperatures.
2. Superplasticity was achieved with maximum elongations to failure of up to >1000% at a strain rate of 1.0×10^{-2} s^{-1}. These results demonstrate the occurrence of high strain rate superplasticity after ECAP.
3. Cavities were nucleated in this alloy during tensile testing at 723 K. Quantitative measurements showed a majority of the cavities had sizes comparable with the as-pressed grain size. These cavities were elliptical in shape and preferentially elongated approximately in the direction of the tensile stress.

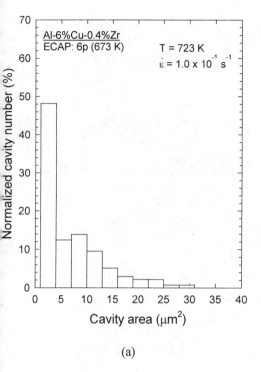

(a)

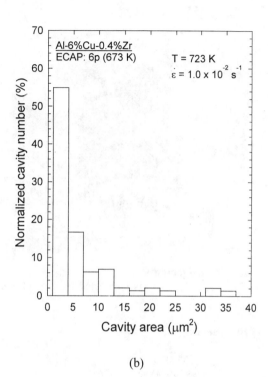

(b)

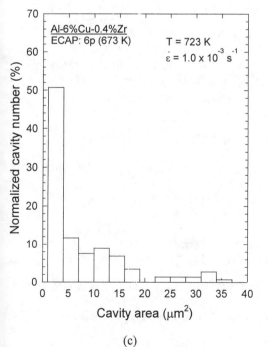

(c)

Fig. 3 Distribution of cavity sizes after ECAP at 673 K and tensile testing at 723 K with strain rates of:
(a) 1.0×10^{-1} s^{-1};
(b) 1.0×10^{-2} s^{-1};
(c) 1.0×10^{-3} s^{-1}.

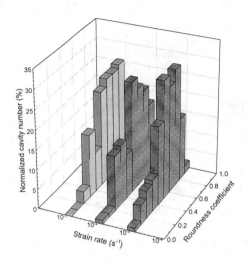

Fig. 4 Distribution of cavity roundness coefficients after ECAP at 673 K and testing at 723 K at three different strain rates.

Acknowledgements

This work was supported by the U.S. Army Research Office under Grant DAAD19-00-1-0488.

References

1. Y. Ma, M. Furukawa, Z. Horita, M. Nemoto, R. Z. Valiev and T. G. Langdon, Mater. Trans. JIM 37 (1996), p. 336.
2. S.X. McFadden, R.S. Mishra, R.Z. Valiev, A.P. Zhilyaev and A.K. Mukherjee, Nature 398 (1999), p. 684.
3. R.Z. Valiev, N.A. Krasilnikov and N.K. Tsenev, Mater. Sci. Eng. A137 (1991), p. 35.
4. R.Z. Valiev, A.V. Korznikov and R.R. Mulyukov, Mater. Sci. Eng. A168 (1993), p. 141.
5. S. Ferrasse, V.M. Segal, K.T. Hartwig and R.E. Goforth, J. Mater. Res. 12 (1997), p. 1253.
6. Y. Iwahashi, Z. Horita, M. Nemoto and T.G. Langdon, Acta Mater. 46 (1998), p. 3317.
7. T. G. Langdon, Metal Sci. 16 (1982), p. 175.
8. R.Z. Valiev, D.A. Salimonenko, N.K. Tsenev, P.B. Berbon and T.G. Langdon, Scripta Mater. 37 (1997), p. 1945.
9. P.B. Berbon, M. Furukawa, Z., Horita, M. Nemoto, N.K. Tsenev, R.Z. Valiev and T.G. Langdon, Phil. Mag. Lett. 78 (1998), p. 313.
10. S. Lee, P.B. Berbon, M. Furukawa, Z. Horita, M. Nemoto, N.K. Tsenev, R.Z. Valiev and T.G. Langdon, Mater. Sci. Eng. A272 (1999), p. 63.
11. R. Grimes, C. Baker, M.J. Stowell and B.M. Watts, Aluminium 51 (1975), p. 3.
12. R. Grimes, M.J. Stowell and B.M. Watts, Metals Tech. 3 (1976), p. 154.
13. A.J. Barnes, Mater. Sci. Forum 170-172 (1994), p. 701.
14. X. Zhao and T.G. Langdon, Mater. Res. Soc. Symp. Proc. 196 (1990), p. 215.
15. H.C. Kim, T.H. Ahn, C.H. So, Y. Ma, X. Zhao and T.G. Langdon, Scripta Metall. Mater. 26 (1992), p. 423.
16. H.C. Kim, T.H. Ahn, C.H. So, Y. Ma, X. Zhao and T.G. Langdon, J. Mater. Res. 9 (1994), p. 2238.
17. M. Furukawa, Y. Iwahashi, Z. Horita, M. Nemoto and T.G. Langdon, Mater. Sci. Eng. A257 (1998), p. 328.
18. S. Lee and T. G. Langdon, Mater. Res. Soc. Sym. Proc. 601 (2000), p. 359.
19. K. Higashi, M. Mabuchi and T.G. Langdon, ISIJ Intl. 36 (1996), p. 1423.

Corresponding author: Cheng Xu, e-mail: chengxu@usc.edu

Industrial Superplastic Forming Research and Application for Commercial Aircraft Components at Israel Aircraft Industries

B. Gershon[1], I. Arbel[1], S. Hevlin[1], Y. Milo[1] and D. Saltoun[2]

[1] Israel Aircraft Industries, Ben Gurion International Airport, Lod 70100, Israel

[2] M.S.I. Engineering Software Ltd., Tel Aviv 61251, Israel

Keywords: Al Alloys, 2195, 6061, Application, Commercial Aircraft, Superplastic Forming, Ti-Alloys, SP 700

Abstract

After research and development of SPF process we turned to a serial production process in Israel Aircraft Industries (IAI). The SPF technology is highly sophisticated and used extensively for commercial and military aircraft components. It is an ideal manufacturing process for complex shaped parts. This paper outlines our research of the SPF technology and the application of this method in the field of shaped parts productions. The main opportunity to use the SPF technology arouse from the needs of a new technology in producing complex-shaped parts for the new model of IAI aircraft "Galaxy". The application of SPF technology can diminish the manufacturing cost by reducing the preparation time, the need for extensive welding or joining methods, by simplifying the manufacturing process. The IAI has launched a program to use new superplastic (SP) alloys such as Ti-4.5Al-3V-2Fe-2Mo (SP-700) and extend research to use the aluminum alloy Weldalit in serial production for civil aircraft components. We were the first to produce by the SPF process production part from the conventional (coarse-grained) 6061 aluminum alloys sheets.

This paper contains results of our research and practical applications with SPF techniques and various materials such as aluminum and titanium alloys. Samples of actual aircraft components and experimental parts, including computer simulation of the manufacturing process. In addition results including illustrations of manufacturing process from a techno-economical point of view.

Introduction

The SPF process can meet the requirements of modern industry and especially production of civil aircrafts having the need for high mobility of production, high changeability of models, short period for organisation, new technologies for tools production.
SPF is a solution to this requirment by opening new opportunities for design of integral parts of aircraft structure.
Resently appears new SP materials such as SP-700 and Weldalit (2195). We investigated the characteristics and the possibilities of the recently developed new alloys for the SPF process in production of aircraft components.

The problems in SPF materials provision and their high cost provoke interest to investigation of the utilization of conventional materials such as 6061and commertial pure titanium in SPF production of deep complex shape parts.

Materials, Method and Equipment

Sheets from titanium and high strength aluminum alloys have advantages for applications in the aerospace industry due to their high strength, low density, heat resistance and other useful properties. Many of the sheet metal structures in airframes have complex shapes and compound curvature with intricate details.

Ti-6Al-4V is still the most popular titanium alloy used in the aerospace industry. However, it suffers from poor workability and associated high susceptibility to cracking, which inevitably results in high processing costs.

Near final-shape Ti-6Al-4V alloy parts are manufactured by superplastic forming (SPF) at temperatures in excess of 900°C. This has several disadvantages including the need for expensive die materials. The dies have their service life shortened by the high working temperatures involved
A new β-rich α+β titanium alloy, Ti-4.5Al-3V-2Fe-2Mo (SP-700), was developed by NKK in Japan, to improve hot workability, hardenability and mechanical properties over Ti-6Al-4V alloy. This alloy has remarkable superplastic formability at temperatures below 800°C, about 130°C lower than those used for Ti-6Al-4V. In addition, it can achieve higher strengths with air cooling after solution heat treatment, than those obtainable from Ti-6Al-4V solution heat-treated and water quenched.

The use of this alloy for the superplastic forming of complex aerospace components will be more cost effective than Ti-6Al-4V. The above mentioned advantages of SP-700, and many others not mentioned here, make this alloy suitable for manufacturing parts by SPF, for our new "Galaxy" business jet, using conventional hot-forming equipment.

The Ti- alloys sheets: Ti-6Al-4V, SP-700, CP-2, CP-3, Ti-3Al-2.5V used in this research and application in manufacturing parts by SPF consisted from 0.5 to 5 mm thick sheets in annealed condition.

For research of industrial application of aluminum alloys: Supral (2004), 7475, 2090, 8090, 2195 (Weldalite) were examined, sheets with thickness 0.5-5 mm.

Due to the high cost of the equipment for SPF and low rentability because of small series, IAI has chosen a way of adapting and modernizing available standard press equipment. This gave IAI an opportunity to realize the SPF technology in manufacture aircraft components on existing equipment. For this purpose there were 2 presses: "MURDOCK" which used for conventional hot forming parts from Ti-alloys, and also the usual hydraulic press with heating of working parts. Press "MURDOCK" allows to make SPF of a parts by size 1400x1100x220 mm of any SP aluminum alloys, a brass and same Ti-alloys however there are some technological restrictions due to the design of the given press.

Specialized (converted) for SPF the hydraulic press allows to make parts from material suitable for SPF (as for example Ti-6Al-4V) in a range of temperatures up to 1000°C with the maximum size 500x500x170mm. On this equipment it is also possible to perform diffusion bonding, however for parts of smaller sizes. Now IAI will carry out work on expansion of available industrial base for SPF with the purpose of manufacture of parts of the large size up to 2.5m in length from various alloys.

IAI use the same various forming methods for aircraft components: female forming, drape forming, back pressure forming, diaphragm forming, forming in conventional die for hot forming with use welded from 2 sheets pack.

Results

The air condition disposal in the tail of aircraft "Galaxy" requires the creation of a fire-resistant and structural parts. The possibility of servicing the engine is also need. Originally the assembled part was designed and constructed from 20 Ti-alloys parts and 600 rivets. Our design of this construction "Frame" (Fig.1) using SPF allows to eliminate 500 rivets and reduce the labour-intensity by 4 times. This integrative design is durable, the weight now is merely half of the previous and increase fire-resistant.

The tensile test results obtained after SPF, and SPF + Annealing, show very little change from those found in the raw SP-700 material (see Table 1).

Thickness measurements revealed a maximum reduction of 50% from that of the original thickness of the raw material.

Table 1: Mechanical Properties SP-700

PROPERTIES	RAW MATERIAL	AFTER SPF	SPF + ANNEAL
UTS (MPa)	1044	1043	1042
YS (MPa)	992	1008	1003
EL(%)	11	10	10

Fig.2 shows a nacelle bulkhead segment from SP700 made by SPF. Previous attempts in conventional hot forming this part from Ti-6Al-4V had failed due to the difficulty in uniformly heating the massive forming tool and inaccurate interface working surfaces between the die parts. The use of a welded two-sheet pack enabled the production of a highly accurate detail part as shown in this example. When SP-700 is SP formed at temperatures of 750-800°C, the flow stress is approximately 0.7 times that of Ti-6Al-4V in SPF temperature range and at a strain rate of 10^{-3} s^{-1}.

For the purpose of exploring of SPF technology of large parts in the universal tool, an experiment was made to produce the half fuel tank with volume of 210 liter, from 2004 Al-alloy (Fig.3).

Shroud detail parts (Fig.4) were made from Ti-alloy by SPF is one of the main fire-resistant parts of "Galaxy" tailcone. The forming was provided from commercial pure titanium CP-2 (Ti-65A) for the reducing of raw material cost.

Virtual manufacturing become an important part of the modern production in IAI and SPF modeling. The Mark (simulation) system is one of its important element. This system allows to avoid the mistakes in design in the very early stages to reduce parts price. The example of such virtual modeling is tailcone part "Astra" (Fig.5, 6). On the section A-A could seen high tolerance of virtual SPF with thickness distribution in an actual part.

Tensile test of Al-alloy 2195 specimens were taken in different condition as follows: raw material, as SPF and air cooled. SPF+T5- same as above with aging for 24 hours at 163°C. SPF+T62-solution at 535°C for 45 minutes and water quench with aging at 163°C for 24 hours.

Table 2: Mechanical Properties of Al-alloy 2195

CONDITION	U.T.S. (MPa)	Y.S. (MPa)	EL. (%)
Raw material	208.3	138.3	14.2
After SPF	466.0	335.8	12.0
SPF+T5	484.2	371.3	10.0
SPF+T62	609.3	527.3	7.2
Al 7475-T76	560.0	460.0	8.0

For the high durability and reliability of connecting details with limited thickness of parts is very important. Therefore we use SPF diaphragm forming for the production of locator (Fig.7).

Due to problems with availability of Al alloys for SPF we used in our research coarse-grained 6061. Part named "Torus" was made (Fig.8). There were good results in thickness distribution, but there was cavitation in small radius areas, that makes using parts from this alloy problematic. Evidently the future of this alloy in SPF is connected with receiving of fine grain structure sheet material.

SP-700, today is about 15% more expensive than Ti-6Al-4V, probably because of the relatively small quantities that are manufactured. However, its use may be easily justified for short-run production of detail parts because the cheaper tooling and easier processing at lower working temperatures. Aircraft parts suitable to SPF are mainly from aluminum alloys but the potential of titanium alloys also high.

Cutting of contour SPF formed parts, on account of intricate shape, uses a waterjet technology. Recently SPF parts were used approximetly 25% Ti alloys and 75% from SP Al alloys in three types of IAI Aircrafts: "Galaxy", "Astra", "Dornier-428" and it will be used in the next projects.

Discussion

In spite of excellent results with SPF and active research, this technology is still not very used in modern industry. We think that the reasons are:
1) Difficulty in acquiring equipment, with high prices and long repay.
2) High exploitation cost.
3) High material cost and problems of them manufacturing, standardisation (especially accordance to requirement of FAI) and acquisition.
4) Problems with control of part thickness.
5) The conservative approach of designers and lack of their level of knowledge in this technology.
6) Difficulties in providing of process isoterming and temperature control especially in large parts.
7) The use of ceramics also has a high cost and problematic in view of lack of accuracy short duration of operating and details fragility.

Significant moments for overcoming of braking factors and increasing utilisation of SPF may be in our opinion:
-design of low cost aluminum alloys suitable to welding;
-increasing of nomenclature complex shaped parts from SP-alloys;
-increase output of known alloys as Supral (2004) for wide use;
-research for utilization of common alloys (for example 6061);
-modification of existing alloys for increase of plastic resources;
  using of universal tools,changeable inserts on stable basis,multiplace tools;
-working out plastic materials for tools stable up to temperature 600°C;
-improvement of tools from ceramics materials and reducing the price of ceramic;
-developing Ti-alloys and steels with low temperature SPF (for example SP-700) to shorten production time and cost;
-developing of new high temperature lubricating materials that facilitate the process of forming and easy removal after forming.

Conclusion

We can see the extensive application SPF and DB in the IAI in the very near future. Using new SP alloys, the combination of its positive characteristics provides us with a new alloy, which has excellent prospects for wide use in aerospace designs. SPF simulation have a good perspective, as integral part of virtual manufacturing. Its use will shorten manufacturing times and reduce costs, subsequently leading to a cheaper aircraft.

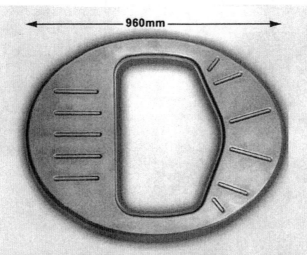

Fig. 1 IAI "Galaxy" fire-wall rear frame (SP-700)

Fig. 2 Expermental nacelle bulkhead segment (SP-700)

Fig. 3 Expermental half fuel tank (2004)   Fig. 4 "Galaxy" shroud detail parts (CP-2)

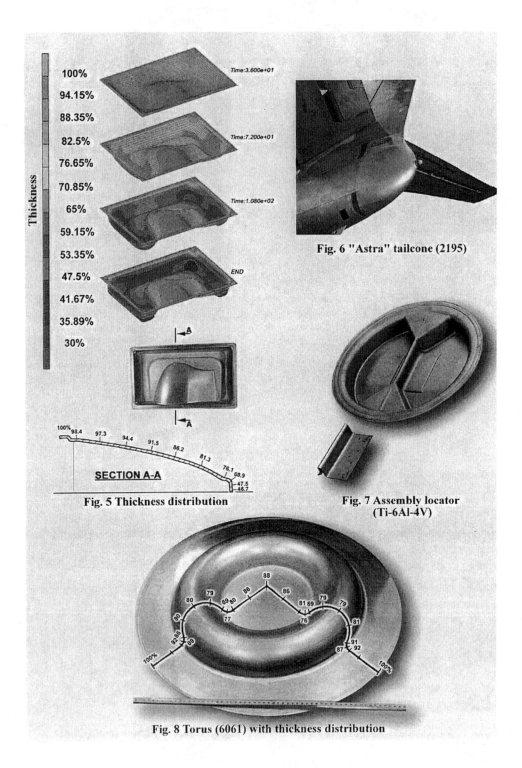

Fig. 5 Thickness distribution

Fig. 6 "Astra" tailcone (2195)

Fig. 7 Assembly locator (Ti-6Al-4V)

Fig. 8 Torus (6061) with thickness distribution

Superplastic Alumina Ceramics Dispersed with Zirconia and Spinel Particles

B.-N. Kim, K. Morita, K. Hiraga and Y. Sakka

National Research Institute for Metals, Mechanical Properties Division,
Sengen 1-2-1, Tsukuba, Ibaraki 305-0047, Japan

Keywords: Alumina, Elongation, Spinel, Strain Rate, Stress, Zirconia

Abstract

Superplastic deformation behavior is examined for Al_2O_3-based ceramics dispersed with 10 vol% ZrO_2 and 10 vol% spinel ($MgO \cdot 1.3Al_2O_3$) particles. The maximum tensile elongation obtained is 850% at a strain rate of 5.0×10^{-4} s^{-1} and at 1500 °C. The static grain growth rate of this material is about 0.6 times the growth rate of a Al_2O_3-10 vol% ZrO_2. The dynamic grain growth rate is considerably decreased by the multiple-phase dispersion, which leads to the enhanced superplasticity; large tensile elongation and high available strain rate. Observations of deformed microstructure show spherical ZrO_2 particles embedded in elongated Al_2O_3 grains, suggesting that the primary deformation mechanism would be grain boundary sliding associated with grain boundary diffusion.

Introduction

High-temperature tensile ductility in fine-grained Al_2O_3 is limited owing to rapid grain growth and resultant severe cavitation during deformation. Accordingly, many trials have been made to suppress the grain growth by use of an additive such as MgO or ZrO_2. The dynamic grain growth of MgO-doped Al_2O_3, however, is active to limit tensile ductility to ~80% at 1350~1500 °C [1]. Although some additional improvement is possible by the codoping of CuO and MgO, the maximum tensile elongation has remained 140% [2]. On the other hand, ZrO_2-particle dispersion is much more effective. Recently, Nakano et al. [3] made 10% ZrO_2-dispersed Al_2O_3 by colloidal processing. The grain size of Al_2O_3 was about 0.45 μm and ZrO_2 particles were uniformly dispersed at the multiple junctions of the matrix grains. The ZrO_2-dispersed Al_2O_3 showed an elongation of 550% at 1500 °C, at 1.7×10^{-4} s^{-1}. However, the strain rates available for superplastic deformation in this material are still low for industrial application. In this study, we try to improve the superplasticity of Al_2O_3 ceramics by the multiple dispersion of ZrO_2 and spinel phases. We expect that the multiple dispersion is more effective on the inihibition of grain growth than the single dispersion, which is expected to lead to enhanced superplasticity.

Experimental

The starting materials were commercial powders of high purity α-Al_2O_3, 3 mol% Y_2O_3-stabilized tetragonal ZrO_2 and MgO with nominal particle diameters of 0.2 μm, 0.07 μm and 0.017 μm, respectively. The amounts of ZrO_2 and MgO powders were determined to make a material of Al_2O_3-10 vol% ZrO_2-10 vol% spinel after sintering. The spinel phase was formed by the chemical interaction between Al_2O_3 and MgO powders during sintering. The powders were mixed in a ball-mill for 24 h, together with 5 mm-diameter Al_2O_3 balls and ethanol. The green compacts were prepared by pressing the mixed powder at 40 MPa and further by cold-isostatic pressing at 200 MPa. The green compacts were sintered at 1450 °C for 1 h in air. The respective crystalline phases in the sintered body were identified by X-ray diffraction (XRD) with Cu K_α.

From the sintered bodies, bone-shaped tensile specimens were machined with a gauge length of 10 mm, a width of 3 mm and a thickness of 2 mm. Constant displacement-rate tensile tests for the specimens were carried out with an Instron-type testing machine at 1500 °C in vacuum. For monitoring strain and stress during deformation, uniform elongation was assumed in the gauge portion.

Bulk density was measured for the machined specimens using the Archimedes method. Microstructural examination was performed for both undeformed and deformed specimens using a scanning electron microscope (SEM) and a high resolution transmission electron microscope (HRTEM). The grain size was determined by measuring the average grain area of each phase A_g on SEM photographs and by calculating the equivalent grain radius from $\sqrt{(A_g/\pi)}$. The molar composition of the spinel phase was analyzed by XRD for both furnace-cooled and quenched samples. The composition ratio between Al_2O_3 and MgO was calculated by $(0.86109-a_0)/(3a_0-2.37195)$, where a_0 (in nm) is the determined lattice parameter from the diffraction pattern for spinel phases [4].

Results and Discussion

Microstructure

A SEM photograph of the etched surface of the as-sintered Al_2O_3-10 vol% ZrO_2-10 vol% spinel is shown in Fig. 1. The measured bulk density is 4.17 g/cm³, which is nearly identical to the ideal density within an experimental error. The measured average radii of Al_2O_3, ZrO_2 and spinel grains are 0.25 μm, 0.09 μm, and 0.18 μm, respectively. The ZrO_2 particles locate at triple and quadruple junctions of matrix grains, while the spinel particles are distributed as a part of matrix grains. The size of spinel particles is comparable to that of Al_2O_3 grains.

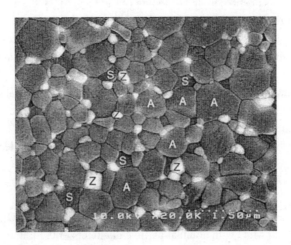

Fig. 1 As-sintered microstructure. 'A' is Al_2O_3, 'Z' is ZrO_2 and 'S' is spinel.

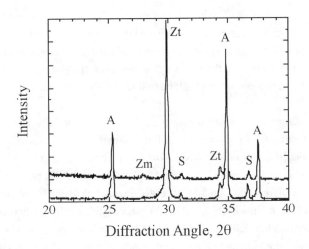

Fig. 2 XRD pattern for deformed (the upper) and undeformed (the lower) samples. 'A' is Al_2O_3, 'S' is spinel, 'Zt' is tetragonal ZrO_2 and 'Zm' is monoclinic ZrO_2.

XRD examinations showed that most ZrO_2 particles are tetragonal in the as-sintered specimens. Remakable diffraction peaks were not detectable for monoclinic ZrO_2. After superplastic tests, however, a diffraction peak appeared for monoclinic phase, as shown in Fig. 2. Due to grain growth during test, the tetragonal ZrO_2 transformed to the monoclinic phase. The ratio of Al_2O_3 to MgO in the spinel phase was determined to be 1.1 at room temperature and 1.3 at 1500 °C. HRTEM observations of the as-sintered Al_2O_3-10 vol% ZrO_2-10 vol% spinel represented clean boundaries without glassy phases. At all grain boundaries and interphase boundaries observed, no boundary phase was found.

Figure 3 represents the microstructure deformed at 1500 °C. The average grain aspect ratios for the respective phases are less than 1.5. The relatively small grain aspect ratio compared with the

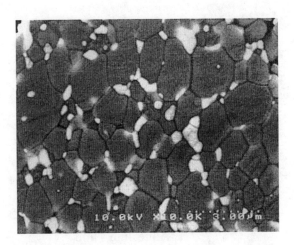

Fig. 3 Deformed microstructure (560%) at 8.33×10^{-4} s^{-1} and at 1500°C. The stress axis is vertical.

macroscopic strain indicates that Rachinger sliding is the primary deformation mechanism. Grain elongation during deformation can also be restricted by grain boundary migration under the deformation of Lifshitz sliding. A higher mobility of grain boundary has a larger effect on restricting grain elongation [5]. As will be shown in the next section, however, the grain growth rate, that is, the grain boundary mobility of the present material is low, so that the effect of boundary migration is not significant.

The accommodation mechanism for Rachinger sliding can be inferred to some extent from the deformed microstructures. Close observation of Fig. 3 shows that the shape of ZrO_2 particles depends on the location. While ZrO_2 particles on Al_2O_3/Al_2O_3 grain boundaries are elongated along the boundaries, the particles within Al_2O_3 grain are equiaxed. The equiaxed ZrO_2 particles observed within the Al_2O_3 grain after deformation indicate that the stress-directed diffusion is not considerable around the ZrO_2 particles. The local diffusion around the ZrO_2 may be limited by the harder surrounding matrix grain, because it requires the deformation of the Al_2O_3 matrix by diffusional or by dislocation processes. Hence, we can conclude that at least, lattice diffusion and dislocation motion in Al_2O_3 grains are not the main accommodation mechanism for Rachinger sliding as well as interphase diffusion between Al_2O_3 and ZrO_2. The elongation of the Al_2O_3 grains along the stress axis seems to be associated with grain boundary diffusion.

Static grain growth

Static grain growth behavior in a polycrystalline aggregate is generally represented by

$$R^m - R_0^m = kt \qquad (1)$$

where R is the average radius of grain, R_0 is the initial radius, m is the grain growth exponent, k is the rate constant, and t is the annealing time. The growth kinetics of Al_2O_3 grains in the present material can be characterized with m=4, as shown in Fig. 4. The slope of Fig. 4 corresponds to k/R_0^4, and an activation energy Q_g of 588 kJ/mol was obtained from its temperature dependence.

The static grain growth behavior of the present material is very similar to those of both mono-

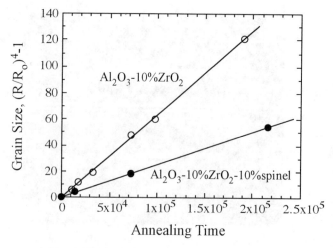

Fig. 4 Static grain growth behavior at 1500 °C.

lithic Al_2O_3 [6] and Al_2O_3-10 vol% ZrO_2 [7], while the value of k for the present material is about 0.6 times the value for Al_2O_3-10 vol% ZrO_2. For Al_2O_3 and Al_2O_3-10 vol% ZrO_2, the value of Q_g was reported to be 564 and 572 kJ/mol, respectively, with the same m-value of 4. From the similar values of Q_g and the same m-value, we consider that the static grain growth behaviors of these three materials are controlled by the same mechanism. In ref. [7], we noted that the lattice diffusion of the oxygen ion controls the static grain growth in Al_2O_3. Although the ZrO_2 and spinel particles grow during annealing, it seems that the basic controlling mechanism of the static grain growth in the present material is identical to that of monolithic Al_2O_3. The dispersion of ZrO_2 and spinel particles is thus supposed to play a role only in grain boundary pinning.

Superplastic deformation

Typical stress-strain curves of Al_2O_3-10 vol% ZrO_2-10 vol% spinel are shown in Fig. 5 for three initial strain rates at 1500 °C. The flow stress increases with increasing strain rate. The strain-hardening after initial yielding is very limited at all strain rates, while the stress-strain curve of pure Al_2O_3 is characterized by extensive strain-hardening followed by rapid stress drop after reaching a peak stress. The strain-hardening for Al_2O_3-10 vol% ZrO_2-10 vol% spinel is also more limited than for Al_2O_3-10 vol% ZrO_2 [3]. Since the strain-hardening mainly results from grain growth, the limited strain-hardening indicates the inhibited grain growth during deformation.

The strain-rate dependence of the tensile elongation is shown in Fig. 6. Largely enhanced superplastic properties are found in the present material. The addition of 10 vol% spinel to Al_2O_3-10 vol% ZrO_2 enhanced the strain rate by a factor of 10. Even at a strain rate of 8.3×10^{-3} s^{-1}, the material exhibited a tensile elongation of 200%.

The maximum elongation obtained is 850% at 1500 °C, which is the largest elongation in Al_2O_3-based ceramics reported so far. The maximum elongation obtained in ceramics is 1038% for ZrO_2 containing 5 wt% SiO_2 [8]. Considering that the present material contains no glassy phase and that the

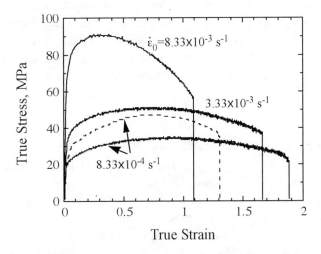

Fig. 5 Typical stress-strain curves. The dotted curve is for Al_2O_3-10%ZrO_2.

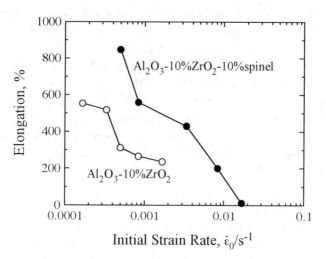

Fig. 6 Strain rate dependence of tensile elongation.

matrix is Al_2O_3 which is harder than ZrO_2, we can say that significant enhancement was achieved in the superplasticity of Al_2O_3.

The stress exponent n determined by using 10% flow stress is 2.7, which is nearly consistent with that for Al_2O_3-10 vol% ZrO_2 (n=2.4). The nearly same n-value indicate that the primary deformation mechanism of Al_2O_3-10 vol% ZrO_2 is not affected significantly by the dispersion of 10% spinel. We consider that the suppressed grain growth by the additional spinel dispersion resulted in the enhanced superplasticity of the present material.

Conclusions

The present Al_2O_3-10 vol% ZrO_2-10 vol% spinel represented enhanced superplasticity; large tensile ductilities and high available strain rates. The maximum tensile elongation obtained was 850% at an initial strain rate of 5.0×10^{-4} s^{-1} and at 1500 °C. The multiple-phase dispersion suppressed the grain growth rate considerably, and the suppressed grain growth led to the enhanced superplasticity. From observations of spherical ZrO_2 particles embedded in elongated Al_2O_3 grains, we consider that the primary deformation mechanism is grain boundary sliding associated with grain boundary diffusion.

References

[1] Y. Yoshizawa and T. Sakuma, Acta Metall. Mater. 40 (1992), p. 2943.
[2] T. J. Davies, A. A. Ogwu, N. Ridley and Z. C. Wang, Acta mater. 44 (1996), p. 2373.
[3] K. Nakano, T. S. Suzuki, K. Hiraga and Y. Sakka, Scripta mater. 38 (1998), p. 33.
[4] H. U. Viertel and F. Seifert, Neues Jahrb. Mineral. Abh. 134 (1979), p. 167.
[5] B. -N. Kim and K. Hiraga, Acta Mater., to be published.
[6] J. Besson and M. Abouaf, Acta Metall. Mater. 39 (1991), p. 2225.
[7] B. -N. Kim, K. Hiraga, Y. Sakka and B. -W. Ahn, Acta Mater. 47 (1999), p. 3433.
[8] K. Kajihara, Y. Toshizawa and T. Sakuma, Scripta Metall. 28 (1993), p. 559.

Grain Refinement and Superplasticity of Reaction Sintered TiC Dispersed Ti Alloy Composites Using Hydrogenation Treatment

N. Machida, K. Funami and M. Kobayashi

Department of Mechanical Engineering, Chiba Institute of Technology,
Tsudanuma, Narashino 275-0016, Japan

ABSTRACT

The purpose of this study is to develop TiC dispersed Ti-6Al-4V alloy composites with useful mechanical properties by using reaction sintered fine TiC dispersion technique in the powder metallurgy method and the hydrogenation treatment due to grain refinement of matrix material. Mo_2C and Ti-6Al-4V prealloyed powder were mechanically blended in a high energy ball mill. The mixed powder was pressed into dies and consolidated by reaction sintering following HIP treatment. The dispersed particles Mo_2C produced reactive TiC in the matrix during sintering. In grain refining treatments, hydrogenation, hot roll forming and heat treatment to cause fine dispersed Ti hydride precipitation, and dehydrogenation process were introduced. The result of this fabrication process for controlled grain refinement demonstrate the following: 1) The microstructure of dehydrogenated and recrystallized specimen indicated very fine grain size, less than 3μ m in diameter. Then, the reacted TiC particles became smaller by hot roll forming after hydrogenation. 2) In Ti-6Al-4V alloy composite with 3vol% Mo_2C, tensile strength and elongation at room temperature were increased more than 14% in comparison with those of non-hydrogenation. 3) Wear resistance of the developed composite had the tendency to increase as blended Mo_2C content increased and its superplasticity was confirmed.

INTRODUCTION

It is well known that titanium alloys are useful materials for their superior mechanical properties, good corrosion resistance, and high strength-to-weight ratio. On the other hand, these alloys have some disadvantages, such as poor wear resistance, low hardness and low strength. Therefore, to overcome these deficiencies, it is more effective to use the composite materials of titanium alloy matrix. From this point of view, the authors have developed fine TiB dispersed Ti-6Al-4V alloy composites utilizing reaction sintering in the powder metallurgy method[1,2]. In this process, Ti-6Al-4V prealloyed powder, which has a fine microstructure with α and β phase, was blended with boride particles, TiB_2, MoB and CrB powder, to precipitate TiB dispersion particles. It was found that reacted fine TiB particles were effective to enhance the mechanical properties such as wear resistance, hardness and superplasticity. These particles were no doubt to increase wear resistance, but they showed needle-shape structure and its dimension was increased with volume fraction of blended boride powder. Obviously, these needle-shape particles have a little effect on tensile strength improvement in comparison with equiaxed particles and, at the same time, it is necessary to make finer the grain size in matrix titanium alloy to improve superplastic behavior.

On the other hand, for reactive TiC particles, it has been examined that there wear resistance was not poor on large scale compared to that of TiB particles and the shape of reacted particles was in equiaxed structure [3]. Furthermore, Mo element should have a characteristic to reverse the matrix structure from α phase to β phase. Hence, Mo_2C powder was selected for secondary blended powder in addition to a new refinement treatment, in which a process utilizing hydrogen for the titanium alloys would prodice fine and equiaxed grain structures homogeneously[4] .

In this paper, Ti-6Al-4V prealloyed powder was mechanically blended with Mo_2C powder, the blended powder was consolidated by reaction sintering in a vacuum, and HIP treatment to disperse

TiC in the matrix and to increase the density of specimens. The details of the fabrication process of grain refinement technique, especially, an effect of dispersed particles TiC on precipitation sites for hydrides and mechanical properties of specimen in which dehydrogenation treatment was applied, were investigated.

EXPERIMENT

Fabrication process of composite materials and hydrogeneration

Matrix alloys were prealloyed from Ti-6Al-4V powder with its average size of 45 μm in diameter. This powder was mechanically blended and alloyed for 600minutes using the planet type ball mill with Mo_2C (3.8 μm in average diameter) powder. The volume fractions of the blended reinforcement Mo_2C powder were 3, 5 and 10vol% respectively. In order to avoid the contamination with oxidation, these operations were performed in a glove box, in which argon gas run out under 10ml/min. The chemical compositions of each powder are shown in Table1. The blended powder was pressed into dies at room temperature under the stress of 431.2MPa and then the green compacts in the size of 10x10x50 mm were consolidated by reaction sintering at 1323K for 50minutes in a vacuum. Each encapsulated sintered sample in steel cylindrical can, in which the sample filled up with BN powder, was HIPed at 1473k for 120minutes under the pressure of 200MPa to reduce pores which remained even in sintering. The densities of sintered and HIPed samples were determined by Archimedes technique. Furthermore, both the upper and lower surfaces of HIPed sample were mechanically ground to remove the roughness after stripping.

For these specimens, hydrogenation was carried out in hydrogen gas under 0.06MPa in condition of 1023K for 0.3ks. The hydrogen content for each specimen is listed in Table 2. An initial hydrogen content of these HIPed specimens is 0.002%

Table 1 Chemical composition of powders (wt%)

Powder	Fe	Si	C	N	O	Al	V	Ti	Mo
Ti-6Al-4V	0.03	0.01	0.01	0.01	0.27	5.92	3.8	bal	
Mo_2C	0.044		5.88		0.34				bal

Table 2 Hydrogen content of various Mo_2C blended specimens (mass%)

Blended Mo_2C volume fraction	0	3	5	10
H	0.03	0.02	0.016	0.015

Hot roll working, dehydrogenation and mechanical test

Finally, these hydrogenated samples packaged by a steel can and insulated by BN powder, were hot-rolled in reduction of about 83% after heating at 1073K for 30minutes, and air-cooled. In the grain refining treatments of matrix material, hydrogenation make fine dispersed Ti hydride precipitation for dislocation pile up sources, hot rolling to induce heavy strain energy, heat treatment for recrystallization and dehydrogenation process to recover the ductility were performed. Microstructures were observed by using SEM. After the hot rolling process, the specimens were heated at 1073K for 115.2ks in a vacuum in order to dehydrogenate and recrystallize.

Tensile strength, Vickers microhardness and wear resistance for these specimens at room temperature were measured. Tensile test was carried out on an Instron type machine in a crosshead speed of 0.5mm/min, using the specimen of a cross section area of 3x0.8 mm^2 and 3mm in height. Wear test was carried out on a pin-on-disk machine loaded 9.6 N with sliding velocity of 63m/min with no lubrication. A circular disk with HT50 steel plate was used as the disk material.

Furthermore, tensile tests were carried out at 1173K in the initial strain rate of 3×10^{-2} and $3\times10^{-3}s^{-1}$ in order to evaluate the elongation and to confirm the superplasticity using an Instron type testing machine.

RESULTS AND DISCUSSION

Reaction sintering and HIP treatment

The precipitation of TiC in Ti-6Al-4V alloy matrix blended with Mo_2C in sintering is based on the following reaction:

$$Ti + Mo_2C \rightarrow TiC + 2Mo - 104.2(KJ/mol) \text{ at } 1473K$$

Fig.1 shows microstructures of HIPed TiC dispersed composites obtained by reaction sintering after being blended with Mo_2C. After the sintering, it was possible to increase the relative densities to over 98% with following HIP treatment. Some spherical pores observed in every part of sintered block were almost eliminated by HIP treatment. The reaction sintered samples show distribution of small size and extremely fine granular TiC, and the dispersed TiC particles increased with increasing blended Mo_2C content. The produced TiC disperses in Ti-6Al-4V alloy matrix like the shape of small equiaxed particles as shown in Figure 1(a), and this TiC particle was confirmed by EPMA as shown in Fig.1 (b), (c). But these TiC particles precipitate at matrix grain boundaries and an appearance of TiC dispersion is not uniform. It seems that the agglomerate of Mo_2C powder occurs in milling process. The matrix showed β rich phase by means of X-ray diffraction, although reactive Mo acted as phase stabilizer element. In the case of Mo_2C 3vol %, Ti-6Al-4V alloy matrix show a typical α and β laminated structure.

Grain growth of matrix of Mo_2C blended composites was fairly arrested by these dispersed TiC particles in comparison with that of non-blended Ti-6Al-4V alloy. These structures are a problem of superplastic deformation as described later. The largest problem in superplasticity is a size of TiC particles. These problems will be solved by the improvement of milling process, using of finer Mo_2C powder and of an isolating solvent, and the improvement of sintering process like the reduction of sintering time.

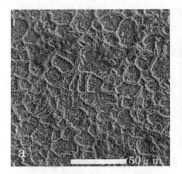 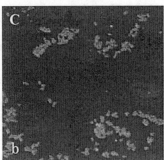 

Fig.1 Microstructure of HIPed TiC dispersed Ti alloy composites blended Mo_2C5vol%(a) SEM image (b), (c) Surface analyzed by EPMA

Hydrogenation treatment

Three dimensional microstructures of hot rolled plates in the reduction of about 83% of TiC dispersed Ti-6Al-4V alloy composites with hydrogenation treatment are shown in Fig.2. In comparison with the microstructures after HIPing in Fig.1, TiC particles dispersed in the matrix are refined and slightly oriented to the rolling direction, and also the grains in the matrix are refined up to below 3 μm in every composite. In these TiC dispersed composites, finer microstructures were obtained in comparison with composites blended with TiB_2, CrB and MoB powder in the same volume fraction. However, the reacted TiC particles were still not dispersed uniformly.

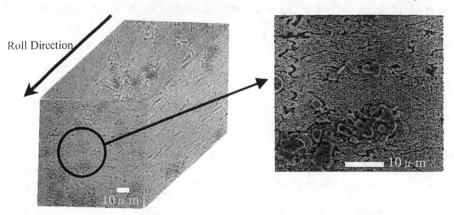

Fig.2 Microstructure of TiC dispersed Ti alloy composites with hydrogenation treatment (Mo_2C 5vol% blended)

Mechanical properties at room temperature

Fig.3 shows the blended Mo_2C content dependence of tensile strength of TiC dispersed Ti alloy composites at room temperature. Tensile strength increases almost linearly with increasing Mo_2C content because of an increment of hard TiC particle produced and of the decrease of matrix grain size. In case of TiC dispersed composites with hydrogenation treatment, an appearance of enhancement of tensile strength shows a similar tendency. This rate of increment with increasing Mo_2C content changes remarkably in TiC dispersed composites with hydrogenation treatment according to TiC dispersion particle and matrix grain size.

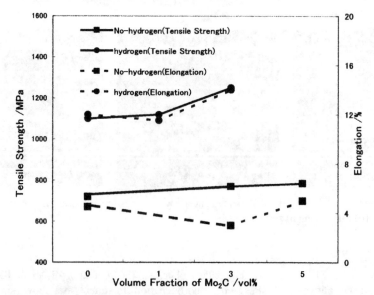

Fig.3 Mo_2C content dependence of tensile strength of TiC dispersed composites with hydrogenation treatment

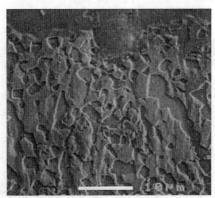

Fig.4 Microstructure of fracture region of 5vol% Mo_2C blended composites with hydrogenation treatment in tensile test at room temperature

Fig.4 shows the microstructure of fracture surface of composites with hydrogenation treatment (Mo_2C 5vol%) by tensile test. The fracture of specimen takes place along a grain boundary which shows typical brittle fracture. Then, it seems that the bonding between precipitated TiC particles and matrix is sufficient from experimental observation that shows the intergranular fracture of TiC particles in itself. An increment of hardness grows higher with increasing Mo_2C content, with the same tendency as tensile strength

The result of wear test of TiC dispersed composites are shown in Fig.5. In this figure, the amount of wear per sliding distance of 5×10^3 m is exhibited together with that of the comparative materials. It was found that the amount of wear of TiC dispersed composites was remarkably affected by dispersed TiC particle content. A large amount of mixed Mo_2C powder especially improves wear resistance. When comparing the results of TiB with those of wear test of TiC, it was found that wear resistance of TiB was far better than that of TiC. However, in the case of tensile strength, this relation was completely contrary to the result of wear resistance

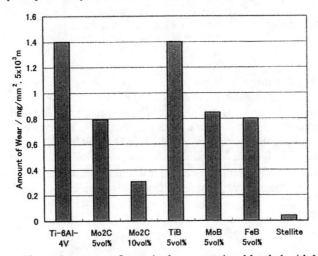

Figure.5 Amount of wear in the composites blended with Mo_2C

Furthermore, tensile tests were conducted at 1173 K to confirm superplasticity in the initial strain rate of 3×10^{-2} and $3 \times 10^{-3} s^{-1}$. As exemplified in 3vol% Mo_2C blended composites with hydrogenation treatment in Fig.6, total elongation of gage length is about 497%. Fig.7 shows fracture surface appearance.

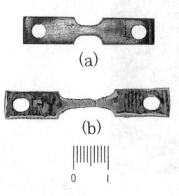 

Fig.6 Superplasticity deformation of 3vol% Mo_2C blended composites with hydrogenation treatment, (a)before deformation, (b)after deformation.

Fig.7 Microstructure of fracture region of 3vol% Mo_2C blended composites with hydrogenation treatment in tensile test at 1173K

CONCLUSION

Ti-6Al-4V alloy matrix composites reinforced with dispersed TiC particles were produced by reaction sintering after blending with Mo_2C powder. In the grain refining treatments of matrix material, hydrogenation to cause fine dispersed Ti hydride precipitation, hot rolling to induce heavy strain energy, heat treatment for recryatallization, and dehydrogenation process to recover the ductility were performed. As a result of this fabrication process for grain refinement, the following conclusions can be obtained:1) The microstructure of dehydrogenated and recrystallized specimen was very fine grain size less than 3 μ m in diameter. 2) In Ti-6Al-4V alloy composites with 3vol% Mo_2C, tensile strength and elongation at room temperature was increased more than 14% in comparison with that of non hydrogenation.3) Wear resistance had the tendency to increase as blended Mo_2C content increased, and its superplasticity was confirmed.

REFERENCES

1. M.Kobayashi, S.Ochiai, K.Funami, C.Ouchi and S.Suzuki, "Superplasticity of fine TiC and TiN dispersed Ti-6Al-4V Alloy Composites," ,Mat. Sci. Forum, 1994,pp. 549-554.
2. K.Funami, M.Kobayashi, S.Suzuki and C.Ouchi, "Superplasticity of fine TiB dispersed Ti-6 Al-4V Alloy Composite obtained by Reaction Sintering," Mater.Sci.Forum,1997, pp.515 -520.
3. W.Backofen,I.R.Turner and D.H.Avery, Trans.ASM,57 ,1964 ,pp.980
4. H.Yoshimura, K.Kimura, M.Hayashi, M.Ishii, T.Hanamura and J.Takamura, "Ultra fine Equiaxed Grain Refinement and Improvement of Mechanical Properties of $\alpha + \beta$ Type Ti Alloy by Hydrogenetion, Hot working, Heat treatment and Dehydrogenation," Mat.Trans.,JIM,1994, pp266-272

On the Activation Energies Observed in Al-based Materials Deformed at Ultrahigh Temperatures

B.Y. Lou[1], T.D. Wang[1], J.C. Huang[1] and T.G. Langdon[2]

[1] Institute of Materials Science and Engineering, National Sun Yat-Sen University, Kaohsiung 804, Taiwan ROC

[2] Departments of Materials Science and Mechanical Engineering, University of Southern California, Los Angeles CA 90089-1453, USA

Keywords: Activation Energy, Al Alloy, Al Composites, Deformation Mechanisms

Abstract

This study examines the exceptionally high activation energies estimated from data obtained from various composites exhibiting high strain rate superplasticity and loaded at ultrahigh temperatures near or slightly above the solidus. It is shown that the activation energies for the 6061 composite and for Al 6061 and 1050 alloys are below 200 kJ/mol at temperatures lower than the incipient melting temperature, T_i, but they become significantly higher at temperatures near or slightly above T_i and this increase is independent of the alloy system, the grain size, whether the material is superplastic and the dominant deformation mechanism at $T<T_i$. There is a transition from plastic flow in a solid state to a viscous-like flow behavior in the Al-6061 systems at $T>610°C$ and in the Al-1050 alloy at $T>650°C$. It is concluded that the high values of Q_t have no physical meaning in terms of the rate-controlling diffusive species.

Introduction

There is evidence suggesting that the liquid phase originating from partial melting at temperatures near the solidus point may contribute to the high tensile elongations observed during high strain rate superplasticity (HSRS) in various aluminum based composites [1-4]. These HSRS composites have solute-rich second phases along the grain boundaries and/or the matrix/reinforcement interfaces which are formed during powder metallurgy or thermomechanical processing and which start to dissolve and then partially melt during high temperature superplastic loading. It is considered that the presence of these small amounts of liquid phases along the grain boundaries or interfaces may lead to a relaxation of the stress concentrations, a sharp strain rate increment or flow stress decrement, unusually high activation energies and the formation of long filaments on the fracture surfaces.

It is instructive therefore to consider whether a similar liquid effect may occur in cast aluminum alloys or cast pure Al systems where it is anticipated that pronounced solute-rich second phases are not present. In this report, the role of the liquid phase was examined during the high temperature deformation of three aluminum alloys or composites and with emphasis on behavior at temperatures near or slightly above their solidus points.

Experimental

The materials examined were the 6061/20%SiC_w HSRS composite and cast Al-6061 (Al-1%Mg-

0.6%Si) and Al-1050 (99.5% industrial grade pure Al) alloys. All of the three systems were subjected to thermomechanical treatments (TMT) or equal channel angular pressing (ECAP) to give a fine grain size in the range of ~0.5-1.0 μm. The materials were then tested to failure over temperatures from 500-650°C at strain rates from 10^{-4}-10^{-1} s^{-1}. The fracture surfaces were examined using a JEOL-6400 scanning electron microscopy (SEM). The incipient partial melting temperature was determined using a Setaram-131 differential scanning calorimeter (DSC).

Results

The grain size of the 6061/SiC$_w$ composite remained fine at high temperatures and the material exhibited HSRS with a tensile elongation over 450% but the grain sizes of the TMT or ECAP Al-6061 and Al-1050 alloys grew significantly to ~50-500 μm. Typical tensile elongation data are summarized in Table 1. The dependence of the flow stress on test temperature and strain rate is summarized in Figs. 1-3 for the 6061/SiC$_w$ composite and the 6061 and 1050 alloys. The optimum test temperature was ~600±10°C for the two 6061 materials and ~630±10°C for the Al-1050 alloy. The incipient partial melting temperature, T_i, was determined from the DSC curves as ~580-585°C for the Al-6061 alloy and ~646°C for the Al-1050 alloy.

Table 1 Typical high rate or low rate tensile failure elongations at their optimum test temperatures for the three materials

Material	T, °C	$\dot{\varepsilon}$, s^{-1}	e, %
6061/SiC$_w$	590-610	10^{-4}	100
		10^{-1}	450
6061	590-610	10^{-4}	400
		10^{-1}	220
1050	620-640	10^{-4}	130
		10^{-1}	200

The values of the apparent strain rate sensitivity, m_a, were determined in the high strain rate ranges from 10^{-2} to 10^{-1} s^{-1} as ~0.3 for the 6061 composite and ~0.2 for the two alloys. Values of the apparent activation energy, Q_a, varied from ~180 to >1000 kJ/mol at temperatures of 500-650°C, as shown in Table 2. There is evidence for a threshold stress, σ_{th}, in the 6061 composite in Fig. 1b. In determining the threshold stresses for this material through a plot of σ against $\dot{\varepsilon}^m$, a best fit was obtained using a true strain rate sensitivity, m_t, of ~0.5. This suggests that grain boundary sliding (GBS) may operate during HSRS.

Figure 4 shows the variation of the flow stress, σ, or the effective flow stress, σ-σ_{th}, as a function of 1/T at fixed strain rates of 1 x 10^{-1} or 5 x 10^{-2} s^{-1} for the 6061/SiC$_w$ composite, the 6061 alloy and the 1050 alloy. The stress exponent n (=1/m_t) is ~2 for the 6061 composite and ~5 for the two alloys. It can be seen from Fig. 4 that the fitted lines bend downward with increasing T so that, as T increases and approaches T_i, the estimated value for Q becomes larger. After incorporating the temperature dependence of the elastic modulus, E(T), the true activation energy, Q_t, decreased only slightly from Q_a, as shown in the last two columns of Table 2.

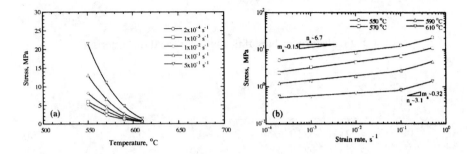

Fig. 1 The dependence of flow stress of the 6061/SiC$_w$ composite as a function of (a) test temperature and (b) strain rate.

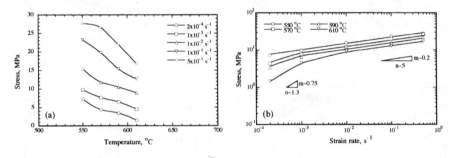

Fig. 2 The dependence of flow stress of the 6061 alloy composite as a function of (a) test temperature and (b) strain rate.

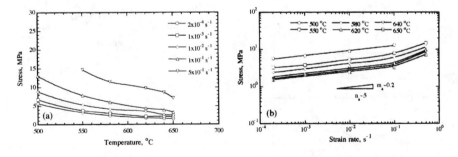

Fig. 3 The dependence of flow stress of the 1050 alloy as a function of (a) test temperature and (b) strain rate.

As mentioned earlier, T_i was determined as ~580-585°C for the Al-6061 alloy and ~646°C for the Al-1050 alloy. The flow stress decreased appreciably near or above T_i and this led to a pronounced increase in the estimated activation energies. The occurrence of partial melting at the grain triple points or matrix/whisker interfaces was supported by the presence of liquid filaments observed on the fracture surfaces of the 6061 composite (Fig. 5a) or Al-6061 alloy loaded at 590°C but not on the Al-1050 specimens at this same temperature. At an ultrahigh temperature of 620°C for the 6061 system or 655°C for the Al-1050, a more severe liquid phase led to liquid decohesion along the grain boundaries and interfaces as shown in Fig. 5b.

Table 2 Estimated true values of m and Q for the three materials

Material	T °C	$\dot{\varepsilon}$ s^{-1}	m_a	m_t	σ_{th} MPa	Q_a kJ/mol	Q_t kJ/mol
6061/SiC$_w$	500-550	1x10^{-2}-5x10^{-1}	~0.3	~0.5	~6.0	195	190
	550-590	1x10^{-2}-5x10^{-1}	~0.3	~0.5	~3.2	600	595
	590-610	1x10^{-2}-5x10^{-1}	~0.3	~0.5	~0.9	1027	1000
6061	500-580	5x10^{-3}-5x10^{-1}	~0.2	~0.2	0	190	185
	580-610	5x10^{-3}-5x10^{-1}	~0.2	~0.2	0	370	360
1050	500-640	5x10^{-3}-1x10^{-1}	~0.2	~0.2	0	195	195
	640-650	5x10^{-3}-1x10^{-1}	~0.2	~0.2	0	1179	1164

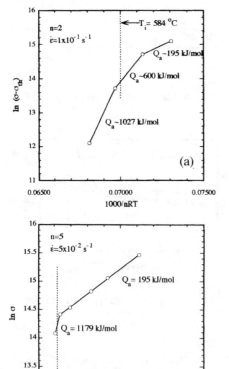

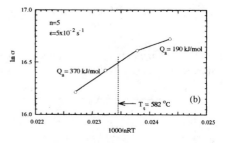

Fig. 4 The dependence of flow stress or effective flow stress at a fixed strain rate as a function of 1/T for (a) 6061/SiC$_w$ composite, (b) Al-6061 alloy and (c) Al-1050 alloy. The T$_i$ temperatures for these three materials are marked by vertical dotted lines.

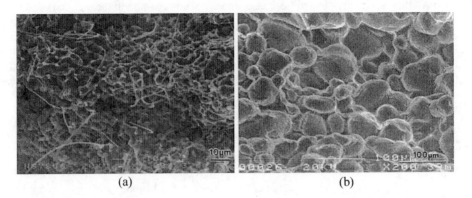

(a) (b)

Fig. 5 SEM micrographs showing (a) long filaments on the fracture surface of the 6061/SiC$_w$ composite loaded at 590 °C, and (b) failure due to liquid decohesion along the grain boundaries in the Al-6061 alloy loaded at 620 °C.

It can be seen from Table 2 that the 6061/SiC$_w$ composite showed an apparent increase in the value of Q_t with increasing temperature, especially at temperatures of $T>T_i$. Similar effects were also observed for the unreinforced Al-6061 alloy although the incremental increase in Q_t at $T>T_i$ was less pronounced than in the composite. Without the alloying in Al-1050, Q_t remained constant over the temperature range of 500-640°C and increased abruptly at $T>T_i$. Thus, at temperatures lower than T_i, the activation energies for all three materials are below 200 kJ/mol. However, when the test temperature is greater than T_i, Q_t increases significantly, and the occurrence of this increase is independent of (i) the alloy system (whether Al-6061 or Al-1050), (ii) the grain size (from 2 μm in the composite to 500 μm in Al-1050), (iii) whether superplastic or not (as indicated by the elongation data in Table 1) and (iv) the dominant deformation mechanism at $T<T_i$ (whether grain boundary sliding for the fine-grained composite or power-law creep for the alloys). This suggests that the flow stress decreases appreciably when the test temperature is close to or slightly above T_i, resulting in a sharp increase in the values recorded for Q. It is suggested that there is a transition from plastic flow in a solid state at the lower temperatures to a viscous-like flow behavior in the presence of a liquid phase. Therefore, the values estimated for Q from the observed temperature dependence of the flow stress or strain rate may not reflect a single deformation mechanism.

Evidence from experiments on the creep behavior of metal matrix composites suggests that load transfer may make a significant contribution in composites and this may lead to exceptionally high values for the measured activation energies [5,6]. The same approach was evaluated in this study by using the load transfer coefficient, α, under conditions of control by either lattice diffusion, $\alpha_{lattice}$, or grain boundary diffusion, α_{gb}. Table 3 summarizes the calculated best-fit values for $\alpha_{lattice}$ and α_{gb}. Since a reasonable value for the load transfer coefficient at such high temperatures should not exceed 0.5, $\alpha_{lattice}$ seems unlikely. Therefore, the corrected values for Q_t listed in the last column were obtained using α_{gb} for control by grain boundary diffusion. At the ultrahigh temperatures above T_i, there can be no load transfer effect and yet it is apparent that the value of Q_t remains exceptionally high.

It is reasonable to conclude that the values of the true activation energies determined in these experiments cannot be explained by incorporating a threshold stress or the effect of load transfer into the analysis since no threshold stress was evident and there was no load transfer in the cast Al-6061 and Al-1050 alloys and, in addition, load transfer is unimportant in the 6061 composite at T>T_i. On the contrary, the increase in Q_t at ultrahigh temperatures near or slightly above the solidus or partial melting temperatures is most readily explained in terms of a transition in the flow mechanism to a liquid viscous-like flow at T>610°C for the Al-6061 systems and T>650°C for the Al-1050 alloy. Similar conclusions have been reached also in another investigation [7]. Thus, the high values estimated for Q_t have no physical meaning in terms of the rate-controlling diffusive species.

Table 3 Corrections incorporating the load transfer effect for the 6061/SiC$_w$ composite

T, °C	$\alpha_{lattice}$	α_{gb}	Q_a, kJ/mol	Q_t, kJ/mol
550	0.87	0.39	195	70
570	0.83	0.14	195	70
590	0.65	0	600	595
610	0.50	0	1027	1000

Summary

(1) The incipient partial melting temperatures, T_i, were measured as ~580-585°C for an Al-6061 composite or alloy and ~646°C for an Al-1050 alloy
(2) Liquid filaments were observed on the fracture surfaces of the 6061 composite and alloy at 590°C but not the Al-1050 alloy. At 620°C for the 6061 materials or at 655°C for the Al-1050 alloy, there was liquid decohesion along the grain boundaries and interfaces.
(3) At temperatures near or above T_i, Q_t increases to high values and this increase is independent of the alloy system, the grain size, whether the material is superplastic and the dominant deformation mechanism at T<T_i. It is concluded there is a transition from plastic flow in a solid state to viscous-like flow in the presence of a liquid phase at T>610°C for the Al-6061 system and T>650°C for the Al-1050 alloy.

References

[1] J. Koike, M. Mabuchi and K. Higashi, Acta Metall. Mater., 43 (1995), p. 199.
[2] K. Higashi, T. G. Nieh and J. Wadsworth, Acta Metall. Mater., 43 (1995), p. 3275.
[3] K. Higashi, M. Mabuchi and T. G. Langdon, ISIJ International, 36 (1996), p. 1423.
[4] H. Iwasaki, M. Mabuchi and K. Higashi, Acta Mater., 45 (1997), p. 2759.
[5] K. T. Park and F. A. Mohamed, Metall. Mater. Trans., 26A (1996), p. 3119.
[6] Y. Li and T. G. Langdon, Acta Mater., 46 (1998), p. 3937.
[7] W.-J. Kim and O. D. Sherby, Acta Mater., 48 (2000), p. 1763.

Cooperative Grain Boundary Sliding at Room Temperature of a Zn-20.2%Al-1.8%Cu Superplastic Alloy

J.D. Muñoz-Andrade[3,4], A. Mendoza-Allende[1], G. Torres-Villaseñor[2] and J.A. Montemayor-Aldrete[1]

[1] Instituto de Física, Universidad Nacional Autónoma de México,
Apartado Postal 20-364, México 01000 D.F., México

[2] Inst. de Investiagción en Materiales, Universidad Nacional Autónoma de México,
Apartado Postal 70-360, México D.F., México

[3] Departamento de Materiales, División de Ciencias Básicas e Ingeniería,
Universidad Autónoma Metropolitana Unidad Azcapotzalco, Av. San Pablo No. 180,
Col. Reynosa Tamaulipas C.P. 02200 México D.F., México

[4] For partial fulfillments of the Program of Individualized Doctorate at the Facultad de Ingeniería, Universidad Central de Venezuela, Caracas, Venezuela

Abstract.- By applying a new technique [1-2] which provides a mesoscopic coordinate system inscribed on the surface of a tensile specimen, with 371 µm gage length for a Zn-20.2%Al-1.8%Cu superplastic alloy deformed at room temperature it is possible to show that: Deformation of the sample it is homogeneous at macroscopic level, but inhomogeneous at mesoscopical level. The inhomogeneity is ascribed to the sliding of grain blocks. For 28.5% of deformation the distribution function for the block sizes is described by: $N(x) = 1.37\, x^3 \exp(-3x/12.2\, \mu m)$, where, $N(x)$ is the number of blocks of size x, inside an area of about $172 \times 244\, (\mu m)^2$.

1. Introduction

In 1971, Raj and Ashby [3] theoretically analyzed the possibility of shear of grain groups as an entity, and also this mechanism has been considered later by other authors [4,5]. However, the actual experimental accurrence of cooperative grain boundary sliding (CGBS), i.e, movement of grain groups as a unit, along with grain boundary sliding of individual grains during superplastic deformation, has been reported only recently [6-10].

Recently a technique has been presented [1-2] that allows to observe the superplastic deformation at three levels: macroscopical, mesoscopical and microscopical. Here, this technique has been used to determine the grain sizes, or of grain block sizes wich contibutes to the macroscopical deformation.

2. Experimental procedure

The material used for this investigation (Zn-20.2%Al-1.8%Cu) was extruded to obtain strips of 20 mm wide and 5 mm. The extruded plates of Zn-20.2%Al-1.8%Cu superplastic alloy were cold rolled into strips of 1 mm thickness. Tension test samples with gage length of 371 µm parallel to the rolling direction were prepared. The geometry and size of the specimen designed for tension test in SEM are shown in the microphotography given in Fig. 1(a), the grain size by using the mean linear intercept was 3.5 µm. The specimen was mechanically polished on successively finer grades of emery paper and diamond paste.

By using of a pyramidal-shaped micro vickers indenter three sets of diamond pyramidal figures were inscribed on the surface of the specimen along on a longitudinal straight line as follow one pyramidal-shaped figure was inscribed approximately at the center of the specimen and a regular trapezoid delimited by pyramidal indention figures in each corner of such figure was inscribed on the surface of the tension sample. Centered on each trapezoid another pyramidal indentation was inscribed see Fig. 1(b). The gage length size of the tension specimen is delimited by the two sets formed each by five pyramidal figures. This gage length has been used for the determination of the macroscopical deformation (371µm).

The experiments were performed at constant cross head velocity, v = 0.1 mm/min, giving a nominal macroscopic initial strain rate of $4.04 \times 10^{-3} s^{-1}$.

3. Experimental results

Figure 2 illustrate the experimental data for the velocity of specific material objects as a function of distance along the tension axis, with origin fixed at the rest end of the specimen (left diamond pyramidal figure on the sample). The straight line is for homogeneous deformation. A good agreement it is found for x-values lowers than 500 µm. For distances larger than such distance the discrepancy between the condition for constant cross head velocity and the experimental data may be due to non-homogeneity of plastic deformation between the central region of the sample and the region near to the shoulders.

See also the Fig. 3 where the non homogeneous character of deformation it is shown by exhibiting a surface material sliding in parallel layers which descend toward to the center of the free surface sample; looking like what would happen with a set of playing cards which after be arranged one above other was subjected to the action of a shear stress that it was applied on the surface of the upper card.

On the other hand, between 5.4% and 28.5% of true deformation the sample surface display an extensive quantity of grain blocks sliding during deformation (see Fig. 4). The gliding zones between grain blocks appear like bright lines. In the Fig. 5, the histogram of the sizes of grains blocks as measured inside an area of 172 x 244 $(\mu m)^2$, it is shown. The area used for these measurements represents 11.4% of the total deformation area.

On the frequency distribution which appears in Fig. 5 the following aspects can be observed the distribution have a maximum in frequency at a size block of 12.2 μm. The blocks have discrete sizes: one, two... up to seven grain sizes length as calculated by using standard procedures [11, 12], 1.74 times the mean linear intercept, L = 3.5 μm. We make note that a natural grouping of block sizes around some values was observed on the raw data before Fig. 5 was build up.

To describe the experimental data on Fig. 5 it is required a distribution function which satisfies the condition that very small and very large blocks are rare. For blocks of sliding grains we have no experimental [6-10] or theoretical evidence [13] of any possible distribution that it corresponds closely with the actual distribution of blocks grain sizes in a Zn-20.2%Al-1.8%Cu superplastic alloy at room temperature deformed.

However, in other areas of plastic deformation of solids certain type of mathematical functions with similar features that here required has been used successfully, and some of these expression [14-17] appear to be the general kind of physical sensible function as needed for our case. Based on such previous work, here we propouse a modified algebraic expression as the following,

$$N(x) = Ax^n \exp(-nx/x_m) \qquad (1)$$

Like the more convenient to describe the experimental data on Fig. 5. A, n, and x_m are constant to be determined. To describe Fig. 5 a value n = 3 was chosen. The maximum on the experimental data occurs for a size x = 12.2 µm, and according to Eq. (1), N(x) has a maximum value of 124 blocks for x = x_m = 12.2 µm. So that A = 1.37 and N(x) appear like,

$$N(x) = 1.37x^3 \exp(-3x/12.2 \text{ µm}) \qquad (2)$$

In Fig. 5 a drawing of the numerical calculations obtained with this expression it is shown as a continuous curve. The black points on the curve are to denote the calculated value at which an experimental measurement has been reported. The total number of blocks as calculated with Eq. (2) at the experimental blocks sizes is 383 which is very close to the experimental number of blocks of 374. From the close agreement between experimental data and the choosed curve we can arrive to the conclusion that Eq. (2) gives a good description to the experimental block size distribution.

4. Discussion

Experiments reported here showed and homogeneous deformation on the macroscopic level. A partially non-homogeneous deformation at mesoscopically scale can be observed. There is a strong experimental evidence that this inhomogeneity in deformation at mesoscopic level is due to blocks of grain sliding as an entity. Apparently one important reason for the occurrence of deformation in such a non-homogeneous manner can be attributed to the fact that the material here deformed, has been tested under stress and temperature conditions where a behavior of grain boundary sliding between individual grains it is not expected.

In our case (Zn-20.2%Al-1.8%Cu) to real strain rate of 1.6×10^{-3} s^{-1} and T = 300K, in a material with a grain size of 6.1 µm; the number of grains in each block has values between 1 to 7 grains. This data means a smaller number of grains for block that those observed for Zn-22%Al to strain rate 10^{-3} s^{-1} and T = 523 K with size of grain of 1.1µm (15 to 30 grains) [8]. In the case of Zn-22%Al a grain boundaring sliding, grain to grain would be expected, however it is not observed at least initially during deformation. This situation has explained by the authors of these report [8], with an analogy which relates the grain boundary sliding processes with the occurring processes during monocristals deformation. According to this analogy the gliding grain blocks could be forming arrangements at mesoscopical level with the grains occupying the place of atoms and forming in this way the pseudo particles called cellular dislocations (according to the Zeling and Mukherjee model) [13, 19]. This last possibility requires more experimental work in order to be supported or rejected.

5. Conclusions

1) Plastic deformation of the specimen it is homogeneous at macroscopic level, but partially inhomogeneous at mesoscopic level. This inhomogeneity is ascribed at the gliding of grain blocks which behaves as entities.

2) After a real deformation of 5.4%, all the sample surface is covered by gliding grain blocks. This blocks have different sizes and their distribution function it is obtained after a true deformation of 28.5%.

3) The technique here used opens the possibility of further studies about some non-homogeneous aspects of superplastic deformation at mesoscopical level. As for instance, the kinetical role of the gliding of grain block during deformation.

Figure captions

Fig. 1(a)   Tension test specimen for SEM after 5.4% of true deformation.

Fig. 1(b)   Mesoscopical aspect of microestructure on the tension specimen after a true deformation of 5.4%.

Fig. 2   Linear relationship between velocity of material points at the surface of the sample and distance along tension axis as measured from the rest side of the sample. After a true deformation of 47%.

Fig. 3   SEM microphotography at mesoscopical level on the tension specimen after a true deformation of 47%.

Fig. 4   Grain blocks sliding during deformation. The gliding zones between grain blocks appear like bright zones where oxide films are absent. After a true deformation of 28.5%.

Fig. 5   The distribution function of sliding blocks as a function of the block sizes. After a true deformation of 28.5%.

References

1. J. D. Muñoz-Andrade: M. Sc. Thesis, Universidad Central de Venezuela, Caracas, Venezuela, 1996.
2. J. D. Muñoz-Andrade, A. Mendoza-Allende, G. Torres-Villaseñor and J. A. Montemayor-Aldrete, J. Mater. Sci. (JMSC3O77-98) accepted for publication 26 June 2000.
3. R. Raj and M. F. Ashby, Metall. Trans. 2 A (1971) 1113.
4. W. R. Cannon and W. D. Nix, Philos. Mag. 27 (1973) 9.
5. J. E. Morral and M. F. Ashby, Acta Metall. 22 (1974) 567.
6. M. G. Zeling, M. R. Dunlap, R. Rosen and A. K. Mukherjee, J. Appl. Phys. 74 (1993) 4972.

7. M. G. Zeling and M. V. Alexsandrova, in Superplasticity in Advanced Materials, edited by S. Hori, M. Tokizane and N. Furushiro, The Japan Society for Research on Superplasticity, Osaka, Japan (1991) p. 95.
8. V. V. Astanin, O. A. Kaibyshev, S. N. Faizova, Scr. Metall. Mater. 25 (1991) 2663.
9. H. S. Yang, M. G. Zeling, R. Z. Valiev and A. K. Mukerjee, Scr. Metall. Mater. 26 (1992) 1707.
10. M. G. Zelin and A. K. Mukherjee, Metall. Mater. Trans. 26 A (1995) 747.
11. H. Ishikawa, D. G. Bhat, F. A. Mohamed and T. G. Langdon, Metall. Trans. 8^a (1997) 523.
12. Y. Ma and T. G. Langdon, Metall. Mater. Trans. 25 A (1994) 2309.
13. M. G. Zelin and A. K. Mukherjee, Phil. Mag. 68 A (1993) 1183.
14. J. S. Koehler in: Imperfections in Nearly Perfect Crystals, W. Shockley et al. Eds. New York, Wiley (1950) p. 197.
15. U. F. Kocks, Phil. Mag. 13 (1966) 541.
16. J. C. M. Li in: Strength and Plasticity, Physics of (A. S. Argon ed) Cambridge, Mass. USA, MIT, (1969) p. 245.
17. R. Gasca-Neri and W. D. Nix, Acta Metall. 22 (1974) 257.
18. A. Orlova, Scripta Metall. 16 (1982) 1133.
19. M. G. Zelin and A. K. Mukherjee, J. Mater. Sci. 28 (1993) 6767.

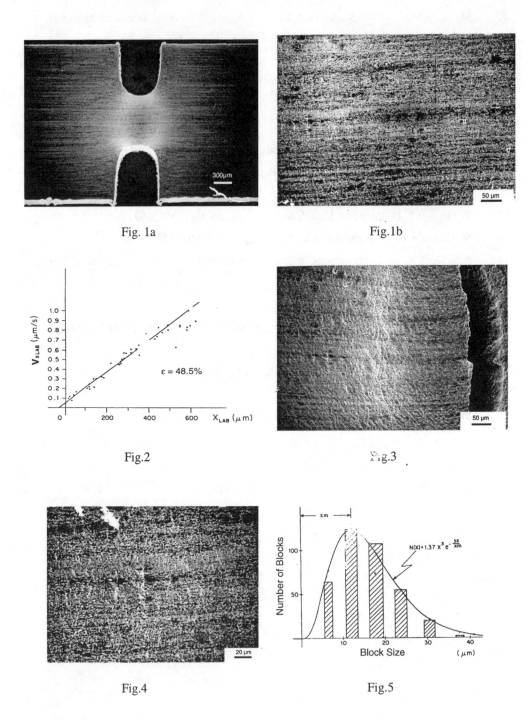

Fig. 1a

Fig.1b

Fig.2

Fig.3

Fig.4

Fig.5

Superplastic Deformation Behavior of the Y-TZP Tested in Torsion

Y. Motohashi[1], S. Sanada[2] and T. Sakuma[1]

[1] Department of Mechanical Engineering, Ibaraki University, Hitachi, Ibaraki 316-8511, Japan
[2] Graduate School of Science and Engineering, Ibaraki University, Hitachi, Ibaraki, Japan

Keywords: 3Y-TZP, Activation Energy, Ceramics, Concurrent Grain Growth, Constitutive Equation, Creep, Grain Size Exponent, Torsion

Abstract

Constant torque torsional creep tests for 3Y-TZP have been conducted in an ambient atmosphere at 1623K to 1773K in the torque range of 0.53 to 1.51Nm. While there are no traces of primary and tertiary creeps, the creep-rate decreases with the increase in strain due to concurrent grain growth. In the shear stress versus shear strain-rate relations obtained in the present experimental range, there exist two regions; one in which m-value is 0.43 to 0.45 and the other in which m-value is 0.21 to 0.22 above and below the shear stress of around 14MPa, respectively. In the former, i.e., high m region, grain size exponent, $p \simeq 2$ and activation energy, Q=443KJ/mol, whereas in the latter, i.e., low m region, $p \simeq 3$ and Q=490KJ/mol. The former region corresponds to the superplastic region II. Regarding the latter region, it seems to be relating to the existence of threshold stress.

1. Introduction

Most of the experiments on the mechanical properties of superplastic ceramics have been examined so far by means of tension, compression and bending tests. The main mode of the superplastic deformation in ceramics is, no doubt, grain boundary sliding. It is, therefore, surely interesting to study superplastic deformation behaviors of ceramics under torsional stresses and to compare the results with those under tensile and compressive stresses. The torsion test may also be an effective measure to clarify the mechanism of ceramics superplasticity. In addition to these merits, the torsion test has following advantages[1]: If the applied torque is kept constant, the stress remains constant, i.e., constant stress creep can be achieved; Material flow is mechanically stable, i.e., a large amounts of strains can be achieved without flow localization. Moreover, from the viewpoint of practical applications such as plastic working of ceramics, the torsional deformation tests at high temperatures will give us many useful informations as high temperature forging-ability and so forth.

Many studies on superplastic behaviors of metallic materials under torsional stresses have been performed so far[2-6], but as long as we know there have been almost no studies on the behaviors of superplastic ceramics subjected to torsional stresses. In this study, torsional creep tests of a typical superplastic ceramic, 3Y-TZP, are carried out at different temperatures with various constant applied torques. The strain-rate sensitivity index, m, grain size exponent, p, and activation energy, Q, observed in different creep conditions are investigated and using these data we derive an empirical constitutive equation for the creep deformation process of 3Y-TZP.

2. Experimental

The material used for this study is a fine-grained 3Y-TZP of which average grain size d_0 is 0.43 μm. The chemical composition of the 3Y-TZP is (in wt%), Y_2O_3:5.15±0.20, $Al_2O_3 \leq 0.10$, $SiO_2 \leq 0.02$, $Fe_2O_3 \leq 0.01$, $Na_2O \leq 0.04$, ZrO_2:bal. Solid cylindrical specimens of gauge length 10mm and radius 3mm were used. The specimen was mounted on the test section, made of SiC, of the apparatus having the liberty to move freely in the longitudinal direction. Thus, the specimen was deformed essentially under pure torsion[7].

The applied torque Γ was kept constant during the test. The shear strain rate and shear stress at the outer radius of the gauge section are readily calculated from the following equations[1]:

$$\dot{\gamma} = \dot{\theta} R / L \qquad \cdots (1)$$
$$\tau = \Gamma (3+n'+m') / 2 \pi R^3 \qquad \cdots (2)$$

where θ is the angle of twist, L and R are the length and radius of the gauge section, and n' and m' are, respectively, the slopes of the plots of Γ versus θ at constant $\dot{\theta}$, and of Γ vs. $\dot{\theta}$ at constant θ, on logarithmic coordinates. The test was carried out in an ambient atmosphere at 1623 to 1773K in the torque range of 0.53 to 1.51 Nm. The specimen was kept 20 minutes at the chosen temperature before testing. The temperature deviation was ±5K. Each specimen was twisted to pre-determined strain. After the twisting, the specimen was air quenched directly in the torsion apparatus. The microstructural evolution during straining was studied using SEM.

All quantities used in this study are those at the surface of the specimen unless otherwise stated.

3. Results and Discussion

3.1 Creep curve

Figure 1 shows the variation of the angle of twist θ with creep time t, and the relation between the θ and the twist-rate $\dot{\theta}$ estimated from Fig.1 is shown in Fig.2. It is seen that the θ vs. t relations, namely creep curves show no trace of tertiary creep. The curves are linear at the beginning stage and then they gradually deviate toward downward with the increase in the creep time. The deviation from the linear relation is showing a tendency to become more evident as the angle of twist and creep time increase. The main cause of this deviation may be the occurrence of the concurrent grain growth as will be stated in the later section. In addition to this, a change in the grain aspect ratio during the creep deformation might have some effect on that. It is quite natural that there appears no tertiary creep, since the creep deformations were being carried out under a condition of constant torques. It seems that the first linear region in the creep curves corresponds to the secondary creep (steady state creep), because non-Newtonian flow such as grain boundary sliding must be active from the very beginning of the deformation.

Since the amount of the concurrent grain growth is different for each of the specimens depending on the applied creep conditions, we have used twist-rates at the points where the curves, obtained

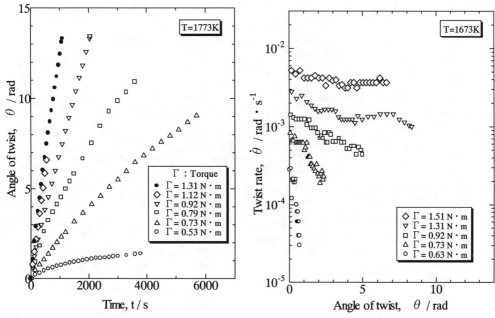

Fig.1 Variation of twist angle with creep time.

Fig.2 Variation of twist rate with the increase in twist angle.

by a least squares method on the data points shown in Fig.2, intersect the vertical axis at $\theta = 0$. This treatment allow us to discuss the stress versus strain-rate relations and so forth with the specimens having the same microstructure. The microstructure under discussion is therefore corresponding to that just before the creep deformations, i.e., the initial microstructure.

3.2 Shear stress versus shear strain-rate relation

Figure 3 is plotted to obtain m'-values which are necessary quantities to estimate stresses from Eq.2. Incidentally, $n'=0$, since the applied torques were kept constant during the creep deformations. The relation between the shear stress τ and the shear strain-rate $\dot{\gamma}$, evaluated from Eqs.1 and 2, is shown in Fig.4. It is seen from the figure that there exist two regions; one is a region above and the other is below the shear stress of around 14MPa. The m-values estimated are 0.43 to 0.45 above the border stress and are 0.21 to 0.22 below it. Hereafter we call the former as high m region and the latter as low m region.

Both regions are also found on the diagram of the equivalent stress $\overline{\sigma}$ versus equivalent strain-rate $\overline{\dot{\varepsilon}}$ relation, since exchange-relationships between τ and $\overline{\sigma}$, and, γ and $\overline{\varepsilon}$ are as follows:

$\overline{\sigma} = \sqrt{3}\ \tau$, and $\overline{\varepsilon} = \gamma / \sqrt{3}$.

Figure 5 is plotted to see the relationships between τ and $\dot{\gamma}^m$ using average m-values of 0.44 and 0.21 for the high and low m regions, respectively. It is evident that each of the data points can be approximated to straight lines and all of them pass through the origin of axes, viz.,

$\tau \propto A(T)\ \dot{\gamma}^{0.44}$ and $\tau \propto A'(T)\ \dot{\gamma}^{0.21}$.

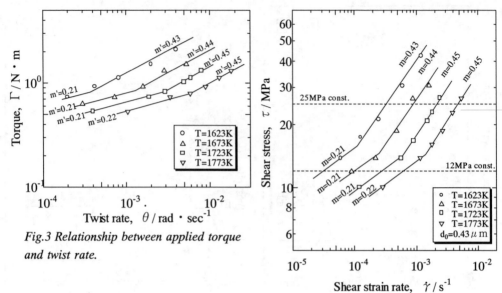

Fig.3 Relationship between applied torque and twist rate.

Fig.4 Relationship between shear stress and shear strain-rate at different temperatures.

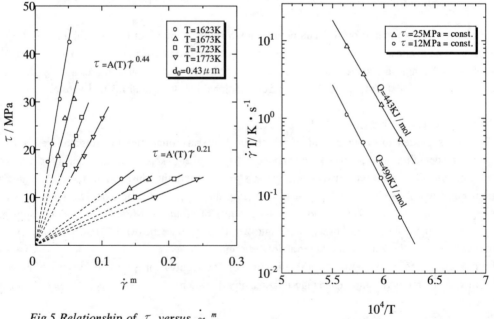

Fig.5 Relationship of τ versus $\dot{\gamma}^m$.

Fig.6 Relationship of temperature compensated shear strain-rate vs. 1/T.

3.3 Activation energy

Relationships between the temperature compensated shear strain-rate and reciprocal of temperature at constant shear stress levels of 25 MPa in the high *m* and 12MPa in the low *m* regions are examined to evaluate apparent activation energies for both regions [see Fig.6]. It is assumed that the shear modulus of the 3Y-TZP has almost no temperature dependence between 1623 and 1773K. The grain size is a constant since, as stated in the previous section, the microstructure under consideration is the initial one. The activation energies obtained are 443KJ/mol and 490KJ/mol in the high m and low m regions respectively. Incidentally, the same activation energy values are obtained when the temperature compensated equivalent strain-rate is used. These values are close to the activation energy, 460KJ/mol, found for the volume self-diffusions of cations, Y^{3+}, Zn^{4+}, evaluated in the temperature range from 1473K to 1773K[8].

3.4 Grain size dependence of strain-rate

Figure 7 shows the movement of the $\ln \tau$ vs. $\ln \dot{\gamma}$ lines with the increase in the shear strain. It is evident that the $\ln \tau$ vs. $\ln \dot{\gamma}$ lines shift to the lower side of the strain-rate as the shear strain increases. The main cause of this must be the concurrent grain growth taking place during the creep deformations, because average grain sizes d_{av} of each specimen after creep deformations have grown much larger than initial one as denoted in the figure.

From Fig.7, the relationship between the shear strain-rate and grain size can be estimated and the result is shown in Fig.8. It is found that in the high m region grain size exponent p is approximately 2 and in the low m region it is approximetly 3.

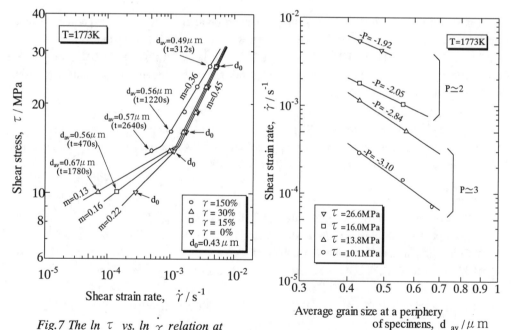

Fig.7 The $\ln \tau$ vs. $\ln \dot{\gamma}$ relation at different shear strains.

Fig.8 Relationship between shear strain-rate and grain size.

3.5 Empirical constitutive equation

From the experimental results obtained in this study, we can deduce empirical constitutive equation for the creep deformation of the 3Y-TZP as follows;

$$\dot{\gamma} = K\, d^{-2}\, \tau^{2.3}\, \exp(-Q/RT) \quad \cdots \quad (3)$$

in the high m region, and
$$\dot{\gamma} = K' \, d^{-3} \, \tau^{4.8} \exp(-Q'/RT) \quad \cdots \quad (4)$$
in the low m region, respectively.

Since, in the high m region, $m=0.44 \simeq 0.5$, and $p \simeq 2$, it is most probable that this region corresponds to the superplastic region II in which grain boundary sliding is the main deformation mode. Although it is uncertain yet what the governing deformation mode is in the low m region, it seems that this region is caused by the existence of the threshold stress from reasons mentioned below: Since the strain-rate values used in this study are based on the initial microstructure, the strain-rates are by no means affected by the microstructural evaluation such as concurrent grain growth; Although, the m-value detected in the low m region is close to that observed for materials having larger grain sizes, it is improbable that a deformation process controlled by slip within grains would be active under such a low stress level. Therefore, if there exists the region 0 at far lower strain-rates[9], then the low m region may be a transition, because of the effect of the threshold stress, from the GBS as the main deformation process to diffusional creep as governing process in the region 0. It is interesting to note that the region I arises even though the 3Y-TZP used in this study is of commercial purity.

Conclusions

There exist two regions on the $\ln \tau$ vs. $\ln \dot{\gamma}$ diagram obtaind in the present experimental range; One is above and the other is below the shear stress of around 14MPa. In the former region, $m \simeq 0.44$, $p \simeq 2$, and Q=443KJ/mol, whereas in the latter region, $m \simeq 0.21$, $p \simeq 3$ and Q=490KJ/mol. From these data, we have derived the following emperical constitutive equations for the creep deformation of 3Y-TZP,
$$\dot{\gamma} = K \, d^{-2} \, \tau^{2.3} \exp[(-443KJ/mol)/RT] \quad \cdots \quad (\tau > 14MPa)$$
$$\dot{\gamma} = K' \, d^{-3} \, \tau^{4.8} \exp[(-490KJ/mol)/RT] \quad \cdots \quad (\tau < 14MPa)$$
The former region corresponds to the superplastic region II, while the latter seems to be relating to the existence of threshold stress.

References

[1] D.S.Fields, Jr and W.A.Backofen, Proc. ASTM, 57 (1957), p.1259.
[2] A.Arieli and A.K.Mukherjee, Acta Metall. 28 (1980), p.1571.
[3] Y.Combres, Ch.Levaillant and F.Montheillet, AGARD Conf. Proc., No.426 (1988), p.101.
[4] M.J.Mayo and W.D.Nix, Acta Metall. 37 (1989), p.1121.
[5] H.J.McQueen, W.Blum and Q.Zhu, Mater. Sci. Forum, 170-172 (1994), p.193.
[6] M.K.Khraisheh, A.E.Bayoumi, C.H.Hamilton, H.M.Zbib and K.Zhang, Scripta Met. Materia. 32 (1995), p.955.
[7] F.Montheillet, M.Cohen and J.J.Jonas, Acta Metall. 32 (1984), p.2077.
[8] H.Solmon et al., Ceramic Trans. 24 (1991), cited from M.J-Melendo, A.D-Rodriguez and A.B-Leon, J.Am. Ceram. Soc. 81 (1998), p.2761.
[9] T.G.Nieh, J.Wadsworth and O.D.Sherby, *Superplasticity in metals and Ceramics*, Cambridge Univ. Press, (1997), p.40.

Corresponding Author: Y.Motohashi, e-mail: motohasi@mech.ibaraki.ac.jp

Control of Thickness Distribution in a Al7475 Axi-Symmetric Cup

N. Suzuki[1], M. Kohzu[2], S. Tanabe[2] and K. Higashi[2]

[1] Production Engineering Section, Manufacturing Department, Japan Aircraft Mfg. Co., Ltd. (NIPPI), 3175, Showa-machi, Kanazawa-ku, Yokohama 236-0001, Japan

[2] Department of Metallurgy and Material Science, College of Engineering, Osaka Prefecture University, 1-1 Gakuen-cho, Sakai 599-8231, Japan

Keywords: Al Alloy, Plane Strain, Sheet Metal Forming, Strain Rate, Superplasticity

Abstract

A numerical analysis of superplastic blowforming under axi-symmetrical deformation with controlling maximum strain rate was performed. This method of analysis is able to account for the strain hardening with regard to strain rate of formed material. The strain rates of each portion in the formed material are estimated by theoretical equations. The equations are based on stress-strain curves on a superplastic tensile test under certain strain rates as far as the magnitude of strain caused in some actual processes. A forming experiment was carried out with the condition calculated by the analysis method on a truncated cone die for 7475 Al-alloy. The deformation process and variation of thickness on the truncated cone formed under analyzed condition was good agreement with the result predicted by the calculation. By using this analysis method, superplastic blow forming of the hemispherical dome which was controlled for the thickness distribution to be in uniform, was planned. The blank sheet was supplied from the bottom of a conical pan designed by the analysis for the preforming. The bottom of the pan was reformed to the hemispherical dome on the final forming. As a result, the dome was formed with uniform thickness distribution successfully, and this verified the validity of the analysis method and the effect of control on thickness of superplastic formed parts.

Introduction

It is important to predict thickness distribution precisely on the forming plan for superplastic formed parts applied to aircraft structures[1-3]. At the same time, it is indispensable to establish suitable forming conditions to keep soundness of mechanical property on formed parts[4, 5]. The optimum strain rate and temperature are the elements of condition to be noted to control on superplastic forming.

In this study, the development of numerical analysis on superplastic forming under axi-symmetrical deformation has been put in operation. The new equation for stress-strain relationship which has fidelity to the result of tensile test has been developed[6]. By using this equation, the analysis method is able to account for the strain hardening with regard to strain rate of the formed material. And it is expected that the variation of stress, strain rate, strain and thickness on the formed material are analyzed precisely by the consideration of contacting between die and material under certain frictional conditions.

Using a blank sheet which has modified distribution of thickness with intended pattern is one of the technical means to control the thickness distribution actively on formed parts. In some case, machining is applied to supply that blank sheet[7]. Two stage forming is the another method without machining to supply that blank sheet. The superplastic forming of hemispherical dome with uniformity of thickness was analyzed and tried out by the application of two stage forming. On this forming, a truncated cone was superplastically formed as the 1st stage, and it's bottom was reformed to hemisphere as the 2nd stage. The figure of truncated cone for preforming and the forming conditions of 2 stages were determined by the numerical analysis method.

Theoretical
Relationship of stress-strain

The stress-strain relationship; $\sigma = k \varepsilon^m$ has been used conventionally on many numerical analysis, where value m is strain rate sensitivity index[1, 2, 7-10]. This equation is not able to include the influence of work hardening. However, most of superplastic materials have confirmed some strain hardening, for example 7475 Al-alloy [11-13]. And it has been experienced that value m is not constant for strain and strain rate. Only by using that equation, it is hard to predict the thickness distribution precisely for the superplastic forming of the materials with work hardening.

In the stress-strain curves shown in Fig.1(a), their tangents intersect in a point (X_p, y_p) except for two curves at high function of strain rate, and each slope f_1 (hardening coefficient) is a function of strain rate $\dot{\varepsilon}$. Since strain rate is controlled within $2.0*10^{-4}$/sec. in this experiments, $f_1(\dot{\varepsilon})$ can approximate by linear function as shown in Fig.1(b). As a result, following constitutive equations are obtained.

$$\sigma = f_1(\dot{\varepsilon})\varepsilon + f_2(\dot{\varepsilon}) \quad (1)$$

$$f_1(\dot{\varepsilon}) = a\dot{\varepsilon} + c \quad (2)$$

$$f_2(\dot{\varepsilon}) = -f_1(\dot{\varepsilon})X_p + y_p \quad (3)$$

Where f_2 is intercept of the tangent, and a, c, X_p and y_p are constants.

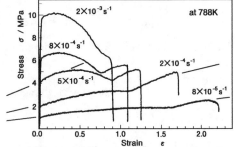

(a) Stress-strain curves for some strain rates

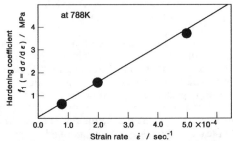

(b) Relationship between strain hardening coefficient and strain rate

Fig. 1 Tensile characteristics of 7475 Al-alloy at 788K and its formularization

Analysis model

Some analysis conditions are held as following. A forming material is deformed under membrane theory, and is formed as a portion of sphere on non-contacting areas before bottom contacting. A blank sheet is constituted with small elements divided on axial section. A contacting areas of forming material are loaded normal stress by forming pressure, at the same time, that areas are loaded by drag forces obtained with Coulomb's friction due to sliding with die. A volume of each element is constant during superplastic forming. An axi-symmetrical deformation elements are established as Fig.2 schematically.

Before bottom contact, the relationship of latitude direction stress $\sigma_{\phi i}$ and circumferential direction stress $\sigma_{\theta i}$ on the element i is shown by equation (4) and (5) at dome apex and periphery respectively. (Refer [9] for example.)

$$\sigma_{\phi i} = \sigma_{\theta i} \quad (4)$$

$$\sigma_{\theta i} = 1/2\, \sigma_{\phi i} \quad (5)$$

A function of stress distribution is assumed throgh equation (6) which will satisfy the boundary conditions given equation (4), (5).

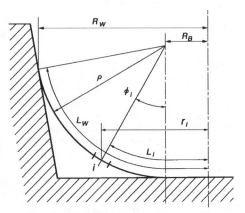

Fig. 2 Symbols of dimension used in analysis

$$K_{Di} = 1 - \frac{1}{2}\left(\frac{L_i}{L_w}\right)^{\frac{\rho}{R_w}} \quad (6)$$

Where K_{Di} is a ratio of stress, $\sigma_{\theta i}/\sigma_{\phi i}$. K_{Di} will vary continuously not only with the location of element i (L_i/L_w; ref. Fig.2) but also with progress of forming.

After bottom contacting, $\sigma_{\phi i}$ is given as equation (7) by the stress balance of vertical direction on small element i using geometrical relationship, $r_i - R_B = \rho \sin\phi_i$.

$$\sigma_{\phi i} = [(1 + \frac{R_B}{r_i})\rho P]/(2\, t_i) \quad (7)$$

$\sigma_{\phi i}$ is given as equation (8) when $R_B=0$, and is given as equation (9) when $\rho \simeq 0$ respectively.

$$\sigma_{\phi i} = \rho P / (2 t_i) \quad (8)$$

$$\sigma_{\phi i} = \rho P / t_i \quad (9)$$

For simplifying equation (7) with satisfying equation (8) and (9), $\sigma_{\phi i}$ is shown as equation (10).

$$\sigma_{\phi i} = \frac{[2 - \frac{1}{\{1 + (R_B / \rho)\}}] \rho P}{2 t_i} \quad (10)$$

In this equation, R_B / ρ is a index representing the progress of forming. Stress of circumferential direction $\sigma_{\theta i}$ is given by a stress ratio K_{Bi} ($= \sigma_{\theta i} / \sigma_{\phi i}$) which is assumed as a function of stress distributions shown by equation (11).

$$K_{Bi} = 1 - \frac{L_i - R_B}{2(L_W - R_B)} \quad (11)$$

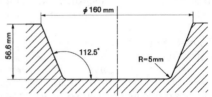

Fig. 3 A cross section of a forming test die

Calculation flow

The calculation is carried out by iteration of time increment. An equivalent strain increment is given by time increment and strain rate shown in equation (12) for each element.

$$\Delta \varepsilon_{eqi} = \dot{\varepsilon}_{eqi} \Delta j \quad (12)$$

At first, time increment and temporary forming pressure are given. To keep the balance for this pressure, stress, strain rate, strain and deformation caused on element i are calculated. By the geometrical relationship between die and deformed blank sheet, contact conditions on each element are adjusted. An element with the maximum strain rate is picked up, and its strain rate is compared with standard strain rate (or optimum strain rate). If the strain rate deviate from standard strain rate, the forming pressure loaded on the model will be adjusted repeatedly before converging. With the determination of forming pressure by that adjustment loop, the deformation of element and shape of blank sheet are determined on the time increment. This calculation will be continued until the completion of forming.

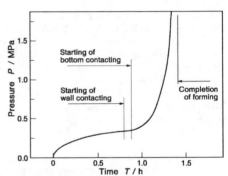

Fig. 4 Pressurization schedule

Experimental

The experiment to form truncated cone was carried out by using a die shown in Fig.3. Al 7475 alloy sheet with thickness of 1.5mm was used as a blank sheet. $2.0*10^{-4}$/sec. of maximum strain rate and 100 elements on the radial direction of blank sheet were employed for calculations of the analysis. The roughness of die surface was $Ra=1.5$. BN powder lubricant was applied to the forming surface of the die and a coefficient of friction $\mu=0.25$ was employed by experiences. The equation for stress-strain relationship at 789K[(1)-(3)] was applied with constants $a=7.72 \times 10^3$ MPa·s, $c=2.45 \times 10^{-2}$ MPa, $X_p=-1.01$, $Y_p=5.61 \times 10^{-1}$ MPa. The

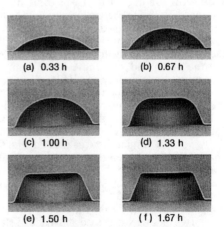

Fig. 5 Change in axial cross section of blank sheet with forming time

truncated cone was formed through the contacting with wall and bottom in the cavity of die under 789K ± 15K.

Results and Discussion
Fig.4 shows the pressurization schedule. The change of forming pressure is similar to one of plane strain type on the contacting condition with forming die[6]. Fig.5 shows the deformation process of blank sheet with the views of cut section. The outline of non-contacting area is kept as an arc of a circle. The forming process was consisted with the timing shown in Fig.4. The distributions of measured thickness are shown in Fig.6 with calculated ones. They are in good agreement. The area of minimum thickness is located along the corner of bottom.

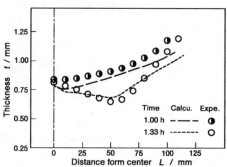

Fig. 6 Thickness distributions in forming process

Control of thickness distribution
Objection
By using the analysis method shown in previous paragraphs, two stage forming method has been proposed to control thickness distribution actively, in superplastic forming of an axi-symmetric cup. A part of preformed cup is prepared as the blank sheet for the final forming(2nd stage) which will have intended distribution of thickness. A shape of preforming die should be designed by the analysis in order to get the distribution of thickness in the area of reforming to control the distribution of thickness on the formed cup. At first, a cup will be formed using blank sheet with uniform thickness as preforming. Then a part of preformed cup will be reformed to the cup with a intended distribution of thickness. This method requires no machining of thickness on blank sheet in advance but superplastic forming twice. For a confirmation of effect on this forming method, a superplastic forming of hemispherical dome with uniform distribution on thickness was considered as experiment.

Experimental
The two stage superplastic forming was planned to get the hemispherical dome which has radius of 70mm and uniform distribution of thickness. The hemispherical dome would be formed in a bottom of a truncated cone which was prepared as a preforming formed on the 1st stage. So the radius of bottom in the truncated cone was 70mm, but the depth and the angle of wall should be determined adequately. Al 7475 alloy, 1.5mm thickness, was used for a blank sheet. The analysis method on axi-symmetric cup was applied to determine the shape of the truncated cone for the 1st stage forming. At the same time, the conditions for two stages forming were calculated with the die shape which had been determined, and with the material property
(relationship of stress-strain, ref. Fig.1) formed under 789K. $2.0*10^{-4}$/sec. was applied as the optimum strain rate for the forming of two stages.
Two dies were designed and made of stainless steel according to the result of analysis. The surface condition of them was prepared to Ra=1.0-1.5 and

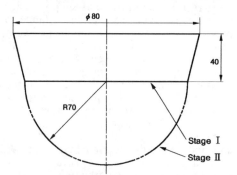

Fig. 7 Figures of axial cross section of pan and dome on the two stage forming

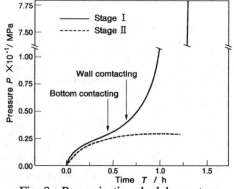

Fig. 8 Pressurization shedules on two stages forming

applied with BN powder. Coefficient of friction was estimated as μ =0.25 on the surface condition. A forming experiment was carried out under the condition same with the analysis.

Result and discussion

The optimum shape of a truncated cone for the 1st stage forming(preforming) was analyzed and determined as shown in Fig.7. At the same time, the shape of hemispherical dome is shown in this Figure, which would be formed in the bottom of the truncated cone for the 2nd stage forming(final forming). Fig.8 shows the pressurization schedules for the superplastic forming of two stages. A high pressure is needed for the 1st stage after the contacting of material and die at the bottom and the wall, but is not needed through the forming for the 2nd stage. Fig.9, 10 and 11 show a change of equivalent stress, a change of equivalent strain rate and a change of equivalent strain respectively calculated for the forming condition. These figures show those distributions along the latitude of dome with forming process in 2nd stage. The equivalent stress and the equivalent strain shown in Fig.9 and 11 increase with forming time. The peak of equivalent strain rate is kept $2.0*10^{-4}$/sec. through the forming process(Fig.10).

In the experiment, the two stage super plastic forming of this condition was completed successfully. Fig.12 shows the truncated cone formed in the 1st stage, and hemispherical dome formed in the 2nd stage which has been made from the bottom of that truncated cone. They have been formed so it perfectly fits with the die and with no necking nor breakage. Fig.13 shows the distributions of thickness in the truncated cone and the hemispherical dome measured on an axial cross section with the calculated ones. They are in good agreement. It shows the uniform distribution of thickness in the hemispherical dome. And it concludes that it is possible to control a distribution of thickness by using the method of two stage superplastic forming. Fig.14 shows the magnified view of axial section on the hemispherical dome. It shows uniform distribution of thickness.

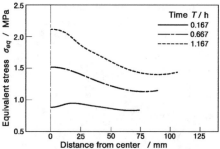

Fig. 9 Change of equivalent stress with forming time

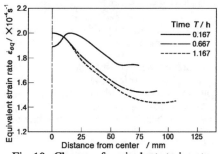

Fig. 10 Change of equivalent strain rate with forming time

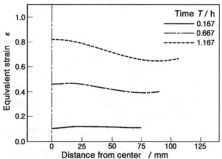

Fig. 11 Change of equivalent strain with forming time

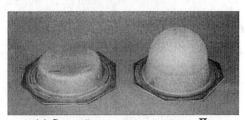

(a) Stage I (b) Stage II

Fig. 12 Views of formed shape on each forming stage

Summary

The method of numerical analysis on superplastic forming under axi-symmetric deformation was developed. A relationship of stress-strain in that method was able to account for the strain hardening with regard to strain rate. The forming experiment was carried out on the conical die for 7475 Al-alloy sheet, and the truncated cone was formed in good agreement to the forming condition calculated by this analytical method. A forming method has been proposed in order to control a distribution of thickness on superplastic formed cup using a blank sheet with uniform thickness. This method is composed of multiple stage forming. A final cup is formed from a part of preformed

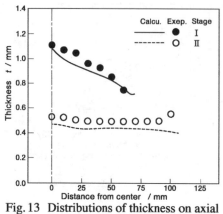

Fig. 13 Distributions of thickness on axial cross section

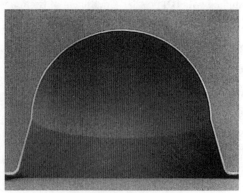

Fig. 14 View of the axial cross section of a hemispherical dome

cup which formed with intended distribution of thickness. The analysis method is applied to determine a shape of preforming and these forming conditions. The challenge of superplastically forming a hemispherical dome with uniform distribution of thickness was handled by using this forming method. A truncated cone was designed for the shape of preforming by the analysis, and its bottom was formed to hemispherical dome. As a result, the uniform distribution of thickness was achieved successfully.

Acknowledgments
This work was supported by Aerospace Division, Japan Aircraft Mfg. Co., Ltd.

References
[1] J.M. Story, "Superplasticity in Aerospace II ", The Minerals, Metals & Materials Society, (1990), 151-166
[2] C.H. Hamilton, INDUSTRIAL RESEARCH & DEVELOPMENT-December(1983), 72-76
[3] R.J. Stracey and R.G. Bultler, Materials Science Forum vols.170-172 (1994), 725-730
[4] Y. Miyagi, M. Hino and T. Eto, KOBELCO Tech. Rev., no.2 (1987), 45-48
[5] G.J. Mahon, D. Warrington, R.G. Butler and R Grimes, Materials Science Forum vols.170-172 (1994), 187-192
[6] N. Suzuki, M. Kohzu, S. Tanabe and K. Higashi, Materials Science Forum vols.304-306 (1999), 777-782
[7] A. Takahashi, S. Shimizu and T. Tuzuku, JSTP vol.31, no.356 (1990), 1128-1134
[8] N. Chandra and K. Chandy, J. Materials Shaping Technology, Vol.9, No.1(1991), 27-37
[9] Z. X. Guo, K. Higashi and N. Ridley, MRS Int'l. Mtg. on Adv. Mats., vol.7 (1989), 141-146
[10] N. Akkus, M. Kawahara and H. Nishimura, Materials Science Forum vol.170 (1994), 633-638
[11] K. Higashi, Japan Inst. of Light Metals, vol.39, no.11, (1989), 751-764
[12] H. Iwasaki, Y. Irie, T. Hayami, K. Higashi, T. Ito, Japan Inst. of Light Metals, vol.39, no.11, (1989), 798-804
[13] H. Iwasaki, T. Hayami, K. Higashi, T. Ito, S. Tanimura, JSTP, vol.32, no.362 (1991), 359-363

For correspondence
Nobuyuki SUZUKI; e-mail < nsuzuki@mail.nippi.co.jp > fax no. 045-773-5129
Masahide KOHZU; e-mail < kohzu@mtl.osakafu-u.ac.jp > fax no. 0722-59-3340
Shigenori TANABE; e-mail < tanabe@mtl.osakafu-u.ac.jp > fax no. 0722-59-3340
Kenji HIGASHI; e-mail < higashi@mtl.osakafu-u.ac.jp > fax no. 0722-59-3340

A Molecular Dynamics Study of Large Deformation of Nanocrystalline Materials

H. Ogawa[1,2], N. Sawaguchi[1] and F. Wakai[3]

[1] National Industrial Research Institute of Nagoya, AIST
1-1, Hirate-cho, Kita-ku, Nagoya 462-8510, Japan

[2] National Institute for Advanced Interdisciplinary Research AIST,
1-1-4, Higashi, Tsukuba, Ibaraki 305-8562, Japan

[3] Material Structure Laboratory, Tokyo Institute of Technology,
Nagatsuta-cho, Midori-ku, Yokohama 226-8503, Japan

Keywords: Deformation, Molecular Dynamics, Neighbor Switching, Polycrystals, Zirconia

Abstract

Molecular dynamics simulations were carried out for model polycrystals of yttria-stabilized zirconia in order to investigate the microscopic mechanism of plastic deformation. The model polycrystals were composed of eight and sixteen grains with two- and three-dimensional grain arrangements, respectively. During the deformation, a neighbor-switching event similar to Ashby and Verrall's model was observed for both model structures. Further complex motions of the grains were observed in the case of the three-dimensional grain arrangement. By analyzing the time variation of the positions of oxide ions, mass transport during the deformation was found to take place mainly at the grain boundary regions.

Introduction

Superplasticity of metallic and ceramic materials is quite interesting phenomena, and many experimental and theoretical studies have been carried out to investigate the deformation mechanisms [1]. In 1970's, Ashby and Verrall [2] considered a model configuration composed of four grains contacting to each other, and proposed a deformation process callded *neighbor-switching event*. In their model, large deformation can be attained by the switching of grain configuration accompanied by the grain boundary slidings. The model has been widely accepted as the main mechanism of superplasticity, but is quite difficult to be observed in the actual deformation process even by the latest experimental techniques, especially for three-dimensional grain arrangements.

In the previous papers [3,4], the present authors carried out molecular dynamics (MD) simulations of polycrystalline matter composed of four grains, and investigated the microscopic processes at the grain boundary during deformation. These studies, however, were limited to not enough number of grains for reproducing the neighbor-switching event. Recently, the authors extended the previous studies by increasing the number of grains and the maximal strain, and simulated the rearrangement of grain configuration in a model polycrystal similar to the Ashby and Verrall's model [5, 6]. In the present paper, the authors also extend the grain arrangement from a planner one to three-dimensional one in order to investigate more realistic, three-dimensional processes in the deformation of polycrystalline matters.

MD simulation

As shown by Ashby and Verrall [2], the minimum number of grains to reproduce the neighbor-switching event is four. However, it is preferable to add several grains surrounding the main grains to more realisticly simulate the process of deformation. The numbers of grains used in this study are eight and sixteen, and the grain centers are arranged on two-dimensional hexagonal and three-dimensional b.c.c.-like lattices, respectively, as shown in Fig. 1. The polycrystalline structure was generated by the weighted Voronoi construction as described in the previous paper [2]. The model structures are composed by the grains of a nanometer size of which diameter is about 5 nm. All grains have the same geometries of hexagonal prism in the case of the 8 grains model, and of similar to the truncated octahedron in the case of the 16 grains model. The crystal orientations of the grains were selected to be different to each other.

The target material of the present simulation is yttria-stabilized zirconia which is the first ceramic material observed the superplasticity [7]. The selected composition is 0.067 Y_2O_3 – 0.933 ZrO_2. The atomic configuration in each grain was constructed by substituting the appropriate

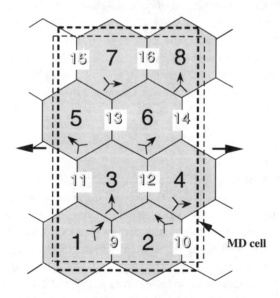

Fig. 1 Two-dimensional projection of the model polycrystalline structure used in this study. The hexagonal regions denote the individual grains and the dotted rectangle (thick) is the MD basic cell. The grain centers are denoted by the numbers for 2D-hexagonal type (1–8) and 3D-b.c.c. type (1–16) model structures. The centers of the grains 9–16 are on the different plane from that for 1–8. Initial orientations of the grains in the 8 grain model are indicated by small arrows. The dotted rectangle (thin) show the MD basic cell after small horizontal tensile.

numbers of yttrium ions and oxygen vacancies for zirconium and oxide ions in the ZrO_2 structure, respectively. Grains were arranged in the rectangular MD basic cells of approximately $8 \times 13 \times 1.5$ nm³ and $9 \times 13 \times 7$ nm³ for the 8 and 16 grains models, respectively. For both models, three-dimensional periodic boundary conditions were applied. The interatomic potentials of Born-Mayer type proposed by Butler et al. [8] and Dwivedi and Cormack [9] were used. Temperature and the deformation of the system were controlled by the Nosé's thermostat and the cell scalings, respectively.

Results and Discussion

The MD calculation was started at 300 K and heated up to 2500K in order to obtain the thermally relaxed structure. This sample was then cooled down to 2000 K and held at this temperature for 50 ps. An initial configuration of grains similar to that of the Ashby and Verrall model is clearly recognized in the simulated 8 grain structure, as shown in Fig. 2 (a). The tensile simulation (horizontal direction) was carried out for this model structure with the strain rate of 0.025 ps⁻¹. In order to avoid an unexpected fracture of the sample, the cell dimensions in the y and z directions were scaled so as to maintain the initial density value. In the beginning of the deformation, an elastic behavior was observed at strains ε lower than 0.03. It turned to a plastic behavior at $\varepsilon \sim 0.04$. The tensile stress reached to the maximum value, 4.7 GPa, at $\varepsilon \sim 0.07$, and then decreased rapidly to about 1 GPa at $\varepsilon > 0.12$. The simulated stress-strain relation is typical for plastic deformation of polycrystalline materials as shown in the previous studies [4, 10].

The time variation of the grain configuration during deformation is analyzed up to $\varepsilon \sim 0.55$.

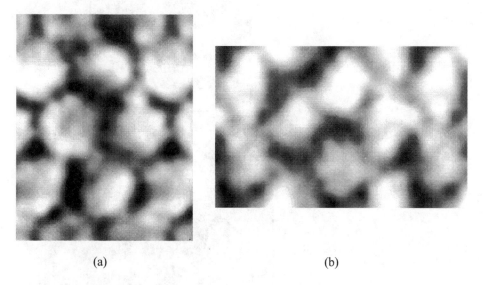

(a)   (b)

Fig. 2 Snapshots of the simulated sample before (a) and after (b) the horizontal elongation. Dark region represents the area where the local potential energy is high (i.e. the grain boundary).

Fig. 3 Three-dimensional images of the morphology of grain core regions of the 16 grains model at the initial (top), intermediate (middle), and final (bottom) stages of deformation. The grain shape was evaluated as the isosurface of the local bond orientational orders.

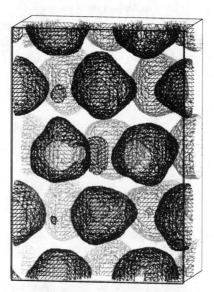
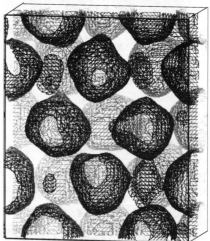
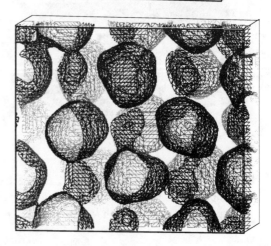

The position of the grain boundary was estimated by the local potential energy distribution. As shown in Fig. 2 (b), the neighbor-switching event was successfully reproduced and the eight original grains can be clearly recognized even at the maximal strain. The shapes of the grains are slightly modified but not elongated extensively. This result is consistent with the Ashby and Verrall model. A small amount of grain rotation is recognized (up to about 5 degrees). The grain boundary layer in the initial structure had a thickness of about 1 nm, and became slightly thicker toward the x direction during deformation.

Similar configurational variation of the grains was also observed in the 16 grains model. In the case of this model, however, visualization of the grain morphology is quite difficult because the grain boundary network spreads three-dimensionally. In this study, the authors try to recognize the grain shapes and configuration by using the local bond orientational orders (BOO) [11]. If we assume the isosurface of the BOO values as the shape of the grain core region, three-dimensional images of the grain morphology can be visualized as shown in Fig. 3. The neighbor-switching event is recognized by comparing the arrangements of the four grains near the center of the MD basic cell at the initial (figure top), intermediate (middle), and final (bottom) stages of the deformation. In the case of the three-dimensional configuration, each grain has more degrees of freedom compared with the two-dimensional model (both translational and rotational). Such complex variation of the grain morphology can be seen, for example, by the variation of cross sections of the grains at the front cell boundary.

Not only the grain morphology, but also the mass transport at the atomistic level is important for understanding the mechanism of superplasticity. Mass transport in polycrystals can be divided into two parts: bulk diffusion and boundary diffusion. In the case of superplasticity, boundary

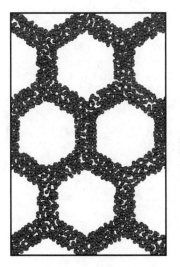
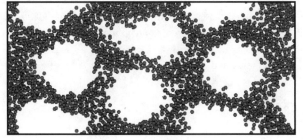

Fig. 4 Variation of the positions of the same oxide ions before (left) and after (right) the deformation of the 8 grains model. Only the ions existed in the initial boundary regions of 1 nm thickness are displayed.

diffusion is considered to be more important. The diffusive motions of oxide ions at the grain boundary during the tensile simulation of the 8 grains model are shown in Fig. 4. In the figure, only the oxide ions initially exist in the boundary layer of 1 nm thickness are displayed. Throughout the deformation, such oxide ions remained close to the boundary even though the boundary topology was changed by the neighbor-switching event. It means that the mass transport occurs mainly at the grain boundary, which is consistent with the Ashby and Verrall's model.

Summary

The present simulation was successful for visualizing what happens during large deformation of a polycrystalline matter. The model polycrystal showed a similar neighbor-switching event to that originally proposed by Ashby and Verrall. Although the model structures used in this study are much simpler, even for the 16 grains model, than the actual polycrystalline materials, simulated results are informative for verifying the theoretical models. In conclusion, the present authors stress that the MD simulation is useful for investigating the deformation process of polycrystalline materials.

References

[1] T.G. Nieh, J. Wadsworth and O.D. Sherby, *"Superplasticity in Metals and Ceramics"*, Cambridge University Press, (1997), 273 pp.
[2] M.F. Ashby and R.A. Verrall, *Acta Metal.*, **21** (1973), p. 149.
[3] H. Ogawa, F. Wakai and Y. Waseda, *Mol. Simul.*, **17** (1996), p. 179.
[4] H. Ogawa, N. Sawaguchi and F. Wakai, *Mater. Sci. Forum*, **243-245** (1997), p. 351.
[5] H. Ogawa, *Mol. Simul.*, **21** (1999), p. 357.
[6] H. Ogawa, N. Sawaguchi and F. Wakai, *Trans. Mater. Res. Soc. Jpn*, **24** (1999), p. 253.
[7] F. Wakai, S. Sakaguchi, and H. Matsuno, *Adv. Ceram. Mater.*, **1**, (1986), p. 259.
[8] V. Bulter, C.R.A. Catlow and B.E.F. Fender, *Solid State Ionics*, **5**, (1981), p. 539.
[9] A. Dwivedi and A.N. Cormack, *J. Solid State Chem.*, **79**, (1989), p. 218.
[10] H. Ogawa, in *"Mesoscopic Dynamics of Fracture"*, ed. by H. Kitagawa *et al.*, Springer-Verlag, (1998), p. 155.
[11] H. Ogawa and Y. Waseda, *Z. Naturforsch.*, **49a**, (1994), p. 987.

Interface Models for GB Sliding and Migration

C. Shet, H. Li and N. Chandra

Department of Mechanical Engineering, FAMU-FSU College of Engineering,
Florida State University, Tallahassee FL 32310, USA

Keywords: Cohesive Zone Model, Crystal Plasticity, Grain Boundary Sliding, Molecular Dynamics

Abstract The predominant strain producing mechanism in superplastic materials is attributed to grain boundary sliding (GBS). Though there have been many attempts to model GBS, most of them are heuristic and not based on the microscopic deformation at the scale of grain boundaries. Lack of such models has hindered the progress of a methodical development of alloying processes. Hence an attempt is made to develop a constitutive model of grain boundary using atomic simulation results. Detailed atomic simulations are carried out on aluminum bicrystals of $\Sigma 3$ and $\Sigma 9$ coincident site lattice boundaries to obtain characteristic behavior of grain boundary sliding/migration. Using the results thus obtained a constitutive model based on cohesive zone approach is developed to predict the behavior of grain boundaries under the external loads and hence GBS. This constitutive model is implemented in a finite element method (FEM) code to analyze the GBS/migration at a continuum scale.

1. Introduction

FEM studies have been carried out to study the grain boundary as an interface within polycrystalline materials using an interface model and accounting for damage, inelastic slip and dilation [1,2]. Here crystalline materials are modeled based on von Mises continuum theory of plasticity or based on specific crystalline properties. The phenomenological models used in these works are effective in predicting the debonding behavior, but lack the grain boundary parameters responsible for grain boundary sliding and migration. These approaches are based on cohesive zone models (CZM) [3], which incorporate a non-linear constitutive behavior at the interface or at a process zone near the crack tip. Fig.1 shows a bimaterial system (material A and B) with an interface between them subjected to traction **T**. The interface cohesive law describes the evolution of tractions and the coupling between them as a function of normal and tangential displacements (Fig.2) and their rate along the interface.

The main objective of this work is to model the grain boundary sliding/migration at a mesoscale based on continuum mechanics, using the data obtained from atomic simulations. This is a new approach of linking atomic level properties and parameters to a mesoscopic grain level computation that can be included within the framework of continuum mechanics. Atomic simulation based on molecular dynamic approach using Embedded Atom Method (EAM) [4,12] is used to extract the cohesive properties of the grain boundaries. Currently there is no direct link between atomic simulation results and continuum mechanics. These cohesive behavior characteristics are employed in CZM to model the interface between the grains. The proposed cohesive zone model is implemented in a general purpose FEM. The grains are modeled as crystalline materials in the framework of crystal plasticity. FEM analyses are carried out on $\Sigma 9$ bicrystals to demonstrate the feasibility of the above approach.

2. The Crystal Model

The general theory on constitutive equations for plastic deformation of ductile single crystals are well described by Rice [5], Hill and Rice [6], Pierce et al [7], Asaro and Rice [8]. A brief outline of the model is described here.

Kinematics:
The total deformation gradient **F** is written as
$$\mathbf{F} = \mathbf{F}^* \cdot \mathbf{F}^P \qquad (1)$$
$\mathbf{F}^*$ denotes stretching and rotation of the lattice, which also includes any rigid body rotations. The inelastic deformation $\mathbf{F}^P$ arises due to plastic shearing on crystallographic slip systems. A particular slip system α is specified by the vectors $(\mathbf{s}, \mathbf{m})$, where **s** and **m** are considered as lattice vector in reference configurations (they are orthogonal to each other), and they are subjected to stretching and rotation as follows
$$\mathbf{s}^* = \mathbf{F}^* \cdot \mathbf{s}, \quad \mathbf{m}^* = \mathbf{m} \cdot \mathbf{F}^{*-1} \qquad (2)$$
The Eulerian velocity gradient in the current configuration is given by
$$\mathbf{L} \equiv \dot{\mathbf{F}} \cdot \mathbf{F}^{-1} = \mathbf{D} + \mathbf{\Omega} \qquad (3)$$
where the rate of stretching **D** and the antisymmetric spin tensor $\mathbf{\Omega}$ may be decomposed into lattice part (superscript *) and plastic parts (superscript P) as
$$\mathbf{D} = \mathbf{D}^* + \mathbf{D}^P, \quad \mathbf{\Omega} = \mathbf{\Omega}^* + \mathbf{\Omega}^P \qquad (4)$$
The plastic part of velocity gradient is given by
$$\dot{\mathbf{F}} \cdot \mathbf{F}^{-1} - \dot{\mathbf{F}}^* \cdot \mathbf{F}^{*-1} = \mathbf{F}^* \dot{\mathbf{F}}^P \cdot \mathbf{F}^{P-1} \mathbf{F}^{*-1} = \mathbf{D}^P + \mathbf{\Omega}^P = \sum_\alpha \dot{\gamma}^{(\alpha)} \mathbf{s}^{*(\alpha)} \mathbf{m}^{*(\alpha)} \qquad (5)$$
where $\dot{\gamma}^{(\alpha)}$ is the rate of slipping on a slip system α. Further the symmetric (stretching) and antisymmetric (spin) part of plastic deformation can be written as
$$\mathbf{D}^P = \sum_\alpha \mathbf{P}^{(\alpha)} \dot{\gamma}^{(\alpha)}, \quad \mathbf{\Omega}^P = \sum_\alpha \mathbf{W}^{(\alpha)} \dot{\gamma}^{(\alpha)} \qquad (6)$$
Where the tensors $\mathbf{P}^{(\alpha)}$ and $\mathbf{W}^{(\alpha)}$ are given by
$$\mathbf{P}^{(\alpha)} = \tfrac{1}{2}(\mathbf{s}^{*(\alpha)} \mathbf{m}^{*(\alpha)} + \mathbf{m}^{*(\alpha)} \mathbf{s}^{*(\alpha)}), \quad \mathbf{W}^{(\alpha)} = \tfrac{1}{2}(\mathbf{s}^{*(\alpha)} \mathbf{m}^{*(\alpha)} - \mathbf{m}^{*(\alpha)} \mathbf{s}^{*(\alpha)}) \qquad (7)$$

The Constitutive Model:
Since the elastic properties are unaffected by slipping, the relation between the symmetric rate of stretching of the lattice $\mathbf{D}^*$ and the Jaumann rate of Kirchhoff stress is given by
$$\overset{\triangledown}{\boldsymbol{\tau}^*} = \mathbf{L} : \mathbf{D}^* \qquad (8)$$
where **L** is the symmetric tensor of elastic moduli and $\overset{\triangledown}{\boldsymbol{\tau}^*}$ is the co rotational stress rate on axes that rotate with crystal lattice, and is related to co rotational stress rate on axes rotating with material $\overset{\triangledown}{\boldsymbol{\tau}}$,
$$\overset{\triangledown}{\boldsymbol{\tau}^*} = \overset{\triangledown}{\boldsymbol{\tau}} + \sum_\alpha (\mathbf{W}^{(\alpha)} \cdot \boldsymbol{\tau} - \boldsymbol{\tau} \cdot \mathbf{W}^{(\alpha)}) \cdot \dot{\gamma}^{(\alpha)} \qquad (9)$$
Rearranging Eq. (9)
$$\overset{\triangledown}{\boldsymbol{\tau}} = \mathbf{L} : \mathbf{D} - \sum_\alpha (\mathbf{L} : \mathbf{P}^{(\alpha)} + \mathbf{W}^{(\alpha)} \cdot \boldsymbol{\tau} - \boldsymbol{\tau} \cdot \mathbf{W}^{(\alpha)}) \cdot \dot{\gamma}^{(\alpha)} \qquad (10)$$

Hardening laws:
The shear rates are developed within the framework of viscoplasticity. Based on the Schmid law, the slipping rate $\dot{\gamma}^{(\alpha)}$ of the slip system in a rate dependent crystalline solid is dependent on the corresponding resolved shear stress $\tau^{(\alpha)}$ as
$$\dot{\gamma}^{(\alpha)} = \dot{a}^{(\alpha)} \left[\tau^{(\alpha)} / g^{(\alpha)} \right] \left[\left| \tau^{(\alpha)} / g^{(\alpha)} \right| \right]^{(1/m)-1} \qquad (11)$$
Where $\dot{a}^{(\alpha)}$ is the reference strain rate on the slip system α, the function $g^{(\alpha)}$ defines the current strain hardening state of the slip system and m is a rate sensitivity exponent. The

strain hardening is characterized by the evolution of the strengths $g^{(\alpha)}$ through the incremental relation

$$\dot{g}^{(\alpha)} = \sum_\beta h_{\alpha\beta} |\dot{\gamma}^{(\beta)}| \qquad (12)$$

where $h_{\alpha\beta}$ are the slip hardening moduli, the sum ranging over all the active slip systems and is taken in the form

$$h_{\alpha\beta} = qh + (1-q)h\delta_{\alpha\beta} \qquad (13)$$

where
$$h = h(\gamma) = h_o \operatorname{sech}^2 (h_o \gamma / \tau_s - \tau_o) \qquad (14)$$

h_o is the initial hardening modulus. τ_o is the yield stress and is equal to the initial value of current yield strength $g^{(\alpha)}(0)$. τ_s is the saturation strength. The parameter q induces the effect of latent hardening as compared to the self-hardening of the slip systems.

3. Interface Constitutive Relations

The constitutive law for the cohesive surface defines mechanical relation between the traction and displacement jump across the surface and is based on phenomenological observations. As the interface separates, the magnitude of traction $T_n = \mathbf{n} \cdot \mathbf{T}$ increases, reaches a maximum and then falls to zero when complete separation occurs. The work of separation is path independent. The interface is characterized by the work of separation, the strength in the normal and the tangential directions, and by coupling parameters. Defining a work potential of the type $\phi(\Delta_n, \Delta_t)$, such that the interfacial traction is given by

$$\mathbf{T} = \partial \phi / \partial \Delta \qquad (15)$$

Following the work of Xu and Needleman [2], the interfacial potential is defined as

$$\phi(\Delta_n, \Delta_t) = \phi_n + \phi_n \exp\left(-\frac{\Delta_n}{\delta_n}\right)\left\{\left[1 - r + \frac{\Delta_n}{\delta_n}\right]\left[\frac{(1-q)}{(r-1)}\right] - \left[q + \left[\frac{(r-q)}{(r-1)}\right]\frac{\Delta_n}{\delta_n}\right]\exp\left(-\frac{\Delta_t^2}{\delta_t^2}\right)\right\} \qquad (16)$$

where $q = \phi_t / \phi_n$, $r = \Delta_n^* / \delta_n$, ϕ_t, ϕ_n = work of tangential separation and normal separation, Δ_n^* = is the value of Δ_n after complete shear separation when $T_n = 0$, and $\phi_n = \sigma_{max} e \delta_n$, $\phi_n = \tau_{max} \sqrt{\frac{e}{2}} \delta_t$. The interfacial traction can be obtained from Eq. (15) as

$$T_n = -\left(\frac{\phi_n}{\delta_n}\right)\exp\left(-\frac{\Delta_n}{\delta_n}\right)\left\{\left[\frac{\Delta_n}{\delta_n}\right]\exp\left(-\frac{\Delta_t^2}{\delta_t^2}\right) + \left[\frac{(1-q)}{(r-1)}\right]\left[r - \frac{\Delta_n}{\delta_n}\right]\left[1 - \exp\left(-\frac{\Delta_t^2}{\delta_t^2}\right)\right]\right\} \qquad (17)$$

$$T_t = -\left(\frac{\phi_n}{\delta_n}\right)\left[\frac{\Delta_t}{\delta_t}\right]\left[\frac{2\delta_n}{\delta_t}\right]\left\{q + \left[\frac{(r-q)}{(r-1)}\right]\frac{\Delta_n}{\delta_n}\right\}\exp\left(-\frac{\Delta_n}{\delta_n}\right)\exp\left(-\frac{\Delta_t^2}{\delta_t^2}\right) \qquad (18)$$

4. Atomistic Simulation for Cohesive Properties

Atomic simulation is based on molecular dynamics, which uses EAM potentials [4] to define the energy of each atom. The total energy of the system can be sum of energy of all atoms, where energy of one atom is given as

$$E_c = \sum_i F_i \left(\sum_{j \neq i} \rho_i^a (r_{ij})\right) + \tfrac{1}{2} \sum_{i,j(j \neq i)} \phi(r_{ij}) \qquad (19)$$

The first term in the Eq.19 is the energy required to embed an atom in a cloud of electron density, and the second part represents pair potential. In molecular statics the total energy is minimized to give equilibrium position for all atoms in a given crystal. In the molecular dynamics, the atomic positions are traced according to the prevailing or applied stress/temperature condition in the context of Newtonian mechanics. A molecular dynamics /statics code 'DYNAMO' developed at Sandia National Laboratory, Livermore is used in the present simulations.

Bicrystals with different CSL boundaries are generated based on the specific orientations of the bicrystal. In our simulations about 6000 atoms are used to simulate the crystal in three dimensions, and the number of atoms so selected is sufficient to predict the grain boundary behavior on an atomic scale. Periodic boundary conditions are imposed in X and Z directions (Fig 3a). Holding a few atomic layers in bottom crystal, the top crystal is subjected to tractions in order to induce the grain boundary sliding/migration.

Fig. 3a shows a portion of $[110]\Sigma9(2\bar{2}1)$ aluminum bicrystal. Fig. 3b and 3c shows the displaced position of atoms when subjected to shear and normal body force applied to all the atoms in top crystal. The loads are applied incrementally and the new positions of atoms are obtained by molecular dynamics. At the end of each incremental step the new position of all atoms are equilibrated using molecular statics. After obtaining equilibrium position the cohesive energy of the system is calculated as the difference in energies between new deformed configurations to that of original configuration of bicrystal (before application of load). Grain boundary energy and other interacting parameters are evaluated as a function of GB sliding/migration. Behavior pattern for different crystals are studied by plotting the energy as a function of sliding/migration.

Fig. 4a and 5a show typical plots showing the variation of cohesive energy with normal and tangential displacement of grain boundary respectively for the $\Sigma9$ bicrystal. The smooth curve fit is obtained by employing a curve of the type given by Eq.16. This equation is chosen in order to obtain the derivatives in an analytical form. Since Eq.16 is force fitted, the shear energy-separation curve is not best for the data set. The derivative of energy-separation curve gives the traction causing separation. Fig 4b and 5b show the plots of traction as a function of separation distance. The area under the traction-separation curve yields the work of separation. From the simulated results, it can be observed that complete debonding occurs when the distance of separation reaches a value of 2 to 3 Å. For $\Sigma9$ bicrystal tangential work of separation along the grain boundary is of the order 3 J/m^2 and normal work of separation is of the order $2.6 J/m^2$. When simulations were carried out for $\Sigma3$-bicrystal, the work of separation ranges from 1.5 to $3.7 J/m^2$. Rose et al. [9] have reported that the adhesive energy (work of separation) for aluminum is of the order 0.5 J/m^2 and the separation distance 2 to 4 Å.

The numerical value of the cohesive energy is very low when compared to the observed experimental results. The measured value of energy for fracture exceeds the cohesive energy computed by our work as well others. For example Wang and Anderson [10] have reported that the measured energy to fracture copper bicrystal with random grain boundary is of the order 54 J/m^2 and for $\Sigma9$ copper bicrystal the energy to fracture ranges from 128 to $970 J/m^2$. It should be noted that the experimental value of fracture energy includes the plastic work in addition to work of separation [11].

It should be noted that when the separation distance is of the order of nanometers (as in our work and that in ref. [9]), the energy values in J/m^2 are in low single digit values. However, the values used in the typical fracture work are in hundreds (54 to $1000 J/m^2$) and the separation distance used is of the order of micrometers. This discrepancy is noteworthy and goes into the heart of what is the scale in the cohesive zone theory and needs further study. As such, higher values are used in FEM simulations to be consistent with the mesoscopic scale (in μm) of the problem.

5. Numerical Results
The separation process in the CZM is idealized as a very thin strip, which is then modeled by interface elements. The virtual work contribution of interface to the total system is

$$\int \delta \dot{\phi} \, dS = \int \left(\dot{T}_n \delta \Delta_n + \dot{T}_t \delta \Delta_t \right) dS \qquad (20)$$

Integrating the constitutive model in the virtual work and using usual procedures finite element equations for rate equilibrium can be developed as

$$[K]\{\dot{U}\} = \{\dot{F}\} \qquad (21)$$

The proposed cohesive zone model (CZM) is implemented in a general purpose FEM code ABAQUS as an USER ELEMENT option. To demonstrate the cohesive model, a bicrystal Σ9 symmetric tilt grain boundary with a slight initial separation along the grain boundary as shown in Fig. 6a is analyzed for normal separation. Because of symmetry only the top crystal is modeled. Around 4000 four node, rectangular elements were used to model the top crystal. The interface is modeled by 46 cohesive elements. Fig. 6b shows the meshwork on the top crystal. Fig. 6c shows the fine mesh around the point of initial separation. The grain separation takes place in [$\bar{1}14$] direction. The grain separation by means of de-cohesion is activated by applying a far end incremental displacement. The material considered is a model anisotropic material available in the literature. The material properties (SI Units) assigned are $C_{11} = 168400$, $C_{12} = 121400$ and $C_{44} = 75400$, $\tau_o = 40$, $\tau_s = 1.8\tau_o$, m=10, $h_o = 8.9\tau_o$, q = 1, $\dot{a} = 0.001$. The cohesive material properties used are $\sigma_{max} = 180$, $\phi_n = \phi_t = 680$.

Inelastic (permanent) strain is represented in the crystal plasticity codes as accumulated slips. Fig. 7a shows the contours of accumulated slip of all the slip systems just before decohesion take place near the point of initial separation. The innermost contour having highest slip value, shows maximum strain occurs near the point of initial separation. Once the cohesive elements near the point of initial separation debond completely, higher slip contour shifts away from the initial point of separation indicating the reduced slip activity at that point as shown in Fig. 7b. Fig. 8 (a-c) shows the slip on each the of active slip systems on (111) planes. Fig. 9 shows the contour plot for stress σ_{22}. With the progressive decohesion of cohesive elements starting from the point of initial separation, the intensity of stress contours also moves, showing the new grain boundary tip positions. For the selected scale range, innermost contour for the stress σ_{22} shown in the Fig. 9 remains constant at 169 MPa, but it has been observed that for few elements near the point of separation the highest value of stress σ_{22} is of the order 180 MPa. This value corresponds to ultimate stress of single crystal under uniaxial tensile loading and also the input cohesive property σ_{max}.

6 Conclusions
Cohesive properties for CZM Σ3 and Σ9 STGB were extracted from atomistic simulations by applying shear and normal tractions on bicrystals. The magnitudes of cohesive properties obtained from atomistic simulation are low when compared to macroscopic experimental observations. FEM simulation using cohesive zone model were conducted at a continuum scale, where the progress of grain boundary separation is demonstrated. The simulations also show the onset of slip and accumulated slip in various slip planes. Thus the preliminary work shows that a methodology of linking atomic scale to continuum scale is feasible.

7 References

1. P. Cannmo, K. Runesson and M. Ristinma, *Inter. J. of Plasticity*, 11(8)(1995) 949-970.
2. X. P. Xu and A. Needleman, *Modelling Simul. Mater. Sci. Eng.* 1, (1993) 111-132.
3. G.I. Barenblatt, The mathematical theory of equilibrum cracks in brittle fracture. Advances in Applied Mechanics, 7(1962) 55-129.
4. M.S. Daw and I. Barka, *Phys. Rev. Lett.*, 50(1983), 1285.
5. J.R.Rice, 1971, *J. Mech. Phys. Solids*, 19, 433
6. R. Hill and J.R Rice, 1972, *J. Mech. Phys. Solids*, 20, 401
7. D. Pierce, R.J.Asaro and A. Needleman, 1983, *Acta Metall*, 31,p1951.
8. R. J. Asaro and J.R Rice, 1977, *J. Mech. Phys. Solids*, 25, 309.
9. J.H Rose, R. S John, John Ferrante, *Physical Reviews B*, Vol.28, 1983.
10. Jian Sheng Wang and P. M Anderson, 1991, *Acta Metall. Mater.*, 39, 779-792.
11. J.R Rice and J. S Wang, *Mater. Sci. Eng.*, A107 (1989), 23-40.
12. N. Chandra, P. Dang, *J. Mat. Sci*, 34 (1999), 655-666.

Note: corresponding author N. Chandra (chandra@eng.fsu.edu)

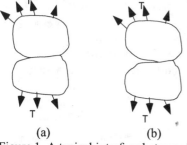

Figure 1. A typical interface between bimateral subjected to traction **T**. (a) before deformation and (b) after deformation

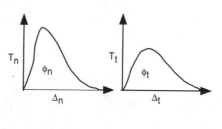

Figure 2. The evolution of traction with displacement jump.

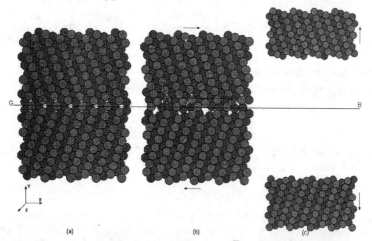

Figure 3 (a) A small portion of bicrystal with $\Sigma 9(2\bar{2}1)$ CSL grain boundary, (b) Displaced position of atoms when top crystal is subjected to shear loading. Atom A has moved to a new position B after shear separation (c) Displaced position of atoms when top crystal is subjected to normal traction showing normal separation

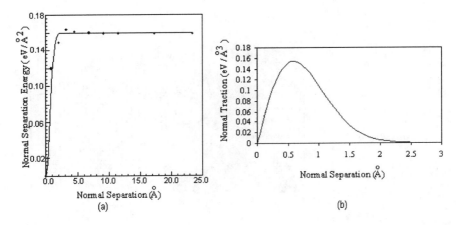

Figure 4 (a) Evolution of normal separation energy with separation distance. (b) Variation of normal traction with normal separation distance

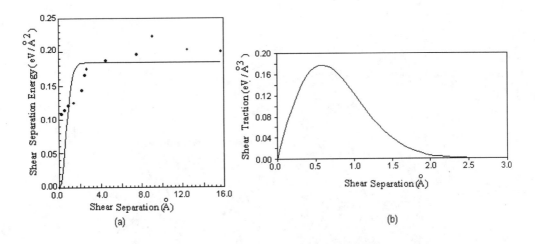

Figure 5 (a) Evolution of shear separation energy with separation distance. (b) Variation of shear traction with normal separation distance

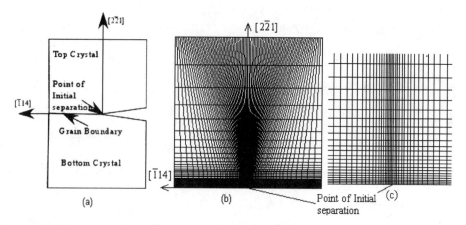

Figure 6 (a) The geometric configuration of $[110]\Sigma 9\,(2\bar{2}1)$ bicrystal with initial separation (b) Coarse mesh of top crystal. (c) Fine mesh near the point of initial grain separation.

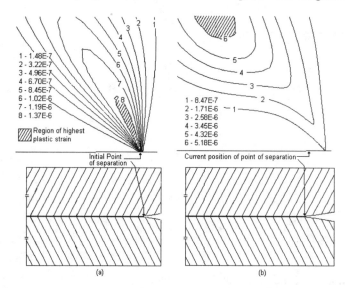

Figure 7 (a) Contours of accumulated slip of all the slip system just before separation of few elements near the point of initial separation. (b) Contours of accumulated slip on all the slip system just after separation of few elements near point of initial separation.

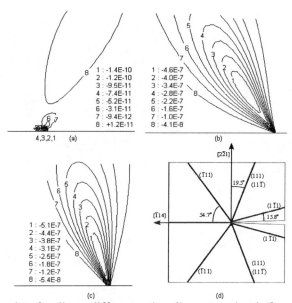

Figure 8. Contour plots for slip on different active slip systems just before separation of cohesive elements near the point of initial separation (a) (111)[$\bar{1}$10] or (11$\bar{1}$)[$\bar{1}$10] (b) ($\bar{1}$11)[101] or ($\bar{1}$11)[0$\bar{1}$1], (c) (1$\bar{1}$1)[011] or (1$\bar{1}$1)[10$\bar{1}$], (d) Orientation of different slip planes on (110) plane

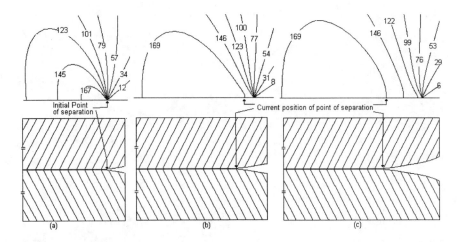

Figure 9 (c) Contours of stress σ_{22} around crack tip just before the decohsion of cohesive elements near the crack tip. (b) and (a) shows the movement of highest stress contour with the progress of decohesion and point of grain separation.

Evaluation of Superplastic Material Characteristics Using Multi-Dome Forming Test

A. El-Morsy, N. Akkus, K. Manabe and H. Nishimura

Graduate School of Engineering, Tokyo Metropolitan University,
1-1 Minami-Osawa, Hachioji-shi, Tokyo 192-0397, Japan

Keywords: Box Forming, Finite Element Method Simulation, Hemispherical Forming, Multi-Dome Forming, Superplasticity

Abstract

Superplastic characteristics of Ti-alloy "Ti-4.5Al-3V-2Fe-2Mo" have been evaluated by multi-domed single sheet forming instead of commonly used tensile test. The flow stress versus strain rate relationship was established from a circular sheet containing four different domes, with diameters of 20, 25, 30, and 35mm. Multi-dome forming experiments were conducted under constant pressure at a superplastic temperature of 800°C. Membrane theory was used to calculate the flow stress and the strain rate at the apex of each dome. Forming processes such as superplastic hemispherical and box forming were simulated by FEM using material characteristics obtained from multi-dome forming and tensile tests. Finally the results of estimated bulging time and final thickness distribution obtained from the simulation were compared to those of experiments. Material characteristics obtained from multi-dome forming experiments showed good agreements with the FEM simulation in the estimated bulging time, whereas the tensile tests had some discrepancies with regard to the estimated bulging time. Almost the same final thickness distribution were obtained in the simulations by using two material characteristics, multi-dome forming and tensile jump test, but the one from multi-dome forming was more close to real experiments.

1-Introduction

Determination of the material characteristics is very important for designing mechanical parts and simulating the new designs. Some difficulties are arisen during the superplastic deformations tests because of difficult testing conditions at the elevated temperature. The limited use of the measurement equipment at the high temperatures makes also the evaluation of the material characteristics less accurate. The design and simulation costs can also be reduced by accurately determined material characteristics. The superplastic characteristics of materials are commonly evaluated by the tensile test [1,2,3]. But the evaluation of the materials characteristics by the tensile test includes certain disadvantages, such as: tensile test is based on a uniaxial stress system; however, the biaxial stress system is effective in real forming processes. It is difficult to measure the actual strain rate and control the temperature distribution uniformly during the test, with long furnace. Recently, however, some researchers reported on the dome-forming method, in which the material characteristics were obtained under a biaxial stress system [4,5]. The dome-forming method offers some advantages over the tensile testing method, such as shorter testing times, less equipment and easy clamping of test pieces during the testing. Furthermore, the same equipment used in the actual superplastic forming can also be used to obtain material characteristics with small modifications. In this study we propose a new test, multi-dome free bulge forming test, to evaluate the superplastic material characteristics.

The aim of this study is an evaluation of superplastic material characteristics obtained through the multi-dome forming on Ti-alloy "SP-700". To confirm the validity of the proposed method, forming processes such as superplastic hemispherical and box forming were simulated by finite element method using the material characteristics obtained from multi-dome forming and tensile tests. The results of estimated bulging time and wall thickness distribution obtained from the simulation were compared to those of experiments.

2- Experimental Procedure

The experiments of multi-dome forming test were performed using the superplastic material α-β Ti-alloy Ti-4.5Al-3V-2Fe-2Mo, commercial name "SP700". The chemical composition of the material is given in **Table 1**. The geometry of the blanks used as specimens was of a 120 mm in diameter and an initial thickness of 0.72 mm.

Figure 1 shows the illustration of experimental set-up for multi-dome bulge-forming test. In the experiments each test piece was clamped between the upper and lower dies. An electric furnace was used to heat the test piece to 800°C. Superplastic temperature was maintained within ± 2°C by using a thermo controller. After waiting five minutes to ensure a uniform temperature distribution throughout the sheet, constant Ar gas pressure was applied to deform the sheet to various heights within the mold, which had four holes whose diameters of 20, 25, 30, and 35mm. **Figure 2** shows the metal mold with the four holes. **Table 2** shows the forming pressure and the forming time of the four experiments from which a strain rate-flow stress curve in the range from 2×10^{-5} to 2×10^{-3} s^{-1} is obtained. As a tensile test, a series of superplastic tensile jump test for Ti-alloy "SP-700" was conducted to examine the relationship between the flow stresses and the strain rates. The details of tensile jump test will not be mentioned here, since it is very commonly used to evaluate the superplastic material characteristics.

Table 1 Chemical composition (%) of Ti-alloy "SP-700".

Fe	Mo	Al	V	Ti
2.0	2.0	4.5	3.0	balance

Table 2 Experimental conditions of multi-dome forming test

Forming pressure / MPa	Bulging time / sec
0.7	300
0.4	600
0.2	600
0.15	1200

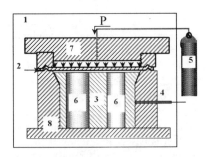

Fig. 1 Illustration of the experimental setup for the multi-dome forming test.

1 Heating furnace 2 Specimen
3 Die   4 Thermocouple
5 Argon gas tank   6 Holes
7 Blank holder   8 Die holder

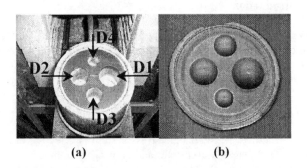

(a) (b)

Fig. 2 Photo of metal mold with the four holes and test piece after multi-dome forming test, respectively.

3 - Analytical Equations

A set of equations was formulated to evaluate the superplastic material characteristics of sheet metal, and to obtain the relationship between the stress and strain state. **Figure 3** shows the illustration of a deformed sheet. A constitutive equation, which defined the relationship between the flow stress and the strain rate, was employed in this study as:

$$\sigma = K \dot{\varepsilon}^m \quad (1)$$

Where σ is the flow stress, K is the strength coefficient, m is the strain rate sensitivity of the material, and $\dot{\varepsilon}$ is the strain rate.

The stresses at the apex of the dome can be calculated using the membrane theory, in which the thickness stress is usually very small and can be ignored ($\sigma_r = 0$). Considering that meridional stress, σ_m, and the circumferential stress, σ_c, are equal to each other at the dome apex, and they can be calculated by:

$$\sigma_c = \sigma_m = \frac{Pr}{2S_f} = \frac{P}{4S_f}\left(\frac{h^2 + R^2}{h}\right) \quad (2)$$

Where P is the forming pressure, r is the radius of curvature, R is the radius of die, h is the bulging height, and S_f is the final thickness at the apex of dome.

Considering the volume constancy during plastic deformation process ($\varepsilon_c + \varepsilon_m + \varepsilon_t = 0$), one can easily obtain the relationship of the strains at the apex of the dome in the three dimensions as:

$$\varepsilon_c = \varepsilon_m = -0.5\varepsilon_t \quad (3)$$

If the above equation is combined with von Mises effective strain equation, then the relationship between effective strain and thickness strain at the apex can be found from [4,6];

$$\varepsilon_e = |\varepsilon_s| = \left|\ln\frac{Sf}{Si}\right| \quad (4)$$

where S_i is the initial thickness of initial sheet. The strain rate sensitivity index m can be obtained from the following equation:

$$m = \frac{d\log\sigma}{d\log\dot{\varepsilon}} = \frac{\log(\sigma_2/\sigma_1)}{\log(\dot{\varepsilon}_2/\dot{\varepsilon}_1)} \quad (5)$$

In this study, the bulge-time is the same for all domes; thus, the strain ratio can be used instead of the strain rate. Hence, the strain rate sensitivity index m can be obtained from the following equation:

$$m = \frac{d\log\sigma}{d\log\varepsilon} = \frac{\log(\sigma_2/\sigma_1)}{\log(\varepsilon_2/\varepsilon_1)} \quad (6)$$

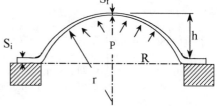

Fig. 3 Illustration of a deformed sheet

4- FEM simulation

Numerical analysis such as Finite Element Method is very effective method to simulate superplastic deformation, since superplasticity is a non-linear and large deformation problem. In the present study, hemispherical dome and box forming were simulated using a commercially available FEM code called MARC-Mentat, in order to compare the experimentally-obtained material characteristics from multi-dome forming test with those obtained from tensile test. The material characteristics m and K, which were obtained from the experiments of the multi-dome forming test and of the tensile test, were included in a special sub-routines, which defined the relationship between stresses and strain rates to characterise the superplasticity of the materials and were linked to FEM solver. The incremental approach was based on the rigid-plastic formulations and elastic strains were neglected in both modelings. The Finite Element mesh of the dome and box were generated using 4-node axi-symmetric and plane stress elements, respectively. The details of the FEM model can be seen in references [7,8]

5- Results and Discussion
5-1 Material characteristics by Multi-dome forming experiments

The flow stress and strain rate values at the apex of the dome are obtained from the well-known dome forming equations. **Figure 4** compares the flow stress versus the strain rate relationship of the superplastic tensile jump test and the multi-dome forming method. A considerable difference is seen between two curves, the one obtained from Multi-dome forming test provided a lower flow stress than the tensile test. Several reason can be thought for this flow stress difference between two tests such as: the tendency toward grain growth in the tensile tests, because of the high temperatures and long times dictated by the low strain rates. Such grain growth would increase the flow stress values in the tensile test pieces. The shape factors and the uniaxial stress system in tensile test and biaxial deformation system in dome forming process, may be, is the another reason for the different flow stress versus strain rate curves. This difference of the flow stress for a certain strain rate between two testings is yet to be investigated. Song et. al reported that higher m values obtained from single dome forming experiments [5]. **Figure 5** shows the relationship between strain rate and strain rate sensitivity index, m, obtained from multi-dome forming and jump tests in the present study. In both experiments, very close m values were obtained at the lower strain rates such as less than 2×10^{-4} s^{-1}. However, multi-dome forming tests revealed higher strain rate sensitivity index for the strain rates higher than 2×10^{-4} s^{-1}. The maximum m was 0.62 for jump test and 0.72 for multi-dome forming test.

5-2 Superplastic Forming Exercises and Comparison with FEM Simulations

Single dome and box forming experiments and their respective FEM simulations were performed to check the accuracy of the material characteristics obtained from both multi-dome and tensile tests. The strain rates, under which the experiments of dome forming and box forming were conducted, were used to obtain m value from **Fig. 5**.

Dome Forming

Figure 6 compares the bulging time of three experimentally deformed domes with the bulging time obtained from their FEM simulations. This figure reveals that there is a notable difference in the estimation of bulging time of the dome, when the m and K values obtained from the jump tests are used. The same simulations give considerable good estimation of the dome bulging time if the same parameters obtained from multi-dome forming tests are used.

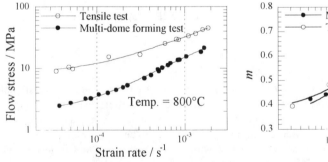

Fig. 4 Flow stress vs. strain rate relationship for the tensile test and the multi-dome forming test

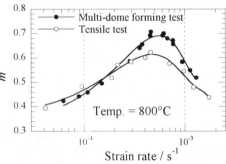

Fig. 5 Relation between the strain rate and strain rate sensitivity index m

Figure 7 compares the thickness distributions of experimentally deformed domes with that of obtained from FEM simulations. These figures indicate that there is no big difference in the estimated wall thickness distribution between experiments and simulations of both material characteristics especially for the smaller bulge high. But for larger deformations, simulation results, obtained from material characteristics of multi-dome forming tests, are in closer agreement with the experiments than the results of estimated wall thickness distribution from material characteristics of the jump tests.

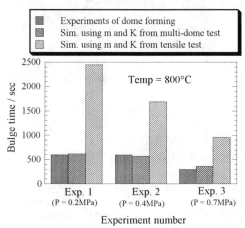

Fig. 6 Comparison of estimated bulging time with the experiments of domes

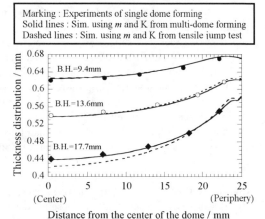

Fig. 7 Comparison of estimated thickness distribution with experiments at various bulging height

Box Forming

Bulge forming of box shapes and their FEM simulations were conducted in the same way as dome forming exercise. **Figures 8** and **9** compare the bulge forming time and the final thickness distribution of box, respectively. The same tendency, which is observed in the single dome forming simulations, were also observed in box forming simulations. As seen in **Fig. 8**, estimation of box bulging time with material characteristics from multi-dome forming test fits better with

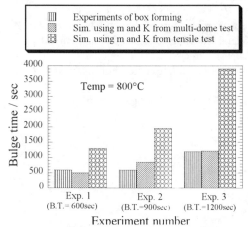

Fig. 8 Comparison of estimated bulging time with the actual experimental time.

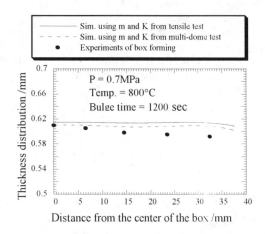

Fig. 9 FEM estimated wall thickness at cross-section A-A compared to that of experiments

experimental one. Almost the same final thickness distribution was obtained in the simulations by using two material characteristics (**Fig. 9**). Otherwise in **Fig. 10**, the one from multi-dome forming was more close to real experimental thickness distribution at the cross-section A-A of the box.

6- Concluding Remarks

Multi-dome forming test is proposed to evaluate the material characteristics of the superplastic sheet. The superplastic characteristics of Ti-alloy "Ti-4.5Al-3V-2Fe-2Mo" have been evaluated by multi-dome forming test and the results were compared with the more commonly used tensile step strain rate test. Some discrepancies with regard to superplastic flow stress and strain rate sensitivity index, m, have been observed between two tests. Multi-dome forming tests revealed lower superplastic flow stress than tensile step strain rate tests. The maximum m value of the Ti-alloy was found to be higher in multi- dome forming test than that of tensile step strain rate test.

Fig. 10 Photo of test piece after box forming.

Single dome and box forming experiments and their FEM simulations were performed to examine the accuracy of the material characteristics obtained from multi-dome and tensile test. Results of the FEM simulations revealed that material characteristics obtained from multi-dome forming test are more suitable to simulate the deformation process especially for the estimation of the forming time. Further studies are needed to understand the shape effects of the test pieces and the effect of the stress system on superplastic material characteristics.

References

1- Y. Takayama, N. Furushiro, T. Tozawa, and H. Kato, Materials Science Forum vol. 170-172 (1994), pp. 561-570.
2- A. Wisbey, B. Geary, D.P. Davies, and C.M. Ward-Close, Materials Science Forum vol. 170-172 (1994), pp. 293-298.
3- H. Iwasaki, K. Higashi, S. Tanimura, T. Komatubara and S. Hayami, Superplasicity in Advanced Materials, (1991), pp. 447-452.
4- F. U. Enikeev and A. A. Kruglov, Int. J. Mech. Sci. Vol. 37, No. 5 (1995), pp. 473-483.
5- Song Yu-Quan and Z. C. Wang, Materials Science and technology, Vol. 9 (1993), pp. 57-60.
6- William F. Hosford and Robert M. Caddell, " Metal Forming : Mechanics and Metallurgy", Prentice-Hall, Inc., Englewood Cliffs, N.J. 07632, 1983, pp. 64.
7- N. Akkus, K. Manabe, M. Kawahara, H. Nishimura, Materials Science Forum vols. 243-245 (1997), pp. 729-734.
8- T. Usugi, N. Akkus, M. Kawahara, H. Nishimura, Materials Science Forum vols. 304-306 (1999), pp. 735-740.

Corresponding author: nihat@mech.metro-u.ac.jp

Superplastic Deformation Mechanism of ZrO$_2$ and Al$_2$O$_3$ with Additives Studied by Electron Microscopy

A. Kumao[1] and Y. Okamoto[2]

[1] Department of Electronics and Information Science, Kyoto Institute of Technology, Matsugasaki, Sakyo-ku, Kyoto 606-8585, Japan

[2] Department of Chemistry and Materials Technology, Kyoto Institute of Technology, Matsugasaki, Sakyo-ku, Kyoto 606-8585, Japan

Keywords: Al$_2$O$_3$, Doping Effect, Dynamic Grain Growth, Electron Microscopy, Grain Boundary Sliding, ZrO$_2$

Abstract

The compressive deformation mechanism of several fine-grained ceramics with additives was investigated using transmission electron microscopy and X-ray spectroscopy. In Y$_2$O$_3$-doped ZrO$_2$, the superplastic deformation is due to the grain boundary sliding although no amorphous phase was observed at grain boundaries. Pure Al$_2$O$_3$ was hardly deformed, however, large strain was obtained in ZrO$_2$-doped Al$_2$O$_3$ ceramics, in which very thin amorphous layers containing Zr atoms existed between Al$_2$O$_3$ grains. Deformation of this material was found to be caused via both grain boundary sliding and dynamic grain growth. In MgO-doped Al$_2$O$_3$, the deformation mechanism was analogous to that of ZrO$_2$-doped Al$_2$O$_3$ except no amorphous phase existed at the boundaries.

1. Introduction

It is generally known that common ceramics are brittle and easily destroyed by applying a large stress. However, it was found that by adding a small amount of metal oxides to fine-grained zirconia and alumina, strength and brittleness of these ceramics were improved and a large strain could be realized [1-4]. For such ceramics, measurement of the stress-strain curve [5] and research of the deformation mechanism [6-8] were performed. Especially, the investigation of the grain boundary structure is important to elucidate the deformation behavior of the ceramics composed small-sized grains. Recently, the development of new superplastic ceramics and the characterization of ceramics prepared with high strain rate are actively carried out together with the applied research of superplastic materials [9-11]. In this paper, zirconia and alumina with the additives were deformed by compressive stress, and the structure before and after deformation was observed using an electron microscope. Based on the structure change, deformation mechanism is discussed.

2. Experimental Procedure

The samples treated here, zirconia containing 3 mol% Y$_2$O$_3$ [Y-TZP], aluminas containing 2500 ppm ZrO$_2$ [AZ] and 10 vol.% ZrO$_2$ [AZ10], and alumina with an additive of 200 ppm MgO [AM], were prepared as follows: each fine powder as mother material and each additive were wet-blended at the fixed proportion. After drying, they were molded under a high pressure and sintered at 1300 °C in air. Each sample was formed into a cylindrical shape of 4 mmϕ × 10 mm. These specimens were called 'original samples'. The second samples were made at 1300 °C by applying a compressive stress to the original sample under a constant strain rate. In order to observe the samples by means of a transmission electron microscope (TEM), both the original and compressed samples were cut to about 0.5 mm thickness and thinned by mechanical polishing. Finally, the samples were etched using an Ar ion beam. To clarify the role of the additives on the plastic deformation, analysis of the local composition was carried out with a JEM-2010 instrument equipped energy dispersive X-ray spectroscopy (EDS).

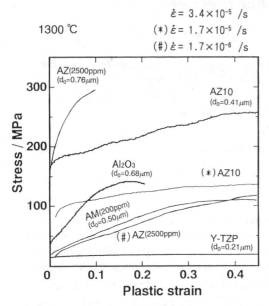

Fig.1. Stress-strain curves for several kinds of fine-grained ceramics. $\dot{\varepsilon}$ means the strain rate. Deformation behavior depends on the strain rate.

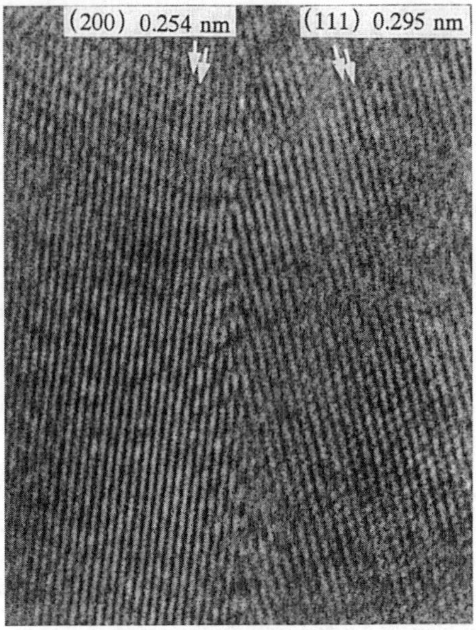

Fig.2. High resolution TEM image of the grain boundary of Y-TZP ceramics. There is no amorphous phase at the boundary and two grains stick closely.

3. Results and Discussion

Fig.1 shows the stress-strain curves for various fine-grained ceramics obtained through the present experiment. It is understood that Y-TZP shows the superplasticity because strain becomes gradually larger under a constant stress of about 20 MPa and a constant strain rate of 3.4×10^{-5} /s at 1300 °C. Pure alumina and alumina containing 2500 ppm ZrO_2 [AZ] were hardly deformed at strain rate of 3.4×10^{-5} /s. However, whether large strain is obtained or not depends on the strain rate. For example, in the case of AZ large strain was obtained without fracture when the strain rate was degreased to 1/20 that of 3.4×10^{-5} /s. Similar phenomenon was observed on AZ10, which meant the alumina containing 10 vol.% ZrO_2. AZ10 deformed at strain rate of 1.7×10^{-5} /s seems to show the superplastic deformation behavior.

(1) ZrO_2 containing 3 mol% Y_2O_3 [Y-TZP, strain 0.53]

As the result of TEM observation, it was found that the average grain sizes of the original sample and the sample after deformation were 0.20 μm and 0.26 μm, respectively [8]. However, no crystallographic structure change occurred during deformation, although a strain remained in the interior of the grains. This fact indicates that stress was relaxed at the grain boundary owing to grain boundary sliding. It was also found that there was no amorphous layer at grain boundaries and that each grain stuck closely as shown in Fig.2. High resolution electron microscope images of the grain boundaries were analyzed on an atomic scale and it was proven that grain boundaries were commonly formed with low index crystal planes such as {111}, {110} and {310}. The results obtained from the EDS analysis show that yttrium atoms added to zirconia distribute uniformly in the material. Y_2O_3 has no effect on superplastic deformation but contributes to the reinforcement of this material. The superplastic deformation of this ceramics is due to the grain boundary sliding without an amorphous boundary layer.

Fig.3. Electron microscope images of pure alumina before(a) and after(b) deformation, showing many cavities in (a) and a great many dislocations in (b). In the original sample of (a), no strain remained in the interior of the grains.

(2) Pure alumina   [strain 0.20]

In this material, large strain was not obtained as shown in Fig.1. Fig.3(a) shows the image of the original sample. It is seen that this material is composed of small and perfect single crystals of about $0.7 \mu m \phi$ and that there are cavities at many grain triple junctions. Fig.3(b) shows the image of the alumina after deformation. It is found that the grains grew dynamically to approximately 3 times that of the original sample during deformation and a great many dislocations arose. This means that grain boundary sliding did not happen and strain hardening occurred.

(3) Al_2O_3 containing 2500 ppm ZrO_2 [AZ, strain 0.50]

Fig.4(a) shows the electron microscope image of the original sample, in which very small particles having black contrast were all over observed among large and white particles. These black and white particles were found to be zirconia and alumina, respectively, by means of EDS analysis. This shows a reasonable result because zirconia and alumina do not dissolve into each other. High resolution electron microscope images revealed that there were amorphous layers of about 1 ~ 1.5 nm width at the grain boundaries between alumina particles as shown in Fig.4(b), where lattice fringes of alumina could be recognized in both grains. Figs.4(c) and 4(d) are EDS spectra obtained from the interior of a grain and grain boundary, respectively, in which the areas of diameter of about 2 nm were analyzed. As might be expected, Zr atoms were not detected in the grain interiors, while they exsisted in the boundary layer. In the deformed sample as shown in Fig.5, the averaged particle size of alumina became about double that of the original sample and the contrast depending on the strain was not observed. Therefore, it is reasonable to consider that the deformation of this material was due to both dynamic grain growth and grain boundary sliding with the help of amorphous layer including Zr atoms. This fact is quite different from the deformation of pure alumina, in which grain boundary sliding did not occur and a great many dislocations originated in the interior of the grains.

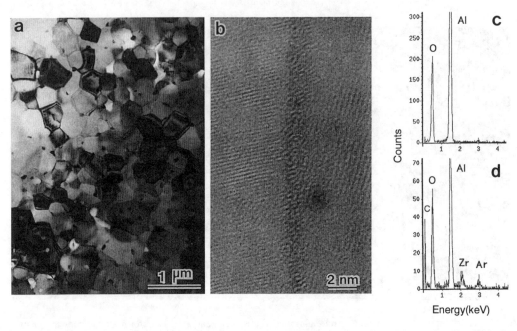

Fig.4. (a) Image of fine-grained Al_2O_3 ceramics containing 2500 ppm ZrO_2 before deformation. Small black particles are ZrO_2. (b) High magnified image of the boundary between Al_2O_3 particles. The layer having black contrast consists of amorphous phase about 1-1.5 nm thick, where Zr atoms are included as shown in Fig.4(d). (c) and (d) show EDS spectra obtained from the interior of a grain and grain boundary, respectively, in which the areas of diameter of about 2 nm was analyzed. The spectrum of Ar is caused by Ar ion beam used for TEM specimen preparation.

Fig.5. Image of alumina containing 2500 ppm ZrO_2 after deformation. Grain growth is recognized.

Fig.6. Image of alumina containing 10 vol.% ZrO_2 before deformation. White and black particles correspond to alumina and zirconia, respectively.

(4) Al_2O_3 containing 10 vol.% ZrO_2 [AZ10, strain 0.49]

Fig.6 shows the electron microscope image of AZ10 before deformation. White particles are alumina, while black particles having complex contrast are zirconia, the size of which is apparently larger in comparison to black particles observed in the sample AZ, which is shown in Fig.4(a). In this case, the sample AZ10 may be considered to be Al_2O_3-ZrO_2 composite. Therefore, large amounts of zirconia are unnecessary to cause superplastic deformation of fine-grained alumina. The reason is that only a very small part of the ZrO_2 added to alumina forms a very thin amorphous layer at the boundaries between alumina particles and facilitates grain boundary sliding. The rest of the ZrO_2 only forms zirconia particles and does not contribute to the plastic deformation. When this sample was deformed under the constant strain rate of 1.7×10^{-5} /s (Fig.1), dynamic grain growth of alumina particles was observed. The deformation mechanism of this material is due to both grain growth and grain boundary sliding.

(5) Al_2O_3 containing 200 ppm MgO   [AM, strain 0.60]

An electron microscope image of the original sample is shown in Fig.7(a), in which cavities are not observed. This is different from the sample of pure alumina. Fig.7(b) shows the high magnified image near the triple junction of this sample. Since the amount of MgO was very small and it is soluble in alumina, Mg atoms were not detected by means of EDS at the grain boundaries(Fig.7(c)) as well as in the interior of the grains. However, since Mg atoms facilitate the rearrangement of Al atoms at the grain boundary and in the crystal lattice, strain originated by the compressive stress was relaxed not only at the grain boundary but also in the interior of the grains. Therefore, Mg atoms seem to help the plastic deformation. After deformation, it was proven that the grain size doubled and there was no strain within the grains. The deformation of this material was found to be due to both dynamic grain growth and grain boundary sliding, although amorphous layers did not exist at the grain boundary.

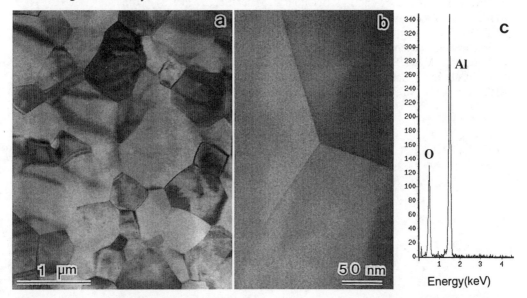

Fig.7. Low(a) and high(b) magnified images of alumina containing 200 ppm MgO before deformation. (c) EDS spectrum at the grain boundary, showing no Mg atoms, because the amount of Mg atoms is very small and they did not segregate at the boundary.

4. Conclusion

The deformation mechanism of zirconia and alumina with additives, which were composed of fine grains with sharp corners and relatively straight edges, was examined using transmission electron microscopy and energy dispersive X-ray spectroscopy. The superplastic deformation of Y-TZP is due to grain boundary sliding, where there was no amorphous layer between zirconia particles. In pure alumina, deformation hardly occurred because grain boundary sliding did not occur and strain hardening originated in the deformed sample. The deformation of aluminas (AZ and AZ10) containing a small amount of zirconia is due to both grain boundary sliding and dynamic grain growth of the alumina grains. Especially, amorphous layers of about 1 ~ 1.5 nm width between alumina particles contribute to grain boundary sliding. In order to get large strain for these ceramics, however, it was necessary to make the strain rate as small as possible. In the case of AM, the deformation mechanism is almost the same as the case of AZ except no amorphous layer existed at the grain boundaries. Thus, in order to generate the superplastic deformation of ceramics composed of small-sized grains, grain boundary sliding is always necessary because the strain in the interior of the grains has to be relaxed.

Acknowledgement

The authors would like to thank Mr. K. Nishio for TEM specimen preparation. This research was supported in part by a Grant-In-Aid for Scientific Research from the Ministry of Education, Science and Culture, Japan.

References

[1] R.C. Garvie, R.H. Hannink and P.T. Pascoe, Nature, **258** (1975), p.703.
[2] T.K. Gupta, J.H. Bechtold, R.C. Kuznicki, L.H. Cadoff and B.R. Rossing, J. Mater. Sci., **12** (1977), p.2421.
[3] T.K. Gupta, F.F. Lange and J.H. Bechtold, J. Mater. Sci., **13** (1978), p.1464.
[4] M. L. Mecartney, W. T. Donlon and A. H. Heuer, J. Mater. Sci., **15** (1980), p.1063.
[5] Y. Okamoto, J. Ieuji, Y. Yamada, K. Hayashi and T. Nishikawa, Advances in Ceramics, Vol.24: Science and Technology of Zirconia 3, (1988), p.565.
[6] T. Sakuma, Mater. Sci. Forum, **233-234** (1997), p.321.
[7] K. Okada and T. Sakuma, J. Am. Ceram. Soc., **79** (1996), p.499.
[8] A. Kumao, N. Nakamura, H. Endoh, Y. Okamoto and M. Suzuki, Mater. Sci. Forum, **304-306** (1999), p.567.
[9] Y. Ikuhara, P. Thavorniti and T. Sakuma, Acta Mater., **45** (1997), p.5257.
[10] G.Sanders, Mater. Sci. Forum, **304-306** (1999), p.805.
[11] Y. Motohashi, T. Sakuma and C.C. Chou, Int. Conf. on Thermomechanic Processing of Steels and Other Materials, Wollongong, Australia, (1997), p.90.

Temperature Effects on the Localization and Mode of Failure of Al 5083

John D. Watts, Xianglei Chen, Anthony Belvin, Z. Chen and N. Chandra

Department of Mechanical Engineering, FAMU-FSU University,
2525 Pottsdammer Street, Tallahassee FL 32310, USA

Keywords: Al Alloy 5083, Cavity, Localization, Mode of Failure, Superplasticity, Temperature Effect

Introduction

Superplastic forming has been accepted as an economical means to produce highly complex, lightweight parts with a single forming operation, when the number of parts is at an optimal level. Where quality and performance are the primary criteria, as in aerospace industry, superplastic forming (SPF) has been widely used. However, where the cost is the predominant factor as in automotive industries the use is rather slow until recently with the advent of inexpensive superplastic materials and processing techniques [1]. One such inexpensive superplastic material is Aluminum Alloy 5083. Because this material's lightweight, moderate strength, good corrosion resistance and weldability, it has been the focus of many studies investigating its properties and applications. [2,3,4,5].

A predominant factor in determining the cost of the SP process is the temperature of formation; the higher the temperature required, the higher the cost becomes (due to lower tool life and the requirement to sometimes use an entirely new tool material). Hence the industry is always seeking to form components using SPF at the lowest possible temperature for any given alloy (and possibly a specific part). Hence the temperature dependence of superplastic property is an important technological issue. Yet another viewpoint of the same problem arises from a purely mechanics consideration. An unanswered question is "At what specific temperature, or a range of temperatures, does the material begin to behave more like a fluid with strain-rate dependent behavior than that of a low temperature solid with a strain-rate independent behavior (usually with a strong strain hardening tendency)?" Such a transition will manifest itself at the macroscopic scale as increased flow stress or tendency to spring back, or at the microscopic scale as a cavitation leading to lowered post-SPF properties.

The purpose of the paper thus is to investigate this scientific and technological issue of transition from plastic to superplastic behavior in materials from mechanics and materials perspectives. Temperature can play an increasingly important role in the deformation, elongation and cavity growth within Al5083, especially at high homologous temperatures. A transition in behavior may occur induced by different deformation mechanisms and evidenceded as different modes of failure. In this research work, simple tension tests are conducted at various temperatures (for a fixed strain-rate) to investigate the transitional effect of temperature on the typical plastic and superplastic properties of Al 5083. The effects of temperature on stress level, mode of failure and cavitation density are studied.

Experiment

Aluminum 5083 alloy sheet used in this study was provided by Sky Aluminum Company and received in the form of a cold rolled sheet with a thickness of 1.5 mm (.0805 in). The chemical compositions are given in Table 1. Complete static recrystallization can be achieved by heating the sheet up to 555°C for 40 minutes to produce an equiaxed grain structure with an average grain size of 17 µm.

Table 1: Chemical composition of Al 5083 alloy (wt%)

Mg	Mn	Cr	Fe	Si	Ti	Al
4.70	0.65	0.13	0.04	0.04	0.03	balance

In order to eliminate the effect of strain rate dependence, a constant true strain rate of 0.02% s^{-1} was used for the entire range of temperatures. It should be noted that to maintain constant true strain-rate, the crosshead velocity has to be continuously increased, which is approximated in our procedure by means of nine different discrete steps. Care should be exercised in comparing our results with that of other published data where a constant velocity (with an initial specified strain rate but subsequently decreased value with deformation) tests are typically used but not stated explicitly.

The temperatures at which the tension tests were conducted are listed in Table 2. Since we are interested in the transition from general plastic deformation to superplastic deformation, more tests were conducted near the superplastic regime where the transition in behavior is more dramatic. The gauge length of the specimen was 6.33mm (0.25in) and the gage width was 7.3mm (0.2 in) with the tensile axis parallel to the rolling direction. An MTS tensile testing machine equipped with an Oxygon vacuum furnace was used to perform all uniaxial tensile tests. All tests were repeated three times.

Table 2: Temperature Range and * Classification

Plastic	Transitional	Superplastic
25	400	510
200	450	520
300	475	530
350	500	540
		550
		560
		570

*- In retrospect, the temperatures selected fell into three distinct categories (plastic, superplastic, and transition). That distinction is maintained throughout this paper

Image analysis was performed with an optical microscope to observe cavitation and failure modes. When measuring cavity volume fraction, the built-in camera first scans the surface to produce a black and white image. Then, using ImagePro ™, cavities can be observed as the dark areas in the digitized picture. The cavity area (number of dark pixels in the observation field) is then divided by total area of the observation field. This will give us the area fraction of cavities for a specific field. If an equal cavity area fraction is assumed in the cross section, then this can be interpreted to cavity volume fraction. A digital camera in conjunction with the image analysis software was used to study the localization/diffusional necking behavior.

Results and Discussion

A. Tensile properties:

The flow stress from the full range of temperatures was plotted. Figure 1 (a) shows the stress –strain behavior of the tested material from room temperature to 475°C, while Figure 1 (b) shows the behavior from 475°C to 560°C at a fixed true strain rate of $2 \times 10^{-4} s^{-1}$. The data was curve-fit to a sixth order polynomial and is shown in the figure below. Fig. 1(c) shows the variation of normalized Young's Modulus as a function of

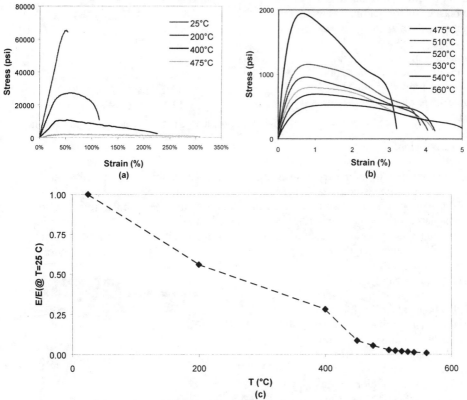

Figure 1: (a) Stress/Strain for the plastic and transitional temperature ranges.
(b) Stress/Strain for the Superplastic temperature range.
(c) Young's Modulus as a Function of Temperature, Normalized with E (T=25°C))

test E_{RT} is the modulus at room temperature and E is the modulus at the test temperature. From room temperature to 400°C from below, Young's Modulus decreases almost linearly. In the transition region (From about 400°C to 500°C), Young's Modulus decreases drastically down to about $1/20^{th}$ of the original stiffness at 25°C. Within the typical superplastic region, Youngs Modulus decreases only slightly as temperature increases.

Elongation data was obtained after performing three tests at each temperature. Figure 2 shows the total elongation obtained in the specimen as a function of temperature as well as the deviations in measurements with the error bars. As the temperature increases from room temperature (25°C) to 400°C, the increase of elongation is not significant when compared with the change in temperature. Between 400°C and 500°C, there is a sharp increase in elongation. Since this region is the transition region between common plastic deformation and superplastic deformation, new mechanisms begin to play an increasingly important role in deformation of Al5083. It seems that in this transitional region, mechanical properties of Al5083 become increasingly temperature sensitive.

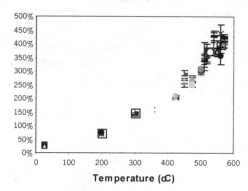

Figure 2: Elongation to failure as a function of temperature

B. Cavitation

For many superplastic materials such as 5083 alloy, cavitation is an important strain limiting process. Many studies have focused on the cavity initiation and cavity growth. In this work we focus on the origin and type of failure (ductile shear, tensile overload or cavity coalescence) by examining the spatial distribution of cavities away from the fractured end. Cavity volume fraction was measured in the failed specimens at different locations along gauge length in the plane parallel to tensile axis (see Figure 3). Figure 3 illustrates the method used to measure the cavity volume fraction. While a higher magnification is useful in capturing small cavities, we may not be able to inquire the entire region. On the other hand, smaller magnification may miss cavities but cover the entire region. In this work a magnification of 100X was used, and the areas were acquired in succession. Hence the volume fraction (based on the cavity area) is accurate with the caveat that small cavities that could not be detected at 100X magnification have a negligible contribution to the volume

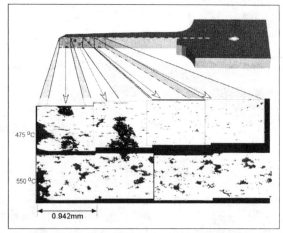

Figure 3: Using Optical microscopy and image analyis, regions of the tensile specimens were analyzed frame by frame to obtain measures of cavitation

fraction and are ignored.

As shown in Fig.4, the distribution of cavities along the gauge length varies with temperatures. When temperature is lower than 500 °C, much less cavity was seen away from the fracture surface. Between 500 °C and 540 °C, the cavity increases significantly. Besides, the difference of cavity volume fraction becomes less, which means the deformation becomes more uniform when temperature increases. It can be seen from Figure 4 that cavity volume fraction decreases when the temperature past 550 °C.

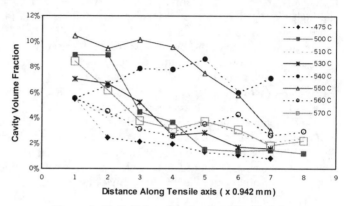

Figure 4: Cavity Volume Fraction Along Tensile Axis

It is to be noted that this study is not focused on the origin of cavities and factors influencing cavity growth. The present analysis relies on the relationship between the spatial distribution of cavities and the eventual mode of failure. Previous study showed that the cavity volume fraction is a function of strain rate and strain at a specific temperature [6]. Besides, cavity growth can also be affected by stress state [6,7]. Effect of temperature on different strain rate and stress state should also be considered when the constitutive equation of cavity growth is formulated.

C. Mode of Failure

Fig. 5 shows a picture of the samples tested at various temperatures. At room temperature, the fracture propagates through the material at a 45° angle relative to the axis of tension and the failure of Al 5083 is observed as a classical, shear induced ductile failure. The region of failure is very localized and necking occurs within 1 or 2 millimeters of the failure surface. As the temperature is raised from 25°C, the strain to failure increases steadily (Fig. 2). The orientation of the failure surface changes with temperature, becoming more normal to the tensile axis. The final failure during this range is generally due to tensile overload at a section by reduced area of cross section. The trend continues up to a temperature of about 400 °C. Both elastic modulus and yield strength show a linear decrease up to this temperature. Profile of section width of uniaxial specimens after deformation in plastic and superplastic ranges Al5083 is shown in Fig. 6. Above 400 °C, the more gradual necking shows that an increase in temperature will most definitely be accompanied by a decrease in the localization of damage leading to failure. Though the strength properties (Young's modulus and yield strength), failure behavior (total elongation, reduction in area), and cavitation behavior (see Figures 3 and 4, cavity volume fraction along tensile axis) all show the same trend, the rate of change with respect to temperature changes from that of the lower temperature tests. It can be seen from Fig. 2 that the region where sharpest change takes place is in transitional region (around 400°C to 500°C), where superplastic deformation starts to play a major role. In superplastic deformation, localized necking is prevented (or prolonged) due to high levels of strain-rate sensitivity factor m. Because of this, the necking is less severe and higher elongation with corresponding uniformity is found [4]. Therefore, when the temperature is increased to above 500 °C (up to 540°C), necking becomes significantly less apparent. Instead of just one obvious crack (break) occurring at the region of failure (as found in the lower temperatures), more cracks are formed further away from the failed area, and sometimes more than one break surface are observed. At elevated temperatures (above 500 °C), the entire fractured surface is serrated, probably due to the linkage of cavities (randomly oriented) and eventual failure.

The optical microscope results show that both number and size of the cavities tend to increase significantly along the tensile axis of the specimen with temperature (see Fig. 3). When the temperature is at, or below 475 °C, cavitation is highest near the failure surface and decreases immediately thereafter. The deformation is less uniform and elongation is mainly controlled by typical necking process. However, when temperature is as high as 550 °C, the deformation is more uniform and necking is not very obvious (see Fig. 6). However, the cavities tend to grow faster and interlink, causing the specimen to fail. This explains the relatively lower elongation of 5083 at 550 °C than 540°C. Therefore, when temperature is above 540°C, strain to failure is considerably affected by cavity growth, interlinkage and coalescence.

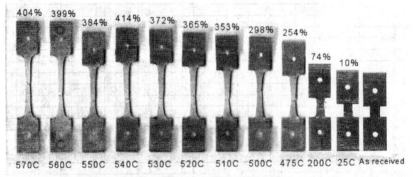

Figure 5: Uniaxial tensile specimens tested at temperatures ranging from room temperature (25C) through the plastic and superplastic ranges up to 570C.

Conclusions

Temperature is shown to have pronounced effects on the deformation behavior and modes of failure on Al5083. Based on the mechanical behavior (properties such as E and total elongation), Fracture behavior (type and mode of failure), and the microstructural characteristics (cavity volume distribution), three different temperature regimes have been identified. Clearly, superplastic behavior is observed above 500 °C and plastic behavior from room temperature up to about 400 °C with a transitional region between 400 °C and 500 °C. In the plastic regime (RT to400 °C) the failure is mainly controlled by typical plastic localization. The actual transition from plastic to superplastic behavior begins at about 400 °C. With increased temperature, the deformation becomes less localized and more cavities with larger size are generated before failure. In the superplastic range, the total elongation is controlled by cavity linkage. The distribution of cavities varies with temperature and is most uniform at 550 °C, although maximum elongation occurs just below at 540 °C.

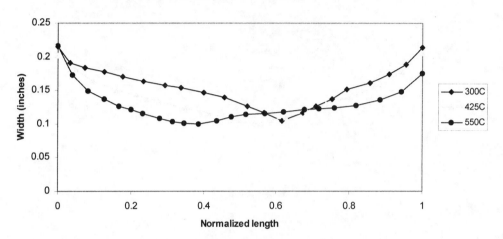

Figure 6: Profile of section width of uniaxial specimens after deformation in plastic and Superplastic ranges

References

1. N. Chandra, S,C, Rama and Z. Chen, "Critical Issues in the Industrial Applications of SPF-Process Modeling and Design Issues", Materials Transactions, JIM, Vol 40, (1999), pp 700-713
2. K.A. Padmanabhan, G.J. Davis, Materials Research and Engineering, Vol.2.(1980), p9-25.
3. E.P.Lautenschlager, J.O. Brittain, Review of Scientific Instruments, Vol.39.(1968), p1563-1565.
4. Dingqiang Li, Dongliang Lin, Yi Liu, Materials Science and Engineering A, Vol.249. (1998), p206-216.
5. Jingsu He, Yanwen Wang, The Superplasticity of Metals, 1986, p6-8.
6. K. Kannan, C.H. Johnson and C. H. Hamilton, Metallurgical and Materials Transactions, Volume 29A, April 1998, p1211-1220
7. Z. Chen, "Experimental and Numerical Analysis of Superplastic Damage in Aluminum Alloys", Dissertation, Florida State University

Superplastic Diffusion Bonding of a Ti-24Al-14Nb-3V-0.5Mo Intermetallic Alloy

Hanliang Zhu, Zhiqiang Li and Chunxiao Wang

Beijing Aeronautical Manufacturing Technology Research Center, Dept 106,
POB 863, Beijing 100024, China P.R.

Keywords: Apparent Activation Energy, Diffusion Bonding, Microstructure, Superplasticity, Ti-24Al-14Nb-3V-0.5Mo Intermetallic Alloy

Abstract

Superplastic diffusion bonding behaviors of a Ti-24Al-14Nb-3V-0.5Mo intermetallic alloy were studied at temperatures ranging from 940 to 1020°C. The investigated alloy exhibits duplex ($\alpha_2 + \beta$) microstructure with an average grain size smaller than 2 μm. The values of the apparent activation energy in the superplastic regime were determined and the values were close to that of the apparent activation energy of the β phase, which indicates that most of the plastic deformation was concentrated in the soft β phase. The strain exponent and the dimensionless constant of the power law creep equation were also determined to the investigated alloy. Based on the studies, the theoretical calculations of superplastic diffusion bonding were carried out. The effects of the diffusion bonding parameters, such as surface finishes of primary sheets, diffusion bonding temperature and pressure, on the predicted bonding time were studied. The experiments of isostatic diffusion bonding were carried out and the calculated results were in agreement well with the experimental results. According to the theoretical calculations and the experimental results, the optimum diffusion bonding process of the investigated alloy was determined. The microstructural transformations during superplastic diffusion bonding process and bonding mechanism were discussed too.

1. Introduction

Titanium aluminides are promising candidates for application in advanced aerospace programs because of their attractive high temperature strength, low density, good creep and oxidation resistance[1]. Since the formability of these materials is very limited, superplastic forming (SPF) appear to be a promising route for fabricating structure parts of aeroengines [2]. The superplastic behavior of titanium aluminides have been demonstrated in the past few years[3-8].
Diffusion bonding (DB) is a solid state joining process in which two clean metallic surfaces are brought into contact at elevated temperature under a low pressure. By selectively bonding only specific areas of two or more thin sheets of alloy and then internally expanding the resulting sandwich into a mold, using a low gas pressure, components with a predefined external shape and an internal cellular structure can be fabricated. Therefore, the DB/SPF process is a suitable technology for fabricating parts with complex shapes. The study on DB process of titanium aluminides just begins [9,10].
The aim of this study was to determine the superplastic deformation parameters of a Ti-24Al-14Nb-3V-0.5Mo intermetallic alloy. Based on the study, the theoretical calculation and experimental work were carried out to determine the optimum process of superplastic diffusion bonding and the microstructure transformations during the process were discussed too.

2 Material and Superplastic Deformation Parameters

The titanium aluminides used in this study were manufactured by Central Iron and Steel Research Institute in the form of rolled sheet of 1.5mm thickness. The alloy has atomic percent compositions Ti-24Al-14Nb-3V-0.5Mo, and the oxygen and nitrogen contents are 560ppm and 150ppm respectively, the hydrogen content is 31ug/g.

The three-dimensional microstructure of the as-received material is shown in Fig.1. The material has an $\alpha_2 + \beta$ microstructure with a grain size of about 2 μm. Fine spheroidal α_2 phase particles(light) are dispersed within the β matrix. The relative volume fraction of α_2 phase is about 55%.

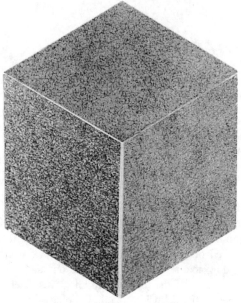

Fig.1 The 3-D microstructure of the as-received material

The investigated alloy exhibits better superplasticity at temperatures from 960℃ to 980℃. At temperature of 960℃ and strain rate of range from 3.5×10^{-4} to 4×10^{-3} s^{-1}, the m value is greater than 0.5. A maximum m value is obtained to be 0.76 at 960℃ and at a strain rate of 1.5×10^{-3} s^{-1}. The relations between the flow stress and the strain rates at three temperatures were studied and the apparent activation energy can be calculated by using the usual power law creep relationship[1]

$$\dot{\varepsilon}_s = A\sigma^n \exp(-\frac{Q}{RT}) \tag{1}$$

Where $\dot{\varepsilon}_s$ is the steady state strain rate, A is a dimensionless constant, n is the stress exponent, Q is the apparent activation energy, R and T have their usual meaning. The average value of the apparent activation energies for three stress level, i.e. 10, 30 and 60 MPa, shown in Fig.2, by a least square curve fitting technique, was 162.5 kJ/mol. This value is close to that of 180±30kJ/mol, obtained in the steady-state flow stress which is reached after a strain of about 0.6 in the temperature range from 930℃ to 980℃ for Ti-25Al-10Nb-3Mo-1V by D.Jobart[2]. Because the grain size of the investigated alloy is much finer, the alloy exposes long-term steady-state flow stress. That the apparent activation energy of the investigated alloy is also close to that of lattice diffusion in β-Ti of 150kJ/mol[11], indicates that most of the plastic deformation is concentrated in the soft β phase. To determine the values of the dimensionless constant (A) and the stress exponent (n) of the investigated alloy, the stress-strain rate data in temperature-compensated form is shown in Fig.3. It can be obtained that the values of A and n are 2818 and 2.39 respectively.

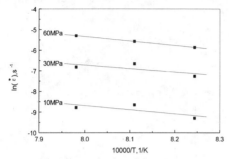

Fig.2 Determination of the apparent activation energy of the investigated alloy

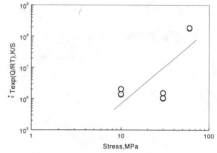

Fig.3 Stress-strain rate data in temperature-compensated form for the investigated alloy

3 Diffusion Bonding Behavior

The theoretical model describing the filling of interfacial voids under isostatic loading has been developed by Pilling [10,12]. The voids are filled by two process: time dependent plastic collapse of the cylinder wall and interfacial transfer of mass to the void by diffusion. Bonding is complete when either the void cross-section or void height becomes zero. The model, which is described in references 10 and 12 in detail, is used to study the effects of various parameters on the bonding time of the investigated alloy by using the superplastic deformation parameters obtained above and the calculation results are discussed to obtain the optimum diffusion bonding parameters.

Under pressure, superplastic flow takes place on the two sheets, which will be bonded. The superplastic deformation is useful in breaking the surface oxide film of the two sheets and leading the surfaces to contact. The calculation relationship between pressure and critical bonding time together with experimental measurements is shown in Fig.4. It is evident that the critical bonding time decreases at a parabolic rate with decreasing the pressure. When the applied pressure is 4.5 MPa, which is gas pressure usually used for SPF process of Ti_3Al intermetallic alloy, the bonding time is 3.99h.

The diffusion bonding temperature is the most important process parameter. The variations of critical bonding time with temperature under pressure of 4.5MPa and 7MPa together with experimental results are shown in Fig.5. With increasing the bonding temperature, the critical bonding time decreases rapidly. The reason is that diffusion and superplastic deformation may be accelerated with the thin layer along the bonding interface by raising the temperature. However, the grain size and α_2 phase volume fraction will increase with temperature, which leads to poor SPF behavior and poor post-SPF mechanical properties. The bonding temperature of superplastic alloys is often selected to be within the temperature region giving optimum superplasticity. Therefore, the bonding temperature of 980℃ is suitable for the investigated alloy.

Isochronal pressure-temperature contours for bonding in 2h for different surface finishes were calculated and shown in Fig.6. It is evident that the finer the surface finish is, the easier the bonding is.

Usually, the diffusion bonding process is divided into two parts dealing with the contacting process through the bonding interface and the subsequent diffusion, and the diffusion includes grain boundary diffusion and volume diffusion[9,12]. These different mechanisms had been considered in the present mathematical model. The rates of void closure by different mechanisms and the overall rate are shown in Fig.7. It is evident that plastic deformation is the dominant mechanism in the initial stages of the bonding process. The contribution from grain boundary diffusion was negligibly small compared with those of other mechanisms. With increasing the area fraction bonded, the void size becomes smaller and smaller, and the contribution from diffusion process becomes bigger and bigger. In the later stages of the bonding process, diffusion plays an important role for interfacial voids become smaller.

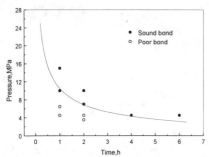
Fig.4 Various of critical bonding time with pressure at 980℃, together with experimental measurements

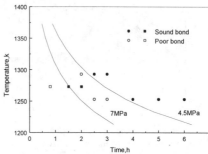
Fig.5 Variations of bonding time with temperature under pressure of 4.5MPa and 7MPa, together with experimental results

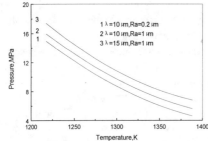
Fig.6 Predicted 2h bonding contour for different surface finishes

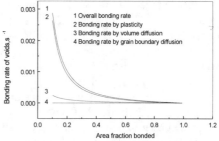
Fig.7 Contribution of different bonding mechanisms

4 Microstructural Transformations

The experiments of isostatic superplastic diffusion bonding of the investigated alloy were carried out at 980℃ under gas pressure and vacuum condition. The sample was cut to be 20×20mm and polished. The wavelength and amplitude of the surface roughness are 10μm and 0.2μm respectively. The sample was located on two stainless steel sheets to make a pack, which was located between a couple of dies, and deflated to obtain the vacuum condition. The argon was put on the pack from the lower die and the gas pressure was 4.5MPa. The experiments were conducted in one hundred-ton SPF/DB press.

The microstructures after isostatic superplastic diffusion bonding are shown in Fig.8. The bonding interface is incontinuous for 2h and sound bonding interface is obtained for 4h and 6h. Fig.9 describes the comparison of experiment and theoretical predictions of area fraction bonded. It is evident from Fig.9 that agreement between the experimental results and the theoretical predictions is excellent at long bonding time. However, the theoretical estimates of area fraction bonded for 2h is higher than experimental value, which is due to the cylindrical approximation, used in the model, having a larger volume than the real void. It can also be found from Fig.9 that there is an evident bonding interface, which is rich in α_2 phase, in each microstructure although the void closure is carried out for 4h. The mechanical properties of the bonding interface being rich in α_2 phase may decrease. However, the size of the bonding interface being rich in α_2 phase becomes smaller and disappears in some point for 6h, which indicates that dynamic recrystallisation takes place during superplastic diffusion bonding process. However, completing the dynamic recrystallisation process is so longer, which may influence the SPF process and the mechanical properties of the alloy.

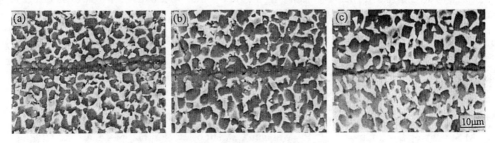

Fig.8 Microstructures after isostatic superplastic diffusion bonding at 980℃, 4.5MPa for (a) 2h (b) 4h (c) 6h

During superplastic diffusion bonding process, the grain size increases. The variation of grain size with time is shown in Fig.10. By fitting technology, the dynamic equation of grain size growth is

$$d = 0.68 + 1.26\exp(\frac{t}{7.3}) \tag{2}$$

Where d and t are grain size and bonding time respectively. The increase in grain size will degraded the SPF behavior and the overall properties of the alloy. The equation is useful to predict the grain size during SPF or DB process.

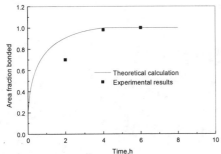

Fig.9 Comparison of experiment and theoretical predictions of area fraction bonded.

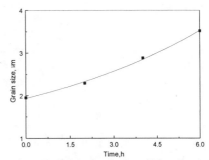

Fig.10 Variation of grain size with bonding time

5 Conclusions

The as-received Ti-24Al-14Nb-3V-0.5Mo intermetallic alloy exhibits duplex ($\alpha_2+\beta$) microstructure and finer grain size of 2μm. The apparent activation energy of the investigated alloy is 162.5kJ/mol, which is close to that of lattice diffusion in β-Ti of 150kJ/mol. The strain exponent and the dimensionless constant for theoretical calculation of superplastic diffusion bonding are 2.39 and 2818 respectively. According to the experimental results and theoretical calculation, the effects of pressure, temperature and surface finishes of primary sheets on the bonding time were studied. The optimum superplastic diffusion bonding process under gas pressure is 980℃, 4.5MPa and 4h for wavelength of 10μm and amplitude of 0.2 μm. During superplastic diffusion bonding process, the grain size increases and the relation between grain size and bonding time is obtained.

Acknowledgments

The authors would like to thank Dr.Shiqong Li of Central Iron and Steel Research Institute for supplying Ti-24Al-14Nb-3V-0.5Mo sheets.

References

[1] H.S.Yang, P.Jin, A.K.Mukherjee. Mat. Sci. Eng., A153(1992), p.457
[2] D.Jobart, J.J.Blandin. Mat. Sci. Eng., A207(1996), p.170
[3] H.S.Yang, M.G.Zelin, R.Z.Valiev A.K.Mukherjee. Scripta Metal. Mater., 26(1992), p.1707
[4] Yi Guo, Peng Liu, Yiwen Wu, Zijing He, Dong Li. Journal of Mater. Sci. Tech. 9 (1993), p. 53
[5] M.Strangwood, A. Gingell, E. R. Wallach C.A.Hippsley. Journal of Mater. Sci. Letters 11 (1992), p. 317
[6] H.S.Yang, P.Jin,E.Dalder et al. Scripta Metal. Mater., 25(1991), p.1223
[7] W.Cho, A.W.Thompson, J.C.Williams. Metall. Trans., 21A(1990), p.641
[8] M.J/Mendiratta, H.A.Lipsit. J.Mtaer.Sci., 15(1980), p.2985
[9] Y.Maehara, Y.komizo, T.G.Langdon. Mater. Sci. Tech. 4(1988), p.669
[10]J.Pilling, N.ridley, M.F.Islam. Mater. Sci. Eng., A205(1996), p.72
[11]H.J.Frost, M.F.Ashby, Deformation-Mechanism Maps, Pergamon, 1982.
[12]J.Pilling. Mater. Sci. Eng., 100(1988), p.137

High Strain Rate Superplasticity in Al-16Si-5Fe Based Alloys with and without SiC Particulates

J.S. Han[1], M.S. Kim[1], H.G. Jeong[2] and H. Yamagata[3]

[1] School of Materials Science and Engineering, Inha University,
#253 Yonghyun-dong, Nam-gu, Inchon 402-751, Korea

[2] Institute for Materials Research, Tohoku University,
Katahira 2-1-1, Aoba-ku, Sendai 980-8577, Japan

[3] R&D Division, Yamaha Motor Co. Ltd,
#2500 Shingai, Iwata, Shizuoka 438-8501, Japan

Keywords: Al-Si-Fe Based Alloy, Cavitation, Grain Boundary Sliding, High Strain Rate, Microstructure, Powder Metallurgy, SiC Particulate, Superplasticity

Abstract Superplasticity at a high strain rate of $1.4 \times 10^{-1} s^{-1}$ was exhibited at 783K in the powder metallurgy processed Al-16Si-5Fe-1Cu-0.5Mg-0.9Zr (wt%) alloys with and without 5 wt% addition of SiC particulates. The maximum elongations of 350% and 270% were obtained for the monolithic and composite material, respectively, where high m values were found. The high strain rate superplasticity in these materials was attributed to the presence of a thermally stable fine microstructure and the operation of grain boundary sliding. The lower superplastic elongation of the composite specimen compared to that of the monolithic specimen was suggested to the result of an increased density of cavities owing to the SiC decohesion at the SiC/matrix interface.

1. Introduction

Powder metallurgy (PM) processed hyper-eutectic Al-Si based alloys containing a higher level of Fe and hard SiC particulates were recently succeeded in applying to mass-production engine pistons for two-stroke-cycle snowmobiles [1]. It was revealed that the commercialized PM piston alloys had enhanced high temperature tensile and fatigue strength in addition to the increased abrasion resistance suitable for the piston operating under severe environment. With accompanying high strength level at elevated temperatures for the PM piston alloys, the forming of the PM alloys to net shape was somewhat difficult via conventional processing procedure [2]. So that, careful study on deformation behavior for this type of alloy is required to find out optimum processing parameters. In this work, high temperature tensile properties including high strain rate superplasticity are investigated using PM Al-16Si-5Fe-1Cu-0.5Mg-0.9Zr (wt%) alloys with and without SiC particulates, which are similar composition to the above-mentioned PM piston alloys.

2. Experimentals

Compositions of Al-16Si-5Fe-1Cu-0.5Mg-0.9Zr (wt%) along with 0 and 5 wt% additions of SiC were used in this study. The starting materials were air atomized Al-16Si-5Fe-1Cu-0.5Mg-0.9Zr alloy powders and SiC particulates. The aluminum alloy powders and the mixtures of aluminum alloy powders and SiC particulates were degassed and canned. The canned powders were extruded to a fully dense rod with the

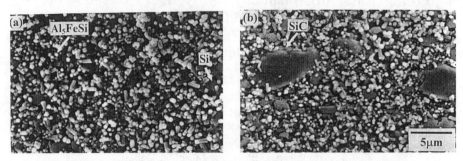

Fig. 1 SEM micrographs of as-extruded (a) monolithic and (b) composite materials.

extrusion ratio of 18 to 1. The extruded bars were machined to tensile specimens having a gage dimension of 12mm length and 3mm diameter. The tensile axis was selected to be parallel to the extrusion direction. Tensile tests were carried out in air with a strain rate range from 1.4×10^{-4} to $7 \times 10^{-1} s^{-1}$ and at a temperature range from 673 to 783K. As soon as deformed to failure, the tensile specimens were cooled by air. Microstructure of as-extruded specimens and deformed specimens was examined by using optical microscope (OM), scanning electron microscope (SEM) and transmission electron microscope (TEM). X-ray diffraction (XRD) and energy dispersive spectrometer (EDS) analyses were performed to characterize constituent phases.

3. Results and Discussion

The SEM micrographs shown in Fig. 1 reveal a fairly uniform distribution of Si particles and small intermetallic constituents in the as-extruded monolithic and composite materials. The small intermetallics are found to be Al_5FeSi by EDS and XRD analyses. In the composite material (Fig. 1(b)), SiC particulates having the mean diameter of about 5 μm are also observed. Fig. 2 shows the TEM micrographs of the as-

Fig. 2. TEM microghraphs of as-extruded (a) monolithic and (b) composite materials.

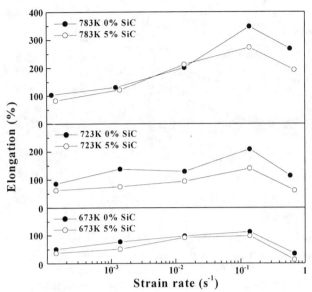

Fig. 3. The variation of elongation as a function of strain rate and temperature for monolithic and composite materials.

extruded monolithic and composite materials. Fine grains with almost equiaxed appearance are visible in both materials. The average grain sizes of the monolithic and composite materials are determined to about 800 and 700 nm, respectively. In both materials very small Al_5FeSi intermetallics (100-400 nm) are observed on grain boundaries and in the matrix.

Fig. 3 shows variations of elongation as a function of strain rate at different test temperatures for both materials. As can be seen from this figure, the elongations of both materials are less than 120% at the lower strain rate of $1.4\times10^{-4}s^{-1}$ for all test temperatures, while the elongations tend to increase with increasing strain rate up to $1.4\times10^{-1}s^{-1}$, and then decrease with a further increase in strain rate. At the strain rate of $1.4\times10^{-1}s^{-1}$ where the peak of elongation appears, the maximum elongations of 350% and 270% are obtained at 783K for the monolithic and composite material, respectively. The relationship between the flow stress at a given strain of 0.1 and strain rate is plotted in Fig. 4 at various testing temperatures for both materials. The flow stresses of both materials tend to increase with increasing strain rate and decreasing test temperature. Also, it can be shown from Fig. 4 that the maximum of the strain rate sensitivity, m, is relatively high at temperature of 783K, namely, 0.4 and 0.35 for monolithic and composite materials, respectively. In contrast, at lower temperature regime (673~723K), the maximum of the strain rate sensitivity is relatively low (<0.3). It is widely accepted that the characteristic feature of superplasticity is a high strain rate sensitivity of the flow stress. In the present study, the maximum elongations over 250% are obtained at 783K and at a high strain rate of $1.4\times10^{-1}s^{-1}$ where high m values appear, indicating that the large elongations in the monolithic and composite materials arise from superplastic deformation.

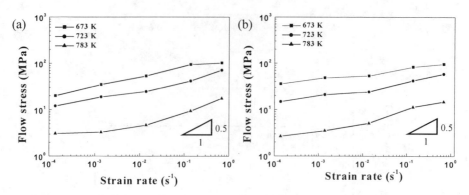

Fig. 4. The variation of flow stress as a function of strain rate and temperature for (a) monolithic and (b) composite materials.

It has been reported that the dominant flow mechanism in high strain rate superplasticity for fine-grained aluminum alloys and aluminum alloy composites is grain boundary sliding [3-7]. In the present investigation, the occurrence of grain boundary sliding was verified by the observations of specimen surfaces after superplastic deformation. One example of the SEM micrograph for the composite specimen deformed superplastically at 783K under the strain rate of $1.4\times10^{-1}s^{-1}$ is shown in Fig. 5, in which the grain boundary sliding and rotation are apparent. A similar surface morphology to this figure is also observed in the superplastically deformed monolithic specimens.

The TEM observations of the inner section of the both specimens after superplastic deformation indicated a limited grain growth during superplastic deformation. A typical TEM micrograph is presented in Fig. 6 for the monolithic specimen deformed superplastically at 783K with the strain rate of $1.4\times10^{-1}s^{-1}$. It can be seen from this figure that the grain size is still small (about 1μm) after superplastic deformation.

From the foregoing results, it was revealed that both Al-Si-Fe alloys and SiC/Al-Si-Fe composites exhibit a high strain rate superplasticity at 783K, although the superplastic elongation is reduced from 350 to 270% with the addition of 5 wt% of a few large (~5μm) SiC particulate. Such a difference in the elongation between two materials was

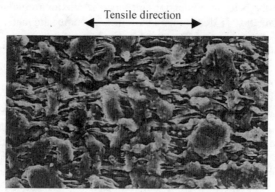

Fig. 5. Surface morphology of the composite material deformed superplastically at 783K with the strain rate of $1.4\times10^{-1}s^{-1}$.

Fig. 6. TEM micrograph of the monolithic material deformed superplastically at 783K with the strain rate of $1.4 \times 10^{-1} s^{-1}$.

reflected in cavitation appearance. The OM micrographs of the polished longitudinal section for monolithic and composite specimens deformed superplastically at 783K under the strain rate of $1.4 \times 10^{-1} s^{-1}$ are shown in Fig. 7. Note that the composite specimen contains a considerably higher density of cavities than the monolithic one does. In order to consider the increased density of cavities in the composite specimen, the longitudinal section near fractured region was observed. A typical SEM micrograph is illustrated in Fig. 8 for the superplastically deformed composite specimen. In this figure, a decohered interface between SiC and the matrix is clearly visible, implying that cavities can be formed in the highly strain-localized areas in the vicinity of the SiC particulates. Such a decohered interface is occasionally observed in the composite specimen after superplastic deformation. The increased density of cavities in the composite specimen is due to particulate decohesion at the particulate/matrix interface may reduce the load bearing areas until total failure of the specimen occurs, leading to lower elongation to failure compared to the case of the monolithic specimen which does not contain the particulates. Therefore, the SiC particulate/matrix decohesion is

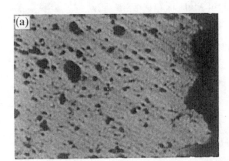 

Fig. 7. Optical micrographs of the polished longitudinal section of (a) monolithic and (b) composite materials deformed superplastically at 783K with the strain rate of $1.4 \times 10^{-1} s^{-1}$.

Fig. 8. The evidence of a decohered interface between SiC and matrix on the surface of the composite material deformed superplastically at 783K with the strain rate of $1.4 \times 10^{-1} s^{-1}$.

suggested as an important controlling factor for the fracture process in the superplastic SiC/Al-Si-Fe composite.

4. Conclusions

(1) The powder metallurgy processed Al-16Si-5Fe-1Cu-0.5Mg-0.9Zr (wt%) alloys with and without 5 wt% addition of SiC particulates exhibited superplasticity at a high strain rate of $1.4 \times 10^{-1} s^{-1}$ at 783K, where high m values were found. The maximum elongations of 350% and 270% were obtained for the monolithic and composite material, respectively.
(2) The high strain rate superplasticity in these materials was attributed to the presence of a thermally stable fine microstructure and the operation of grain boundary sliding during deformation.
(3) The increased density of cavities in the composite specimen owing to the SiC decohesion at the SiC/matrix interface is suggested to cause the lower superplastic elongation compared to the case of the monolithic specimen.

References

[1] H. Yamagata and T. Koike, J. Japan Inst. Light Met. 48 (1998), p. 52.

[2] H. Yamagata, Mater. Sci. Forum, 304-306 (1999), p. 797.

[3] T. G. Neih, J.Wadsworth and T. Imai, Scripta Metall. 26 (1992), p. 703.

[4] M. Mabuchi and K. Higashi, Philo. Mag. Lett. 70 (1994), p. 1.

[5] M. Mabuchi, T. Imai, K. Kubo, K. Higashi, Y. Okada and S. Tanimura, Mater. Lett.11 (1991), p. 339.

[6] M. Mabuchi and K. Higashi, J. Mater. Res. 10 (1995), p. 2494.

[7] M. Kawazoe, T. Shibata, T. Mukai and K. Higashi, Scrip. Mater. 36 (1997), p. 699.

Corresponding Author
M.S.Kim : mskim@inha.ac.kr

Author Index

A

Adam, Laurent 295
Afonichev, D.D. 315
Aida, T. 309
Aizawa, Tatsuhiko 35, 327
Akimoto, H. 309
Akkus, N. 587
Arbel, I. 527
Asahina, T. 327

B

Bai, B.Z. 105
Bang, W. 171
Barnes, A.J. 3
Bate, Peter S. 477
Bekki, Yoichiro 165
Belvin, Anthony 599
Berdin, V.K. 225
Bernault, E. 301
Bewlay, B.P. 417
Blandin, J.J. 301, 437
Bruemmer, S.M. 93

C

Carrino, L. 219
Chandra, N. 261, 387, 577, 599
Chang, Young Won 159, 171, 231
Chen, Chilong 333
Chen, S.H. 205
Chen, W.C. 425
Chen, Xianglei 599
Chen, Z. 599
Cheng, Jung-Ho 411
Cho, H.S. 443
Chokshi, Atul H. 123
Chu, J.P. 205
Chung, Li Chuan 411
Comley, P.N. 41
Curtis, R.V. 47, 213

D

Daniel, B.S.S. 371
Dashwood, R.J. 339, 357
Dendievel, R. 301
Doi, Y. 309
Dougherty, L.M. 93

E

Dunand, D.C. 177
Dunwoody, B.J. 59
Dupuy, L. 437
Dutkiewicz, J. 111

El-Morsy, A. 587

F

Ferguson, D.E. 425
Ferguson, H.S. 425
Flower, H.M. 339, 357
Flower, Terry 23
Funami, K. 539
Furukawa, Minoru 431, 489
Furushiro, N. 249

G

Garmestani, H. 393
Garriga-Majo, D. 47, 213
Gershon, B. 527
Ghosh, A.K. 237
Gigliotti, M.F.X. 417
Giuliano, G. 219
González-Doncel, G. 255
Grimes, R. 339, 357
Gu, J.L. 105
Guillard, Stephane 283

H

Ha, Tae Kwon 159, 171, 231
Han, J.S. 443, 613
Hänninen, H. 277
Hao, Hongqi 363
Harrison, A.W. 339
Herling, Darrell R. 465
Hevlin, S. 527
Higashi, Kenji 147, 153, 327, 345, 459, 565
Hill, P.S. 99
Hiraga, K. 187, 193, 533
Horita, Zenji 431, 471, 489
Hosaka, F. 117
Hsiao, I.C. 381
Huang, J.C. 381, 515, 545
Humphreys, F.John 477

I

Ikuhara, Yuichi 129, 141, 399
Inoue, K. 205
Ishikawa, K. 459
Islamgaliev, R.K. 449
Iwasaki, H. 153, 327

J

Jeong, H.G. 443, 613
Jin, Zhe 425
Jocelyn, Alan 23
Juszczyk, A.S. 47

K

Kaibyshev, Oscar A. 73, 225, 417
Karavaeva, M.V. 225
Kashaev, R.M. 225
Kato, H. 165
Kim, B.-N. 533
Kim, M.S. 443, 613
Kitazono, K. 289, 405
Kobayashi, M. 539
Kohzu, M. 147, 565
Komura, S. 431
Kondo, Naoki 183
Kumao, A. 593
Kuribayashi, K. 117, 289, 405
Kuśnierz, J. 111
Kuwabara, A. 399

L

Langdon, T. G. 135, 321, 431, 471,
  489, 521, 545
Lee, S. 471
Lee, Sang-Mok 321
Lee, Sungwon 521
Li, H. 577
Li, Zhiqiang 53, 607
Lim, Hyun Tae 231
Liu, Yi 363
Lou, B.Y. 545

M

Mabuchi, M. 35, 153
Mabuchi, Mamoru 327
Machida, N. 539
Mahoney, M.W. 507
Malysheva, S.P. 315
Manabe, K. 587
Matsuki, K. 309

Matsushita, J. 117
McFadden, S.X. 499
McNelley, T.R. 255
Medrano, R.E. 199
Mendoza-Allende, A. 551
Milo, Y. 527
Mimurada, Junpei 129, 141
Mishra, R.S. 425, 507
Mohamed, Farghalli A. 83
Montemayor-Aldrete, J.A. 551
Mori, T. 153
Morita, K. 187, 533
Moriwaki, K. 459
Motohashi, Y. 559
Mukai, T. 147, 459
Mukherjee, Amiya K. 283, 499
Mukhtarov, Sh.Kh. 417
Muñoz-Andrade, J.D. 551
Murzinova, M.A. 315

N

Nakamura, M. 327
Nakano, K. 193
Nakano, Manabu 129, 141
Namilae, S. 387
Nash, Doug 23
Nazar, A.M.M. 199
Neishi, K. 471
Nemoto, Minoru 431, 489
Nieh, T.G. 111, 387
Niikura, Akio 165
Nishimura, H. 587

O

Ogawa, H. 571
Ohji, Tatsuki 183
Okamoto, Y. 593
Okanda, Y. 459
Ota, S. 471
Otsuka, M. 117

P

Padmanabhan, K.A. 371
Pagliaria, D. 47
Pérez-Prado, M.T. 255
Pimenoff, J. 277
Ponthot, Jean-Philippe 295
Pulino-Sagradi, D. 199

R

Rajagopal, K.R. 261
Rauch, E.F. 437
Ridley, N. 99
Robertson, I.M. 93
Rojo, Marco A. Hernandez 65
Romu, J. 277
Ruano, O.A. 255

S

Sabelkin, V.P. 65
Sadeghi, R.S. 273
Sakka, Y. 187, 193, 533
Sakuma, T. 559
Sakuma, Taketo 129, 141, 399
Salishchev, G.A. 315
Saltoun, D. 527
Sanada, S. 559
Sanders, Daniel G. 17
Sasaki, Kazutaka 129, 141
Sasaki, S. 165
Sato, E. 117, 289, 405
Satou, T. 117
Sawaguchi, N. 571
Schuh, C. 177
Sergueeva, A.V. 499
Sheikh-Ali, A.D. 393
Shet, C. 387, 577
Shimojima, K. 327
Smith, Mark T. 465
Song, Jin Hwa 231
Soo, S. 47
Sosa, Siari S. 135
Sudhir, B. 123
Sun, X.J. 105
Sundar, R.S. 405
Suzuki, N. 565
Suzuki, T.S 193
Suzuki, Yoshikazu 183
Syutina, L.A. 225

T

Tagata, T. 153
Takayama, Y. 165
Tan, Ming Jen 333
Tanabe, S. 565
Todd, R.I. 99
Torres-Villaseñor, G. 551
Tsutsui, H. 147

U

Umakoshi, Y. 205, 249
Utsunomiya, A. 431

V

Valiev, Ruslan Z. 449
Valitov, V.A. 417
Vetrano, J.S. 93

W

Wakai, F. 571
Walter, J.D. 47
Wang, Chunxiao 607
Wang, T.D. 515, 545
Warashina, K. 249
Watanabe, H. 147, 165, 459
Watts, John D. 599
Wen, C.E. 327
Whittingham, Roy 29
Wu, Xin 363

X

Xu, Cheng 521

Y

Yagodzinskyy, Y. 277
Yamada, Y. 327
Yamagata, H. 443, 613
Yang, L.Y. 105
Yasuda, H.Y. 205
Yokota, S. 399
Yunusova, N.F. 449

Z

Zelin, Michael 283
Zherebtsov, S.V. 315
Zhu, Hanliang 53, 607

Keyword Index

A
Accommodation 345
Activation Energy 147, 183, 545, 559
AlCuAgMgZr Alloys 111
Al-Cu-Zr Alloy 249
Alloying Additions 357
Alloying Costs 357
Al-Mg Alloy 153, 165
Al-Mg-Sc Alloy 93, 431
Al-Mg-Zr 339
Alpha case 41
Al-Si Alloy 443
Al-Si-Fe Based Alloy 613
Alumina 193, 533, 593
Alumina Reinforcement 123
Aluminium 59, 93, 499
Aluminium Alloys 255, 381, 437, 471, 507, 521, 527, 545, 565
Aluminium Alloy 2195 .... 277, 465, 527
Aluminium Alloy 5083 599
Aluminium Alloy 6061 .... 515, 527
Aluminium Alloy 7475 333
Aluminium Composites 545
Aluminium Products 29
Aluminosilicate Glass 123
Analytical Model 219
Anisotropy 41
Apparatus for Impulsive Forming 65
Apparent Activation Energy 607
Application 527
Atomistic Mechanisms 371
Automotive 59
AZ91 35

B
Boundary Sliding 83
Box Forming 587

C
Casting Investments 47
Cavitation 83, 153, 345, 521, 613
Cavity 599
Cavity Formation 193
Cavity Growth 193
Cavity Size Distribution 193

Ceramics 23, 73, 135, 559
CGBS Bands 73
Chemical Bonding State 399
Coarse-Grained Iron Aluminides 231
Co-Doping 129
Cohesive Zone Model 577
Commercial Aircraft 527
Commercial Alloy 3
Competitiveness 3
Complex Analysis 213
Complex Loading Conditions 225
Composites 135
Compression 123
Computer Control 35
Concurrent Grain Growth 559
Constitutive Equation 261, 559
Contact 295
Continuous 93
Cooperative GBS 301
Cooperative Grain Boundary Sliding .. 73, 371
Cooperative Processes 283
Creep 135, 187, 301, 405, 559
Crystal Plasticity 577
CSL Boundaries 381

D
Damage 437
Deformation 571
Deformation Banding 255
Deformation Mechanism Map 117
Deformation Mechanisms 105, 449, 545
Deformation States 283
Dental Implant Superstructure 47
Dental prostheses 47
Dies 47
Die-Set Selection 35
Diffusion 135
Diffusion Bonding 23, 29, 99, 607
Diffusional Creep 117
Dislocation 187
Dopant 141
Doping Effect 593
Dual Phase 129
Ductility 83, 141
Dynamic 93

Dynamic Grain Growth 593
Dynamic Recovery 205
Dynamic Recrystallization 205, 231, 315, 363

E
EBSD 381
ECAE 437, 465
Effective Diffusivity 147
Electron Microscopy 593
Elongation 153, 533
Equal-Channel Angular Pressing 321, 431, 471, 489, 521
Equal-Channel-Angular-Extrusion 465
Explosive Forming 65

F
Fine-Grained Structure 345
Finite Element Method 219, 295
Finite Element Method Simulation 587
Finite Element Modelling 225, 273, 411
Flow Localization Factor 411
Former Alpha Boundaries 83
Fracture Surface 165
Friction Stir Processing 507

G
Geometric Simulation 213
Grain Boundary 255, 381, 399
Grain Boundary Character Distribution (GBCD) 333
Grain Boundary Energy 387
Grain Boundary Plasticity 345
Grain Boundary Sliding 153, 249, 345, 371, 393, 443, 577, 593, 613
Grain Growth 443, 477
Grain Misorientation 507
Grain Refinement 111, 225, 309, 477, 507
Grain Size 459, 477
Grain Size Exponent 147, 559

H
Hemispherical Forming 587
High Strain Rate 443, 613
High Strain Rate Superplasticity 309, 321, 327, 339, 345, 357, 431, 449, 459, 489, 507, 515, 521
Hydrogen Alloying 315
Hydrostatic Stress 387

I
Impurity Segregation 83
Induction Heater 47
Industrial Applications 53
Inhomogeneous Cavitation 333
Injection Forming 35
In-situ Composite 405
In-situ TEM 93
Instability 411
Intermetallic Particles 277
Intermetallics 405
Internal Stress 289
Internal Stress Superplasticity 405
Internal Variable Theory 159, 171
Interphase Boundary Sliding 371
Intragranular Slip 393
Iron Aluminide 205

L
Large Strain 477
Lattice Dislocations 393
Liquid Phase 73, 443
Load Relaxation 171
Load Relaxation Test 231
Localization 599
Low Temperature Superplasticity 345, 381, 449, 459
Low-cost 23

M
Magnesium 35
Magnesium Alloy 147, 327, 363, 459
Markets 3
MAXStrain Technology 425
Mechanical Properties 199
Mechanisms of Deformation 73
Mesoscopic Grain Boundary Sliding 371
Metal Matrix Composite 289
Metallic Glass 371
Metals 73
Microduplex Pb-Sn Alloys 159
Micrograin Superplasticity 83
Microstructure 183, 363, 443, 607, 613
Misorientation Angle 381
Misorientation of Grain Boundaries 249
Mode of Failure 599
Modeling 99, 301
Molecular Dynamics 571, 577
Molecular Orbital Calculations 399
Monolithic Polycrystal 289

Multidimensional Space 65
Multi-Dome Forming 587

N
Nano Particle 515
Nanocrystalline Structure 371, 417, 499
Nanomaterials... 499
Natural Configurations 261
Near-Net Shaping 35
Neighbor Switching.................................. 571
NiAl 405
Nickel... 499
Nickel Aluminides.................................... 309
Nickel Based Alloy................................... 417
Novel 23
Numerical Simulation............................... 295

O
Optimal Parameters 65
Orientation Imaging Microscopy OIM....... 93
Oxygen Rich Layer.................................... 41

P
Percolation 123
Phase Boundary Sliding............................ 159
Phase Transformation 177
Plane Strain.. 565
Plastic Deformation 199
Polycrystals.. 571
Powder Metallurgy 327, 443, 515, 613
Precipitate Free Zone................................ 171
Press Design for Production 29
Pressure-Time Profile............................... 213
Prismatic Die Shape................................. 213
Processing Techniques 3
Production Costs...................................... 357

Q
Quality 23

R
R&D Initiatives.. 3
Rapid Solidification.................................. 339
Rate Controlling Process 371
Recrystallization 93, 477
Rigid-Plastic 273
Role of Solute Atoms 249

S
Scandium 357
Segregation.. 387
Sever Straining 165
Severe Plastic Deformation 425, 431,
    437, 449, 471, 489, 499
Sheet Metal Forming 565
Shells 65
SiC Particulate... 613
Silicon Nitride 183
Simulation 273
Solid State Bonding................................... 99
SP 700 .... 41, 527
SPF 59, 465
SPF/DB.. 53
Spinel... 533
Stability 477
Stability Criterion 411
Stainless Steel.. 199
Sticking Contact 213
Stoke Control... 35
Strain Concentration................................ 411
Strain Rate 205, 533, 565
Strain Rate Sensitivity............................. 153
Stress 533
Stress Exponent 147, 187
Stress Sensitivity 183
Structural Superplasticity 159, 231, 371
Submicrocrystalline Structure 315, 417
Superplastic Deformation............. 23, 111, 183
Superplastic Forging 459
Superplastic Forming ..29, 219, 277, 489, 527
Superplastic Response 261
Superplasticity............. 35, 47, 53, 65, 73, 93,
    99, 105, 117, 129, 141, 165, 171,
    177, 199, 205, 225, 283, 315, 363,
    393, 399, 411, 417, 443, 465, 499,
    565, 587, 599, 607, 613

T
Techno-Economics....................................... 3
Temperature 443
Temperature Effect.................................. 599
Texture 165, 255
Thermal Cycle... 289
Thermodynamics..................................... 261
Thermomechanical Treatment.................. 515
Thermomechanics 295
Thickness Profile..................................... 213
Ti-24Al-14Nb-3V-0.5Mo
    Intermetallic Alloy 607
Titanium 23, 499

Titanium Alloys 47, 65, 105, 177,
    225, 315, 527
Titanium Composites 177
Titanium Products 29
Tool Design 59
Torsion 559
Transmission Electron Microscopy
    TEM 93, 187
TZP 399

U
Ultrafine Grain Size 105, 321, 425,
    431, 471, 489
Ultrafine Grained Materials 449
Unstable Plastic Flow 411

V
Viscoplasticity 295
Volume Control 273

W
Warm Rolling 277
Wet Mechanical Alloying 309

Y
YSZ 141
Yttria-Stabilized Zirconia 135
3Y-TZP 559

Z
Zircaloy 4 99
Zirconia 117, 129, 187, 533, 559, 571
Zirconium 357
Zn-Al Alloy 321
ZrO_2 593

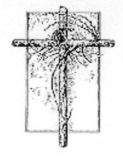

A HISTORY OF EASTCOTE PARISH CHURCH (SAINT LAWRENCE)

JEAN A GIBSON

An environmentally friendly book printed and bound in England by
www.printondemand-worldwide.com

This book is made entirely of chain-of-custody materials

www.fast-print.net/store.php

A History of Eastcote Parish Church (Saint Lawrence)
Copyright © Jean A Gibson 2013

All rights reserved

No part of this book may be reproduced in any form by photocopying
or any electronic or mechanical means, including information storage
or retrieval systems, without permission in writing from both the
copyright owner and the publisher of the book.

The right of Jean A Gibson to be identified as the author of this work has
been asserted by her in accordance with the Copyright, Designs and
Patents Act 1988 and any subsequent amendments thereto.

A catalogue record for this book is available from the British Library

ISBN 978-178035-607-5

First published 2013 by
FASTPRINT PUBLISHING
Peterborough, England,
on behalf of St Lawrence Church, Eastcote.

DEDICATION

To the past, present and future community of

St Lawrence Church,

Eastcote

To Brian and Jo Radford

With thanks for your encouragement in this project which has been completed after many years.

Jean R. Gibson
December 2013

FOREWORD
By The Reverend Stephen Dando

This history of St Lawrence's has opened to us a vision of the growth and development of the Parish Church in Eastcote. It has enabled us to see how worship, Christian community and wider community have related to one another as well as how society has changed since the arrival of a church in Eastcote. We see that the five incumbents all come from a catholic tradition in which the Eucharist has played, and continues to play a central part. The five Incumbents have encouraged Eucharistic worship to be a springboard for outreach and care for the wider community, as well as ensuring that the worship itself is open to all and that everyone has a part to play in it.

As years have passed, the place of the church in society has become less obvious, with a powerful move towards secularisation. Sunday opening and Sunday football were not the challenges faced by The Revd Rupert Godwin, and social activities for the majority of the population no longer hinge on what the church has to offer. The changes have been chronicled with great care, so that we are able to see how life in Eastcote has changed and how the church and the Christian community has responded to those changes.

We now live in a much more diverse and multicultural society and this is reflected in the worshipping community, but despite all the changes, the themes of outreach to the wider community, fellowship, mission, collaborative ministry and worldwide concerns all continue to be important today. However the centrality of the Eucharist has always been the core of what St Lawrence Church has stood for and it is through being fed in this way, that the people of this parish have been enabled to become a community that looks outwards and reaches out in love and service to all.

Jean has captured this theme and shown us the detail of how it has been worked out through the various incumbencies and changes in society. We are most grateful to her for her diligence, her attention to detail and for her tireless work. We thank her for such a wonderful piece of social and ecclesiastical history.

Once you have read this book you will echo my thanks to Jean and those who have helped her, and will come to understand why there is such a marvellous sense of prayer and worship at St Lawrence, Eastcote.

Fr. Stephen

July 2012

INTRODUCTION

When I undertook to write the history of St Lawrence Church Eastcote the original plan was to have it ready for the 75th Anniversary of the present permanent church, consecrated in 1933. I had not realised the mammoth task I had undertaken and 2008 passed without publication. My aim then was to complete the book by for the 90th Anniversary of the Dedication of the Mission Church. Without the tremendous efforts of the parishioners of those early days we might not have the present building. This Anniversary in 2010 also passed without publication.

St Lawrence, to whom the church is consecrated, was Archdeacon of Rome in the third century. During the persecution of Christians by the Emperor Valerian, it was learned that Lawrence was the keeper of the treasures of the Church. He was arrested and ordered to produce them. He asked for a day in which to make the collection. All night he hurried about the poorest streets of Rome. The next day he appeared at the tribunal, followed by a crowd of poor and sick people. 'These are the treasures of the Church' he said, for it was not gold, but the humble folk befriended by the Church, who constituted her true wealth. Lawrence was condemned to death. According to tradition, Lawrence was martyred by being roasted on iron bars, hence the gridiron as the symbol associated with him. His martyrdom certainly is historical, probably the year AD 258. The Church celebrates St Lawrence Day on 10 August.

Some comment on the spelling of Lawrence is necessary. When a short history of the parish was being written for the 25th anniversary in 1958, Father Bill Hitchinson did some research. He was struck by the fact that the name St Lawrence appeared on the foundation stone and on the plaque in the Lady Chapel, whereas the name was spelt with a 'u' on the notice board. He could find no trace of any official change in the spelling. He thought it might have been felt that the 'u' was correct, as the name was Laurentius in Latin, which did not use a 'w'. As the Order in Council separating the parish from Ruislip spelt it with a 'w' he reverted to the original spelling. It will be seen in reproductions from the *Leaflet*, forerunner of a parish magazine, that Laurence was usually spelt with a 'u' during Father Godwin's incumbency. A plaque recording the completion of the 'east end' of the present church hall has a 'u'. For consistency I have used 'w' for Lawrence throughout the text.

Father David Coleman wrote in the Foreword to *A Guide to St Lawrence Church Eastcote* (2004), that church buildings are important for three reasons. Firstly they are sacraments in stone, silent witnesses to God's truth. Secondly they are consecrated and set apart from secular use as a reminder and an assurance of the presence of God. Thirdly a church building is 'home' to the Christian community. For these reasons it is right to discuss the church building and furnishings, together with other property belonging to the church. They are an essential part of the mission of the Church in Eastcote.

However it is the church community of St Lawrence's, who are the Church, witnessing to their Christian faith not only by their attendance at their 'home', but also by their fellowship and love for each other, and for those with whom they come in contact. Their role in various aspects of church life, and their mission in the wider community is all part of the history of St Lawrence Church.

From the beginning it seemed logical to have chapters on the different incumbencies with the idea of showing how each had its own ethos, reflecting to some extent the viewpoint of successive Vicars, and taking into account how changes in society influenced the mission of the Church in the parish of Eastcote. The first chapter is more of an introduction, dealing with the creation of a new parish, and the events up to 1933. The remaining chapters follow the church's history up to the end of 2010.

I have included quotes from source material as the language used seemed to reflect the particular period in which it was written. I have referred to some members of the St Lawrence community, and others in the wider community, by name. Some readers may feel that many other people contributed as much to the life of St Lawrence's. However sources are not always consistent as to the information recorded and I hope that 'naming names' may help others to recollect with pleasure their association with the church. It also shows that St Lawrence Church is a community of individual members, each able to play a part in the mission of the Church.

Again some may think that I have gone into too much detail regarding various organisations associated with the church. Some groups no longer exist but they met the needs of a particular time. My reason for details, such as the topics of speakers at meetings, and destinations of outings, was to show how, through the different incumbencies, the church organisations reflect the influences of contemporary society and the wider community.

I trust that this history will give some idea of how the parish church of Eastcote has adapted to challenges posed by changes in society, and yet maintained its role of Christian mission.

ACKNOWLEDGEMENTS

Most of the information has been taken from St Lawrence Church magazines, Minutes of Annual Parochial Church Meetings, reports of various organisations, *Record of Services* books, and printed copies of special services. I have also had access to the Minutes of the Parochial Church Council, Mothers' Union, Church of England Men's Society, and Women's Guild. A scrapbook of newspaper cuttings belonging to Father Godwin was also of interest.

For the early chapters I had access to magazines and other documents of St Martin's Church, Ruislip that gave interesting background on the years leading to the Consecration of the present church in 1933. I appreciate the help I had from St Martin's in this respect.

I am grateful for the assistance given to me by staff at Uxbridge, Ruislip, and Eastcote libraries, for reminiscences from Norman Parker and Beryl Orders, and for information resulting from correspondence with the personal assistant of the Bishop of Willesden, the Records and Archives Centre of The Children's Society, The Girls' Friendly Society, and The Boys' Brigade.

My thanks go to the Ruislip, Northwood and the Eastcote Local History Society (RNELHS) and to Uxbridge Library for permission to use photographs from their archives. I am also grateful to Eileen Bowlt for permission to reproduce the map of the parishes in the Deanery of Hillingdon from her book *The Goodliest Place in Middlesex*, also for information in that book on population figures. Geographers' A-Z Map Co Ltd has given permission to reproduce a section of a modern street map on which the parish boundaries are shown, and Photopro Photography has kindly given permission to reproduce photographs in the possession of St Lawrence Church, which were taken by F. & J. Hare. Other photographs have been taken or by members of the church community, many by Arthur Plummer. Fairtrade and the Women's World Day of Prayer have allowed the reproduction of their respective logos.

It is impossible to give thanks to all those who have helped me to complete this project but some must be acknowledged by name. First is Dorothy Reile. Little did she realise when she offered to proofread my efforts just what she had let herself in for. She has read several drafts of the text, including the later ones with illustrative material. She has been a wonderful support throughout and I might have given up had she not been there to encourage me. Dorothy was not well enough to help with the final proofread. Others who have read the text, or parts of it, with a different approach have given pertinent comments and suggestions for which I am most grateful. They include Father

Stephen, the Revd Rachel Phillips, Father David Coleman, Sylvia Hooper, Barbara Plummer, Karen Spink and a few MU members.

My thanks go to Geoff Higgs who provided the appendix listing productions of the St Lawrence Players. I also express my gratitude to Andrew Bedford who compiled several of the appendices, some rather complicated and entailing numerous revisions, particularly that of the Table of Communicants 1933 to 2010. My thanks also go to Brian Beeston for writing the scouting sections in the book. I am also grateful to the many people who I have contacted to confirm details about different aspects of church life or activities of organisations,

Finally my thanks go to Arthur Plummer who came to my aid when my computer went 'on the blink' and has been responsible for the presentation of many of the appendices. Arthur scanned photos and other illustrative material and inserted them in the text in preparation for publication and he helped compile the extensive index. His patience with these and my numerous revisions has been remarkable. Arthur's contribution in the last months before publication has been immeasurable. I am most grateful.

Publications referred to include:-
Bradshaw Paul *A Companion to Common Worship* SPCK 2002

Bowlt Eileen *The Goodliest Place in Middlesex* Hillingdon Borough Libraries 1989

Edwards Ron *Eastcote From Village to Suburb; A Short Social History 1900-1945* Hillingdon Borough Libraries 1987

Kemp W.A.G. *The History of Eastcote Middlesex* 1963.

Eastcote a pictorial history Ruislip, Northwood and Eastcote Local History 1984

CONTENTS

Foreword iv
Introduction v
Acknowledgements vii
List of illustrative material x

Chapter I	The Early Years - moves towards a separate parish	1
Chapter II	The Reverend R. F. Godwin - Vicar 1933 to 1956	15
Chapter III	The Reverend W. H. Hitchinson - Vicar 1956 to 1980	71
Chapter IV	The Reverend D. M. H. Hayes - Vicar 1980 to 1990	156
Chapter V	The Reverend D. Coleman Vicar -1990 to 2004	216
Chapter VI	The Reverend S. Dando - Vicar from 2005	276

Appendices
1 Clergy and other members of the team ministry 313
2 Subsequent appointments of St Lawrence clergy 315
3 Bishops of the London Diocese and Archdeacons of Northolt 317
4 Church Officials 320
5 Table of Communicants during the years of the Mission Church 324
6 Table of Communicants from October 1933 to December 2010 325
7 Table showing numbers of Baptisms and Marriages 332
8 Confirmation numbers 333
9 Ecumenical Services - numbers attending 338
10 Electoral Roll, Stewardship and Quota/Common Fund Figures 340
11 Fêtes, Fairs and other Fund Raising events from 1956 343
12 Accounts Summaries 2008, 2009, 2010 346
13 St Lawrence Players' Productions 348
14 Parish boundaries shown on a modem street map 353

Index 354

References are listed at the end of chapters in which footnotes have been used.

Photograph work had been credited where appropriate in the captions. Other photographs have come from St Lawrence archives or have been lent by friends.

If any copyright material has been used without permission being given, the author wishes to apologise for the inadvertence.

Photographs, Illustrations and Maps

Eastcote Halt station	1
Map of Eastcote 1916	2
Temporary Church	3
Invitation to Dedication	4
Record of Dedication	4
Accounts for 1927	5
Wooden Shrine	9
War Memorial (1925)	10
Page from *Register of Services* 1922	12
The Revd R. F. Godwin	15
Devon Parade Field End Road	16
Map of modem parishes	18
The Church Interior - from an early print	19
The Church Exterior - from an early print	19
The Foundation Stone 10 December 1932	20
Contemporary description of the church	21
Consecration 1933, Order of Procession	23
Plaque in Lady Chapel	24
Page from *Register of Services* 1933	24
The Aumbry	25
Our Lady and the Holy Child	25
Scale of Charges for Hall Hire 1936	27
Plaque in the hall 1939	28
The Methodist Chapel, Field End Road	29
Plaque on front wall of completed hall	29
Daisy-bell	30
Copies of the *Leaflet* 1934 and 1955	32
Confirmation Certificate - Tony Stamp 1950	35
Page from Tony Stamp's *Before The Altar*	36
Kerswell's Restaurant 1946	38
Youth Fellowship programmes 1955	52
Players' *Jack and the Beanstalk* 1952	55
London Diocesan Mission 1949	62
The Reverend William Hitchinson	71
Father Bill with his Curates	74
Cover Designs of *Gridiron* in the 1960s	76
Statue of St Lawrence	77
Detail from the Great Rood	78
Two of the Stations of the Cross	78
The East Window	79
The Lady Chapel Window	80
Public Auction catalogue 1973	81
The Chancel 1954	82
Garden of Remembrance	84
Memorial Book	84
Group under the tree 1978	85
Services 1974 and 1975	87
Weekday Service in the Lady Chapel 1961	88
Lighting the Paschal (Easter) Candle 1961	90
Confirmation 1964	91
View towards the West End 2000	96
Highgrove House	98
Infant Sunday School Class 1962	99
Scale model of St Lawrence Church 1967	100
Parish Outing to Cambridge 1962	107
Summer Fete 1962	112
Mission Exhibition 1962	115
No Small Change 1965	117
Open House Floral and Art Display 1972	119
Chapel of St Francis House, Hemingford Grey	120
Christian Aid Week 1963	123
Stewardship Campaign Father Bill's letter	126
Parish Supper 1963	127
Time and Talents	128
Players' *Murder on the Nile* 1959	140
Players' *Dry Rot* 1961	141
Women's World Day of Prayer	152
Fr Bill	154
Fr David Hayes' Institution and Induction	156
Fr Alan Body and Fr Peter Day	158
East End prior to changes in Sanctuary	160
East End after 1988	160
The Bronze Crucifix	161
The Parish Hall Building Project Brochure	163

What's On and Who's Who 1986	177	Holiday Club 1994	249
Cover of Jubilee Booklet 1983	180	Sunday Service -SLICK 2003	250
Father David's Introductory Letter	181	Stewardship Campaign 2004	255
		May Fair 2002	257
Golden Jubilee Programme	182	MU Garden Party	260
The Sanctuary Golden Jubilee Festival	183	Percy Pether Clock	261
		Players' *The Wizard of Oz* 1998	263
Review of Summer Concert 1985	186	Flower Festival photos	265
		Introduction to Diamond Jubilee Booklet	270
Stewardship Renewal 1985	190	Diamond Jubilee Programme	271
Charitable Giving 1981	193	Three Vicars	272
Christingle Service 1984	195	Fr David and Ann –Anniversary 2004	273
May Fair 1985	198	The Revd Stephen Dando	276
Women's Guild	205	The Revd Rachel Phillips with Revds Stephen Dando and Wendy Brooker	278
Players' *Playing For Time* 1984	207		
Players' *The Railway Children* 1987	208	Captain Donald Woodhouse	279
		Alan Wright	279
Golden Jubilee Festival Brochure cover	210	Finding Faith	286
		Confirmation 2010	287
The Revd David and the Revd Ann Coleman	216	Children's Corner in the Church	288
The Revd Michael and Mrs Bolley	218	Announcement of Fun Afternoon 2010	291
The Revd Sue Groom with others	219	Understanding Islam	292
		Fairtrade Logo	292
The Revd Ruth Lampard	220	Cristo Redentor Church, Luanda	293
The Lady Chapel 2008	224		
Decorators of the hall rooms 1992	225	Our Charities for 2009	294
		The Choir in Dublin	296
Sunday Evenings February 1997	230	Autumn Concerts 2007	297
		The Anniversary Cake, Dedication Service	298
Ministry of Healing 1997	231		
House Eucharists 1997	232	Diary Dates for 75th Anniversary	299
Year of Prayer 1995	233	The Ball Committee	300
Teaching and Nurture MAP	235	Family Fun Day 2010	301
Study Groups and other events Lent 1996	236	May Fair 2007	302
		Christmas Market 2005	303
Christians for Life 1997	238	Guides arrangement - Flower Festival 2008	305
The Font – Flower festival 2000	239		
London Challenge 2002	240	MU banner Flower Festival 2008	308
Call-in Centre	242		
Magazine Covers 1987 and 1996	244	Players' *75 Glorious Years*	310
		Fr Stephen with the Rt Revd Wilfred Wood	311
Free for All 2000	245		

CHAPTER I

THE EARLY YEARS 1920 to 1933

Moves towards a separate parish
While researching material for this history, it soon became obvious that a chapter on the years prior to the Consecration of St Lawrence Church, on 21 October 1933, was necessary. Some background as to why and how the parish came into being, and of the problems of the Mission Church in the 1920s, may help to illustrate how many of the traditions associated with the parish church of today had their beginnings in those years.

In the early years of the 20th century Eastcote was little more than a hamlet, ecclesiastically within the parish of St Martin's, Ruislip. According to Ron Edwards in *Eastcote, From Village to Suburb* the population of some 600 lived in 120 scattered farms and cottages. There were four large houses of varying ages: Eastcote House (demolished 1964), Haydon Hall (demolished 1967), Highgrove House and Eastcote Place (now residential flats). In addition there were a few Victorian villas, a smithy, four beer houses or public houses, and a Methodist Chapel.[1] The St Lawrence Church *Silver Jubilee Programme,* of 1958, states that as recently as 1910 the hamlet had 'some 150 houses, two shops, a couple of cars, no pavements and not a few ponds at inconvenient spots'.

The opening of the station in 1906 led to the centre of Eastcote moving away from the old village (RNELHS)

Eastcote Village Institute was the centre of social life. Built on the High Road in 1893, in a meadow opposite what is now Flag Walk, it was supported financially by the more affluent families in the area. Activities at the Institute included a fruit and flower show, choral and instrumental concerts, and treats for the village children.

The extension of the Metropolitan Railway from Harrow to Uxbridge in 1904, and the opening of Eastcote station as a halt in 1906, heralded changes which were to occur after World War I, when the countryside between Harrow and Uxbridge was gradually transformed by developers and Eastcote became more populated.

However the major developments were in the 1930s, when between 1931 and 1939 there was a big increase in the population of Northwood and Ruislip (which included Eastcote).

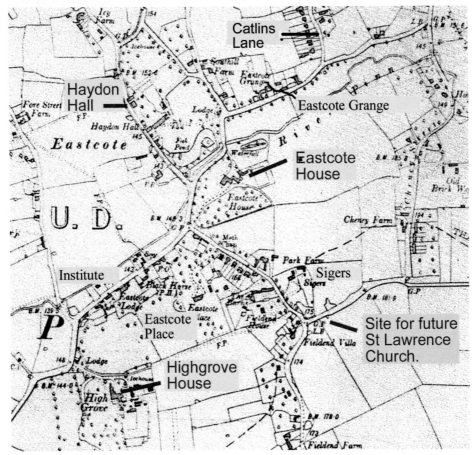

Section of Ordnance Survey map 1916; indicates rural nature of area and places mentioned in the text

The Mission Church

The idea of providing Eastcote with a parish church had been mooted before World War I, and in 1913 an acre of land had been bought for this purpose by the Bishop of London's Fund, from Mr Ralph Hawtrey Deane (1888-1924), owner of Eastcote House. He never lived in Eastcote. He was one of the owners of large estates in the area, who were to sell land to developers and whose houses were to be inhabited by future parishioners.

Led by laymen such as Kenneth Goschen of Sigers, Field End Road and Tom (T.G.) Cross, of Catlins Lane (from 1930 at St Catherine's Farm), Sunday evening services were held in the Eastcote Village Institute. When the First World War ended in 1918 the prospect of a church in Eastcote was revived and these Sunday services seem to have provided the stimulus for building the temporary church. With financial support from the diocesan authorities the temporary corrugated Mission Church was dedicated on Saturday 18 December 1920, on the site of the present Parish Hall. Large numbers attending Sunday evening services soon led to the erection of an additional ten feet to the original building. When the extension was made four of the original windows were altered 'for better ventilation'.

Temporary Church, dedicated 18 December 1920. Picture shows the extension to the original church   (RNELHS 1922)

Early in 1921 Ralph Hawtrey Deane gave a further acre of land to the church. The first Missioner, the Revd P.D. Ellis (December 1920 to October 1921), had lodgings in a cottage at Eastcote House. With this additional land the members of the Mission Church in Eastcote were optimistic and hopes rose of having a permanent church in the not too distant future. When the Revd Ellis left Eastcote, such was the optimism of some members that a separate Permanent Church Building Fund had been set up and proposals for a parsonage house had been put forward.

The Status of the Mission Church
From the start there was some confusion as to the status of the Mission Church in Eastcote, for at the first church meeting in January 1921 the Revd Ellis told the church members that the Mission was 'responsible only to the Bishop' but he also emphasised that Mr Goschen and Mr Whaley, whom he had nominated, were to act as wardens but 'with no legal status'. At this meeting elections to committees also took place.

Those early years were difficult, for while the members of the Mission Church, for periods without their own priest, were anxious to be independent of the mother church, the vicar of St Martin's was concerned that as legal head of the parish he was responsible for what went on in the Eastcote area. This led to some misunderstanding and tension between the two parties, in which the London Diocesan Home Mission was inevitably much involved. *Minutes* of church meetings of the Mission Church indicate an enthusiastic membership, while the St Martin's Church magazines *Ruislip Outlook* shed light on the issues from the viewpoint of the Revd Edward Cornwall Jones, (Vicar 1923-38). Matters were amicably resolved, but referring to the issues may give some understanding of the challenges faced when the Church embarks on the creation of any new parish, and also show how great were the achievements of the last Missioner and first Vicar of St Lawrence Church, the Revd R. F. Godwin.

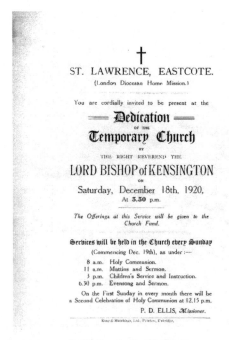 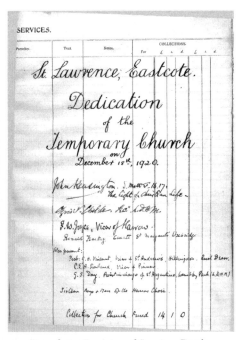

Invitation to the Dedication   Page from *Register of Services* Book.

The Revd J. McIntyre was Missioner from November 1921 to September 1922. During the following eleven weeks, when visiting priests conducted services, the 'Parochial Church Council', unofficial as it was, had further discussions regarding 'the building of a Permanent Church'. After the Revd J. G. Dale's arrival in December 1922 these discussions continued and various plans were considered. By 1924 the Vicar of Ruislip saw the daughter church at Eastcote as pushing forward plans for a permanent building rather too quickly. While acknowledging the role of the Revd Dale, he also paid tribute to the Revd William Arthur Gordon Gray, the previous Vicar of St Martin's (1900-23), without whose 'foresight' and 'initiative' Eastcote would not be 'on the verge of a desirable independence' (*Ruislip Outlook*). However the Vicar was concerned about the formation of a separate new parish while building developments in the area were uncertain and was not agreeable to a general appeal for funds. After a rather tense few months, and after consultation with the London Diocesan Home Mission, he agreed to the formation of a Permanent Church Building Committee under his chairmanship. At the Vestry and Annual Meeting in 1925, Mr Cross, at that time treasurer of the daughter church of St Lawrence, reported that the debt of the parsonage had been cleared and that the Permanent Building Fund exceeded £1000, which was regarded as a 'fine achievement'.

The Revd Dale left Eastcote in November 1925. Eastcote was again without its own priest, this time for ten weeks, and had visiting priests conducting services through Advent and Christmas, and into February. There was no service on Ash Wednesday 1926, for though the Revd D. Lewis was appointed in December 1925, he married in January 1926 and did not conduct his first service until the first Sunday in Lent. Relations with the vicar of Ruislip improved with the Revd Lewis's acceptance of the document, known as the Memorandum, which balanced the rights of the incumbent and the Missioner, and acknowledged the Vicar of Ruislip as the legal head in the transition from district to parish.

Accounts for 1927 (still part of Ruislip parish)

The Revd Lewis conducted his last service at Eastcote on 15 December 1929. When the Revd H.E. Nuttall conducted his first service the following Sunday, Eastcote had been created a Conventional District, a step in the transition to independence. The Revd Nuttall conducted his last service on 29 September 1931. The Revd R.F. Godwin, the last Priest Missioner conducted his first service on 3 October 1931. The final step to independence was the creation of Eastcote as a 'Peel' District in 1931, agreed to by the Bishop of London, the Vicar of Ruislip, and the Revd Godwin.

A 'Peel' district had authority by the Peel Act of 1843, passed when Sir Robert Peel was the Prime Minister (1841-46). The Act was a result of the Industrial Revolution and the growth of manufacturing towns. It allowed for the sub-division of populous places and the creation of separate parishes, relieving incumbents of their vested obligations. The district was divided from the mother parish before the church was consecrated, and the Missioner was licensed to the charge of the future parish.

Really the Revd Godwin was Priest-in-Charge, appointed to become the first Vicar. In the *Register of Services* the Revd Godwin wrote, 'District made a "Peel District" as from 9/11/31', just a month after his first service in Eastcote.

Eastcote was one of forty-five new parishes created at this time to meet the needs of the expanding urban population on the outskirts of London in the 20th century, especially north of the River Thames. Another local 'forty-five' church is All Saints', Hillingdon.

The Church building
Minutes of early meetings of the Mission Church make several references to improvements, repairs and additions to the building. In February 1921 a harmonium had been purchased for £26, to replace the one on loan, with the promise that Mr J.D. Marshall (of The Grange, in the High Road), a member of the church finance committee, would 'cart the instrument from London'. This in turn was part-exchanged for a second-hand French harmonium in 1928. The harmonium seems to have been at the back of the church, as was the choir, and although no specific reference has come to light as to its role in the services, in 1926 a 'cupboard for choir music was to be erected in the church'. In 1924 the church roof had to be painted with red oxide paint to combat rust, the work entrusted to Mr Tapping at a cost of about £9.15s. That same year 24 new hassocks were bought at 2/- *(10p)* each.

The Mission Church was soon a centre of village activities. Arrangements were made for hiring the temporary building, with charges of 10/6 *(55p)* per night for 'local entertainment' and 21/- *(£1.10p)* for 'outsiders'. In 1924, when a new brick Eastcote Institute was built in Fore Street, the stage and supports from the 1893 tin hut in the High Road were purchased for £5. Proceeds from various entertainments paid for this and for the redecoration of the inside of the building. Further redecoration was necessary in 1928, by which time the heating system had been up-dated with the installation of a New Tortoise stove.

During the years of the Mission Church a number of plans for the permanent church were considered, and although changes of Priest Missioner delayed matters to some extent, by 1931 the prospect of having a permanent church was a reality. By then the sum of £2,000 had been deposited with the Church Commissioners from the Permanent Church Fund.

The London Gazette of 13 November 1931 published the scheme, later ratified by an Order in Council, which 'constituted a separate district for spiritual purposes and that same shall be named the parish of St Lawrence'. Progress was accelerated, Sir Charles Nicholson was commissioned to draw up plans for a permanent church, and the foundation stone of the permanent church was laid on 10 December 1932, by Lt Commander Ralph Hawtrey Deane, son of the previous owner of the land.

The Parsonage
A loan of £500 from the Ecclesiastical Commissioners was granted in 1921 for the purpose of having a clergy house. In 1922 when it was felt that the parsonage was urgently needed, and after much local discussion and consultation with the London Diocesan Fund, it was agreed to have a bungalow built by the Universal Housing Co. Ltd of Rickmansworth, at a cost of £1,320. Reference to a further loan of £200 was agreed. The Revd McIntyre seems to have been a tenant in a cottage belonging to Mr Marshall the whole time he was Priest Missioner from November 1921 to September 1922. By the end of 1922 the Parsonage was built, and the Revd Dale, Missioner from December 1922 to November 1925, lived there. In 1923 arrangements were made to install a hot water system at a cost of £30.15s.0d. In December 1927 a report stated the water supply in the parsonage to be 'completely disorganised' and 'needing expert advice', but by the end of January 1928 the water system in the parsonage 'was now satisfactory'.

The Accounts for 1927 presented at the Ruislip Vestry and Annual Meeting, give some indication of the financial situation in the Eastcote district. The Vestry and Annual Meeting in April 1929 record the Electoral Roll having 270 Ruislip names and 211 Eastcote names (See Appendix 10).

Church Services in the Mission Years
As is the practice in the Church of England, worship focused on Sunday services and the Church's major festivals. However the Eucharist was not the principal Sunday service as it is today. Emphasis was on Morning and Evening Prayer, usually referred to as Matins and Evensong, as set out in *The Book of Common Prayer (BCP),* (first published in 1549, following the Reformation but with later revisions). 'The Parish Communion movement' which started in the mid 1930s, and aimed to restore the Eucharist as the main act of worship on a Sunday morning, made little progress until the 1960s.

Information has been extracted from the *Register of Services* Books, which give times of services, and numbers of communicants at Holy Communion services. However total congregational figures are not recorded, except on rare occasions. Even this has proved to be frustrating as there are entries where times of Communion services have been recorded with no figures but no comments to indicate whether or not there were communicants. While not consistent as to information recorded, these books do indicate a willingness to adapt, as Eastcote was growing in population and changing in character.

It is not surprising that changes and additions usually followed the appointment of a new incumbent. It must have been difficult for the people of this new District, for as well as having six different Priest Ministers, there were periods in between when visiting priests conducted services.

Sunday Services and Major Festivals
The pattern of weekly Sunday services was shown on the invitation to the Dedication of the Temporary Church on 18 December 1920: Holy Communion at 8am, Matins and Sermon at 11am, Children's Service and Instruction at 3pm, and Evensong and Sermon at 6.30pm. A second celebration of Holy Communion was held on the first Sunday of each month at 12.15pm, but was soon moved to 12noon.

At the major feasts of Christmas and Easter, and on Whitsunday from 1923, there were three celebrations at 7am, 8am and 12noon. Beginning on Christmas Day 1929, a few days after the Revd H. E. Nuttall became the fifth Priest Missioner, the later celebration was brought forward yet again to follow on after Matins and Sermon. It is interesting to note that throughout these years the number of communicants on Easter Sunday was greater than on Christmas Day, even in the years when Christmas was on a Sunday (see Appendix 5). One has to remember that transport was limited and the weather in December seems to have been more severe than in recent years. It should also be borne in mind that stress was placed on *BCP* instruction that parishioners should communicate at least three times a year, and that Easter is the one obligatory occasion. Whitsunday services seem to have been well attended with greater communicant numbers than the total of all four Sundays in a normal four-week month.

Weekday Services
The chart with footnotes (see Appendix 5), which shows numbers of Communicants may help to clarify points made in this section.

For much of this period services in 'ordinary' weeks were inconsistent and seemed to depend on the incumbent. By 1932 regular Holy Communion services were established at 7am on Wednesday and 10.30am on Thursday. On Ash Wednesday and Ascension Day, Holy Communion celebration was at 7am, with Evensong at 8pm. From 1930 Revd Nuttall introduced an additional Holy Communion at 10.30am on these special days of the Church.

During Lent in some years there were two Holy Communion celebrations in the week, and sometimes three; most years there was one weekday Evensong. The number of services in Holy Week gradually increased and, from 1930 on Monday to Thursday there was Holy Communion at 7am and 10.30am and an evening service. Services on Good Friday varied from year to year.

All Saints' Day and All Souls' Day had Holy Communion celebrations most years. St Lawrence Day, 10 August, passed without mention in *Register of*

Services books in some years.

Baptism and Confirmation

Although St Lawrence Church was not a Peel District in its own right until 1931 and unable to hold marriages until the permanent church was consecrated in 1933, the Priest Missioners baptised at the church, and also prepared candidates for Confirmation but at other churches in the Deanery. There was one baptism the day following the Dedication in 1920 and an annual average of around 15, the names being recorded in a separate *Baptism* Book. Until Eastcote became a 'Peel District' and the church had its own *Confirmation Register,* the names of candidates, where and by whom they were confirmed, were recorded in the relevant *Register of Services* Book. Annual numbers ranged from 2 to 18 but in 1926 the entry merely stated 'Confirmation Service at Ruislip', and there are no references to Confirmations in the years 1927-9. The First Communion of the newly confirmed was usually on the Sunday following Confirmation.

Other Annual Services

In addition to the Church's Holy Days, other annual services held by the Mission Church continue today, each of them witnessing to the wider community of Eastcote in a special way. Harvest Thanksgiving Sunday was of particular significance in an area where there were still a number of farmers and farm workers and in many years a Harvest Festival Evensong was held one evening in the preceding week.

Armistice Sunday was marked by a service at the War Memorial Cross, now held in the War Memorial Gardens in Field End Road. Occasionally this started at 11.15am on 11 November but for unknown reasons it was often held in the afternoon of the Sunday nearest to Armistice Day.

Eastcote War Memorial

It is relevant at this point to give some background to the Eastcote War Memorial Cross for the temporary wooden shrine it replaced is now in the north aisle of St Lawrence Church. This wooden shrine was erected on a patch of grass, at the junction of Field End Road and Bridle Road in 1917 to commemorate the local men who had fallen in the War, at that time numbering six. The shrine, dedicated

Wooden Shrine in the church

The War Memorial, Field End Road 1925
(RNELHS)

by the Right Revd Bishop of Kensington, was a gift from Mr and Mrs B.J. Hall of Field End Lodge, who also gave one to Ruislip Parish Church. Field End Lodge, used as a Voluntary Aid Detachment Hospital (VAD) during World War I, is now part of the Tudor Lodge Hotel. In 1921 the wooden shrine was replaced with the official War Memorial Cross. The shrine was put first in the temporary church and finally in its present position. The central panel has the names of the 16 Eastcote men who gave their lives in the 1914-18 War, and the wings record the other 118 Eastcote men who joined H. M. Forces in that War and survived. The names of nine Eastcote men who died in the 1939-45 War have been added to the Memorial.

The site of the War Memorial Cross proved unsuitable for the growing area and after being damaged by one of the buses that ran from Uxbridge along Bridle Road it was repaired and moved to the War Memorial Gardens in Field End Road in 1929.

Committees
All parish communities have committees, organisations and clubs connected with their local church, some having close ties with the church and others witness to the role of the Church in serving the community at large. While the community in Eastcote was growing and facilities at St Lawrence Church were limited, one would not expect a large number of groups, and as it remained an area within St Martin's and had no church magazine of its own the information is sketchy. However that information shows that fellowship, and concern for young people, were as important in those early years, as they are today.

At the first meeting of the church members in 1921, as well as the appointing of 'Acting Wardens', and the election of Sidesmen, members were also elected to serve on three committees: Entertainments, Grounds and Church Works, and Finance. As the years went by, other issues led to the formation of different committees such as Social, Fête, very important for fund raising, and Young People's Welfare, showing concern for the young in the parish.

The Summer Fête
In 1921 the Finance Committee discussed a proposal for a combined Summer Fête and Agricultural Show. A committee was appointed to organise the fête, and it was decided 'to ascertain the opinion of the local farmers'. Like the Harvest Festival held during the week, this is a further illustration that Eastcote in the 1920s was very different from the Eastcote of today. The annual summer garden fête, held for some years in the grounds of Mr and Mrs Kenneth Goschen of Sigers in Field End Road (now the area of Robarts Close) was a happy social occasion as well as being financially successful. A Ladies Sewing Circle was set up to make articles for sale each year. Jumble sales were held to raise money for the materials for these ladies and for other fête expenses. These annual events had support from tenants of the large properties in the area, including Lady Anderson who lived at Eastcote Place, and who appears to have had a stall in at least two of these fêtes. After Mr Goschen became a Director of the Bank of England in 1922, and moved from the area in 1923, the fête or fair was sometimes held in the grounds of Field End Hall. This was a community hall in the Telling Brothers' development of houses and shops in the area of Morford Way, Morford Close and Field End Road. In the early 1930s it was converted into the Ideal Cinema and is at present the site of Initial House, near to the Manor House pub.

Other Social Occasions
Although particular reference to the celebration of St Lawrence's Day is not found every year in the *Register of Services*, there is reference in *Minutes* to the 'annual St Lawrence Party on 13 August 1924 in the Parsonage gardens', so one may assume this was another regular summer social event on the calendar. Other social occasions mentioned in some *Minutes* books which seem to have been regular events, included 'theatricals', a New Year's Party or Social, and a 'Children's Party and Social for all the village children'. This was presumably a Christmas party for those who attended the Children's Service on Sundays but open to others as a form of out-reach to the community at large. The 'theatricals' could have been the seeds of what was to become the St Lawrence Players, for there were requests for Mrs Harrison Smith to 'get up some theatricals again' in aid of church funds at fêtes. As is the case today, people in the area supported events held to raise money for St Lawrence's Church.

Children's Services
The sources of information for these years have been of little help as regards Church teaching for children and young people, in what we think of today as Sunday School. The *Register of Services* books note Children's Service and Instruction, or Catechism.

Uniformed Groups
Evidence that the Mission Church saw the need to encourage the children of the community can be seen in the uniformed organisations associated with it.

One organisation referred to for a short period was the Church Lads' Brigade under Capt. Parker. From 1921-1922 weekly Sunday services were held for the Lads. *Minutes* refer to 'the purchase of 12 vests and one set of boxing gloves for £3', to helping to finance a camp in 1922 and making a grant of £10 to the group in 1924. No further information is available. It seems that the Girls' Friendly Society (GFS), founded in 1875, was active in Eastcote rather longer than the Lads' Brigade, for it was listed as an organisation in the first Eastcote *Leaflet* of 1932. With the aims of developing the whole person, girls were expected to participate in 'worship and service to others', as well as through study and recreation. There is an entry in the *Register of Services* of a GFS Corporate Communion in 1928. In 1929 certificates were awarded for ten years' membership and that same year the Eastcote Branch won the silver cup for Advanced Country Dancing.

Page from *Register of Services* Book February to March 1922

There are very few references to events involving any section of the Scout movement in these years. The Service books record the Dedication of Girl Guide Colours in July 1923, Rededication of Guide and Brownie Colours in 1930, and very occasional Parade Services, including a Scout, Cub, Guide and Brownie Parade in March 1932. Apart from these there is one reference in *Minutes* to the Girl Guides Operetta 'being a success' in May 1926.

The Mothers' Union
One organisation to become important in the parish was the Mothers' Union. Soon after Eastcote became a 'Peel' District in November 1931, moves were made to set up a separate Eastcote branch. An inaugural meeting was held in February 1932, and by the end of that year monthly Corporate Communions and meetings were routine features of the Eastcote Branch in the Uxbridge Rural Deanery.

St Lawrence Church in the Local Community and Beyond
This chapter ends with a few more references to St Lawrence's connections with the local community and further away, to illustrate that a Christian community witnesses beyond the local church membership. The temporary building was, of course, used by the church members for social and fund raising events including socials, New Year and children's parties, meetings, whist drives, sales, and stage entertainments. In arranging for the hire of the building for 21/- per night for 'outsiders', as well as 10/6 per night for 'local entertainment', the church provided facilities for many village activities not directly associated with the church. The Women's Institute (WI) had use of the hall the first Monday each month for £1 when the Eastcote Branch was formed in 1924, which points to the WI being an 'outsider' but having some special concession. In 1928 the 'fee for the Hall for theatrical performances', in support of this group was 10/- for three nights. The Branch transferred to the new brick-built Eastcote Institute building in Fore Street when it replaced the 'tin hut' in the High Road. The Institute building was purchased by the Middlesex County Council to open as an Infants School in 1927 and was, at that time, the only school in Eastcote.

Arranging for the use of church facilities is one way to connect with the local community but the people of St Lawrence Church from early days, have given support in other ways. Records show that some of the Priest Missioners prepared boys for Confirmation who were at St Vincent's Home and School for Defective Children, as it was known when it opened in 1912, and that the boys occasionally attended Holy Communion services at St Lawrence's. Another local establishment to be supported by the Eastcote Church was the Northwood and Pinner Cottage (Community) Hospital in Pinner Road, which formally opened in December 1924. Until 1926 proceeds from special collections were sent St Thomas's Hospital in London but from 1926 the appeal of supporting a local institution led to these annual donations going to the local Hospital. In some years produce given at the Harvest Festivals was taken to this Hospital.

The Permanent Church Fund was the major cause for which money was raised in the Mission years but the annual Diocesan Quota, then as now, led to appeals for commitment to regular giving by parishioners. Most years a 'Week of Prayer and Self Denial' held in the autumn, also raised funds for the Diocese. Collections for specific charities became regular features in the 1920s. The Christmas collections were for the 'Waifs and Strays' (now The Children's Society) and the Good Friday collections for The Church Mission to the East London Jews. For several years collections were taken at the War Memorial services for the Earl Haig Fund (now The Royal British Legion). Support for overseas mission was demonstrated in special collections, taken on alternate years, for the Society of the Propagation of the Gospel and the Church Missionary Society. As happens today, St Lawrence Church responded to special appeals such as St Paul's Restoration Fund in 1925.

Perhaps the increase of population in the area and in the membership of St Lawrence Church is reflected in the few figures available for the Diocesan Quota, £15 in 1921, £26 in 1927 and £40 in 1931. The years which followed, when the Revd Godwin was Priest-in-Charge and Vicar, saw the growth and influence of the Church of St Lawrence in Eastcote.

References
1. Edwards Ron *Eastcote From Village to Suburb A Short Social History 1900-1945;* Hillingdon Libraries 1987 p 9

CHAPTER II

THE REVEREND RUPERT FREDERICK GODWIN VICAR 1933 to 1956

This chapter considers the period of the Revd Rupert Frederick Godwin's incumbency and his influence as Vicar from 1933-1956, years during which the community of St Lawrence Church established its presence in Eastcote. Although the year 2008 marked the 75th anniversary of the consecration of the permanent church with the Revd Godwin as the first vicar, his contribution really began before 1933. He came to Eastcote in October 1931 as the last Priest Missioner, but with the knowledge that he was to be the first Vicar. It seems appropriate to digress and explain why there was not a special service at which he was instituted and inducted as Vicar, as is the practice when a vicar is appointed to an existing parish.

The Revd R F Godwin
(Photo from 1966 magazine)

In the case of a 'Peel District' where the district is divided from the mother parish before the church is consecrated, the Missioner of the district is licensed to the charge of the future parish before the church is built. When that has been done and the church has been consecrated, the Minister automatically becomes the first Vicar of the parish without any further ceremony.

Eastcote district became a 'Peel District', as from 9 November 1931, and therefore a separate parish. The Revd Godwin was then licensed to St Lawrence Church as Minister or Priest-in-Charge on 8 February 1932, by the Bishop of London at Fulham Palace, and so became the first Vicar immediately after the consecration on 21 October 1933.

The Development of Eastcote

This book is not the place to give a detailed history of Eastcote, but some reference to the growth of the area is necessary, in order to appreciate the challenges faced by the new parish church. St Lawrence Church, Eastcote had been created as a new and separate parish specifically to meet the demands of a growing population. In the early days of the Mission Church, the Vicar of Ruislip may have had some misgivings about the formation of a separate parish, while building development in the area had been uncertain, but the 1920s saw more sustained growth.

Devon Parade, Field End Road.
On the left the Telling shops (see page 11) and the Manor House in the centre distance
(Uxbridge Library *The Gazette Series 1935*)

It is impossible to give separate and accurate population figures for Eastcote, but figures from local history books give evidence of considerable growth. Eileen Bowlt's *The Goodliest Place in Middlesex*, states the population figures for Northwood and Ruislip (which included Eastcote) as 9,113 in 1921, as 16,035 in 1931, and, then a big increase, to 47,000 in 1939. The last increase was nearly all in Ruislip and Eastcote.[1] Ron Edwards' *Eastcote, From Village to Suburb* gives the population of Eastcote, which at the beginning of the 20th Century was just under 600, as close to 15,000 by 1939.[2] The Vicar reported in the April 1949 *Leaflet* that he had been 'advised by a kind Councillor, that the population of Eastcote was between 16,000 and 18,000'. In 1955 the population was thought to be 20,000.

Moves to Establishing the Parish of Eastcote

That the Revd Godwin successfully guided the parish through years in which, added to the many difficulties facing a new parish with an expanding population, were the demanding times of a major war and its consequences, will be made evident in the following pages. Building projects were delayed and some church activities were suspended or curtailed during the war, and it was many years after the war before the potential of St Lawrence Church was fulfilled.

The *Uxbridge Gazette* in October 1943 reported, that after a decade of 'setbacks and disappointments', the Church of St Lawrence had 'quite a reputation for its well-ordered services'. By the time the Revd Godwin retired in 1956, the problems of the 1939-45 War and its aftermath had been overcome and the activities and influence of the 'new parish' was established in Eastcote. At the Revd Godwin's retirement presentation in 1956, the Vicar's Warden reminded parishioners of how Father Godwin had 'assisted and supervised the building of our splendid church, the vicarage and this fine hall, of which we can be justly proud'. As these buildings, together with the St Lawrence Centre, originally the parsonage bungalow, are the material or outward and visible signs of the presence of a Christian community it seems appropriate at this stage to discuss why and how these buildings came into being. Obviously the 'why' and 'how' will touch on other aspects of the life of the church which will be gone into in some detail in later sections.

The Permanent Church

By November 1931 the prospect of having a permanent parish church was a reality. Eastcote had become 'the parish of St Lawrence', £2,000 had been deposited with the Church Commissioners from the Permanent Church Fund, and a man had been appointed, who was to lead the church community from being a mission hall gathering, to a parish church congregation.

Various plans for the permanent church had been considered in the 1920s, and though nothing came of these, they may have helped to clarify the type of building that would be suitable for the expanding urban population of Eastcote. In June 1932 the Vicar was able to write in the parish *Leaflet*, a publication he had started in the February, that the parish was 'within sight of a permanent Church'. The London Diocesan Fund had commissioned Sir Charles Nicholson to draw up plans for the permanent church. The sale of the site of St Stephen's Spitalfields, provided £5,500, *via* the London Diocesan Fund, towards the total cost which, was not to exceed £8,000, including fittings and furniture. The link with St Stephen's is commemorated by a plaque in the Lady Chapel, and is a reminder to those who worship at St Lawrence Church to *be faithful to the trust we inherit*. In August 1932 the plans of the new church were on view.

In September notice was given that the parish boundaries had been defined and weather permitting 'beating the bounds' would take place on 8 October. In October 1932 the Vicar reported that Sir Charles Nicholson's plans had been approved and the site of the church fixed. The PCC chose Messrs Wooldridge & Simpson, of Frenchay Road, Oxford, as the main contractors for the building; they were also to furnish the interior. The contract was for £6,719, which did not include the organ.

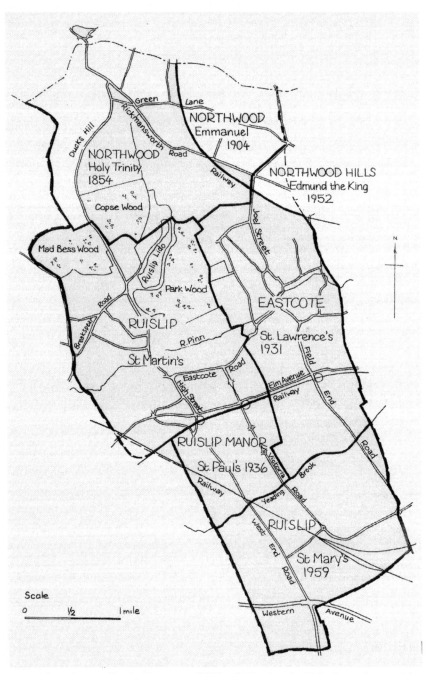

Parishes (now within the Deanery of Hillingdon) created within the ancient parish of St. Martin's, Ruislip during the 20th century
(from Eileen Bowlt's *The Goodliest Place in Middlesex* – P 30)

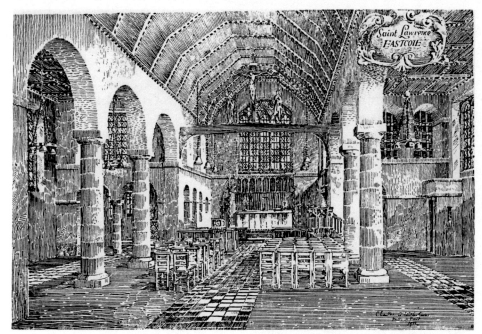
The Church Interior – from an early print

The Church Exterior – from an early print

From the time Eastcote was declared a 'Peel District', in November 1931, efforts to raise money for the Building Fund were intensified. By September 1932 this Fund stood at £2,600 (including the £2,000 deposited with the Church Commissioners). The autumn sale of work remained the main fund raising event, with the 1932 autumn sale being particularly successful, raising £156. More modest sums were raised by whist drives and jumble sales, organised by various church groups. In November 1932 a special St Lawrence Appeal Fund was launched, and literature was circulated throughout the parish so that most residents of Eastcote were aware that, at last, there was to be a permanent church. For some reason this was not a great success. Perhaps it was too early for the new parish church to expect a response from the community at large, when there were no signs of building activity.

Foundation Stone laid on 10 December 1932

Lt Commander Ralph Hawtrey Deane, son of the last owner of the land, laid the foundation stone of the permanent church on 10 December 1932, in the presence of the new Rural Dean of Uxbridge, the Revd H.W. Reindorp, on his first visit to the parish. The Bishop of Kensington, the Rt Revd Bertram Fitzgerald Simpson, conducted the service. The Vicar of St Martin's was prevented from attending but had loaned the church's processional cross; St Martin's Choir joined that of St Lawrence. Scouts and Guides provided a guard of honour.

The Vicar's letters in 1933 issues of the *Leaflet* reported on the progress of the building of the church. 'Bricklayers will have finished their splendid work by Easter then it's the roof'. 'Clearance of the scaffolding has revealed a very fine building...hope church may be completed by August'. 'Great topic of conversation is the Consecration on 21 October'

Father Godwin seemed aware that establishing a new parish was not going to be easy, and in his September letter also reminded readers that 'a church is intended to house a congregation...all should develop a keenness to become regular members' and he went on to warn 'changes there must be, and we must all be equal to the enlarged status which will be ours as the congregation of a Church possessed of full parochial rights'.

ST. LAURENCE CHURCH, EASTCOTE

THE BUILDING

The surroundings of the new church seem to suggest a building treated more as a country church than as a town one. The church has no clerestory and is roofed in three parallel spans, the roof over the organ loft being carried up to form a small belfry with saddleback roof; three bells are provided for here.

The material of the building is brick with tiled roofs and dressings of Weldon and Clipsham stone; the floors are of red tile with stone steps and the block flooring is Australian jarrah.

The interior is simple in treatment. The walls are plastered, the columns are of Weldon stone, and the roofs have boarded ceilings decorated in colour. Heating is effected on the pipeless system, and the lighting is by electricity in shaded lamps. The church is furnished with oak stalls, pulpit, etc., and a rood beam marks the entrance to the Chancel.

The organ loft is over the choir vestry, the organ being played from a console on the opposite side of the church. There is a chapel on the north side of the chancel, and good vestry accommodation on the south side. The font stands at the west end and has an oak cover. The seating is with chairs of special design with folding kneelers. The glazing is in clear white glass throughout, and the flat parts of the roofs (over vestry, etc.) are laid with copper.

The following are the principal contractors employed:—
General work: WOOLDRIDGE & SIMPSON, of Frenchay Road, Oxford.
Heating: THE LONDON PIPELESS HEATING Co.
Electric Lighting: W. FRANKLIN & SON, 45, Buchanan Gardens, Willesden.
Bells: JOHN TAYLOR & Co., Loughborough.
Organ: THE JOHN COMPTON ORGAN Co., Willesden.
Chairs: MEALING BROS., High Wycombe, Bucks.
Block Flooring: THE ACME FLOORING & PAVING Co., Barking.

CHARLES A. NICHOLSON,
Architect.

THE BELLS

A. M. Walrond, Esq., through whose kind offices the church is equipped with three bells, supplies the following regarding the Barron Bell Trust.

The Barron Bell Trust was founded and endowed by the late Miss Emma Barron of Veseys, Great Holland, Essex, in 1925, to provide, retune, and repair the bells of churches situated in Great Britain (Church of England). The Trustees have since that date provided or repaired bells in the following counties: Essex, Yorkshire, Dorset, Hants, Middlesex, and Kent—the largest repairs being at the Parish Church of Portsea and Christchurch Abbey, Hants., whilst they have provided a full set of 10 bells to Great Holland Church in memory of the Founder.

THE ORGAN

The organ has been built by the John Compton Organ Company, Ltd., of Willesden, N.W.10. It is a two-manual instrument, designed and constructed on the extension system, with electric action. The pipes are enclosed in a swell chamber and located in an elevated position on the north side of the Chancel. The keyboards of the organ are placed on the south side of the Chancel at floor level, so that the organist can hear organ, choir, and congregation to advantage. Blowing is by Discus apparatus and electric motor.

Description of the church from
The Architect and Building News dated 1933.
Note the spelling of Lawrence

Although the 1932 St Lawrence Appeal Fund had not been a success, a separate Organ Fund had had a good response. A two-manual organ supplied by John Compton Organ Company Ltd. of Willesden, at the cost of around £900, was installed in time for the Consecration.

Throughout its history many material gifts have been made to the Church of St Lawrence, Eastcote, and while it is impossible to list them all it is perhaps acceptable to mention two of the early ones at this point. One gift, still used in the church, was a pair of carved chairs, given in March 1932, for use in the sanctuary 'when we get our church'. A second gift was that of the three bells for the small belfry.

Mr Norman Parker, who had been a choirboy from 1942-1946, made a special effort to attend Evensong in the summer of 2006, when he was on holiday from South Africa. He had not been in the church for 60 years, and his first comment was that the pulpit had been moved. To date, no one else remembers the pulpit being other than where it is today.

The Consecration of St Lawrence Church
Saturday 21 October 1933
As 21 October 1933 is such an important day in the history of St Lawrence Church, Eastcote, and as the Consecration of a church is an event few people experience, it seems right to give some idea of this memorable day. A perfect October day favoured the proceedings. The Bishop of London, the Right Revd Arthur Foley Winnington-Ingram consecrated the church, accompanied by the Bishop of Kensington and a number of other clergy and other dignitaries, including the Registrar. One of the clergy was the Revd P. D. Ellis, at that time Vicar of Pinner, who had been the first Missioner priest of Eastcote.

A large number of people gathered for the occasion, not only from the surrounding area but also from other parts of the Diocese. Many people came from the Revd R. F. Godwin's former parishes of Tottenham and Paddington. The church was packed to overflowing and some visitors were unable to get into the church. A procession headed by the Cross bearer, made the time-honoured circuit of the church singing the hymn *Blessed City, Heavenly Salem*, and halted at the south-west door. Here the Bishop of London was presented with the petition praying for the consecration of the church. Having assented, the Bishop knocked three times on the closed door, the doors were opened and the keys handed to the Bishop. He then gave the salutation of Peace and offered prayer.

The Revd Godwin, the Vicar, recited a short Litany. Then the Bishop, attended by the Archdeacon of Middlesex, his chaplain and the Vicar, proceeded around the church and in turn asked blessing on the font, chancel, lectern, pulpit, clergy and choir stalls, organ, chapel and Holy Table. Prayers were offered for benefactors and others. On returning to the chancel the sentence of consecration was read by the Registrar, and the Bishop performed the act of Consecration.

> **ORDER OF PROCESSION FOR THE CONSECRATION OF**
> St. LAWRENCE CHURCH, 21st OCTOBER 1933
>
> *At 3.55 p.m., the Clergy and Choir being robed in the vestry,*
>
> *there shall be formed a procession in the following order;—*
>
> The Crossbearer.
> The Choir.
> The Clergy.
> The Rural Dean.
> The Foreman.
> The Builder.
> The Architect
> The Churchwardens of St. Martin's, Ruislip.
> The Vicar of Ruislip.
> The Director of the Forty-five Churches Fund.
> The Secretary of the London Diocesan Fund.
> The Incumbent of St. Laurence, Eastcote.
> The Registrar.
> The Archdeacon of Middlesex.
> The Bishop's Chaplain.
> THE BISHOP OF KENSINGTON.
> The Bishop's Chaplain.
> THE BISHOP OF LONDON.
> --- oo ---

Extract from *Leaflet* in 1933.
Note the different spelling of Lawrence in the same article

In his address the Bishop said it was a great day for him, for it was a vindication of his policy of pulling down old churches in the City, which had lost their usefulness. The proceeds from the site of St Stephen's, Spitalfields had provided a considerable sum towards the cost of the Eastcote permanent church.

The Bishop of Stepney conducted Holy Communion on the Sunday morning, when 166 people received the sacrament. In the evening the address was given by the Revd R. W. Odell, Director of the Forty-five Churches Fund.

Plaque in the Lady Chapel

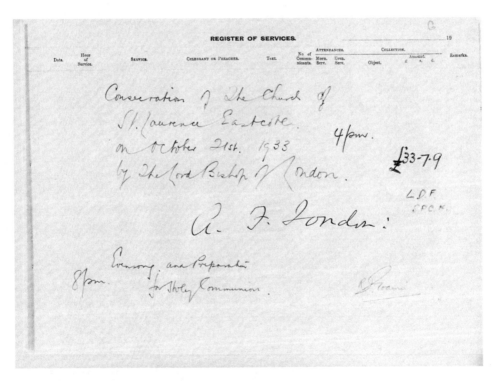

Page from *Register of Services*

Some additions and some alterations to the Church

It is hardly surprising that there were few additions or alterations in the new church during the incumbency of the Revd Godwin. However as is the case with all additions or alterations in the church a faculty, or authorisation, from the

The Aumbry

diocese had to be obtained. In 1935, the balcony arch overlooking the chancel was filled in with a painted wooden screen. It was given by the widow of Harry Cubitt, the Quantities Surveyor for the church, to complete the scheme of Sir Charles Nicholson. When in 1941 the Churchwardens and Deputy Churchwardens wanted the congregation to mark his tenth anniversary at St Lawrence, Eastcote, the Vicar wanted a frame and door to complete the aumbry.* at the High Altar. A fund was set up, a design by Sir Charles Nicholson submitted, but doubt was expressed as to when the work would be done owing to war conditions. In fact the blessing of the aumbry took place the following year.

During these formative years of the new parish, many gifts were made which supported the High Anglican Church tradition associated with St Lawrence Church, and the great importance attached to attendance of the congregation at Holy Communion. Gifts included the churchwardens' wands, which are carried on ceremonial occasions, altar silver, dossal curtains, (the hanging cloth behind the altar), and altar frontals and vestments, in seasonal liturgical colours.

In 1954 a statue of Our Lady and the Holy Child, which stands at the entrance to the Lady Chapel, was presented in memory of William Collins and his family.

Parochial Church Council (PCC) *Minutes* record that there were some mishaps. In 1942 strikers for the gongs in church were stolen. In 1943 it was reported that the Home Guard damage to the church roof had been repaired. There are no details on either of these incidents. 1948 *Leaflets* refer to 'faults' on the organ resulting from a rotten main cable. The replacement and instalment of a new one, to run along the sanctuary step rather than under ground, cost £100. The *Leaflet* of October 1950 reports 'The ravages of the fire in the Lady Chapel have now been removed.'

Our Lady and the Holy Child

---

* The Aumbry is used to securely store consecrated sacraments. Normally sufficient is stored for use when communion is taken to the homes of those who are sick.

The Vicarage

In 1922 the bungalow had been built for the priest at St Lawrence Church. From 1932, his first summer at Eastcote, the Revd Godwin held an annual summer 'At Home' there for friends and supporters of the church, as one way of getting to know his parishioners. Once he became the Vicar of the parish the building was known as the vicarage. From the beginning the building seems to have had drawbacks, some referred to in the first chapter. In 1935 some improvement was made, at least to the exterior appearance, when in the Vicar's words 'the woodwork is now a pleasing green'.

Few details are available as to the planning of a new vicarage. *Minutes* refer to a controversial PCC meeting in September 1936, chaired by the Archdeacon of Middlesex. The *Leaflet* of that month records 'great progress has been made with the new vicarage...very soon the roof timbers will be on'. The Vicar moved into the new vicarage, designed by the architect of the church, Sir Charles Nicholson, and built behind the church, at the end of March 1937.

The bungalow was redecorated and the Standing Committee was 'empowered to find a suitable tenant'. The Standing Committee of a church is a small group of members of the PCC, who have the responsibility to decide certain church issues. The revenue from the property was to be divided between the parish and the London Diocesan Fund, for the Vicar's stipend. The first tenants were Mr and Mrs Munson. For many years, from 1941 the tenants were Mrs and Miss Angell.

What was the future of the Mission Church building and its contents after the Consecration of the Church? The War Memorial was transferred to the permanent church. The altar from the Mission Church was sent to the Bishop of Buckingham, for use at Stoke Poges. The harmonium went to St Paul's, Roxeth. The lectern was given to the newly established mission church at Pinner Green. St Lawrence Church was one of three parishes from which the Parish of St Edmund the King was created, and lost part of the Joel Street area. When the permanent church was consecrated in 1964 St Lawrence Church presented a Processional Cross to the new parish. The church hall, according to the Vicar in January 1934, 'sadly needs a thorough restoration...to make the hall comfortable for our own and outside needs'. With the increased annual budget that came in having the new church, only minor improvements could be made for the time being.

In 1935 a Permanent Hall Fund (PHF) was set up, and in March 1936 *PCC Minutes* record a balance of £50, towards the estimated cost of £1,500-£2000. Fund raising events for this project included the annual autumn sale, whist drives and jumble sales, but other means were also tried. One novel scheme, organised by Miss Warren from December 1935 until 1949, was a birthday guild. Participants paid a shilling or sixpence into the Fund on *their* birthday, and though large amounts were not raised the proceeds were welcome additions. The Vicar introduced PHF boxes in July 1936, suggesting every member of the congregation took one and returned it at the autumn sale of work. In 1938 the Vicar's annual summer 'At Home' was enlarged into a Vicarage Fête, making it a means of assisting the Fund as well as being an enjoyable social occasion. Profits from these fêtes were never as much as from the annual sale of work but they were a different type of event.

> **ST. LAWRENCE, EASTCOTE, CHURCH HALL.**
>
> Scale of Charges for Hire.
>
> Socials, Whist Drives, Meetings, etc.—Evenings : 7 to 11 p.m., 15/- May to September, 21/- October to April. Afternoons : 2 to 5.30 p.m., 12/6 May to September, 17/6 October to April.
>
> Classes.—Mornings : 10 a.m. to 12 noon, 4/- May to September, 7/6 October to April. One hour : 4/- May to September, 5/- October to April. Afternoons : 2.30 to 5 p.m., 2/6 Summer Sessions, 7/6 Winter Sessions.
>
> Brownies and Guides.—1/- hour. Church Hall only. In the grounds of the Church only under strict supervision.
>
> The Council reserve the right to refuse lettings.
>
> By order of the St. Lawrence Parochial Church Council : F. Adamson (Hon. Sec.), 81, Abbotsbury Gardens, from whom particulars can be obtained.

Scale of Charges for Hall Hire 1936

The Vicar made several references to the old hall in 1939 *Leaflets*, emphasising the need for improved facilities not only for church events, but those of the wider community. In March the Fund stood at nearly £600. By May, with a loan of £500 from the Diocese (free of interest), and help from other sources, the decision had been made to build the 'east end' of the permanent hall. This was to include a stage, two committee rooms, cloakrooms, lavatories and a kitchen. It was hoped that it would then be possible to have 'a full licence from Middlesex County Council, so that 'facilities for letting will be increased'.

Plans were shown to 'congregation and friends' at a meeting in July 1939. It was also revealed that, for the first time in the history of the parish, there would be 'for our small numbers, a considerable debt,' of £800. In spite of the outbreak of war the work was to go ahead. The architects, Messrs Seely & Paget, advised that the first portion would be finished in ten weeks. War conditions prevented a foundation stone laying ceremony but a small plaque was inserted in one of the interior walls. Although the permanent section was unfinished at the time of the Annual Sale on 2 December 1939, and there was 'general confusion in the hall', the profits were 'much better than last year' (1938 over £100). One can imagine that seeing some material improvement in the hall encouraged people to be generous. After Evensong on 4 February 1940 the Venerable Archdeacon of Middlesex officially declared the new section open.

In 1940 the problem of permanent black-out for the hall was solved by courtesy of the Home Guard who had use of the premises as a 'convenient and central quarters to carry on their important work'. The war meant that the anticipated increase in hall lettings did not occur, in fact there were no lettings, and there was no prospect of completing the permanent hall. Fund raising activities continued, to meet the commitment of repaying £100 annually to the diocese, and to build up the PHF. The whist drives held in one of the new committee rooms were particularly popular and successful. A big effort was made for the annual sale in 1940, when it was held on two days and raised over £130. It was to be many years before a two-day sale was to be repeated, but from 1941 until 1946 the sale started on the Saturday morning.

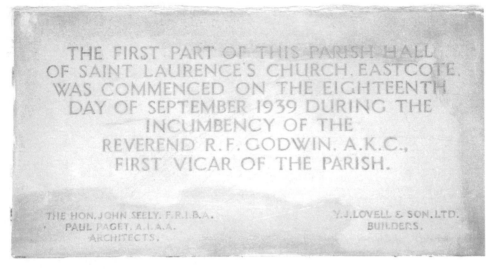

Plaque in the first part of the present hall, September 1939.
Note spelling of Saint Lawrence

By 1941 it had become increasingly difficult to stock stalls and the number of side shows increased. However by August 1942 the Fund was in the black, and at the Annual Parochial Church Meeting (APCM), held after Evensong on 6 February 1944, it was reported that a 'highly creditable' £350 had been raised in 1943. At the 1945 APCM the Fund had nearly £1,000.

Although times were difficult after the war, and several years were to pass before there was any prospect of completing the hall, efforts were maintained to build up the PHF. The annual sale, always a successful event, attracted support from the wider community, and the one in December 1947 brought the total in the Fund to nearly £2,000. The Vicar always acknowledged the contribution of those involved, but often singled out the efforts of the ladies, as in the January 1948 *Leaflet*, 'Once again we must thank the ladies for working so hard ...'. With the growth in congregations there was a more optimistic outlook. In January 1952 the Vicar once again asked for people to take PHF boxes, claiming 'every parishioner should be personally interested in this fund which means so much to the future working of the Parish'. The possibility of procuring, a 'permit' to build, saw the completion of the permanent hall a little nearer. The architect, Mr Beddall-Smith, approached the relevant authorities and a 'permit' was finally granted in 1954.

In May 1954 it was reported that to complete the hall 'with cloak room, etc., as required by the County Council to obtain a licence for plays, dancing, etc., will cost at least £6,300'. At that time the PHF stood at £3,800. The diocesan authorities indicated that if St Lawrence Church raised another £400 by the end of June, they might lend £2,000, at interest repayable over some years. This was quite a challenge, and though some were pessimistic, a meeting of representatives of all the church organisations found 'everyone willing to do their utmost'. The first big appeal since the Consecration of the Church was launched. Circulars were delivered at almost every house in the parish and, when collectors called for donations, most residents were sympathetic to the cause. This effort alone raised £140. A parochial meeting of members raised another £150, and together with the proceeds of the summer fête and other events, the Vicar was able to report, in July, that almost £650 had been raised.

The loan from the diocese was granted and a building licence from the Ministry of Supply was issued. The contractor was Messrs Prowting, Ltd. By November building was under way. With no parish hall the Christmas bazaar 1954, a less ambitious event than usual, was held at the Methodist Chapel, Chapel Hill, by kind permission of the Trustees of Eastcote Methodist Church. The building of 6a and 6b Field End Road, next to Georgian Lodge, now stands on this site.

The Methodist Chapel, Field End Road.
This part of Field End Road is known locally as Chapel Hill (RNELHS 1940)

Plaque at the front wall of the completed hall, commemorating the blessing of the hall
4 May 1955

At last after many years of delay and much effort, the permanent hall was a reality. On 4 May 1955, The Lord Bishop of London, the Right Revd Dr J.W. Wand, blessed the new hall on his way from the vicarage to the church, for the confirmation of 'a record number of candidates'. The hall was officially opened on 8 June in the presence of Mr T.G. Cross, Chairman, and several members of the Ruislip-Northwood Urban District Council. (Mr Cross had been present at the opening of the Mission Church in 1920). A series of events to celebrate, what the vicar referred to as a 'red letter' day in the annals of St Lawrence, Eastcote,' followed, ending with a Carnival Dance on the Saturday, enjoyed by over a hundred young people.

The above account clearly shows how apt were the words of the Vicar's Warden, in 1956, when he reminded the parishioners of how the Vicar had 'assisted and supervised the building of our splendid church, the vicarage and this fine hall, of which we can be justly proud'.

The Grounds
Changes in the actual grounds of the church, as against the buildings within the grounds, have taken place over the years. The fence has 'survived' for many years. The yew-tree hedge was planted to commemorate King George V's Silver Jubilee in 1935.

Directly concerned with the church grounds was, and remains, the problem of parking. In the early 1950s more people became owners of cars and some seemed unaware that their 'parking' showed a lack of consideration for others. Comments in the *Leaflet* include, in June 1952 'the present practice of 'depositing' a car on the drives in the churchyard presents some difficulties…will owners kindly 'park' their cars right and left of the main entrance gate…the ground has been built up and there is no risk of 'bogging'. Worshippers needed at least one reminder, for, in February 1954, 'the 'parking' of cars on the drive…now that they are more numerous is inconvenient…kindly use the 'car park' or the ground to the side of the parish hall'.

Although some parishioners had cars, for the majority a walk was needed to get to church. The report of a new bus service, the 225, introduced in January 1944 was really of no help in getting people to church. The route between Eastcote station and Northwood Hills roundabout via Field End Road was ideal for the area of the parish, but it did not run on Sundays. Hope was expressed that other services would follow.

St Lawrence Church is fortunate to have spacious grounds and some of

Daisy-bell, beside the 'North' wall of the Lady Chapel
(Thought to be *Gazette*.)

the organisations associated with the church have made and continue to make use of them in their varying activities. From 1947 the Scout Group has had its own meeting place and headquarters in the grounds. There had been some discussion that the Scouts could make use of one of the air raid shelters in the church grounds but these were obviously demolished. Mr Parker recalled these shelters, one where the car park is now, and the other in front of the hall.

Mr Parker also recalled the Vicar allowing the choirboys to play cricket, on the grass alongside the church before choir practice and before Evensong. Presumably it was not at the same time as the pedigree Guernsey cow, Daisy-bell, grazed there to keep the grass down!

Church Services
Obviously the Revd Rupert Godwin's contribution to the parish of Eastcote went far beyond 'over-seeing' the development of the buildings as described earlier. The local paper's report in 1943, that the church had 'quite a reputation for its well-ordered services', owed much to the conviction and guidance of the Vicar. From the beginning he encouraged members of the congregation to bring friends and neighbours to church activities, and this included attending church services. He saw the expanding population of Eastcote as presenting a great opportunity for increasing the size of the congregation.

It is sometimes said that the strength of a church community is seen, to some extent, through the witness of its members in their church attendance. In his letters in the monthly *Leaflet,* the Vicar often stressed the importance of church attendance, and frequent Communion, as obligations of Church members. In April 1948, when referring to the 'larger number of candidates' for Confirmation, he wrote that 'Communicant status is the one that counts for Church membership'.

Father Godwin's strong views on the role of Sunday School and the fellowship of Church membership, for the growth of the church, are discussed in later sections on Church organisations.

Information regarding services has been taken mainly from issues of the *Leaflet* and from the relevant *Register of Services* books. As with the years of the Mission Church, the latter rarely record attendance numbers. These two sources show that, though there were some changes in the services by 1956, *The Book of Common Prayer (BCP)* containing services unknown by many in the twenty-first century, continued to be the handbook of the Church. Comparison of the services listed in the January 1934 and December 1955 *Leaflet* may be of interest.

Moves were made towards liturgical revision in the mid 1930s but it was not until after the war that these made any progress. Proposals were put forward to experiment with an acceptable revision of the services in the *BCP*. However it was not until the 1960s, years after the Revd Godwin retired, that significant changes were made.

One big change was made when the *English Hymnal* replaced the *Ancient and Modern* Hymn Book in 1955. The Vicar thought 'it is about the best collection of hymns that express true devotion, fine poetry and a virile Christianity more suited to our age than the mawkish sentimentality of much Victorian hymnody'. It was used until 2001.

Copies of the *Leaflet* for 1934 and 1955

Numbers of Communicants

When looking at the Table of Communicants, it should be borne in mind that during World War II, routine was disrupted by air raids and the involvement of many people in war-service of some kind. The war had a profound effect on church life in Eastcote as elsewhere, bringing some people back to their church while uprooting many for several years. With post-war building there was further growth in the population of Eastcote, and that, together with the London

Mission of 1949, to which the Revd Godwin was deeply committed, contributed to the increased Communicant figures of the last years of his incumbency. *Service Books* covering the years when the Revd Godwin was Vicar give evidence of the increase in the number of Communicants and although this does not in itself prove an increase in Church membership, it is surely an indication that this was the case. Certainly there was a considerable increase in the number of Communicants at the major Church Festivals of Easter Day, Whitsunday and Christmas from 1933 to 1956 (see Appendix 6).

Sunday Services
Sunday services in these years developed from those established during the years of the Mission Church. At first Holy Communion was celebrated at 8am and 11.30am (after Matins). Evensong (evening Prayer) at 6.30 pm, remained a regular service, apart from a few years during the war, when it was changed because of the black-out. Sunday School was normally at 3pm. However changes were made to the other morning services, for it seems that the Revd Godwin was a supporter of the 'Parish Communion Movement' referred to in the first chapter, with its aim to restore the Eucharist as the main act of worship on a Sunday morning. One senses that the Deanery Mission of 1934, and the Vicar's recognition of the need to adapt services for the growing population, may also have influenced the changes made in 1935. From June 1935, both Matins and Sung Eucharist were brought forward by half an hour, to 10.30am and 11am respectively. The Vicar expressed the hope that the change to 11 o'clock 'will enable many to come who have been hitherto debarred from attendance owing to domestic ties'. The fact that this change was maintained suggests that it was a successful move.

As attendance figures were not recorded it is impossible to compare the size of congregations. Records show that the number of communicants at 8 am was always far greater than that at the later service. It should be remembered that during these years, people usually fasted before receiving Communion and greater emphasis was placed on personal preparation than is the case today.

Two other services were held at 4pm on Sundays by arrangement with the Vicar. The first was Baptism, and the second was Churching: a service of thanksgiving attended by women after the birth of a child.

Weekday Services
Although attendance was low, regular weekday Holy Communion services were held at 7am on Wednesdays and 10.30 am on Thursdays. The first Thursday of each month was established as the Mothers' Union Corporate Communion from the time their branch was formed in 1932.

Holy Communion was also celebrated on the Saints' Days, selected in *The Book of Common Prayer,* often with two celebrations, at 7am and 10.30am, to accommodate both the working members of the congregations and others.

Sometimes the service of Evensong was held on Saints' Days. Most years at least one Holy Communion service was held on St Lawrence Day, and some years there was Evensong. There was no celebration of the Patronal Festival on a Sunday at this time.

Two important days of the Church's Year, which fall during the week, are Ascension Day and Corpus Christi. On each of these days there were two Holy Communion celebrations and in most years there was Evensong on Ascension Day. There were also celebrations on All Saints' and All Souls' Days.

Church Festivals or Feasts
As would be expected the number of communicants at Christmas, Easter and Whitsunday, were considerably higher than on ordinary Sundays. From 1933 there was the addition of the 11.30pm Sung Eucharist on Christmas Eve until the war years brought an end to this from 1939 to 1943. With the added problem of the 'black-out', from 1940 to 1943, only two Christmas services were held at 8.15am and 11am. The 7am service was re-instated in 1946.

Lent and Easter
On more than one occasion the Vicar commended members of the church to use the opportunity of Lent 'to improve our membership in our Church' or 'improve our religious life'. In his letter of March 1941 he said 'Lent does not mean enforced gloominess…it means that we become more active in searching out our true motives for accepting or rejecting our particular way of life'.

On Ash Wednesday, in addition to the weekly 7am service a second Holy Communion was held at 10.30am. This was stopped in 1940, the year after the outbreak of war, and was not re-instated. Other Ash Wednesday services varied during these years. Some years Evensong was held either in the afternoon or in the evening. Some services seem particularly austere: the Litany (General Supplication) and A Commination (a recital of divine threats against sinners).

Lent and Holy Week were marked by additional services. On a few Palm Sundays the ceremony of the Blessing and Distribution of Palms took place at morning services, and some years Evensong was followed by the singing of Stainer's *Crucifixion*.

Until 1940 services on Good Friday were: 10am Children's Service; and between 12 and 3pm Matins, Litany, Ante-Communion, Readings and Meditation. From 1941 the practice of having a visiting preacher to conduct the Three Hours' Devotion replaced these services. Ante-Communion was at 8am. Most years there was also an evening service.

The usual routine for Holy Saturday was 8am Ante-Communion and 8pm Compline. The ceremony of the Blessing of the Holy Fire took place for the first time in 1950.

Other Annual Services
Although some services mentioned in this section did occur each year, many of them were for only a few consecutive years and others were intermittent. It seems right to refer to them as they show, once more, how the Revd Godwin responded to the needs of the times, and how the parish church witnessed to the wider community of Eastcote and beyond.

Regular services included the celebration of the Dedication each October, and an annual Harvest Festival in September or October. Confirmation services were held most years at St Lawrence's, the first in July 1934 when fourteen candidates were presented. The number of St Lawrence candidates is shown in Appendix 8.

Each year there were Sundays 'of special intention' for a variety of charities, including hospitals and overseas missions, some are referred to in the 'Mission in the Local Community and Beyond' section of this chapter.

Confirmation Certificate 1950

One service of public witness, which started in the days of the Mission Church, and continued until the outbreak of World War II, was the service held at the War Memorial Cross in the Memorial Gardens in Field End Road, on the Sunday afternoon nearest Armistice Day. During the war the 'two minutes silence' was not observed publicly. In 1947 the 'newly-formed Eastcote Branch of the British Legion decided to revive this public expression'. A service, at 3pm on Sunday 9 November, at St Lawrence Church was followed by a procession to the War Memorial. Since 1952 the War Memorial service has been held to enable the 'two minutes silence' to be observed at 11am.

In addition to the carols sung at Sunday services during the season of Christmas, there was the annual service of lessons and carols. One imagines the content to be similar but it was referred to as Nine Chapters from 1933-8, then Nine Lessons until 1953 and finally, as the more familiar Nine Lessons and Carols. Toy services, to which Sunday school scholars took gifts, which were later sent for distribution to poor children elsewhere, were referred to most years. From 1933-36 a Watch-night service was held. It is not everyone's way of bringing in the New Year, so perhaps this was discontinued because of low numbers.

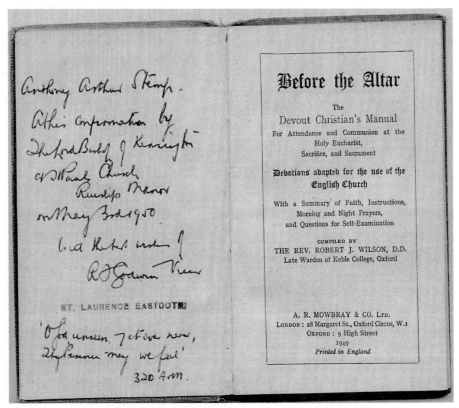

Page from Tony Stamp's copy of *Before The Altar*.
He was one of ten confirmed at St Paul's, Ruislip Manor in 1950

Inevitably some of the services related to the war. King George VI ordered the first National Day of Prayer in May 1940. Commenting on the high church attendance on that day, the Vicar commented in his letter '…it shows us that as a nation, in spite of a sad falling off...[the nation] is still Christian at heart'. Other National Days of Prayer, observed each September of the war, were also well attended. In addition the first anniversary of the 'Battle of Britain' was kept, on 21 September 1941 as RAF Sunday, and as Battle of Britain Sunday in several Septembers after the war.

There were also special services to mark the death of George V, accession, Coronation and death of George VI, and accession and coronation of Elizabeth II.

On occasions there was a Children's Eucharist. The *Record of Services* also refers to Parade Services, at 9.45am Sunday, almost weekly for a period from May 1940 to February 1942, but they are not mentioned in the *Leaflet* so it may be assumed these were for HM Forces posted in the area. Perhaps this is an example of the church meeting the needs of the community at a particular time.

Music at St Lawrence Church
A passing reference was made to the choir of the Mission Church in the previous chapter but from 1933, with a proper church organ and choir stalls, music played a more prominent role in Sunday services. In the early 1932 St Lawrence *Leaflets*, in the days of the 'tin hut', appeals were made for choir members, for 'boys and men but also more ladies'. With the permanent church there was a change, for in November 1933 the Vicar wanted applications, 'from gentlemen and boys', declaring the 'mixed choir is dissolved, with grateful thanks'. In December 1933 there were '11 gentlemen in the choir and a host of boys waiting to be trained'. Practices were held on a Friday, boys at 7pm, gentlemen at 8pm. From May 1936 there was a second weekly practice for boys on Tuesdays, and this seems to have continued through the war years and for the remainder of the Revd Godwin's time as Vicar. By March 1934 the choir were robed, the boys' surplices and ruffs being approved by the School of English Church Music (SECM) to which the choir was soon affiliated.

The SECM was inaugurated on 6 December 1929, the feast of St Nicolas, on the initiative of Sir Sydney Nicholson, at that time organist at Westminster Abbey. The SECM consisted of a training college for church musicians (the College of St Nicolas), and an association of affiliated churches committed to improve the singing at church services, particularly that of the canticles and psalms. Choral festivals were held triennially, in 1930 at the Royal Albert Hall, and 1933 and 1936 at the Crystal Palace. This building was burned down during the night of 30 November and 1 December 1936. From 1936 representatives of the St Lawrence choir attended the SECM festivals. Norman Parker recalled attending summer schools at Oxford and Canterbury, run by Sir Sydney for top choristers. Perhaps the fact that Sir Sydney was the brother of Sir Charles Nicholson, (architect of St Lawrence Church and the vicarage led to his being responsible for a festival of the combined choirs of the area at St Lawrence's in 1937. In 1945, by command of King George VI, the SECM became the Royal School of Church Music (RSCM).

The Revd Godwin expressed appreciation to the organist and choir, among others, for their part in the services, particularly at Festivals. He referred, for example, to renditions of Bach's Passion music on Palm Sunday 1936, Stainer's *Crucifixion* in 1938 and 1944, and Stanford's *Te Deum* at the Dedication service in 1945. Stainer's *Crucifixion* was said to be particularly suitable for church choirs, as it had congregational participation. According to PCC *Minutes* of 1934 the organist was paid £30 per annum. Though the Vicar often made appeals for more boys and tenor voices to join the choir the only reference to female voices since 1933 was in 1947, when he acknowledged the contribution of 'a ladies' choir in the Service of Nine Lessons.

The Revd Godwin was always anxious to have a 'well-ordered service' and must have been upset to report that at the Dedication services in 1939 the servers and choir 'suffered in some measure owing to the non-appearance of albs and surplices from the laundry'. It is hardly surprising that by 1952 the Vicar was of the opinion that renewal of the old garments 'cannot long be deferred'.

Like other church organisations, the choir had various social events. There are accounts of Christmas outings, with the servers, to pantomimes at Daly's Theatre, *Charley's Aunt* (1935), followed by tea at Slaters' restaurant, and at the King's Theatre, Hammersmith, *Jack and the Beanstalk,* (1936), *Robinson Crusoe* (1937), *Jack and Jill* (1938), *Sleeping Beauty* (1939), followed by enjoyable teas at either the vicarage or in the church hall. For the remaining years of the war the choirboys had to be content with the 'enjoyable teas', and some years they joined in the parish social evenings. The January 1944 L*eaflet* reported the choirboys

Kerswell's Restaurant – Old Peoples victory party 1946 (RNELHS)

having a successful carol party. During their carol singing around Eastcote, they were invited to sing at Messrs Kerswell's restaurant, which had opened in 1943. Their 'takings' were considerably increased from patrons dining there. The restaurant was on the first floor presumably over Kerswell Ltd (Bakers). It was where Budgens is today.

Other references to music at St Lawrence include organ recitals, and recitals by the St Lawrence Orchestra, Quartet or Trio playing after Evensong, and at summer fêtes. The choir also contributed to the successful annual Mystery plays, which were produced from 1935 onwards.

It may interest readers to learn that in 1937 members of the church choir felt the need for a local choral society. Issues of the *Leaflet* refer to the inauguration of the Eastcote Choral Society in the church hall on 6 October 1937, its first public effort, a sacred concert in the church on 29 December 1937, and further concerts in the hall on 18 May 1938 and 25 April 1939. Meetings were held on Wednesdays in the church hall. The advent of war seems to have led to the closure of the Society. The present Eastcote Choral Society was formed in 1970 as successor to the Eastcote and District Choir (founded in 1946). Archives of the present Society have nothing about the earlier group. There is an entry, in a *Record of Services* book, of an Eastcote Choral Society Recital on 21 December 1955, presumably held in St Lawrence Church.

Revd R. F. Godwin's Sociable Community
Once Eastcote had become a 'Peel' District in November 1931, St Lawrence's Church was able to have official Churchwardens, a Parochial Church Council (PCC), with a Standing Committee and various Sub-committees. The Churchwardens and PCC were appointed or elected at the Annual Parochial Church Meeting (APCM). Mr T.G. Cross and Capt. E.T. Weddell were the first Churchwardens. (the list of Churchwardens from 1933 to 2010 is in Appendix 4). There were also church organisations and clubs. The Revd Godwin did much personally, both to establish many of these groups and influence their ethos. Frequently in his monthly letter, the Vicar referred to the importance of committed Church membership and on occasion this led to the closure or re-organisation of a church group.

Some social events that will be referred to when writing about specific church groups were shared by the congregation as a whole, and a number by the wider community of Eastcote. In fact the attendance of the latter at public events such as fêtes, sales, music recitals and St Lawrence Players' productions was important. These occasions provided an opportunity for encouraging local residents to become involved with their parish church, and their financial support was much appreciated.

Soon after his arrival in Eastcote in October 1931, the Vicar launched the *Leaflet* as one means of making the presence of St Lawrence known to the community at large. In his monthly letters he asked new residents to 'make themselves known to him' and offered to call on them, as he was keen to get them involved in parish affairs. In the early days he was also 'at home', at certain times during the month, for those who preferred to call on him.

It has been shown that Father Godwin regarded regular church attendance and frequent Communion as vital for full members of the Church. In addition he was aware from his first days, that much hard work was needed by members of the church to build up some 'cohesion', if it was to move from being a 'new' to becoming an 'established' parish. In order for the church to grow, individual members of the congregation, and various church groups, had to support each other for the benefit of all. The Vicar was of the opinion that this would happen only if members of St Lawrence became a 'sociable' community. With this in mind he promoted various social events to help members of the congregation 'to get to know one another better.'

The Summer Fête
In 1932 the Revd Godwin held his first summer 'At Home' in the vicarage garden. In 1934 the *Leaflet* reported the annual event 'a great success' with an attendance of about 180. As well as refreshments 'a programme of dancing by Madame Sturmer's pupils gave much enjoyment'. From 1939 it was the Cheney School of Dancing under Madame Barnes who provided this regular and popular entertainment.

In 1938 the annual event was enlarged into a fête with various sideshows. These fêtes were not only enjoyable social occasions, but also helped to swell the Permanent Hall Fund (PHF), set up in 1935. The fêtes entailed a concerted effort by members of the congregation both before and on the day, and helped to build up some of the 'cohesion' within the church community so desired by the Vicar. They could also be regarded as a part of the Church's mission, as stallholders, providers of refreshments and of entertainments welcomed members of the wider community. Over the years, church members, both as individuals and organisations or groups, provided an extensive programme of entertainment. The first fête, of 1938, was a fairly modest venture compared with later ones, as regards sideshows. They included a fortune-teller, a bran tub, a lemonade bar, and a stall run by the St Lawrence Needlework Guild. In spite of the outbreak of war in 1939, or perhaps because of it, the numbers of stalls and sideshows increased. The Vicar's account of the 1941 Fête reported the 'side-shows presented an attractive picture and were keeping people amused throughout the afternoon'. Darts, ball-in-bucket, mop throwing, and ringing bottles were among the many stalls that extracted the pennies, and the ankle competition was an innovation. The St Lawrence Trio gave a musical performance during teatime.

In 1941 another new feature, that of a baby show was tried. It attracted a number of entrants, and became so popular, that in its last three years, from 1946 until 1948, it was held on a separate day. In 1947 the Vicar reported an entry of around 90 babies, placed in different classes; most of them had been baptised at St Lawrence Church. In 1945 the St Lawrence Juvenile Players made their debut with scenes from *Alice in Wonderland*. From 1948 productions by the St Lawrence Players became a regular part of the summer fête day, usually in the evening. Items that seem to have occurred only once were a Punch and Judy show (1944), pony rides (1946), a fancy dress parade (1948), and a coconut shy (1950). The apparatus for this which had been recently unearthed, was used at fêtes in the early years of the Mission, probably 1921-1922.

When giving his account of the day's proceedings in the *Leaflet,* the Vicar always expressed his thanks to all who had contributed to the success of the day. Mention was made of the organisers of the fête and holders of stalls or sideshows, and the part played by specific church organisations or groups. The dancing displays and St Lawrence Needlework Guild stall had special reference, as did the teams of ladies who were responsible for the refreshments, which were led first by Mrs King-Palmer, and later by Mrs Dalton. Some may recall that food rationing continued for several years after the end of the war, when catering for events such as the fête could not have been easy.

Reports in the local paper show this summer social event was popular with Eastcote residents, and gate numbers appear to support this view, for example in 1944 nearly 400 paid for admission. The fête was not a fund raising event on the scale of the autumn sale, but the takings rose from £25 in 1938 to nearly £100 in 1954, when a big effort was made to swell the PHF.

Parochial Socials
The first parochial social, known at this time as the 'Vicar's Social', as against the socials of individual church organisations or groups, was held in January 1934. In his January *Leaflet* letter the Vicar expressed regret that some members of the Mission Church had left, but he looked forward to an increase in membership from the growing population of Eastcote. He recognised that 'in a new parish the problem of the old and new residents is an ever present one', and he hoped a social would be a means through which 'old and new residents will make themselves known to one another'. A party from the Revd Godwin's previous parish of St Michael's, Paddington was to have given a concert, but owing to foggy weather, they and a number of St Lawrence's members were unable to attend. Well over a hundred were at the second Vicar's Social in the February, when the Paddington visitors did attend and perform. The Vicar reported 'an encouraging number of fresh faces, many wishing to be associated with St Lawrence now that they had settled in'.

That first Vicar's Social of January 1934 became an annual event for several years. In January 1937 the Vicar's Christmas Social entertainment was in modern 'revue' style. The social in January 1938 was 'a most successful evening' with nearly 100 attending, and it was 'pleasing to see most of our new members present'. The entertainment on that occasion was a performance of the short play *The Wrong Flat*, which led to the formation of the St Lawrence Mummers. The Christmas social in December 1938 included amusements, games and the singing of carols. The *Leaflet* of February 1939 announced a Military Whist Drive instead of a social for that month, so it would seem a social of some kind had become a more regular affair.

With the coming of war in September 1939 the parish socials were difficult to arrange, though some of the church organisations continued to have their separate Christmas socials throughout the war. There was one 'all-in' social in January 1941, consisting of Sunday school scholars, Cubs and Scouts, choirboys and the congregation generally. Mrs King-Palmer 'worked another miracle and provided all with a most sumptuous tea' and a concert entertained everyone. That was the last of parochial Christmas socials for a number of years.

After the war the Vicar encouraged the congregation to support St Lawrence Players' productions, the Scouts and Cubs Gang Shows, whist drives and other fund raising efforts for the PHF fund, but the parish Christmas social of pre-war days was a thing of the past.

However in January 1952 the Communicants' Guild, seen by the Vicar as a means of 'keeping us together in the fellowship of the church', organised the first Guild supper. This was a rather more formal social evening, with a sit-down meal. By 1956, just before the Vicar retired, the supper seemed established as an annual event.

Dedication Celebrations
Annual services to celebrate the Dedication of St Lawrence Church on 21 October 1933 have been referred to, but from 1935 social events also took place. It is likely that the Deanery Mission of Help held in October 1934 (see later section in this chapter), coming just a year after the Dedication, together with preparations for the annual sale of work was sufficient activity for that year. From 1935, up to and during the war, socials were held following Evensong on St Lawrence Day, whichever day of the week it was celebrated. At these socials there was some form of entertainment and refreshments, but comments in the *Leaflet* give the impression that Father Godwin's real purpose was for parishioners to get to know fellow worshippers. In 1936 the third Birthday Social gave 'evidence of the definite building up of congregational sociability'.

From 1946 the Festival was emphasised with a Dedication Festival Week. In 1946 there were four special preachers on the two Sundays. The Revd Godwin regretted there was not a better response to the events of the week, an organ recital, a whist drive, a social, and a Saturday concert by the Ruislip Orchestra. He was keen that, following the upheavals of war, the parishioners should 'recover, as much as we can, our former Church life'. All could help if 'we are more regular worshippers' and 'more sociable'.

For the remaining years of the Revd Godwin's incumbency the Entertainment Committee seems to have been responsible for co-ordinating regular socials as well as the social events of Dedication Week.

Church Organisations

Mothers' Union

One of the early organisations at St Lawrence Church was the Mothers' Union (MU), with its aim 'to strengthen and preserve marriage and family life through Christianity'. The MU movement developed from small beginnings in 1875, when Mary Sumner, wife of a country parson, gathered together the women of her parish to teach them how to 'make and keep their homes Christian and happy'.

The Objectives of the MU were -

> **To promote and support marriage and its wider understanding.**
> **To encourage parents to bring up their children in the faith and life of the Church.**
> **To maintain a worldwide fellowship of Christians united in prayer, worship and service.**
> **To promote conditions in society favourable to stable family life and the protection of children.**
> **To help those whose family life has met with adversity.**

With access to MU *Enrolling Members' Report Forms* and *Minutes* of various MU meetings, together with news of MU activities recorded in issues of the *Leaflet,* there has been a vast amount of detail from which to select information. In view of the importance the Revd Godwin attached to the MU in the life of the parish, referring to it as 'the backbone of the Church and parish' and 'one of the parish's greatest assets', it seems fitting to write at some length about the MU during this formative period of Eastcote parish church. Reference has been made elsewhere to difficulties faced by the Vicar in establishing a sound tradition in the newly formed parish. There is evidence that the MU, in which the Vicar took a keen interest, played a part in this 'sound tradition'. He chaired the AGMs, encouraged members to become involved in all aspects of parish life and suggested ways in which MU members could reach out to the wider local community.

The St Lawrence Branch, in the Uxbridge Rural Deanery, had its inaugural meeting in February 1932. *Report Forms* show a membership of 30 by September 1932, with 49 in 1934, fluctuating lower numbers during war and the late 1940s, but rising again to 44 in 1956. Membership was on a communicant basis, with a Corporate Communion service on the first Thursday of each month at 10.30am. Apparently some members found Thursdays inconvenient and, from

1934 until 1950, the 8am service on the first Sunday of each month was also regarded as a Corporate Communion for members of the MU. In some years, the Thursday services were moved to another day to celebrate Candlemas or Ash Wednesday, and additional services were held on other Church festivals. However the Thursday Corporate Communion was the focus of the St Lawrence MU.

As well as attending services at their own church, representatives of the Branch attended annual MU Lady Day and festival services at other churches in the Deanery. St Lawrence Church obviously took its turn in hosting these events. On one occasion when the Lady Day service was held at St Lawrence's over 250 members attended (1954). Members were encouraged to attend the annual London Diocesan Festival Service at St Paul's Cathedral when branch banners were carried in procession: in the case of St Lawrence this was often done by a male member of the congregation.

There seems to have been more than one St Lawrence's banner Soon after the branch was formed members had raised money for a branch banner, which the Vicar blessed in September 1933, at the opening meeting of the 1933-4 year. The banner was carried in processions in following years, at Deanery and Diocesan services. A further banner was given specifically to St Lawrence Church and dedicated in 1943. The old one was sent to Mary Sumner House (the headquarters of the MU), in the hopes that it would eventually go to 'some badly damaged East End Branch'. Several St Lawrence members were present at the dedication of the banner after it was given to the MU St Matthew Branch in Stepney, which had lost its church in an air raid. From 1944, for a few years the two branches exchanged social visits. There are later references to yet another new banner in 1979.

Meetings were held at 3pm on the third Wednesday of the month, though these were not on a regular basis until the early 1950s. Often there were guest speakers at these meetings but members also spoke on many occasions. Examples will show that the talks were usually connected, either directly or indirectly, with the aims and objectives of the MU. Sometimes they were linked with parish, deanery or diocesan projects. There were talks about Sunday observance, prayer, marriage and divorce, Christianity in the home, and religious education. Some of the Vicar's talks were on the teaching of the Church, when he showed vestments and ornaments (sacred vessels etc.), and explained their history and meaning. Other talks were on the work related with moral-welfare in Uxbridge deanery, and several were associated with the 1949 'Mission to London'. Sometimes speakers were from national organisations, such as Dr Barnardo's, or the Waifs and Strays (The Children's Society since 1982).

Addresses on the work of the MU in India, Africa, China and elsewhere, give evidence of MU work overseas. There are references to deanery support of the Church of England Zenana Missionary Society in N.W. India during the 1930s,

1940s and early 1950s. Annual sewing days were organised when MU members, supplied with materials, made up garments for the Mission. In 1937 when this was held in the parish hall nearly 70 garments were completed. From 1950 there are references to the St Lawrence branch exchanging letters and photographs with a branch in South Africa at Port Elizabeth.

The 'worldwide fellowship' of the MU is also shown in references to the Wave of Prayer, which each January brought in turn before the throne of God, the needs of every branch of the Mothers Union throughout the world. The Dioceses of Accra and Sierra Leone in West Africa were associated with London on the days this continuous Wave was devoted to the London Diocese. Once a year St Lawrence MU members were urged to play their part 'by spending a specific 15 minutes in their parish church asking God's blessing on the work of the MU in the London Diocese and also on their fellow members'.

In keeping with the aims of the MU, members, both as a body and as individuals, were involved in the wider community as well as parish life. With other ladies in the parish they sold flags on the annual Church Army Violet Days, and they organised flag days for charitable causes such as the United Services in 1943. Apart from attending the regular MU services, members made the effort to attend the special services in Lent and were involved in St Lawrence projects such as the 1949 Mission to London. In 1955 they seem to have been particularly active. They helped in the house-to-house collection for the new church hall and visited young mothers in the hope of their joining the MU. They also supported the new Sunday School, started at Newnham Avenue School in January 1955, by visiting parents of the many children who were on the register there, and tried to interest these newcomers to the area in St Lawrence Church activities.

From their first year the St Lawrence MU undertook to have stalls at Christmas bazaars and summer fêtes, which raised money, first for the permanent church, and then for the parish hall. Sewing parties, which met to make goods to sell at these annual events, also gave an opportunity for members to get to know each other. Their Christmas 'Stock Exchange' stalls in the 1940s proved popular and financially successful. In 1945 the MU stall contributed £16.8.9d to the total takings of nearly £120. The December 1954 Sale of Work stall raised nearly £25 out of the total of around £100.

Obviously the branch was affected by the war and its aftermath in several ways. For one thing it was more difficult to get outside speakers and, from September 1939, meetings held in the church began at 2.15pm instead of 3.00pm, as the church was not allowed to have lights. In 1940 it was recorded that 'despite a year of terrible war...this Branch has been more than able to maintain its activities'. 1942 was 'a quiet and difficult year...many members being ill, on war work or evacuated'. After the war ended in 1945 it took several years for the branch to recover. Some interest was shown in a renewal campaign of 1946-7 but there was some hostility to the standard set up by the MU, particularly on the subject of divorce. In 1948 members were reminded of 'the spiritual character of the MU'. Following the 1949 'Mission to London' there were encouraging signs with increasing membership, partly the result of post-war house building in the area, and greater commitment.

There was of course a social side to the MU. One annual event was the Christmas party or social to which MU branches of other churches were often invited. Early guests were Ruislip members in 1934. Before the war the MU funds bore the cost of the food. For 30 people in 1936 it was estimated that 'the cost of the catering would be under £1'. From 1941, during the difficult days of rationing, members took their own food and pooled it. In 1950 *Minutes* record that an allowance of tea, sugar and milk had been obtained from the Food Office. Entertainment at these socials ranged from songs and sketches, to games and competitions and whist drives. Outings were a regular feature of the MU year from 1933, when a visit to Brighton was 'very much enjoyed'. Over the years outings were to Clacton, Windsor, Eastbourne, Hastings, Bournemouth, Brighton and Folkestone. In 1937 there was one outing to see the illuminations for the Coronation of George VI and a second to Bognor Regis. In 1940 the trip to Hastings (5/- return) was cancelled and replaced by a garden party at the vicarage and the next outing to the sea was in 1946, when the members went to Folkestone. In 1950 the return coach fare to Worthing was 9/- per person.

During these formative years of the new parish the MU had its 'ups and downs', but the early 1950s saw signs of its increasing influence in the life of St Lawrence. In 1952, for the first time, a committee member represented the MU on the PCC and all members were encouraged to join the Electoral Roll. By the end of the Revd Godwin's incumbency the St Lawrence MU was well established, with good attendances at corporate communions and meetings, and commitment to duties and responsibilities as members of the parish church.

Church of England Men's Society
The Vicar referred to the MU as 'the backbone of the Church and parish' and 'one of the parish's greatest assets' but was of the opinion that there was always plenty of work in the parish for the men to do. Conscious of their important role in the years of the Mission Church, the Vicar 'wanted to extend the activities in the parish for the men of the congregation', and felt that a branch of the Church of England Men's Society (CEMS) would help in this matter. By March 1935 a branch of the CEMS had been formed. As with other St Lawrence organisations membership was on a Communicant basis. Although reference was made to attendance at an annual conference in 1936, membership was low and with the coming of war and all that that entailed, the branch never really got off the ground.

Possibly, related to the 1949 London Mission, and certainly encouraged by the Bishop of London, the Vicar made moves in 1951 to re-form a branch of the CEMS at St Lawrence's. He saw this as a means of bringing 'men of the congregation…into a closer fellowship'. The first meeting had 'a record attendance of men of the congregation', the next 'if sparse in numbers, (fourteen) made up for it in enthusiasm and a business-like despatch'. There was a Service of Admission in April and more members came forward. By 1952 the Society was recognised as 'a real force in the parish' playing its part in the annual garden fête by providing and manning side-shows, and helping to make the church grounds more attractive. The *Leaflet* reported interesting meetings and an enjoyable walk to Harefield.

In 1953 members were urged to attend a meeting and social of the Wembley, Harrow and District Federation of the CEMS, which had been established. Branch meetings included lectures with titles such as 'The Ornaments of the Church' and 'The Relationship between Church and State'. Discussions took place in autumn 1955 on 'ways and means of helping the Vicar and of developing other aspects of Church work'.

One outcome of these discussions, in the spring of 1956, was a series of 'visiting teams' to call on the parents of Sunday school scholars, to encourage them to 'join up', especially those from the Newnham Avenue area. As may be read elsewhere, the Sunday school, which started in Newnham Avenue School in 1955, had by November around 180 children attending. Here was an opportunity to get their parents, fathers as well as mothers, involved in St Lawrence Church activities.

By the time Revd Godwin retired the St Lawrence Branch of the CEMS seemed a permanent and outward-looking organisation.

Sunday School
In addition to the weekly Sunday School and Children's Service, the *Record of Services* books and issues of the *Leaflet* refer to services specifically for children on special days of the Church year such as Good Friday and Easter Day. Bringing young people into the family of the Church is a vital part of the Church's mission and Father Godwin recognised, that the new parish of Eastcote, with its growing population had opportunities to extend this ministry. The one obstacle to achieving this end was the perennial one, the lack of teachers. In his September 1934 letter, the Vicar wrote at length on the matter. The last of successive expert teachers, loaned by the Council of Youth to the new parish, to instruct young teachers, had left. He appealed for more teachers, writing that 'a new parish and a new congregation means hard work in building up, and we are not even in sight of that cohesion which enables the "established" parish to pursue its work from year to year with the benefit to all concerned'.

From 1933 there were three groups of children, meeting at 3pm on Sunday afternoons. Children over eleven met in the church for Catechism with the Vicar, the infants were in the choir vestry, and the juniors met in the hall. The war may have necessitated some changes as to times and places, but reports in the *Leaflet* refer to increasing numbers, and inevitably to the need for more teachers. In 1943 the possibility of a room 'over the station bridge', to meet the needs of that area, made no headway at this time owing to lack of teachers.

Again and again appeals for teachers, in the Vicar's letters, and from succeeding Sunday School superintendents, had little or no response. In the *Leaflet* of September 1946 the Vicar reported that 'under Miss Gray's enthusiastic leadership the Junior Sunday School is flourishing…but…Senior School … very great loss…in resignation of Miss Troughton'. There were instances of crises when teachers left, to marry, do National Service or to go to college, and their places being taken by another. Occasionally there was an optimistic comment, such as May 1948, when following the Confirmation of 80 candidates, 'several had offered their services as Sunday School Teachers'. Perhaps it was less daunting to undertake this service in the company of others, who would together attend the instruction classes for new teachers.

Possibly connected with the large number of Confirmation candidates in 1948, the PCC raised the subject of how to involve 'some of the older children in the parish'. A short 10am Children's Service started in October 1948. Members of the laity conducted this regular addition to Sunday worship.

The association of Sunday School with Nativity plays and Christmas parties is inevitable, and these did indeed take place. Christmas parties for the different age groups were annual events and enjoyable social occasions. Few details are recorded, but in 1946 the infants had a successful fancy dress parade, which was repeated by them and by the seniors in 1947. No doubt much depended on the teachers. From 1951 the highlight of the Christmas party was a special performance of the St Lawrence Players' pantomime.

The tradition of having a Christmas tree in church dates back to 1933. From 1934 there had been a crib, though a crib service was not introduced until later. As part of the St Lawrence Christian community members of the Sunday School were made aware of the needs of others. For many years the Sunday School scholars contributed toys, some years at a special toy service, which were placed under the tree, for later distribution to poor children in Plaistow, Bethnal Green or Rotherhithe. After the war the toys went to the Church of England Children's Society to distribute to their Homes.

The *Leaflet* of February 1945 reported 'The Sunday School scholars gave a very successful Nativity Play on Sunday, Jan.14th'. At the time it could not have been recognised as a thing of major importance, but it sowed the seeds not only of the future St Lawrence Players, but also the Youth Fellowship.

Summer treats do not appear to have been regular events, but reports of some are worth mentioning here. In 1934 the infants had a party and games in the Vicarage garden, while the senior and junior children went to Regents' Park Zoo.

In 1937 the 'treat took place in the church field, the weather was kind, the children enjoyed themselves thoroughly in hay-making and running races'. In 1947 the seniors 'treat was held in the grounds of Eastcote House'. There were, at least two, more ambitious summer treats. The first summer outing was in 1950 when 90 children went to Windsor. The St Lawrence Players began to finance the outings with profits from their annual pantomime, a sign of a move towards the Vicar's ideal of 'cohesion' in the parish. In 1952 the destination was California, near Wokingham where they 'spent an enjoyable day, plenty of amusements and a good tea, amid lovely surroundings'. This last comment indicates how much Eastcote had been developed since the days of the Mission Church.

The more optimistic comment of 1948 was followed by further pleas for teachers, especially male, in 1949. Writing in the *Leaflet* in April 1949, the year of the London Diocesan Mission held in May, the Vicar expressed the hope, that among other things it would result in more church members offering to become teachers, for 'these young people should be our future congregation and we ought to be able to do our best for them'.

In December 1949 the report that 'numbers have grown too large' for a summer party, led to that first 'outing' of summer 1950. One assumes there had been offers to join the band of teachers at Bridle Road, but not enough to have a Sunday school 'over the station bridge', something the Vicar had wanted since 1943. In 1952 he expressed regret that the church had been unable to accept facilities offered by Ruislip and Northwood Urban District Council, 'especially since other religious bodies have been able to do so'. This was owing to a lack of teachers.

Newnham Avenue Sunday School
Just before the Vicar's retirement, after years of frustration regarding Sunday School, there was a big step forward. With the support of the PCC, arrangements were made to start a new Sunday School, in January 1955, at Newnham Avenue School, which had opened in 1952. In February 1955, Mr Lane, the Superintendent reported having 139 children on the registers, with 15 teachers. His March report was even more encouraging, with the school going from strength to strength. In spite of particularly severe weather 'we now have 160 children on the register…last Sunday 21 new pupils arrived in the snow'. Classes were held at 10am to enable teachers to attend Sung Eucharist at St Lawrence church.

In 1955, instead of an outing, parties, catering for the three different age groups, were held in the church grounds. This, not only enabled 'many of the children "from over the bridge" to make a closer acquaintance with their Parish church', but allowed them to meet those who attended classes at the church. The children from 'over the bridge' also attended the Dedication Festival that October, 'an imposing sight'. Hopes were expressed that there would be sufficient teachers 'to instruct and hold these children together through the "teen-age" years'. Eight members of the Youth Fellowship had become teachers, an example of one organisation supporting another, though there was still a need for other teachers for the senior classes.

Building on the presence of the Church in the Newnham Avenue area of Eastcote, various moves were made to get the parents of the Sunday School children interested in their parish church. Mr Lane arranged a social to meet the parents, the Mothers' Union promised to visit parents and, in 1956, members of the Church of England Men's Society undertook to visit parents with the particular purpose of involving more men in parish activities.

The progress of the Sunday school, supported by other church groups, seems to indicate that in the last years of Revd Godwin's incumbency, 'hard work in building up' some 'cohesion' in the new parish had been achieved.

Uniformed Organisations
The early issues of the *Leaflet* listed several uniformed organisations; Scouts and Cubs, Guides, Brownies and Girls' Friendly Society. The Guides and Brownies were designated as 'open' groups. The fact that they were referred to as 'open' groups suggests that members' families did not have to belong to the church, and that the organisations merely used the church premises for their meetings. Perhaps that was why the Vicar decided, that from May 1936, they, together with the Girls' Friendly Society, should be removed from the organisations list.

4th Eastcote St Lawrence Scout Group
In the mid-1930s the choir had become well established with a good number of boys. Consequently there was a ready-made group of 8 – 11 year olds to establish a Wolf Cub Pack. Thus in May 1938 the St Lawrence Cub Pack was formed under the leadership of Harry Rann, a friend of the Revd R. F. Godwin and a server and Sunday School teacher at the Church. As this was the fourth Scout Group to be established in Eastcote, it became registered as the '4th Eastcote', although in its early years, the Group was known as St Laurence Eastcote, the "u" being used in the official title. As the Group was sponsored by the Church, it became part of the 'London Diocesan Boy Scouts Association' (LDBSA) and as such was a constituent member of the LDBSA District, and not part of the Eastcote Northwood District. The decision was made to use the Diocese of London colours of red and black for the scarf worn by the groups members. These colours have remained in the Group scarf.

Within a year or so, a Scout Troop section was formed, to cater for the 11 - 18 year olds. Weekly meetings were held in the 'old' Parish Hall and leaders of both sections were drawn from St Lawrence Church members. With the outbreak of war in 1939, a number of the younger leaders were called up for military service and replacement leaders and helpers were recruited from 'older' fathers, such as Jack Ellis, Harold Smith and Harold Poole. Harry Rann served in the Army and subsequently was ordained, eventually being appointed a Prebendary of Exeter Cathedral. Despite those early wartime years the numbers grew and the Group flourished such that in August 1942, the first Scout Summer Camp was held in the grounds of the 1st Chesham Bois Scout Group at Chesham. The Scout Troop possessed very little camping equipment at that time and new gear was difficult to obtain in wartime Britain. However, those attending all survived despite having a very rainy week in borrowed tents, covered with grass and leaves, acting as the required camouflage! The camping equipment situation slowly improved and the next three Summer Camps were held at Hampden Bottom, near Great Missenden.

With the war years over, many of the former leaders such as Gerald Collins and Stan Peterken returned from military service and took up active roles. With growing numbers of boys and the need to store an increasing amount of equipment, the decision was made that the Group should strive to have its own meeting place and Headquarters. In the summer months of the mid-forties Annual Garden Fêtes were held in the grounds of Eastcote House, organised by a very active Parents' Committee under the Chairmanship of Mr P. J. Beere. Other fund-raising activities were also held and funds were accumulated such that by 1947 there was enough to commence building. With the agreement of the Church and Diocese, land at the back of the Vicarage was made available, a pre-fabricated building was purchased and the new Headquarters was completed and opened by the Bishop of Kensington in October 1947. At this time the Group comprised a Cub Pack, a Scout Troop, a Senior Scout Troop and a Rover Crew. The LDBSA was disbanded and the Group was moved into the Ruislip District (rather than Eastcote/Northwood District) to make the two Districts more evenly balanced.

By the early fifties, numbers had grown so much that a second Cub Pack and Scout Troop had been formed. Membership of the Sections had been extended to all, not solely Church members. This resulted in some 'misunderstanding' of the purpose of a sponsored Group and in the spring of 1950 a decision was made to divide the membership and establish an 'open' Group, which was named 6th Eastcote. Numbers in the 4th Eastcote were more than halved, but Cub, Scout and Rover sections continued, under the then Group Scoutmaster Stan Peterken.

Brownies and Guides
In April 1939 as well as St Lawrence, Eastcote Scout Group, which included Cubs, the *Leaflet* listed 6th Eastcote Brownie Pack. Reports of a social and 'an expanding pack', is all the information available about this group. As with other organisations the outbreak of war is likely to have influenced matters and the pack closed. However in 1941 the Vicar reported the possibility of having a new pack and this started in 1942. Meeting on Saturday afternoons, with Miss Fisher as Brown Owl until May 1949, this seemed a successful group. The pack continued under other leaders until the middle of 1953.

From April 1943 attempts to start a Guide company were unsuccessful, until May 1950 when Miss Piper became leader. When she left in April 1952 there was no one to take her place.

While the Scout Group went from strength to strength, the girls' sections seemed unable to get established at this time, owing to the lack of leaders.

St Lawrence Youth Fellowship
Between 1942 and 1956 reports of differently named church youth groups appeared in the *Leaflet*. In December 1942 the St Lawrence Youth Club 'had an interesting lecture on Holland…propose to start a P.E. class…possibly a drama class'. In 1943 the Youth Circle 'started to bring together the younger people in the parish'. Activities included play reading, tennis, socials, cycling and walks. By May 1946 it had become the Youth Fellowship. It was open to any over 14, who were members of the congregation, and had a weekly subscription of 2d.

THE ST. LAURENCE YOUTH FELLOWSHIP
Tuesday, August 2nd.—" Nuts Nite."
Tuesday, August 9th.—Games Evening.
Tuesday, August 16th.—Miss West : " Work of a Welfare Officer."
Tuesday, August 23rd.—Square Dance.
Saturday, August 27th.—Visit to Coliseum.
Tuesday, August 30th.—Mr. Morgan on " Geneva."
Saturday, September 17th.—Youth Fellowship Dance.

ST. LAURENCE YOUTH FELLOWSHIP
Programme for October
Tuesday, 4th.—Dancing Evening.
Tuesday, 11th.—Communicants' Guild.
Tuesday, 18th.—A Talk by Geoffrey Williams.
Tuesday, 25th.—A Talk by Father Dobb of Dr. Barnardo's.

ST. LAURENCE YOUTH FELLOWSHIP
Programme for December :—
Dec. 6.—From Bach to Bop : A " record " evening.
Dec. 13.—Communicants Guild.
Dec. 20.—Carol Singing.
Dec. 27.—Christmas Social.

Youth Fellowship programmes from 1955 issues of the *Leaflet*

When Junior Players started a club in August 1948 (with Miss Angell as leader until 1953) the 'history' of the church youth club becomes even more confusing. Until July 1948 the Youth Fellowship met on Tuesdays at 7pm, yet from August the St Lawrence Players Club is listed as meeting at the same time. In January 1950 an appeal for a games leader, stated the club 'is a group of young people growing up into the Church life'…open to new members from 12 years of age'. The need for adults to help and to encourage young people was as pressing then as it is now. If, for a while, there were two groups catering for different aged children, the November 1953 *Leaflet* reported, the Junior Players have been re-organised as 'the Youth Fellowship with their own committee, under the watchful care of a committee of parents'.

There were other short-lived groups. Issues of the *Leaflet* give evidence that from 1941, in addition to the Brownie venture, efforts were made to meet the needs of 'older' young members in the parish. In 1941 a Thursday Night Club was opened for 'young people and the Forces stationed in the neighbourhood of Eastcote' and 'had many pleasant evenings'. In March 1942 the Table Tennis Club 'held a very pleasant social evening and dance recently'. By the end of 1942 with declining numbers the two groups were forced to close. Several years after the war, in June 1953, a table tennis club was on the Organisation list but there are no reports on the success or otherwise of this club.

Another group that ran only from 1941 to 1942 was a club for boys. Cricket, football and other outdoor games were arranged, and in November 1942 the club had two football teams. In March 1942 the boys presented a 'very successful' concert.

St Lawrence Players
(see Appendix 13 for list of productions)
Readers familiar with the St Lawrence Players of recent years may be surprised to learn that, in connection with the Epiphany Mystery play, the January 1935 *Leaflet* referred to a group by that name, who 'have worked hard at rehearsals'. In the same *Leaflet* notice was given of the St Lawrence Players Corporate Communion to be held on the Feast of the Epiphany 1935. This indicated the importance the Vicar attached to members of church organisations attending Holy Communion; however no further reference has been found regarding this aspect of a drama group for the years when the Revd Godwin was Vicar. In both 1935 and 1936 thanks were expressed to the contribution of the choir to the success of the St Lawrence Players in the performances of the Mystery plays. There is no further mention of the name St Lawrence Players for a number of years.

The success of a short play, *The Wrong Flat*, 'on a large but very temporary stage', as part of the parish social of January 1938, led to the formation of a dramatic society, known as the St Lawrence Mummers. Two plays, *The Man on the Sofa* and *The Old Geyser,* were performed on Wednesday 27 April 1938; admission was free with a silver collection. The Mummers held a social evening in September, in aid of stage fixtures, inviting all to come and 'spend a jolly evening'. Admission was by ticket and the price of 9d included refreshments. Tickets were obtainable from the Vicar or any of the Mummers. There were further successful productions in December 1938 and in April 1939, when the silver collection made a profit £2 for the Permanent Hall Fund. The group had 'fervent hopes' that the April show was the last 'on a temporary platform, with all the attendant worries'. The new section of the hall, with a permanent stage, was officially opened in February 1940. The Mummers, like other organisations, faced problems when the war started, one being the 'difficulty in filling male parts', and in May 1940 decided to 'suspend activities for the present'.

The formation of a new dramatic society followed the success of Sunday School scholars in the Christmas Play of January 1945. Several young people were keen to do more in the way of acting and the first appearance of the St Lawrence Juvenile [sic] Players was at the summer fête on 23 June 1945, when they presented scenes from *Alice in Wonderland* arranged by Miss Angell. In July the Players raised money for the 'Save the Children' Fund when they gave a performance of two scenes from *Alice in Wonderland,* together with concert items. Other productions took place in November and January, and at their first birthday in March 1946, it was reported that 'the year's work is very satisfactory'. In addition the young people had sent £20 to charities, mainly for children.

In July 1946 an appeal was made for older members. There seem to have been joint efforts by junior and senior Players, in Mystery Plays at Christmas in 1946 and 1947, in two performances of two plays in May 1947, and in three performances of *The Sleeping Beauty* in December 1947. By this time the St Lawrence Senior Players was established as a separate group and had become more ambitious. There was an encouraging review in the local newspaper following two performances of *The School for Scandal* in February 1948. During Dedication Festival Week in October 1948 there were two performances of *The Man who Stayed at Home*. In both 1949 and 1950 the group produced two short plays on the day of the summer fête, and *I Have Five Daughters* (an adaptation of *Pride and Prejudice*) in November 1949.

Meanwhile the Junior Players had started the St Lawrence Players Club in 1948. In October 1950 they presented two short plays, and in January 1950 their carol party raised funds for the Permanent Hall Fund. By 1954 the Junior Players Club had evolved into the St Lawrence Youth Fellowship.

It is likely that members from both groups took part in some productions for a number of years, and almost certainly in the Nativity Plays of 1948, 1949 and 1950, and the Easter Play of 1949. In January 1950 Miss Angell, on behalf of 'Players, senior and junior,' thanked people for their support. Miss Angell contributed much to establishing successful drama at the church, and appears to have remained responsible for the annual Nativity Play until the end of the Revd Godwin's incumbency.

The two performances of *Candied Peel* by the senior St Lawrence Players, in May 1950, was the first of many successful productions directed by Hylda Darby. The tradition of a pantomime continued, with afternoon and evening performances on the Saturdays from 1951 to 1953, and an additional Friday performance in 1954 and 1956. The Players also presented the pantomime at Sunday School parties each year, and often gave the profits from the public performances towards the cost of Sunday School summer outings. In 1956 the profits from *Dick Whittington* amounted to £24.11s.6d.

The earlier practice of Friday and Saturday performances continued, with productions each May and autumn, usually in November. The Players also presented an item for the annual summer fête, busy days on which many members helped in other ways to make them both profitable and happy social occasions.

Jack and the Beanstalk 1952

The dramatic efforts of the Players were curtailed during the building of the new hall. There was no 1954 autumn production and no 1955 pantomime. The first production of the St Lawrence Players in the new hall was *The Blue Goose* in October 1955.

By the time the Revd Godwin retired in 1956 the annual pattern of four productions had been established. They were the pantomime, a play in the spring, an item at the summer fête, and another play in the autumn. These entertainments were anticipated, one could almost say expected, from this successful and popular senior drama group.

District Visitors
This group had its origins way back in 1877. Before there was any idea of a separate parish in Eastcote, a few ladies' of St Martin's parish agreed to become District Visitors for the purpose of visiting poorer neighbours, giving sacramental alms and distributing tracts. Some ladies were responsible for particular roads in the 'hamlet' of Eastcote. After Eastcote became a 'Peel District' in November 1931, the Revd Godwin maintained District Visitors as a system of communication.

In February 1932 the first *Leaflet* was issued. A 'modest venture to begin with', it was the parish magazine of its day. The Revd Godwin wanted the *Leaflet* to be, not only a publication for members of St Lawrence Church, but also a means of making the presence of St Lawrence Church known to the community at large. The lady District Visitors were responsible for monthly delivery and the collection of subscriptions; 1/- a year.

As Eastcote became more developed as a suburb, the Vicar was anxious 'to open up new areas for distribution' of the *Leaflet,* and called for more ladies to undertake this 'great work'. In February 1945 thanking 'our faithful band of District Visitors, who so willingly distribute…monthly…700', he added 'there are many residents who look forward to their call with the *Leaflet*'. With further development the Vicar regarded the Visitors as doing 'vital work', helping a large number to renew their church membership, when arriving in a new parish.

Following the opening of the Newnham Sunday school, and building on the presence of St Lawrence Church in that area, the District Visitors supported Mothers' Union and Church of England Men's Society members in making the activities of the church more widely known, by delivering the *Leaflet.*

The Flower Guild
From the early days of the *Leaflet* the Vicar often expressed thanks to the ladies who contributed and arranged the flowers in church. Needless to say, there were also requests for others, 'to provide flowers for the altars and the War Memorial'. For a few years from May in 1936 when the *Leaflet* listing was re-organised, the 'flower ladies' were a part of the Sanctuary Guild, but by the end of the war were known as the Flower Guild. Their efforts throughout the year, and particularly at Church festivals, were much appreciated.

Members of the congregation were invited to subscribe to the flower fund to help with the cost, when flowers were scarce or expensive, and to provide for festival flowers. At Easter 1949, the idea of giving a lily, or the money for one, as a memorial for a loved one 'found a very ready response'. Another precedent, that of providing 'in memoriam' flowers, for a particular Sunday remains a tradition at St Lawrence.

St Lawrence Needlework Guild
In the days of the Mission Church a Sewing Circle was set up to make articles for the annual Sale. In 1935 its name was changed to the St Lawrence Needlework Guild. This church group met regularly, either in the choir vestry, or in private houses. As well as making gifts for the summer fêtes and autumn sales, the ladies took orders 'for plain or fancy needlework'. Meetings were times of encouragement and co-operation about the practical work, and particularly helpful during the war and post-war years, when materials were in short supply.
The contribution of this Guild towards the Permanent Hall Fund (PHF) was immense. In February 1944 the Vicar reported 'the main support of the PHF is

the St Lawrence Needlework Guild, which has performed wonders under Mrs Batty's able guidance'. By 1950 it had been responsible for raising nearly £1,000.

The rising cost of materials led to the Guild's first social concert in October 1950. The success of this venture led to a second concert in April 1951, when members of the congregation provided the entertainment. In 1952 the Guild organised dances in April and October which proved a popular means of fund raising. Tickets were 1/6d, which included refreshments.

The history of this Guild shows that, as has been implied about other church groups, the parish of Eastcote was moving to that 'cohesion' of an established church so desired by the Vicar.

The Communicants' Guild
The Vicar formed the Communicants' Guild in 1950 (after the London Diocesan Mission), though a Communicants' Fellowship had been listed in the *Leaflet* from 1932 until March 1943. A service of renewal was held on 19 September, 'for all communicants who have been confirmed at or from St Lawrence parish', and was followed by a social hour in the hall. Thereafter the Vicar hoped to have a service of preparation on the third Tuesday of each month, followed by the social hour. He hoped this would be a means of 'keeping us together in the fellowship of the Church'.

By January 1951 the Guild was said to be 'shaping well'. The Vicar hoped that 'through its members we can retrieve some of the many "lapsed" Communicants'. In May 1951 most of the twenty-six newly confirmed attended the Guilds' preparation service before their first Communion on Whit Sunday.

The first anniversary service and meeting in September 1951 was well attended. The Vicar hoped members would 'have the advance of St Lawrence's Church at heart, and offer themselves for service wheresoever needed'. In the December *Leaflet* the Guild was regarded as 'a very necessary feature in the family life of the parish', and by September 1952 as 'the heart of the Church life'.

In addition to the regular 'social hour' after the monthly service, other social events were arranged. In August 1951 members had a country walk in the area of Amersham, and in the summer of 1952 two afternoons of picnics and sports at King's College Fields in Ruislip. In 1954 a Reunion Dance was held at the Cavendish Rooms, Field End Road. Tickets were 1/6d, with refreshments extra. What was to become an enjoyable annual event was a Guild supper, first held in January 1952. The one in 1953 was 'voted one of the most enjoyable social features we have had'. About 80 had a splendid supper, provided by Miss Angell and her helpers. Nearly 100 sat down 'to a delightful meal' in 1954, when it was 'pleasing to see some fresh faces'. In February 1956, thanks were expressed to the 14 ladies who made the fifth 'supper the success it was'. The fact that it was held in 'our lovely new hall' possibly played its part in seeing this as 'an encouraging event for the future'.

Mission in the Local Community and Beyond
Since it became a separate parish, St Lawrence Church has been involved in many aspects of what might be called 'the mission of the Church'. Though not overtly referred to as such, mission has been at the heart of much that has been written in the preceding pages. Here discussion is about charitable organisations, which were financially supported by the church, with Deanery or Diocesan Missions in which St Lawrence was involved, and with support given, both pastoral and material, to groups in the Eastcote community between 1933 and 1956.

Support of Charitable Organisations
Three societies, the Church Army, the Church of England Children's Society (Waifs and Strays), and the Society for the Propagation of the Gospel had regular support. Several other societies had financial support when the parish responded to specific appeals.

The Church Army
St Lawrence Church maintains a tradition of supporting the work of the Church Army, with its widespread ministry of turning 'Concern into Action'. During the years the Revd Godwin was at Eastcote, financial support was mainly through the annual 'flag' collection in April or May, known as Violet Day, when ladies of the parish undertook to sell, what were in fact, lovely little paper violets.

As with any such project, one person was responsible for organising Violet Day, and up to and including 1936, it seems to have been a different lady each year. Appeals for help were made in the *Leaflet*, some years having a better response than others. The number of helpers had some influence in the sum of money collected. However, one has to remember, when looking at the sums raised, that there was a considerable growth of population in Eastcote during the period covered by this chapter. In 1932 the amount collected was £4/6/1 and in 1936 it was £6/6-. From 1937 until 1948 Miss Warren was in charge. It is possible that some ladies gained experience and confidence helping year by year, for from 1943 to 1947 the annual sum was over £30, with £39 being collected in 1944. Mrs Smith took over the organisation in 1948, and after a couple of disappointing years, collections increased. In 1955 she was congratulated in the *Leaflet* on her record total of £60.

Over these years, representatives from the Church Army, such as Capt. Clementson in 1935, and Capt. G. Gardham in 1943, were welcomed to services at St Lawrence, when congregations heard of the many activities of the Church Army. Several issues of the *Leaflet,* especially during the early 1940s, refer to these activities. For example, in April 1941 there is reference to 'the society's work for the men and women of the Forces', at home and abroad, and to the 'mobile canteens at home' which gave vital help to the civilian population, particularly 'where towns and cities have been severely bombed'.

The Waifs and Strays
(now The Children's Society)
The custom, begun in the days of the Mission Church, of sending collections from the Christmas services to the Waifs and Strays Society continued. Money was also raised through members of the church having Society boxes, collected once a year by the Branch Secretary, and the contents sent to headquarters. The proceeds in 1934 were £4/19/8 and in 1954 £54. Secretaries seem to have held the position for a number of years, Miss Russell up to 1946, Miss Dalton from 1947-53 and the Misses Gibbeson and Williams from 1954. Support for the Society was encouraged when visiting preachers came to the church.

For several years the Society seems to have had special links with Eastcote. In 1935 its Fête was held in the grounds of Haydon Hall, by kind permission of Mrs Bennett Edwards. There was entertainment, stalls and dancing… Admission 6d for adults, 3d for children…teas were 9d a head. Members of the church assisted in many ways to make the event a success. During the war, from 1940 to 1945, the Society moved from its headquarters in Kennington Old Town Hall to St Michael's School in Joel Street. In 1942 the Founder's Day service was held in St Lawrence Church instead of at the usual venue, Westminster Abbey.

Foreign Mission Societies

Society for the Propagation of the Gospel
St Lawrence Church consistently supported the Society for the Propagation of the Gospel (SPG), throughout the Revd Godwin's time in Eastcote. Money for this Society, with its own St Lawrence Secretary, (Miss Warren until 1949 and then Miss B. Orders), was mainly raised from missionary boxes issued to members of the congregation. These were returned at an annual meeting when most years there was a guest speaker who described some aspect of work on the mission field. Subscribers to the national publication, *The Mission Field,* were invited to these meetings. Few figures are available as to the amounts collected through the boxes, or from the collections on Missionary Sundays. One part of a Missionary Week-End, in 1947, had Sunday School children 'in Indian costumes portray the every-day life of Indians'.

Some members of the congregation attended the annual Diocesan Missionary Festival at St Paul's Cathedral. A few went to the annual SPG Rallies held at the Royal Albert Hall. In 1935 a party from the church went to a special pageant there, *From Sea to Sea*, with over 2,000 performers, which told the story of the Society's work over the 234 years of its existence. After the war more seemed to attend the Rally, for in 1955 a coach was hired 'so as to take a larger number of our people'.

Some Other Societies
It is always beneficial for a church congregation to hear the views of visiting preachers, as well as those of their own clergy. Among the many visiting preachers were representatives from charitable organisations. Some came to strengthen links already made with the parish, such as the Church Army, the Waifs and Strays Society, and the SPG. The Vicar also invited representatives from other organisations, such as Dr Barnardo's Homes, the British and Foreign Bible Society (BFBS), and the Missions to Seamen, to place before the congregation the work of their particular society. Normally the response was a donation to the society, but there were occasions when the commitment was over a longer period, such as in 1941, 1947 and 1954 when the proceeds of Lent boxes were sent to the Society for the Promotion of Christian Knowledge (SPCK). In 1954 a representative of the Universities Mission to Central Africa (UMCA), 'was so persuasive' that a branch of the Society was formed in the parish.

During these years the Good Friday collections were given to the East London Fund for work among the Jews. The purpose of this organisation was to 'serve the Church in her bridging of the gap between Christ and the Jewish people'.

Deanery and Diocesan Missions
In addition to regular support of mission work and annual Missionary services, Church of England parishes have organised local missions, either to re-vitalise a church community or as outreach to the wider community of that parish. During the years under discussion St Lawrence Church was involved in one Deanery Mission and one Diocesan Mission which were more ambitious schemes, each set up to meet the challenge of a particular circumstance.

Rural Deanery of Uxbridge Mission of Help October 1934
The 1934 Deanery Mission of Help was related to the significant growth in population in the area of Northwood and Ruislip (including Eastcote), which rose from 9,114 in 1921 to 16,035 in 1931. The Mission was to meet the challenge of how to bring those new to the area into the life of the Church. From the early issues of the *Leaflet* the Vicar had asked newcomers to make themselves known to him as he felt that it was from them 'that we shall build up our congregation'.

In the months before the Mission 'much labour' had gone into preparation as to the 'best possible way to reach all Church-people in the Deanery', and the Vicar expected all members of St Lawrence Church to support the venture, by telling neighbours, and others about the Mission. If members took greater interest 'in our Church' it would help others to renew their interest.

Circulars were delivered to each house and the Mission started with a public meeting in the Central Hall, Uxbridge on Friday 5 October. The centre for the parishes of Ruislip, Ickenham and Eastcote was St Martin's Ruislip, where meetings for residents were held on Mondays and Wednesdays. The Revd O. H. Gibbs-Smith was the Mission Preacher on four Sunday evenings at St Lawrence

Church, when there were large congregations. The Mission ended with a Foreign Missionary Demonstration at the Hall in Uxbridge, on 8 November, with speakers from the Church Missionary Society (CMS), the SPG and the Student Christian Movement (SCM). A 'char-a-banc' took Eastcote people to and from the Uxbridge meetings, at the cost of 1/3d a head.

The Vicar reported in November that the Mission had had a rousing effect on people generally. After the Mission, in response to many requests, he gave an instructional course of addresses on Sunday evenings. 1935 saw some growth at St Lawrence's, both in attendance at weekday services, and in the size of the Sunday congregations. The latter influenced the change in time of the Sung Eucharist in May, referred to earlier in this chapter. Whether this growth, and the formation of a branch of the Church of England Men's Society (CEMS) was a direct result of the Mission it is impossible to say.

A year after the Mission, in the autumn of 1935, the Vicar expecting a large number of new residents to Eastcote, wrote after the second Consecration Festival, that St Lawrence was 'slowly but surely acquiring a congregation of keen church-people'.

The 'Week of Prayer and Self-denial' certainly had better than usual responses in 1934 and 1935.

The Diocesan Week of Prayer and Self-denial
From the *Register of Service* books and *Leaflets* these special weeks took place each autumn, at least up to the Diocesan Mission of 1949. The purpose was to 'forward the work of the Church in the London Diocese'. Parishioners were expected to help meet the Diocesan Quota through donations in self-denial envelopes. As its name suggests, this annual Week was not restricted to the financial side of Church life. The Vicar appealed for attendance at the extra services during the Week and for improved attendance at Sunday services on a regular basis. He also reminded parishioners that each member of the Church had a responsibility to its own parish, and should offer service in the form of time and talents.

The London Diocesan Mission of May 1949
When Father Godwin had remarked on the high attendance at St Lawrence's, on one of the National Days of Prayer during the war, he expressed the feeling that the nation 'is still Christian at heart'. It is relevant to refer to a series of articles in the *Church Times* of Lent 2008 that looked back 60 years, to the 1948 report of a committee of the Church Assembly (the precursor of the General Synod), entitled *Report on the Spiritual Discipline of the Laity*. The war had ended in 1945 and leaders of the Church of England were much concerned with how much people's lives had been changed by the experiences of war. Faith in a loving God had been challenged, and the routines of spiritual life had been disrupted.

St. Laurence, Eastcote.

MISSION TO LONDON—LOCAL FIXTURES

Conferences.

Monday, March 7th: The Communist Challenge, S. Laurence Church Hall at 8.15.

Monday, April 4th: Evangelisation, S. Mary's Church Hall, 8.15.

Monday May 2nd: Methods of "Following-up." Details later.

Quiet Afternoons.

Saturday, March 12th: For all church people of the three parishes, at S. Paul's Church.

Saturday, May 14th: For all helpers in the three parishes separately.

Processions of Witness.

All church people are asked to make an effort to make these processions a great demonstration of our strength and our faith.

Good Friday, 15th April, Saturday, 23rd April, Saturday, May 7th. Details will be announced later, but the idea is for separate processions to be organised in each parish, meeting for a joint service. A different area will be covered each time.

POSTER PUBLICITY HELPS THE MISSION

The Mission Executive Committee has decided on the most modern methods of publicity to attract the attention of Londoners to the Mission.

Among the methods of publicity chosen are posters. Some of these posters are appearing on hoardings and in Underground stations throughout the Diocese soon; newspaper advertisements will appear in the London Press nearer the time of the Mission itself.

Well-wishers of the Church have spaces, such as windows, available for use. For this purpose, two posters have been produced. These posters feature the Mission slogan, "Recovery Starts Within," and the date of the Mission.

It is important to remember that these posters are not intended to do the Mission's work. They are intended to pave the way, to introduce the Mission and the fact of its existence, to a wide public. When these posters appear throughout London, we shall be making our first approach to what have been described as the "fringers." But as the Bishop of Kensington said at the Royal Albert Hall on May 24th, 1948, "There will be a lot of advertisement, but don't put your trust in that. The advertisement that counts, that does the trick, that strikes home to the ear and eye and heart of the wayfaring man is your own testimony."

H. H. Greaves Limited, 106/110 Lordship Lane, S.E.22.

London Diocesan Mission 1949
from February 1949 issue of the *Leaflet*

It seems likely that the research, which led to the report of 1948, influenced the Diocese of London to organise a campaign with the aim of a renewal of Church life. The Diocesan Mission of 1949 was on a much bigger scale and involved the laity of St Lawrence Church far more than the Deanery one of 1934.

Obviously much planning had gone on before news of the Mission ever reached the congregation in May 1948. It was a 'great venture, never before attempted over so large an area as a diocese'. Its success would depend on individual efforts in every parish, for as the Vicar wrote 'renewal starts within'. It was important that people within the diocese, whether churchgoers or not, should know the Mission was to be held, hence the many preparations in the months leading up to it.

The first area meeting, of representatives from St Mary's, South Ruislip, St Paul's, Ruislip Manor, and St Lawrence's, took place in September 1948 (see map page 18). In November a meeting was held in St Lawrence's hall, when details were given, regarding the letters and circulars that were to go to every house. The success of 'personal canvassing' was stressed. It was up to the laity to encourage 'lapsed' members to renew their Church life. A personal invitation, to take an interest in the Mission, and to attend the meetings in the Mission Week proper, Sunday 15 May to Sunday 22 May, was more likely to result in success, than merely using letter boxes. St Lawrence's District Visitors were quite used to knocking at people's doors, but it was an entirely new experience for many.

In February and March 1949 the *Leaflet* announced the many fixtures connected with the Mission. There were prayers and intercessions and study groups, conferences and quiet afternoons during March, April, and the first two weeks of May, and Good Friday processions in each of the three parishes. 'Modern' methods of publicity were used to attract the attention of Londoners to the Mission, such as posters on hoardings and in Underground stations.

According to the *Leaflet* all the preparations had been worthwhile. Both the opening meeting at St Paul's, and the closing one at St Lawrence's, were packed. Attendance at the weekday meetings at the Bourne School, in Southbourne Gardens, improved throughout the week. The Missioner, Canon R.R. Roseveare, was said to have 'held his hearers from start to finish of his most arresting and inspiring addresses'. A Church Army sister lived in the parish for the Mission Week.

As to the results of the Mission, the Vicar was concerned that there should be a 'keener sense of membership in the Church'. There was a 'follow-up' meeting at St Paul's in October, which coincided with Dedication Festival Week. The Vicar must have been referring to the recent Mission when he questioned 'just how far we are witnessing for the Church of Christ in the parish of St Lawrence'. The growth in Sunday School numbers, which has been mentioned earlier, suggests that some had offered their services as teachers. However, reporting on the Diocesan Anniversary Meeting at the Royal Albert Hall, the Revd Godwin was disappointed not to have a much larger congregation at St Lawrence's.

For a few years after the London Mission the diocesan publication, *The London Churchman,* was distributed in the parish with the St Lawrence *Leaflet* as a centre insert. This change was not a success and was discontinued.

Mission in the Local Community
St Lawrence's has regarded support of charitable bodies, and appeals, as part of the Church's mission, and this has inevitably included many different local organisations. The following have been chosen because they illustrate some of the history of the wider community in these early years, for none of them exist today.

St Michael's School, Joel Street
In November 1934 the Vicar reported the opening of St Michael's School, Joel Street, 'built on the latest approved pattern…accommodates boarders, some of whom attend St Lawrence Church'. He took a scripture lesson every week at the school. As with so many things the war caused disruption, and in 1939 the school was evacuated to Devon.

From 1940 to 1945 the building was the temporary headquarters of the Waifs and Strays during the period of their evacuation from Kennington, London (the Society became the Church of England Children's Society in 1946, and The Children's Society in 1982). In 1945 the building was sold to the Middlesex County Council who decided to adapt it for use, as a 'special needs' school. The premises were unoccupied for some time, mainly because of the difficulties in obtaining building materials in the post war years. It opened, in 1950, as a boarding school for thirty physically disabled boys, retaining the name St Michael's. The boys received medical attention, and at the same time continued their education.

The Vicar of St Lawrence Church was appointed as Chaplain to the school. The Revd Godwin prepared several boys for confirmation. The Lord Bishop of Kensington visited the school to confirm boys in 1951 and 1955. The *Leaflet* in 1951 reported that 'four newly-confirmed lads …received their first Communion' at St Lawrence's 'two of them in invalid chairs'.

Perhaps it was as a result of the 1949 Diocesan Mission that, from 1950, a few men from St Lawrence's were responsible for leading a service for the boys each Sunday morning. This is one example of service in the Eastcote community.

As the years went by there were changes, but St Michael's remained a special needs school until the 1980s when pupils were transferred in phases to other local schools. The premises were demolished in 1985 and Deerings Drive was built on the site.

The Retreat
In the *Leaflet* of 1951 which referred to the mission of the church at St Michael's

school, the Vicar also noted the work of St Lawrence people at the Retreat, 'one of Eastcote's old houses, standing in a peaceful garden on Chapel Hill' (North end of Field End Road near the Methodist Chapel). Referring to it as an old house is a little misleading, for it was built in the early 1900s and was not in the same category as the really old houses such as Eastcote House or Highgrove House.

The Middlesex County Council opened the Retreat, in the grounds of what is now Alison Close, Field End Road, as a residence for old people in the late 1940s. The Vicar had responded to a request for spiritual help from St Lawrence Church. It seems that a monthly weekday Holy Communion celebration was provided, conducted by the Vicar. In addition to this, with the authority of the Vicar, men from the church congregation led a regular Sunday Evening service. The Vicar reported that seeing those old people, who were able to gather for their evening service 'no matter what their denomination …one realises how much this opportunity of an act of worship means to them all'. Two ladies shared the work as pianists for several years. The piano was a gift from one of the worshippers at St Lawrence's. When one pianist left in 1952, there was an appeal to provide music requiring, 'the ability to play hymn tunes at sight, and a desire to aid the Church in bringing spiritual help and comfort to the aged'. Four people, two men and two ladies, answered the appeal to provide music.

In some years produce from the annual Harvest Festival was given to the Retreat home.

St Vincent's Home, Wiltshire Lane
A link with what was called St Vincent's Cripples Home and School for Defective Children, made in the early days of the Mission Church through the visits of ladies who were District Visitors, was re-enforced when the Revd Godwin prepared boys at the Home for Confirmation. In addition a member of the congregation arranged for parties of boys to be taken from the Home to church by car. Some years the Home had a share of fruit and vegetables from the Harvest Festival. The buildings of St Vincent's Nursing Home now stand on the site.

Northwood and Pinner Community Hospital
Another place that had a share of produce from St Lawrence's annual Harvest Festival during the 1930s was the Community Hospital at Northwood Hills. What had started as a Voluntary Aid Detachment Hospital (VAD) in temporary buildings in World War I, was by 1931, a well-loved local cottage hospital, in permanent accommodation. As it was on the borders of the original St Lawrence parish the Vicar referred to it as 'our own hospital'. Collections on Hospital Sundays often went to this local hospital. In 1941 and 1942 appeals were made for eggs, particularly from those who kept poultry, or for money to buy eggs for the hospital. One may assume that as it does today, St Lawrence Church responded to other local appeals.

Finances and the Freewill Offering Fund
Reference has been made already to fund raising efforts in Father Godwin's incumbency, for the building of the parish church and the hall, but also for charitable causes, and different 'in house' church organisations.

It is almost inevitable that raising sufficient income to meet that of the expenditure of a parish has been, and remains, a challenge. From the first *Leaflet* in February 1932, the Revd Godwin made periodic appeals for members of the congregation to join the Freewill Offering Scheme (FWO). This was very like the Stewardship scheme of today. With a commitment to regular giving on the part of members of the congregation, whether one is at church or prevented from attending, the finances of a parish are more easily and efficiently run.

Naturally with the Consecration of the church, in October 1933, the expenses rose considerably. In June 1934 the Vicar reported that £150 'new money' was needed 'now that we have a Church to maintain, and a Parish Hall'. Although in February 1935 the FWO had been 'a great success' and thanks were expressed to Mrs Rye for 'creating enthusiasm for the Fund', the Secretary of the Parochial Church Council (PCC), Mr Adamson, had a letter in the October issue of the Leaflet imploring others to join the FWO. Lists of PCC secretaries and FWO officers are in Appendix 4.

The Diocesan Quota
(For figures throughout the history of the church see Appendix 10)
Then as now, as well as meeting the running costs of St Lawrence Church, the parish had to contribute to the Diocesan Quota each year. From the Quota a diocese pays the stipends of the clergy and meets other diocesan expenses. Few Quota figures are available for these years and it is impossible to make real comparisons with the Diocesan Quotas of the twenty-first century. However the following may interest some readers. In 1927 when Eastcote was a Mission Church the Quota was £26. In 1932 when it was a 'Peel' district it had risen to £40. In 1951 the Quota was £85. In 1952 two figures were given, the Standard contribution of £82 and a Quota of Ambition of £107. In 1954 the Standard, or Contribution of Obligation, had risen to £116, and in 1956 it was £170. In 2010 the Quota (now known as the Common Fund contribution) was £77,108.

The challenge of meeting the Quota each year was obviously not peculiar to the Eastcote parish for, as mentioned previously, each autumn there was a Diocesan Week of Prayer and Self-denial. Although parishioners were frequently urged to contribute to the Quota through the FWO scheme, and other moneys also went towards this parish commitment, members of the congregation were encouraged to make additional donations in this special Week in order to meet the annual Quota. Each December a service took place in St Paul's Cathedral when the Bishop of London received the gifts from parish representatives. In 1934, the year of the Deanery Mission, and in 1935, there was a better that usual response but in most years the Vicar expressed disappointment at the number of envelopes returned. In 1936 and 1937 the Lent appeals were on behalf of the Diocese. As was the case with other fund-raising efforts at St Lawrence's in Father Godwin's incumbency 'boxes were available' as a means of supporting the cause.

The Vicar wrote in the *Leaflet* of May 1937, that with hundreds of new houses being built in the area, it was 'becoming increasingly impossible to work the parish'. The solution would be an Assistant Priest. The prospect of that happening receded in 1939, when the parish had a loan from the Diocese to help meet the cost of building the east end of the permanent hall. In the opinion of the Vicar 'the importance of paying our Quota to the Diocese becomes even more prominent'.

With the outbreak of war finance became an increasing challenge. However the Vicar was able to report in 1942 that the sum collected in self-denial envelopes, returned after the Week of Prayer and Self-Denial, had enabled the parish to meet the Quota. In fact every year the parish had sent more than the Quota of Obligation, 'thanks to the painstaking efforts of the faithful few'. It was hoped that after the war the parish would be able to send the Quota of Ambition, a step towards having an Assistant Priest.

The Bishop of London's Reconstruction Fund
In 1946 the chance of sending the Quota of Ambition was dashed with the launch of the Bishop of London's Reconstruction Fund appeal in 1946. A large number of churches in the Diocese of London had been badly damaged or destroyed during the war and the appeal was for £750,000 towards the re-building of these churches. The War Damage Commission also allocated money for this purpose. In the October 1946 *Leaflet* the Vicar informed his parishioners that the PCC had decided 'we should find at least £30 as a minimum per annum for the next five years'. This was 'new money', and had nothing to do with the Quota. All existing obligations still had to be met.

As Mr A.C. Riddle, the first secretary for the Fund, reported early in 1947, a few personal subscriptions, the proceeds of the St Lawrence Players' Mystery Play, and returned Postage Stamp Cards, had made 'little headway with our subscriptions' for the fund. Whether the stamps affixed to the Cards were used or unused ones is not made clear. The Sunday Schools (Junior and Senior) and the MU were the main supporters of these cards.

The poor response to the appeal is likely to have been influenced by the fact that it coincided with efforts to raise money for the Permanent Hall Fund (PHF). Possibly this led the PCC to decide in June that, instead of dividing the profits of the 1947 summer fête between the PHF and the Bishop's Fund as agreed earlier, the whole of the proceeds should go to the latter. In the event the weather, 'very doubtful at the start, ended with a thunderstorm and heavy rain' and with 'only fair attendance' there was only a small profit of £15.

In July 1947 the Vicar wrote at length of the importance of the Bishop's Fund. He referred to the many Visitations made by the Bishop to gain support for the Fund, but added that the purpose of these visits was 'to arouse in church people a renewed sense of membership in the Church...worship to Almighty God and Almsgiving'. Perhaps this was a hint of the Diocesan Mission that was to come in 1949.

While appeals for their own parish church had support from the wider community of Eastcote, the Bishop's Reconstruction Fund was a different matter and depended largely on committed Church people. The years following the war made many financial demands on the congregation, but an appeal for the Fund, in November 1948, stressed how fortunate St Lawrence Church had been. It was one of only 77 churches in the Diocese of London to escape war damage. A great number were bombed out of use and 91 were completely destroyed.

As well as the proceeds of the Cards, the ubiquitous boxes, collections at some weddings, recitals and other performances, the sale of annual blotters also raised some money for the Fund. These blotters produced each year from 1947 to 1952, had a piece of blotting paper in the centre of a small booklet. The one for 1952 included the Church's calendar, information about St Lawrence Church services and organisations, and a number of advertisements for local shops and businesses, and raised the odd sum of £5/0/7d.

The Reconstruction Fund had raised £75,000 by November 1952. The special appeal was discontinued at the end of 1952 so that, in the words of the Bishop, 'parishes may concentrate on attaining their Quota of Ambition'. The secretary of the FWO stated that the main reason for concentrating on the Quota of Ambition was 'to bring up to a reasonable level, the utterly inadequate stipends of many of the clergy'. For the last years of the Revd Godwin's incumbency the target was £450 per annum, but the highest achieved was £353 in 1954, when the Quota of Obligation was £116. Unfortunately a number of generous subscribers were lost in 1955, either because they had left the parish or they had died. In view of the commitments in respect of the loans on the building of the hall, the increased cost of church expenses generally, including the aim to submit the Quota of Ambition, there was the inevitable appeal for fresh subscribers to the join the FWO.

The Growth of Church Numbers
Trying to make any assessment as to growth in numbers of the St Lawrence congregation during these years is difficult. It has been shown that Father Godwin considered regular church attendance, and frequent Communion, as obligations of Church members. Numbers of communicants have been discussed and shown to have increased during his incumbency, but no comparisons can be made regarding church attendance as the *Register of Services* books rarely recorded attendance figures. The Revd Godwin may have supported the 'Parish Communion Movement', which aimed to restore the Eucharist as the main act of

worship on a Sunday morning, but it made little progress until the 1960s. Many people did not receive Communion each week but may have regularly attended one of the Sunday morning services.

The Electoral Roll and the Annual Parochial Church Meeting
(For numbers throughout the history of the church see Appendix 10)
Turning to the numbers on the Electoral Roll is of some help. The figure for Eastcote in 1929, before it became a separate 'Peel' District, was 211. One may assume that the considerable population growth in the area between 1931 and 1939 contributed to the increase in Electoral Roll numbers, from 327 in 1932 to 428 in 1939. One has to remember the difficulties associated with the war and its aftermath, when looking at the Roll figures for those years. They ranged between 428 in 1939 to 395 in 1947. A thorough revision of the Roll in 1948 reduced the figure to 245. Newcomers were called upon to complete the necessary form for admission to the Roll. The Vicar considered that all regular members of the congregation over 18 should enrol. Probably the London Diocesan Mission of 1949, together with a further population growth in the area, contributed to the increase in Roll numbers to 344 in 1956. Issues of the *Leaflet* give estimates of Eastcote population in 1949 as between 16,000 and 18,000, and in 1955 as 20,000. In 2010 there were 282 names on the Electoral Roll.

Having one's name on the Roll enables a member to vote at the Annual Parochial Church Meeting (APCM). The APCM should be of interest to all church members, for the agenda covers more than electing and agreeing to various church officials and representatives. The audited accounts are presented, and reports on the church fabric and main events of the past year are given. Members are able to ask questions. Attendance at the APCM compared with Roll numbers has seldom been high, and this was the case in the years the Revd Godwin was at Eastcote. Though in 1936 he reported that the hall was 'well filled', most years he commented on the poor attendance. Actual figures are available only from 1952 when 40 attended, regarded by the Vicar as 'poorly attended'. In 1954 there were 36 and in 1956 there were 48, the highest figure recorded to that date. In 2010 of the 282 on the Roll 33 were at the APCM.

Revd Rupert Frederick Godwin's Legacy
As has been remarked at several places in this chapter the first Vicar of Eastcote faced many problems establishing a new parish in uncertain times. The Revd Godwin had anticipated that with the great increase of population in Eastcote and the various efforts and missions, particularly those of the Deanery and Diocesan Missions of 1934 and 1949, there would be a much larger congregation at St Lawrence Church. Actual growth in numbers attending church, or how many of those attending were on the Electoral Roll is one thing but numbers in themselves do not constitute the witness of a Christian community.

The Vicar seems to have been involved and to have had influence in all the church organisations and groups encouraging members both, to bring those from the wider community of Eastcote into the Christian community of St Lawrence Church, but also to serve the local and the wider community.

In his last letter of April 1956 the Vicar wrote of completing the 'trio of endeavours' that he had hoped to accomplish, the building of the church, vicarage, and lovely church hall. He did much more than this, for he built up the 'cohesion' he had regarded as necessary for an established church to pursue its work.

He was pleased to remain in London when he retired in 1956 to Jersey House, a clergy home, in The Bishop's Avenue, N2. The Tube which had led, at the beginning of the century, to the development of Eastcote meant that he had 'easy access to London Town' from Hampstead. He died in November 1965. The fact that the church's East window was dedicated, in 1969, as a memorial to Revd Godwin is a tribute to his twenty-five years' service to the Eastcote parish.

An excerpt from his obituary in the January 1966 *Gridiron* seems a fitting conclusion to this chapter.

> Perhaps what impressed most in Father Godwin was his steadfast adherence to principle. A convinced 'Prayer Book Catholic' he nailed his colours to the mast from the first and would tolerate neither watering-down nor what he called unwarranted additions. He was punctilious in observing a dignified and restrained ritual. Though he always disclaimed any oratorical gifts, he had a talent for building up a band of devoted workers and for winning young people without playing down to them. Those who knew him best came to value him as a true friend. May he rest in peace.

References
1. Bowlt Eileen M. *The Goodliest Place in Middlesex,* Hillingdon Bor. Lib. 1989 p261
2. Edwards Ron *Eastcote, From Village to Suburb,* Hillingdon Bor. Lib. 1987 p47

CHAPTER III

THE REVEREND WILLIAM HITCHINSON
VICAR 1956 to 1980

The Revd W Hitchinson, formerly Assistant Priest at St Mary's, Twickenham, came to St Lawrence Church with his wife and two children in May 1956. Most people who knew the Revd Hitchinson, and are alive today, remember him as Father Bill, and although it wasn't until November 1975 that he signed his magazine letters in that form, he will usually be referred to by that title, or Vicar, in most of this chapter.

The Reverend William Hitchinson

As the previous chapters have shown the St Lawrence Church congregation did not exist in isolation from the local community of Eastcote. Local housing developments put on hold during World War II, together with the baby boom which had started in the mid 1940s, necessitated a number of new schools being built in Eastcote. Bourne School was opened in 1946, Field End Junior and Infants in 1947 and 1952, Newnham Junior and Infants in 1952, St Nicholas' in 1955, and St Mary's in 1957. Pupils and their parents were potential members of the congregation. Other developments in the immediate area of the church included Winslow Close, since redeveloped as The Forresters, The Sigers and Farthings Close.

It is relevant at this point to look briefly beyond the local scene, for this period saw much in the way of social and political changes which affected the life of people in general and led to questioning the role of the Church. Influential events of the 1960s and 1970s included the Campaign for Nuclear Disarmament (CND), Anti-Vietnam War demonstrations and the Miners' Strike. In addition there was student unrest both in this country and abroad. It was also the time of Beatlemania, of Mods and Rockers, and of Flower Power, miniskirts and the contraceptive pill. For the young being a hippie was the thing, flowers, psychedelic drugs and 'love' were seen as the answer to the problems of the modern world.

The changes connected with the development of the area, together with the changes in society as a whole, presented both challenges and opportunities to St Lawrence Church. The growing population is reflected in figures recorded in the church registers, showing increased numbers of communicants, baptisms, marriages and confirmations. Just as Father Godwin had had to guide the St Lawrence Church community through the problems associated with a newly formed parish, then World War II and its aftermath, so the Revd Hitchinson had to deal at parish level with the role of the Church in changing times. This chapter will show how he tried to adapt church services to meet the needs of the local community. He was particularly concerned about young people and the relevance of the Church in their lives. In his Vicar's Remarks on the year of 1971, delivered at the Annual Parochial Church Meeting (APCM) in 1972, he accepted that the downward trend in attendance at church services and fewer numbers of confirmations were part of a national trend. As part of the mission of the Church he was wholeheartedly behind an ambitious new magazine, which was widely distributed in the parish from 1961 to 1971.

As in the previous chapter, there are sections devoted to specific activities or aspects of the community life of the parish and official church organisations, in order to show that in many respects the traditions of the earlier years were maintained. One important example is that the form of worship continued, on the whole, to be in the High Anglican tradition, though it will be shown that changes were made by major revisions to the Church of England liturgy during Father Bill's incumbency. In following developments in the separate aspects of life at St Lawrence Church it will be seen that there was both continuity and change to meet the challenges of those particular times. This was, and remains, inevitable if the Church is to have any meaningful role to play.

In his first magazine letter of June 1956, the Revd W. H. Hitchinson paid tribute to the Revd R. F. Godwin and all that he had done to establish the parish of Eastcote. He also wrote of the future, and of what he hoped to achieve as Vicar.

As he repeated these words in his March 1974 letter, they are worth quoting as they express what he saw as guiding principles of the Church and they lay at the heart of his ministry.

> *There are two sides to parish life – Worship and Fellowship – and these, I am sure should be the mainspring of all our activities if we are to be alive to ourselves, and also to extend the influence of the Church. Worship must, of necessity, come first, as our Blessed Lord has commanded, but from our worship together there should also emerge a fellowship which, in its outpouring, will embrace all those with whom we come into contact.*

Father Bill's Assistant Curates (see Appendix 1)
Although there had been hopes that the church would have an assistant priest during the Revd Godwin's incumbency, it was not until 1958 that one was appointed. The Revd Peter Goodridge came to Eastcote as a newly ordained Deacon in May 1958, and was ordained Priest in 1959. Like succeeding assistant curates Father Goodridge did good work with the Junior Church, and the youth of the parish. He was involved in the successful Youth Fellowship, and with the Scout Group. Much of the success of the new parish magazine, the *Gridiron* that started in 1961, was the result of Father Goodridge's enthusiasm. He maintained the spiritual life of the parish during Father Bill's temporary absence through illness, from October 1963 to February 1964, at the time the parish was experiencing the early weeks of a Christian Stewardship Programme. Father Goodridge left Eastcote in April 1964 when he was appointed Vicar of St Philip's, Tottenham. From there he moved to St Martin's, West Drayton and then to Truro where he served as Diocesan Director of Education and Canon Librarian of Truro Cathedral. He died in 2005.

The possibility of church owned property as accommodation for an assistant priest had been under discussion when Father Goodridge was in Eastcote, but he spent his six years living in accommodation at different parishioners' homes. Some two years after Father Goodridge left the parish the Revd W. P. van Zyl was appointed as assistant priest in 1966 and he lived at 79 The Sigers, one of the two newly built maisonettes which the church had recently bought. Mr Youngs, Verger/Caretaker, and his wife lived at 77. Father van Zyl celebrated his first Holy Eucharist on 26 June, was licensed on 6 October, and preached his farewell sermon on Sunday 18 June 1967. What Father Bill had hoped would be a stay of two or three years was barely a year when Father van Zyl returned with his family to his native Rhodesia.

Fortunately after a few months Father Bill had a replacement in the person of the Revd Clive Pearce who had been ordained Priest in 1966. He was Assistant Priest from November 1967 until October 1973, when he left to become the Vicar of St Anselm's, Hatch End. Like Father Goodridge Father Pearce was much involved with the young people in the parish, among other things helping to launch the Sunday evening Youth Fellowship in 1971.

Frs. Huw Chiplin, Peter Goodrich, Bill, Clive Pearce
At Father Bill's 'farewell service' October 1979
(F & J Hare/Photopro Photography)

Father Bill's last assistant priest was the Revd Huw Chiplin, ordained priest in 1974, who came to the parish in November 1975. He lived at 26 Woodlands Avenue, the first of the houses that the church has provided for assistant clergy over the years. Father Huw left in January 1979 to be Curate at St Michael's and Christ Church, Notting Hill and Priest in Charge of St Francis, North Kensington. He became Vicar of St Matthew's, Hammersmith in 1984. Once again it is in references to the work he did with the youth of the parish that Father Huw's name frequently occurs.

During the last year of Father Bill's incumbency the church was again without an Assistant Priest. When Father Bill left Eastcote after Easter 1980 the parish had to rely on visiting clergy until Father David Hayes came in June 1980.

The Church Magazine
It may seem odd to include at this point in the history of the Parish Church of Eastcote, a section specifically on the church magazine but the magazines, particularly those of 1961 to 1971, have been a valuable and interesting source of information.

From the first issue of the *Leaflet* in 1932 (printed by Messrs Greaves), 'a modest venture to begin with', the Revd Rupert Godwin had wanted it to be, not just a publication for members of St Lawrence Church, but also a means of making the presence of the parish church known to the community at large. In this it had success, partly as a result of the faithful District Visitors' monthly delivery and the collection of subscriptions. The first two Vicars of Eastcote were in turn editors of the *Leaflet*. By the late 1950s, with the local population growth and with the Revd Peter Goodridge as an Assistant Curate it was felt the time had come for a new style magazine.

The mission of the Church is always to spread the gospel and in April 1961 the new magazine was one aspect of that mission. The leading article in the first issue stated that the intention was to appeal to those 'outside the Church' and be a 'vehicle of Christian propaganda in a largely indifferent society'. The magazine had a new name *Gridiron*, the traditional emblem of St Lawrence, was much bigger than the *Leaflet*, having eight pages (including advertisements), and had an inset, *The Sign* a Church of England publication. The intention was for each issue of *Gridiron* to have a leading article concerning Christian principles, a teaching article dealing with an important aspect of the Christian faith, news of parish organisations, and a photograph. Particulars of relevant local events, and Christian action both at home and abroad, with possibly a correspondence column, were to be included in this ambitious venture. It cost 6d, the price of a cup of tea, and remained at that price throughout the years of its publication, (2½ p for the last two months after the decimal system was introduced).

Dr D. Harrison, who had submitted articles under the heading of 'S. Lawrence Calling' in the *Leaflet* from 1953, was editor of *Gridiron* from 1961 to 1966 when Mrs. P. Beeston took his place (see Appendix 4 for list of editors). From August 1961 until May 1970, the magazine also featured a monthly 'Children's Corner' with a story, and often a prayer or puzzle, for younger readers. It was written by Mrs R. F. Edwards, mother of Mrs R. Chrzczonowicz and grandmother of John Chrzczonowicz, who are both members of St Lawrence Church today.

The magazine was a vast improvement compared with the *Leaflet* that had served the Parish since 1932, and from July 1961 increased in size to twelve pages. For a time the circulation reached 1,150 and much of the success of this enterprise resulted from the enthusiasm of Fr Goodridge (curate from 1958-64), and to the distributors of the magazines, the successors of the earlier District Visitors. Changes were made to the cover in January 1963, when a small drawing of the church tower was included, and again in January 1964, when the drawing was replaced with an unsatisfactory coloured picture of the statue of St Lawrence taking up a third of the cover.

The wide circulation of the early issues was short lived. By April 1964 the magazine was not selling well and financial losses necessitated action. In September the size returned to eight pages and in the November to the cover of January 1963, to give room for more text. Feed back from those who helped to distribute the magazines, favoured retaining *The Sign* and having a more

Cover Designs of *Gridiron* in the 1960s

conventional cover with a picture of the Parish Church. A half-page drawing of the church on the May 1965 issue, and modified in June, remained the cover design until March 1971. Sales continued to decline. In September 1970 the decision was made that it was impractical to carry on with the current format. Once the contracts with advertisers had expired the Magazine Committee were to produce a duplicated magazine until a more permanent solution could be found. D. S. Martell Ltd. of Ramsgate were to supply printed covers having a reproduction of a photograph of the church, and advertisements which they obtained and for which they paid the church £6 per page or pro rata. Volunteers were to assist with the assembling of the magazine each month. In response to a request, at a distributors meeting, Fr Clive Pearce (Curate from 1968-73) agreed to write a series of articles on Anglican beliefs and practices for the do-it-yourself magazine.

In his 'Vicar's Letter' of March 1971 Father Bill expressed regret at the 'change to a cheaper form of supplying parishioners with their monthly news', and gave special thanks to the printers, Messrs Greaves, with whom the church had been associated since 1932. He could not speak for earlier years, but since 1956 only once had the magazine failed to appear and that was owing to a printers' strike, in August 1959, and on that occasion Fr. Goodridge produced a 'Strike Special'.

The do-it-yourself magazine, *The Parish Magazine of S. Lawrence Eastcote*, remained at 2½p until April 1978 when the price was raised to 5p. Most of the features of the ambitious magazine of 1961 to 1971 were included, and the reports of the various church organisations indicate a persuasive magazine committee.

Maintenance and Additions to The Church Building

In 1962 a review of the first five years of Father Bill's incumbency reported that part of the exterior brickwork had been refaced, the interior decorated, the old coal-fired heating replaced by 21 electric night-storage heaters, and the lighting re-wired. The belfry had been strengthened, a new bell hammer fitted and the organ had been overhauled. Problems associated with the roof occurred and included 'rain in at the west end of the church and in the Lady Chapel' in 1959. There were also references to damp walls in the Lady Chapel and the sacristy. These have all caused trouble on more than one occasion in the history of St Lawrence Church.

The review of 1962 referred to the statue of St Lawrence, which had been presented to the church in 1961 in memory of Thomas Grandison (Churchwarden 1951-1959). The statue was designed, made and donated by

Statue of St Lawrence carrying a money bag and book, symbols of his office. The gridiron, showing rear left, indicates the manner of his death

Sidney Starling (Churchwarden 1965-1972). He also carved the statues of St Lawrence and Mary with the Child Jesus, which are in outside niches over the west door and the east wall of the Lady Chapel.

After the use of temporary Stations of the Cross from Holy week in 1959, a donation of £50 was given towards permanent ones. Parishioners were also invited to purchase one in memory of a loved one. For Holy Week 1962 the 14 permanent Stations of the Cross were in place on the walls around the church. At the end of 1962, altar gates were fitted to the original rails at both altars, at a cost of £112.

Station of the Cross 9
Jesus Falls for the Third Time

Detail from the Great Rood taken at the restoration in 1962

Station of the Cross 12
Jesus Dies on the Cross

A big undertaking in 1962 was the restoration of the Great Rood above the entrance to the chancel. It had previously been cleaned *in situ* but on this occasion the figures of the Blessed Virgin and St John, and the crucifix were removed, cleaned, skilfully repainted and re-erected. Apparently one or two very anxious moments were experienced when the rood was being re-hung, as the hooks came away from the ceiling, but finally all went well.

One of the striking features of St Lawrence Church is the stained glass. The concept of a window to enrich the east end of the church had been discussed for a number of years and with the death of the Revd Rupert Godwin in 1966, interest in the project was revived. The idea of relating a memorial to Father Godwin, and his twenty-five years' service to Eastcote Parish, with the high calling of a person to the priesthood was the guiding principle. A design, a complicated one, developed from sketches by Sidney Starling (a Churchwarden and amateur artist), was submitted and a faculty was granted in April 1968. An appeal was launched in June 1968 and by early August, £1500, the cost of the window, had been received. Money came from donations and many fund-raising efforts. The Right Revd Robert Stopford, Lord Bishop of London, 1961-73, dedicated the window on 2 June 1969.

A description of the East Window

The importance of the Eucharist in the life of the Church is portrayed by a priest in vestments raising the Host, against the cruciform halo of our Lord. The priest's chasuble is red the 'proper' colour of St Lawrence. Golden rays extend over the whole world on which the priest stands, thus symbolizing Christ as the Bread of Life and the Light of the World. The two male figures on the left represent Europe and Africa, and two females on the right Asia and America. God the Father is evoked by the beams of light streaming forth, God the Son by the elevated bread and suggestion of a cross, and God the Holy Spirit by the dove and tongues of fire in the traceries. The window also has seven stars representing the seven Spirits of God (I Cor. 12:8-10). The shafts of crystals are symbolic of the rising prayers of the faithful, and the ribbons the bonds of hope and love (I Cor.13:13).

The stained glass window above the entrance to the Lady Chapel, depicting the Annunciation of our Lady was dedicated on 29 January 1975 by the Right Revd Graham Leonard, Bishop of Truro, 1973-81 (Bishop of Willesden 1964-1973; Bishop of London 1981-1991). A parishioner, who funded all the costs, remained anonymous until after her death in 1979, when her name was revealed and a plaque (back-dated) was put in the Lady chapel

Description of the Lady Chapel Window
Apparently the three sections of glass panes led to the rather unusual design, which include several symbols associated with Mary. The central pane has a heart and a sword (Luke 2:35) and a moon (Rev. 12:1). Balancing the figure of Gabriel there are a dove representing the Holy Spirit, and a lily representing purity.

Just prior to the dedication of the Annunciation window, the January 1975 magazine reported that major renovation of the church had almost been completed during 1974.

The re-decoration included the repainting of the interior roofs by Messrs Campbell, Smith & Co., who had been doing this sort of work for 100 years. There were not sufficient funds for the cross members to be re-stencilled, but they were carefully cleaned and had a coat of varnish. All the stonework, including the font and the window surrounds, was cleaned, together with the Stations of the Cross and statue of St Lawrence. It was hoped that the fixing of radiator shelves would disperse the heat dust and help to prevent future discolouring of the stonework.

One contribution to the special Church Decorating Fund in 1973 was from a Public Auction, organised by Mr and Mrs Christopher Lally and held in November. Mr Rowland Clay conducted the auction in a professional and humorous manner. It was both an enjoyable and financially successful event, making a profit of £585. It was a new means of fund-raising for St Lawrence's and one that was repeated in the 1980s when Father David Hayes was Vicar.

> **PUBLIC AUCTION**
>
> S. LAWRENCE CHURCH HALL
> BRIDLE ROAD
> EASTCOTE, MIDDX.
>
> **CATALOGUE OF LOTS**
>
> including
>
> BRASS, POTTERY, GLASS, PLATE, COINS
>
> PICTURES, BOOKS, CIGARETTE CARDS
>
> POSTCARDS, ELECTRICAL ITEMS
>
> CURIOS, ETC.
>
> which will be Sold by Auction
>
> on SATURDAY, 3rd NOVEMBER, 1973
>
> Sale to commence 2.0 p.m.
>
> On view in morning 10 a.m. - 12 noon
>
> PLEASE BRING THIS CATALOGUE WITH YOU
> PRICE 10p

Cover of Public Auction Catalogue
3 November 1973

As early as 1973 the Mothers' Union (MU) had donated £10 towards a new carpet for the sanctuary, which was then more than 40 years old. The carpets in the chancel were even older. Various church groups had fund raising events for the new carpets, but it was not until the end of 1976, with the laying of red carpets in the sanctuary and chancel that the renovation of the church was finally completed.

It is not possible to mention all the gifts that were made during Father Bill's incumbency but it is appropriate to refer to many that reflect the High Anglican tradition of St Lawrence Church. Some were gifts from individuals, often in memory of a loved one, others were from one of the church organisations and yet others as a result of fund-raising efforts by the congregation as a whole. By 1960 a purple cope and sets of vestments in white, red, green and gold were added to those that had already been in use. On Whitsunday 1965 a new set of red vestments* were used for the first time. They matched curtains and altar frontals given the previous year. All were hand made by members of the MU, assisted by two non-MU parishioners. The cost of the materials was £62.10s.0d; the money was raised over a period of two years. A professionally made frontal for the high

---

* The handmade chasuble is shown on page 71

The Chancel, before the Altar Gates were added in 1962
(*Diocesan Magazine* October 1954)

altar alone would have cost £90. By 1970 a set of silver candlesticks and crucifix and new altar linen were in use. By 1980 items given for use at Holy Communion services included a solid silver and gilt chalice and paten, a silver flagon, and a silver lavabo bowl and jug. Candelabra (a pair of branched candlesticks) were used for the first time at the Altar of Repose on Maundy Thursday 1976.

The Church Hall

Even in the days of the Mission Church, the temporary building was used not only by members of the congregation, but also by the wider community of Eastcote. The hall lettings were a source of income and following the completion of the permanent hall in 1955 it was important that the facilities were kept up-to-date. In 1960 and 1961 there were discussions as to the possibility of extending the hall beyond the committee rooms, with another small hall having flats above for the curate and the caretaker. The proposed plans were recognised as impracticable by the PCC Hall Committee and further consideration was given to repairing the bungalow.

Improvements to the hall went ahead. During the 1960s it was re-decorated and had new curtains. In the summer of 1968 the kitchen was fitted with a new sink, strip lighting and a tiled floor, but the prospect of a larger kitchen was something for the future. At the end of 1971 a team of the Players re-painted the walls of the stage and the proscenium arch. In February 1972 the Vicar's Letter reported that the old, unsatisfactory, overhead heaters in the hall had been replaced with a new heating system (by Emery Installations) at a cost of £866.

One important regular hall 'let', which started in February 1963, is an example of the church's outreach to the Eastcote community. The Eastcote Nursery School, for children up to five years of age, was organised for many years by Mrs Edna Edwards. Another group that still uses the hall, and has been associated with St Lawrence's from 1958, is the Phoenix Ladies' Club.

The Vicarage

A change of Vicar always necessitates some work on the vicarage, and with the arrival of the Hitchinsons in 1956 the vicarage was adapted to the needs of a family. Later in 1956 the Finance and General Purposes Committee of the PCC recommended that £300 should be spent 'on a second hand car for the Vicar's use', licensed and insured by the PCC. A car was purchased in February 1957, and in 1962 a garage was erected in the vicarage grounds.

Accommodation and Church Grounds
Miss Angell, who had been a tenant since 1941, was still living in the bungalow in 1956 but the building was in a poor state. Although some repairs were made after her tenancy ended in 1961, a report revealed that a great deal of money would be required if it was to provide accommodation for a curate and a caretaker. It would need £4,000 to be converted into two dwellings and £2,000 to keep it as a single dwelling. It was at this time that the possibility of extending the hall was discussed, and the decision was made to 'make the bungalow habitable' for a caretaker. The Revd Peter Goodridge (June 1958-April 1964), the first Assistant Curate, had accommodation at different parishioners' homes. Mrs Borer, caretaker from 1945-1961, had accommodation locally. Mr W. Youngs, the new caretaker, moved into the bungalow in 1962.

The Vicar's Notes in the June 1964 magazine reported that the PCC were discussing plans 'for the erection of two houses for curate and verger' (caretaker). They were to be built on the land behind the bungalow and when completed the bungalow would be demolished. This ambitious scheme came to nothing. In October 1965 the Vicar reported that the church had obtained a loan to purchase two maisonettes, numbers 77 and 79, The Sigers. Mr Youngs occupied number 77 until he retired in 1970 and the property was sold. In turn both the Revd W.P. van Zyl and the Revd Clive Pearce lived at 79. After Father Clive left the parish in 1973 this property was also sold. The conversion of the bungalow to provide extra accommodation for the Junior Church was started in 1968.

By 1975 it was felt the church should provide a house for the assistant priest and 26 Woodlands Avenue was purchased. The Revd Huw Chiplin was the only curate to live there and after he left the parish in January 1979 the house was put up for sale and was sold at the end of 1979.

During the 1960s and early 1970s there were discussions about selling the land to the 'liturgical' north of the church, (the land between the church and 4 Bridle Road). It was referred to as 'the Vicar's Close' or 'surplus land', but by 1973 it was finally agreed not to take the matter any further. A long magazine article on the church gardens and grounds in November 1973, reported that while the future of the land to the 'north' of the church was uncertain the area had remained, 'if not somewhat neglected, to say the least a little fallow'. In the summer of 1973, through the 'Herculean efforts' of the Scouts and their supporters, great steps had been taken to clear much of the undergrowth and to cut the grass. Summer fêtes and other church activities were to make use of this area.

As early as 1961 the PCC Minutes refer to a Garden of Remembrance to be laid out by professionals and then maintained by the garden group. With uncertainty over the 'Vicar's Close', it wasn't until October 1974 that a decision was made as to the site of the Garden. The annual Dedication Gift Day in 1974, together with other gifts and promises raised £1,000 towards the final cost of £1,752. 28p.

Garden of Remembrance

The Garden of Rest for the interment of the ashes of parishioners and their loved ones, was finally blessed and hallowed on 16 April 1978 by the Archdeacon of Northolt, the Venerable Roy Southwell. A Memorial Book, recording the names of loved ones who have died, was placed in the church at the end of 1977. The book was given in loving memory of William James Lane, 'devoted worker at this church' who died in January 1977.

That long magazine article also paid tribute to the 'gardening group' that had started more than 15 years previously with 'a team of more than twenty', and had done so much to establish the church gardens. In 1973 there were only four regular helpers. There was a plea for more assistance, in particular for help in clearing the ground behind the bungalow for use by the Youth Club. Even in those days, it was the 'efforts of a few stalwarts', who turned out regularly, who kept the grounds in order. It seems that Sidney Gover, Tony Cattle, Ron Carratt and John Styles were the mainstay of the team over a long period though others must have been involved, including Mr Cornell. At one time there were over 100 rose bushes to care for, as well as grass to cut and bushes and trees to keep under control.

Memorial Book,
Flower Festival 2008

Group with Fr Bill under the tree 1978 (Uxbridge Library)

In concluding this section a link to the present day seems relevant. Since 2008 the seat around the tree in the car park is a popular place to chat. Some extracts from a 1978 magazine article indicate that the tree was also a focal point in the days of Father Bill.

> ...*we stand in its shade and chat about holiday plans or make last minute decisions about the Fête...gather waiting for a parish ramble to get under way...for the youth group it is a meeting place after service...parish news is exchanged... we need somewhere to gather, irrespective of the reason.*

Worship
Before any discussion of church attendance and the range of services it seems appropriate to refer again to Father Bill's guiding principles of 'Worship and Fellowship'. His letters of July to December 1956 give some indication of just how much importance he attached to worship, in one letter reminding his readers of 'the primary aim of the church ... to build its members into a worshipping community', and in another that 'prayer and worship are the most important aspects of our life together'. Father Bill, as other vicars have done and still do, used his magazine letters as one means of reminding parishioners of the teachings of the Church. From 1961 to 1971, the years of the large ambitious magazine, the 'teaching articles' included topics such as the use of vestments, ornaments of the altar, the ceremonial cleansing of the chalice as part of the Eucharistic rite (1964), and the use of incense (1970). Father Bill felt very strongly that the church organisations owed their existence to the Church and had a part to play in its evangelistic mission. Later sections will show that worship was an important feature of these organisations.

The Vicar's views coincided with the Church of England Children's Council's 1956 campaign, *Operation Firm Faith for Christ and His Church,* which stressed the importance of worshipping together as a family. The purpose of this campaign was 'the revival of Christian family life and worship in every parish'. In one 1956 letter, while acknowledging the 'excellent work in Sunday School...and other parochial organisations', Father Bill wanted greater co-ordination 'so that family worship may once again be the rule of our parish church'. There was considerable growth in the Junior Church during this period, but the Vicar frequently stressed the need for parents to 'come to church with their children'.

On the whole services at St Lawrence's continued to be in the High Anglican tradition, but as is right there were alterations and additions. Some changes the Vicar made were in response to the needs of the parishioners and some were introduced to enhance worship at St Lawrence's. Others were the outcome of revision by the Church of England of its liturgy. Why and how this revision of the 1960s and early 1970s took place, and how Father Bill was able to gain the support of his congregation to accept what were quite radical changes, is discussed later in this chapter.

Sunday Services
When Father Bill became Vicar the regular Sunday services were Holy Communion at 8am and 11am, Matins at 10.15am, and Evensong at 6.30pm. There are several examples of his concern to provide services to meet particular needs of the community and he encouraged people to make their views known to him. As early as the summer of 1956, he asked for comments as to 'whether the services were at the best times' for the parishioners. Possibly as a result of this an additional Holy Communion service was held at 9am on Dedication Sunday, with the primary objective of catering for young communicants. From Advent Sunday 1956 the 9am service was permanent. Whether the addition of the 9am service had, in itself, any significant effect on the total number of communicants it is impossible to say.

It was not until the middle of 1960 that any <u>attendance</u> figures were recorded in the *Record of Services* books. Matins had provided a reasonable length service for families until children were confirmed and then attended a Holy Communion service. Attendance at Matins in 1965 was usually over 80, with 165 at Easter and 145 for Harvest Festival, but in 1970 the average was well down to between 20 and 30, with only 70 at Easter and 23 at Harvest. This decline was partly related to the changes referred to already. As numbers at Matins declined members of the choir were transferred to the 9am or 11am choirs. When only seven people were in church for Matins on Whitsunday 1972 (and that included the Vicar, the organist and two sidesmen!), Father Bill decided that 'the general need for this service no longer existed'. The last Matins was held on 30 July 1972. It was appreciated that this would disappoint some, but pointed out that the clergy would be enabled to visit branches of St Lawrence Junior Church more regularly, a vital part of the ministry of the Church.

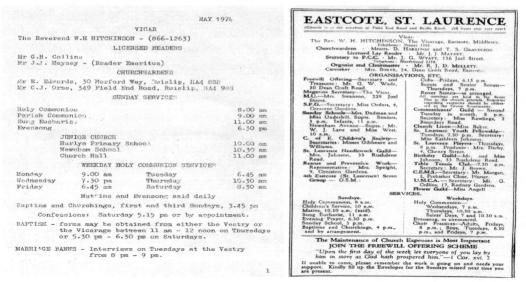
List of services in 1974 compared with those in 1975

Attendance at Evensong in 1965 averaged around 70, with 105 at Easter and 136 at Harvest, but by 1970 the average was down to between 30 and 40, with only 53 at Easter and 65 at Harvest. In November 1972 the Vicar reported that 'a mere 60 turned up for Evensong' at Harvest Thanksgiving. Declining numbers at Evensong, with certain notable exceptions, was in keeping with the national trend. The notable exceptions were when there were visiting preachers, 'farewells' for assistant priests, and for Father Bill's farewell when some 460 attended Evensong. By the end of the 1970s the average attendance was under 30. By then the 9am Parish Communion was established as the main act of worship on a Sunday at St Lawrence's.

The two other Sunday services held, as in the past, by arrangement with the Vicar were Baptism and Churching, though the latter was no longer included with other services listed in the magazine after March 1978. In his magazine letters, Father Bill often referred to meeting former parishioners and, on one occasion in 1974, a lady reminded him of the baptism of her daughter, which had taken place some seventeen years previously. There were three babies to be baptised. The Vicar took the first and then the second but when he approached the godparent of the third baby she looked blank and turned to the mother, who look bewildered and turned to the grandmother. She shook her head. There was a buzz among the relatives as they tried to recall where they had last seen the baby. The race for home discovered the baby sleeping peacefully in her pram in the garden! After some delay the Baptism took place.

In 1970 Father and Mrs Clive Pearce's son was baptised at the 11am service on Easter Sunday, the first time that a Baptism had taken place within the context of a Solemn Eucharist at St Lawrence Church.

Weekday Services
For the first few months of Father Bill's incumbency, weekday Holy Communion services were held as they had been in Father Godwin's time, that is at 7am Wednesday, 10.30am Thursday and at varying times on Saints' Days. By the end of 1957 there were Holy Communion services each day, Monday to Saturday, and these were largely maintained until Father Bill's retirement in October 1979. From September 1977 a change, from an early morning to a late afternoon celebration on Tuesdays, appears to have been made to accommodate young communicants.

A Weekday Service in the Lady Chapel 1961 – before Altar Gates were added in 1962

A further example of the Vicar's response to the needs of another group of parishioners was a monthly afternoon pram Eucharist, which was first held in October 1971. He wanted mothers to 'bring their babies and not worry if the children wandered about'. There was opportunity for a cup of tea and a chat. In 1974 Father Bill thought the service provided a focal point in the month for some mothers and the services were held until 1978.

Advent and Christmas

On the Saturday before Advent 1957, Father Bill conducted a Quiet Afternoon in church to help people prepare for the penitential season of Advent, which he regarded as 'a time for taking spiritual stock of ourselves'. These afternoons seem to have been appreciated for they continued throughout his incumbency, often led by visiting clergy, on two occasions by Father Goodridge when he was Vicar of St Philip's, Tottenham.

The Churches Together in Eastcote Carol Service, which attracts people from the wider community of Eastcote, was initiated in 1965 when a Carol Service was held on Monday 21 December at St Lawrence Church, led by the combined choirs of the Eastcote Anglican, Methodist and Roman Catholic churches. By 1967 the Presbyterian (now the United Reformed) had joined the other three and this Carol Service became an annual event. The service of Nine Lessons and Carols, established in Father Godwin's time, remained a feature of the Christmas season, usually taking place on the Sunday after Christmas. Although there had been a tradition of having a crib in church for many years, a *Record of Service* book refers for the first time to a 'Blessing of the Crib' service at 6.30pm on Christmas Eve 1961. By the end of the 1960s the Crib Service was a regular fixture of the calendar and remains a popular service today.

Numbers of Communicants at Christmas were very high, with over 600 from 1957 to 1974. There were 904 in 1960, over 800 in 1961, 1962, and 1964 to 1966. From 1959 to 1967 (apart from 1963) there were over 400 Communicants at Midnight Mass. From 1975 numbers declined and in 1979 had fallen to a total of under 500.

Ash Wednesday and Lent

Each year parishioners were reminded of the importance of observing the great penitential season of Lent by magazine articles by Father Bill. These included guide lines on making Lenten resolutions, a special Lent prayer, and meditations for Holy Week. The Vicar's wish to have services at times convenient to parishioners possibly influenced the times of some Ash Wednesday services. Throughout this period there was one at 10.30am, but the time of the earlier morning service varied. From 1959 there was an 8pm Sung Mass, ('said' in 1963 because of the severe weather which some may recall). Most years from 1960 there was a fourth service at 6pm and in 1973 a fifth, the monthly Pram service. The significant numbers of Communicants at Ash Wednesday services, for many years reaching over 200, seem to reflect Father Bill's teaching about this Day of Obligation.

In Lent, in addition to the normal daily morning Holy Communion services, there was at least one other service during the week. Over the years these varied and included Compline and Address, Stations of the Cross, and Holy Eucharist (some years followed by a talk and discussion). In 1977 and 1978 a period of conducted and silent prayer before the Blessed Sacrament was held on Saturday afternoons. Several years a special 3pm Mothering Sunday service was held.

Holy Week

On Palm Sunday 1957 the Vicar introduced the Blessing and Distribution of Palms at all Communion services for the first time. In the years 1957 to 1962 there were several 'firsts' at St Lawrence's. A section on liturgical revision 1981-2000, in *A Companion to Common Worship* refers to the number of parishes using any 'liturgical possibilities' for Holy week as being in 'a minority' as recently as the late 1970s.[1] It seems Father Bill wanted to use the liturgical forms available in the 1950s to enhance worship in the most solemn week of the Church's

Lighting the Paschal (Easter) Candle 1961

year, and gradually brought in forms new to St Lawrence's. When, in 1958 he gave notice of a 'first' 8pm Maundy Thursday Sung Eucharist, he added, not surprisingly, that 'every parishioner should regard it as an obligation to be present' at one of the three Holy Communion services (four from 1971) held on that day. In 1959 the Maundy Thursday Sung Eucharist was followed by the 'first' Vigil at the Altar of Repose. For a few years it was from 9pm to 9am (men only from 10pm to 6am!), but the Vigil was soon only from 9pm to midnight, as today. The Wednesday of Holy Week 1959 also witnessed the 'first' Stations of the Cross service.

On Good Friday 1960 as well as the children's morning service and the Three Hours' Devotion established in Father Godwin's time, Father Bill conducted the 'first' Liturgy and Holy Communion. There was no consecration on Good Friday. The hosts used were from the reserved sacrament, consecrated on the Maundy Thursday.

Although Vigil Ceremonies had been held on Holy Saturdays from 1950 it was not until 1962 that the 'first' Vigil Ceremonies and Midnight Mass of Easter was celebrated, with the service starting at 11pm. From 1967 the Vigil Ceremonies and first Eucharist of Easter was at the earlier time of 8pm. The number of communicants was never high, 72 in 1964, but averaging around 40. The chart of Communicants shows these 'firsts' of Holy Week (see Appendix 6).

Additional acts of devotion in Holy Weeks included performances by the church choirs of Stainer's *Crucifixion*, Maunder's *Olivet to Calvary*, and Faure's *Requiem*.

In the earlier years there were several presentations, by the local Studio School of Speech and Drama, of scenes from Dorothy L. Sayers' *A Man Born to be King*.

By 1957, Father Bill's first Easter Day in the parish, the 9am was a regular Sunday service and, with the traditional 7am service held at festivals, it meant he celebrated four services. By Easter 1960 he had the assistance of Father Goodridge and for four years until 1964 there was an extra 6am service. As at Christmas, total numbers of Easter communicants were high with over 800 from 1960 to 1962. From 1964 numbers fell and by Easter 1980 had fallen to around 400. Figures suggest that for the occasional communicant, Easter was less important than Christmas.

Other Annual Services
Apart from Palm Sunday, and of course Easter Day, other Sundays when numbers of communicants were higher than on normal Sundays were Whit Sunday, Harvest Festival and, until 1968, the Sunday when the newly confirmed received their first Communion. This was usually the Sunday after

Confirmation 1964

Confirmation. It was not until 1968 that Confirmation was within the context of the Eucharist. The number of St Lawrence Church candidates was high from 1957 to 1963, with a total of 108 in 1957 and 69 in 1963. In two years there were as many as eight preparation classes.

Celebrations of Holy Communion on important days of the Church's Year continued as in the past. While some of these might fall at a weekend, those of Ascension and Corpus Christi were always on a Thursday. In 1957 and 1958 there were three morning services on those days, including one at 6.30am (Sung Eucharist on Corpus Christi in 1957). The Vicar reminded parents that their children could by law be absent from school on Ascension Day (as on Ash Wednesday) 'for the purpose of attending their church for worship'. An 8pm Sung Eucharist, held for the first time on Ascension Day 1959, was a permanent feature from 1960, for both Ascension Day and Corpus Christi. Several years there was a 6pm celebration as well, which suggests the Vicar again responding to the wishes of some parishioners. The chart of Communicant numbers shows that numbers of communicants particularly on Ascension Day were high for many years, over 200 from 1959 to 1964.

Two other important days of the Church's Year were All Saint' Day and All Souls' Day. Until 1960 if these fell on a weekday there were three morning services but that year, in keeping with the other 'firsts', there was a 'first' 8pm Sung Eucharist on both days, and from 1971 to 1978 a 6pm as well. If either fell on a Saturday there was just the normal 8.30am service.

Two days in the year, special to St Lawrence's, were the Patronal Festival (10 August) and the anniversary of the Consecration of the Church (21 October), often called the Dedication Festival. It is obvious that Father Bill wanted parishioners to attach more importance to worship on these days than had been the case in the past. From 1959 there was an 8pm Sung Eucharist on the Patronal Festival and from 1971 to 1973, what the Vicar referred to as the 'usual four services of weekday festivals'. The tradition of celebrating the Patronal Festival on a Sunday began in 1974.

Father Bill's principle of 'Worship and Fellowship' is clearly seen in the way the Dedication Festival was celebrated. Apart from 1963, when Stewardship started, there was a Day of Prayer and Gift Day. The Vicar regarded it as 'an obligation of every parishioner to be in church at some time during the day to offer up thanksgiving and prayers for the needs of the whole Church'. In the years up to 1963 he was at the church door from 9am to 5pm to receive gifts of money. In 1958, Silver Jubilee Year, over 200 people spent time in church and £330 was given. From 1973 the money was for specific purposes, such as the re-decoration of the interior of the church, or the Garden of Rest.

In 1964 the decision was made to celebrate 'this most important event in the life of the parish on the most convenient Saturday nearest to the date so as to combine our worship with fellowship'. The Sung Eucharist followed by a social was not a great success that year but in 1967, when there was a reunion for those confirmed in the last seven years, there were more communicants at the Sung Eucharist. The practice of celebrating the Dedication in the week with 'the usual four services of weekday festivals', with light refreshments following the 8pm appears to have been a more acceptable and successful combination of 'worship and fellowship', until 1974. In 1958 the Silver Jubilee celebrations extended well beyond the day itself with a number of visiting clergy at many services. From 1974, like the Patronal Festival, the Feast of the Dedication was celebrated on the Sunday nearest to 21 October.

Alternative Services
Changes made by Father Bill, not only to meet the particular needs of the Eastcote community, but also to underline the importance of worshipping together as a family were all in place before major changes initiated elsewhere were introduced regarding the format of services,

As mentioned earlier the 'Parish Communion movement' of the 1930s wanted to restore the Eucharist as the main act of worship on a Sunday morning but it was not until 1965, some twenty years after World War II had ended, that any real progress was made to bring this about. For those familiar only with the Parish Eucharist as the main Sunday service it may be of interest to learn why and how the changes of the 1960s and 1970s took place.

An article in the June 1965 magazine entitled 'A Landmark' pointed out that the Church of England had not revised its liturgy for 300 years, though some provinces of the Anglican Communion (in other countries) had an up-to-date Prayer Book. St Lawrence's had adhered to the Prayer Book of 1662, 'with such amendments from that of 1928 as were permitted' by the Bishop of London. Briefly the changes in the 1928 Prayer Book had been rejected by Parliament, and though in widespread use, did not have 'lawful authority'. Only a constitutional change regarding the relationship between Church and State would enable the Church of England to make certain variations and experiments in worship without having to present them for parliamentary approval.

The 'Landmark' referred to the Prayer Book (Alternative and Other Services) Measure, which Parliament had passed in March 1965, and which became operative in May 1966. The purpose was 'to authorise the use by way of experiment of alternative forms of service deviating from *The Book of Common Prayer' (BCP)* and the use of forms of service for use on special occasions; and to authorise minor variations in public prayer'. The Measure would allow the Convocations and House of Laity to authorise, by two-thirds majorities, services differing from those in the *BCP* for a seven-year period (which could be renewed for a second seven years), but all such services would require the consent of the Parochial Church Council (PCC) before they could be used in that particular parish church.

As far as St Lawrence Church was concerned the Vicar was anxious to give members of the congregation the opportunity of experiencing each of the new rites of the Service of Holy Communion (Series I, II, and III), as they were authorised, prior to discussion and decision by the PCC. On Sunday 11 June

1967, instead of Evensong, Series I Eucharist was celebrated and following the consent of the PCC later in the month was adopted at St Lawrence's. By January 1968 The Series I 'little red book', published by the Church Union, had been in use for six months and though Father Bill agreed that it was 'by no means perfect... its value lies in its clarity', unlike the official version published by the of Society for Promoting Christian Knowledge (SPCK) with its numerous alternatives. In 1968 the PCC agreed to the use of Series II Service of the Eucharist for mid-week services and by 1970 its use had been extended to Sundays. Alternative Series II rites of High Mass and Confirmation and of Holy Baptism were also used from 1968.

Commenting on the changes on Alternative Services in February 1970 Father Bill made it clear that St Lawrence Church aimed to preserve 'the dignity of our services whilst relating the changes to what is happening throughout the Church at large'. It was important to remember that 'our common purpose in worship, the Mission of the Church to the world and our own part in it' should enable all to adapt to the inevitable changes at a time of transition.

In November 1971 parishioners were able to participate in a provisional draft of a new Order for the Holy Communion known as Alternative Service Series III, which was used experimentally on weekdays. After amendments its use was finally authorised from February 1973. Once again after demonstration of the service, discussion at the Annual Parochial Church Meeting (APCM) and the consent of the PCC, the last and 'definitive' rite of the Series came into use at St Lawrence's from 24 June 1973.

Each of these new rites for Holy Communion gave the opportunity of emphasising the Ministry of both Word and Sacrament. However it may surprise readers that the use of contemporary language only came with Series III, when forms such as 'thee', 'thine', 'wilt' and 'hast' disappeared. A modern translation of the Bible, the Jerusalem Bible, had been used at most services at St Lawrence's since 1968, the exception being some major festivals.

A further constitutional change took place in the Church of England in 1970, which simplified the procedure of liturgical revision. This was the advent of Synodical Government when three Houses of Bishops, Clergy and Laity considered proposals together, although they voted independently. One of Synod's first major achievements was through the Church of England (Worship and Doctrine) Measure passed by Parliament in 1974. Under its terms Synod 'could produce and authorise forms of service as and when it was considered necessary, provided that the *BCP* remained available and unaltered'. The Measure of 1965 was redundant and Series III forms of service, all in contemporary English, were produced. *The Alternative Service Book 1980 (ASB)*, published in November 1980, was authorised for ten years, but later extended until December 2000.

Choirs and Organists
There are few references to the church choir in the magazines of this period so one assumes the choir played its traditional role in church services as it had done during Father Godwin's time as Vicar. It may be remembered that from 1933 the choir was exclusively male. There was a magazine appeal from Mr Mullet, the organist and choirmaster from 1954 to 1958, (see Appendix 4 for list of Organists) 'for boys of course but also for older men', and one in 1962 for 'four young men (over 18) …and after training, to sing bass and tenor parts in the 9am choir'. Mrs Dedman, who was an organist at St Lawrence's for over 20 years from 1959, instructed the 9am choir and successive 'head' organists were responsible for the other choirs. There is no reference as to when female voices were again in the choir, but a notice in the September 1963 issue of the *Gridiron* stating 'there are vacancies for both adults and boys' could mean that female adults were again acceptable. There had been ladies as well as men and boys in the days of the Mission Church. By 1971 at the latest there were girls in the junior choir.

During the 1960s there appear to have been many wanting to join the three choirs, who sang at three (until the last Matins in July 1972) morning services and one evening service each Sunday. According to a Vicar's Notes of 1966 future applications were to be made to him 'so that a balance between the choirs may be maintained'. Members of all three choirs sang new carols as well as old favourites at a successful 'Three Choirs' Carol Service in December 1964. From 1965 members of the choirs participated in the 'Interfaith' Carol Service, and referred to as such for a number of years. This was an ecumenical venture and likely to have been the responsibility of the misnamed Interfaith Committee, consisting of members from St Thomas More and St Lawrence churches.

After nearly nine years as head organist and choirmaster Mr Edwards resigned in 1969. He had played the organ at nearly all of the weddings and extra services at Christmastide and during Holy Week. He was succeeded by a series of organists and choirmasters in each case for relatively short periods. In his May 1978 Vicar's Letter Father Bill expressed his appreciation of 'our organists and choirs' and thanked Mrs Dedman, Mrs Lumsden, Mrs Holt, Mr B. Wall and Mr John Brion 'for their help on the musical side' during the months without a regular organist for the 11am and 6.30pm services. From 1979 Mrs Dedman was the sole official organist.

Extensive Magazine Choir Notes of 1970 indicate a lively interest in the musical activities of the choirs at that time. Apart from the holiday month of August an anthem was sung at Evensong each Sunday and occasionally there were recitals after Evensong. Every effort was made for members of the three choirs to join together for special occasions. Musical events during 1970 included performances of part of Berlioz's *The Childhood of Christ*, Stainer's *Crucifixion* and Faure's *Requiem*. For the *Requiem* the combined choirs of the church together with the Ickenham Strings, under their conductor Mr N. Lane, 'gave a moving performance'. At the end of 1970 in addition to being part of a choir of around 60 for the Interfaith Carol Service, some 50 members of the three St Lawrence choirs took part in the church's own Nine Lessons and Carols Service.

A magazine article in July 1971 reminded readers of the purpose of the Royal School of Church Music (RSCM) to which the choirs of St Lawrence were affiliated. 'The main object of the RSCM is to stimulate interest in the

Looking towards the West End with Choir Stalls in the foreground.
Flower Festival 2000

promotion of church music and the maintenance of a high standard of choral singing'. As the Holy Communion became the important service on a Sunday the choir played an important part in introducing the congregation to various settings of the Creed and Gloria. Settings by Vaughan Williams, Merbecke and Martin Shaw are three examples sung in the early 1970s.

In February 1972 the choir 'excelled themselves in their performance of Handel's *Messiah*', Parts I and III. This was a performance with a difference and did not mean months of rehearsal, a type of performance often undertaken by choirs in recent years. The choir was in fact an augmented choir. There was a rehearsal on the Saturday afternoon to which 'anyone who enjoys singing was invited' provided they had a copy of the work. This was a time of power cuts and the performance had been timed to finish before one scheduled to start at 9pm. In the event the power cut was early, just as the choir were singing the Amens. There was a slight delay before, accompanied by the piano, the performance was completed by candlelight.

A full performance of the *Messiah* took place in December 1973, when despite the cold weather those who attended had a very enjoyable evening. David Lacey, organist and choirmaster at the time, conducted the augmented church choirs and 'the end result was excellent'. Special thanks were expressed to the soloists and orchestra, 'who, in spite of the petrol shortages had come from as far afield as High Wycombe, Slough and Wraysbury and most of them gave their services free'. It was regretted that there had not been a larger audience.

Choir News in May 1974 intimated that the 9am choir would be unable to continue singing in four parts unless there was a response to the appeal for tenors and basses. On the brighter side the same report referred to the fund raising for new choir robes. Whist drives run by the choir and other fund raising efforts of church organisations were well supported. Much time and thought was given, by Mrs Dedman and others, to the design and colour of the robes. The bright red robes were worn for the first time for the Christmas services in 1975. The following year new red sanctuary and chancel carpets completed a major renovation of the church, so no doubt some thought had been given to choice of colour for the robes with this in mind.

In September 1979 there was a further appeal for choir members. It seems likely that many churches had, and continue to have, difficulty in maintaining a full complement of people in their choirs that play such an important role in leading the singing in church services.

Junior Church
Although Father Bill was in sympathy with the campaign of 1956, *Operation Firm Faith for Christ and His Church* and its purpose, 'the revival of the Christian family life and worship in every parish', he was realistic enough to realise that Sunday School, or Junior Church remained a way of bringing many children into the family of the Church.

The Sunday School 'over the bridge' at Newnham Avenue School, which had opened in January 1955 with Mr Lane as Superintendent, was well established when the Revd William Hitchinson became Vicar in 1956, but there were other areas of growing population in the parish that needed similar centres. The August 1957 *Leaflet* reported that the church had 25 enthusiastic teachers, 16 of whom were either at school or just started on business careers. There was also an appeal for 20 or more who would 'make time' to prepare their lessons and give an hour of time on Sunday mornings, for 'this work is the most important side of the worshipping community'. Members of the church were urged to ponder on the words, "Freely ye received, freely give". Mr Lane opened new branches of the Junior Church at Harlyn Drive Primary School, Northwood Hills in 1957 and St Mary's (now Haydon School), Wiltshire Lane in 1958.

The success of the four branches of the Junior Church is reflected in the numbers of candidates who were confirmed. One or two Confirmations in this period are of interest. In 1957, the first full year of Father Bill's incumbency more than 100 children were confirmed at St Lawrence Church, and in 1959, in addition to the

93 confirmed at the church, 14 children from the Harlyn Drive branch of the Junior Church were confirmed at St Michael's School, Joel Street with five St Michael's boys. The link with this school for children with special needs, which had opened in 1950, had been maintained, an example of the church serving the wider community. Another example is included here to show how Father Bill met the needs of small sections of the local community for a short time, regarding it as one aspect of the mission of the Church. In 1958 the Vicar responded to the request to run a Sunday School at Highgrove House, where there were 21 boys and 7 girls between the ages of 7 and 14. Highgrove House was at that time used by Middlesex County Council to house homeless families. PCC *Minutes* of February 1959 refer to the Highgrove House Sunday School 'now in being' and to children from there 'being brought to church' one Sunday, but there are no further references.

Highgrove House. One of the two remaining large houses referred to at the beginning of Chapter 1 (RNELHS)

In the November 1959 magazine the Vicar referred to 50 teachers for Junior Church at the four centres in the parish but said there was a need for a constant stream of volunteers to help with young people. A special Golden Jubilee publication in 1983 refers to attendance at Sunday schools in 1959 as '600 at four centres throughout the parish'. *Minutes* of many PCC meetings indicate that there was great concern about the lack of teachers for the Junior Church. For example, in October 1960 the discussion referred to 'a need for older members of the congregation to help...appeals had been met with almost complete indifference ...many confirmed had now lapsed'. At the Annual Parochial Church Meeting in 1962 it was reported that the Sunday Schools were doing well but there was 'a tremendous turnover of teachers'.

The fact that the Sunday Schools were doing well owed much to Mr Lane. When Mr and Mrs Lane left the area in 1962 the Vicar paid tribute, in the July *Gridiron*, to Mr Lane's work as secretary of the Free Will Offering (FWO) scheme, and as Junior Church Organiser. In the latter he was responsible for the whole of its administration. He would be 'much missed for his enthusiasm for this part of the Church's witness in this place'. Mr Bob Davies replaced Mr Lane as Junior Church Organiser until 1966 when Mr E. (Ted) Edwards took over. Mr Christopher Smith was Superintendent of the Newnham Avenue branch from 1962 to 1972, when Mr Reg Horsfall took his place.

Infant Class in the Parish Hall 1962

An example of the keen interest taken in the work of the Junior Church in Eastcote was recorded in the September 1967 issue of the *Gridiron*. An "open Sunday" held at the Newnham School centre reported 'the welcome presence of a very large number of parents'. As well as a service conducted by the children there was an exhibition of practical work done by the scholars from all the centres during the year. The teaching they had received was 'imaginatively translated into pictures, maps, models etc'. Pride of place was given to a beautiful scale model of St Lawrence Church, Eastcote, constructed by three of the senior boys, Peter Johnson, Geoffrey Mills and Peter Moore.

The social changes of the 1960s, and possibly the importance Father Bill attached to the family worshipping together, contributed to a decline in Junior Church numbers. From May 1966 St Mary's was no longer listed as a centre for Junior Church. In the September 1974 magazine the Vicar wrote of the decision to close the branches at Newnham Avenue and Harlyn Drive, and to concentrate on one centre of the parish – the parish hall. Parents could bring their children to the hall and it could 'ease their own problem of church-going… Junior Church and the Sung Eucharist are at the same time'. At the time of this change in the organisation of the Junior Church Mr Edwards, who had been superintendent at the parish hall as well as Junior Church Organiser, resigned and his place was taken by Mr Horsfall, to be replaced in turn by Mr Ian Mitton and Miss Judith Brion.

For a number of years the Junior Church had been in the habit of attending the 11am Eucharist on the first Sunday of each month. A further move towards family worship came in 1976, when the decision was made that most Sundays the Junior Church (excluding the Kindergarten) should meet in the hall for the first part of the Liturgy, join the congregation at the Offertory and stay for the rest of the service. In September this was revised with the children returning to the hall after the Consecration.

Scale model of St Lawrence Church 1967

The season of Christmas saw the Junior Church involved in Nativity productions. The 1973 production 'was a moving experience'. In 1974 after the Nativity play there was a short Toy Service, the toys being distributed later to children at St Vincent's Hospital and under-privileged children of the district. Although there are few reports of Junior Church Christmas parties it is likely they were held each January. Some years, thanks were expressed in March issues of the magazines, to organisers, helpers and caterers for the parties held on three consecutive Saturdays.

Father Bill regarded the work of the Junior Church 'as being of the utmost importance to the Church of tomorrow'. With his guiding principles of 'Worship and Fellowship', his concern was, that having brought young people into the family of the Church as worshippers, he wanted to foster fellowship and keep them in that family.

Youth Groups
When Father Bill became Vicar there was a club, the Youth Fellowship, for 14 to 18 year-olds. However the new Vicar wanted to embrace other younger members of the parish in social groups suitable for their age. As early as February 1957 he was anxious to start a club for the 11 year-olds, as he had 'a

nucleus of 75 now being prepared for Confirmation' and he wanted a social club to encourage these young people to remain within the family of the Church. He also recognised the need for something for those over 18. Following discussions in 1957 and 1958 the formation of new groups was decided upon. There was to be a Junior Guild for the 11 to 14 year-olds and the St Lawrence Association for those over 18.

The Junior Guild
Either late 1958 or early 1959 the Junior Guild, led by Mrs Keeler, was started for the young newly confirmed. There are few details about this group, but the Vicar referred to it in 1961 'as a small, compact and thriving little body'. In the April 1962 *Gridiron* there was praise for an excellent and varied film show, when those present were 'impressed by the expertise of the projectionists', two 13 year-olds, Brian Aldred and David Carrington. The audience was reported to have 'half expected the usual usherettes to appear complete with a tray and torch…instead Mrs Keeler provided welcome refreshments'.

As so often happens the problem of leaders had much to do with the short life of this group. Father Bill regretted its closure in May 1964, saying 'it was a valiant attempt to deal with the 11 to 15 year-olds of the parish'.

The St Lawrence Association
In keeping with the Vicar's guiding principles of 'Worship and Fellowship', another short-lived organisation was formed in 1959. The purpose of the St Lawrence Association was to provide an opportunity for those over 18 to have fellowship through a social group suited to their interests. It was aimed at young married people (many were married in their late teens or early twenties), those returning from the Forces (National Service still operated for some until 1962 or 1963) and 'mothers who get little break from household chores' (many no longer worked when they had a family). According to the May 1959 to March 1961 magazines, the Association met on Fridays. In spite of its short life, and apparent failure, it seems right to mention the Association for it is another example of Father Bill's attempts to connect with sections of the church community.

The Youth Fellowship
The age range of this church organisation for young people fluctuated between 1956 and 1980, and was closed for periods in 1967, 1970 and 1973. The Youth Fellowship which had been functioning with its own committee under the guidance of a few adults since 1953, maintained many features of its early years until 1967, when the problem of leaders, both adults to guide and young committee members to organise club activities, led to its closure. The parish magazines reveal some of the people involved in the 1960s. Thanks were expressed to Mrs Onions and Mr and Mrs Davies. Some of the young committee members who worked hard in these years were John Sturman, Stephen Lee, Susan Robinson and David Bowden.

Needless to say Father Bill was much concerned that 'Worship and Fellowship' should be at the heart of the Youth Fellowship. In the early years of his incumbency the Fellowship was actively involved in parish affairs through the Communicants' Guild and the Vicar's Social evenings, and helping at the annual summer fair and Christmas market. He was of the opinion that the Youth Fellowship was a responsibility of the parish. The Minutes of the Annual Parochial Church Meeting (APCM) of 1962 include in the Vicar's Remarks, that the 'potential of youth should be harnessed and guided not only in church but also during the week'. He always wanted leaders to be regular communicant members of the Church and he wanted the Fellowship, as an organisation, to be involved in parish life.

Over the years the Fellowship hosted a number of speakers, who addressed the Fellowship on a wide range of subjects. An evening centred on Christian Stewardship and entitled 'Chat, Food and Dance' was held on 30 October 1963, two days after the parish supper, which had launched Stewardship at St Lawrence Church. It indicated the importance attached to the Fellowship's involvement, as an organisation, in parish life. A series of talks by local representatives of other Christian denominations and one on Dr Beeching's Railway Plan, showed interest beyond the community of St Lawrence (1963), as did one on the Samaritans (1966). As well as 'outside' speakers several members of the congregation supported the group by giving talks on their career or holiday experiences.

Successive committees tried to arrange varied programmes that would appeal to the membership and reflect the current trends. Club evenings usually allowed time for various games and socialising. The *Gridiron* of January 1966 encouraged new membership. 'In addition, every week you can join in table tennis, darts, billiards, cards, chess, etc., etc., whilst listening to a variety of records. Coffee and "coke" may be bought during the evening'. Some evenings there was dancing (occasionally with instruction), or a social to which members of other local youth clubs or members of the congregation were sometimes invited. A film on the building and consecration of Coventry Cathedral (1963) was part of the autumn 1966 programme.

Apart from Tuesday club nights, events were arranged in which members could participate at weekends or in the school holidays. Current trends were reflected in the changing names given to the regular and popular Saturday dances, ranging from a Pancake Dance and a Farmers' Dance in 1957, a New Year's Dance in 1962, a Rock Dance in 1964, and Beat Balls in 1964 and 1965.

Most years in the 1960s the Youth Fellowship organised hikes for Easter Monday and August Bank Holiday, but these were not always well supported. Some 20 members joined by Father Goodridge and a few other adults, 'braved the English climate for their usual Easter hike' in 1963, but in 1964 only 11 braved the Easter Monday rain, in a walk from Great Missenden to Wendover.

While other events such as inter-club quizzes and table tennis matches did not take place every year, carol singing in Eastcote for charitable causes, was a consistent feature of the Fellowship's programme. An example of practical help for others in the 1960s was the club's financial support to individual boys suffering from leprosy at special homes in Ghana.

In July 1967 Susan Robinson and David Bowden, who were by then assistant leaders, resigned prior to their marriage. It was decided to close the club temporarily. When the Fellowship re-opened in October1968 Stephen Lee and Sally Rees were the leaders and it appeared to run on similar lines to that of the earlier 1960s, including the annual carol singing. One venture was 'a successful midnight hike' in June 1969, which replaced the usual August hike. Unfortunately, owing to lack of leaders the club was closed again in 1970.

Mrs Carratt and Mr Horsfall helped to restart a group in 1971, and Karen Lloyd, Andrew Lee, Ian Mitton and Angela Dew were hard-working committee members. Club evenings were held after Evensong on Sundays for young people between 13 and 21. Activities on club evenings were much as they had been in the past but with tennis tournaments and beetle drives, as well as listening to records and general socialising. There were talks on first aid, one on hot-air ballooning, another on Eastcote by Ron Edwards (a local historian), and one on photography by Joe Orme (Churchwarden). A successful Youth Fellowship Dance was held at the Vagabonds clubhouse in Pinner, in December 1971. In August 1972 around 25 of the membership of 30 were regularly attending club evenings. To begin with a number of them attended Corporate Communion on the first Wednesday of each month.

Early outings included those to the Ordnance Survey Offices in Southampton and the Middlesex County Press at Hillingdon in 1971. During 1972 to 1973 the Youth Fellowship took part in inter-club quizzes. In 1972 and 1973 several members went to Sunday evening services at St Martin-in-the-Fields and St Paul's Cathedral. A visit to the crypt of St Martin's for *Alive* in March 1972, a modern folk-version of St Luke's Gospel in which there was audience participation, was 'much enjoyed'.

In spite of the efforts of the Youth Fellowship committee, magazine reports in 1972 indicate a lack of backing from the all-important general membership. The committee expressed disappointment that few now attended the Corporate Communion services. They were discouraged that their efforts to provide something different had poor response. Only six took part in a planned walk along the Grand Union Canal from Uxbridge to Cowley on Easter Monday. Only 12 members went on the trip to Oxford and Blenheim, which they had organised for the end of the summer holidays. Even the attempt in the winter to make record sessions more modern, with discos having DJs, failed to get much support. Inevitably all this led to yet another closure in 1973.

However there were some young parishioners who were keen to have their own organisation. A new Youth Discussion Group had its first meeting in December 1975. The aim was 'to meet fortnightly on Sunday evenings at 8pm in members' homes to discuss topical issues of particular interest to young people'. 'Young people' were those who were over 16, though this was reduced to 15 by July 1977. From September 1976 meetings were weekly, and were usually held in a room in the bungalow, now the St Lawrence Centre. From the beginning Bible Study was an important feature of the club's programme. For a time these were twice a month, at first on Sundays, and then also on Wednesdays after the 7.30pm service. By 1980 Bible Study had dropped to once a month on a Sunday, and a short study or prayer meeting after each Wednesday evening service during Lent.

A digression is relevant here for in 1976 the PCC Fellowship Committee became actively involved in providing discos, something many of the young people of the church wanted. The fellowship created by these popular discos when adults and young people supported each other, no doubt played a part in encouraging the latter to establish the Group as a worthwhile organisation of the St Lawrence community.

Discussions, the purpose of the Youth Discussion Group when it was formed, were included in programmes arranged by successive committees, but as the club branched out to include other events they became less frequent. Although the subject of the discussion was not often recorded, some of the early ones were on Euthanasia, the Shroud of Turin, the Power of Prophecy, and Relationships.

Programmes soon included traditions of earlier youth groups. There were quizzes and evenings of table tennis, darts etc in the church hall, but with the addition of rounders on Cheney Fields, presumably the area now known as the Long Meadow. New events, in which guests sometimes participated, were the club's own version of popular radio and TV programmes, such as *The Brains Trust* and *Just a Minute* (1976, 1978 and 1979), *Desert Island Discs* and *Call my Bluff* (1978 and 1979), and *University Challenge* (1979).

From 1978, encouraged by Father Huw, members again went carol singing, this time in the wards at St Vincent's Hospital, renewing a former association with that establishment. It must be put on record that each of the Assistant Priests had in turn supported the youth groups of their time. A 'one off' and 'out of parish' visit was made in 1978, when several members went to a service at Emmanuel Church, Northwood where Cliff Richard was the attraction playing gospel music and talking about his faith.

Pancake parties were once again annual events, but there were many other occasions when members gathered together to enjoy food. Barbecues, and evenings sampling Chinese, Italian and Greek foods were popular. There was a special party to celebrate the Queen's Jubilee in 1977, when everyone had to wear red, white and blue.

The young people contributed to parish life in many ways. In November 1976 club members provided the first of the parish breakfasts consisting of rolls and marmalade, and tea or coffee. They were also responsible for painting the kitchen in the church hall in 1978, and in decorating 'their room' in the bungalow, and painting white lines in the car park in 1979. As a group they were at both the parish harvest buffet and Father Bill's leaving party in 1979. When Father Bill left Eastcote it was evident that the Youth Club was, as he had wished, 'involved in parish life'.

The group's success owed much to the hard work of the young people themselves. This is well supported by Angela Dew's regular secretary's reports in the parish magazines. Ian Mitton was chairman until his marriage in 1979, when Jim Rhymer succeeded him. Perhaps the persuasive powers of the magazine committee had something to do with more frequent and fuller reports than in earlier years, but *PCC Minutes* refer to the Youth Group's involvement in parish affairs, suggesting an enthusiastic and outward-looking group. It is possibly towards the end of this period that the group became known as St Lawrence Young People or SLYP.

Fellowship
Father Bill's guiding principles of 'Worship and Fellowship', which lay at the heart of his ministry, may be seen in much that went on at St Lawrence's. He referred to them frequently in magazine letters, reminding his readers of their importance. From the beginning he regarded regular communicant worship as an essential quality of full Church membership and expected members of church organisations to abide by this. However although worship came first, he saw fellowship as the other side of parish life. He believed that from 'our worship together there should also emerge a fellowship which, in its outpouring, will embrace all those with whom we come in contact'.

In rather the same way that the Revd Godwin had stressed the importance of members of St Lawrence Church becoming a 'sociable' community, so Father Bill emphasised the importance of 'fellowship'. In order to achieve his 'social community' Father Godwin had hosted 'at home' days, which developed into the summer fêtes. He also instituted 'Vicar's Socials', which developed into parish socials, until the war brought these to an end. After the war Father Godwin formed the Communicants' Guild in 1950, with its social hour following the service on the second Tuesday of each month, hoping this would be a means of 'keeping us together in the fellowship of the Church'. Father Bill's 'Worship and Fellowship' was to a large extent a continuation in the tradition of Father Godwin's 'sociable community'.

Parish Nights
The Communicants' Guild evening continued for a few months after Father Bill came to Eastcote but in the December 1956 he wrote in his magazine letter, that the 'second Tuesday in each month will in future be known as "Parish Night" when, after our service in church, we can meet as a parish over a cup of tea and share our problems together'. Until the end of 1960 the Vicar made these evenings opportunities to discuss various aspects of the Christian faith. It is likely that articles concerning Christian principles and Christian faith, which appeared in issues of the *Gridiron* from 1961, were the outcome of these parish nights. Following meetings with the Hall Committee it was agreed that, beginning in April 1961 the Hall should be free on Wednesday evenings, 'to be used at the Vicar's discretion for parish purposes'. These 'purposes' included meetings, Scout Whist Drives and socials. There are magazine references to the whist drives in 1961, 1962 and 1964, and the occasional mention of meetings but 'fellowship' came more openly in other social activities.

Many social events were annual features but as is always the case, much depended on people being willing to organise and help in making these viable. There is also the point that certain social activities are popular with people for a few years and then fall out of favour for new ventures to take their place.

In keeping with his belief that there were two sides of parish life, 'Worship and Fellowship', it is hardly surprising that a number of opportunities for fellowship were connected with Church celebrations. It has been shown that Father Bill was keen that parishioners should attach more importance to worship on two days of the year which were special to St Lawrence's, the Patronal Festival (10 August) and the anniversary of the Consecration, or Dedication, of the church (21 October). With many parishioners away on holiday during August, social events to celebrate the Patronal Festival were held only in 1967, 1968, 1970, 1971, 1973 and took the form of light refreshments after the evening Eucharist service. From 1974 both the Patronal and Dedication Festivals were always celebrated on a Sunday. In 1956, as well as introducing a Day of Prayer and Gift Day on 20 October to celebrate the Dedication Festival, there was a social in the evening. On 5 October 1957 a more formal Birthday Supper was held, rather on the lines of the Communicants' Guild suppers, which had taken place so successfully from 1952 to 1956. A parish social was one of the many events that marked the Silver Jubilee celebrations of October 1958. It seems that these particular socials were intermittent and it was not until 1964, when the Dedication was celebrated on Saturday 25 October, that the 7pm Sung Eucharist was again followed by 'a social, fun and games'. Similar social gatherings took place in 1967 to 1969, 1972 to 1973 (40th Anniversary), 1976 (presentation to Mr G. Collins, Licensed Reader), and 1979 (Father Bill's Official Leaving).

Harvest Suppers
Harvest Suppers were popular social events, that of 1960 was 'well attended and enjoyed', and was repeated in 1961 when numbers were restricted to 130 and

tickets (price 5/- a head), issued 'in strict rotation on receipt of application forms'. On that occasion items of light entertainment were presented. There were no Harvest Suppers in 1963 (the year of the launch of Stewardship) or 1964, and from 1965 the supper was of a 'more informal character'. In 1967 there was a re-organisation of Parochial Church Council (PCC) sub-committees. One new one was the Fellowship Committee, which was responsible for organising socials, entertainments and parish gatherings, mainly work covered previously by the Entertainments Committee. By 1970 this committee had, in the words of the Vicar, 'got into their stride', and there was a glowing report in the *Gridiron* of the Harvest Supper they had organised. 'The stage was transformed into an old world country scene…decorations on the tables featured clever little centre motifs symbolizing the harvest theme'. The meal consisted of roast beef, potatoes, carrots and peas, followed by fresh fruit salad with cream, and coffee and biscuits. After the meal there were items of entertainment, and then the tables were cleared for folk dancing. Reports in issues of the *Gridiron* suggest that similar Harvest Suppers were held most years in the 1970s. In 1979, instead of having the Harvest Supper on the day before the Festival a buffet supper was held after the Sunday evening service.

The Annual Ball

The 1970s saw a number of new ventures on the part of the Fellowship Committee each fostering fellowship within the parish community. In his January 1971 letter the Vicar looked forward to a parish dance to be held at Greenford Hall, Greenford on Friday 12 February 1971, he wanted 'the support of lots of parishioners to make it a success', and hoped it would be 'a most happy parish party'. It was indeed a success and the first of five Annual Balls organised by Mervyn Shepherdly with a supportive team, and ladies of the parish who provided buffet suppers. Dancing was to the music of Ron Hill and his band. The price of tickets (including refreshments) gradually increased from £1/1/0d in 1971, to £2 in 1975, but on this last occasion the buffet supper was prepared and served by students of the Ealing Technical College School of Catering. In 1973 there were a number of guests including the Bishop and Mrs Leonard, the Rural Dean and Mrs Connor and Father and Mrs. R. Ames. (From St Edmund's Northwood Hills). There were also representatives from the parish of St Martin's, St Edmund's, St Thomas More and Eastcote Methodist churches.

Church Outing to Cambridge 1962

Parish Outings

A brief notice in a *Leaflet* refers to a parish outing to Canterbury on 24 October 1959. More detail was given in issues of the *Gridiron* of coach outings in the 1960s. The first was in May 1961, when two 33-seater coaches (cost 13/- per person) took parishioners to Coventry to see the new Cathedral (consecrated in 1963). At the time of the visit the only part of the interior the party was able to see was the crypt, as work was still in progress on the remainder. The journey back to Eastcote 'was enlivened by the formation of the St Lawrence "Glee Club" ably conducted by a member of our choir'. There were other visits to Cambridge (1962), Lancing College and Chichester (1963), Bath, Downside Abbey and Wells (1964), Brighton (1965), and Winchester (1967). Perhaps it was because by the 1960s more people had cars that no further outings were arranged until 1977, when 52 parishioners went on a coach trip to the Cotswolds, ending a 'successful day' with Evensong at Christ Church, Oxford.

Parish Rambles

1975 may have seen the last of the Parish Balls but the Fellowship Committee had new ideas of events with 'the emphasis on fellowship and a chance to get to know each other even better.' In September 1975 the first parish ramble took place, when a 'gallant band of nineteen adults (including Father Huw Chiplin) and seven children set out by car for Great Missenden, waved off by the Vicar'. The hope that was expressed in the magazine report, that the parish ramble 'could be repeated and perhaps become part of the St Lawrence calendar', was certainly fulfilled. Usually over 30 people took part and sometimes there were over 40, with children as young as six and occasionally the odd dog or two. Reports show that the rambles were popular and real family events, times of fellowship and fun, when parishioners were 'part of the family of God together in a truly enjoyable experience'. In each of the years 1976 to 1979 at least three rambles took place, led by several different people from various starting points, including Dunsmore near Wendover, Berkhamstead, Wheathampstead, and Chesham. From 1977 there were rambles during the Christmas period, two in 1979, one locally around Ruislip Woods and one further afield in the area of Burnham Beeches.

The following from a report of a walk in the Hambleden area, in September 1976, seems to express how many people felt about the rambles. 'Apart from the sheer delight of walking through fields and woods unhampered by motorists and petrol fumes, the best thing about a ramble is that you have time to talk to people'.

Discos

The Youth Fellowship had problems of leadership, and at times, of support from their own members. This meant there were periods when there was no church youth group. The Youth Discussion Group, which had started in December 1975, consisted of a number of staunch members of the Church, and possibly they influenced forward looking members of the PCC Fellowship Committee to provide something the young people of the parish really wanted, that was discos.

In February 1976 the first disco was held. Although several adults were involved, John Pearson, a member of the Fellowship Committee, was largely responsible for organising these popular events. The first discos had DJs but later progressed to bands, two mentioned were Mike Pawley and Jem Fleming. The price of an entrance ticket, which included a soft drink and crisps, was 25p for that first disco but had risen to 50p by 1980. The discos were aimed at young people from 11 to 16, Guides, Scouts, choir members and the Youth Fellowship, and their personal friends. The tickets were sold by the young people and the rule was, no ticket no entry.

Writing about the ninth disco of June 1978, John reported an average of 85 people and an average of £1.53½ p profit, adding 'we have achieved the object of not making money'. Reminiscing in his report about the 15th disco, of February 1980 when 98 came, he wrote that 'the mood and feeling of the disco reflects the company present (and that includes the DJ), the music and perhaps the season of the year. Winter discos are warm enclosed affairs but in the summer with light evenings and open windows everything is more relaxed'. These popular discos, with adults and young people supporting each other, certainly illustrated fellowship within the community of St Lawrence.

New Types of Parish Social Evenings
From the mid 1970s in addition to rambles and discos, the Fellowship Committee was involved in a number of other events, which contributed to fellowship. In January 1976 they organised a parish social evening which took the form of a square dance, when 'about a hundred people of all ages gathered…to among other things, honour their partners…look for the Northern Lights, doh-se-doh, and do the Bingo Waltz'. Ignorance of square dancing was no excuse as there was a caller, John Smith. Tickets were 50p including refreshments 'provided by Mrs Barker and her band of helpers'. This 'enjoyable family evening' was followed by other square dance evenings, one later in 1976 and others in 1977, 1978 and 1979. The last was reported to have been 'poorly attended but enjoyed by those who were there'.

In the first week of January 1977 there were two opportunities for fellowship. The first was when light refreshments were served after the Solemn Eucharist on the Feast of the Epiphany, the last of the four services of that day. By this time the practice of having light refreshments on such occasions seemed to be routine. Two days later the Fellowship Committee arranged a family social evening, at the especially low price of 10p per ticket, in which 'everyone joined with great gusto'. There were quizzes, games and Old Time Dancing, and after coffee and mince pies during the interval, much laughter was had watching an old 'Keystone Kops' film.

At the end of 1977, when 31 December fell on Saturday, a New Year's Eve social with a Scottish flavour was held. It was specifically for those who might be at home and an opportunity for families of St Lawrence's to have a get-together. There were games, dancing and refreshments. The evening rounded off with a Watch Night Service, conducted by Captain Donald Woodhouse. For the few families who went it was obviously a happy occasion.

One further 'one off' event was the 'Do-it-yourself' barbecue of July 1976, that long hot summer. It was not as the name implies, that people had to cook their own food, or even bring their own food, but that people had to bring their own chairs, glasses and wine. Some one else did the cooking. There was a magnificent spread of mountains of sausages and chicken and many varieties of salads, followed by fruit salad and ice cream. The Ruislip Banjo Band provided music in the hall. It was another 'happy and enjoyable evening'.

The Annual Sale

Although the November event was variously referred to as the autumn sale, the autumn fair, the Christmas market, or the Christmas bazaar, it was traditionally an annual effort to raise money, as well as a time of fellowship. Members of the congregation, both as individuals and as members of church organisations, had from the early days contributed to the success of these events. In addition the sales were also opportunities of mission to the wider community of Eastcote who came to spend money.

During the first years of Father Bill's incumbency, the *Leaflet* had only brief references to the bazaar or market, mainly the fact that it would be held on a Saturday afternoon. After the *Gridiron* started in 1961 more details were recorded. Until the re-organisation of the PCC committees in 1967, the Entertainments Committee was responsible for this and other social events.

The Christmas market of 1961 was certainly a success for 'over 600 persons paid for admission and the total receipts at £276 were greater than in 1960'. The Needlework Guild took £71.10s. 0d and the 'colossal amount of sweets' given by children of the Junior Church raised £18. When Stewardship was launched in 1963 the decision was made that proceeds from the annual fêtes and markets would be given to charities. The coffee market of 1963 'was low key'. It raised £48, of which the £35 raised by the Needlework Guild was sent to a charity to help mentally handicapped children. The proceeds of the 1964 market (£150) went to the Save the Children Fund, via the Chairman of the Council's Christmas Appeal, and that of 1965 was sent to help the needs of the Church in Mauritius (see p114). There was no sale in 1966 but influenced to some extent by the persuasion of the new Fellowship Committee and the backing of church organisations a bazaar was held in 1967. The stalls were well stocked and the proceeds of a 'most enjoyable social occasion', about £180, were divided between the Building Fund and the Memorial Fund to Father Godwin. The optimism that Stewardship would bring in sufficient funds to enable the proceeds from fêtes and markets to be given to charities was short-lived. In the remaining years of Father Bill's incumbency, apart from 1973, St Lawrence Church's 40th anniversary, when the proceeds went to the Assistant Clergy Fund, the annual sale again became an occasion for raising funds for the church.

From 1969, when for the first time the Christmas market opened at 11am, it started to become a more ambitious affair. Different opening times were tried. In 1970 the Fellowship Committee were keen to make the event a resounding success and the market was open on two days. People were urged to help, to 'come and join in the fun and enjoy making a real contribution'. The hall was packed on the Friday evening and there was a 'steady flow of customers' throughout the Saturday. The takings reached nearly £700. In 1971 when 'Mrs Barker and her ladies smoothed the passage of the workers with suppers, lunches and teas', £750 was raised towards the cost of new heating in the hall. Then for a number of years the market was held on the Friday evening and Saturday morning, with refreshments for both workers and customers, in the form of sausage and mash, coffee and sandwiches. From 1977 it was Saturday afternoon instead of the morning, with suppers on Fridays and teas on Saturdays.

On the whole the goods on the stalls were what one would expect but there were the occasional novel 'service' stalls. In 1972 and 1973 there was a stall providing a sharpening, grinding and repair service for small hand held tools. One year at least Joe Orme, a jeweller and member of St Lawrence's, undertook jewellery repairs for a small price. The first Grand Draw was held in 1976. Then as now local trades people and businesses, as well as churchgoers, were generous with their gifts to the draw and tombola. These brought in nearly £300 of the total of £841. A report on this market pointed out that apart from the financial success the 'social aspect of the efforts undertaken together should not be overlooked'.

Summer Fêtes
By 1956 the summer fêtes were popular social events both for members of St Lawrence Church and the wider community of Eastcote. The work that went on beforehand, the variety of stalls and sideshows on the day, together with the evening entertainment provided by the St Lawrence Players all indicated the 'sociable community' wanted by Father Godwin. As has been suggested Father Bill's 'Worship and Fellowship' was a different way of expressing the same sentiments.

In his letter in the July 1958 *Leaflet* Father Bill linked together the preparations that were going on for two annual events, Confirmation and the Garden Fête. He wanted his readers to encourage the newly confirmed to join in all our activities to 'find their joy in Worship in, and in service to, their Parish church from which comes their Fellowship in such events as garden fêtes'. After the 1960 fête he wrote ' I regard this occasion as one of our main opportunities of enjoying the fellowship which emerges from our worship together'.

A full report in the August issue of *Gridiron* 1961 regretted that 'the attendance fell short of our expectations' at the fête on 1 July. It had been the hottest day for

Summer Fête 1962

14 years, and there was a rival attraction 'over the road' at the church of St Thomas More. Among the sideshows were darts, putting, treasure hunt and the shooting gallery, the chance of throwing wet sponges at cheeky choirboys or trying one's luck on the coconut shies. A new attraction was the balloon race and for the second year a model railway. The hot weather meant that ice cream and soft drinks did a brisk trade, and even 'provided the crew of a local bus with cold drinks, while the passengers waited patiently in the bus which had stopped outside the gates'.

In September 1964, following the decision in 1963 that the proceeds of future fêtes and markets would be given to charities, the Anglican and Roman Catholic churches in Eastcote worked together and held a fête in the grounds of St Thomas More Church. The proceeds went to the local Freedom from Hunger Campaign. The combined committee set up to organise that fête, and rather oddly referred to as the Interfaith Committee, was soon enlarged to include members from the local Methodist and Presbyterian (United Reformed since 1972) churches. It seems likely that it was through this committee that the first Eastcote ecumenical carol service referred to earlier was held in 1965.

The Interfaith Committee was certainly responsible for a most successful Grand Fête in June 1965, held in the grounds of St Lawrence Church and which raised over £700 for the local Old Folks' Association. The fête was opened by Kenneth Kendal of TV and radio fame. It was a real community effort with helpers from the four churches, trades people and local organisations such as the Rotary Club and the Women's Institute, and many others. In the evening there was a choice of entertainment between a dance, held in the church hall of St Thomas More, and the *Fête Accompli*, presented by Methodists, Roman Catholics and the St Lawrence Players in the hall of St Lawrence Church.

For the remainder of the 1960s, although church organisations took part in manning stalls and sideshows, and in providing tea at the annual Summer Fête, these summer events appear to have been relatively subdued affairs. A notice in the *Gridiron* of the Garden Fête on 6 June 1970 promised 'a more active one ...due largely to the needs of the financial budget which requires the boost of a good Fête'. By this time it was realised 'that the needs for fêtes and bazaars still remains despite the valiant attempts which parishes are making in Stewardship programmes'. It certainly was 'a more active one' including children's fancy dress competition, Punch and Judy show and lucky programme competition. Many features were in the fêtes of subsequent years, but one unusual stall sold fruit and vegetables from Brentford market (some of the committee members collected the produce at 4.30am). Exhibits in the bungalow in 1970 were floral arrangements, in which some members of the Flower Guild were involved, pottery by Mrs Jean Jones, and sketch portraits by Mr Surridge. It was a great social occasion and even guests at the only wedding of the day made purchases at the fête. The day rounded off with a barbecue hop in the evening.

The fêtes of the 1970s had many stalls, sideshows and events, which were repeated year by year. However different things were tried, some for a single year, others for two or more. One example is the successful dog show in 1974. Also in 1974 the magnum-sized champagne bottle full of halfpennies, which had stood in the church for people to donate their coins, was used for a guessing competition. When the coins were counted and bagged there were 12,829!! The nearest guess was quite a good one, 12,961. Several years there were enjoyable dancing displays, either by pupils from Newnham School or members of St Lawrence Guides or Brownies.

In 1976, 1977 and 1978 the wider community was made fully aware that the parish church was having its annual fête. A band led a procession from 26 Woodlands Avenue, home of the curate, Father Huw Chiplin, through Eastcote. The procession in 1977, Queen Elizabeth's Jubilee Year, must have been particularly impressive, with a 'town crier' (Captain Don Woodhouse), the Vicar in his Canon's robes, Father Huw in his cassock, the two Churchwardens, the Jubilee Queen, her attendants, cubs, brownies, scouts and guides, decorated bicycles, and mums and dads. The procession set off to the 'sound of drums and fifes...out into Field End Road, over the station bridge, through the shopping centre and then into the church grounds'. The police helped the procession at the traffic lights and the crossing of roads.

Although in 1979 there was no procession through the town, Father Bill reported that it had not detracted from the enjoyment of the fête. 'There were plenty of stalls, tea, fun and sideshows and lots of people came to share the afternoon with us'. This is surely a fitting quote to conclude this section, for it implies that his guiding principles set out in his first magazine letter had been fulfilled.

The Mission of the Church
Much that is written in this chapter relates to 'the mission of the Church' in its widest sense. Father Bill often included 'service' or 'mission' when referring to parish life and his guiding principles of 'Worship and Fellowship'. In 1957, soon

after he came to Eastcote, he expressed the view that it was the responsibility of the Parochial Church Council (PCC) to be positively involved in the mission of the Church both at home and overseas. Among other things a new sub-committee of the PCC, the Missionary Committee, was asked to bring the overseas work of the Church into focus. One way of gaining support from parishioners might be to have a link between the parish and a particular overseas diocese.

In his magazine letter of October 1960 Father Bill, while emphasising the need for the support of the Church overseas, said 'we have to bear in mind the work of the Church here at home'. He referred to support at Deanery and Diocesan level, through Moral Welfare, Ordination Candidates, and the London Diocesan Fund. He mentioned The Children's Society and other societies claiming our attention. Finally he stressed the 'spread of the gospel in the local parish and one of the best ways in which we can help is to offer our services to the youth and children'.

During Father Bill's time in Eastcote the parish continued to support many of the charitable organisations that St Lawrence's had supported in the past. It also responded to local, national and international appeals and crises. Personal contact or concern on the part of one or a few enthusiastic people can encourage support for a particular cause. Visiting clergy, or other speakers, who can talk of their own experiences help to gain interest in a project.

Overseas: Korea and Mauritius
One link between the parish and a particular area of the world seems to have started in January 1961, when Father Archer Torrey, of the Korean Mission visited St Lawrence's to speak of the work of the Church overseas and particularly in Korea. The September magazine reported that the Korean Mission had received £51.9s.7d, part of the total proceeds of £137.4s.4d from the 1961 Lent boxes. The Junior Church had up to that time, sent their Lent savings and most of their Sunday collections, nearly £100, to help orphans in Korea. St Lawrence Church supported the Mission up to 1976.

St Lawrence Church's aid to Mauritius is an example of a parish link with a particular overseas diocese and one that lasted for twenty years or more. From Father Bill's magazine letters it would appear that he was a personal friend of Alan Rogers, who was the Lord Bishop of Mauritius from 1959-1966. Father Bill had been curate at St Mary's Twickenham when Alan Rogers was Vicar. The Bishop made several visits to Eastcote. When he visited on Dedication Sunday 1961, at each of the services the Bishop told of the devastation caused by cyclones in 1960, and of the needs of the Church in Mauritius. This led to the parish making a special commitment to raise sufficient money to rebuild the church at Riviere des Anguilles. The first payment of £100 was made before the Bishop returned to Mauritius, and it was estimated that another £1400 would be needed. When the Bishop spent the day in Eastcote in February 1965, he brought photographs for the Missionary Exhibition held in March, and was handed a cheque for a further £300. The parish had news regarding progress towards the rebuilding of the church through the magazine and the visits of the Bishop. By 1968 it was known that the church was to be dedicated in the name of St Lawrence and it was hoped that the 'Foundation stone will be laid reasonably soon and the building completed during 1969'.

This was not to be and in December 1971 Father Bill informed the parishioners that owing to the closure of the sugar factory at Riviere des Anguilles the population decreased to such an extent that it was decided that it would be wrong to replace the building destroyed in the cyclone. With the agreement of the PCC, it was arranged that the money donated by the parish, around £1,200, would be transferred to help the small island of Rodrigues where it would be used at the Bishop's discretion. In 1974 parishioners learned that the money was used to build accommodation for male teachers at the only secondary school on the island. It was to be called St Lawrence House in recognition of the help the Eastcote parish had given to the area over many years. The parish continued to support the cause through interest, prayer and annual donations to The Friends of Mauritius into the 1980s.

Missionary Exhibitions
The Missionary Committee certainly 'brought the overseas work of the Church into focus' in 1961. The first Missionary Exhibition took place in the parish hall

---

THE MAGAZINE OF EASTCOTE PARISH CHURCH   PRICE SIXPENCE

THE CHURCH OVERSEAS

AN EXHIBITION

THE PARISH HALL

Wednesday, November 14th, at 8 p.m.

An exhibition of photographs; book stalls; refreshments

THE CURTAIN LINE, a play by Cyril J. Davey

Speaker: The Rev. JOHN F. CAMPBELL, Metropolitan Sec. of S.P.G.

Missionary Exhibition 1962

---

in November and had exhibits about the two parish links. In addition there were speakers from the Universities Mission to Central Africa (UMCA) and the Korean Mission, as well as photographs and bookstalls. The Society for the Propagation of the Gospel (SPG) film *While There is Time* urged the necessity for praying and acting *today*.

The Exhibition aroused interest and more parishioners attended the second Missionary Exhibition in November 1962. Once again there were photographs, bookstalls and a speaker, this time from SPG. On the UMCA stall, pencils were popular while the SPG stall had a good trade in Christmas cards. The occasion also showed the involvement of younger members of the congregation as the Youth Club performed *The Curtain Line*, a play about missionaries, produced by Miss Angell.

In 1963 the parish was busy with the launch of Stewardship and it was not until Lent 1965 that what was to be the last Missionary Exhibition was held. The Exhibition was associated with parish study groups and the 1965 Lent course *No Small Change*. The Vicar referred to the need for a big change 'to take place in our whole outlook to the mission of the Church both at home and overseas'.

In addition to the Korean Mission and the commitment to Mauritius, St Lawrence Church contributed regularly to the Leprosy Mission, the Society for Promoting Christian Knowledge (SPCK), SPG and UMCA. As in the past some of the money was raised through Lent and Advent savings, some by collections in church, some from specific events and some from collection boxes issued to members of the congregation. In many years members of the parish attended the Annual Rally of the SPG at the Royal Albert Hall. One of the 21 who went in 1964 found it a 'most moving and inspiring occasion'. In May 1965 a Sung Eucharist and Rally at Alexandra Palace marked the union of the SPG and the UMCA into one society, the United Society for the Propagation of the Gospel (USPG). Over 40 parishioners attended and choir members from St Lawrence sang in the choir of 1,000 voices.

As is the case today the parish responded to specific crises, such as that of the Nicaraguan Earthquake Appeal in 1972. Response to a personal connection can be seen in the support of the International Nepal Fellowship. From 1977 donations were given towards the work Alison Craven was doing in Nepal. The fact that the Craven family were members of St Lawrence Church for many years encouraged people to take an interest in the cause.

No Small Change
It has been established already that the 1960s were years of particular challenge for the Church. Changes in church services, which have been discussed, was one response, and the Prayer Book (Alternative and Other Services) Measure of 1965 regarded as a 'landmark' towards those changes. The Lent study groups *No Small Change* of 1965 could also be regarded as a landmark in connection with the Mission of the Church. St Lawrence Church was one of 50 parishes in the London Diocese selected to take part in the pilot scheme. The text of Father Bill's March 1965 magazine article gives some idea of the purpose of *No Small Change* (the entire article is reproduced on the next page). Articles in later magazines of 1965 imply that parish participation in the discussions was good

and should lead to 'changes that ought to be made to make our witness for Christ and His Church more effective'.

> Printed in Gridiron, March 1965
> ## NO SMALL CHANGE
>
> On Saturday. February 6th, representatives from the parish attended a study conference in London on the subject of "Mutual Responsibility and Interdependence", which is a statement drawn up by the leaders of the Anglican Communion and presented to the Anglican Congress at Toronto in August, 1963. In this statement we are called upon as a parish prayerfully to examine our life and work that we may be better fitted to serve God and our fellow men and women: it is a challenge to all who profess and call themselves Christian.
>
> The rapidly changing situation in the world today needs a new approach to missionary thinking in every part of the world so that all God's people can see their true responsibilities and the ways in which they can learn from, and help, each other. So vast is the change in this new thinking that an old saying has been given a new meaning—"No Small Change"— and clearly a big change has to take place in our whole outlook to the mission of the Church both at home and overseas. We need to ask ourselves some searching questions on our membership of the Church, its services, its organisations, our personal part in the worship and our personal share in the problems and tasks of the local Parish; and prayerfully reach the best answers of which we are capable, so that God's Kingdom may be advanced. Gone is the time when communities or countries could exist without reference to what is happening in the world around them. The advance in science, technology, medicine, industry, education and, most of all perhaps, in transport and communication have drawn us all closer together and perhaps we are beginning to see, albeit painfully slowly, that the problems and tasks in one area of the world are not dissimilar to those in many other areas. In the same way, gone is the time when an organisation of a parish, the parish itself or the diocese can ignore the problems and tasks facing the Church in every quarter of the globe and this applies equally to the Church in Africa, Asia, America and Australia, as to the Church here at home. Each has a part to play in the great theme of mutual responsibility and interdependence.
>
> Whilst we have to "keep the machinery going" in our parochial life with all its joys, problems and difficulties, we also need to see how far we can take a fuller part in the "give- and-take of Christian family life", both in our own backyard and across the world. With this end in view I propose to initiate study groups during Lent on the lines—much modified—of the suggestions made at the study conference early in February: I hope to receive full co-operation from all the organisations in the parish at the various meetings to be held in this connection. There has not been time to work out the details for printing here, but everyone will be told, in church, in due time, how we shall try to tackle the course. There are five parts, viz. :—1. The Parish looks at its purpose. 2. The Parish looks at its neighbours. 3. The Parish looks at the world. 4. The Parish looks at God's people, and 5. The Parish plans for action. In addition, a course of sermons will be preached on Sunday evenings in Lent.
>
> I hope you will all endeavour to come to the sessions, whether afternoon or evenings I must leave to your convenience, and to attend each Sunday, first for your Communion and then again in the evening. Lent is not only a time for giving up things!!
> W. H. H.

In a long July 1966 magazine article Father Bill referred back to *No Small Change* and commented on how the pattern of parish life had been changing, with the increase in car ownership and television viewing. Among the many changes he felt there was a 'growing consciousness that we are all involved in the mission of the Church'.

Mission in the Local Community
It is now right to look at another aspect of mission, that of 'service' or 'mission' in the local community. The importance Father Bill attached to the 'spread of the gospel in the local parish' has been shown in the Worship section in which the discussion refers to a number of changes he initiated as regards church services. Here it is appropriate to mention a few services, which may have come about as a result of discussions of the PCC Missionary Committee and seem to illustrate Father Bill's guiding principles of 'Worship and Fellowship'. Afternoon services of reunion were held on consecutive Sunday afternoons in the autumn of 1958 (Silver Jubilee Year), 1961 and 1962 for people who had had contact with the parish church, through Baptism, Confirmation or Marriage services in the recent past. These were a positive attempt to keep those people within the fellowship of the Church. Perhaps the launch of Stewardship made it impossible to hold similar services in 1963, and there are only occasional references to these reunion services in later years.

The church may not have been successful in having these reunion services as annual events but successive *Register of Services* books give evidence of other services embracing the wider community of Eastcote, some of which remain popular today. Ones which began in Father Bill's incumbency included the Crib Service (from 1961), the Old Folks' Association Service (from 1965) and the Ecumenical (Interfaith) Carol Service (from 1965).

Two further examples from the 1960s indicate the new approach to mission. The St Lawrence Branch of the Mothers' Union (MU) had been an important organisation since 1932, but it was recognised that another women's organisation was needed and in 1965 the Women's Guild was formed. Another example of fostering fellowship was the Evening Gathering in March 1966, when 58 people whose names had been added to the Electoral Roll during the past year, were invited to sherry and refreshments with members of the PCC. These days wine would be more likely than sherry.

Some suggestions of the Missionary Committee in 1960 were to make the presence of the parish church known to the wider local community. One suggestion led to the church services being advertised on notice boards put up by the local council near the War Memorial and near Cavendish Hall in 1961, for which the church paid an annual rent. Another saw the publication of a new style of parish magazine in 1961. In his Vicar's Remarks at the 1961 Annual Parochial Church Meeting (APCM) Father Bill referred to the new style magazine resulting from the efforts of the Missionary Committee and 'hoped it would receive enthusiastic support from the Parish'. By this he implied the area of Eastcote and not just members of St Lawrence Church.

Some aspects of what might be called 'mission to the local community' had been established in Father Godwin's time and continue to the present day. Although the church's fêtes, markets, St Lawrence Players' productions and other church events may be considered to be social as well as fund raising occasions, they also offer opportunities for mission. Once again words from Father Bill's first magazine letter are relevant here. 'From our worship together there should also emerge a fellowship which, in its outpourings, will embrace all those with whom we come in contact'. Traditionally the local community has been generous in its support of these church events.

Open House ;
Floral and Art Display 1972

There was certainly plenty of support from the local community when St Lawrence Church was involved in the Greater London Festival and held 'Open House' from Thursday 29 June to Sunday 2 July 1972. As well as floral displays, from individuals and organisations, the church had exhibits of different types of works of art, an exhibition of altar vessels and vestments, and musical items by the choir and organists. In fact there were so many works of art that the overflow was exhibited in the hall on the Saturday. Light refreshments were available in the bungalow (now the St Lawrence Centre) Thursday to Saturday mornings and afternoons. As far as church members were concerned it was a busy time, for in addition to the Floral and Art Display the summer fête was held on the Saturday. The August magazine recorded the Open House, Floral and Art Display 'a tremendous success…very rewarding to see so many people visiting'. The original supply of programmes had run out by lunchtime on Friday and 600 more were printed. The article concluded with thanks to local shopkeepers and all who had helped to make 'such a successful and enjoyable event'.

Another aspect of the church's mission to the local community was through the links St Lawrence Church and its members had with local schools, hospitals and old people's homes. For many years there were the Sunday Schools at

Newnham, Harlyn Drive and St Mary's School. Contact with St Michael's School, Joel Street and St Vincent's Hospital, Wiltshire Lane was maintained. During these years new additions to the work of the parish included visiting people in Whitby Dene Old Folks' Home (opened 1966), Frank Welch sheltered accommodation (opened 1968) and residents in the flats at Missouri Court (opened 1976).

Quiet Days, Retreats and Pilgrimages
While Father Bill was rightly concerned about the Mission of the Church he was, as is every priest, concerned about the spiritual growth of members of his own parish church. Church services, study groups, parish meetings and teaching through magazine letters and articles were some of the ways in which he was able to do this. During his time as Vicar of Eastcote several new opportunities for spiritual growth were introduced.

In the September 1965 issue of the *Gridiron* Father Bill hoped to 'resume our normal practice' of holding a Quiet Afternoon on the Saturday before Advent. The first had been in 1960, and these afternoons, which were seen as 'a useful preparation for the penitential season of Advent', were held most years. Often they were conducted by visiting clergy, on two occasions (1968 and 1977) by Father Peter Goodridge who had been Assistant Priest at St Lawrence from 1958 to 1964.

The Chapel at St Francis House,
Hemingford Grey

Many members of St Lawrence Church have experienced the benefits of a Parish Retreat but the retreat movement is a comparatively modern organisation association. The Association for Promoting Retreats was founded in 1911, and by 1962 there were over 70 retreat houses in the British Isles. In the 1960s, at a time when the role of the Church was being questioned, the retreat movement was seen 'as a reaction against the rush and turmoil of modern life'. The first Eastcote Parish Retreat took place at St Francis House, Hemingford Grey, Huntingdonshire in October 1962. Throughout Father Bill's time at Eastcote, and for many years after, Michael Bedford (now Assistant Priest at St Martin's,

Ruislip) arranged these annual retreats. Nearly all of them were at St Francis House in late September or early October. In 1968 and 1970 the venue was the Anglican Convent of All Saints, London Colney. The cost of the two-day retreat in 1962, excluding transport, was 35/- (£1.75p) but by 1979 had risen to £7.50p. Those who have been to Hemingford Grey have been impressed by the atmosphere of St Francis House and appreciated the delightful village and surrounding countryside.

A small party represented St Lawrence Parish in the few years of the Bishop's annual Pilgrimage to St Mary's Church, Willesden. St Mary's housed a mediaeval shrine believed to have been the scene of a mystical appearance of the Blessed Virgin. Blackened by age, the wooden statue, known as the 'Black Virgin of Willesden', was destroyed during the reign of Henry VIII. A new statue was made though not in place for the first pilgrimage in 1971 and no mention is made of the pilgrimages after 1974. Father Bill encouraged parishioners to take part in 'what undoubtedly is an inspiring afternoon'.

On a rather more ambitious scale than Quiet Afternoons, Retreats and Pilgrimages was the possibility of attending performances of the Passion Plays at Oberammergau. The July 1966 magazine referred to 1960 when 'this parish sent two parties' and feelers were put out to see if parishioners were interested in another 'parish outing' during the performances in 1970. The cost of the 1960 holiday of 12 days had been about £34 and it was realised that it would be rather more in 1970. For whatever reason it seems there was not sufficient support for this venture.

St Lawrence Church and Ecumenical Activity in Eastcote
At the beginning of this chapter mention was made of the development of the Eastcote area and how the increase in the population affected the St Lawrence community. Reference was also made to the changes in society and to the questioning of the role of the Church. During the 1960s efforts were made to try to bring about closer relations between the churches of Eastcote.

The Week of Prayer for Christian Unity
According to a *Register of Services* book the Week of Prayer for Christian Unity was first observed at St Lawrence Church in January 1957, shortly after Father Bill became Vicar. It may be of interest to give a little background. It seems to have been initiated by Anglicans in 1908 or 1909 with the aim of re-uniting separated Christians with the Roman Catholic Communion. At the time it appears to have had limited appeal, and it was not until the 1930s that Abbé Paul Couturier, of Lyons, broadened the appeal and united Christians all over the world with the universal prayer of our Lord 'that they may all be one'. It was realised by many Christians that they could use this special week that includes 25 January, the Conversion of St Paul, to pray for unity in their own churches and chapels.

In 1963 the *Gridiron* printed a message from Father Langdale, Priest in charge of St Thomas More Roman Catholic Church, in which he wrote 'we have begun to realise what great scandal is caused by the spectacle of Christian disunity... we can also love one another better, not in a spirit of shallow compromise, but with deep charity'. The January 1964 *Gridiron* had an article by the Revd J. D. Cardew, Minister of Eastcote Methodist Church in which he asserted that 'God is at present occupied in stirring up people to think about uniting the Churches'.

The first of exchange visits between members of Eastcote churches seems to have taken place during the Week of Prayer in 1964. St Lawrence members were invited to a service at the Methodist church on the Monday and members of the Methodist Church accepted the invitation to join the St Lawrence community at a Sung Eucharist on the Friday evening. After each of these services there was a social gathering. In 1965 members of all four local churches, Anglican, Methodist, Presbyterian and Roman Catholic held a united service in St Lawrence Church Hall. From then on an ecumenical service was held each Week of Prayer at one of the four churches.

Each January Father Bill drew attention to the Week of Prayer. In 1965 he asked that people should pray that 'God's will be done in us, and through us... that the unity of the Church may come about in the way God wills and when He wills'. In 1979 the message was the same but words seem more in keeping with the times, 'that the unity of the Church may come about in God's own way and in His own time'.

Anglican-Methodist Union
Attending each other's services of worship was one thing, but in the 1960s Anglican-Methodist Unity Proposals were on the agenda and might have led to unity between these two branches of the Christian faith. In 1963 the Report of the Conversations between representatives of the two churches was published and was the fruit of six years of discussion. There would be a period of 'testing the ground'. In Eastcote, as in all parts of the country, there were meetings and discussions on the matter. In 1964, just before the Week of Prayer, the Eastcote Methodist Minister, the Revd J. D. Cardew, addressed members of St Lawrence's congregation to explain the issues involved. The exchange church visits in the 1964 Week of Prayer for Christian Unity were part of the 'testing'. The March *Gridiron* articles, *Looking back on Unity Week*, by the Revd J. David Cardew and the Revd Peter Goodridge, indicate that, while there was appreciation by both Anglicans and Methodists of the exchange visits, unity was unlikely in the foreseeable future. For the remainder of the 1960s discussions went on at local and national level, culminating in the Convocations of Canterbury and York and the Methodist Conference in July 1969.

In a long August *Gridiron* article Father Bill commented on the outcome of the July meetings. There had been an adequate majority for union at the Methodist Conference, which indicated a willingness to go forward, but the Convocations had 'hesitated to give the required majority for the present proposals, bound up as they are with doctrinal ambiguities'. Father Bill realised that some would be disappointed at the outcome but he also referred to the great strides that had been made in the ecumenical field and that we 'have much for which to thank God in the real co-operation that now exists between churches of different denominations'.

It was reported at the 1970 Annual Parochial Church Meeting (APCM) that in spite of the voting at national level regarding the Anglican-Methodist Unity Proposals, co-operation with the other Eastcote churches would continue through the Interfaith Committee. The few years of activity over the Proposals may have contributed to the friendship that developed between the Eastcote churches during the 1970s.

Christian Aid

Christian Aid has been raising money to help alleviate problems arising from poverty for over eighty years. Readers will be aware that special appeals are made in times of emergency in addition to the annual Christian Aid Week in May. It was during Father Bill's time in Eastcote that members of St Lawrence Church started to co-operate with members of Eastcote Methodist Church in an effort to cover all the roads in the area in the annual door-to-door collection. Appeals for volunteers were made through the parish magazines but sadly, then as now, the ideal never seems to have been accomplished. The first reference to the two churches having 'combined for some years to raise more than £500' was in the *Gridiron* of April 1965, when the appeal for volunteers referred to the disappointing total of £435 collected in 1965. By 1980 St Andrew's United Reformed (formerly Presbyterian) Church members were also involved.

Great Britain had set a target of £2million to be raised by May 1960 for a special World Refugee Year. The chairman of the organising committee for Northwood

> **CHRISTIAN AID WEEK**
>
> **MAY 13–18**
>
> LORD, WHEN SAW WE THEE AN HUNGRED, OR ATHIRST........?
> INASMUCH AS YE DID IT NOT TO ONE OF THE LEAST OF THESE, YE DID IT NOT TO ME.
> *(S. Matthew's Gospel, chapter 25)*

Christian Aid Week 1963

and Ruislip (which included Eastcote) hoped the area would raise £5,000. In his magazine letter of October 1959 the Vicar appealed to parishioners to 'set aside between now and the end of May 1960 a sum of not less than £1 per family'. Envelopes were issued with magazines in the spring and returned and presented at the altar at Evensong on 29 May 1960. The parish total was around £620 and the wider local area exceeded £13,000. The success of this World Refugee Year may have been the incentive that led to the Eastcote Anglican and Methodist churches combining their efforts.

Each year there was an appeal for more helpers and some years additional efforts were made to bring home to people the role of Christian Aid in meeting the needs of those less fortunate than ourselves. In 1965 there was a presentation by professional actors of *No Man is an Island* on the aims and objectives of Christian Aid. In 1971 a church bell was tolled at 2pm on Sunday 16 May 'to remind us all of those dying every second for want of food'. The co-operation between the Eastcote churches appears to have led to higher collections in the 1970s, with over £650 in 1978 and 1979.

The Interfaith Committee
The rather oddly named Interfaith Committee has been referred to already in connection with the combined Summer Fête of September 1964. Originally made up of members from St Thomas More and St Lawrence churches, by 1965 it also had members from the Eastcote Methodist and St Andrew's Presbyterian churches. That year, in addition to the successful Grand Fête and *Fête Accompli*, the Committee arranged the first, of what has become a popular annual event, the Ecumenical Carol Service. Up to 1970 it was referred to as the Interfaith Carol Service. Some form of refreshment has always followed these services giving opportunity for congregations of the different churches to socialise.

Father Bill reported 1965 as –

> something of a vintage year in ecumenical relations between the Anglicans, Methodists, Presbyterians and Roman Catholics in Eastcote, first by the united service in St Lawrence Parish Hall in January, then the joint summer Fête, when we raised over £700 for the local Old Folks' Association; followed by a joint thanksgiving service in the Methodist Hall; and finally, three days before Christmas, the joint Carol Service in St Lawrence Church.

The magazines of the 1970s and Parochial Church Council (PCC) *Minutes* occasionally refer to events arranged by the Interfaith Committee. There were summer outings in 1972 and 1973. From 1971 there were 'monthly' prayer services at one of the churches but they seem to have been less frequent as the years went by and the last reference to them was in the Secretary's Report at the 1977 APCM. At this time St Andrew's United Reformed and the Methodist churches seemed anxious to disband the Committee, by then known as the Eastcote Association of Christian Churches. Father Bill and Father Anglian, of St Thomas More, wanted the prayer meetings to continue and the Committee to have further meetings regarding the future. Obviously the Committee had successful 'further meetings' as they at least arranged the Ecumenical Carol Services, which remained an annual event of the Eastcote churches.

Freewill Offering, Stewardship and the Diocesan Quota
The section on the Fabric of St Lawrence Church gives some idea of major financial projects during Father Bill's incumbency. Elsewhere is information about several fund-raising efforts, many of which were social occasions as well. Like any private person or institution the church tries to work to a budget, and knowing that there is a regular income helps in this matter. The major source of income is from parishioners in the form of contributions, either in church collections or through schemes where donors make commitments to regular giving. Up until 1963 this was through the Freewill Offering Scheme (FWO). Mr G. W. Wade was Secretary and Treasurer from 1956 until 1960 when Mrs Onions succeeded him.

An appeal in the December 1958 issue of the *Leaflet* is typical of the periodic pleas for more subscribers. 'I know we call it free-will, but those who love our Lord should regard it as an obligation to assess what they can give and then give whether they are able to be present each week or not'. Father Bill's letter in the *Leaflet* of July 1959 indicated that a move towards a more formal planned-giving scheme was being considered, for he referred to 'leaflets at the back of the church dealing with the Christian stewardship of money'. The question of launching a Christian Stewardship campaign, which is designed 'to quicken the life of the parish', was discussed by the PCC over several meetings but the 1960 November *Leaflet* reported that it had been decided 'to postpone consideration of such a scheme for the time being'.

In fact it was not until 1963 that the decision was made that St Lawrence Church should embark on its first Christian Stewardship Programme. Father Bill described the purpose of the programme in the September 1963 *Gridiron*. The Lambeth Conference of 1958 had considered the Christian Stewardship Movement, which had begun as a scheme for increasing the income of the Church, but the emphasis had changed and broadened out. He added that it had many of the characteristics associated with previous Missions and regarded it 'as one of the most important events which has ever occurred in the life of our parish'. Although Christian Stewardship is often thought of in terms of financial giving it is far more than this, as Father Bill made clear in the October 1963 magazine. Stewardship involves the use of our time and our talents as well as our money. However, the need for a regular income to meet the expenses of a church means that emphasis is, in the first instance, on financial giving. This is usually referred to as Phase I of a Stewardship Campaign, and the other two aspects of time and talents as Phase II.

Leading up to this 'major event' preliminary work had been done by a Stewardship Steering Committee, who were members of St Lawrence Church. Mr Rex Clarke of the Council for Christian Stewardship directed the Stewardship Programme, which ran from 13 October to 17 November 1963. The Programme opened with special preachers at the Sunday services on 13 October and during the following two weeks there were training sessions for Lady Visitors and Men Visitors (meeting separately!). During the training sessions emphasis was placed on the proper teaching of Christian Stewardship. 'The principle of the Stewardship of the whole of life was the theme of the Programme'.

> THE VICARAGE,
> EASTCOTE,
> MIDDLESEX.
>
> Dear Friends,
>
> Our Christian Stewardship Programme will begin on Sunday, October 13th. There will be special preachers at all services, and I hope that you will all do your very best to be in church on that day.
>
> Mr. Clarke of the Council for Christian Stewardship is coming to our parish to take charge of the whole Programme, the highlight of which will be the Parish Supper on Monday, October 28th.
>
> May I again direct your attention to this quotation from the Lambeth Conference Report?
>
> "There can be no forward steps without a full acceptance of Christian Stewardship. By Stewardship, we mean the regarding of ourselves—our time, our talents, and our money—as a trust from God to be utilised in His service. This teaching is an urgent need in every congregation; a parish without a sense of Stewardship has within it the seeds of decay."
>
> I ask you for your prayers and for your enthusiastic support for the Christian Stewardship Programme. It will undoubtedly be a full and exciting time, and I hope and pray that it will lead us to a better and more exciting life dedicated to the glory of God and the good of His Church.
>
> Yours sincerely,
>
> William H Hitchin

Fr. Bill's October 1963 letter

A Commissioning service was held on Sunday 27 October. On the following evening 424 individuals representing 288 families attended the Parish Supper. This was held in the Exhibition Hall of the Queen's Building at London Airport, which had been opened by the Queen on 16 December 1955. Nine coaches transported guests from all parts of the parish. Unfortunately the Vicar was unable to be present as he was in hospital after major surgery. He did not return to parish duties until February 1964. The report paid special thanks to the ladies who had contributed to the success of the evening. They had made the supper and transport arrangements, persuaded so many families to attend and been admirable hostesses. Several ladies also helped with the buffet supper at the Stewardship 'Chat, Feed and Dance' of the Youth Fellowship held two evenings later, when Mr Clarke provided the 'Chat'.

Over the course of the next few weeks the visitors met members of the parish in their homes, and gradually the response cards were collected. At the close of the Programme out of the 554 people contacted just over 400 responded with 150 cards outstanding. Over 100 were negative responses and the positive responses promised £4,973.3s.3d. The Stewardship system came into operation on 1 December. Envelopes and money offerings were put in a box and offered at the altar during the services. No collections were taken at any service. Sunday 8 December was a day of thanksgiving for the Programme and the promises made were offered to God at Evensong. The summary report had made it clear that 'there is no point at which the Parish may be said to have "done" Stewardship'.

The Parish Stewardship Supper 1963
(F&J Hare/Photopro Photography)

A Stewardship Committee was appointed, with Mr W. A. Henley as chairman, Mr Starling as Recorder and Mrs Onions as Correspondent. It was to report on the initial campaign with regard to money, and to prepare for the second part of the Programme concerning the use of time and talents.

In connection with fund raising, as mentioned earlier, it was decided that proceeds from the annual fêtes and markets would no longer be needed to boost the parish purse but would be given to charities. This policy was not followed for long as the income from Stewardship and church collections failed to bring in sufficient funds. By 1970 fêtes and markets were once again looked upon as important events for raising funds for the church.

In February 1964 the Stewardship Committee reported that promises had been made amounting to £6,804 and preparations were being made for Phase II of the

---

Printed in Gridiron, *March 1965*

VICAR'S NOTES
Christian Stewardship.

In the recent news-letter two parish groups were unfortunately omitted— apologies to the Players and to the ladies who so nobly help out with refreshments.
Here is the complete list:

Baptismal card distribution	Office work
Book repairing	Players
Guides and Brownies	Road Wardens
Choir	Scouts and Cubs
Church cleaning	Servers
Crèche	Sewing Guild
Flower Guild	Sidesmen
Gardening	Sunday School teachers and pianists
General repairs	
Hospitality—clergy	visiting Visitors
Intercession groups	Working party
Laundry	Refreshment group
Magazine distribution	Transport
Youth Fellowship	C.E-M.S.
Home help	Mothers' Union
	Artists for posters

Requests for help from any of these groups should, in the first instance be given to Mr. Henley, 41 Chamberlain Way, Pinner (PINner 3700), who will then get into touch with the appropriate group. Similarly, if you are in doubt about the identity of the group or section leader and wish to offer your help, again please let Mr. Henley know and he will put you in touch with the group concerned.

Gardens. The grounds of the church are looking very ship-shape and much credit is due to the gardening group who have been working hard since the spring. In addition to the routine work, they have tackled the long grass and shrubs on the North side of the church in readiness for the fete and have levelled off the ground near to the front entrance of the hall. This portion has now been concreted by Mr. Youngs, and we shall now have a little more parking space, particularly when the Blood Transfusion Unit takes over the hall.

Programme. This was launched on 21 October, and by the end of the year members of the congregation had made offers of time and talents. The aim was to cover all the work of the parish that could be covered by laity, and for the church to play a greater part in serving the community at large. A newsletter listed groups working for the parish needing help. It included book repairing, gardening, laundry, refreshments, transport, and the various branches of youth work so important for the future of the church.

By 1967 the scheme was well established and *Gridiron*'s report of the Annual Parochial Church Meeting (APCM) referred to Mr Starling, reporting on Phase I, pleading for more promises and explaining the advantage of covenanting, whereby the Church could recover the Income Tax paid by the donor. As is the case today only the Recorder had knowledge of amounts promised. Mr Henley reporting on Phase II listed the various items of work carried out by different groups, mentioning that the Road Warden Scheme had been revived. This was the scheme that had been set up by the Church of England Men's Society but had been abandoned in 1962. Towards the end of 1968 the special Stewardship Committee was disbanded and the work taken over by other committees.

Starting in 1968, 'Stir-Up' Sunday (so called because of the Collect) was set aside as Stewardship Sunday. The opening words of the Collect for that day, the last Sunday before Advent show it to be an appropriate choice. 'Stir up, we beseech thee, O Lord, the wills of thy faithful people'. Each year Father Bill preached on the subject of Christian Stewardship. In his *Gridiron* letter of December 1977 commenting that Stewardship Sunday was also the celebration of Christ the King, he wrote that the day is 'set aside for personal assessment and for re-examination of our lives and of our commitment to the Christian Faith'.

As far as Phase I was concerned there was cause for anxiety in the late 1960s, for at the APCM in 1969 it was reported that while at the commencement of the Stewardship scheme there were over 300 subscribers the numbers had fallen to just under 200. At the APCM in 1970 the Treasurer, Mr Dale, 'explained the seriousness of the financial situation' in a fair amount of detail. A few details will show just how serious. The *total* income (not just from Stewardship) for 1964 was £6,900 but the estimated income for 1970 was £5,300, a decrease of £1,600. The estimated deficit for 1970 was £1,900. The Stewardship Recorder, Mr Lane, then reported that while 'costs had been constantly increasing the income from Christian Stewardship had continued to fall'. He referred to the booklet which had been sent to parishioners giving reasons for the fall, a loss of around 25 members a year, leading to a fall of about a third of the income.

The year 1971 saw not only a further fall in the number of people in the scheme to 162, and a further reduction in the income from Stewardship to £3,317, but, for those days, a large increase in the Diocesan Quota, from £750 in 1970 to £1,011 in 1971. Details of Stewardship income, together with amounts of Tax Refund, numbers in the scheme and other figures are in Appendix 10.

As has been explained in the previous chapter as well as meeting the running costs of St Lawrence Church, the parish has to contribute to the Diocesan Quota. As always it is impossible to make real comparisons regarding the actual amounts but Appendix 10 gives figures for some years in this period. In 1956 the Quota was £170, had risen to £750 in 1970 and to £4,694 in 1980.

In the Vicar's magazine letter of November 1972 he asked his readers to 'take our Christian discipleship more seriously…give more of our time, our abilities and of our money not indeed only for Church activities but to serve the whole community'. His *Gridiron* letter of December 1975 contained much on the problem of clergy stipends but once again appealed for parishioners to give serious consideration to Christian Stewardship. 'There are some who just use the Church for what they want. Christian Stewardship means being used by God in His church for his purposes'.

The reduction of Income Tax, twice in the 1970s, had an adverse effect on the church's income as it meant a decrease in the amount of Tax that could be recoverable by the church. The rise in clergy stipends and the 'questioning of the role of the Church' in this period didn't help the parish's financial situation. It should be remembered that it was in the 1970s that the parish also took on more expense in buying property to accommodate clergy assistants, first in The Sigers and then Woodlands Avenue. By 1979 the Quota had increased to £3,881 and records show that by 1979 the number of parishioners in the Stewardship scheme was down to 102. A Renewal Programme planned for the autumn of 1979 was postponed, perhaps because Father Bill was to retire after the Dedication Festival.

Gift Day
In Father Bill's first year in Eastcote, and long before Stewardship was introduced to the parish, the September *Leaflet* 1956 announced that the first Gift Day was to take place on Saturday 20 October, the day before Dedication Sunday. It would provide an opportunity for each parishioner to give 'thanks to God in a tangible form by presenting a gift in thanksgiving for many blessings received'. Father Bill 'agreed to sit outside the church all day', (from 9am to 5pm) to receive these gifts personally. These would be offered at the altar the following day. That first year the total was nearly £114. The *Register of Services* Books covering the years of Father Bill's incumbency record the Day of Prayer and Gift Day in connection with Dedication Festival events, from 1956 to 1962. In the years up to 1962 Father Bill continued to 'receive the gifts personally'. In 1958, Silver Jubilee Year, it was suggested that 25 coins or notes, pennies or pounds should be given and the large sum of £330 was donated on this special occasion. In 1959 a respectable sum of around £200 was donated. No records of amounts are available for the following three years.

With the start of Stewardship in the parish in 1963 this 'opportunity' to give 'thanks to God in a tangible form' was regarded as no longer necessary.

However Stewardship was not as successful as had been hoped and, just as from 1970 the fêtes and markets were once again important fund-raising events, so in 1973 a Gift Day was re-introduced on the Saturday of the Dedication celebrations. In his Vicar's magazine letter Father Bill reminded parishioners of the need to 'thank God' for the church building, for the ministry of faithful priests and for the devoted souls who had worshipped and worked in the parish in the past. He wanted the period of celebration to be 'one of real gratitude for the past by showing our concern for the future'. As before Father Bill sat outside the church to receive the gifts. The money donated, nearly £400, was allocated to the re-decoration of the interior of the church. In 1974 the Gift Day produced around £901, which went towards the cost of the Garden of Rest for the interment of ashes.

In 1975 when the problem of clergy stipends had become more pressing the £400 contributed on Gift Day went to the Assistant Clergy Fund. In 1976 the sum contributed was much less, and half of the £200 'was earmarked' for the work of the Church in Mauritius and the remainder for the Garden of Rest.

Since 1974 Dedication Festival services had been celebrated on the Sunday nearest to 21 October. In 1977 and 1978 the Gift Day was also transferred to the Sunday. Envelopes were issued to parishioners prior to the Day. The Secretary's Report at the 1979 APCM indicated that the PCC were considering reverting to a Saturday and a possible wider coverage of the parish for Gift Day envelopes. On 21 October 1979 Father Bill preached what was thought to be his farewell sermon at the Dedication Festival and there was no Gift Day in 1979.

Mothers Union
With Father Bill's emphasis on 'Worship and Fellowship' as being at the heart of the Christian life it is hardly surprising that he maintained a close association with the group that the Revd Rupert Godwin had regarded as 'one of the parish's greatest assets'. From the August committee meeting of 1956, which welcomed her to the parish, Mrs Hitchinson also took a keen interest in the MU.

The regular monthly Thursday Corporate Communion services remained the focus of the Branch. Coffee in the hall after the Corporate Communion services gave an opportunity for 'fellowship'. Most years the annual report records these as 'well attended' by members, but membership numbers seemed to reflect the changes taking place in society as a whole. Membership increased from 44 in 1956 to reach 70 in 1961, and was in the upper 70s from 1962 to 1967. From 1969 numbers began to decline and by 1980 there were only 33 members. In 1973 it seemed possible that the Branch might close when Mrs Worden, the Enrolling Member responsible for returning *Report Forms* to Mary Sumner House, resigned for health reasons. The committee with its secretary Mrs Pearmain, 'worked extremely well', and at the Triennial Meeting of 1976 Mrs Pearmain became Enrolling Member (see Appendix 4 for list of MU Officials).

Branch members continued to attend and to host meetings and services of other MU branches in the Deanery. In 1965 some 170 members attended the Overseas Deanery Meeting in the church hall. The Deanery services had followed a familiar structure for years but the *Gridiron* report of the Deanery Festival Service held at All Saints', Hillingdon in 1969, remarked on a 'service with a difference - for the first time members of the MU reading the lessons and leading the prayers'. Figures for attendance, available only when St Lawrence branch hosted this annual Festival Service, indicate that other branches were also suffering from declining numbers. While in 1962 around 240 attended, in 1970 eight parishes were represented by only 100 members.

The proportion of St Lawrence MU members attending the London Diocesan Festival Service at St Paul's Cathedral was never very great. In 1964 *Gridiron* reported 'that only three members out of 80 attended…when Gerald Collins kindly carried the banner'. However the following year 24 attended. Mr Holmes carried the banner, as Gerald Collins had become a Lay Reader. In 1970 the Bishop of Coventry, The Right Reverend C. K .N. Bardsley, gave the address, and 'banners from 90 branches were carried in procession, adding to the splendour of this thrilling occasion'. In 1977 the report referred to Father Bill who 'looked splendid in his robe', taking his place in his Prebendal stall. Father Bill had been appointed a Prebend in 1976 (see p 152-153).

Father Bill's last Deanery Lady Day Service at St Lawrence's in 1979 was the occasion of the dedication of the last banner of the Branch. As early as 1961 there had been discussions that a new one was needed. In the later 1970s two members, Mrs Olive Bernard and Mrs Edna Pearmain, attended ecclesiastical embroidery classes and, with the contribution of others, both in the design and sewing of the banner, a new banner was finished in 1979. The banner with its representation of the Madonna and Child appliquéd in several colours remains in the church to this day.

As in the past the subjects of the regular monthly Wednesday afternoon meetings were usually associated with the aims and objectives of the MU. Sometimes members gave illustrated talks about their holidays and most years the Vicar spoke on some aspect of the Church. In 1977 'with the dramatic rise in prices', it was necessary to raise the price of tea and biscuits that followed the talk or presentation to 5d (Coffee after Corporate Communion was raised to 8d).

Changes in society in the 1960s meant changes in the community at St Lawrence's, and the MU, like other church organisations, had to move with the times to some extent while still keeping the role of mission and 'handmaid of the Church'. In the May 1964 *Gridiron* there was a suggestion that some MU meetings should be open ones on the grounds that the MU should 'be outward-looking enough to knock down the walls of exclusiveness and, without any lowering of standards, share what we have with others who want it'. By 1969 there were those who agreed with the views of the Lord Bishop of Norwich, quoted in the September *Gridiron* that 'the Mothers' Union should open its membership to all women within the Church whose purpose it is to strengthen, safeguard and promote Christian life'.

The August 1973 parish magazine recorded an item that had made the 6pm BBC TV News on 4 July. The MU Council had 'by an overwhelming majority, decided to admit divorcees and unwed mothers into full membership of the Mothers' Union'.

The Branch maintained the 'worldwide fellowship' of the MU by joining the annual January Wave of Prayer when the Dioceses of London and Sierra Leone were remembered. It kept up its links with the branch at Port Elizabeth until that closed in 1973. In 1978 it was reported that the new link with a branch in Burma 'was flourishing'.

References to a MU Prayer Group meeting once a month were made in AGM reports of 1957, 1958 and 1959 there is nothing in magazines or MU *Report Forms* in later years. However there are references to the Prayer Group and Prayer Groups in several issues of the *Gridiron* from 1960 to 1970, either listed as the intention of the day under *This Month's Prayers,* or in notices of fortnightly meetings at 8pm on Wednesdays. Members met in different houses where, after reading a passage from the Bible, time was spent in meditation and prayer, with special intercession for the sick. It seems likely that the early MU groups may have inspired the later parish ones and bear out Father Bill's concept that church organisations were handmaids of the Church, each contributing to the whole.

Although by the 1960s many social occasions were organised for the 'fellowship' of the parish as a whole, the MU still had their own social events. Branch visits to cathedral cities and stately homes seem largely to have replaced outings to seaside resorts, though there are reports of trips to Brighton (1964, 8/6d return), Bournemouth and Eastbourne. In 1961 there was 'a most enjoyable outing to Guildford's new cathedral', in 1967 it was Winchester and in 1977 Salisbury. Penshurst Place (1966), Syon House (1973) and Waddesdon Manor (1978) were among other destinations. Several years there were theatre visits. One to the Westminster Theatre to see *Give a Dog a Bone* in 1965 cost 7/6d (including the coach fare).

Coffee mornings at members' houses were all part of the fellowship side of the MU, as were fund raising efforts of jumble sales, beetle drives and bring and buy events. Some of the money raised was for MU funds but some was used for gifts to the church. The earlier section about the fabric of the church referred to the red vestments, curtains and altar frontals given in 1964 and 1965, and this was followed by a green set in the next year or so. By 1973 members felt that it had been some time since they had provided anything for the church and gave £10 to open the Carpet Fund, for the new sanctuary carpet.

The tradition of having a summer garden party at the home of one of the MU members seems to have started in 1968 and became a popular annual event. The magazine report on the August 1974 party is typical and includes, 'We had some gentleman guests, friends and several tiny tots, it all helped to make it a Home and Family time'.

MU members were very much involved in parish life, supporting church projects such as Stewardship in 1963, *No Small Change* in 1965, and the Open House Floral Art Display in 1972. They helped at summer fêtes, autumn fairs, and fund raising events for the Memorial window and the St Paul's Cathedral Restoration Fund (1971). During this period they provided refreshments at the Confirmation and Pram services, catered for Harvest suppers and helped to make posies for Mothering Sunday. Members also helped as individuals in the wider community of Eastcote.

Shortly before he left Eastcote Father Bill thanked the Branch for its help to the church, and in view of his forthcoming retirement asked for the same loyal support to the new Vicar and his family.

Church of England Men's Society
While the St Lawrence Branch of the MU was long established when Father Bill became Vicar, the same could not be said of the Church of England Men's Society (CEMS). The first attempt in 1935 'to extend the activities in the parish for the men of the congregation' had little success, for it had barely started when WWII broke out. The second attempt in 1951 was more successful, possibly because of the more settled times following the end of the war, and the fact that there was already a fellowship between several men in the congregation. Members had to agree to the Rule of Life of the Society:

> **'In the power of the Holy Spirit, to pray to God every day, to be a faithful Communicant, and by active witness, fellowship and service to help forward the Kingdom of God'.**

By 1956 the Society was energetic in parish affairs and played its part as a 'handmaid of the Church'. One of the outcomes of the 1956 *Operation Firm Faith* campaign, in which the CEMS was involved, was the setting up of the Road Warden Scheme in 1957. The idea was that each road should have a Warden to act as liaison officers between the church and the people of the parish. The Society made small notice boards for the Wardens to erect in their front gardens to display information about St Lawrence's. The CEMS was also responsible for organising a small church library and a car service to get the elderly to church.

In 1961 the St Lawrence Branch was active in bringing about the Uxbridge Rural Deanery Federation, formed to strengthen the work of the Society for the Church.

At the inaugural meeting two members from St Lawrence's were elected into office, Mr L. J. Lally as President and Mr S. G. Peterken as Chairman. The first Federation Service was held at St Martin's, Ruislip on 8 June. In keeping with the church magazine's intention to be 'a vehicle of propaganda in a largely indifferent society', an article in the September issue of the *Gridiron* appealed for men of the parish to join the St Lawrence branch of the Society in which Communicant Churchmen were bound together by the 'Rule of Life'.

Meetings were held twice a month. The main meeting was on the first Monday of the month, usually with a speaker, and study group evening. The talks covered many aspects of Christian life including Church Unity, Christian Education in State Schools, Racial Discrimination and Atomic Warfare. The study group evenings included study and discussion of the General Secretary's book *Learning and Living*.

At the 1961 Annual Parochial Church Meeting the Vicar referred to the CEMS branch that had just completed ten years, as a small but enthusiastic body of members.

In 1962 Mr Ford gave up the chairmanship of the branch after seven years and was succeeded by Mr Pennington. The new committee 'predicted a change in the Society's approach to matters in which it was involved'. At this time appeals for more volunteers for the Warden scheme were not successful. The scheme was revived in 1965 as part of Phase II of Stewardship.

One new approach to increasing CEMS membership was a 'men only' supper in February 1963, following a celebration of Holy Communion, to which all men in the congregation were invited. An item in the 1963 April *Gridiron* stressed that the main monthly meetings were open to all men who had been confirmed with no lower age limit. At one of these meetings the Vicar gave an illustrated talk on his training at Chichester Theological College. The annual Uxbridge Federation Service was held at St Lawrence's in 1963.

In 1963 a demonstration of the working of the church organ and a visit to Mr Bossanyi's studio in Field End Road to see how he made his stained glass windows for churches, were unusual and interesting evenings for members. Examples of Mr Bossanyi's work may be seen in Uxbridge Station and in York, Canterbury and Washington Cathedrals. As a resident in Field End Road Mr Bossanyi did not take kindly to the development of Farthings Close, referring in a letter to 'the ugly monotonous row of houses revealed by the wintry bare trees'. An evening visit to Chenies, when the Rector showed them round St Michael's Church was another interesting outing. The Bedford Chapel, contains what the late architectural historian Nickolas Pevsner, described as 'a rich a store of funeral monuments as any in the country'.

Later in 1963 the Branch was well represented at the annual National Conference at Durham in September, where among the many services and discussions of 'the stimulating weekend', the Archbishop of York's sermon had a message with relevance in the twenty-first century. He made three points, 'the Christian church is now a minority movement in the world, it is disunited, and it has its priorities wrong'.

This message was not an encouraging one. Society was changing and in spite of the 'new approach' of the St Lawrence Branch, men were not attracted to join the CEMS. By the end of 1964, what had been an active society was in decline, and the Parochial Church Council learned that out of a membership of 25, only six attended the meetings. Although an account of the Society's activities in March 1965 recorded a 'most interesting and informative visit to the London College of Divinity in Northwood (now the London School of Theology), and that the branch supported the Uxbridge and Ealing federations in providing teams to lead services of divine worship at some wards at St Bernard's Hospital, Southall, the Branch closed that year.

The last reference to the CEMS, in the last issue of the *Gridiron*, was to a series of lectures at Acton Town Hall in 1971 being 'well attended by clergy and people of the Willesden jurisdiction'. Whether any were from St Lawrence's was not recorded.

St Lawrence Women's Guild
The idea that a women's fellowship should be formed is likely to have been connected with the 1965 *No Small Change* project, in which St Lawrence Church was a pioneer parish. Obviously the Vicar was involved in preliminary moves for this project at a diocesan level before 1965 and in January 1965 the possibility of a women's group was discussed at a Parochial Church Council meeting. Reporting on the inaugural meeting in the June *Gridiron,* the Vicar was pleased so many ladies had met to discuss the formation of an organisation 'that will meet the need for which it was formed and be a source of strength to the church and the community'. Whether it was the exclusive attitude of the MU at that time, or that many women were not free during the day, the fact remains that there were those in the parish for whom the MU did not cater.

After some discussion the name was chosen, an annual subscription of 5/- was agreed and a programme of meetings, to be held at 8pm on the second Wednesday of each month, was planned. What Father Bill regarded as guiding principles of the Church 'Worship and Fellowship', must have influenced the Women's Guild when drawing up the ethos of this new group. The aims were 'to worship, to give service and help, and to have fellowship'. Guild reports in the *Gridiron* show that members carried out these aims, and were a lively and active organisation and true 'handmaids' of the Church.

From September 1966 an annual Rededication Service was held and in November 1966 the first of regular three-monthly Corporate Communion services took place. Members also attended the anniversary services of the Women's Fellowship groups at the Eastcote Methodist and St Andrew's United Reformed churches. These links were all part of the ecumenical moves being made with local churches.

The monthly meetings covered an amazingly wide range of interests, were often of a topical nature, and frequently open to those outside the Guild. The subjects chosen show that through the years the committees worked hard to achieve full and varied programmes with speakers from many organisations, as well as calling on their own members, and their relations, for presentations. The subjects of visiting speakers included, the Work of a Probation Officer, Marriage Guidance, the Samaritans, Crime Prevention, the National Trust, the Church Army, the British and Foreign Bible Society, the Citizen's Advice Bureau and the Ramblers' Association. Topical films were shown such as *Father Borelli's Work with Children in Naples* (1966) and *Cathy Come Home* (1967). There were evenings when visitors or members showed slides of holidays or hobbies and included ones of Oberammergau shown by Vera Herbert and Marjorie Jones in 1979 (the year before the ten-yearly Passion Play), and of Orchids by Mollie Pottinger.

Sometimes subjects of meetings were of local interest, for example, Old Eastcote (Mr Morris), St Edmund the King, Northwood Hills (Father Ames), Community Relations (member of the American Base), Northwick Park Hospital which opened in 1969 (Chief Nursing Officer), Whitby Dene Home which opened in 1957 (Matron) and St Vincent's Home. Other evenings were more practical; one in 1974 was a Tupperware party, a means of purchasing this brand of plastic containers. More women were interested in home dressmaking than in this century, and at one of the several dressmaking demonstration evenings, thanks were expressed to Mr and Mrs Read of the local fabric shop (alas no longer in Eastcote) for giving lengths of material and sewing books to raffle. Other demonstrations were on make-up, flower arranging, cake icing (Barbara Plummer), yoga (Iris Castles), making the most of your camera (Joe Orme), wine making (Ted Edwards), car maintenance (Joe Orme and Ted Edwards) and one, of a different nature, on decimalisation which came into effect on 15 February 1971 (Geoffrey Sykes). These last are examples of family and friends getting involved and of the 'co-ordination' Father Bill desired between St Lawrence members.

There are many examples of the Guild carrying out the aim 'to give service and help', both within the church community and beyond. Like other church organisations they ran stalls at fêtes and fairs and held jumble sales and beetle drives to raise money for church projects and other charities. One of the charities was the local hostel in Queen's Walk, which cared for unmarried mothers and their children. For a few years in the 1970s there appears to have been a close link with the hostel. Guild members took birthday presents and cards, and Easter eggs to the children. They also arranged outings 'as a family unit' to pantomimes at Watford (1971), Ickenham (1973), and in 1972 the children attended a performance of the St Lawrence Players' production of *Salad Days*. More recently Queen's Walk was connected with *Welcare*, a Deanery association, caring for unmarried mothers and babies. Until 2008 St Lawrence Church supported Queen's Walk by sending an annual donation, mainly the proceeds from the inter-house Christmas post.

Another form of service and help was that of providing refreshments at the Pram Services of the 1970s, at Confirmations and other church events, and by making posies for the children at Mothering Sunday services. One tangible gift of the Guild to the church was a silver and gilt chalice and paten. A designer's set made by Mr J. Johns in the last year of his apprenticeship as a silversmith it was consecrated by the Vicar at the July 1970 Corporate Communion.

Through these years Father Bill, and Fathers Pearce and Chiplin, gave talks and led question times and discussions on Church matters, covering topics such as the Church's Year, Baptism and Confirmation, the history of Holy Thursday and church vestments. One lively discussion evening was about the new rites of Holy Communion in 1972.

The aim of the Women's Guild 'to have fellowship' was experienced not only through worship and service as already described but also in specific social activities.

There were simple and relaxed coffee evenings in members' homes and social evenings in the church hall to which family and friends were invited. The magazine has accounts of the Guild co-hosting a cheese and wine party with the Magazine Distributors (1972) and a social with the Youth Fellowship (1974), organising a jumble sale with the Mothers' Union (1975), and records 'a delightful Nativity Play given by the Tuesday evening Brownies' at the Christmas meeting of 1978.

From the first year committees organised a variety of outings. Some visits were the result of interest in recently opened buildings, Brunel University (1968) and Northwick Park Hospital. Trips included those to *Alibi for a Judge* at the Savoy Theatre, with Andrew Cruikshank (1966), a Passion play at The Empire Pool Wembley (1967), and several to the theatre at Windsor. Windsor was also the destination of a number of summer outings in the late 1970s. The one in1976 seems to have been particularly successful family affair. The party travelled by coach past London Airport and 'were thrilled to see a Jumbo Jet hover like an enormous bird before landing…on to Runnymede…a Salters river launch to Windsor…light refreshments on board…some walked by the river… some visited the apartments in the Castle…seats in the choir stalls for Evensong…a perfect finish to a day to remember'. The magazine account ended with thanks to

Enid Lumsden and the Guild committee for the idea and Pat Beeston 'for a flawless organisation'.

The Guild fulfilled 'the need for which it was formed' and was 'a source of strength to the church and the community'. Some women became members of the St Lawrence community through the Guild. The reports of the AGMs in the magazines did not always name the committees but the following appear to have been chairpersons Mrs McKellar, Mrs Joan Edwards, Mrs Enid Lumsden and Mrs Mollie Wilson.

The Society of Mary
A small group with more limited appeal than either the Mothers' Union or the Women's Guild was the Society of Mary. A Ward of this Society was formed in 1965 and included members from other churches in the area. The Society of Mary was founded just before World War II, but had its origins in the 19th century Oxford Movement, which emphasised the Catholic heritage of the Church of England, and had a growing devotion to Our Lady. Members pledged themselves to follow a simple devotional rule. Monthly devotional meetings were held in church. In April 1969 a few members of the St Lawrence Ward joined a coach party from the parish of Holy Innocents, Hammersmith, for a pilgrimage to Walsingham.

The Needlework Guild
For many years, this church group, which had done so much to raise money for the Permanent Hall Fund, continued to meet most Thursday afternoons in the Committee Room. Members enjoyed a friendly chat over a cup of tea and made items for the fêtes and bazaars. They were grateful for 'gifts of suitable materials for aprons and peg bags', and members also held tea parties and coffee mornings in aid of funds to purchase other material. For at least a couple of years in the early 1960s, members undertook to dress dolls for Christmas presents. Although there was some response to appeals for new members with fresh ideas, by the mid 1970s the Guild no longer fulfilled a useful purpose and reluctantly decided to close down. In his May 1975 *Gridiron* letter Father Bill thanked the group who had 'met weekly for the benefit of the parish'. Times had changed and the Guild had been both a good handmaid of the Church and one of fellowship.

The Flower Guild
The Flower Guild, unlike the Needlework Guild, has remained a handmaid of the Church right up to the present day. The Vicar's Remarks at the Annual Parochial Church Meetings refer to the Guild's contribution to the parish. The members supplied and arranged flowers for the vases on the two altars and at the statues of Our Lady and St Lawrence. Members of the congregation could supply flowers or give money for flowers for members to arrange. The cost of flowers for festivals was met from donations from the congregation, and for several years in the 1970s, from the proceeds of jumble sales organised by the Guild. A particularly successful jumble sale was that in 1978 when the profit was £78.

Items in the magazines thanking present members and appealing for new volunteers who were asked to contact Mrs Hitchinson suggest that she was very involved with the Guild, at least up to 1972. Later Mrs Mary Baker and Mrs Beryl Orme were mentioned.

As with other organisations the Guild played its part at fêtes and fairs, providing goods and manning stalls. Among other new features at the 1970 summer fête was the room in the bungalow set aside for floral arrangements in which some members were involved. This successful venture may have influenced the decision that St Lawrence Church would take part in the Open House and Floral Display held in the church from 29 June to 2 July 1972, referred to earlier. In 1973 the Guild, together with some members from the Women's Guild and the Mothers' Union, 'made 150-200lb' of marmalade for sale at the summer fête. This marmalade contribution continued for a number of years.

St Lawrence Players
The St Lawrence Players was by 1956 a well-established and popular organisation of the church. The annual pattern of three major productions continued, often presenting three performances of a production. For many years there was also a revue on the evening of the annual fête. In 1971 the Players' Theatre Club was formed to get round the new Greater London regulations requiring a licence for stage performances in the Church Hall. Each member, for an annual subscription of 10p, received a membership card, notification of productions and tickets by post, and was able to take up to four guests.

Murder on the Nile 1959

The Players' New Year production was usually a pantomime and the practice of inviting children to the Saturday matinee performance was maintained for a number of years. There are magazine reports of boys and staff from St Michael's School, boys from St Vincent's Hospital and children from Queen's Walk Hostel enjoying these outings. Naturally children from St Lawrence Junior Church were also in the audience. Another tradition during these years was to donate the profits from the New Year production to children, often it was to the Junior Church, but in 1965 and 1966 it was the Lady Hoare Thalidomide Appeal (for children born with malformation of limbs as a result of mothers taking the drug in early pregnancy).

In 1964 the comedy *The Geese are Getting Fat* was regarded as a 'break from the usual pantomime', and appreciated by both audience and Players. There were other years when the pantomime tradition was broken. From 1967 to 1969 there were 'a farce', and two amusing plays. In 1970 the announcement of *A Revue for You* was described as 'a sort of musical warm-up to the rest of the year'. It was back to pantomime with *Sinbad the Sailor* in 1971, followed by the musicals *Salad Days,* 'an entirely successful evening' (1972), *Oliver*, 'Congratulations...on a big undertaking...sets were first class' (1973) and *Where's Charlie,* 'a light pleasant show' (1974). Reviews of the rest of the 1970s new year productions were on much the same lines, a typical one being that on *Babes in the Wood* (1975). 'It was fun right from the start...we hissed, we booed, we shouted and sang...the set and costumes were well up to the Players' standard'.

Magazine reviews of the spring and autumn productions often referred to the choice or suitability of the play, and to audience reaction. Examples of comments are those on *Man Alive*, 'an hilarious and enjoyable romp' (1965), *This Happy Breed*, 'an ambitious choice' (1967), *The House by the Lake*, 'they did what they could with what was quite poor material (1972), *Off the Hook,* 'a very enjoyable and worthwhile production' (1976), and *The Same Sky,* 'credit is due to all the Players for this thought-provoking play' (1979).

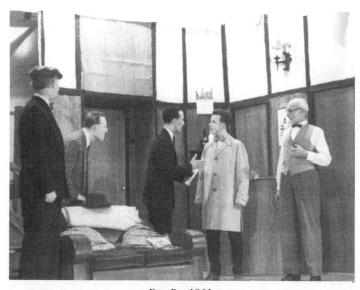

Dry Rot 1961

The tradition of the Players, which had started in Father Godwin's time, of presenting a revue on the evening of the fête continued for some years, and they also presented a revue after the Christmas Market in 1956, Father Bill's first in Eastcote. From 1957 to 1963 there was the 'normal' form of revue after the fête, but in 1962 the programme was a little different. Other church organisations (Scouts, Cubs, Guides, Youth Fellowship, Junior Guild and the M.U.) showed their talents before the interval, and the second half of the programme was presented by the Players 'in their usual competent and relaxed manner'. The contribution of the Players in entertainment that followed the churches of Eastcote fêtes of 1964 and 1965 has been referred to in an earlier section.

The Players, like other church organisations, did their bit to raise money for Father Godwin's Memorial, producing a Summer Show in July 1966 and an Old Time Musical in 1968. The cast and 'back-room boys' were congratulated on the Summer Show which they had presented on Friday 12 and Saturday 13 June 1970, the week after the fête, but there had been a disappointing number of people in the audiences. This seems to have been the end of these annual revues. Perhaps these popular evenings had run their course? In any case three major productions a year is a commendable output for this organisation. One must also take into account the fact that the Fellowship Committee of the Parochial Church Council (PCC) were organising a variety of new events in the 1970s.

Naturally reviews commented on the success of the directors of the pantomimes, plays and revues, and on actors' performances and their ability to portray different types of characters. Hylda Darby directed the majority of the productions but from 1972 Leigh Smith, Ann Morgans, and Anton and Jean Jungreuthmayer were each responsible for more than one production (see Appendix 12 for full list of productions).

Magazine reviews gave praise to several people who were members of the Players for a relatively short time. However it seems right to name those members who were mentioned for outstanding performances over many years, and to assume it will not give offence to others who also contributed to the enjoyment and popularity of the productions. Those whose names occurred frequently were, in no particular order, Jean Jungreuthmayer, Leigh Smith, Beryl Orders, Marjorie Howard, Barbara Williams, and Wally Durston.

Reviews often had comments on the scenery, and when relevant on costumes, and here too the Players had the reputation of a high standard. A few examples indicate the importance of this aspect of the productions. 'Wally Durston is to be congratulated on his reproduction of the drawing-room of the Fordyce family' in *Goodnight Mrs Puffin* (1967). 'The design of the Great Hall of the Convent of our Lady of Rheims was superb' in *Bonaventure* (1971). In 1973 the sets in *Oliver* 'were first class…the Players have some excellent designers…and talented members who built [it]'. Several reviews commended the designers on dealing with the limitations of the small stage, and thanked the 'back-room boys and girls' for their contribution to a production.

By strange coincidence the last production of the St Lawrence Players, while Father Bill was Vicar, was that of the pantomime *Alice in Wonderland* (1980), scenes from which had been performed by the St Lawrence Juvenile (*sic*) Players way back in 1945, when the seeds of the church's dramatic society had been sown. The subsequent history of the St Lawrence Players, both in Father Godwin's time and in Father Bill's time in Eastcote, indicate a flourishing organisation that contributed to fellowship within the community life of St Lawrence Church.

4th Eastcote Scout Group

With the arrival of the first Curate at St Lawrence Church in 1958, The Revd Peter Goodridge, (who in his youth had been an active Scout) he was appointed as the Group's first Chaplain. He attended many of the activities and camps and his involvement with the youth of the Parish no doubt resulted in the slow but steady growth in numbers in all Sections. Father Peter subsequently held various appointments and was finally appointed Canon Librarian of Truro Cathedral. In 1962 a Senior Scout Troop was re-established to cater for the Scouts on reaching the age of 16 years. Probably the most notable activity in the mid-sixties was the Senior Scout Hiking Expedition in Germany, from the Rhine to the Moselle over the Hunsruck.

In order to have the money for the purchase of equipment, various fund-raising activities were arranged in the sixties and seventies, including Whist Drives, Jumble Sales, Car Washes and Waste Paper Collections. Even some of the annual Group Parties held in January each year became very successful Gang Shows to add funds to the coffers. With the introduction of Venture Scouting to the movement, our own Venture Scout Unit (VSU) was established to carry on the activities of the Senior Scout Troop and this Unit was called the 'Gryphonhurst Unit', named after the old mansion situated on the corner of Field End Road and Bridle Road. The Unit flourished for many years and was noted for its production of the Group's news magazine *Interface*, produced twice a year for almost ten years. During this period the Group was under the leadership of Mr C K Lawrence (Lawrie), who was then followed by Gerald Collins until his appointment as District Commissioner in 1976.

Brownies and Guides

During the Revd Godwin's incumbency attempts to establish Brownie and Guide units with children in families belonging to St Lawrence Church had not been very successful, owing to the lack of people willing to be leaders. The Guide Company had closed in 1952 and the Brownie Pack in 1953. Like Father Godwin, Father Bill wanted leaders of the uniformed organisations to be communicant members of the Church, and between 1956 and 1980 there were times when urgent appeals were made for suitable people to lead, or to help with, either a Brownie Pack or Guide Company. On each occasion the group managed to keep going. By 1980 there were two Brownie Packs and two Guide Companies.

Over the years a number of people helped as Brown Owls, Guide Captains, and in other capacities, and without them the groups would not have survived. Nevertheless it is right to mention a few by name for they were each involved with the Guide movement for many years. Mrs Hazel Martin and Mrs Harris Davies ran the 6th Eastcote Brownie Pack for ten years (1959-1969). Mrs Doreen Murrell then became Brown Owl for a number of years before becoming District Commissioner in 1979. Mrs Yvonne Horsfall ran the 3rd Pack from 1965, when the Pack started, until 1975. In 1970 she offered to help run 6th Guide Company 'for the time being' but remained as leader until 1978. A number of older Guides and mothers of girls either in the Brownies or the Guides started by helping in some way and then later became leaders.

Brownies
It must have been shortly after Father Bill came to the parish that the 6th Eastcote Brownie Pack, which met on Thursdays, was re-started. The December 1959 *Leaflet* referred to its 'third birthday' and reported that 58 Brownies had been trained, six Brownies had gained First Class badges, the Pack had the maximum of 24, and there was a waiting list of 12. By 1959 the 6th Eastcote Guide Company, which met on Fridays, had been running for two years. The number of young girls wanting to become Brownies led in 1965, to the opening of a second group, the 3rd Eastcote Pack, which met on Tuesdays. Although there was obviously a need for a second Guide Company, if the enthusiasm of the girls was to be maintained beyond the Brownie age, the problem of Guide leaders meant it was not until 1979 that the 3rd Eastcote Company was started.

An account, by a Sixer (leader of each group of six girls) of the 6th Pack, in the *Gridiron* (February 1962) seems typical of a weekly Brownie evening. 'We start off by having our Brownie ring, in which we sing our Six rhymes' (one for each Six)…we play games and have work groups in which we learn knots, flags and semaphore, and many other things'. One imagines that some evenings were spent making the most of church grounds with its 'on site' facility for 'tracking' and other outdoor activities, such as inter-six sports evenings.

The Packs visited the local woods, Ruislip, Bayhurst and Mad Bess, for rambles, tracking, tree identification etc., and at least once for Brownies Revels with other packs from the Eastcote District (1971). In 1976 the 3rd Pack had a particularly enjoyable evening, when several mums took the girls, suitably dressed, to Bayhurst Woods for 'a tramp's supper'. Two 'dutiful husbands' prepared a sausage and beans supper cooked on a barbecue, and organised tracking for the brownies.

Outings, some by one of the packs, some by each in turn, and some the two together, included local ones to Rayner's Lane ambulance station, Pinner fire station, Ruislip police station and the Civic Centre at Uxbridge. In 1969 there was a joint visit to the Ice Show at Wembley. One London day outing was a visit to Girl Guide Headquarters, Changing the Guard at Buckingham Palace, a picnic in St James's Park, and a tour of the Royal Mews. Another London trip was to St Paul's Cathedral and the Mansion House.

Brownie Revels have been referred to but there were other occasions when St Lawrence Church Brownies met those from other packs. One annual event was the Eastcote District Inter-pack Shield. Six Brownies represented each pack, a Sixer, a Seconder, two Brownies with 2nd Class badges, and two without. The 6th Pack won the shield in 1960, 1963 and 1968. In 1975 both church packs took part in the District Swimming Gala at Highgrove Swimming Pool (opened in 1964), with the 6th winning the Gala Shield and the 3rd coming a close second. In 1977 the 3rd came first and the 6th second.

The various visits and outings organised by the leaders all seemed to have been enjoyed by the Brownies. One popular event that was particularly difficult to arrange was a pack holiday, which among other things required finding suitable accommodation such as a church hall. The 6th had their first pack holiday at Felpham, Sussex in 1974. It was followed by ones at Cowley, near Cheltenham (1975), at Penn, Buckinghamshire (1976) and at Bognor (1978). The 3rd Pack had their first pack holiday at Flackwell Heath, Buckinghamshire in 1978.

As with several other church organisations the Brownies held fund-raising events, some large and some small. A few were for their own funds such as in 1974 when the 6th Pack made almost £100 profit from a Jumble Sale, to buy equipment for their first pack holiday. In 1978 the 3rd Pack raised money for their first pack holiday by collecting 4,000 wine bottles for which a bottle recovery company paid 1p per bottle. 1974 was special, for it was the Diamond Jubilee year of the Brownie Guides. The 6th having had a very successful jumble sale decided to do 60 'good turns' by helping people to shop or clean shoes etc., while the 3rd Pack marked the special year by having a sponsored 'sing in', which raised more than £50, with which they bought groceries for distribution 'to old folk in the district'. In almost opposite types of event the 3rd Pack raised £62 from a sponsored 'silence' in March 1976 while the 6th raised £62 from a sponsored 'song-a-minute'. The girls sang 38 songs and Christmas carols in half an hour.

Both packs contributed to the church community by making gifts and helping to man stalls at fêtes and bazaars. In 1974 and 1975 individual brownies donated some of their pocket money to the Vicar's Carpet Fund.

In June 1974, together with the 6th Guide Company, the two packs put on a 'Guides Own' revue which included a Brownie 'Wombles' gymnastic display. Many years the Brownies and Guides joined together, usually with other packs and companies in the Eastcote District, to celebrate Thinking Day, 22 February, the Founder's birthday. The day started with a service in one of the Eastcote churches and was usually followed by a party or entertainment. From early 1968 a combined Church Parade of the uniformed organisations attached to St Lawrence's was part of the 9am service on the second Sunday of the month.

Girl Guides
The 6th Eastcote Guides started in 1957, only a few months after the 6th Brownies. The *Gridiron* of April 1967 congratulated the Girl Guide Company on reaching its tenth birthday in March, adding 'being Lent, its celebrations were subdued, but adequate'. Although people seemed willing and able to help with the running of the company, the problem of appointing a suitable person as Captain was another matter, and there were periods when the future of the Company was in doubt. In 1963 there were several months when the company was without a Captain and there was concern on behalf of the brownies, keen to move up into guides, and on the guides themselves. From 1964 to 1968 a number of people took on the role of Captain, but in 1969 the Company was again without a Captain. In April 1970 Mrs Yvonne Horsfall came forward and by 1974, under her leadership, the Company was flourishing and had 39 Guides. After she left at Easter 1978 there was a further difficult period, particularly as there were a large number of girls of guide age eager to 'go up into Guides'. By January 1979 the crisis had passed and with 'suitable' help a second company, the 3rd Eastcote, was started, with Mrs Enid Lumsden as Captain, having helped with 3rd Brownies since 1976.

Typical Guide evenings followed a similar pattern to that of the Brownies, but more mature as to content, for example learning more difficult knots and Morse code rather than semaphore. Time was spent on working towards a variety of proficiency badges. The June 1971 *Gridiron* reported Karen Lloyd as the first member of the Company to become a Queen's Guide. Later in the year Barbara Horsfall was congratulated on being 'the third this year' to attain this high award. PCC *Minutes* reveal Claire Shepherdly to have been the 'second' Queen's Guide.

The companies attached to St Lawrence Church had contact with other guides apart from Thinking Day celebrations. Like the brownies representative patrols took part in various District Competitions and Swimming Galas. The 6th won the District Shield on at least one occasion in 1965. In 1965 the 6th Company was also chosen to take part in a pageant presented by the North-West Division at Field End School.

A few guides from St Andrew's Church joined those of the 6th Company for the Company's first camp at Plumpton Green, Sussex in 1964. Other 6th Company camps were on the Duke of Norfolk's Estate (1965), Brightwell Park, Watlington (1972) and in the grounds of Woodrow House, near Amersham (1979). In 1969 when the guides were unable to camp as a Company, because they had no Captain, 23 guides 'through the generosity of other Guiders' joined local companies who camped at East Grinstead and Horsham. In 1971 six guides took part in a District weekend camp at Seer Green, Buckinghamshire. In 1979 five guides from the newly formed 3rd Company enjoyed their first experience of camp at Seer Green at a similar District camp.

Jumble sales seem to have been the favourite and successful events held by the guides as a means of raising money, most of which was spent on camp equipment. In 1964 and 1965 the 6th made profits of £41.2s.4d and £50. Later ones of 1972, 1975 and 1979 were equally rewarding, with the one in April 1979 raising nearly £100. In 1979 the 3rd raised nearly £80 at their first Jumble Sale. Of this about £40 went to the purchase of a flag for the newly formed company, and some of the rest was spent on camp equipment. It is likely that many of the wider Eastcote community, as well as those of the congregation, supported these financial efforts, for jumble sales were a popular means of fund-raising in the 1960s and 1970s.

In these years the Guides were involved in the 'fellowship' of the St Lawrence community in several ways. They helped at church social and fund-raising events, manning stalls and running sideshows. In 1971 the 6th are recorded as 'busy knitting and sewing for the Christmas Market' and in 1975 gave displays of Scottish and country dancing at the church fête. Like the Youth Club members and other young people of the St Lawrence community, the guides appreciated the discos that the Fellowship Committee organised.

Badminton Club
The possibility of a church badminton club was discussed at the June 1956 meeting of the Parochial Church Council and started that autumn. Those interested had to apply for membership to the Vicar for, in keeping with other church organisations, membership was restricted to regular communicant members of the Church. Originally the club met on Wednesdays, juniors 14-18 from 6.30pm-8.00pm (annual subscription 10s.6d) and seniors over 18 from 8.15pm-10.30pm (annual subscription £1.1s.0d.). Racquets were provided. From January 1961, if not earlier, the evening had changed to its present evening, Friday, and had only one senior section.

The only references to matches against other clubs are in two issues of the *Gridiron* in 1963, losing to the Radio Chemical Centre, Amersham in February, and drawing with All Saints' Church, Hillingdon in March.

The *Minutes* of PCC Meetings in 1966 record the Vicar's concern that many members of the St Lawrence Church Badminton Club were not Church members. Obviously the decision was made to close the Club, for in the *Gridiron* of December the Vicar's Notes included an invitation to 'regular weekly Communicants…to start 'a new Badminton Club for Church members'. The April 1967 *Gridiron* reported that there was a waiting list to join the Club. The

Vicar was 'willing to issue application forms on the understanding that they will be considered'.

Perhaps it was the closure of the Youth Fellowship in July 1967 that led the Vicar to ask members of the badminton club if they would give instruction to younger members of the Church. Seven people were willing to help and by early 1968 a Junior Badminton Club was meeting on Saturday mornings. By 1969 the younger members met on Sunday afternoons, and there was a waiting list. The Junior Badminton Club was on the list of Church organisations until May 1974.

In 1969, in contrast to the junior section, the Senior Club needed members. There were periodic appeals for new Church members. In October 1974 the appeal offered 'recreation and fellowship to all ages and both sexes - one of the few parish organisations to do so'. In February 1977 the appeal had a different approach. 'Why not do a good turn to your heart and lungs? Switch off the "box" and play badminton'.

In spite of times when membership has been low the St Lawrence Badminton Club has kept going.

MATTERS OF INTEREST
The London Borough of Hillingdon
St Lawrence Church does not exist in isolation from the wider community. It is situated geographically in a particular area of local government. When Mr Len Lally, a longstanding member of St Lawrence Church, was appointed Chairman of the Ruislip-Northwood Urban District Council in 1961, he wrote a long article in *Gridiron* about local government referring to the fact that decisions of the Council affected all residents. Mr Lally was of the opinion that the Church and the District Council were 'working towards the same end - a better and fuller life for us all'. He gave information about building applications, the proposed new swimming bath on the Highgrove site, food hygiene, vandalism and the flowerbeds in the shopping area of Field End Road.

The report of a Royal Commission had been published in 1960 and recommended replacing the Middlesex County Council and the London County Council with the newly constituted Greater London Council (GLC) of 52 Boroughs. Each borough would be large enough to be responsible for practically all the local services and make local government more 'local'. Needless to say much discussion took place and alterations made before the GLC came into being on 1 April 1965. The number of boroughs was reduced to 32 and the minimum population of each increased from 100,000 to 200,000. Ruislip-Northwood was combined with Uxbridge, Yiewsley-and-West Drayton, and Hayes-and-Harlington to form the London Borough of Hillingdon. A Civic Service was held at St Lawrence Church to mark the occasion. In 1979, his last year as Vicar, Father Bill served as Chaplain to Len Lally when he was Mayor of Hillingdon.

There have been more recent occasions when the Vicar of St Lawrence Church has served as Chaplain to the Mayor of Hillingdon. A notice about the new Hillingdon Hospital appeared in the early 1964 issues of the *Gridiron*.

The Royal British Legion Eastcote Branch (created 'Royal' 1971)
For many years there was a close connection between the Eastcote Branch of the British Legion and St Lawrence Church. In 1975 the Branch's old Standard was 'laid up' in the church and hangs above the choir stalls on the south side of the chancel. New colours were dedicated in June 1976. In November 1976 Father Bill and the Revd Huw Chiplin attended the Eastcote Headquarters of the British Legion to receive the Branch's gift to the church of a lectern edition of the Jerusalem Bible (first published 1966). When he retired in 1979 Father Bill, who had been chaplain to the Men's section for 23 years and also a vice-president, was presented with a camera in appreciation of his services to the Branch.

The Church of St Edmund the King
With the opening of Northwood Hills Station in 1934 and local housing development in that area St Edmund the King was established as a mission church, in rather the same way that St Lawrence's had started in 1920. St Lawrence Church was one of three parishes from which this new parish was created and lost part of the Joel Street area in the early 1950s (see map on page 18). It was not until October 1964 that the permanent church at Northwood Hills was consecrated. Father Bill and the PCC agreed to 'show our joy with them by presenting them with a Processional Cross'.

Synodical Government and St Lawrence's
The challenges and changes to the Church in the 1960s led to a number of changes in Church administration. One of these was the introduction of Synodical Government. Father Bill described it in his March 1970 letter, as a system of government to include parochial clergy and lay-people at all levels of government in the Church from parochial level through to the General Synod of the Church of England.

At a PCC meeting in 1969 Father Bill explained that this would give the PCC more power than in the past. He was of the opinion this was what had been done in the Eastcote parish for a number of years, though in many parishes most of the decisions had been that of the parish priest. It may be recalled that he started a Missionary Committee in 1957. With the challenges of the 1960s there were several re-organisations of PCC sub-committees, reflecting Father Bill's view that the laity should take greater part in decisions and activities of the St Lawrence Church community. From 1967 the committees were Finance and General Purposes, Education (a new one), Stewardship, Hall and Planning and Fellowship (replacing Entertainments). In addition to PCC members, representatives of all the organisations were co-opted to some sub-committees in the hope that the organisations would strengthen the evangelistic mission of the Church and become more concerned with the affairs of the parish.

One change at St Lawrence's, under the new system of Synodical Government, was the election by the PCC of Mr C. Meese Tyrer as the first Electoral Roll Officer in 1970. The Roll would be revised annually as in the past but in 1972 and every sixth year thereafter a new Roll would be compiled. Messrs Bedford, Darby and Orme were elected to the new Deanery Synod at the 1970 Annual Parochial Church Meeting (APCM).

The Episcopal Areas of the Diocese of London

The year 1970 saw another change as regards organisation, for a new Episcopal area, Edmonton, was created in the London Diocese, making five areas or jurisdictions. These are each under the care and oversight of a bishop, that of the Bishop of London and the Area Bishops of Stepney, Kensington, Willesden and Edmonton. From the days of the Mission Church in 1920 St Lawrence Church had been under the authority of the Bishop of Kensington. In October 1970 the parish was transferred to the new Archdeaconry of Northolt, in the jurisdiction of the Bishop of Willesden (at that time Graham Leonard). A sixth area bishop in the Diocese of London, the Bishop of Fulham, had at that time the pastoral care for Church of England parishes on the Continent of Europe. Since the Ordination of women to the priesthood in 1994 he has pastoral care of parishes operating under the London Plan. This ensures that those unable to accept other bishops because of their involvement with the ordination of women have proper episcopal care.

A further change took place in the Diocese of London in July 1979, when Gerald (Ellison) the Lord Bishop of London legally delegated a range of duties and powers to the area Bishops, thus giving them autonomy over various activities within the Diocese. Each area was to have its own Synod. At that time Willesden was to have a Synod of 105 clergy and 105 laity. A service was held in St Paul's Cathedral on Sunday 27 January 1980 to formally inaugurate the re-organisation of the Diocese of London into Areas.

Bishops' Visitations

The Bishop of Kensington's Visitation to the Deanery of Hillingdon in 1967 involved the PCC in a special meeting in September 1967, in preparation for a Deanery Conference, which was held in the evening of 13 October. In connection with the challenges to the Church in the 1960s, the Bishop had submitted a series of seven questions for the PCC to discuss regarding the attitude of people who felt no need to 'participate in the corporate worship of the Church'. The conclusions of the individual PCC findings were summarised and discussed at the Deanery Conference, which ended a busy day for the Bishop in the Hillingdon Deanery. The reports of the findings of the Conference were the subject of leading articles in the *Gridiron* for several months in 1968. This seems to be a further example of Father Bill's attempts to get the ordinary parishioner involved in the Mission of the Church as he asked for, and received, comments on the articles.

One Bishop's Visitation that was possibly of more importance for the parish as a whole was the one in 1972. The Vicar's Letter of September 1972 gave a full account of the Bishop of Willesden's Official Visitation of the parish on 20 July. The 'of' is significant, for in addition to church services at 10.30am and 6 and 8pm he travelled round the parish. Visits were made to the old people's home at Whitby Dene, where he administered the Blessed Sacrament, to the Retreat Home, and also to St Vincent's Hospital. He also administered the Blessed Sacrament in private homes and visited housebound parishioners, some of whom had known the original church before 1933. The Bishop was 'particularly impressed with the great number of youngsters who attended the services'.

Bishops of London have attended St Lawrence Church on some of the major occasions in its history. The Right Revd Arthur Foley Winnington-Ingram consecrated the church in October 1933. The Right Revd John William Charles Wand dedicated the church hall in 1955, when he also confirmed 43 candidates. The Right Revd Henry Colville Montgomery Campbell attended the Jubilee Celebrations in October 1958 and confirmed 93 candidates in St Lawrence Church in 1959.

Licensed Lay Reader (today known as Reader)
A 1964 magazine article by Mr J. J. Masey, Licensed Lay Reader at St Lawrence Church, explained that the ministry of laity went back to the early days of the Christian Church. As the Church grew and its organisation developed, those who assisted the threefold ministry of bishops, priests and deacons were admitted to minor orders of which the chief was the Reader. There had been periods when the office had fallen into abeyance but had been revived in 1866. A form of admission was drawn up by which Readers were licensed to perform nearly all the duties performed by deacons. Mr Masey had 'his call to the Readership during the war' and had served the Church for 28 years when he resigned in 1969. He was listed in the magazine as Reader Emeritus until June 1980 at least.

When in 1965 Mr Gerald Collins was admitted to the Order of Readers and licensed to act in the Parish of St Lawrence, Father Bill wrote that he looked forward 'to sharing some of the spiritual work of the parish with Mr Masey and Mr Collins'.

From 1979 Readers were able to conduct funerals under certain conditions, these were the authorisation of the Bishop, the relative's goodwill and at the invitation of the incumbent.

Lay Chalice Assistants
Until 1970 the Bishop's permission to assist with the chalice at the administration of the Holy Communion had usually been confined to Licensed Readers. In 1970 this permission was extended to 'men of good standing' in the parish. Just before St Lawrence Church left the jurisdiction of the Bishop of Kensington, in addition to Mr Collins six men, Robert Davies, John Dedman, William Henley, Lloyd

Jones, Joseph Orme, and Frank Reeve, were the first Eastcote parishioners to have this privilege. Father Bill wrote that 'they will greatly assist the clergy at 8am, 9am, and 11am and at other services as and when they are required'.

Women's World Day of Prayer

The February 1969 *Gridiron* gave notice, that the observance of the annual Women's World Day of Prayer (WWDP) would change from the first Friday of Lent to the first Friday of March, the day on which it is still held. The WWDP should not be confused with the January Wave of Prayer of the Mothers' Union. The WWDP really started in 1919 when two separate 'days of prayer' were united. One had started in 1887 as a day of prayer for the women of one particular Presbyterian Church in New York. The other was started in 1891 when Baptist missionaries, appalled by the deprivation of women in many parts of the world, called for a day of prayer for overseas missions. It was after World War II that the WWDP really grew and by 1968 there were 128 countries taking part. An International Committee was set up to oversee the work of the Movement worldwide.

Women's World Day of Prayer Logo
In the centre is the cross, formed by praying figures from the four corners of the earth. All are joined together within the circle of the world and enfolded in God's love. (WWDP)

Today, meetings at International level choose the themes of future services from suggestions received from all over the world. The themes are then allocated to countries willing to prepare an Order of Service. Once a draft order has been agreed it is sent to National Committees, who may amend the service to suit their own country. More than 170 countries and islands take part and the International Order of Service is translated into over 60 languages and 1,000 dialects.

Bishop Ramsey School

As early as 1973 the Secretary's Report at the APCM referred to the suggestion of having a large Church of England Comprehensive School in the area of the Manor School, Ruislip, the main part of which would fall within the parish of Eastcote.

The report of Deanery Synod meetings given at the APCM in 1976 referred to four governors being elected to the new school. In addition Father Bill was appointed a Governor representing the London Diocesan Board of Education. The school officially opened in September 1977.

Father Bill's Retirement

Before trying to evaluate Father Bill's enormous contribution to the parish, two of his personal celebrations deserve mention. In 1976 the Bishop of London nominated Father Bill to the Prebendal Stall of Weldland in St Paul's Cathedral, to take effect from 12 May. Father Bill attended St Paul's for the Collation and Installation into the Prebend. The dictionary definition of prebend is 'part of the revenue of cathedral or collegiate church granted to a canon or member of the chapter as a stipend; portion of land or tithe from which this stipend is drawn'. The income had long been transferred to the Church Commissioners. The title Prebendary and Canon are one and the same title. The appointment to a Prebend or Canonry is in the nature of an ecclesiastical knighthood. Although his official title was The Revd Prebendary etc., Father Bill continued to sign himself as Father Bill.

On 10 June 1979, Trinity Sunday, Father Bill celebrated 25 years in the Priesthood, and following an earlier Sung Eucharist (10.30am and not 11am) there was a social gathering for sherry and cake. Father Bill was 'deeply touched' by the numbers who attended the service and had 'emotional moments during the procession'. In replying to the speech of the Churchwarden, Mr Darby, Father Bill paid tribute to his wife 'who has been a tower of strength to me all through my ministry'.

At the end of June 1979 it was known that Father Bill was to retire from the parish. He planned to have a holiday in September and have his farewell service on 21 October, Dedication Sunday. Through no fault of his he was unable to move to his new home in Northwood Hills until after Easter 1980. This was a gain for the parish for it meant a shorter interregnum. Father Bill remained at the vicarage and officiated at the Lenten and Easter services in 1980.

21 October 1979 was a memorable occasion. The text for Father Bill's farewell sermon was I Corinthians 3:11. 'For the foundation, nobody can lay any other than the one which has already been laid, that is Jesus Christ'. Father Peter Goodridge, Father Clive Pearce and Father Huw Chiplin, three of his former colleagues were able to be present on this important evening. Among the many people who attended were Mr and Mrs Lally (Mayor and Mayoress) and representatives from various organisations. Among those from the wider community of Eastcote were the Matron and some sisters from St Vincent's Hospital, a place that had links with St Lawrence's from the early days of the Mission Church.

In reply to the tribute given by Miss Margaret Case, St Lawrence's first female Churchwarden, Father Bill said he was 'deeply touched by the many expressions of love and affection' and grateful for the splendid arrangements the Churchwardens had made for his farewell sermon, presentation and reception. He noted that from the time he became a Lay Reader in 1935, he had 'served under six Bishops of London, five Bishops of Kensington and two Bishops of Willesden and one wife'. He said that Mrs Hitchinson had done so much for the parish and that 'it was only through her loving co-operation in the home and every department of parish life that I have been able to do my job'.

He mentioned the two Masters of Ceremony, Mr Gerald Collins and Mr John Martin, who had been responsible for the dignity of the church services during his incumbency. Father Bill's reference to the 'dignity of the church services' brings to mind what he wrote in his magazine letter of June 1956, about the two sides to parish life - Worship and Fellowship, part of which is worth repeating here. 'Worship must, of necessity, come first, as our Blessed Lord has commanded, but from our worship together there should also emerge a fellowship which, in its outpouring, will embrace all those with whom we come in contact'. Father Bill referred to 'a number of hiccups over the years', but felt that from the time he came to Eastcote he had been 'surrounded by a host of most loyal people'.

Fr. Bill's 'farewell service' in October 1979
(F & J Hare/Photopro Photography)

Father Bill's Legacy
Today it is more or less taken for granted that the laity has an important role in the decisions and activities of St Lawrence Church. Father Bill seems to have been in the vanguard as regards vicars who involved their congregations in many

of the changes in church life. This chapter has shown many aspects of how Father Bill guided the parish through difficult years. In his first years at Eastcote he, with support from his wife, seemed personally responsible for much that went on in the parish, even to editing the magazine. By the time he retired there was greater involvement from members of the congregation.

Throughout all the changes Father Bill's guiding principles of Worship and Fellowship may be seen. While maintaining the High Anglican tradition of worship of the earlier years he gained the support of the congregation for substantial changes in church services.

He left the parish in 1980 but returned to St Lawrence Church on a number of occasions. He died on 15 November 2003. A paragraph from Father Clive Pearce's appreciation of Father Bill, printed in the April 2004 issue of the church magazine, is an appropriate end to the chapter.

> *Father Bill deftly managed his four assistant priests and three lay Readers John Masey, Gerald Collins and Norman Rogers. He would send his staff what were affectionately known as his 'kindly notes' to bring order to worship. He took on board Eucharist reform in the Church of England from 1965 onwards, moving from Series I to Series III in a matter of eight years with new music and, with what was thought odd and new-fangled in those days, the sign of peace. There was always dignity and reverence in the way he presided at the liturgy. This he always celebrated in his measured and clear way of speaking, with feeling, reverence and humility. His life was always under girded by regular daily Holy Communion and the recitation of the offices. If there was ever any disagreement with one of his curates the discussion always ended with a glass of scotch and then an adjournment to church to say Evensong together.*

1   Bradshaw Paul *A Companion to Common Worship* SPCK 2002 p24.

CHAPTER IV

THE REVEREND DAVID HAYES
VICAR 1980 to 1990

Fr David Hayes' Institution and Induction June 1980
Fr Ken Toovey, Fr David, Bp Hewlett Thompson,
flanked by Churchwardens Joe Orme and Margaret Case
(F & J Hare/Photopro Photography)

In February 1980 the parish knew that the Revd David Hayes had been appointed as the third Vicar of St Lawrence, Eastcote. As Father Bill remained in the parish until Easter 1980 the interregnum was relatively short, and the Institution and Induction of Father David took place on 24 June. He was familiar with the area as he had been Curate at St Martin's, Ruislip from 1970 to 1975, during the incumbency of the Revd Kenneth Toovey. At the time of his appointment Father David was Priest-in-Charge of Ludford in the Diocese of Hereford (1975-1980).

With hindsight the magazine notice of 'An Evening of Special Prayer' on Friday 6 June for the parish, for Father David, and for Monica, his wife, and for Timothy and Rachel, his children, may be seen as an indicator of the importance Father David gave to the life of prayer. In addition to the chain of silent prayer, from 5.30pm to 9pm, Father Kenneth Toovey, then Dean of Hillingdon, celebrated at the 7.30pm Eucharist.

At his first PCC meeting Father David and the Council agreed that 1981 would be not only a good time for a Stewardship Renewal Campaign, but also for 'taking in a wider field of mission'. Father David saw 1981 as the planning year for the Golden Jubilee celebrations in 1983. In his April 1981 magazine letter he asked readers to consider the role of St Lawrence's in the future. 'To what ministry is God calling us in the years ahead? What is our responsibility as the Church within society given our location here in Eastcote?' As will be shown, during Father David Hayes' incumbency much of the tradition of the past was maintained but new and sometimes challenging things were introduced.

Father David Hayes' Assistant Curates
During his time at St Lawrence's Father David had three assistant priests, the Revd Masaki Narusawa (July 1981 to May 1984), the Revd Alan Boddy (July 1984 to April 1987), and the Revd Peter Day (June 1987 to January 1991).

Father Huw Chiplin had left in January 1979 and St Lawrence Church was without an assistant priest until July 1981. With the prospect of a Stewardship Renewal Campaign and his desire for 'taking in a wider field of mission', it is hardly surprising that Father David was delighted when an experienced priest was appointed and took up his duties on 26 July. In a letter in the August magazine Father Masaki Narusawa introduced himself. Born in Tokyo, Japan, like most of his countrymen he was not brought up in the Christian faith. From his early childhood he had travelled widely, living in the U.S.A. and Germany as well as Japan and, for the last ten years, in England. While working in the City as a commodity/metal trader for an American company he became a Christian. He was ordained in 1978 and served his first curacy at St Alphage, Edgware.

At the 1983 APCM Father David recalled the memorable occasion of the marriage of Father Masaki to Alicia Turner in October 1982. It was indeed unusual for a priest to marry in his parish, at a service attended by 370 guests, and presided over by a Bishop with eight concelebrating priests. Like most assistant priests at St Lawrence's Father Masaki was involved in all aspects of pastoral care in the parish. After nearly three years in the parish Father Masaki, Alicia and their daughter Ester moved to the parish of St Mark with Christ Church, Glodwick, Oldham. Their final Sunday was 6 May 1984. Father David 'thanked God for Father Masaki's priestly ministry here in this parish, and for his practical contribution to the life of the Parish'. Among the variety of his contributions were the Weekly Bulletin editing (and from January 1984 the Magazine editing), the House Group and the Walsingham Pilgrimages. Alicia was also thanked for what she had done, especially as a Sunday School teacher and in starting and overseeing the Crèche at 11am.

St Lawrence's next assistant priest, Father Alan Boddy, was ordained Deacon on 1 July 1984. Father David's June 1984 letter pointed out that the parish was fortunate to be 'allowed a deacon as recent house-keeping [St Lawrence's financial contribution through the Diocesan Quota] had been below the level required for a parish with two clergy'. He asked parishioners to give some thought to this. The parish was also fortunate in that Father Alan was 'a mature and lively Christian who had gained a wealth of experience through his years of secular employment'. Father Alan had been for many years in the Civil Service working in a variety of Home Office Departments, on such matters as explosives legislation, fire service training and parole for long sentence prisoners. He said he had a number of interests but pride of place went to music. Choral music added to the splendid occasion of Father Alan's First Mass on Wednesday 3 July 1985, after his ordination as priest the previous Sunday.

Fr Alan Boddy and Fr Peter Day

Father Alan left St Lawrence's in April 1987. Paying tribute to Father Alan in his May magazine letter Father David wrote, 'Father Alan's departure after Solemn Evensong on Easter Day is completely in character, namely choosing to leave us on the greatest Christian Festival...he had exercised his ministry with sensitivity and great care of people and we have all benefitted from the skills of his previous career'. From 1987 to 1990 Father Alan was Curate at St Mary Abbots with St George, Kensington. From 1990 he was Chaplain in HM Prison Service.

Once again the parish was not long without an assistant priest, for on 28 June 1987 Peter Day had his first day as Deacon at St Lawrence Church. Like Father Alan, Father Peter's theological training came after years in another profession, pharmacy, in which he had had wide experience. He was ordained priest in 1988 and celebrated his First Mass on Wednesday 6 July. Father Peter was involved in many of the new enterprises of the 1980s. He was responsible for the spiritual and pastoral care of the parish during the nine months interregnum of 1990, until Father David Coleman became Vicar of Eastcote in October. Father Peter had done much for the parish. In Father David Coleman's January 1991 letter he wrote that the parish was sad at Father Peter's going but happy that he was moving on to a post of greater responsibility. Father Peter was Curate at St Augustine's, Wembley Park from 1991 until 1994, when he became Vicar at Glen Parva and South Wigston, in the Leicester Diocese.

When in his 1986 Vicar's Report, Father David Hayes referred to the increased support he was being given by Father Alan, he also mentioned Captain Donald Woodhouse, whose family had worshipped at St Lawrence's since 1972. Captain Don became a Lay Minister after his 'retirement' from the Church Army in 1985 and has done, and still does, a great deal for parishioners and the wider community. He has, for example, taken a keen interest in the Scouts but also in work with other young people, the Drop-in Centre and other ecumenical activities of the Eastcote churches. Father David was also grateful to the occasional help he had from Father Wilfred Beale, a retired priest, who with his wife Joan, worshipped at St Lawrence's from 1985.

The Church Fabric
As in the past, changes or improvements of property of St Lawrence's church were inter-related with the needs of the contemporary situation. During Father David's incumbency these reflected the vision he referred to at his first PCC meeting, of 'taking in a wider field of mission'. He was much concerned with the 'responsibility of the Church within society given our location in Eastcote'. On several occasions when the problem of financing projects seemed insurmountable Father David expressed his belief that the money would come if the scheme was right.

The major undertaking, which took place over a number of years, was the upgrading of the church hall. This was an ambitious and worthy scheme but it is right that changes in the church itself should be discussed first.

The Church
Worshippers at St Lawrence Church today are familiar with the priest facing the congregation when celebrating the Eucharist and some may be surprised to learn that this is a relatively recent practice. Although when Father David Hayes came to Eastcote the 'definitive' rite of Alternative Service Series III had been used for a number of years, the priest still celebrated in the east-facing position, with his back to the congregation. During his first year at St Lawrence's Father David introduced the moveable nave altar as the focal point at the Church Parade Masses, so by 1982 many people were familiar with the westward position. In the June 1982 Vicar's Letter Father David explained that the 'purpose of the priest facing the congregation is to bring everyone closer to the drama of the Eucharist…a pastoral help to draw us all closer to the dynamic presence of Christ in the Eucharist'. Starting that month, for an experimental period of six months, the Lady Chapel Altar was brought forward in order that the priest could celebrate facing the congregation at weekday celebrations. The intention was to 'get the feel of the westward position' and eventually to re-order the Sanctuary with a freestanding altar.

East End prior to the changes in the Sanctuary
Flower Festival 1983

The 'eventually' was to be in 1986. At a PCC meeting in July 1982 it was agreed that from October until July 1983 some changes would be made in the Sanctuary, which could suggest ideas for a permanent re-ordering of the Sanctuary. At the 1983 APCM there was unanimous consent that the westward position of the altar was preferable. Initially the PCC was responsible for considering specific proposals but parishioners were later invited to a meeting for their comments and discussion. A faculty was granted in 1984. The original plans to adjust the stonework in the Sanctuary were found to be very expensive (£4,000 to £6,000), and revised plans were drawn up by the middle of 1985. The stonework was to be left intact but the existing levels extended forwards by using wooden joists, so creating a larger top step for a freestanding altar. The whole area of the Sanctuary and the Chancel was to be carpeted and the alternative plan had the merit of being less expensive! A new faculty was applied for and granted. Some of the work was started in summer 1985, including the removal of the reredos, plastering and making good the east wall, and making the wooden joists.

East End after the Re-ordering of Sanctuary and
installation of the Bronze Crucifix 1988

By September the steps had been completed (£1,018) and an order placed for the carpeting (£1,436). Donations to a special St Lawrence PCC Sanctuary Fund were generous, but without the expertise of Mr Tony Cattle who constructed the steps the total cost would have been greater. By July 1986 the new eight-foot wide freestanding altar was in position. The altar had been specially made and given by Tony Cattle as an offering to God for thirty years of worship at St Lawrence's. Sadly he died in 1987, before the re-ordering was completed.

In the spring of 1986 the contrast between the newly painted east wall and the remaining walls of the church was obvious and, under the guidance of Arthur Plummer, the major task of painting the high walls of the church began. One decorating tower had been bought in 1982 and two others were borrowed. In his December 1986 magazine report Arthur expressed sincere thanks to the small but dedicated band of 12 parishioners who had worked so hard to complete the task for the beginning of Advent. The actual cost of decorating the walls was around £200 and if they had not adopted a DIY strategy the church would have remained dirty. To have it done professionally would have run into several thousands of pounds. Arthur also thanked Mrs Jean Jones for giving the Stations of the Cross a wash and brush up, and Mrs Beryl Dew 'who keeps our church so wonderfully clean' and coped with the extra work caused by the decorators. Beryl still keeps the church wonderfully clean. One important remark in Arthur's report was 'it amazes me how often our Lord allows us to just meet a target, be it time or money, so that a job can be done in his name'.

The bronze crucifix on the east wall is a unique work of art. It was the result of much consultation by the PCC, with the sculptor Andrew Cooper, and with the Diocesan Advisory Committee, leading to the granting of a faculty in 1987.

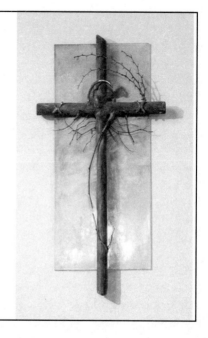

Description of Crucifix.
Mounted on glass the crucifix represents the living wood of the cross and the tree of life. The head of Christ is seen as a knuckle on the trunk and the body and feet are portrayed as branches almost touching each other. The sharp thorns around the halo represent the crucifixion, whilst the branches, grouped beneath, the resurrection. These create a circular frame focusing the onlooker's attention onto the head and halo. Crosses indicate the position of the hands.

After the crucifix was mounted on the wall in April 1988 minor adjustments were made to it and a spotlight installed. The cost of £3,350 was met from the special Sanctuary Fund.

The Bishop of Willesden, the Rt Revd Tom Butler, visited St Lawrence on the Feast of Christ the King (20 November 1988) to celebrate the Eucharist and bless the re-ordered Sanctuary and Crucifix. Immediately after the service the Bishop blessed the Parish Hall developments.

A new furnishing to the church was the Incumbents' Board in 1982. Tony Cattle made it and Sidney Starling painted and lettered the board with the names of former priests of the parish. On 21 October 1983 the plaster statue of Mary and the Child Jesus, made by Sidney Starling, was blessed during the Golden Jubilee Eucharist of Consecration and later placed on the east outside wall of the Lady Chapel, in a niche that has been there since the church was built. In 1984 a small statue of Our Lady of Walsingham, given in the memory of Gerald Collins by his family, was fixed on the north wall of the Lady Chapel. In addition to the re-decorating of the walls of the church in 1986 various work of a maintenance nature went on in the 1980s, including a restoration of the organ and repair to the church roof.

The Parish Hall Building Project
In January 1983 Father David discussed with the PCC, and made public to the congregation, a project in keeping with his vision of St Lawrence's having a 'responsibility as the Church' within the parish at large. Father David saw the church buildings as playing a 'definite commitment to serve others in the wider community', and felt that 1983, Golden year, was the right time to embark on a fresh challenge. Many discussions had taken place before January 1983 and many more followed. The general idea was to develop the church hall and bungalow site to include a second hall to cater for specific needs in the Eastcote community. A sub committee of the PCC, the Project Buildings Committee was appointed and did much hard work, studying and revising several schemes submitted by the architect, Mr Geoffrey Miller. Mrs Ann Sykes, Chairman of the original Planning Committee, played an important role during the whole building project, liaising with various organisations using the Hall as well as being trouble-shooter on many occasions. The congregation were kept up-to-date about the Building Project through reports in the magazine, progress bulletins by Arthur Plummer, and occasional parish meetings, some with the church Architect, Mr Emil Pym. Father David frequently asked for everyone to pray about the Project and reminded parishioners that all should be involved in 'this exciting venture as God's people'.

Nigel Boardman, the Appeal Director, Barbara Plummer and Captain Donald Woodhouse produced a brochure, published by the Church Army Press, explaining the £190,000 Building Appeal. By May 1984 a copy had been sent to grant-making trusts and charities. Among those supporting the Appeal were the

Bishop of Willesden and John Wilkinson, the then MP for Ruislip and Northwood.

Parish Hall building Project Brochure - Artist Impression of the Scheme

The original ideas were modified and in September 1984 it was formally agreed that the project be divided, as far as possible, into self-contained stages, and that work should begin as soon as possible with the funds raised so far. At that time the Fund stood at nearly £30,000. It was not until October 1985 that the first stage of the Building Project was under way. This comprised a new kitchen opening into the Hall followed by the conversion of the existing kitchen into toilets, accessible to the disabled. By the summer of 1986 this had been completed. Work on the second stage of the building development began in February 1987. The two 1939 committee rooms were made into one, with access for the disabled. A first floor room was built over this. A small party of parishioners, once again under the guidance of Arthur Plummer, decorated the new areas between July and December 1987.

There was a social evening in January 1988 to offer the parish's sincere thanks to George Guest and those who supported his fund-raising coach trips to places in the British Isles and the Continent. The new upstairs room was named the George Guest Room in public acknowledgement of the considerable amount of money he had given to the Development Fund, without which the first floor could not have been built at this time. The April 1988 magazine article reported that the Church Hall was beginning to pay its way and there were many more bookings. In November 1988 there was further celebration of the Building Project when the Bishop of Willesden blessed the changes and additions to the Hall. By the end of 1989 changes had been made to cloakroom and storage areas at the foyer area of the Hall, and new house lights were a great improvement to stage productions.

By October 1989 it was known that Father David was taking up another appointment. He made it clear that as the parish prepared for a new incumbent with a constantly changing PCC membership the Fund would remain open as a separate account thereby preserving its charitable status. The 1984 Appeal was launched on the public understanding that the whole scheme, as outlined in the brochure, was to be achieved. Donations from the Diocese (£5,000) and from various grant making trusts (approx £25,000) were made on that understanding. Further developments would be left to the new incumbent and the PCC, in the light of the parish's future needs and responsibilities as the Church within the local community.

Many fund raising efforts took place between 1983 and 1988. To launch the project, and as an example and encouragement, Father David and Richard Darby had run in the London Marathon in April 1983 and through sponsorship raised £4,500. The following year Father David ran in the London Marathon to raise funds to help people who would benefit from the provisions of the new buildings. These funds went to Action for Research into Multiple Sclerosis and also Hillingdon Welcare, the organisation responsible for the Church's social work among families with difficulties.

As has been noted already George Guest, someone not a member of the congregation, contributed a great deal to the Building Fund. Fund raising efforts by members of the congregation were many and various, some producing large sums, others relatively small amounts. All efforts were important, for as Father David had said the project was one in which all should be involved. The first of several Village Auctions organised by Jeannie (Genie) Chivers was held in April 1983 and raised £575. Profits from fêtes and markets sometimes went to the Fund. Among other events raising smaller amounts were the monthly raffles, the sale of second-hand books after the Sunday morning services, jumble sales, the first St Lawrence quiz supper (February 1984), and the sociable pancake evenings after Evensong.

While some people were involved in the building plans and others in fund raising, yet others were getting information and gaining insight as to the best way St Lawrence's could use the improved facilities to serve the local community. In fact two schemes were running using the bungalow, before actual work on the Project began. From autumn 1984 Hillingdon Welcare workers ran the Rainbow Group. This was for mothers and their handicapped babies or toddlers, and the aim was to provide a place for the mums to find support, friendship and comfort from others facing similar problems. The Drop-in Centre started in the following spring and was a further attempt to 'express our ministry of Christian caring within the wider community'. The intention was to befriend anyone who might drop in, and if appropriate to refer those with problems to other agencies. From the beginning lay members of St Lawrence and other Eastcote churches ran this scheme, which in reality mainly attracted those who lived on their own.

The Bungalow and the Church Grounds
The bungalow site had been considered as part of the Building Project and the last of the self-contained stages of that scheme. Major decisions regarding the bungalow were to be left until after the arrival of a new Vicar. However it was important that the best use should be made of the building. At the February 1985 APCM Miss Margaret Case, Churchwarden, reported that the bungalow was being temporarily up-graded to meet the needs of the parish. Repairs and re-decorating were done enabling church organisations and others, such as the Rainbow Group and the Drop-in Centre, to make more use of the building than had been the case in the recent past.

The church and the other buildings, which are all part of the property of St Lawrence's, are on a large site and the grounds require maintenance. Appeals for help to join the 'gardening group' have never brought forth a great number of volunteers but fortunately there have been those who have given of their time and their talents in serving the Church in this way. Over the course of years the personnel in the group changed but some of the stalwarts in the 1980s were Messrs Murrell, Cattle, Cornell, Gough and Priddle, and Mr and Mrs Higgs, Mr and Mrs Lumsden, and Mr and Mrs Towers.

The Vicarage
Early in 1986 members of the congregation were invited to have a look at the new vicarage study in an extension at the front of the house. The work included the re-location of the kitchen in the old study and boiler room area, with the old kitchen becoming a utility room. Father David wrote that having the study at the front of the house was of 'great benefit to all concerned'. There was a major teething problem as, with a sound proof inner door, the front door bell which rang in the utility room couldn't easily be heard when the study door was closed! The Vicar assured his readers that this would be remedied.

The Curate's House
26, Woodlands Avenue had been sold at the end of 1979. By the end of 1980 the PCC agreed that the time was right for looking for a house for a possible curate. At the Annual Parochial Church Meeting in April 1981 Father David reported that 33 Sunningdale Avenue had been purchased and that he was actively 'looking for an assistant priest'. Father Masaki joined the parish in the July and was the first Curate to live in the Sunningdale property. A small band of volunteers re-decorated the house in 1984 for the arrival of Father Alan Boddy and in 1987 for the arrival of Father Peter Day.

Worship

Alongside his belief that the members of the congregation of St Lawrence Church had a responsibility to further the mission of the Church in Eastcote, Father David felt this was only possible if the parish went 'forward together' as a worshipping community.

On the whole the traditional pattern of services already established at St Lawrence's continued during his incumbency. The High Anglican tradition was maintained at St Lawrence's but Father David not only responded to the needs of particular groups within the wider community, but initiated a number of what might be called 'ways of worship' that were new to his parishioners. Father David's frequent references to the importance of prayer in his magazine letters underline the importance he attached to this essential part of the Christian life, and several of the changes and additions reflect this view.

In keeping with the national trend the total number of communicants had fallen by the late 1970s but during the 1980s remained fairly steady (see Appendix 6). Numbers were obviously affected during interregnums or when there was no Assistant Priest, as there were fewer weekday services.

Sunday Services

For most of the 1980s the times and types of Sunday services remained largely as they had been in Father Bill's incumbency, that is a 8am Holy Communion, 9am Parish Communion and 11am Sung Eucharist, and 6.30pm Evensong. In 1981 the 'Rite A' Final Revision Series III of the *Alternative Service Book 1980 (ASB)*, was introduced to congregations at weekday services in January and then at Sunday services in October. Father David hoped the alterations would 'help us to deepen our spiritual life as God's people in this parish'. He saw the *ASB* as an attempt to return to the simplicity of structure of early forms of Christian worship.

Following the introduction of the 'Rite A' service Father David decided to use a new Lectionary starting on Advent Sunday 1981. This was a new scheme of readings at the Eucharist on Sundays and had a three-year cycle. Michael Bedford, Secretary of the St Lawrence Branch of the Bible Reading Fellowship (BRF) at this time, recommended books as 'companions to the new scheme'. It is worth noting that there had been a BRF branch at the church since 1959 and this is still active today.

Another attempt of Father David to deepen the spiritual life of God's people is seen when, from the beginning of 1981, he encouraged the practice of having Holy Baptism within the Mass at 9am or 11am. The Baptism Service in the *ASB* emphasised the importance of this Sacrament taking place within the main worship of the congregation, when the whole congregation may take part and welcome the newly baptised into the Church. When a baptism took place on a

Sunday afternoon the newly baptised was welcomed into the Church at one of the services on the following Sunday. Closely associated with this practice and starting in January 1981 Father David invited mothers of newly baptised children to a monthly Pram Eucharist on the last Thursday morning of every month. The monthly Pram Eucharist had been popular from 1971 to 1978 and when restarted, as a morning rather than an afternoon event, was well attended until the last ones in the summer of 1984. Mothers had the opportunity to chat over a cup of coffee in the Vicarage after the service.

It has been shown already that during Father Bill's incumbency members of the congregation had become more involved in decisions and activities of the parish, and that a few members of the congregation were, with the Bishop's authorisation, assisting clergy at St Lawrence's as Lay Readers and Lay Chalice Administrators. In keeping with the great importance Father David gave to the life of prayer, both as a parish and as an individual, it is relevant to refer to two additions he introduced at St Lawrence's, neither needing the Bishop's authorisation, and both still practised today. One of benefit to corporate worship began in Advent 1985 when some members of the laity were asked to lead the Prayers of Intercession and Thanksgiving during the Eucharist. The other, of a more personal nature, was the Prayer Partner scheme, which started in February 1985. Each 'prayer partner' draws the name of another person participating in the scheme and remembers that person in daily prayer and during Sunday worship throughout the year of a 'prayer partnership'. The name of the partner is a secret between each person and God, and at the end of the year cards are sent to reveal identities to each other. It is one way of showing our love and concern for one another.

A major change took place in February 1989, when the 10am Sung Eucharist replaced the 9am and 11am services. This change had been suggested when Father Bill was Vicar but had not been favourably received. In his Vicar's Report at the 1988 APCM, Father David had said there would be a parish consultation after Easter 1988 regarding Sunday morning services. He made it clear that his mind was not already made up and that after consultation and prayer he, as the incumbent, would decide what was right for the parish. A Worship Questionnaire was distributed to all members of the congregation, asking whether a combined service would strengthen the parish, whether they would favour a combined Eucharist, and if so what adjustments they would be prepared to make. Father Peter and Captain Donald Woodhouse shared Father David's assessment of the 158 replies, a few being joint husband and wife replies. A presentation of the results was made to the PCC who offered their comments. After meetings with those who would have to implement any changes such as Churchwardens, Head Server, Choirmaster, Sunday School Superintendent and others, Father David was of the opinion that having a combined Sung Eucharist at 10am, attended by the majority of the congregation, would strengthen the corporate life and worship at St Lawrence's.

There had been combined 10am services before February 1989. These were occasions when Father David felt that the majority of the congregation should be at one service as the community of St Lawrence's. Two were when visiting clergy preached about Stewardship before the start of Stewardship Renewal Campaigns, in 1981 and 1985, and another was the Sung Eucharist on Dedication Sunday 1983, Golden Jubilee Year. On 5 April 1987, for the launch of the Pastoral Care Scheme, and on 20 November 1988, when the Right Revd Tom Butler, Bishop of Willesden, celebrated the Eucharist and blessed the re-ordered Sanctuary the 10am was the only morning service. Father David was of the opinion that as these special services were concerned with the corporate life of the parish, it was right that 'we should worship as the ONE Body of Christ here at St Lawrence's.

Writing in the April 1989 magazine Father David commented that 'in our worship we are experiencing the death and new life principal – no more 9am, no more 11am celebrations, but a new beginning with fresh opportunities for us individually and corporately'. He recommended using the five minutes before the service in prayerful preparation and at the time of the Communion bell to remain kneeling, unless physically unable, until one came forward to receive Holy Communion. Later in the year he announced that there would be no 8am celebration on Stewardship Sunday 1989 'in order that for one occasion this year we may worship as ONE family of God's people at St Lawrence's'.

By 1980 the average weekly attendance at Evensong had declined to under 30; by 1990, in keeping with the national trend, it was down to under 20. On days of special celebration and with Solemn Evensong, numbers were greater but seldom reached above 40. Strangely enough it was Harvest Festival, not strictly a Church Festival, which had the greatest attendance, often over 80, and as late as 1989 reached 74. On Easter Day 1987 when the parish bade Farewell to Father Alan Boddy 123 attended Solemn Evensong.

Other Sunday evenings which drew larger congregations were Songs of Praise services, Eastcote Churches Ecumenical Services, the London Borough of Hillingdon Civic Services and ones in which St Lawrence Young People (SLYP) took part.

Weekday Services
After Father Bill's official retirement in 1979, and during the months of the interregnum, the weekday Holy Communion services had been reduced to two; 7.30pm Wednesday and 10.30am Thursday. From September 1980 there were Holy Communion services each day except Tuesday (the Vicar's day off), and once Father David had an Assistant Priest in 1981 there were normally Holy Communion services each day. There were seldom many people at these weekday services but they were important for those who attended.

The Table of Communicants (Appendix 6) shows lower numbers of communicants on Ash Wednesday, Ascension Day and Corpus Christi. Perhaps fewer children were withdrawn from school on these days 'for the purpose of attending their church for worship', as had been the case in Father Bill's time.

Advent and Christmas
Two annual events that remained popular were the Ecumenical Carol Service and the Crib Service. In 1984, another service appealing to the wider community of Eastcote was the first Christingle Service at St Lawrence Church. This service, adopted by the Church of England Children's Society, is one in which members of the parish come together to give their gifts to the Society in exchange for a Christingle orange, symbolising the sharing of what we have as individuals.

Other examples of St Lawrence's mission in the local community were the visits of local schools to the parish church. From the autumn of 1981 a Mass for Bishop Ramsey Upper School was held most terms. In the course of the 1980s Bishop Ramsey, Grangewood, Reddiford and Newnham schools had Carol Services at the church.

Lent, Holy Week and Easter
The keeping of Lent followed more or less the pattern of previous years with Mass and Address on Wednesdays and Stations of the Cross each Friday but with Benediction at 5pm on Saturdays. Most years visiting priests gave addresses on Wednesdays. In 1983, the Golden Jubilee Year, the parish was fortunate to have Dr Martin Israel, consultant pathologist and priest, who gave a series of addresses on Christian life.

After the Wednesday services in 1986 members of the congregation gave talks on the theme 'On being a Christian in the world', presented from the viewpoint of the daily life and work of each of the speakers. In January Father David had urged church members to 'to make it a top priority to attend...I am confident that these talks will be a significant stage of development in our growth as God's people in the parish'. The people who gave the 'highly successful' talks helped Father David to organise a series of talks in Lent 1987 on a wide range of subjects of Christian concern. Two were the Christian Challenge of the Inner City and of Coping with Illness in the Family. There were visiting speakers and as in 1983 members of churches in the Deanery were invited.

The start of Holy Week saw the traditional Blessing and Distribution of Palms at morning services. From 1981 to 1987 the 9am service was a Church Parade Mass and in 1989, the first year of the 10am Sung Eucharist, and 1990, there was a Youth Service and Parade in the afternoon. On most Palm Sundays the choir sang a short choral work after Evensong.

For theological reasons from 1982 only one Mass was celebrated on Maundy Thursday. Father David explained in his Vicar's Letter of April that it was customary to have just the Mass of the Last Supper, as it is the particular celebration of the first Maundy Thursday. The Bishop of London expected all the clergy in the diocese to be in St Paul's Cathedral in the morning, to share with him in the Mass of Oils, when the holy oils are blessed for distribution to the parishes and the priests renew the vows of their ordination. In 1987 the *Register of Services* refers to the Mass of the Last Supper 'including the ceremony of the washing of feet', but whether this became a regular feature at this time is uncertain as there are no further references to it, either in the magazines or the *Record of Services* books.

Other Services
The routine of other regular services continued, including those associated with St Lawrence Church organisations, those commemorating Saints of the Church, and the Remembrance Day service at the Eastcote War Memorial. During the two periods of interregnum Confirmation services took place at other churches in the area.

Quiet Days, Retreats and Pilgrimages
Although the practice of holding a Quiet Afternoon before Advent, usually led by visiting clergy, continued for a few years, with so few people taking this opportunity of quiet reflection before the pre-Christmas busyness 1985 was the last one. One of the many events of the Golden Jubilee programme was a special Quiet Afternoon on 15 October 1983, led by Father Ray Phillips, Vicar of All Saints' Church, Hillingdon. All Saints, it may be recalled, was a local '45' church. It was consecrated in the same year as St Lawrence and had the same architect, Sir Charles Nicholson, and so there is a special link with that church.

Annual Retreats were usually well supported and most were at St Francis House, Hemingford Grey in the autumn. One exception was 1985 when St Francis House was closed for renovation and the retreat was held at the modern purpose-built Retreat House of St Columba, near Woking. The conductor was Father Tom Butler, Archdeacon of Northolt at that time and soon to be Bishop of Willesden. The following year it was back to St Francis House in May, when because it was May Day Bank Holiday weekend, it was possible to have a longer retreat – Friday evening to Monday teatime. The cost, excluding travel was £40. Unfortunately, on this occasion the Conductor, the Bishop of Edmonton, was ill. The Warden, Dorothy Glass, managed during the course of Friday evening to obtain the services of two local clergy who were joined by Father Alan Boddy from Saturday afternoon for 24 hours. With the help of Dorothy and Michael Bedford, who organised the retreats in these years, a revised programme was completed. The three clerics apparently 'contrived to maintain a thread in their thought and ideas' having not met and not knowing what each was going to talk about!

The parish was involved in a number of pilgrimages in the 1980s. In May 1981 there was a parish pilgrimage to the Shrine of Our Lady at St Mary's, Willesden.

In October 1982 some members of the congregation joined the London Diocesan Pilgrimage to Walsingham. The first parish pilgrimage to the Shrine of Our Lady at Walsingham took place in May 1983. Another day pilgrimage in 1984 was followed by weekend ones in May 1985 and September 1986.

Young People
From 1981 the magazines no longer referred to the Junior Church, but to the Sunday School, with Junior and Kindergarten sections. At the end of the 1980s a new Junior Church was formed. Possibly because it was a routine part of Sunday worship, the magazines have little information about the Sunday School. However, from this information, together with reports of Church Organisations for the Annual Parochial Church Meeting (APCM), available from 1984, it is possible to show the importance of this Sunday worship, and of social activities associated with these groups.

As in the past more teachers were often needed, and as in the past some were teachers for many years while others, for various reasons, for a relatively short period. Mrs Sylvia Bullock (now the Revd Sylvia Lafford) was Superintendent, at least from 1984 until 1990, when she had a break from teaching after a total of 14 years. Her annual reports echo views of the clergy, saying in 1987 that 'unless we take care to nurture our children in the Christian faith, our "church" of tomorrow will be sadly lacking'. In her 1988 APCM report, when the two sections had to combine because there were only six teachers for the 60 children on the register, Sylvia appealed for more help and wrote 'we must treat this as a priority as there is an increasing lack of Christian education in our schools'. Appreciation was accorded to all those who served the Church by teaching and helping in the Sunday School. Three who had taught for many years in the Kindergarten were Mrs Alice Knott (25 years) and Mrs Jane Morris (8 years) who both retired during 1984, and Mrs Gail Daniels in 1988 after 8 years.

The Sunday School, as part of the community of St Lawrence's, is naturally involved in some of the celebrations and changes that take place. There are a few references to preparations for a Nativity Play and to members of the Sunday School enjoying the annual Christingle Service. Some years there were displays in church for the benefit of the congregation. One Lent it took the form of posters. In 1983, Golden Jubilee Year, the children from both departments took part in making the rainbow exhibition for the Flower Festival, and in 1985 they had a harvest display on the old 'high altar', which had been moved to what is known at present as 'Tot Zone '. On Palm Sunday 1985 the Juniors performed a short mime to narrate the Easter story for the Kindergarten section. In 1986 an afternoon Mothering Sunday Service was followed by a celebration tea and provided an opportunity for teachers and clergy to chat to parents of Sunday School children. Like other organisations, the Sunday School held a fund-raising sale in aid of the 1980s Building Project, and while building work was carried out in the area of the Committee Room the Junior section met in the Scout Group's HQ.

Two popular annual social events, when the children largely provided their own amusements, were the summer picnic and the Christmas disco/party. In October 1984 the Juniors had 'a wonderful day out' when they visited St Paul's Cathedral for an event organised by the London Diocesan Board. Among the activities was a tour of the Cathedral that included the Whispering Gallery and Crypt, and the day ended with a short service under the dome. The Juniors also enjoyed attending the Saturday matinee productions of the St Lawrence Players, that of *The Railway Children* having special mention in the 1987 report.

Crèche
Alicia Narusawa, Father Masaki's wife, started the crèche scheme in September 1983 to encourage those families with young children who didn't go to church, and make it easier for those whose families already attended. The crèche was available for those under the age of three, with parents working a rota system to look after the children. Once a child was old enough to attend Kindergarten the parent of that child was likely to 'drop out' of the rota, but its success or otherwise largely depended on families making use of the facility. By 1990 the scheme seemed to have run its course and helpers were inclined to feel despondent when no one turned up.

Junior Church
A new Junior Church commenced on 24 April 1988 and grew out of an expressed wish of the Sunday School Support Group that there should be a kind of 'sixth form' Sunday School for young people of secondary school age. During the rest of Father David's incumbency Joan Beale, Barbara Plummer and Doreen Plummer provided reports for the APCM. A team of adults and clergy led the weekly sessions, which were held in the George Guest Room. A short service preceded a varied programme of teaching and young people joined the congregation in church, at the time of the Peace. Over the three years average attendance fluctuated between seven and 12, as members moved on to higher education or entered more fully into the worship of the Church. In her 1990 report Barbara wrote that 'the young people …need all our encouragement to grow in their Christian faith'.

In 1988 there were two outings, to St Alban's Abbey and to Westminster Abbey, and in 1990 there was one to the Royal Foundation of St Katherine in Limehouse, a Retreat House and Conference Centre. These outings and the visits to the Eastcote Methodist, St Andrew's United Reformed and St Thomas More Roman Catholic churches helped the young people to understand more about their Christian faith.

St Lawrence Young People (SLYP)
When Father David became Vicar in 1980 the Youth Group, formed in 1975 as a discussion group, had proved itself a worthwhile church organisation for young people from the age of 15, and contributed a great deal to the parish as a whole. Increasingly known as St Lawrence Young People (SLYP it went through a

difficult time in the early 1980s, but by 1987 it had recovered with the whole group taking responsibility for organising club events rather than leaving it to a few committee members. Members who appear to have given much time to the group include Angela Dew, Andrew Pearson, Rob Squires, Chris Baker, Neil Horchover and Mark Kimsey.

The spiritual and social aspects of the club were much as in the past. Father Masaki, Father Alan, and Father Peter in turn, and Captain Don, encouraged the group in a number of ways, often attending their meetings and sometimes going with them on outings. On the spiritual side the clergy involved the group, not only in the Evensong services referred to already, but also occasionally at Sunday morning services. Magazine reports show that Bible study and prayer meetings featured in their programmes. One outstanding evening was that of 11 November 1981, when Father Masaki invited members to his home in Sunningdale Avenue to the first House Eucharist in the parish. There was a long magazine article on what was obviously a moving occasion and the group left 'feeling much closer to God and one another, and will certainly look upon the Eucharist with much greater appreciation'.

The social side of club evenings included talks on a range of subjects, quizzes and *Desert Island Discs*. Physical activities included the usual games, the odd keep-fit, circuit training and country dancing evenings. There were many evenings of cooking, some more successful than others, with pancake events remaining popular. Bonfire evening appears to have been an annual event. Away from Eastcote, members went on rambles, organised car rallies, visited St Alban's (1981), the canal museum at Northampton and St Katherine's Dock (1984) and went on a narrow boat week-end on the Grand Union Canal (1987). The traditional carol singing at St Vincent's Hospital took place most years, sometimes joined by members of the Junior Youth Club.

Reports show SLYP took an active part in parish life. For some years they continued to serve breakfasts after the Sunday services. In 1983 they provided tea, coffee and 'a large selection of biscuits' after the Golden Jubilee Service on the morning of 23 October and they served the cider at the Parish Supper in the evening. They had a few quizzes and debates against teams from the congregation and helped at the annual fêtes and markets.

In 1981 the first of many cricket matches against other members of the congregation took place, usually in Bishop Ramsey Upper School grounds. In 1988 Father David wrote of 'the youth of the parish challenging those in the parish of advancing years to a cricket match'. It turned out to be a really exciting match starting with the older team scoring 53 runs. SLYP were confident they would win as they had a new fast bowler Timothy Hayes, the Vicar's son. In three balls in his first over he claimed a hat trick. However stout resistance from Lionel Williams and the Vicar meant the outcome was not certain and in the end it was a draw. In 1989 SLYP won for the first time. The 'almost' annual parish cricket match was and remains a popular parish event for participants and spectators.

According to the 1990 report, highlights of the year included, winning the cricket match and the building and running of the Father Christmas Grotto. Although numbers attending meetings fluctuated between 8 and 12 it was felt 'essential to continue to provide some form of fellowship for youngsters within the parish'.

Junior Youth Club/Young Communicants
The Junior Youth Club was launched by SLYP in January 1981 and in the May issue of the magazine Father David praised them for 'a quite splendid job'. Rather like the Junior Guild, which had existed from 1958 or 1959 to 1964, it was for the 11 to 13 year-olds recently confirmed or being prepared for confirmation. Its purpose was to make these young people feel that they really did belong to the Church, to provide a setting for continuing Christian instruction, and to be somewhere to 'enjoy one another's company and let off steam'.

Several events in which this group took part helped to make them feel they 'did belong to the Church'. In 1981 they helped decorate the church for Harvest Thanksgiving and helped at the Christmas market, raising £13 running a Treasure Hunt. Some members attended the Players' production of *View from Below*, a few joined members of SLYP carol singing at St Vincent's Hospital, and others served coffee and tea after the Ecumenical Carol Service. Talks and instruction by Father David and Father Masaki, and visits by Iris Castles to introduce them to the mysteries of yoga, all helped to establish this club as a church organisation.

Social activities consisted mainly of table tennis and badminton and, in the lighter evenings, rounders on Cheney Fields. They also took part in a Five-a-Side Football Competition organised by the Metropolitan Police. In 1982 they enjoyed Valentine, pancake and Hallowe'en parties.

Much of the success of the first year or so of the club was the wholehearted way in which SLYP members encouraged the younger people. Those responsible included Angela Dew, Judith Brion, Sue Mitton and Rob Squires. However, by the end of 1982 Angela had left and there was an appeal for other leaders and the club seemed likely to fail.

From September 1983 the club was called The Young Communicants Club and was led by the clergy and a rota of parents. In 1985 Mrs Pat Dunhill reported benefits to the parents as well as the children, as they had met other members of the congregation who were just faces before and had become friends. The club was 'proving a happy link between the young people and the church'. There was a steady attendance of 12 at the monthly Sunday meetings and a variety of activities prevented these becoming 'samey'. In 1988 Mrs Anne Barnicoat reported taking a flexible approach, providing games on a regular basis but also a coffee bar 'where some of the girls are quite happy to meet and have a chat and listen to tapes'. There were competitions and discussions and a disco party at Christmas. Father David, Father Peter or Captain Don went to each session, which ended with Compline in church, where all seemed to gain something from this 'wind down'.

A number of members then moved up to SLYP and there is no further reference to the club. It had been a worthwhile venture and an example of the parish responding to the needs of a particular group at a particular time.

'Forward Together' - New Enterprises
It has been noted already that Father David believed members of the St Lawrence congregation had a responsibility to further the mission of the Church, and that this would be possible only if the parish went 'Forward Together' as a worshipping community. The Worship section of this chapter referred to the new approach to Baptism, the revival of Pram Eucharist Services, and the start of the Prayer Partner Scheme. The Parish Hall Building Project was important as part of a 'definite commitment to serve others in the wider community' with better facilities. Discussion of Church Organisations and other church activities will show that Father David's 'Forward Together' influenced much that was done in this incumbency.

Father David's 'Forward Together' ethos, with what could be called its twin ideals of wider mission and a worshipping community, will be recognised as inter-related in many of the enterprises discussed in the following paragraphs. It should be remembered that much of the tradition of St Lawrence Church of earlier days remained but, as in Father Bill's time, the Church met the challenges of a changing society. It could be said that Father Bill's 'Worship and Fellowship' had taken a step forward. The majority of the 1980s initiatives, some in a revised form, have an important role in the life of the parish in the 21st century.

Some initiatives are likely to have come about as a result of the Bishop's Visitation to Eastcote in October 1981, following the successful Phase I of the Stewardship Renewal Campaign. During his visit groups discussed what was most valued at St Lawrence's. When what was done as a Parish was listed under the headings, Care, Community and District, it was felt that much more was needed. Subsequently the PCC discussed a number of ways of putting into practice the 'Forward Together' policy. Father Masaki's coming to Eastcote in July 1981 enabled new areas of mission to be opened up.

At the end of 1981 Father Masaki was responsible for organising accommodation, in the homes of parishioners and the church hall, for around 50 young people who were part of several thousand taking part in the London Taizé Pilgrimage. About 150 from France, Spain and Portugal were in the area of Pinner, North Harrow and Eastcote from 27 December to 1 January. The programme for the visitors included services, talks, visits, discussions and afternoon workshops. One morning the Taizé Office was said in St Lawrence Church with readings and prayers in four languages. Part of a report about the pilgrimage, said that the melodic and meditative chants in the service would not be forgotten by those who were present.

In his Vicar's Report at the 1982 APCM Father David commented on his appointment as Chairman of the Governors of Bishop Ramsey School in November 1981, seeing it as 'a great opportunity of teaching and commending the Christian faith and hopefully, a valuable link between the school and the parish'. One link was the start of a Mass at St Lawrence Church each term for members of the Upper School, while another was the Bishop Ramsey Carol Service. This comprehensive Church of England school had been officially opened in July 1978, on two sites in Eastcote Road and Warrender Way. In 2008, with the completion of new buildings, the whole school is now accommodated on the Eastcote Road site. The connection with St Lawrence Church remains.

The Golden Jubilee Year events of 1983 are discussed in a separate section but Father David's January 1983 letter shows the concern he had about the infrequent worship of some church members. He recognised that 'these days, as we all know, there are many calls on our commitment and loyalties', but regarded worship and receiving of the Sacrament on Sundays and other major days as a priority. He said 'our Golden Rule needs to be to put our worship of God first on a Sunday', and so related the start of the Golden Jubilee to his 'Forward Together' ethos. The Golden Jubilee celebrations seem to have given 'Forward Together' fresh impetus.

A day conference, for PCC members and leaders of Church Organisations, was held on 2 June 1984 at St Andrew's Church Hall, Uxbridge, to 'consider our spiritual needs and commitment'. Father Ian Stanes, the Willesden Area Officer for Mission, Ministry and Evangelism directed the day's events. Groups discussed biblical passages concerned with the early Church's life and fellowship as the Body of Christ. This was related to life at St Lawrence's under the theme 'Forward All Together'. In Father David's July letter he pointed out that the strengthening of corporate life and development of our mission in the parish was the concern of ALL.

Well before the discussions of that 1984 conference, a February 1984 magazine article explained the purpose of the parish House Group meetings, which were started by Father Masaki early in 1983. The idea came from the time of the very early Church, when people met in each other's homes because they had nowhere else to go. Although members of St Lawrence's House Group attended Holy Communion each week, meeting less formally in a small group enabled them to exchange views and was regarded as a very valuable part of spiritual life. Those who attended felt a bond growing between them, and a strengthening of commitment to the Lord and to their church. It is impossible to say how long this House Group scheme survived. Members of the congregation appear to have been involved for a number of years for the magazine's 'What's On Who's Who' page had Mrs R. (Regula) Sharp as the contact person as late as September 1988.

```
                    WHAT'S ON and WHO'S WHO
PARISH BREAKFAST & BOOK SALE after 9am Service, in the Hall
CRECHE            During 11am Service    Mrs Anne Thiel      R.632079
YOUNG FAMILIES GROUP                     Mrs Anne Thiel      R.632079
SOCIETY OF MARY    1st Sat 4.30pm  Mrs Madeline Gardner R.674986
HOUSE GROUP as announced in the Bulletin Mrs R Sharp      866 7612
DROP-IN CENTRE Mon 2-3.30pm Thur 7.30-9pm Mrs R Sharp     866 7612
YOUNG COMMUNICANTS CLUB 1st Wed  8pm  Mrs Pat Dunhill    R.634071
CHOIR PRACTICE Thurs 6.30-8.30pm  Mrs Frances Bradford    427 2576
MOTHER & BABY/TODDLER GROUP Thu 1.30pm Mrs Gloria Lewi   868 7657
MOTHERS' UNION 1st Thurs    10.30am     Corporate Communion
               3rd Wed       2.30pm     Mrs Monica Hayes  866 1263
WOMEN'S GUILD  2nd Wed       8.00pm     Mrs M Dedman      866 7634
ST LAWRENCE YOUNG PEOPLE    Sun 8pm     Mr Neil Horchover 868 6339
ST LAWRENCE PLAYERS Rehearsals Thu 8pm Mrs Win Brion      868 9037
BADMINTON       Friday      8.45pm      Mr Michael Wray   866 8032
BROWNIES 3rd Eastcote Tues  6.00pm      Mrs Liz Overall   R.637634
         6th Eastcote Thurs 6.00pm      Mrs Jean Weeks    868 5046
GUIDES   3rd Eastcote Wed   6.00pm      Mrs Caroline Dann 868 9144
         6th Eastcote Fri   6.30pm      Mrs Jean Weeks    868 5046
CUBS     4th Eastcote Tues  6.45pm      MrsShirley Horchover 868 6339
SCOUTS   4th Eastcote Fri   7.30pm      Mr Brian Beeston  868 9335
VENTURE SCOUTS 4th Eastcote Mon 8pm     Mr Richard Darby  868 8930
-------------------
HALL SECRETARY       Mrs Pam Gough    50 The Chase        866 2852
PCC SECRETARY        Miss Beryl Orders1 Middleton Drive   429 1298
STEWARDSHIP RECORDER Mrs Sheila Aleong 24a Lime Grove     866 4906
SUNDAY BULLETIN EDITOR Fr David, Fr Alan (Deadline 6pm Wed)
ORGANIST & CHOIRMASTER         Mrs Frances Bradford       427 2576
ELECTORAL ROLL OFFICER & CHRISTIAN AID Miss Doreen Plummer
MASTER OF CEREMONIES Mr John Martin    31 Lowlands Road 868 9002
HONORARY SACRISTAN             Mr Michael Bedford         866 4332
                               Mrs Madeline Gardner       R.674986
BIBLE READING FELLOWSHIP       Mr Michael Bedford         866 4332
LIFTS TO CHURCH etc.           Miss Margaret Case         866 6326
FLOWERS IN CHURCH              Mrs Margaret King          R.635610
FLOWERS FOR WEDDINGS           Mrs Barbara Leeson         868 9185
MAGAZINE EDITORIAL TEAM: Mrs Barbara Higgs 28 The Chase 866 2311
Mesdames Doreen Murrell, Barbara Plummer, Dr Jean Yates
```

What's On and Who's Who - Magazine August 1986

At the 1986 APCM Father David reported on a new enterprise in the way of mission in the wider Eastcote community. Four parishioners had taken part in a bereavement-counselling course to enable them, in appropriate cases, to visit the bereaved after funerals. This lay ministry remains an important part of mission by the parish church. Since November 1996 there has been an annual Sunday afternoon Bereavement Service to which the recently bereaved are invited, but all are welcome. The service is followed by tea in the hall.

Sunday 5 April 1987 saw the launch of the Pastoral Care Scheme. This was a venture 'to strengthen and complement the pastoral care that <u>already</u> exists within the life of the parish'. As the present Parish Link Scheme is similar an account of the 1987 scheme seems relevant.

At his institution as vicar of the parish Father David was authorised to share with the Bishop of Willesden, the cure of the souls in the parish of Eastcote. By 1987 he shared this pastoral care with Father Alan Boddy and Captain Donald Woodhouse, with occasional help from Father Wilfred Beale. At that time the parish had a population of about 24,000 people, with a congregation represented by around 230 households. There could be a ten per cent change in congregation over a year and it was not possible for the clergy to offer direct adequate pastoral care to all. Unfortunately some needing the services of a priest slipped through the net and it was decided that the existing structure could be strengthened. In writing about a proposed means of meeting this challenge, Father David pointed out that the sharing of the cup at every Eucharist is a reminder that each member of a congregation belongs, through Christ, to other members of that congregation, and so pastoral care is the concern of all.

After Pastoral Care slips, completed by most members of the congregation, had been received, the names were grouped together in localities of six to eight households. Members of the congregation living outside the parish boundary were of course incorporated in the scheme. Regula Sharp was to co-ordinate the venture and after two years hand it over to someone else. One person from each group was appointed by Father David to act as lay pastor for two years and then another person in the group would take over. These changes would in Father David's words 'dispel the idea of anyone being the Vicar's MI5 contact!' Each lay pastor was to know and keep in touch with members in their group, and this would require sensitivity and discretion. They were to refer to the Vicar or his staff in the case of specific pastoral need, for instance in the case of serious illness. They were also to help in promoting the mission of the Church in the neighbourhood. The scheme was likened to a 'Spiritual Neighbourhood Watch'. In his 1988 Vicar's Report, Father David regarded the launching of the Pastoral Care Scheme as the most outstanding feature of the past year. He said that it was remarkably successful and very real pastoral work had been done. Mrs Gill Tuffin became co-ordinator of the scheme in 1989.

One initiative is likely to have resulted from the early days of the Pastoral Care Scheme and again illustrates Father David's 'Forward Together'. In August 1987 Father David wrote about a new monthly meeting at which members of the congregation were invited to share with the clergy in praying for those in need. The first meeting was held in September 1987 in the Lady Chapel before the Wednesday evening 7.30pm Mass. The names of those to be prayed for were announced. Any of those present, for whom the prayers were offered, were then invited to go forward to receive the laying on of hands by the priest and any others wishing to share in the laying on of hands. He said that 'the healing power of Christ resides within his Body the Church of which we are all members'. After the laying on of hands there was a period of quiet until the beginning of the Mass.

In January 1988, following several requests from the congregation, the parish embarked on a course of Bible Study. Each course was to be self-contained, meet at different venues and on different evenings to enable as many people as possible to attend. Participants were expected to have read the relevant passages in advance, and attend all sessions, in order to explore the Bible instead of having a series of lectures. Father Peter led the first course. There were no courses during Lent because of the usual Wednesday evening Lent addresses. The subsequent self-contained courses were each led by a different person. From 1989 the study courses usually took place in the George Guest Room. Bible Study evenings remain a feature of parish life.

In October 1988 Father Peter Day introduced the Tuesday House Eucharist into the weekday pattern of worship. Father Masaki had invited St Lawrence Young People to his home in 1981 for the first House Eucharist in the parish, but in 1988 the Tuesday Eucharist was held at the home of a different parishioner, week by week. Rather like the House Group and the 'laying on of hands' meetings, the number of people involved was never high, but Father David saw it as 'a significant contribution to the corporate life of the Church in this parish'. The House Eucharist as a weekly fixture ended at the end of 1989 but was revived on a monthly basis in the early 1990s.

The 1980s could be regarded as a fruitful period as regards Father David's 'Forward Together' but the findings of a Parish Appraisal revealed possibilities for further development. In July 1988 all parishes within the Willesden Area had been asked to undertake this self-assessment, which proved to be very time-consuming. Miss Margaret Case, one of the Churchwardens at that time, co-ordinated the project. Information from the inevitable forms was collated by a small working party in time for an Extraordinary Meeting of the PCC with Father Michael Colclough, the Area Dean on 26 May 1989. In June Father David thanked 'all those who had contributed information to the Parish Appraisal giving a profile of the life our Parish church within the context of Eastcote as a place'. He saw it as 'an excellent tool to enable us to tackle the mission of the Church in Eastcote in the 1990s'. As it turned out this task fell to his successor as Father David left the parish in January 1990.

St Lawrence's Golden Jubilee 1983
There are references to events of 1983 throughout this chapter, but as the year was a landmark in the history of St Lawrence Church the Golden Jubilee Year merits a section of its own.

As early as April 1981 Father David saw 1981 as the planning year for the Golden Jubilee, which 'would be a time for giving thanks for the last 50 years and looking forward to the next 50 years'. The parish would need to ask, 'to what ministry is God calling us in the years ahead? What is our responsibility as the Church within society given our location here in Eastcote?' Members of the congregation were invited to submit their ideas, in writing, to the Vicar and a special Golden Jubilee Committee was set up.

The Golden Jubilee Programme (see page 183) outlines the year's events from 3 April, Easter Day and comments on some of them are appropriate. However, the special events really began with the Wednesday evening Lent addresses given by Dr Martin Israel. He was both a consultant pathologist and a priest, and one of the most sought after retreat conductors and guides to the Christian life in this country at that time. The series of addresses was entitled 'Bursting the Egg of Modern Life'. St Lawrence's was fortunate to have someone of Dr Israel's calibre and renown and members of other churches in the Deanery were invited to the addresses. There was a record attendance of 121 on Ash Wednesday and in the following weeks the attendance was always over 80. Those who heard Dr Israel remember him as an inspiring and powerful speaker.

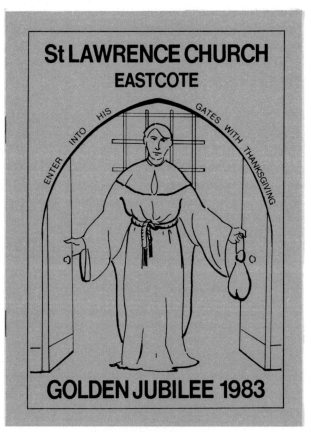

Cover of Jubilee Booklet

> "Truly the Lord is in this place.
> This is nothing less than a house of God.
> This is the gate of heaven."
>
> *Genesis 28 vv 17, 19*

Dear Friends,

This Golden Jubilee booklet contains extracts from past magazines recalling some of the events in the parish of Saint Lawrence, Eastcote since 1920. These extracts make fascinating reading as they touch on the human side of our parish's history. We can truly give thanks for fifty years in which God has been active through the members of the Church here since the consecration of the church building. But let us not forget the years before in which a small band of worshippers worked so hard with the priest—missioners in order to provide the spiritual basis for the erection of the present church buildings.

These extracts are a salutary reminder that a parish can never remain static. In this area of north—west London we are conscious of the continual movement of people, and of rapid change over the years. In 1983 we cannot stand still. As we look forward to the next fifty years, it is right for us to ask, and earnestly seek to answer, searching questions: Into what new patterns of Christian life as the Church here is God calling us? What is our mission to a parish of twenty three thousand souls? Are we using our buildings to the glory of God, serving not only our own needs but the community around us? What are our responsibilities beyond the borders of this parish?

May this Golden Jubilee year be one in which we can rejoice, and enjoy the various events celebrating our first fifty years, in glad expectation of the next fifty years.

May God Bless You All,

DAVID HAYES

Vicar

Introductory Letter – Golden Jubilee Booklet

Father John Weeks, Area Rural Dean, who on 24 April was the first of several visiting preachers in 1983, was also the Vicar of St Matthew's, Yiewsley. The first social event of Jubilee Year was an enjoyable Barn Dance held on the last day of April.

By 1983 St Lawrence Church had links with Bishop Ramsey School and members of the Upper School had visited the church for Mass once a term since 1981. According to the March magazine the concert by the School Choir scheduled for May, was brought forward to March. The choir sang Easter music, including Faure's *Requiem*, *Missa Brevis* by Britten, *Easter Sequence* by

Kenneth Leighton and a setting of Psalm 148 by Holt. Light refreshments were served in the hall after the concert and the whole evening was a great success.

Golden Jubilee Programme 1983

April 3rd	Easter Day – Start of Festival
April 24th	Preacher at 11am – Fr John Weeks, Area Dean
April 30th	Barn Dance
May 6th	Bishop Ramsey School Choir at 8pm
May 14th	Pilgrimage to Walsingham
May 18th–21st	'The Vigil' by Ladislas Fodor, an Easter Play – 'The Players' directed by John Woodnutt
May 29th	Preacher at 11am – Mr Gerald Collins
June 5th	Preacher at 9am – Fr Brian Copus
June 11th	Garden Party
June 18th	Sine Nomine Singers – Concert in Church at 8pm
June 23rd–26th	Flower Festival & Art and Craft Exhibition
June 26th	Preacher at 11am – Fr Ray Philips, Vicar of All Saints, Hillingdon
July 3rd	Preacher at 9am – Fr Anthony Ball
July 10th	Preacher at 11am – Fr Huw Chiplin
July 18th	Cricket Match – St Lawrence v All Saints at Hillingdon
July 24th	Car Rally
July 31st	Preacher at 11am – Fr Clive Pearce
August 7th	Preacher at 9am – Canon Christopher Mutukisna; Preacher at 11am – Fr Peter Goodridge
September 19th	Call My Bluff – St Lawrence v All Saints at Eastcote
October 2nd	Harvest Thanksgiving; Preacher at 9am – Capt Donald Woodhouse; Celebrant at 11am – Canon Bill Hitchinson
October 15th	Quiet Afternoon
October 21st–23rd	JUBILEE WEEKEND
Friday 21st	8pm Mass – The Bishop of Willesden – followed by Sherry in the Hall
Saturday 22nd	Concert – Eastcote Choral Society at 8pm
Sunday 23rd	10am Combined Service – Preacher, The Ven Tom Butler, Archdeacon; 6.30pm Evensong – Preacher, Canon Bill Hitchinson; 7.30pm Parish Supper – in the Hall
December 17th	Nativity Play and Toy Service 3pm in Church

Programme of Golden Jubilee Events

The May Pilgrimage to the Shrine of Our Lady at Walsingham, the first parish pilgrimage, was organised by Father Masaki. Walsingham had been founded as a centre of pilgrimage in The Middle Ages, destroyed by Henry VIII, and restored in the 1920s. Father Masaki wrote in the February magazine that 'every year thousands of Christians visit this beautiful Norfolk village to pray, to offer thanksgiving and petitions and to experience renewal in their faith and commitment'. Father David wrote 'In our Golden Jubilee year it is right that we, as a parish, should make this act of pilgrimage, reminding us of our need of a deeper devotion to God, that we are a pilgrim Church, and that the Church is greater than the church at parish level'. Father David and Father Masaki and over 50 people went by coach, leaving Eastcote at 8am and returning from Walsingham at 5.30pm. The cost, including afternoon tea, was £6 for adults, with reductions for younger people.

In May the St Lawrence Players' *The Vigil*, an Easter play by Cadislas Fodor based on the mystery of the empty tomb, was presented in the church as a Deanery event. John Woodnutt, a practising Christian and a professional actor who lived locally, directed the play. It was a memorable production. A review referred to the church setting enhancing the performance. 'For two hours we were indeed a jury confronted by the mystery of the Empty Tomb, and the question of fraud or truth. The Paschal Candle carried to the central position on the stage, after the exit of all the performers in silence, was most effective'.

Mr Gerald Collins, the preacher at the 11am Service on 29 May, had been a Licensed Reader at St Lawrence's since 1965. He died in August 1983. He had contributed a great deal to the life of the parish. He was a devout churchman, a respected Scouter, and for a time a member of the St Lawrence Players, remembered by some for his performances as the Dame in pantomimes.

Comments on the visiting preachers of June, July and August may be of interest. Father Brian Copus was Vicar of St Mary's, South Ruislip. Father Ray Phillips is of particular interest as he was Vicar of All Saints' Hillingdon, our 'twin' parish. Father Anthony Ball was Vicar of St Paul's, Ruislip Manor. Father Huw Chiplin, Father Clive Pearce and Father Peter Goodridge had all been Assistant Priests at St Lawrence's.

On successive Saturdays in June parishioners enjoyed a Garden Party and a concert by the Sine Nomine Singers. However the highlight of that month was the ambitious Flower Festival and Arts and Crafts Exhibition, which took place from 23 to 26 June. Comments in the August magazine by Father David and John Pearson (member of the Golden Jubilee Committee) indicate that the corporate effort of those who produced the displays of flowers in the church and the exhibits of a variety of arts and crafts in the hall, together with those who gave of their time to be on duty and help with refreshments, all contributed to a most enjoyable and successful Golden Jubilee celebration. The Flower Guild

The Sanctuary, Golden Jubilee Festival

was responsible for many of the flower displays but other church organisations, neighbouring churches, local schools, and other local organisations all expressed themselves creatively with their arrangements and exhibits. The Arts and Crafts Exhibition appears to have surprised people by a remarkable display of talent.

The Friday evening concert given by the Figaro Players, from the Royal Academy of Music, was another form of artistic talent, which gave much pleasure. The Saturday Barbecue was thoroughly enjoyed, apart from the midges! The eventful week ended at Evensong on the Sunday with a presentation by the St Lawrence Young People, based on the Genesis version of the Creation. As John Pearson remarked 'It was perhaps apt that this weekend was brought to a close by the folk who may well be organising our 75th anniversary celebrations!'

Two Jubilee events in July were outdoors. The July cricket match against All Saints', Hillingdon was arranged when it was known that Father David was to preach at All Saints' 10am Service on 17 July, the Sunday of their Golden Jubilee Consecration celebrations. As the parish priests exchanged pulpits it was thought a good idea that the congregations of the churches should meet 'and take the sporting field against each other'. On Monday 18 July some members of St Lawrence's went with the team to Hillingdon to give moral and vocal support. They were a little taken aback to find that All Saints' had two or three players who could 'play a bit'. St Lawrence batted first and made 84 for the loss of nine wickets in the allotted 20 overs. All Saints' scored 86 for two in only 12 overs to win the match 'quite handsomely'. Nevertheless the game between the two Jubilee parishes was considered 'a lot of fun'. The other outdoor event was also enjoyed. The Car Rally saw 15 car drivers, with navigators and crews, driving through parts of Hertfordshire and Buckinghamshire scanning a list of questions and directions. Everyone got back to Eastcote to be met with tea and cakes and a chance to find out 'where that elusive clue was after all.'

August and September were quiet months as regards the Golden Jubilee programme. There was one evening of fun in September, when teams from All Saints', Hillingdon and St Lawrence's once more took 'the sporting field against each other'. This time the 'contest' was at Eastcote in the form of the panel game *Call My Bluff*. Some of the words the teams were asked to define, either truthfully or otherwise as tactics demanded, were 'grimthorpe, quoiler, straddlebob, shoyhoy and zurf'. Once again All Saints' were the winners.

Quite rightly, October was THE month of Golden Jubilee Year. It began with Harvest Thanksgiving when the preacher at 9am Church Parade and 11 am services was Captain Donald Woodhouse. 1983 was a special year for Captain Don personally as in June he was awarded the O.B.E. in the Queen's Birthday Honours, for his service in the Church Army. The celebrant at 11am, Canon Bill Hitchinson, was welcomed as a visitor and he needs no comment here. On 9 October Father Richard Fenwick, the new incumbent of St Martin's, Ruislip preached at 11am on his first visit to St Lawrence's. On Saturday 15 October Father Ray Phillips led the Quiet Afternoon.

The St Lawrence Players' production of *Dear Octopus*, a play in which members of the family are together to celebrate the Golden Wedding of Charles and Dora

Randolph. The grand toast to 'the Family, that Dear Octopus, whose tentacles we never quite escape, nor in our inmost hearts ever quite wish to', was an apt choice.

During the year a number of past parishioners returned to take part in anniversary celebrations.

The Jubilee Weekend of Friday, Saturday and Sunday, 21 to 23 October began on Friday 21 when the Bishop of Willesden, the Right Revd Hewlett Thompson, celebrated at the Dedication Eucharist. Over 150 people attended this 'official birthday' Service. Appropriately, the sherry party in the hall after the Service had a family atmosphere. On the Saturday evening the Eastcote Choral Society presented a concert. Readers may recall that in 1937 members of the church choir started a choral society with that name, so the concert possibly brought back memories of those early days to some of the concertgoers. In 1983 there were normally three Sunday morning services, but on Dedication Sunday the two sung services of 9 and 11am were combined in a 10am Eucharist at which the Archdeacon, the Venerable Tom Butler was the preacher. There was a congregation of 215. Solemn Evensong, when Canon Bill Hitchinson was again the visiting preacher, had a high attendance of 119. The Parish Supper after the Service was another relaxed and happy occasion of fellowship.

The Golden Jubilee Committee had arranged a varied and very ambitious programme. For parishioners, members of the wider community, and a number of ex-St Lawrence parish friends, Jubilee Year had been one of rejoicing and enjoyment in celebrating the first fifty years of St Lawrence Church.

Music at St Lawrence's
During Father David's incumbency the organists were Miss Joanna Paul (1981 to 1983), Mr Tony Smith (1983 to 1984), and Mrs Frances Bradford (from October 1984). When there was no 'official' organist Mrs Millicent Dedman was organist and Mrs Anne Holt and Mrs Enid Lumsden assistant organists. Having sufficient numbers in the choir to lead the singing in church services remained a problem and there were periodic appeals for new voices. One in 1983 was for 'boys and girls from 8 years…and ladies and gentlemen [to make a] commitment to one hour practice, one of the two Sunday morning services and an occasional Evensong'. The two morning services meant the choir seldom sang as one body. Towards the end of 1985, for a trial of three months, on the first and third Sundays members combined to sing either at 9am or 11 am but it was not possible to get the full potential of the choir at either service. By 1987 Father David's concern about the choir being weakened by dividing for the two services, led to the decision that the choir should be at the service attended by the largest number of worshippers, which was the 11am. The situation improved to some extent in 1989 when the 10am Sung Eucharist was introduced. At the APCM Millicent Dedman was presented with a gift token in appreciation of her 33 years as organist for the 9am service.

Throughout the 1980s some of the choir sang at Evensong each week and at choral weddings. The choir also sang at weekday Sung Eucharist services. One memorable occasion was that of Father Alan Boddy's first mass in July 1985 when the choir, joined by members of choirs that Father Alan had sung with in the past, sang Vaughan Williams *Mass in G minor*. At the Ecumenical Carol Services the choir sang with members from local churches. From 1982 to 1988 Palm Sunday Evensong was followed by the choir singing music suitable for Holy Week (*The Cross of Christ* 1982, *St John's Passion* 1987), and was well supported.

Affiliation to the Royal School of Church Music (RSCM) means the choir is not an isolated group and sometimes leads to participation with other churches for special services and 'choir days'. Frances Bradford seems to have given the choir encouragement in a number of ways and the choir took part in the Hillingdon Deanery Festival at St Giles', Ickenham in 1985, a Festival Service at Westminster Cathedral (RC) in 1986, and one at St Paul's in 1988, when she reported there had been 'a good deal of preparation beforehand and an afternoon rehearsal, but the result was appreciated by all'.

EVEN THE VICAR DIDN'T REALISE...........

The Choir announced their 'Entertainment' well in advance but were extremely secretive about the programme. On the night they surprised everyone with the tremendous range of their talents. If you came to the Summer Concert a few weeks ago you would have been delighted to hear again Hazel and Michelle Vincent's adaption of the Pie Jesu, part of Andrew Lloyd-Webber's arrangement of the Mass. Jean Jones and James Beeston gave a lovely rendering of the Nunc Dimittis which would be familiar to those who watched Smiley's People on television. It would take too long to describe in detail all the solo pieces but thanks must go to Iris Castles for her monologue - how often have we also realised, when sitting in Church, that we have left the soup on - or in my case the immersion heater? The first half of the concert ended with an amusing version of Onward Christian Soldiers - the Choir and Clergy must be irritated by all those coughs and sneezes.

The second half of the programme was a performance of Captain Noah and His Floating Zoo, a compilation of ten songs with words by Michael Flanders which tells the story of the Flood from the Old Testament. The final song tells of the rainbow which came after the rains - a joyous note on which to end. The only way in which to round off such a successful evening was loud applause for the choir and a bashful bow by Frances Bradford and, of course, the Vicar's admission that he hadn't realised what a wealth of talent the members of our Choir possess.

Review of Summer Concert in November 1985 Magazine

The first of the 'Choir Entertains' evenings, for which Frances Bradford was responsible, was on 29 June 1985 and was a rousing success. In October 1987, in addition to a variety of vocal and instrumental items, the St Lawrence Orchestra made its 'first' appearance. It may be recalled that there was a St Lawrence Orchestra during Father Bill Hitchinson's time as Vicar and it seems likely when there were sufficient instrumentalists to play together as a group it was given this title. The review of the October 1987 concert referred to 'the volume of applause and cheers usually associated with the last night of the Proms'. The proceeds from these two concerts were for the Building Fund, while that of June 1989 was for the Choir Fund. The programme on that occasion included *Pilgrim* (a musical version of Bunyan's *Pilgrim's Progress*), the St Lawrence Orchestra and individual vocal and instrumental items.

Over the years the church has been host to a number of outside musical groups. The one that has been most consistent is the Sine Nomine Singers. The earliest visit of which there is a record is March 1982 when their concert included Faure's *Requiem*. In December they sang Charpentier's *Midnight Mass*, based on French carols and were joined by the choir of Newnham Junior School in the singing of other carols.

In June 1988 the Sine Nomine Singers presented a light-hearted programme of part-songs, madrigals and nonsense songs, including a setting of three Ben Jonson poems, specially written for this their 30th anniversary year by founder-conductor, Michael Rose. The group still give concerts at St Lawrence's today.

Stewardship
The first Christian Stewardship Campaign had been launched in October 1963 and although Stewardship Sundays had been held in the following years, the number of people with financial commitments in the scheme had dropped from over 300 in 1963 to 102 in 1979. Numbers on the Electoral Roll had risen from 377 in 1963 to 415 in 1964 but by 1979 had dropped to 285. Meanwhile the expenses of the parish had risen considerably, including its contribution to the Diocesan Quota.

As has been noted in the previous chapter a Renewal Campaign for autumn 1979 had been postponed, when it was known that Father Bill intended to retire in the October.

Shortly after Father David became Vicar, in June 1980, preparations were made for a Renewal Campaign to begin in May 1981. Mr John Watson of the London Diocese Christian Stewardship Committee directed the Campaign. The January 1981 magazine contained an outline Campaign and Father David hoped people would avoid these dates when planning their holidays. Successive issues of the magazine had articles by Father David and other members of the Stewardship Steering Committee, which was chaired by Captain Donald Woodhouse. As with the first campaign Phase I concentrated on the financial aspect of Christian Stewardship.

In one letter Father David reminded his readers that Christian Stewardship was at the heart of the Gospel and that our own life and all we have is a gift from God. In another letter he wanted 'all to accept the principle of Christian Stewardship'. He regarded the campaign as a major event in the life of the parish and wrote that if successful 'our spiritual life as a parish and as individuals will be deepened and strengthened as men and women of faith'. In February Father David asked ALL to make use of a Bible Study offered by Captain Don, 'for our deeper understanding of what God says to his people concerning discipline and use of money'. There were challenging articles by Euan Lumsden, the Parish Treasurer, and Tom Kirkley, the Stewardship Recorder. These outlined the parish's obligations, including mortgage payments on a curate's house, and referred to a promising response with new Covenants and increased giving, before the Renewal Campaign had officially begun.

May saw the completion of Campaign preparations. Leading up to the opening Service at 10am on 31 May, Father David, Captain Don and some of the young people of the parish helped to present the challenge of the Gospel at the 9am and 11am Services on the first four Sundays in May. In preparation for visiting homes in the parish after the Parish Supper, a number of parishioners attended three training sessions led by John Watson. They were commissioned at the 10am Service on 7 June, Whitsunday. Father David regarded it as an appropriate day for them to receive the authority of the Holy Spirit 'to go out and talk to others about the Stewardship Campaign'. He made it clear that visits would only be made to those who were agreeable and asked for parishioners to pray for the person visiting them.

At the Stewardship Supper, held on Friday 12 June in the Winston Churchill Hall (opened 1965), the principles and practice of Christian Stewardship were explained by John Watson. Mrs Ann Sykes was the Supper hostess chairman, and in the August magazine Father David thanked the many ladies of the parish who, working as a team, provided 'a truly excellent supper, on a shoe-string budget'. As well as being a Campaign evening it was seen as an enjoyable social occasion, 'with much laughter'. John Pearson, a member of the Campaign Committee, paid tribute to the hostesses, and to the visitors, who welcomed guests on arrival. He regarded the Supper as 'one of the most outstanding and successful events in the history of the Parish Church of Eastcote'. He felt the Campaign had brought a lot of people together and really united the church community.

Details in the Recorder's Report at the 1982 APCM show Phase I of the Campaign was even more successful than had been hoped. Overall membership had increased from 98 at the end of 1980, to 209 at the end of 1981. More importantly, the number of those who were able to sign covenants went up from

48 to 142. With the benefit of new and increased membership, income from Christian Stewardship totalled £11,208 (compared with £4,973 in 1980), and covenanted giving included in that total, enabled the Recorder to claim an Income Tax rebate of £2,141 for the 1981 financial year. The Diocesan Quota had also risen from £4,694 in 1980 to £7,081 in 1981 (see Appendix 10).

The Bishop of Willesden (The Rt Revd Hewlett Thompson)'s Visitation to St Lawrence Church on 11 October seemed to point the way to the other aspects of Stewardship, the use of our time and our talents. In his sermon at the combined 10am Service the Bishop said that St Lawrence was a rich parish, not just financially but in natural and spiritual gifts, and these energies must flow out rather than stagnate. During the remainder of his Visitation the Bishop met members of the congregation, had lunch with the PCC and stimulated group discussions about the needs of the parish. Father David felt the Bishop's visit had 'helped us to look into our potential talents as a parish'.

In 1982 Stewardship Sunday was kept on 23 May. Father David's letter emphasised Christian Stewardship was rooted in the Gospel and reminded parishioners to respond to God's generosity in practical ways. Stewardship was not only about financial giving but is also about the liberal giving of our time and talents. After Evensong on 23 May, St Lawrence Young People (SLYP) gave a presentation on this aspect of Stewardship. A July magazine article reminded readers that one way SLYP did 'their time and talents bit' was in providing coffee and rolls after the 9am Sunday Service, and felt parishioners should support this effort rather than go home to have their coffee. Since the 1981 Renewal Campaign efforts had been made to get more people involved in helping with different aspects of parish life, and parishioners were encouraged to come forward and become part of an existing team. A number of people made use of a new enterprise started in the autumn of 1981, which organised lifts for those having difficulty getting to church.

In 1983 Stewardship Sunday was held on 30 October, a busy month in St Lawrence's Golden Jubilee year. Father David's letter implored parishioners to read the first-class leaflet produced by the Stewardship Committee as it 'concerns a central part of our personal and corporate Christian life'. In 1984 there was only one service on Stewardship Sunday, 28 October, as Father David felt that parishioners 'should together as God's people here, share in worship as we consider our personal and practical response to God's generosity'. Captain Don preached at this service and as Chairman of the Stewardship Committee made it known that it seemed right and proper to have another campaign in 1985.

Income from Stewardship rose in the 1980s but so did in the Diocesan Quota. Details of the 1985 Christian Stewardship Renewal Programme were given in the early 1985 magazines. An article, by Captain Don, entitled 'Honour the Lord with Generosity', recalled the message of Prebendary Michael Saward's February sermon, in which he had compared Eastcote with his own parish of similar size in Ealing, and which had seen a tremendous growth in giving in recent years. He challenged the pattern and practice of Stewardship in Eastcote declaring that the church had set its sights too low.

> <u>Sunday 17th February</u> - The Rev. Michael Saward, Vicar of Ealing, preaches at 9 and 11 a.m.
>
> <u>Sunday 19th May</u> - Commissioning of Stewardship visitors at 6.30 p.m. The Rev. Michael Saward preaches.
>
> <u>Friday Evening 24th May</u> - Stewardship Supper at the Winston Churchill Hall.
>
> <u>Whitsunday 26th May</u> - The Rev. Tom Butler preaches at 10 am.
>
> "Where your treasure is, there will your heart be also".
> Matthew 6 v 21

Christian Stewardship Renewal Campaign 1985 (from Magazine)

As in previous campaigns much preliminary work was done before the commissioning of Visitors. Alan Boddy chaired the renewal Programme Steering committee, Arthur Plummer was chairman of the Visitors, Mary Raper of the Hostesses, and Hazel Martin co-ordinated the catering for the Supper. Once again the supper was a social success.

Financially the Campaign does not appear to have been as successful as that of 1981. Mrs Sheila Aleong who had succeeded Tom Kirkley in 1985, after seven years as Stewardship Recorder, presented details at the 1986 APCM. There were 162 members in the scheme compared with 209 in 1981, of whom 122 had Covenants compared with 142 in 1981. The Electoral Roll was 240 in 1985 compared with 302 in 1981.

Special Overdraft and Thanksgiving Sunday appeals in 1986, 1987 and 1988 helped towards balancing the books but what was really needed were commitments of more generous giving by an increased number of people. Obviously there are losses in church membership for various reasons, and other people become new members of the congregation. For old and new members the financial needs of the parish need to be explained on a regular basis. A Parish Meeting held in September 1986 to discuss the life and finances of the parish does not appear to have been well attended, in spite of a three-line whip from the Vicar. Among the presentations one by John Watson and Roger Clayton Pearce of the Diocesan Stewardship Committee compared the results of the 1981 and 1985 campaigns with those of other parishes, pointing out that the average giving of £2.65p.each week was below the national average of £3.54. It was accepted that with the recent Overdraft Sunday together with the parish undertaking the re-ordering of the Sanctuary and an ambitious Building Project 'people were tired of such calls on their pockets'.

At the 1989 APCM the Recorder's Report, on the year ending 31 December 1988, showed an increase on 1987, even though there had been a decrease in the amount of tax reclaimed. Father David said it was encouraging that there had been 7 new and 11 increased covenants. He also commented that the size of contributions did not necessarily reflect the spirituality of a person's giving. He also thanked Sheila Aleong for her work, which was 'carried out meticulously, going beyond the facts and figures to the exercise of pastoral care'.

Throughout the 1980s articles by the clergy and laity on all aspects of Stewardship appeared from time to time in the magazines, though the emphasis seemed to be on the need for more generous giving of money. Annual Stewardship Sundays continued, and were held from 1986 to 1989. The visiting preacher in 1987 was the Venerable Eddie Shirras, the Archdeacon of Northolt. In 1989, with his many years of experience of the Church in a part of the Anglican Communion very different to that of Eastcote, Bishop William Herd, formerly Bishop of Karamoja in the Province of Uganda, brought a fresh challenge to our thinking about Christian Stewardship.

At the 1990 APCM Geoffrey Andrews, member of the Stewardship Committee, stated that the Diocesan Quota, or Common Fund as it was sometimes known, was the main cause of our financial problems, and had risen by 50% over the last three years whereas our giving had risen by only 20%. Figures show that numbers on the Electoral Roll fell from 302 in 1981 to 225 by 1989, the last full year of Father David's incumbency. During the same period Stewardship giving had gone up from £11,208 to £27,905, but the Quota had risen considerably, from £7,081 to £31,008. There was some reduction in the Quota while the parish was without a Vicar (from the end of January to the end of October in 1990), but would rise when there was a new Vicar of Eastcote. A Stewardship Renewal Campaign was inevitable in the not too distant future.

What about the time and talents aspects of Stewardship? The work of sub-committees of the PCC, the involvement of parishioners in the fêtes and markets and other church activities, together with the names of people on the 'What's On and Who's Who' pages of magazines, indicate that many gave of their time and talents. A number were routine offerings and as such not often mentioned in magazine articles. However, there were some references. For example in June 1986 an article appealed for help with the mammoth task of decorating the interior of the church, wanted volunteers to join the scheme which gave people lifts to church, invited help with making items for the Autumn Fayre, and hoped members of the congregation would, 'swell the ranks of our teams for assembling the magazine'.

The Drop-in Centre started in April 1985, is one example of the giving of time and talents for the benefit of the wider community of Eastcote. The Pastoral Care Scheme set up in 1987 was another illustration of caring for others.

Dealing with the Deficit in the Interregnum
During the interregnum of 1990 the budget for the year again showed an expected deficit. The Stewardship Committee suggested that, in place of a gift day to make up the shortfall, members of the congregation should give of their time and talents to raise funds. In the July magazine Alan Wright, Chairman of the Committee, referred to two of the proposals. One was to hold a Bring and Buy sale after the 10am Service on the first Sunday in each month, starting on 1 July. People were encouraged to 'bring something along and drop in for a coffee or tea and a chat'. This proved successful and similar 'after service' sales have been held over the years. The second proposal was to hold an Antiques Fair, possibly in October. In the event it was held on 8 December but being nearer Christmas, it was hoped that it would attract people looking for unusual presents. The income from letting stalls to people who deal in antiques, together with the proceeds from stalls manned by members of St Lawrence's, the sale of light lunches and teas, plus the inevitable raffle, all contributed to make the event a success, both financially and socially. It was the first of many annual Antiques and Bric-a-brac Fairs efficiently organised by Mrs Joan Sherriff-Gibbons (see Appendix 11).

By the beginning of September the response to the appeal for fundraising had been very encouraging and over £700 had been raised. At the 1991 APCM the Stewardship Committee was able to report that the fund raising efforts in 1990 had been worthwhile. During the year there had been 'two jumble sales, the Antiques Fair, coffee mornings, open gardens, Sunday lunch and Christmas lunch, taste-ins, a beetle drive and bring and buy sales'. These events had not only helped to clear the deficit but everyone had benefitted from these times of fellowship.

Charitable Giving
In the course of its history St Lawrence Church has given financial support to numerous charitable causes. Among societies and organisations receiving support have been a variety of overseas missions, those associated with emergency relief both at home and abroad, and national and local projects. Support for some has been for a short period while for others it has spanned over many years. Each chapter in this book refers to only a few charities but sources evidence a church community always conscious of, and committed to, the custom of tithing. An article in the April 1982 magazine gives an explanation of this policy, and gives details of giving for 1981.

In addition to this tithing from income there were, and still are, separate appeals for specific causes, when money is raised through special collections, self-denial boxes, and a variety of fund-raising activities. Sometimes part or all the profits from a fête, fair, flower festival, or other church event has been allocated to a charity.

To some extent the success of any particular cause depends on the response of the individual but the manner in which information is presented can be influential. Visiting speakers with personal experience and parish connections, together with the enthusiasm of a few, can encourage support as has been shown in the previous chapter concerning work in Mauritius and Nepal.

```
                              TITHE

     In 1981 the parish gave away 10% of its income as is its
custom.  This amounted to £2,131, as against £1,390 in 1980.

     People sometimes ask what our various money-raising events
are "in aid of".  One could answer that they help towards our
donations to 21 good causes.  The Christmass Market and Summer
Fete/Plant sales raised £1,698, that is 80% of the donations'
total.  Or we could say they help to pay the electricity bill,
or the mortgage interest on the curate's house.  The point
really is that we are tithing our total income to help other
people.  Here are the details of our giving:

OVERSEAS                                                £
     Friends of Mauritius                              200
     International Nepal Fellowship                     50
     Leprosy Mission                                    25
     S.P.C.K.                                           40
     U.S.P.G.                                          255
                                                       570
HOME
     Additional Curates Society                         50
     Bishop Ramsey School                               50
     Chichester Theological College                      2
     C.E.Childrens Society Centenary Appeal             20
     Church of England Pension Board                    70
     Church Army                                       200
     Clergy Orphan Corporation                          30
     Fellowship of S.Nicholas Childrens Homes          200
     Helen House                                       200
     Hillingdon Deanery Wel-Care                       304
     Kings College, London                              50
     Order of S.Paul, Alton                            100
     Ruislip-Northwood Old Folks Association            25
     S.Lukes Hospital for the Clergy                   160
     S.Francis House, Hemingford Grey                   50
     S.Paul's, Ruislip Manor, Appeal                    50
                                                     2,131

               Euan Lumsden - Honorary Treasurer
```

Charitable Giving.
Page from April 1982 Magazine

The Church Army

It is likely that support for the Church Army during these years was related to having Church Army personnel worshipping at St Lawrence's, although the tradition of supporting the Church Army goes back to the days of Father Godwin.

Until they retired to Folkestone in the early 1980s, Captain Frank Collier and his wife Freda had been members of the congregation for many years. Captain Don and his wife Norma, and son Ian, have been at St Lawrence's since 1972.

In this period there are several magazine articles about the Church Army, which celebrated its Centenary from autumn 1981 and throughout 1982. In January 1981 the launch of a Centenary Appeal for £2¼ million was reported. In June 1981 Captain Frank led an Act of Remembrance at the tomb of Wilson Carlile (Founder of the Army) in the crypt of St Paul's Cathedral, just before the Annual Founder's Day Service. On 9 June 1982 HM the Queen attended a Thanksgiving Service in Westminster Abbey, the next day the Archbishop of Canterbury Commissioned new Officers in Southwark Cathedral, and an evening Open-air Rally took place in the Jubilee Gardens. A paragraph of one magazine article is worth quoting. 'After 100 years, the name is still the same, its Officers are still Captains and Sisters, the emphasis still on the Gospel that is relevant and the proclaiming of that Good News by word and showing it forth through practical concern for people in need'.

Throughout the 1980s appeals were made for collectors on Church Army Flag Day in the Eastcote shopping area, or house-to-house collecting during the preceding week. For many years this was organised by Mrs Mollie Wilson. In 1983, Jubilee Year and the year Captain Don was awarded the O.B.E., some of the proceeds from the Flower Festival and Arts and Crafts Exhibition went to the Church Army (£95.50). By 1989 Captain Don had taken on the role of organiser and in his appeal for help suggested people might hold a coffee morning, instead of collecting. At least one coffee morning, with a Bring and Buy Stall was held in 1990.

Helen House
One new cause supported by St Lawrence's during the 1980s and beyond, was Helen House. In August 1980 the Anglican Community of All Saints' Sisters at Oxford launched an appeal for £400,000 to establish the first hospice in the country for children suffering from incurable disease. In order to give her parents a short break, Mother Frances Dominica and the nuns had on several occasions nursed a little girl called Helen, who was suffering from irreversible brain damage. Their experiences led to the appeal, to which there was good response. A magazine item in July 1981 reported that £350,000 had been given 'by a vast number of friends throughout the country'. By the autumn of 1982 Helen House had had its first patients.

In his February 1981 letter Father David referred to the £603.43 raised from the 1980 Advent Self-Denial Box, and collections at the Guides and Scouts Carol Service and Midnight Mass, saying 'it was a marvellous result and reflects a real spirit of giving in our parish for which we thank God'. Fund-raising in 1981 was through the Lent Self-Denial Box (£229.82), a coffee evening (£75.32), the 'Wishing Well' run by the 6th Eastcote Guides at the May Fair (£8), and part of the proceeds of a sponsored walk by five members of the Junior Youth Club (£20).

OUR FIRST CHRISTINGLE SERVICE

This beautiful candlelight service has been adopted by the Church of England Childrens Society as a way in which the members of a parish can come together to exchange their gifts to the Society for a Christingle orange, symbolising the sharing of what we have as individuals.

The Christingle is an orange with four sticks of fruit wrapped around with a red ribbon and topped by a candle. The orange represents the world; the red ribbon, the blood of Christ and the candle, the Light of the World, Our Saviour Jesus Christ.

The Christingle Service is a family service, where children, parents and grandparents can come together to share and to give. Thousands of children and their families have already benefitted from Christingle Services through the work of The Children's Society. It is the voluntary giving of talent, time and money as well as the support of prayer all over the country which helps the Society to carry on its vital work.

The Children's Society works with its family and community projects, specialist residential homes, centres for children with mental and physical handicaps and through adoption and fostering.

It believes in families and that is what Christingle Services are all about: family services to share the fruits of family life, that many may reap its rewards.

A Christingle will be held in this parish on Saturday, 15th December at 5.00 p.m. when a Christingle will be given to each Sunday School child when they present their gifts. Do join us for this special service.

Page from Magazine December 1984

On 3 May 1982, Bank Holiday Monday, the Fellowship Committee organised a Ploughman's Lunch and sale of goods. Guest of honour was Mother Frances, who told the assembled company of the purpose and aims of Helen House. Following this highly successful event, which raised around £350, a visit to All Saints' Convent was arranged and the party had a guided tour of the building that would house the children and the people who would be caring for them.

In 1983, like the Church Army, Helen House benefitted from the profits of the Flower Festival and Arts and Crafts Exhibition (£341.60). In a long letter in the October 1983 magazine Mother Frances said that over 50 children had stayed at Helen House in its first nine months, some returning for many visits. Parents, a few brothers and sisters had also stayed, and even one golden Labrador whose young mistress was a frequent visitor. She also reported that through the generosity of many thousands of people they had an Endowment Fund of £1 million to meet the annual running costs which had risen to £125,000.

The Children's Society
Like the Church Army support for The Children's Society goes back to the days of the Mission Church. In 1984 a new association with the Society began when the first Christingle Service (see previous page) was held.

ST LAWRENCE CHURCH AND ECUMENICAL ACTIVITY IN EASTCOTE
Christian Aid
By the time Father David came to the parish support for Christian Aid Week, each May, was established as a combined effort of St Thomas More, St Andrew's, Eastcote Methodist and St Lawrence churches. As now, each church was allocated certain roads with the intention of having sufficient volunteers for a house-to-house collection to cover all the roads in Eastcote. During the 1980s Miss Doreen Plummer was the co-ordinator for the four churches until 1988, when Mrs Susan Everitt of St Andrew's took over, and Mrs Sadie Wright became the organiser for St Lawrence's. Each year there was a plea in the magazine for collectors. Each year the necessity for supporting Christian Aid seemed to become more urgent. In 1986 Members of St Lawrence's covered only 20 of the 50 roads allocated but by 1988 only five were not done. Results of the annual collection are not available for every year but support grew, rising from £715 in 1980 to £1,757 in 1988. Some years the collection had a special focus with catchy phrases, such as that of 1985 on The Homeless, with *Charity begins with the Homeless*.

In February 1987 Father David asked magazine readers to attend two meetings at St Thomas More Church, on the worldwide and varied range of Christian Aid work. Captain Don emphasised that support for this cause was a joint venture of all the Eastcote churches and it should 'be seen that we care together'. Reporting on these meetings Captain Don was in no doubt that 'Christian Aid demands our ongoing and increasing support'. Collections at these meetings enabled the churches to send £56.40p to the famine appeal for Mozambique. In March a Coffee Morning and Bring and Buy Sale was held at the Methodist Church Hall as part of the Eastcote churches combined support for Christian Aid.

In addition to supporting the Eastcote churches' efforts, in 1984 the St Lawrence Harvest collection and the profits from the Harvest supper were sent to the Christian Aid Famine Relief Fund. In Lent 1985 parishioners were encouraged to give up a meal a week and to put the money saved into the Lenten Austerity box at the back of the church, with the proceeds going to Christian Aid-Ethiopia.

Other Ecumenical Activities
Another example, which like Christian Aid Week had become a joint venture in the mid 1960s, was the Eastcote churches' observance of the annual Week of Prayer for Christian Unity in January. Each year a united service was held in one of the four Eastcote churches. Some progress towards unity appears to have been made since Father Bill's 1979 letter 'that the unity of the Church may come about in God's own way and in His own time'. In his January 1983 letter Father David wrote 'We give thanks to God for the understanding, trust and working together already achieved by the churches worldwide, and we pray for that Unity that our Lord wills for His Body'.

It has been shown that the Eastcote churches were 'working together', and in January 1985 the annual united service developed into what became known as Tea Table Conferences. The format of these events has been maintained to the present time. A talk with an opportunity for questions was followed by tea, allowing further discussion in a less formal context, and the proceedings concluded with a service. The talks were on a variety of themes including, The Role of the Church in Argentina (1985), the Roman Catholic View Point, when Father Gerard Hughes author of *God of Surprises* was the speaker (1986), and Towards the Year 2000 - reconciling the global and the local (1989). Since 1985, when the first was held at St Lawrence Church, these sociable Tea Table Conferences have been held at each of the local churches and have led to greater 'understanding' and 'trust' between members of different Christian denominations in Eastcote.

Another local inter church experiment deserves mention. As part of the scheme to experience worship in each other's churches, one church hosted the only Evening Service on fifth Sundays of the month. Taking out festivals and summer holiday months, in practice this amounted to two occasions a year at the most. The first was held in June 1985 and the last seems to have been in April 1989.

One annual service that was always held at St Lawrence Church during these years was the well-attended Eastcote Churches' Carol Service. Local schools that intermittently held Carol Services at St Lawrence's in the 1980s were Bishop Ramsey, School Junior, Grangewood (in Fore Street, for children with severe learning difficulties), and Reddiford (in Cecil Park Pinner, a private school). A District Girl Guide Carol Service was held one Sunday afternoon in 1986 but this does not appear to have been repeated.

Fellowship

One dictionary's definition of fellowship is, participation, sharing, community of interests, companionship, intercourse, friendliness. Throughout its history the fellowship of the community of St Lawrence Church has been of great importance. Father Godwin's 'sociable' community and Father Bill's 'Worship and Fellowship' have illustrated this in the years up to 1980. Father David Hayes was convinced that fellowship through various social events played its part in strengthening the corporate life of the church community, which in turn would help to further the mission of the Church, all part of his 'Forward Together'.

Fêtes and Fairs

Page from April 1985 Magazine

The Fellowship Committee was responsible for organising many of the fund-raising and social events, some of which had become fixtures in the annual calendar. Most notable were those referred to variously as Spring/May fêtes, fairs, or fayres, and Christmas or autumn fairs, fayres or markets, which have always involved the wider community of Eastcote. Father David's comments in numerous magazine letters indicate the value he placed on such events. Expressing thanks to all who in any way contributed to the success of the 1981 Christmas Market, planners, helpers and those who came and spent their money, he wrote 'As Vicar, I believe strongly that we need such events as an expression of our life as the Body of Christ here. It is important for members of our different Sunday Services to meet on a social level…from a clergy point of view such events are golden pastoral opportunities, in that one is able to meet so many people'.

Reference has been made elsewhere to fund-raising efforts, many in connection with the Building Project. The Appeal led to greater efforts to raise money for St Lawrence Church and from 1984 some of the profits from the fairs and fêtes were allocated to Building Funds, but not all. Coffee and Bring and Buy mornings, Beetle Drives and other well-tried means raised money which went towards stocking stalls for the autumn fairs. One Grand Jumble Sale in October 1984 raised £200, which helped to bring about a profit of just over £2,000 from the Christmas Market that year. In 1989 half the £1,801 from the 1988 Christmas Market went to the Building Appeal, and half to purchasing *Celebration Hymnal*, new hymnbooks to supplement the *English Hymnal*, which had been in use since 1979.

From its early beginnings St Lawrence Church fêtes were always looked upon as enjoyable social occasions with less emphasis on fund-raising, than was the case with autumn events. In June 1982 in place of the usual type of fête the Fellowship Committee decided to turn the clock back to serve afternoon tea on the Vicarage lawn, if the weather was suitable, and have games for the children. The weather was not suitable but the 'indoor' garden party had 'smashing teas', and the rain held off long enough for games and competitions for the children. In June 1983 there was the celebratory Jubilee Garden Party. After the official launch of the Building Appeal in 1984 the annual fête seemed to be on a more ambitious scale. Greater corporate effort from the parish, a wide variety of stalls and sideshows, plus the popular ploughman's lunch had good support from the wider community of Eastcote, and brought increased profits (see Appendix 11). However Father David's May 1986 letter reminded parishioners that the May Fair was a 'valuable pastoral opportunity to meet the wider community of our parish whom we are here to serve'. In 1989 the profit of £1,237 went to the Church Urban Fund, which had started that year.

Harvest Suppers
The practice of having Harvest Supper after Solemn Evensong on Harvest Thanksgiving Sunday, which had been started in 1979, continued throughout the 1980s, except for 1983, when there were Jubilee celebrations. Perhaps people were encouraged to attend Solemn Evensong with the prospect of a good supper and an enjoyable social evening after the service. Usually attendance was over 80. For several years Mrs Beryl Clements and her team organised these suppers.

For a few years the profits from the Harvest Suppers went to SOS Sahel (Trees for Sahara).

Other Social Events
As in the past the type of social events varied during the 1980s. Several were intermittent, popular for a time, and then not so well supported. This may have been because people wanted something different, or no one was willing to organise that particular event. There were also changes in congregation membership with different age groups having different interests that affected social events. The Fellowship Committee worked hard, and in organising a variety of events did much to strengthen the corporate life of the church community.

New Year Socials were held in 1981, 1982 and 1983, in the form of the traditional mixture of fun and games. In 1984 there was a different form of entertainment when St Lawrence's had its first Quiz Supper, described as 'a sort of team Mastermind with a break for fish and chips'. The cost was £2.50p per person (supper inclusive). This successful venture was organised by John Martin and his family in aid of the Building Project. It seems the next Quiz Supper was one in 1987 organised by the St Lawrence Players.

In 1984 and 1985, after Evensong on the Sunday before Shrove Tuesday, many people enjoyed a variety of savoury and sweet pancakes served by Regula Sharp and others. There were a number of Coffee and Bring and Buy Sales, always regarded as occasions of fellowship as well as fund-raising events. There was the occasional Barbecue or Ploughman's Lunch. In 1988 and 1989 there were opportunities to enjoy food and fellowship at the Friday Luncheon Club for which Marie Gunson was largely responsible. The magazine invitation to the Christmas Lunch in 1989 shows what good value these meals were. 'Come and enjoy a turkey with all the trimmings and Christmas pudding…£3.75p'.

On the more energetic side there were square and barn dances, which were thoroughly enjoyed by the relatively small numbers who supported them. Parish rambles remained a fairly regular feature on the social calendar, and were popular as occasions when people had time to talk and get to know each other better. In 1981 St Lawrence Young People (SLYP), challenged 'those of advancing years', St Lawrence Old People (SLOP), to a cricket match to be held at Bishop Ramsey Upper School. From then on the SLYP versus SLOP cricket match was a summer fixture, with one exception, that of 1983 when the match was between St Lawrence Church and All Saints' Church, Hillingdon.

In June 1982 about 30 people took a trip down the River Thames in a boat named *New Princess of Wales*. Despite the 'vagaries of the weather' it was voted a 'super afternoon out'. Thanks were given to Joe Orme for the organisation and to everyone who made it such fun. The other major outing of 1982 was to Helen

House, a visit resulting from the successful Ploughman's Lunch held in aid of the hospice.

The wording of the invitation to an event in 1985 was in keeping with Father David's 'Forward Together'. 'Are you an 8 o'clocker? Are you a 9 o'clocker? Are you an 11o'clocker? Are you an Evensonger? Then all come and be a 12 o'clocker at the Parish Sherry Party on Sunday, 21 April'. A more ambitious event was planned for the following year, when ex-clergy and ex-parishioners were invited to a Parish Re-union afternoon tea on Saturday 21 June 1986. Parishioners were asked to mention the re-union when sending Christmas cards in 1985, and nearer the day they were asked to make a special effort to attend. It was seen as a good opportunity to renew old friendships and show off the work already begun in connection with the Building Project.

A new venture initiated in 1989 was the St Lawrence Social Club. At the inaugural meeting on Wednesday 26 April ideas were put forward as to what people would like to do. The aim was to provide an opportunity for members of the congregation to meet together socially and strengthen fellowship within the church; another manifestation of Father David's 'Forward Together'. Members of the first committee were Enid Lumsden, Derek Allcock, Sadie and Alan Wright, and Ann Sykes as treasurer. During its first year there was an average attendance of 30 members and friends. Among the varied programme of activities were a mystery coach trip, a speaker on The Samaritans, a theatre outing to Watford Palace to see Brian Murphy in *Roll on Friday*, an Any Questions evening, and a Christmas dinner and dance (£4 including a glass of wine). This new venture proved to be successful.

Mothers' Union
Although the membership of the MU never reached the high numbers of the 1950s and 1960s, the Branch was still of importance. Father David Hayes regarded it as not 'merely another parish organisation at branch level' but part of a worldwide organisation within the communicant life of the Church of England. Monica Hayes, in a long article in the November 1986 magazine, reflected on various aspects of the MU. Among other points she maintained that it was nationally, 'a positive force for influence on social trends'. She also felt that locally, its purpose 'is to be informed about our world, to form opinions on issues relevant to family life, to hear what other people do in their various walks of life, to support MU endeavours overseas, and above all to pray'.

The monthly Corporate Communion services, followed by coffee and fellowship, remained the focus of MU activities, and keeping the annual Wave of Prayer each January witnessed to the worldwide nature of the MU. The pattern of attending and hosting services and meetings with other MU branches within in the Deanery continued. Each year a few members attended the Diocesan Festival Service at St Paul's Cathedral, usually obtaining the services of a willing male to carry the banner in the procession.

The 50th anniversary of St Lawrence's as a fully communicant Branch was celebrated on 31 October 1985, with a special Corporate Communion. Members were pleased that Father Bill Hitchinson returned to give the address, and Father Clive Pearce also attended the service. Mrs Joan Hitchinson was among a number of past and present members who were welcomed to the sherry party that followed. The celebratory cake, iced in blue and gold with the MU badge, was made by Mrs Barbara Plummer. There was much reminiscing over old photographs, and the list of Enrolling Members.

The Enrolling Member was the official link with MU headquarters, Mary Sumner House. At the November 1982 AGM and Triennial Elections Mrs Edna Pearmain agreed to hold this office for a further three years, and Mrs Edith Bedford remained as Treasurer. At the Triennial General meeting in January 1986 they both resigned, each having served nine years, but remained valuable members of the Branch. Monica Hayes was elected Branch Leader (the designation 'Enrolling Member' had been discontinued), a position she held until just before Father Hayes and his family left the parish. In October 1989 Mrs Marion Hayman became Branch Leader.

Meetings showed that members were 'informed about our world' and fulfilled the other objectives to which Monica Hayes referred. There were a number of talks about different areas of the world, with many meetings being open to visitors. Among them were talks on Israel, Burma, and Bulgaria. One, given by the Rev Peter Goodridge, St Lawrence's former first curate, was about a visit to his sister in India. Another, by Mrs Jane Arden in 1983, was on the Province of Central Africa, where her husband Donald had been Bishop from 1971 to 1980. As Assistant Bishop of Willesden (1981 to 1994) Bishop Donald visited St Lawrence's to confirm candidates in 1984, 1987 and 1989. His most recent visits were in Holy Week 2009 when he preached at the Monday, Tuesday and Wednesday Eucharist Services, and December 2010 when he confirmed eleven adults.

Among the wide range of topics of speakers 'on issues relevant to family life' were those on the MU and the single parent, Traidcraft, drugs and AIDS. There were speakers on The Children's Society, the Salvation Army, the RNLI, and other charities. Talks about 'what other people do in their various walks of life' included Christian Education and Church Schools given by Michael Bedford, at present a Priest at St Martin's, Ruislip, and Acting and Christianity by John Woodnutt, a local actor. Mr Lloyd Jones, a member of the St Lawrence congregation, gave a talk about Westminster Abbey and a few months later he conducted a tour of the Abbey.

Bring and buy sales and raffles, at both Branch and Deanery meetings, raised money to support the work of the MU both at home and overseas; this became increasingly difficult with a small membership. The sale of Webb Ivory cards and gifts at monthly meetings was another small but useful source of revenue.

However with the proceeds from jumble sales the Branch was able to make gifts to the church. One example was the pair of wedding stools, used for the first time at the wedding of Father Masaki and Alicia in October 1982. As with other church organisations the MU played their part at the annual fêtes and markets.

Of course MU members participated in the various social functions of the church community, but they also had their own social events. The annual Garden Party remained a popular event. The one in 1981, when MU members from St Andrew's, Uxbridge, St Anselm's, Hatch End, and St Paul's, South Ruislip were present, as well as members from St Lawrence's congregation, had a record attendance and appears to have been a particularly happy afternoon. It was also an opportunity to chat with the new curate, the Revd Masaki Narusawa.

Apart from one Branch outing to Ely in 1989 it was through Deanery organised outings that members visited Tewkesbury, Bristol, Bath, Bury St Edmund's, Worcester and Canterbury during the early 1980s.

In addition to the MU January Wave of Prayer members were encouraged to support the Women's World Day of Prayer, held at different Eastcote churches each year. In keeping with Monica Hayes' reflections that one of the purposes of the Branch was 'above all to pray' a revitalised monthly Wednesday afternoon Prayer Group started in May 1988. Meetings, which were held at different homes in the parish, had a pattern of a Bible reading, followed by vocal prayer and a period of silence. In 1989 all members in the London Diocese were asked to keep Wednesday, 24 May as a day of prayer, bearing in mind their objective 'to maintain a worldwide fellowship of Christians, united in prayer, worship and service'.

St Lawrence Women's Guild
The aims of the Women's Guild 'to worship, to give service and help, and to have fellowship' were maintained, though with increasing difficulty as the organisation failed to attract many new members. The three-monthly Corporate Communion services continued and the group did much to 'give service and help', as well as having a great loyalty to the Guild and real fellowship with each other.

Each year Guild members gave service and help in a number of ways. As in the past, members provided posies for Mothering Sunday, in 1982 and 1983 as many as 200. They were also responsible for refreshments at Confirmations, some years a buffet supper other years just coffee and biscuits; it rather depended on the time and day of the week. In 1981 they also provided food after the Civic Service, apple tarts for the Stewardship Renewal Campaign supper and helped with refreshments on a number of other occasions. The Guild provided welcome packs of groceries to Father Alan Boddy (1984) and Father Peter Day (1987) when they arrived in the parish. The Guild was involved in the 50th Anniversary celebrations in 1983 including the Flower Festival.

In addition to their contributions at fêtes and markets the Guild held a variety of fund raising events. They needed money in order to finance some of their service and help, such as flowers for the Mothering Sunday posies. They seem to have had particularly successful jumble sales in the early 1980s, raising around £65-70 each year, and this enabled them to pay for more experienced speakers at their meetings. During Father David's incumbency the Guild made many donations both to church funds and outside charities. Among other gifts it sponsored Father David and Richard Darby, in their London Marathon for the Building Project.

Arrangements were made for many outside speakers. Among the more unusual ones were a Display of Costumes from the Crinoline to the Mini-skirt, Court Dancing from the 13th to the 18th Centuries, Mrs Eileen Sheridan's cycle ride from Land's End to John O'Groats (872 miles in 2 days 11 hours and 7 minutes), Windmills, Well Dressing, and Norfolk Village Signs. In 1981 and 1987 Belltones from Denham gave programmes of 'off the table' hand bell ringing. In 1984 there was an evening of working with glass. One of topical interest was on the Thames Barrier in 1985, a meeting having a large audience with visitors from local churches. Apparently the work on the Barrier, which began in 1974, took much longer than expected with 2,000 men working on the site day and night, depending on the tides, and costing £450 million. It was finished in 1982 and officially opened by the Queen in May 1984.

Guild members and others associated with St Lawrence's often gave talks or demonstrations. Topics ranged from cookery demonstrations, flower arranging, dressmaking, macramé and keep-fit, to showing slides and talking about holidays. St Lawrence clergy not only chaired Annual General Meetings of the Guild but also gave talks to members; Father Masaki on the history of vestments, Father David one on the Holy Land and another comparing the life of a priest in a county diocese with that of a London suburb, Father Alan on Ancient Egypt and Father Peter on why and how he came from being a pharmacist to Assistant Priest at St Lawrence's. In 1987 the Revd Reg Ames, from the adjoining parish of St Edmund the King, Northwood Hills, gave a talk about pantomimes. He had played dame for many years at that church's annual pantomime and members enjoyed an entertaining evening.

Outings included those to the Royal Mews and the Queen's Gallery (1980) and Kensington Palace (1985). They had 'a lovely warm and sunny day' for the trip to Gloucester Cathedral and Painswick House in 1981. There were visits to the theatre in Windsor and the more local Beck Theatre. The only outing in 1983 was to Piccotts End Medieval Wall Paintings at Hemel Hempstead. The last outing of the Guild was in 1988 when members, and Captain and Mrs Woodhouse, went to Bekonscot Model Village at Beaconsfield. This visit was connected with a talk Captain Don had given in 1987 on many aspects of the Church Army's work in general, and in which he referred to its association with the village, and the fact that profits from the entry fees were donated to the Church Army. In spite of it being a chilly evening with heavy rain the group appreciated the escorted tour and the helper who 'operated the splendid model trains from the Electronic Room'.

WOMEN'S GUILD

Last year the Women's Guild went into abeyance while seeking God's guidance on their future. At their Annual General Meeting in February the members of the Guild decided that the time had come for the Guild to be disbanded - a decision reached with much sadness.

The Women's Guild has been an important part of the social fabric of our parish for around twenty-five years. Under the chairmanship of Topsy Dedman they have held regular monthly meetings of great variety and interest. Through the work and prayer of the Guild many women have been drawn into the life of the church. The Guild has also supported the parish in practical ways. In years past they have given us a Chalice and Paten and new Cruets, and each year have provided the posies for Mothering Sunday. They have also made many donations to charity.

The changing world in which we live means that it has become difficult to recruit new members and to find new officers, so it is right to disband rather than go on with no new life coming in. The Social Club, which meets once a month on Wednesday evenings, has taken over some of the functions of the Guild and the specific needs of women within the church are catered for through the Mothers' Union.

I know that the clergy have found the Women's Guild a great support and I would like to thank the members for all that they have done in the past and to wish them God's blessing on their lives.

Fr. Peter

From the Magazine, April 1990

Of course there were some evenings when the 'to have fellowship' took the form of coffee and chat, games or a social. However the 20th Anniversary of the Guild, in 1985, was marked by a special evening. Invitations were sent to past members and to Father Bill and Mrs Hitchinson. The evening started with a Thanksgiving Eucharist service conducted by Father Bill. Mrs McKellar (founder member of the Guild) read the lesson. After the service Father David officially welcomed everyone to the buffet supper in the hall. The party food, including the birthday cake made by Mrs Barbara Plummer, was much enjoyed. Needless to say there was a great deal of talk and 'reminiscences of earlier days'.

The Report of the Chairwoman (Mrs Dedman) at the 1987 AGM seems to capture the depth of fellowship in this organisation. She thanked members for 'their loyalty and support and the bond of closeness and friendliness which is always apparent when we meet'. Unfortunately although meetings were often open to non-members, the Guild failed to attract many new members. Most members had contributed as office holders or committee members over 22 years and had hoped that new and younger members with fresh ideas would take over. After the September 1988 committee meeting a letter was sent to Father David suggesting that, unless they could find a Chairwoman, Treasurer and a new committee at the 1989 AGM, the Guild should be discontinued. At the 1989 AGM Father David's suggestion that the Guild should 'go into abeyance for a year' was accepted. It was with much sadness that in February 1990, shortly after Father David had left the parish, the decision was made to disband the Guild. It was typical of the group that during the 'abeyance' they had stalls at both the fête and the market in 1989, and made posies for Mothering Sunday in 1989 and 1990. Father Peter's article in the April 1990 magazine paid tribute to the part played by the Guild in the life of the parish for nearly 25 years (see previous page).

The Society of Mary
For the first few years of Father David's incumbency the monthly devotional meeting was held, as in the past, on 1st Wednesday evening, but from March 1985 it was moved to 1st Saturday afternoon at 4.30pm. During the season of Lent it was, therefore, held prior to the weekly Benediction.

Society reports refer to attendance at celebrations of the festivals of Our Lady; notably the Solemnity of Mary, Mother of God (1 January), The Purification of Saint Mary the Virgin, usually called Candlemas (2 February), The Annunciation (25 March), and The Assumption of Our Lady (15 August). For most of the celebrations there was an evening Sung Eucharist but some services were at 10.30am. The Society's May Devotion and AGM, held at a London church, was an opportunity to meet people from other groups of the Society.

Society members were grateful to those who organised pilgrimages to shrines of Our Lady at Willesden (1981) and Walsingham (1982 to 1986). The pilgrimage of 1985 'was indeed a happy one, with services and devotions led by Father David, and the smooth running travel and accommodation arranged by Michael Bedford'.

Gerald Collins had been secretary of the Ward for two years when he died in August 1983. Madeline Gardner, his successor, commenting on his death in the 1984 APCM Report stated that he 'always set an example of the highest in devotion to Our Lady'. The small statue of Our Lady in the Lady Chapel was given in his memory.

The Society's meetings were open to any member of the congregation and quite a number became members in the course of the years under discussion. The support that Father David and his successive Assistant Curates gave to the Society was much appreciated. The fellowship of the St Lawrence community may be recognised in the help given by Anne Holt, organist, and Margaret King and the Flower Guild for arranging flowers on behalf of the Society at the Statue of Our Lady at the entrance to the Lady Chapel.

St Lawrence Players
Most years during the period 1980 to 1990 the Players' productions followed the pattern of two plays, in May and October, and a musical in February. There were Saturday matinee performances of the musicals, and in 1980 to 1982 and 1984 an additional performance on the Thursday (see Appendix 12 for list of productions).

The Players contributed to the celebrations of the Golden Jubilee year in 1983 with two special productions. In May *The Vigil*, an Easter play by Cadislas Fodo, was based on the mystery of the empty tomb. Directed by John Woodnutt, a local professional actor, and assisted by Judith Howe (née Brion), it was a great success. The July magazine review referred to an impressive performance being 'enhanced by the church setting'. In October the production was *Dear Octopus*, a play celebrating the Golden Wedding of Charles and Dora Randolph and was another successful production.

The October 1984 production was a revue entitled *Playing for Time*, a nostalgic trip in words and music, of the social and economic changes between 1914 and 1956, ending with *My Fair Lady*. There was a glowing review of the revue, 'the costumes were incredible – the casting so apt – the timing quite superb…well done Players, another success'.

Playing For Time 1984

A fire in the hall in May 1985 meant the cancellation of the Saturday performance of *A Murder is Announced*. Among the losses were relatively new curtains (1981) and items lent by members for the production. The 1986 APCM Report referred to the hard work that goes into putting on a show, adding that 'our reward of course is the actual production'. Having, for the first time ever, to cancel a performance meant 'utter disappointment and sense of anti-climax'. In spite of this they had their usual after show party and started a sing - song around the piano. This apparently inspired them to put on an Old Time Musical for their October production, which 'appeared to be universally popular'. In the October 1985 magazine Father David thanked those who coped so efficiently when the hall was set on fire, offering a special thank you to the St Lawrence Players who undertook the first big clean-up.

The February musicals appear to have been particularly ambitious and successful, most of them admirably produced by Judith Howe. In 1986 *Scrooge*, a musical version of Dickens *A Christmas Carol* was highly praised. In 1987 *The Railway Children*, a UK first amateur production was regarded as 'quite a scoop' for the Players. The 1988 production of *Camelot* was seen as 'one of the more ambitious undertaken by the Players, with a large cast and many changes of scenery and costume', and having a number of younger members in the show. *Hans Anderson* in 1990 again had many children including ones from Choir, Cubs, Scouts, Brownies and Guides; hopefully some Players of the future.

The Railway Children 1987

In May 1987 the Players were really forced to 'rest' as far as a stage production was concerned, as the hall was out of action owing to the rebuilding work. However they took the opportunity to meet their audience in a more social way and arranged a Supper Quiz. After a fish, or chicken, and chips supper from Pisces Restaurant the evening was spent answering quiz questions devised and presented by Shirley and David Horchover. The evening was declared a great success and Supper Quizzes became annual events.

The October 1988 production of *A Pack of Lies*, based on the story of the Communist spies Peter and Helen Kroger, and their time in Ruislip in 1960-1961, aroused great interest. The review praised many of the cast, and concluded with…'seldom has there been a production here to match the drama that was unfolded before us'. As the play was about a local event a few words about it are appropriate.   The Jacksons are horrified to learn that the Krogers, their 'Canadian' friends across the road, are suspected of being spies. The play exposes the strengths and weaknesses of the Jacksons as they come to terms with feeling they have betrayed old friends, by allowing their house to be used for 'viewing' the Krogers' house. Alan Hooper and Janet Ford were commended for outstanding performances as the Jacksons, and Barbara Williams and Derek Allcock were 'convincing' as the 'Canadians'.

Few reviews mentioned the names of individuals; most referred to the joint efforts of the Players. One of 1981 is typical. 'The whole production, including the scenery, was a great team effort on behalf of the cast and all those who work so hard behind the scenes'. Mrs Win Brion had been chairman of the Players for six years when she resigned in 1989 and Mrs Ann Sykes was elected to this office.

The Players remained a popular organisation of the church with different generations of some families as members. As well as the pleasure the Players gave through their productions, they played their part in other church events, and like other church organisations they continued to contribute at fêtes and markets by running stalls.

The Flower Guild
Apart from the seasons of Advent and Lent, when there are no flowers in church, members of this group, continued to supply and arrange flowers about three times a year. On these occasions two members shared the cost as well as being responsible for the arrangements. Flowers for Church Festivals were not the financial responsibility of the Guild but could give members one or two further opportunities to arrange flowers during the course of the year. Members were also responsible for arranging flowers for weddings. Of course the Golden Jubilee was a special year for the Guild, who organised the exhibits in church in connection with the Flower Festival and Arts and Craft Exhibition of June 1983.

The cost of flowers for Church Festivals was partly met by donations from members of the congregation, but the Guild also held fund raising events. Until 1984 there was an annual Jumble Sale but these seemed to have fallen out of favour. On at least one occasion the Guild combined with the Scout Group, the boys washing cars in the car park, and the ladies providing refreshments and running a cake stall in the Hall. It is likely that as a result of this more enjoyable way of raising money, from 1986 the Guild began their popular Coffee and Bring and Buy Days. For a number of years these were held at the home of Barbara Leeson, a long-standing member of the Guild. Some years these were morning and evening events, others morning and afternoon, but always they were well-supported and happy events.

As with other organisations some ladies have given of their time, talents and money over many years. Information available for this period indicates that Mrs Margaret King and Mrs Sylvia Gayler were very active members of the Guild.

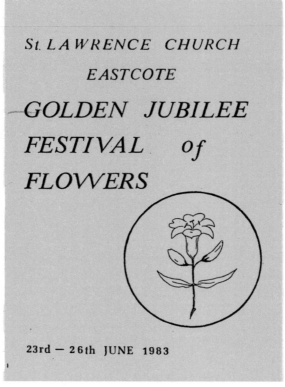

Golden Jubilee Festival Brochure Cover

4th Eastcote Scout Group

With the Group now well-established and adequately funded, numbers grew steadily and the amount of equipment owned could no longer be accommodated in the Headquarters. Not only did the Group have tentage, marquee etc. but seven canoes and a trailer! So planning permission was sought and obtained and a new Store was erected at the end of the existing HQ. The foundations were dug out, concrete laid and the steel structure was put together and built by volunteer parents and Leaders under the supervision of the then Group Chairman, Mr Alan Brooke. Equipment was now stored away from the meeting area, which was just as well, for Beaver Scouts had started in the UK for 6–8 year olds in 1986.

The demand for membership was such that two Beaver Colonies were soon flourishing and this created the need for two Cub Scout Packs – the Comets and Meteors. The numbers of Gryphonhurst Venture Scout Unit (VSU) had slowly diminished but when numbers built up again, a resurrected VSU was started named the Meinzapint Unit – taken from the name of a narrow boat seen on the Canal! District Father and Son Camps were bi-annual events (later named Parent and Scout Camps – for obvious reasons), and in the intervening years a number of Family Camps were held, to which entire families were invited to camp for the weekend.

Each Section, Beaver, Cub Scout and Venture, continued with its own respective outings, activities, camps and expeditions under the Group Scout Leadership of Ann Sykes and later Euan Lumsden. Captain Donald Woodhouse became Group Chaplain. Euan retired in 1995 with John Giles taking on his responsibilities for a couple of years until Mark Winn was appointed Group Scout Leader. The Scout Movement has never been static in developing and changing structures, activities and even age groups, so it was not unexpected when Venture Scouts Units were replaced with Explorer Units, for the 14-18 year olds shortly into the new millennium. A new Unit has been formed at the 4th Eastcote, open to other Groups in the area who may not have their own Unit, known as the 'Diehards', after the nickname of the Middlesex Regiment.

It was pleasing to see that long service to the Scout Movement was marked when in April 2004 the Scout Association's Award of the Silver Acorn was awarded to Brian Beeston ('Bob') "in recognition of specially distinguished service" as a Leader with the 4th Eastcote for a period of 50 years, the Group's first 'home-grown' Silver Acorn. Not all Groups continue to survive in the area and, with the decline in numbers, the two Districts were combined in 2006 to form the new Ruislip Eastcote Northwood Scout District. Leaders for the Group and its Sections have not always been 'plentiful', but certainly adequate. The Leaders we have had, have been of high quality and thus the Group has existed successfully for almost 75 years. Indeed, with Sections covering all age groups and offering a good and varied programme of activities, no wonder the 4th Eastcote is currently the second biggest Group in the county of Greater London Middlesex West.

Throughout its history the Scout Group has always been a full and active member of the St Lawrence Church family, with sections being represented at the Uniformed Organisation Parade services. It contributes in all the activities such as fêtes, markets, bazaars, family days etc. and in particular in providing the outdoor cooking for barbecues! The Group remains ever grateful to the church for the use of the land on which its headquarters is built, and the surrounding open area for its outdoor activities.

Brownies and Guides
At the beginning of 1980 there were two Brownie packs and two Guide companies. The 6th Brownies and the 6th Guides had been functioning since the late 1950s. The 3rd Brownie Pack had started in 1965 and the 3rd Guide Company in 1979. At his first APCM meeting Father David referred to Mrs Enid Lumsden, starting and building up the 3rd Company to 26 members. Unfortunately the problem of leadership, and of helpers, meant that by 1989 there was only one pack and one company.

The 3rd Brownie Pack nearly closed down in 1987 when the pack said goodbye to Mrs Liz Overall who had been Brown Owl for seven years. Although a few people tried to keep the pack going it had closed by the 1989 APCM. The 6th Brownie Pack seems to have been led for many years by Mrs Jean Weeks until 1987, when Mrs Kay Drew took over and remained Brown Owl until 1992.

When Jean Weeks left the area in 1987 she also had been Captain of the 6th Guide Company but once again lacking sufficient leaders or helpers the company closed in 1988 and joined with the 3rd, who were more fortunate with leaders and helpers. Miss Caroline Dann had become Captain in April 1986 when Enid Lumsden moved to be Assistant Guider with the Eastcote District Ranger Guides.

It can be assumed that weekly meetings followed the traditional pattern suited to the different age groups, which included girls working towards, and gaining, various badges. They took part in inter-pack and inter-company events, such as District Challenge Cups and Swimming Galas, though these seem to have been less frequent than in the 1960s and 1970s. In October 1983 brownies and guides took part in a Gang Show in which they 'had a lot of fun and companionship' and raised about £400 for the Paul Strickland Scanner Appeal, at Mount Vernon Hospital.

As church organisations brownies and guides were involved in several church activities. They attended the uniformed Parade Services and helped to decorate the church for annual Harvest Festivals, and the Golden Jubilee Flower Festival in 1983. They also helped at the annual fêtes, markets and other events of the church. As in the past, the girls responded to specific appeals. In February 1985 they had an enjoyable Disco raising £64 for the Building Fund. The 3rd Guides had already raised £70 from a sponsored silence. The following paragraphs give some idea of the activities of the four groups.

Brownies
The 6th and 3rd Packs enjoyed Brownie Revels with other packs in the District, that of 1984 celebrating the 70th anniversary of the formation of the Brownies. In 1981 the brownies started the tradition of Maypole Dancing at the church fête. To some extent the out-of-meeting activities are bound to be influenced by the Brown Owl and other helpers of the pack, and this in turn depends on what other commitments they each have, so the two packs have a slightly different history.

The 1984 APCM 6th Pack Report notes that 'bulbs were again planted in September and cared for by the brownies for Christmas for the residents at Missouri Court', so the tradition of visiting Missouri Court goes back at least to 1983. Most years the brownies took bowls of bulbs but one year it was bowls of fruit and another gift boxes, but they always sang carols. That was until January 1990 when they took calendars they had made and sang brownie songs. The residents had said they would prefer a visit in January, as it was something they could look forward to in the quiet period after Christmas. Some years the brownies enjoyed jacket potatoes on their return to the hall.

The challenge of finding suitable accommodation such as a church hall, and getting sufficient helpers, for a pack holiday appears to have been met most years in the 1980s. The 6th Pack had holidays at Datchet (1983), Tring (1984), Frimley Green (1986 and 1987), Farnham Common (1988), Potters Bar (1989), and Cobham (1990). In her 1989 report Kay Drew (Brown Owl) referred to the hard work in preparation for the Pack Holiday at Farnham Common being worthwhile, as she had gained her Pack Holiday Licence. During 1988-89 Captain Donald Woodhouse 'adopted' the Pack and his visits to their meetings were 'very welcome and enjoyed by all'.

The APCM 3rd Pack Reports usually referred to 'busy years' with Brownie activities, fulfilling their commitment to the church, having a waiting list, and thanking those who had helped with the Pack. During 1983-84 the 3rd Pack Brownies reached the front page of the local press, after they had knitted hundreds of squares and made up four beautiful blankets for Save the Children Fund. For a few years the girls saved some of their pocket money and gave donations to The Children's Society. The year 1986-87 was a good year for the Pack as among other things they won the Eastcote District competition and '18 brownies worked very hard with the Red Cross' to gain their First Aid badges. When the Pack was forced to close from lack of leaders a few of the brownies joined the 6th Pack.

Girl Guides
The Guide Movement as such was older than the Brownies and celebrated its 75th anniversary in 1985. Representatives from St Lawrence's took part in many of the District and Division activities, including a multi-faith act of worship at RAF Stanmore, a Division service at Emmanuel Church, Northwood, a Regional Rally at Crystal Palace, and a Division Camp at Stoke Poges. A magazine article by Doreen Murrell about the camp recalls being persuaded to transport a six-foot diameter ball to the site after it had been inflated at the local garage. Needless to say the inflated ball would not fit into a car but the Stoke Poges garage owners were resourceful and the ball eventually arrived strapped between two open van doors.

Something many Guides look forward to is the opportunity to camp 'under canvas'. At Whitsun 1981 some of the 6th Company camped at Seer Green, Buckinghamshire. In spite of bad weather, it rained every day but the last, when it was fine to take down the tents, the girls 'had a good time and received some useful camping experience', especially those camping for the first time under canvas. Most years some Guides went to camps organised at District level. In 1983 12 Guides from the 6th company enjoyed a weekend 'learners' camp at Whitsun, and in the summer holidays 3 from the 6th and 3 from the 3rd made up a patrol to camp with another company at Sarrat. Other camp venues included

Cliveden (1986), Chalfont Heights Scout Camp (1987), and Blacklands Farm near East Grinstead (1988 and 1990). The 1991 APCM Report makes it sound an exciting camp-site where activities included swimming, canoeing, rock-climbing, archery and abseiling.

During the 1980s the 3rd Company went on a number of rambles or hikes. An evening hike in October 1981 started at St Vincent's Hospital and the girls, 'clad in woolly hats and anoraks walked to the Lido', using the moon and stars to light their way. In February 1982 patrol leaders and 'seconders', under the supervision of Guiders and helpers, had a ramble starting from Moor House at Chenies, Buckinghamshire. They map-read their way to Latimer, where they had their packed lunch, returning to Chenies by a different route. They had all enjoyed the ramble and 'arrived home with rosy cheeks and muddy boots'. In 1985, the 75th celebratory year, some went on a night hike with 'a few voluntary [*sic*] Dads' (from Chenies to Latimer), and also had a more local ramble along the River Pinn.

Outings by the 3rd Company, 'of interest' rather than specifically connected with Guiding, included one with Captain Don for a tour of Church Army work in parts of London (1981), a visit to The Commonwealth Institute (1982), and a tour of the Opera House, Covent Garden with the Eastcote Rangers (1986). There was a theatre outing to *The Lion, the Witch and the Wardrobe* (1984) at the Westminster Theatre. From 1987 when 6th Company began to have problems with leadership, Guides first from both companies, then from the combined unit enjoyed visits to pantomimes, *Jack and the Beanstalk* (1987) at RAF Uxbridge, *Aladdin at Richmond* (1988) and *Babes in the Wood* at The Beck Theatre (1990).

Caroline Dann began her 1990 APCM Report stating that she had only one regular unit helper with 28 girls and that she required additional help. Over the years parents of girls of brownie or guide age had given invaluable help in the four units associated with St Lawrence's but the sad fact is when Father David left the parish there were only the 3rd Brownie Pack and the 6th Guide Company.

Father David Hayes Leaves Eastcote
Unlike Father Godwin and Father Bill, who had retired when they left St Lawrence's, Father David had another appointment. His letter to the parish giving details of his appointment was reproduced in the October 1989 magazine. He was to be instituted as Rector to the City Centre Parish of Canterbury in January 1990. The full dedication of the parish is St Peter with St Alphege and St Margaret, and St Mildred with St Mary de Castro. The parish covered the whole of Canterbury within the old city wall, with the exception, of course, of the Cathedral and its precincts. Father David also had the responsibility for St Thomas' Eastbridge Pilgrims' Hospital, which provided accommodation for a small number of pensioners. From 1997 he was Guardian of the Greyfriars. Father David retired in 2007. He has visited the parish on a few occasions,

including the church's 75th Anniversary celebrations in 2008, when he preached at 8 and 10 am on Sunday 10 August, the Patronal Festival of St Lawrence.

On Friday 12 January 1990 well over 200 people gathered in St Lawrence Church for the Eucharist, sung by Father David, which was one of Thanksgiving for his ministry in Eastcote. After the final hymn the vicarage family were 'put on show', in full view of the congregation, and Margaret Case (Churchwarden) expressed appreciation of all that Father David had undertaken in his years as Vicar. Frank Reeve (Churchwarden) made presentations to Father David, Monica, Timothy and Rachel. The evening ended with happy fellowship in the hall where the catering committee served refreshments. Father Peter proposed a toast to Father David, who later cut a cake, made and beautifully decorated by Barbara Plummer with various symbols of Christian life. On 27 January around 30 members of St Lawrence's attended Father David's Induction (by the Archbishop of Canterbury) and Institution (by the Archdeacon of Canterbury) in St Mildred's Church.

During his years as Vicar Father David built on the work of his predecessor, Father Bill, in that the laity was increasingly involved in what can be regarded as the mission of the Church in Eastcote. The number of new enterprises that were undertaken bears witness to this. Most of the initiatives are now well-established aspects of St Lawrence's in Eastcote, and although some have not survived it is fitting that reference has been made to them in this chapter. Changing society means that members of the Church have to adapt to change and sometimes try new ways to reach out to the wider community. Father David often reminded his parishioners that the mission of the Church was possible only if the parish was a worshipping community. His 'Forward Together' helped St Lawrence Church to meet the challenges of the 1990s.

CHAPTER V

THE REVEREND DAVID COLEMAN

VICAR 1990 to 2004

The Revd David and the Revd Ann Colman 2004

By July 1990 it was known that the new vicar was to be Father David Coleman. He had been ordained Deacon in 1973 and Priest in 1974. He had served his title at Christ and St John with St Luke, Isle of Dogs, and his second curacy was at Holy Cross, Greenford (1977 to 1980). He became Vicar of St Peter's, Cricklewood in 1980, and Vicar of St Alban the Martyr and St Michael, Golders Green in 1985. He was married with three children, Kate 17, Ben 12, and Christopher 7. His wife Ann had been ordained Deacon in 1987. Father David's Induction and Institution as the fourth Vicar of the parish took place on 29 October 1990.

From the beginning Father David Coleman made it known that he had been nurtured as a Christian in the catholic tradition of the Church of England, and had been sustained by that tradition and its spirituality. He was therefore in sympathy with the High Anglican tradition established at St Lawrence's by Father Godwin, and maintained by both Father Bill Hitchinson and Father David Hayes. Like them he saw the need to relate the mission of the Church to the contemporary situation. Father Bill's 'Worship and Fellowship' had moved on to Father David (Hayes)'s 'Forward Together', with parishioners taking a more positive role in the mission of the Church. Also like his predecessors Father David emphasised that the mission of the Church was only possible if the parish was a worshipping community.

At his first APCM in 1991 Father David referred to the Decade of Evangelism that had been launched recently in the Diocese of London. He said 'evangelism must always begin with ourselves, and our conformity to the Gospel. Everything must be related to our worship, but if it was vital worship it would reflect a concern for those within our community and beyond'. The Parish Appraisal, which had taken place in 1989, is likely to have aided the PCC sub-committee in their deliberations concerning the mission of the Church in Eastcote in the 1990s, and their proposals were in keeping with Father David's vision for the church.

Towards the end of 1992 the Bishop of London's *Agenda for Action* had turned the Decade of Evangelism into a positive challenge to all parishes in the Diocese. At St Lawrence's six sub-committees of the PCC were formed to consider questions posed in the *Agenda for Action*. They were to deal with Care and Service, Finance, Buildings, Worship and Prayer, Teaching and Nurture, and Ministry. During 1993 extended consideration by the sub-committees and PCC resulted in a Mission Action Plan (MAP), a series of very definite proposals for enhancing the mission of the Church within the church community and beyond. The MAP, reviewed each year, influenced many aspects of life associated with St Lawrence's.

Father David Coleman's Ministerial Colleagues
The role of the vicar of a church had changed considerably since St Lawrence Church was established. Source material for Father David Coleman's incumbency clearly gives a very different picture from that of the time of Father Godwin, or even Father Bill, when the Vicar seemed to organise so much of church affairs. Throughout his incumbency Father David acknowledged the support of his ministerial colleagues with whom he shared 'in the cure of souls'.

During earlier interregnums the parish had relied on help from local clergy, but during the interregnum of January to October 1990 the parish was fortunate to have Father Peter Day who, as resident priest, was responsible for the spiritual and pastoral care of the parish until the new Vicar arrived. Although they only worked together for a relatively short time, Father David was grateful to Father Peter for 'his kindness to me on my advent as Vicar'. Father Peter presided for the last time at the Parish Eucharist on Sunday 29 January 1991.

A few months later the parish was able to welcome Father John Foulds as Assistant Curate. A single man in his late twenties he was ordained Deacon in 1991 and Priest in 1992. From then Father John officiated at the Eucharist and at Baptisms, Weddings and Funerals. Like other curates he did rewarding work among young people and was involved in new ventures in the parish, including the monthly House Eucharist and Prayer Groups, each held in the homes of different parishioners, as well as the more informal ways of worship on Sunday evenings.

In his 1994 Vicar's Remarks, delivered for the first time during the Parish Eucharist preceding the APCM, Father David referred to Father John as a 'colleague who was unfailingly loyal and helpful' and was delighted that Father John was to spend an extra year at St Lawrence's. His last Sunday was 19 February 1995. Father John went on to be Curate-in-charge of the Church of the Annunciation, Kenton from 1995 to 1999 when he became Priest-in-Charge of Holy Trinity, Meir, in the Diocese of Lichfield.

At the 1995 APCM Father David announced that the parish would be getting another minister, probably an experienced priest. By July it was known that that person was the Revd Michael Bolley. He and his wife Monica had two children, Ruth who was 11 and Simon who was 8. Father Michael had been ordained Deacon in 1992 and Priest in 1993 and had been Curate at St John the Baptist, Pinner since 1992. He was licensed at a Solemn Eucharist on Thursday 7 September. He had responsibilities greater than was usual for a curate, and was appointed for a period of five years, and had the official title of Associate Vicar Father Michael's particular role as Associate Vicar, set out in a Specification of Ministry agreed by the Bishop and the PCC, was to promote the mission of the

Revd Michael Bolley and his wife Monica

Church. During October 1995 the PCC reviewed and updated the MAP, and Father Michael put forward his 'ideas about ways in which we might reach out to people beyond the Church'. During his years at St Lawrence's he initiated several new projects. These included the *'Christians for Life'* course, and the parish-wide newspaper *Free for All*. In 1999 the Bishop requested that Father Michael be released before the five years were up, in order to take up his appointment as Priest-in-Charge at Holy Trinity, Southall. His last Sunday at St Lawrence's was 11 April 1999.

The new Assistant Curate at St Lawrence's was the Revd Sue Groom who had been ordained Deacon in 1996 and Priest in 1997. She had been at St Mary the Virgin, Harefield before coming to Eastcote in October 1999 with her husband Phil. She was licensed as Assistant Curate at the 10am Eucharist on Sunday 10 October. At this service the Bishop of Willesden also administered the sacrament of Confirmation and re-hallowed the recently re-ordered Lady Chapel. Just over a week later the parish was left in the good hands of the Revd Sue, assisted by the Revd Ann Coleman, Captain Don and others, while Father David led a parish pilgrimage to the Holy Land. From the beginning of May to the middle of August 2001 she again had the 'care of the spiritualities of the parish' when Father David had his first Sabbatical in 28 years. The Revd Sue was in the parish for only two years but many people were deeply appreciative of her ministry. She was appointed Priest-in-Charge of St Matthew, Yiewsley and had her last Sunday at St Lawrence's on Dedication Sunday, 21 October 2001. She left Yiewsley in 2007 to take up a new post as Director of Deanery Licensed Ministers for Kensington.

The Revd Sue Groom with Fr David Coleman and
Churchwardens Andrew Bedford (left) and Lionel Williams

Father David's last full time stipendiary priest was the Revd Ruth Lampard. She was ordained Deacon in 2000 and Priest in 2001 and had been Curate at St Peter Mount Park, Ealing, before coming to St Lawrence's in November 2001. In his 2002 Vicar's Remarks Father David commented that the Revd Ruth had already made her mark. In 2003, commenting on the 'quality of assistant curates we have been pleased to welcome to St Lawrence's' he referred to 'Ruth's particular charisma for nurturing young people…and for her sense of fun'. In March 2004 she announced her engagement to the Revd Ian Tattum, Priest-in-Charge at St Ippolyts, Hitchin in Hertfordshire. The Revd Ruth's last service at St Lawrence's was 19 September 2004. In 2008 she was appointed Associate Vicar of St Mary the Boltons, near West Brompton, with special care for the St Jude's part of the parish, the children and young people, and environmental concerns.

The Revd Ruth Lampard
at her leaving party 2004

In addition to the full time stipendiary priests at St Lawrence's there were, over the years, several other people who shared 'in the cure of souls'. Foremost among these was Father David's wife, the Revd Ann Coleman. She had been ordained Deacon in 1987 and when the Colemans came to Eastcote she was with the Advisory Council for the Church's Ministry. From 1991 to 1993 the Revd Ann taught at Bishop Ramsey School and in 1993 was appointed its Chaplain. At that time the Revd Ann was also licensed as a curate to the parish and officially joined the ministerial team. In 1994 she was one of the first women to be ordained to the priesthood. She celebrated the Eucharist for the first time on Wednesday 20 April and thereafter took her place on the rota of celebrants. Although her ministry had its main focus at the school, St Lawrence's had a share in her liturgical, preaching and pastoral ministry. Even after she became Course Leader at the North Thames Ministerial Training College in 1999 the Revd Ann still contributed much to life at St Lawrence's. In 2004 it was known that she was to become Director of Wydale Hall, the York diocesan retreat and conference centre. The parish of Eastcote benefitted a great deal from the Revd Ann's priestly ministry throughout her husband's incumbency.

After Father Michael left the parish in April 1999 it was difficult to cover all the regular Masses. Father David was grateful for the assistance of the Revd Wendy Brooker, Chaplain at the Royal Orthopaedic Hospital in Stanmore. She and her husband Derek had returned to live in the locality and the Revd Wendy was able to celebrate the Eucharist at St Lawrence's from time to time.

Another member of the team was the Revd Laurence Hillel who was appointed Chaplain of Bishop Ramsey School in 2001. The Bishop felt that Father Laurence should be attached to a parish church and he joined St Lawrence's as an honorary Curate. Although the main focus of Father Laurence's ministry was the School, Father David was appreciative of his thoughtful teaching ministry at St Lawrence's, both in preaching and study groups. Although Father Laurence moved to St Anne's Brondesbury in June 2004 in his capacity as Chaplain at Bishop Ramsey School he still celebrates the Eucharist when members of the School attend St Lawrence's for a School Mass, and he also helps to familiarise pupils with the church.

One lay member of the ministerial team frequently mentioned by Father David is Captain Donald Woodhouse OBE. He and his wife Norma, and son Ian, have worshipped at St Lawrence's since 1972. Captain Don joined the ministerial team when he became a Lay Minister on his 'retirement' from the Church Army in 1985. Over the years he has contributed a great deal to St Lawrence Church. In his Vicar's Remarks in 2004 Father David said 'Captain Don continues indefatigably with a preaching and pastoral ministry, as well as inspiring our children's holiday clubs in the summer, and at Easter and Christmas. This is in addition to continuing work with the Church Army and the Scout movement'. He is also a keen supporter of the ecumenical moves made by local Eastcote churches. Many members of St Lawrence's and the wider community of Eastcote have reason to be grateful for his ministry.

Another member of the laity who joined the team in 1998 was Alan Wright, licensed to the office of Reader by the Bishop of London in St Paul's Cathedral on 1 December. This was the culmination of several years of study. In addition to the usual functions of a Reader, such as preaching and assisting in the taking of services, Alan continued to co-ordinate the Parish Care Scheme and was responsible for overseeing the provision of Bible study. He thus helped the 'worshipping community' of St Lawrence by his contribution to worship, teaching and pastoral care.

Elaine Garrish joined the team towards the end of 1999 when, after a year's training, she was commissioned as a Pastoral Assistant. In 2002 Father David thanked Elaine for her valuable work in visiting not only individuals, but also Whitby Dene Residential Home, and Missouri Court, and for energising the ministry of healing and prayer support. In 2004 Elaine began training for the ordained ministry and was ordained priest in 2008.

Over the years St Lawrence's has fostered many vocations to the priesthood and Elaine was the last of several during Father David's incumbency. The others were Gill Tuffin ordained priest in 1996, Marion Smith ordained priest in 1999, Michael Bedford ordained priest in 2002, and Sylvia Lafford ordained priest in 2005. In their time at St Lawrence's they had each contributed a great deal to the parish.

In concluding this section on Father David's ministerial colleagues it is right to point out that while Father David had support in worship, teaching and pastoral care in the parish, he also had extra-parochial commitments that could be very time-consuming. From 1992 to 2002 he was Chairman of Governors of Bishop Ramsey School. While Father Coleman did not think it took as much of his time as it had that of Father David Hayes, it was a sizeable commitment. He was also Chaplain at St Vincent's Hospital. Another extra-parochial ministry was in Post Ordination Training where he was one of three tutors in the Willesden Area. Yet another aspect of his extra-parochial ministry was, for a few years, that of Area Chaplain of the Mothers' Union. In addition, on at least two occasions he was Chaplain to the Mayor of Hillingdon. In each of these roles Father David gave advice and support to those outside the St Lawrence Church community.

St Lawrence Church Fabric
Fabric Reports on the years 1990 to 2004 show that being responsible for the property of St Lawrence Church imposed a considerable financial burden on the parish at a time when the income of the church, for various reasons, was not keeping up with expenditure. While fund-raising efforts, donations from parishioners and friends, and grants, were sources of much of the money, without various legacies, some of the work could not have been done. It should also be remembered that any work undertaken had to be approved by the Diocesan Advisory Council and a Faculty obtained, which could take a long time.

The Church
The major expenditure during these years was for work on the church roof. The 1992 APCM Churchwardens' Fabric Report referred to work on guttering and roof tiles costing over £3,300. During 1992 there was a leak in the Lady Chapel ceiling. Further work was done on the main roof but the Vicar's Letter of September 1994 revealed there were severe problems and by the 1995 APCM the full extent was made known. All the nails holding the battens supporting the roof had to be replaced – a major job costing in the region of £25,000. Through the generosity of benefactors there was £5,000 in the Roof Fund and, as the Common Fund (Quota) payments were up to date, there was the possibility of a grant from the Diocese, perhaps of £5,000. This still left a sum of at least £15,000 to be raised, and if this was not raised by the time the work was completed the church would have to take out a loan.

In his May 1995 magazine letter Father David appealed for generosity to meet the cost of this very necessary and urgent work, pointing out the responsibility of parishioners to maintain the building. 'Our church building is an important resource and asset for our Christian ministry here in Eastcote. It is wise and proper stewardship for us to keep it in good order'.

The September 1995 magazine revealed that the total cost had risen and was likely to be in the region of £33,000, as some 5,000 tiles needed to be replaced, about a tenth of the total number covering the entire building. Among the fund-raising efforts was one to 'sponsor' a tile or tiles. Barbara and Lionel Williams 'built' a replica scaled model of St Lawrence's with the roof divided into over 2800 numbered squares, each representing two actual tiles. For every £1 given the donor would fix a sticker on the square of their choice. Obviously this 'tiling' of the roof was just a bit of fun as the actual cost of the real tiles was a fraction of the overall repair bill.

In January 1996 Father David was able to write that, by 'a tremendous effort – of which we may be justifiably proud', the sum of over £33,000 was raised in nine months, and thanks were expressed at the Eucharist services on 14 January for the success of the appeal.

Before the 1995 crisis relating to the church roof was known, other problems associated with the church building had to be dealt with. By the 1993 APCM it was obvious that the central heating should be replaced. At that time the Common Fund or Quota had not been fully paid so there was no possibility of a grant from the Diocese. However over £2,500 was raised by generous contributions from parishioners and a new gas convector system installed which, by 1995, proved to be cheaper to run than the old electric heaters. The organ presented a further problem in 1993. Nearly as old as the church, the Compton organ needed a thorough overhaul at a cost of nearly £8,000, or be replaced with an Allen organ. Choir members started to raise money for the Organ Fund with, among other efforts, a sponsored 'Organ Play'. By 1995 the organ had been repaired, the cost having been met from the Organ Fund, donations and other fund-raising efforts. In 1994 a minor but helpful addition to the fabric of the church were the handrails on the steps at the main entrance.

Changes were made in the Lady Chapel in 1999. The Vicar's Remarks for the 1998 APCM referred to the re-ordering of the Lady Chapel, made possible by a portion of the generous legacy of Elsie Fischer. When the high altar sanctuary area was re-ordered in the 1980s to accommodate modern liturgical rites, with the priest in the westward facing position, the Lady Chapel was left more or less as it had been since 1933, with the exception that the altar had been brought forward. This was felt to be unsatisfactory, both in regard to the celebration of the weekday Mass, and as a place for private devotion. The re-ordering included the extension and carpeting of the altar step, the provision of a new altar similar in style to the high altar, and the installation on the east wall of a figure of the risen Christ incorporated in a new reredos of blue damask brocade The work was carried out in July and August 1999 at a cost of just under £8,000. The Lady

Chapel was re-hallowed on 10 October when the Bishop of Willesden (The Rt Revd Graham Dow) visited St Lawrence's to licence the Revd Sue Groom and administer the sacrament of Confirmation.

The re-ordered Lady Chapel 2008 Flower Festival

A new sound system was installed in December 1997, much of the cabling being carried out by Andrew Bedford, thus reducing the installation costs. It replaced the small loop and address system of 1990. Work entirely 'Do It Yourself' was the re-decoration of the inside of the church in 2000. As in 1986 the work was done by a small band of helpers under the guidance of Arthur Plummer, and at great saving to parish finances. Soon after this, new notice boards were installed and the re-ordering of the bookstall area was started. The children's corner was moved from the bookstall area to the northwest corner of the nave of the church.

Other work in connection with the church building included attempts to deal with the recurring problem of the damp east wall in the Sacristy. It is hoped that this was finally solved when the Sacristy was refurbished early in 2005. This, together with the refurbishment of the church toilet, was paid out of a legacy from Eva Starling.

Among church 'furnishings' added during Father David's incumbency were new sedilia (seats for clergy and servers) in the Sanctuary (1992), and new bronze-finished candlesticks for both the high altar and the altar in the Lady Chapel (1993). By 2003 altar frontals and lectern falls in colours of the liturgical seasons were in use. Other additions were a stand for the Advent wreath or candles (1996), a new hymnbook, *Complete Anglican Hymns Old and New* (2001), and new candelabra (2004).

The Hall and the St Lawrence Centre
The 1984 Appeal for the Parish Hall Building Project had been launched on the understanding that the whole scheme, including some development of the bungalow site, would eventually be achieved. Towards the end of Father David Hayes' incumbency it had been agreed that decisions regarding the bungalow, the last of the self-contained stages of the Building Project, should be left to the new incumbent and the PCC in the light of the contemporary situation. Although when Father David Coleman came to St Lawrence's the bungalow was reported to be 'in a bad state', it was treated as an annex to the hall, and for that reason it seems logical to discuss both the hall and the bungalow in this section.

Careful maintenance of the hall, by the Hall Management Committee, was vital as there were a number of 'outside' bookings in addition to the regular use by church organisations and church functions. This was very satisfying as part of the Building Project had been to cater for the needs of the Eastcote community. During 1992 the main hall was painted professionally for reasons of safety but painting of other rooms was done by a 'SMALL group of trusty volunteers'. In contrast to the problem of the church roof, it was the hall floor that caused concern for a number of years and was replaced in 1997 at the cost of £9,124.

The decorators of the hall rooms 1992
(Uxbridge Library)

The future of the bungalow remained a problem, and it was not until 1996 that, as part of the Mission Action Plan, a small group from the newly elected PCC was set up to consider a way forward. The findings of the sub-committee were reported to the PCC in September. The options ranged from doing nothing to demolishing and replacing the building. While the latter would be the ideal it was agreed that the most cost effective option was for a full refurbishment. In

December 1996 a structural survey was made and a positive report issued. Progress reports on the project were given during 1997, including the fact that a portion of a generous legacy from Elsie Fischer was to be allocated towards the refurbishment. The building contract was awarded to Hamlet & Johnson of Northwood and the replacement doors and windows to Silver Glass of Northwood Hills.

In June 1998 a magazine article entitled 'St Lawrence Centre Appeal', gave some details as to what the Centre would provide. Among these were a parish office, enabling church business and enquiries to be dealt with more effectively, space for printing facilities, and attractive meeting rooms. The aim was to 'transform a semi-derelict embarrassment into a valuable and useful resource of which we can be proud'. The total cost was in the region of £67,000 and was financed to a great extent – some £37,000 - from the Elsie Fischer bequest. This meant a shortfall of £30,000 that had to be raised. The Appeal was officially launched on Sunday 21 June 1998.

Work commenced in July 1998 and the building was handed back to the parish in November when the furnishing of the interior began. The Churchwardens, and those they recruited, worked hard to make sure the Centre was ready for the official opening. This took place on 21 February 1999 when the Archdeacon of Northolt, the Venerable Pete Broadbent (Bishop of Willesden from 2001) blessed and officially opened the Centre, after preaching at the 10am Parish Eucharist. At the 1999 APCM, Father David regarded the 'transformation of the old bungalow' into the St Lawrence Centre as 'the major development in our life here at St Lawrence over the past 12 months'.

The area near the Centre was soon landscaped to complete the 'transformation'. The large area behind the Centre was gradually cleared, and the old Portacabin that had been used to store items from the 'old bungalow', was replaced with three metal storage containers, designated for use by the St Lawrence Players, the Guides and Brownies, and the church and others as agreed by the PCC. Problems with the heating system in the Centre were finally sorted out by 2002.

Meanwhile during 2001 and 2002 quite a large sum of money had to be spent on the Hall. Among major expenditure were new units in the kitchen (£8,300), and work for the main hall. In the main hall as well as a total decoration (£2,470), there were new cupboards (£2,215), and new hall and stage curtains (£3,555). The Players contributed £1,100 towards the cost of the stage curtains. As with any building, maintaining the Hall and the Centre at a good standard requires constant attention on a day-to-day basis. This is the responsibility of the Hall Management Committee, but Churchwardens inevitably get involved.

The Church Grounds
In April 1991 a magazine article by the Vicar and the Churchwardens drew attention to the fact that the church was fortunate in having extensive grounds.

These provided an appropriate setting for the church, as well as affording space for reflection and - in a designated area - a place for the interment of ashes of the departed. The article reported that a faculty had been applied for, so that certain work could be undertaken in the Garden of Remembrance. When granted, it led to the exchange of positions of bench and crucifix to their present positions, with a new seat facing the west end of the church. People were invited to place tributes in memory of loved ones, around the foot of the crucifix on the new paved area, rather than on the grass. It was hoped this would enable the grass to be kept in a better condition than previously.

The long article referred to changes already made in the area adjacent to the west end of the church and along the boundary of the church grounds. In appealing for help with the upkeep of the grounds, readers were reminded that this was part of our responsible Christian stewardship. In 1999 a distinctive feature of the church grounds, the concrete cones around the central grass island, with its popular tree, were put in place to preserve the grass.

Annual Fabric Reports express thanks to the work of the Hall Management Committee, and to various groups of volunteers. These included the dedicated team of ladies, affectionately known as 'the parish scrubbers' who keep the interior of the Centre clean, the 'maintenance team' who do various jobs in the buildings and the grounds of St Lawrence's, the small but 'dedicated gardeners', for many years under the guidance of Euan Lumsden, and others with no special group title. This giving of time and talents as part of responsible Stewardship saves the parish money, and also helps to build up the fellowship of the church community.

The Curate's House
During Father John's time at St Lawrence's (1991-1995) work on the Assistant Curate's house at 33 Sunningdale Avenue included external decoration and the refurbishment of the bathroom. The house was not considered suitable when Father Michael was appointed in 1995 and the Sunningdale house was sold for £95,000, less £8,834 legal and survey fees and repayment of mortgage. The new clergy house at 163 Pine Gardens cost £122,831 and with additional expenses it meant having to borrow around £43,000. With Father Michael's departure in 1999 the PCC decided to change the assistant clergy housing and purchase a smaller house, possibly within walking distance of the church. The sale of 163 Pine Gardens realised £147,368 and the cost of 1 Farthings Close, the new property, was £136,062 including fees. The Revd Sue Groom was the first of St Lawrence's Assistant Curates to live there.

Worship and Prayer
It is relevant to refer again to Father David's comments made at the 1991 APCM, with regard to the Decade of Evangelism, when he said that 'evangelism must always begin with ourselves, and our conformity to the Gospel', and that 'everything must be related to our worship'. A Worship and Prayer sub-

committee was one of six, set up to consider the challenge of the Bishop of London's *Agenda for Action*. The importance Father David attached to 'vital worship' reflecting concern for those within the church community and beyond will be evident in the following.

The Mission Action Plan (MAP), formulated by the PCC in response to the Bishop's *Agenda for Action,* included definite objectives under the heading of Worship and Prayer. In an article on the MAP a paragraph from the December 1993 magazine stated-:

> *In the crucial area of prayer and worship, our priorities are; building on the strong foundation of liturgical worship we already enjoy; enabling people to understand the liturgy better, and thus enter into it more fully; making sure our worship is attractive to all age groups; increasing the diversity of styles of worship available at St Lawrence.*

A worship group was formed to assist the clergy in the planning of services and possible developments in the worshipping life of the parish. From 1998 this became a formal Worship sub-committee of the PCC.

Services - Traditional and New
Like his predecessors Father David introduced new types of services to meet the needs of the times but he made it clear in his Vicar's Remarks in 1994 that St Lawrence's would remain a parish in the sacramental tradition of Anglicanism. While he was Vicar the 10am Sung Eucharist, celebrated with the traditional accoutrements, would be the principal Sunday service. However, together with well-loved music and ceremonial there must also be a place for newer forms to make worship more accessible. With some exceptions the times and frequency of Sunday services remained as they had been since February 1989. However as early as December 1990 a magazine article invited people to offer suggestions to Father David for improvements in the timing of weekday services, and also welcomed anyone to join the clergy in Morning Prayer and Evening Prayer on weekdays, an ancient practice of the Church.

The chapter on Father Bill's incumbency gave some background to changes in forms of worship culminating in *The Alternative Service Book 1980 (ASB)*. Authorisation for its use for ten years was later extended to December 2000. In November 2000 Father David wrote about the considerable growth over the past 20 years in liturgical thought and understanding. There had also been a desire in the Church as a whole for a richer and more flexible rite and this was coupled with developments in linguistic usage, such as inclusive language. From 2001 the *ASB* would cease to be authorised. The new Eucharistic rite of *Common Worship*, which was similar to the *ASB* Rite A used at St Lawrence's since 1981, was adopted from Advent 2000.

One of the objectives put forward by the PCC in 1993 regarding Worship and Prayer was to 'increase the diversity of styles of worship available', so there obviously was already some diversity. Father David's adherence to the sacramental tradition of Anglicanism regarding the Sunday 10am Sung Eucharist meant that any diversity of style of worship would be in Sunday evening worship. As far as the 10am service was concerned the Parade Services continued to be held on a monthly basis during term time, and involved young people who were not necessarily regular churchgoers. Sometimes Father David had young people making a more active contribution, such as in April 1991 when St Lawrence Young People (SLYP) took a prominent part in the service. From 2000 a special Holiday Club Eucharist allowed the regular congregation to share in the outcome of the previous week's Children's Holiday Club, a worthwhile venture that had been running since 1992.

The normal Sunday evening service was, as in the past, the Order of Evening Prayer according to *The Book of Common Prayer (BCP)*. The average attendance had declined at the end of the 1970s to under 30, and by the 1990s was rarely above 20. Numbers were higher when Solemn Evensong was celebrated and there were four servers and more 'ceremonial'.

One 'diversity of styles of worship' was Songs of Praise. Several of these services were held. In the early 1990s at Eastertide and Harvest numbers attending were over 60 but by the turn of the century were down to around 30. In 1993 Father John was involved with a new venture. This was the Praise Service, a time of informal worship, with drama, singing, prayer and a music group. The first of these appears to have been in 1993 when it was incorporated with the Christingle Service, which from that year has been an annual and popular December event (see Appendix 9). Marion Smith was much involved with the Praise Services that continued until she was ordained deacon in 1998. Others associated with them were Phil Williams and Father Michael.

An entirely different kind of service became a regular form of Sunday evening worship when, in 1997 the Guild of Our Lady ceased to meet. Father David hoped that the laudable object of the Guild, to promote devotion to Our Lady, would find wider appreciation on Sunday evenings. The form of service varied, sometimes it was Vespers and Benediction and sometimes Solemn Evensong followed by Benediction. It was in the latter form that Father David conducted his last service at St Lawrence's on 24 October (Dedication Sunday) 2004. Other types of evening services held occasionally included reflective Celtic or Taizé ones, seasonal meditations at Evensong during Lent, and special services on Palm Sunday, some years with members of other Eastcote churches.

One last type of Sunday evening service led by the young people must be mentioned. Under the enthusiastic guidance of the Revd Ruth the first *On Trial* in March 2002, was followed by *Joseph* (October 2002), *Moses* (July 2003), and *David and Goliath* (December 2003). By 2003 the St Lawrence Intelligent Cool

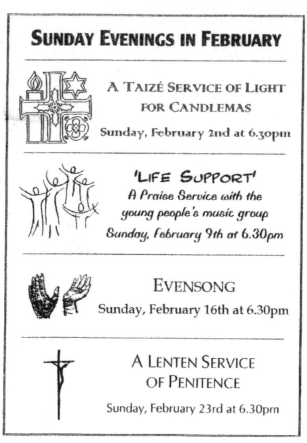

Sunday Evenings – February 1997

Kids (SLICK), as they were by then known, were joined by Joyful Chorus, a group for young singers aged 8 or over that had existed for about 18 months. These services were well attended by the members of the congregation.

During 1999 the Millennium was celebrated by *The God Who Comes*, an evening of drama, music and reflection (27 and 28 November). On 31 December a Said Mass at 11pm was followed by a time of prayer, and ended with Reflection and Lighting of the Millennium Candle.

In October 2001 *The Complete Anglican Hymns Old and New*, replaced the *Celebration Hymnal* (in use for the past 12 years), and *The English Hymnal*, which had been used since 1955. From February 2003 the evening service was at 6pm.

The most important work of the 1981-86 Liturgical Commission of the Church of England was *Lent, Holy Week, Easter: Services and Prayers (LHWE)*. This had a fresh approach to the use of Scripture, and richer use of symbolic action than previous official Church of England texts. Writing on the observance of Lent in the February 1991 magazine, Father David pointed out that the quiet of a weekday celebration could be a worthwhile aid to growth in the spiritual life. On Wednesdays there was opportunity for silent prayer before the Blessed Sacrament immediately after the 7.30pm Eucharist that lasted to 8.30pm. From 1991 the Stations of the Cross devotion was on Saturdays at 5pm (not Fridays as in the past) and seen by Father David as a suitable preparation for the Sunday Eucharist. A maximum attendance of 13 in 1991 fell to six in 1997 and was discontinued. Since 1999 the Stations of the Cross devotion has preceded the Wednesday Eucharist, which begins at the Offertory and is followed with silent prayer before the Blessed Sacrament. In 1991, at his first Easter at St Lawrence's, Father David replaced the previous Good Friday services with the 2pm Solemn Liturgy of the Day.

The Ministry of Healing
The first publication of the 1981-86 Liturgical Commission was *Ministry to the Sick*, which provided for the laying on of hands with prayer in public worship. Father David Hayes had introduced 'prayer for those in need and the laying on of hands' in 1987, when once a month at 6.45pm it preceded the 7.30pm Wednesday Eucharist.

In 1993 Father David Coleman recognised that this was an inconvenient time for many people. He also wanted to bring the healing ministry more into the mainstream of the parish's worshipping life, and for it to be 'placed within the context of the ultimate healing service, which is the Eucharist'. From 19 May 1993 the ministry of healing took place within the Wednesday Eucharist on a monthly basis. On that first occasion the Bishop of Willesden celebrated the Eucharist and exercised the ministry of healing in the course of a Parish Visitation. Over the years the ministry of healing has become an integral part of the worshipping life of the parish. From November 1994 the ministry of healing was available following the 10am Parish Eucharist on a regular basis. More recently this ministry is regularly offered within this main Sunday Parish Eucharist, something Father David had hoped for in 1993. Healing is also available after the 8am Holy Eucharist on 'healing' Sundays. Members of the laity have taken an active part in the ministry of healing from the 1980s.

THE MINISTRY OF HEALING

The ministry of laying on of hands and anointing with prayer for healing will be available following the 10.00am Parish Eucharist on

SUNDAY, FEBRUARY 9TH

Notice in February 1997 Magazine

By 2000 Father David decided there should be some revision of the way the sick and needy were prayed for in Sunday services. He felt that the long prayer list had become rather large and unwieldy and was of limited use in enabling people to pray meaningfully for others. From June prayer requests were pinned on an Intercession Board at the entrance to the Lady Chapel. As before the names were printed in the weekly bulletin. Although the names were read out particular needs were not mentioned at the Sunday Eucharist services, but were voiced at the monthly 7.30pm Wednesday Eucharist for the Sick. Elaine Garrish was responsible for the intercession board bulletin list. In September she helped launch a new venture in the form of a Prayer Ministry Team. Members of the team were available in the Baptistry to talk with, or pray for anyone with a personal need or concern, during the monthly healing ministry held after the 10am Sunday Eucharist. It was seen as an outreach of support and prayer, and people were invited to speak to members of the team at any time. This is another example of greater involvement of the laity during these years.

Other Prayer Meetings
After Father John's Ordination to the Priesthood in June 1992 it was possible to have daily celebration of the Eucharist. In July he also re-instated the House Eucharist on the last Friday of every month, rather than the weekly basis of the 1980s. During Father Michael's time as Associate Vicar, these House Masses

HOUSE EUCHARISTS

CONTINUING A SERIES ON
'PATHWAYS OF PRAYER'

THURSDAY, FEBRUARY 6TH AT 8.00PM
'PRAYING THROUGH SCRIPTURE'

at   the home of Percy Pether

FRIDAY, FEBRUARY 21ST AT 8.00PM
'PRAYING THROUGH SILENCE'

at   the home of Gwen Bedminster

THURSDAY, MARCH 6TH AT 8.00PM
'PRAYING THROUGH LIFE'

venue to be arranged

Notice in February 1997 Magazine

were more frequent and in 1997 included a series of studies on 'Pathway of Prayer'.

It is impossible to say exactly when prayer groups first met in the homes of different members of the congregation but it seems Father John was responsible for starting these some time in 1992. He saw these prayer groups as 'opportunities to meet with each other in the informality of a home to pray together and to get to know other members of the church family at St Lawrence's even better'. Alan Wright and Marion Smith were involved from autumn 1994,

from

The **Bishop** *of* **Willesden**

A Good Year of Prayer

How do you tell whether or not the 'Year For Prayer' has been a success?

- 700 people attended one of the 13 evenings in different localities when I came to pray with the churches. People went away excited, having stood and prayed in silence, or aloud, with candles and by writing their prayers.
- 70 came to 'Prayer for Men', with cooked breakfast. Another 20 were turned away. "It's the stomach!" said my wife, Molly.
- 4,000 copies of 'Pathways of Prayer' were used. Many have spoken positively to us of the course.
- Nearly 600 copies of the course 'Praying as a Church' have been ordered by parishes.
- 31 people are enjoying the 8 week course PALS (Prayer and Listening Support), learning how to help others in the journey of prayer.
- 5 teaching days and a weekend in Eastbourne were all well attended, from 'Praying as Families' to 'Leading Prayers in Church'.

These are days of increasing darkness in this country. There is so much confusion and so little hope. Young people do not know where to look to for meaning. The answer to darkness is light. Light is Jesus Christ. Light is God himself. We need to keep our attention on him who is light. We need to pray.

Let us build on our year for prayer. Keep your daily rhythm of time for God and let it grow. Encourage prayer in your church. Many of the teaching days will be repeated in 1996 in case you missed them first time round.

+ *Graham*

Prayer for Men 2

Finding your own style for prayer

- Cooked breakfast and prayer
- Saturday 23rd March
- 8 - 10am
- All Saints, Waltham Drive, Edgware

Call 0181 451 0189 to book

The Bishop's review of the Year of Prayer

when these group meetings combined a time of study and discussion with a time of prayer. In Lent 1995, the Prayer Group followed Bishop Graham's Lent Course on Personal Prayer. This was one part of his *'A Year for Prayer'*, which included a number of teaching days on different aspects of prayer at different locations in his Area of Willesden.

Reference has been made to Father Michael's specific role to promote the mission of the Church. In a November 1995 magazine article he wrote about one of his early initiatives, stressing the importance of prayer when 'considering together the best ways to go about drawing others to Christ and his Church'. He started monthly gatherings of Prayer for the Mission of the Church, usually held after the Wednesday evening Eucharist. He saw them as a time when 'we pray together about our common life and witness'. In March 1997 Father Michael discontinued them, for although he felt that much good had come from them, they were generally attended by a faithful few and he looked for other ways of involving more people.

In September 2001 the Revd Ann Coleman and Deirdre (wife of the Revd Laurence Hillel) started a contemplative prayer group called Sacred Space, which met monthly after the Wednesday evening Eucharist, usually in the St Lawrence Centre. As its title suggests it was an opportunity to find time and space to be still before God. A short time of listening to readings or music was followed by a time of silent prayer together. Although meetings were held after the Revd Ann and Deirdre left St Lawrence's, like Father Michael's Prayer for the Mission of the Church, Sacred Space was not supported by many people, and this, together with the problem of having someone to organise the evening, led to it being discontinued.

Words of Father David's written in 1999 at the time of the 25th anniversary of his ordination to the priesthood seem appropriate to end this section. Commenting on the numerical decline of the Church since the early 1970s, which were largely sociological and outside its control, he acknowledged good things to celebrate. 'I think, for example, of the deepening and enrichment of the liturgy; of the growth of lay ministry; and the rediscovery of aspects of our life, such as the ministry of healing, in which the activity of the Spirit has been evident'.

Teaching and Nurture
In 1993, in response to the Bishop of London's *Agenda for Action*, one area of church life identified as needing more positive action was Teaching and Nurture. Just as earlier chapters have shown that different aspects of church life are inter-related, so the different areas of church life identified in 1993 cannot be discussed in complete isolation from each other. For example the area of Teaching and Nurture inevitably involves Ministry, and Ministry involves Care and Service. Among the Teaching and Nurture objectives was the provision of more in the way of opportunities for study, and increasing the awareness of what Eastcote Parish Church had to offer to the people of the parish. There was to be a

> ## TEACHING AND NURTURE
>
> *Our priorities are:*
>
> ❖ *to provide a varied pattern of Christian Education which will enable people of differing ages, and at different stages of the journey of faith, to learn more about Christian faith and practice, and to grow in Christian maturity;*
>
> ❖ *to cater specifically for those who know little or nothing of Christian faith and who want to find out more, and who may in due time come to full Christian commitment.*
>
> *We will seek to further these priorities by:*
>
> ❑ *supporting the effective work of the Sunday School;*
>
> ❑ *re-establishing the Junior Church for those aged 11 and over;*
>
> ❑ *continuing to provide annual Confirmation preparation courses for young people and adults, together with opportunities for 'refresher' courses in basic Christianity;*
>
> ❑ *providing at least two in-depth study courses per year;*
>
> ❑ *establishing within the next few months a group with a structured programme for those enquiring about the Christian faith;*
>
> ❑ *establishing study and suppport groups throughout the parish within the next four years;*
>
> ❑ *continuing and developing the work of the Baptism Visitor Scheme in bringing more knowledge of Christianity and the local church to those who seek baptism for their children.*

One section of the 1995 Mission Action Plan

programme of parish study including, initially, two parish based courses a year, and within a five-year time scale it was hoped to establish study groups throughout the parish.

The 'Forward Together' of Father David Hayes, together with the 1989 Parish Appraisal, played a part in preparing St Lawrence Church for the Decade of Evangelism and the *Agenda for Action* with its various challenges of mission. In 1991 Father David had said that evangelism always had to begin with us, and our conformity to the Gospel, and he encouraged members of the congregation to take advantage of opportunities to widen and deepen their faith.

Study Courses
As implied by 'more positive action', teaching and nurture had been part of parish life well before 1993. From at least as early as 1986, study groups had supplemented the traditional Lent observances of prayer, fasting and almsgiving.

Members met in several homes around the parish to study and pray together, sometimes in ecumenical groups, as is the case today.

Prayer before the Blessed Sacrament
on Wednesdays in Lent following the 7.30 pm Mass

On March 6th this will be Prayer for the Mission of the Church

Stations of the Cross
on Saturdays in Lent at 4.30 pm

Except on March 2nd, when there will be Vespers and Benediction

Morning and Evening Prayer on weekdays
Morning Prayer takes place in Church at 9.00 am Tuesday to Friday

Evening Prayer takes place in Church at 5.00 pm Monday to Friday

You are most welcome to join the clergy and others for

Morning or Evening Prayer on a regular basis or occasionally

Ecumenical Lent Study Groups
organized by Churches Together in Eastcote, on the theme:

'Building Bridges'
An opportunity to share and reflect upon experiences of conflict

between people and upon the healing of relationships

The course is in five sessions:

Reconciliation and Creation	Families and Gender
Race and Religions	Rich and Poor

Histories of Conflict

This is an enjoyable way of meeting people from other churches and sharing our faith with one another. Groups meet at various times.

For more information, and to take part in the course,

please take a form from the back of the Church

and return it to the Rev'd Robin Pagan by February 12th.

Notice of Study Groups and other events in Lent 1996

In the early 1990s the clergy were largely responsible for St Lawrence's programmes of study. Occasionally it was just a single evening, such as in February 1992 when Father David talked, and invited questions, about the significance of various things used in worship and often taken for granted, such as vestments, bells, incense, and candles. After Easter he held a series of study evenings in the George Guest Room on the synoptic Gospels. These were well attended, and in 1993 a study course was held on St John's Gospel. 1995 seemed a particularly busy year. After Easter Father David had a Confirmation refresher course, in the summer, following on from the Bishop's 'Year of Prayer', Ann Coleman led a course on Great Teachers of Prayers, and in the autumn the course was 'Introduction to the Old Testament'. In summer 1996 there was a major course of teaching sermons on the subject of the Eucharist at both the 8am and 10am services. Members of the clergy continued to lead study groups throughout Father David's incumbency. One of the last conducted by Father David was an in-depth study of St Luke's Gospel in autumn 2003. This was good preparation as 2003 to 2004 was the year of Luke, in the three-year cycle of the new Sunday Lectionary.

During the years under discussion several members of the laity took a positive part in study courses. The ambition to establish study groups throughout the parish took a step forward in 1994 when the prayer group meetings embarked on a programme of study and discussion on topics such as Ethics, Evangelism, Environment, and Eternity. Marion Smith and Alan Wright as well as Father John led these particular meetings. Since 1998, when Marion became a deacon and Alan a reader, the regular Bible Study meetings have been held at Alan's house.

During these years Gill Tuffin, Michael Bedford, and Sylvia Lafford were also each involved in the 'Teaching and Nurture' aspect of parish life as part of their training for ordination. In 2000 Michael presented an evening entitled 'One feast, Three Celebrations in Church', in which he explained that the liturgies of the Lord's Supper, the Lord's Passion and the Lord's Resurrection were not attempts at historical re-enactment, but 'are a calling **into the present** the saving work of Jesus who brings the past, the present and the future into unity'. In 2003 Sylvia was responsible for organising a major course 'Faith and Work', in which members of the congregation shared their experiences of how being a Christian helps or challenges us as we try to live out our faith.

Christians for Life
In an up-dated Mission Action Plan (MAP) at the end of 1995 one of the priorities agreed upon, as regards Teaching and Nurture, was to establish a 'structured programme for those enquiring about the Christian faith'. In his Vicar's Remarks at the 1996 APCM Father David referred to the fact that in a society that was becoming more and more 'post-Christian' a basic working knowledge of Christianity could no longer be assumed. During 1996 Father Michael, used *Emmaus: the Way of Faith*, a relatively new resource programme, which was rooted in the need for evangelism, nurture and growth, to establish a 'nurture' programme for people who wanted to explore the basics of the Christian faith. The first *Christians for Life* course started in February 1997. Father David saw the launch of *Christians for Life* as a very significant development. These courses were very successful. They were held for many years and led to a number of people being presented for Confirmation.

CHRISTIANS FOR LIFE

at Saint Lawrence Eastcote

A course for anyone who wants to find out more about the Christian faith.

It is particularly suitable for:

people interested in investigating Christianity

those who want to brush up on the basics

newcomers to the Church

anyone thinking about becoming a Confirmed member of the Church

Welcome evening
Come to a simple supper and find out more
Wednesday, January 22nd at 8.15pm in the Church Hall

For more information, pick up a leaflet

January 1997 magazine

Cursillo

In addition to study courses, not all mentioned here, members of the congregation had other opportunities to widen and deepen their faith. In 1996 a few experienced a Cursillo Weekend. Cursillo – a Spanish word meaning 'short course' - is a movement that was formed in Spain after the Spanish Civil War to help the spiritual life of war-ravaged young men and has since spread around the world. The three days long weekend consisted of talks, discussions and times for meditation, with a daily Eucharist and service of Compline. Some of those who attended Cursillo weekends in 1996 became members of Reunion Groups, a key part of what Cursillo offers. The purpose of these small local groups is to provide encouragement, support, and mutual accountability amongst committed Christians. Members of the congregation still go on Cursillo weekends and periodically larger numbers of people from the Diocese of London gather together for a Eucharist, lunch and Ultreya (a structured meeting). Father Michael who introduced Cursillo to St Lawrence's is still very much involved with the movement.

Other Examples of Nurture

Though not always noted in the annual MAP traditional ways of nurture such as work with young people, Confirmation preparation courses, quiet days and retreats, were maintained. At the beginning of Lent 2001 there was a retreat with a difference when members of the congregation were invited to take a retreat but stay at home. Father David organised a Week of Guided Prayer, in which individuals met regularly with a member of the ministerial team as guide, and sought to listen to God in prayer through using the Scriptures. A rather more ambitious venture took place in October 1999 when Father David led a parish pilgrimage to the Holy Land in October, as part of the Millennium celebrations. Although aware that relatively few people would actually undertake the pilgrimage, Father David felt it could be a journey of prayer for the whole parish. To this end he held several meetings and wrote a number of magazine articles on different aspects of the Holy Land. Accounts in later magazines show that the pilgrimage was highly successful and was followed, in 2001, by one to Rome and Assisi. In response to interest for another parish journey, Father David arranged a parish holiday in Vienna and Prague in September 2003.

Baptism Visitors

Among other things *Agenda for Action* had identified the laity as a great resource of the Church. In 1994 lay members of St Lawrence's congregation offered something new to the people of the parish in their role as Baptism Visitors. There had been contact after Baptism in the past but this was a more positive step. Following the initial contact of parents with the clergy, the visitor became involved and in addition to explaining the baptism service, told the family something about the Church. The visitor attended the baptism service, stood with the baptism party around the font, and handed them the baptismal candle. The emphasis was not so much on theological expertise as upon a friendly welcome, and this remains an important part of St Lawrence's mission in the wider community of Eastcote.

The Font – Flower Festival 2000

The London Challenge

'We're to be expressions of God's love for all the citizens of London'.

A further example of the laity as a resource of the Church in sharing the Gospel was highlighted as a result of the London Challenge of 2002, which the Bishop of London pointed out was not his challenge – it came from Christ Jesus himself. The Good News of God's love, which we are asked to share, was not new but 2000 years old. The Revd Ann Coleman wrote about the Challenge. 'What we have to do is to interpret it afresh for our post modern culture – one which seems more concerned with individualism than community, more concerned with rights than with responsibilities'. The London Challenge video was shown to the congregation at both morning services on 5 May and, after the 10am service, the congregation was split into groups, each group responding to a series of questions drawn up by the clergy and churchwardens. The feedback from the discussions was encouraging, and the revised MAP was based largely upon the responses from the congregation. Father David wrote 'The Challenge had enabled us as a parish to reflect on current good practice in our ministry and mission, and has encouraged us to seek appropriate and sustainable developments'. Ann Coleman masterminded the excellent displays in the church, which were an effective focus for the imaginative and well-attended London Challenge Celebration held with the Bishop of Willesden (The Rt Revd Pete Broadbent) on 17 September. At this service he was presented with the 2002 MAP, revised in the light of the Challenge.

London Challenge Celebration 17 September 2002

Care and Service
In the December 1993 magazine Father David made clear the overlap of different aspects of church life, and the need for greater involvement of the laity in the 'ministry' of the parish. As far as Care and Service was concerned the priority was ecumenical co-operation, regarded as 'vital in identifying areas of community significance, and addressing them'. There was also the desire to establish a network of contacts to 'be responsible for disseminating information about the church in their locality'.

From Father Bill's time as Vicar, there had been 'ecumenical co-operation' with the other churches in Eastcote. The joint fêtes and concerts of the 1960s had not been repeated, but the January observance of the Week of Prayer for Christian Unity, co-operation over Christian Aid Week each May, and the Carol Service became part of the annual calendar. In Father David Hayes' time other annual events were added, including the January Tea Table Conference, further occasional services, and some Lent study groups. In June 1993 Father David (Coleman) expressed the hope that members of St Lawrence's would support a closer relationship, resulting from the inauguration of the Churches Together in Eastcote (CTE). A formal council, consisting of the clergy and ministers, and three representatives from each of St Lawrence, St Thomas More, Eastcote Methodist, Northwood Hills and St Andrew's United Reformed churches, aimed to build upon the foundations of the many years of informal friendship. St Lawrence's first representatives were Nicholas Metcalfe, Regula Sharp and Jean Yates. From the beginning Captain Don was an active member of the council. From June 1994 a short formal AGM followed by tea and a service became another annual event. Christmas cards with information about Christmas services at all the churches, have been delivered annually throughout the parish.

The Drop-in Centre, the ecumenical venture started in 1985, had the objective of providing a venue for people with problems to drop in and find sympathy and possible help. By the time Father David became Vicar in 1990 it was really a Monday club, yet still serving the local community. A number of people, mostly living on their own, attended regularly, some weekly, for seven or eight years. Bill Maxted was a weekly visitor for 15 years. Christmas parties were enjoyable occasions, but the tea, chat and laughter of the weekly gatherings were probably more important. In 2000, with falling numbers, those running the joint venture decided that people might be put off by the name and changed it to the Call-In Centre. By 2004 some 33 members from the local churches had helped on a rota basis about once a month. Each have played their part in making the venture a success, but it seems right to name Pam Gough, Marion Hayman, Regula Sharp, and for many years Joan Reeve, who have been involved from the start. Captain Don has called in most weeks and his quizzes are much appreciated. The fact that in 2010 there are several visitors each week shows that the Call-In Centre still fulfils a need in the local community.

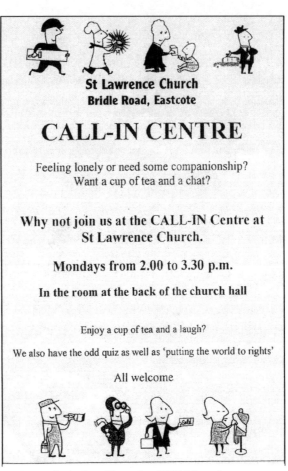
Invitation, December 2003 Magazine

When the MAP was reviewed in 1996, the PCC agreed that the focus should be on one specific item in each section of the Plan. For example in the area of Teaching and Nurture it was the launch of the *Christians for Life* course, which has been discussed earlier. In the area of Care and Service it was the challenge of pastoral care. The Pastoral Care Scheme of 1987 had been successful but by 1997 the scheme needed re-organising. The aim was to improve the level of pastoral care. With Alan Wright as co-ordinator, a new Parish Care Scheme was launched on Dedication Sunday 19 October 1997. The main change was that the parish was divided into eight geographical areas, each with between 30 and 40 people and two 'parish carers'. As before the carers were the links between clergy and members of the congregation. When further re-organisation took place in 2006 the name was changed to the Parish Link Scheme with more 'links' and fewer people in each group. One pleasant feature is the sending of Christmas and Easter cards via the links to all members of St Lawrence Church.

Churches Together in Eastcote and Council of Christians and Jews
Co-operation between members from St Thomas More and St Lawrence churches for the Summer Fête of 1964 had been through a committee confusingly called

the Interfaith Committee. Subsequently there was increasing 'ecumenical' co-operation between the local churches leading to the inauguration of the Churches Together in Eastcote (CTE) on Trinity Sunday, 6 June 1993. In the 21st century there is much discussion on true 'interfaith', as well as 'ecumenical' issues, and it seems relevant to mention the launch of the Hillingdon Branch of the Council of Christians and Jews (CCJ) on Monday 5 July 1993. This took place at the Winston Churchill Hall, just a month after the inauguration of the CTE. The Right Revd Richard Harries, then Bishop of Oxford, was the guest speaker. From that time members of St Lawrence's have shown interest in CCJ, several by attending meetings and a few serving on committees.

Outreach and Communication
The challenge of the Decade of Evangelism with its subsequent *Agenda for Action* and resulting annual MAP, led to the PCC formulating a more active approach for enhancing the mission of the Church both within the church community and outwards to the wider community. Father Michael came to the parish in 1995 with the particular responsibility of promoting the mission of the Church and some projects he initiated have been discussed already. This chapter has already considered many aspects of mission through personal contact but better communication through the printed word was also needed.

In the revised MAP of 1995 a new category, Outreach and Communication, was added to those identified in 1993. One objective was 'to improve the quality of material produced for internal communication within our church'. During 1996 the Publicity and Publications Committee, usually known Pub and Pubs, was set up to replace the old Magazine Committee. It reflected the more active approach to the mission of the Church. In the July 1996 magazine Father David referred to the importance of using modern technology to disseminate information and to the progress that had been made over the last year. The Parish Magazine was largely produced on computer, and the PCC decision to invest in a new copy printer meant the finished article had a more professional look. During 1996 Sadie Wright became editor when she succeeded Father Michael who had stepped in temporarily when Mrs P. Brown resigned as she was leaving the district. She had been editor since 1993, following Barbara Higgs who had been editor for the previous ten years (see Appendix 4).

This was a period when many companies adopted a logo to promote themselves to the public, with the chosen logo reflecting something significant about the organisation it represented. In keeping with modern trends St Lawrence's adopted a logo as one way of trying to get the Church more widely known. The logo chosen was based on the crucifix behind the high altar, something unique to our church. The logo first appeared in the new cover design of the September 1996 magazine, replacing a design of symbols of the Church, which had been used since 1987. With minor modifications the magazine cover has been the same up to the present time. However as regards actual production, which had been 'in house' since April 1971, there were changes. A number of people had

 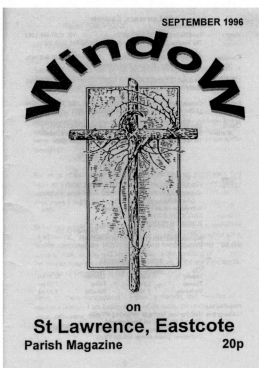

Magazine Covers for 1987 (left) and 1996

been involved in the printing, collating, folding, stapling, and distribution but as various members of the team withdrew production became increasingly difficult for the few dedicated ones remaining. In 2001 arrangements were made for Bishop Ramsey School to print and collate the prepared text, which was then collected and bundled up ready for distribution. Although many people had given of their time and talents over the years, special tribute was paid to Ron Towers who had been responsible for much of the organisation and practical work associated with the production and distribution for well over 20 years. The price of the monthly magazine went up from 10p to 20p in 1990 and to 30p in 1999.

From September 1996 the logo began to be used on church literature of all kinds, including weekly bulletins, notice boards, and letterheads. By 2001 The Pub and Pub Committee had redesigned the Sunday bulletin, erected new internal and external notice boards, produced car stickers and bookmarks, and a general information booklet for use within the church. In 2004 a Historical Guide to St Lawrence Church was published. To these mainly 'in house' means of disseminating information must be added the physical upgrading of the bookstall, with its increased variety of cards and books, mostly to buy but some to borrow, which has been managed by Doreen Plummer for many years, helped for a long time by Margaret Case.

A second objective of the 1995 MAP was 'to make local people more aware of our church and what it has to offer, and of the relevance and importance of Christian faith'. One effective way, and needing the all-important personal contact, was through approaching local shops to display church posters about fêtes and other church events. Fortunately, as is the case today, local tradespeople were supportive.

Newspaper of the Parish Church of St. Lawrence, Eastcote
Number 10   Easter 2000

Brownies for the Future!

On Sunday February 20th the beginning of the Morning Service at St.Lawrence was taken over by the 6th Eastcote Brownies as they presented a Time Capsule to Father David for safekeeping in the Church. The Capsule contains a collection of memorabilia that provide a unique snapshot of everyday life in Eastcote at the start of the 21st Century — items that include a supermarket till receipt, coins, a daily paper from January 3rd and some Brownie badges, to list but some. The twenty four Brownies, all aged between 7 and 10 years, filled in reports detailing their ages, interests, impressions and experiences of life at the turn of the millennium — as well as their hopes and aspirations for the future — and these accounts have been sealed in the capsule too.

Do not open until 2030AD

Impact of Technology
Finally, Marion Glynn, Tawny Owl, wrote about the issues of the day, discussing the impact of the latest technology, such as the internet and mobile phones, on our lives. The capsule has been sealed until the year 2030, and an official record has been made in the Church's archives for it to be opened when that year comes around. By that time, of course, many of the present Brownies will be mums themselves, hopefully with their own daughters in the Brownies.

Future of the Brownies
But what will the Brownies of the future be like? Will we see our first generation of Space Brownies camping out on some far away planet, sending hypercards home via the EtherNet? Buzz Lightyear watch out! On a more serious note, however, the Brownies will only be there in 2030 if we have Brownie Leaders — there are 80 girls on the waiting list to join, far more than the 6th Eastcote pack can hope to accommodate. If you would like to find out how to get involved, please contact Mrs Marion Glynn, Tawny Owl, on 020 8866 9269.

Watch this Space!
Digital diary users can book the date for the Capsule's opening now; the rest of us will have to watch this space in *Free For All* — online, in colour and fully interactive by then, naturally...

Services during Holy Week and Easter

Holy Week, which begins on Palm Sunday (April 16th) and reaches its culmination on Easter Day (April 23rd), is the most important week of the year for Christians. We recall and celebrate the events at the very heart of our faith—the suffering and death of Jesus and his resurrection to new life. Visitors are welcome at any of our services: please join us if you can.

Palm Sunday (April 16th)
8.00am......................Eucharist
10.00am....................Blessing of Palms in the Parish Hall and Procession to Church for the Parish Eucharist with dramatic reading of the Passion
6.30pm.....................*Churches Together in Eastcote* United Service

Monday to Wednesday
8.00pm......................Eucharist & Address

Maundy Thursday (April 20th)
8.00pm......................Solemn Eucharist of the Lord's Supper followed by the Watch of Prayer until midnight

Good Friday (April 21st)
10.30am - 12.30pm.. Activities for children aged 4 to 11. Charge £1, including hot cross buns. *Ends with:*
12.00 noon...............Children's Worship for Good Friday
2.00pm.....................Liturgy of the Passion and Death of the Lord

Easter Eve (April 22nd)
9.00pm.....................The Celebration of the Resurrection of the Lord: Easter Vigil and First Eucharist of Easter

Easter Day (April 23rd)
8.00am......................Eucharist
10.00am....................Easter Festival Parish Eucharist
6.30pm.....................Evensong and Benediction

In This Issue...

Break a leg, Vicar?
  Everything you ever wanted to know about Pulpit Falls.......2

2000 Years of Flowers
  The Ladies of St. Lawrence Say it With Flowers..............3

The May Fair
  Don't miss this year's fun!....3

Mothering Sunday..................4

Free for All – Easter 2000

A far more ambitious plan was to produce a parish newspaper three or four times a year, and distribute it to every household in the parish. Father Michael was responsible for producing the early issues, the first of which was in December 1996. The first issue of *Free for All* had a variety of articles about the life and activities of St Lawrence Church, including details of the special Christmas services and of the first *Christians for Life* course. In 1999, when Father Michael left the parish, Arthur Plummer became editor and, apart from one or two issues, he has been responsible until the present time. Father Michael printed the first issue, but Brian Beeston was involved from the beginning and has been, and remains, responsible for printing most of the now twice-yearly issues of *Free for All*. The newspaper continues to keep the local community up-to-date with old and new events and activities at St Lawrence's. The ambitious project has been highly successful. In addition to evidence that people come to events at the church as a result of reading a *Free for All*, a good thing in itself, a number have joined St Lawrence's congregation. Distribution of *Free for All* was and remains an enormous task, and depends on members of the congregation and their friends to deliver the newspaper. There were well over 8,000 households in 1996 and by Easter 2010, issue 30, the number had increased to nearly 8,900.

By 2000 the church had made use of modern technology to set up the St Lawrence Church web site, *www.st-lawrence-eastcote.org.uk*, as another means of disseminating information, largely to people in the wider community. Among other information the Weekly Bulletin section is up-dated each week.

Young People
Bringing children and young people into the Church is always an important part of the mission of any church. During Father David's incumbency, as in the past, providing suitable forms of worship, and opportunities for social activities, were adapted in response to the needs of different age groups at the time. Again, as in the past, there were often requests for more adult help.

Sunday School
During these years the numbers on the Sunday School register fluctuated, as inevitably did regular attendance. APCM reports show changes of the age range of those in Sunday School, and of other church groups for young people, so direct numerical comparisons are impossible. There were 30 of 4-12 year olds on the register in 1991, 60 of 4-10 year olds in 1998, and 46 of 5-10 year olds in 2005. In 1995 the children were divided into three groups Seekers 4-5, Discoverers 6-7, and Followers 8-10.

Attempts to have up-to-date teaching methods led to the introduction in 1999 of the Scripture Union 'Salt' teaching resources for lessons. Once a term the children met as one large group and the session was led by one of the clergy or Captain Don. This gave an opportunity to reinforce the children's importance as part of the church family. The tradition of the children joining the 10am congregation at the Offertory and returning to the hall after the Consecration continued, and sometimes they showed their work to the congregation during the service. In 2003 a new venture called 'Godly Play' was introduced. This scheme aimed to build on the children's natural spirituality as they responded to a Bible story that was enacted using simple cut out figures. One group of children each week joined these sessions that were held in the George Guest Room. These sessions seem to have inspired a quieter and more reflective response and were an effective supplement to the Scripture Union resources. The children were reported to have 'reacted to the sessions very positively'.

Although there were often pleas for more helpers and teachers fortunately there was always someone willing to take on the role of planning the Sunday sessions. During these years the organisers were Sylvia Gayler (1990 to 1993), Jane Coan (1993 to 1997), Marion Glynn (1997 to 1999), Anne Hand (1999), and Sylvia Hooper (1999 to 2006). A number of people were involved either as teachers or helpers for short or long periods. Usually this was on a rota basis, and their contribution was much appreciated. Rosemary Metcalfe was one who in 1995 decided to have a well-earned break after ten years. From 2000 as a result of the requirements of the Church's Child Protection Policy a number of parents helped on a rota basis after the appropriate official checks had been carried out.

The children appreciated visits from Father David and his ministerial team. APCM reports indicate that there were regular displays in church of the work produced by the children. The children learned the music for the sung parts of the service, and on occasions they made a positive contribution to church services, including Mothering Sunday and Harvest Services. Each year they took part in the popular Christingle Service when many people from the wider community attended. In 1993 they 'sang, read poems and prayers and led the congregation in a dance!' In 1998 they joined with members of the Junior Church in a play *The Bethlehem Herald*. All these activities helped the children feel part of the St Lawrence community.

As part of the church community the children took part in the annual fairs and markets by running sideshows and manning stalls. The APCM reports record enjoyable summer picnics and games in Eastcote House grounds in the 1990s, and successful Epiphany parties from 2002.

The Crèche
In 1990 Jane Williams and her band of willing helpers may have been somewhat despondent when so few parents made use of the crèche but this situation improved. In January 1992 the venue was moved from the rather dingy facilities of the bungalow to the committee room of the hall. In 1994 Jane reported that the scheme had attracted new parents with young children and in 1997 there were often as many as ten children, always supervised by two adults. By this time Jane's own children had moved up to Sunday School and she was replaced by Sally Leakey until 2003 when Helen Britter took over. The Crèche operated each

week except Parade Sundays. The children played games, sang songs, made friends and joined the congregation in church at the Peace.

Junior Church
While very young children moved from crèche to Sunday School, others moved from Sunday School to Junior Church, which was for the 11-14 year olds. Some APCM Reports indicate that with this age in particular, membership and attendance may have been influenced by friends. During 1991 some 7 members left to take part in the whole of the 10am Sunday Eucharist. In 1992 there were 8 who attended regularly. In 1994 there were 10 but numbers reduced to 6 when the older members reached the age of 16 and moved on to join the main congregation. In 1998 a Young People's 14+ group was started to cater for the previous Junior Church members, but several were involved in the life of the church in other ways such as serving or giving occasional assistance at Sunday School and this group did not survive. In 2005 things looked better for the Junior Church and it was reported that between 10 and 16 young people were attending on a regular basis. They met in the George Guest Room and joined the church service at the Peace. While Reports on the Junior Church were written in turn by Doreen Plummer, Marion Smith, Monica Bolley, Val Heaney and Adrienne Williams, there were several other adults who gave of their time and talents to help Junior Church members.

In 1994 the Junior Church 'adopted' a 10 year-old Indian boy through Tear Fund and their child support unit and, in spite of their low membership, were able to fund his Christian education and health care for a few years. In order to do this the young people had various fund-raising efforts such as car washing. For several years their main social event was a visit to Slough for a Pizza Hut lunch and ice-skating.

The Holiday Club
In keeping with the mission of the Church a new venture to attract children from the wider community of Eastcote was tried in 1992. This was the Holiday Club organised by the Revd Ann Coleman, which took place the first full week after the schools had broken up for the summer holidays. Each morning from Monday to Friday there were opportunities for children aged 6 to 12 years to take part in sport and craft activities. It was a great success and Holiday Clubs have taken place every year since 1992, each year having a theme, often making use of the church grounds as well as the buildings. Over these years a number of people were responsible for organising these highly successful events and they had support from many willing helpers. Most years Captain Don played an important part in the success of these mornings. Several years over 100 children took part in these weeks and in 1994 there were 140.

From 1994, at the latest, there was a pre Christmas Children's Afternoon but these pre Christmas clubs are now morning events. In 1995 there was a morning of activities on the Tuesday of Holy Week, which was linked with the traditional

10am Good Friday Children's Service. From 1997 a morning of activities started at 10am on Good Friday and ended with a service in church. The second issue of the parish newspaper *Free for All* may have contributed to the success of the 1997 Good Friday event when 130 children attended. There was usually some display of the children's handiwork in church on the Sunday following holiday activity events.

```
HOLIDAY CLUB '94

140 children filled the car park on Monday
1st August, waiting to join the Holiday Club.
Eventually everyone was registered, and there
began a week of activities culminating with
Captain Don's play, 'The Pied Piper of
Eastcote.' This involved a cast of thousands.

Each morning we 'visited' a different part of
the world, and enjoyed craft activities
relating to that country.

Sweetmaking, jewellery making, painting,
glitter and glue proved to be popular once
again. Football, snooker, and table tennis
were also high priorities for the children.

We heard stories from different parts of the
world, including news of the St. Lawrence
Children's Hospice in Romania, and the young
boy in India supported by the Junior Church.
The children were eager to join in the songs,
games, and prayers, and about twenty came to
church on St. Lawrence's Sunday to help to
take our 'Chinese Dragon' to church.

My thanks to all the helpers whose ideas and
enthusiasm enabled the week to be such a
great success. We look forward to our
Christmas Children's Afternoon - I think!

Ann Coleman
```

1994 Magazine Report – Holiday Club

SLICK

Reference was made early in this chapter to Father David's recognition of Ruth's 'particular charisma for nurturing young people'. Ruth was largely responsible for the success of a new social group, which emerged in 2002 in response to the enthusiasm for 'something for the young people expressed in the London Challenge consultation'. The group led a few Sunday evening services, starting with *On Trial* in March 2002. By 2003 the group had a name: St Lawrence Intelligent Cool Kids (SLICK). SLICK challenged the PCC to an afternoon of activities on Sunday 9 March 2003. It was not an afternoon of activities such as experienced in the children's Holiday Club sessions, but a series of games on a Star Wars theme. Spectators paid to witness this afternoon of fun, which was enjoyed by all. Other social events enjoyed by SLICK were videos, pizza nights and a Halloween party. After Ruth left in September 2004 Adrienne Williams took over the running of this group and organised events, one of which was a coffee morning after the 10am Eucharist on 20 March 2005 in aid of Cancer Research.

MOSES

*I'm an Israelite:
Get me out of here!*

The story of a chosen man with a big stick

**An act of worship exploring faith
through the life of Moses**

Led by
SLICK
&
Featuring
Joyful Chorus

**at St Lawrence Church
Sunday 6th July at 6.00pm**

Refreshments follow service:
e.g. frog spawn, red sea cocktails,
boil sweets & dead fly biscuits
(and some murky brown hot drinks)

Sunday Service led by SLICK 2003

St Lawrence Church Pray and Play
In February 2003 Ruth started a Friday afternoon service for parents and toddlers. Rather on the lines of the Pram Services of Father Bill's time it was a short service with interaction stories, prayer and play followed by coffee in the St Lawrence Centre. These services seem to have been a response to specific needs and the relevant *Record of Services* book shows they enabled a few parents to receive Holy Communion during the week. They were discontinued when Ruth left the parish and Father David no longer had an Assistant Curate.

St Lawrence Mother and Toddler Group
One example of St Lawrence's mission to the wider community of Eastcote was

the Mother and Toddler Group that opened in April 1997. Many of those who attended these Thursday afternoon sessions were already associated with St Lawrence's but the majority were newcomers. Money left over after the closure of St Lawrence Young People (SLYP) was put towards buying play equipment for this new group. A magazine report of August 1997 showed this venture was successful with 60 children on the register and around 20 children attending each week. This meant a lively but safe environment for the variety of activities that were available. Each family paid £1 each week to attend and soon the group was able to buy more equipment, which the organisers felt played a large part in the success of this venture. Over the years a number of parents were involved in running the group and as children moved on to playgroups their mums handed over to others. They were always grateful for the support given by a few members of the MU who helped with refreshments. Later reports show that this group fulfilled a local need. Summer and Christmas parties and a mum's night out, were all enjoyable events. In 2001 their first fun afternoon in aid of charity was organised and raised £86. The one in 2002, at which the Mayor of Hillingdon (Mrs Josephine Barrett) was guest, raised £120 for The Shooting Star Trust.

Music at St Lawrence's
Numbers in the choir were never very high in these years, a maximum of 14 for a time, but members were conscientious and their contribution greatly enhanced the 10am Sunday and weekday Sung Eucharists. A number sang at choral weddings, (21 in 1993) and funerals, including that of 'Topsy' Millicent Dedman in 1991 (former organist). A small but much appreciated group continued to sing at Sunday evening services. Frances Bradford (appointed in 1984) remained organist until her retirement in 1994. Anne Holt, assistant organist, expressed 'gratitude for her enthusiasm and encouragement over ten years', and Father David wrote of her 'musical skill and dedication, but above all her Christian sensitivity to the liturgy'. Enid Lumsden, another assistant organist stepped in to take choir practice. With no official organist the church 'was grateful to Enid and Anne for holding the fort until a new organist was appointed', which unfortunately didn't take place in this incumbency. When Enid retired in 2003 Anne took over, though Enid still played the organ when needed.

There were times when the choir was augmented, such as at Father David's Institution and Induction in October 1990, and various Church Festivals. Several annual reports expressed gratitude to members of the congregation who joined them for special occasions but also appealed for people to become regular members. The choir helped congregations to familiarise themselves with new settings for the Eucharist. The Choir Report of 2001 wrote of them 'still getting to grips with the new harmonies and word changes' of the new hymnbook – *Complete Anglican Hymns Old and New*.

On Palm Sundays in the early 1990s, while Frances Bradford was still organist, in place of Evensong there were devotions for choir and congregation, *God so loved the world* (1991), *Behold your King* (1992), and *The Way of the Cross* (1993). During these years the choir took part in the Royal School of Church Music Festivals at St Paul's Cathedral. Later, in 2001 some members enjoyed taking part in the Harrow Deanery Choirs Festival at St Andrew's Church, Roxbourne.

At least two of the popular 'Choir Entertains' evenings, started in 1985, took place in the early 1990s. These social occasions were also fund-raising events. One in June 1991 was in aid of Church Funds and one in June 1993 (Jubilee Year) was in aid of the Organ Fund. Another effort for the Organ Fund was the 'Organ Play' in May 1993 when Frances, Anne, and Enid played through all the 656 hymns in the *English Hymnal*. Other fund-raising in 1992 and 1993 took the form of coffee mornings and beetle drives, some for the Roof Repair and some for the purchase of new choir robes which they had in 1993.

As in the past and as happens today concerts were given in church by visiting musicians. In May 1998 and 2001 concerts in support of Christian Aid featured 18th century music played on period instruments. One visiting group was the Amici Chamber Choir who presented a Summer Concert in July 1999, a concert of Sacred Choral Music in April 2000, and Music for Holy Week on Palm Sunday 2004.

Stewardship
At Father David's first APCM in 1991 the reports of the Treasurer and the Stewardship Recorder showed that in real terms stewardship giving had decreased and the parish was heading for a deficit of around £11,500. The Diocesan Quota (Common Fund payment) for 1990 had been £33,107 (reduced from some £40,000 because of the interregnum) but the 1991 Quota was likely to rise to £46,500. In his June magazine letter Father David made it clear that 'if the work of the Church is not to be debilitated by a chronic financial crisis' there had to be a radical increase in our giving. To this end a Stewardship Renewal Campaign was held from 6 to 27 October 1991.

As with previous campaigns much work was done before October. Michael Dyer, Diocesan Finance and Stewardship Adviser, directed the Campaign and Alan Wright was Chairman of the Stewardship Committee. The Revd Eddie Shirras, Archdeacon of Northolt, preached at the 10am Eucharist on the day of the launch of the campaign 6 October (Harvest Sunday). Father David considered the day highly appropriate as 'Harvest is about thanking God for his bounty towards us'. He also reminded people of the significance of Dedication Sunday, on 20 October, and the need for members of St Lawrence to be generous, 'not only to maintain the building and the ministry, but so that we can extend the work of the Church'. The social side of the campaign was less ambitious than some in the past, and consisted of informal gatherings when volunteer hosts were 'At Home' to members of the congregation.

The Recorder's Report for the 1992 APCM showed that membership had increased from 137 in 1990 to 183, and that 57 existing members had increased their giving. Father David felt this was a justification for the campaign. Accounts for 1992 show an increase in income from Stewardship and Tax Refund on Covenants. However the Treasurer reported that the Quota for 1992 would be £55,802, and even with an increase in income it was feared the total could not be paid. Figures show that this was the case for the three years from 1992 to 1994 (see Appendix 10).

A Stewardship Sunday was held in November 1992, when church members were asked to review their giving. In connection with this Father David wrote a great deal in the magazine about the financial situation of the Church of England in general, and that at St Lawrence's in particular. It seems relevant to include some of this contemporary information. Churches need maintaining and clergy need paying. On the whole, parishes are responsible for maintaining church buildings while dioceses meet the cost of incumbents from the Common Fund. As the Church is, in theological terms, the body of Christ there is a fundamental unity, and we are bound to mutual support. This means more affluent parishes subsidise those in inner cities or remote rural areas. At Eastcote the 1992 Common Fund assessment was £56,000. What was to be spent on maintenance, worship, heating etc was estimated to be £18,000, making a total of £74,000. Income from giving (Stewardship, alms etc) was £58,000. Other income (fees, fundraising etc) was £10,000. All this made for an estimated deficit of £6,000.

The Financial Report at the 1993 APCM revealed that the parish had paid only £46,000 of the £55,802 Common Fund. The Treasurer referred to the fact that St Lawrence Church was considered an affluent parish and was expected to contribute towards the cost of the more needy parishes in the Diocese. If we did not pay our quota it would be written off, but we might not get another curate and would be unlikely to get any grant for such things as heating. In his Vicar's Remarks of 1993 Father David referred to the extended consideration being given by the various sub-committees to the *Agenda for Action* and expressed the hope that the discussions would lead to a relevant and realisable set of objectives. 'They would, however, be dependent on getting our finances on a sounder footing'. Father David made it clear that although the Diamond Jubilee was a time for celebration and enjoyment as part of our thanksgiving, the Jubilee Fund would not be used towards the Common Fund. It would be divided equally between the domestic needs of St Lawrence – a major item such as the organ, or heating replacement, - and an outside charity, SOS Sahel. 'We must look outwards, it was to those who were generous that God was generous; it was what we measured out that we would receive'.

In his Vicar's Remarks at the 1994 APCM Father David spoke about that year's Stewardship Campaign (17 April to 22 May), and emphasised how important stewardship was if St Lawrence's was to meet its financial obligations to the diocese. He hoped people would pledge 5% of take home pay. The parish also needed the time and talents of parishioners – 'it was the greatest asset we had'. Income from Stewardship and Tax Refund on Covenant went up and at the 1995 APCM the Treasurer reported that a new system had been introduced and the Diocese had managed to reduce the Quota from £61,000 in 1993 and £59,000 in 1994 to £53,000 for 1995. The parish was able to pay in full once more.

There were Stewardship Sundays in June 1995 and October 1996 when people were again asked to review their giving but the response was poor. In 1997 as part of the Mission Action Plan (MAP) a Stewardship Renewal Programme, 'The Responsibility is Ours', was organised for 18 May to 15 June. The Bishop of Willesden (The Rt Revd Graham Dow) officiated and preached at the Eucharist of Thanksgiving at the end of the Programme. The increased revenue enabled the church to meet the quota to the Common Fund. In 1998, in view of the St Lawrence Centre Appeal it was decided that it would not be appropriate to have a renewal programme. Letters were sent to all members in October thanking them for past response and asking for continued support. At the 1999 APCM concern was voiced about a decrease in income from stewardship of £2,000 but an increase in expenditure of £3,500. Membership for 1998 was 139 compared with 157 in 1997. Fortunately, as had happened in the past, a legacy helped the financial situation, but a full-scale renewal campaign was essential.

The Stewardship Committee ran a very successful campaign from 4 June to 2 July 2000, with membership rising to 179. In his 2001 Vicar's Remarks Father David commented on how the parish had marked the millennium. He referred to the special services and flower festival, but regarded the year 2000 important for the stewardship campaign which was 'a resounding success in raising the whole profile of Stewardship in the parish, and attracted several new members, as well as encouraging many existing members to make revised commitments'. Father David also mentioned that the time and talents side of stewardship in 2000 had included the redecoration of the interior of the church building.

The 2003 Vicar's Remarks noted that the main focus of parish life in 2002 had been the 'London Challenge' and that this had enabled the parish to celebrate the enormous amount of ministry already going on. It also posed questions for the future, one of which was the matter of finance. Father David expressed the view that the parish had 'made real strides in trying to address this sometimes delicate subject in a proper manner', but warned that 'the pressures on the diocese continue to grow, in an uncertain economic climate'. The role of the Stewardship Committee as set out in the annual MAP was

(a) to monitor parish giving and advise the PCC as to any necessary action to maintain an adequate level of stewardship income,
(b) to keep before the parish the vital need for stewardship in the terms of time and talents, and to encourage fund raising where appropriate (other than that of the Fellowship Committee [sic]).

Each July from 2001 to 2003 Andrew Bedford gave a presentation of the Committee's work in church, with varying degrees of success as regards increased giving. Andrew also organised the two-pronged approach in the Stewardship Campaign of 13 June to 11 July 2004. People were asked to look at how they worshipped as well as at their financial giving. An interim report on the campaign in the September magazine showed positive responses to both 'prongs'. The results indicated the necessity for having periodic full-scale campaigns.

People have committed to:

- *Sunday worship*
  - *65% said that they would endeavour to attend every week;*
  - *13% at least twice a month, and*
  - *13% at least once a month.*

- *Weekday Eucharists*
  - *19% aim to attend one of our 6 weekday Eucharists.*

- *Morning or evening prayer*
  - *4% said that they would attend.*

- *Offering a short prayer at 9am or 5pm – the start times of morning and evening prayer*
- *35% said that they would stop what they were doing and join with others around the parish in offering up prayers.*

On the financial side the outcome was also positive:

- *30 people have joined the stewardship scheme, and*
- *52 people have increased their regular giving.*

Extracts from the 2004 Stewardship Campaign Report

Charitable Giving
Annual accounts give details of donations given by St Lawrence Church to various charities some of which, as in the past, were from the church's income and some from special collections or events. A few examples may be of interest.

Two charities supported by St Lawrence's from the time of the Mission Church, and which continue to receive donations now, were The Children's Society and the Church Army. A donation was sent to the United Society for the Propagation of the Gospel (USPG) most years. Helen House, Oxford, supported for a few years in the 1980s had further donations from 1999. The local Queen's Walk Hostel also received donations. The congregation responded to Lent appeals and emergency appeals such as Tear Fund (Rwanda) 1994, Mozambique Flood Relief (2000), and the DEC Tsunami Appeal (2005) which raised £2,831.

Fellowship
Previous chapters have shown the fellowship of the community at St Lawrence's has been strengthened through social events and this in turn has helped to further the mission of the church in Eastcote. When, in 1997, terms of reference were agreed for the various PCC committees, the role of the Fellowship Committee was

(a) to promote Christian fellowship within the parish by organising social functions and entertainments and to assist in the mission of the Church in Eastcote by these activities,
(b) to assist in the raising of funds by organising two annual Fêtes and other occasional events.

The May fairs or fêtes, and Christmas markets or bazaars, were the responsibility of the Fellowship Committee but Pat Lomax seems to have been the organiser of these events up to the 1994 May Fair. The Committee was then collectively responsible until 2002 when a small sub-committee took over the running of these two highly successful annual events. From 1995 the market was held on the Friday evening as well as on the Saturday. Many of the long established and popular stalls, sideshows and events continued to be regular features of the fairs and markets in these years, but as time went on there were added attractions. St Lawrence's has always been grateful for donations from local businesses and shopkeepers, and for the attendance of members of the wider community of Eastcote, each playing a vital part in making these events both happy and financially rewarding occasions (see Appendix 11).

As both the fair and the market were fund raising as well as social events it is relevant to refer to another successful fund raising and social event though it was not the responsibility of the Fellowship Committee. In 1990 Joan Sherriff-Gibbons organised the first Antique and Bric-a-Brac Fair to help deal with the deficit during the interregnum. She continued to organise or be involved in these Fairs until the last one in 2004.

Among the annual functions and entertainments for which the Fellowship Committee was responsible was a Harvest meal. Up to 1993 this was a supper following Harvest Festival Solemn Evensong but this was changed in 1994 to a Harvest lunch following a later, 11am, Parade and Parish Mass. Another popular

Handbill for 2002 May Fair

event was the supper quiz organised by Ann Sykes from 1989, with the help first of her husband and then of others, until 2010. It may be remembered that the St Lawrence Players had a successful supper quiz in 1987. A summer event was the cricket match held on Bishop Ramsey School Field, apart from one year (1994). Most years it was a Vicar's XI versus a Wardens' XI, but in 1993 it was between St Lawrence, Eastcote and All Saints' Hillingdon, at home, with a return match at Hillingdon in 1994. In 2000 and 2001 the match was between St Lawrence's and St Paul's, Ruislip Manor.

Some of the occasional or one off events were a barn dance (1990,1992, 1996), Line Dancing (1999, 2001), a Race Night (1997), a Donkey Derby (2002), a Murder Mystery Evening (1998). Needless to say these functions always had refreshments of some kind. As part of the Millennium Celebrations the Committee was responsible for providing refreshments and arranging the Art and Crafts Exhibition the weekend of the Flower Festival, that jointly raised £1,025 for Jeel Al Amal Children's Home in Bethany. Other events in 2000 were a Family Fun Evening and a Casino Night. Another well-supported new venture of 2000 was a parish meal at the Taste of the Mediterranean restaurant in Eastcote. A party from the parish returned there in 2002 and 2004. In 2001 it was held at The Tudor Lodge Hotel another local venue. In 2003 the party went further afield for Dinner and Dog Racing at Walthamstow. The Committee also organised parish outings to Southend (2002 and 2004), and to Clacton-on-Sea (2003).

As well as providing refreshments for events that they organised, the Committee were responsible for refreshments after the annual Ecumenical Carol Service, always held at St Lawrence's until fairly recently, and for the CTE Tea Table Conference when St Lawrence's hosted this annual event. Some years they arranged refreshments for special occasions such as Confirmations, Father David's Induction and Father Peter's Farewell reception (1990), and the Bishop of Willesden's visit at the end of the Stewardship Renewal Programme (1997).

St Lawrence Social Club
The St Lawrence Social Club was formed in 1989 (in Father David Hayes' time) to 'provide an opportunity for members of the congregation to meet together socially and strengthen the fellowship within the church'. Writing on the closure of the Women's Guild in April 1989 Father Peter referred to the Social Club having taken over some of the functions of the Guild, and Social Club reports show that many of the activities were similar. The biggest difference was the inclusion of men as members, and although the Club was formed for 'members of the congregation', it has always been open for others to join. By the time Father David Coleman became Vicar the Social Club was well established and had an average attendance of 30 members and friends. This average soon went up to between 40 and 50. Club evenings were usually held, as they are today, on the fourth Wednesday of each month. It was not until June 1991 that the Social Club was listed on the 'Who's Who and What's On' page of the magazine.

In the course of each year there were both 'home' and 'away' meetings. As with the social activities of other church organisations the Club had many outside speakers. Talks on local or relatively local topics included those on Metroland (1992), Some Remarkable Women of Pinner and Old Houses of Middlesex (1994), the Compass Arts Theatre, Ickenham and the History of Northolt Aerodrome (1996). Among the members who gave talks were Geoff Higgs on Putting a Television Programme Together (1998), Lionel and Barbara Williams on their experiences in East Africa in their younger days, (2000), and David Horchover on Sales Promotion (2003).

Some of the 'home' evenings had entertainment as well as information. One from the 'Bell Tones', a hand bell-ringing group from Denham, who played 'off the table' as opposed to in the hand, treated the Club to 'a feast of music throughout the evening' (1995). A visit from David Ball who described himself as 'A Magical Entertainer', gave an account of how the Magic Circle was formed in 1905, and treated his audience to a variety of tricks (1996).

In addition to providing speakers for some of the evenings, members frequently organised their own entertainment. A few examples must suffice. There was an enjoyable evening when they recalled the old but popular TV and radio parlour games of *What's My Line*, *Twenty Questions* and *Call My Bluff*. (2001). Members seem to have had a great time at a Night at The Races (2004). Then there were the summer barbecues when the success depended to some extent on the weather.

There were a variety of 'away' activities. Walks, some local, some in London were usually led by a member of the Club, but guided walks included Jack the Ripper Haunts, from Tower Hill to Spitalfields (1995), and places in St Alban's with 'ghostly' connections (1999). Members spent some pleasant summer evenings on canal trips (1994, 2000, 2002). Other outings were to the Organ Museum at St Alban's (1999) and Bletchley Park (2002). Starting in 2001 there were summer visits to the Coach and Horses Petanque Stadium, Croxley Green, where teams from the Club battled it out to become Boules Champions of St Lawrence Church.

From 1991 there were regular outings to the Six Bells in Ruislip for the annual skittles evening, when members and guests had fun themselves and also provided entertainment for other customers at the pub. The other annual outing was to the theatre. Usually this was to the Palace Theatre, Watford, but in 2002 and 2003 it was to the James Theatre at Watford Grammar School for Boys, as the Watford theatre was closed for extensive renovation. In 2003 there was a second theatre visit to a 'stunning production' in Regent's Park Open Air Theatre of Cole Porter's *High Society*.

A very popular annual event was the seasonal and festive dinner in December. Each year it had a similar pattern of pre-prandials and chat in the Committee Room, an enjoyable meal, quizzes and games, and ended with Enid Lumsden playing the piano for carol singing.

The Club enabled members of the congregation, and others, to enjoy a variety of social occasions and strengthen fellowship. As part of the St Lawrence community the Social Club supported the annual fairs, markets and other fund-raising efforts, and other church events.

Mothers Union
When Father David became Vicar in 1990 the membership of the St Lawrence MU Branch was 30. Chairing the 1991 Triennial Meeting, Father David expressed his delight in finding an active MU membership at St Lawrence's. However there was concern that although the membership had remained stable, the MU did not attract younger members. Like his predecessors Father David

considered the MU an important branch of the body of the Church. In 1992 he became MU Willesden Area Chaplain as part of his extra-parochial ministry.

Numbers declined largely owing to increasing age and frail health. Some members died while others moved to be near their families. In spite of this the Branch maintained the traditional programme of services and meetings. The Thursday morning Corporate Communion, followed by coffee and fellowship, remained a monthly feature. The first meeting of the calendar year was a service, again followed by fellowship in the hall. The regular January Wave of Prayer was sometimes held in the Lady Chapel but more often it was at the home, first of Edna Pearmain, and from 1996 of Marion Hayman, who also provided lunch.

As numbers declined there were fewer 'outside' speakers. Members learned about organisations such as the Citizens' Advice Bureau, and heard about the work of a Magistrate, of a Hillingdon Councillor (from Catherine Dann a member of the congregation), and of Chaplains at London Airport and Mount Vernon Hospital. Among other talks were those on the mother and baby unit at Holloway Prison, the History of the Ickenham Compass Theatre, and early years at St Lawrence Church by Barbara Williams (a long-standing member of the congregation). Over the years several MU members gave talks, usually on their holiday experiences. As well as Father David, who attended most of the meetings, his successive ministerial colleagues all took an active interest in the MU. Father David's talks included one on his sabbatical in America in 2001.

The ministerial team also joined members and their guests at two popular annual social events. One was the summer garden party held at Marion Hayman's home. In addition to the delightful garden, appetising meal and pleasant company there was usually a raffle, and bring and buy stall. Each December the monthly meeting was the Christmas party, which normally consisted of carols,

Mother's Union Garden Party

entertainment and food. Marion Hayman and Nancy Ashenden appear to have contributed readings and poems for several years, while Hazel Vincent and Frances Bradford and, from 1995, Anne Holt provided the musical entertainment. Don Woodhouse frequently amused everyone with his contributions. His 'Albert goes to Church' (1998), a tribute to Stanley Holloway, was much appreciated.

Most years a few St Lawrence representatives attended Deanery, Area and Diocesan events, including the Diocesan Festival Service at St Paul's Cathedral. MU reports expressed thanks to the male members of the congregation, John Martin, Percy Pether, Alan Braithwaite and Arthur Plummer who carried the Branch's banner in the procession at this annual service. Percy considered himself an Honorary Member of the Branch mainly because of the number of occasions he had carried the banner! He attended Thursday

The Percy Pether Clock

morning services and always joined the members for coffee after the monthly Corporate Communion   He also added his own brand of humour during MU Deanery outings, and at Branch garden parties and Christmas festivities. In 2001 the Pether family gave a clock to the MU in memory of Percy (1916-2000). It is on a wall in the Elsie Fischer Room of the St Lawrence Centre

Annual APCM reports record members going on enjoyable Deanery outings to Winchester, Portsmouth, Wells and Tewkesbury and other cathedral cities. In 1995 there were a number of guests in the party of 18 who went on the outing to Canterbury, where they were entertained to tea by Father David Hayes and Monica in the refectory of the Eastbridge Hospital, no longer a hospital but part of Father David's ministry, incorporating his house and eight almshouses.

Members of the MU continued to take part in church events, and on at least one occasion in 1994, they were responsible for the teas at the annual cricket match. On 15 September 2002 MU members hosted a coffee morning in the church hall following the 10am service in which Jan Wilson, a former London Diocesan

President, preached about the MU. Articles on display boards in the hall gave information on the history of the MU that had celebrated 125 years in 2001.

As early as the 1994 Triennial Meeting, concern about the future of the MU at St Lawrence's had become a real problem. Father David, with his high regard for the work of the MU, and knowing that Marion Hayman was only prepared to serve as Branch Leader for three more years, was anxious that the Branch should continue beyond 1997. The problem of leadership and decline in numbers was not confined to St Lawrence's. During 1995 the Willesden Area piloted a new scheme that would enable all Christians to become members of the MU without feeling they must attend branch meetings. Divorcees and unmarried mothers had been admitted as full members since 1973, and during the 1980s membership was extended again so that those married with no children and single persons of both sexes could join. In 1997 there were only 23 members and attendance at meetings was as low as 12-15. At the end of 1997 it looked as though the Branch would close for lack of a leader but Father David prevailed upon the committee to keep it open, and Joyce Boot and Susan Miller agreed to be Joint Branch Leaders.

With further reduction in numbers the decision was made at the 2003 Triennial Meeting to stay open, but meet only once a month after the Thursday Corporate Communion, and to discontinue the monthly Wednesday afternoon meetings from January 2004. Father David hoped the Branch would continue with the garden party and social gathering at Christmas, and the Branch managed to do this. They also kept the Wave of Prayer, and a faithful few still attended some outside MU events.

The Society of Mary
For several years members of the St Lawrence 'ward' of the Society of Mary continued to attend the monthly 4.30pm Saturday afternoon devotional service of the 'Office' and Benediction, and the major annual festivals of Our Lady. Some members attended the May Devotion and AGM of the Society.

Annual reports expressed thanks to Margaret King and Joan Reeve for arranging flowers at the statue of Our Lady, and to Anne Holt for playing the organ. As in the past the Vicar and his successive Assistant Curates were appreciated for the time and support given to the Society.

The year 1994 was one of change. Reference has been made earlier in this chapter to the Ordination of the Revd Ann Coleman as Priest in April 1994. In his 1994 APCM Vicar's Remarks Father David referred to the subject of the ordination of women as priests and said the decision had been welcomed by many and deplored by some. The Society of Mary at national level had been against the ordination of women and while members of the St Lawrence ward held differing views it was felt inappropriate for the Society of Mary to act as though all its members were of one mind. The decision was made to disband the ward and to maintain its activities under the title of the Guild of Our Lady.

The Guild had a membership of about a dozen for the next couple of years but increasing commitments made it difficult to maintain reasonable numbers on Saturdays. In 1997 it was felt right for the Guild to cease to meet. Father David paid tribute to Madeline Gardner, who had been a pillar of strength to the Guild and to the earlier Society. Some members remained individual members of the Society of Mary. The object of the Guild had been to promote devotion to Our Lady and this had found expression through the Saturday services. Father David wanted to provide the opportunity for this devotion to Our Lady, and so Vespers and Benediction became one regular form of Sunday evening worship.

St Lawrence Players
The reputation of the St Lawrence Players was such that a high standard was expected of all their productions. Annual reports and magazine reviews certainly indicate that audiences were always appreciative. Reviews commended the cast, gave praise to the sets, and paid tribute to back stage crews.

Apart from 1994 and 2000 the practice of having three productions each year was maintained. In 1994 for a variety of reasons there was no October production and members spent the summer renovating scenery, clearing out their storage

The Wizard of Oz – 1998

space and enjoying various social events. Among these was a Quiz Supper. In 2000 there was no May play as members were busy meeting requirements connected with the decision that, after many years as a private Theatre Club, the Players should be licensed by the London Borough of Hillingdon for public performances. This meant they had to replace all the old scenery with new, made from fire retardant ply-wood/hardboard/timber, and to treat almost everything used on the stage with a fireproofing liquid - a time consuming and expensive occupation. Fortunately the Players had enough members with the ability to undertake this work. One advantage of being licensed was that the Players could advertise to the public. During 2002 new hall and stage curtains were purchased costing £3,555. The Players contributed £1,100 towards the cost of the stage curtains, which dramatically transformed the appearance of the hall on the evenings of performance. From 2003 'front of house' members dressed in black and white for easy identification.

While from 1981 to 1990 there had been an annual musical in the New Year, between 1991 and 2005 there were musicals, pantomimes, plays, and a few rather different evenings of entertainment. In 1991 the Players presented an *Edwardian Parlour Party* - a miscellany of Christmas readings, music and supper. The two sell-out evenings were thoroughly enjoyed with 90 people partaking of supper each evening. In 1997 *Song, Style and Supper* was a musical entertainment with melodrama and nostalgia and another successful evening.

For the Players 1997 was a special year in which they celebrated their own 50th anniversary, for although the first production was in May 1948 this drama group had been formed in 1947. The May 1997 production of *When We Are Married*, which began their Golden season, was the third staged by the Players, the others being in 1964 and 1979. Some of the 1997 cast had been in the 1979 production. The magazine review referred to 'a thoroughly enjoyable play, well acted and produced and full of fun'. It also made special mention of the set itself, with its attention to detail and complementing the character of the play. In addition to the stage productions of their special year the Players found time to socialise with an outing to 'The Mill' at Sonning, and an anniversary party in the church hall to which they invited past members. Around 70 past and present members, including two founder members Hylda Darby and Beryl Orders, spent an enjoyable evening reminiscing and looking at old photographs, programmes and posters.

The Players continued their involvement in the annual fairs and markets and like other church organisations played an active part in celebratory church events. In June 1993 as part of the Diamond Jubilee Flower Festival they organised the Arts and Crafts Exhibition in the hall, and in October they presented *60 Glorious Years* - a miscellany of music and nostalgia covering, of course, 1933-1993. In November 1999 as part of the Millennium celebration the Players presented in church T*he God Who Comes*, an evening of drama, music and reflection. This 2000 year-old story with its modern interpretation was seen as 'an excellent choice', and special mention was made of the 'lighting effects and music, which added real atmosphere to a very moving production'.

The Flower Guild
Under the guidance of Margaret King, members of the Flower Guild continued to play their part in the community of St Lawrence by

Flower Festival 2008

Flower Festival 2008

Flower Festival 2000

Flower Festival 2000

Examples of Flower Festival arrangements

contributing time, talent, and money for floral decorations in the church. New members were and are welcome. For most of the 1990s members held an annual coffee and bring and buy morning, to raise funds towards the cost of flowers for major events in the Church's year. From 1991 making the posies for Mothering Sunday was an opportunity for members to get together.

In 1998 the festivals of Christmas and Easter were linked in a new way. The branches of the Christmas tree were made into a cross to which different symbols were added during Lent. On Easter Day individual blooms brought by the congregation were, with the assistance of members of the Guild, attached to the cross and it was transformed into one of flowers. The result was much admired and the cross of flowers was repeated in 1999 and 2000.

Reference has been made to the Flower Festival and Arts and Crafts Exhibition of both the Diamond Jubilee and Millennium celebrations.

The theme of the Diamond Jubilee was 'Celebration' and that of the Millennium '2000 Years in Flowers'. The Flower Guild was responsible for organising the floral displays in the church in which many individuals, churches, and organisations participated.

Guides, Brownies and Rainbow Guides
The 6th Eastcote Brownie Pack and the 3rd Eastcote Guide Company remained an important part of St Lawrence's for both church members and those of the wider local community. In 1996 the 1st Eastcote Rainbow Guide unit opened for girls below Brownie age.

3rd Eastcote Guides
Caroline Dann held the position of Captain throughout this period and appears to have had successive willing helpers. APCM Reports indicate a traditional pattern of activities for much of this period, and the award of proficiency badges seems to have been routine. In addition to the links with the 6th Brownies through District and Area activities, and church events, the guides invited the younger girls for barbecues and games. One evening in 1994 there were games, campfire songs, toasting marshmallows over candles, and an enrolment. From 1999, at the latest, some guides helped at Brownie meetings.

For several years a number of guides had great fun at a 'water dabble day' at the Hillingdon Outdoor Activity Centre where they tried canoeing, sailing, windsurfing, and raft building. Reports also tell of a variety of camping experiences including one at Blacklands, near East Grinstead in Sussex, where there was swimming, archery, canoeing and rock climbing. The girls took part in hikes, mainly night ones, and several sponsored walks. A sponsored walk in 1991, in aid of *Guide Dogs for the Blind*, meant 40 times round the Field End School playing fields. In 1997 they raised money for the Willow Tree site at Harefield. In 1998 the guides attended a campfire there, and some of the 3rds were among the first to camp on the site. The County obtained lottery funding to build a shower block and a pack holiday house, officially opened by the Mayors of Harrow and Hillingdon in 2000. The Middlesex North West Guides' Willow

Tree Centre was the venue for many Guiding events, including one to mark the Queen's Jubilee, when over 1,000 attended.

In 2000 the girls took part in events to celebrate the Millennium. As well as two activity days at Willow Tree, they had a wonderful day at the Millennium Dome on a special Guide Association Day when the attendance of 20,000 was 'an amazing sight'. In 2001 and 2002 they enjoyed special concerts at Wembley Arena organised by the Guide Association, solely for members.

The guides continued to play their part at church events such as the fair and market, helped at the Harvest lunch (from 1994), and assisted at Parade Services. The Company was responsible for attractive displays at the Diamond Jubilee and Millennium Flower Festivals, and organised a wide game for the Family Fun Day in 2000.

In 2002 as part of their 'guiding service' the girls, with the support of members of the Eastcote Village Conservation Panel, planted 3,000 daffodil bulbs and 3,000 crocus bulbs beside Eastcote Road near Highgrove Pool. They were awarded the 'Hinman Conservation Shield' for the year's most outstanding contribution to conservation in the Eastcote area.

In 2001 the new Guide Programme was seen as 'an exciting step forward'. Reports show that members of the 3rd Company responded well to the challenge of the theme's initiative, called 'Go For Its', where they plan and organise many of the activities themselves. In 2002 the Guide Association was re-named Girl Guiding.

6th Eastcote Brownies

Lack of sufficient adult help continued to be a problem at times. Visits from Captain Don were particularly welcome and the brownies were always delighted to see him. Kay Drew who had been Brown Owl since 1987, continued until 1992 when she married and moved from the area. Elaine Wigington who had been involved from the mid 1980s took over the running of the Pack but with no permanent help forthcoming it was a difficult time. She had support from a few parents on a rota basis but had to do the planning, record keeping, etc. herself. When in 1993 Elaine decided to relinquish her role there was a danger that the Pack would have to close. After many unsuccessful appeals for help Miss Katharine Maines came forward and was Brown Owl until she moved in 1997. Several people had the position of Brown Owl for two to three years each, until Jackie Smith took over in 2004.

On the whole the weekly meetings followed the traditional pattern until the new Brownie Programme was introduced in 2003, giving more flexibility and fun, and allowed the brownies to help with planning the activities. APCM Reports refer to 'the gradual introduction of a new yellow and brown uniform which was very cheerful and more comfortable' (1990) and 'updated uniforms' (2003).

The girls worked towards gaining a variety of badges including crime prevention, safety in the home, signallers, scientist, needlecraft, road safety, world cultures, jesters, and entertainers. Their outings ranged from visits to local fire and ambulance stations and Eastcote House walled garden, to District and Area Brownie Revels, from trips to the Science Museum and the Planetarium, to Ten Pin Bowling and visits to pantomimes at different venues. The brownies considered the 1998 St Lawrence Players' pantomime *The Wizard of Oz* 'wonderful and thoroughly enjoyed by all'.

Reports refer to the brownies' continued involvement with church functions including the traditional maypole dancing at the annual fair, helping in various ways at both fair and market, and contributing to the decoration of the church at each Harvest Festival. The Pack also had displays at the Diamond Jubilee (1993) and Millennium Flower Festivals. During 1999, with the approach of the Millennium, the girls put together a time capsule which was sealed and placed with church records and is to be opened in 2030. A *Free for All* item about this is shown on page 245. For several years they arranged table decorations for the Harvest meal. Among a number of their fund-raising events, usually for charity, was a sponsored Kim's Game in aid of the St Lawrence Centre.

In 1991, four leaders and 17 brownies enjoyed a Pack Holiday at Ellesborough, near Wendover. From 1992 when the Pack had no one with an Overnight Licence a few brownies went on Pack Holidays with other local packs. Following the opening of the Willow Tree Centre in 2000 brownie visits to the site included the event celebrating the Queen's Jubilee. According to the 2003 Report the highlight of the year was an overnight stay for 18 brownies at the Centre.

Another successful activity of 2003 was a Carnival Creations day of arts, craft, music and dance at Hatch End Art Centre. The girls learnt how to use African drums and made a Chinese banner. They also performed a Justin Timberlake dance routine on stage to an audience of 500.

In 2004 the Revd Ann wrote a play for the brownies to celebrate 90 years of Brownies. It told of the start of Brownies and the different things they did and the 6th Pack acted it out in church at a Parade Service after Easter. In September the girls provided a Guard of Honour at the wedding of Jackie Smith (née Herbert).

One event of the interregnum must be mentioned. The girls decided to hold a sponsored silence to remember the children who had been affected by the 2004 Boxing Day tsunami. They raised over £290 for the Tsunami Relief Appeal.

1st Eastcote Rainbow Guides
The Rainbow Guides section of the Guide movement, for girls aged 5-7, started in 1987 but it was not until October 1996 that a unit was opened at St Lawrence's

with Mrs Pauline McDonald as Rainbow Guider. On 20 November all 15 recruits made their Rainbow Promise in front of parents and invited guests. The occasion was reported in the *Ruislip and Northwood Gazette* and the church's *Free for All*. With support from successive unit helpers Pauline led this popular group throughout Father David's incumbency and beyond.

The regular meetings were based on 'themes', and annual reports indicate that the girls had a great deal of fun. During summer months the meetings were often held outdoors. Over the years the themes included butterflies, bubbles, buildings and homes, printing, movement and sound, creepy crawlies, elephants, music and songs, flower arranging, leaf and bark rubbing, and stained glass windows. Various experiments in a science evening included taking fingerprints with talcum powder and testing static electricity by sticking balloons to the hall walls.

From 1997 there were regular summer walks to Eastcote House walled garden. In alternate years there were visits to Rayners Lane Ambulance Station and Ruislip Fire Station. The girls also went to Tesco's at Pinner Green for a 'behind the scenes' look at how a supermarket operates, when the girls thoroughly enjoyed being allowed a go on the till. They made several visits to the Willow Tree site. In 2000 they took part in the National Guide Day, the District Fun Day and the Division Rainbow Party, and in 2002 they enjoyed the Queen's Jubilee celebration day.

As well as sharing the occasional meeting with the 6th Brownies, the rainbows joined the brownies for the annual visit to a local pantomime, always enjoyed by everyone.

Like other uniformed organisations the rainbows helped at fairs and markets, and other church events, and they attended Parade Services.

Special Celebrations
In the course of this chapter there have been many references to happy social events organised by the Fellowship Committee, special sub-committees, or one of the St Lawrence Church organisations. A number of them were regular or annual events but there were two special years in that the events marked significant or 'historic' anniversaries, one local and one universal. 1993 was the Diamond Jubilee of St Lawrence Church, and 2000 the beginning of the third Christian Millennium.

St Lawrence Church Diamond Jubilee
As far as the parish church of Eastcote was concerned 1993 was a year of many celebrations associated with its Diamond Jubilee and organised by a special Jubilee Committee. It was indeed an eventful year and it should be remembered that while all the celebrations were going on, much was being done regarding the *Agenda for Action*. Although several of the 1993 celebrations are mentioned elsewhere a few comments on the printed Diamond Jubilee Programme may be

of interest. Among named visiting preachers and celebrants the Revd Clive Pearce and the Revd Peter Day were past Assistant Curates. Father Ray Phillips was Vicar of All Saints' Hillingdon, our 'sister' church. One important visiting preacher, not shown on the Programme, was Father Bill, St Lawrence's second Vicar. He preached at Evensong following the Re-Union Tea at the end of the Flower Festival and Arts and Crafts Festival.

Dear Friends,

Welcome to our Diamond Jubilee booklet. Here you will find details of some of the special events we have planned to enable us to celebrate this happy year.

As a part of our 'thank-offering', we have decided to establish a 'Jubilee Fund', to which contributions are invited. At the end of the year this fund will be divided equally between the domestic needs of the Church, and an external charity.

For sixty years Eastcote Parish Church, dedicated in honour of St. Lawrence, has stood as a witness to the eternal truths of the Christian Faith.

In this building the sacraments of Christ have been celebrated – Baptism administered and the Eucharist offered. Here couples have pledged their love in marriage, and the departed have been commended to God's mercy. Here the Word of God has been proclaimed and preached.

For all of this – for the vision and enthusiasm of those who built this Church, and for the faithfulness of those who have ministered and worshipped here since – it is right to give God thanks and praise.

But our Diamond Jubilee is not just about looking back. It is also an opportunity for us to consecrate ourselves anew to God's service, and to look to the future. Into what new ways of 'being the Church' is God calling us? That is a question we must face with honesty as well as confidence, if we would be true to the inheritance we have received.

May our Jubilee year be one in which we rejoice – but rejoice with expectation.

May God bless you all.

DAVID COLEMAN
Vicar of Eastcote

Introduction to the Diamond Jubilee Booklet

Diamond Jubilee Programme 1993

April 11th	Easter Day. Opening of Jubilee Programme
April 25th	Fr. Clive Pearce preaching at 10am
May 3rd	May Fair
May 16th to May 23rd	Exhibition in Church by Ruislip, Northwood and Eastcote Local History Society
May 19th	Bishop of Willesden's parish visitation
May 23rd	Bishop of Willesden celebrating and preaching at 10am
June 6th	Trinity. Archdeacon of Northolt preaching at 10am Inauguration of 'Churches Together in Eastcote. 3.00pm, at St. Thomas More Church
June 9th	Confirmation at 8pm. Bishop Donald Arden presiding
June 20th	'Songs of Praise' at 6.30pm
June 24th to June 27th	Flower Festival and Arts & Crafts Festival
June 27th	Re-Union Tea (3pm) and Evensong (5pm)
July 4th	Cricket Match v. All Saints' Hillingdon, 2pm. Solemn Evensong 6.30pm. Preacher Fr. Ray Phillips
July 18th	Fr. Peter Day preaching at 10am
August 2nd to August 6th	Children's Festival Play Scheme, 10am–12.30pm each day
August 8th	St. Lawrence Sunday. Fr. F. Giles preaching at 10am. Wine afterwards in the Vicarage garden
Sept. 11th	Garden Party
Sept. 19th	'Sing-in' Messiah
Sept. 26th	Harvest Festival and Supper
Oct. 1st/3rd	Parish Retreat
Oct.. 8th/9th	Players production 'Sixty Glorious Years'
Oct. 24th	Dedication Festival. Bishop of London celebrating and preaching at 11am, followed by parish lunch.

Diamond Jubilee Programme 1993

Writing in the August 1993 magazine Father David regarded that event as one of the highlights of the year so far. He thanked all who had contributed to making the Festival so successful, and expressed pleasure in being able to welcome so many former members of St Lawrence Church to the Re-Union Tea that was blessed with fine weather.

The 'Sing-in Messiah' in September gave an opportunity for anyone who wanted to take a choral part in a performance of Handel's *Messiah*. Under the guidance of Richard Bourne, a professional conductor, and with Derek Stevens as organist, there was an afternoon rehearsal from 4pm-5.15pm. After a break for refreshments the singers joined four professional soloists for the evening performance. A review said that the event was a resounding success. Both singers and audience enjoyed themselves and all acknowledged that in such a short time, wonders could be achieved with this much-loved work.

The Jubilee Committee was responsible for a variety of successful celebratory events that strengthened the fellowship of the St Lawrence community. The Secretary's Report for the 1994 APCM referred to the successful Jubilee Fund that resulted in the presentation of £2,593.17 to SOS Sahel, and £3,100 being passed to the Organ and Heating Funds. The Jubilee Fund consisted of donations, most of the money from special events, and profits from the sale of Jubilee items. These items included St Lawrence designed mugs (£2.50), printed cotton tea-towels, one design of St Lawrence's Church and one of places in Old Eastcote (£2.50 each), and notelets with the same designs (£1 for a packet of 10). Father David thanked Margaret Case for all her work in administering the Jubilee Fund.

The Reverends David Coleman, Bill Hitchinson and David Hayes
on the afternoon of the Re-Union Tea

The Millennium
References to parish events in the millennium year of 2000 are in other sections of this chapter but it seems appropriate to re-iterate views made by Father David in his 2001 Vicar's Remarks on how the parish had marked the millennium. 'We had special services, and we had enjoyed a marvellous flower festival. But the year 2000 was when we took a serious handle on Stewardship'.

Father David and the Revd Ann
This section ends with celebrations remembered particularly for their association with Father David and the Revd Ann. Their almost annual invitation to wine in the vicarage garden after the St Lawrence Sunday 10am Eucharist was much appreciated for its fellowship and relaxed atmosphere. However, in 1995 the invitation was for tea before Solemn Evensong as there was a service at the Eastcote War Memorial at 11.30am to commemorate VJ Day and the cessation of hostilities in 1945.

One of the first personal Coleman celebrations in which the St Lawrence community and others had a share was the Revd Ann's first celebration of the Eucharist on Wednesday 20 April 1994, at which 158 of the 171 attending the service received Communion. On 12 July 1997 members of St Lawrence were invited to a Bring and Share Supper to celebrate Father David and Ann's Silver Wedding Anniversary. For their 30th Wedding Anniversary on 14 July 2002 everyone at St Lawrence's was invited to drinks and nibbles after the 10am Parish Eucharist

On Wednesday, 30 June 1999 over 140 members of St Lawrence's joined Father David in a Solemn Service of Thanksgiving for the 25th Anniversary of his Ordination to the Priesthood. On Saturday 15 May 2004 Father David and Ann celebrated significant anniversaries of their priestly ordinations, 30th and 10th respectively, with an 11am Solemn Mass, at which the preacher was Father Ken Leech. Some 178 friends from St Lawrence and further afield attended the service which was followed by a buffet lunch.

Fr David Coleman and Ann after celebrating their anniversaries on 15 May 2004

Father David goes North

In July 2004 it was known that Father David and Ann would be leaving the parish and relocating to the North. In the August magazine Father David wrote of his time as Vicar of Eastcote as 'years of great fulfilment in what may justifiably be called one of the best parishes in the diocese of London'. He referred to the Archbishop's invitation to Ann to become the Director of Wydale Hall, the York diocesan retreat and conference centre. Father David had been invited by the Bishop of Whitby to become the Continuing Ministerial Education Officer (CME) for the Cleveland Archdeaconry (the North Riding of Yorkshire). This post entailed providing training for both clergy and laity, with a special responsibility for Post Ordination Training (POT), a role he had fulfilled in Willesden for the last nine years. The post of CME entailed being the priest in charge of a small parish within the North Yorkshire National Park, and near to Wydale. Although both Father David and Ann 'felt upheld and cherished' by St Lawrence's they accepted the 'promptings of the Holy Spirit' that it was time for them to undertake fresh challenges. Father David's last Sunday as Vicar of Eastcote was Dedication Sunday, 24 October 2004. In 2008 the Colemans returned south when Father David became Vicar of St Leonard's, Heston.

Father David's Legacy
Like his predecessors, Father Bill and Father David Hayes, Father David Coleman maintained the traditions established by Father Godwin, and like them he adapted the mission of the church to the contemporary situation. Father David Hayes' 'Forward Together' and the Parish Appraisal of 1989 helped the parish to make a positive response to the *Agenda for Action* in the MAP of 1993. In his introduction to the Diamond Jubilee booklet Father David wrote of the Jubilee as an opportunity for us to consecrate ourselves anew to God's service, to look to the future, and to ask into what new ways of 'being the Church' was calling us?

Father David was of the opinion that the London Challenge of 2002 had enabled the parish, as a Christian community, to celebrate what had been achieved. However it left the parish with the problem of how to share the good news of Jesus Christ with a wider community that often seemed to have little feeling for the spiritual depths of life.

In his Vicar's Remarks before the 2004 APCM Father David reported that all the clergy of the Willesden Episcopal area had been on a conference, the theme of which was 'new ways of being Church'. While agreeing that the Church needed to find new ways of presenting the message of Jesus Christ to the present generation, he was of the opinion that such ways could only flow 'from what we have already received'. He was appreciative of the ways of being Church that had evolved at Eastcote, expressed at its most profound level in worship. 'A way of being church that is both traditional and inspired by our heritage, while at the same time being flexible, open and innovative'.

In his last magazine letter Father David referred to the Eucharist, rendered in a traditional Catholic, yet unfussy style, as being at the heart of the parish. Throughout his time as Vicar he had sought to complement this ethos with an emphasis on shared and collaborative ministry. He saw the ordained ministry as absolutely necessary to the integrity of a Christian community but went on to say that the ordained priesthood existed to 'inspire, facilitate, and enable the priesthood of all believers'. He felt this was expressed in various schemes in the parish, and above all in the Sunday Eucharist where the clergy were assisted by altar servers and lay administrators, completely inclusive in age and gender.

For Father David collaborative ministry gave the priest space to perform the role so necessary today, namely reflective leadership. He certainly showed 'reflective ministry' and guided the parish through the challenges of the 1990s and into the

21st century. He hoped our quest for a new Vicar would be motivated not so much by 'Who shall care for us?' as by 'Who shall bring us new insight and vision?'

Shortly before Father David left the parish it was agreed that St Lawrence Church should have a Parish Prayer, and it was hoped that it would unite and strengthen the community during the interregnum and into the future. It was simple, easy to memorise and relevant to all.

**Grant us Lord faith to believe
and strength to do your will.
Amen.**

CHAPTER VI

THE REVEREND STEPHEN DANDO
VICAR FROM 2005

The Revd Stephen Dando

By March 2005 parishioners knew that the Revd Stephen Dando would be the new Vicar. At that time he was Vicar of St Mary's Illingworth in the Diocese of Wakefield. Before training for the priesthood at Mirfield, he taught in South London primary schools. He was ordained Deacon in 1983, and Priest in 1984. He served his curacy in Wandsworth and then returned to Yorkshire where he was Vicar successively in two parishes. Father Stephen is married to Elaine, who is also a priest and was licensed as chaplain to University College, London at the end of February 2004. They have six children whose ages at that time ranged from 19 to 28, a rather different age range from Father David Coleman and Ann's children on his appointment. Father Stephen's Collation and Induction as the 5th Vicar of Eastcote took place on Monday 13 June 2005.

Father Stephen took over traditions established during earlier incumbencies but just as each of his predecessors had adapted to their contemporary society Father

Stephen seeks to make the mission of the Church appropriate for the 21st century. In his magazine message to parishioners in June 2005 Father Stephen referred to his first visit to St Lawrence Church in the January when he had a sense of stillness, of prayer and worship. He went on to say that he regarded St Lawrence's as the place where we are fed by God through word and sacrament, so that we can then go out to share God's love with all people and together reach out to the whole community.

During the previous incumbency there had been moves to greater involvement of the laity and Father David had referred to the 'priesthood of all believers' as being important. Father Stephen took up this theme in his review for the 2006 APCM acknowledging what had been achieved in the past and predicting the need 'to adapt to new ways, to the laity doing much that was previously done by clergy'.

Possibly with the aim of getting more laity involved, and to a greater depth, Father Stephen arranged an Away Day in September 2006 for members of the PCC. This took place at St John's, Great Stanmore where the Rector, the Revd Alison Christian, asked the PCC what their 'dreams' were for St Lawrence's. Through group and plenary discussion the Rector took up on what was good and positive about St Lawrence's and stimulated the PCC to consider steps to forward St Lawrence's mission in the parish. The Away Day was followed in December by a congregational consultation, where in groups they discussed three questions. Why did you first come to St Lawrence's? Why did you keep coming and what keeps you here? How would you like to see St Lawrence's grow? The feedback from the consultation and from the PCC Away Day contributed to a new Mission Action Plan (MAP), drafted early in 2007. The MAP focused on three areas which people felt important. These were Welcome and Nurture, Work with Young People, and Our Relationship with the Local Community. In connection with the MAP new sub-committees of the PCC were formed, the Children and Youth Committee in 2007 and the Mission Committee in 2008, the latter incorporating the responsibilities of the former Publicity and Publications Committee. Obviously there is overlapping in the three areas with decisions made in any area influencing many aspects of church life as will be seen in this chapter.

Father Stephen's Ministry Team
In his Vicar's Remarks for the 2006 APCM Father Stephen compared the situation, as regards clergy assistance, with that of 2004 when Father David had the support not only of an Assistant Curate but also of other clergy living or worshipping in the parish. He thought it unlikely that St Lawrence would have another curate in the near future and hoped that the Revd Wendy's contribution would continue. Although Elaine's direct participation in the life of the parish was limited by her commitments to University College and by this time St Pancras, Euston Road, he valued her support and encouragement.

In October 2006 it was announced that Rachel Phillips was to become Assistant Curate at St Lawrence Church after her ordination as Deacon on 30 June 2007. Before her ordination training she had been a solicitor for 15 years and she had been organist and choir director at St John's, Great Stanmore. Incidentally Rachel met the St Lawrence PCC there at their Away Day. On 28 June 2008 she was one of nine, including Elaine Garrish, (former parishioner and pastoral assistant at St Lawrence's) ordained Priest at St Andrew's, Roxbourne. Rachel's inspiration, dedication and enthusiasm have been invaluable, and include work with children and young people, new forms of Sunday evening services, furthering the relationship of the church with the local community, and St Lawrence's link with Cristo Redentor, Luanda. Her last Sunday at St Lawrence's was on 11 July 2010 when the well-attended 10am Eucharist was followed by refreshments in the hall. In September she was licensed as Priest in charge of the parish of Northaw and Cuffley in the St Alban's Diocese.

The Revd Rachel Phillips
presiding at her first Eucharist assisted by
Revd Stephen Dando (left) and Revd Wendy Brooker

As Father Stephen had hoped the Revd Wendy's contribution to the parish continues. In grateful recognition of that contribution she was formally acknowledged as an Assistant Priest of St Lawrence Church in 2008. She became a Deaconess in 1985, and although she was ordained Deacon in 1987 it was another seven years before she was among the first women to be ordained Priest in 1994. On 27 June 2010 she celebrated 25 years of Church ministry. There was a special Festival Eucharist of St Peter when the preacher was a former member of the congregation, the Revd Sylvia Lafford. After the service the congregation enjoyed refreshments as part of this joyful occasion.

Captain Don is another member of Father Stephen's team who has contributed, and continues to contribute, to the life of the parish in many ways. At the end of the 2008 summer Holiday Club Eucharist a small presentation was made to him to celebrate his 60 years as a Church Army Captain.

Another member of the team who, together with the Revd Wendy and Captain Don, did so much to ensure that the worship and pastoral care continued to a high standard during the 2004-2005 interregnum was the Reader, Alan Wright. Like them he continues to play an important part in the team.

Captain Don Woodhouse

Alan Wright

In 2007 Jane Coan and Tricia Hughes joined the Ministry Team after they were licensed as Pastoral Assistants. As part of their course of training they had become more deeply involved in parish affairs. Among other things they helped start the Film Club in February 2007, compiled a Daily Prayer Cycle (which in the course of a month remembers every road in the parish), and soon took over co-ordinating the Parish Links Scheme from Alan. Tricia left the parish in 2008. Jane remains a valued member of the team.

Though not members of the Ministry Team it is appropriate at this point to refer to the churchwardens who, in these days of Health and Safety, Child Protection and other regulations, and a general increase in paper work, have increasing responsibility in the running of the parish. Andrew Bedford and Lionel Williams, the Churchwardens who managed the interregnum so well, remained in office and provided some continuity for Father Stephen until 2006. Andrew had been in office since 1997 and Lionel since 1995. They had each given unsparingly of their time and talents and they continue to do a great deal for the parish (see Appendix 4 for list of Churchwardens).

With increasing administrative work the decision had been made to create a new part-time paid post, and Sharon Evans had been appointed Parish Administrator early in 2004. After 18 years as Hall Bookings Secretary Ann Sykes had stepped down and Sharon became responsible for all bookings for the hall and the St Lawrence Centre. Among other duties she soon took over the weekly bulletin (previously done on a rota basis by Judith Howe, Alan Hooper, Geoff Higgs and Arthur Plummer). Sharon left in 2009 and was succeeded by Tessa Maude.

Inevitably in this chapter there is no attempt at any assessment of the incumbency as a whole, as has been possible in previous chapters, for Father Stephen is still Vicar of St Lawrence Church and no one knows what the future holds for the mission of the Church in this area. However it is hoped that the following pages will show how Father Stephen and his Ministry Team, and members of the St Lawrence community are committed to that mission.

The Fabric of St Lawrence Church
The church buildings are the outward and visible signs of the presence of the Church in Eastcote and as part of its mission St Lawrence's tries to make everyone feel welcome to those buildings, particularly to the church. Since the 2007 MAP with its three areas of Welcome and Nurture, Work with Young Children, and Our Relationship with the Local Community, this has been expressed more positively.

During the interregnum the PCC discussed draft plans submitted by the church's architect regarding 'disabled access' to the church, hall and St Lawrence Centre. Of course no major decisions could be made until Father Stephen became Vicar but it was important to consider how the church could comply with the Disability Discrimination Act, which came into force in October 2004. While access to the hall and centre was dealt with relatively easily, that of access to the church has been the subject of many meetings.

The West End of The Church
By 2006 none of the early proposals for disabled access to the church via the South Door seemed feasible. Blackwoods Architects, based in Aylesbury, were asked to consider an access for all via a West Door. As Father Stephen wrote in the August 2006 *Window* this would give everyone a fine view of the altar and East Window when they entered the church. It would physically contribute to making a more welcoming atmosphere. Part of this more ambitious proposal was to have a toilet and small kitchen in the area. There would be a 'gathering' space for people to enjoy coffee after Mass, and would be a retreat for mothers and toddlers when necessary during a service. The 'gathering' space would be partitioned off with glass panels and doors and a gallery roof. Optimistically it was hoped that the re-ordering of the west end would be achieved by the 75th anniversary in 2008.

Over the following months, and years, various proposals were discussed by the PCC and working parties set up by the PCC, and members of the congregation had the opportunity to express their views after a presentation in church. In early 2009 after consultations with the architect and the Diocesan Advisory Council (DAC), involving revisions and much discussion, the time had come for the architect to go out to tender to interested builders. By this time the DAC had agreed that keeping the existing font was not practical and that there should be a substantial moveable wooden font that could take the lid of the original font. The subject of etchings for the glass panels was under discussion. It was thought that if the work could start early in October (depending on funding), the majority of it should be finished by Christmas. The quotes received were much higher (by some 30% to 50%) than anticipated and in November 2009 the PCC agreed the work should be done in phases.

Phase 1 was to enhance the current 'disabled access', near the choir vestry on the south side of the church. It was to have a permanent ramp and an automatic opening inner door (the outer door to be fixed open when the church is in use). As far as the West End development was concerned the gardeners' store in the north west corner of the church was to be converted to provide toilet and kitchen facilities. Phase 2 was to install the glass partitions and gallery to form the foyer. Phase 3 was to create the new door in the west wall and the associated external work for access, and to close the existing main entrance. The PCC felt that Phases 2 and 3 should be done together.

It should be remembered that it took many years for the 1984 Parish Hall Building Project to reach completion in 1999, with the transformation of the old bungalow into the St Lawrence Centre, and although it is hoped that it will not take so long to achieve the West End development, the decision for the phased approach as regards the West End of the church is appropriate. While the object is to make the church more accessible and welcoming, with the 2008 commitment of St Lawrence's to look 'onwards in faith and outwards in love', careful and prayerful consideration as to what is right for the mission of the Church in the community of Eastcote and beyond may influence future decisions. At the time of writing no major changes have been made.

Maintenance of the Church
Prior to the refurbishment of the sacristy and church toilet in 2005, the copper roof was stripped off and replaced with three layers of roofing felt. In 2008 the lead centre guttering between the south aisle and the nave was repaired with T-pen (a lead substitute) to deal with the intermittent leaks that had occurred in the south aisle. In 2009, following the attempted theft of lead from the north aisle valley gutter, the damaged bays were repaired incorporating new T-pen expansion joints. The copper roofing of the small north door porch was removed and replaced by a new high performance felt roofing system.

In September 2008 Mannering & Sons Ltd (Electrical Contractors) carried out the complete renewal of the electrical distribution equipment at a cost of £5,517.19, which included work in the hall.

During 2008 the Children's Corner in the church was made more welcoming to young children and their carers by decorating it with a Noah's Ark theme and providing new furniture. The area is well used and much appreciated.

The Hall and the St Lawrence Centre
From 2006 a Premises Committee of the PCC became responsible for the day-to-day maintenance of the car park, St Lawrence Centre, the hall and the shipping containers; work that had previously been done by a Hall Committee

In both the hall and the St Lawrence Centre work was mainly of a routine nature, but in 2007 and 2009 there were problems of tree roots blocking drains from the toilets at the front of the hall. With the expanding use of the hall by the church, parish and public the Premises Committee felt the hall needed a face-lift. During 2008 the Committee Room was redecorated and the carpet replaced with laminated flooring and the disabled toilet was refurbished. By the 2010 APCM the hall ceiling, walls and pelmets had been cleaned by the maintenance team. Other work included the redecoration of the hall kitchen, the refurbishment of the hall toilets and the resurfacing of the hall floor.

Apart from improvement for disability access in the form of a ramp and railings leading up to the front door of the Centre in 2006, the annual reports refer to general maintenance and the willing band of 'parish scrubbers'.

The Church Grounds
As in the past the relatively small band of helpers who tend the grounds, cutting the lawns and trying to keep the flowerbeds tidy and attractive, keep up the tradition of volunteers who, throughout the history of St Lawrence Church, have given their time to maintain the extensive grounds of the church. There have been a few recent changes in the church grounds. One noticeable improvement is the circular seat round the weeping ash tree, given in 2006 by Elizabeth and Andrew Bedford. As in the past this tree remains a focal point for meeting and all the more enjoyable for the seat.

In 2008 the Phoenix Ladies who have used the hall since 1958, planted a rosemary bush with a plaque outside the Garden of Rest, to commemorate their 50th birthday. Unfortunately during 2008 the Garden of Rest was vandalised twice. The damaged crucifix was replaced and dedicated and standard roses that were torn down were replaced in 2009.

In November 2008 another small team organised by Heather Chamberlain began work in the woodland area behind the Centre, clearing paths and creating new ones. The plan is to make it an area to be enjoyed by everyone, especially the youth organisations and the holiday club. In the summer of 2009 the Holiday Club enjoyed using the woodland for the Rainforest Adventure and planted a wayfarer tree (Viburnum Lady Eve), one of 75 trees to be planted in the parish to commemorate the church's 75th Anniversary. The first tree, an acer, was planted on 12 October 2008 in front of the Centre, after the 10am Parish Eucharist.

1 Farthings Close
After the Revd Ruth left the parish in 2004 the prospect of having another curate in the near future was unlikely. The decision was made to keep the property and let it furnished until such time as another curate was appointed. The house was refurbished with a new kitchen and bathroom, new carpets and curtains, and the utilities were brought up to standard. From May 2005 it was let unfurnished until 2007 when the Revd Rachel Phillips moved in. The amount received from the letting was £19,490, which almost equalled the sum expended to refurbish the property. When the Revd Rachel left in 2010 the property was again let unfurnished until such time as the parish has another curate.

Worship
Father Stephen maintains the catholic tradition of worship at St Lawrence's but just as his predecessors made changes so Father Stephen has responded, and continues to respond to the circumstances of the 21st century. Some changes were made during 2006 and the 2007 MAP has naturally influenced the style of a few services since that time. Bearing in mind that the formal brief of the Worship Committee is 'to develop the worshipping life of the Parish in consultation with the incumbent and to advise the PCC on matters pertaining to the liturgy', many of the changes follow discussions between Father Stephen and members of the laity and reflect the need to adapt and try new ways of worship.

Continuity and Change - Sunday Services
In the October 2006 issue of *Window* Father Stephen wrote that while he accepted Sunday evening services offered a different way of responding to God, he stressed 'that must not be at the cost of abandoning our staple diet, The Holy Eucharist, in which we are fed and strengthened each week, so that we can truly be Christ's disciples'. At that time he explained a change in the presentation of the Holy Eucharist, made to draw attention to the two parts of the Eucharist. As a reminder 'that Christ is in our midst' the first part, that of being fed by the Ministry of the Word (our readings, prayers and teaching), was moved to the front of the Chancel. At the 10am Sunday Eucharist the Gospel is read from the middle of the church. After the Peace the action moves from the Chancel to the Altar, for the Sacrament of being fed by the Body and Blood of Christ.

The 10am Parish Mass remains the principal Sunday service. At this service the celebrant is assisted by other clergy, by lay administrators of the chalice, and by altar servers. The organ and choir lead the music, and during Festival seasons the use of incense adds to the ceremonial aspect of the service. In 2009 the role of the laity in the distribution of the Communion was emphasised by the decision that Chalice Administrators from the congregation were no longer required to wear robes.

Since 1994 The Ministry of Healing had been offered after the 10am Eucharist but on 4 June 2006 it was offered within the service for the first time, something Father David Coleman had hoped would happen. The day was appropriate as it was The Feast of Pentecost, when the Church gives thanks for the gift of The Holy Spirit that enables Christians to do God's work, and that gift includes the work of healing. The Laying On Of Hands, within the 10am and after the 8am Eucharist, is offered on a regular basis.

As in the past a Parade Service usually takes place once a month during term time, and since October 2006 there has been the occasional 'All Age' Eucharist on those months when there is no Parade Service. Members of the different uniformed and younger groups of St Lawrence's make a positive contribution to the service. These services are conducted in a manner that emphasises the joy of the Resurrection and the tone of these services is somehow more relaxed. Attendance has increased and is usually well over 200, with figures in 2010 of 237 at Candlemas, 299 on Mothering Sunday, and 281 on 20 June when there was a special World Cup Parade Service, during the period of the World Cup in South Africa. Young people were encouraged to wear their football kit to church and invited to play football in the church grounds after the service.

Another special Sunday Service in 2010 was on Valentine's Day, when couples were given the opportunity to renew their marriage vows.

The sacrament of Holy Baptism taking place within the context of the 10am Mass is now more frequent, with a few taking place in a Parade Service. These occasions give an opportunity for the congregation to welcome the newly baptised into the Church and for the baptism party, or parties, to meet members of the community of St Lawrence's. Recently there have been Holy Baptism services at 11.45am and some of the 10am congregation attend to welcome new Church members.

Although a number of churches no longer have Sunday evening worship St Lawrence's still has a service most Sundays of the year. The Order of Evening Prayer from *The Book of Common Prayer* *(BCP)* remains the routine Sunday evening service when attendance usually just reaches double figures. For this service the small congregation sits in the choir stalls which helps to create a feeling of united worship. The number and styles of other Sunday evening worship has increased when attendance is higher, actual numbers depending on the type of service. Some people prefer quiet reflective services while others are more at home with more congregational participation. Taizé and Songs of Praise are examples of different styles of service that have been maintained. One evening service, Solemn Evensong and Benediction, a service that had originally been associated with Guild of Our Lady, was discontinued in 2007, an example of accepting that this type of service was no longer appropriate.

The first Choral Evensong was held in 2006 a few months after the appointment of James Mooney-Dutton as Organist and Choirmaster in January 2006. In 2008, the 75th Anniversary Year, there were five such evenings when around 40 people attended.

Annual Services in the Church's Year
The pattern of annual Advent and Lent services was established before Father Stephen came to the parish and changes and additions aim to 'develop the worshipping life of the Parish'. One addition during Advent is the opportunity for silent prayer before The Blessed Sacrament following the Wednesday evening Eucharist. In 2008 the decision was made for St Lawrence's to once again hold its own Festival of Carols, as well as being involved in the Ecumenical Carol Service. From 2008 this has been held at different Eastcote churches, rather than always at St Lawrence's as had happened in the past. In December 2008 and 2009 the congregation benefitted from the Revd Rachel's research, into how a traditional Christmas carol service might help people to *know* God, when she gave informative illustrated talks entitled *Curious about Carols?* By 2009, Posada, a custom started at St Lawrence's in Father David Coleman's time, was a regular feature of the season of Advent. Figures representing Mary and Joseph travelled around the parish staying a night in different homes until brought to the crib in church on Christmas Eve at the well-attended Crib service. A new venture in Advent 2009 was the opportunity for people to bring their own crib and crib figures to the 10am Eucharist to be blessed.

Sunday evenings in Lent in the last few years have had a diversity of style of worship. In 2007 there were short talks on prayer followed by a time of prayer and concluding with the quiet reflective service of Compline. Each Sunday in Lent 2009 and 2010 a member of the congregation shared with others their personal journey of faith before the evening closed with Compline.

Changes have been made in other annual Sunday services that enable members of the congregation to take a more active part. For example at the Bereavement Service in November members of the congregation may light candles in memory of loved ones.

Fun and Faith
When Father Stephen became Vicar of Eastcote there was already a diversity of styles of worship. He has introduced others some of which have been discussed already. Though services in the past had often been preceded or followed by some form of social or fellowship, at the beginning of 2006 Father Stephen set the tone for Fun and Faith to go together when a Solemn Eucharist to celebrate the Epiphany was followed by 'a hugely popular' Curry Party. The Fun and Faith theme continued with champagne and strawberries at Corpus Christi, and a cream tea and Choral Evensong rounded off a heat-wave weekend that included a parish coach trip to Bournemouth. Since then there have been more instances of Fun and Faith and these have been popular with many who attend St Lawrence

Church. At the Easter services there are opportunities to bang drums, clash cymbals and wave banners as outward expressions of joy at the Resurrection. Since 2008 young members of the congregation have enjoyed the face painting and balloons at the 10am Easter Day Eucharist. The 2010 World Cup Parade Service has been mentioned already. The July Family Fun Day that attracts people from the wider community ends with an informal service.

Some aspects of worship at St Lawrence's in 2010 may differ from that of the early years of the church and this is right for the Church is the Body of Christ and as such must adapt to meet the challenges of the 21st century. While some traditions including Bible Study meetings and Lent Quiet Days remain, owing to the closure of St Francis House the last Parish Retreat at Hemingford Grey took place in 2009. The Oberammergau Passion Play was visited by ten members of the congregation in 2010.

In 2010 the decision was made that instead of running a *Christians for Life* there would be two shorter courses each lasting about six weeks. The first is an introduction to the Christian Faith and the second to look more deeply at how Christians are called to live and worship. The new approach seems to have been successful as 11 adults (the highest number in recent years) were confirmed by Bishop Arden on 5 December 2010. In March Father Stephen reported that Bishop Pete had given official permission to allow children from Year 4 (aged about 8), to receive Communion before being confirmed. This was, of course, after proper preparation and with parental permission, and only allowed for those children who have been baptised. On 25 July the Archdeacon The Revd Rachel Treweek came to St Lawrence's and 18 children were formally admitted and able to share in Communion.

Finding Faith

Explore what Christians believe

Six Monday evenings starting
Monday 14th June
at 7.45 pm

Elsie Fischer Room in the St Lawrence Centre

All welcome

You may use this as part of the preparation
for Baptism or Confirmation

Finding Faith item from Magazine, May 2010

Confirmation, December 2010

Young People
Work with young people is always important in the life of a church for those young people are the Church of the future. The success of such work depends as it has always done on the willingness of adults to provide leadership and to encourage children and young people to become members of the Church community.

In connection with the commitment of nurturing and supporting children there has been a Children's Advocate for many years. In December 2001, a Parish Child Protection Policy was adopted by the PCC with Barbara Plummer as the church's first Children's Advocate. Since she became a Churchwarden in 2006 Emma Kimsey and Maria Smith have jointly held that position. The Children's Advocate tries to ensure that all children's groups within the church provide a safe and caring environment by ensuring that all leaders and helpers have up to date declaration forms and Criminal Record Bureau (CRB) Disclosures, where appropriate.

Following the PCC Away Day in September 2006 and the congregational consultation in December, the challenge of how to encourage young people, and to keep them as active members of the Church, was recognised as of major importance. It was seen as so important that 'improving our work with Young People' was one of the three areas of focus in the 2007 Mission Action Plan (MAP). In May 2007 the Children and Youth Committee (a sub-committee of the PCC) was set up to support work with children and young people, and 'to encourage young people to participate in the life of the church and to grow in faith in Christ'. As Father Stephen acknowledged in the July 2010 magazine, since coming to the parish in 2007 Rachel has been behind many of the new ventures, but the support of members of the laity has made such ventures possible and successful.

Between 2005 and 2010 there have been launches of new groups and re-launches of others. The groups that take place during the 10am Sunday Eucharist may be described as spiritual, while those that meet at other times are more social. Obviously many individual young people are involved in two groups.

Crèche/Tots Club
The crèche, catering for the 0-5s continues to operate on most Sundays and remains a worthwhile facility for the few who take advantage of it. The children's corner in the church had a major 'make-over' in 2008. Now known as the TotZone it remains popular, especially since the furniture and toys have been 'soundproofed'.

Following a pilot on three Sundays in June 2009, Tots Club for the 0-5s was launched in September and has proved a successful new venture and an excellent way of introducing Baptism families to church life. Initially the group met in the committee room of the hall but with increasing numbers it moved to the Elsie Fischer room in the Centre. Meeting once a month children and their parents or carers enjoy games, stories, songs and craft, before moving into church during the Eucharist at the time of the Peace.

Children's Corner in the Church

Sunday School
The Sunday School meets on the same basis as previously. From 2006 it has had two classes, Years 1-3 (ages 5 to 7) and 4-6 (ages 8 to 11). John Blanchard became Co-ordinator in 2006, taking over from Sylvia Hooper. Attendance still varies, averaging 15-20 each week. When they leave at the age of 11 the

children are each given a youth Bible. The teachers continue to use the Scripture Union material and once a term each group enjoys a session of Godly Play. Parties at the end of the summer term and at Epiphany, when children are invited to bring a friend, remain popular. Their contribution at the Christingle Service is always enjoyed.

Junior Church/Explorers
From 2005 to 2008 the Junior Church continued as in the past, but with dwindling numbers, a new arrangement was made for young people aged 11+. Instead of meeting in the Centre they joined the main congregation until after the Gospel, then went to the choir vestry for teaching, and returned to the church at the Peace. This enabled young people in the choir or serving to join the Junior Church. From 2008 the group became known as Explorers.

5-2-7
This was a group launched in September 2009. It is a social club for children in Years 5 to7 (ages 9 to 12), an age group particularly well represented in Sunday School. With Philippa Cooper as leader, 5-2-7 immediately proved popular with the children, who have enjoyed pizza-making, DVDs, fireworks, craft, games and much more, as well as decorating the tea-lights which were used in church over Christmas and at Candlemas. The group meets from 5 to 7pm on Sunday evenings.

SLICK
St Lawrence Intelligent Cool Kids (SLICK), the social group for young people of 11+ had been formed in Father David Coleman's incumbency. SLICK was for members of the St Lawrence community attending church on Sundays either as members of Junior Church, Joyful Chorus or serving at the altar.

As members of that community they helped at fairs and markets and raised money for Cancer Research and Ovarian Cancer. Adrienne's Williams' 2006 Annual Report of SLICK recorded regular 'get togethers' mainly on Saturday mornings. Table tennis provided a lot of fun, and as a result of a successful Bingo Night that raised £80 they were able to buy a CD player. In 2008 they organised a quiz that raised money for Christian Aid. A change in leadership resulted in a new format for the group. In September 2009 SLICK was re-launched for those of 12 and over and meets on Sunday evenings about once a month. Under Richard Maude's enthusiastic leadership the young people have enjoyed activities ranging from big-screen Wii competitions to street dance and a memorable Scalextric evening with the young people of Thomas More Church. Future plans include more joint ventures with Thomas More and other churches, to enable the young people to meet new friends.

The Children's Holiday Club and Workshops continue to attract children from the wider community of Eastcote as well as those who are directly associated with St Lawrence Church.

Outreach to the Local Community
Throughout its history each of the Vicars have reminded parishioners of the importance of the mission of the Church to the wider community of Eastcote. This has been done through church services and social events, personal contact, and various forms of publicity. Father Godwin had wanted the *Leaflet* to make 'the presence of the parish church known to the community at large'. Father Bill hoped the *Gridiron* would 'appeal to those outside the Church', and be a 'vehicle of Christian propaganda' in what he considered a largely indifferent society. Since 1971 the church magazine has not had such a wide circulation and has been largely for internal communication. However the launch of *Free for All* in 1996, with distribution to every household in the parish, has proved a successful means of reaching the wider community.

Until the latter part of the 20th century a large proportion of the population in the area was familiar with various aspects of the church community. For some it was through baptism, marriage, and funeral services, for others through attendance at annual Easter, Christmas and harvest services, for others through young people's involvement with church organisations and for many through support of the church's fêtes, markets and flower festivals.

While much had been done in the area of Outreach and Communication since the MAP of 1995, the need for a more positive approach to forward St Lawrence's mission in the post-Christian society of the 21st century was recognised. This was highlighted in the MAP of 2007 when one of the three areas of focus was 'outreach to the local community'. The aim was to improve disability access and facilities in both the church and hall, establish premises for outreach and service in Field End Road, and generally raise the profile of the church in the local community. Earlier sections of this chapter have discussed efforts to improve disability access and to offer less formal styles of worship. In May 2008 the Mission Committee was set up 'to ensure the church is outward-looking and mission-focused in all its literature, publicity and activities…'

Among many aspects of church life to which Father Stephen referred in his review of the 75th anniversary year was the first Family Fun Day on 19 July 2008. This successful event to 'offer God's gift of joy to the community' has been repeated in 2009 and 2010. Another example of 'outreach to the local community' was a joint venture with other local churches. In December 2009 the long awaited ecumenical outreach in Field End Road was achieved at the inaugural evening at Costa Coffee in Eastcote, appropriately called Carols @ Costa. The aim was to provide an opportunity for people who do not normally attend church to meet together to enjoy some fun, fellowship and a 'subtle' faith input. The feedback was positive and further sessions took place during 2010, known as Grounded @ Costa.

The Costa venture is a recent outcome of the close relationship between the Eastcote churches, through Churches Together in Eastcote (CTE). Local ecumenical activity has developed since it started in the 1960s and many earlier initiatives remain today. These include the various annual meetings and services, and the Monday Call In, celebrating its 25th anniversary in 2010. One venture that had been tried in the past and now seems to be successfully established is the CTE Lent groups.

Announcement of Fun Afternoon 2010

In 2008 an opportunity for members of the church to be 'outward-looking' was to attend the *Understanding Islam* course, sponsored by the CTE. A magazine article introducing the course stated that so many Muslims had moved to Britain that they now number by far the biggest religious group after British Christians. Muslims, Jews and Christians are cousins in faith, for they each trace their ancestry through Abraham. There was a need for Christians to appreciate the 'inner dimensions of Islam' get to know and understand their cousins. The course led by Dr Chris Hewer, the St Ethelburga Fellow in Christian-Muslim relations, consisted of ten two-hour sessions, and included a visit to a mosque. Many people from churches and synagogues benefitted from his excellent teaching.

A rather different way of reaching outwards was the planting of the first of '75 trees for 75 years'. The aim was to plant 75 trees within the parish boundaries as a reminder of the church's 75th anniversary year in 2008, and as a contribution to the care of God's creation in this area. The first tree was planted on 12 October 2008 in the grounds of the church outside the St Lawrence Centre. Many more have been planted since then by individuals and local groups, including Eastcote Residents Association, local schools, those living in sheltered accommodation, the 2009 Children's Holiday Club, and those who attend the Monday Call In. By July 2010 the number of trees planted had reached 60.

The outline of the course is as follows:

1.	17 April	A Muslim-eye view of creation
2.	24 April	The Prophet Muhammad
3.	1 May	The Qur'an
4.	8 May	An overview of Muslim history
5.	15 May	Principal beliefs of Islam
6.	22 May	The ritual practice of Islam
7.	5 June	Living constantly remembering God
8.	12 June	Visit to a mosque
9.	19 June	Islam and other faiths
10.	26 June	Muslims in Europe and especially in Britain.

Understanding Islam course outline March 2008 magazine

Another aspect of 'outreach to the local community' may be seen in the fortnightly visits to Whitby Dene, one of the local residential homes. The aim of the visits is to express commitment as a church to 'love one another'. Lay administers of the chalice, and others of the St Lawrence community, share with some of the residents a service of praise and thanksgiving and administer Holy Communion.

Outreach Beyond the Local Community
As in the past the community of St Lawrence Church shows concern for people in many areas of the world, through prayer and practical help. Two examples illustrate this aspect of outreach.

Fairtrade
St Lawrence Church has supported Fairtrade since 2002. In March 2005 Alan and Sylvia Hooper organised a Fairtrade evening to sample wine and snacks. In 2006 the church was registered as a Fairtrade Church. Each annual Fairtrade Fortnight (late February to March) has a different theme but the goal is always to get people talking and buying Fairtrade. Parishioners are encouraged to purchase

Fairtrade Logo

goods with the Fairtrade logo. The foundation works to ensure that farmers in developing countries are given a guaranteed fair price for the commodities they produce – one way to fight poverty. The church uses Fairtrade wine for Holy Communion and Fairtrade coffee is used by all church organisations. There is also a stall after the 10am Sunday service every two months when there is the opportunity to buy a range of products including chocolate, biscuits, dried fruit, pasta and preserves.

ALMA
As well as being the 75th anniversary of St Lawrence Church, 2008 was the tenth anniversary of the establishment of the Angola, London and Mozambique Association (ALMA). Over the years St Lawrence's has supported this mission link between the diocese of London and the Anglican churches of Angola and Mozambique through various appeals, such as the Bishop of London's annual Lent Appeals and the Mozambique Flood Relief in 2000. Moves were made to establish a link with a parish in Angola, that of Cristo Redentor (Christ the Redeemer) in Luanda – the capital of Angola, and in 2008 the link was formerly established. In his Review of 2009 Father Stephen reported that Rachel had visited Cristo Redentor when on a holiday visit to her sister, and that this had been followed by the visit of Father Nunes Pedro (ALMA link officer in Angola) to St Lawrence's. Members of St Lawrence's were able to see pictures and films, and hear at first-hand about Cristo Redentor and its members. Some money has been raised to help fund a roof for their church, but now the challenge is how to communicate on a regular basis and how to share our hopes and concerns with them in prayer.

Cristo Redentor Church, Luanda

Charitable Giving
Charitable giving is an essential way of ensuring that the church is 'outward-looking' and each year the church aims to give 5% of its income to charities.

Members of the congregation are invited to nominate charities they would like to benefit. The PCC Finance Committee considers the merits of these bearing in mind certain criteria and once the selected ones are agreed upon by the PCC the list is published.

Our charities for 2009

We had many more suggestions for charities that St Lawrence could support in 2009 than we have had in previous years. This gave the Finance Committee a challenge to whittle them down to a number we could realistically help.

They selected the following, which the Standing Committee has now endorsed:

January/February	Michael Sobell House
Lent	Diocesan Lent Appeal
April/May	Angola Link Parish
June/July	USPG
September/Harvest	Church Urban Fund and Upper Room
October	Church Army
Christmas Services	Let the Children Live
Christmas Post	London Haven – Breast Cancer
Christingle	Children's Society

Many of these are new to St Lawrence, and that was one of the considerations that influenced the committee in making their selection. The committee also decided to go for fewer charities to which we could give more, rather than donating smaller amounts to many charities.

Smaller, local charities which have had an impact on one or more of our parishioners' lives were favoured by the committee, though they also chose at least one from each of the six categories that we aim to focus on, specifically:

- World mission agencies
- Christian development and relief organisation
- Other church organisations
- Charitable agencies working overseas
- National charitable organisations
- Local charitable causes

As 2009 progresses, there will be regular articles in the magazine and displays at the back of church giving more information about our chosen charities.

Article from November 2008 Magazine

Most years the charities vary, with some new to St Lawrence's but the Church Army and The Children's Society continue to be supported. The money from the Christmas card postage went to the Queen's Walk hostel until 2009 when a donation was sent to London Breast Haven. The church has also responded to emergency appeals such as that of the Haiti Earthquake in January 2010.

Music at St Lawrence's
When Father Stephen became Vicar in 2005 St Lawrence's had been without an official organist since 1994 and it was not until January 2006 that a new one was appointed. James Mooney-Dutton was welcomed as the official organist initially for a 'six-month trial period'. Anne Holt had held the fort since taking over from Enid Lumsden in 2003 but she resigned at the end of 2005 after she moved from the area. The church community was grateful to Anne for all that she had done. Incidentally Enid and a few other members of the congregation play the organ when needed. The appointment of James was significant for among other experience he had spent a year at Norwich Cathedral as Organ Scholar, a post that included training the main cathedral choir. Under James's guidance music at St Lawrence's has gone from strength to strength though there is always an appeal for new members in the annual choir report. Until 2010 junior choir rehearsed on Tuesday and the adults on Friday evenings. Choir practice is now on Thursdays for both choirs with an overlap of time to allow for them to practise together.

Music plays an integral part in the 10am Eucharist, the principal Sunday service since 1989, and the choir has built up a repertoire of psalms and anthems to enhance the worship at this service. Although there is no longer a choir at the majority of Sunday evening services, the occasional Choral Evensong services that started in 2006 show the quality and commitment of choir members. Of course the choir are also in evidence at Sung Eucharist services during the week, at some weddings and funerals, and at the annual CTE Carol Service.

Members of the choir have participated in several events with other choirs. During 2006 the choir joined choirs of the Harrow Deanery to sing Evensong at St Mary's Church, Harrow on the Hill, and at St Paul's Cathedral. In 2007 and 2009 some of St Lawrence choristers sang Choral Evensong at St Paul's as part of the London Diocese Choir. On Good Friday 2009 some members took part in a scratch performance of Stainer's *Crucifixion* at St John's Greenhill. The choir have also sung carols at local residential homes, in the shopping centre of Eastcote, and under the tree in Trafalgar Square.

The year 2008 was particularly full and enjoyable for the choir. In addition to all the preparation and singing at the many special services of the church's 75th anniversary, and their participation in regular 'outside' events, in October the choir hosted the Harrow Deanery Choirs' Festival. This was the first time this annual event had been held outside the Harrow Deanery, and was considered an honour and privilege. The highlight was the trip to Dublin in May when all the adults and members of the junior choir, augmented by James' friends from other choirs, sang at the 11am service and Choral Evensong at Christ Church Cathedral while their own choir were on holiday. The Dean complimented the choir 'for achieving a standard as high as the resident choir's'.

The Choir in Christ Church Cathedral, Dublin.
James Mooney-Dutton (left) and the Revd Rachel Phillips (right)

Concerts by visiting musicians have long been a feature at St Lawrence's but James was responsible for organising a series of monthly concerts that started on 17th February 2007 with the purpose of raising money for the organ fund. The first of these concerts was an organ recital when James entertained an appreciative audience of nearly 60 people to a light-hearted selection of pieces spanning five centuries. At that time the church did not have a piano of a good enough standard for use in concerts but by September the church had purchased a baby grand piano and this increased the variety of concerts. In addition to fund-raising for the newly designated Church Music Fund there have been concerts for charities in November 2007 for the Cancer Unit at Mount Vernon Hospital and January 2009 for Breast Cancer Haven London. As may be seen in diary dates for 2008 (see page 299) ten concerts were arranged during the year. The Carlo Curley Organ Concert in October attracted an audience of 250. This was hardly surprising for this American concert organ virtuoso has been named the 'Pavarotti of the Organ'. A 3 manual Makin digital organ was hired for his recital, and was used at the Harrow Deanery Choirs' Festival the following day. A magazine appreciation of the concert referred to two pieces performed by Carlo and James on two organs as making 'your toes curl and your spine tingle'. Owing to small audiences at some of the other concerts in 2008 the decision was made to hold them every two months.

**Autumn Concerts in aid of
the Church Music Fund**
(All start at 7.30 pm)

Saturday 15th September 2007

**Opening Recital on the new Reid-Sohn piano
by Ivan Yanakov**
*winner of ARTISTS INTERNATIONAL DEBUT
WINNERS SERIES in New York City*

Saturday 20th October 2007

MorrisLenson Guitar Duo

Saturday 17th November 2007

The Walbrook Singers
*Directed by Paul Ayres
James Mooney-Dutton - piano and organ*

Notice of Autumn Concerts in Magazine

2008 Onwards and Outwards

References have been made already to the 75th anniversary of St Lawrence's and to some aspects of the life of the church community in 2008. While the year's programme is outlined in the '2008 diary dates' it seems appropriate to discuss some of the events and activities, most not mentioned elsewhere, in what was a notable year in the history of the present church building.

In the January 2008 magazine Father Stephen wrote of a history of the church being published at the end of the year, and of former St Lawrence clergy being invited to preach during the course of the year. The time scale for the former was optimistic! Father Stephen felt that 'understanding our past is important because it will lay the foundations for the next steps we must take'. Visiting preachers were invited to reflect on '2008 Onwards and Outwards', the theme chosen for the year and in keeping with the Anniversary Prayer that replaced the 2004 parish prayer.

Anniversary Prayer

Loving God, as we look back in thanksgiving, we look forward in hope.
May you lead us onwards in faith and outwards in love.
Amen.

The anniversary celebrations were launched on the Feast of the Epiphany when after the Festival Eucharist three members of the congregation cut a celebratory cake. They were the Mayor, Councillor Catherine Dann, and two who had been at the church in 1933, Joyce Boot and Beryl Orders.

In addition to monthly concerts, a relatively new venture, the '2008 diary dates' includes annual events such as the Supper Quiz, the May Fair, the three St Lawrence Players' productions, the Parish Picnic and Cricket Match, and the Holiday Club. Dates of annual services such as Confirmation, St Lawrence and Dedication Days are shown but in this anniversary year they were particularly joyful occasions. Father David Hayes was welcomed on the Feast of St Lawrence when he preached at the 10am Eucharist, and joined the congregation to share good company at the parish barbecue in the vicarage garden after the service. The Dedication Festival Mass on Tuesday 21 October was another enjoyable celebration, with many guests including clergy and members of local churches. Father David Coleman, other past clergy and members of St Lawrence, and the new Mayor of Hillingdon, Councillor Brian Crowe were also present. Joyce Boot, Beryl Orders and Norman Suckley together cut a birthday cake. All had been associated with St Lawrence Church at least as far back as 1933 and Norman had been a server at the Dedication Service.

Joyce Boot, Norman Suckley & Beryl Orders cutting the Anniversary Cake. The Revds Wendy Brooker, Stephen Dando and David Coleman in the background
Dedication Service 21st October 2008

There were other events of 2008 that gave opportunities for members of the parish to grow in faith. In February there was a Religious Speed Questioning session when the ministerial team attempted to answer any questions 'thrown at them'. In June there was a Gardens Quiet Day that enabled people to visit someone else's garden and spend time quietly and reflectively and wonder at the beauty of God's creation.

2008	diary dates		
January		July	
Sunday 6	Official Launch at 10am service Cake and wine afterwards	Monday 5	Concert – Frida Backman - violin
		Sunday 6	Parish Picnic and Cricket Match against All Saints Hillingdon (also celebration their 75th Anniversary)
Saturday 19	Concert – Zsuzsa Vamosi-Nagy & Atsuko Kawakami – flute and piano	Saturday 19	Family Fun Day
February		Monday 28 to Friday 1 August	Holiday Club
Saturday 2	Supper Quiz	August	
Thursday 7 To Saturday 9	Players production	Sunday 10	St Lawrence Day Barbecue after 10 am service
Saturday 16	Concert – Trio Anima – flute, viola & harp	Tuesday 26	Parish Pilgrimage with accommodation at Wyedale Hall, Yorkshire
Sunday 24	Religious speed questioning		
March		September	
Saturday 15	Concert –St James Ensemble - mezzo – soprano, clarinet, viola and Piano	Saturday 13	Concert – LaurineRochut – Violin & Piano
April		October	
Saturday 5	Concert - Zalas Trio – Clarinet, violin & piano	Thursday 2 To Saturday 4	Players Production – 75 Glorious Years
Sunday 20	Songs of Praise style Evensong	Friday 10	Concert – Carlo Curly – organ – Tickets on sale from James Mooney-Dutton at £8 each.
Sunday 27	Confirmation at 10 am service		
May		Saturday 11	Harrow Deanery Choir Festival
Monday 5	May Fair	Friday 17	Dedication Festival Ball
Thursday 8 to Saturday 10	Players production	Tuesday 21	Dedication Festival Service at 8 pm Birthday cake and wine after the service.
Saturday 17	Concert – Mary Callanan – piano	November	
June		Saturday 15	Concert – Valerie Welbanks – cello
Thursday 5 To Monday 9	Flower Festival		All concerts are in Church and start at 7.30 pm. Performers subject to change
Saturday 14	Lunar Saxophone Quartet		Please check weekly bulletin and monthly Window magazine for details
Sunday 22	Garden quiet day, 2 – 5 pm.		
Saturday 28	Rachel's Ordination as a Priest.		

Diary Dates for 75th Anniversary

June was also the month of the highly successful Flower Festival when once again one could marvel at the beauty of the flowers but also appreciate the skill of the flower arrangers.

At the end of August 32 parishioners set out on the Parish Pilgrimage to Yorkshire, which was superbly organised by Lionel Williams. The party was based at Wydale Hall, North Yorkshire, where Ann Coleman had been the Director. A magazine account reports 'visits to ancient monasteries and places of local interest interspersed with a daily diet of prayer and thanksgiving at some rather special sites'. The Eucharist was celebrated in many places including Whitby, Fountains Abbey, York Minster, and a chapel in the crypt of St Mary, Lastingham. While Father Stephen and parishioners were away the priestly ministry of the parish was left in the capable hands of the Revd Rachel, who had been ordained priest in June.

One event of 2008 that deserves special comment was the Dedication Festival Ball, held on Friday 17 October. Initially there may have been some concern about holding the event in the church but it was an opportunity to draw on medieval traditions; to celebrate the 75th anniversary of the dedication of the church building within the building itself. Meticulous planning by the Ball Committee regarding the setting, decoration and four-course meal, together with a 'most masterly master of ceremonies', and the music of Perry Parsons Big Band made for 'a fantastically successful evening'.

The Ball Committee

Christian Fellowship

Christian fellowship was very much in evidence at the Dedication Festival Ball and has been at the heart of much that has taken place at St Lawrence's throughout its history. While a special Ball Committee was responsible for the Ball, and since 2002 two or three members of the congregation have run the annual fêtes and markets, it is the responsibility of the Fellowship Committee 'to promote Christian fellowship within the parish by organising social functions and to assist in the raising of funds by organising occasional events'.

Summaries of the Fellowship Committee's activities in reports for Annual Parochial Church Meetings give some indication of just how much the committee does, particularly with catering. Annual events that have been the Committee's responsibility for many years include the Parish Supper Quiz (£6 per person in 2005 and £8 in 2010 - the 22nd annual quiz), the Cricket Match, and the Harvest Lunch (£7 and £3.50 for under 12s in 2005, £8 and £4 in 2009), and various other church festivals and events.

In 2006 three new enterprises were added to the calendar. The first was the simple bread and soup lunch on a Saturday in Lent, when proceeds go to the Lent Appeal. The second was the St Lawrence Sunday Bring and Share Lunch after the 10am Eucharist. However in 2010 the celebration took the form of Choral Evensong followed by drinks and refreshments at the Vicarage. July 2006 a third new 'social function' undertaken by the Fellowship was 'Fun, Games, Cream Teas and Choral Evensong, in 2007 it was Cream Teas and Sponsored Hymn Sing and in 2008 the July event had developed into a Family Fun Day referred to earlier. By 2010 it seems to have been established as another popular annual event, now ending with informal worship.

Several 'one off' events deserve mention for, like the more routine ones, they were occasions of Christian fellowship at which refreshments contributed an important part. The earliest was Father Stephen's Collation and Induction in 2005. The year 2008 was a very busy year for the Fellowship Committee, for as

Family Fun Day 2010 – Informal Worship

well as the annual and many 75th Anniversary events there were the Mayor's Civic Service (Councillor Catherine Dann was Mayor for the second time and Father Stephen was her Chaplain), and the Revd Rachel's first Eucharist following her ordination as Priest. In 2009 Father Stephen celebrated the 25th anniversary of his ordination as a priest and after the Festival Eucharist there was a 'feast of music and international food'. In 2010 there were events for two members of Father Stephen's ministry team. The first was in celebration of the Revd Wendy's 25 years of church ministry and the second for the Revd Rachel's Farewell, and on both occasions there were refreshments after the 10am Eucharist.

Many 'social functions' are quite ambitious but it is worth mentioning a week-by-week opportunity to enjoy fellowship after the 10am Sunday Eucharist, when coffee is served either in the hall or in church. It may be recalled that in 1976 members of the contemporary Youth Club provided breakfast after the 9am Holy Communion service. Since 1989 the main service has been at the later time and the refreshment provided is normally more simple and has been organised by Sheila Aleong for some time.

Between 2005 and 2010 the Fellowship Committee organised a number of events 'away' from St Lawrence's. These include Parish Meals at the Tudor Lodge Hotel (2005 and 2006), Dinner and Dog Racing at Walthamstow (2007). Parish coach outings in December have a particular focus of Advent worship in a cathedral city - usually a Communion service at the beginning of the visit. Visits, linked to Christmas Markets in the cities, have been to Winchester (2007), Canterbury (2008), Bury St Edmunds (2009), and Bath (November 2010).

The monthly ramble, revived in July 2009, is a further opportunity for Christian fellowship. Starting points have included Ruislip Lido, Bayhurst Woods, Rickmansworth Aquadrome, Roxbourne Park and Horsenden Hill. Though not as well supported as those of the 1970s, these Saturday morning circular rambles in the local countryside attract from a dozen to 25 people.

Fairs and Bazaars
Successive groups of two or three people have been responsible for co-ordinating the May Fair and Christmas Bazaar, two very important annual occasions in the life of the parish. These events not only raise essential funds for the church, but also make a contribution to the church's mission in the wider community. Over the years both events have accumulated a variety of traditional stalls and amusements, while some are part of the programme for only a few years. Many things are common to both the fair and the bazaar but each also has some that are suited to the season, and in the case of the fair to being mainly outdoors (weather permitting). The fact that those responsible for these events are called 'co-ordinators' is important, for it is vital that, both before the event and on the day, they have the support of a large team of people. Much depends on having people to run the stalls or amusements.

As in the past the local community continues to support these events. Local shops and businesses are generous, donating to advertise in the programme, providing prizes, and sponsoring stalls. Just as important is the attendance of the residents of Eastcote and beyond who come year after year, spend money and enjoy the social occasion.

Brownies Maypole dancing – May Fair 2007

After the May Fair in 2007 (when £5,500 was raised) an article by Sue Cobb, one of the three co-ordinators, thanked all for their different contributions and ended by referring to the team effort 'clearly demonstrating that vast warmth of spirit that is always so evident at St Lawrence's'. That year one of the sponsored stalls, the plant stall, raised £639. In 2009 shortly after the Fair, Enid and Euan Lumsden, after 20 years of running the popular and profitable plant stall, felt it was time to take a well-earned retirement. The parish has always been good at contributing plants for the stall and the practice seems to be continuing.

Christmas Market 2005

In an article about the 2008 Christmas Bazaar (when £5,200 was raised), Richard Green one of two co-ordinators, recalled 'fond memories of St Lawrence's bazaar' some 40 years previously. 'The tradition of holding the Bazaar on Friday evening and Saturday morning has not lost its magic. There is always a great atmosphere on the Friday evening and Saturday gives an opportunity for the younger children to visit and Father Christmas was kept very busy'. Among his thanks to the many people who had helped he referred to Captain Don and his latest book that raised over £300.

The fairs and bazaars continue to be rewarding fund-raising events and happy social occasions for helpers and visitors and remain firm fixtures on the annual calendar.

Income and Expenditure (see Appendices 10 and 11)
Before commenting on a few items of income and expenditure it is appropriate to mention Euan Lumsden, who was Honorary Treasurer for well over 30 years (1975 to 2008, with a short gap in 1983). At the end of his Vicar's Remarks in 2008, delivered during the 10am service, Father Stephen thanked Euan for his hard work year after year. The congregation echoed Father Stephen's thanks and Euan was presented with a retirement gift. At the APCM when he presented the Financial Report for the year ended 31 December 2007, Euan said that when he

took it over the late Father Hitchinson had commented 'it was the first time he had understood the accounts'. Father Stephen remarked, 'Euan takes all our figures and makes sense of them for which we are extremely grateful'.

Euan's successor as Treasurer, Mark Britter, is experienced in IT and from 2008 the Annual Accounts Summary show income and expenditure as pie charts as well as in tabulated form (see Appendix 12). In looking at these figures it should be borne in mind that 2008 was the 75th anniversary year, a year of special services and events. Since that time the nation has been in economic crisis, and this is reflected in a decrease in some sources of income, for example from Fund Raising and Events, and from the Interest on reserve funds.

As far as expenditure is concerned (£156,027 in 2008; £146,870 in 2009; £148,935 in 2010) the largest item is the Diocesan Quota or Common Fund (54.8% (£79,000) in 2008, 56.1% (£75,108) in 2009, and 48.0% (£77,108) in 2010.

Having referred earlier to the fairs and bazaars 'raising essential funds for the church' it can be seen, that together with other fund raising events, they accounted for 11.1 % in 2008, 10.3% in 2009, and 8.2% in 2010, of the gross income; quite a significant proportion. One gross income that has increased is Hall and Centre Lettings, rising (13% in 2008, 14% in 2009, and 17.3% in 2010). Of course Planned Giving, together with the tax recoverable through Gift Aid, and monies from General Collections has always been the main source of income. The figures for the years 2008 to 2010 show the importance of the total of these three sources of income; 64.7% (£93,358) in 2008, 65.8% (£88,139) in 2009, and 56.9% (£91,383) in 2010.

The Planned Giving relates to the Stewardship Scheme and includes taxable pledges, non-taxable pledges and promises. There have been presentations in church encouraging members of the congregation to join the scheme, or review their giving, and there was a fairly successful mini Stewardship Campaign in 2009. There is likely to be a major Stewardship Campaign in the near future. However, it is important that members of the St. Lawrence community are committed to Christian Stewardship of their time and talents as well as their money if St Lawrence Church is to fulfil its mission in Eastcote.

Uniformed Organisation
All units of the uniformed organisations continue to be involved in church services and other church events as well as participating in Scouting and Girl Guiding activities.

3rd Eastcote Guides
One notable event in 2005 was the marriage of their Captain, Caroline Dann. As part of Caroline's 'hen party' the girls rose to the challenge of making her a wedding dress out of newspaper. She commented that unbeknown to them it was

very like the real thing'. As Caroline Scott she continued to lead the Company until the autumn of 2010. The company had been under her able leadership for over 20 years.

Among the wide range of activities the guides have had craft, beauty, cooking (indoors and outdoors), fitness and DVD evenings. In November 2007 the guides visited the Mayor of the London Borough of Hillingdon, at the Civic Centre in Uxbridge, and found out about the role of Mayor. The Mayor was Catherine Dann, Caroline's mother, and a member of the St Lawrence congregation. Five guides were enrolled in the Mayor's Parlour. Visits to the Hillingdon Outdoor Activity Centre remained popular where in 2009 the girls tried the low ropes course. In 2009 they took part in planet watching at Ruislip Lido.

The 3rds have been to the Willow Tree site in Harefield many times. On one occasion they took part in 'Walks [*sic*] for the World', a sponsored walk around

Guides' arrangement - Flower Festival 2008

the site raising money for guides in other countries. In connection with the 'Go for it!' challenge badges, in 2006 they completed 'Go for it! Camp' and 'Go for it! London'. In 2008 (Beijing Olympics year) the girls were at the 'Going for Gold' County camp.

Outings more in the nature of entertainment included a Guide Association 'Big Gig concert' (2006) and *Holiday on Ice* (2007), both at Wembley. In November 2008 they went to Chessington World of Adventure where they had a great time going on rides in the dark, 'even though it rained all day'.

As part of their guiding service the girls had fundraising events for, among other charities, UNICEF and Breakthrough Breast Cancer. They also planted bulbs in the green area near the Tudor Lodge Hotel in 2009. This was done, as in 2002, with the encouragement of the Eastcote Conservation Panel.

In 2010 the Guiding Association (Girl Guiding) celebrated the 100th anniversary of its foundation, and among other special events, members of the Company attended the closing ceremony in Leicester Square. Since Baden-Powell started the movement it has quite rightly adapted to changes in society and it continues to attract girls to become members. Towards the end of 2010 Joanne Ellingham took over the leadership of the 3rd Eastcote Guides.

6th Eastcote Brownies

It seems likely that the 'new' Brownie Programme of 2003 played a part in re-vitalising all Brownie packs. Jackie Smith's annual reports as Brown Owl of the 6th Pack convey a real spirit of enthusiasm on the part of the brownies as they helped to plan activities, keep busy and have fun.

Among the range of badges the girls gained were Brownie Traditions and Disability Awareness. In 2007 the brownies put on a Christmas show for their parents to gain their Entertainer's badge. In addition to solo performances and a group performance, they made programmes and designed their outfits, and had a raffle and refreshment stall. It was a good example of the girls working together as a team. The evening was also a financial success raising £190 towards the unit's census money. This is an annual levy on all members of the Girl Guiding movement and benefits guiding at both local and national levels. In 2006 and 2009 the girls took part in 'Kids on the Catwalk' to raise funds towards their group census.

Visits to the Willow Tree site were reported each year, some with rainbow guides and guides from other units. In May 2008 (Beijing Olympics year) they took part in the 'Going for Gold' event. It opened with a Chinese display team and two very large Chinese dancing dragons. Each year from 2007 the 6th Pack have had a Pack Holiday at Willow Tree, each with a theme. In 2007 it was 'The Circus', and in 2010 Wild Wild West when the activities included line dancing and High Noon party. During 2010 the site was also the venue for Guiding Centenary celebrations. At one event in June, called Hip Hip Hooray the girls took part in different activities and had fun learning about the history of the movement.

Apart from their visits to the Willow Tree site, the brownies have enjoyed outings to local pantomimes. In 2007, like the guides, they visited the Mayor's

Parlour at the Civic Centre when Catherine Dann was Mayor. They had a tour of the chambers, saw the Mayor in her robes and Chain of Office, and enjoyed squash and as many cakes as they wanted.

The Pack continues to hold fund-raising events for charities. In 2007 they completed the 'Cheeky Monkey Challenge' by walking round the church 24 times three-legged and raised £420.50p for Children with Leukaemia.

In 2008 the 75th Anniversary of the church, in addition to their ongoing contributions to the church community, the brownies made a collage as an altar frontal for Harvest Festival.

1st Eastcote Rainbow Guides
The Annual Reports of Pauline Mc Donald, who continues as enthusiastic leader of the 1st Rainbow Guides, refer to having 'another busy year' and to the girls' enjoyment as members of this unit, still the only one in the area. There always seems to be a long waiting list of youngsters eager to join.

The practice of 'theme evenings' has been maintained with many craft activities. However, to be in line with the more up-to-date approaches of the other branches of the Guiding movement, in 2005 a new Rainbow programme was introduced. In the course of the year every rainbow achieved their first Roundabout badge. In order to achieve a Roundabout badge the girls had to meet certain challenges posed by four criteria, look, learn, laugh and love. During the Mothering Sunday Parade Service in 2009 Father Stephen presented 15 rainbows with Rainbow Roundabout badges.

Rainbow outings included ones to Tesco's and Ruislip Fire Station. Enjoyable summer activities included walks to Eastcote House walled garden where new rainbows made their promise, an outing to Ruislip Lido, a mini Olympics and outdoor games. Visits were made to the Willow Tree site to take part in various guiding events, including the 'Going for Gold' in 2008. That year the rainbows also had a most enjoyable day celebrating the Rainbow 21st Birthday in a county party. In 2010 the girls took part in the Hip Hip Hooray event at the Willow Tree site.

Mothers Union
Sadly by the time Father Stephen became Vicar the MU, the organisation that Father Godwin had looked upon as 'the backbone of the Church and parish' and 'one of the parish's greatest assets', had declined to a membership of 10. The small group met only once a month for Corporate Communion. Visitors were always welcome to join members for coffee and fellowship after the service. At the annual Christmas meeting in December 2005 members and guests enjoyed carols, readings, quizzes, and seasonal refreshments. In January 2006 the Wave of Prayer was again held at Marion Hayman's home and was followed by a light lunch.

The MU report for the 2006 APCM made it known that the current Joint Branch Leaders, Susan Miller and Joyce Boot, were to stand down at the November Triennial Meeting, and it was anticipated that the Branch would close, not through lack of enthusiasm on the part of members, but through the size of membership. A few members continued to attend Deanery events, including an outing in 2006 to Old Arlesford, the home of Mary Sumner in Hampshire, and the annual Diocesan Festival Service.

The October 2006 magazine confirmed the closure of the Branch. During the All Saints' Sunday Festival Eucharist on 5 November there was a thanksgiving for all that the MU had contributed to the life of the parish for well over 70 years. The inaugural meeting of the branch had been in February 1932, shortly after Eastcote became a 'Peel' district. Father Stephen reminded parishioners that the MU is a missionary organisation and that their work continues, for while the MU has declined in this country it is very active in Third World countries. The formal closure of the St Lawrence Branch of the MU took place at the Triennial Meeting on Wednesday 15 November. At this meeting Father Stephen said that all could retain a role in the MU by joining another group or become a Diocesan member. He urged all to keep praying, as prayer plays a vital role in the MU. The St Lawrence MU banner is kept in the church as a reminder of the church's long association with the MU.

Mothers Union banner
Flower Festival 2008

Since the formal closure, a few people, usually Susan Miller and Joyce Boot, have provided coffee and fellowship in the Committee Room after the 10.30am Thursday Eucharist each week.

Social Club

By the time Father Stephen became Vicar the programme of the Social Club had settled into a pattern of outdoor events from May to August and indoor events from September to April. Weather and other unforeseen events necessitated the odd last minute change. The skittle alley at the Six Bells was not available in 2006 and so brought to an end one annual event. The year 2006 saw the last of the boule evenings at Croxley Green as the ground was 'decked' over. In 2008 the Club visited the Hillingdon Sports Centre at Gatting Way (near Swakeleys Roundabout), for a boules session at the relatively recently installed facility there. In 2009 when a repeat visit was rained off, over half the members of the boules party spent part of the evening at The Waters Edge at the Lido. Over the years there were walks in Ruislip Woods and in the area of the Lido, Old Redding, and the Chess valley at Chenies, and more local ones around old Eastcote, as well as a guided London Walk around Piccadilly and Westminster (2007). A narrow boat canal trip from Horsenden Hill Willowtree Marina in July 2006 was enjoyed by 30 club members in spite of the thunder and lightning and heavy rain at the start of the trip.

As in the past the Club had presentations in the church hall on a wide variety of topics. These included the role of a Metropolitan Police Negotiator (2005), the History of Eastcote and the history of the Mary Rose (2006), Life as a member of the Guild of Professional Toastmasters and the work of a Tour Leader (2007), The Royal Botanical Gardens at Kew, Life at the Bar, and Mercy Ships (2008), the RNLI and Victorian Silhouettes (2009) and Whitefriars Glass (2010).

A couple of fun evenings organised by Club members were the St Lawrence Donkey Derby (2006), a repeat of a similar event in 2004, and an evening of Trivial Pursuit (2007). Activities in the annual programme are the short AGM, followed a quiz or some other enjoyable pastime, the summer barbecue, the theatre trip, and the Christmas dinner. The Social Club remains a source of enjoyment and fellowship.

Flower Guild

Throughout its history successive vicars have expressed thanks to the ladies who contribute and arrange the flowers in church. Currently the ladies of the Flower Guild contribute to making the church beautiful by giving of their time, talents and money. Again as in the past there are periodic appeals for new members. The Flower Festival of 2008, which they organised as part of the 75th anniversary celebrations was particularly rewarding. A cheese and wine lunch and a flower demonstration raised money towards the cost of the displays. The Festival itself was blessed with good weather and it was well supported. The profit of £1,500 was shared between two children's hospices – Helen House in Oxford and the Shooting Star Hospice at Twickenham.

St Lawrence Players

Annual reports and magazine reviews show that the Players maintain the tradition of three productions each year. Among productions between 2005 and 2010 was one of interest for its local setting, *A Pack of Lies* (first performed at St Lawrence's in 1988), programmes of short plays, and *Jane Eyre* a play with a large cast and a large number of period costumes. As in the past the reviews comment on the cast, sets, directors and producers, and all those behind-the-scenes people who contribute to entertaining and enjoyable evenings.

Scene from *75 Glorious Years* 2008

As part of the St Lawrence Church community the Players take an active part in the annual fairs and markets and other aspects of church life. For example in 2009, in addition to their stage productions in the hall, the Players presented a fictional drama in church on Palm Sunday evening 2009 entitled *A New Light Dawns*.

During 2007 and 2008 The St Lawrence Players celebrated their own 60th anniversary at the same time as the church celebrated its 75th anniversary. The joint celebration was marked in October 2008 by a lavish entertainment entitled *75 Glorious Years*. The show with its 17 items of solo turns, drama, singing, dance, and musical sequences was thoroughly enjoyed by cast and audience. The opening and closing musical sequences each had around 30 performers, including three generations of the Williams/Worker family and involving three generations of the Sykes/Kimsey family. The opening selection from *My Fair Lady* soon had the audience in the right mood, and *Mamma Mia* had a number of new youngsters, which suggests the Players has a bright future.

It seems relevant to refer to the Players' entertainment to the wider community of Eastcote on 17 July 2010, when Eastcote residents celebrated the restoration of the walled garden of Eastcote House (demolished in 1964). The St Lawrence Players gave two performances of a special Eastcote Mummers Play to appreciative audiences in the grounds of the house. It may be remembered that in 1913 an acre of land was purchased from Mr Ralph Hawtrey Deane, the owner of Eastcote House, with the idea of providing Eastcote with a parish church. Shortly after the Dedication of the Mission Church in 1920 Ralph Hawtrey Deane gave an additional acre. This 2010 link between St Lawrence Church,

Eastcote House, and the wider community of Eastcote is a further reminder of the importance of the Church's mission in this particular area at this particular time in the history of the Church of St Lawrence, Eastcote.

Each of St Lawrence's incumbents have been rightly concerned with the mission of the Church in Eastcote and each have emphasised that worship is at the heart of the church community. Each, in ways suitable to their particular time, expressed the importance of fellowship within that community if it is to witness in the wider community. In the early days of the parish church Father Godwin had been concerned to promote a 'sociable' community, Father Bill's guiding principles had been 'Worship and Fellowship, Father David Hayes' 'Forward Together', while Father David Coleman's took the mission of the church forward with greater involvement of the laity in the ministry of the church.

Fr Stephen with the Rt Revd Wilfred Wood,
a 'valued guide for over 30 years'
25th anniversary of Fr Stephen's Ordination as a Priest
2009

Into a New Decade
Father Stephen has built on the work of his predecessors to meet the challenges of the 21st century. In his 2008 APCM Vicar's Remarks he seemed to carry forward the sentiments of his predecessors regarding the St Lawrence community. 'Only when we regularly meet together, spend time together and worship together will we be a community that can flourish and grow and that can reach out to serve the wider community in which God has placed us'. Outreach to the Eastcote community, and beyond, has been more focused in the last few years. In his Review of 2009 Father Stephen said this would only continue to bear fruit if it is built on a solid foundation of prayer and worship. He concluded by saying that as the church entered a new decade it was ready to open its doors and hearts to welcome whoever God sends us.

APPENDICES

Appendix 1

Clergy and Other Members of the Team Ministry

Page 1 of 2

Priest Missioners	Dates
P. D. Ellis	18 Feb 1920 – 9 Oct 1921
J. McIntyre	6 Nov 1921 – 30 Sept 1922
J.G. Dale	10 Dec 1922 – 29 Nov 1925
D. Lewis	21 Feb 1926 – 15 Dec 1929
H.E. Nuttall	22 Dec 1929 – 29 Sept 1931
R.F. Godwin	3 Oct 31 licensed as Minister-in-Charge 8 Feb 1932. Automatically Vicar immediately after Consecration

Vicars		
1933 - 1956	Rupert Frederick Godwin	21 Oct 1933 – ? Apr 1956 (RIP 1 Dec 1965)
1956 - 1980	William Henry Hitchinson	30 May 1956 – 6 Apr 1980 (RIP 15 Nov 2003)
1980 - 1990	David M.H. Hayes	24 Jun 1980 – 28 Jan 1990
1990 - 2004	David Coleman	29 Oct 1990 – 24 Oct 2004
2005 -	Stephen Dando	13 Jun 2005 –

Assistant Curates		
1958-1964	Peter Goodridge	1 Jun 1958 – 26 Apr 1964 (RIP 13 Dec 2005)
1966-1967	W.P. van Zyl	26 Jun 1966 – 18 Jun 1967
1967-1973	Clive Pearce	12 Nov 1967 – 30 Oct 1973
1975-1979	Huw Chiplin	30 Jul 1975 – 21 Jan 1979
1981-1984	Masaki Narusawa	26 Jul 1981 – 6 May 1984
1984-1987	Alan Boddy	15 Jul 1984 – 19 Apr 1987
1987-1991	Peter Day	28 Jun 1987 – 27 Jan 1991
1991-1995	John Foulds	26 Jun 1991 – 19 Feb 1995
1995-1999	Michael Bolley	7 Sept 1995 – 11 Apr 1999 (Associate Vicar)
1999-2001	Sue Groom	10 Oct 1999 – 21 Oct 2001
2001-2004	Ruth Lampard	10 Nov 2001 – 19 Sept 2004
2007-2010	Rachel Phillips	30 Jun 2007 – 11 Jul 2010

Appendix 1

Clergy and Other Members of the Team Ministry

Page 2 of 2

Other Clergy		
1993-2004	Ann Coleman	1993-24October 2004
2001-2004	Laurence Hillel	5 Apr 2001 – 20Jun 2004
1985 -	Captain Donald Woodhouse	Lay Minister from 1985
After 1999	Wendy Brooker	1999 -

Lay Readers/ Readers	
1939-1956	Dr Fraser A.W. Luckin
1941-1969	J.J. Masey (Emeritus until1980 at least)
1965-1983	Gerald H. Collins (RIP July 1983)
1998-	Alan Wright

Pastoral Assistants	
1999-2007	Elaine Garrish
2007-	Jane Coan
2007-8	Tricia Hughes

Appendix 2
Subsequent Appointments of St Lawrence Church Clergy

Name & Time at St Lawrence	New appointment Dates	Subsequent appointments
Vicars		
Rupert Frederick Godwin 1933-1956	1956	Retired RIP 1 Dec 1965
William Henry Hitchinson 1956-1970	1980	Retired RIP 15 Nov 2003
David Hayes 1980-1990	1990-2007 1990-2007 1997-2007 2003-2007 2007	Rector Cant St Peter w St Alphege Master Eastbridge Hosp. Guardian of the Greyfriars, Canterbury P-in-c, Blean Retired
David Coleman 1990-2004	2004-2008 2008	P-in-c Upper Ryedale and Continuing Ministerial Education Officer, Cleveland Archdeaconry Vicar St Leonards, Heston

Name & Time at St Lawrence	New appointment Dates	Subsequent appointments
Curates		
Peter Goodridge 1958-1964	1964 1971 1985-1997 1987 1996	Vicar St Philip, Tottenham Vicar St Martin, West Drayton Diocesan Director Education, Truro Hon Canon Truro Cathedral Canon Librarian Truro Cathedral RIP 13 Dec 2005
W.P. van Zyl 1966-1967	1967	Returned to Rhodesia
Clive Pearce 1967-1973	1973-	Vicar St Anselm, Hatch End
Huw Chiplin 1975-1979	1979 1984	Curate St Michael & Christ Church & P-in-c St Francis, N Kensington Vicar St Matthew, Hammersmith
Masaki Narusawa 1981-1984	1984	Curate, St Mark with Christ Church, Glodwick, Oldham

Appendix 2
Subsequent Appointments of St Lawrence Church Clergy

Name & Time at St Lawrence	New appointment Dates	Subsequent appointments
Alan Boddy 1984-1987	1987 1990	Curate, St Mary Abbots with St George, Kensington Chaplain, HM Prison Service Retired 2002
Peter Day 1987-1991	1991 1994	C St Augustine, Wembley Park Vicar, Glen Parva and South Wigston, Leics.
John Foulds 1991-1995	1995 1999	Curate in Charge, Church of the Annunciation, Wembley Park P-in-c Holy Trinity, Meir, Lichfield
Michael Bolley 1995-1999	1999 2006	P-in-c, Holy Trinity, Southall Vicar, Holy Trinity, Southall
Sue Groom 1999-2001	2001 2003 2007 2009	P-in-c, St Matthew, Yiewsley Vicar, St Matthew, Yiewsley Dir. Lic. Min., Kensington Area Priest in Charge, Henlow & Longford also Asst. Diocesan Dir. Of Ordinands
Ruth Lampard 2001-2004	2005 2006-2008 2008	Hon Curate, St George, Letchworth Chaplain to Bishop of Kensington Assoc. Vicar, St Mary the Boltons
Rachel Phillips 2007-2010	2010	Vicar, Northaw & Cuffley, St Albans
Other Clergy		
Ann Coleman 2001-2004	2004-2008 2005-2008 2008	Director, Wyedale Hall, Yorks. Dioc Moderato Reader Training Tutor, St Mellitus College, London
Laurence Hillel 2001-2004	2001 2004	Chaplain, Bishop Ramsey School Brondesbury St Anne with Kilburn Holy Trinity (Non-stipendiary)

Appendix 3

Bishops associated with St Lawrence Church
(in addition to Confirmation - see Appendix 8)

Year appointed	Name	St Lawrence association
Bishops of London		
1901	Arthur Foley Winnington-Ingram	Consecration 1933
1939	Geoffrey Francis Fisher	
1945	John William Charles Wand	Parish hall 1955
1956	Henry Colville Montgomery Campbell	Silver Jubilee 1958
1961	Robert Wright Stopford	East Window 1969
1973	Gerald Alexander Ellison	
1981	Graham Douglas Leonard	Bp Willesden 1964-73 Bp Truro 1973-81 Lady Chapel Window 1975
1991	David Michael Hope	
1995	Richard John Carew Chartres	
Edmonton (new Area)		
1970	Alan Francis Bright Rogers	Friend of Father Bill Bp Mauritius 1959-66 Bp Fulham 1966-70 Bp Edmonton 1970-75
1975	William John Westwood	
1985	Brian John Masters	
1999	Peter William Wheatley	
Fulham		
1926	Basil Staunton Batty	
1947	William Marshall Selwyn	
1948	George Ernest Ingle	
1955	Robert Wright Stopford	
1957	Roderic Norman Coote	
1966	Alan Francis Bright Rogers	

Appendix 3

Bishops associated with St Lawrence Church
(in addition to Confirmation - see Appendix 8)

Page 2 of 3

Year Appointed	Name	St Lawrence association
Fulham (Cont)		
1970	John Richard Satterthwaite	
1980-1982	No appointment	
1982	Brian John Masters	
1985	Charles John Klyberg	
1996	John Charles Broadhurst	
Kensington		
1911	John Primatt Maud	
1932	Bertram Fitzgerald Simpson	
1942	Henry Colville Montgomery Campbell	
1949	Cyril Eastaugh	
1962	Edward James Keymer Roberts	
1964	Ronald Cedric Osbourne Goodchild	
later bishops not relevant as St Lawrence under Willesden from 1970		
Willesden St Lawrence transferred 1970		
1964	Graham Douglas Leonard	Parish Visitation 1972
1974	Geoffrey Hewlett Thompson	Stewardship 1981 Golden Jubilee 1983
1985	Thomas Frederick Butler	Bp 1991 Leicester Bp 1998 Southwark Sanctuary and Hall (George Guest Room) 1988
1992	Graham Dow	Bp Carlisle 2000 Stewardship 1997 Lady Chapel 1999
2001	Pete Alan Broadbent	The London Challenge 2002

Appendix 3

Archdeacons of Northolt associated with St Lawrence Church

Year Appointed	Name	St Lawrence association
1970	Roy Southwell	Garden of Rest 1978 Retired 1980
1980	Thomas Frederick Butler	Golden Jubilee 1983 Stewardship 1985
1985	Edward Scott Shirras	Stewardship 1987 Stewardship 1991
1992	Michael Colclough	
1995	Pete Broadbent	St Lawrence Centre 1999
2001	Christopher Chessup	
2005	Rachel Treweek	Communion of Young People 2010

Appendix 4
CHURCH OFFICIALS

Year	Church Wardens	
1933		
1934	Capt E.T. Weddell	T.G. Cross
1935	A.E.G. Hillier	T.G. Cross
1936	A.E.G. Hillier	S Morris Scott
1937	W.H. Morgan	G.W. Young
1938	W.H. Morgan	G.W. Young
1939	J.J. Masey	G.W. Young
1940	J.J. Masey	G.W. Young
1941	J.J. Masey	G.W. Young
1942	J.J. Masey	G.W. Young
1943	J.J. Masey	S.G. Peterken
1944	J.J. Masey	S.G. Peterken
1945	J.J. Masey	S.G. Fisher (took over Peterken in Forces)
1946	J.J. Masey	S.G. Fisher
1947	J.J. Masey	S.G. Fisher
1948	J.J. Masey	S.G. Fisher
1949	J.J. Masey	S.G. Fisher
1950	S.G. Fisher	T.S. Grandison
1951	D. Harrison	T.S. Grandison
1952	D. Harrison	T.S. Grandison
1953	D. Harrison	T.S. Grandison
1954	D. Harrison	T.S. Grandison
1955	D. Harrison	T.S. Grandison
1956	D. Harrison	T.S. Grandison
1957	D. Harrison	T.S. Grandison
1958	D. Harrison	J.G. Wyatt
1959	P.W. Darby	J.G. Wyatt
1960	P.W. Darby	J.G. Wyatt
1961	P.W. Darby	J.G. Wyatt
1962	P.W. Darby	J.G. Wyatt
1963	P.W. Darby	G.H. Collins
1964	P.W. Darby	G.H. Collins
1965	P.W. Darby	S.W. Starling
1966	P.W. Darby	S.W. Starling
1967	P.W. Darby	S.W. Starling
1968	P.W. Darby	S.W. Starling
1969	P.W. Darby	S.W. Starling (A.B. Dale . acted for part of year)
1970	P.W. Darby	S.W. Starling
1971	P.W. Darby	S.W. Starling
1972	E. Edwards	C.J. Orme
1973	E. Edwards	C.J. Orme
1974	E. Edwards	C.J. Orme
1975	E. Edwards*	C. J. Orme (*Edwards left A. Plummer elected until APCM)
1976	A. Plummer	M. Bedford
1977	A. Plummer	M. Bedford
1978	A. Plummer	Miss M.M. Case
1979	Miss M.M. Case	C.J. Orme
1980	Miss M.M. Case	C.J. Orme
1981	Miss M.M. Case	C.J. Orme
1982	Miss M.M. Case	C.J. Orme
1983	Miss M.M. Case	F. Reeve
1984	Miss M.M. Case	F. Reeve
1985	Miss M.M. Case	F. Reeve
1986	Miss M.M. Case	F. Reeve
1987	Miss M.M. Case	F. Reeve
1988	Miss M.M. Case	F. Reeve
1989	Miss M.M. Case	F. Reeve
1990	Miss M.M. Case	F. Reeve
1991	Miss M.M. Case	F. Reeve

Appendix 4
CHURCH OFFICIALS

Church Wardens (Cont)	
1992	Miss M.M. Case / A. Braithwaite
1993	Miss M.M. Case / A. Braithwaite
1994	Miss M.M. Case / F. Reeve
1995	Miss M.M. Case / L.L.S. Williams
1996	Miss M.M. Case / L.L.S. Williams
1997	A.J. Bedford / L.L.S. Williams
1998	A.J. Bedford / L.L.S. Williams
1999	A.J. Bedford / L.L.S. Williams
2000	A.J. Bedford / L.L.S. Williams
2001	A.J. Bedford / L.L.S. Williams
2002	A.J. Bedford / L.L.S. Williams
2003	A.J. Bedford / L.L.S. Williams
2004	A.J. Bedford / L.L.S. Williams
2005	A.J. Bedford / L.L.S. Williams
2006	Mrs S. Hooper / Mrs B. Plummer
2007	Mrs S. Hooper / Mrs B. Plummer
2008	Mrs S. Hooper / Mrs B. Plummer
2009	Mrs S. Cobb / Mrs B. Plummer
2010	Mrs S. Cobb / Mrs B. Plummer

PCC Secretary	
1932	T.G. Cross
1934	L.C. Brown
1935	F. Adamson
1936	F. Adamson
1937	G.A. Bolton
1938	G. Collins
1941	The Vicar
1945	Miss Lamprell
1951	S.G. Peterken
1955	J.G. Wyatt
1958	J.F. Dedman
1962	Miss J Hanson
1966	Mrs D McDonald
1974	Mrs Ann Sykes
1983	Miss Beryl Orders
1998	Mrs Jane Williams
2008	Mrs Elaine Wigington

Freewill Offering Secretary	
1932	T.M. Cross
1933	A. Ellis
1934	Mrs Rye
1939	T. Muggeridge
1942	G. Young
1943	Mrs Fisher
1952	W.J. Lane
1956	G.W. Wade
1960	Mrs Onions

Stewardship Recorder	
1963	S. Starling
1970	W.J. Lane
1977	D.K.D. McGregor. On sudden death of Lane
1978	T. Kirkley
1985	Mrs Sheila Aleong

Appendix 4
CHURCH OFFICIALS
Page 3 of 4

Treasurer	
No PCC Minute book for years prior to 1941	
1941	Young & Luckin (no initials but elsewhere A W Luckin)
1943	Vicar, Churchwardens & Luckin
1944	Vicar & Churchwardens
1945	Vicar, Churchwardens, & Luckin
1947	Vicar & Churchwardens
1957	Vicar, Churchwardens & G. Collins
1958	G. Collins & J Manger
1961	S.W. Starling & J Manger
1963	A.B. Dale & S.W. Starling
1965	A.B. Dale & G Collins
1969	A.B. Dale
1974	A.B. Dale
1975	Euan Lumsden
1983	Colin Goodier
1984	Euan Lumsden
2008	Mark Britter

Children's Advocate	
2001	Mrs B Plummer
2006	Mrs Emma Kimsey Mrs Maria Smith

Electoral Roll Officer	
1971	C. Tyrer (First at St Lawrence)
1973	Mrs Doris McDonald
1974	Miss Doreen Plummer
1998	Mrs Jean Gibson

Magazine Editors	
1961	Dr D. Harrison
1966	Mrs Pat Beeston
1979	Mrs Doris McDonald
1984	Fr Masaki Narusawa Mrs Ann Oke Mrs Barbara Higgs
1993	Mrs P. Brown
1996	Fr. Michael Bolley Mrs Sadie Wright

MU Officials	
Enrolling Member	
(Responsible for returning *Report Forms* to Mary Sumner House)	
1932-1935	Mrs Taylor
1935	Mrs Cross
1936-1940	Mrs Batty
1941-1949	Mrs Milne
1950-1955	Mrs Conisbee
1956-1961	Mrs Swanson
1962-1969	Mrs O. Bernard
1970-1972	Mrs Worden
1976-1984	Mrs Pearmain
Branch Leader	
1985-1988	Mrs Monica Hayes
1989-1997	Mrs Marion Hayman
1998-2006	Miss Joyce Boot Mrs Susan Miller (Joint)

Appendix 4

CHURCH OFFICIALS

Organist	
1932-39	Mr M.A. MacFarlane
1941-43	F.A. Lark
1944-47	D. May
1947-54?	S.B. Beck
1954-58	R.J.D. Mullet
1959	J.W.C. Ravenshill & Mrs Dedman
1960	D. A. Edwards, Mrs Dedman E. Glover
1961	D. A. Edwards Mrs Dedman Mrs Timberlake
1969	D. Brown Mrs Dedman
1971	D. Brown Mrs Dedman Mrs Timberlake
1972	Mrs Dedman Mrs Timberlake
1973	David Lacey and Mrs Dedman
1974	Last time Lacey listed. Mrs Dedman seems in charge
1978	from April Mr A. Edwards and Mrs Dedman
1979	Mrs Dedman
1981	Miss Joanna Paul
1983	Tony Smith
1984	(Oct) Mrs Frances Bradford to Christmas 1994
1995	Mrs Ann Holt and Mrs Enid Lumsden (Both voluntary)
2006	James Mooney-Dutton

Parish Administrator.	
2004	Mrs Sharon Evans
2009	Mrs Tessa Maude

Appendix 5

Table of Communicants during the Years of the Mission Church – 18 December 1920 to 20 October 1933

Notes

Information has been taken from the Register of Services books. In a few cases no figures recorded for some services, or are difficult to decipher, and this is reflected particularly in the total Sunday numbers of 1922 and 1929 and 1930.

1922 No figures recorded for Ascension Day.
1926 No priest Ash Wed. (Dale in parish the following Sunday).
1930 August figures difficult to decipher.
1933 No priest Ash Wed. (Godwin in a nursing home).

Figures in red after the numbers indicate the number of services for that day and are given only when there is a change from the previous year.

Total Sunday numbers include Christmas when on a Sunday.

Total Weekday numbers exclude Ash, Ascension, and Christmas (if on a weekday). The higher Weekday totals are years when there were many communicants in Lent, Holy Week and the Week of Prayer.

	Ash Wednesday	Easter	Ascension	Whitsun	St Lawrence	Christmas	Total Sunday	Total Weekday	Annual (approx)
1920						Sat 71 (2)	31		102
1921	7 (1)	108 (3)	16 (1)	64 (2)	-	Sun 87 (3)	954	17	994
1922	5	104	-	42	-	Mon 80	738	51	874
1923	4	112	10	84	Fri 9 (1)	Tue 107	1022	376	1519
1924	8	150	13	66	Sun 40	Thu 134	1184	340	1679
1925	9	152	13	90	Mon 10	Fri 75	1171	287	1555
1926	-	156	10	71	Tue 5	Sat 93	1106	16	1225
1927	7	154	5	82	-	Sun 94	1077	13	1102
1928	6	157	9	73	-	Tue 112	1176	12	1315
1929	3	127	7	82	Sat 6	Wed 93	898	14	1015
1930	16 (2)	161	21 (2)	64	Sun 20	Thu 140	1462	268	1907
1931	7	197	25	102		Fri 150	1685	220	2087
1932	14	198	19	106		Sun 182	2074	340	2447
1933	-	219	13	89			1537	198	1748

Appendix 6

Table of Communicants – 21 October 1933 to 31 December 2010

Page 1 of 7

Notes

The table needs to be read across the two page spread to obtain the total figures for a full year.

The table shows the number of communicants on some of the special days of the Church's Year. During the history of the church there have been additions and changes and mostly these have been referred to in the relevant chapters. Christ the King celebrated by the Church of England only from 1970 was a festival instituted by the Pope in 1925

The figures in red after the numbers indicate the number of services for that day and are only given when there was a change from the previous year.

Christmas numbers are given as follows, using 1933 as an example:-
65 at Midnight, plus 145 at 3 other services, total 210.
Number of services is 4 i.e. Midnight plus 3.
From 1962 Easter numbers are shown in a similar way:-
49 at Vigil, plus 801 at 5 other services, total 850. 6 Vigil plus 5

The last column gives the approximate annual total number at all church Holy Communion services plus those receiving in home or in hospital.

From 1974 to 2004 both St Lawrence and Dedication celebrated on Sundays.
1988 only one 10am service for Christ the King, (Pontifical Sung Mass Blessing Altar, Crucifix and Sanctuary).
From 19 February 1989 Sunday services at 8am and 10am 2 instead of 8am, 9am and 11am. So only 2 at Pentecost and Lawrence but at Dedication and Christ the King 3 at 8am, 10am and 5pm.

Appendix 6
Numbers of Communicants – 21 October 1933 to 31 December 2010
Page 2 of 7

Year	Ash Wednesday		Maundy Thursda		Good Friday	Easter				Ascension		Whitsun/ Pentecost		Corpus Christi	
						Vigil	Day		Total						
1933															
1934	13	2	14	2			250		3	28	2	105	3		
1935	28		16				276			34		121		11	1
1936	16		19				279			23		103		15	2
1937	11		14				224			32		108		13	
1938	14		24				259			30		133		18	
1939	15		21				304			27		139		-	
1940	4	1	14				270			31		155		23	
1941	20		16				339			44		136		12	
1942	10		28				258			29		136		17	
1943	8		15				279			27		127		12	
1944	5		13				287			23		164		24	
1945	6		14				295			16		115		18	
1946	3		12				284			23		125		11	
1947	4		23				278			21		145		11	
1948	15		17				329			27		204		16	
1949	11		21				374			29		171		-	
1950	8		14				385			24		168		18	
1951	9	2	24				323			52		239		10	
1952	7	1	24				373			47		166		29	
1953	13		22				339			46		215		13	1
1954	7		20				413			62		226		18	2
1955	13		27				421			52		243		23	
1956	6		26	2			432			27	1	214	3	24	
1957	74	3	64				582		4	157	2	302	4	75	
1958	107		179				671			150	3	373		89	
1959	213	2	169				771			293	4	454		114	
1960	346	4	179	3	43		801		5	239		449		82	3
1961	243		216		53		806			247	3	434		114	
1962	260	3	191		81	49	801	850	6	237		491		84	
1963	254		192		95	56	753	809		245		379		90	
1964	223		195		84	72	663	735		231		487		85	
1965	211	3	186	3	92	53	642	695	5	198	3	446	4	77	3
1966	192		178		97	41	662	703		191		393		98	

Appendix 6
Numbers of Communicants – 21 October 1933 to 31 December 2010

Year	St Lawrence		Dedication		All Saints		All Souls		Christ the King	Christmas				Annual Total
										Mid-night	Day	Total		
1933					18	2	15			65	M 145	210	4	800
1934	F 11	2	89	3	22		8			74	Tu 117	191		3,142
1935	Sa 9		56		21		6			75	W 119	194		3,418
1936	M 4		34		Su 65		7	1		111	F 119	230		3,543
1937	Tu 7		26		11		2			114	Sa 137	251		2,952
1938	W 9		72		9	1	4			72	Su 142	214		3,361
1939			46		8	2	17			War	M 192	192	3	3,218
1940			55		9		4			War	W 219	219	2	3,617
1941			53		9	1	7			War	Th 230	230		4,004
1942	M 8	1	57		Su 53		9			War	F 211	211		3,860
1943	Tu 6		40		15	2	4			War	Sa 210	210		3,405
1944	Th 11	2	51		5	1	18	2		117	M 102	219	3	3,412
1945	F 3	1	53		18	2	Su 4	1		160	Tu 102	262		3,284
1946	Sa 5		60		6	1	1			147	W 108	255	4	3,222
1947	Su 59	2	50		8		Su 66	2		166	Th 158	324		3,213
1948	-		81		5		5	1		200	Sa 135	335		3,912
1949	W 3	1	63		10		5			214	Su 179	393		4,466
1950	Th 14	2	60		4		13			191	M 158	349		3,945
1951	F 4	1	85		30	2	2			210	Tu 170	380		4,340
1952	Su 71	3	59		4	1	6	1		191	Th 161	352		4,712
1953	M 4	1	98		Su 59		4			230	F 199	429		5,595
1954	Tu 15		115		9	1	5			206	Sa 177	383		5,916
1955	W 8		99		9		4			214	Su 228	442		5,846
1956	F 78	2	197		51	2	10	2		262	Tu 267	529	5	7,059
1957	Sa 50		234		54	3	24			298	W 359	657		11,626
1958	Su 204	3	25th Tu 128 / Su 340	2	Sa 32	1	M 30	3		334	Th 399	733		15,217
1959	M 107	2	W 299	4	Su 274		29			407	F 385	792		19,132
1960	W 99	3	F 147		102	3	50			401	Su 503	904		21,381
1961	Th 137		Su 388	3	91		118			442	M 382	824		22,161
1962	F 74		Su 374		130		73			431	Tu 427	858		21,899
1963	Sa 36	1	Su 301		98		54	1		398	W 374	772		21,643
1964	M 53	3	Sa 92	2	Su 320		76	3		420	F 421	841		21,253
1965	Tu 74	3	Th 127	3	81	3	58	3		415	Sa 438	853	5	20,329
1966	W 97		F 115		114		89			420	Su 443	863		19,743

Appendix 6
Numbers of Communicants – 21 October 1933 to 31 December 2010

Year	Ash Wednesday		Maundy Thursda		Good Friday		Easter				Ascension		Whitsun/ Pentecost		Corpus Christi	
							Vigil		Day	Total						
1967	211	4	152		124		43		619	662	166		455		87	
1968	183	3	158		103		15		615	630	182		374		99	
1969	131	4	131		85		36		585	621	174		365		111	
1970	167		126		104		41		573	614	168		345		79	
1971	185		125	4	103		43		570	613	197	4	318		100	4
1972	157		130		94		41		564	605	156		368		117	
1973	161	5	133		99		31		459	490	157		345		81	
1974	154	4	117		85		42		525	567	134		290		79	
1975	153		149		85		32		412	444	148		274	3	96	
1976	165		111		87		36		377	413	117		270	4	76	
1977	144		119		90		38		418	456	117		295		77	
1978	139		119		84		33		375	408	150		301	3	60	
1979	119		110		81		31		349	380	4	95	212		67	3
1980	105	2	107	2	77		31		379	410	83	2	196		65	2
1981	109		111		84		38		324	362	81		208	2	65	
1982	99		83	1	83		47		325	372	79		203	3	54	
1983	150		80		77		25		315	340	83		207		59	
1984	97		77		79		33		305	338	74		228		44	
1985	88		72		79		32		293	325	68		163	2	61	
1986	99		87		95		37		274	301	76		191	3	48	
1987	105		78		86		38		310	348	67		183		54	
1988	104		87		84		40		249	289	61		174		73	
1989	91		75		79		40		254	294	3	72	196	2	43	
1990	89		65		69	1	42		240	282	63		179		62	
1991	70		73		89		42		226	268	75		158		52	
1992	72		65		77		41		228	269	68		152		44	
1993	83		59		80		38		219	257	60		132		31	
1994	71		62		74		38		200	238	66		139		53	
1995	80		63		71		43		200	243	51		136		49	
1996	70		71		72		41		208	249	61		142		51	
1997	58		59		71		37		187	224	51		148		37	
1998	60		59		63		35		194	229	63		128		45	
1999	77		57		69		37		198	235	56		139		44	

Appendix 6
Numbers of Communicants – 21 October 1933 to 31 December 2010

Year	St Lawrence		Dedication		All Saints		All Souls		Christ the King		Christmas				Annual Total
											Mid-night	Day		Total	
1967	Th 89		Sa 181	2	97		126				420	M 353		773	19,827
1968	Sa 66	2	M 82	3	83		Sa 49	1			350	W 378		728	18,975
1969	Su 269	3	Tu 102		Sa 25	1	106	3			310	Th 386		696	18,008
1970	M 97		W 151		Su 249		95		210	3	292	F 453		745	18,206
1971	Tu 104	4	Th 150	4	84	4	113	4	207		370	Sa 447		817	19,311
1972	Th 95		Sa 70	3	80	5	151		252		378	M 416		794	18,884
1973	F 76		40th F 76 / Su 328	4 / 3	102	4	94		220		309	Tu 417		726	17,416
1974	Su 214	3	182		64		74	2	204		281	W 334		615	16,065
1975	165		233		Sa 45	2	Su 230		196		234	Th 298		532	15,664
1976	199		225		69	4	89	4	190		287	Sa 262		549	14,993
1977	174		230		76		95	5	210		291	Su 234		525	15,660
1978	170		208		60		105	4	198		302	M 205		507	4	15,703
1979	152		251		68	2	60	2	180		246	Tu 242		488	14,313
1980	180		180		34		Su 199		179		253	Th 209		462	13,405
1981	182		187		Su 186		60	3	217		279	F 174		453	14,622
1982	189		190		36		65	2	181		260	Sa 171		431	14,630
1983	173		50th 193		35	2	70	3	185		262	Su 184		446	14,405
1984	165		181		66		59	2	190		257	Tu 177		434	14,125
1985	214		179		42		57		196		291	W 138		429	3	13,638
1986	157		171		Su 195		58		164		297	Th 142		439	13,790
1987	156		170		Su 183		58		179		294	F 119		413	13,784
1988	171		188		39	2	54		181	1	235	Su 143		378	13,863
1989	159	2	166	3	53		75		178	3	239	M 111		350	13,618
1990	130		140	2	72		59		169	2	260	Tu 121		381	12,638
1991	164		152		36	1	50		175		231	W 110		341	12,027
1992	136		Su 147		Su 147		55		157		234	F 82		316	11,317
1993	151		Su 191		Su 146		59	3	117		185	Sa 70		255	11,610
1994	145		146		40		49	2	149		170	Su 89		259	11,181
1995	125		161		42		65		161		171	M 75		246	10,953
1996	136		145		41		Sa 38	1	135		151	W 86		237	10,893
1997	133		152		Su 142		37	2	154		180	Th 100		280	10,528
1998	128		131		Su 148		43		139		139	F 90		229	9,925
1999	113		143		Su 136		51		140		140	Sa 74		214	9,990

Appendix 6

Numbers of Communicants – 21 October 1933 to 31 December 2010

Page 6 of 7

Year	Ash Wednesday	Maundy Thursday	Good Friday	Easter			Ascension	Whitsun/ Pentecost	Corpus Christi
				Vigil	Day	Total			
2000	68	61	68	42	198	240	51	135	42
2001	67	55	66	36	211	247	52	129	33
2002	54	68	47	42	179	221	48	134	13 Conf / 8pm 95
2003	74	69	52	42	174	216	49	132	33
2004	60	72	54	45	193	238	40	133	34
2005	58 2	58	51	34	164	198	37	132	37
2006	77	63	60	55	186	241	49	136	51
2007	41	57	57	89	170	259	14	89	49
2008	68	59	65	41	182	223	44	120	33
2009	68	54	62	36	177	213	43	103	48
2010	48	56	57	36	217	253	55	109	29

Appendix 6

Numbers of Communicants – 21 October 1933 to 31 December 2010

Page 7 of 7

Year	St Lawrence	Dedication	All Saints	All Souls		Christ the King	Christmas			Annual Total
							Mid-night	Day	Total	
2000	125	120	30	56		123	160	M 60	220	9,969
2001	103	145	48	49		139	179	Tu 84	263	10115
2002	101	131	33	Sa 39	1	130	174	W 74	248	9,819
2003	104	138	136	45	2	124	179	Th 79	258	10,136
2004	110	210	128	25	1	138	160	Sa 80	240	10,085
2005	W 24 2	123	136	49		131	175	Su 81	256	9,127
2006	Th 36 / Su 104	100	W 10 / Su 138	52	2	133	162	M 64	226	9,199
2007	F 24 / Su 102	109	Th 18 / Su 123	34	1	112	118	Tu 75	193	9,008
2008	145	75th / Tu 121	Sa 4 / Su 120	37	2	94	135	Th 80	215	9,071
2009	Su 95 / M 15	W 18 / Su 104	Su 123	40		124	131	no	203	8,639
2010	Su 102 / Tu 20	112	Su 102	32		140	108	Sa 82	190	8,563

Appendix 7
Table showing numbers of Baptisms and Marriages

Year	Baptisms	Marriages	Year	Baptisms	Marriages
1920	1		1965	91	59
1921	18		1966	88	59
1922	11		1968	93	60
1923	6		1969	94	68
1924	16		1970	82	52
1925	24		1971	61	57
1926	13		1972	63	58
1927	18		1973	88	47
1928	14		1974	83	34
1929	6		1975	84	36
1930	13		1976	56	28
1931	27		1977	75	24
1932	20		1978	70	20
1933	25	1	1979	63	33
1934	43	5	1980	65	18
1935	28	12	1981	50	33
1936	47	14	1982	41	23
1937	47	8	1983	48	34
1938	76	19	1984	38	33
1939	28	28	1985	46	34
1940	32	32	1986	33	27
1941	83	31	1987	42	32
1942	29	29	1988	45	24
1943	129	16	1989	38	19
1944	111	20	1990	38	33
1945	114	31	1991	47	18
1946	141	20	1992	55	26
1947	145	31	1993	65	23
1948	128	25	1994	66	28
1949	107	26	1995	60	19
1950	95	23	1996	72	19
1951	95	36	1997	56	11
1952	89	22	1998	51	10
1953	97	29	1999	72	12
1954	69	24	2000	61	11
1955	74	31	2001	60	12
1956	82	28	2002	59	9
1957	99	50	2003	60	3
1958	92	40	2004	32	8
1959	90	51	2005	22	10
1960	101	45	2006	32	11
1961	110	52	2007	29	6
1962	100	34	2008	19	6
1963	110	49	2009	16	4
1964	92	44	2010	31	11

Appendix 8
Number of candidates prepared for Confirmation by St Lawrence clergy
See Appendix 3 for full names of Bishops in the Diocese of London

Year	Date	Venue	Bishop	Bishop Name	No.	Note
1931	8 Dec	St Peter's Harrow	LB London	Arthur W-Ingram	1	
1932	29 May	St John's Hillingdon	Bury (St Edmunds?)	Walter Whittingham	13	
1933	8 Apr	Westminster Abbey	LB London	Arthur W-Ingram	1	
1934	24 Mar	Westminster Abbey	LB London	Arthur W-Ingram	2	
	9 June	St Paul's Cath.	LB London	Arthur W-Ingram	1	
	4 Jul	St Lawrence	Kensington	Bertram Simpson	14	
	3 Nov	St Paul's Cath.	Stepney	Charles Curzon	1	
1935	2 Nov	St Paul's Cath.	LB London	Arthur W-Ingram	1	
	8 Dec	St Margaret's Uxbridge	Kensington	Bertram Simpson	13	
1936	22 Nov	St Martin's Ruislip	Kensington	Bertram Simpson	11	
1937	21 Nov	St Lawrence	Kensington	Bertram Simpson	16	
1938	20 Nov	St Martin's Ruislip	Kensington	Bertram Simpson	10	
1939	4 June	St John's Hillingdon	Kensington	Bertram Simpson	1	
	19 Nov	St Lawrence	Kensington	Bertram Simpson	7	
1940	17 Nov	St Martin's Ruislip	Kensington	Bertram Simpson	4	
1941	14 Jan	St Lawrence	Kensington	Bertram Simpson	3	
	22 Nov	St Lawrence	Kensington	Bertram Simpson	32	
1942	7 Mar	St Paul's Cath.	LB London	Geoffrey Fisher	1	
	4 Jul	St Paul's Cath.	LB London	Geoffrey Fisher	1	
	25 Oct	St Paul's Ruislip Manor	Kensington	Henry Campbell	7	
1943	6 Mar	All Saints' Hillingdon	Kensington	Henry Campbell	3	
	12 Dec	St Margaret's Uxbridge	Kensington	Henry Campbell	3	
1944	30 Apr	St Lawrence	Kensington	Henry Campbell	20	
	5 Sept	Privately (RIP 12 Sept 1944)	Kensington	Henry Campbell	1	
	15 Oct	St Paul's Ruislip Manor	Kensington	Henry Campbell	2	
1945	14 Oct	St Martin's Ruislip	Kensington	Henry Campbell	22	
	17 Nov	St Paul's Cath.	LB London	William Wand	2	
1946	23 Mar	Westminster Abbey	LB London	William Wand		
	13 Oct Sun	St Lawrence	Kensington	Henry Campbell	13	
1947	23 Apr	St Paul's Ruislip Manor	Kensington	Henry Campbell	8	
	11 Oct	St Paul's Cath.	LB London	William Wand	1	

Appendix 8
Number of candidates prepared for Confirmation by St Lawrence clergy
Page 2 of 5

Year	Date	Venue	Bishop	Bishop Name	No.	Note
1948	10 Ap	St Lawrence	Kensington	Henry Campbell	17	
	14 Ap	St Lawrence	Kensington	Henry Campbell	11	
	8 May	St Paul's Cath.	LB London	William Wand	2	
	11 Dec	St Paul's Cath.	LB London	William Wand	2	
1949	9 Apr	Westminster Abbey	LB London	William Wand	2	
	4 May	St Lawrence	Kensington	Henry Campbell	45	Includes 16 Roxeth & Harrow Church Lads Brigade
	8 Oct	St Paul's Cath.	LB London	William Wand	2	
	10 Dec	St Paul's Cath.	LB London	William Wand	1	
1950	3 May	St Paul's Ruislip Manor	Kensington	Cyril Eastaugh	10	
1951	25 Apr	St Lawrence	Kensington	Cyril Eastaugh	22	
		St Michael's School, Joel St.	Kensington	Cyril Eastaugh	4	
	14 Jul	St Paul's Cath.	LB London	William Wand	1	
1952	30 Apr	St Paul's Ruislip Manor	Kensington	Cyril Eastaugh	25	
1953	6 May	St Lawrence	Kensington	Cyril Eastaugh	26	
	23 May	St Paul's Cath.	LB London	William Wand	1	
	3 Oct	St Paul's Cath.	Fulham	George Ingle	1	
1954	31 May	St Paul's Ruislip Manor	Kensington	Cyril Eastaugh	38	
1955	6 Mar	St Michael's School, Joel St.	Kensington	Cyril Eastaugh	4	
	4 May	St Lawrence	LB London	William Wand	43	
1956	6 Mar	St Paul's Ruislip Manor	Kensington	Cyril Eastaugh	22	
	17 Mar	St Paul's Cath.	LB London	Henry Campbell	2	
	24 Mar	St Michael's School, Joel S.	Kensington	Cyril Eastaugh	6	
1957	28 Mar	St Mary's Hayes	LB London	Henry Campbell	1	
	26 Jun	St Lawrence	Kensington	Cyril Eastaugh	104	
	10 Jul	St Martin's Ruislip	Kensington	Cyril Eastaugh	1	
	14 Dec	St Paul's Cath.	LB London	Henry Campbell	2	
1958	20 Jun	St Lawrence	Kensington	Cyril Eastaugh	64	
	29 Jun	St Paul's Ruislip Manor	Kensington	Cyril Eastaugh	1	
	11 Oct	St Paul's Cath.	LB London	Henry Campbell	5	
1959	15 Feb	St Michael's School, Joel St.	Kensington	Cyril Eastaugh	19	
	21 Jun	St Lawrence	LB London	Henry Campbell	93	
	11 Jul	St Paul's Cath.	LB London	Henry Campbell	3	
	12 Jul	St Barnabas Northolt	Kensington	Cyril Eastaugh	1	
1960	13 Feb	St Paul's Cath.	LB London	Henry Campbell	7	
	19 Jun	St Lawrence	Kensington	Cyril Eastaugh	72	

Appendix 8
Number of candidates prepared for Confirmation by St Lawrence clergy
Page 3 of 5

Year	Date	Venue	Bishop	Bishop Name	No.	Comment
1961	19 Mar	St Michael's School, Joel St.	Kensington	Cyril Eastaugh	4	
	18 Jun	St Lawrence	Bishop of Antigua	Newnham Davis	79	deputy for our bishop
	2 Jul	St Nicholas Perivale	Kensington	Cyril Eastaugh	1	
	9 Dec	St Paul's Cath.	Kensington	Cyril Eastaugh	2	
1962	23 Mar	St Lawrence	Kensington	Edward Roberts	53	
1963	22 Feb	St Lawrence	Kensington	Edward Roberts	65	
	24 Mar	St Michael's School, Joel St	Kensington	Edward Roberts	3	
	18 May	St Paul's Cath.	LB London	Robert Stopford	1	
1964	31 May	St Lawrence	Kensington	Ronald Goodchild	36	
	31 May	St Paul's Cath.	LB London	Robert Stopford	2	
1965	9 May	St Lawrence	Bishop of Antigua	Newnham Davis	33	deputy for our bishop
1966	15 May	St Lawrence	Kensington	Ronald Goodchild	50	
	18 Jun	St Paul's Cath.	LB London	Robert Stopford	2	
1967	5 Mar	Emmanuel Northwood	Willesden	Graham Leonard	3	
	18 Mar	Westminster Abbey	LB London	Robert Stopford	1	
	28 May	St Lawrence	Kensington	Ronald Goodchild	53	
	10 Jun	St Paul's Cath.	Ass Bp London	Thomas Craske	2	
	28 Jun	St Richard Northolt	Kensington	Ronald Goodchild	1	
1968	23 Jun	St Lawrence	Kensington	Ronald Goodchild	36	First time within context of a Eucharist
1969	11 May	St Lawrence	Ass Bp London	Cyril Sainsbury	40	
	26 Jun	St Mary's South Ruislip	Kensington	Ronald Goodchild	3	
	29 Jun	St Paul's Ruislip Manor	Kensington	Ronald Goodchild	1	
	25 Oct	St Paul's Cath.	Fulham	Alan Rogers	1	
1970	3 May	St Andrew's Roxbourne	Willesden	Graham Leonard	2	
	7 Jun	St Hilda's, Ashford	Kensington	Ronald Goodchild	2	
	28 Jun	St Lawrence	Kensington	Ronald Goodchild	31	
1971	2 May	St Lawrence	Willesden	Graham Leonard	33	
	2 Jun	St Edmund of Canterbury, Yeading	Willesden	Graham Leonard	1	
1972	14 May	St Lawrence	Willesden	Graham Leonard	23	
1973	6 May	St Lawrence	Ass Bp London	Cyril Sainsbury	38	
	3 Jun	St Andrew's Roxbourne	Willesden	Graham Leonard	1	

Appendix 8
Number of candidates prepared for Confirmation by St Lawrence clergy
Page 4 of 5

Year	Date	Venue	Bishop	Bishop Name	No.	Comment
1974	31 Mar	John Keble Ch. Mill Hill	Edmonton	Alan Rogers	1	
	12 May	St Lawrence	Willesden	Hewlett Thompson	22	
1975	22 Mar	Westminster Abbey	LB London	Gerald Ellison	1	
	4 May	St Lawrence	Willesden	Hewlett Thompson	35	
1976	22 Feb	St Mary's Kenton	Willesden	Hewlett Thompson	4	
	4 Apr	St Martin's Ruislip	Willesden	Hewlett Thompson	1	
	2 May	St Lawrence	Willesden	Hewlett Thompson	22	
1977	1 May	St Paul's Ruislip Manor	Willesden	Hewlett Thompson	1	
	8 May	St Lawrence	Willesden	Hewlett Thompson	36	
1978	28 May	St Lawrence	Willesden	Hewlett Thompson	19	
	3 Oct	St Martin's Ruislip	Willesden	Hewlett Thompson	3	
1979	10 Feb	St John's Hillingdon	Willesden	Hewlett Thompson	1	
	20 May	St Lawrence	Willesden	Hewlett Thompson	17	
1980	23 Mar	St Martin's Ruislip	Willesden	Hewlett Thompson	4	Interregnum (prepared by Hitchinson)
	19 Jun	St John's Pinner	Willesden	Hewlett Thompson	11	
1981	7 May	St Lawrence	Willesden	Hewlett Thompson	19	
1982	2 May	St Lawrence	Willesden	Hewlett Thompson	15	
1983	27 Apr	St Lawrence	Willesden	Hewlett Thompson	21	
	6 Nov	St Anselm's Hatch End	Willesden	Hewlett Thompson	3	
1984	23 May	St Lawrence	Ass Bp Willesden	Donald Arden	7	
1985	10 Feb	St Mary's South Ruislip	Willesden	Hewlett Thompson	1	
	29 Apr	St Lawrence	Willesden	Hewlett Thompson	19	
1986	27 Apr	St Lawrence	Willesden	Thomas Butler	14	
	10 Jul	Christ the Saviour Ealing	Willesden	Thomas Butler	1	
1987	13 May	St Lawrence	Ass Bp Willesden	Donald Arden	11	
1988	27 Apr	St Lawrence	Willesden	Thomas Butler	8	
1989	16 Apr	St Lawrence	Ass Bp Willesden	Donald Arden	11	
1990	2 May	St Mary's South Ruislip	Willesden	Thomas Butler	13	Interregnum
1991	22 May	St Lawrence	Willesden	Thomas Butler	10	
1992	11 Jun	St Mary's South Ruislip	Willesden	Graham Dow	8	
	5 Jul	All Saints' Queensbury	Willesden	Graham Dow	1	
1993	9 Jun	St Lawrence	Fulham	John Klyberg	10	
1994	26 May	St Lawrence	Willesden	Graham Dow	10	
1995	8 Jun	St Edmund the King Northwood Hills	Willesden	Graham Dow	3	

Appendix 8
Number of candidates prepared for Confirmation by St Lawrence clergy

Year	Date	Venue	Bishop	Bishop Name	No.	Comment
1996	19 Jun	St Lawrence	Willesden	Graham Dow	7	
1997	18 Jun	St Edmund the King Northwood Hills	Willesden	Graham Dow	8	
1998	16 Jun	St Lawrence	Willesden	Graham Dow	9	
1999	28 Feb	St Augustine's Wembley	Willesden	Graham Dow	6	
	10 Oct	St Lawrence	Willesden	Graham Dow	12	
2000	20 Jun	St Edmund the King Northwood Hills	Willesden	Graham Dow	1	
2001	12 Jun	St Lawrence	Willesden	Pete Broadbent	12	
2002	30 May	St Lawrence	Willesden	Pete Broadbent	11	
2003	11 May	St Lawrence	Willesden	Pete Broadbent	16	
	1 June	St John's Stanmore	Willesden	Pete Broadbent	1	
2004	16 May	St Lawrence	Willesden	Pete Broadbent	18	
2005	-	-	-	-	-	Interregnum
2006	14 May	St Lawrence	Willesden	Pete Broadbent	26	
2007		St Lawrence	Willesden	Pete Broadbent	10	
2008	27 Apr	St Lawrence	Willesden	Pete Broadbent	9	
2009	10 May	St Lawrence	Willesden	Pete Broadbent	4	
2010	5 Dec	St Lawrence	Ass Bp London	Donald Arden	11	

Appendix 9

Ecumenical Services – Numbers Attending

Page 1 of 2

Year	Carols	Christingle	Crib	Bereavement
	no figs until1973. Held at St L. from 1965 (not 1968 or 1969) venue varied from 2008	1984, 1986, 1990, on Sat; Sun from 1993	started from 1961 at latest but no figs. until 1986	from 1996 On Sun
1973	253			
1974	201			
1975	253			
1976	No figs			
1977	No figs			
1978	174			
1979	182			
1980	227			
1981	174			
1982	293			
1983	219			
1984	No figs	No figs		
1985	218			
1986	278	No figs	208	
1987	350		177	
1988	247		122	
1989	252		158	
1990	229	100	141	
1991	250		175	
1992	213		215	
1993	146	131	186	
1994	195	169	197	
1995	124	166	231	
1996	168	238	297	86
1997	214	220	347	57
1998	256	246	333	89
1999	213	204	275	86
2000	224	234	340	61
2001	241	241	346	64
2002	212	253	257	114
2003	207	215	278	95
2004	No figs	247	320	103

Appendix 9
Ecumenical Services – Numbers Attending
Page 2 of 2

Year	Carols	Christingle	Crib	Bereavement
2005	255	212	263	93
2006	219	273	232	97
2007	229	183	275	109
2008	St Thomas More	191	238	135
2009	Methodist	232	248	88
2010	St Andrew's	182	213	95

Appendix 10
Electoral Roll, Stewardship and Quota/Common Fund Figures
Page 1 of 3

Notes
From 1963 the Stewardship scheme was introduced.
From 1972 official Electoral Roll Officer required and new Electoral Roll every 6th year.
From 1994 APCM held after 10am Service Sunday instead of weekday evening.

NB. Quota		Due £		Underpaid £
	1992	55,802	Under paid	9,802
	1993	60,859	Under paid	13,859
	1994	58,946	Under paid	8,946

Year	Elect Roll	Attend. APCM	Stewardship Income £	Tax relief £	No. in scheme	Quota £
1927						26
1928						
1929	211					
1930						
1931						
1932	327					
1933	340					
1934	411					
1935	434					
1936	422					
1937	450					
1938	443					
1939	428					
1940	?					
1941	429	32				
1942	414					
1943	406					
1944	403					
1945	402 New	25				
1946	404					
1947	395					
1948	245 Revd					

Appendix 10
Electoral Roll, Stewardship and Quota/Common Fund Figures
Page 2 of 3

Year	Elect Roll	Attend APCM	Stewardship Income £	Tax Relief £	No in Scheme	Quota £
1949	264					Poss 84
1950	274					
1951	276					85
1952	278	40				Standard 82 Ambition 107
1953	306	41				
1954	314	36				
1955	330	37				
1956	344	48				
1957		82				
1958						198 with Clergy Widows Fund
1959						174/192
1960						337
1961						364
1962					160 covenants	
1963	377	62			> 300	313
1964	415	76	5,259	667		
1965	345	72	4,651	731		384
1966	336	64	4,171	735		
1967	332	64	3,866	721		
1968	324	42	3,450	670		480
1969	328	52	3,413	663	<200	750
1970	328	76	3,399	852		750
1971	338	69	3,317	853		1,011
1972	314 New	71	3,521	875	134	1,154
1973	310	58	3,372	925	?	1,236
1974	313	52	3,892	1,120	351	1,438
1075	327	49	4,121	1,325	147	2,129
1976	339	75	4,073	1,445	139	2,054
1977	348	63	3,672	1,308		1,353
1978	252 New	54	3,996	1,258		3,531

Appendix 10
Electoral Roll, Stewardship and Quota/Common Fund Figures

Year	Elect Roll	Attend APCM	Stewardship Income £	Tax Relief £	No in Scheme	Quota £
1979	285	90	3,984	1,047	102	3,881
1980	288	55	4,973	1,293	98	4,694
1981	302	60	11,208	2,141	209	7,081
1982	298	62	15,583	5,764	202	9,685
1983	304	60	15,483	5,712	196	12,888
1984	238 New	66	18,283	6,242	183	18,074
1985	240	43	19,311	6,284	162	20,960
1986	244	54	20,106	7,053	153	22,536
1987	238	62	23,234	7,309	155	25,161
1988	225	59	24,778	6,510	142	28,229
1989	225	59	27,905	7,170	134	31,008
1990	241 New	59	28,884	7,889	137	33,107
1991	244	63	32,376	7,699	182	47,787
1992	243	52	40,488	10,971	183	46,000
1993	243	42	38,615	10,401	182	47,000
1994	244	62	41,359	11,425	170	50,000
1995	249	51	41,072	11116	169	53,058
1996	238 New	61	42,950	10,788	154	52,572
1997	241	64	44,450	11,053	157	55,196
1998	252	46	43,040	11,050	139	57,459
1999	274	45	40,882	10,473	138	60,500
2000	282	53	49,349	12,843	179	54,500
2001	274	52	51,480	12,940	166	59,000
2002	260 New	54	51,459	13,194	165	50,154
2003	266	38	51,826	13,533	167	54,166
2004	273	37	52,937	13,870	161	58,499
2005	263	46	56,007	14,816	173	62,496
2006	283	45	54,404	14,284	162	65,621
2007	243 New	45	55,337	13,316	166	70,178
2008	257	44	60,684	16,852	160	79,000
2009	264	51	56,859	15,391	156	75,108
2010	282	33	58,723	16,200	154	77,108

Appendix 11

Fêtes, Fairs and other Fund Raising events from 1956

Figures were not always available

Notes

Most figures are taken from annual Statement of Financial Activities presented at APCM, copies of which are in PCC Minutes Books. No copy of 1982 available and 1983 made no comparisons with 1982 fete and fair figures.

Table only shows figures for events that took place annually for many years.

Proceeds went to church funds unless otherwise indicated.

Reference is made to some other fund-raising in the text (NB George Guest's contribution to the Building Project) and efforts for specific charities (e.g. 1982 when Ploughman's Lunch raised funds for Helen House in May), and 'afternoon tea' was the social occasion in June.

The amounts shown in the accounts in Appendix 12 are these figures plus other fund-raising events so the figures do not tally.

* Info from magazines.

Interfaith organised event.

+ gross all fund-raising costs shown as one figure.

~ Fair total is included in the amount shown in the Fete column.

1990 and 1991 Antique Fair...figs in Accounts seem to include amount in 'Fund raising other functions' ie. No separate figs until 1992.

Regular concerts started in Feb. 2007 in aid of the Music Fund.

Year	Fête £	Fair £	Gift Day £
1956	-		c 114 *
1957	-		Over 100 *
1958	-	190 *	330 (Silver Jubilee)
1959	-		183 *
1960	150 *		170 *
1961	126 *	276 (600 paid admission) *	?
1962	150 *	232 *	120 *
1963	125	48 (coffee market low key - Start of Stewardship)	
1964	400 Joint with St Thomas More*# Freedom from Hunger Campaign	150 Save the Children Fund *	

Appendix 11
Fêtes, Fairs and other Fund Raising events from 1956

Year	Fête £	Fair £	Gift Day £
1965	400 + Eastcote churches *# local Old Folks Assoc.	Proceeds to Mauritius.	
1966	Ruislip Northwood Round Table Fête *	No sale	
1967	Vicarage Garden Fête	c 180	
1968	Vicarage Garden Fête	?	
1969	Vicarage Garden Fête	c 300 *	
1970	880 Total Fête Fair & others	700 ~ less expenses (Fri. Sat.)	
1971	1,005 (Total Fête Fair & others)	750 ~	
1972	307 (Fête and other functions)	690	
1973	831 (Total Fêtes Fair & others)	507 ~ Asst. Clergy Fund	378 (40th Anniv.)
1974	228	632	901
1975	332	797	408
1976	441	841 Ist Grand Draw	209
1977	438	1,003	299
1978	556	1,010	290
1979	629	834	-
1980	834 incl. Plant Sale	1,050	934
1981	783 incl. Plant Sale	915	Stewardship Renewal
1982	afternoon tea instead of fête.		Not recorded
1983	417	1,042	
1984	590	2,023	
1985	1,170	1,927	
1986	308 (Acs state half of proceeds)	1,808	
1987	812	1,921	
1988	646	1,801	
1989	1,237 Church Urban Fund	1,938	
1990	1,091	1,446 (excl. Draw)	

Appendix 11

Fêtes, Fairs and other Fund Raising events from 1956

Year	Fête £	Fair £	Antique Fair £
1991	1,149	1,595	Antique Fair from 1992
1992	1,496	1,634	1,135
1993	1,548	1,157	1,150
1994	1,841	1,370	1,360
1995	1,540	1,385	1,380
1996	2,243	2,000	1,687
1997	2,086	2,291	1,881
1998	1,510	1,775	1,650
1999	2,085	2,110	2,072
2000	1,927	2,380	1,192
2001	2,132	4,116	2,783
2002	3,413	4,197	1,342
2003	3,293	3,558	1,230
2004	3,581	3,692	1,087 (last one)
2005	4,032	4,400	
2006	4,425	4,823	
2007	6,378 +	5,284 +	
2008	6,517 +	4,702 +	
2009	6,335 +	5,651 +	
2010	5,143 +	3,735 +	

Appendix 12

Accounts 2008, 2009 and 2010

Page 1 of 2

ST LAWRENCE CHURCH - ACCOUNTS SUMMARY 2009

What we received.....

Our income for 2009 was £133,984 (2008 : £144,192) and came from the following sources. Planned giving includes weekly/monthly envelopes, Bankers orders donations, and donations under Gift Aid. The Income Tax recovered is what we get back under the Gift Aid Scheme.

	2009		Increase/(Decrease) from 2008 to 2009		2008	
	% of total	£	%	£	£	% of income
Planned Giving	42.4%	56,859	(6.3%)	(3,825)	60,684	42.1%
Income Tax Recovered	11.5%	15,391	(8.7%)	(1,461)	16,852	11.7%
General Collections	11.9%	15,889	.4%	66	15,823	11.0%
Giving Subtotal	65.8%	88,139	(5.6%)	(5,220)	93,359	64.7%
Fund Raising & Events	10.3%	13,774	(14.0%)	(2,251)	16,025	11.1%
Hall & Centre Letting	14.0%	18,749	.0%	8	18,741	13.0%
Interest	2.1%	2,846	(72.0%)	(7,321)	10,167	7.1%
Wedding/Funeral Fees	2.1%	2,841	30.7%	667	2,174	1.5%
Magazine	1.3%	1,712	.7%	12	1,700	1.2%
Other	0.7%	923	(54.4%)	(1,103)	2,026	1.4%
Legacy	3.7%	5,000		5000	0	0.0%
Rent	0.0%	0		0	0	0.0%
	100.0%	133,984	(7.1%)	(10,208)	144,192	100.0%

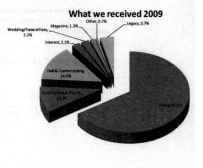

What we spent.....

Our expenses for 2009 were £146,870 (2008 : £156,027), with the largest item being our Diocesan Parish Contribution. This is shown in 2 parts; the actual cost to the diocese of providing our incumbent (including housing, pension training and support), and the contribution we are asked to make to help support other parishes, the diocese, and the Church nationally.

	2009		Increase/(Decrease) from 2008 to 2009		2008	
	% of income	£	%	£	£	% of income
Actual cost of Incumbent	46.7%	62,590	5.2%	3,114	59,476	41.2%
Diocesan contribution	9.3%	12,518	(35.9%)	(7,006)	19,524	13.5%
Common Fund Subtotal	56.1%	75,108	(4.9%)	(3,892)	79,000	54.8%
Church Services	6.0%	8,092	(8.6%)	(766)	8,858	6.1%
Outreach and Youth activities	1.7%	2,329	(29.9%)	(993)	3,322	2.3%
Clergy and Vicarage costs	3.7%	5,004	2.5%	123	4,881	3.4%
Charitable Giving	5.4%	7,247	(6.6%)	(515)	7,762	5.4%
Church Utilities & Cleaning	12.9%	17,309	67.4%	6,971	10,338	7.2%
Parish Administration	7.2%	9,671	11.4%	991	8,680	6.0%
Hall and Centre Expenses	6.9%	9,285	(36.2%)	(5,275)	14,560	10.1%
Church Maintenance	1.2%	1,661	(79.9%)	(6,609)	8,270	5.7%
West End Project	4.3%	5,763	201.7%	3,853	1,910	1.3%
Magazine costs	1.0%	1,351	18.5%	211	1,140	0.8%
Fund Raising & Events costs	2.5%	3,310	(35.2%)	(1,800)	5,110	3.5%
Other	0.6%	740	(66.3%)	(1,456)	2,196	1.5%
	109.6%	146,870	(5.9%)	(9,157)	156,027	108.2%
Deficit in 2009	(9.6%)	(12,886)			(11,835)	(8.2%)

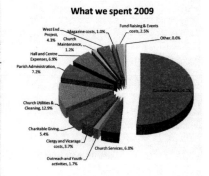

346

Appendix 12

Accounts 2008, 2009 and 2010

Page 2 of 2

ST LAWRENCE CHURCH - ACCOUNTS SUMMARY 2010

What we received.....

Our income for 2010 was £160,562 (2009 : £133,984) and came from the following sources. Planned giving includes weekly/monthly envelopes, Bankers orders donations, and donations under Gift Aid. The Income Tax recovered is what we get back under the Gift Aid Scheme.

	2010 % of total	2010 £	Increase/(Decrease) from 2008 to 2009 %	Increase/(Decrease) from 2008 to 2009 £	2009 £	2009 % of income
Planned Giving	36.6%	58,740	3.3%	1,881	56,859	42.4%
Income Tax Recovered	10.1%	16,200	5.3%	809	15,391	11.5%
General Collections	10.2%	16,443	3.5%	554	15,889	11.9%
Giving Subtotal	56.9%	91,383	3.7%	3,244	88,139	65.8%
Fund Raising & Events	8.2%	13,189	(4.2%)	(585)	13,774	10.3%
Hall & Centre Letting	17.3%	27,725	47.9%	8,976	18,749	14.0%
Interest	0.6%	951	(66.6%)	(1,895)	2,846	2.1%
Wedding/Funeral Fees	4.7%	7,620	168.2%	4,779	2,841	2.1%
Magazine	1.0%	1,593	(7.0%)	(119)	1,712	1.3%
Other	1.1%	1,790	93.9%	867	923	0.7%
Legacy	3.1%	5,000		0	5,000	3.7%
Rent	4.3%	6,932		6,932	0	0.0%
Insurance Claim	2.7%	4,379				
	100.0%	160,562	19.8%	26,578	133,984	100.0%

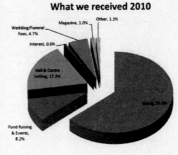

What we spent.....

Our expenses for 2010 were £148,935 (2009 : £146,870), with the largest item being our Diocesan Parish Contribution. This is shown in 2 parts; the actual cost to the diocese of providing our incumbent (including housing, pension training and support), and the contribution we are asked to make to help support other parishes, the diocese, and the Church nationally.

	2010 % of income	2010 £	Increase/(Decrease) from 2008 to 2009 %	Increase/(Decrease) from 2008 to 2009 £	2009 £	2009 % of income
Actual cost of Incumbent	40.0%	64,257	2.7%	1,667	62,590	46.7%
Diocesan contribution (20%)	8.0%	12,851	2.7%	333	12,518	9.3%
Common Fund Subtotal	48.0%	77,108	2.7%	2,000	75,108	56.1%
Church Services	7.4%	11,932	47.5%	3,840	8,092	6.0%
Outreach and Youth activities	1.5%	2,350	.9%	21	2,329	1.7%
Clergy and Vicarage costs	2.6%	4,100	(18.1%)	(904)	5,004	3.7%
Charitable Giving	4.2%	6,751	(6.8%)	(496)	7,247	5.4%
Church Utilities & Cleaning	10.1%	16,242	(6.2%)	(1,067)	17,309	12.9%
Parish Administration	6.7%	10,794	11.6%	1,123	9,671	7.2%
Hall and Centre Expenses	5.7%	9,181	(1.1%)	(104)	9,285	6.9%
Church Maintenance	1.3%	2,047	23.2%	386	1,661	1.2%
West End Project	0.4%	675	(88.3%)	(5,088)	5,763	4.3%
Magazine	0.6%	1,034	(23.5%)	(317)	1,351	1.0%
Fund Raising & Events costs	3.7%	5,873	77.4%	2,563	3,310	2.5%
Other	0.5%	848	14.6%	108	740	0.6%
	92.8%	148,935	1.4%	2,065	146,870	109.6%
Surplus (deficit in 2009)	7.2%	11,627			(12,886)	(9.6%)

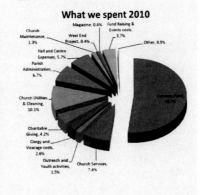

Appendix 13
St Lawrence Players Productions.

Page 1 of 5

Date	Production	Dates	Director
1948	School for Scandal	Feb 6, 7	Miss L Angell
	The Man Who Stayed At Home	Oct 23	Miss L Angell & Muriel Ferris
1949	Murder By Request / A Flat & A Sharp	Jun 18	
	I Have Five Daughters	Nov 26	L. D. Angell
1950	Dick Whittington	Feb 3	L. D. Angell
	Candied Peel	May 20	Hylda Darby
	Distinguished Gathering	Dec 9	Hylda Darby
1951	Cinderella	Feb 3	Hylda Darby
	Baa Baa Black Sheep	May 19	Hylda Darby
	St. Laurence Minstrels	Jun	
	This Happy Breed	Nov 10	Hylda Darby & Fred Conisbee
1952	Jack and the Beanstalk	Feb 2	Hylda Darby
	Nothing But The Truth	Apr 28	Hylda Darby
	St. Laurence Minstrels	Jun	
	The Poltergeist	Nov 15	Hylda Darby & Gerald Collins
1953	Aladdin	Jan 24	Hylda Darby
	Hay Fever	May 2	Hylda Darby
	Something Different	Jun	
	The Shop At Sly Corner	Nov 14	Hylda Darby
1954	Cinderella	Jan 30	Hylda Darby
	The White Sheep of the Family	May 1	Hylda Darby
	Nothing But The Truth (Act II)	Jun	
1955	A Programme of Entertainment	Jun 18	
	The Blue Goose	Oct 21, 22	Hylda Darby
1956	Dick Whittington	Feb 3, 4	Hylda Darby
	A Lady Mislaid	May 4, 5	Hylda Darby
	Waters of the Moon	Oct 5, 6	Hylda Darby
	Christmas show inc. Harlinquinade	Nov 24	
1957	Aladdin	Jan 18, 19, 26	Hylda Darby
	Queen Elizabeth Slept Here	May 3, 4	Hylda Darby
	Sit Down a Minute Adrian	Nov 8, 9	Hylda Darby
1958	The Babes in the Wood	Jan 25	Hylda Darby
	The Hollow	May 2, 3	Hylda Darby
	Baa Baa Black Sheep	Oct 30, 31, Nov 1	Hylda Darby
1959	Sinbad The Sailor	Jan 23, 24	Hylda Darby
	Murder On The Nile	Apr 23, 24, 25	Hylda Darby
	The Middle Watch	Nov 5, 6, 7	Hylda Darby
1960	Puss In Boots	Jan 23, 29, 30	Hylda Darby
	Hawk Island	Apr 29, 30	Hylda Darby
	Ten Little Niggers	Oct 28, 29	Hylda Darby

Appendix 13
St Lawrence Players Productions.

Page 2 of 5

Date	Production	Dates	Director
1961	Jack & the Beanstalk	Jan 27, 28	Hylda Darby
	Dry Rot	Apr 28, 29	Hylda Darby
	Haul for the Shore	Oct 27, 28	Hylda Darby
1962	Dick Whittington	Feb 2, 3	Hylda Darby
	To Kill A Cat	May 4, 5	Hylda Darby
	Distinguished Gathering	Oct 26, 27	Hylda Darby
1963	Aladdin	Feb 1, 2	Hylda Darby
	Emma	May 3, 4	Hylda Darby
	Fete's Finale	Jun 29	Hylda Darby
1964	The Geese Are Getting Fat	Feb 7, 8	Hylda Darby
	When We Are Married	May 1, 2	Hylda Darby
	A Show For You	Sep 12	Hylda Darby
	Mansfield Park	Nov 13, 14	Hylda Darby
1965	Cinderella	Feb 5, 6	Hylda Darby
	Man Alive	May 7, 8	Hylda Darby
	Fete Accompli	Jun 19	
	Waters Of The Moon	Oct 22, 23	Hylda Darby
1966	Little Red Riding Hood	Feb 11, 12	Hylda Darby
	Watch It Sailor	May 6, 7	Hylda Darby
1967	Post Horn Gallop	Jan 27, 28	Hylda Darby
	This Happy Breed	May 5, 6	Hylda Darby
	Goodnight Mrs Puffin	Oct 13, 14	Hylda Darby
1968	Running Riot	Feb 2, 3	Hylda Darby
	A Letter From The General	May 17, 18	Hylda Darby
	A Revue	Jun 22	
	Rock-a-Bye, Sailor	Oct 25, 26	Hylda Darby
1969	Haul For The Shore	Feb 1	Hylda Darby
	Blithe Spirit	May 2, 3	Hylda Darby
	A Lady Mislaid	Oct 24, 25	Hylda Darby
1970	A Revue For You	Jan 30, 31	
	Mystery At Blackwater	May 8, 9	Hylda Darby
	A Murder Has Been Arranged	Oct 23, 24	Hylda Darby
1971	Sinbad The Sailor	Feb 5, 6	Hylda Darby
	Bonaventure	May 7, 8	Hylda Darby
	The Bride And The Bachelor	Oct 29, 30	Hylda Darby
1972	Salad Days	Feb 4, 5	Hylda Darby
	Happiest Days of Your Life	May 5, 6	Jean Jungreuthmayer
	Soiree Musicale	Jul 8	A Leda production
	The House By The Lake	Oct 13, 14	Hylda Darby
1973	Oliver!	Feb 1, 2, 3	Hylda Darby
	Ladies In Retirement	May 4, 5	Hylda Darby
	Quiet Wedding	Oct 26, 27	Hylda Darby
1974	Where's Charlie	Feb 7, 8, 9	Hylda Darby
	Midsummer Mink	May 17, 18	Hylda Darby
	Dear Octopus	Oct 18, 19	Leigh Smith

St Lawrence Players Productions.

Date	Production	Dates	Director
1975	Babes In The Wood	Jan 30, 31	Hylda Darby
	The Orchard Walls	May 9, 10	Leigh Smith
	Music Hall	Oct 9, 10, 11	Come Into The Parlour / A Victorian Entertainment
1976	Where The Rainbow Ends	Feb 5, 6, 7	Hylda Darby
	Off The Hook	May 7, 8	Jean Jungreuthmayer
	Triple Bill	Oct 8, 9	Iris C, Jean J, Ann M
1977	The Wizard of Oz	Feb 3, 4, 5	Jean Jungreuthmayer
	Man Alive	May 6, 7	Hylda Darby
	Music Hall		
1978	The Heartless Princess	Feb 3, 4	Jean Jungreuthmayer
	A Letter From The General	May 5, 6	Hylda Darby
	A Voyage Round My Father	Oct 6, 7	Anton Jungreuthmayer
1979	Story of Cinderella	Feb 2, 3	Jean Jungreuthmayer
	The Same Sky	May 4, 5	Anton Jungreuthmayer
	When We Are Married	Oct 5, 6	Ann Morgans
1980	Alice in Wonderland	Jan 31, Feb 1, 2	Judith Brion & Anton Jungreuthmayer
	The Shop At Sly Corner	May 9, 10	Hylda Darby
	Music Hall	Oct 3, 4	
1981	Toad of Toad Hall	Feb 5, 6, 7	Judith Brion, Anton Jungreuthmayer
	Celebration	May	Jean Jungreuthmayer
	Suddenly at Home	Oct 9, 10	Judith Brion
1982	Salad Days	Feb 4, 5, 6	Win Brion
	Pygmalion	May 13, 14, 15	Linda Cannon
	Confusions	Oct 8, 9	Ann S, Jean J, Anton J, Judith H, Dereck A
1983	Viva Mexico	Feb 11, 12	Judith Howe
	The Vigil	May 18, 19	John Woodnutt
	Dear Octopus	Oct 13, 14, 15	Judith Howe
1984	Oliver!	Feb 9, 10, 11	Jean Jungreuthmayer
	On Monday Next	May 18, 19	Win Brion
	Playing for Time	Oct 12, 13	Derek Allcock
1985	Half a Sixpence	Feb 8, 9	Judith Howe
	A Murder is Announced	May 17, 18	Judith Howe
	Old Time Music Hall	Oct 11, 12	
1986	Scrooge	Feb 14, 15	Judith Howe
	How the Other Half Loves	May 16, 17	Judith Howe
	Three Into One Will Go	Oct 10, 11	Judith Howe
1987	The Railway Children	Feb 6, 7	Judith Howe
	Outside Edge	Oct 9, 10	Judith Howe
1988	Camelot	Feb 5, 6	Judith Howe
	Murder Mistaken	May 13, 14	Dorothy Bentote
	Pack of Lies	Oct 7, 8	Judith Howe

Appendix 13
St Lawrence Players Productions.

Page 4 of 5

Date	Production	Dates	Director
1989	Finian's Rainbow	Feb 3, 4	Judith Howe
	Brush With a Body	May 12, 13	Sue Worker & Barbara Williams
	Music Hall Miscellany	Oct 6, 7	Derek Allcock & Ann Sykes
1990	Hans Andersen	Feb 9, 10	Judith Howe
	Murder at Deem House	May 11, 12	Sue Worker & Barbara Williams
	Stepping Out	Oct 5, 6	Judith Howe
1991	Edwardian Evening	Jan 11, 12	
	My Fair Lady	May 16, 17, 18	Judith Howe
	Pride and Prejudice	Oct 4, 5	Dorothy & Barbara
1992	Dick Whittington	Jan 10, 11	Mark Kimsey & Ann Sykes
	Table Manners	May 8, 9	Judith Howe
	Lord Arthur Savile's Crime	Oct 9, 10	Stephen Kimsey
1993	Pickwick	Feb 11, 12, 13	Judith Howe
	Darling I'm Home	May 14, 15	Alison Higgs & Emma Sykes
	Sixty Glorious Years	Oct 7, 8, 9	Derek Allcock, Ann Sykes
1994	Sweeney Todd	Feb. 10, 11, 12	Stephen Kimsey
	The Anniversary	Jun 17, 18	Duncan Sykes
1995	Carousel	Feb 9, 10, 11	Mark Kimsey & Emma Kimsey
	Party Piece	May 12, 13	Judith Howe
	Blithe Spirit	Oct 6, 7	Dorothy Bentote
1996	Aladdin	Jan 18, 19, 20	Judith Howe
	Play On	May 10, 11	Alison Higgs, Emma Sykes
	Absent Friends	Oct 11, 12	Alan Hooper
1997	Song, Style & Supper	Feb 7, 8	Ann Sykes
	When We Are Married	May 15, 16, 17	Judith Howe
	Wild Goose Chase	Oct 10, 11	Dorothy Bentote
1998	The Wizard of Oz	Feb 12, 13, 14	Judith Howe
	The Forsyte Saga	May 14, 15, 16	Valerie Clarke
	Man Alive	Oct 9, 10	Barbara Williams
1999	Steel Magnolias	Feb 12, 13	Emma Kimsey
	I'll Get My Man	May 14, 15	Dorothy Bentote
	The God Who Comes	Nov 27, 28	Judith Howe, Alan Hooper & Win Brion
2000	Cinderella	Feb 20, 21, 22	Judith Howe
	The Day After The Fair	Oct 6, 7	Win Brion
2001	Alarms and Excursions	Feb 9, 10	Dorothy Bentote
	It Runs In The Family	May 10, 11, 12	Valerie Clarke
	Murdered to Death	Oct 11, 12, 13	Malcolm Bentote
2002	Season's Greetings	Feb 7, 8, 9	Mark Kimsey
	Rebecca	May 9, 10, 11	Valerie Clarke
	Run For Your Wife	Oct 10, 11, 12	Ritchard Tysoe

Appendix 13
St Lawrence Players Productions.
Page 5 of 5

Date	Production	Dates	Director
2003	Happy Families	Feb 6, 7, 8	Alan Hooper
	The Odd Couple	May 8, 9, 10	Valerie Clarke
	Time and Time Again	Oct 9, 10, 11	Dorothy Bentote
2004	Hobson's Choice	Feb 5, 6, 7	Valerie Clarke
	Confusions	May 13, 14, 15	Malcolm Bentote
	The Dancing Years	Oct 7, 8, 9	Graeme Gibaut
2005	Murder By Misadventure	Feb 10, 11, 12	Valerie Clarke
	Pack of Lies	May 12, 13, 14	Mark Kimsey
	Plaza Suite	Oct 6, 7, 8	Dorothy Bentote
2006	After September	Feb 9, 10 11	Valerie Clarke
	Breath of Spring	May 11, 12, 13	Graeme Gibaut
	Just Between Ourselves	Oct 12, 13, 14	Dorothy Bentote
2007	Proscenophobia	Feb 8, 9, 10	Valerie Clarke
	Day of Reckoning	May 10, 11, 12	Mark Kimsey
	Hidden Truths	Oct 11, 12, 13	Graeme Gibaut & Valerie Clarke
2008	Din Dins	Feb 7, 8, 9	Valerie Clarke
	London Suite	May 8, 9, 10	Dorothy Bentote & Estelle Dunham
	Sevety Five Gloroius years	Oct 2, 3, 4	Ann Sykes
2009	Deckchairs 1	Feb 5, 6, 7	Valerie Clarke
	Anybody for Murder?	May 7, 8, 9	Dorothy Bentote
	Jane Eyre	Oct 8, 9, 10	Valerie Clarke
2010	Curtain Up on Murder	Feb 4, 5, 6	Mark Kimsey & Ritchard Tysoe
	Out Of Focus	May 13, 14, 15	Katherine Plummer
	Cranford	Oct 7, 8, 9	Sue Worker

Appendix 14

The boundaries of the parish shown on a modern street map

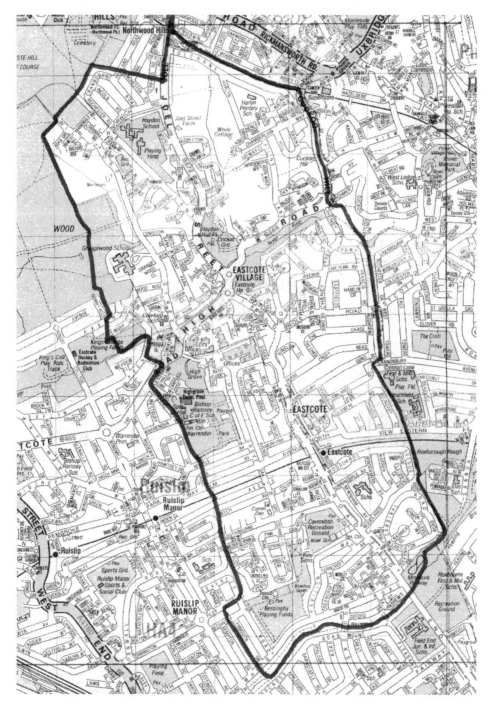

Reproduced by permission of Geographers' A-Z Map Co Ltd. © Crown Copyright 2011.
All rights reserved. Licence number 100017302

INDEX

Pages in brackets are Appendix entries.

Adamson, Mr, 66
Advent, 89, 169, 224, 228, 285, 301
Aleong, Sheila, 190, 191, 301, (321)
All Saints' Church, Hillingdon, 6, 132, 147, 183, 184, 270
All Saints' Day, 8, 34, 92, 308, (327, 329, 331)
All Souls' Day, 8, 34, 92, (327, 329, 331)
Allcock, Derek, 201, 209
altar gates, 78
Altar of Repose, 82, 90
Alternative Service Book (ASB), 94, 166, 228
Alternative Services, 93, 94
Ames, Revd Reg., 107, 137, 204
Amici Chamber Choir, 252
Ancient and Modern Hymn Book, 32
Anderson, Lady, 11
Andrews, Geoffrey, 191
Angell, Miss, 26, 52, 54, 57, 83, 115
Anglian, Father, 124
Anglican Convent of All Saints, 120
Angola, London & Mozambique Association (ALMA), 293
Anniversary, 75th, 297–99
Annual Balls, 107
Annual Parochial Church Meeting (APCM), 28, 39, 69, 72, 94, 102, 123
annual sale, 27, 28, 56, 110
Antiques Fair, 192
Arden, Rt Revd Donald, 202, 286
Armistice Day, 9, 36
Ascension Day, 8, 34, 92, 169, (324, 326, 328, 330)
Ash Wednesday, 5, 8, 34, 44, 89, 92, 169, 180, (324, 326, 328, 330)
Ashenden, Nancy, 261
Assistant Clergy Fund, 110, 131
Associate Vicar, 218
aumbry, 25
badminton club, 147, 148
Baker, Chris, 173
Baker, Mary, 140
Ball, David, 258
Ball, Revd Anthony, 183
Baptism, 9, 33, 87, 94, 118, 166, 284, 288
Baptism Visitors, 239
barbecues, 104, 184
Bardsley, Rt Revd C K N ., 132
Barker, Mrs., 109, 111

Barnardo's, 44, 60
Barnes, Madame, 40
Barnicoat, Anne, 174
Barrett, Mrs Josephine, 251
Battle of Britain Sunday, 37
Batty, Mrs, 57
bazaars, 302
Beale, Joan, 172
Beale, Revd Wilfred, 159, 178
Beddall-Smith, Mr, 28
Bedford, Andrew J., 224, 255, 279, 282, (321)
Bedford, Edith, 202
Bedford, Elizabeth, 282
Bedford, Revd Michael, 120, 166, 170, 202, 206, 222, 237, (320)
Beeston, Brian, 211, 246
Beeston, Pat, 75, 138, (322)
bereavement, 177, 285
Bernard, Olive, 132, (322)
Bible Reading Fellowship, 166
Bible Study, 104, 179, 188, 221, 237, 286
Bishop of London's Reconstruction Fund, 67–68
Bishop Ramsey School, 152, 176, 181, 220, 221, 222, 244, 257
Blanchard, John, 288
Blessing of the Holy Fire, 34
Boardman, Nigel, 162
Boddy, Revd Alan, 157–58, (313, 316)
Bolley, Monica, 218, 248
Bolley, Revd Michael, 218, (313, 316)
Book of Common Prayer (BCP), 7, 31, 33, 93, 229, 284
Boot, Joyce, 262, 298, 308, (322)
Borer, Mrs, 83
Bossanyi, Mr, 135
Bourne, Richard, 271
Bowden, David, 101, 103
Bowlt, Eileen, 16
Bradford, Frances, 185, 186, 187, 251, 261, (323)
Braithwaite, Alan, 261, (321)
bring and buy, 133, 192, 260
Brion, John, 95
Brion, Judith, 100, 174, *see also* Howe, Judith
Brion, Win, 209
Britter, Helen, 247
Britter, Mark, 304, 322

Broadbent, The Rt Revd Peter Alan, 240, 318
Broadbent, Ven Peter Alan, 226, 319
Brooke, Alan, 210
Brooker, Revd Wendy, 221, 277, 278, 301, (314)
Brown, Mrs P, 243, (322)
Brownies, 50, 51, 113, 143–46, 211–13, 226, 267–68, 306–7
bungalow, 7, 26, 83, 165, 225, 226
Butler, The Rt Revd Thomas Frederick, 162, 168, 170, 185, (318)
Butler, Ven Thomas Frederick, (319)
Call-In Centre, 241, *see also* Drop-in Centre
Campbell, Smith & Co, 80
Candelabra, 82, 225
Candlemas, 44, 206, 284, 289
Cardew, Revd J.D., 121, 122
Care and Service, 217, 234, 241, 242
Carlile, Wilson, 194
Carol Service, 89, 96, 118, 124, 169, 241, 285, 295
carols, 36, 42, 95, 145, 187, 212, 260, 295, 307
Carratt, Mrs, 103
Carratt, Ron, 84
Case, Margaret, 153, 165, 179, 215, 244, 271, (320, 321)
Castles, Iris, 137, 174
Catechism, 11, 47
Cattle, Tony, 84, 161, 162, 165
Cavendish Hall, 118
Celebration Hymnal, 199, 230
Central Hall, Uxbridge, 60
Chalice Administrators/assistants, 167, 283
Chamberlain, Heather, 282
chancel, 22, 25, 78, 81, 97, 149
Chapel Hill, 29, 65
Charitable giving, 192, 255, 293
Cheney Fields, 104, 174
Cheney School of Dancing, 40
Children and Youth Committee, 277, 287
Children's Advocate, 287
Children's corner, 224, 282, 288
Children's Holiday Club, 221, 229, 248, 282, 289
Children's Service and Instruction, 8, 11
Children's Society (The), 48, 114, 169, 196, 202, 213, 256, 294
    Waifs and Strays, 14, 44, 58, 59, 60, 64
Chiplin, Revd Huw, 74, 83, 108, 113, 138, 149, 153, 157, 183, (313, 315)
Chivers, Jeanie/Genie, 164

choir, 6, 37–39, 50, 53, 86, 89, 95–97, 109, 116, 119, 169, 185–87, 223, 251, 295
Choral Evensong, 284, 285, 295, 300
Christ Church Cathedral, Dublin, 295
Christ the King, 129, 162, (325, 327, 329, 331)
Christian Aid, 123, 124, 196, 197, 241, 252, 289
Christian, Revd Alison, 277
Christians for Life, 218, 237, 242, 246, 286
Christingle, 169, 171, 196, 229, 247, 289
Christmas, 8, 34, 36, 100, 169, 266, 285, 290, (324, 325, 327, 329, 331)
Chrzczonowicz, John, 75
Chrzczonowicz, Rosemary, 75
Church Army, 45, 58, 60, 63, 162, 193–94, 204, 214, 221, 256, 294
Church Assembly, 61
church grounds, 30–31, 47, 226–27, 282–83
church hall. *see* Permanent Hall Fund
Church Hall Committee Room, 163, 282
Church Lads' Brigade, 12
church logo, 243–44
Church Magazine, 10, 74, 135, 155, 243–44
    Gridiron, 75, 290
    Leaflet, 12, 17, 39–40, 56, 74–77, 290
    The Parish magazine of St Lawrence Eastcote, 77
Church of England (Worship and Doctrine) Measure 1974, 94
Church of England Men's Society, 46, 50, 129, 134–36
Church Parade, 146, 159, 169, 184
church roof, 77, 162, 222–23
Church Urban Fund, 199
Church web site, 246
Church's Child Protection Policy, 247
Churches Together in Eastcote, 89, 197, 241–43, 258, 290, 291
Churching, 33, 87
Civic Centre, 145, 305, 307
Civic Service, 148, 203, 301
Clarke, Rex, 125
Clay, Rowland, 80
Clements, Beryl, 199
Clementson, Capt. Church Army, 58
Coan, Jane, 247, 279, (314)
Cobb, Sue, 303, (321)
Colclough, Revd Michael, 179, (319)
Coleman, Revd Ann, 216, 219, 220, 234, 237, 240, 248, 262, 272–74, 299, (314, 316)

Coleman, Revd David, 216–17, 222, 231, 241, 272–75, 285, 298, 311, (313, 315)
Collier, Capt and Mrs Frank, 194
Collins, Gerald H, 51, 132, 143, 151, 154, 155, 162, 183, 206, (314, 320, 321, 322)
Collins, William, 25
Common Fund. *see* Diocesan Quota
Common Worship, 90, 228
communicants, 8, 33, 57, 68, 72, 92, 166, 169, (324–31)
Communicants' Guild, 57
Communion, 7, 8, 9, 13, 31, 33, 34, 40, 44, 53, 57, 64, 65, 68, 86, 87, 88, 90, 91, 93, 135, 151, 155, 168, 272, 283, 292, 293, 302, *see also* Eucharist and Mass
Community of Eastcote, 9, 70, 82, 89, 199, 248, 256, 281, 290
Complete Anglican Hymns Old and New, 225, 230
Compline, 34, 89, 175, 238, 285
Confirmation, 9, 29, 64, 65, 91, 94, 174, 237, 238, (333–37)
Consecration. *see* Dedication
Cooper, Andrew, 161
Cooper, Philippa, 289
Copus, Revd Brian, 183
Cornell, Mr, 84, 165
Corpus Christi, 34, 92, 169, 285, (326, 328, 330)
Costa Coffee, 290–91
Council of Christians and Jews, 243
Couturier, Abbé Paul, 121
Coventry Cathedral, 102
Craven, Alison, 116
Crèche, 157, 172, 247, 288
Crib service, 48, 89, 118, 169, 285
cricket match, 173, 200, 257
Criminal Record Bureau (CRB), 287
Cristo Redentor, 278, 293
Cross, T.G. (Tom), 3, 5, 29, 39
Cub Scouts, 13, 42, 50–51, 113, 142, 208, 210–11
Cubitt, Henry, 25
Curley, Carlo, 296
Cursillo, 238
Daily Prayer Cycle, 279
Dale, A.B., 129, (320, 322)
Dale, Revd J.G., 4, 5, 7, (313)
Dalton, Miss, 59
Dalton, Mrs, 41
Dando, Revd Elaine, 276, 277
Dando, Revd Stephen, 276–77, 283, 285, 293, 311, (313)
Daniels, Mrs Gail, 171

Dann, Caroline (Mrs Scott), 212, 214, 266, 304
Dann, Catherine, 260, 298, 301
Darby, Hylda, 54, 142, 264
Darby, P.W., 150, 153, (320)
Darby, Richard, 164
Davies, Bob/Robert, 99, 151
Davies, Mr & Mrs, 101
Davies, Mrs Harris, 144
Day of Prayer and Gift Day, 92, 106, 130
Day, Revd Peter, 157, 158, 165, 179, 203, 217, 270, (313, 316)
Deane, Ralph Hawtrey, 2, 3, 310
Deane, Ralph Hawtrey, Com.R.N., 7, 20
Deanery Mission of 1934, 33, 60
Decade of Evangelism, 217, 227, 235, 243
Dedication, 8, 35, 42–43, 84, 92, 106, 153, 185, 298, (325, 327, 329, 331)
  Consecration, 22–24
Dedman, John, 151, (321)
Dedman, Millicent (Topsy), 95, 97, 185, 205, 206, 251, (323)
Dew, Angela, 103, 105, 173, 174
Dew, Beryl, 161
Diamond Jubilee, 269–71
Diamond Jubilee, Brownies, 145
Diocesan Advisory Council (DAC), 222, 281
Diocesan Mission of 1949. *see* London Mission 1949
Diocesan Quota, 14, 61, 66, 125, 129, 157, 187, 189, 191, 252, 304
  Common Fund, 222, 223, 253, 254, 304
  Quota, 223
Diocesan Week of Prayer and Self-denial, 14, 61, 66
Disability Discrimination Act 2004, 280
discos, 103, 104, 108–9, 147
District Visitors, 55–56, 56, 63, 65, 75
Dow, The Rt Revd Graham, 224, 254, (318)
Drew, Kay, 212, 213, 267
Drop-in Centre, 159, 164–65, 191, 241, *see also* Call-in Centre
Dunhill, Pat, 174
Durston, Wally, 142
Dyer, Michael, 252
East London Fund, 60
East Window, 70, 79, 280
Eastcote Choral Society, 39, 185
Eastcote House, 1, 2, 49, 51, 247, 268, 269, 307, 310
Eastcote Methodist Church, 29, 89, 121, 123
Eastcote Nursery School, 82

Eastcote Place, 1, 11
Eastcote station, 2, 30
Eastcote Village Conservation Panel, 267
Eastcote Village Institute, 1, 3
Eastcote War Memorial, 9, 170, 272
Easter, 8, 34, 56, 87, 90, 91, 158, 169, 180, 266, 286, 290, (324–26, 328, 330)
Easter Vigil, 90, (326, 328, 330)
Edwards, D.A., 95, (323)
Edwards, E. (Ted), 99, 100, 137, (320)
Edwards, Mrs Bennett, 59
Edwards, Mrs Edna, 82
Edwards, Mrs Joan, 139
Edwards, Mrs R.F., 75
Edwards, Ron, 1, 16, 103
Electoral Roll, 7, 46, 69, 150, 187, 190, 191, (322)
Ellingham, Joanne, 306
Ellis, Jack, 50
Ellis, Revd P.D., 3, 22, (313)
Emmanuel Church, Northwood, 104, 213
English Hymnal, 32, 230, 252
Entertainments Committee, 107, 110
Epiphany, 53, 109, 247, 285, 289, 297
Episcopal Areas, 150
Eucharist, 7, 33, 34, 37, 49, 79, 87, 89, 90, 91, 92, 94, 100, 155, 159, 166, 173, 179, 228, 231, 238, 252, 274, 283, 284, 308, see also Communion and Mass
Evans, Sharon, 280, (323)
Evening Prayer, 7, 229, 284, *see also* Evensong
evening worship. *see* Evensong
Evensong, 7, 8, 33, 34, 86, 87, 168, 229, 284
Everitt Susan, 196
Faculty, 24, 222, 227
Fairtrade, 292
Family Fun Day, 286, 290, 300
Farthings Close property, 72, 135, 227, 283
Fellowship Committee, 107, 108, 200, 256–58, 300–302
Field End Junior and Infants Schools, 71, 146
Field End Lodge, 10
Finance and General Purposes Committee, 83
Finance Committee, 11, 294
Fischer, Elsie, 223, 226, 261, 288
Fisher Miss, 51
Fleming, Jem, 108
Flower Festival, 183, 264–66, 309
Flower Guild, 56, 113, 139–40, 209–10, 264–66, 309
Ford, Janet, 209

Ford, Mr, 135
Forty-five Churches, 6, 23
Forward Together, 166, 175–76, 235, 311
Foulds, Revd John, 217–18, (313, 316)
Free for All, 245–46, 249, 290
Freewill Offering, 66, 125, (321)
Gang Shows, 42, 143
Garden of Remembrance/Rest, 84, 131, 227
gardeners, 227, 281
gardening group, 84, 165
Gardham Capt G, 58
Gardner, Madeline, 206, 263
Garrish, Revd Elaine, 221, 232, 278, (314)
Gayler, Sylvia, 210
Gibbeson, Miss B, 59, *see also* Williams, Barbara
Gibbs-Smith, Revd O.H., 60
Gift Day, 84, 92, 130–31
Giles, John, 211
Girl Guides, 13, 50, 51, 146–47, 211, 213–14, 266–67, 304–6
Girls' Friendly Society, 12, 50
Glass, Dorothy, 170
Glynn, Marion, 247
Godwin, Revd Rupert Frederick, 5, 6, 15, 17, 20–22, 31, 33, 39, 43, 46, 61, 63–65, 69–70, 79, 217, (313, 315)
Golden Jubilee, 162, 168, 179–85, 199
Good Friday, 8, 34, 63, 90, 231, 249, (326, 328, 330)
Goodridge, Revd Peter, 73, 77, 83, 91, 122, 143, 153, 183, (313, 315)
Goschen, Kenneth, 3, 11
Gough, Pam, 241
Gough, Peter, 165
Gover, Sydney, 84
Grand Draw, 111
Grandison, Thomas, 77, (320)
Grange, The, 6
Grangewood School, 197
Gray, Miss, 48
Gray, Revd W.A.G., 4
Great Rood, 78
Greater London Festival 1972, 119
Greaves, Messrs, 74, 77
Green, Richard, 303
Groom, Revd Sue, 219, 224, (313, 316)
Guest, George, 163
Guild of Our Lady, 229, 263, see *also* Society of Mary
Gunson, Marie, 200
Haiti Earthquake Appeal (2010), 295
Hall Committee, 106
Hall Management Committee, 225, 227
Hall, Mr and Mrs B.J., 10

Harlyn Drive Primary School, 97, 98, 99
harmonium, 6, 26
Harries, Rt Revd Richard, 243
Harrison Smith, Mrs, 11
Harrison, Dr D, 75, 320, (322)
Harrow Deanery Choirs' Festival, 295, 296
Harvest, 9, 13, 35, 86, 87, 105, 107, 290
Haydon Hall, 1, 59
Hayes, Monica, 156, 201, 202, 203, (322)
Hayes, Rachel, 156, 215
Hayes, Revd David M.H., 156, 159, 167, 178, 198, 214–15, 261, 298, (313, 315)
Hayes, Timothy, 156, 173
Hayman, Marion, 202, 241, 260, 262, 307, (322)
Heaney, Val, 248
Helen House, 194, 196, 201, 256, 309
Henley, Mr W.A., 127, 129, 151
Herbert, Vera, 137
Herd, The Rt Revd William, 191
Hewer, Dr Chris, 291
Higgs, Barbara, 165, 243, (322)
Higgs, Geoff, 165, 258, 280
High Anglican tradition, 72, 155, 166, 216
Highgrove House, 1, 65, 98
Highgrove Swimming Pool, 145
Hillel, Deirdre, 234
Hillel, Revd Laurence, 221, 234, (314, 316)
Hillingdon Hospital, 149
Hillingdon Outdoor Activity/Sports Centre, 266, 305, 309
Hillingdon Welcare, 164
Hillingdon, London Borough of, 148, 168, 263
Hillingdon, Mayor of, 148, 222, 266, 298, 305
Hinman Conservation Shield, 267
Historical Guide to St Lawrence Church, 244
Hitchinson, Canon/Revd William Henry/Father Bill, 71–73, 97, 130, 131, 152–54, 184, 185, 202, 304, (313, 315)
Hitchinson, Joan, 139, 153, 202, 205
Holmes, Mr, 132
Holt, Anne, 95, 185, 207, 251, 261, 262, 295, (323)
Holy Communion. *see* Communion
Holy Communion (Series I, II, and III), 93
Holy Week, 8, 34, 90, 169, 248
Home Guard, 25, 27
Hooper, Alan, 209, 280, 292
Hooper, Sylvia, 247, 288, 292, 321
Horchover, David, 208, 258
Horchover, Neil, 173
Horchover, Shirley, 208
Horsfall, Barbara, 146
Horsfall, Reg, 99, 100, 103
Horsfall, Yvonne, 144, 146
House Eucharist, 173, 179, 217, 232
House Group, 176–77
Howard, Marjorie, 142
Howe, Judith (née Brion), 207, 208, 280
Hughes, Father Gerard, 197
Hughes, Tricia, 279, (314)
Interfaith Committee, 124, 243
International Nepal Fellowship, 116
interregnum, 156, 170, 217, 275
Islam, course, 291
Israel, Dr Martin, 180
Jeel Al Amal Children's Home, 258
Jones, Jean, 113, 161
Jones, Lloyd, 152, 202
Jones, Marjorie, 137
Jones, Revd Edward Cornwall, 4
Jubilee, The Queen's, 104, 113, 267, 268, 269
jumble sales, 20, 26, 133, 137, 139, 147, 164, 192, 203, 204
Jungreuthmayer, Anton, 142
Jungreuthmayer, Jean, 142
Junior Church. *see* Young People
Junior Guild. *see* Young People
Junior Players, 52, 54
Junior Youth Club. *see* Young People
Keeler, Mrs, 101
Kendal, Kenneth, 112
Kerswell's Restaurant, 39
Kimsey, Emma, 287, (322)
Kimsey, Mark, 173
Kindergarten. *see* Young People
King, Margaret, 207, 210, 262, 264
King, Palmer Mrs, 41, 42
Kirkley, Tom, 188, 190, (321)
Knott, Alice, 171
Korean Mission, 114, 115, 116
Lacey, David, 96, (323)
Lady Chapel, 17, 25, 80, 159, 162, 223–24
Lafford, Revd Sylvia, 171, 222, 237, 278
laity, 48, 63, 128, 149, 151, 154, 167, 215, 231, 237, 277, 283, 287, 311
Lally, Len, 134, 148, 153
Lally, Mr & Mrs Christopher, 80
Lampard, Revd Ruth, 219, 249–50, (313, 316)
Lane, William James, 49, 50, 84, 99, 129, (321)
Langdale, Father, 121
Lawrence, C K ('Lawrie'), 143

Laying on of Hands
  Ministry of Healing, 178, 221, 231, 234, 284
Leakey, Sally, 247
Lectionary, 166
Lee, Andrew, 103
Lee, Stephen, 101
Leech, Father Ken, 273
Leeson, Barbara, 209
Lent, 8, 34, 60, 67, 89, 116, 169, 209, 230–31, 239, 241, 285, 291, 293, 300
Lent, Holy Week, Easter Services and Prayers (LHWE), 230
Leonard, The Rt Revd Graham Douglas, 80, 150, (317, 318)
Lewis, Revd D., 5, (313)
Litany, 34
Lloyd, Karen, 103, 146
local community. *see* Community of Eastcote
Lomax, Pat, 256
London Challenge, The, of 2002, 240, 249, 254, 274
London Diocesan Home Mission, 4, 5
London Gazette, The, 7
London Mission 1949, 33, 46, 61–64, 69
London, Bishop of, 5, 22, 29, 67, 79, 93, 150, 217, 240, 293
Lumsden, Enid, 95, 138, 139, 146, 165, 185, 201, 211, 212, 251, 259, 295, (323)
Lumsden, Euan, 165, 188, 211, 227, 303, (322)
magazine. *see* Church Magazine
Maines, Katherine, 267
maintenance team, 227, 282
Mannering & Sons Ltd, 281
markets, 110, 118, 127, 130, 191, 198, 256, 290, 300
marriages, 9, 72, (332)
Marshall, J.D., 6, 7
Martell, D.S., 77
Martin, Hazel, 144, 190
Martin, John, 154, 200, 261
Masey, J.J. (John), 151, 155, 314
Mass, 89, 90, 166, 169, 176, 223, 283, 284, *see also* Communion and Eucharist
Matins, 7, 8, 33, 34, 86, 95
Maude, Richard, 289
Maude, Tessa, 280, (323)
Maundy Thursday, 82, 90, 169, (326, 328, 330)
Mauritius, 110, 114, 115, 131
McDonald, Pauline, 269
McIntyre, Revd J., 4, 7, (313)

McKellar, Mrs, 139, 205
Memorial Book, 84
Metcalfe, Nicholas, 241
Metcalfe, Rosemary, 247
Methodist Chapel, 1, 29
Millennium, 230, 239, 264, 272
Miller, Geoffrey, 162
Miller, Susan, 262, 308, (322)
ministry of healing. *see* Laying on of Hands
Mission, 14, 58–61, 64–65, 75, 113–19, 217, 234, 290, 311
Mission Action Plan (MAP), 217, 225, 228, 237, 254, 287
Mission Church, 1–14, 16, 26, 35, 46, 150, (324)
Mission Committee, 277
Missouri Court, 119, 212, 221
Mitton, Ian, 100, 103, 105
Mitton, Sue, 174
Mooney-Dutton, James, 285, 295, (323)
Morgans, Ann, 142
Morning Prayer, 7, 228, *see also* Matins
Morris, Jane, 171
Mother and Toddler Group, 250–51
Mother Frances Dominica, 194
Mothering Sunday, 90, 171, 247, 307
Mothers' Union, 13, 43–46, 131–34, 201–3, 259–62, 307–8
Mothers' Union banner, 44, 132, 261, 308
Mount Vernon Hospital, 212, 260, 296
Mozambique, 196, 256, 293
Mullet, Mr, 95, (323)
Munson, Mr & Mrs, 26
Murrell, Doreen, 144, 213
Murrell, Eric, 165
Music, 37, 185–87, 251–52, 295–96
Narusawa, Alicia, 157, 172
Narusawa, Revd Masaki, 157, 179, 203, (313, 315)
National Day of Prayer, 36–37
Needlework Guild/Sewing Circle, 11, 56–57, 110, 139
Newnham School, 49, 50, 71, 97, 99, 169, 187, 197
Nicaraguan Earthquake Appeal 1972, 116
Nicholson, Sir Charles, 7, 17, 25, 26, 37, 170
Nicholson, Sir Sydney, 37
Nine Lessons and Carols, 36
No Small Change, 116–17, 136
Northwick Park Hospital, 137, 138
Northwood and Pinner Community Hospital, 13, 65
Nuttall, Revd H.E., 5, 8, (313)
Oberammergau, 121, 137, 286

Odell, Revd R.W., 23
Old Folks' Association, 112, 118
Onions, Mrs, 101, 125, 127, (321)
Operation Firm Faith, 85, 134
Orders, Beryl, 59, 142, 264, 298, (321)
Ordination of women, 150
organ, 18, 22, 77, 162, 223, 252, 296
organist, 38, 95, 97, 185, 251, 295, (323)
Orme, Beryl, 140
Orme, Joe/Joseph, 103, 111, 137, 150, 152, 200, (320)
Outreach and Communication, 243, 290
Overall, Liz, 212
Palm Sunday, 34, 90, 169, 229, 310
Parade Services, 13, 37, 211, 212, 229, 267, 269
Parish Administrator, 280, (323)
Parish Appraisal (1989), 179, 217, 235
Parish Balls, 108
Parish Care Scheme, 221, 242, *see also* Parish Links and Pastoral Care
Parish Communion. *See* Communion
Parish Communion Movement, 7, 33, 93
parish hall, 29, 30, 45, 99, 115
Parish Hall Building Project, 162–65, 175, 225, 281
Parish Links, 279, *see also* Parish Care and Pastoral Care
Parish Magazine. *see* Church Magazine
Parish Outings, 107
Parish Prayer, 275, 297
Parish Rambles, 108, 302
parish scrubbers, 227, 282
Parker, Norman, 22, 31, 37
Parsonage, 3, 5, 7, *see also* St Lawrence Centre
Paschal Candle, 90, 182
Pastoral Assistant, 278, (314)
Pastoral Care Scheme, 168, 178, 191, 242, *see also* Parish Care and Parish Links
Patronal Festival, 34, 92, 93, 106, 215
Paul Strickland Scanner Appeal, 212
Paul, Joanna, 185, (323)
Pawley, Mike, 108
Pearce, Revd Clive, 73, 77, 83, 87, 138, 153, 155, 183, 202, 270, (313, 315)
Pearce, Roger Clayton, 190
Pearmain, Edna, 131, 132, 202, 260, (322)
Pearson, Andrew, 173
Pearson, John, 108, 183, 184, 188
Peel District, 5–6, 15
Pennington, Mr, 135
Pentecost, 284, *see also* Whitsun
Permanent Church (Building) Fund, 3, 14, 17
Permanent Hall Fund, 26–30, 56, 151

Perry Parsons Big Band, 299
Peterken, S.G./Stan, 51, 134, (320, 321)
Pether, Percy, 261
Phillips, Revd Rachel, 278, 285, 287, 293, 299, (313, 316)
Phillips, Revd Ray, 170, 183, 184, 270
Phoenix Ladies, 82, 282
Pilgrimages, 120, 121, 157, 170, 182, 239
Pine Gardens Property, 227
Piper, Miss, 52
Planned Giving, 304
Plummer, Arthur, 161, 162, 163, 190, 224, 246, 261, 280, (320)
Plummer, Barbara, 137, 162, 172, 202, 206, 215, 287, (321, 322)
Plummer, Doreen, 172, 196, 244, 248, (322)
Poole, Harold, 50
population, 1, 6, 16, 31, 41, 60, 69, 72, 148, 178, 290
Posada, 285
Pottinger, Mollie, 137
Praise Service, 229
Pram Eucharists/Services, 88, 167, 250
Pray and Play, 250
prayer and worship, 85, 228, 277, 311
Prayer before The Blessed Sacrament, 231, 285
Prayer Book (Alternative and Other Services) Measure 1965, 93, 116
Prayer for the Mission of the Church, 234
Prayer Partner scheme, 167
Prayers of Intercession and Thanksgiving, 167
Premises Committee, 282
Priddle, Mr, 165
Priest Missioner, 5, 6, 7, 8, 15
Prowting, Messrs, 29
Publicity and Publications Committee (Pub & Pubs), 243, 277
Pym, Emil, 162
Queen's Walk Hostel, 137, 140, 256, 294
quiet afternoons/days, 63, 89, 120, 170, 239, 286, 298
Quiz Supper, 164, 200, 257
Quota. *see* Diocesan Quota
Rainbow Group, 164, 165
Rainbow Guides, 268, 307
Rann, Revd Harry, 50
Raper, Mary, 190
Reader, 151, 155, 221, (314)
Recorder's Report, 188
Reddiford School, 197
Rees, Sally, 103
Reeve, Frank, 152, 215, (320)
Reeve, Joan, 241, 262

Reindorp, Revd H.W., 20
Remembrance Sunday, 170
Retreat House of St Columba, 170
Retreat, The, 64
retreats, 120, 170, 239, 286
Rhymer, Jim, 105
Richard, Cliff, 104
Riddle, A.C., 67
Road Warden Scheme, 129, 134
Robinson, Susan, 101, 103
Rogers, Norman, 155
Rogers, The Rt Revd Alan, 114, 317
Roseveare, Canon R.R., 63
Royal Albert Hall, 37, 59, 63, 116
Royal British Legion, 14, 149
Royal School of Church Music, 37, 186
Ruislip-Northwood Urban District Council, 29, 49, 148
Russell, Miss, 59
Rye, Mrs, 66, (321)
Sacred Space, 234
Sacristy, 77, 224, 281
Sanctuary, 22, 159–62
Saward, Preb. Michael, 189
Scout Group, 31, 50–51, 143, 210–11, 304
Scouts, 13, 20, 31, 42
Seely and Paget, 27
Sharp, Regula, 177, 178, 200, 241
Shepherdly, Claire, 146
Shepherdly, Mervyn, 107
Sherriff-Gibbons, Joan, 192, 256
Shirras, Ven. Eddie, 191, 252, (319)
Shooting Star Hospice, 309
Sigers (House), 3, 11
Silver Glass of Northwood Hills, 226
Silver Jubilee, 30, 92, 106, 130, 151
Simpson, Rt Revd Bertram F., 20, (318)
Sine Nomine Singers, 183, 187
Sing-in Messiah, 271
Smith, Christopher, 99
Smith, Harold, 50
Smith, Jackie, 267, 268, 306
Smith, John, 109
Smith, Leigh, 142
Smith, Maria, 287, (322)
Smith, Mrs, 58
Smith, Revd Marion, 222, 229, 233, 237, 248
Smith, Tony, 185, (323)
Sociable Community, 39–40, 105
Society of Mary, 139, 206, 262, *see also* Guild of Our Lady
Songs of Praise, 168, 229, 284
SOS Sahel, 200, 253, 271
Southwell, Ven Roy, 84, (319)
SPCK, 60, 94, 116, 155

SPG, 59, 60, 61, 115, 116
Squires, Rob, 173, 174
St Andrew's Church, Roxbourne, 252, 278
St Andrew's Church, Uxbridge, 176
St Andrew's United Reformed Church, 123, 137, 241
St Bernard's Hospital, 136
St Edmund the King, Northwood Hills, 26, 137, 149, 204
St Francis House, Hemingford Grey, 120, 170, 286
St Lawrence Association, 101
St Lawrence Centre, 225–26, 261, 282, 292
St Lawrence Day, 8, 34, 42, 300, (324, 325, 327, 329, 331), *see also* Patronal Festival
St Lawrence Intelligent Cool Kids (SLICK). *see* Young People
St Lawrence Juvenile Players, 41
St Lawrence Mummers, 42, 53
St Lawrence Orchestra, 39, 187
St Lawrence Players, 11, 41, 48, 49, 53–55, 112, 140–43, 207–9, 226, 263–64, 309–11, (348–52)
St Lawrence Players' Theatre Club, 140
St Lawrence Social Club, 201, 258–59, 309
St Lawrence Women's Guild, 136–39, 203–6
St Lawrence Young People (SLYP). *see* Young People
St Martin's Church, Ruislip, 1, 4, 20, 60, 156
St Martin-in-the-Field, 103
St Mary's Church, Harrow on the Hill, 295
St Mary's Church, South Ruislip, 63, 183
St Mary's Church, Twickenham, 71, 114
St Mary's Church, Willesden, 121, 170
St Mary's School, 71, 97, 99, 119
St Michael's School, 64
St Nicholas School, 71
St Paul's Cathedral, 44, 59, 66, 150, 170
St Paul's Church, Roxeth, 26
St Paul's Church, Ruislip Manor, 63, 183
St Stephen's Church, Spitalfields, 17
St Thomas More, R.C. Church, 121, 124, 196, 241, 242–43
St Vincent's, 13, 65, 100, 104, 151, 173
Stanes, Father Ian, 176
Starling, Eva, 224
Starling, Sydney, 78, 79, 127, 129, 162, (320, 321, 322)
Stations of the Cross, 78, 89, 90, 169, 231
Stevens, Derek, 271

Stewardship, 125–30, 168, 175, 187–92, 227, 252–55, 272, 304, (321)
Stir-Up Sunday, 129
Stopford, Rt Revd Robert, 79, (317)
Sturman, John, 101
Styles, John, 84
Suckley, Norman, 298
Sumner, Mary, 43, 44, 131, 202, 308
Sunday School. *see* Young People
Sunday services, 8, 33, 86–87, 166–68, 228–30, 283–85
Sunningdale Avenue, church property in, 165, 227
Surridge, Mr, 113
Sykes, Ann, 162, 188, 201, 209, 211, 257, 280, 310, (321)
Sykes, Geoffrey, 137
Synod, 61, 94, 149–50
Taizé, 175, 229, 284
Tea Table Conference, 241, 258
Teaching and Nurture, 217, 234, 237
Tear Fund, 248, 256
Telling Brothers, 11
The London Churchman, 64
The Mission Field, 59
The Sigers, Church property in, 83, 130
The Sign, 75
The Vigil, 182, 207
Thompson, The Rt Revd Hewlett, 185, (318)
Toovey, Revd Kenneth, 156
Torrey, Father Arthur (Korean Mission), 114
Tots Club, 288
Towers, Mavis, 165
Towers, Ron, 165, 244
Treweek, Revd Rachel (Archdeacon), 286, (319)
Trinity Sunday, 153, 243
Troughton, Miss, 48
Tsunami Appeal (2005), 256
Tudor Lodge Hotel, 10, 258, 301, 306
Tuffin, Revd Gill, 178, 222, 237
Tyrer, Chris, 150, (322)
UMCA, 60, 115
Universal Housing Co. Ltd, Rickmansworth, 7
Uxbridge Gazette, 17
van Zyl, Revd W.P., 73, 83, (313, 315)
Venture Scouts, 211
Vespers and Benediction, 229, 263, 284
Vestry and Annual Meeting, 5, 7
Vicar's Close, 83–84
Vicar's Remarks, 72, 102, 228, 237, 254, 262, 272, 274, 277, 311
Vicar's Social evenings, 102

vicarage, 26, 40, 82, 153, 165
Vincent, Hazel, 261
Voluntary Aid Detachment Hospital (VAD), 10, 65
Wade, G.W., 125, (321)
Waifs and Strays. *see* Children's Society (The)
Walsingham, 139, 157, 162, 171, 182, 206
Wand, The Rt Revd John William Charles, 29, 151, (317)
Warren, Miss, 26, 58, 59
Watch Night, 109
Watford theatre, 259
Watson, John, 187, 188, 190
Wave of Prayer, 45, 133, 152, 201, 203, 260, 262, 307
Weddell, Capt E.T., 39, (320)
weddings, 68, 95, 186, 209, 251, 295
Week of Guided Prayer, 239
Week of Prayer for Christian Unity, 121–22, 197, 241
Weekly Bulletin, 157, 246
Weeks, Jean, 212
Weeks, Revd John, 181
Welch, Frank, 119
Welcome and Nurture, 277, 280
West End of The Church, 280–81
Whaley, Mr, 3
whist drives, 13, 20, 26, 28, 42, 46, 106
Whitby Dene, 119, 137, 151, 221, 292
Whitsun, 8, 33, (324, 326, 328, 330)
Wigington, Elaine, 267, (321)
Wilkinson, John MP, 163
Williams Phil, 229
Williams, Adrienne, 248, 250, 289
Williams, Barbara (née Gibbeson), 142, 209, 258, 260
Williams, Jane, 247, (321)
Williams, Lionel, 173, 223, 258, 279, 299, (321)
Willow Tree Centre, 266–67, 268
Wilson, Jan, 261
Wilson, Mollie, 139
Windsor, 46, 49, 138, 204
Winn, Mark, 211
Winnington-Ingram, The Rt Revd Arthur Foley, 22, 151, (317)
Winslow Close, 72
Women's World Day of Prayer (WWDP), 152
Woodhouse, Capt Donald, 109, 113, 159, 162, 167, 178, 184, 187, 204, 211, 213, 221, 261, (314)
Woodlands Avenue, church property in, 74, 83, 113, 130, 165
Woodnutt, John, 182, 202, 207

Wooldridge & Simpson, Messrs, 18
Worden, Mrs, 131, (322)
World War II, 32, 35, 71, 72, 93, 139, 152
Worship and Fellowship, 73, 85, 92, 100, 101, 105, 113, 118, 136, 155, 311
Worship sub-committee, 228
Wright, Alan, 192, 201, 221, 233, 237, 242, 252, 279, (314)
Wright, Sadie, 196, 243, (322)
Wydale Hall, 220, 273, 299
Yates, Jean, 241
Young People, 10, 11, 72, 100, 188, 287 5-2-7, 289
- Junior Church, 97–100, 171–72, 247, 248, 289
- Junior Guild, 101
- Junior Youth Club, 174–75

Kindergarten, 171–72
St Lawrence Intelligent Cool Kids (SLICK), 229–30, 249–50, 289
St Lawrence Young People (SLYP), 168, 172–74, 184, 189
Sunday School, 11, 47–50, 55, 59, 86, 97, 98, 171, 172, 246–47, 288–89
Youth Circle, 52
Youth Club, 52, 84
Youth Discussion Group, 103–5, 108
Youth Fellowship, 48, 50, 52–53, 54, 101–3
Youngs, Mr, 73, 83
Youth Circle. *see* Young People
Youth Discussion Group. *see* Young People
Youth Fellowship. *see* Young People